T0213422

Lecture Notes in Artificial Intelligence   10534

Subseries of Lecture Notes in Computer Science

LNAI Series Editors

Randy Goebel
University of Alberta, Edmonton, Canada
Yuzuru Tanaka
Hokkaido University, Sapporo, Japan
Wolfgang Wahlster
DFKI and Saarland University, Saarbrücken, Germany

LNAI Founding Series Editor

Joerg Siekmann
DFKI and Saarland University, Saarbrücken, Germany

More information about this series at http://www.springer.com/series/1244

Michelangelo Ceci · Jaakko Hollmén
Ljupčo Todorovski · Celine Vens
Sašo Džeroski (Eds.)

Machine Learning and Knowledge Discovery in Databases

European Conference, ECML PKDD 2017
Skopje, Macedonia, September 18–22, 2017
Proceedings, Part I

 Springer

Editors
Michelangelo Ceci 🆔
Università degli Studi di Bari Aldo Moro
Bari
Italy

Jaakko Hollmén
Aalto University School of Science
Espoo
Finland

Ljupčo Todorovski
University of Ljubljana
Ljubljana
Slovenia

Celine Vens
KU Leuven Kulak
Kortrijk
Belgium

Sašo Džeroski
Jožef Stefan Institute
Ljubljana
Slovenia

ISSN 0302-9743   ISSN 1611-3349   (electronic)
Lecture Notes in Artificial Intelligence
ISBN 978-3-319-71248-2   ISBN 978-3-319-71249-9   (eBook)
https://doi.org/10.1007/978-3-319-71249-9

Library of Congress Control Number: 2017961799

LNCS Sublibrary: SL7 – Artificial Intelligence

Printed on acid-free paper

This Springer imprint is published by Springer Nature
The registered company is Springer International Publishing AG
The registered company address is: Gewerbestrasse 11, 6330 Cham, Switzerland

Preface

We are delighted to introduce the proceedings of the 2017 edition of the European Conference on Machine Learning and Principles and Practice of Knowledge Discovery in Databases (ECML PKDD 2017). This year the conference was held in Skopje, Macedonia, during September 18–22, 2017. ECML PKDD is an annual conference that provides an international forum for the discussion of the latest high-quality research results in all areas related to machine learning and knowledge discovery in databases as well as innovative application domains. This event is the premier European machine learning and data mining conference and builds upon a very successful series of 18 ECML, ten PKDD (until 2007, when they merged), and nine ECML PKDD (from 2008). Therefore, this was the tenth edition of ECML PKDD as a single conference.

The scientific program was very rich and consisted of technical presentations of accepted papers, plenary talks by distinguished keynote speakers, workshops, and tutorials. Accepted papers were organized in five different tracks:

- The conference track, featuring regular conference papers, published in the proceedings
- The journal track, featuring papers that satisfy the quality criteria of journal papers and at the same time lend themselves to conference talks (these papers are published separately in the journals *Machine Learning* and *Data Mining and Knowledge Discovery*)
- The applied data science track (formerly industrial track), aiming to bring together participants from academia, industry, government, and non-governmental organizations in a venue that highlights practical and real-world studies of machine learning, knowledge discovery, and data mining
- The demo track, presenting innovative prototype implementations or mature systems that use machine learning techniques and knowledge discovery processes in a real setting
- The nectar track, offering conference attendees a compact overview of recent scientific advances at the frontier of machine learning and data mining with other disciplines, as published in related conferences and journals

In addition, the PhD Forum provided a friendly environment for junior PhD students to exchange ideas and experiences with peers in an interactive atmosphere and to get constructive feedback from senior researchers. This year we also introduced the EU Projects Forum with the purpose of disseminating EU projects and their results to the targeted scientific audience of the conference participants. Moreover, two discovery challenges, six tutorials, and 17 co-located workshops were organized on related research topics.

Following the successful experience of last year, we stimulated the practices of reproducible research (RR). Authors were encouraged to adhere to such practices by making available data and software tools for reproducing the results reported in their papers. In total, 57% of the accepted papers in the conference have accompanying software and/or data and are flagged as RR papers on the conference website. In the proceedings, each RR paper provides links to such additional material. This year, we took a further step toward data publishing and facilitated the authors in making data and software available on a public repository in a dedicated area branded with the conference name (e.g., figshare, where there is an ECML PKDD area).

The response to our call for papers was very good. We received 364 papers for the main conference track, of which 104 were accepted, yielding an acceptance rate of about 28%. This allowed us to define a very rich program with 101 presentations in the main conference track: The remaining three accepted papers were not presented at the conference and are not included in the proceedings. The program also included six plenary keynotes by invited speakers: Inderjit Dhillon (University of Texas at Austin and Amazon), Alex Graves (Google DeepMind), Frank Hutter (University of Freiburg), Pierre-Philippe Mathieu (ESA/ESRIN), John Quackenbush (Dana-Farber Cancer Institute and Harvard TH Chan School of Public Health), and Cordelia Schmid (Inria).

This year, ECML PKDD attracted over 600 participants from 49 countries. It also attracted substantial attention from industry and end users, both through sponsorship and submission/participation at the applied data science track. This is confirmation that the ECML PKDD community is healthy, strong, and continuously growing.

Many people worked hard together as a superb dedicated team to ensure the successful organization of this conference: the general chairs (Michelangelo Ceci and Sašo Džeroski), the PC chairs (Michelangelo Ceci, Jaakko Hollmén, Ljupčo Todorovski, Celine Vens), the journal track chairs (Kurt Driessens, Dragi Kocev, Marko Robnik-Šikonja, Myra Spiliopoulou), the applied data science chairs (Yasemin Altun, Kamalika Das, Taneli Mielikäinen), the nectar track chairs (Donato Malerba, Jerzy Stefanowski), the demonstration track chairs (Jesse Read, Marinka Žitnik), the workshops and tutorials chairs (Nathalie Japkowicz, Panče Panov), the EU projects forum chairs (Petra Kralj Novak, Nada Lavrač), the PhD forum chairs (Tomislav Šmuc, Bernard Ženko), the Awards Committee (Peter Flach, Rosa Meo, Indrė Žliobaitė), the discovery challenge chair (Dino Ienco), the production and public relations chairs (Dragi Kocev, Nikola Simidjievski), the local organizers (Ivica Dimitrovski, Tina Anžič, Mili Bauer, Gjorgji Madjarov), the sponsorship chairs (Albert Bifet, Panče Panov), the proceedings chairs (Jurica Levatić, Gianvito Pio), the area chairs (listed in this book), the Program Committee members (listed in this book), and members of the MLJ and DMKD Guest Editorial Boards. They made tremendous effort in guaranteeing the quality of the reviewing process and, therefore, the scientific quality of the conference, which is certainly beneficial for the whole community. Our sincere thanks go to all of them.

We would also like to thank the Cankarjev Dom congress agency and the student volunteers. Thanks to Springer for their continuous support and Microsoft for allowing us to use their CMT software for conference management. Special thanks to the sponsors (Deutsche Post DHL Group, Google, AGT, ASML, Deloitte, NEC, Siemens, Cambridge University Press, *IEEE/CAA Journal of Automatica Sinica*, Springer, IBM Research, *Data Mining and Knowledge Discovery*, *Machine Learning*, EurAi, and

GrabIT) and the European project MAESTRA (ICT-2013-612944), as well as the ECML PKDD Steering Committee (for their suggestions and advice). Finally, we would like to thank the organizing institutions: the Jožef Stefan Institute (Slovenia), the Ss. Cyril and Methodius University in Skopje (Macedonia), and the University of Bari Aldo Moro (Italy).

September 2017

Michelangelo Ceci
Jaakko Hollmén
Ljupčo Todorovski
Celine Vens
Sašo Džeroski

Organization

ECML PKDD 2017 Organization

Conference Chairs

Michelangelo Ceci	University of Bari Aldo Moro, Italy
Sašo Džeroski	Jožef Stefan Institute, Slovenia

Program Chairs

Michelangelo Ceci	University of Bari Aldo Moro, Italy
Jaakko Hollmén	Aalto University, Finland
Ljupčo Todorovski	University of Ljubljana, Slovenia
Celine Vens	KU Leuven Kulak, Belgium

Journal Track Chairs

Kurt Driessens	Maastricht University, The Netherlands
Dragi Kocev	Jožef Stefan Institute, Slovenia
Marko Robnik-Šikonja	University of Ljubljana, Slovenia
Myra Spiliopoulou	Magdeburg University, Germany

Applied Data Science Track Chairs

Yasemin Altun	Google Research, Switzerland
Kamalika Das	NASA Ames Research Center, USA
Taneli Mielikäinen	Yahoo! USA

Local Organization Chairs

Ivica Dimitrovski	Ss. Cyril and Methodius University, Macedonia
Tina Anžič	Jožef Stefan Institute, Slovenia
Mili Bauer	Jožef Stefan Institute, Slovenia
Gjorgji Madjarov	Ss. Cyril and Methodius University, Macedonia

Workshops and Tutorials Chairs

Nathalie Japkowicz	American University, USA
Panče Panov	Jožef Stefan Institute, Slovenia

Awards Committee

Peter Flach   University of Bristol, UK
Rosa Meo   University of Turin, Italy
Indrė Žliobaitė   University of Helsinki, Finland

Nectar Track Chairs

Donato Malerba   University of Bari Aldo Moro, Italy
Jerzy Stefanowski   Poznan University of Technology, Poland

Demo Track Chairs

Jesse Read   École Polytechnique, France
Marinka Žitnik   Stanford University, USA

PhD Forum Chairs

Tomislav Šmuc   Rudjer Bošković Institute, Croatia
Bernard Ženko   Jožef Stefan Institute, Slovenia

EU Projects Forum Chairs

Petra Kralj Novak   Jožef Stefan Institute, Slovenia
Nada Lavrač   Jožef Stefan Institute, Slovenia

Proceedings Chairs

Jurica Levatić   Jožef Stefan Institute, Slovenia
Gianvito Pio   University of Bari Aldo Moro, Italy

Discovery Challenge Chair

Dino Ienco   IRSTEA - UMR TETIS, France

Sponsorship Chairs

Albert Bifet   Télécom ParisTech, France
Pance Panov   Jožef Stefan Institute, Slovenia

Production and Public Relations Chairs

Dragi Kocev   Jožef Stefan Institute, Slovenia
Nikola Simidjievski   Jožef Stefan Institute, Slovenia

ECML PKDD Steering Committee

Michele Sebag	Université Paris Sud, France
Francesco Bonchi	ISI Foundation, Italy
Albert Bifet	Télécom ParisTech, France
Hendrik Blockeel	KU Leuven, Belgium and Leiden University, The Netherlands
Katharina Morik	University of Dortmund, Germany
Arno Siebes	Utrecht University, The Netherlands
Siegfried Nijssen	LIACS, Leiden University, The Netherlands
Chedy Raïssi	Inria Nancy Grand-Est, France
Rosa Meo	Università di Torino, Italy
Toon Calders	Eindhoven University of Technology, The Netherlands
João Gama	FCUP, University of Porto/LIAAD, INESC Porto L.A., Portugal
Annalisa Appice	University of Bari Aldo Moro, Italy
Indré Žliobaité	University of Helsinki, Finland
Andrea Passerini	University of Trento, Italy
Paolo Frasconi	University of Florence, Italy
Céline Robardet	National Institute of Applied Science in Lyon, France
Jilles Vreeken	Saarland University, Max Planck Institute for Informatics, Germany

Area Chairs

Michael Berthold	Universität Konstanz, Germany
Hendrik Blockeel	KU Leuven, Belgium
Ulf Brefeld	TU Darmstadt, Germany
Toon Calders	Universiteit Antwerpen, Belgium
Bruno Cremilleux	Universite de Caen Normandie, France
Tapio Elomaa	Tampere University of Technology, Finland
Johannes Fuernkranz	TU Darmstadt, Germany
João Gama	FCUP, University of Porto/LIAAD, INESC Porto L.A., Portugal
Alipio Mario Jorge	FCUP, University of Porto/LIAAD, INESC Porto L.A., Portugal
Arno Knobbe	Leiden University, The Netherlands
Stefan Kramer	Johannes Gutenberg University of Mainz, Germany
Ernestina Menasalvas	Universidad Politecnica de Madrid, Spain
Siegfried Nijssen	Leiden University, The Netherlands
Bernhard Pfahringer	University of Waikato, New Zealand
Jesse Read	Ecole Polytechnique, France
Arno Siebes	Universiteit Utrecht, The Netherlands
Carlos Soares	University of Porto, Portugal
Einoshin Suzuki	Kyushu University, Japan
Luis Torgo	University of Porto, Portugal

Matthijs Van Leeuwen   Leiden Institute of Advanced Computer Science,
                                 The Netherlands
Indré Žliobaité   University of Helsinki, Finland

Program Committee

Niall Adams
Mohammad Al Hasan
Carlos Alzate
Aijun An
Aris Anagnostopoulos
Fabrizio Angiulli
Annalisa Appice
Ira Assent
Martin Atzmueller
Antonio Bahamonde
Jose Balcazar
Nicola Barberi
Christian Bauckhage
Roberto Bayardo
Martin Becker
Srikanta Bedathur
Jessa Bekker
Vaishak Belle
Andras Benczur
Petr Berka
Michele Berlingerio
Cuissart Bertrand
Marenglen Biba
Silvio Bicciato
Albert Bifet
Paul Blomstedt
Mario Boley
Gianluca Bontempi
Henrik Boström
Marc Boulle
Pavel Brazdil
Dariusz Brzezinski
Rui Camacho
Longbing Cao
Francisco Casacuberta
Peggy Cellier
Loic Cerf
Tania Cerquitelli
Edward Chang

Thierry Charnois
Duen Horng Chau
Keke Chen Wright
Silvia Chiusano
Arthur Choi
Frans Coenen
Roberto Corizzo
Vitor Santos Costa
Boris Cule
Tomaz Curk
James Cussens
Alfredo Cuzzocrea
Claudia d'Amato
Maria Damiani
Stepanova Daria
Jesse Davis
Tijl De Bie
Martine De Cock
Juan del Coz
Anne Denton
Christian Desrosiers
Nicola Di Mauro
Claudia Diamantini
Tom Diethe
Ivica Dimitrovski
Ying Ding
Stephan Doerfel
Carlotta Domeniconi
Frank Dondelinger
Madalina Drugan
Wouter Duivesteijn
Robert Durrant
Ines Dutra
Dora Erdos
Floriana Esposito
Nicola Fanizzi
Fabio Fassetti
Ad Feelders
Stefano Ferilli

Carlos Ferreira
Cesar Ferri
Maurizio Filippone
Asja Fischer
Peter Flach
Eibe Frank
Elisa Fromont
Fabio Fumarola
Esther Galbrun
Patrick Gallinari
Dragan Gamberger
Byron Gao
Paolo Garza
Eric Gaussier
Rainer Gemulla
Konstantinos Georgatzis
Pierre Geurts
Dorota Glowacka
Michael Granitzer
Caglar Gulcehre
Francesco Gullo
Stephan Gunnemann
Tias Guns
Sara Hajian
Maria Halkidi
Jiawei Han
Xiao He
Denis Helic
Daniel Hernandez
Jose Hernandez-Orallo
Thanh Lam Hoang
Arjen Hommersom
Frank Höppner
Tamas Horvath
Andreas Hotho
Yuanhua Huang
Eyke Huellermeier
Dino Ienco
Georgiana Ifrim

Bhattacharya Indrajit
Szymon Jaroszewicz
Klema Jiri
Giuseppe Jurman
Toshihiro Kamishima
Bo Kang
U Kang
Andreas Karwath
George Karypis
Samuel Kaski
Mehdi Kaytoue
Latifur Khan
Frank Klawonn
Dragi Kocev
Levente Kocsis
Yun Sing Koh
Alek Kolcz
Irena Koprinska
Frederic Koriche
Walter Kosters
Lars Kotthoff
Danai Koutra
Georg Krempl
Tomas Krilavicius
Matjaz Kukar
Meelis Kull
Jorma Laaksonen
Nicolas Lachiche
Leo Lahti
Helge Langseth
Thomas Lansdall-Welfare
Christine Largeron
Pedro Larranaga
Silvio Lattanzi
Niklas Lavesson
Nada Lavrac
Florian Lemmerich
Jiuyong Li
Limin Li
Jefrey Lijffijt
Tony Lindgren
Corrado Loglisci
Peter Lucas
Gjorgji Madjarov
Donato Malerba

Giuseppe Manco
Elio Masciari
Andres Masegosa
Wannes Meert
Rosa Meo
Pauli Miettinen
Mehdi Mirza
Dunja Mladenic
Karthika Mohan
Anna Monreale
Joao Moreira
Mohamed Nadif
Ndapa Nakashole
Mirco Nanni
Amedeo Napoli
Sriraam Natarajan
Benjamin Nguyen
Thomas Nielsen
Xia Ning
Kjetil Norvag
Eirini Ntoutsi
Andreas Nuernberger
Francesco Orsini
Nikunj Oza
Pance Panov
Apostolos Papadopoulos
Evangelos Papalexakis
Panagiotis Papapetrou
Ioannis Partalas
Gabriella Pasi
Andrea Passerini
Dino Pedreschi
Jaakko Peltonen
Jing Peng
Ruggero Pensa
Nico Piatkowksi
Andrea Pietracaprina
Gianvito Pio
Susanna Pirttikangas
Marc Plantevit
Pascal Poncelet
Miguel Prada
L. A. Prashanth
Philippe Preux
Kai Puolamaki

Buyue Qian
Chedy Raissi
Jan Ramon
Huzefa Rangwala
Zbigniew Ras
Chotirat Ratanamahatana
Jan Rauch
Chiara Renso
Achim Rettinger
Fabrizio Riguzzi
Matteo Riondato
Celine Robardet
Pedro Rodrigues
Juan Rodriguez
Fabrice Rossi
Celine Rouveirol
Stefan Rueping
Salvatore Ruggieri
Yvan Saeys
Alan Said
Lorenza Saitta
Ansaf Salleb-Aouissi
Claudio Sartori
Christoph Schommer
Matthias Schubert
Konstantinos Sechidis
Giovanni Semeraro
Sohan Seth
Vinay Setty
Junming Shao
Nikola Simidjievski
Sameer Singh
Andrzej Skowron
Dominik Slezak
Kevin Small
Tomislav Smuc
Yangqiu Song
Mauro Sozio
Papadimitriou Spiros
Jerzy Stefanowski
Gerd Stumme
Mahito Sugiyama
Mika Sulkava
Stephen Swift
Sandor Szedmak

Andrea Tagarelli
Domenico Talia
Letizia Tanca
Jovan Tanevski
Nikolaj Tatti
Maguelonne Teisseire
Aika Terada
Georgios Theocharous
Hannu Toivonen
Roberto Trasarti
Volker Tresp
Isaac Triguero
Panagiotis Tsaparas
Karl Tuyls
Niall Twomey
Nikolaos Tziortziotis
Theodoros Tzouramanis

Antti Ukkonen
Jan Van Haaren
Martijn Van Otterlo
Iraklis Varlamis
Julien Velcin
Shankar Vembu
Deepak Venugopal
Vassilios Verykios
Ricardo Vigário
Herna Viktor
Christel Vrain
Jilles Vreeken
Willem Waegeman
Jianyong Wang
Ding Wei
Cheng Weiwei
Zheng Wen

Joerg Wicker
Marco Wiering
Pengtao Xie
Makoto Yamada
Philip Yu
Bianca Zadrozny
Gerson Zaverucha
Filip Zelezny
Bernard Zenko
Junping Zhang
Min-Ling Zhang
Shichao Zhang
Ying Zhao
Mingjun Zhong
Arthur Zimek
Albrecht Zimmermann
Marinka Zitnik

Additional Reviewers

Anes Bendimerad
Asim Karim
Antonis Matakos
Anderson Nascimento
Andrea Pagliarani
Ricardo Ñanculef
Caio Corro
Emanuele Frandi
Emanuele Rabosio
Fábio Pinto
Fernando Martinez-Pluned
Francesca Pratesi
Fangbo Tao

Kata Gabor
Golnoosh Farnadi
Tom Hanika
Heri Ramampiaro
Vasileios Iosifidis
Jeremiah Deng
Johannes Jurgovsky
Kemilly Dearo
Mark Kibanov
Konstantin Ziegler
Ling Luo
Liyuan Liu
Lorenzo Gabrielli

Marcos de Paula
Martin Ringsquandl
Letizia Milli
Christian Poelitz
Riccardo Guidotti
Andreas Schmidt
Shi Zhi
Yujia Shen
Thomas Niebler
Albin Zehe

Sponsors

Gold Sponsors

Deutsche Post DHL Group	http://www.dpdhl.com/
Google	https://research.google.com/

Silver Sponsors

AGT	http://www.agtinternational.com/
ASML	https://www.workingatasml.com/
Deloitte	https://www2.deloitte.com/global/en.html
NEC Europe Ltd.	http://www.neclab.eu/
Siemens	https://www.siemens.com/

Bronze Sponsors

Cambridge University Press	http://www.cambridge.org/wm-ecommerce-web/academic/landingPage/KDD17
IEEE/CAA Journal of Automatica Sinica	http://www.ieee-jas.org/

Awards Sponsors

Machine Learning	http://link.springer.com/journal/10994
Data Mining and Knowledge Discovery	http://link.springer.com/journal/10618
Deloitte	http://www2.deloitte.com/

Lanyards Sponsor

KNIME	http://www.knime.org/

Publishing Partner and Sponsor

Springer	http://www.springer.com/gp/

PhD Forum Sponsor

IBM Research	http://researchweb.watson.ibm.com/

Invited Talk Sponsors

EurAi	https://www.eurai.org/
GrabIT	https://www.grabit.mk/

Invited Talks Abstracts

Towards End-to-End Learning and Optimization

Frank Hutter

University of Freiburg

Abstract. Deep learning has recently helped AI systems to achieve human-level performance in several domains, including speech recognition, object classification, and playing several types of games. The major benefit of deep learning is that it enables end-to-end learning of representations of the data on several levels of abstraction. However, the overall network architecture and the learning algorithms' sensitive hyperparameters still need to be set manually by human experts. In this talk, I will discuss extensions of Bayesian optimization for handling this problem effectively, thereby paving the way to fully automated end-to-end learning. I will focus on speeding up Bayesian optimization by reasoning over data subsets and initial learning curves, sometimes resulting in 100-fold speedups in finding good hyperparameter settings. I will also show competition-winning practical systems for automated machine learning (AutoML) and briefly show related applications to the end-to-end optimization of algorithms for solving hard combinatorial problems.

Bio. Frank Hutter is an Emmy Noether Research Group Lead (eq. Asst. Prof.) at the Computer Science Department of the University of Freiburg (Germany). He received his PhD from the University of British Columbia (2009). Frank's main research interests span artificial intelligence, machine learning, combinatorial optimization, and automated algorithm design. He received a doctoral dissertation award from the Canadian Artificial Intelligence Association and, with his coauthors, several best paper awards (including from JAIR and IJCAI) and prizes in international competitions on machine learning, SAT solving, and AI planning. In 2016 he received an ERC Starting Grant for a project on automating deep learning based on Bayesian optimization, Bayesian neural networks, and deep reinforcement learning.

Frontiers in Recurrent Neural Network Research

Alex Graves

Google DeepMind, UK

Abstract. In the last few years, recurrent neural networks (RNNs) have become the Swiss army knife of sequence processing for machine learning. Problems involving long and complex data streams, such as speech recognition, machine translation and reinforcement learning from raw video are now routinely tackled with RNNs. However, significant limitations still exist for such systems, such as their ability to retain large amounts of information in memory, and the challenges of gradient-based training on very long sequences. My talk will review some of the new architectures and training strategies currently being developed to extend the frontiers of this exciting field.

Bio. Alex Graves completed a BSc in Theoretical Physics at the University of Edinburgh, Part III Maths at the University of Cambridge, a PhD in artificial intelligence at IDSIA with Jürgen Schmidhuber, followed by postdocs at the Technical University of Munich and with Geoff Hinton at the University of Toronto. He is now a research scientist at DeepMind. His contributions include the Connectionist Temporal Classification algorithm for sequence labelling (now widely used for commercial speech and handwriting recognition), stochastic gradient variational inference, and the Neural Turing Machine/Differentiable Neural Computer architectures.

Using Networks to Link Genotype to Phenotype

John Quackenbush

Dana-Farber Cancer Institute, Harvard TH Chan School of Public Health

Abstract. We know that genotype influences phenotype, but aside from a few highly penetrant Mendelian disorders, the link between genotype and phenotype is not well understood. We have used gene expression and genetic data to explore gene regulatory networks, to study phenotypic state transitions, and to analyze the connections between genotype and phenotype. I will describe how networks and their structure provide unique insight into how small effect variants influence phenotype.

Bio. John Quackenbush received his PhD in theoretical particle physics from UCLA in 1990. Following a postdoctoral fellowship in experimental high-energy physics, he received an NIH research award to work on the Human Genome Project and helped map chromosome 11 and sequence chromosomes 21 and 4. After four years of working in genomics and computational biology at the Salk Institute and then at Stanford University, John joined The Institute for Genomic Research (TIGR), pioneering microarray expression technologies and analytical methods. In 2005, he joined the Dana-Farber Cancer Institute and the Harvard T. H. Chan School of Public Health, where he uses computational and systems biology methods to explore the complexities of human disease, including cancer. In 2011, he cofounded Genospace, a precision-medicine software company acquired by Hospital Corporation of America in 2017. John's many awards include recognition as a White House Champion of Change for making genomic data useful and widely accessible.

Multi-target Prediction via Low-Rank Embeddings

Inderjit Dhillon

University of Texas at Austin

Abstract. Linear prediction methods, such as linear regression and classification, form the bread-and-butter of modern machine learning. The classical scenario is the presence of data with multiple features and a single target variable. However, there are many recent scenarios where there are multiple target variables. For example, recommender systems, predicting bid words for a web page (where each bid word acts as a target variable), or predicting diseases linked to a gene. In many of these scenarios, the target variables might themselves be associated with features. In these scenarios, bilinear and nonlinear prediction via low-rank embeddings have been shown to be extremely powerful. The low-rank embeddings serve a dual purpose: (i) they enable tractable computation even in the face of millions of data points as well as target variables, and (ii) they exploit correlations among the target variables, even when there are many missing observations. We illustrate our methodology on various modern machine learning problems: recommender systems, multi-label learning and inductive matrix completion, and present results on some standard benchmarks as well as an application that involves prediction of gene-disease associations.

Bio. Inderjit Dhillon is the Gottesman Family Centennial Professor of Computer Science and Mathematics at UT Austin, where he is also the Director of the ICES Center for Big Data Analytics. Currently he is on leave from UT Austin and works as Amazon Fellow at A9/Amazon, where he is developing and deploying state-of-the-art machine learning methods for Amazon search. His main research interests are in big data, machine learning, network analysis, linear algebra and optimization. He received his B. Tech. degree from IIT Bombay, and Ph.D. from UC Berkeley. Inderjit has received several awards, including the ICES Distinguished Research Award, the SIAM Outstanding Paper Prize, the Moncrief Grand Challenge Award, the SIAM Linear Algebra Prize, the University Research Excellence Award, and the NSF Career Award. He has published over 175 journal and conference papers, and has served on the Editorial Board of the Journal of Machine Learning Research, the IEEE Transactions of Pattern Analysis and Machine Intelligence, Foundations and Trends in Machine Learning and the SIAM Journal for Matrix Analysis and Applications. Inderjit is an ACM Fellow, an IEEE Fellow, a SIAM Fellow and an AAAS Fellow.

Enabling a Smarter Planet with Earth Observation

Pierre-Philippe Mathieu

ESA/ESRIN, EO Science, Applications and New Technologies

Abstract. Nowadays, teams of researchers around the world can easily access a wide range of open data across disciplines and remotely process them on the Cloud, combining them with their own data to generate knowledge, develop information products for societal applications, and tackle complex integrative complex problems that could not be addressed a few years ago. Such rapid exchange of digital data is fostering a new world of data-intensive research, characterized by openness, transparency, and scrutiny and traceability of results, access to large volume of complex data, availability of community open tools, unprecedented level of computing power, and new collaboration among researchers and new actors such as citizen scientists. The EO scientific community is now facing the challenge of responding to this new paradigm in science 2.0 in order to make the most of the large volume of complex and diverse data delivered by the new generation of EO missions, and in particular the Sentinels. In this context, ESA is supporting a variety of activities in partnership with research communities to ease the transition and make the most of the data. These include the generation of new open tools and exploitation platforms, exploring new ways to disseminate data, building new partnership with citizen scientists, and training the new generation of data scientists. The talk will give a brief overview of some of ESA activities aiming to facilitate the exploitation of large amounts of data from EO missions in a collaborative, cross-disciplinary, and open way, for uses ranging from science to applications and education.

Bio. Pierre-Philippe Mathieu is an Earth Observation Data Scientist at the European Space Agency in ESRIN (Frascati, Italy). He has spent 20+ years working in the field of environmental and ocean modelling, weather risk management and remote sensing. He has a degree in mechanical engineering and an M.Sc. from the University of Liege (Belgium), a Ph.D. in oceanography from the University of Louvain (Belgium), and a Management degree from the University of Reading Business School (UK).

Automatic Understanding of the Visual World

Cordelia Schmid

Inria

Abstract. One of the central problems of artificial intelligence is machine perception, i.e., the ability to understand the visual world based on input from sensors, such as cameras. Computer vision is the area which analyzes visual input. In this talk, I will present recent progress in visual understanding. It is for the most part due to the design of robust visual representations and learned models capturing the variability of the visual world based on state-of-the-art machine learning techniques, including convolutional neural networks. Progress has resulted in technology for a variety of applications. I will present in particular results for human action recognition.

Bio. Cordelia Schmid holds an M.S. degree in Computer Science from the University of Karlsruhe and a Doctorate, also in Computer Science, from the Institut National Polytechnique de Grenoble (INPG). Her doctoral thesis received the best thesis award from INPG in 1996. Dr. Schmid was a postdoctoral research assistant in the Robotics Research Group of Oxford University in 1996–1997. Since 1997 she has held a permanent research position at INRIA Grenoble Rhone-Alpes, where she is a research director and directs an INRIA team. Dr. Schmid has been an Associate Editor for IEEE PAMI (2001–2005) and for IJCV (2004–2012), editor-in-chief for IJCV (2013—), a program chair of IEEE CVPR 2005 and ECCV 2012 as well as a general chair of IEEE CVPR 2015 and ECCV 2020. In 2006, 2014 and 2016, she was awarded the Longuet-Higgins prize for fundamental contributions in computer vision that have withstood the test of time. She is a fellow of IEEE. She was awarded an ERC advanced grant in 2013, the Humboldt research award in 2015 and the Inria & French Academy of Science Grand Prix in 2016. She was elected to the German National Academy of Sciences, Leopoldina, in 2017.

Abstracts of Journal
Track Articles

A Constrained l1 Minimization Approach for Estimating Multiple Sparse Gaussian or Nonparanormal Graphical Models

Beilun Wang, Ritambhara Singh, Yanjun Qi
Machine Learning
http://doi.org/10.1007/s10994-017-5635-7

Identifying context-specific entity networks from aggregated data is an important task, arising often in bioinformatics and neuroimaging applications. Computationally, this task can be formulated as jointly estimating multiple different, but related, sparse undirected graphical models (UGM) from aggregated samples across several contexts. Previous joint-UGM studies have mostly focused on sparse Gaussian graphical models (sGGMs) and can't identify context-specific edge patterns directly. We, therefore, propose a novel approach, SIMULE (detecting Shared and Individual parts of MULtiple graphs Explicitly) to learn multi-UGM via a constrained l1 minimization. SIMULE automatically infers both specific edge patterns that are unique to each context and shared interactions preserved among all the contexts. Through the l1 constrained formulation, this problem is cast as multiple independent subtasks of linear programming that can be solved efficiently in parallel. In addition to Gaussian data, SIMULE can also handle multivariate Nonparanormal data that greatly relaxes the normality assumption that many real-world applications do not follow. We provide a novel theoretical proof showing that SIMULE achieves a consistent result at the rate O $(\log(Kp)/n_{tot})$. On multiple synthetic datasets and two biomedical datasets, SIMULE shows significant improvement over state-of-the-art multi-sGGM and single-UGM baselines (SIMULE implementation and the used datasets @https://github.com/QData/SIMULE).

Adaptive Random Forests for Evolving Data Stream Classification

Heitor M. Gomes, Albert Bifet, Jesse Read, Jean Paul Barddal, Fabrício Enembreck, Bernhard Pfharinger, Geoff Holmes, Talel Abdessalem
Machine Learning
http://doi.org/10.1007/s10994-017-5642-8

Random forests is currently one of the most used machine learning algorithms in the non-streaming (batch) setting. This preference is attributable to its high learning performance and low demands with respect to input preparation and hyper-parameter tuning. However, in the challenging context of evolving data streams, there is no random forests algorithm that can be considered state-of-the-art in comparison to bagging and boosting based algorithms. In this work, we present the adaptive random forest (ARF) algorithm for classification of evolving data streams. In contrast to previous attempts of replicating random forests for data stream learning, ARF includes an effective resampling method and adaptive operators that can cope with different types of concept drifts without complex optimizations for different data sets. We present experiments with a parallel implementation of ARF which has no degradation in terms of classification performance in comparison to a serial implementation, since trees and adaptive operators are independent from one another. Finally, we compare ARF with state-of-the-art algorithms in a traditional test-then-train evaluation and a novel delayed labelling evaluation, and show that ARF is accurate and uses a feasible amount of resources.

An Expressive Dissimilarity Measure for Relational Clustering Using Neighbourhood Trees

Sebastijan Dumančić, Hendrik Blockeel
Machine Learning
http://doi.org/10.1007/s10994-017-5644-6

Clustering is an underspecified task: there are no universal criteria for what makes a good clustering. This is especially true for relational data, where similarity can be based on the features of individuals, the relationships between them, or a mix of both. Existing methods for relational clustering have strong and often implicit biases in this respect. In this paper, we introduce a novel dissimilarity measure for relational data. It is the first approach to incorporate a wide variety of types of similarity, including similarity of attributes, similarity of relational context, and proximity in a hypergraph. We experimentally evaluate the proposed dissimilarity measure on both clustering and classification tasks using data sets of very different types. Considering the quality of the obtained clustering, the experiments demonstrate that (a) using this dissimilarity in standard clustering methods consistently gives good results, whereas other measures work well only on data sets that match their bias; and (b) on most data sets, the novel dissimilarity outperforms even the best among the existing ones. On the classification tasks, the proposed method outperforms the competitors on the majority of data sets, often by a large margin. Moreover, we show that learning the appropriate bias in an unsupervised way is a very challenging task, and that the existing methods offer a marginal gain compared to the proposed similarity method, and can even hurt performance. Finally, we show that the asymptotic complexity of the proposed dissimilarity measure is similar to the existing state-of-the-art approaches. The results confirm that the proposed dissimilarity measure is indeed versatile enough to capture relevant information, regardless of whether that comes from the attributes of vertices, their proximity, or connectedness of vertices, even without parameter tuning.

Constraint-Based Clustering Selection

Toon Van Craenendonck, Hendrik Blockeel
Machine Learning
http://doi.org/10.1007/s10994-017-5643-7

Clustering requires the user to define a distance metric, select a clustering algorithm, and set the hyperparameters of that algorithm. Getting these right, so that a clustering is obtained that meets the users subjective criteria, can be difficult and tedious. Semi-supervised clustering methods make this easier by letting the user provide must-link or cannot-link constraints. These are then used to automatically tune the similarity measure and/or the optimization criterion. In this paper, we investigate a complementary way of using the constraints: they are used to select an unsupervised clustering method and tune its hyperparameters. It turns out that this very simple approach outperforms all existing semi-supervised methods. This implies that choosing the right algorithm and hyperparameter values is more important than modifying an individual algorithm to take constraints into account. In addition, the proposed approach allows for active constraint selection in a more effective manner than other methods.

Cost-Sensitive Label Embedding for Multi-label Classification

Kuan-Hao Huang, Hsuan-Tien Lin
Machine Learning
http://doi.org/10.1007/s10994-017-5659-z

Label embedding (LE) is an important family of multi-label classification algorithms that digest the label information jointly for better performance. Different real-world applications evaluate performance by different cost functions of interest. Current LE algorithms often aim to optimize one specific cost function, but they can suffer from bad performance with respect to other cost functions. In this paper, we resolve the performance issue by proposing a novel cost-sensitive LE algorithm that takes the cost function of interest into account. The proposed algorithm, cost-sensitive label embedding with multidimensional scaling (CLEMS), approximates the cost information with the distances of the embedded vectors by using the classic multidimensional scaling approach for manifold learning. CLEMS is able to deal with both symmetric and asymmetric cost functions, and effectively makes cost-sensitive decisions by nearest-neighbor decoding within the embedded vectors. We derive theoretical results that justify how CLEMS achieves the desired cost-sensitivity. Furthermore, extensive experimental results demonstrate that CLEMS is significantly better than a wide spectrum of existing LE algorithms and state-of-the-art cost-sensitive algorithms across different cost functions.

Efficient Parameter Learning of Bayesian Network Classifiers

Nayyar A. Zaidi, Geoffrey I. Webb, Mark J. Carman,
François Petitjean, Wray Buntine, Mike Hynes, Hans De Sterck
Machine Learning
http://doi.org/10.1007/s10994-016-5619-z

Recent advances have demonstrated substantial benefits from learning with both generative and discriminative parameters. On the one hand, generative approaches address the estimation of the parameters of the joint distribution—P(y, x), which for most network types is very computationally efficient (a notable exception to this are Markov networks) and on the other hand, discriminative approaches address the estimation of the parameters of the posterior distribution—and, are more effective for classification, since they fit P(y—x) directly. However, discriminative approaches are less computationally efficient as the normalization factor in the conditional log-likelihood precludes the derivation of closed-form estimation of parameters. This paper introduces a new discriminative parameter learning method for Bayesian network classifiers that combines in an elegant fashion parameters learned using both generative and discriminative methods. The proposed method is discriminative in nature, but uses estimates of generative probabilities to speed-up the optimization process. A second contribution is to propose a simple framework to characterize the parameter learning task for Bayesian network classifiers. We conduct an extensive set of experiments on 72 standard datasets and demonstrate that our proposed discriminative parameterization provides an efficient alternative to other state-of-the-art parameterizations.

Gaussian Conditional Random Fields Extended for Directed Graphs

Tijana Vujicic, Jesse Glass, Fang Zhou, Zoran Obradovic
Machine Learning
http://doi.org/10.1007/s10994-016-5611-7

For many real-world applications, structured regression is commonly used for predicting output variables that have some internal structure. Gaussian conditional random fields (GCRF) are a widely used type of structured regression model that incorporates the outputs of unstructured predictors and the correlation between objects in order to achieve higher accuracy. However, applications of this model are limited to objects that are symmetrically correlated, while interaction between objects is asymmetric in many cases. In this work we propose a new model, called Directed Gaussian conditional random fields (DirGCRF), which extends GCRF to allow modeling asymmetric relationships (e.g. friendship, influence, love, solidarity, etc.). The DirGCRF models the response variable as a function of both the outputs of unstructured predictors and the asymmetric structure. The effectiveness of the proposed model is characterized on six types of synthetic datasets and four real-world applications where DirGCRF was consistently more accurate than the standard GCRF model and baseline unstructured models.

Generalized Exploration in Policy Search

Herke van Hoof, Daniel Tanneberg, Jan Peters
Machine Learning
http://doi.org/10.1007/s10994-017-5657-1

To learn control policies in unknown environments, learning agents need to explore by trying actions deemed suboptimal. In prior work, such exploration is performed by either perturbing the actions at each time-step independently, or by perturbing policy parameters over an entire episode. Since both of these strategies have certain advantages, a more balanced trade-off could be beneficial. We introduce a unifying view on step-based and episode-based exploration that allows for such balanced trade-offs. This trade-off strategy can be used with various reinforcement learning algorithms. In this paper, we study this generalized exploration strategy in a policy gradient method and in relative entropy policy search. We evaluate the exploration strategy on four dynamical systems and compare the results to the established step-based and episode-based exploration strategies. Our results show that a more balanced trade-off can yield faster learning and better final policies, and illustrate some of the effects that cause these performance differences.

Graph-Based Predictable Feature Analysis

Björn Weghenkel, Asja Fischer, Laurenz Wiskott
Machine Learning
http://doi.org/10.1007/s10994-017-5632-x

We propose graph-based predictable feature analysis (GPFA), a new method for unsupervised learning of predictable features from high-dimensional time series, where high predictability is understood very generically as low variance in the distribution of the next data point given the previous ones. We show how this measure of predictability can be understood in terms of graph embedding as well as how it relates to the information-theoretic measure of predictive information in special cases. We confirm the effectiveness of GPFA on different datasets, comparing it to three existing algorithms with similar objectives—namely slow feature analysis, forecastable component analysis, and predictable feature analysis—to which GPFA shows very competitive results.

Group Online Adaptive Learning

Alon Zweig, Gal Chechik
Machine Learning
http://doi.org/10.1007/s10994-017-5661-5

Sharing information among multiple learning agents can accelerate learning. It could be particularly useful if learners operate in continuously changing environments, because a learner could benefit from previous experience of another learner to adapt to their new environment. Such group-adaptive learning has numerous applications, from predicting financial time-series, through content recommendation systems, to visual understanding for adaptive autonomous agents. Here we address the problem in the context of online adaptive learning. We formally define the learning settings of Group Online Adaptive Learning and derive an algorithm named Shared Online Adaptive Learning (SOAL) to address it. SOAL avoids explicitly modeling changes or their dynamics, and instead shares information continuously. The key idea is that learners share a common small pool of experts, which they can use in a weighted adaptive way. We define group adaptive regret and prove that SOAL maintains known bounds on the adaptive regret obtained for single adaptive learners. Furthermore, it quickly adapts when learning tasks are related to each other. We demonstrate the benefits of the approach for two domains: vision and text. First, in the visual domain, we study a visual navigation task where a robot learns to navigate based on outdoor video scenes. We show how navigation can improve when knowledge from other robots in related scenes is available. Second, in the text domain, we create a new dataset for the task of assigning submitted papers to relevant editors. This is, inherently, an adaptive learning task due to the dynamic nature of research fields evolving in time. We show how learning to assign editors improves when knowledge from other editors is available. Together, these results demonstrate the benefits for sharing information across learners in concurrently changing environments.

Knowledge Elicitation via Sequential Probabilistic Inference for High-Dimensional Prediction

Pedram Daee, Tomi Peltola, Marta Soare, Samuel Kaski
Machine Learning
http://doi.org/10.1007/s10994-017-5651-7

Prediction in a small-sized sample with a large number of covariates, the "small n, large p" problem, is challenging. This setting is encountered in multiple applications, such as in precision medicine, where obtaining additional data can be extremely costly or even impossible, and extensive research effort has recently been dedicated to finding principled solutions for accurate prediction. However, a valuable source of additional information, domain experts, has not yet been efficiently exploited. We formulate knowledge elicitation generally as a probabilistic inference process, where expert knowledge is sequentially queried to improve predictions. In the specific case of sparse linear regression, where we assume the expert has knowledge about the relevance of the covariates, or of values of the regression coefficients, we propose an algorithm and computational approximation for fast and efficient interaction, which sequentially identifies the most informative features on which to query expert knowledge. Evaluations of the proposed method in experiments with simulated and real users show improved prediction accuracy already with a small effort from the expert.

Learning Constraints in Spreadsheets and Tabular Data

Samuel Kolb, Sergey Paramonov, Tias Guns, Luc De Raedt
Machine Learning
http://doi.org/10.1007/s10994-017-5640-x

Spreadsheets, comma separated value files and other tabular data representations are in wide use today. However, writing, maintaining and identifying good formulas for tabular data and spreadsheets can be time-consuming and error-prone. We investigate the automatic learning of constraints (formulas and relations) in raw tabular data in an unsupervised way. We represent common spreadsheet formulas and relations through predicates and expressions whose arguments must satisfy the inherent properties of the constraint. The challenge is to automatically infer the set of constraints present in the data, without labeled examples or user feedback. We propose a two-stage generate and test method where the first stage uses constraint solving techniques to efficiently reduce the number of candidates, based on the predicate signatures. Our approach takes inspiration from inductive logic programming, constraint learning and constraint satisfaction. We show that we are able to accurately discover constraints in spreadsheets from various sources.

Learning Deep Kernels in the Space of Dot Product Polynomials

Michele Donini, Fabio Aiolli
Machine Learning
http://doi.org/10.1007/s10994-016-5590-8

Recent literature has shown the merits of having deep representations in the context of neural networks. An emerging challenge in kernel learning is the definition of similar deep representations. In this paper, we propose a general methodology to define a hierarchy of base kernels with increasing expressiveness and combine them via multiple kernel learning (MKL) with the aim to generate overall deeper kernels. As a leading example, this methodology is applied to learning the kernel in the space of Dot-Product Polynomials (DPPs), that is a positive combination of homogeneous polynomial kernels (HPKs). We show theoretical properties about the expressiveness of HPKs that make their combination empirically very effective. This can also be seen as learning the coefficients of the Maclaurin expansion of any definite positive dot product kernel thus making our proposed method generally applicable. We empirically show the merits of our approach comparing the effectiveness of the kernel generated by our method against baseline kernels (including homogeneous and non homogeneous polynomials, RBF, etc...) and against another hierarchical approach on several benchmark datasets.

Offline Reinforcement Learning with Task Hierarchies

Devin Schwab, Soumya Ray
Machine Learning
http://doi.org/10.1007/s10994-017-5650-8

In this work, we build upon the observation that offline reinforcement learning (RL) is synergistic with task hierarchies that decompose large Markov decision processes (MDPs). Task hierarchies can allow more efficient sample collection from large MDPs, while offline algorithms can learn better policies than the so-called "recursively optimal" or even hierarchically optimal policies learned by standard hierarchical RL algorithms. To enable this synergy, we study sample collection strategies for offline RL that are consistent with a provided task hierarchy while still providing good exploration of the state-action space. We show that naïve extensions of uniform random sampling do not work well in this case and design a strategy that has provably good convergence properties. We also augment the initial set of samples using additional information from the task hierarchy, such as state abstraction. We use the augmented set of samples to learn a policy offline. Given a capable offline RL algorithm, this policy is then guaranteed to have a value greater than or equal to the value of the hierarchically optimal policy. We evaluate our approach on several domains and show that samples generated using a task hierarchy with a suitable strategy allow significantly more sample-efficient convergence than standard offline RL. Further, our approach also shows more sample-efficient convergence to policies with value greater than or equal to hierarchically optimal policies found through an online hierarchical RL approach.

Preserving Differential Privacy in Convolutional Deep Belief Networks

NhatHai Phan, Xintao Wu, Dejing Dou
Machine Learning
http://doi.org/10.1007/s10994-017-5656-2

The remarkable development of deep learning in medicine and healthcare domain presents obvious privacy issues, when deep neural networks are built on users' personal and highly sensitive data, e.g., clinical records, user profiles, biomedical images, etc. However, only a few scientific studies on preserving privacy in deep learning have been conducted. In this paper, we focus on developing a private convolutional deep belief network (pCDBN), which essentially is a convolutional deep belief network (CDBN) under differential privacy. Our main idea of enforcing ε-differential privacy is to leverage the functional mechanism to perturb the energy-based objective functions of traditional CDBNs, rather than their results. One key contribution of this work is that we propose the use of Chebyshev expansion to derive the approximate polynomial representation of objective functions. Our theoretical analysis shows that we can further derive the sensitivity and error bounds of the approximate polynomial representation. As a result, preserving differential privacy in CDBNs is feasible. We applied our model in a health social network, i.e., YesiWell data, and in a handwriting digit dataset, i.e., MNIST data, for human behavior prediction, human behavior classification, and handwriting digit recognition tasks. Theoretical analysis and rigorous experimental evaluations show that the pCDBN is highly effective. It significantly outperforms existing solutions.

Robust Regression Using Biased Objectives

Matthew J. Holland, Kazushi Ikeda
Machine Learning
http://doi.org/10.1007/s10994-017-5653-5

For the regression task in a non-parametric setting, designing the objective function to be minimized by the learner is a critical task. In this paper we propose a principled method for constructing and minimizing robust losses, which are resilient to errant observations even under small samples. Existing proposals typically utilize very strong estimates of the true risk, but in doing so require a priori information that is not available in practice. As we abandon direct approximation of the risk, this lets us enjoy substantial gains in stability at a tolerable price in terms of bias, all while circumventing the computational issues of existing procedures. We analyze existence and convergence conditions, provide practical computational routines, and also show empirically that the proposed method realizes superior robustness over wide data classes with no prior knowledge assumptions.

Sparse Probit Linear Mixed Model

Stephan Mandt, Florian Wenzel, Shinichi Nakajima,
John Cunningham, Christoph Lippert, Marius Kloft
Machine Learning
http://doi.org/10.1007/s10994-017-5652-6

Linear mixed models (LMMs) are important tools in statistical genetics. When used for feature selection, they allow to find a sparse set of genetic traits that best predict a continuous phenotype of interest, while simultaneously correcting for various confounding factors such as age, ethnicity and population structure. Formulated as models for linear regression, LMMs have been restricted to continuous phenotypes. We introduce the sparse probit linear mixed model (Probit-LMM), where we generalize the LMM modeling paradigm to binary phenotypes. As a technical challenge, the model no longer possesses a closed-form likelihood function. In this paper, we present a scalable approximate inference algorithm that lets us fit the model to high-dimensional data sets. We show on three real-world examples from different domains that in the setup of binary labels, our algorithm leads to better prediction accuracies and also selects features which show less correlation with the confounding factors.

Varying-Coefficient Models for Geospatial Transfer Learning

Matthias Bussas, Christoph Sawade, Nicolas Kühn,
Tobias Scheffer, Niels Landwehr
Machine Learning
http://doi.org/10.1007/s10994-017-5639-3

We study prediction problems in which the conditional distribution of the output given the input varies as a function of task variables which, in our applications, represent space and time. In varying-coefficient models, the coefficients of this conditional are allowed to change smoothly in space and time; the strength of the correlations between neighboring points is determined by the data. This is achieved by placing a Gaussian process (GP) prior on the coefficients. Bayesian inference in varying-coefficient models is generally intractable. We show that with an isotropic GP prior, inference in varying-coefficient models resolves to standard inference for a GP that can be solved efficiently. MAP inference in this model resolves to multitask learning using task and instance kernels. We clarify the relationship between varying-coefficient models and the hierarchical Bayesian multitask model and show that inference for hierarchical Bayesian multitask models can be carried out efficiently using graph-Laplacian kernels. We explore the model empirically for the problems of predicting rent and real-estate prices, and predicting the ground motion during seismic events. We find that varying-coefficient models with GP priors excel at predicting rents and real-estate prices. The ground-motion model predicts seismic hazards in the State of California more accurately than the previous state of the art.

Vine Copulas for Mixed Data: Multi-view Clustering for Mixed Data Beyond Meta-Gaussian Dependencies

Lavanya Sita Tekumalla, Vaibhav Rajan, Chiranjib Bhattacharyya
Machine Learning
http://doi.org/10.1007/s10994-016-5624-2

Copulas enable flexible parameterization of multivariate distributions in terms of constituent marginals and dependence families. Vine copulas, hierarchical collections of bivariate copulas, can model a wide variety of dependencies in multivariate data including asymmetric and tail dependencies which the more widely used Gaussian copulas, used in Meta-Gaussian distributions, cannot. However, current inference algorithms for vines cannot fit data with mixed—a combination of continuous, binary and ordinal—features that are common in many domains. We design a new inference algorithm to fit vines on mixed data thereby extending their use to several applications. We illustrate our algorithm by developing a dependency-seeking multi-view clustering model based on Dirichlet Process mixture of vines that generalizes previous models to arbitrary dependencies as well as to mixed marginals. Empirical results on synthetic and real datasets demonstrate the performance on clustering single-view and multi-view data with asymmetric and tail dependencies and with mixed marginals.

Weightless Neural Networks for Open Set Recognition

Douglas O. Cardoso, João Gama, Felipe M. G. França
Machine Learning
http://doi.org/10.1007/s10994-017-5646-4

Open set recognition is a classification-like task.It is accomplished not only by the identification of observations which belong to targeted classes (i.e., the classes among those represented in the training sample which should be later recognized) but also by the rejection of inputs from other classes in the problem domain. The need for proper handling of elements of classes beyond those of interest is frequently ignored, even in works found in the literature. This leads to the improper development of learning systems, which may obtain misleading results when evaluated in their test beds, consequently failing to keep the performance level while facing some real challenge. The adaptation of a classifier for open set recognition is not always possible: the probabilistic premises most of them are built upon are not valid in a open-set setting. Still, this paper details how this was realized for WiSARD a weightless artificial neural network model. Such achievement was based on an elaborate distance-like computation this model provides and the definition of rejection thresholds during training. The proposed methodology was tested through a collection of experiments, with distinct backgrounds and goals. The results obtained confirm the usefulness of this tool for open set recognition.

Classification of High-Dimensional Evolving Data Streams via a Resource-Efficient Online Ensemble

Tingting Zhai, Yang Gao, Hao Wang, Longbing Cao
Data Mining and Knowledge Discovery
http://doi.org/10.1007/s10618-017-0500-7

A novel online ensemble strategy, ensemble BPegasos(EBPegasos), is proposed to solve the problems simultaneously caused by concept drifting and the curse of dimensionality in classifying high-dimensional evolving data streams, which has not been addressed in the literature. First, EBPegasos uses BPegasos, an online kernelized SVM-based algorithm, as the component classifier to address the scalability and sparsity of high-dimensional data. Second, EBPegasos takes full advantage of the characteristics of BPegasos to cope with various types of concept drifts. Specifically, EBPegasos constructs diverse component classifiers by controlling the budget size of BPegasos; it also equips each component with a drift detector to monitor and evaluate its performance, and modifies the ensemble structure only when large performance degradation occurs. Such conditional structural modification strategy makes EBPegasos strike a good balance between exploiting and forgetting old knowledge. Lastly, we first prove experimentally that EBPegasos is more effective and resource-efficient than the tree ensembles on high-dimensional data. Then comprehensive experiments on synthetic and real-life datasets also show that EBPegasos can cope with various types of concept drifts significantly better than the state-of-the-art ensemble frameworks when all ensembles use BPegasos as the base learner.

Differentially Private Nearest Neighbor Classification

Mehmet Emre Gursoy, Ali Inan, Mehmet Ercan Nergiz, Yucel Saygin
Data Mining and Knowledge Discovery
http://doi.org/10.1007/s10618-017-0532-z

Instance-based learning, and the k-nearest neighbors algorithm (k-NN) in particular, provide simple yet effective classification algorithms for data mining. Classifiers are often executed on sensitive information such as medical or personal data. Differential privacy has recently emerged as the accepted standard for privacy protection in sensitive data. However, straightforward applications of differential privacy to k-NN classification yield rather inaccurate results. Motivated by this, we develop algorithms to increase the accuracy of private instance-based classification. We first describe the radius neighbors classifier (r-N) and show that its accuracy under differential privacy can be greatly improved by a non-trivial sensitivity analysis. Then, for k-NN classification, we build algorithms that convert k-NN classifiers to r-N classifiers. We experimentally evaluate the accuracy of both classifiers using various datasets. Experiments show that our proposed classifiers significantly outperform baseline private classifiers (i.e., straightforward applications of differential privacy) and executing the classifiers on a dataset published using differential privacy. In addition, the accuracy of our proposed k-NN classifiers are at least comparable to, and in many cases better than, the other differentially private machine learning techniques.

Ensemble-Based Community Detection in Multilayer Networks

Andrea Tagarelli, Alessia Amelio, Francesco Gullo
Data Mining and Knowledge Discovery
http://doi.org/10.1007/s10618-017-0528-8

The problem of community detection in a multilayer network can effectively be addressed by aggregating the community structures separately generated for each network layer, in order to infer a consensus solution for the input network. To this purpose, clustering ensemble methods developed in the data clustering field are naturally of great support. Bringing these methods into a community detection framework would in principle represent a powerful and versatile approach to reach more stable and reliable community structures. Surprisingly, research on consensus community detection is still in its infancy. In this paper, we propose a novel modularity-driven ensemble-based approach to multilayer community detection. A key aspect is that it finds consensus community structures that not only capture prototypical community memberships of nodes, but also preserve the multilayer topology information and optimize the edge connectivity in the consensus via modularity analysis. Empirical evidence obtained on seven real-world multilayer networks sheds light on the effectiveness and efficiency of our proposed modularity-driven ensemble-based approach, which has shown to outperform state-of-the-art multilayer methods in terms of modularity, silhouette of community memberships, and redundancy assessment criteria, and also in terms of execution times.

Flexible Constrained Sampling with Guarantees for Pattern Mining

Vladimir Dzyuba, Matthijs van Leeuwen, Luc De Raedt
Data Mining and Knowledge Discovery
http://doi.org/10.1007/s10618-017-0501-6

Pattern sampling has been proposed as a potential solution to the infamous pattern explosion. Instead of enumerating all patterns that satisfy the constraints, individual patterns are sampled proportional to a given quality measure. Several sampling algorithms have been proposed, but each of them has its limitations when it comes to (1) flexibility in terms of quality measures and constraints that can be used, and/or (2) guarantees with respect to sampling accuracy. We therefore present Flexics, the first flexible pattern sampler that supports a broad class of quality measures and constraints, while providing strong guarantees regarding sampling accuracy. To achieve this, we leverage the perspective on pattern mining as a constraint satisfaction problem and build upon the latest advances in sampling solutions in SAT as well as existing pattern mining algorithms. Furthermore, the proposed algorithm is applicable to a variety of pattern languages, which allows us to introduce and tackle the novel task of sampling sets of patterns. We introduce and empirically evaluate two variants of Flexics: (1) a generic variant that addresses the well-known itemset sampling task and the novel pattern set sampling task as well as a wide range of expressive constraints within these tasks, and (2) a specialized variant that exploits existing frequent itemset techniques to achieve substantial speed-ups. Experiments show that Flexics is both accurate and efficient, making it a useful tool for pattern-based data exploration.

Identifying Consistent Statements About Numerical Data with Dispersion-Corrected Subgroup Discovery

Mario Boley, Bryan R. Goldsmith, Luca M. Ghiringhelli,
Jilles Vreeken
Data Mining and Knowledge Discovery
http://doi.org/10.1007/s10618-017-0520-3

Existing algorithms for subgroup discovery with numerical targets do not optimize the error or target variable dispersion of the groups they find. This often leads to unreliable or inconsistent statements about the data, rendering practical applications, especially in scientific domains, futile. Therefore, we here extend the optimistic estimator framework for optimal subgroup discovery to a new class of objective functions: we show how tight estimators can be computed efficiently for all functions that are determined by subgroup size (non-decreasing dependence), the subgroup median value, and a dispersion measure around the median (non-increasing dependence). In the important special case when dispersion is measured using the mean absolute deviation from the median, this novel approach yields a linear time algorithm. Empirical evaluation on a wide range of datasets shows that, when used within branch-and-bound search, this approach is highly efficient and indeed discovers subgroups with much smaller errors.

Lagrangian Relaxations for Multiple Network Alignment

Eric Malmi, Sanjay Chawla, Aristides Gionis
Data Mining and Knowledge Discovery
http://doi.org/10.1007/s10618-017-0505-2

We propose a principled approach for the problem of aligning multiple partially overlapping networks. The objective is to map multiple graphs into a single graph while preserving vertex and edge similarities. The problem is inspired by the task of integrating partial views of a family tree (genealogical network) into one unified network, but it also has applications, for example, in social and biological networks. Our approach, called Flan, introduces the idea of generalizing the facility location problem by adding a non-linear term to capture edge similarities and to infer the underlying entity network. The problem is solved using an alternating optimization procedure with a Lagrangian relaxation. Flan has the advantage of being able to leverage prior information on the number of entities, so that when this information is available, Flan is shown to work robustly without the need to use any ground truth data for fine-tuning method parameters. Additionally, we present three multiple-network extensions to an existing state-of-the-art pairwise alignment method called Natalie. Extensive experiments on synthetic, as well as real-world datasets on social networks and genealogical networks, attest to the effectiveness of the proposed approaches which clearly outperform a popular multiple network alignment method called IsoRankN.

Local Community Detection in Multilayer Networks

Roberto Interdonato, Andrea Tagarelli, Dino Ienco,
Arnaud Sallaberry, Pascal Poncelet
Data Mining and Knowledge Discovery
http://doi.org/10.1007/s10618-017-0525-y

The problem of local community detection in graphs refers to the identification of a community that is specific to a query node and relies on limited information about the network structure. Existing approaches for this problem are defined to work in dynamic network scenarios, however they are not designed to deal with complex real-world networks, in which multiple types of connectivity might be considered. In this work, we fill this gap in the literature by introducing the first framework for local community detection in multilayer networks (ML-LCD). We formalize the ML-LCD optimization problem and provide three definitions of the associated objective function, which correspond to different ways to incorporate within-layer and across-layer topological features. We also exploit our framework to generate multilayer global community structures. We conduct an extensive experimentation using seven real-world multilayer networks, which also includes comparison with state-of-the-art methods for single-layer local community detection and for multilayer global community detection. Results show the significance of our proposed methods in discovering local communities over multiple layers, and also highlight their ability in producing global community structures that are better in modularity than those produced by native global community detection approaches.

Measuring and Moderating Opinion Polarization in Social Networks

Antonis Matakos, Evimaria Terzi, Panayiotis Tsaparas
Data Mining and Knowledge Discovery
http://doi.org/10.1007/s10618-017-0527-9

The polarization of society over controversial social issues has been the subject of study in social sciences for decades (Isenberg in J Personal Soc Psychol 50(6):1141–1151, 1986, Sunstein in J Polit Philos 10(2):175–195, 2002). The widespread usage of online social networks and social media, and the tendency of people to connect and interact with like-minded individuals has only intensified the phenomenon of polarization (Bakshy et al. in Science 348(6239):1130–1132, 2015). In this paper, we consider the problem of measuring and reducing polarization of opinions in a social network. Using a standard opinion formation model (Friedkin and Johnsen in J Math Soc 15(3–4):193–206, 1990), we define the polarization index, which, given a network and the opinions of the individuals in the network, it quantifies the polarization observed in the network. Our measure captures the tendency of opinions to concentrate in network communities, creating echo-chambers. Given this numeric measure of polarization, we then consider the problem of reducing polarization in the network by convincing individuals (e.g., through education, exposure to diverse viewpoints, or incentives) to adopt a more neutral stand towards controversial issues. We formally define the ModerateInternal and ModerateExpressed problems, and we prove that both our problems are NP-hard. By exploiting the linear- algebraic characteristics of the opinion formation model we design polynomial-time algorithms for both problems. Our experiments with real-world datasets demonstrate the validity of our metric, and the efficiency and the effectiveness of our algorithms in practice.

Micro-Review Synthesis for Multi-entity Summarization

Thanh-Son Nguyen, Hady W. Lauw, Panayiotis Tsaparas
Data Mining and Knowledge Discovery
http://doi.org/10.1007/s10618-017-0491-4

Location-based social networks (LBSNs), exemplified by Foursquare, are fast gaining popularity. One important feature of LBSNs is micro-review. Upon check-in at a particular venue, a user may leave a short review (up to 200 characters long), also known as a tip. These tips are an important source of information for others to know more about various aspects of an entity (e.g., restaurant), such as food, waiting time, or service. However, a user is often interested not in one particular entity, but rather in several entities collectively, for instance within a neighborhood or a category. In this paper, we address the problem of summarizing the tips of multiple entities in a collection, by way of synthesizing new micro-reviews that pertain to the collection, rather than to the individual entities per se. We formulate this problem in terms of first finding a representation of the collection, by identifying a number of "aspects" that link common threads across two or more entities within the collection. We express these aspects as dense subgraphs in a graph of sentences derived from the multi-entity corpora. This leads to a formulation of maximal multi-entity quasi-cliques, as well as a heuristic algorithm to find K such quasi-cliques maximizing the coverage over the multi-entity corpora. To synthesize a summary tip for each aspect, we select a small number of sentences from the corresponding quasi-clique, balancing conciseness and representativeness in terms of a facility location problem. Our approach performs well on collections of Foursquare entities based on localities and categories, producing more representative and diverse summaries than the baselines.

MixedTrails: Bayesian Hypothesis Comparison on Heterogeneous Sequential Data

Martin Becker, Florian Lemmerich, Philipp Singer, Markus Strohmaier, Andreas Hotho
Data Mining and Knowledge Discovery
http://doi.org/10.1007/s10618-017-0518-x

Sequential traces of user data are frequently observed online and offline, e.g., as sequences of visited websites or as sequences of locations captured by GPS. However, understanding factors explaining the production of sequence data is a challenging task, especially since the data generation is often not homogeneous. For example, navigation behavior might change in different phases of browsing a website or movement behavior may vary between groups of users. In this work, we tackle this task and propose MixedTrails, a Bayesian approach for comparing the plausibility of hypotheses regarding the generative processes of heterogeneous sequence data. Each hypothesis is derived from existing literature, theory, or intuition and represents a belief about transition probabilities between a set of states that can vary between groups of observed transitions. For example, when trying to understand human movement in a city and given some data, a hypothesis assuming tourists to be more likely to move towards points of interests than locals can be shown to be more plausible than a hypothesis assuming the opposite. Our approach incorporates such hypotheses as Bayesian priors in a generative mixed transition Markov chain model, and compares their plausibility utilizing Bayes factors. We discuss analytical and approximate inference methods for calculating the marginal likelihoods for Bayes factors, give guidance on interpreting the results, and illustrate our approach with several experiments on synthetic and empirical data from Wikipedia and Flickr. Thus, this work enables a novel kind of analysis for studying sequential data in many application areas.

On Temporal-Constrained Sub-trajectory Cluster Analysis

Nikos Pelekis, Panagiotis Tampakis, Marios Vodas,
Christos Doulkeridis, Yannis Theodoridis
Data Mining and Knowledge Discovery
http://doi.org/10.1007/s10618-017-0503-4

Cluster analysis over Moving Object Databases (MODs) is a challenging research topic that has attracted the attention of the mobility data mining community. In this paper, we study the temporal-constrained sub-trajectory cluster analysis problem, where the aim is to discover clusters of sub-trajectories given an ad-hoc, user-specified temporal constraint within the dataset's lifetime. The problem is challenging because: (a) the time window is not known in advance, instead it is specified at query time, and (b) the MOD is continuously updated with new trajectories. Existing solutions first filter the trajectory database according to the temporal constraint, and then apply a clustering algorithm from scratch on the filtered data. However, this approach is extremely inefficient, when considering explorative data analysis where multiple clustering tasks need to be performed over different temporal subsets of the database, while the database is updated with new trajectories. To address this problem, we propose an incremental and scalable solution to the problem, which is built upon a novel indexing structure, called Representative Trajectory Tree (ReTraTree). ReTraTree acts as an effective spatio-temporal partitioning technique; partitions in ReTraTree correspond to groupings of sub-trajectories, which are incrementally maintained and assigned to representative (sub-)trajectories. Due to the proposed organization of sub-trajectories, the problem under study can be efficiently solved as simply as executing a query operator on ReTraTree, while insertion of new trajectories is supported. Our extensive experimental study performed on real and synthetic datasets shows that our approach outperforms a state-of-the-art in-DBMS solution supported by PostgreSQL by orders of magnitude.

Social Regularized von Mises–Fisher Mixture Model for Item Recommendation

Aghiles Salah, Mohamed Nadif
Data Mining and Knowledge Discovery
http://doi.org/10.1007/s10618-017-0499-9

Collaborative filtering (CF) is a widely used technique to guide the users of web applications towards items that might interest them. CF approaches are severely challenged by the characteristics of user-item preference matrices, which are often high dimensional and extremely sparse. Recently, several works have shown that incorporating information from social networks—such as friendship and trust relationships—into traditional CF alleviates the sparsity related issues and yields a better recommendation quality, in most cases. More interestingly, even with comparable performances, social-based CF is more beneficial than traditional CF; the former makes it possible to provide recommendations for cold start users. In this paper, we propose a novel model that leverages information from social networks to improve recommendations. While existing social CF models are based on popular modelling assumptions such as Gaussian or Multinomial, our model builds on the von Mises–Fisher assumption which turns out to be more adequate, than the aforementioned assumptions, for high dimensional sparse data. Setting the estimate of the model parameters under the maximum likelihood approach, we derive a scalable learning algorithm for analyzing data with our model. Empirical results on several real-world datasets provide strong support for the advantages of the proposed model.

The Best Privacy Defense Is a Good Privacy Offense: Obfuscating a Search Engine User's Profile

Jörg Wicker, Stefan Kramer
Data Mining and Knowledge Discovery
http://doi.org/10.1007/s10618-017-0524-z

User privacy on the internet is an important and unsolved problem. So far, no sufficient and comprehensive solution has been proposed that helps a user to protect his or her privacy while using the internet. Data are collected and assembled by numerous service providers. Solutions so far focused on the side of the service providers to store encrypted or transformed data that can be still used for analysis. This has a major flaw, as it relies on the service providers to do this. The user has no chance of actively protecting his or her privacy. In this work, we suggest a new approach, empowering the user to take advantage of the same tool the other side has, namely data mining to produce data which obfuscates the user's profile. We apply this approach to search engine queries and use feedback of the search engines in terms of personalized advertisements in an algorithm similar to reinforcement learning to generate new queries potentially confusing the search engine. We evaluated the approach using a real-world data set. While evaluation is hard, we achieve results that indicate that it is possible to influence the user's profile that the search engine generates. This shows that it is feasible to defend a user's privacy from a new and more practical perspective.

Tour Recommendation for Groups

Aris Anagnostopoulos, Reem Atassi, Luca Becchetti,
Adriano Fazzone, Fabrizio Silvestri
Data Mining and Knowledge Discovery
http://doi.org/10.1007/s10618-016-0477-7

Consider a group of people who are visiting a major touristic city, such as NY, Paris, or Rome. It is reasonable to assume that each member of the group has his or her own interests or preferences about places to visit, which in general may differ from those of other members. Still, people almost always want to hang out together and so the following question naturally arises: What is the best tour that the group could perform together in the city? This problem underpins several challenges, ranging from understanding people's expected attitudes towards potential points of interest, to modeling and providing good and viable solutions. Formulating this problem is challenging because of multiple competing objectives. For example, making the entire group as happy as possible in general conflicts with the objective that no member becomes disappointed. In this paper, we address the algorithmic implications of the above problem, by providing various formulations that take into account the overall group as well as the individual satisfaction and the length of the tour. We then study the computational complexity of these formulations, we provide effective and efficient practical algorithms, and, finally, we evaluate them on datasets constructed from real city data.

Contents – Part I

Ensembles and Meta Learning

Feature Selection and Extraction

Kernel Methods

Learning and Optimization

Matrix and Tensor Factorization

Networks and Graphs

Neural Networks and Deep Learning

Contents – Part II

Recommendation

Regression

Transfer and Multi-task Learning

Unsupervised and Semisupervised Learning

Contents – Part III

Demo Track

Anomaly Detection

Concentration Free Outlier Detection

Fabrizio Angiulli[✉]

University of Calabria, 87036 Rende, CS, Italy
fabrizio.angiulli@unical.it
http://siloe.dimes.unical.it/angiulli

Abstract. We present a novel notion of outlier, called Concentration Free Outlier Factor (CFOF), having the peculiarity to resist concentration phenomena that affect other scores when the dimensionality of the feature space increases. Indeed we formally prove that CFOF does not concentrate in intrinsically high-dimensional spaces. Moreover, CFOF is adaptive to different local density levels and it does not require the computation of exact neighbors in order to be reliably computed. We present a very efficient technique, named *fast*-CFOF, for detecting outliers in very large high-dimensional datasets. The technique is efficiently parallelizable, and we provide a MIMD-SIMD implementation. Experimental results witness for scalability and effectiveness of the technique and highlight that CFOF exhibits state of the art detection performances.

Keywords: Outlier detection · Curse of dimensionality

1 Introduction

Outlier detection is a prominent data mining task, whose goal is to single out anomalous observations, also called outliers [2]. While the other data mining approaches consider outliers as noise that must be eliminated, as pointed out in [11] "one person's noise could be another person's signal", thus outliers themselves are of great interest in different settings (e.g. fraud detection, ecosystem disturbances, intrusion detection, cybersecurity, medical analysis, to cite a few).

Data mining outlier approaches can be supervised, semi-supervised, and unsupervised [8,13]. Supervised methods take in input data labeled as normal and abnormal and build a classifier. The challenge there is posed by the fact that abnormal data form a rare class. Semi-supervised methods, also called one-class classifiers or domain description techniques, take in input only normal examples and use them to identify anomalies. Unsupervised methods detect outliers in an input dataset by assigning a score or anomaly degree to each object.

Unsupervised outlier detection methods can be categorized in several approaches, each of which assumes a specific concept of outlier. Among the most popular families there are distance-based [4,5,16,23], density-based [7,15,20], angle-based [18], isolation-forest [19], subspace methods [1,14], and others [2,9,25].

© Springer International Publishing AG 2017
M. Ceci et al. (Eds.): ECML PKDD 2017, Part I, LNAI 10534, pp. 3–19, 2017.
https://doi.org/10.1007/978-3-319-71249-9_1

This work focuses on unsupervised outlier detection problem in the full feature space. In particular, we introduce a novel notion of outlier, the Concentration Free Outlier Factor (CFOF), having the peculiarity to resist concentration phenomena affecting other measures. Informally, the CFOF score measures how many neighbors have to be taken into account in order for the object to be considered close by an appreciable fraction of the population. The term distance concentration refers to the tendency of distances to become almost indiscernible as dimensionality increases, and is part of the so called curse of dimensionality problem [6,10]. And, indeed, the concentration problem also affects outlier scores of different families due to the specific role played by distances in their formulation [17,25]. Moreover, a special kind of concentration phenomenon, known as hubness, concerns scores based on reverse nearest neighbor counts [12,22], that is the concentration of the scores towards the values associated with anomalies, which results in almost all the dataset composed of outliers.

The contributions of the work within this scenario are summarized next:

- As a major peculiarity, we formally show that, differently from the practical totality of existing outlier scores, the CFOF score distribution is not affected by concentration phenomena arising when the dimensionality of the space increases.
- The CFOF score is adaptive to different local density levels. Despite local methods usually require to know the exact nearest neighbors in order to compare the neighborhood of each object with the neighborhood of its neighbors, this is not the case for CFOF, which can be reliably computed through sampling. This characteristics is favored by the separation between inliers and outliers guaranteed by the absence of concentration.
- We describe the *fast*-CFOF technique, which from the computational point of view does not suffer of the problems affecting (reverse) nearest neighbor search techniques. The cost of *fast*-CFOF is linear both in the dataset size and dimensionality. Moreover, we provide a multi-core (MIMD) vectorized (SIMD) implementation.
- The applicability of the technique is not limited to Euclidean or vector spaces. It can be applied both in metric and non-metric spaces equipped with a distance function.
- Experimental results highlight that CFOF exhibits state of the art detection performances.

The rest of the work is organized as follows. Section 2 introduces the CFOF score and its properties. Section 3 describes the *fast*-CFOF algorithm. Section 4 presents experiments. Finally, section draws conclusions.

2 The Concentration Free Outlier Factor

2.1 Definition

We assume that a dataset $\mathbf{DS} = \{x_1, x_2, \ldots, x_n\}$ of n objects belonging to an object space $\mathbb{U}$, on which a distance function dist is defined, is given in input.

We assume that $\mathbb{U} = \mathbb{D}^d$ (where $\mathbb{D}$ is usually the set $\mathbb{R}$ of real numbers), with $d \in \mathbb{N}^+$, but the method can be applied in any object space equipped with a distance function (not necessarily a metric).

Given an object x and a positive integer k, the *k-th nearest neighbor* of x is the object $nn_k(x)$ such that there exists exactly $k-1$ objects lying at distance smaller than $\text{dist}(x, nn_k(x))$ from x. It always holds that $x = nn_1(x)$. We assume that ties are non-deterministically ordered. The *k nearest neighbors set* $\text{NN}_k(x)$ of x, where k is also called the *neighborhood width*, consists of the objects $\{nn_i(x) \mid 1 \leq i \leq k\}$.

By $\text{N}_k(x)$ we denote the number of objects having x among their k nearest neighbors:

$$\text{N}_k(x) = |\{y : x \in \text{NN}_k(y)\}|,$$

also referred to as *reverse k nearest neighbor count* or *reverse neighborhood size*.

Given a parameter $\varrho \in (0,1)$ (or equivalently a parameter $k_\varrho \in [1,n]$ such that $k_\varrho = n\varrho$), the *Concentration Free Outlier Score*, also referred to as CFOF, is defined as:

$$\text{CFOF}(x) = \min\{k/n : \text{N}_k(x) \geq n\varrho\}, \tag{1}$$

that is to say, the score returns the smallest neighborhood width (normalized with respect to n) for which the object x exhibits a reverse neighborhood of size at least $n\varrho$ (or k_ϱ).[1]

Intuitively, the CFOF score measures how many neighbors have to be taken into account in order for the object to be considered close by an appreciable fraction of the dataset objects. We notice that this kind of notion of perceiving the abnormality of an observation is completely different from any other notion so far introduced in the literature.

The CFOF score is adaptive to different density levels. This characteristics is also influenced by the fact that actual distance values are not employed in its computation. Thus, CFOF is invariant to all of the transformations the leave unchanged the nearest neighbor ranking, such as translation or scaling. Also, duplicating the data in a way that avoids to affect the original neighborhood order (e.g. by creating a separate, possibly scaled, cluster from each copy of the original data) will preserve original scores.

Consider Fig. 1 showing a dataset consisting of two normally distributed clusters, each consisting of 250 points. The cluster centered in $(4,4)$ is obtained by translating and scaling (by a factor 0.5) the cluster centered in the origin. The top 25 CFOF outliers for $k_\varrho = 20$ are highlighted (objects within small circles). It can be seen that the outliers are the "same" objects of the two clusters.

---

[1] Notice that k (or k/n), representing a neighborhood width, denotes the output of CFOF, while the other outlier definitions employ k as an input parameter. We warn the reader that, in order to make more intelligible the comparison of CFOF with other outlier techniques, sometimes we will refer to k as an input parameter (the use will be clear from the context). Moreover, in order to avoid confusion and to maintain analogy with the input parameter ϱ, we also refer to ϱ as k_ϱ.

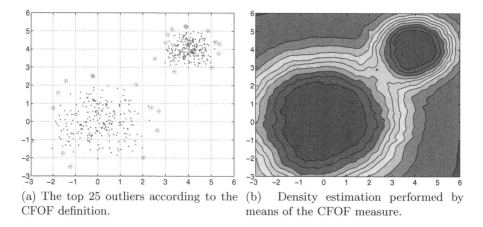

(a) The top 25 outliers according to the CFOF definition.

(b) Density estimation performed by means of the CFOF measure.

Fig. 1. Two normal clusters with different standard deviation.

2.2 Relationship with the Distance Concentration Phenomenon

The term distance concentration, which is part of the so called curse of dimensionality problem [6], refers to the tendency of distances to become almost indiscernible as dimensionality increases. In a more quantitative way this phenomenon is measured through the ratio between a quantity related to the mean μ and a quantity related to the standard deviation σ of the distance distribution of interest. E.g., in [10] the intrinsic dimensionality ρ of a metric space is defined as $\rho = \mu_d^2/(2\sigma_d^2)$, where μ_d is the mean of the pairwise distance distribution and σ_d the associated standard deviation. The intrinsic dimensionality intends to quantify the expected difficulty of performing a nearest neighbor search: the smaller the ratio the larger the difficulty to search on an arbitrary metric space.

In general, it is said that we have concentration when this kind of ratio tends to zero as dimensionality goes to infinity, as it is the case for objects with i.i.d. attributes.

The concentration problem also affects different families of outlier scores, due to the specific role played by distances in their formulation.

Figure 2 reports the sorted scores of different outlier detection techniques, that are aKNN [5], LOF [7], ABOF [18], and CFOF (the parameters k of aKNN, LOF, and ABOF, and k_ϱ of CFOF, are held fixed to 50 for all the scores), associated with a family of uniformly distributed datasets having fixed size ($n = 1000$) and increasing dimensionality $d \in [10^0, 10^4]$. The figure highlights that, except for CFOF, the other scores exhibit a concentration effect. For aKNN (Fig. 2a) the mean score value raises while the spread stay limited. For LOF (Fig. 2b) all the values tend to 1 as the dimensionality increases. For ABOF (Fig. 2c) both the mean and the standard deviation decrease of various orders of magnitude with the latter term varying at a faster rate than the former one. As for CFOF the score distributions for $d > 100$ are very close and exhibit only slight changes. Notably, the separation between scores associated with outliers and inliers is always ample.

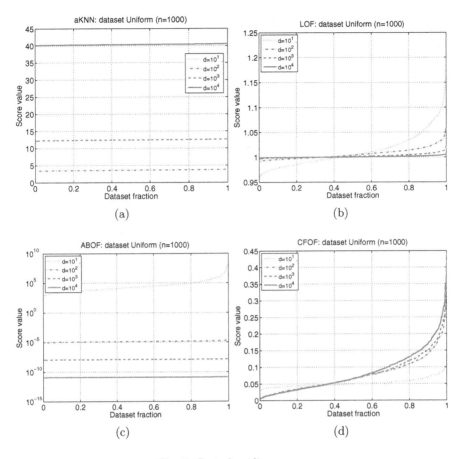

Fig. 2. Sorted outlier scores.

2.3   Relationship with the Hubness Phenomenon

CFOF has connections with the reverse neighborhood size, a tool which has been also used for characterizing outliers. In [12], the authors proposed the use of the reverse neighborhood size $N_k(\cdot)$ as an outlier score, which we refer to as RNN count (RNNc for short). Outliers are those objects associated with the lowest RNN counts. However, RNNc suffers of a peculiar problem known as *hubness* [21]. As the dimensionality of the space increases, the number of *anti-hubs*, that are objects appearing in a much lower number of k nearest neighbors sets (possibly they are neighbors only of themselves), overcomes the number of *hubs*, that are objects that appear in many more k nearest neighbor sets than other points, and, according to the RNNc score, the vast majority of the dataset objects become outliers with identical scores.

We provide evidence that CFOF does not present the hubness problem. Figure 3 reports the distribution of the $N_k(\cdot)$ value and of the CFOF absolute score for a ten thousand dimensional normal dataset (a very similar behavior has

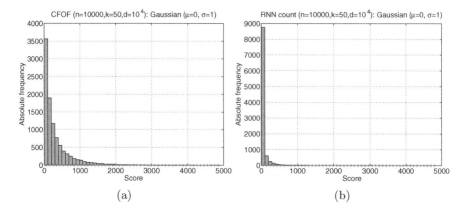

Fig. 3. Distribution of CFOF and RNN counts.

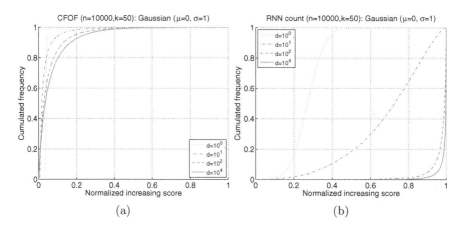

Fig. 4. Comparison between CFOF and RNN counts.

been observed also for uniform data). Notice that CFOF outliers are associated with the largest score values, hence to the tails of the corresponding distribution, while RNNc outliers are associated with the smallest score values, hence with the largely populated region of the associated score distribution, a completely opposite behavior. To illustrate the impact of the hubness problem with the dimensionality, Fig. 4 shows the cumulative frequency associated with the normalized, between 0 and 1, increasing score. This transformation has been implemented here in order to make the comparison much more interpretable. Original scores have been mapped to $[0,1]$. CFOF scores have been divided by their maximum value. The mapping for $N_k(\cdot)$ has been obtained as $1 - \frac{N_k(x)}{\max_y N_k(y)}$, since outliers are those objects associated with the lowest counts. The plots make evident the deep difference between the two approaches. Here both n and k for RNNc (k_ϱ for CFOF, resp.) are held fixed, while d is increased. As for RNNc, the hubness problem is already evident for $d = 10$ (where objects with a normalized score ≥ 0.8 corresponds to about the 40% of the dataset), while the curve for $d = 10^2$ closely resembles that for $d = 10^4$ (where almost all the dataset objects have a normalized score ≥ 0.8). As far as CFOF is concerned, the two curves for $d = 10^4$ closely resemble each other and the number of objects associated with a large score value always correspond to a very small fraction of the dataset population.

2.4   Concentration Free Property of CFOF

In this section we formally prove that the CFOF score is concentration-free. Specifically, the following theorem shows that the separation between the scores associated with outliers and the rest of the scores is guaranteed in any arbitrary large dimensionality.

Before going into the details, we recall that the concept of intrinsic dimensionality of a space is identified as the minimum number of variables needed to represent the data, which corresponds in a linear space to the number of linearly independent vectors needed to describe each point.

Theorem 1. *Let $\mathbf{DS}^{(d)}$ be a d-dimensional dataset consisting of realizations of a d-dimensional independent and (non-necessarily) identically distributed random vector $\mathbf{X}$ having distribution function f. Then, as $d \to \infty$, the CFOF scores of the points of $\mathbf{DS}^{(d)}$ do not concentrate.*

Proof. Consider the squared norm $\|\mathbf{X}\|^2 = \sum_{i=1}^{d} X_i^2$ of the random vector $\mathbf{X}$. As $d \to \infty$, by the Central Limit Theorem, the standard score of $\sum_{i=1}^{d} X_i^2$ tends to a standard normal distribution. This implies that $\|\mathbf{X}\|^2$ approaches a normal distribution with mean $\mu_{\|\mathbf{X}\|^2} = \mathbf{E}[X_i^2] = d\mu_2$ and variance $\sigma_{\|\mathbf{X}\|^2}^2 = d(\mathbf{E}[(X_i^2)^2] - \mathbf{E}[X_i^2]) = d(\mu_4 - \mu_2^2)$, where μ_2 and μ_4 are the 2nd- and 4th-order central moments of the univariate probability distribution f.

In the case that the components X_i of $\mathbf{X}$ are non-identically distributed according to the distributions f_i $(1 \leq i \leq d)$, the result still holds by considering the average of the central moments of the f_i functions.

Let x be an element of $\mathbf{DS}^{(d)}$ and define the zeta score z_x of the squared norm of x as

$$z_x = \frac{\|x\|^2 - \mu_{\|\mathbf{X}\|^2}}{\sigma_{\|\mathbf{X}\|^2}}.$$

It can be shown [3] w.l.o.g. assume that $\mathbf{E}[X] = 0$, for large values of d, the number of k-occurrences of x is given by

$$N_k(x) = n \cdot Pr[x \in \mathrm{NN}_k(\mathbf{X})] \approx n\Phi\left(\frac{\Phi^{-1}(\frac{k}{n})\sqrt{\mu_4 + 3\mu_2^2} - z_x\sqrt{\mu_4 - \mu_2^2}}{2\mu_2}\right).$$

Let $t(z_x)$ denote the smallest integer k such that $N_k(x) \geq n\varrho$. By exploiting the equation above it can be concluded that

$$t(z_x) \approx n\Phi\left(\frac{z_x\sqrt{\mu_4 - \mu_2^2} + 2\mu_2\Phi^{-1}(\varrho)}{\sqrt{\mu_4 + 3\mu_2^2}}\right).$$

Since $\mathrm{CFOF}(x) = k/n$ implies that k is the smallest integer such that $\mathrm{N}_k(x) \geq n\varrho$, it also follows that $\mathrm{CFOF}(x) \approx t(z_x)/n = \hat{t}(z_x)$.

Moreover, since as stated above the $\|\mathbf{X}\|^2$ random variable is normally distributed, it also holds that for each $z \geq 0$

$$Pr\left[\frac{\|\mathbf{X}\|^2 - \mu_{\|\mathbf{X}\|^2}}{\sigma_{\|\mathbf{X}\|^2}} \leq z\right] = \Phi(z),$$

where $\Phi(\cdot)$ denotes the cdf of the normal distribution.

Thus, for arbitrarily large values of d and for any standard score value $z \geq 0$

$$Pr\left[CFOF(\mathbf{X}) \geq \hat{t}(z)\right] = 1 - \Phi(z),$$

irrespective of the actual data dimensionality value d.

3   Score Computation

CFOF scores can be determined in time $O(n^2 d)$, where d denotes the dimensionality of the feature space (or the cost of computing a distance), after computing all pairwise dataset distances.[2] Next we introduce a technique, named *fast*-CFOF which does not require the computation of the exact nearest neighbor sets and, from the computational point of view, does not suffer of the curse of dimensionality affecting nearest neighbor search techniques.

The technique builds on the following probabilistic formulation of the CFOF score. Assume that the dataset consists of n i.i.d. samples drawn according to an unknown probability law $p(\cdot)$. Given a parameter $\varrho \in (0, 1)$, the (*Probabilistic*) *Concentration Free Outlier Factor* CFOF is defined as follows:

$$\mathrm{CFOF}(x) = \min\left\{k/n : \mathbf{E}\left[Pr[x \in \mathrm{NN}_k(y)]\right] \geq \varrho\right\}. \tag{2}$$

To differentiate the two definitions reported in Eqs. (1) and (2), we also refer to the former as *hard*-CFOF and to the latter as *soft*-CFOF. Intuitively, the *soft*-CFOF score measures how many neighbors have to be taken into account in order for the expected number of dataset objects having it among their neighbors correspond to the fraction ϱ of the overall population.

3.1   The *fast*-CFOF Technique

Given a dataset **DS** and two objects x and y from **DS**, the building block of the algorithm is the computation of $Pr[x \in \mathrm{NN}_k(y)]$. Consider the boolean function $B_{x,y}(z)$ defined on instances z of **DS** such that $B_{x,y}(z) = 1$ if z lies within the region $\mathcal{I}_{\mathrm{dist}(x,y)}(y)$, and 0 otherwise. We want to estimate the average value $\overline{B}_{x,y}$ of $B_{x,y}$ in **DS**, which corresponds to the probability $p(x,y)$ that a randomly picked dataset object z is at distance not grater than $\mathrm{dist}(x,y)$ from y.

---

[2] It is generally recognized that this cost can be reduced to $O(dn \log n)$ in low dimensional spaces.

It is enough to compute $\overline{B}_{x,y}$ within a certain error bound. Thus, we resort to *batch sampling*, consisting in picking up s elements of **DS** randomly and estimating $p(x,y) = \overline{B}_{x,y}$ as the fraction $\hat{p}(x,y)$ of the elements of the sample satisfying $B_{x,y}$ [24]. Given $\delta > 0$ (an *error probability*) and $\epsilon, 0 < \epsilon < 1$ (an *absolute error*), if the sample size s satisfies certain conditions [24] then

$$Pr[|\hat{p}(x,y) - p(x,y)| \leq \epsilon] > 1 - \delta. \tag{3}$$

For large values of n, since the variance of the Binomial distribution becomes negligible with respect to the mean, the cdf $binocdf(k;p,n)$ tends to the step function $H(k - np)$, where $H(k) = 0$ for $k < 0$ and $H(k) = 1$ for $k > 0$. Thus, we can approximate the value $Pr[x \in \mathrm{NN}_k(y)] = binocdf(k;p(x,y),n)$ with the boolean function $H(k - k_{up}(x,y))$, with $k_{up}(x,y) = n\hat{p}(x,y)$.[3] It then follows that we can obtain $\mathbf{E}[Pr[x \in \mathrm{NN}_k(y)]]$ as the average value of the boolean function $H(k - n\hat{p}(x,y))$, whose estimate can be again obtained by exploiting batch sampling. Specifically, *fast*-CFOF exploits the one single sample in order to perform the two estimates above described.

The algorithm *fast*-CFOF receives in input a list $\varrho = \varrho_1, \ldots, \varrho_\ell$ of values for the parameter ϱ, since it is able to perform a *multi-resolution analysis*, that is to compute scores associated with different values of the parameter ϱ with no additional cost. Both ϱ and parameters ϵ, δ can be conveniently left at the default value ($\varrho = 0.001, 0.005, 0.01, 0.05, 0.1$ and $\epsilon, \delta = 0.01$; see later for details).

First, the algorithm determines the size $s = \lceil \frac{1}{2\epsilon^2} \log\left(\frac{1}{\delta}\right) \rceil$ of the *sample* (or *partition*) of the dataset needed in order to guarantee the bound reported in Eq. (3). We notice that the algorithm does not require the dataset to be entirely loaded in main memory, since only a partition at a time is needed to carry out the computation. Thus, the technique is suitable also for disk resident datasets. We assume that dataset objects are randomly ordered and, hence, partitions can be contiguous. Otherwise, randomization can be done in linear time and constant space by disk-based shuffling. Each partition, consisting of a group of s consecutive objects, is processed by the subroutine *fast*-CFOF_part (see Algorithm 1), which estimates CFOF scores of the objects within the partition through batch sampling.

The matrix hst, consisting of $s \times B$ counters, is employed by *fast*-CFOF_part. The entry $hst(i,k)$ of hst is used to estimate how many times the sample object x'_i is the kth nearest neighbor of a generic object dataset. Values of k, ranging from 1 to n, are partitioned into B log-spaced bins. The function k_bin maps original k values to the corresponding bin, while k_bin^{-1} implements the reverse mapping (by returning a certain value within the corresponding bin).

For each sample object x'_i, the distance $dst(j)$ from any other sample object x'_j is computed (lines 3–4) and, then, distances are ordered (line 5) obtaining the list ord of sample identifiers such that $dst(ord(1)) \leq dst(ord(2)) \leq \ldots \leq dst(ord(s))$.

---

[3] Alternatively, by exploiting the Normal approximation of the Binomial distribution, a suitable value for $k_{up}(x,y)$ is given by $k_{up}(x,y) = n\hat{p}(x,y) + c\sqrt{n\hat{p}(x,y)(1 - \hat{p}(x,y))}$ with $c \in [0,3]$.

Algorithm 1. *fast*-CFOF_*part*

Input: Dataset sample $\langle x'_1, \ldots, x'_s \rangle$ of size s, parameters $\varrho_1, \ldots, \varrho_\ell \in (0,1)$,
     dataset size n
Output: CFOF scores $\langle sc'_{1,\varrho}, \ldots, sc'_{s,\varrho} \rangle$

1   initialize matrix hst of $s \times B$ elements to 0;
   // Nearest neighbor count estimation
2   **foreach** $i = 1$ *to* s **do**
     | // Distances computation
3 | **foreach** $j = 1$ *to* s **do**
4 | $dst(j) = \mathrm{dist}(x'_i, x'_j)$;
     | // Count update
5 | $ord = sort(dst)$;
6 | **foreach** $j = 1$ *to* s **do**
7 | $p = j/s$;
8 | $k_{up} = \lfloor np + c\sqrt{np(1-p)} + 0.5 \rfloor$;
9 | $k_{pos} = k\_bin(k_{up})$;
10 | $hst(ord(j), k_{pos}) = hst(ord(j), k_{pos}) + 1$;

   // Scores computation
11 **foreach** $i = 1$ *to* s .**do**
12 | $count = 0$;
13 | $k_{pos} = 0$;
14 | $l = 1$;
15 | **while** $l \leq \ell$ **do**
16 | **while** $count < s\varrho_l$ **do**
17 | | $k_{pos} = k_{pos} + 1$;
18 | | $count = count + hst(i, k_{pos})$;
19 | $sc'_{i,\varrho_l} = k\_bin^{-1}(k_{pos})/n$;
20 | $l = l + 1$;

Moreover, for each element $ord(j)$ of ord, the variable p is set to j/s (line 7), representing the probability $p(x'_{ord(j)}, x'_i)$, estimated through the sample, that a randomly picked dataset object is located within the region of radius $dst(ord(j)) = dist(x'_i, x'_{ord(j)})$ centered in x'_i. The value k_{up} (line 8) represents the point of transition from 0 to 1 of the step function $H(k - k_{up})$ employed to approximate the probability $Pr[x'_{ord(j)} \in \mathrm{NN}_k(y)]$ when $y = x'_i$. Thus, before concluding each cycle of the inner loop (lines 6–10), the $k\_bin(k_{up})$-th entry of hst associated with the sample $x'_{ord(j)}$ is incremented.

The last step consists in the computation of the scores. For each sample x'_i the associated counts are accumulated till their sum goes over the value ϱs and the associated value of k is employed to obtain the score.

The temporal cost of the technique is $O(s \cdot n \cdot d)$, where s is independent of the number n of dataset objects and can be considered a constant, and $n \cdot d$ is the size of the input, hence *the temporal cost is linear in the size of the input*.

As for the spatial cost, $O(Bs)$ space is needed for storing counters hst, $O(2s)$ for distances dst and the ordering ord, $O(\ell s)$ for storing scores, and $O(sd)$ for the buffer maintaining the sample, hence *the spatial cost is linear in the sample size*.

Before concluding, we notice that *fast*-CFOF is an embarrassingly parallel algorithm, since partition computations do not need to communicate intermediate results. Thus, it is readily suitable for multi-processor/computer system. We implemented a version for multi-core processors (using `gcc`, `OpenMP`, and the `AVX x86-64` instruction set extensions) that elaborates partitions sequentially, but employs both MIMD (cores) and SIMD (vector registers) parallelism to elaborate each single partition.

4   Experimental Results

Experiments are performed on a Intel Core i7 2.40 GHz CPU (having 4 cores with 8 hardware threads, and SIMD registers accommodating 8 single-precision floating-point numbers) based PC with 8 GB of main memory, under the Linux operating system. As for the implementation parameters, the number B of hst bins is set to 100 and the constant c used to compute k_{up} is set to 2. We assume 0.01 as the default value for the parameters ϱ, ϵ, and δ.

Some of the dataset employed are described next. *Clust2* is a dataset family (with $n \in [10^4, 10^6]$ and $d \in [2, 10^3]$) consisting of two normally distributed clusters centered in the origin and in $(4, \ldots, 4)$, with standard deviation 1.0 and 0.5 along each dimension, respectively. *MNIST* is a dataset consisting of handwritten digits[4] composed of $n = 60000$ vectors and $d = 784$ dimensions.

4.1   Accuracy

The goal of this experiment is to assess the quality of the result of *fast*-CFOF for different sample sizes, that is different combinations of the parameters ϵ and δ. We notice that the default sample size is $s = 26624$. With this aim we first computed the exact dataset scores by setting the sample size s to n.

Figure 5 compares the exact scores with those obtained for the standard sample size on the *Clust2* (for $n = 10^5$ and $d = 100$) and *MNIST* datsets. The blue curve is associated with the exact scores sorted in descending order and the x-axis represents the outlier rank position of the dataset objects. As for the red curve, it shows the approximate scores associated with the objects at each rank position. The curves highlight that the ranking position tends to be preserved and that in both cases top outliers are associated with the largest scores.

We can justify the accuracy of the method by noticing that the larger the CFOF score of x and, for any y, the larger the probability $p(x, y)$ that a dataset object will lie in between x and y and, moreover, the smaller the impact of the

---

[4] http://yann.lecun.com/exdb/mnist/.

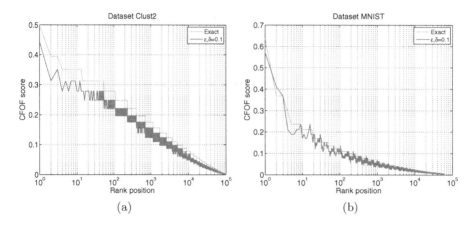

Fig. 5. Accuracy analysis of *fast*-CFOF.

Table 1. Spearman correlation between the exact and approximate outlier rankings computed by *fast*-CFOF.

Clust2 ($n = 100000, d = 100$)							
ϵ	δ	s	ϱ_1	ϱ_2	ϱ_3	ϱ_4	ϱ_5
			0.001	0.005	0.01	0.05	0.1
0.1	0.1	512	–	0.874	0.943	0.981	0.986
0.025	0.025	3584	0.933	0.985	0.991	0.996	0.996
0.01	0.1	15360	0.988	0.996	0.997	0.998	0.997
0.01	0.01	26624	0.994	0.998	0.998	0.998	0.997
MNIST ($n = 60000, d = 784$)							
ϵ	δ	s	ϱ_1	ϱ_2	ϱ_3	ϱ_4	ϱ_5
			0.001	0.005	0.01	0.05	0.1
0.1	0.1	512	–	0.526	0.679	0.886	0.939
0.025	0.025	3584	0.683	0.899	0.938	0.979	0.988
0.01	0.1	15360	0.929	0.978	0.985	0.993	0.995
0.01	0.01	26624	0.965	0.989	0.992	0.996	0.997

error ϵ on the estimated value $\widehat{p}(x, y)$. Intuitively, the objects we are interested in, that are the outliers, are precisely the one least prone to bad estimations.

We employ the Spearman's rank correlation coefficient to assesses relationship between the two rankings. This coefficient is high (close to 1) when observations have a similar rank. Table 1 reports Spearman's coefficients for different combinations of ϵ, δ, and ϱ. The coefficient ameliorates for increasing samples (very high values are reached for the default sample) and larger ϱ values (that exhibit high coefficient values also for small samples).

4.2   Scalability

Figure 6 shows the execution time on the *Clust2* and *MNIST* datasets.

Figure 6a shows the execution time on *Clust2* for the default sample size, $n \in [10^4, 10^6]$ and $d \in [2, 10^3]$. The largest dataset considered ($n = 10^6$ and $d = 10^3$, occupying 4 GB of disk space) required about 44 min. *fast*-CFOF exhibits a sublinear dependence from the dimensionality, due to the exploitation of the SIMD parallelism. As for the dashed curves, they are obtained by disabling MIMD parallelism. The performance ratio between the two versions is about 7.6, thus confirming the effectiveness of the parallelization schema.

Figure 6b shows the execution time on *Clust2* ($n = 10^6, d = 10^3$) and *MNIST* (180 MB of disk space) for different sample sizes. As for *Clust2*, the execution time drops from 44 min, for the default sample, to about 24 min, for $s = 15360$ ($\epsilon = 0.01, \delta = 0.1$). Finally, as for *MNIST*, the whole dataset ($s = n$) required less than 6 min, while about 3 min are required with the default sample.

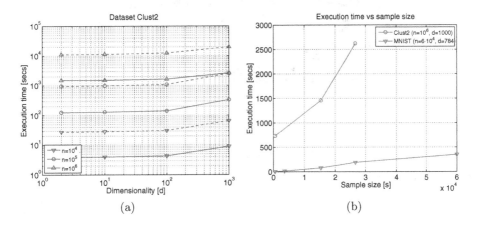

Fig. 6. Scalability analysis of *fast*-CFOF.

4.3   Effectiveness

On *Clust2*, we used the distance to cluster centers as the ground truth. Specifically, for each dataset object, the distance R from the closest cluster center has been determined and distances associated with the same cluster have been normalized as $R' = \frac{R - \mu_R}{\sigma_R}$. Table 2 reports the Spearman's correlation between normalized distances R' and CFOF scores. The high correlation values witness for both the meaningfulness of the definition and its behavior as a local outlier measure even in high dimensions.

Figure 7 shows the height top outliers of *MNIST*. It appears that these digits are deformed, quite difficult to recognize, and possibly misaligned within the 28×28 cell grid.

Table 2. Spearman correlation between the normalized distance to the object's cluster center and the score computed by *fast*-CFOF.

Clust2 ($n = 100000, d = 100$)							
ϵ	δ	s	ϱ_1	ϱ_2	ϱ_3	ϱ_4	ϱ_5
			0.001	0.005	0.01	0.05	0.1
0.1	0.1	512	–	0.874	0.943	0.981	0.987
0.025	0.025	3584	0.932	0.985	0.992	0.997	0.998
0.01	0.1	15360	0.987	0.997	0.998	0.999	0.999
0.01	0.01	26624	0.993	0.998	0.999	0.999	0.999

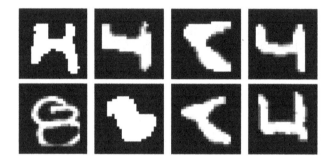

Fig. 7. Top CFOF outliers of *MNIST*.

4.4 Comparison with Other Approaches

We compared CFOF with *a*KNN, LOF, and ABOD, by using some labelled datasets as ground truth. The datasets, randomly selected at the UCI ML Repository[5], are: *Breast Cancer Wisconsin Diagnostic* ($n = 569, d = 32$), *Image segmentation* ($n = 2310, d = 19$), *Ozone Level Detection* ($n = 2536, d = 73$), *Pima indians diabetes* ($n = 768, d = 8$), *QSAR biodegradation* ($n = 1055, d = 41$), *Yeast* ($n = 1484, d = 8$). Each class in turn is marked as abnormal, and a dataset composed of all the objects of the other classes plus 10 randomly selected objects of the abnormal class is considered. Table 3 reports the Area Under the ROC Curve (AUC) obtained by CFOF (*hard*-CFOF has been used), *a*KNN, LOF, and ABOD. As for the parameters k_ϱ and k, for all the methods the corresponding parameter has been varied between 2 and 100, and the best result has been reported in the table. Notice that the wins are 16 for CFOF, 4 for *a*KNN, 2 for LOF, and 4 for ABOD. The comparison points out that CFOF represents an outlier detection definition with its own peculiarities, since the other methods behaved differently, and state of the art detection performances.

---

[5] https://archive.ics.uci.edu/ml/index.html.

Table 3. AUCs for the labelled datasets.

Dataset	Class	CFOF	aKNN	LOF	ABOD
Breast	0	0.929	*0.936*	**0.952**	0.914
	1	**0.805**	0.685	*0.780*	0.404
Image	1	**0.942**	0.812	*0.846*	0.649
	2	**0.990**	0.988	0.987	*0.989*
	3	**0.956**	0.817	*0.919*	0.713
	4	0.936	**0.971**	*0.949*	0.941
	5	**0.933**	0.688	*0.884*	0.688
	6	0.979	*0.979*	0.968	**0.982**
	7	**0.993**	0.973	*0.982*	0.976
Ozone	0	**0.728**	0.677	0.662	*0.680*
	1	**0.656**	0.429	*0.591*	0.426
Pima	0	**0.736**	0.509	*0.626*	0.454
	1	*0.677*	**0.700**	0.670	0.626
QSAR	0	**0.692**	0.503	0.444	*0.545*
	1	**0.818**	0.706	0.706	*0.757*
Yeast	0	**0.769**	0.526	*0.568*	0.487
	1	**0.743**	0.678	*0.729*	0.629
	2	**0.788**	0.327	*0.437*	0.313
	3	0.772	**0.832**	0.700	*0.820*
	4	0.721	**0.808**	0.695	*0.803*
	5	**0.735**	*0.728*	*0.728*	**0.735**
	6	*0.766*	0.613	**0.783**	0.636
	7	**0.794**	0.550	*0.587*	0.543
	8	0.814	*0.881*	0.850	**0.892**
	9	0.980	0.993	*0.981*	**1.000**

5   Conclusions

We presented the Concentration Free Outlier Factor, a novel density estimation measure whose main characteristic is to resist concentration phenomena usually arising in high dimensional spaces and to allow very efficient and reliable outlier detection through the use of sampling. We are extending the study of the theoretical properties of the definition, assessing guarantees of the *fast*-CFOF algorithm, and extending the experimental activity. We believe that the CFOF score can offer insights also in the context of other data mining tasks. We are currently investigating its application in other classification scenarios.

References

1. Aggarwal, C.C., Yu, P.S.: Outlier detection for high dimensional data. In: Proceedings of the International Conference on Managment of Data (SIGMOD) (2001)
2. Aggarwal, C.C.: Outlier Analysis. Springer, New York (2013). https://doi.org/10.1007/978-3-319-47578-3
3. Angiulli, F.: On the behavior of intrinsically high-dimensional spaces: distance distributions, direct and reverse nearest neighbors, and hubness. Manuscript submitted for publication to an international journal (2017). Available at the author's site
4. Angiulli, F., Fassetti, F.: Dolphin: an efficient algorithm for mining distance-based outliers in very large datasets. ACM Trans. Knowl. Disc. Data, **3**(1), Article 4 (2009)
5. Angiulli, F., Pizzuti, C.: Outlier mining in large high-dimensional data sets. IEEE Trans. Knowl. Data Eng. **2**(17), 203–215 (2005)
6. Beyer, K., Goldstein, J., Ramakrishnan, R., Shaft, U.: When is "nearest neighbor" meaningful? In: Beeri, C., Buneman, P. (eds.) ICDT 1999. LNCS, vol. 1540, pp. 217–235. Springer, Heidelberg (1999). https://doi.org/10.1007/3-540-49257-7_15
7. Breunig, M.M., Kriegel, H., Ng, R.T., Sander, J.: LOF: identifying density-based local outliers. In: Proceedings of the International Conference on Managment of Data (SIGMOD) (2000)
8. Chandola, V., Banerjee, A., Kumar, V.: Anomaly detection: a survey. ACM Comput. Surv. **41**(3), 15:1–15:58 (2009)
9. Chandola, V., Banerjee, A., Kumar, V.: Anomaly detection for discrete sequences: a survey. IEEE Trans. Knowl. Data Eng. **24**(5), 823–839 (2012)
10. Chávez, E., Navarro, G., Baeza-Yates, R., Marroquín, J.L.: Searching in metric spaces. ACM Comput. Surv. **33**(3), 273–321 (2001)
11. Han, J., Kamber, M.: Data Mining, Concepts and Technique. Morgan Kaufmann, San Francisco (2001)
12. Hautamaki, V., Karkkainen, I., Franti, P.: Outlier detection using k-nearest neighbour graph. In: Proceedings of the International Conference on Pattern Recognition (ICPR), pp. 430–433 (2004)
13. Hodge, V., Austin, J.: A survey of outlier detection methodologies. Artif. Intell. Rev. **22**(2), 85–126 (2004)
14. Kriegel, H.-P., Kröger, P., Schubert, E., Zimek, A.: Outlier detection in axis-parallel subspaces of high dimensional data. In: Theeramunkong, T., Kijsirikul, B., Cercone, N., Ho, T.-B. (eds.) PAKDD 2009. LNCS (LNAI), vol. 5476, pp. 831–838. Springer, Heidelberg (2009). https://doi.org/10.1007/978-3-642-01307-2_86
15. Jin, W., Tung, A.K.H., Han, J.: Mining top-n local outliers in large databases. In: Proceedings of the International Conference on Knowledge Discovery and Data Mining (KDD) (2001)
16. Knorr, E., Ng, R.: Algorithms for mining distance-based outliers in large datasets. In: Proceedings of the International Conference on Very Large Databases (VLDB), pp. 392–403 (1998)
17. Kriegel, H.-P., Kröger, P., Schubert, E., Zimek, A.: Interpreting and unifying outlier scores. In: Proceedings of the SIAM International Conference on Data Mining (SDM), pp. 13–24 (2011)
18. Kriegel, H.-P., Schubert, M., Zimek, A.: Angle-based outlier detection in high-dimensional data. In: Proceedings of the International Confernce on Knowledge Discovery and Data Mining (KDD), pp. 444–452 (2008)

19. Liu, F.T., Ting, K.M., Zhou, Z.-H.: Isolation-based anomaly detection. TKDD **6**(1), 3:1–3:39 (2012)
20. Papadimitriou, S., Kitagawa, H., Gibbons, P.B., Faloutsos, C.: LOCI: fast outlier detection using the local correlation integral. In: Proceedings of the International Conference on Data Enginnering (ICDE), pp. 315–326 (2003)
21. Radovanovic, M., Nanopoulos, A., Ivanovic, M.: Hubs in space: popular nearest neighbors in high-dimensional data. J. Mach. Learn. Res. **11**, 2487–2531 (2010)
22. Radovanovic, M., Nanopoulos, A., Ivanovic, M.: Reverse nearest neighbors in unsupervised distance-based outlier detection. IEEE Trans. Knowl. Data Eng. **27**(5), 1369–1382 (2015)
23. Ramaswamy, S., Rastogi, R., Shim, K.: Efficient algorithms for mining outliers from large data sets. In: Proceedings of the International Conference on Management of Data (SIGMOD), pp. 427–438 (2000)
24. Watanabe, O.: Sequential sampling techniques for algorithmic learning theory. Theor. Comput. Sci. **348**(1), 3–14 (2005)
25. Zimek, A., Schubert, E., Kriegel, H.-P.: A survey on unsupervised outlier detection in high-dimensional numerical data. Stat. Anal. Data Min. **5**(5), 363–387 (2012)

Efficient Top Rank Optimization with Gradient Boosting for Supervised Anomaly Detection

Jordan Frery[1,2(✉)], Amaury Habrard[1], Marc Sebban[1], Olivier Caelen[2], and Liyun He-Guelton[2]

[1] Univ. Lyon, Univ. St-Etienne, UMR CNRS 5516, Laboratoire Hubert-Curien, 42000 Saint-Étienne, France
{jordan.frery,amaury.habrard,marc.sebban}@univ-st-etienne.fr
[2] Worldline, 95870 Bezons, France
{jordan.frery,olivier.caelen,liyun.he-guelton}@worldline.com

Abstract. In this paper we address the anomaly detection problem in a supervised setting where positive examples might be very sparse. We tackle this task with a learning to rank strategy by optimizing a differentiable smoothed surrogate of the so-called *Average Precision* (AP). Despite its non-convexity, we show how to use it efficiently in a stochastic gradient boosting framework. We show that using *AP* is much better to optimize the top rank alerts than the state of the art measures. We demonstrate on anomaly detection tasks that the interest of our method is even reinforced in highly unbalanced scenarios.

1 Introduction

Anomaly detection in DNA sequences, credit card transactions or cyber security are some illustrations of supervised learning settings where data is often highly unbalanced (i.e. a few anomalies versus a huge amount of genuine/normal data). A naive approach to tackle this binary problem would consist in applying a standard classification, such as SVM, boosting or logistic regression, using classic margin-based surrogate loss functions, like the hinge loss, the exponential loss or the logistic loss. However, since abnormal instances are often very sparse in the feature space using such algorithms cannot be directly appropriate [1]. There exist several methods to get rid of the issues due to unbalanced datasets. The most famous are sampling-based strategies, either by undersampling or oversampling the data [2,3]. The former aims at removing instance from the majority class while the latter creates synthetic data from the minority class. Other hybrid methods such as SMOTEBoost [4], RUSBoost [5] and Adacost [6] combine a learning algorithm with a sampling or cost sensitive method. However, it turns out that these approaches have been shown to be hard to use when facing highly unbalanced situations [7] leading to either insufficient generated diversity (by oversampling) or too drastic reduction of the dataset size (by undersampling). In addition, sampling methods induce a bias in the posterior probabilities [8,9].

© Springer International Publishing AG 2017
M. Ceci et al. (Eds.): ECML PKDD 2017, Part I, LNAI 10534, pp. 20–35, 2017.
https://doi.org/10.1007/978-3-319-71249-9_2

On the other hand, it is worth noticing that a peculiarity of the use cases mentioned above is the need to resort to a (often limited) number of human experts to assess the potential anomalies found by the learned model. Actually, our contribution stands in a context where the number of false positives (FP) may be significantly larger than the false negative (FN) due to the high class imbalance and where the impact of FP is very penalizing. For example, in fraud detection for credit card transactions, it is out of the question to automatically block a credit card without the expert approval (which may risk the confidence of customers having their credit card falsely blocked). In this context, the goal of the automatic system is more to give the shortest list of alerts preventing the expert from going through thousands of transactions. In other words, one aims at maximizing the number of true positives in the top rank alerts (i.e. the so-called *precision*) rather than discriminating between abnormal and normal cases.

This is the reason why we tackle in this paper the supervised anomaly detection task with a *learning to rank* approach. This strategy has gained a lot of interest in the information retrieval community [10]. Given a query, the goal is to give the most relevant links to the user in a small set of top ranked items. It turns out that apart the notion of query, the anomaly detection task can relate to this setting aiming at finding the anomalies with the highest precision without giving too many genuine examples to the experts.

In such settings, different machine learning algorithms have been efficiently used such as SVMs (e.g. SVM-Rank [11], SVM-AP [12]) or ensemble methods (e.g. random forest [13], boosting [14]). It turns out that gradient boosting has shown to be a powerful method on real life datasets to address learning to rank problems [15]. Its popularity comes from two main features: (i) it performs the optimization in function space [16] (rather than in parameter space) which makes the use of custom loss functions much easier; (ii) boosting focuses step by step on difficult examples that gives a nice strategy to deal with unbalanced datasets by strengthening the impact of the positive class. In order to be efficient in learning to rank problems, gradient boosting needs to be fed with a loss function leading to a good precision in the top ranked examples.

In the literature, many approaches resort to pairwise loss functions [17–19], typically checking that every positive example is ranked before any negative instance. Note that all those methods implicitly optimize the area under the ROC curve. Therefore they aim at minimizing the number of incorrectly ranked pairs but do not directly optimize the precision of top ranked items as shown in [20]. To overcome this issue, recent works in learning to rank suggested to optimize other criteria like the Average Precision (AP) or the Normalized Discounted Cumulative Gain (NDCG) such as in Adarank [21], LambdaMART [22] or LambdaRank [23]. It has been shown that both AP and $NDCG$ are much more suited for enhancing ranking methods. However, due to the non convexity and non differentiability of those criteria, the previous methods rather work on standard surrogate convex objective functions (such as the pairwise cross-entropy or the exponential loss) and take into account the AP and $NDCG$ in the form of weighting coefficients only. In other words, the gradients are not computed as derivatives of AP and $NDCG$. Therefore, used in this way, these criteria only tend to guide the optimization process in the right direction. We claim here that there is room for doing much better and directly considering the analytical expressions of those criteria in a gradient boosting method.

In this paper, our contribution is three-fold: (i) focusing on AP, we show how to optimize a loss function based on a surrogate of this criterion; (ii) unlike the state of the art learning to rank methods requiring a quadratic complexity to minimize the ranking measures, we show that AP can be handled linearly in gradient boosting without penalizing the quality of the solution; (iii) compared to the state of the art, we show that our method allows us to highly improve the quality of the top-ranked items. We even show that this advantage is much larger when the imbalance of the datasets is very important. This is a particularly interesting feature when addressing anomaly detection problems where the positive examples are very sparse.

The rest of this paper is organized as follows: In Sect. 2 we first introduce our notations, then describe our performance measures and present an approximation to AP. We then describe our method in a boosting framework and define a more suitable smoothed AP as the loss function in Sect. 3. We demonstrate the effectiveness of our work in the experiments section where we compare several state of the art machine learning models in Sect. 4.

2   Evaluation Criteria and Related Work

We consider a binary supervised learning setting with a training set $\mathcal{S} = \{z_i = (x_i, y_i)\}_{i=1}^{M}$ composed of M labeled data, where $x_i \in \mathcal{X}$ is a feature vector and $y_i \in \{-1, 1\}$ is the label. In unbalanced scenarios, $y = 1$ often describes the minority (positive) class while $y = -1$ represents the majority (negative) class. Let P (resp. N) be the number of positive (resp. negative) examples such that $P + N = M$. We also define $\mathcal{S}^+ = \{z_i^+ = (x_i^+, y_i^+) | y_i = +1\}_{i=1}^{P}$ and $\mathcal{S}^- = \{z_i^- = (x_i^-, y_i^-) | y_i = -1\}_{i=1}^{N}$ where $\mathcal{S}^+ \cup \mathcal{S}^- = \mathcal{S}$. We assume that the training data $z_i = (x_i, y_i)$ is independently and identically distributed according to an unknown joint distribution $\mathcal{D}_{\mathcal{Z}}$ over $\mathcal{Z} = \mathcal{X} \times \{-1, 1\}$.

In this work, we aim at learning from S a function (or hypothesis) $f : \mathcal{X} \to \mathbb{R}$ that gives a real value to any new $x \in \mathcal{X}$. Assessing the quality of f requires the use of an evaluation criterion. It is worth noticing that most of the criteria are based on the true positive TP, true negative TN, false positive FP and false negative FN quantities. For example, the accuracy is defined as $\frac{TP+TN}{M}$. It is known that optimizing the accuracy is NP-hard due to the non convexity and non differentiability of TP and TN. Therefore, classification algorithms resort to surrogates like the hinge loss, the logistic loss or the exponential loss which are convex functions used in Support Vector Machines, logistic regression and boosting, respectively. However, when S is highly unbalanced, optimizing such losses may lead to a classifier which always predict the negative class. A solution to overcome this issue may consist in addressing unbalanced scenario from a *learning to rank* point of view. Rather that discriminating examples belonging to the positive and negative classes, we rather aim at ranking the data with a maximal number of TP in the top ranked examples.

In this context, two measures are well used in the literature: the *pairwise* $AUCROC$ measure and the *listwise* average precision AP. From a statistical point of view, the $AUCROC$ represents the probability that a classifier ranks a randomly drawn positive instance higher than a randomly chosen negative one. The expression of this measure is equivalent to the Wilcoxon-Mann-Whitney statistic [24]:

$$AUCROC = \frac{1}{PN} \sum_{i=1}^{P} \sum_{j=1}^{N} I_{0.5}(f(x_i^+) - f(x_j^-)), \tag{1}$$

where $I_{0.5}$, is a special indicator function that yields 1 if $f(x_i^+) - f(x_j^-) > 0, 0.5$ if $f(x_i^+) - f(x_j^-) = 0$ and 0 otherwise. In the following we will use the classic indicator function $I(*)$ that yields 1 if $*$ is true, 0 otherwise.

$1 - AUCROC$ has been exploited in Rankboost algorithm [17] as an objective function where the authors use the exponential as a surrogate to the indicator function. Let $\ell_{roc}(z_i, f) = \frac{1}{N} \sum_{j=1}^{N} e^{(f(z_j^-) - f(z_i^+))}$ be the loss suffered by f at z_i. We get the following upper bound on $1 - AUCROC$:

$$1 - AUCROC \leq \frac{1}{P} \sum_{i=1}^{P} \frac{1}{N} \sum_{j=1}^{N} e^{(f(z_j^-) - f(z_i^+))} = \frac{1}{P} \sum_{i=1}^{P} \ell_{roc}(z_i, f) = \mathbb{E}_{z_i \in S^+} \ell_{roc}(z_i, f) \tag{2}$$

We can notice that this objective is a pairwise function inducing an algorithmic complexity $\mathcal{O}(PN)$. Moreover, as illustrated later in this section and shown in [20], ℓ_{roc} is not well suited to maximize the precision in the top ranked items.

A better strategy consists in using an alternative criterion based on the average precision AP, defined as follows:

$$AP = \frac{1}{P} \sum_{i=1}^{P} p(k_i), \tag{3}$$

where $p(k_i)$ is the precision with respect to the rank k_i of the ith positive example. Since the rank depends on the outputs of the model f, we get:

$$p(k_i) = \frac{1}{k_i} \sum_{j=1}^{P} I(f(x_i^+) \leq f(x_j^+)) \tag{4}$$

with

$$k_i = \sum_{j=1}^{M} I(f(x_i^+) \leq f(x_j)). \tag{5}$$

Plugging Eqs. (4) and (5) in Eq. (3) we get:

Fig. 1. Two rankings (with two positives and eight negatives examples) ordered from the highest score (at the top) to the lowest. On the left, we get $AUCROC = 0.63$ and $AP = 0.33$. On the right, $AUCROC = 0.56$ and $AP = 0.38$. Therefore, the two criteria disagree on the best ranked list. (Color figure online)

$$AP = \frac{1}{P} \sum_{i=1}^{P} \frac{1}{\sum_{j=1}^{M} \mathrm{I}(f(x_i^+) \le f(x_j))} \sum_{j=1}^{M} \mathrm{I}(y_j = 1)\mathrm{I}(f(x_i^+) \le f(x_j^+)). \quad (6)$$

AP has been used in recent papers to enhance learning to rank algorithms.

In [20,23], the authors introduce a new objective function, called LambdaRank, which can be used with different criteria, including AP. This function depends on the criterion of interest without requiring to compute the derivatives of that measure. This specificity allows them to bypass the issues due to the non differentiability of the criterion. The objective function takes the following form:

$$\frac{1}{N} \sum_{i=1}^{P} \ell_{\lambda Rank}(z_i^+, f) \quad (7)$$

with $\ell_{\lambda Rank}(z_i^+, f) = \frac{1}{N} \sum_{j=1}^{N} \log(1 + e^{-(f(x_i^+) - f(x_j^-))})|AP_{ij}|$ the loss suffered by f at z_i. Here, $|AP_{ij}|$ is the absolute difference in AP when one swaps, in the ranking, example x_i with x_j. LambdaMART [22] made use of LambdaRank in a gradient boosting method and got good results as reported in [15]. However, it is worth noticing that in this algorithm, the analytical expression of AP as defined in Eq. (6) is not involved in the calculation of the gradient. $|AP_{ij}|$ can be viewed as a weighting coefficient which hopefully tends to guide the optimization process towards a good solution. One objective of this paper is to directly use AP in the algorithm and therefore to use the same criterion at both training and test time.

Let us before focus on the effect of $AUCROC$ and AP in terms of quality of top ranked items. Figure 1 compares these criteria in two different situations according to the location of two positive (in dark color) and eight negative (in light color) examples that are ordered according to their predicted scores (highest score at the top). The key point of this figure is to show that $AUCROC$ and AP disagree on which list is the best. $AUCROC$ prefers the list on the left because the positive examples are rather well ranked even though the first three

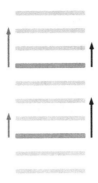

Fig. 2. Comparison of the emphasis given by AP (arrows on the left) and the emphasis given $AUCROC$ (arrows on the right) [20]. One can compare this emphasis to the intensity of gradient w.r.t. the examples if AP and $AUCROC$ were continuous functions. (Color figure online)

items are negative. Therefore, we can note that this criterion is very relevant if we are interested in classifying examples into two classes, for example, the classifier being based on a threshold (likely after the fifth rank, here) splitting the items into two parts. AP is in favor of the list on the right because it prefers to champion the top list accepting to pay the prize to miss some positives. This criterion is thus very relevant to deal with anomaly and fraud detection where the goal is to provide the shortest list of alerts (here, typically the first two items) with the largest precision.

Figure 2 (inspired from [20]) illustrates graphically how the emphasis is done while computing gradients from pairwise loss function such as $AUCROC$ (black arrows on the right) or a listwise loss function such as AP (red arrows on the left) respectively. We can notice that a learning algorithm optimizing the $AUCROC$ would champion first the worst positive to get a good classifier (w.r.t. an appropriate threshold) while the AP would promote first the best positive to get a good top rank.

The previous analysis shows the advantage of optimizing AP in a learning to rank algorithm. This is the objective of the next section where we introduce a differentiable expression of AP in a gradient boosting algorithm.

3   Stochastic Gradient Boosting with AP

In this section, we present the stochastic gradient boosting framework as introduced in [25]. Then we instantiate the loss function in two different ways: first, we introduce a differentiable version of AP using the sigmoid function. Then, in order to reduce the algorithmic complexity, we suggest to use a rough approximation based on the exponential function. We show that this second strategy allows us not only to drastically reduce the complexity but also, to get similar or even better results than the sigmoid-based loss. We give some explanations about this behavior at the end of the section.

3.1   Stochastic Gradient Boosting

Like other boosting methods, gradient boosting is based on a sequential and adaptive learning over weak learners that are linearly combined. However, instead of setting a weight for every example, gradient boosting builds each new weak learner on the residuals of the previous linear combination. We can see gradient boosting as gradient descent in functional space. The linear combination at step t is defined as follows:

$$f_t(x) = f_{t-1}(x) + \gamma_t h_t(x),$$

with $h_t \in \mathcal{H}$ an hypothesis belonging to a class of models $\mathcal{H}$ (typically, regression trees) and γ_t the weight underlying the performance of h_t in the linear combination. Residuals are defined by the negative gradients of the loss function computed w.r.t. the previous linear combination of weak learners:

$$g_t(x) = -\left[\frac{\partial \ell(z_i, f_{t-1}(x_i))}{\partial f_{t-1}(x_i)}\right], i = 1 \ldots M.$$

As in standard boosting, hard examples get more importance along the iterations of gradient boosting. Note that a mini-batch strategy is usually used to speed-up the procedure by randomly selecting a proportion $\lambda \in [0,1]$ of examples at each iteration. Additionally, this stochastic approach allows us to avoid falling in a local optima. A generic version of the stochastic gradient boosting is presented in Algorithm 1.

Algorithm 1. Stochastic gradient boosting

INPUT: a training set $S = \{z_i = (x_i, y_i)\}_{i=1}^{M}$, a parameter $\lambda \in [0,1]$, a weak learner
Require: Initialize $f_0 = argmin_h \ell(y, h)$
  for $t = 1$ to T **do**
    Select randomly from S a set $S' = \{x_i, y_i\}_{i=1}^{\lambda M}$

$$g_t(x) = -\left[\frac{\partial \ell(z, f_{t-1}(x))}{\partial f_{t-1}(x)}\right], \forall z = (x, y) \in S' \qquad (8)$$

    Fit a weak classifier (e.g. a regression tree) $h_t(x)$ to predict the targets $g_t(x)$
    Find $\gamma_t = argmin_\gamma \ell(z, f_{t-1}(x) + \gamma h_t(x))$
    Update $f_t(x)$ such that $f_t(x) = f_{t-1}(x) + \gamma_t h_t(x)$
  end for

The key step of this algorithm takes place in Eq. (8). It requires the definition of a differentiable loss function with its associated gradients. Unlike the state of the art ranking methods which make use of gradient boosting, we aim at directly optimizing in the loss function ℓ a surrogate of AP.

3.2   Sigmoid-Based Surrogate of AP

To define a loss function ℓ based on AP, we need to transform the non differentiable Eq. (6) into an expression for which one will be able to compute gradients on AP. Therefore, we need to get rid of the indicator function. A standard way consists in replacing $I(f(x_i) \leq f(x_j))$ by the sigmoid function:

$$I(f(x_i) \leq f(x_j)) \approx \frac{1}{1 + e^{-\alpha(f(x_j) - f(x_i))}} = \sigma(f(x_j) - f(x_i)) \quad ,$$

with α a smoothing parameter. As α grows the approximation gets closer to the true AP. Considering that $\sum_{j=1}^{M} I(y_j = 1) = P$, we get the following differentiable surrogate of AP:

$$\begin{aligned}
\hat{AP}_{sig} &= \frac{1}{P} \sum_{i=1}^{P} \frac{1}{\sum_{j=1}^{M} \frac{1}{1 + e^{-\alpha(f(x_j) - f(x_i^+))}}} \sum_{j=1}^{P} \frac{1}{1 + e^{-\alpha(f(x_j^+) - f(x_i^+))}} \\
&= \frac{1}{P} \sum_{i=1}^{P} \frac{\sum_{j=1}^{P} \sigma(f(x_j^+) - f(x_i^+))}{\sum_{h=1}^{M} \sigma(f(x_h) - f(x_i^+))} \\
&= \frac{1}{P} \sum_{i=1}^{P} \hat{p}(k_i) \approx \frac{1}{P} \sum_{i=1}^{P} p(k_i).
\end{aligned} \tag{9}$$

From $\hat{AP}_{sig}$, we get the following objective function:

$$1 - \hat{AP}_{sig} = \mathbb{E}_{z_i \in S^+} \ell_{ap}^{sig}(z_i, f),$$

where $\ell_{ap}^{sig}(z_i, f) = 1 - \hat{p}(k_i)$ is the loss suffered by f in terms of precision at z_i (let us remind that k_i is the rank (predicted by f) of the ith positive example z_i). In fact, we can simply rewrite our objective function as:

$$1 - \hat{AP}_{sig} = \frac{1}{P} \sum_{i=1}^{P} \frac{\sum_{j=1}^{N} \sigma(f(x_j^-) - f(x_i^+))}{\sum_{h=1}^{M} \sigma(f(x_h) - f(x_i^+))}$$

For the sake of simplicity, let us use the following notations: $\sigma(f(x_j) - f(x_i)) = \sigma_{ji}$ and we have

$$\frac{\partial \sigma_{ji}}{\partial f_t(x_j)} = -\sigma_{ji}(1 - \sigma_{ji}) = -\sigma'_{ji},$$

$$\frac{\partial \sigma_{ji}}{\partial f_t(x_i)} = \sigma_{ji}(1 - \sigma_{ji}) = \sigma'_{ji}.$$

The gradient w.r.t. $f_t(x_p^+)$ or $f_t(x_p^-)$, for positive and negative examples respectively, are given by:

$$
\begin{aligned}
\frac{\partial(1 - \hat{AP}_{sig})}{\partial f_t(x_p^+)} &= \frac{\partial(1 - \hat{AP}_{sig})}{\partial \sigma_{jp}} \frac{\partial \sigma_{jp}}{\partial f_t(x_p^+)} + \frac{\partial(1 - \hat{AP}_{sig})}{\partial \sigma_{pi}} \frac{\partial \sigma_{pi}}{\partial f_t(x_p^+)} \\
&= \sum_{j=1}^{P} \frac{(\sigma'_{jp} \sum_{h=1}^{M} \sigma_{hp} - \sigma_{jp} \sum_{h=1}^{N} \sigma'_{hp})}{(\sum_{h=1}^{N} \sigma_{hp})^2} \\
&\quad + \sum_{i=1}^{P} \frac{(\sigma'_{pi} \sum_{h=1}^{N} \sigma_{hi} - \sigma_{jp}\sigma'_{pi})}{(\sum_{h=1}^{N} \sigma_{hi})^2},
\end{aligned}
\tag{10}
$$

$$
\begin{aligned}
\frac{\partial(1 - \hat{AP}_{sig})}{\partial f_t(x_p^-)} &= \frac{\partial(1 - \hat{AP}_{sig})}{\partial \sigma_{pi}} \frac{\partial \sigma_{pi}}{\partial f_t(x_p^-)} \\
&= \sum_{i=1}^{P} \sum_{j=1}^{P} \frac{-\sigma_{ji}\sigma'_{pi}}{(\sum_{h=1}^{N} \sigma_{hi})^2},
\end{aligned}
$$

As, $\frac{\partial(1 - \hat{AP}_{sig})}{\partial \sigma_{jp}} \frac{\partial \sigma_{jp}}{\partial f_t(x_p^-)} = 0$, since the example x_p from the previous formulation will always be positive in $1 - \hat{AP}$.

In the following, we call $SGBAP_{sig}$, Stochastic Gradient Boosting algorithm using this sigmoid-based approximation $1 - \hat{AP}_{sig}$.

3.3 Exponential-Based Surrogate of AP

It is worth noticing that by approximating the indicator function by the sigmoid function, the computation of the gradients as stated above is performed in quadratic time. This can be a too strong algorithmic constraint to deal with real world applications like fraud detection in credit card transactions (see the experimental section where the dataset contains 2,000,000 transactions). To overcome this issue, we suggest here to resort to a less costly surrogate of AP using the exponential function as an approximation of the indicator function.

$$
I(f(x_i) \leq f(x_j)) \approx e^{(f(x_j) - f(x_i))}.
$$

As already done in Rankboost [17], we can show that the use of this exponential function allows us to reduce the time complexity for binary datasets to $\mathcal{O}(P + N)$.

Using the new approximation, AP takes the following form:

$$
\begin{aligned}
\hat{AP}_{exp} &= \frac{1}{P} \sum_{i=1}^{P} \frac{\sum_{j=1}^{P} e^{f(x_j^+)} e^{-f(x_i^+)}}{\sum_{h=1}^{M} e^{f(x_h)} e^{-f(x_i^+)}} = \frac{1}{P} \sum_{i=1}^{P} \frac{e^{-f(x_i^+)} \sum_{j=1}^{P} e^{f(x_j^+)}}{e^{-f(x_i^+)} \sum_{h=1}^{M} e^{f(x_h)}} \\
&= \frac{\sum_{j=1}^{P} e^{f(x_j^+)}}{\sum_{h=1}^{M} e^{f(x_h)}}
\end{aligned}
$$

As for the sigmoid approximation, we rather use $1 - \hat{AP}_{exp}$ to minimize it.

$$1 - \hat{AP}_{exp} = \frac{\sum_{h=1}^{M} e^{f(x_h)} - \sum_{j=1}^{P} e^{f(x_j^+)}}{\sum_{h=1}^{N} e^{f(x_h)}} = \frac{\sum_{n=1}^{N} e^{f(x_n^-)}}{\sum_{h=1}^{M} e^{f(x_h)}} \qquad (11)$$

Finally, finding the gradients of this new objective function is straightforward.

$$\frac{\partial 1 - \hat{AP}_{exp}}{\partial f(x_p^+)} = \frac{-e^{f(x_p^+)} \sum_{n=1}^{N} e^{f(x_n^-)}}{(\sum_{h=1}^{M} e^{f(x_h)})^2} \qquad (12)$$

$$\frac{\partial 1 - \hat{AP}_{exp}}{\partial f(x_p^-)} = \frac{e^{f(x_p^-)} \sum_{i=1}^{M} e^{f(x_h)} - e^{f(x_p^-)} \sum_{n=1}^{N} e^{f(x_n^-)}}{(\sum_{h=1}^{M} e^{f(x_h)})^2}$$

In the following, we call our method SGBAP, the stochastic gradient boosting based on our approximation $1 - \hat{AP}_{exp}$.

Note that, in Eq. 12, one can see an adverse effect brought by the exponential approximation of the indicator function. Indeed, if an $f(x_i)$ is first in the ranking, the gradient of $x_i, g(x_i)$, should decrease as there is no other position in which it will improve the overall AP. However, in our approximation, when $f(x_i)$ is significantly high, the gradient for this example will be the highest. Assume $\forall j \in \mathcal{S} \setminus x_i, f(x_i) \gg f(x_j)$, we have $g(x_i) \approx 1$ and $g(x_k) \approx 0 \quad \forall k \in \mathcal{S}^- \setminus x_i$. In fact, this effect is limited with stochastic gradient boosting. Indeed, since $g(x_i)$ is not computed during all the iteration thanks to the random mini-batches, the gradient is then automatically regularized. However, running the gradient boosting algorithm instead of the stochastic version would raise the previous effect. The same holds for any basic gradient descent based algorithm.

3.4 Comparison Between the Approximations of AP

In this section, we compare experimentally the approximations used in this paper - $\hat{AP}_{exp}$ and $\hat{AP}_{sig}$ - with a simple one-dimensional sample described in Table 1. For this experiment, we use a simple linear model $f(x) = \theta_0 + \theta_1 x$. The toy dataset has been made such that the model has three ranking choices: (i) rank the examples in descending order from $x = +7$ to $x = -6$ (when $\theta_1 > 0$), (ii) rank the examples in descending order from $x = -6$ to $x = +7$ ($\theta_1 < 0$) or (iii) give the same rank to every example ($\theta_1 = 0$). We give the AP and $AUCROC$ measures in each case: $AP = 0.29, AUCROC = 0.52$ when $\theta_1 < 0$, $AP = 0.33, AUCROC = 0.49$ when $\theta_1 > 0$ and $AP = 0.22, AUCROC = 0.5$ when $\theta_1 = 0$.

Figure 3, shows that the two objective functions considered are obviously not convex. However, they both find their minimum in $\theta_1 > 0$ which yields the best AP. In comparison, we show in the supplementary material a pairwise based and an accuracy based objective function that find their minimum in $\theta_1 < 0$.

Note that, on Fig. 3, $1 - \hat{AP}_{exp}$ has another advantage than the time complexity over the sigmoid approximation. Indeed, for negative examples with high

Table 1. Toy dataset constituted of 14 examples on the real line with their associated labels. x correspond to the feature value and y the class.

x	−6	−5	−4	−3	−2	−1	0	1	2	3	4	5	6	7
y	−1	−1	−1	+1	+1	−1	−1	−1	−1	−1	−1	−1	+1	−1

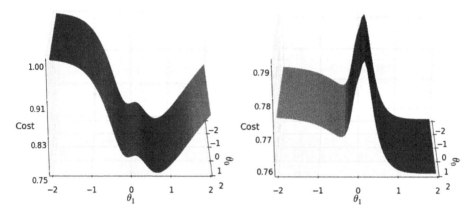

Fig. 3. $1 - \hat{AP}_{exp}$ (on the left) and $1 - \hat{AP}_{sig}$ (on the right) costs in function of the two model parameters θ_0 and θ_1. (Better with color)

scores (e.g. when $\theta_1 > 1$), $1 - \hat{AP}_{sig}$ tends to have vanishing gradients while, for $1 - \hat{AP}_{exp}$, they tend to increase exponentially. Indeed, on Fig. 3, the cost increases for the exponential approximation while it decreases for the sigmoid approximation.

4   Experiments

In this section, we present an experimental evaluation of our approach in two parts. In a first setup, we provide a comparative study with different state-of-the-art methods and various evaluation measures on 5 unbalanced public datasets coming either from the UCI Irvine Machine Learning repository or the LIBSVM datasets[1] and on a real dataset of credit card transactions provided by the private company ATOS/Worldline[2]. In a second experiment, we study the robustness of our method to undersampling of positives instances.

4.1   Top-Rank Quality over Unbalanced Datasets

In this experiment, we use the public datasets Pima, Breast cancer, HIV, Heart cleveland, w8a and the real fraud detection dataset over credit card transactions provided by ATOS/Worldline. This dataset contains 2 millions transactions

---

[1] http://archive.ics.uci.edu/ml/   and   https://www.csie.ntu.edu.tw/~cjlin/libsvm tools/.

[2] ATOS/Wolrdline is leader in e-transaction payments http://worldline.com/.

labelled as 1 fraudulent or −1 genuine where 0.2% are fraudulent. It is constituted of 2 subsets of transactions of 3 consecutive days each. The first one is
fixed as the training set and the second as the test. Each subset being separated
by one week in order to have the same week days (e.g. Tuesday to Thursday)
in train and test. This setting models a realistic scenario where the feedback for
every transactions is obtained only few days after the transaction was performed.
The properties of the different datasets are summarized in Table 2.

Table 2. Properties of the 6 datasets used in the experiments.

	#examples	Positives ratio	#features
Pima	767	34%	8
Breast cancer	286	30%	9
HIV	3,272	13.3%	8
Heart cleveland (4 vs all)	303	4.3%	13
w8a	64000	3%	300
Fraud	2,000,000	0.2%	40

We now describe our experimental setup. For the public datasets where the
training/testing sets are not available directly, we randomly generate 2/3-1/3
splits of the data to obtain training and test sets respectively. Hyperparameters
are tuned thanks to a 5-fold cross-validation over the training set, keeping the
values offering the best AP. We repeat the process over 30 runs and average the
results.

We compare our method, named SGBAP, to 4 other baselines[3]: SGBAP$_{sig}$
as defined previously, GB-Logistic which is the basic gradient boosting with
a negative binomial log-likelihood loss function [16] (pointwise and accuracy
based for binary datasets), LambdaMART-AP [22] a version of gradient boosting
that optimizes the average precision and RankBoost [17], a pairwise version of
AdaBoost for ranking. For each method, we fix a time limit to 86,000 s.

We evaluate the previous methods according to 4 criteria measured on the
test sets. First, we use the classic average precision (AP) and $AUCROC$. Additionally, we also consider 2 measures to best assess the quality of the approaches
for top-rank precision. For this purpose, we use the performance $Pos@Top$,
defined in [26], that gives the percentage of positive example retrieved before
a negative appears in the ranking. In other words, it corresponds to the recall
before the precision drops under 100%. We also evaluate the $P@k$ from Eq. 4. In
our setup, we set k to be the number of positive examples, which makes sense
in our context of highly unbalanced data when the objective is to provide a
short list of alerts to an expert and where the number of positive examples is

---

[3] Note that we did not use Adarank in our evaluation because the weights updates
rely on a notion of query that is not adapted to our framework.

much smaller than the negative examples. In fact, the latter measure is both precision and recall at rank k. This measure is also equivalent to the F_1 score since the latter is an harmonic mean between precision and recall. Note that we show experimentally in the supplementary material that AP is actually highly correlated to the F_1 score.

The results obtained are reported on Table 3. First, we can remark, that except for the Pima dataset that has the highest positive ratio, our approach is always better in terms of AP. SGBAP is also better than other baselines in terms of $Pos@top$ which is the hardest measure for evaluating the top-rank performance. Additionally, we see that for all datasets with a significantly low positive ratio (less than 15%), our approach is always better according to the

Table 3. Results obtained for the different evaluation criteria used in the paper. We indicate in bold font the best method with respect to each dataset and each evaluation measure. A – indicates that the method did not finish before the time limit.

Dataset	Algorithm	AUCROC	AP	Pos@Top	P@k
Pima	GB-Logistic	0.8279 ± 0.0352	0.7125 ± 0.0267	0.0388 ± 0.0379	$\mathbf{0.6608 \pm 0.0296}$
	RankBoost	$\mathbf{0.8352 \pm 0.0359}$	0.7281 ± 0.0621	$\mathbf{0.0620 \pm 0.0546}$	0.6586 ± 0.0298
	LambdaMART-AP	0.8177 ± 0.0304	$\mathbf{0.7338 \pm 0.0528}$	0.0407 ± 0.0443	0.6559 ± 0.0257
	SGBAP	0.8276 ± 0.0418	0.7119 ± 0.0486	0.0579 ± 0.0577	0.6455 ± 0.0356
	SGBAP$_{sig}$	0.8215 ± 0.0215	$0.7091\pm, 0.0328$	0.0388 ± 0.0346	0.6514 ± 0.0325
Breast cancer	GB-Logistic	0.6821 ± 0.0756	0.5089 ± 0.0562	0.0931 ± 0.0561	0.4457 ± 0.0739
	RankBoost	0.6492 ± 0.0562	0.4838 ± 0.0632	0.0461 ± 0.0513	0.4626 ± 0.0629
	LambdaMART-AP	0.6733 ± 0.0419	0.5280 ± 0.0680	0.0859 ± 0.0828	$\mathbf{0.5196 \pm 0.0624}$
	SGBAP	0.7124 ± 0.0596	$\mathbf{0.5602 \pm 0.0830}$	$\mathbf{0.1019 \pm 0.1018}$	0.4980 ± 0.0612
	SGBAP$_{sig}$	$\mathbf{0.7131 \pm 0.0521}$	0.5503 ± 0.0443	0.0729 ± 0.0693	0.5061 ± 0.0574
HIV	GB-Logistic	0.8598 ± 0.0155	0.5557 ± 0.0376	0.0303 ± 0.0284	0.5391 ± 0.0364
	RankBoost	0.8599 ± 0.0127	0.5464 ± 0.0276	0.0401 ± 0.0363	0.5309 ± 0.0254
	LambdaMART-AP	0.8222 ± 0.0466	0.4286 ± 0.0887	0.0075 ± 0.0176	0.4874 ± 0.0814
	SGBAP	$\mathbf{0.8661 \pm 0.0150}$	$\mathbf{0.5737 \pm 0.0347}$	$\mathbf{0.0536 \pm 0.0410}$	$\mathbf{0.5445 \pm 0.0351}$
	SGBAP$_{sig}$	0.7578 ± 0.0231	0.3928 ± 0.0434	0.041 ± 0.0250	0.3902 ± 0.0439
Heart cleveland	GB-Logistic	0.7544 ± 0.1020	0.1638 ± 0.0931	0.0133 ± 0.0498	0.1 ± 0.1420
	RankBoost	$\mathbf{0.8109 \pm 0.0515}$	0.1739 ± 0.0638	0.0150 ± 0.0565	0.0967 ± 0.1335
	LambdaMART-AP	0.7277 ± 0.1225	0.1809 ± 0.1011	0.0383 ± 0.0863	0.1333 ± 0.1287
	SGBAP	0.7789 ± 0.1178	$\mathbf{0.2188 \pm 0.1103}$	0.0483 ± 0.0970	$\mathbf{0.2017 \pm 0.1044}$
	SGBAP$_{sig}$	0.7983 ± 0.0638	0.2136 ± 0.0964	0.045 ± 0.0906	0.1566 ± 0.1295
w8a	GB-Logistic	0.9544 ± 0.0039	0.7385 ± 0.0154	0.0534 ± 0.0529	0.7091 ± 0.0152
	RankBoost	$\mathbf{0.9712 \pm 0.0028}$	0.7649 ± 0.0135	0.0392 ± 0.0451	0.7277 ± 0.008
	LambdaMART-AP	–	–	–	–
	SGBAP	0.9701 ± 0.0029	$\mathbf{0.8351 \pm 0.0100}$	$\mathbf{0.1779 \pm 0.0978}$	$\mathbf{0.7972 \pm 0.0132}$
	SGBAP$_{sig}$	–	–	–	–
Fraud	GB-Logistic	0.8808	0.1477	0.0009	0.2411
	RankBoost	$\mathbf{0.8829}$	0.1560	0.0005	0.2449
	LambdaMART-AP	–	–	–	–
	SGBAP	0.6878	$\mathbf{0.1747}$	$\mathbf{0.0059}$	$\mathbf{0.3203}$
	SGBAP$_{sig}$	–	–	–	–

P@k measure. Overall, we can remark that when the imbalance is high, our app-roach is always significantly better than other baselines according to 3 criteria: $AP, Pos@top$ and $P@k$ which clearly confirms that our method performs bet-ter for optimizing top-rank results. Note that, for the dataset HIV, SGBAP$_{sig}$ performed quite poorly. We believe that this is because of the early vanishing gradient due to the imbalance in the dataset. This effect does not appear in heart cleveland dataset most likely because of the small dataset size.

4.2   Top Rank Capability for a Decreasing Positive Ratio

In this section, we present an experiment showing the robustness of our approach when the ratio of positives decreases. We consider the Pima dataset because it has the highest ratio of positive instances and because our approach did not perform the best for all criteria. We aim at under-sampling the positive class (i.e. to decrease the positive ratio $\frac{P}{M}$). We start from the original positive ratio (34%) and go down to 3% by steps of ~ 0.05. For every new dataset, we follow the same experimental setup as described previously. At the end of the 30 runs for a given positive ratio dataset, we compute the average rank obtained by the examples in the test set and remove the top k positive instances such that $\frac{P-k}{M}$ is equal to the next positive ratio to evaluate. We repeat the previous set up until we reach 3% of positive examples in the dataset. We repeat this process independently for each method. The objective is to remove from the current dataset the easiest positive examples for each approach to evaluate its capability to move at the top new positive examples. Note that this makes harder the problem of ranking correctly in the top positive instances. Thus, the top rank performance measures should globally decrease.

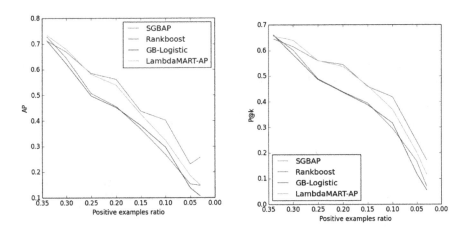

Fig. 4. The average precision and P@k at different positive example ratio for pima dataset.

The results with respect to the AP criterion and $P@k$ are presented on Fig. 4. From this experiment, we see that SGBAP outperforms the other models as the imbalance ratio increases and notably when the ratio of positives becomes smaller than 15% which confirms that our approach behaves clearly the best when the level of imbalance is high in comparison to other state of the art approaches.

5   Conclusion and Perspectives

In this paper, we presented SGBAP, a novel Stochastic Gradient Boosting based approach for optimizing directly a surrogate of the average precision measure. Our approximation is based on an exponential surrogate allowing us to compute our criterion in linear time which is crucial for dealing with large scale datasets such as for fraud detection tasks. We claim that this approach is well adapted for supervised anomaly detection in the context of highly unbalanced settings. Indeed, our criterion focuses specifically on the top-rank yielding a better *precision* in the top k positions.

A perspective of this work would be to optimize other interesting measures for learning to rank such as NDCG by means of a stochastic gradient descent approach. Another direction, would be to adapt the optimization of the surrogate of average precision to other learning models such as neural networks where we could take benefit from recent results in non-convex optimization.

References

1. Cortes, C., Mohri, M.: AUC optimization vs. error rate minimization. In: NIPS, pp. 313–320 (2003)
2. Chawla, N.V., Bowyer, K.W., Hall, L.O., Kegelmeyer, W.P.: SMOTE: synthetic minority over-sampling technique. JAIR **16**, 321–357 (2002)
3. Ramentol, E., Caballero, Y., Bello, R., Herrera, F.: SMOTE-RSB*: a hybrid pre-processing approach based on oversampling and undersampling for high imbalanced data-sets using SMOTE and rough sets theory. Knowl. Inf. Syst. **33**, 245–265 (2012)
4. Chawla, N.V., Lazarevic, A., Hall, L.O., Bowyer, K.W.: SMOTEBoost: improving prediction of the minority class in boosting. In: Lavrač, N., Gamberger, D., Todorovski, L., Blockeel, H. (eds.) PKDD 2003. LNCS (LNAI), vol. 2838, pp. 107–119. Springer, Heidelberg (2003). https://doi.org/10.1007/978-3-540-39804-2_12
5. Seiffert, C., Khoshgoftaar, T.M., Hulse, J.V., Napolitano, A.: RUSBoost: a hybrid approach to alleviating class imbalance. IEEE Trans. Syst. Man Cybern. **40**(1), 185–197 (2010)
6. Fan, W., Stolfo, S.J., Zhang, J., Chan, P.K.: AdaCost: misclassification cost-sensitive boosting. In: ICML, pp. 97–105 (1999)
7. Dal Pozzolo, A., Caelen, O., Bontempi, G.: When is undersampling effective in unbalanced classification tasks? In: Appice, A., Rodrigues, P.P., Santos Costa, V., Soares, C., Gama, J., Jorge, A. (eds.) ECML PKDD 2015. LNCS (LNAI), vol. 9284, pp. 200–215. Springer, Cham (2015). https://doi.org/10.1007/978-3-319-23528-8_13
8. Niculescu-Mizil, A., Caruana, R.: Predicting good probabilities with supervised learning. In: ICML, pp. 625–632 (2005)

9. Dal Pozzolo, A., Caelen, O., Johnson, R.A., Bontempi, G.: Calibrating probability with undersampling for unbalanced classification. In: SSCI, pp. 159–166 (2015)
10. Liu, T.Y.: Learning to Rank for Information Retrieval. Springer, Heidelberg (2011). https://doi.org/10.1007/978-3-642-14267-3
11. Joachims, T.: Optimizing search engines using clickthrough data. In: SIGKDD, pp. 133–142 (2002)
12. Yue, Y., Finley, T., Radlinski, F., Joachims, T.: A support vector method for optimizing average precision. In: SIGIR, pp. 271–278 (2007)
13. Breiman, L.: Random forests. Mach. Learn. **45**(1), 5–32 (2001)
14. Freund, Y., Schapire, R., Abe, N.: A short introduction to boosting. J.-Jpn Soc. Artif. Intell. **14**(771–780), 1612 (1999)
15. Chapelle, O., Chang, Y.: Yahoo! learning to rank challenge overview, pp. 1–24 (2011)
16. Friedman, J.H.: Greedy function approximation: a gradient boosting machine. Ann. Stat. **29**, 1189–1232 (2001)
17. Freund, Y., Iyer, R., Schapire, R.E., Singer, Y.: An efficient boosting algorithm for combining preferences. J. Mach. Learn. Res. **4**, 933–969 (2003)
18. Burges, C., Shaked, T., Renshaw, E., Lazier, A., Deeds, M., Hamilton, N., Hullender, G.: Learning to rank using gradient descent. In: ICML, pp. 89–96 (2005)
19. Herschtal, A., Raskutti, B.: Optimising area under the ROC curve using gradient descent. In: Proceedings of the Twenty-First International Conference on Machine learning, p. 49. ACM (2004)
20. Burges, C.J.: From RankNet to LambdaRank to LambdaMART: an overview. Learning **11**, 23–581 (2010)
21. Xu, J., Li, H.: AdaRank: a boosting algorithm for information retrieval. In: SIGIR, pp. 391–398. ACM (2007)
22. Wu, Q., Burges, C.J., Svore, K.M., Gao, J.: Adapting boosting for information retrieval measures. Inf. Retr. **13**(3), 254–270 (2010)
23. Burges, C.J., Ragno, R., Le, Q.V.: Learning to rank with nonsmooth cost functions. In: Schölkopf, P.B., Platt, J.C., Hoffman, T., (eds.) NIPS, pp. 193–200 (2007)
24. Hanley, J.A., McNeil, B.J.: The meaning and use of the area under a receiver operating characteristic (ROC) curve. Radiology **143**(1), 29–36 (1982)
25. Friedman, J.H.: Stochastic gradient boosting. Comput. Stat. Data Anal. **38**(4), 367–378 (2002)
26. Li, N., Jin, R., Zhou, Z.H.: Top rank optimization in linear time. In: NIPS, pp. 1502–1510 (2014)

Robust, Deep and Inductive Anomaly Detection

Raghavendra Chalapathy[1], Aditya Krishna Menon[2(✉)], and Sanjay Chawla[3]

[1] University of Sydney and Capital Markets Cooperative Research Centre
(CMCRC), Sydney, Australia
rcha9612@uni.sydney.edu.au
[2] Data61/CSIRO, Australian National University, Canberra, Australia
aditya.menon@data61.csiro.au
[3] Qatar Computing Research Institute, Al Rayyan, Qatar
schawla@qf.org.qa

Abstract. PCA is a classical statistical technique whose simplicity and
maturity has seen it find widespread use for anomaly detection. However,
it is limited in this regard by being sensitive to gross perturbations of the
input, and by seeking a linear subspace that captures normal behaviour.
The first issue has been dealt with by *robust PCA*, a variant of PCA
that explicitly allows for some data points to be arbitrarily corrupted;
however, this does not resolve the second issue, and indeed introduces
the new issue that one can no longer inductively find anomalies on a
test set. This paper addresses both issues in a single model, the *robust
autoencoder*. This method learns a nonlinear subspace that captures the
majority of data points, while allowing for some data to have arbitrary
corruption. The model is simple to train and leverages recent advances
in the optimisation of deep neural networks. Experiments on a range of
real-world datasets highlight the model's effectiveness.

Keywords: Anomaly detection · Outlier detection · Robust PCA
Autoencoders · Deep learning

1   Anomaly Detection: Motivation and Challenges

A common need when analysing real-world datasets is determining which
instances stand out as being dramatically dissimilar to all others. Such instances
are known as *anomalies*, and the goal of *anomaly detection* (also known as *outlier detection*) is to determine all such instances in a data-driven fashion [9].
Anomalies can be caused by errors in the data but sometimes are indicative of
a new, previously unknown, underlying process; in fact Hawkins [14] defines an
outlier as an observation that *deviates so significantly from other observations
as to arouse suspicion that it was generated by a different mechanism.*

Principal Component Analysis (PCA) [15] is a core method for a range of
statistical inference tasks, including anomaly detection. The basic idea of PCA
is that while many data sets are high-dimensional, they tend to inhabit a low-
dimensional manifold. PCA thus operates by (linearly) projecting data into a

© Springer International Publishing AG 2017
M. Ceci et al. (Eds.): ECML PKDD 2017, Part I, LNAI 10534, pp. 36–51, 2017.
https://doi.org/10.1007/978-3-319-71249-9_3

lower-dimensional space, so as to separate the *signal* from the *noise*; a data point which is far away from its projection is deemed as anomalous.

While intuitive and popular, PCA has limitations as an anomaly detection method. Notably, it is highly sensitive to data perturbation: one extreme data point can completely change the orientation of the projection, often leading to the masking of anomalies. A variant of PCA, known as a *robust* PCA (RPCA) limits the impact of anomalies by using a clever decomposition of the data matrix [8]. We will discuss RPCA in detail in Sect. 2, but note here that it still carries out a linear projection, and further cannot be used to make predictions on test instances; that is, we cannot perform *inductive* anomaly detection.

In this paper, we will relax the linear projection limitation of RPCA by using a deep and robust autoencoder [13,30]. The difference between RPCA and a deep autoencoder will be the use of a nonlinear activation function and the potential use of several hidden layers in the autoencoder. While this modification is conceptually simple, we show it yields noticeable improvements in anomaly detection performance on complex real-world image data, where a linear projection cannot capture sufficient structure in the data. Further, the robust autoencoder is capable of performing inductive anomaly detection, unlike RPCA.

In the sequel, we provide an overview of anomaly detection methods (Sect. 2), with a specific emphasis on matrix decomposition techniques such as PCA and its robust extensions. We then proceed to describe our proposed model based on autoencoders (Sect. 3), and present our experiment setup and results (Sects. 4 and 5). Finally, we describe directions for future work (Sect. 6).

2   Background and Related Work on Anomaly Detection

Consider a feature matrix $\mathbf{X} \in \mathbb{R}^{N \times D}$, where N denotes the number of data points and D the number of features for each point. For example, N could be the number of images in some photo collection, and D the number of pixels used to represent each image. The goal of anomaly detection is to determine which rows of $\mathbf{X}$ are anomalous, in the sense of being dissimilar to all other rows. We will use $\mathbf{X}_{i:}$ to denote the ith row of $\mathbf{X}$.

2.1   A Tour of Anomaly Detection Methods

Anomaly detection is a widely researched topic in the data mining and machine learning community [2,9]. The two primary strands of research have been the design of novel algorithms to detect anomalies, and the design *efficient* means of discovering all anomalies in a large dataset. In the latter strand, starting from the work of Bay and Schwabacher [4], several optimisations have been proposed to discover anomalies in near linear time [12].

In the former strand, which is our primary focus, most emphasis has been on non-parametric methods like distance and density based outliers [7,21]. For example, distance-based methods define a domain-dependent dissimilarity metric, and deem a point to be anomalous if it is relatively far away from its neighbours [35]. Another popular approach is the one-class SVM, which learns a smooth boundary that captures the majority of probability mass of the data [27].

In recent years, matrix factorization methods for anomaly detection have become popular. These methods provide a *reconstruction matrix* $\hat{\mathbf{X}} \in \mathbb{R}^{N \times D}$ of the input $\mathbf{X}$, and use the norm $\|\mathbf{X}_{i:} - \hat{\mathbf{X}}_{i:}\|_2^2$ as a measure of how anomalous a particular point $\mathbf{X}_{i:}$ is; if the reconstruction is close to the input, then it is deemed normal; else, anomalous. We describe several popular examples of this approach, beginning with principal component analysis (PCA).

2.2   PCA for Anomaly Detection

PCA finds the directions of maximal variance of the data. Supposing without loss of generality that the data matrix $\mathbf{X}$ has zero mean, this may be understood as the result of a matrix factorisation [6]:

$$\min_{\mathbf{W}^T \mathbf{W} = \mathbf{I}, \mathbf{Z}} \|\mathbf{X} - \mathbf{W}\mathbf{Z}\|_F^2 = \min_{\mathbf{U}} \|\mathbf{X} - \mathbf{X}\mathbf{U}\mathbf{U}^T\|_F^2. \tag{1}$$

Here, the reconstruction matrix is $\hat{\mathbf{X}} = \mathbf{X}\mathbf{U}\mathbf{U}^T$, where $\mathbf{U} \in \mathbb{R}^{D \times K}$ for some number of *latent dimensions* $K \ll D$. We can interpret $\mathbf{X}\mathbf{U}$ as a projection (or encoding) of $\mathbf{X}$ into a K-dimensional subspace, with the application of $\mathbf{U}^T$ as an inverse projection (or decoding) back into the original D dimensional space.

2.3   Autoencoders for Anomaly Detection

PCA assumes a linear subspace explains the data. To relax this assumption, consider instead

$$\min_{\mathbf{U}, \mathbf{V}} \|\mathbf{X} - f(\mathbf{X}\mathbf{U})\mathbf{V}\|_F^2 \tag{2}$$

for some non-decreasing *activation function* $f \colon \mathbb{R} \to \mathbb{R}$, and $\mathbf{U} \in \mathbb{R}^{D \times K}, \mathbf{V} \in \mathbb{R}^{K \times D}$. For the purposes of anomaly detection, one can define the reconstruction matrix as $\hat{\mathbf{X}} = f(\mathbf{X}\mathbf{U})\mathbf{V}$.

Equation 2 corresponds to an autoencoder with a single hidden layer [13]. Popular choices of $f(\cdot)$ include the sigmoid $f(a) = (1 + \exp(-a))^{-1}$ and the rectified linear unit or ReLU $f(x) = \max(0, a)$. As before, we can interpret $\mathbf{X}\mathbf{U}$ as an encoding of $\mathbf{X}$ into a K-dimensional subspace; however, by applying a nonlinear $f(\cdot)$, the projection is implicitly onto a nonlinear manifold.

2.4   Robust PCA

Another way to generalise PCA is to solve, for a tuning parameter $\lambda > 0$,

$$\min_{\mathbf{S}, \mathbf{N}} \|\mathbf{S}\|_* + \lambda \cdot \|\mathbf{N}\|_1 : \mathbf{X} = \mathbf{S} + \mathbf{N}, \tag{3}$$

where $\|\cdot\|_*$ denotes the trace or nuclear norm $\|\mathbf{X}\|_* = \mathrm{tr}((\mathbf{X}^T\mathbf{X})^{1/2})$, and $\|\cdot\|_1$ the elementwise ℓ_1 norm. For the purposes of anomaly detection, one can define the reconstruction matrix $\hat{\mathbf{X}} = \mathbf{X} - \mathbf{N} = \mathbf{S}$.

Intuitively, Eq. 3 separates $\mathbf{X}$ into a signal matrix $\mathbf{S}$ and a noise matrix $\mathbf{N}$, where the signal matrix has low-rank structure, and the noise is assumed to not overwhelm the signal for most of the matrix entries. The trace norm may be seen as a convex relaxation of the rank function; thus, this objective can be understood as a relaxed version of PCA.

Equation 3 corresponds to robust PCA (RPCA) [8]. Unlike standard PCA, this objective can effortlessly deal with a single entry perturbed arbitrarily. When $\lambda \to +\infty$, we will end up with $\mathbf{N} = \mathbf{0}, \mathbf{S} = \mathbf{X}$, i.e. we will claim that there is no noise in the data, and so all points are deemed normal. On the other hand, when $\lambda \to 0$, we will end up with $\mathbf{N} = \mathbf{X}, \mathbf{S} = \mathbf{0}$, i.e. we will claim that there is no signal in the data, and so points with high norm are deemed anomalous.

2.5 Direct Robust Matrix Factorization

Building upon RPCA, Xiong et al. [32] introduced the direct robust matrix factorization method (DRMF), where for tuning parameters K, e one solves:

$$\min_{\mathbf{S},\mathbf{N}} \quad \|\mathbf{X} - (\mathbf{N} + \mathbf{S})\|_F^2 : \operatorname{rank}(\mathbf{S}) \leq K, \|\mathbf{N}\|_0 \leq e. \tag{4}$$

As before, the matrix $\mathbf{N}$ captures the anomalies and $\mathbf{S}$ captures the signal. Unlike RPCA, one explicitly constraints $\mathbf{S}$ to be low-rank, rather than merely having low trace norm; and one explicitly constraints $\mathbf{N}$ to have a maximal number of nonzeros, rather than merely having bounded ℓ_1 norm. The lack of convexity of the objective requires a bespoke algorithm for the optimisation.

2.6 Robust Kernel PCA

Another way to overcome the linear assumption of PCA is the robust kernel PCA (RKPCA) approach of [25]. For a feature mapping Φ into a reproducing kernel Hilbert space, and projection operator $\mathbf{P}$ of a point into the KPCA subspace, it is proposed to reconstruct an input $\mathbf{x} \in \mathbb{R}^D$ by solving the pre-image problem

$$\hat{\mathbf{x}} = \operatorname*{argmin}_{\mathbf{z} \in \mathbb{R}^D} E_0(\mathbf{x}, \mathbf{z}) + C \cdot \|\Phi(\mathbf{z}) - \mathbf{P}\Phi(\mathbf{z})\|^2, \tag{5}$$

where E_0 is a robust measure of reconstruction error (i.e. not merely the Euclidean norm), and $C > 0$ is a tuning parameter. RKPCA does not explicitly handle gross outliers, unlike RPCA; however, by choosing a rich feature mapping Φ, one can capture nonlinear anomalies. This choice of feature mapping must be pre-specified, whereas autoencoder methods implicitly *learn* a good mapping.

3   From Robust PCA to Robust Autoencoders

We now present our robust (convolutional) autoencoder model for anomaly detection. The method can be seen as an extension of robust PCA to allow for a nonlinear manifold that explains most of the data.

3.1   Robust (Convolutional) Autoencoders

Let $f\colon \mathbb{R} \to \mathbb{R}$ be some non-decreasing activation function. Now consider the following objective, which combines the salient elements of Eqs. 2 and 3:

$$\min_{\mathbf{U},\mathbf{V},\mathbf{N}} \|\mathbf{X} - (f(\mathbf{X}\mathbf{U})\mathbf{V} + \mathbf{N})\|_F^2 + \frac{\mu}{2} \cdot (\|\mathbf{U}\|_F^2 + \|\mathbf{V}\|_F^2) + \lambda \cdot \|\mathbf{N}\|_1, \qquad (6)$$

where $f(\cdot)$ is understood to act elementwise, and $\lambda, \mu > 0$ are tuning parameters. This is a form of *robust autoencoder*: one encodes the input into the latent representation $\mathbf{Z} = f(\mathbf{X}\mathbf{U})$, which is then decoded via $\mathbf{V}$. The additional $\mathbf{N}$ term captures gross outliers in the data, as with robust PCA. For the purposes of anomaly detection, we have reconstruction matrix $\hat{\mathbf{X}} = f(\mathbf{X}\mathbf{U})\mathbf{V}$.

When $\lambda \to +\infty$, we get $\mathbf{N} = \mathbf{0}$, and the model reduces to a standard autoencoder (Eq. 2). When $\lambda \to 0$, then one possible solution is $\mathbf{N} = \mathbf{X}$ and $\mathbf{U} = \mathbf{V} = \mathbf{0}$, so that the model memorises the training data. For intermediate λ, the model augments a standard autoencoder with a noise absorption term that endows robustness.

More generally, Eq. 6 can be seen as an instance of

$$\min_{\theta,\mathbf{N}} \|\mathbf{X} - (\hat{\mathbf{X}}(\theta) + \mathbf{N})\|_F^2 + \frac{\mu}{2} \cdot \Omega(\theta) + \lambda \cdot \|\mathbf{N}\|_1, \qquad (7)$$

where $\hat{\mathbf{X}}(\theta)$ is some generic predictor with parameters θ, and $\Omega(\cdot)$ a regularisation function. Observe that we could pick $\hat{\mathbf{X}}(\theta)$ to be a convolutional autoencoder [19, 30], which would be suitable when dealing with image data; such a model will be studied extensively in our experiments. Further, the regulariser Ω could involve more general matrix norms, such as the $\ell_{1,2}$ norm [16].

3.2   Training the Model

The objective function of the model of Eqs. 6, 7 is non-convex, but unconstrained and sub-differentiable. There are several ways of performing optimisation. For example, for differentiable activation f, one could compute sub-gradients with respect to all model parameters and apply backpropagation. However, to leverage existing advances in training deep networks, we observe that:

- For fixed $\mathbf{N}$, the objective is equivalent to that of a standard (convolutional) autoencoder on the matrix $\mathbf{X} - \mathbf{N}$. Thus, one can optimise the parameters θ using any modern (stochastic) optimisation tool for deep learning that exploits gradients, such as Adam [20].
- For fixed θ (i.e. $\mathbf{U}, \mathbf{V}$ in the standard autoencoder case), the objective is

$$\min_{\theta,\mathbf{N}} \|\mathbf{N} - (\mathbf{X} - \hat{\mathbf{X}}(\theta))\|_F^2 + \lambda \cdot \|\mathbf{N}\|_1,$$

which trivially solvable via the soft thresholding operator on the matrix $\mathbf{X} - \hat{\mathbf{X}}(\theta)$ [3], with solution

$$\mathbf{N}_{ij} = \begin{cases} (\mathbf{X} - \hat{\mathbf{X}}(\theta))_{ij} - \frac{\lambda}{2} & \text{if } (\mathbf{X} - \hat{\mathbf{X}}(\theta))_{ij} > \frac{\lambda}{2} \\ (\mathbf{X} - \hat{\mathbf{X}}(\theta))_{ij} + \frac{\lambda}{2} & \text{if } (\mathbf{X} - \hat{\mathbf{X}}(\theta))_{ij} < -\frac{\lambda}{2} \\ 0 & \text{else.} \end{cases}$$

We thus alternately optimise $\mathbf{N}$ and θ until the change in the overall objective is below some threshold. The use of stochastic optimisation for the first step, and the simplicity of the optimisation for the second step, means that we can easily train the model where data arrives in an online or streaming fashion.

3.3   Predicting with the Model

One convenient property of our model is that the anomaly detector will be inductive, i.e. it can generalise to unseen data points. One can interpret the model as learning a robust representation of the input, which is unaffected by gross noise; such a representation should thus be able to accurately model any unseen points that lie on the same manifold as the data used to train the model.

Formally, given a new $\mathbf{x}_* \in \mathbb{R}^D$, one simply computes $f(\mathbf{x}_*^T \mathbf{U})\mathbf{V}$ to score this point. The larger $\|\mathbf{x}_* - \mathbf{V}^T f(\mathbf{U}^T \mathbf{x}_*)\|_2^2$ is, the more likely the point is deemed to be anomalous. We emphasise that such inductive predictions are simply not possible with the robust PCA method, as it estimates parameters for the $N \times D$ observations present in $\mathbf{X}$, with no means of generalising to unseen data.

3.4   Connection to Robust PCA

While the robust autoencoder of Eq. 6 has clear conceptual similarity to robust PCA, it may seem that choices such as the ℓ_2 penalty on $\mathbf{U}, \mathbf{V}$ are somewhat arbitrarily used in place of the trace norm. We now show how the objective can in fact be naturally derived as an extension of RPCA.

The trace norm can be represented in the variational form [26] $\|\mathbf{S}\|_* = \min_{\mathbf{W}\mathbf{V}=\mathbf{S}} \frac{1}{2} \cdot (\|\mathbf{W}\|_F^2 + \|\mathbf{V}\|_F^2)$. The robust PCA objective is thus equivalently

$$\min_{\mathbf{W},\mathbf{V},\mathbf{N}} \frac{1}{2} \cdot (\|\mathbf{W}\|_F^2 + \|\mathbf{V}\|_F^2) + \lambda \cdot \|\mathbf{N}\|_1 : \mathbf{X} = \mathbf{W}\mathbf{V} + \mathbf{N}.$$

This objective has the disadvantage of being non-convex, but the advantage of being amenable to extensions. Pick some $\mu > 0$, and consider a relaxed version of the robust PCA objective:

$$\min_{\mathbf{W},\mathbf{V},\mathbf{N},\mathbf{E}} \|\mathbf{E}\|_F^2 + \frac{\mu}{2} \cdot (\|\mathbf{W}\|_F^2 + \|\mathbf{V}\|_F^2) + \lambda \cdot \|\mathbf{N}\|_1 : \mathbf{X} = \mathbf{W}\mathbf{V} + \mathbf{N} + \mathbf{E}.$$

Here, we allow for further systematic errors $\mathbf{E}$ which have low average magnitude. We can equally consider the unconstrained objective

$$\min_{\mathbf{W},\mathbf{V},\mathbf{N}} \|\mathbf{X} - (\mathbf{W}\mathbf{V} + \mathbf{N})\|_F^2 + \frac{\mu}{2} \cdot (\|\mathbf{W}\|_F^2 + \|\mathbf{V}\|_F^2) + \lambda \cdot \|\mathbf{N}\|_1 \tag{8}$$

This re-expression of robust PCA has been previously noted, for example in Sprechmann et al. [29]. To derive the robust autoencoder from Eq. 8, suppose now that we constrain $\mathbf{W} = \mathbf{X}\mathbf{U}$. This is a natural constraint in light of Eq. 1, since for standard PCA we factorise $\mathbf{X}$ into $\hat{\mathbf{X}} = \mathbf{X}\mathbf{U}\mathbf{U}^T$. Then, we have the objective

$$\min_{\mathbf{U},\mathbf{V},\mathbf{N}} \|\mathbf{X} - (\mathbf{XUV} + \mathbf{N})\|_F^2 + \frac{\mu}{2} \cdot (\|\mathbf{XU}\|_F^2 + \|\mathbf{V}\|_F^2) + \lambda \cdot \|\mathbf{N}\|_1.$$

Now suppose we modify the regulariser to only operate on $\mathbf{U}$ rather than $\mathbf{XU}$:

$$\min_{\mathbf{U},\mathbf{V},\mathbf{N}} \|\mathbf{X} - (\mathbf{XUV} + \mathbf{N})\|_F^2 + \frac{\mu}{2} \cdot (\|\mathbf{U}\|_F^2 + \|\mathbf{V}\|_F^2) + \lambda \cdot \|\mathbf{N}\|_1.$$

This is again natural in the context of standard PCA, since there we have $\mathbf{W} = \mathbf{XU}$ satisfying $\mathbf{W}^T\mathbf{W} = \mathbf{I}$. Observe now that we have derived Eq. 6 for a linear activation function $f(x) = x$. The robust autoencoder thus extends this model by employing a nonlinear activation.

3.5 Relation to Existing Models

Our contribution is a nonlinear extension of RPCA for anomaly detection. As noted above, the key advantages over RPCA are the ability to capture nonlinear structure in the data, as well as the ability to detect anomalies in an inductive setting. The price we have to pay is the lack of convexity of the objective function, unlike RPCA; nonetheless, we shall demonstrate that the model can be effectively trained using the procedure described in Sect. 3.2.

Some works have employed deep networks for anomaly detection [31,34], but without explicitly accounting for gross anomalies. For example, the recent work of [34] employed an autoencoder-inspired objective to train a probabilistic neural network, with extensions to structured data; the use of an RPCA-style noise matrix $\mathbf{N}$ may be useful to explore in conjunction with such methods.

Our method is also distinct to denoising autoencoders (DNA), wherein noise is explicitly added to instances [30], whereas we *infer* the noise automatically. The approaches have slightly different goals: DNAs aim to extract good features from the data, while our aim is to identify anomalies.

Finally, while nonlinear extensions of PCA-style matrix factorisation (including via autoencoders) have been explored in contexts such as collaborative filtering [23,28], we are unaware of prior usage for anomaly detection.

4 Experimental Setup

In this section we show the empirical effectiveness of Robust Convolutional Autoencoder over the state-of-the-art methods on real-world data. Our primary focus will be on non-trivial image datasets, although our method is applicable in any context where autoencoders are useful e.g. speech.

4.1 Methods Compared

We compare our proposed Robust Convolutional Autoencoder (RCAE) with the following state-of-the art methods for anomaly detection:

- **Truncated SVD**, which for zero-mean features is equivalent to PCA.
- **Robust PCA (RPCA)** [8], as per Eq. 3.

- **Robust kernel PCA (RKPCA)** [25], as per Eq. 5.
- **Autoencoder (AE)** [5], as per Eq. 2.
- **Convolutional Autoencoder (CAE)**, a convolutional autoencoder without any accounting for gross anomalies i.e. Eq. 7 where $\lambda = +\infty$.
- **Robust Convolutional Autoencoder (RCAE)**, our proposed model as per Eq. 7.

We used TensorFlow [1] for the implementation of AE, CAE and RCAE[1]. For RPCA and RKPCA, we used publicly available implementations[2,3].

4.2   Datasets

We compare all methods on three real-world datasets:

- `restaurant`, comprising video background modelling and activity detection consisting of snapshots of restaurant activities [32].
- `usps`, comprising the USPS handwritten digits [17].
- `cifar-10` consisting of 60000 32×32 colour images in 10 classes, with 6000 images per class [22].

For each dataset, we perform further processing to create a well-posed anomaly detection task, as described in the next section.

4.3   Evaluation Methodology

As anomaly detection is an unsupervised learning problem, model evaluation is challenging. For the `restaurant` dataset, there are no ground truth anomalies, and so we perform a qualitative analysis by visually comparing the anomalies flagged by various methods, as done in the original robust PCA paper [8].

For the other two datasets, we follow a standard protocol (see e.g. [32]) wherein anomalies are explicitly identified in the training set. We can then evaluate the predictive performance of each method as measured against the ground truth anomaly labels, using three standard metrics:

- the area under the precision-recall curve (AUPRC)
- the area under the ROC curve (AUROC)
- the precision at 10 (P@10).

AUPRC and AUROC measure ranking performance, with the former being preferred under class imbalance [11]. P@10 measures classification performance, being the fraction of the top 10 scored instances which are actually anomalous.

For `CIFAR − 10`, the labelled dataset is created by combining 5000 images of dog and 50 images of cat; a good anomaly detection method should thus flag the cats to be anomalous. Similarly, for `usps`, the dataset is created by a mixture

---

[1] https://github.com/raghavchalapathy/rcae.

[2] http://perception.csl.illinois.edu/matrix-rank/sample_code.html.

[3] http://www3.cs.stonybrook.edu/~minhhoai/downloads.html.

Table 1. Summary of datasets used in experiments.

Dataset	# instances	# anomalies	# features
restaurant	200	Unknown (foreground)	19200
usps	231	11 ('7')	256
cifar-10	5000	50 (cats)	1024

of 220 images of '1's, and 11 images of '7'as in [33]. Details of the datasets are summarised in Table 1.

Additionally, we also test the ability of our model to perform denoising of images, as well as detecting inductive anomalies.

4.4   Network Parameters

Although we have observed that deeper RCAE networks tend to achieve better image reconstruction performance, there exist four fold options related to network parameters to be chosen: (a) number of convolutional filters, (b) filter size, (c) strides of convolution operation and (d) activation applied. We tuned via grid search additional hyper-parameters, including the number of hidden-layer nodes $H \in \{3, 64, 128\}$, and regularisation λ within range $[0, 100]$. The learning, drop-out rates and regularization parameter μ were sampled from a uniform distribution in the range $[0.05, 0.1]$. The embedding and initial weight matrices were all sampled from the uniform distribution within range $[-1, 1]$.

5   Experimental Results

In this section, we present experiments for three scenarios:

(a) non-inductive anomaly detection,
(b) inductive anomaly detection, and
(c) image denoising.

5.1   Non-inductive Anomaly Detection Results

We present results on the three datasets described in Sect. 4.

(1) restaurant dataset. We work with the restaurant video activity detection dataset [32], and consider the problem of inferring the background of videos via removal of (anomalous) foreground pixels. Estimating the background in videos is important for tasks such as anomalous activity detection. It is however difficult because of the variability of the background (e.g. due to lighting conditions) and the presence of foreground objects such as moving objects and people.

For this experiment, we only compare the RPCA and RCAE methods, owing to a lack of ground truth labels.

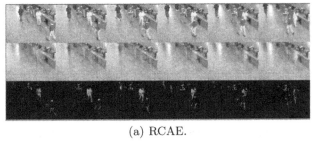

(a) RCAE.

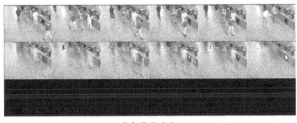

(b) RPCA.

Fig. 1. Top anomalous images containing original image (people walking in the lobby) decomposed into background (lobby) and foreground (people), **restaurant** dataset.

Parameter settings. For RPCA, rank $K = 64$ is used. Per the success of the Batch Normalization architecture [18] and Exponential Linear Units [10], we have found that convolutional+batch-normalization+elu layers provide a better representation of convolutional filters. Hence, in this experiment the RCAE adopts four layers of (conv-batch-normalization-elu) in the encoder part and four layers of (conv-batch-normalization-elu) in the decoder portion of the network. RCAE network parameters such as (number of filter, filter size, strides) are chosen to be (16, 3, 1) for first and second layers and (32, 3, 1) for third and fourth layers of both encoder and decoder layers.

Results. While there are no ground truth anomalies in this dataset, a qualitative analysis reveals RCAE to outperforms its counterparts in capturing the foreground objects. Figure 1 compares the top 6 most anomalous images for RCAE and RPCA. We see that the most anomalous images contain high foreground activity (which are recognised as anomalous). Visually, we see that the background reconstruction produced by RPCA contains a few blemishes in some cases, while for RCAE the backgrounds are smooth.

(2) usps dataset. From the **usps** handwritten digit dataset, we create a dataset with a mixture of 220 images of '1's, and 11 images of '7', as in [33]. Intuitively, the latter images are treated as being anomalous, as the corresponding images have different characteristics to the majority of the training data. Each image is flattened as a row vector, yielding a 231×256 training matrix.

Table 2. Comparison between the baseline (bottom four rows) and state-of-the-art systems (top three rows). Results are the mean and standard error of performance metrics over 20 random training set draws. Highlighted cells indicate best performer.

(a) usps				(b) cifar-10		
Methods	AUPRC	AUROC	P@10	AUPRC	AUROC	P@10
RCAE	0.9614 ± 0.0025	0.9988± 0.0243	0.9108 ± 0.0113	0.9934 ± 0.0003	0.6255 ± 0.0055	0.8716 ± 0.0005
CAE	0.7003 ± 0.0105	0.9712 ± 0.0002	0.8730 ± 0.0023	0.9011 ± 0.0000	0.6191 ± 0.0000	0.0000 ± 0.0000
AE	0.8533 ± 0.0023	0.9927 ± 0.0022	0.8108 ± 0.0003	0.9341 ± 0.0029	0.5260 ± 0.0003	0.2000 ± 0.0003
RKPCA	0.5340 ± 0.0262	0.9717 ± 0.0024	0.5250 ± 0.0307	0.0557 ± 0.0037	0.5026 ± 0.0123	0.0550 ± 0.0185
DRMF	0.7737 ± 0.0351	0.9928 ± 0.0027	0.7150 ± 0.0342	0.0034 ± 0.0000	0.4847 ± 0.0000	0.0000 ± 0.0000
RPCA	0.7893 ± 0.0195	0.9942 ± 0.0012	0.7250 ± 0.0323	0.0036 ± 0.0000	0.5211 ± 0.0000	0.0000 ± 0.0000
SVD	0.6091 ± 0.1263	0.9800 ± 0.0105	0.5600 ± 0.0249	0.0024 ± 0.0000	0.5299 ± 0.0000	0.0000 ± 0.0000

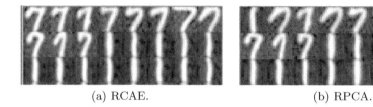

(a) RCAE.   (b) RPCA.

Fig. 2. Top anomalous images, usps dataset.

Parameter settings. For SVD and RPCA methods, rank $K = 64$ is used. For AE, the inputs are flattened images as a column vector of size 256, and the hidden layer is a column vector of size 64 (matching the rank K).

For DRMF, we follow the settings of [33]. For RKPCA, we used a Gaussian kernel with bandwidth 0.01, a cost parameter $C = 1$, and requested 60% of the KPCA spectrum (which roughly selects 64 principal components).

For RCAE, we set two layers of convolution layers with the filter number to be 32, filter size to be 3×3, with number of strides as 1 and rectified linear unit (ReLU) as activation with max-pooling layer of dimension 2×2.

Results. From Table 2, we see that it is a near certainty for all '7' are accurately identified as outliers. Figure 2 shows the top anomalous images for RCAE, where indeed the '7's are correctly placed at the top of the list. By contrast, for RPCA there are also some '1's placed at the top.

(3) cifar-10 dataset. We create a dataset with anomalies by combining 5000 random images of dog and 50 images of cat, as illustrated in Fig. 3. In this scenario the cats are anomalies, and the goal is to detect all the cats in an unsupervised manner.

Parameter settings. For SVD and RPCA methods, rank $K = 64$ is used. We trained a three-hidden-layer autoencoder (AE) (1024-256-64-256-1024 neurons).

(a) RCAE.

(b) RPCA.

Fig. 3. Top anomalous images, `cifar-10` dataset.

The middle hidden layer size is set to be same as rank $K = 64$, and the model is trained using Adam [20]. The decoding layer uses sigmoid function in order to capture the nonlinearity characteristics from latent representations produced by the hidden layer. Finally, we obtain the feature vector for each image by obtaining the latent representation from the hidden layer.

For RKPCA, we used a Gaussian kernel with bandwidth $5 \cdot 10^{-8}$, a cost parameter $C = 0.1$, and requested 55% of the KPCA spectrum (which roughly selects 64 principal components). The RKPCA runtime was prohibitive on the full sample (see Sect. 5.4), so we resorted to a subsample of 1000 dogs and 50 cats.

The RCAE architecture in this experiment is same as for **restaurant**, containing four layers of (conv-batch-normalization-elu) in the encoder part and four layers of (conv-batch-normalization-elu) in the decoder portion of the network. RCAE network parameters such as (number of filter, filter size, strides) are chosen to be (16, 3, 1) for first and second layers and (32, 3, 1) for third and fourth layers of both encoder and decoder.

Results. From Table 2, RCAE clearly outperforms all existing state-of-the art methods in anomaly detection. Note that basic CAE, with no robustness (effectively $\lambda = \infty$), is also outperformed by our method, indicating that it is crucial to explicitly handle anomalies with the $\mathbf{N}$ term.

Figure 3 illustrates the most anomalous images for our RCAE method, compared to RPCA. Owing to the latter involving learning a linear subspace, the model is unable to effectively distinguish cats from dogs; by contrast, RCAE can effectively determine the manifold characterising most dogs, and identifies cats to be anomalous with respect to this.

5.2 Inductive Anomaly Detection Results

We conduct an experiment to assess the detection of *inductive* anomalies. Recall that this is a capability of our RCAE model, but not e.g. RPCA. We consider the following setup: we train our model on 5000 dog images, and then evaluate

Table 3. Inductive anomaly detection results on `cifar-10`. Note that RPCA and DRMF are inapplicable here. Highlighted cells indicate best performer.

	SVD	RKPCA	AE	CAE	RCAE
AUPRC	0.1752 ± 0.0051	0.1006 ± 0.0045	0.6200 ± 0.0005	0.6423 ± 0.0005	0.6908 ± 0.0001
AUROC	0.4997 ± 0.0066	0.4988 ± 0.0125	0.5007 ± 0.0010	0.4708 ± 0.0003	0.5576 ± 0.0005
P@10	0.2150 ± 0.0310	0.0900 ± 0.0228	0.1086 ± 0.0001	0.2908 ± 0.0001	0.5986 ± 0.0001

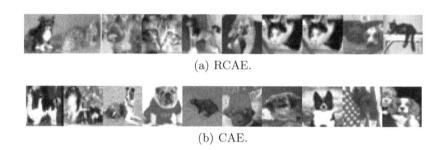

(a) RCAE.

(b) CAE.

Fig. 4. Top inductive anomalous images, `cifar-10` dataset.

it on a test set comprising 500 dogs and 50 cat images. As before, we wish all methods to accurately determine the cats to be anomalies.

Table 3 summarises the detection performance for all the methods on this inductive task. The lower values compared to Table 2 are indicative that the problem here is more challenging than anomaly detection on a single dataset; nonetheless, we see that our RCAE method manages to convincingly outperform both the SVD and AE baselines. This is confirmed qualitatively in Fig. 4, where we see that RCAE correctly identifies many cats in the test set as anomalous, while the basic CAE method suffers.

5.3   Image Denoising Results

Finally, we test the ability of the model to de-noise images, which is a form of anomaly detection on individual pixels (or more generally, features). In this experiment, we train all models on a set of 5000 images of dog from `cifar-10`. For each image, we then add salt-and-pepper noise at a rate of 10%. Our goal is to recover the original image as accurately as possible.

Figure 5 illustrates that the most anomalous images in the presence of noise contain images of the variations of dog class images (e.g. containing person's face). Further, Fig. 6 illustrates for various methods the mean square error between the reconstructed and original images. RCAE effectively suppresses the noise as evident from the low error. The improvement over raw CAE is modest, but suggests that there is benefit to explicitly accounting for noise.

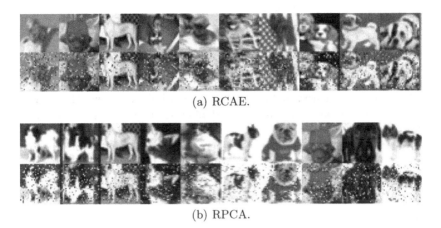

(a) RCAE.

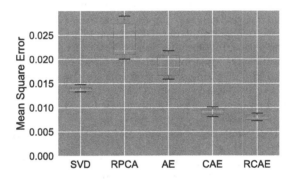

(b) RPCA.

Fig. 5. Top anomalous images in original form (first row), noisy form (second row), image denoising task on `cifar-10`.

Fig. 6. Illustration of the mean square error boxplots obtained for various models on image denoising task, `cifar-10` dataset. In this setting, RCAE suppresses the noise and detects the background and foreground images effectively.

5.4   Comparison of Training Times

We remark finally that our RCAE method is comparable in training efficiency to existing methods. For example, on the small-scale `restaurant` dataset, it takes 1 min to train RPCA, and 8.5 min to train RKPCA, compared with 10 min for our RCAE method. The ability to leverage recent advances in deep learning as part of our optimisation (e.g. training models on a GPU) is we believe a salient feature of our approach.

We note that while the RKPCA method is fast to train on smaller datasets, on larger datasets it suffers from the $O(n^2)$ complexity of kernel methods; for example, it takes over an hour to train on the `cifar-10` dataset. It is plausible that one could leverage recent advances in fast approximations of kernel methods [24], and studying these would be of interest in future work. Note that the issue of using a fixed kernel function would remain, however.

6 Conclusion

We have extended the robust PCA model to the nonlinear autoencoder setting. To the best of our knowledge, ours is the first approach which is *robust, nonlinear* and *inductive*. The robustness ensures that the model is not over-sensitive to anomalies; the nonlinearity helps discover potentially more subtle anomalies; and being inductive makes it possible to deploy our model in a live setting.

While autoencoders are a powerful mechansim for data representation they suffer from their "black-box" nature. There is a growing body of research on outlier description, i.e., explain the reason why a data point is anomalous. A direction of future reason is to extend deep autoencoders for outlier *description*.

References

1. Abadi, M., Agarwal, A., Barham, P., Brevdo, E., Chen, Z., Citro, C., Corrado, G.S., Davis, A., Dean, J., Devin, M., et al.: TensorFlow: large-scale machine learning on heterogeneous distributed systems. arXiv preprint arXiv:1603.04467 (2016)
2. Aggarwal, C.C.: Outlier Analysis. Springer, New York (2016)
3. Bach, F., Jenatton, R., Mairal, J., Obozinski, G.: Convex optimization with sparsity-inducing norms. In: Optimization for Machine Learning. MIT Press (2011)
4. Bay, S.D., Schwabacher, M.: Mining distance-based outliers in near linear time with randomization and a simple pruning rule. In: International Conference on Knowledge Discovery and Data Mining (KDD) (2003)
5. Bengio, Y., et al.: Learning deep architectures for AI. Found. Trends® Mach. Learn. **2**(1), 1–127 (2009)
6. Bishop, C.M.: Pattern Recognition and Machine Learning. Springer, New York (2006)
7. Breunig, M.M., Kriegel, H.P., Ng, R.T., Sander, J.: LOF: identifying density-based local outliers. In: ACM SIGMOD record, vol. 29, pp. 93–104. ACM (2000)
8. Candés, E., Li, X., Ma, Y., Wright, J.: Robust principal component analysis?: recovering low-rank matrices from sparse errors. In: 2010 IEEE Sensor Array and Multichannel Signal Processing Workshop (SAM), pp. 201–204. IEEE (2010)
9. Chandola, V., Banerjee, A., Kumar, V.: Outlier detection: a survey. ACM Comput. Surv. (2007)
10. Clevert, D.A., Unterthiner, T., Hochreiter, S.: Fast and accurate deep network learning by exponential linear units (ELUs). arXiv preprint arXiv:1511.07289 (2015)
11. Davis, J., Goadrich, M.: The relationship between Precision-Recall and ROC curves. In: International Conference on Machine Learning (ICML) (2006)
12. Ghoting, A., Parthasarathy, S., Otey, M., Ghoting, A., Parthasarathy, S., Otey, M.E.: Fast mining of distance-based outliers in high-dimensional datasets. Data Min. Knowl. Disc. **16**(3), 349–364 (2008)
13. Goodfellow, I., Bengio, Y., Courville, A.: Deep Learning. MIT Press, Cambridge (2016). http://www.deeplearningbook.org
14. Hawkins, D.: Identification of Outliers. Chapman and Hall, London (1980)
15. Hotelling, H.: Analysis of a complex of statistical variables into principal components. J. Educ. Psychol. **24**, 417–441 (1933)
16. Huang, J., Zhang, T.: The benefit of group sparsity. Ann. Stat. **38**(4), 1978–2004 (2010)

17. Hull, J.J.: A database for handwritten text recognition research. IEEE Trans. Pattern Anal. Mach. Intell. **16**(5), 550–554 (1994)
18. Ioffe, S., Szegedy, C.: Batch normalization: accelerating deep network training by reducing internal covariate shift. arXiv preprint arXiv:1502.03167 (2015)
19. Jain, V., Seung, S.: Natural image denoising with convolutional networks. Adv. Neural Inf. Process. Syst. **21**, 769–776 (2008)
20. Kingma, D., Ba, J.: Adam: a method for stochastic optimization. arXiv preprint arXiv:1412.6980 (2014)
21. Knorr, E.M., Ng, R.T.: A unified notion of outliers: properties and computation. In: KDD, pp. 219–222 (1997)
22. Krizhevsky, A., Hinton, G.: Learning multiple layers of features from tiny images. Technical report (2009)
23. Lawrence, N.D., Urtasun, R.: Non-linear matrix factorization with Gaussian processes. In: International Conference on Machine Learning (ICML) (2009)
24. Lopez-Paz, D., Sra, S., Smola, A.J., Ghahramani, Z., Schölkopf, B.: Randomized nonlinear component analysis. In: International Conference on Machine Learning (ICML) (2014)
25. Nguyen, M.H., Torre, F.: Robust kernel principal component analysis. In: Advances in Neural Information Processing Systems (NIPS) (2009)
26. Recht, B., Fazel, M., Parrilo, P.A.: Guaranteed minimum-rank solutions of linear matrix equations via nuclear norm minimization. SIAM Rev. **52**(3), 471–501 (2010)
27. Schlkopf, B., Platt, J.C., Shawe-Taylor, J.C., Smola, A.J., Williamson, R.C.: Estimating the support of a high-dimensional distribution. Neural Comput. **13**(7), 1443–1471 (2001)
28. Sedhain, S., Menon, A.K., Sanner, S., Xie, L.: AutoREC: autoencoders meet collaborative filtering. In: International Conference on World Wide Web (WWW) (2015)
29. Sprechmann, P., Bronstein, A.M., Sapiro, G.: Learning efficient sparse and low rank models. IEEE Trans. Pattern Anal. Mach. Intell. **37**(9), 1821–1833 (2015)
30. Vincent, P., Larochelle, H., Lajoie, I., Bengio, Y., Manzagol, P.A.: Stacked denoising autoencoders: learning useful representations in a deep network with a local denoising criterion. JMLR **11**, 3371–3408 (2010)
31. Williams, G., Baxter, R., He, H., Hawkins, S., Gu, L.: A comparative study of RNN for outlier detection in data mining. In: International Conference on Data Mining (ICDM) (2002)
32. Xiong, L., Chen, X., Schneider, J.: Direct robust matrix factorizatoin for anomaly detection. In: International Conference on Data Mining (ICDM). IEEE (2011)
33. Xu, H., Caramanis, C., Sanghavi, S.: Robust PCA via outlier pursuit. In: Advances in Neural Information Processing Systems, pp. 2496–2504 (2010)
34. Zhai, S., Cheng, Y., Lu, W., Zhang, Z.: Deep structured energy based models for anomaly detection. In: International Conference on Machine Learning (ICML) (2016)
35. Zhao, M., Saligrama, V.: Anomaly detection with score functions based on nearest neighbor graphs. In: Advances in Neural Information Processing Systems (NIPS), pp. 2250–2258 (2009)

Sentiment Informed Cyberbullying Detection in Social Media

Harsh Dani, Jundong Li[✉], and Huan Liu

Computer Science and Engineering, Arizona State University, Tempe, AZ, USA
{hdani,jundongl,huanliu}@asu.edu

Abstract. Cyberbullying is a phenomenon which negatively affects the individuals, the victims suffer from various mental issues, ranging from depression, loneliness, anxiety to low self-esteem. In parallel with the pervasive use of social media, cyberbullying is becoming more and more prevalent. Traditional mechanisms to fight against cyberbullying include the use of standards and guidelines, human moderators, and blacklists based on the profane words. However, these mechanisms fall short in social media and cannot scale well. Therefore, it is necessary to develop a principled learning framework to automatically detect cyberbullying behaviors. However, it is a challenging task due to short, noisy and unstructured content information and intentional obfuscation of the abusive words or phrases by social media users. Motivated by sociological and psychological findings on bullying behaviors and the correlation with emotions, we propose to leverage sentiment information to detect cyberbullying behaviors in social media by proposing a sentiment informed cyberbullying detection framework. Experimental results on two real-world, publicly available social media datasets show the superiority of the proposed framework. Further studies validate the effectiveness of leveraging sentiment information for cyberbullying detection.

Keywords: Cyberbullying detection · Social media
Sentiment information

1 Introduction

Cyberbullying is an increasingly important and serious social problem, which can negatively affect the individuals. It is defined as the phenomena of using the internet, cell phones and other electronic devices to willfully hurt or harass others. Due to the recent popularity and growth of social media platforms such as Facebook and Twitter, cyberbullying is becoming more and more prevalent. It has been identified as a serious national health concern by the American Psychological Association and the White House. In addition to that, according to the recent report by National Crime Prevention Council, more than 40% of teens in the US have been bullied on social media platforms [1]. The victims of cyberbullying often suffer from depression, loneliness, anxiety and low self-esteem [2]. In more tragic scenarios, the victims might attempt suicide or suffer

© Springer International Publishing AG 2017
M. Ceci et al. (Eds.): ECML PKDD 2017, Part I, LNAI 10534, pp. 52–67, 2017.
https://doi.org/10.1007/978-3-319-71249-9_4

from interpersonal problems. Since cyberbullying is not restricted by time and place, it has more insidious effects than traditional forms of bullying [3].

Traditional mechanisms to combat cyberbullying behaviors include the development of standards and guidelines that all users must adhere to, employment of human editors to manually check the bullying behavior, the use of profane word lists, and the use of regular expressions, etc. However, these mechanisms fall short in social media since social media data is naturally dynamic [4]. As a result, the maintenance of these mechanisms is time and labor consuming. Also, they cannot scale well. Therefore, it demands the use of a principled learning framework to accurately detect new cyberbullying behaviors automatically.

The detection of cyberbullying in social media is a far more challenging task than one can expect due to the following two reasons: First, the content information in social media is short, noisy and unstructured [5]. The short and unstructured texts make traditional text representation techniques, i.e., bag-of-words very sparse and high-dimensional. As a result, traditional machine learning classifiers often cannot work well due to the curse of dimensionality [6]. Second, the users in social media intentionally obfuscate the words or phrases in the sentence to evade the manual and automatic checking. Obfuscation such as "n00b" makes it impossible for traditional mechanisms to accurately detect abusive words or phrases, leading to more false positives.

Previous psychological and sociological studies on the bullying behaviors and emotional intelligence suggest that emotional information can be used to better understand the bullying behaviors [7]. Emotional intelligence refers to the ability of an individual to accurately perceive emotion, use emotions to facilitate thought, understand and manage the emotion [8]. The lower the emotional intelligence of the user, the more likely an individual will be involved in the bullying behaviors [9]. Motivated from this insight, we investigate if the use of sentiment information of the post content could help better understand and accurately detect cyberbullying behaviors in social media.

In this paper, we attempt to perform cyberbullying detection in a supervised way by proposing a principled learning framework. More specifically, we first investigate whether sentiment information is particularly correlated with cyberbullying behaviors. Then, we discuss how to deal with short, noisy, unstructured content and how to properly leverage sentiment information for cyberbullying detection. Methodologically, we present a novel learning framework called Sentiment Informed Cyberbullying Detection (SICD). Experiments on two real-world social media datasets validate the effectiveness of the proposed framework. To summarize, we make the following contributions:

- We formally define the problem of sentiment informed cyberbullying detection in social media;
- We verify the sentiment difference between normal posts and bullying posts by comparing their sentiment score distributions;
- We present a principled learning framework which leverages sentiment information of user posts to detect cyberbullying in social media; and

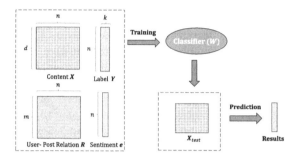

Fig. 1. Proposed sentiment informed cyberbullying detection framework.

– We perform empirical experimental studies on two real-world, publicly available social media datasets to verify the efficacy of the proposed framework.

2 Problem Definition

We first introduce the notations used in this paper. We use boldface uppercase letters (e.g., $\mathbf{A}$) to denote matrices, boldface lowercase letters to denote vectors (e.g., $\mathbf{a}$) and lowercase letters (e.g., a) to denote scalars. We denote the transpose of matrix $\mathbf{A}$ as $\mathbf{A}^T$ and the transpose of vector $\mathbf{a}$ as $\mathbf{a}^T$. $\mathrm{Tr}(\mathbf{A})$ denotes the trace of matrix $\mathbf{A}$ if it is square. The entry of matrix $\mathbf{A}$ at the row i and column j is denoted as $\mathbf{A}_{ij}$. We denote the i-th row of matrix $\mathbf{A}$ as $\mathbf{A}_{i*}$ and the j-th column as $\mathbf{A}_{*j}$. $||\mathbf{A}||_{2,1}$ denotes the $\ell_{2,1}$-norm such that $||\mathbf{A}||_{2,1} = \sum_i \sqrt{\sum_j \mathbf{A}_{ij}^2}$.

Let $\mathbf{C} = [\mathbf{X}, \mathbf{Y}]$ denote the corpus of social media posts, where $\mathbf{X} \in \mathbb{R}^{d \times n}$ is the content matrix of these posts, $\mathbf{Y} \in \mathbb{R}^{n \times k}$ is a one-hot label matrix, n is the number of posts, d is the number of features, k is the number of classes. In this work, we set $k = 2$, indicating that a post is either normal or bullying. The social media corpus $\mathbf{C}$ is generated by a set of m users, i.e., $\{u_1, u_2, \ldots, u_m\}$, $\mathbf{R} \in \mathbb{R}^{m \times n}$ denotes the user-post relationships (as shown in Fig. 1), $\mathbf{R}_{ij} = 1$ if post j is posted by u_i and $\mathbf{R}_{ij} = 0$ otherwise. Meanwhile, each post in $\mathbf{C}$ is associated with a sentiment score in the range of $[-1, 1]$, -1 denotes the most negative sentiment score and 1 denotes the most positive sentiment, and $\mathbf{e}$ represents the sentiment score vector for all n posts. With these notations, we now formally define the problem of sentiment informed cyberbullying detection as follows:

Given a corpus of social media posts with the content information $\mathbf{X}$ and the label information $\mathbf{Y}$, the user-post relationships $\mathbf{R}$ and the sentiment score of posts $\mathbf{e}$, we aim to learn a classifier $\mathbf{W}$ to automatically detect whether the unseen social media posts (i.e., test data) are normal posts or bullying posts.

3   Exploratory Data Analysis

One important motivation of the problem we study is to investigate the correlation between sentiment information and cyberbullying behaviors. We first introduce two real-world social media datasets and then present our observations from these two datasets.

3.1   Datasets

We use two publicly available social media datasets, both datasets contain labeled social media posts, i.e., the post is either labeled as normal or bullying.

Twitter is a microblogging website which allows users to post 140 characters messages. The posts in this dataset have been manually labeled as bullying or normal. This dataset has been kindly provided by Xu et al. [2].

MySpace is a social networking website which allows its registered users to view pictures, read chat and check other users' profile information. Also, each post in the dataset is manually labeled as normal or bullying. This dataset has been kindly provided by Bayzick et al. [10].

Detailed statistics of these two datasets are summarized in Table 1.

Table 1. Statistics of Twitter and MySpace datasets.

	Twitter	MySpace
# of posts	7,321	3,245
# of features	3,709	4,236
# of positive posts	2,102	950
# of negative posts	5,219	2,295
# of users	7,043	1,053
Avg. posts per user	1.04	2.98

3.2   Verifying the Sentiment Score Distribution Difference

We conduct an empirical study to verify if the sentiment distribution of the normal posts is different from the bullying posts. Particularly, we learn a distant supervision based sentiment classification model [11] on Stanford Twitter Sentiment dataset. Then we employ it to obtain the sentiment score of each post in the Twitter and MySpace datasets. The sentiment score of each post is normalized in the range of $[-1, 1]$. Figure 2(a) and (b) show the sentiment score distribution of the normal and the bullying posts in both Twitter and MySpace datasets, respectively. In these figures, X-axis shows the sentiment polarity score and Y-axis shows the density of posts. We can observe that two distributions are

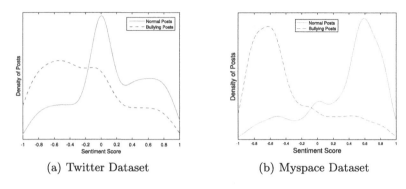

(a) Twitter Dataset   (b) Myspace Dataset

Fig. 2. Sentiment distribution of normal and bullying posts.

centered with different mean values. This suggests that there is a clear difference between the sentiment score distribution of the normal posts and the bullying posts, and the sentiment of bullying posts are more negative than normal posts.

3.3 Verifying Sentiment Consistency

In this subsection, we aim to investigate whether the sentiment scores of two posts with the same class labels, i.e., both posts are normal or bullying, are more similar than two randomly chosen posts. We use two-sampled t-test to verify the statistical significance of the above-stated hypothesis.

Suppose $d(p_i, p_j)$ denotes the sentiment similarity of two social media posts p_i and p_j, which can be computed by the RBF kernel. Let s_c and s_d be two vectors of the same length, each element in s_c denotes the sentiment similarity of two posts p_i and p_j with the same class label, and each element in s_d denotes the sentiment similarity of two randomly selected posts. We then use two-sampled t-test to investigate whether the sentiment similarity of two posts with the same class label is higher than two randomly chosen posts. The null hypothesis is as follows: $H_0 : \tau_c \leq \tau_d$ and the alternative hypothesis is as follows: $H_1 : \tau_c > \tau_d$, where τ_c and τ_d represent the sample means of s_c and s_d, respectively.

The result of t-test, i.e., p-values obtained on Twitter and MySpace are $1.09e^{-11}$ and $1.028e^{-7}$, respectively. It suggests that there is a strong statistical evidence (with a significance level $\alpha = 0.01$) to reject the null hypothesis. In other words, we validate the alternative hypothesis that the sentiment scores of two posts with the same class label are more similar than two randomly chosen posts. The two-sampled t-test results further pave the way to incorporate sentiment information for cyberbullying detection.

4   The Proposed Framework - SICD

In this section, we introduce the proposed SICD framework in detail. First, we present how to model the short, noisy and unstructured user post content.

Then we discuss how to model the user-post relationships and how to model sentiment information for cyberbullying detection.

4.1   Modeling Content of Social Media Posts

In order to find better text representation for cyberbullying detection, we employ unigram model with TF-IDF as feature values due to its success in cyberbullying detection [12]. Also, we perform stopwords removal and stemming.

In social media, the posts made by users are often short, noisy and unstructured. Also, these posts are not necessarily about the same topic which causes the vocabulary size to be extremely large. Hence, traditional text representation techniques such as n-grams and bag-of-words become extremely high-dimensional. Also, short text content of posts causes these feature representations to be extremely sparse. Such high-dimensional and sparse feature representations often cause poor prediction performance of traditional machine classifiers.

In recent years, sparse learning has been widely used to alleviate the negative effects of high-dimensional features to improve the prediction performance. Hence, we employ sparse learning techniques to deal with sparse, noisy and unstructured posts. More specifically, we use $\ell_{2,1}$-norm regularization term to seek a more compact feature space. The $\ell_{2,1}$-norm regularization selects a subset of relevant features across all the data instances with joint sparsity [13,14]. The classification problem then can be formulated as follows:

$$\min_{\mathbf{W}} \frac{1}{2}||\mathbf{X}^T\mathbf{W} - \mathbf{Y}||_F^2 + \lambda||\mathbf{W}||_{2,1}, \tag{1}$$

where λ is a parameter to control the feature sparsity. In the above formulation, the first term minimizes the least squared loss between post content and class labels and the second term seeks a more compact feature representation.

4.2   Modeling User-Post Relationships

Text data in social media is often linked due to the presence of various social relations, and these correlations can be explained by well-received social science theories such as Homophily [15] and Social Influence [16]. In particular, we hypothesize that if two social media posts are from the same user, they are more likely to have the same class label than two randomly chosen posts. In order to test this hypothesis, we create two equally sized vectors $\mathbf{up_c}$ and $\mathbf{up_d}$, where each element of the first vector denotes the label difference (Euclidean distance) of two posts by the same user and each element of the second vector denotes the label difference of two randomly chosen posts. We perform two-sampled t-test to investigate the above hypothesis. We form the null hypothesis as H_0: $m_c \geq m_d$ and the alternative hypothesis as H_1: $m_c < m_d$, where m_c and m_d represent the sample means of $\mathbf{up_c}$ and $\mathbf{up_d}$, respectively. The results show that there is a strong evidence (with a significance level $\alpha = 0.01$) to reject the null hypothesis.

In order to model the above mentioned user-post relationships, we propose to add a regularization term to minimize the label difference of the two posts if they are from the same user. Specifically, we first construct an affinity matrix $\mathbf{A} \in \mathbb{R}^{n \times n}$ from matrix $\mathbf{R}$ as follows: $\mathbf{A} = \mathbf{R}^T\mathbf{R}$, such that $\mathbf{A}_{ij} = 1$ denotes that two social media posts are by the same user and $\mathbf{A}_{ij} = 0$ otherwise. With this, the user-post relationships can be modeled by minimizing the following term:

$$\frac{1}{2}\sum_{i=1}^{n}\sum_{j=1}^{n}\mathbf{A}_{ij}||\hat{\mathbf{Y}}_{i*} - \hat{\mathbf{Y}}_{j*}||_2^2 = \mathrm{Tr}(\mathbf{W}^T\mathbf{X}\mathbf{L_A}\mathbf{X}^T\mathbf{W}), \tag{2}$$

where $\hat{\mathbf{Y}} = \mathbf{X}^T\mathbf{W}$ is the predicted value of the class label $\mathbf{Y}$. $\mathbf{L_A} = \mathbf{D_A} - \mathbf{A}$ is the Laplacian matrix, $\mathbf{D_A}$ is a diagonal matrix with $\mathbf{D_A} = \sum_i \mathbf{A}_{ij}$.

4.3 Modeling Sentiment Information

Motivated by psychological and sociological findings on the correlation of emotions and bullying behaviors, we propose to incorporate sentiment information to detect cyberbullying behaviors. From Sect. 3, we have an observation that sentiment score distributions of normal posts and bullying posts are different and posts with the same label are more likely to have similar sentiment scores than two randomly chosen posts. Now we discuss how to leverage these observations to perform cyberbullying detection.

To model sentiment information of posts, we construct an undirected affinity graph $\mathbf{S} \in \mathbb{R}^{n \times n}$ where each node denotes a social media post and edge weight denotes the sentiment similarity. In this paper, we propose to construct the k-nearest neighbor graph to model the sentiment affinity between different posts. More specifically, the matrix $\mathbf{S}$ can be defined as:

$$\mathbf{S}_{ij} = \begin{cases} \exp(-\frac{||e_i - e_j||_2^2}{\sigma^2}) & \text{if } e_i \in \mathcal{N}_k(e_j) \text{ or } e_j \in \mathcal{N}_k(e_i) \\ 0 & \text{otherwise,} \end{cases}$$

where $\mathcal{N}_k(e_i)$ denotes the k-nearest neighbors of post p_i in terms of sentiment score. Then, we propose to model sentiment information with another Graph Laplacian [17]. The key idea is that if the sentiment scores of two posts are close to each other, their labels are similar. We formulate the above idea by minimizing the following term:

$$\frac{1}{2}\sum_{i=1}^{n}\sum_{j=1}^{n}\mathbf{S}_{ij}||\hat{\mathbf{Y}}_{i*} - \hat{\mathbf{Y}}_{j*}||_2^2 = \mathrm{Tr}(\mathbf{W}^T\mathbf{X}\mathbf{L_S}\mathbf{X}^T\mathbf{W}), \tag{3}$$

where $\mathbf{L_S} = \mathbf{D_S} - \mathbf{S}$ is the Laplacian matrix of the sentiment affinity matrix $\mathbf{S}$. Here, $\mathbf{D_S}$ denotes the diagonal degree matrix with $\mathbf{D_S} = \sum_i \mathbf{S}_{ij}$.

4.4 Sentiment Informed Cyberbullying Detection (SICD)

As illustrated from the previous sections, we employ sparse learning to model the content of the social media post. Also, we model user-post relationships and

sentiment information. By considering all the types of the information, the task
of sentiment informed cyberbullying detection can be formulated as:

$$\min_{\mathbf{W}} F(\mathbf{W}) = \frac{1}{2}||\mathbf{X}^T\mathbf{W} - \mathbf{Y}||_F^2 + \lambda||\mathbf{W}||_{2,1}$$
$$+ \frac{\alpha}{2}\text{Tr}(\mathbf{W}^T\mathbf{XL_A}\mathbf{X}^T\mathbf{W}) + \frac{\beta}{2}\text{Tr}(\mathbf{W}^T\mathbf{XL_S}\mathbf{X}^T\mathbf{W}), \qquad (4)$$

where α and β are parameters to control the contribution of user-post relationships and sentiment information, respectively. By solving the objective function
in Eq. (4), we can get $\mathbf{W}$ as the learned classifier. To detect the cyberbullying
behaviors on unseen social media post $\mathbf{x}$, we can use the following formulation:
$\arg\max_{i \in \{bully, normal\}} \mathbf{x}^T\mathbf{W}_{*i}$.

5   Algorithmic Details

Due to the presence of the $\ell_{2,1}$-norm, the optimization problem in Eq. (4) is non-
smooth but convex. Now we introduce how to solve the optimization problem
along with the time complexity analysis.

5.1   Optimization Algorithm for SICD

A natural choice to solve the optimization problem in Eq. (4) is to use subgradient descent method [18]. However, it has a very slow convergence rate, i.e.,
$O(\frac{1}{\epsilon^2})$ where ϵ denotes the desired accuracy, which makes it not suitable for
real-world applications. In recent years, proximal gradient descent [19,20] has
been widely used to solve large-scale non-smooth convex optimization problems,
where the objective function can be separated into both smooth and non-smooth
parts. In our scenario, $||\mathbf{W}||_{2,1}$ is the non-smooth part and the other terms form
the smooth part $f(\mathbf{W})$. In each iteration of proximal gradient descent, $F(\mathbf{W})$ is
linearized around the current estimate $\mathbf{W}_t$, where t indicates the t-th iteration.
In particular, $\mathbf{W}$ is updated by solving the following optimization problem:

$$\mathbf{W}_{t+1} = \arg\min_{\mathbf{W}} \mathcal{G}_{\eta_t}(\mathbf{W}, \mathbf{W}_t). \qquad (5)$$

$\mathcal{G}_{\eta_t}(\mathbf{W}, \mathbf{W}_t)$ is defined as:

$$\mathcal{G}_{\eta_t}(\mathbf{W}, \mathbf{W}_t) = f(\mathbf{W}_t) + \langle \nabla f(\mathbf{W}_t), \mathbf{W} - \mathbf{W}_t \rangle + \frac{\eta_t}{2}||\mathbf{W} - \mathbf{W}_t||_F^2 + \lambda||\mathbf{W}||_{2,1} \quad (6)$$

where η_t is the step size that can be determined by the backtracking line search
algorithm. $\langle \mathbf{A}, \mathbf{B} \rangle$ denotes the dot product between two matrices $\mathbf{A}$ and $\mathbf{B}$:
$\langle \mathbf{A}, \mathbf{B} \rangle = \text{Tr}(\mathbf{A}^T\mathbf{B})$. The gradient of the smooth part $f(\mathbf{W})$ is formulated as:

$$\nabla f(\mathbf{W}_t) = \mathbf{X}\mathbf{X}^T\mathbf{W}_t - \mathbf{X}\mathbf{Y} + \alpha\mathbf{XL_A}\mathbf{X}^T\mathbf{W}_t + \beta\mathbf{XL_S}\mathbf{X}^T\mathbf{W}_t. \qquad (7)$$

In Eq. (6), we ignore the terms that are not related to $\mathbf{W}$ and the objective function boils down to the following optimization problem:

$$\mathbf{W}_{t+1} = \pi_{\eta_t}(\mathbf{W}_t) = \arg\min_{\mathbf{W}} \frac{1}{2}||\mathbf{W} - \mathbf{U}_t||_F^2 + \frac{\lambda}{\eta_t}||\mathbf{W}||_{2,1}, \tag{8}$$

where $\mathbf{U}_t = \mathbf{W}_t - \frac{1}{\eta_t}\nabla f(\mathbf{W}_t)$. The above problem can be further decomposed into k sub-problems. Each sub-problem can be formally formulated as follows:

$$\mathbf{w}_{t+1}^i = \arg\min_{\mathbf{w}^i} ||\mathbf{w}^i - \mathbf{u}_t^i||_2^2 + \frac{\lambda}{\eta_t}||\mathbf{w}^i||_2, \tag{9}$$

where the $\mathbf{w}_{t+1}^i$, $\mathbf{w}^i$ and $\mathbf{u}_t^i$ are the i-th row of the matrix $\mathbf{W}_{t+1}$, $\mathbf{W}$ and $\mathbf{U}_t$, respectively. Given the value of λ, the Euclidean projection of the above optimization problem has a closed-form solution, which can be formulated as:

$$\mathbf{w}_{t+1}^i = \begin{cases} (1 - \frac{\lambda}{\eta_t||\mathbf{u}_t^j||_2})\mathbf{u}_t^j; & \text{if } ||\mathbf{u}_t^j||_2 \geq \frac{\lambda}{\eta_t}, \\ \mathbf{0}; & \text{otherwise.} \end{cases} \tag{10}$$

Since the algorithm described above has closed-form Euclidean projection [20], hence it has the same convergence rate (i.e., $\frac{1}{\epsilon}$) as traditional gradient descent algorithms for smooth convex optimization problems. As discussed in [20], the proximal algorithm can be further accelerated to achieve the optimal convergence rate of $O(\frac{1}{\sqrt{\epsilon}})$ by employing Nestrov's method [21].

5.2   Time Complexity Analysis

Given a corpus of $\mathbf{C}$ with n social media posts and a feature dimension of d, it requires $O(nd)$ operations to obtain the gradient of the least squared formulation. The Euclidean projection for the $\ell_{2,1}$-norm according to Eq. (10) requires $O(2n)$ operations [20]. Third, the Laplacian regularization for the modeling of user-post relationships requires $O(nd)$. Similarly, the Laplacian regularization for the modeling of sentiment information also requires $O(nd)$. Also, by employing the Nestrov's accelerated method, we can achieve the optimal convergence rate of $O(\frac{1}{\sqrt{\epsilon}})$. Hence, the total time complexity of the proposed Algorithm is $O(\frac{1}{\sqrt{\epsilon}}(nd + 2n + nd + nd)) = O(\frac{1}{\sqrt{\epsilon}}(nd))$.

6   Experiments

In this section, we perform experiments to evaluate the effectiveness of the proposed SICD framework. After introducing the experimental settings, we present the detailed experimental results.

6.1 Experimental Settings

We follow standard experimental settings [12] to evaluate the performance of the proposed SICD framework. To avoid the bias brought by imbalanced class distributions, we use AUC and F1-measure as the classification metrics.

There are three positive parameters involved in our framework. λ controls the contribution of the sparse regularization. α controls the contribution of user-post relationships and β controls the contribution of sentiment information modeling. In the experiments, we set these parameters as $\lambda = 0.1$, $\alpha = 0.1$, $\beta = 0.05$, and $k = 20$ for the k-nearest neighbor in Eq. (3) by using grid search strategies.

6.2 Performance Evaluation

We compare our proposed SICD framework with the following baseline methods:

- **LS:** Traditional linear classification method with least squared loss [22].
- **Lasso:** This is a supervised sparse learning method [22] which uses ℓ_1-norm sparse regularization on the basis of least squared loss.
- **MF:** We perform NMF [23] on the content information for a compact representation and then apply SVM.
- **POS:** This method uses TF-IDF features, POS-tags of the bigrams, and the list of profane words as feature sets and then classifies posts using SVM [24].
- **USER:** This method uses TF-IDF features, and user related features such as gender and age as feature sets and then classifies posts using SVM [12].

For the *USER* baseline, if the user did not provide age or gender information, we impute the age information by the mean value and gender information by the most frequent value of others. For all methods, we perform five-fold cross-validation and report the average results. Particularly, we first divide 80% of the data as training data and the remaining 20% as the test set. The Tables 2 and 3 summarize the results on the Twitter dataset and MySpace dataset, respectively. It should be noted that in these tables, the training ratio is varied among the 80% training data.

We draw the following observations from these two tables:

- *SICD* consistently outperforms other baseline methods on both datasets with the varied training ratio. We also perform pairwise Wilcoxon signed-rank test [25] between *SICD* and other baselines, the results show that SICD is statistically significant better (with significance level $\alpha = 0.01$).
- *Lasso* and *MF* both achieves better performance than *LS*. It indicates that performing dimensionality reduction on the original content matrix can reduce the noisy information contained and helps improve the performance.
- As we increase the training ratio from 10% to 90%, the performance of *SICD* tends to increase gradually. It shows that more training data helps achieve better performance on the cyberbullying detection problem.
- *POS* outperforms *LS*, *Lasso* and *MF*, which indicates that POS tags of the frequent bi-grams and the list of profane words are often good indicators of cyberbullying behaviors.

Table 2. Classification evaluation of different methods on Twitter data.

	Train ratio	10%	20%	30%	40%	50%	60%	70%	80%	90%
F1	LS	0.4057	0.4105	0.4128	0.4264	0.4454	0.4519	0.4586	0.4662	0.4724
	Lasso	0.4635	0.5254	0.5734	0.5783	0.5870	0.5927	0.6039	0.6120	0.6187
	MF	0.5090	0.5197	0.5785	0.5819	0.5882	0.5916	0.5974	0.6008	0.6225
	POS	0.4934	0.5279	0.5812	0.5864	0.5985	0.6023	0.6104	0.6191	0.6247
	USER	0.4789	0.5190	0.5797	0.5805	0.5820	0.5939	0.6089	0.6178	0.6235
	SICD	**0.5601**	**0.5965**	**0.6127**	**0.6265**	**0.6354**	**0.6445**	**0.6697**	**0.6894**	**0.7056**
AUC	LS	0.6103	0.6142	0.6168	0.6259	0.6309	0.6338	0.6392	0.6435	0.6519
	Lasso	0.6419	0.6934	0.7219	0.7234	0.7297	0.7318	0.7532	0.7617	0.7698
	MF	0.6567	0.6745	0.7261	0.7281	0.7309	0.7335	0.7368	0.7397	0.7446
	POS	0.6497	0.6915	0.7245	0.7310	0.7391	0.7426	0.7469	0.7583	0.7624
	USER	0.6389	0.6867	0.7236	0.7295	0.7338	0.7431	0.7446	0.7516	0.7603
	SICD	**0.7049**	**0.7369**	**0.7567**	**0.7684**	**0.7869**	**0.7934**	**0.7977**	**0.8051**	**0.8169**

Table 3. Classification evaluation of different methods on MySpace data.

	Train ratio	10%	20%	30%	40%	50%	60%	70%	80%	90%
F1	LS	0.3937	0.4112	0.4370	0.4432	0.4691	0.4807	0.4916	0.5018	0.5075
	Lasso	0.3960	0.4338	0.4625	0.4759	0.4811	0.4925	0.5046	0.5120	0.5235
	MF	0.4145	0.4427	0.4861	0.4917	0.5109	0.5164	0.5254	0.5364	0.5446
	POS	0.4390	0.4502	0.4745	0.4818	0.4897	0.5012	0.5191	0.5286	0.5341
	USER	0.4267	0.4478	0.4723	0.4789	0.4842	0.4981	0.5085	0.5256	0.5320
	SICD	**0.4928**	**0.5086**	**0.5301**	**0.5572**	**0.5691**	**0.5791**	**0.5886**	**0.6071**	**0.6105**
AUC	LS	0.6096	0.6138	0.6236	0.6284	0.6348	0.6408	0.6485	0.6516	0.6547
	Lasso	0.6106	0.6219	0.6331	0.6378	0.6412	0.6501	0.6521	0.6587	0.6625
	MF	0.6147	0.6276	0.6487	0.6495	0.6569	0.6573	0.6639	0.6748	0.6837
	POS	0.6248	0.6314	0.6385	0.6418	0.6479	0.6509	0.6576	0.6657	0.6517
	USER	0.6210	0.6303	0.6349	0.6392	0.6441	0.6487	0.6461	0.6625	0.6658
	SICD	**0.6509**	**0.6549**	**0.6681**	**0.6915**	**0.6920**	**0.7224**	**0.7296**	**0.7404**	**0.7539**

- Similarly, *USER* outperforms *LS*, *Lasso* and *MF* which indicates that adding user based features such as gender and age helps improve the classification performance. However, it is inferior to the proposed *SICD*, one potential reason is that the age or gender information is often scarcely available in social media due to privacy reasons [26,27].
- Figure 3 demonstrates the prediction by *SICD* visually for the Twitter dataset. The top words for the bullying posts according to the ground truth and prediction by *SICD* are described in the top-left and bottom-left part of Fig. 3. Similarly, the top words for the normal post according to the ground truth and SICD are described in the top-right and bottom-right part of the Fig. 3. As we can observe, there is a significant overlap of the words in ground truth and prediction by *SICD* which visually demonstrates the effectiveness of the proposed framework.

Fig. 3. Prominent words for Twitter dataset.

6.3   Impact of Sentiment Information

In order to investigate the impact of sentiment information on cyberbullying detection, we assess the effectiveness of different types of information in SICD. In particular, we compare our proposed method with the following methods:

Table 4. Impact of sentiment information in Twitter and MySpace datasets.

Methods	Twitter		MySpace	
	F1	AUC	F1	AUC
Sentiment	0.2014	0.5214	0.2106	0.5197
Content	0.4724	0.6519	0.5075	0.6547
Content+UPR	0.6544	0.7921	0.5908	0.7032
Content+SENT	0.6298	0.7846	0.5894	0.6891
SICD	**0.7056**	**0.8169**	**0.6105**	**0.7539**

- **Content:** This is the traditional Least Squared classification model where only content information **X** is used.
- **Sentiment:** We first compute the sentiment score of the each post and then calculate its distance from the mean of the bullying and normal posts groups. The post is classified into the group with the shorter distance.
- **Content+UPR:** This method is a variant of our method, where the sentiment regularization is removed.
- **Content+SENT:** This method is a variant of our method, where the user-post relationship regularization is removed.

The experimental results on Twitter and MySpace are in Table 4. We have the following observations:

- With all the types of the information considered, *SICD* achieves the best cyberbullying detection performance.
- The *Sentiment* method achieves the worst performance. It indicates that although we observe the difference in the sentiment scores of the normal posts and the bullying posts, we cannot just use this information to detect cyberbullying. *Content* achieves better performance compared to *Sentiment*. It suggests that content information is the most effective source of information to perform cyberbullying detection.
- The *Content+UPR* and the *Content+SENT* achieves better performance than *Content* and *Sentiment*. It indicates that integration of either user-post relationships or sentiment information helps achieve better performance compared to traditional text-based cyberbullying detection methods.

6.4 Parameter Sensitivity

Our proposed SICD has two important parameters: α and β. The parameter α and β control the contribution of user-post relationships and sentiment information, respectively. To better understand the effects of these two parameters, we vary the values of α and β as {0.001, 0.01, 0.1, 0.2, 0.4, 0.8, 1.0, 10.0, 100.0} and report the classification performance on both datasets. The classification results w.r.t. AUC are shown in Fig. 4(a) and (b).

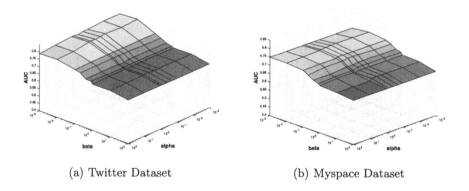

(a) Twitter Dataset   (b) Myspace Dataset

Fig. 4. Impact of the parameters α and β on the proposed framework

It can be observed that when β is around 0.01, SICD achieves the best performance. When we gradually increase the value of β, the cyberbullying detection performance first increases and then keeps stable. We can also observe that SICD is not very sensitive to the parameter α, thus we can tune it in a safe range. In usual practice, the parameters α and β should be in the range of $[0.001, 1]$.

7   Related Work

In this section, we briefly introduce the related work in cyberbullying detection and sentiment analysis in social media.

We first briefly review the related literature of detecting cyberbullying behaviors in social media. Dinakar et al. [24] proposed the problem of modeling textual information to detect cyberbullying behaviors on the web. They used concatenation of several feature sets, such as TF-IDF features, POS tags of frequent bigrams and list of profane words to predict the presence of bullying. Dadvar et al. [12] used the user related features such as gender, age to show that such user based features can be used to improve the prediction performance. Xu et al. [2] proposed several models such as BoW based models, LSA (Latent Semantic Analysis) based and LDA based models to predict the bullying behaviors in social media. However, most of them presented an exploratory study rather than providing a principled learning framework. Dinakar et al. [1] presented a common sense based reasoning approach to construct the bullying knowledge base and incorporated it into the cyberbullying detection framework. However, the construction of such knowledge base for each dataset is a labor intensive work. Also, real-world social networks often evolve over time which makes the development of this knowledge base even more difficult and time-consuming. In the later work, Squicciarini et al. [3] presented an approach based on pairwise interactions between users in social networks to identify the bullying users. Particularly, the authors considered interactions of the cyberbullies with normal users in addition to the bag-of-words text analysis.

Another research area related to our work is sentiment analysis in social media. Traditional sentiment analysis has been extensively studied in literature. It has been applied to different corpus such as product reviews [28–30], movie reviews [31,32] and newspaper articles [33]. Recently, the sentiment analysis in social media has received increasing attention since social media is an opinion-rich resource. Sentiment analysis finds many applications in social media realm [34–36] such as poll-rating prediction [37], event detection and prediction [38]. However, the use of sentiment analysis to detect malicious behaviors in social media is limited. One particular use of sentiment analysis to detect malicious posts from social media is done by Cambria et al. [39]. [40] used sentiment analysis to identify various emotions from the bullying behaviors. More specifically, the authors used a trained model and applied it to the Twitter dataset to discover various emotional patterns. However, this work is different from ours as we leverage the sentiment score difference between normal posts and bullying posts and proposed a principle learning framework.

8   Conclusion and Future Work

In this paper, we study the problem of sentiment informed cyberbullying detection in social media. The unique characteristics of the social media data and intentional obfuscation of the abusive words present unique challenges for cyberbullying detection. Motivated by the psychological and sociological findings, we propose to leverage sentiment information to help detect cyberbullying behaviors in social media. First, we conduct an exploratory data analysis on the

Twitter and MySpace datasets and observe that sentiment information can be potentially useful for cyberbullying detection. Methodologically, we propose a principled sparse learning framework by incorporating sentiment information and user-post relationships. Finally, we conduct extensive experiments on two real-world datasets. The experimental results show the effectiveness of the proposed model as well as the impact of sentiment information.

There are many future directions. Most of the work done so far in cyberbullying detection has been found in the English language. However, it is important to develop methods to handle other languages as well. Another future work is to investigate the impact of the sarcasm information hidden in the posts for cyberbullying detection.

Acknowledgements. This material is based upon work supported by, or in part by, the NSF grant 1614576, and the ONR grant N00014-16-1-2257.

References

1. Dinakar, K., Jones, B., Havasi, C., Lieberman, H., Picard, R.: Common sense reasoning for detection, prevention, and mitigation of cyberbullying. TiiS **2**(3), 18 (2012)
2. Xu, J.M., Jun, K.S., Zhu, X., Bellmore, A.: Learning from bullying traces in social media. In: NAACL-HLT, pp. 656–666 (2012)
3. Squicciarini, A., Rajtmajer, S., Liu, Y., Griffin, C.: Identification and characterization of cyberbullying dynamics in an online social network. In: ASONAM, pp. 280–285 (2015)
4. Li, J., Hu, X., Jian, L., Liu, H.: Toward time-evolving feature selection on dynamic networks. In: ICDM, pp. 1003–1008 (2016)
5. Baldwin, T., Cook, P., Lui, M., MacKinlay, A., Wang, L.: How noisy social media text, how diffrnt social media sources? In: IJCNLP, pp. 356–364 (2013)
6. Li, J., Cheng, K., Wang, S., Morstatter, F., Trevino, R.P., Tang, J., Liu, H.: Feature selection: a data perspective. arXiv preprint arXiv:1601.07996 (2016)
7. Kokkinos, C.M., Kipritsi, E.: The relationship between bullying, victimization, trait emotional intelligence, self-efficacy and empathy among preadolescents. Soc. Psychol. Educ. **15**(1), 41–58 (2012)
8. Mayer, J.D., Roberts, R.D., Barsade, S.G.: Human abilities: emotional intelligence. Annu. Rev. Psychol. **59**, 507–536 (2008)
9. Mckenna, J., Webb, J.A.: Emotional intelligence. Br. J. Occup. Ther. **76**(12), 560–561 (2013)
10. Bayzick, J., Kontostathis, A., Edwards, L.: Detecting the presence of cyberbullying using computer software. In: WebSci (2011)
11. Go, A., Bhayani, R., Huang, L.: Twitter sentiment classification using distant supervision. CS224N Project Report, Stanford, vol. 1, p. 12 (2009)
12. Dadvar, M., De Jong, F.: Cyberbullying detection: a step toward a safer internet yard. In: WWW, pp. 121–126 (2012)
13. Li, J., Wu, L., Zaïane, O.R., Liu, H.: Toward personalized relational learning. In: SDM, pp. 444–452 (2017)
14. Li, J., Dani, H., Hu, X., Liu, H.: Radar: residual analysis for anomaly detection in attributed networks. In: IJCAI (2017)

15. McPherson, M., Smith-Lovin, L., Cook, J.M.: Birds of a feather: homophily in social networks. Annu. Rev. Sociol. **27**, 415–444 (2001)
16. Marsden, P.V., Friedkin, N.E.: Network studies of social influence. Sociol. Methods Res. **22**(1), 127–151 (1993)
17. Chung, F.R.: Spectral Graph Theory, vol. 92. American Mathematical Society, Providence (1997)
18. Boyd, S., Vandenberghe, L.: Convex Optimization. Cambridge University Press, Cambridge (2004)
19. Ji, S., Ye, J.: An accelerated gradient method for trace norm minimization. In: ICML, pp. 457–464 (2009)
20. Liu, J., Ji, S., Ye, J.: Multi-task feature learning via efficient l2, 1-norm minimization. In: UAI, pp. 339–348 (2009)
21. Nesterov, Y.: Introductory Lectures on Convex Optimization: A Basic Course, vol. 87. Springer, New York (2013). https://doi.org/10.1007/978-1-4419-8853-9
22. Hastie, T., Tibshirani, R., Friedman, J.: The elements of statistical learning
23. Lee, D.D., Seung, H.S.: Algorithms for non-negative matrix factorization. In: NIPS, pp. 556–562 (2001)
24. Dinakar, K., Reichart, R., Lieberman, H.: Modeling the detection of textual cyberbullying. In: The Social Mobile Web, pp. 11–17 (2011)
25. Demšar, J.: Statistical comparisons of classifiers over multiple data sets. JMLR **7**, 1–30 (2006)
26. Burger, J.D., Henderson, J., Kim, G., Zarrella, G.: Discriminating gender on Twitter. In: EMNLP, pp. 1301–1309 (2011)
27. Mislove, A., Viswanath, B., Gummadi, K.P., Druschel, P.: You are who you know: inferring user profiles in online social networks. In: WSDM, pp. 251–260 (2010)
28. Ding, X., Liu, B., Yu, P.S.: A holistic lexicon-based approach to opinion mining. In: WSDM, pp. 231–240 (2008)
29. Hu, M., Liu, B.: Mining and summarizing customer reviews. In: KDD, pp. 168–177 (2004)
30. Liu, B.: Sentiment analysis and opinion mining. Synth. Lect. Hum. Lang. Technol. **5**(1), 1–167 (2012)
31. Pang, B., Lee, L.: A sentimental education: sentiment analysis using subjectivity summarization based on minimum cuts. In: ACL, p. 271 (2004)
32. Zhuang, L., Jing, F., Zhu, X.Y.: Movie review mining and summarization. In: CIKM, pp. 43–50 (2006)
33. Pang, B., Lee, L., Vaithyanathan, S.: Thumbs up?: sentiment classification using machine learning techniques. In: EMNLP, pp. 79–86 (2002)
34. Hu, X., Tang, L., Tang, J., Liu, H.: Exploiting social relations for sentiment analysis in microblogging. In: WSDM, pp. 537–546 (2013)
35. Hu, X., Tang, J., Gao, H., Liu, H.: Unsupervised sentiment analysis with emotional signals. In: WWW, pp. 607–618 (2013)
36. Cheng, K., Li, J., Tang, J., Liu, H.: Unsupervised sentiment analysis with signed social networks. In: AAAI, pp. 3429–3435 (2017)
37. O'Connor, B., Balasubramanyan, R., Routledge, B.R., Smith, N.A.: From tweets to polls: linking text sentiment to public opinion time series. In: ICWSM, pp. 1–2 (2010)
38. Bollen, J., Mao, H., Pepe, A.: Modeling public mood and emotion: Twitter sentiment and socio-economic phenomena. In: ICWSM, pp. 450–453 (2011)
39. Cambria, E., Chandra, P., Sharma, A., Hussain, A.: Do not feel the trolls. ISWC (2010)
40. Xu, J.M., Zhu, X., Bellmore, A.: Fast learning for sentiment analysis on bullying. In: Workshop on Issues of Sentiment Discovery and Opinion Mining (2012)

zooRank: Ranking Suspicious Entities in Time-Evolving Tensors

Hemank Lamba[1(✉)], Bryan Hooi[1], Kijung Shin[1], Christos Faloutsos[1], and Jürgen Pfeffer[2]

[1] Carnegie Mellon University, Pittsburgh, USA
{hlamba,kijungs,christos}@cs.cmu.edu, bhooi@andrew.cmu.edu
[2] TU Munich, Munich, Germany
juergen.pfeffer@tum.de

Abstract. Most user-based websites such as social networks (Twitter, Facebook) and e-commerce websites (Amazon) have been targets of group fraud (multiple users working together for malicious purposes). How can we better rank malicious entities in such cases of group-fraud? Most of the existing work in group anomaly detection detects lock-step behavior by detecting dense blocks in matrices, and recently, in tensors. However, there is no principled way of scoring the users based on their participation in these dense blocks. In addition, existing methods do not take into account temporal features while detecting dense blocks, which are crucial to uncover bot-like behaviors. In this paper (a) we propose a systematic way of handling temporal information; (b) we give a list of axioms that any individual suspiciousness metric should satisfy; (c) we propose zooRank, an algorithm that finds and ranks suspicious entities (users, targeted products, days, etc.) effectively in real-world datasets. Experimental results on multiple real-world datasets show that zooRank detected and ranked the suspicious entities with high accuracy, while outperforming the baseline approach.

1 Introduction

User-based systems, such as web-services like Amazon, Twitter or corporate IT networks, have become popular targets of fraud or attacks. A popular research problem is to detect the spammers/fraudsters/attackers that are trying to attack a given system [3,11,13,21]. Similarly, in the social networks setting, there are multiple websites where anyone can buy fake Facebook page-likes or Twitter followers. In all these cases, such fraudulent activities take the form of "lockstep" or highly synchronized behavior: such as, multiple users liking the same set of pages on Facebook, or multiple users following the same users almost at the same time on Twitter [3]. Such behavior results in dense blocks in matrices/tensors. The reason behind these blocks is intuitive, as most of the fraudsters have constrained resources (accounts, IP addresses, time, etc.) and they reuse their resources to add as many fraudulent activities as possible to maximize their profits.

© Springer International Publishing AG 2017
M. Ceci et al. (Eds.): ECML PKDD 2017, Part I, LNAI 10534, pp. 68–84, 2017.
https://doi.org/10.1007/978-3-319-71249-9_5

Various methods have been proposed to identify users exhibiting such behavior, which involve finding dense blocks in tensors [11,21] or clustering in subgraphs [3,23]. However, for security experts monitoring the systems, it is imperative to know which users are more suspicious than other users, since it directs their attention to such users for further analysis or actions. In this paper we propose a method that ranks entities effectively (see Fig. 1) for a security analyst to view. Consider Fig. 2; all three users, A, B and C are participating in dense blocks (as they are part of the 2 rectangles), however their contribution towards the suspiciousness of each block is different. A core question we answer in our paper is as follows:

Informal Problem 1 (Individual Suspiciousness Metric). *Given multimodal temporal data in the form of (userId, productId, ..., timestamp), how can we find and score suspicious entities (e.g. users, activities, products, days,etc.)?*

In addition, almost all the social networking websites and services have timestamps associated with every user activity. However, few approaches in the literature consider temporal features [3]. These timestamps can be useful for detecting fraudsters. However, it is not clear in dense block detection literature, in what ways we can incorporate the temporal information available to us. In this paper we answer the following question:

Informal Problem 2 (Temporal data handling). *Given data in the form of (cat 1, cat 2, ..., timestamp), how can we generate features from timestamps useful for detecting fraudsters? Here cat 1,cat 2 are any categorical features (generally userId, productId, activityId, ratings, etc.).*

We propose ZOORANK, a novel approach for successfully scoring entities based on their participation in suspicious dense blocks. We introduce a set of axioms that any ideal individual scoring metric should satisfy. We show theoretically, that our proposed scoring function satisfies the proposed axioms. Additionally, ZOORANK also provides a framework to make good use of temporal information that generally exists in all the real-world datasets. As shown in Fig. 1, ZOORANK successfully finds suspicious users in multiple real-world datasets (Software Marketplace data and Reddit data) with high accuracy. Additionally, the suspicious users found by our method showed clearly anomalous patterns. In Fig. 1 (Bottom Left), we see that multiple users are working in groups to target certain products. Similarly, in Fig. 1 (Bottom Right), the suspicious users detected by our method show extremely regular and bot-like behavior resulting in spikes in the inter-arrival time distribution (difference in seconds between consecutive posts).

Our main contributions are as follows:

- **Theory**
  - *Axioms*: We propose a set of axioms that an individual scoring metric for measuring contribution of a user towards a suspicious block should follow.

Problem Setup (User×Product×Rating×Days   (User × InterArrival Time)
                  ×InterArrival Time)

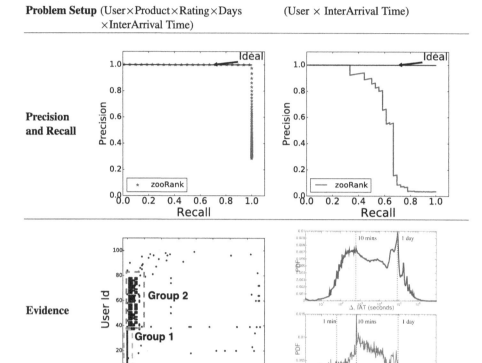

Fig. 1. Effectiveness of ZOORANK on real world datasets. (Top Left) Perfect precision-recall on software marketplace dataset. (Top Right) ZOORANK obtains good precision recall on Reddit dataset. (Bottom Left) Top 100 suspicious users found by ZOORANK show high synchronicity (formed groups) in rating and reviewing top suspicious products. (Bottom Right) The suspicious users (bottom; red) detected by ZOORANK for Reddit dataset show irregular spikes in inter-arrival time distribution, as compared to all the users (top; blue). (Color figure online)

- *Metric*: We propose an individual suspiciousness scoring metric.
- *Proofs*: We further prove that our proposed individual metric follows all the proposed axioms.
- **Temporal Features:** We provide a way of creating temporal features from the timestamp information present in the data.
- **Multimodality and Effectiveness:** The proposed approach ZOORANK can take into account various features, including temporal features. The approach detects suspicious entities in all modes of the data. We tested ZOORANK on various real-world datasets and were able to find suspicious entities with high accuracy, revealing interesting fraud patterns.

Reproducability: Our code and link to the datasets used is available at https://goo.gl/rrvDTx.

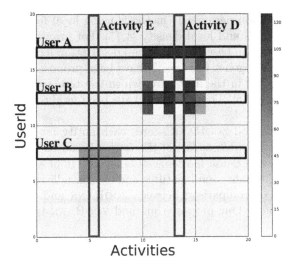

Fig. 2. How to rank users based on their suspiciousness, matching human intuition (A > B > C)?

2 Background and Related Work

A lot of work exists in the literature which aims at finding dense blocks, but none of the methods present a way of scoring the individual entities in dense blocks.

Detecting dense blocks: Densest-subgraph identification (i.e., the problem of finding a subgraph with maximum average degree) has been broadly studied in theory [5,8]. These theoretical results have been extended and applied to anomaly and fraud detection [11,20] since dense subgraphs (dense blocks) in real-world graph data tend to indicate fraudulent lock-step behavior, such as follower-buying services in Twitter. Spectral methods, which make use of eigen and singular value decomposition, also have been used for detecting dense subgraphs corresponding to 'cut-and-paste' bibliography in patent graphs [18], lock-step followers [13] and small-scale stealthy attacks [19] in social networks. Other approaches for dense-subgraph detection include co-clustering [3] and belief propagation [17]. Recently, dense-block detection in multi-aspect data also has been researched [12,21] for spotting groups synchronized in multiple aspects, such as IPs, review scores and review keywords. For our experiments, we use the best performing dense subgraph detection method M-Zoom [21]. The existing methods, however aim at only finding blocks, and do not provide a rank-list of users to inspect according to their suspiciousness.

Scoring Anomalies: Evaluating the anomalousness or suspiciousness of individuals is complementary to detecting dense blocks, which correspond to group activities. A widely-used approach is to detect outliers. Outlier detection methods are divided into parametric methods assuming underlying data distribution [10] and non-parametric methods using local features, such as distances to neighbors [14] and local density [4,15]. For graph data, on the other hand, various approaches, based on minimum description length [6,9], neighborhood information [22], egonet features [2] have been proposed for scoring nodes. Many methods do exist in the literature, which use temporal information such as inter-arrival time [7,16]. These features have been used to successfully detect bot-like behavior [7].

Our proposed method ZOORANK scores each entity (*individual-scoring*) in any of the dimensions (*multimodal*) of the tensor based on the entity's participation in the suspicious dense blocks (*dense-blocks*). It provides ways of transforming temporal data into useful features and thus handles both *numerical and categorical features*. A comparison between ZOORANK and other algorithms is summarized in Table 1. Our proposed method ZOORANK is the only one that matches all specifications.

Table 1. Comparison of other methods and their features

	N-dim Outlier Methods			Point Processes		Graph/Tensor based Methods					
	GFADD /FADD [15]	LoF [4]	Robust Random Cut [9]	RSC [7]	Self Feeding [16]	SPOKen [18]	CopyCatch [3]	CrossSpot [12]	M-Zoom [21]	FRAUDAR [11]	ZOORANK
Dense Blocks						✓	✓	✓	✓	✓	✓
Individual Scoring	✓	✓	✓								✓
Numerical & Categorical Features								✓			✓
Multimodal and Extensible									✓	✓	✓
Temporal Features			✓	✓	✓			✓			✓

3   Preliminaries and Problem Definition

3.1   Problem Definition

Definition 1 (K-way timed tensor). *A K-way timed tensor is a higher-order matrix containing entries of the form (category 1, category 2, ..., category K, timestamp).*

Many types of data including "like" data from Facebook (UserId, PageId, Timestamp), "follow" data from Twitter (UserId, FolloweeId, Timestamp), activity log from an organization (UserId, OperationId, Timestamp) or network data (Source IP, Source Port, Destination IP, Destination Port, Timestamp) all can be formulated as a K-way timed tensor. We now give a precise definition of the problem statements.

Problem 1 (Temporal Features Handling). *Given a K-way timed tensor $\mathcal{A}$, how can we effectively transform the temporal features associated with $\mathcal{A}$ to generate a categorical tensor $\mathcal{X}$?*

Problem 2 (Individual-Suspiciousness). *Given a L-way categorical tensor $\mathcal{X}$ of size $N_1 \times N_2 \times \cdots \times N_L$ with non-negative entries, compute a **score function** $f_{\mathcal{X}}(i)$, which defines the suspiciousness of entity i in the $m(i)^{th}$ mode of $\mathcal{X}$ with respect to the overall tensor $\mathcal{X}$.*

3.2   Block Level Suspiciousness Metrics

In this paper, we consider three block-level suspiciousness metrics although our proposed method is not restricted to them. The metrics are Arithmetic (g_{ari}), Geometric (g_{geom}) and Density (g_{susp}). Arithmetic computes the arithmetic average mass of a sub-block $\mathcal{Y}$ of a tensor $\mathcal{X}$. Similarly, Geometric metric is the geometric average mass of the block. The Density metric is the KL-divergence (Kullback Leibler) between the distribution of the mass in the sub-block with respect to the distribution of the mass in the tensor. These metrics are explained in the following sections.

3.3   Axioms

In this sub-section, we establish axioms that a good score function $f = f_{\mathcal{X}}(i)$ should satisfy. The suspiciousness of an entity should be based on its participation in dense blocks $\mathcal{B}$. Hence, our first two axioms govern the scores with respect to a single block $\mathcal{Y} \in \mathcal{B}$: our third axiom then governs how the single-block scores are combined to form $f_{\mathcal{X}}(i)$.

Let $\rho_{\mathcal{Y}}$ be the density (i.e. mass divided by volume) of $\mathcal{Y}$, and $\rho_{\mathcal{Y}}(i)$ be the density of the slice of $\mathcal{Y}$ defined by entity i. Similarly, let $\mathcal{C}_{\mathcal{Y}}(i)$ denote the mass of that same slice. The entire list of symbols is shown in Table 2.

Axiom 1 (Mass). *If an entity a has more mass than entity b in a block and given the fixed size of block in both the modes $m(a)$ and $m(b)$, then entity a is more suspicious. Formally*

$$IF \; \mathcal{C}_{\mathcal{Y}}(a) > \mathcal{C}_{\mathcal{Y}}(b), \quad AND \quad \mathcal{N}_{\mathcal{Y}}^{m(a)} = \mathcal{N}_{\mathcal{Y}}^{m(b)}, THEN \; \delta_{\mathcal{Y}}(a) > \delta_{\mathcal{Y}}(b)$$

This is represented in Fig. 2. See how entities are ranked by suspiciousness in the top right block (User A > User B > Activity D).

Table 2. Symbols and definitions

Symbol	Definition
$\mathcal{X}$	Input categorical L-way tensor
$\mathcal{Y}$	Dense block within tensor $\mathcal{X}$
$N_{\mathcal{Y}}^i$	Size of ith mode of block $\mathcal{Y}$
$m(i)$	Mode of entity i
$\rho_{\mathcal{Y}}$	Density of block $\mathcal{Y}$
$C_{\mathcal{X}}$	Sum of the entries in $\mathcal{X}$
$C_{\mathcal{Y}}$	Sum of the entries in $\mathcal{Y}$
$C_{\mathcal{Y}}(i)$	Mass of entity i in $\mathcal{Y}$
$V_{\mathcal{Y}}$	Volume of the block $\mathcal{Y}$
$g()$	Block suspiciousness scoring function
$f()$	Individual-suspiciousness scoring function
$\delta_{\mathcal{Y}}(i)$	Block level suspiciousness of entity i in block $\mathcal{Y}$
$\mathcal{B}$	List of suspicious dense blocks
M	Number of suspicious blocks to be considered

Axiom 2 (Concentration). *Given two entities a, b in different modes $m(a), m(b)$, where number of entities in one mode $(N_{\mathcal{Y}}^{(m(a))})$ is less than the number of entities in the second mode $(N_{\mathcal{Y}}^{(m(b))})$, then for fixed density, entity a is more suspicious than entity b.*
Formally,

$$IF \ N_{\mathcal{Y}}^{m(a)} < N_{\mathcal{Y}}^{m(b)} \quad AND \quad \rho_{\mathcal{Y}} = \rho_{\mathcal{Y}}(a) = \rho_{\mathcal{Y}}(b),$$
$$THEN \quad \delta_{\mathcal{Y}}(a) > \delta_{\mathcal{Y}}(b)$$

This is represented in Fig. 2. See how entities are ranked by suspiciousness in the lower left block (User C > Activity E).

Axiom 3 (Monotonocity). *If for every block, entity a has higher suspiciousness than entity b, then entity a has higher overall suspiciousness.*
Formally,

$$IF \ \delta_{\mathcal{Y}}(a) > \delta_{\mathcal{Y}}(b) \quad \forall \mathcal{Y} \in \mathcal{B}, THEN \ f_{\mathcal{X}}(a) > f_{\mathcal{X}}(b)$$

3.4   Shortcomings of Other Metrics

While these axioms are simple and intuitive, many other candidate metrics are not able to satisfy them. We consider some of them, and show why they fail.

Block Score: One simple metric to consider is the block suspiciousness score itself. The metric is to assign each individual the maximum block suspiciousness score out of all the blocks it is part of. The metric doesn't change if the two entities have different contributions to the block, and hence fails Axiom 1 (*Mass*) and Axiom 2 (*Concentration*).

SVD-score: Any matrix $\mathbf{A}$ can be decomposed using SVD decomposition as follows: $\mathbf{A} = \mathbf{U}\mathbf{\Sigma}\mathbf{V^T}$. Each of the singular values in $\mathbf{\Sigma}$ represents the singular value related to a dense block that exists in the dataset. The metric here is the score of the maximum component for each user. This metric would again fail Axiom 1 (*Mass*) and Axiom 2 (*Concentration*).

Average δ-Block Score: Another proposed metric could be the average of all the contributions by the given entity to all the suspicious blocks. The contribution to a block is computed as the difference in the suspiciousness between the block and the block after removing the specified entity. This statistic fails to satisfy Axiom 3 (*Monotonocity*) as if entity 1 has higher suspiciousness in 2 blocks than entity 2, but entity 2 exists only in one of the blocks, then the mean statistic is ambiguous.

As we show above, the metrics based on aggregation of block statistics do not work. In the following section, we propose ZOORANK, a scalable and effective method for finding and scoring suspicious individual entities in multimodal temporal data.

4 Proposed Approach: ZOORANK

4.1 Temporal Feature Handling

As mentioned, any data from a social networking website or a web service can be represented as a K-way timed tensor. We propose a way to handle such tensors by converting the numerical timestamp mode into interpretable categorical features. We propose to generate 0th-order, 1st-order, and temporal folding features.

- 0th-**order features:** The 0th order features bucketize the timestamp into number of days, hours, minutes, etc. passed since the first observation was made. The temporal resolution can be chosen by practitioners based on the typical level of temporal variation present in their dataset.
- 1st-**order features:** Inter-arrival time is defined as the time interval between 2 consecutive timestamps of the same user. [7] found that bots tend to display regular inter-arrival time behavior such as performing an activity every exactly 5 min, due to automated scripts. To capture this pattern, we propose 1st-order features, which is the log-bucketized inter-arrival time between 2 consecutive operations of a user (generalizable to any entity).
- **Temporal folding features:** We propose another way to detect fraudsters showing periodic behavior, which are common in bot-like behavior. For instance, a group of anomalous users might try to perform multiple login activities only from Wednesday 10 PM to 11 PM, or only on a specific day of the week. We work with 3 such features: (1) day of the week, (2) hour of the day and (3) hour of the week. We call these features temporal folding features.

4.2 Proposed Metric

Our metric is based on the δ-contribution of each entity towards the block suspiciousness score. We first define the δ-contribution for a given entity i in mode $m(i)$ of a specific block $\mathcal{Y} \in \mathcal{B}$, where $\mathcal{B}$ is a list of blocks. We denote this by $\delta_{\mathcal{Y}}(i)$.

Definition 2 (Entity's Block-level Suspiciousness ($\delta_{\mathcal{Y}}(i)$)). *We define $\delta_{\mathcal{Y}}(i)$ as the difference between the suspiciousness score of block $\mathcal{Y}$ and block $\mathcal{Y}$ after removing entity i from the block i.e., $\delta_{\mathcal{Y}}(i) = g(\mathcal{Y}) - g(\mathcal{Y} \setminus i)$.*

We need to aggregate the δ-metric over the entire list of blocks $\mathcal{B}$, in such a way that the given axioms are satisfied. We propose two metrics both of which satisfy the given axioms. The first metric is the sum of the δ-contributions, and the second is the maximum of the δ-contributions. We define the maximum metric as follows:

$$f_{\mathcal{X}}(i) = \max_{\mathcal{Y} \in \mathcal{B}}(\delta_{\mathcal{Y}}(i))$$

We empirically found that the maximum metric performs the best on the real-world datasets, and hence for the rest of the paper, all references to the proposed metric is for the maximum version of the metric.

4.3 Algorithm

After handling the temporal features, we produce a categorical tensor $\mathcal{X}$. Algorithm 1 defines the outline of ZOORANK. The first step is to compute suspicious blocks for the given tensor $\mathcal{X}$. To compute suspicious blocks, any existing method for block detection can be used.

We first find the M top suspicious dense blocks as determined by g (Line 1), where g is one of the metrics defined in Sect. 3.2. These top M suspicious blocks are stored in the list $\mathcal{B}$. For every entity i that has occurred at least once in any of the blocks in $\mathcal{B}$, we compute the individual suspiciousness score function f. This score function captures the contribution of a particular entity towards making the block suspicious. To do this, we compute the marginal contribution of each entity towards that block. This is equivalent to removing the entity i from the block, and re-computing the suspiciousness score (Lines 6–7). The difference between the new suspiciousness score and the original suspiciousness score is the marginal contribution of entity i. We compute the marginal contribution of each entity i over all the blocks (Lines 4–8). We define the individual suspiciousness score of the entity i as the maximum of the marginal contributions of entity i (Line 9). Another potential metric is to replace the maximization in Line 9 by the sum function. We conduct experiments with that metric as well.

This formulation of the scores $f_{\mathcal{X}}(i)$ satisfies intuitively reasonable properties, namely our axioms defined in Sect. 3.3:

Theorem 1. *The scores $f_{\mathcal{X}}(i)$ computed by Algorithm 1, using any of the metrics g_{ari}, g_{geo}, or g_{susp}, satisfies Axioms 1 to 3.*

Data: Tensor $\mathcal{X}$, block scoring function g, number of blocks to consider M,
      mode j to consider
Result: Individual scores for each entity i over the entire tensor : $f_{\mathcal{X}}(i)$

1 $\mathcal{B} = \text{ComputeDenseBlocks}(\mathcal{X}, M, g)$
2 **for** *each entity* $i \in N_j$ **do**
3      $\delta_i = []$
4   **for** $\mathcal{Y} \in \mathcal{B}$ **do**
5   **if** $i \in \mathcal{Y}$ **then**
6   Create new block $\mathcal{Y}'$ by removing the entries of entity i
7   Append $(g(\mathcal{Y}) - g(\mathcal{Y}'))$ to δ_i
8   **end**
9      $f_{\mathcal{X}}(i) = \max(\delta_i)$
10 **end**
11 Sort and output $f_{\mathcal{X}}(i)$

Algorithm 1. zooRank: Detecting Suspiciousness Individuals

Proof. We first start by defining some of the standard block suspiciousness methods as follows:

$$g_{ari}(\mathcal{Y}, \mathcal{X}) = C_{\mathcal{Y}}/(\sum_j N_{\mathcal{Y}}^j/L)$$

$$g_{geo}(\mathcal{Y}, \mathcal{X}) = C_{\mathcal{Y}}/(V_{\mathcal{Y}}^{1/L})$$

$$g_{susp}(\mathcal{Y}, \mathcal{X}) = V_{\mathcal{Y}} \cdot D(\rho_{\mathcal{Y}}||\rho_{\mathcal{X}})$$

where $D(\rho_{\mathcal{Y}}||\rho_{\mathcal{X}}) = \rho_{\mathcal{X}} - \rho_{\mathcal{Y}} + \rho_{\mathcal{Y}} \log \frac{\rho_{\mathcal{Y}}}{\rho_{\mathcal{X}}}$.

zooRank satisfies Axiom 1 (Mass)
If we fix the block's dimensions $N_{\mathcal{Y}}^1, \ldots, N_{\mathcal{Y}}^L$, all three metrics above strictly increase the mass of the block (i.e. $C_{\mathcal{Y}}$); this can be inferred directly from the form of g_{ari} and g_{geo}, and for g_{susp}.

As $C_{\mathcal{Y}}(a) > C_{\mathcal{Y}}(b)$, thus $\mathcal{Y} \setminus a$ has lower mass than $\mathcal{Y} \setminus b$, and since g is strictly increasing in mass (for fixed block dimensions), we get $g(\mathcal{Y} \setminus a) < g(\mathcal{Y} \setminus b)$. Therefore:

$$\delta_{\mathcal{Y}}(a) = g(\mathcal{Y}) - g(\mathcal{Y} \setminus a) > g(\mathcal{Y}) - g(\mathcal{Y} \setminus b)$$
$$= \delta_{\mathcal{Y}}(b)$$

zooRank satisfies Axiom 2 (Concentration)
Using the same reasoning as above, it suffices to show $g(\mathcal{Y} \setminus a) < g(\mathcal{Y} \setminus b)$. Note that $N_{\mathcal{Y}}^{m(a)} < N_{\mathcal{Y}}^{m(b)} \Rightarrow V_{\mathcal{Y} \setminus a} < V_{\mathcal{Y} \setminus b}$ (since removing from a smaller mode decreases the volume more). Consider each metric g_{ari}, g_{geo}, and g_{susp} separately:

– **case 1:** g_{ari}.
  Here $\mathcal{Y} \setminus a$ and $\mathcal{Y} \setminus b$ have the same sum of block dimensions, and $C_{\mathcal{Y} \setminus a} = \rho_{\mathcal{Y}} \cdot V_{\mathcal{Y} \setminus a} < \rho_{\mathcal{Y}} \cdot V_{\mathcal{Y} \setminus b} = C_{\mathcal{Y} \setminus b}$ so that $g_{ari}(\mathcal{Y} \setminus a) < g_{ari}(\mathcal{Y} \setminus b)$.

- **case 2:** g_{geo}.

    Note that $g_{geo}(\mathcal{Y}) = C_{\mathcal{Y}}/(V_{\mathcal{Y}}^{1/L}) = \rho_{\mathcal{Y}} \cdot V_{\mathcal{Y}}/(V_{\mathcal{Y}}^{1/L}) = \rho_{\mathcal{Y}} \cdot V_{\mathcal{Y}}^{\frac{L-1}{L}}$. Thus:

$$g_{geo}(\mathcal{Y} \setminus a) = \rho_{\mathcal{Y}} \cdot (V_{\mathcal{Y} \setminus a})^{\frac{L-1}{L}} < \rho_{\mathcal{Y}} \cdot (V_{\mathcal{Y} \setminus b})^{\frac{L-1}{L}} = g_{geo}(\mathcal{Y} \setminus b).$$

- **case 3:** g_{susp}.

    $g_{susp}(\mathcal{Y} \setminus a) = V_{\mathcal{Y} \setminus a} \cdot D(\rho_{\mathcal{Y}} \| \rho_{\mathcal{X}}) < V_{\mathcal{Y} \setminus b} \cdot D(\rho_{\mathcal{Y}} \| \rho_{\mathcal{X}}) = g_{susp}(\mathcal{Y} \setminus b).$

ZOORANK satisfies Axiom 3 (Monotonocity)

$$f_{\mathcal{X}}(a) = \max_{\mathcal{Y} \in \mathcal{B}} \delta_{\mathcal{Y}}(a) > \max_{\mathcal{Y} \in \mathcal{B}} \delta_{\mathcal{Y}}(b) = f_{\mathcal{X}}(b).$$

5   Experiments

In this section, we conducted experiments to answer the following questions:

- **Q1:** How effectively does ZOORANK find suspicious entities across all modes?
- **Q2:** How generalizable is ZOORANK over different datasets?
- **Q3:** Does ZOORANK scale linearly with size of the data?

5.1   Datasets

We used various real-world datasets including a software marketplace dataset, a dataset from a popular social news aggregation website (Reddit), a dataset about Indian elections from Twitter, and a research lab's intrusion detection dataset.

- **Software Marketplace Dataset (SWM):** We used the SWM dataset that was used previously by [1]. The dataset contains the reviews for all the products (software) under the entertainment category of the marketplace. The dataset contains 1,132,373 reviews from 966,839 unique users for 15,094 products. Each review has a rating from 1 to 5, and the timestamp on which the review was posted. The dataset, thus is in the format (UserId, ProductId, Rating, Timestamp). Previous studies [1,23] manually annotated ground truth labels for suspicious users, which we considered as our ground truth.
- **Reddit Dataset:** Reddit is a social news aggregator website, which allows users to post, comment on, upvote and downvote stories. The dataset was collected and analyzed by [7]. The dataset contains 1,020,834 user comments for 1,036 users. The Reddit dataset is in the form (UserId, #Upvotes, #Downvotes, Length, Timestamp). The dataset has information about ground truth suspicious user accounts.
- **DARPA Intrusion Detection:** The DARPA intrusion detection dataset contains a sample of network data for the US Air Force laboratory. The dataset contains records in the format (Source IP, Destination IP, Timestamp). Further, it also contains labels for anomalous connections. For ground truth, we considered any source IP address that participates in at least 10 such anomalous connections, and any destination IP address that participates in at least 400 such connections. We altered this definition for ground truth thresholds and still achieved similar results as mentioned in the paper.

- **Indian Elections 2014 Dataset:** We collected tweets from 2014 Indian Elections. We crawled all the tweets from the 10% Sample API (Decahose). All the tweets contain the top 5 hashtags on Indian Elections per week. We further considered only those users who have at least 2 tweets in our dataset. This led us to a dataset of tweets from March, 2014 to May, 2014 consisting of 10,786 users.
- **Simulated Dataset:** We also tested our approach on a simulated dataset. For simulation, we used a realistic way of generating user-timestamps [7], then for each of the timestamp, we added activities based on a Poisson distribution. We simulated 3 blocks, comprising of 300, 400 and 200 genuine users respectively, where each block has different parameters for the activity Poisson distribution. For the suspicious blocks, we simulated three blocks for 50, 25 and 25 users respectively. The first block does the most popular activity over the entire duration of the simulation and with random inter-arrival times. The second and third block do the second most and third most popular activities, respectively, at a steady inter-arrival time of 1 min on a single day.

Experimental Settings: All our experiments were conducted on a machine on Intel(R) Xeon(R) CPU W3530 @ 2.80 GHz and 24 GB RAM. For all our experiments, we chose $M = 30$ and used M-Zoom [21] for dense block detection. We created multiple tensors based on different resolutions of time features (such as day of week, hour of the day, Inter-arrival time (in seconds, bucketized), etc.). However, we reported only the best accuracy obtained. The choice of what tensor to use, what block-level metric to use, and what value of M is appropriate, is for the practitioner to decide and depends on the type of data, on which the method is being applied.

5.2   Q1. Effectiveness of ZOORANK

To test the effectiveness of ZOORANK, we compare our ranking of the suspicious entities with the ground truth suspicious users in our datasets. We further test the accuracy of our method on the SWM dataset. For software marketplace, we experimented with different versions of temporal features. Note that our algorithm achieves 100% accuracy in identifying suspicious users in the SWM dataset. From Fig. 3a, we observed that adding the inter-arrival time feature increased the accuracy of the method. Our algorithm can rank entities in multiple modes; hence, we also tried to rank the products on basis of their suspiciousness. Though we do not have ground truth for which products were suspicious, we analyzed the top 5 suspicious products in Table 3b. We used the number of reviews by ground truth fraudsters as an indicator for suspiciousness. It can be observed that all the suspicious products are popular (high number of total reviews) and have also been targeted significantly from fraudsters (high number of fraud users). We also noticed that most of the reviews by fraudsters were highly synchronized and a large majority came on a single day (Fig. 3c).

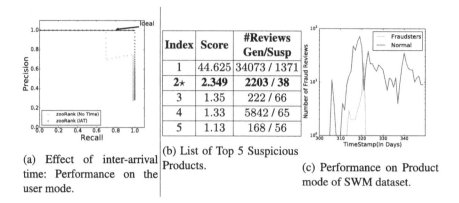

(a) Effect of inter-arrival time: Performance on the user mode.

(b) List of Top 5 Suspicious Products.

(c) Performance on Product mode of SWM dataset.

Fig. 3. ZOORANK is effective. (a) It gives nearly 100% accuracy while identifying suspicious users in the SWM dataset. (b) ZOORANK marks products reviewed by known fraudsters as suspicious. (c) Product #2 received nearly all of it's reviews by fraud users on one single day.

Table 3. ZOORANK is generalizable over multiple datasets, and multiple modes that exist in the datasets.

Dataset	F1-score (SUM)	F1-score (MAX)	Tensor
Reddit	0.62	**0.67**	User × Inter-Arrival Time (IAT)
SWM	0.98	**1.0**	User × Product × Rating × Day × IAT
DARPA (SrcIP Mode)	0.97	**0.988**	SrcIP × DstIP × Day × IAT
DARPA (DstIP Mode)	0.29	**0.37**	SrcIP × DstIP × Hour × IAT

5.3 Q2. Generalizability of ZOORANK

We tested our method on multiple real-world datasets. In Table 3, we present our accuracy on each dataset. We observed that using maximum of the marginal contributions is better than using sum for all of the cases. Further, we also compared our method with a baseline approach. We define the following baseline: **Block Score:** defined as the maximum of all block suspiciousness scores a block is part of. From Fig. 4, it can be observed that our approach clearly is better than the mentioned approach.

For Indian elections data, we did not have any ground truth. We extracted the top 100 suspicious users and evaluated them manually. The results for top 100 suspicious users are shown in Fig. 5. The user ids are sorted by their suspiciousness score, and plotted on the scatter plot along with top suspicious hashtags. Figure 5 clearly shows groups of suspicious users. It is evident that the first two users are "hashtag hijackers". These two users tweeted spam messages with other hashtags but also focussed on generic hashtags related to the Indian elections. Both of these users have an identical behavior, which imply they do follow "lock-step" behavior. The second group of users were tweeting hashtags

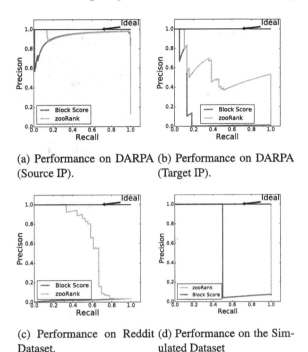

(a) Performance on DARPA (Source IP).

(b) Performance on DARPA (Target IP).

(c) Performance on Reddit Dataset.

(d) Performance on the Simulated Dataset

Fig. 4. ZOORANK is generalizable. ZOORANK outperforms the baseline across different modes (see (a) and (b)) and across multiple datasets (see (c) and (d))

related to themselves and also generic hashtags related to the elections ("self-promoters"). We also spotted the user who tweets out all the trending topics at regular intervals, possibly through automated scripts ("trending topic aggregator"). We believe that the remaining users are users who were discussing indian elections a lot and were influencers in the political discussion. On further analysis, 20 users out of the 100 users were already suspended by Twitter. Thus, our algorithm was able to identify users that were considered spam by Twitter but also users that were missed by Twitter algorithm ("self-promoters") but were clearly malicious.

5.4 Q3. Scalability of ZOORANK

In this section, we evaluate the scalability of the ZOORANK. We measure the effects of the number of blocks and the number of records on the runtime of ZOORANK. To study the effect of the number of records, we generated the dataset with given number of entries in 3 dimensions, where cardinality of each dimension is 10^6. For all our results, we used arithmetic metric and operated on the most suspicious 30 blocks. The results are shown in Fig. 6a, showing that our method scales linearly both in the data size and the number of blocks searched for. For the effect of the number of blocks, we generated a dataset with 10^4 records with a similar number of entries in each dimension.

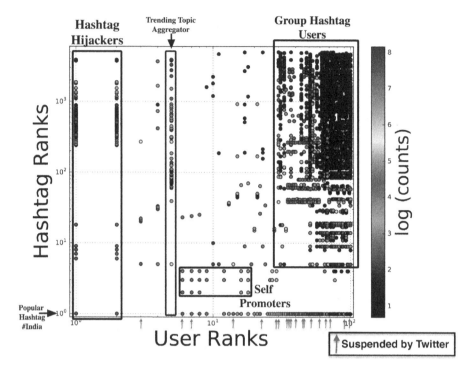

Fig. 5. ZOORANK identifies fraudulent suspicious behavior in Twitter: Top 100 suspicious users, and top hashtags as identified by ZOORANK.

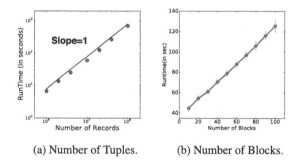

(a) Number of Tuples.   (b) Number of Blocks.

Fig. 6. Scalability of ZOORANK (a) ZOORANK scales linearly with the number of records. (b) ZOORANK scales linearly with the number of blocks we want to find.

6   Conclusions

In this paper, we proposed a set of axioms that a given individual suspiciousness scoring metric should follow. We presented such a metric that satisfies all the proposed axioms. Specifically, our contributions are as follows:

– **Individual-Suspiciousness Metric:** We propose a suspiciousness metric which scores each entity participating in dense blocks. The proposed criteria $f_\chi(i)$ satisfies intuitive axioms.

- **Temporal Features:** The proposed method provides ways to transform the numerical timestamp mode to information rich categorical temporal features.
- **Effectiveness:** The proposed method ZOORANK was successfully tested on various real-world datasets. It scored the suspicious entities with high accuracy, and also uncovered interesting fraud patterns.
- **Scalability:** The method is linearly scalable with the size of the data and can be used for *big-data* problems (see Fig. 6).

Acknowledgement. This material is based upon work supported by the National Science Foundation under Grants Nos. CNS-1314632, IIS-1408924, and by the Army Research Laboratory under Cooperative Agreement Number W911NF-09-2-0053. Any opinions, findings, and conclusions or recommendations expressed in this material are those of the author(s) and do not necessarily reflect the views of the National Science Foundation, or other funding parties. The U.S. Government is authorized to reproduce and distribute reprints for Government purposes notwithstanding any copyright notation here on.

References

1. Akoglu, L., Chandy, R., Faloutsos, C.: Opinion fraud detection in online reviews by network effects. In: ICWSM (2013)
2. Akoglu, L., McGlohon, M., Faloutsos, C.: Oddball: spotting anomalies in weighted graphs. In: Zaki, M.J., Yu, J.X., Ravindran, B., Pudi, V. (eds.) PAKDD 2010. LNCS (LNAI), vol. 6119, pp. 410–421. Springer, Heidelberg (2010). https://doi.org/10.1007/978-3-642-13672-6_40
3. Beutel, A., et al.: Copycatch: stopping group attacks by spotting lockstep behavior in social networks. In: WWW (2013)
4. Breunig, M.M., Kriegel, H.P., Ng, R.T., Sander, J.: LOF: identifying density-based local outliers. In: ACM Sigmod Record (2000)
5. Charikar, M.: Greedy approximation algorithms for finding dense components in a graph. In: Jansen, K., Khuller, S. (eds.) APPROX 2000. LNCS, vol. 1913, pp. 84–95. Springer, Heidelberg (2000). https://doi.org/10.1007/3-540-44436-X_10
6. Eberle, W., et al.: Discovering structural anomalies in graph-based data. In: ICDMW (2007)
7. Ferraz Costa, A., et al.: RSC: mining and modeling temporal activity in social media. In: KDD (2015)
8. Goldberg, A.V.: Finding a maximum density subgraph. Technical report (1984)
9. Guha, S., et al.: Robust random cut forest based anomaly detection on streams. In: ICML (2016)
10. Hawkins, D.M.: Identification of Outliers, vol. 11. Springer, Dordrecht (1980). https://doi.org/10.1007/978-94-015-3994-4
11. Hooi, B., Song, H.A., Beutel, A., Shah, N., Shin, K., Faloutsos, C.: Fraudar: bounding graph fraud in the face of camouflage. In: KDD (2016)
12. Jiang, M., Beutel, A., Cui, P., Hooi, B., Yang, S., Faloutsos, C.: A general suspiciousness metric for dense blocks in multimodal data. In: ICDM (2015)
13. Jiang, M., Cui, P., Beutel, A., Faloutsos, C., Yang, S.: Inferring strange behavior from connectivity pattern in social networks. In: Tseng, V.S., Ho, T.B., Zhou, Z.-H., Chen, A.L.P., Kao, H.-Y. (eds.) PAKDD 2014. LNCS (LNAI), vol. 8443, pp. 126–138. Springer, Cham (2014). https://doi.org/10.1007/978-3-319-06608-0_11

14. Knox, E.M., Ng, R.T.: Algorithms for mining distancebased outliers in large datasets. In: PVLDB (1998)
15. Lee, J.Y., Kang, U., Koutra, D., Faloutsos, C.: Fast anomaly detection despite the duplicates. In: WWW Companion (2013)
16. Vaz de Melo, P.O.S., Faloutsos, C., Assunção, R., Loureiro, A.: The self-feeding process: a unifying model for communication dynamics in the web. In: WWW, pp. 1319–1330
17. Pandit, S., et al.: Netprobe: a fast and scalable system for fraud detection in online auction networks. In: WWW (2007)
18. Prakash, B., et al.: Eigenspokes: surprising patterns and community structure in large graphs. In: PAKDD (2010)
19. Shah, N., Beutel, A., Gallagher, B., Faloutsos, C.: Spotting suspicious link behavior with fbox: an adversarial perspective. In: ICDM (2014)
20. Shin, K., Eliassi-Rad, T., Faloutsos, C.: CoreScope: graph mining using k-core analysis - patterns, anomalies and algorithms. In: ICDM (2016)
21. Shin, K., Hooi, B., Faloutsos, C.: M-zoom: fast dense-block detection in tensors with quality guarantees. In: Frasconi, P., Landwehr, N., Manco, G., Vreeken, J. (eds.) ECML PKDD 2016. LNCS (LNAI), vol. 9851, pp. 264–280. Springer, Cham (2016). https://doi.org/10.1007/978-3-319-46128-1_17
22. Sun, J., Qu, H., Chakrabarti, D., Faloutsos, C.: Neighborhood formation and anomaly detection in bipartite graphs. In: ICDM (2005)
23. Ye, J., et al.: Discovering opinion spammer groups by network footprints. In: COSN (2015)

Computer Vision

Alternative Semantic Representations for Zero-Shot Human Action Recognition

Qian Wang$^{(\boxtimes)}$ and Ke Chen

School of Computer Science, The University of Manchester,
Manchester M13 9PL, UK
{qian.wang,ke.chen}@manchester.ac.uk

Abstract. A proper semantic representation for encoding side information is key to the success of zero-shot learning. In this paper, we explore two alternative semantic representations especially for zero-shot human action recognition: textual descriptions of human actions and deep features extracted from still images relevant to human actions. Such side information are accessible on Web with little cost, which paves a new way in gaining side information for large-scale zero-shot human action recognition. We investigate different encoding methods to generate semantic representations for human actions from such side information. Based on our zero-shot visual recognition method, we conducted experiments on UCF101 and HMDB51 to evaluate two proposed semantic representations. The results suggest that our proposed text- and image-based semantic representations outperform traditional attributes and word vectors considerably for zero-shot human action recognition. In particular, the image-based semantic representations yield the favourable performance even though the representation is extracted from a small number of images per class.
Code related to this chapter is available at:
http://staff.cs.manchester.ac.uk/~kechen/BiDiLEL/
Data related to this chapter are available at:
http://staff.cs.manchester.ac.uk/~kechen/ASRHAR/

Keywords: Zero-shot learning · Semantic representation
Human action recognition · Image deep representation
Textual description representation · Fisher Vector

1 Introduction

Zero-Shot Learning (ZSL) aims to recognize instances from new classes which are not seen in the training data. It is a promising alternative to the traditional supervised learning which requires labour-intensive annotation work on all the classes involved. As shown in Fig. 1, in ZSL, the knowledge learned from training data is transferred to recognise unseen classes through the side information which can usually be acquired with less effort. Although most existing works in ZSL focus on the development of novel recognition models, the side information for

© Springer International Publishing AG 2017
M. Ceci et al. (Eds.): ECML PKDD 2017, Part I, LNAI 10534, pp. 87–102, 2017.
https://doi.org/10.1007/978-3-319-71249-9_6

knowledge transfer plays an equally important role in the success of ZSL. The most popular side information used in ZSL literature are attributes and word vectors. Although they have been widely used in ZSL [9,12,14,26,27], both of them have obvious drawbacks as well, especially for zero-shot human action recognition in video data.

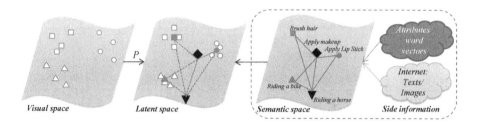

Fig. 1. A schematic diagram of zero-shot learning framework. The work in this paper is highlighted in the dashed box. Human action classes are denoted by coloured markers (blue and black for training and unseen classes respectively) with different shapes. The training data are used to learn the mapping P and training class embedding (blue filled markers in the latent space), then the unseen class embedding (black filled markers in the latent space) is achieved by preserving the semantic distances (red lines). See Sect. 4.2 for more details of our ZSL method. (Color figure online)

The definition and annotation of attributes for human actions (e.g., the attributes defined for UCF101 [10] include "bodyparts-visible: face, fullbody, onehand", "body-motion: flipping, walking, diving, bending", etc.) are subjective and labour-intensive. When a large number of human actions are involved, more attributes are needed to distinguish one human action from the other. As a result, attributes based semantic representations are inappropriate for large scale zero-shot human action recognition. On the other hand, as stated in [2], using a word vector of the class label to represent a human action is far from adequate to illustrate the rich appearance variations. In addition, the word vectors are learned from textual corpus, thus suffering from the catastrophic semantic gap problem (i.e., the difference of information conveyed by visual media and texts).

To address the limitations of existing semantic representations for ZSL, we attempt to explore alternative side information towards enhanced zero-shot human action recognition. The essentials of side information for ZSL are twofold. Firstly, it should be achievable for a large number of human actions without much effort. More importantly, the side information should be able to capture the visually discriminative semantics thus benefiting the ZSL by easily bridging the semantic gap. To this end, we employ action relevant images as the side information resources to extract the semantics of human actions. With the aid of search engines, it is effortless to collect a set of action relevant images by using the action name as the key words. Although still images lack of temporal information in human actions, they provide abundant visually discriminative information which can be exploited to extract high-level semantic representations for human actions. On the other hand, we aim to enhance the word vectors by collecting and encoding textual descriptions of human actions. We believe that the contextual information in the action relevant texts (e.g., description articles of human actions from the web) will remove the ambiguity of the semantics in the original action word vectors which are based solely on the action labels.

To summarise, the contributions of this paper include:

- We propose and implement the idea of using textual descriptions to enhance the word vector representations of human actions in ZSL.
- We propose and implement the idea of using action related still images to represent semantics for video based human actions in ZSL.
- Experiments are conducted to evaluate the effectiveness of the proposed semantic representations in zero-shot human action recognition, and significant performance improvement has been achieved.

2 Related Work

The semantic representation is key for the success of ZSL. Recently, attempts have been made to explore more effective semantic representations for objects/actions towards improved ZSL performance. In this section, we will review the prevailing semantic representations used in ZSL (Table 1), including a variety of extensions of attributes and word vectors, as well as many other less popular approaches proposed in literature.

Table 1. A survey on semantic representations in ZSL

Authors and year	Semantic representation
Lampert et al. [12]	Attributes, annotated manually
Sharmanska et al. [21]	Attributes, enhanced by learning from visual data
Liu et al. [14]	Attributes, enhanced by learning from visual data
Qin et al. [18]	Attributes, enhanced by learning from visual data
Fu et al. [8]	Attributes, enhanced by learning from visual data
Inoue and Shinoda [9]	Word vector, enhanced by a weighted combination of related word (from *WordNet*) vectors
Alexiou et al. [2]	Word vector, enhanced by the synonyms of labels (from Internet dictionaries)
Mukherjee and Hospedales [16]	Word Gaussian distribution
Sandouk and Chen [20]	Word vector, enhanced by contexts (from tags)
Elhoseiny et al. [6]	Tf-idf, based on *Wikipedia* articles
Akata et al. [1]	BOW, based on *Wikipedia* articles
Rohrbach et al. [19]	*WordNet* path length, based on *WordNet* ontology Hit-counts, based on web search results
Chuang Gan et al. [5]	*WordNet* path length, based on *WordNet* ontology

Attributes based semantic representations were firstly proposed for ZSL in [12], thereafter, attributes have been employed for ZSL in many works [3, 26–28].

A set of binary attributes need to be manually defined to represent the semantic properties of objects. As a result, each object class can be represented by a binary attribute vector in which the value of one and zero indicates the presence and absence of each attribute respectively. Since the attributes are shared by seen and unseen classes, the knowledge transfer is enabled. However, as mentioned above, the definition of attributes require experts with domain knowledge to discriminate different classes, and the attribute annotation for a large number of classes could be subjective and labour-intensive.

Alternatively, attributes can be mined automatically from visual features by discriminative mid-level feature learning [7,8,14,18,21], but their semantic meanings are unknown, thus inappropriate for direct use in ZSL. To enhance the attributes' discriminative power and semantic meaningfulness, the manually defined attributes and the ones automatically learned from training data are usually combined. However, the data-driven attributes are usually dataset specific and probably fail on a different dataset.

The other kind of prevailing side information used in ZSL is derived from text resources. One of the most popular semantic representations is word vector (e.g., the ones generated by the *word2vec* tool [15]) due to its convenience and effectiveness. A class label can be easily represented with the vector representation of the corresponding word or phrase. However, word vectors are deficient to discriminate different classes from the visual perspective due to the semantic gap, i.e., the gap between visual and semantic information. As a result, word vectors are usually outperformed by attributes in ZSL.

To alleviate the semantic gap problem, some attempts have been made to enhance the word vectors [2,9,16,20]. Inoue and Shinoda [9] aim to adapt the original word vectors to make two visually similar concepts close to each other in the adapted word vector space by representing a concept with a weighted sum of its original word vector and its hypernym (based on *WordNet*) word vectors. And the weights are learned from visual resources. Alexiou *et al.* [2] enrich the word vector representation by mining and considering synonyms of the action class labels from multiple *Internet dictionaries.* Mukherjee and Hospedales [16] use *Gaussian distribution* instead of a single word vector to model the class labels so that the intra-class variability can be expressed properly in the semantic representations. To address the issue of polysemy, Sandouk and Chen [20] learn a specific vector representation for a word together with its context. That is to say, the same word could have different vector representations when it is in different contexts. Inspired by these works, our work further investigates the possible side information and enabling techniques to enhance the word vectors for ZSL.

Other than attributes and word vectors, other side information has also been investigated for knowledge transfer in ZSL, only if they are able to model the relationships among different classes and relatively easy to obtain. For example, *WordNet* path length is used to measure the semantic correlations between two concepts in [5,19]. The Internet together with search engines provides a natural opportunity to get side information to measure between-class semantic relationships based on hit-count on search results [19]. Textual descriptions of a class rather than the single class name are employed to represent a class in [1,6]. Concept related textual descriptions (e.g., *Wikipedia* page) can be readily obtained from the Internet and then processed with techniques in natural language processing (NLP). Considering our focus on zero-shot human action recognition based on video data, images from the Internet can be alternative side information to texts which have been a typical choice for zero-shot image classification.

3   Method

In this section, we propose our methods of generating semantic representations for zero-shot human action recognition from text and image resources respectively. Firstly, we use search engines to collect action relevant texts and images as the side information. Some typical examples are shown in Fig. 2. Once the side information are collected, we use different encoding approaches to generate the semantic representations for human actions.

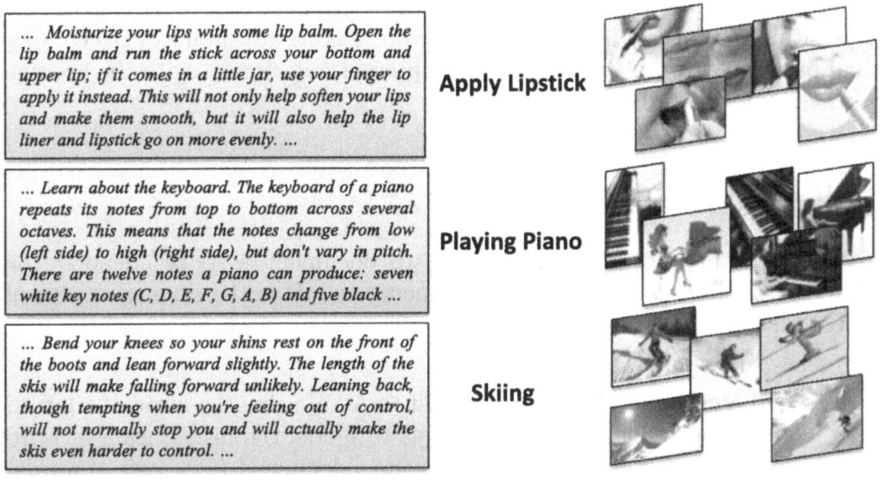

> ... Moisturize your lips with some lip balm. Open the lip balm and run the stick across your bottom and upper lip; if it comes in a little jar, use your finger to apply it instead. This will not only help soften your lips and make them smooth, but it will also help the lip liner and lipstick go on more evenly. ...

Apply Lipstick

> ... Learn about the keyboard. The keyboard of a piano repeats its notes from top to bottom across several octaves. This means that the notes change from low (left side) to high (right side), but don't vary in pitch. There are twelve notes a piano can produce: seven white key notes (C, D, E, F, G, A, B) and five black ...

Playing Piano

> ... Bend your knees so your shins rest on the front of the boots and lean forward slightly. The length of the skis will make falling forward unlikely. Leaning back, though tempting when you're feeling out of control, will not normally stop you and will actually make the skis even harder to control. ...

Skiing

Fig. 2. Examples of collected description texts and images of three human actions from UCF101 (i.e., "Apply Lipstick", "Playing Piano" and "Skiing").

3.1   Text-Based Semantic Representation

Texts Collection. Motivated by the fact a class label is insufficient to depict the complex concepts in the human action, we try to collect textual descriptions from the web to represent each human action. Textual descriptions of human actions can be derived from *WikiHow*, a website teaching people "how to do anything". Inevitably, the description texts for some actions (e.g., "pick", "sit") are not available from *WikiHow*, for which we turn to alternative sources including *Wikipedia* and *Online dictionary*.

Pre-processing. Once the textual descriptions for all the human actions are collected, we end up with a document for each human action class. We use natural language processing techniques to pre-process the unstructured textual data before encoding them into semantic representations. In the first step, we tokenize the documents to get all the words appearing in the documents. After removing the stop words (i.e., the words carrying little semantic meanings such as "is", "you", "of"), we have a dictionary containing d words.

Term-Document Matrix (TD). Given the documents and the dictionary containing all the terms/words in the documents, a term-document matrix M is constructed to represent the term frequency in all documents. M_{ij} denotes the frequency of term i in document j, where $i = 1, 2, \ldots, d$ and $j = 1, 2, \ldots, C$, C is the number of documents, i.e., the number of human actions in a specific dataset. Thus the column vectors in M can be used to represent the semantic representations of human actions. We denote this approach as **TD** in the following sections.

Average Word Vector (AWV). We aim to enhance the word vectors by incorporating the collected textual information. Taking advantage of the compositional property of word vectors, we can represent a document with the average of all the included word vectors.

$$AWV(j) = \frac{1}{n_j} \sum_{i=1}^{n_j} v_i \tag{1}$$

where n_j is the number of terms in the j-th document, $v_i \in \mathbb{R}^D$ denotes the word vector of the i-th term in the document, and D is the dimensionality of word vectors.

Fisher Word Vector (FWV). In contrast to AWV using the mean of all word vectors to represent a document, FWV aims to model the distribution of word vectors in a document. Fisher Vector represents a document (i.e., a set of words) by the gradient of log likelihood with respect to the parameters of a pre-learned probabilistic model (i.e., Gaussian Mixture Model) [17,25]. A Gaussian Mixture Model (GMM) is used to fit the distribution of the word vectors involved in all documents, where the parameters $\Theta = \{\mu_k, \Sigma_k, \pi_k\}, k = 1, \ldots, K$. Let $V^j = \{v_1, \ldots, v_{n_j}\}$ be a set of word vectors from the j-th human action description document. Then the Fisher Vector of j-th document can be denoted by:

$$FWV(j) = [\mathcal{G}_{\mu,1}^{V^j}, \ldots, \mathcal{G}_{\mu,K}^{V^j}, \mathcal{G}_{\sigma,1}^{V^j}, \ldots, \mathcal{G}_{\sigma,K}^{V^j}], \tag{2}$$

where

$$\mathcal{G}_{\mu,k}^{V^j} = \frac{1}{\sqrt{\pi_k}} \sum_{v_i \in V^j} \gamma_{k,i} \left(\frac{v_i - \mu_k}{\sigma_k} \right), \tag{3}$$

$$\mathcal{G}_{\sigma,k}^{V^j} = \frac{1}{\sqrt{2\pi_k}} \sum_{v_i \in V^j} \gamma_{k,i} (\frac{(v_i - \mu_k)^2}{\sigma_k^2} - 1), \tag{4}$$

$$\gamma_{k,i} = \frac{exp[-\frac{1}{2}(v_i - \mu_k)^T \Sigma_k^{-1}(v_i - \mu_k)]}{\Sigma_{t=1}^{K} exp[-\frac{1}{2}(v_i - \mu_t)^T \Sigma_k^{-1}(v_i - \mu_t)]}. \tag{5}$$

The dimension of the Fisher Vector is $2DK$, where D and K are the dimensionality of word vectors and the number of components in the GMM respectively.

3.2   Image-Based Semantic Representation

Human actions are difficult to describe with texts due to the complexity and intra-class variations. Although they lack temporal information, still images can provide abundant information for the understanding of human actions. Compared to the video examples, still images are much easier to collect, annotate and store. Thus we hold the view that still images are a proper kind of side information which can benefit modelling human action relationships with little effort.

Image Collection. Given a human action, we use the label as the key word and search relevant images with search engines. For most human actions we can get a collection of images each of which gives a view of the action. However, for some action names which could have multiple meanings, the additional explaining key words are needed to get reasonable searching results. For example, we use "salsa spin + dancing" and "playing + hula hoop" for the actions "salsa spin" and "hula hoop" respectively. For each human action, we get different numbers of relevant images after removing the ones of poor quality (e.g., irrelevant ones and the ones smaller than 10 Kb) from the returned results. The image collection and filtering can be processed automatically without many human interventions[1].

Feature Extraction. We aim to extract useful information from a set of images to represent a human action. Recently, deep convolution neural networks have been used to extract image features carrying high-level conceptual information. By feeding the images into a pre-trained CNN model, the deep image features can be obtained easily. Then each human action is represented with a set of image feature vectors $F^j = \{f_1, \ldots, f_{n_j}\}$. In the next two sections, we use two approaches to encode the set of image features into the action-level semantic representation.

Average Feature Vector (AFV). Similar to Eq. (1), we can use the average of multiple image features as the human action semantic representation.

$$AFV(j) = \frac{1}{n_j} \sum_{i=1}^{n_j} f_i \tag{6}$$

---

[1] The image scraper tool is available: http://staff.cs.manchester.ac.uk/kechen/ ASRHAR/.

Fisher Feature Vector (FFV). Similar to the processing applied on word vectors in Sect. 3.1, we use Fisher Vector to encode a set of image feature vectors relevant to a specific human action.

$$FFV(j) = [\mathcal{G}_{\mu,1}^{F^j}, \ldots, \mathcal{G}_{\mu,K}^{F^j}, \mathcal{G}_{\sigma,1}^{F^j}, \ldots, \mathcal{G}_{\sigma,K}^{F^j}], \tag{7}$$

where $\mathcal{G}_{\mu,i}^{F^j}$ and $\mathcal{G}_{\sigma,i}^{F^j}$ can be calculated in the same way as Eqs. (3–5).

4 Experimental Settings

4.1 Dataset

We use two human action datasets to evaluate the proposed approaches for zero-shot recognition, i.e., UCF101 [22] and HMDB51 [11]. **UCF101** is a human action recognition dataset collected from YouTube. There are 13,320 real action video clips falling into 101 action categories. In our experiments, we use 5 randomly generated 51/50 (seen/unseen) class-wise data splits. **HMDB51** contains 6,766 video clips from 51 human action classes. Similarly, we use 5 randomly generated 26/25 splits in all experiments.

4.2 Zero-Shot Recognition Method

We employ our recently developed ZSL method, bidirectional latent embedding learning (BiDiLEL) [26], as a test bed in our experiments[2]. To make the paper self-contained, we will briefly describe the main idea of BiDiLEL in this section.

The method employs a two-stage latent embedding algorithm to learn a latent space in which the semantic gap is bridged and zero-shot recognition can be done (see Fig. 1). In bottom-up stage, we learn a projection matrix P by supervised locality preserving projection (SLPP) [4], such that the examples close to each other in the original visual space will still be close in the latent space. By exploiting the local structures and labelling information in the training data, the learned latent space preserves the data distribution and is more discriminative. The properties are expected to generalise well for test examples from unseen classes.

In the top-down embedding, the latent embedding of each seen class can be calculated by averaging the projections of all the training examples from the class and then serve as landmarks guiding the learning of latent embedding of unseen classes. We use the landmarks based Sammon mapping (LSM) [26] which aims to preserve the inter-class semantic distances (measured in the semantic space). As a result, the semantic distances between seen and unseen classes as well as between any pair of unseen classes will be preserved in the latent space.

Once the latent embedding of both seen and unseen classes are obtained, we can do the zero-shot learning in the latent space using the nearest neighbour

---

[2] Like attributes and word vectors, our proposed semantic representations may be directly deployed in all the existing zero-shot human action recognition methods.

method. Specifically, given a test example, we use projection matrix P to map it into the latent space, where its distances to all the class embedding can be calculated, and it will be assigned to the closest class label. For more details, we refer the readers to [26].

4.3 Video Representation

C3D was proposed in [24] for human action recognition. It utilizes 3D ConvNets to learn spatio-temporal features for video streams. According to [26], the C3D video representation outperforms its counterparts in zero-shot human action recognition. We use the model pre-trained on Sports-1M dataset and follow the setting in [24, 26] to extract spatio-temporal deep features (i.e., the 4096-dimensional "fc6" activations of the deep neural network) from 16-frame segments. Finally, the visual representation of a video stream is calculated by averaging the features of all the segments from the video.

4.4 Evaluation

In most existing ZSL works, the evaluations are based on the assumption that test examples are only from unseen classes, which is often referred as to conventional zero-short learning (cZSL). In practice, however, the test examples can be from either training classes or unseen classes. To evaluate ZSL methods in a more practical scenario, the problem of generalised ZSL has been formulated and investigated in [3, 27]. In gZSL, given a test example, the label search space consists of both seen and unseen classes. In our experiments, we follow the protocols in [27] and report both conventional and generalised ZSL (cZSL and gZSL) results using per-class accuracy. In the generalised ZSL scenarios, except the examples from test classes, we also reserve 20% examples from each training class for testing and the rest 80% examples from each training class for training.

Concretely, we report the recognition accuracy of test examples from unseen classes by setting the search space in the unseen label set $\mathcal{U}$ for the cZSL; the accuracy is denoted by $A_{\mathcal{U} \rightarrow \mathcal{U}}$. For gZSL, we set the search space in the whole label set $\mathcal{T} = \mathcal{S} \cup \mathcal{U}$ and report three types of per-class accuracies, i.e., the recognition accuracy of test examples from unseen classes $A_{\mathcal{U} \rightarrow \mathcal{T}}$, the recognition accuracy of test examples from seen classes $A_{\mathcal{S} \rightarrow \mathcal{T}}$ and the harmonic mean,

$$H = 2 * A_{\mathcal{U} \rightarrow \mathcal{T}} * A_{\mathcal{S} \rightarrow \mathcal{T}} / (A_{\mathcal{U} \rightarrow \mathcal{T}} + A_{\mathcal{S} \rightarrow \mathcal{T}}). \tag{8}$$

The ZSL method employed in our experiments works in the inductive setting (i.e., the test example is processed individually), but can be extended to the transductive setting (i.e., all the test examples are assumed to be available as a collection when doing the recognition) easily by using the structured prediction method [26, 28]. The method of structure prediction uses Kmeans to group all the test examples into clusters (the number of clusters is set to be the number of unseen classes) and find a one-to-one map from the clusters to unseen classes. In our experiments, we will report the results of cZSL in both inductive and transductive settings.

5   Experimental Results

In this section, we present the designed experiments and the results to evaluate the effectiveness of proposed semantic representations[3].

5.1   Text-Based Representation

We conduct experiments of zero-shot human action recognition by utilising the proposed text-based semantic representations in Sect. 3.1, i.e., TD, AWV and FWV. We use the 300-dimensional word vectors pre-trained with *word2vec* on Google News dataset (about 100 billion words)[4]. For FWV, we set the value of K in Eq. (2) to be $\{1, 2, 3, 4, 5\}$. The experiments aim to investigate how different text-based semantic representations perform in zero-shot human action recognition. In our experiments, we follow the protocols in [26] using class-wise cross validation to find the optimal values of hyper-parameters. According to the performance on the validation data, cosine distances are employed to calculate the semantic distances for FWV, and Euclidean distances are employed for AWV.

We report the results of conventional ZSL in both inductive and transductive settings in Table 2. With only the textual description sources, the simple encoding method TD can achieve the accuracy of 19.54% and 15.26% respectively on UCF101 and HMDB51, which indicates the textual descriptions collected by search engines are useful for modelling the inter-class relationships. By incorporating the pre-trained word vectors, AWV improves the accuracy to 24.38% and 21.80% respectively on UCF101 and HMDB51. On the other hand, by comparing FWV with different K values, we know that $K = 1$ gives the best results with an accuracy of 23.76% on UCF101 and 19.57% on HMDB51; however, it is still outperformed by AWV on both datasets regardless of inductive or transductive settings. To conclude, AWV performs the best among different text-based semantic representations.

5.2   Image-Based Representation

In our experiments, we collect variant numbers of relevant images for different human actions. The average number of relevant images per class is around 200 and 100 for UCF101 and HMDB51 respectively. To extract the image features, we use the GoogLeNet [23] model pre-trained on ImageNet dataset[5]. The activations of top fully connected layer of GoogLeNet of 1024 dimensions are used as the deep image features. We evaluate the image-based semantic representations encoded with different approaches described in Sect. 3.2, i.e., AFV and FFV. Again, we set the values of K in Eq. (7) to be $\{1, 2, 3, 4, 5\}$. We employ the same experiment protocols as those used in the previous experiments (Sect. 5.1). According to the

---

[3] The scripts and data used in our experiments can be available on our project page: http://staff.cs.manchester.ac.uk/kechen/ASRHAR/.

[4] https://code.google.com/p/word2vec/.

[5] http://www.vlfeat.org/matconvnet/.

Table 2. Results of different text-based semantic representations (mean ± standard error of recognition accuracy %) on UCF101 and HMDB51 datasets. (Sem. Rep.– Semantic Representation, Att–Attributes, WV–Word vector)

Sem. Rep.	UCF101 (51/50)		HMDB51 (26/25)	
	Inductive	Transductive	Inductive	Transductive
Random	2.00	2.00	4.00	4.00
Att	21.54 ± 0.72	**32.00 ± 2.30**	–	–
WV	19.42 ± 0.69	22.05 ± 1.74	21.53 ± 1.75	24.14 ± 3.43
TD	19.54 ± 0.75	24.29 ± 0.65	15.26 ± 0.57	15.33 ± 1.72
AWV	**24.38 ± 1.00**	30.60 ± 2.67	**21.80 ± 0.87**	**26.13 ± 1.29**
FWV (K = 1)	23.76 ± 0.72	28.54 ± 0.70	19.57 ± 1.21	20.41 ± 1.74
FWV (K = 2)	23.61 ± 1.08	28.64 ± 1.45	18.80 ± 1.22	20.01 ± 1.74
FWV (K = 3)	22.21 ± 0.96	24.33 ± 2.34	17.35 ± 1.93	21.37 ± 3.16
FWV (K = 4)	22.11 ± 0.62	28.76 ± 1.03	17.07 ± 1.41	18.80 ± 2.95
FWV (K = 5)	21.50 ± 0.67	27.56 ± 2.43	16.95 ± 1.19	17.20 ± 1.92

performance on the validation data, cosine distances are employed to model the semantic distances for FFV, and Euclidean distances are employed for AFV.

The experimental results are shown in Table 3. Apparently, $K = 1$ again gives the best performance of FFV, achieving 40.12% and 25.82% respectively on UCF101 and HMDB51 in the inductive setting, 50.67% and 31.51% respectively on UCF101 and HMDB51 in the transductive setting. Different from the text-based semantic representations, image-based semantic representations FFV encoded by Fisher Vector outperforms the AFV on both datasets.

Table 3. Results of different image-based semantic representations (mean ± standard error of recognition accuracy %) on UCF101 and HMDB51 datasets.

Sem. Rep.	UCF101 (51/50)		HMDB51 (26/25)	
	Inductive	Transductive	Inductive	Transductive
Random	2.00	2.00	4.00	4.00
AFV	37.24 ± 0.89	50.48 ± 1.35	25.55 ± 1.66	30.77 ± 3.23
FFV (K = 1)	**40.12 ± 1.30**	**50.67 ± 2.45**	**25.82 ± 1.19**	**31.51 ± 1.67**
FFV (K = 2)	38.01 ± 1.58	49.60 ± 1.82	25.50 ± 0.95	28.98 ± 1.94
FFV (K = 3)	36.52 ± 1.38	45.48 ± 0.73	24.27 ± 1.10	26.95 ± 3.38
FFV (K = 4)	35.31 ± 1.17	44.76 ± 2.40	23.22 ± 1.25	25.26 ± 2.32
FFV (K = 5)	34.98 ± 0.68	45.08 ± 1.82	23.09 ± 1.12	23.93 ± 2.06

5.3   Comparison with Other Semantic Representations

In this experiment, we compare the proposed semantic representations with other popular ones. From Tables 2 and 3, we know that AWV and FFV (K = 1) perform the best among the text- and image-based semantic representations respectively. So we consider AWV and FFV (K = 1) as the representatives of the proposed text- and image-based semantic representations. As described in Sect. 4.4, we conduct the experiments in both conventional and generalised ZSL scenarios in our experiments.

We present the experimental results in Table 4. Clearly, the proposed two semantic representations (i.e., AWV and FFV (K = 1)) outperform word vectors and attributes consistently in terms of the conventional ZSL evaluation metric. On UCF101, the use of textual information enhances the word vectors based solely on the action labels by lifting the accuracy from 19.42% to 24.38%, even higher than that of labour-intensive attributes (21.54%). The image-based semantic representation FFV encoded with Fisher Vector gives the best accuracy of 40.12%, significantly higher than its counterparts. This is attributed to the narrower semantic gap between video representation space and image-based semantic space. The still images contain abundant visually discriminative information which can be further encoded into high-level semantic representations of human actions. On HMDB51, the same conclusions can be drawn. It is noteworthy that AWV is only slightly better than WV for HMDB51 dataset. The reason might be the existence of actions which are difficult to describe with texts in this dataset, such as, "sit", "talk", "turn", "stand", "pick", "catch", and etc.

Table 4. A comparison of different semantic representations on UCF101 and HMDB51 datasets (mean $\pm$ standard error)%.

Dataset	Sem. Rep.	cZSL	gZSL		
		$A_{\mathcal{U}\to\mathcal{U}}$	$A_{\mathcal{U}\to\mathcal{T}}$	$A_{\mathcal{S}\to\mathcal{T}}$	H
UCF101	Random	2.00	1.00	1.00	1.00
	WV	19.42 $\pm$ 0.69	4.54 $\pm$ 0.64	84.79 $\pm$ 0.91	8.59 $\pm$ 1.17
	Att	21.54 $\pm$ 0.72	2.48 $\pm$ 0.62	86.39 $\pm$ 1.37	4.78 $\pm$ 1.18
	AWV	24.38 $\pm$ 1.00	5.32 $\pm$ 1.53	**86.43 $\pm$ 1.06**	9.85 $\pm$ 2.66
	FFV	**40.12 $\pm$ 1.30**	**16.55 $\pm$ 1.30**	82.38 $\pm$ 1.17	**27.49 $\pm$ 1.86**
HMDB51	Random	4.00	2.00	2.00	2.00
	WV	21.53 $\pm$ 1.75	2.64 $\pm$ 0.33	58.70 $\pm$ 1.40	5.05 $\pm$ 0.61
	AWV	21.80 $\pm$ 0.87	2.99 $\pm$ 0.35	**62.00 $\pm$ 2.57**	5.69 $\pm$ 0.64
	FFV	**25.68 $\pm$ 1.07**	**5.91 $\pm$ 0.90**	58.57 $\pm$ 1.50	**10.65 $\pm$ 1.48**

Regarding the generalised ZSL scenario, the proposed AWV and FFV perform better on the test examples from unseen classes (with 5.32% and 16.55% respectively on UCF101, 2.99% and 5.91% respectively on HMDB51), outperforming the attributes and word vectors. We also notice that FFV does not perform the best on test examples from seen classes (i.e., $A_{\mathcal{S} \rightarrow \mathcal{T}}$), although it is significantly better than others in terms of harmonic mean (H). This is reasonable and practically preferable with the trade-off between recognition accuracy of examples from seen and unseen classes.

5.4   How Many Images Are Enough?

In the previous experiments, we use all the collected images to encode the image-based semantic representations. In this experiment, we investigate how the number of images affects the encoded semantic representations. We use AFV and FFV ($K = 1$) as the encoding methods and generate the semantic representations for each human action with the number of relevant images to be 5, 10, 20, 30, 40, 50, 60, 70, 80, 90 and 100 respectively (For the case when the total number of collected images for one human action is less than the expected number, we simply use all the collected images of that action in the experiment). The experiments are conducted on two human action datasets in conventional ZSL scenario under both inductive and transductive settings.

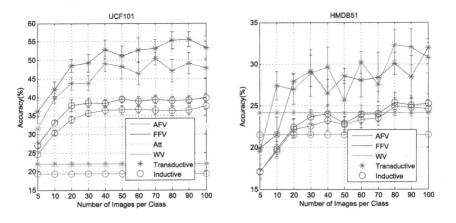

Fig. 3. Effects of number of images on the performance of AFV and FFV ($K = 1$).

The performances of two types of image-based semantic representations with different numbers of images are shown in Fig. 3. For a direct comparison, we display the baseline performance of attributes and word vectors in the figure as well. Using more images usually benefits the performance of AFV and FFV on both datasets. In specific, we can see a dramatic performance boost with the number of images increased from 5 to 40 per class for UCF101. A further increase of images does not improve the performance significantly, which is especially true in the inductive setting. For HMDB51 dataset, the similar trend of performance improvement can be observed from Fig. 3, and the performance improvement stops until the number of images per class increases to around 80. In addition, the proposed image-based semantic representations using only 5 images per class can achieve better performance on UCF101 than attributes and word vectors, and the number rises to 20 for HMDB51 to beat word vectors. To summarise, we are able to use a small number of relevant images to encode the semantic representations of human actions, yet boosting the zero-shot human action recognition accuracy to a large extent.

6   Conclusions and Future Work

We explore the alternative side information to the existing attributes and word vectors towards improved zero-shot human action recognition. The textual descriptions of human actions from the Internet can be used as side information for knowledge transfer in ZSL. In addition, the combination with pre-trained word vectors can further improve the power of text-based semantic representations, even better than the manually annotated attributes. On the other hand, the image-based semantic representations achieve dramatic performance improvement compared with the ones based on other side information (e.g., texts and human annotations), due to the narrower semantic gap. Our experiments also show that a small number of images are enough to gain significant performance improvement.

There are quite a few directions we can follow in our future work. Firstly, we only use a very simple encoding method (TD) for text-based semantic representations in this paper, which results in an extremely high dimensionality and sparse vector representation per document. It has been chosen in this work as a proof of concept, but could be optimised by using alternative techniques such as latent Dirichlet allocation (LDA), latent semantic indexing (LSI), etc. Besides, in our methods of text-based representation encoding, only the occurrences of different words in a given document are considered, and the word orders which play an important role in text understanding have been ignored. Thus the meaning of sentences containing "not" and "but" would be destroyed. To overcome this limitation, some potential techniques recently developed in NLP (e.g., *document2vec* [13]) would be investigated. Currently, we extract image features with deep CNN models pre-trained on large scale object classification dataset (i.e., ImageNet). Although the pre-trained models have already shown great generalization and transferability to other visual recognition tasks, better performance can be expected by fine-tuning the models with our specific human action image data. We have done some preliminary experiments on the combination of two different types of semantic representations, but only get results no better than the use of image-based semantic representation alone. We do not want to rush to the conclusion that the image- and text-based semantic representations are not complementary before further studying the combination methods in our future work.

Acknowledgments. The authors would like to thank Ubai Sandouk from MLO group at The University of Manchester for personal communication and the anonymous reviewers for their valuable comments and suggestions.

References

1. Akata, Z., Malinowski, M., Fritz, M., Schiele, B.: Multi-cue zero-shot learning with strong supervision. In: IEEE Conference on Computer Vision and Pattern Recognition (CVPR), pp. 59–68 (2016)
2. Alexiou, I., Xiang, T., Gong, S.: Exploring synonyms as context in zero-shot action recognition. In: IEEE International Conference on Image Processing (ICIP), pp. 4190–4194. IEEE (2016)
3. Chao, W.-L., Changpinyo, S., Gong, B., Sha, F.: An empirical study and analysis of generalized zero-shot learning for object recognition in the wild. In: Leibe, B., Matas, J., Sebe, N., Welling, M. (eds.) ECCV 2016. LNCS, vol. 9906, pp. 52–68. Springer, Cham (2016). https://doi.org/10.1007/978-3-319-46475-6_4
4. Cheng, J., Liu, Q., Lu, H., Chen, Y.W.: Supervised Kernel locality preserving projections for face recognition. Neurocomputing **67**, 443–449 (2005)
5. Chuang Gan, M.L., Yang, Y., Zhuang, Y., Hauptmann, A.G.: Exploring semantic interclass relationships (SIR) for zero-shot action recognition. In: Twenty-Ninth AAAI Conference on Artificial Intelligence, pp. 3769–3775 (2015)
6. Elhoseiny, M., Saleh, B., Elgammal, A.: Write a classifier: zero-shot learning using purely textual descriptions. In: IEEE International Conference on Computer Vision (ICCV), pp. 2584–2591 (2013)
7. Farhadi, A., Endres, I., Hoiem, D., Forsyth, D.: Describing objects by their attributes. In: IEEE Conference on Computer Vision and Pattern Recognition (CVPR), pp. 1778–1785. IEEE (2009)
8. Fu, Y., Hospedales, T.M., Xiang, T., Gong, S.: Learning multimodal latent attributes. IEEE Trans. Pattern Anal. Mach. Intell. **36**(2), 303–316 (2014)
9. Inoue, N., Shinoda, K.: Adaptation of word vectors using tree structure for visual semantics. In: ACM on Multimedia Conference, pp. 277–281. ACM (2016)
10. Jiang, Y., Liu, J., Zamir, A.R., Toderici, G., Laptev, I., Shah, M., Sukthankar, R.: Thumos challenge: action recognition with a large number of classes (2014)
11. Kuehne, H., Jhuang, H., Garrote, E., Poggio, T., Serre, T.: HMDB: a large video database for human motion recognition. In: IEEE International Conference on Computer Vision (ICCV), pp. 2556–2563. IEEE (2011)
12. Lampert, C.H., Nickisch, H., Harmeling, S.: Learning to detect unseen object classes by between-class attribute transfer. In: IEEE Conference on Computer Vision and Pattern Recognition (CVPR), pp. 951–958. IEEE (2009)
13. Le, Q.V., Mikolov, T.: Distributed representations of sentences and documents. In: International Conference on Machine Learning (ICML), vol. 14, pp. 1188–1196 (2014)
14. Liu, J., Kuipers, B., Savarese, S.: Recognizing human actions by attributes. In: IEEE Conference on Computer Vision and Pattern Recognition (CVPR), pp. 3337–3344. IEEE (2011)
15. Mikolov, T., Sutskever, I., Chen, K., Corrado, G.S., Dean, J.: Distributed representations of words and phrases and their compositionality. In: Advances in Neural Information Processing Systems, pp. 3111–3119 (2013)
16. Mukherjee, T., Hospedales, T.: Gaussian visual-linguistic embedding for zero-shot recognition. In: Conference on Empirical Methods on Natural Language Processing (EMNLP) (2016)
17. Perronnin, F., Sánchez, J., Mensink, T.: Improving the Fisher Kernel for large-scale image classification. In: Daniilidis, K., Maragos, P., Paragios, N. (eds.) ECCV 2010. LNCS, vol. 6314, pp. 143–156. Springer, Heidelberg (2010). https://doi.org/10.1007/978-3-642-15561-1_11

18. Qin, J., Wang, Y., Liu, L., Chen, J., Shao, L.: Beyond semantic attributes: discrete latent attributes learning for zero-shot recognition. IEEE Sig. Process. Lett. **23**(11), 1667–1671 (2016)
19. Rohrbach, M., Stark, M., Szarvas, G., Gurevych, I., Schiele, B.: What helps where- and why? Semantic relatedness for knowledge transfer. In: IEEE Conference on Computer Vision and Pattern Recognition (CVPR), pp. 910–917. IEEE (2010)
20. Sandouk, U., Chen, K.: Multi-label zero-shot learning via concept embedding. arXiv preprint arXiv:1606.00282 (2016)
21. Sharmanska, V., Quadrianto, N., Lampert, C.H.: Augmented attribute representa- tions. In: Fitzgibbon, A., Lazebnik, S., Perona, P., Sato, Y., Schmid, C. (eds.) ECCV 2012. LNCS, vol. 7576, pp. 242–255. Springer, Heidelberg (2012). https:// doi.org/10.1007/978-3-642-33715-4_18
22. Soomro, K., Zamir, A.R., Shah, M.: UCF101: a dataset of 101 human actions classes from videos in the wild. arXiv preprint arXiv:1212.0402 (2012)
23. Szegedy, C., Liu, W., Jia, Y., Sermanet, P., Reed, S., Anguelov, D., Erhan, D., Vanhoucke, V., Rabinovich, A.: Going deeper with convolutions. In: IEEE Confer- ence on Computer Vision and Pattern Recognition (CVPR), pp. 1–9 (2015)
24. Tran, D., Bourdev, L., Fergus, R., Torresani, L., Paluri, M.: Learning spatiotem- poral features with 3D convolutional networks. In: IEEE International Conference on Computer Vision (ICCV), pp. 4489–4497 (2015)
25. Vedaldi, A., Fulkerson, B.: VLFeat: an open and portable library of computer vision algorithms (2008). http://www.vlfeat.org/
26. Wang, Q., Chen, K.: Zero-shot visual recognition via bidirectional latent embed- ding. Int. J. Comput. Vis. (2017)
27. Xian, Y., Schiele, B., Akata, Z.: Zero-shot learning-the good, the bad and the ugly. In: IEEE Conference on Computer Vision and Pattern Recognition (CVPR) (2017)
28. Zhang, Z., Saligrama, V.: Zero-shot recognition via structured prediction. In: Leibe, B., Matas, J., Sebe, N., Welling, M. (eds.) ECCV 2016. LNCS, vol. 9911, pp. 533– 548. Springer, Cham (2016). https://doi.org/10.1007/978-3-319-46478-7_33

Early Active Learning with Pairwise Constraint for Person Re-identification

Wenhe Liu[1]([⊠]), Xiaojun Chang[2], Ling Chen[1], and Yi Yang[1]

[1] CAI, University of Technology Sydney, Sydney, Australia
allenlwh@gmail.com, Ling.Chen@uts.edu.au, yeeiyang@gmail.com
[2] LTI, Carnegie Mellon University, Pittsburgh, USA
cxj273@gmail.com

Abstract. Research on person re-identification (re-id) has attached much attention in the machine learning field in recent years. With sufficient labeled training data, supervised re-id algorithm can obtain promising performance. However, producing labeled data for training supervised re-id models is an extremely challenging and time-consuming task because it requires every pair of images across no-overlapping camera views to be labeled. Moreover, in the early stage of experiments, when labor resources are limited, only a small number of data can be labeled. Thus, it is essential to design an effective algorithm to select the most representative samples. This is referred as *early active learning* or *early stage experimental design* problem. The pairwise relationship plays a vital role in the re-id problem, but most of the existing early active learning algorithms fail to consider this relationship. To overcome this limitation, we propose a novel and efficient early active learning algorithm with a pairwise constraint for person re-identification in this paper. By introducing the pairwise constraint, the closeness of similar representations of instances is enforced in active learning. This benefits the performance of active learning for re-id. Extensive experimental results on four benchmark datasets confirm the superiority of the proposed algorithm.

Keywords: Early active learning · Person re-identification

1 Introduction

The primary target of person re-identification (re-id) is to identify a person from camera shots across pairs of non-overlapping camera views, and research on this topic has attracted considerable attention in recent years [8–10,15,29]. In the field of computer vision, re-id can be formed as an image *retrieval* task. Given a *probe* image of a person from one camera view, the difficulty is to identify images of the same person from a *gallery* of images taken by other non-overlapping camera views. Despite the encouraging results reported in previous works, re-id remains a challenge in several respects. The accuracy of identification is often degrades as a result of the uncontrollable and/or unpredictable variation of appearance changes across camera views, such as body pose, view angle, occlusion and illumination conditions [7,20,23].

© Springer International Publishing AG 2017
M. Ceci et al. (Eds.): ECML PKDD 2017, Part I, LNAI 10534, pp. 103–118, 2017.
https://doi.org/10.1007/978-3-319-71249-9_7

Supervised re-id methods can achieve promising results if there are sufficient labeled training data. Unfortunately, the human labor necessary for labeling training data is sometimes inadequate. This problem becomes extremely severe in the re-id scenario, since labeling for re-id is difficult to achieve. Unlike other recognition tasks which only requires each image to be labeled, re-id requires all pairs of images across camera views to be labeled. It is a tough task even for humans to identify the same person in different camera views among a potentially huge number of imposters [9,20]. At the same time, pairwise labeled data is required for each pair of camera views in the camera network in re-id, thus the labeling cost will become prohibitively high numbers of cameras in today's world. For example, there might be more than over a hundred in one underground train station [20].

To save labor costs, it is essential to design an effective algorithm that can select a subset of samples that are the most representative and/or informative for training. Active learning is widely studied to solve this kind of sample selection problem. As discussed in [18], active learning methods can be divided into two categories. The first category of algorithms select the most informative samples for labeling when there are already some labeled samples. They include uncertainty sampling methods [1,6,11,22] query by committee methods [3,21]. Most of these active learning methods prefer to select uncertainty data, or data that is difficult to analyze. They thus require a certain number of labeled samples to evaluate the uncertainty of the unlabeled data or sampling bias [18] will result. It is therefore recommended that such methods are only applied in the mid-stage of experiments when there are sufficient labeled data. For the purpose of distinguishing between the two categories, we refer to the first category of active learning methods as *traditional active learning*. The second category of active learning methods is considered for application in the early stage of experiments, when there are limited resources for labeling data. In this case, there are no labeled samples, thus labeling a small number of representative data is desirable for training reliable supervised models. In the category of early active learning, there are clustering-based methods [16,19] and transductive experimental design methods [27]. These kinds of active learning algorithms are referred to as *early active learning* or *early stage experimental design* [18]. We illustrate the procedures of and example of the traditional active learning algorithm, QUIRE [6], and our early active learning algorithm with pairwise constraint (abbreviated as EALPC) in Fig. 1.

In the rest of this paper, we focus on the early active learning methods for person re-identification applications. As mentioned, labeling re-id data is extremely labor-consuming and time-consuming. It is therefore highly desirable to enhance the learning performance in re-id applications by early active learning. Unfortunately, early active learning methods currently merely consider analyzing representative samples with pairwise relationships. Therefore, directly applying them for re-id may be not appropriate.

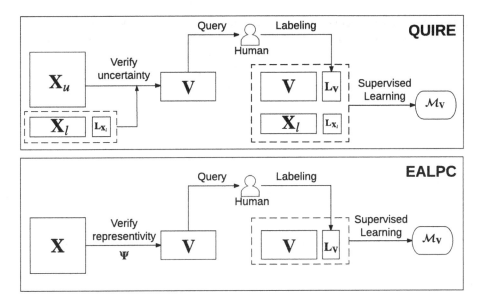

Fig. 1. Procedures of QUIRE [6] (*upper*) and our Early Active Learning with Pairwise Constraint (EALPC) (*lower*). In QUIRE, pre-labeled samples $\mathbf{X}_l$ are used for the uncertainty evaluation on the unlabeled samples $\mathbf{X}_u$. Then, it selects a subset samples $\mathbf{V} \subset \mathbf{X}_u$ for labeling. At last, both $\mathbf{X}_u$ and $\mathbf{V}$ along with their labels are used for supervised learning. In EALPC, unlabeled data $\mathbf{X}$ is analyzed without pre-labeled data. Meanwhile, pairwise constraint $\mathbf{\Psi}$ is introduced to enhance the performance of early active learning for re-id. More details are in Sect. 2.

To overcome the limitations described above, we propose a novel algorithm for person re-identification, Early Active Learning with Pairwise Constraint, abbreviated as EALPC. The main contributions of our work are as follows:

1. We propose a novel Early Active Learning with Pairwise Constraint algorithm for person re-identification. To the best of our knowledge, this is the first method considers to consider both (a) applying early active learning for the re-id application, and (b) extending early active learning schema with pairwise constraint.
2. We introduce the $\ell_{2,1}$-norm to our objective function, which improves the robustness of our methods and suppresses the effects of outliers.
3. We propose an efficient algorithm to optimize the proposed problem. Our optimization algorithm also provides a closed form solution and guarantees to reach the global optimum in the convergence.

2   The Proposed Framework

In this section, we first revisit the early active learning algorithm and then propose our early active learning with pairwise constraint for re-id.

Notation. Let the superscript $^\mathsf{T}$ denote the transpose of a vector/matrix, $\mathbf{0}$ be a vector/matrix with all zeros, $\mathbf{I}$ be an identity matrix. Let $\mathrm{Tr}(\mathbf{A})$ be the trace of matrix $\mathbf{A}$. Let $\mathbf{a}_i$ and $\mathbf{a}^j$ be the i-th column vector and j-th row vector of matrix $\mathbf{A}$ respectively. Let $\langle \mathbf{A}, \mathbf{B} \rangle = \mathrm{Tr}(\mathbf{A}\mathbf{B}^\mathsf{T})$ be the inner product of $\mathbf{A}$ and $\mathbf{B}$, and $\|\mathbf{v}\|_p$ be the ℓ_p-norm of a vector $\mathbf{v}$. Then, the Frobenius norm of an arbitrary matrix $\mathbf{A}$ is defined as $\|\mathbf{A}\|_F = \sqrt{\langle \mathbf{A}, \mathbf{A} \rangle}$. The ℓ_2-norm of a vector $\mathbf{a}$ is denoted as $\|\mathbf{a}\|_2 = \sqrt{\mathbf{a}^T\mathbf{a}}$ and the $\ell_{2,1}$-norm of matrix $\mathbf{A} \in \mathbb{R}^{n \times m}$ is denoted as $\|\mathbf{A}\|_{2,1} = \sum_{i=1}^{n} \sqrt{\sum_{j=1}^{m} a_{ij}^2} = \sum_{i=1}^{n} \|\mathbf{a}^i\|_2$, where a_{ij} is the (i,j)-th element of $\mathbf{A}$ and $\mathbf{a}^i$ is the i-th row vector of $\mathbf{A}$. For analytical consistency, the $\ell_{2,0}$-norm of a matrix $\mathbf{A}$ is denoted as the number of the nonzero rows of $\mathbf{A}$. For any convex function $f(\mathbf{A})$, let $\partial f(\mathbf{A})/\partial \mathbf{A}$ denote its subdifferential at $\mathbf{A}$. We denote $\mathcal{G}$ as a weighted graph with a vertex set $\mathcal{X}$ and an affinity matrix $\mathbf{S} \in \mathbb{R}^{n \times n}$ constructed on $\mathcal{X}$. The (unnormalized) Laplacian matrix associated with $\mathcal{G}$ is defined as $\mathbf{L} = \mathbf{D} - \mathbf{S}$, where $\mathbf{D}$ is a degree matrix with $\mathbf{D}(i,i) = \sum_j S(i,j)$.

2.1   Early Active Learning

We first revisit the early active learning algorithm. Given a set of unlabeled samples $\mathbf{X} \in \mathbb{R}^{d \times n}$, the task of active learning is to select a subset of $m < n$ most representative samples $\mathbf{V} \in \mathbb{R}^{d \times m}$. Then, the selected samples are queried labeling for supervised learning. The labeled subset of data is expected to maximize the potential performance of the supervised learning in the early stage of experiment, when the available resource for labeling data is limited, i.e. only a small number of data can be labeled for supervised learning. Generally, we can define the optimization problem of early active learning as follows:

$$\min_{\mathbf{V},\mathbf{A}} \mathbf{R}(\mathbf{X}, \mathbf{V}, \mathbf{A}) + \alpha \mathbf{\Omega}(\mathbf{A}), \ s.t. \ \mathbf{V} \subset \mathbf{X}, \ |\mathbf{V}| = m. \tag{1}$$

where $\mathbf{V}$ is a subset of $\mathbf{X}$, $\mathbf{A}$ is a transformation matrix. In Eq. (1), the first term $\mathbf{R}(\cdot)$ is the reconstruction loss, the second term $\mathbf{\Omega}(\cdot)$ is the regularization term and $\alpha > 0$ is a leverage parameter. The major purpose of early active learning is to select a subset $\mathbf{V} \subset \mathbf{X}$ with size $m < n$ that can best represent the whole data $\mathbf{X}$ through the linear transformation matrix $\mathbf{A}$. The selected samples are therefore considered to be the most representative.

In [27], an early active learning via a Transduction Experimental Design algorithm (TED) is proposed with the aim of finding the subset $\mathbf{V} \subset \mathbf{X}$ and a project matrix $\mathbf{A}$ that minimizes the least squared reconstruction error:

$$\min_{\mathbf{V},\mathbf{A}} \sum_{i=1}^{n} (\|\mathbf{x}_i - \mathbf{V}\mathbf{a}_i\|_2^2 + \alpha \|\mathbf{a}_i\|_2^2)$$

$$s.t. \ \mathbf{A} = [\mathbf{a}_1, \cdots, \mathbf{a}_n] \in \mathbb{R}^{m \times n}, \ \mathbf{V} \subset \mathbf{X}, \ |\mathbf{V}| = m. \tag{2}$$

where $\mathbf{V}\mathbf{a}_i$ is the representation item of $\mathbf{x}_i$. However, Eq. (2) is an NP-hard problem to solve, thus an approximate solution by a sequential optimization problem is proposed in [27].

2.2 Early Active Learning with Pairwise Constraint

In this work, we focus on early active learning in the person re-id problem. As mentioned previously, person re-id is formed as an image *retrieval* task which aims to re-identify the same person across non-overlapping camera views given a probe image of the person. The analysis of pairwise relationships of images in different camera views is therefore required. For this purpose, we introduce a pairwise constraint to early active learning:

$$\boldsymbol{\Psi_V}(\mathbf{A}) = \sum_{i,j=1}^{n} \|\mathbf{Va}_i - \mathbf{Va}_j\|_2^2 S_{\mathbf{V}}(i,j), \tag{3}$$

where $\mathbf{Va}_i$ is the representation item of $\mathbf{x}_i$ and $S_{\mathbf{V}}(i,j)$ is the (i,j)-th element of similarity matrix $\mathbf{S}$. It is the similarity between the i-th and the j-th representations. In this work we define $S_{\mathbf{V}}(i,j)$ as a Gaussian similarity:

$$S_{\mathbf{V}}(i,j) = \begin{cases} \exp(-\frac{\|\mathbf{Va}_i - \mathbf{Va}_j\|^2}{\sigma^2}), & if \ \mathbf{Va}_i \in \mathcal{N}_k(\mathbf{Va}_j) \ and \ \mathbf{Va}_j \in \mathcal{N}_k(\mathbf{Va}_i) \\ 0 & , \quad otherwise, \end{cases} \tag{4}$$

where $\mathcal{N}_k(\mathbf{x})$ denotes the set of k-nearest neighbors of $\mathbf{x}$. We can then reformulate the pairwise constraint in Eq. (3) by inducing a Laplacian matrix:

$$\boldsymbol{\Psi_V}(\mathbf{A}) = \sum_{i,j=1}^{n} \|\mathbf{Va}_i - \mathbf{Va}_j\|_2^2 S_{\mathbf{V}}(i,j) = \mathrm{Tr}((\mathbf{VA})\mathbf{L_V}(\mathbf{VA})^T), \tag{5}$$

where $\mathbf{L_V} = \mathbf{D} - \mathbf{S_V}$ is the Laplacian matrix and $\mathbf{D}$ is the degree matrix with each element $\mathbf{D}_{ii} = \sum_j S_{\mathbf{V}}(i,j)$. As discussed in [9], minimizing the pairwise constraint will force the similar representations to be close to each other. Following the assumption that visually similar images of a person have a high probability of sharing the similar representation features in re-id [9], this will make early active learning schema more suitable for re-id applications.

After introducing the pairwise constraint, the early active learning for person re-identification can be formulated as:

$$\min_{\mathbf{V},\mathbf{A}} \mathbf{R}(\mathbf{X}, \mathbf{V}, \mathbf{A}) + \alpha \boldsymbol{\Omega}(\mathbf{A}) + \beta \boldsymbol{\Psi_V}(\mathbf{A})$$
$$s.t. \ \mathbf{A} = [\mathbf{a}_1, \cdots, \mathbf{a}_n] \in \mathbb{R}^{m \times n}, \mathbf{V} \subset \mathbf{X}, \ |\mathbf{V}| = m. \tag{6}$$

where $\alpha > 0$ and $\beta > 0$ are leverage parameters of regularization terms. After substituting Eqs. (2) and (5) into Eq. (6) we obtain:

$$\min_{\mathbf{V},\mathbf{A}} \sum_{i=1}^{n} (\|\mathbf{x}_i - \mathbf{Va}_i\|_2^2 + \alpha \|\mathbf{a}_i\|_2^2) + \beta \mathrm{Tr}((\mathbf{VA})\mathbf{L_V}(\mathbf{VA})^T)$$
$$s.t. \ \mathbf{A} = [\mathbf{a}_1, \cdots, \mathbf{a}_n] \in \mathbb{R}^{m \times n}, \mathbf{V} \subset \mathbf{X}, \ |\mathbf{V}| = m. \tag{7}$$

Finding the optimal subset $\mathbf{V} \subset \mathbf{X}$ in Eq. (7) is NP-hard. Inspired by [18], we relax the problem to the following problem by introducing the $\ell_{2,0}$-norm for structure sparsity:

$$\min_{\mathbf{A}} \sum_{i=1}^{n} \|\mathbf{x}_i - \mathbf{X}\mathbf{a}_i\|_2^2 + \alpha\|\mathbf{A}\|_{2,0} + \beta\mathrm{Tr}((\mathbf{X}\mathbf{A})\mathbf{L}_{\mathbf{X}}(\mathbf{X}\mathbf{A})^T) \tag{8}$$

$$s.t. \ \mathbf{A} = [\mathbf{a}_1, \cdots, \mathbf{a}_n] \in \mathbb{R}^{n \times n}, \ \|\mathbf{A}\|_{2,0} = m.$$

However, the $\ell_{2,0}$-norm makes Eq. (8) a non-convex problem. At the same time, the least squared loss used in Eq. (8) is sensitive to the outliers [18], which makes the algorithm not robust.

We note that in previous researches [17,18,26], the $\ell_{2,1}$-norm is used instead of the $\ell_{2,0}$-norm. It is shown in [18] that the $\ell_{2,1}$-norm is the minimum convex hull of the $\ell_{2,0}$-norm when row-sparsity is required. In other words, minimization of $\|\mathbf{A}\|_{2,1}$ will achieve the same result as $\|\mathbf{A}\|_{2,0}$ when $\mathbf{A}$ is row-sparse. As analyzed in [18,30], the $\ell_{2,1}$-norm can suppress the effect of outlying samples. We therefore reformulate Eq. (8) as a relaxed convex optimization problem:

$$\min_{\mathbf{A}} \sum_{i=1}^{n} \|\mathbf{x}_i - \mathbf{X}\mathbf{a}_i\|_{2,1} + \alpha\|\mathbf{A}\|_{2,1} + \beta\mathrm{Tr}((\mathbf{X}\mathbf{A})\mathbf{L}_{\mathbf{X}}(\mathbf{X}\mathbf{A})^T). \tag{9}$$

In Eq. (9), we adopt the $\ell_{2,1}$-norm instead of both the least square reconstruction loss term and the $\ell_{2,0}$-norm structure sparsity term for robustness and suppression of outliers. By inducing the matrix formulation, Eq. (9) is rewritten as follows:

$$\min_{\mathbf{A}} \|(\mathbf{X} - \mathbf{X}\mathbf{A})^T\|_{2,1} + \alpha\|\mathbf{A}\|_{2,1} + \beta\mathrm{Tr}((\mathbf{X}\mathbf{A})\mathbf{L}_{\mathbf{X}}(\mathbf{X}\mathbf{A})^T). \tag{10}$$

After obtaining the optimal solution of $\mathbf{A}$, the importances of samples can be ranked by sorting the absolute row-sum values of $\mathbf{A}$ in the decreasing order. A subset of the representative samples then can be selected corresponding to the top m largest values and query labeling.

Kernelization. The proposed algorithm can be extended to the kernel version for non-linear high dimensional space. We define $\boldsymbol{\Phi}: \mathbb{R}^d \rightarrow \mathcal{H}$ as a mapping from the Euclidian space to a Reproducing Kernel Hilbert Space (RKHS) as $\mathcal{H}$. It can be induced by a kernel function $\mathcal{K}(\mathbf{x}, \mathbf{y}) = \boldsymbol{\Phi}(\mathbf{x})^T\boldsymbol{\Phi}(\mathbf{y})$. Then we can project $\mathbf{X}$ to RKHS space as $\boldsymbol{\Phi}(\mathbf{X}) = [\boldsymbol{\Phi}(\mathbf{x}_1), \cdots, \boldsymbol{\Phi}(\mathbf{x}_n)]$. The proposed problem thus becomes:

$$\min_{\mathbf{A}} \|(\boldsymbol{\Phi}(\mathbf{X}) - \boldsymbol{\Phi}(\mathbf{X})\mathbf{A})^T\|_{2,1} + \alpha\|\mathbf{A}\|_{2,1} + \beta\mathrm{Tr}((\boldsymbol{\Phi}(\mathbf{X})\mathbf{A})\mathbf{L}_{\mathbf{X}}(\boldsymbol{\Phi}(\mathbf{X})\mathbf{A})^T). \tag{11}$$

We denote our Early Active Learning with Pairwise Constraint algorithm in Eq. (10) as EALPC and the kenerlized version of our algorithm in Eq. (11) as EALPC_K.

3   Optimization

We provide an efficient algorithm for optimizing the proposed objective function. Taking the derivative w.r.t. $\mathbf{A}$ in Eq. (10) and setting it to zero, we obtain[1]:

$$\mathbf{X}^T\mathbf{X}\mathbf{A}\mathbf{P} - \mathbf{X}^T\mathbf{X}\mathbf{P} + \alpha\mathbf{Q}\mathbf{A} + \beta\mathbf{X}^T\mathbf{X}\mathbf{A}\mathbf{L}_{\mathbf{X}} = 0, \tag{12}$$

where $\mathbf{P}$ is a diagonal matrix and its i-th diagonal element is $p_{ii} = \frac{1}{2\|\mathbf{x}_i - \mathbf{X}\mathbf{a}_i\|_2}$. $\mathbf{Q}$ is a diagonal matrix and its i-th diagonal element is $q_{ii} = \frac{1}{2\|\mathbf{a}^i\|_2}$. Then by setting the derivative of Eq. (12) w.r.t. $\mathbf{a}_i$ to zero for each i, we obtain:

$$p_{ii}\mathbf{X}^T\mathbf{X}\mathbf{a}_i - p_{ii}\mathbf{X}^T\mathbf{x}_i + \alpha\mathbf{Q}\mathbf{a}_i + \beta\mathbf{X}^T\mathbf{X}\mathbf{A}\mathbf{L}_i = 0, \tag{13}$$

where $\mathbf{L}_i$ is the i-th column vector of $\mathbf{L}_{\mathbf{X}}$. It is sample to verify that $\mathbf{A}\mathbf{L}_i = l_{ii}\mathbf{a}_i + \sum_{k \neq i} l_{ki}\mathbf{a}_k$, where l_{ii} and l_{ki} are the (i,i)-th and (k,i)-th element of $\mathbf{L}_{\mathbf{X}}$ respectively and $\mathbf{a}_k$ is the k-th column vector of $\mathbf{A}$. Therefore, the optimal solution $\mathbf{a}_i^*$ can be calculated by the closed form solution:

$$\mathbf{a}_i^* = (p_{ii}\mathbf{X}^T\mathbf{X} + \alpha\mathbf{Q} + \beta\mathbf{X}^T\mathbf{X}l_{ii})^{-1}(p_{ii}\mathbf{X}^T\mathbf{x}_i - \beta\mathbf{X}^T\mathbf{X}\sum_{k \neq i}\mathbf{a}_k l_{ki}). \tag{14}$$

In Eq. (12), $\mathbf{P}$ and $\mathbf{Q}$ are dependent on $\mathbf{A}$, thus they also need to be determined in each iteration. We propose an iterative algorithm to solve this problem. The detailed algorithm is described in Algorithm 1. In the next section, we will prove that Algorithm 1 converges to the global optimal solution of Eq. (10).

4   Convergence Analysis

We first introduce a lemma proposed in [17]:

Lemma 1. *For any arbitrary vector* $\mathbf{m}$ *and* $\mathbf{n}$ *there is*

$$\|\mathbf{m}\|_2 - \frac{\|\mathbf{m}\|_2^2}{2\|\mathbf{n}\|_2} \leq \|\mathbf{n}\|_2 - \frac{\|\mathbf{n}\|_2^2}{2\|\mathbf{n}\|_2}. \tag{15}$$

Next, in the following theorem we prove the convergence of our algorithm:

Theorem 1. *Algorithm 1 monotonically decreases the objective function value of Eq. (10) in each iteration.*

---

[1] In practice, when $\mathbf{x}_i - \mathbf{X}\mathbf{a}_i = 0$, p_{ii} can be regularized as $p_{ii} = \frac{1}{2\sqrt{\|\mathbf{x}_i - \mathbf{X}\mathbf{a}_i\|_2^2 + \eta}}$. Similarly when $\mathbf{a}_i = 0$, we set $q_{ii} = \frac{1}{2\sqrt{\|\mathbf{a}^i\|_2^2 + \eta}}$. η is a very small constant. It can be verified that when $\eta \to 0$ the problem with η reduces to the original problem in Eq. (12).

Algorithm 1. Algorithm for solving problem in Eq. (10)

Input: The data matrix $\mathbf{X} \in \mathbb{R}^{d \times n}$, parameters α and β.

1 Initialize $\mathbf{A} \in \mathbb{R}^{n \times n}$.

2 **while** *not converge* **do**

3 $\quad$ Compute the diagonal matrix $\mathbf{P}$, where the i-th diagonal element is $p_{ii} = \frac{1}{2\|\mathbf{x}_i - \mathbf{X}\mathbf{a}_i\|_2}$.

4 $\quad$ Compute the diagonal matrix $\mathbf{Q}$, where the i-th diagonal element is $q_{ii} = \frac{1}{2\|\mathbf{a}^i\|_2}$.

5 $\quad$ Update $\mathbf{A}$ by each column $\mathbf{a}_i$ as in Eq. (14):

$$\mathbf{a}_i^* = (p_{ii}\mathbf{X}^T\mathbf{X} + \alpha\mathbf{Q} + \beta\mathbf{X}^T\mathbf{X}l_{ii})^{-1}(p_{ii}\mathbf{X}^T\mathbf{x}_i - \beta\mathbf{X}^T\mathbf{X}\sum_{k \neq i}\mathbf{a}_k l_{ki}).$$

6 **end**

Output: The matrix $\mathbf{A} \in \mathbb{R}^{n \times n}$.

Proof. Suppose in an iteration the updated $\mathbf{A}$ is $\mathbf{A}^+$. According to Step 5 in Algorithm 1 we know that:

$$\mathbf{A}^+ = \underset{\mathbf{F}}{\arg\min}\, f(\mathbf{F}), \tag{16}$$

where we denote the function

$$f(\mathbf{F}) = \mathrm{Tr}((\mathbf{X} - \mathbf{X}\mathbf{F})\mathbf{P}(\mathbf{X} - \mathbf{X}\mathbf{F})^T) + \alpha\mathrm{Tr}(\mathbf{F}\mathbf{Q}\mathbf{F}^T) + \beta\mathrm{Tr}((\mathbf{X}\mathbf{F})\mathbf{L_X}(\mathbf{X}\mathbf{F})^T).$$

Thus, in each iteration when updating $\mathbf{A}$ to $\mathbf{A}^+$ we have

$$\mathrm{Tr}((\mathbf{X} - \mathbf{X}\mathbf{A}^+)\mathbf{P}(\mathbf{X} - \mathbf{X}\mathbf{A}^+)^T) + \alpha\mathrm{Tr}((\mathbf{A}^+)\mathbf{Q}(\mathbf{A}^+)^T) + \beta\mathrm{Tr}((\mathbf{X}\mathbf{A}^+)\mathbf{L_X}(\mathbf{X}\mathbf{A}^+)^T)$$
$$\leq \mathrm{Tr}((\mathbf{X} - \mathbf{X}\mathbf{A})\mathbf{P}(\mathbf{X} - \mathbf{X}\mathbf{A})^T) + \alpha\mathrm{Tr}(\mathbf{A}\mathbf{Q}\mathbf{A}^T) + \beta\mathrm{Tr}((\mathbf{X}\mathbf{A})\mathbf{L_X}(\mathbf{X}\mathbf{A})^T). \tag{17}$$

According to the definition of $\mathbf{P}$ and $\mathbf{Q}$, we thus obtain:

$$\sum_{i=1}^{n}\left(\frac{\|\mathbf{x}_i - \mathbf{X}\mathbf{a}_i^+\|_2^2}{2\|\mathbf{x}_i - \mathbf{X}\mathbf{a}_i\|_2} + \alpha\frac{\|\mathbf{a}^{i+}\|_2^2}{2\|\mathbf{a}^i\|_2}\right) + \beta\mathrm{Tr}((\mathbf{X}\mathbf{A}^+)\mathbf{L_X}(\mathbf{X}\mathbf{A}^+)^T)$$
$$\leq \sum_{i=1}^{n}\left(\frac{\|\mathbf{x}_i - \mathbf{X}\mathbf{a}_i\|_2^2}{2\|\mathbf{x}_i - \mathbf{X}\mathbf{a}_i\|_2} + \alpha\frac{\|\mathbf{a}^i\|_2^2}{2\|\mathbf{a}^i\|_2}\right) + \beta\mathrm{Tr}((\mathbf{X}\mathbf{A})\mathbf{L_X}(\mathbf{X}\mathbf{A})^T). \tag{18}$$

Meanwhile, according to Lemma 1, we can induce the following inequalities:

$$\sum_{i=1}^{n}\left(\|\mathbf{x}_i - \mathbf{X}\mathbf{a}_i^+\|_2 - \frac{\|\mathbf{x}_i - \mathbf{X}\mathbf{a}_i^+\|_2^2}{2\|\mathbf{x}_i - \mathbf{X}\mathbf{a}_i\|_2}\right) \leq \sum_{i=1}^{n}\left(\|\mathbf{x}_i - \mathbf{X}\mathbf{a}_i\|_2 - \frac{\|\mathbf{x}_i - \mathbf{X}\mathbf{a}_i\|_2^2}{2\|\mathbf{x}_i - \mathbf{X}\mathbf{a}_i\|_2}\right), \tag{19}$$

and

$$\sum_{i=1}^{n}\left(\|\mathbf{a}^{i+}\|_2 - \frac{\|\mathbf{a}_i^+\|_2^2}{2\|\mathbf{a}_i\|_2}\right) \leq \sum_{i=1}^{n}\left(\|\mathbf{a}^i\|_2 - \frac{\|\mathbf{a}^i\|_2^2}{2\|\mathbf{a}^i\|_2}\right). \tag{20}$$

After summing Eqs. (19) and (20) in the both sides of Eq. (18), we conclude that:

$$\sum_{i=1}^{n}(\|\mathbf{x}_i - \mathbf{X}\mathbf{a}_i^+\|_2 + \alpha\|\mathbf{a}^{i^+}\|_2) + \beta\mathrm{Tr}((\mathbf{X}\mathbf{A}^+)\mathbf{L}_{\mathbf{X}}(\mathbf{X}\mathbf{A}^+)^T)$$

$$\leq \sum_{i=1}^{n}\left(\|\mathbf{x}_i - \mathbf{X}\mathbf{a}_i\|_2 + \alpha\|\mathbf{a}^i\|_2\right) + \beta\mathrm{Tr}((\mathbf{X}\mathbf{A})\mathbf{L}_{\mathbf{X}}(\mathbf{X}\mathbf{A})^T).$$

(21)

The above inequality indicates that the objective function value of Eq. (10) monotonically decreases in Algorithm 1. □

Meanwhile, let $\partial f(\mathbf{A})/\partial \mathbf{A} = 0$ is equal to solving Eq. (12), thus in convergence, $\mathbf{A}$ will satisfy Eq. (10). As Eq. (10) is a convex problem, $\mathbf{A}$ is the global optimum solution to our problem. Overall, Algorithm 1 will converge to the global optimum solution of Eq. (10).

5   Experimental Study

In the experiments, we compare our proposed EALPC algorithm with five state-of-the-art and classic active learning algorithms. After determining and labeling the most representative samples, we train the re-id models with these samples using five popular re-id algorithms. All experiments are operated on four widely referenced re-id benchmark datasets. We report the average performance of 10 trials of independent experiments on each dataset.

5.1   Datasets and Settings

Datasets. We analyze performance of active learning for re-id on four widely referred benchmark datasets for person re-identification.

1. **VIPeR** [4]. The VIPeR dataset contains 1,264 images of 632 persons from two non-overlapping camera views. Two images are taken for each person, each from a different camera. Variations in viewpoint and illumination conditions occur frequently in VIPeR.
2. **PRID** [5]. The PRID dataset contains images of 385 individuals from two distinct cameras. Camera B records 749 persons and Camera A records 385 persons, 200 of whom are same persons.
3. **i-LID** [30]. The i-LID dataset records 119 individuals captured by three different cameras in an airport terminal. It contains 476 images with large occlusions caused by luggage and viewpoint changes.
4. **CAVIAR** [2]. The CAVIAR dataset contains 72 individuals captured by two cameras in a shopping mall. The number of the images is 1,220, with 10 to 20 images for each individual. The size of the images in the CAVIAR dataset varies significantly from 39×17 to 141×72.

In the experiments, we use the recently proposed Local Maximal Occurrence (LOMO) features for person image representation [12]. As in [14,20], all person images are scaled to 128×48 pixels. We then use the default setting in [12] to produce a 29,960 dimension feature for each image.

Active Learning Algorithms. We choose five active learning algorithms and compare them with our proposed algorithm.

1. **Random.** As a baseline algorithm, we randomly select samples and query labeling.
2. **K-means.** We use the K-means algorithm as another baseline algorithm as in [18]. In each experiment, samples are ranked by their distances from the K cluster centers in ascending order.
3. **QUIRE** [6]. Active learning by Querying Informative and Representative Examples is an algorithm which queries the most informative and representative examples for labeling using the min-max margin-based approach.
4. **TED** [27]. Active learning via Transduction Experimental Design is an algorithm that selects a subset of informative samples from a candidate dataset. It formulates a regularized linear regression problem which minimizes reconstruction error.
5. **RRSS** [18]. Early active learning via Robust Representation and Structured Sparsity is a early active learning algorithm. It uses the $\ell_{2,1}$-norm to introduce structured sparsity for sample selection and robustness. However, RRSS does not consider the pairwise relations in re-id. We also introduce the kernelized RRSS denoted as **RRSS_K**.
6. **EALPC.** Our proposed early active learning with pairwise constraint algorithm is denoted as EALPC. We also use a kernelized version of our algorithm denoted as **EALPC_K**. For kernelization, we construct a Gaussian kernel for the candidate dataset, i.e. $\mathcal{K}(x_i, x_j) = \exp(-\alpha \|x_i - x_j\|^2)$.

To seek the optimal parameters (if any), we apply a grid search in a region of $\{10^{-4}, 10^{-3}, \cdots, 1, \cdots, 10^3, 10^4\}$ with a five-fold cross validation strategy to determine the best parameters.

Re-identification Algorithms. Five state-of-the-art supervised re-id algorithms are chosen for the performance analysis of the proposed early active learning algorithms on person re-id.

1. **NFST** [28]. Null Foley-Sammon Transform space learning is a re-id algorithm for learning a discriminative subspace where the training data points of each of the classes are collapsed to a single point.
2. **KCCA** [14]. Kernel Canonical Correlation Analysis algorithm seeks a common subspace between the proposed images extracted from disjoint cameras and projects them into a new space.
3. **XQDA** [12]. Cross-view Quadratic Discriminant Analysis learns a discriminant low dimensional subspace by cross-view quadratic discriminant analysis for metric learning.
4. **kLFDA** [24]. Kernelized Local Fisher Discriminant Classifier is a closed form method that uses a kernelized method to handle large dimensional feature vectors while maximizing a Fischer optimization criterion.
5. **MFA** [25]. Marginal Fisher Analysis method is introduced for dimensionality reduction by designing two graphs that characterize the intra-class compactness and interclass separability.

Settings. We report the average performance of 10 independent trials. In each trial, we divide each dataset into two equal-sized subsets as training and test sets, with no overlapping of person identities. Following the setting in [20], we divide the probe and gallery sets for re-id as follows: for datasets recording two camera views, e.g. VIPeR and PRID, images of one view are randomly selected for the probe sets, and images from the other view are chosen for the gallery sets. For a multi-view dataset, e.g. i-LID, images of one view are randomly selected as gallery sets and others are chosen as probe images. For the training set, we apply active learning methods to select a subset of training samples and query human labeling. The supervised re-id algorithms are then trained with the labeled samples. For evaluation measurement, we evaluate the performance of re-id by Cumulative Matching Characteristic (CMC) curve, which is the most commonly used performance measure for person re-id algorithms [7,12,13]. CMC calculates the probability that there exists a candidate image in the rank k gallery set that appears to match the prob image. In the experimental study, we also report the Rank One Matching Accuracy from CMC for simplicity.

5.2 Experimental Result Analysis

Performance of Re-id. We illustrate the performance of the active learning algorithms for re-id application in Table 1. In Table 1, each row corresponds to an active learning algorithm, and each column corresponds to a supervised re-id method. On each benchmark dataset, we select 20% of training samples via active learning algorithms and query labeling. The labeled subsets of samples are then adopted by supervised re-id algorithms for training models. We report the rank one matching accuracy in Table 1.

As shown in Table 1, we observe that: (1) All active learning algorithms perform better than Random selection. This indicates that active learning algorithms can select useful samples to improve the performance of re-id. (2) Our algorithms consistently outperform the other active learning algorithms. The table also confirms that our algorithms are better than the RRSS and TED method by around 5% on rank one matching accuracy. RRSS and TED have a similar optimization target to our algorithm but without pairwise constraint. This implies that our method is much suitable for re-id applications as a result of introducing the pairwise constraint. (3) The performance of the kernelized methods is better than the performance of the linear methods with our algorithm. This is consistent with the mathematical analysis in [18] that kernelization produces more discriminative representation by mapping data into high-dimensional feature space. (4) The active learning algorithms with XQDA method for report better rank one matching accuracy than those with LOMO features.

In Fig. 2, we illustrate the performance via CMC curves of active learning methods with XQDA as the re-id algorithm. The percentage of the labeled training sample is set to only 10% to present a more challenging task. We choose XQDA as it returned the best re-id results in the previous experiments. As shown in Fig. 2, we can observe that: (1) Our algorithms outperforms other algorithms consistently on all four benchmark datasets. (2) Compared to the results

Table 1. Rank one matching accuracy (%) on four benchmarks. Percentage of selected instances for labeling is 20% of all samples. Each column is an active learning algorithm and each row is a re-id algorithm. The best result of each re-id algorithm is marked in bold numbers. The best result of the algorithms overall is marked with an asterisk (∗).

Dataset	CAVIAR					VIPeR				
Algorithm	NFST	KCCA	XQDA	kLFDA	MFA	NFST	KCCA	XQDA	kLFDA	MFA
Random	23.65	23.47	21.38	27.55	25.87	26.65	23.01	27.23	22.78	23.64
K-means	26.90	25.99	22.05	27.74	27.40	27.59	26.16	27.59	23.15	24.39
TED	29.78	28.70	29,42	27.94	28.08	27.45	28.53	28.43	25.75	26.09
QUIRE	30.66	30.87	31.56	28.18	26.16	28.39	27.43	28.54	26.25	25.13
RRSS	31.87	30.69	33.57	30.95	29.01	31.56	28.54	30.71	27.34	28.04
RRSS_K	31.69	33.03	35.56	31.41	31.13	31.61	28.73	31.46	28.51	29.40
EALPC	34.12	33.57	37.45	33.09	31.16	32.61	29.45	31.82	28.54	29.56
EALPC_K	**35.00**	**35.20**	**38.75**∗	**33.29**	**31.91**	**33.66**	**30.44**	**34.29**∗	**29.18**	**30.03**
Dataset	PRID					iLIDS				
Algorithm	NFST	KCCA	XQDA	kLFDA	MFA	NFST	KCCA	XQDA	kLFDA	MFA
Random	24.49	25.47	24.00	23.50	20.00	25.96	23.40	25.00	23.35	25.00
K-means	26.16	27.54	27.01	24.70	21.20	27.02	23.94	27.00	25.57	25.20
TED	27.72	27.71	29.32	24.33	22.11	29.15	25.33	28.13	27.33	29.20
QUIRE	27.24	26.90	29.33	24.40	22.50	28.72	25.74	28.03	29.48	30.20
RRSS	29.21	28.44	30.00	25.09	23.97	28.11	27.66	30.82	30.08	30.55
RRSS_K	30.33	29.03	31.05	25.30	24.10	29.17	27.37	32.00	30.30	31.10
EALPC	32.22	30.63	31.03	25.90	25.60	29.26	27.66	32.34	30.43	31.60
EALPC_K	**32.70**	**31.50**	**33.40**∗	**26.06**	**25.70**	**31.19**	**28.72**	**34.00**∗	**31.60**	**32.47**

(a) CAVIAR   (b) VIPeR   (c) PRID   (d) i-LID

Fig. 2. CMC performance comparison of active learning algorithms. XQDA is chosen as the re-id algorithm. The percentage of selected samples is set to 10% of all samples.

in Table 1, all algorithms suffer a decrease in the rank one matching accuracy when the percentage of labeled samples is halved from 20% to 10%. However, our algorithm only decreases around by 5% on rank one matching accuracy whereas the accuracy of others, e.g. Random and K-means, reduces approximately 10%. This indicates that our algorithm is more robust. (3) The matching accuracy of our algorithm is the only one to reach 90% with rank 15 on CAVIAR and

VIPeR, and the only one to reach 90% on rank 20 on PRID and i-LID. This implies that our algorithm is more effective on re-id.

Effects on the Number of Selected Instances. Figure 3 illustrates the performance of re-id when the number of instances that selected by active learning methods varies. As displayed in Fig. 3, we observe that: (1) Generally, rank one matching accuracy of all re-id algorithms increases gradually when the number of selected instances increases. (2) All active learning methods report better performances than Random selection. This indicates that active learning algorithms can improve the performance of re-id applications. (3) Our algorithm consistently performs better than the other active learning algorithms when the number of selected instance increases. More specifically, for our algorithm, kernelized methods is better than the linear methods.

Convergence. In Fig. 4, we draw the objective value of the first 50 iterations of our algorithm on benchmark datasets. In the experiments, we fix the leverage

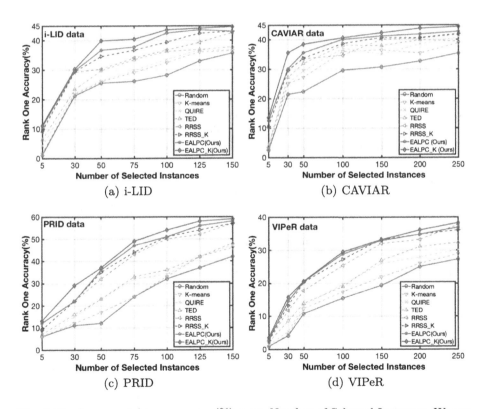

(a) i-LID

(b) CAVIAR

(c) PRID

(d) VIPeR

Fig. 3. Rank one matching accuracy (%) w.r.t. Number of Selected Instances. We use XQDA as the re-id algorithm and train it with varying numbers of samples selected by the active learning methods.

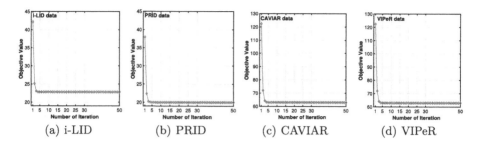

(a) i-LID   (b) PRID   (c) CAVIAR   (d) VIPeR

Fig. 4. Convergence analysis of EALPC on benchmark datasets. The parameters are set as $\alpha = 0.1$ and $\beta = 1$. The percentage of selected samples is 20%.

parameters as $\alpha = 0.1$ and $\beta = 1$ and set the percentage of selected samples to 20%. As shown in Fig. 4, the object values of our algorithm decrease dramatically and barely change after the first five iterations on all the benchmark datasets. This indicates that our algorithm converges very rapidly on all the datasets, which is consistent with our theoretical analysis of convergence.

6 Conclusion

In this work, we have proposed a novel early active learning algorithm with a pairwise constraint for person re-identification. The proposed method is designed for the early stage of supervised re-id experiments when there are limited labor resources for labeling data. Our algorithm introduces a pairwise constant for analyzing graph structures specifically for re-identification. A closed form solution is provided to efficiently weight and select the candidate samples. Extensive experimental studies on four benchmark datasets validate the effectiveness of the proposed algorithm. The experimental results demonstrate that our methods achieve encouraging performance against the state-of-the art algorithms in the filed of early active learning for person re-identification. In future work, our algorithm can be applied to other applications that consider the pairwise relatedness, such as in social network analysis, etc.

Acknowledgements. This work was partially supported by the Data to Decisions Cooperative Research Centre www.d2dcrc.com.au and partially supported by the National Science Foundation under Grant No. IIS-1638429.

References

1. Balcan, M.-F., Broder, A., Zhang, T.: Margin based active learning. In: Bshouty, N.H., Gentile, C. (eds.) COLT 2007. LNCS (LNAI), vol. 4539, pp. 35–50. Springer, Heidelberg (2007). https://doi.org/10.1007/978-3-540-72927-3_5
2. Cheng, D.S., Cristani, M., Stoppa, M., Bazzani, L., Murino, V.: Custom pictorial structures for re-identification. In: Proceedings of the British Machine Vision Conference (BMVC) (2011)

3. Freund, Y., Seung, H.S., Shamir, E., Tishby, N.: Selective sampling using the query by committee algorithm. Mach. Learn. **28**(2), 133–168 (1997)
4. Gray, D., Brennan, S., Tao, H.: Evaluating appearance models for recognition, reacquisition, and tracking. In: Proceedings of IEEE International Workshop on Performance Evaluation for Tracking and Surveillance (PETS), vol. 3 (2007)
5. Hirzer, M., Beleznai, C., Roth, P.M., Bischof, H.: Person re-identification by descriptive and discriminative classification. In: Heyden, A., Kahl, F. (eds.) SCIA 2011. LNCS, vol. 6688, pp. 91–102. Springer, Heidelberg (2011). https://doi.org/10.1007/978-3-642-21227-7_9
6. Huang, S.J., Jin, R., Zhou, Z.H.: Active learning by querying informative and representative examples. In: Advances in Neural Information Processing Systems (NIPS), pp. 892–900 (2010)
7. Karanam, S., Li, Y., Radke, R.J.: Person re-identification with discriminatively trained viewpoint invariant dictionaries. In: Proceedings of the IEEE International Conference on Computer Vision, pp. 4516–4524 (2015)
8. Karanam, S., Li, Y., Radke, R.J.: Sparse re-id: block sparsity for person re-identification. In: Proceedings of the IEEE Conference on Computer Vision and Pattern Recognition Workshops, pp. 33–40 (2015)
9. Kodirov, E., Xiang, T., Fu, Z., Gong, S.: Person re-identification by unsupervised ℓ_1 graph learning. In: Leibe, B., Matas, J., Sebe, N., Welling, M. (eds.) ECCV 2016. LNCS, vol. 9905, pp. 178–195. Springer, Cham (2016). https://doi.org/10.1007/978-3-319-46448-0_11
10. Kodirov, E., Xiang, T., Gong, S.: Dictionary learning with iterative Laplacian regularisation for unsupervised person re-identification. In: Proceedings of the British Machine Vision Conference (BMVC), vol. 3, p. 8 (2015)
11. Lewis, D.D., Catlett, J.: Heterogeneous uncertainty sampling for supervised learning. In: Proceedings of the Eleventh International Conference on Machine Learning, pp. 148–156 (1994)
12. Liao, S., Hu, Y., Zhu, X., Li, S.Z.: Person re-identification by local maximal occurrence representation and metric learning. In: Proceedings of the IEEE Conference on Computer Vision and Pattern Recognition, pp. 2197–2206 (2015)
13. Lisanti, G., Masi, I., Bagdanov, A.D., Del Bimbo, A.: Person re-identification by iterative re-weighted sparse ranking. IEEE Trans. Pattern Anal. Mach. Intell. **37**(8), 1629–1642 (2015)
14. Lisanti, G., Masi, I., Del Bimbo, A.: Matching people across camera views using kernel canonical correlation analysis. In: Proceedings of the International Conference on Distributed Smart Cameras, p. 10. ACM (2014)
15. Ma, A.J., Li, P.: Semi-supervised ranking for re-identification with few labeled image pairs. In: Cremers, D., Reid, I., Saito, H., Yang, M.-H. (eds.) ACCV 2014. LNCS, vol. 9006, pp. 598–613. Springer, Cham (2015). https://doi.org/10.1007/978-3-319-16817-3_39
16. Nguyen, H.T., Smeulders, A.: Active learning using pre-clustering. In: Proceedings of the Twenty-First International Conference on Machine Learning. ACM (2004)
17. Nie, F., Huang, H., Cai, X., Ding, C.H.: Efficient and robust feature selection via joint $\ell_{2,1}$-norms minimization. In: Advances in Neural Information Processing Systems (NIPS), pp. 1813–1821 (2010)
18. Nie, F., Wang, H., Huang, H., Ding, C.H.: Early active learning via robust representation and structured sparsity. In: International Joint Conference on Artificial Intelligence (IJCAI) (2013)

19. Nie, F., Xu, D., Li, X.: Initialization independent clustering with actively self-training method. IEEE Trans. Syst. Man Cybern. Part B (Cybern.) **42**(1), 17–27 (2012)

20. Peng, P., Xiang, T., Wang, Y., Pontil, M., Gong, S., Huang, T., Tian, Y.: Unsupervised cross-dataset transfer learning for person re-identification. In: Proceedings of the IEEE Conference on Computer Vision and Pattern Recognition, pp. 1306–1315 (2016)

21. Seung, H.S., Opper, M., Sompolinsky, H.: Query by committee. In: Proceedings of the Fifth Annual Workshop on Computational Learning Theory, pp. 287–294. ACM (1992)

22. Twomey, N., Diethe, T., Flach, P.: Bayesian active learning with evidence-based instance selection. In: Workshop on Learning over Multiple Contexts, European Conference on Machine Learning (ECML 2015) (2015)

23. Xiao, T., Li, H., Ouyang, W., Wang, X.: Learning deep feature representations with domain guided dropout for person re-identification. In: Proceedings of the IEEE Conference on Computer Vision and Pattern Recognition, pp. 1249–1258 (2016)

24. Xiong, F., Gou, M., Camps, O., Sznaier, M.: Person re-identification using kernel-based metric learning methods. In: Fleet, D., Pajdla, T., Schiele, B., Tuytelaars, T. (eds.) ECCV 2014. LNCS, vol. 8695, pp. 1–16. Springer, Cham (2014). https://doi.org/10.1007/978-3-319-10584-0_1

25. Yan, S., Xu, D., Zhang, B., Zhang, H.J., Yang, Q., Lin, S.: Graph embedding and extensions: a general framework for dimensionality reduction. IEEE Trans. Pattern Anal. Mach. Intell. 29(1), 40-51 (2007)

26. Yang, Y., Shen, H.T., Ma, Z., Huang, Z., Zhou, X.: $\ell_{2,1}$-norm regularized discriminative feature selection for unsupervised learning. In: International Joint Conference on Artificial Intelligence (IJCAI) (2011)

27. Yu, K., Bi, J., Tresp, V.: Active learning via transductive experimental design. In: Proceedings of the 23rd International Conference on Machine Learning, pp. 1081–1088. ACM (2006)

28. Zhang, L., Xiang, T., Gong, S.: Learning a discriminative null space for person re-identification. In: Proceedings of the IEEE Conference on Computer Vision and Pattern Recognition, pp. 1239–1248 (2016)

29. Zheng, L., Yang, Y., Hauptmann, A.G.: Person re-identification: past, present and future. arXiv preprint arXiv:1610.02984 (2016)

30. Zheng, M., Bu, J., Chen, C., Wang, C., Zhang, L., Qiu, G., Cai, D.: Graph regularized sparse coding for image representation. IEEE Trans. Image Process. **20**(5), 1327–1336 (2011)

Guiding InfoGAN with Semi-supervision

Adrian Spurr$^{(\boxtimes)}$, Emre Aksan, and Otmar Hilliges

Advanced Interactive Technologies, ETH Zurich, Zurich, Switzerland
{adrian.spurr,emre.aksan,otmar.hilliges}@inf.ethz.ch

Abstract. In this paper we propose a new semi-supervised GAN architecture (ss-InfoGAN) for image synthesis that leverages information from *few* labels (as little as 0.22%, max. 10% of the dataset) to learn semantically meaningful and controllable data representations where latent variables correspond to label categories. The architecture builds on Information Maximizing Generative Adversarial Networks (InfoGAN) and is shown to learn both continuous and categorical codes and achieves higher quality of synthetic samples compared to fully unsupervised settings. Furthermore, we show that using *small* amounts of labeled data speeds-up training convergence. The architecture maintains the ability to disentangle latent variables for which no labels are available. Finally, we contribute an information-theoretic reasoning on how introducing semi-supervision increases mutual information between synthetic and real data. Code related to this chapter is available at: https://github.com/spurra/ss-infogan.

1 Introduction

In many machine learning tasks it is assumed that the data originates from a generative process involving complex interaction of multiple independent factors, each accounting for a source of variability in the data. Generative models are then motivated by the intuition that in order to create realistic data a model must have "understood" these underlying factors. For example, images of handwritten characters are defined by many properties such as character type, orientation, width, curvature and so forth.

Recent models that attempt to extract these factors are either completely supervised [18,20,23] or entirely unsupervised [3,5]. Supervised approaches allow for extraction of the desired parameters but require fully labeled datasets and a priori knowledge about which factors underlie the data. However, factors not corresponding to labels will not be discovered. In contrast, unsupervised approaches require neither labels nor a priori knowledge about the underlying factors but this flexibility comes at a cost: such models provide no means of exerting control on what kind of features are found. For example, Information Maximizing Generative Adversarial Networks (InfoGAN) have recently been shown to

---

Electronic supplementary material The online version of this chapter (https://doi.org/10.1007/978-3-319-71249-9_8) contains supplementary material, which is available to authorized users.

© Springer International Publishing AG 2017
M. Ceci et al. (Eds.): ECML PKDD 2017, Part I, LNAI 10534, pp. 119–134, 2017.
https://doi.org/10.1007/978-3-319-71249-9_8

learn disentangled data representations. Yet the extracted representations are not always directly interpretable by humans and lack direct measures of control due to the unsupervised training scheme. Many application scenarios however require control over *specific* features.

Embracing this challenge, we present a new semi-supervised generative architecture that requires only few labels to provide control over which factors are identified. Our approach can exploit already existing labels or use datasets that are augmented with easily collectible labels (but are not fully labeled). The model, based on the related InfoGAN [3] is dubbed semi-supervised InfoGAN (ss-InfoGAN). In our approach we maximize two mutual information terms: (i) The mutual information between a code vector and real labeled samples, guiding the corresponding codes to represent the information contained in the labeling, (ii) and the mutual information between the code vector and the synthetic samples. By doing so ss-InfoGAN can find representations that unsupervised methods such as InfoGAN fail to find, for example the category of digits of the SVHN dataset. Notably our approach requires only 10% of labeled data for the hardest dataset we tested and for simpler datasets only 132 labeled samples (0.22%) were necessary.

We discuss our method in full, provide an information theoretical rationale for the chosen architecture and demonstrate its utility in a number of experiments on the MNIST [13], SVHN [19], CelebA [14] and CIFAR-10 [10] datasets. We show that our method improves results over the state-of-the-art, combining advantages of supervised and unsupervised approaches.

2   Related Work

Many approaches to modeling the data generating process and identifying the underlying factors by learning to synthesize samples from disentangled representations exist. An example of an early approach is supervised bi-linear models [27], separating style from the content. Zhu et al. [29] use a multi-view deep perceptron model to untangle the identity and viewpoint of face images. Weakly supervised methods based on supervised clustering, have been proposed such as high-order Boltzman machines [22] applied on face images.

Variational Autoencoders (VAEs) [9] and Generative Adversarial Networks (GANs) [6] have recently seen a lot of interest in generative modeling problems. In both approaches a deep neural network is trained as a generative model by using standard backpropagation, enabling synthesis of novel samples without explicitly learning the underlying data distribution. VAEs maximize a lower bound on the marginal likelihood which is expected to be tight for accurate modeling [2,8,24,26]. In contrast, GANs optimize a minimax game objective via a discriminative adversary. However, they have been shown to be unstable and fragile [17,23].

Employing semi-supervised learning, Kingma et al. [7] use VAEs to isolate content from other variations, and achieve competitive recognition performance in addition to high-quality synthetic samples. Deep Convolutional Inverse Graphics Network (DC-IGN) [11], which uses a VAE architecture and a specially tailored training scheme is capable of learning a disentangled latent space in fully supervised manner. Since the model is evaluated by using images of 3D models, labels for the underlying factors are cheap to attain. However, this type of dense supervision is unfeasible for most non-synthetic datasets.

Adversarial Autoencoders [15] combine the VAE and GAN frameworks in using an adversarial loss on the latent space. Similarly, Mathieu et al. [16] introduces an adversarial loss on the reconstructions of VAE, that is, on the pixel space. Both models are shown to learn both discrete and continuous latent representations and to disentangle style and content in images. However, these hybrid architectures have conceptually different designs as opposed to GANs. While the former learns the data distribution via Autoencoder training and employ the adversarial loss as a regularizer, the latter directly relies on an adversarial objective. Despite the robust and stable training, VAEs have tendency to generate blurry images [12].

Conditional GANs [18,20,23] augment the GAN framework by using class labels. Mirza and Osindero [18] train a class-conditional discriminator while [20,23] use auxiliary loss terms for the labels. Salimans et al. [23] use conditional GANs for pre-training, aiming to improve semi-supervised classification accuracy of the discriminator. Similarly, the AC-GAN model [20] introduces an additional classification task in the discriminator to provide class-conditional training and inference of the generator in order to be able to synthesize higher resolution images than previous architectures. Our work is similar to the above in that it provides class-conditional generation of images. However, due to MI loss terms our architecture can (i) be employed in both supervised *and* semi-supervised settings, (ii) can learn interpretable representations in addition to smooth manifolds and (iii) can exploit continuous supervision signals if such labels are available.

Comparatively fewer works treat the subject of fully unsupervised generative models to retrieve interpretable latent representations. Desjardins et al. [5] introduced a higher-order RBM for recognition of facial expressions. However, it can only disentangle discrete latent factors and the computational complexity rises exponentially in the number of features. More recently, Chen et al. [3] developed an extension to GANs, called Information Maximizing Generative Adversarial Networks (InfoGAN). It enforces the generator to learn disentangled representations through increasing the mutual information between the synthetic samples and a newly introduced latent code. Our work extends InfoGAN such that additional information can be used. Supervision can be a necessity if the model struggles in learning desirable representations or if *specific* features need to be controlled by the user. Our model provides a framework for semi-supervision in InfoGANs. We find that leveraging few labeled samples brings improvements on the convergence rate, quality of representations and synthetic samples. Moreover, semi-supervision helps the model in capturing otherwise difficult to capture representations.

3   Method

3.1   Preliminaries: GAN and InfoGAN

In the GAN framework, a generator G producing synthetic samples is pitted against a discriminator D that attempts to discriminate between real data and samples created by G. The goal of the generator is to match the distribution of generated samples P_G with the real distribution P_{data}. Instead of explicitly estimating $P_G(x)$, G learns to transform noise variables $z \sim P_{noise}$ into synthetic samples $\tilde{x} \sim P_G$. The discriminator D outputs a single scalar $D(x)$ representing the probability of a sample x coming from the true data distribution. Both $G(z; \theta_g)$ and $D(x; \theta_d)$ are differentiable functions parametrized by neural networks. We typically omit the parameters θ_g and θ_d for brevity. G and D are simultaneously trained by using the minimax game objective $V_{GAN}(D, G)$:

$$\min_G \max_D V_{GAN}(D,G) = \mathbb{E}_{x \sim P_{data}}[\log D(x)] + \mathbb{E}_{z \sim P_{noise}}[\log(1 - D(G(z)))] \quad (1)$$

GANs map from the noise space to data space without imposing any restrictions. This allows G to produce arbitrary mappings and to learn highly dependent factors that are hard to interpret. Therefore, variations of z in any dimension often yields entangled effects on the synthetic samples $\tilde{x}$. InfoGANs [3] are capable of learning disentangled representations. InfoGAN extends the unstructured noise z by introducing a latent code c. While z represents the incompressible noise, c describes semantic features of the data. In order to prevent G from ignoring the latent codes c, InfoGAN regularizes learning via an additional cost term penalizing low mutual information between c and $\tilde{x} = G(z, c)$:

$$\min_{G,Q} \max_D V_{InfoGAN}(D, G, Q, \lambda_1) = V_{GAN}(D, G) - \lambda_1 L_I(G, Q),$$
$$I(C; \tilde{X}) \geq L_I(G, Q) = \mathbb{E}_{c \sim P_c, \tilde{x} \sim P_G}[\log Q(c|\tilde{x})] + H(c), \quad (2)$$

where Q is an auxiliary parametric distribution approximating the posterior $P(c|x)$, L_I corresponds to the lower bound of the mutual information $I(C; \tilde{X})$ and λ_1 is the weighting coefficient.

3.2   Semi-supervised InfoGAN

Although InfoGAN can learn to disentangle representations in an unsupervised manner for simple datasets, it struggles to do so on more complicated datasets such as CelebA or CIFAR-10. In particular, capturing categorical codes is challenging and hence InfoGAN yields poorer performance in class-conditional generation task than competing methods. Moreover, depending on the initialization, the learned latent codes may differ between training sessions further reducing interpretability.

In Semi-Supervised InfoGAN (ss-InfoGAN) we introduce available or easily acquired labels to address these issues. Figure 1 schematically illustrates our architecture. To make use of label information we decompose the latent code

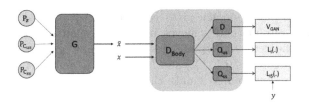

Fig. 1. Schematic overview of the ss-InfoGAN network architecture. P_z, $P_{C_{us}}$ and $P_{C_{ss}}$ are the distributions of the noise and latent variables z, c_{us} and c_{ss}, respectively.

c into a set of semi-supervised codes, c_{ss}, and unsupervised codes, c_{us}, where $c_{ss} \cup c_{us} = c$. The semi-supervised codes encode the same information as the labels y, whereas c_{us} are free to encode potential remaining semantic factors.

We seek to increase the mutual information $I(C_{ss}; X)$ between the latent codes c_{ss} and the labeled real samples x, by interpreting labels y as the latent codes c_{ss}, (i.e. $y = c_{ss}$). Note that not all samples need to be labeled for the generator to learn the inherent semantic meaning of y. We additionally want to increase the mutual information $I(C_{ss}; \tilde{X})$ between the semi-supervised latent codes and the synthetic samples $\tilde{x}$ so that information can flow back to the generator. This is accomplished via Variational Information Maximization [1] in deriving lower bounds for both MI terms. For the lower bounds of $I(C_{ss}; \cdot)$ we utilize the same derivation as InfoGAN:

$$I(C_{ss}; X) \geq \mathbb{E}_{c \sim P_{C_{ss}}, x \sim P_X}[\log Q_1(c_{ss}|x)] + H(C_{ss}) = L_{IS}^1(Q_1), \qquad (3)$$

$$I(C_{ss}; \tilde{X}) \geq \mathbb{E}_{c \sim P_{C_{ss}}, \tilde{x} \sim P_G}[\log Q_2(c_{ss}|\tilde{x})] + H(C_{ss}) = L_{IS}^2(Q_2, G), \qquad (4)$$

where Q_1 and Q_2 are again auxiliary distributions to approximate posteriors and are parametrized by neural networks. With $Q_1 = Q_2 = Q_{ss}$ we attain the MI cost term:

$$L_{IS}(Q_{ss}, G) = L_{IS}^1(Q_{ss}) + L_{IS}^2(Q_{ss}, G) \qquad (5)$$

Since we would like to encode the labels y via latent codes c_{ss}, we optimize $L_{IS}^1(Q_{ss})$ with respect to Q_{ss} and $L_{IS}^2(Q_{ss}, G)$ only with respect to G. The final objective function is then:

$$\min_{G, Q_{us}, Q_{ss}} \max_D \quad V_{ss\text{-}InfoGAN}(D, G, Q_{us}, Q_{ss}, \lambda_1, \lambda_2) \qquad (6)$$

$$= V_{InfoGAN}(D, G, Q_{us}, \lambda_1) - \lambda_2 L_{IS}(G, Q_{ss}) \qquad (7)$$

Training Q_{ss} on labeled real data (x, y) enables Q_{ss} to encode the semantic meaning of y via c_{ss} by means of increasing the mutual information $I(C_{ss}; X)$. Simultaneously, the generator G acquires the information of y indirectly by increasing $I(C_{ss}; \tilde{X})$ and learns to utilize the semi-supervised representations in synthetic samples. In our experiments we find that a small subset of labeled samples is enough to observe significant effects.

We show that our approach gives control over discovered properties and factors and that our method achieves better image quality. Here we provide an information theoretic underpinning shedding light on the reason for these gains. By increasing both $I(C_{ss}; X)$ and $I(C_{ss}; \tilde{X})$, the mutual information term $I(X; \tilde{X})$ is increased as well. We make the following assumptions:

$$X \leftarrow C_{ss} \rightarrow \tilde{X}, \tag{8}$$

$$I(X; \tilde{X}) = 0 \text{ initially}, \tag{9}$$

$$H(C_{ss}) = \mathsf{C}, \tag{10}$$

where C is a constant and $\rightarrow$ are dependency relations. Assumption (8) follows the intuition that the data is hypothesized to arise from the interaction of independent factors. While latent factors consist of z, C_{us} and C_{ss}, we abstract for the sake of simplicity. Assumption (9) formulates the initial state of our model where the synthetic data distribution P_G and the data distribution P_{data} are independent. Finally we can assume that labels follow a fixed distribution and hence have a fixed entropy $H(C_{ss})$, giving rise to (10).

We decompose $H(C_{ss})$ and reformulate $I(X; \tilde{X})$ in the following way:

$$H(C_{ss}) = I(C_{ss}; X) + I(C_{ss}; \tilde{X}) + H(C_{ss}|X, \tilde{X}) - I(C_{ss}; X; \tilde{X}), \tag{11}$$

$$I(C_{ss}; X; \tilde{X}) = I(X; \tilde{X}) - I(X; \tilde{X}|C_{ss}) \tag{12}$$

where $I(C_{ss}; X; \tilde{X})$ is the multivariate mutual information term. While pointwise MI is per definition non-negative, in the multivariate case negative values are possible if two variables are coupled via the third. By using the conditional independence assumption (8), we have

$$I(C_{ss}; X; \tilde{X}) = I(X; \tilde{X}) - I(X; \tilde{X}|C_{ss}) = I(X; \tilde{X}) \geq 0. \tag{13}$$

Thus the entropy term $H(C_{ss})$ in Eq. (11) takes the form

$$H(C_{ss}) = I(C_{ss}; X) + I(C_{ss}; \tilde{X}) + H(C_{ss}|X, \tilde{X}) - I(X; \tilde{X}) \tag{14}$$

Let Δ symbolize the change in value of a term. According to assumption (10), the following must hold:

$$\Delta_{I(C_{ss};X)} + \Delta_{I(C_{ss};\tilde{X})} + \Delta_{H(C_{ss}|X,\tilde{X})} - \Delta_{I(X;\tilde{X})} = 0 \tag{15}$$

Note that $\Delta_{I(C_{ss};X)}$ and $\Delta_{I(C_{ss};\tilde{X})}$ increase during training since we directly optimize these terms, leading to the following cases:

$$\begin{aligned}
\Delta_{I(C_{ss};X)} + \Delta_{I(C_{ss};\tilde{X})} &\geq -\Delta_{H(C_{ss}|X,\tilde{X})} \implies \Delta_{I(X;\tilde{X})} \geq 0 \\
\Delta_{I(C_{ss};X)} + \Delta_{I(C_{ss};\tilde{X})} &< -\Delta_{H(C_{ss}|X,\tilde{X})} \implies \Delta_{I(X;\tilde{X})} < 0
\end{aligned} \tag{16}$$

The first case results in the desired behavior. However the latter case cannot occur, as it would result in negative mutual information $I(X; \tilde{X})$. Hence, based on our assumptions, increasing both $I(C_{ss}; X)$ and $I(C_{ss}; \tilde{X})$ leads to an increase in $I(X; \tilde{X})$.

4   Implementation

For both D and G of ss-InfoGAN we use a similar architecture with DCGAN [21], which is reported to stabilize training. The networks for the parametric distributions Q_{us} and Q_{ss} share all the layers with D except the last layers. This is similar to [3], which models Q as an extension to D. This approach has the disadvantage of negligibly higher computational cost for Q_{ss} in comparison to InfoGAN. However, this is offset by a faster convergence rate in return.

In our experiments with low amount of labeled data, we initially favor drawing labeled samples, which improves convergence rate of the supervised latent codes significantly. During training the probability of drawing a labeled sample is annealed until the actual labeled sample ratio in the data is reached. The loss function used to calculate L_I and L_{IS} is the cross-entropy for categorical latent codes and the mean squared error for continuous latent codes. The unsupervised categorical codes are sampled from a uniform categorical distribution whereas the continuous codes are sampled from a uniform distribution. All the experimental details are listed in the supplementary document. In the interest of reproducible research, we provide the source code on GitHub.[1]

For comparison we re-implement the original InfoGAN architecture in the Torch framework [4] with minor modifications. Note that there may be differences in results due to the unstable nature of GANs, possibly amplified by using a different framework and different initial conditions. In our implementation the loss function for continuous latent codes are not treated as a factored Gaussian, but approximated with the mean squared error, which leads to a slight adjustment in the architecture of Q.

5   Experiments

In our study we focus on interpretability of the representations and quality of synthetic images under different amount of labeling. The closest related work to that of ours is InfoGAN, and the aim was to directly improve upon that architecture. The existing semi-supervised generative modeling studies on the other hand, aim to learn discriminative representations for classification. Therefore we make a direct comparison with InfoGAN.

We evaluate our model on the MNIST [13], SVHN [19], CelebA [14] and CIFAR-10 [10] datasets. First, we inspect how well ss-InfoGAN learns the representations as defined by existing labels. Second, we qualitatively evaluate the representations learned by ss-InfoGAN. Finally, we analyze how much labeled data is required for the model to encode semantic meaning of the labels y via c_{ss}.

We hypothesize that the quality of the generator in class-conditional sample synthesis can be quantitatively assessed by a separate classifier trained to recognize class labels. The class labels of the synthetic samples (i.e. the class conditional inputs of the generator) are regarded as true targets and compared with the classifier's predictions. In order to prevent biased results due to the

---

[1] Implementation can be found at https://github.com/spurra/ss-infogan.

generator overfitting, we train the classifier C by using the *test set*, and validate on the *training set* for each dataset. Despite the *test set* consisting fewer samples, the classifier C generally performs well on the unseen *training set*. In our experiments, we use a standard CNN (architecture described in the supplementary file) for the MNIST, CelebA and SVHN datasets and Wide Residual Networks [28] for CIFAR-10 dataset.

In order to evaluate how well the model separates types of semantic variation, we generate synthetic images by varying only one latent factor by means of linear interpolation while keeping the remaining latent codes fixed.

To evaluate the necessary amount of supervision we perform quantitative analysis of the classifier accuracy and qualitative analysis by examining synthetic samples. To do so, we discard increasingly bigger sets of labels from the data. Note that Q_{ss} is trained only by using labeled samples and hence sees less data, whereas the rest of the architecture, namely the generator and the shared layers of the discriminator, uses the entire training samples in unsupervised manner. The minimum amount of labeled samples required to learn the representation of labels y varies depending on the dataset. However, for all our experiments it never exceeded 10%.

5.1   MNIST

MNIST is a standard dataset used to evaluate many generative models. It consists of handwritten digits, and is labeled with the digit category. Figure 2

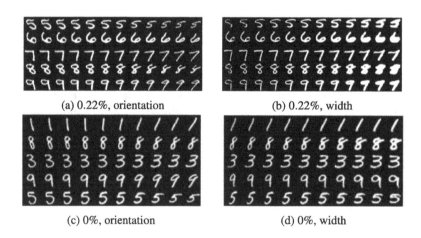

(a) 0.22%, orientation

(b) 0.22%, width

(c) 0% , orientation

(d) 0%, width

Fig. 2. Manipulating latent code on MNIST: in all figures of latent code manipulation we use the convention that a latent code varies from left to right (x-axis) while the remaining codes and the noise are kept fixed. Each row along the y-axis corresponds to a categorical latent code encoding a class label unless otherwise stated. The interpretation of the varying latent code is provided under the image. Synthetic images generated by interpolating the latent codes, encoding the digit "orientation" and "width", between -2 and 2. (a, b) ss-InfoGAN with 0.22% supervision. (c, d) InfoGAN (taken from [3]).

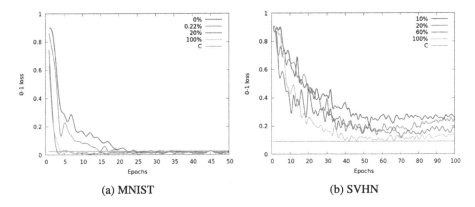

(a) MNIST   (b) SVHN

Fig. 3. 0–1 loss on synthetic samples: in all 0–1 loss figures we plot the classification accuracy on synthetic samples of the respective dataset. During training of ss-InfoGAN, a batch of synthetic samples are randomly generated, and evaluated by the independent classifier C. Colors represent the GAN models trained with different amount of supervision and classifier performance (C) on real validation samples. (Color figure online)

presents the synthetic samples generated with our model and InfoGAN by varying the latent code. Due to lower complexity of the dataset, InfoGAN is capable of learning the digit representation unsupervised. However, using just 0.22% of the available data has a two-fold benefit. First, semi-supervision provides additional fine-grained control (e.g., digits are already sorted in ascending manner in Fig. 2a, b). Second, we experimentally verified that the additional information increases convergence speed of the generator, illustrated in Fig. 3a. The 0–1 loss of the classifier C decreases faster as more labeled samples are introduced while the fully unsupervised setting (i.e. InfoGAN) is the slowest. The smallest amount of labeled samples for which the effect of supervision is observable is 0.22% of the dataset, which corresponds to 132 labeled samples out of 60'000.

5.2   SVHN

Next, we run ss-InfoGAN on the SVHN dataset which consists of color images, hence includes more noise and natural effects such as illumination. Similar to MNIST, this dataset is labeled with respect to the digit category. In Fig. 4, latent codes with various interpretation are presented. In this experiment different amount of supervision result in different unsupervised representations retrieved.

The SVHN dataset is perturbed by various noise factors such as blur and ambiguity in the digit categories. Figure 5 compares real samples with randomly generated synthetic samples by varying digit categories. The InfoGAN configuration (0% supervision) fails to encode a categorical latent code for the digit category. Leveraging some labeled information, our model becomes more robust to perturbations in the images. Through the introduction of labeled samples we are capable of exerting control over the latent space, encoding the digit labels in

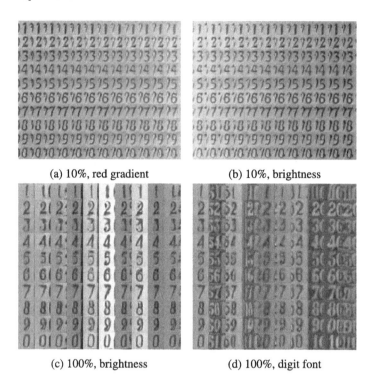

(a) 10%, red gradient

(b) 10%, brightness

(c) 100%, brightness

(d) 100%, digit font

Fig. 4. Manipulating latent code on SVHN: latent codes encoding the "brightness", "digit font" and "red gradient" are interpolated between -2 and 2 for each semi-supervised categorical code. (a, b) ss-InfoGAN with 10% supervision. (c, d) ss-InfoGAN with 100% supervision. (Color figure online)

the categorical latent code c_{ss}. The smallest fraction of labeled data needed to achieve a notable effect is 10% (i.e. 7'326 labels out of 73'257 samples).

In Fig. 3b we assess the performance of ss-InfoGAN with respect to C. The unsupervised configuration is left out since it is not able to control digit categories. As ss-InfoGAN exploits more label information, the generator converges faster and synthesizes more accurate images in terms of digit recognizability.

5.3   CelebA

The CelebA dataset contains a rich variety of binary labels. We pre-process the data by extracting the faces via a face detector and then resize the extracted faces to 32×32. From the set of binary labels provided in the data we select the following attributes: "presence of smile", "mouth open", "attractive" and "gender".

Figure 6 shows synthetic images generated with ss-InfoGAN by varying certain latent codes. Although we experiment by using various hyper-parameters, InfoGAN is not able to learn an equivalent representation to these attributes.

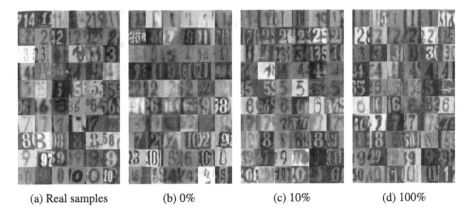

| (a) Real samples | (b) 0% | (c) 10% | (d) 100% |

Fig. 5. Random synthetic SVHN samples: in all figures of randomly synthesized images we present examples of real samples from the dataset and synthetic images. Models are trained with different amount of supervision which is noted under the images, where the 0% supervision corresponds to InfoGAN. In each row, the semi-supervised categorical code encoding the digits is kept fixed while rest of the input vector, (i.e. z, c_{us}) and the remaining codes in c_{ss}, is randomly drawn from the latent distribution. Although each row represents a digit the fully unsupervised model (b) (i.e. InfoGAN) lacks control on the digit category.

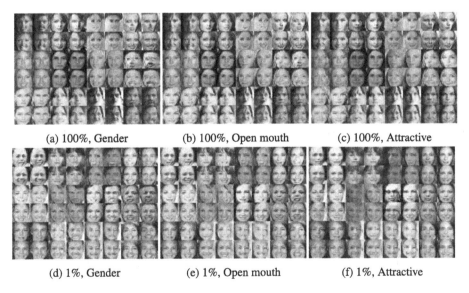

| (a) 100%, Gender | (b) 100%, Open mouth | (c) 100%, Attractive |
| (d) 1%, Gender | (e) 1%, Open mouth | (f) 1%, Attractive |

Fig. 6. Manipulating latent code on CelebA: synthetic samples generated by varying semi-supervised latent codes for the binary attributes "gender", "open mouth" and "attractive". Each 2×2 block corresponds to synthetic samples generated by keeping the input vector (c_{ss}, c_{us}, z) fixed except the first semi-supervised categorical code encoding "smile" attribute (varied across the y-axis) and the other semi-supervised categorical latent code, whose interpretation is given in the caption (varied across the x-axis).

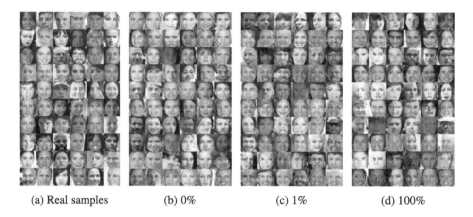

| (a) Real samples | (b) 0% | (c) 1% | (d) 100% |

Fig. 7. Random synthetic CelebA samples: for each synthesized image the latent and noise variables are randomly drawn from their respective distribution.

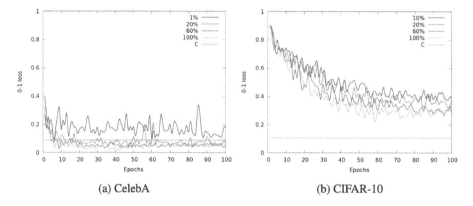

(a) CelebA   (b) CIFAR-10

Fig. 8. 0–1 loss on synthetic samples: (a) the model trained on CelebA dataset by leveraging the minimum amount of supervision that is sufficient to encode label (1%) information shows unstable behavior. (b) None of the models trained on CIFAR-10 dataset achieve the quality enough to reach real sample classification accuracy.

We see that for as low as 1%, c_{ss} acquires the semantic meaning of y. This corresponds to 1'511 labeled samples out of 151'162. Figure 7 presents a batch of real samples from the dataset alongside with randomly synthesized samples from generators trained on various labeled percentages, with 0% corresponding again to InfoGAN.

The performance of ss-InfoGAN on the independent classifier C is shown in Fig. 8a. For the lowest amount of labeling some instability can be observed. We believe this is due to the differences between the positives and the negatives of each binary label being more subtle than in other datasets. In addition, synthetic data generation exhibits certain variability which can obfuscate important parts of the image. However, using 20% of labeled samples ensures a stable training performance.

5.4   CIFAR-10

Finally we evaluate our model on CIFAR-10 dataset consisting of natural images. The data is labeled with the object category, which we use for the first semi-supervised categorical code. In order to stabilize training we apply instance noise [25].

On this dataset the unsupervised latent codes are not interpretable. An example is presented in Fig. 9 where the synthetic samples are generated by varying one of the unsupervised latent codes. Despite the fact that ss-InfoGAN model is trained by using *all* label information, the semantic meaning of this unsupervised representation is not clear. The randomness of the natural images prevent models from learning interpretable representations in the absence of guidance.

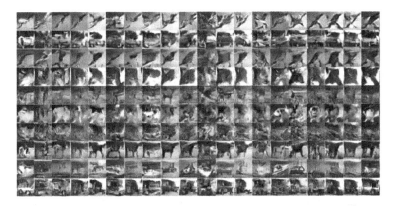

Fig. 9. Manipulating latent code on CIFAR: an example of varying an unsupervised latent code (x-axis) on CIFAR-10. Each row corresponds to a fixed code and represents class labels. The unsupervised latent code is not clearly interpretable.

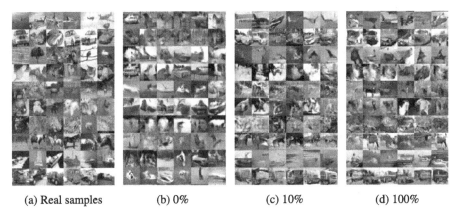

(a) Real samples   (b) 0%   (c) 10%   (d) 100%

Fig. 10. Random synthetic CIFAR-10 samples: real samples and synthetic samples generated by the models trained with different amount of supervision. In each row, the semi-supervised categorical code encoding the image categories is kept fixed while rest of the input vector (i.e. z, c_{us}) and the remaining codes in c_{ss}, is randomly drawn from the latent distribution.

Figure 10 shows synthetic samples generated by models with different supervision configurations. InfoGAN has difficulties in learning the object category (see Fig. 10b) and hence in generating class-conditional synthetic images. For this dataset we find that labeling 10% of the training data (corresponding to 5'000 images out of 50'000) is sufficient for ss-InfoGAN to encode class category (see Fig. 10c).

In Fig. 8b classification accuracy of C on the synthetic samples is plotted, again displaying the similar behavior of having better performance as more labels are available. It is evident that the additional information provided by the labels is fundamental to control *what* the image depicts. We argue that attaining such low amounts of labels is feasible even for large and complicated datasets.

5.5   Convergence Speed of Sample Quality

During the course of the experiments, it is observed that the convergence of synthetic sample quality is faster in comparison to InfoGAN. Figure 11 shows synthetic SVHN samples from a fully supervised ss-InfoGAN and InfoGAN at training epoch 26 and 47. The training epochs are chosen by inspection so that each model starts producing recognizable images. Therefore we can quantitatively say that ss-InfoGAN converges faster than InfoGAN.

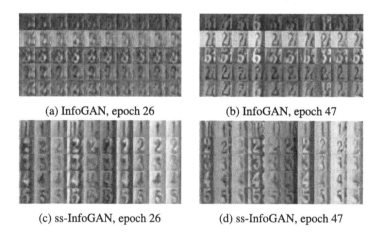

(a) InfoGAN, epoch 26   (b) InfoGAN, epoch 47

(c) ss-InfoGAN, epoch 26   (d) ss-InfoGAN, epoch 47

Fig. 11. Samples from InfoGAN and ss-InfoGAN trained on SVHN at two different epochs

6   Conclusion

We have introduced ss-InfoGAN a novel semi-supervised generative model. We have shown that including few labels increases the convergence speed of the latent codes c_{ss} and that these represent the same meaning as the labels y. This speed-up increases as more data samples are labeled. Although in theory this only improves convergence speed of c_{ss}, we have shown empirically that the sample quality convergence speed has improved as well.

In addition, it was shown that using labeling information is useful in cases where InfoGAN fails to find a *specific* representation, such as in the case of SVHN, CelebA and CIFAR-10. To successfully guide a latent code to the desired representation, it is sufficient that the dataset contains only a minimal subset of labeled data. The amount of required labels ranges from 0.22% for the simplest datasets (MNIST) to a maximum of 10% for the most complex datasets (CIFAR-10). We argue that acquiring such low percentages of labels is cost effective and makes the proposed architecture an attractive choice if control over *specific* latent codes is required and full supervision is not an option.

Acknowledgements. This work was supported in parts by the ERC grant OPTINT (StG-2016-717054).

References

1. Barber, D., Agakov, F.V.: The IM algorithm: a variational approach to information maximization. In: NIPS, pp. 201–208 (2003)
2. Burda, Y., Grosse, R., Salakhutdinov, R.: Importance weighted autoencoders. arXiv preprint arXiv:1509.00519 (2015)
3. Chen, X., Duan, Y., Houthooft, R., Schulman, J., Sutskever, I., Abbeel, P.: Info-GAN: interpretable representation learning by information maximizing generative adversarial nets. ArXiv e-prints, June 2016
4. Collobert, R., Kavukcuoglu, K., Farabet, C.: Torch7: a matlab-like environment for machine learning. In: BigLearn, NIPS Workshop (2011)
5. Desjardins, G., Courville, A., Bengio, Y.: Disentangling factors of variation via generative entangling. ArXiv e-prints, October 2012
6. Goodfellow, I.J., Pouget-Abadie, J., Mirza, M., Xu, B., Warde-Farley, D., Ozair, S., Courville, A., Bengio, Y.: Generative adversarial networks. ArXiv e-prints, June 2014
7. Kingma, D.P., Rezende, D.J., Mohamed, S., Welling, M.: Semi-supervised learning with deep generative models. ArXiv e-prints, June 2014
8. Kingma, D.P., Salimans, T., Welling, M.: Improving variational inference with inverse autoregressive flow. arXiv preprint arXiv:1606.04934 (2016)
9. Kingma, D.P., Welling, M.: Auto-encoding variational bayes. arXiv preprint arXiv:1312.6114 (2013)
10. Krizhevsky, A., Hinton, G.: Learning multiple layers of features from tiny images. Technical report, University of Toronto (2009)
11. Kulkarni, T.D., Whitney, W., Kohli, P., Tenenbaum, J.B.: Deep convolutional inverse graphics network. ArXiv e-prints, March 2015
12. Lamb, A., Dumoulin, V., Courville, A.: Discriminative regularization for generative models. arXiv preprint arXiv:1602.03220 (2016)
13. Lecun, Y., Bottou, L., Bengio, Y., Haffner, P.: Gradient-based learning applied to document recognition. Proc. IEEE **86**(11), 2278–2324 (1998)
14. Liu, Z., Luo, P., Wang, X., Tang, X.: Deep learning face attributes in the wild. In: Proceedings of International Conference on Computer Vision (ICCV), December 2015

15. Makhzani, A., Shlens, J., Jaitly, N., Goodfellow, I., Frey, B.: Adversarial autoencoders. ArXiv e-prints, November 2015
16. Mathieu, M.F., Zhao, J.J., Zhao, J., Ramesh, A., Sprechmann, P., LeCun, Y.: Disentangling factors of variation in deep representation using adversarial training. In: Advances in Neural Information Processing Systems, pp. 5041–5049 (2016)
17. Metz, L., Poole, B., Pfau, D., Sohl-Dickstein, J.: Unrolled generative adversarial networks. arXiv preprint arXiv:1611.02163 (2016)
18. Mirza, M., Osindero, S.: Conditional generative adversarial nets. ArXiv e-prints, November 2014
19. Netzer, Y., Wang, T., Coates, A., Bissacco, A., Wu, B., Ng, A.Y.: Reading digits in natural images with unsupervised feature learning. In: NIPS Workshop on Deep Learning and Unsupervised Feature Learning 2011 (2011)
20. Odena, A., Olah, C., Shlens, J.: Conditional image synthesis with auxiliary classifier GANs. ArXiv e-prints, October 2016
21. Radford, A., Metz, L., Chintala, S.: Unsupervised representation learning with deep convolutional generative adversarial networks. ArXiv e-prints, November 2015
22. Reed, S., Sohn, K., Zhang, Y., Lee, H.: Learning to disentangle factors of variation with manifold interaction. In: Proceedings of the 31st International Conference on Machine Learning (ICML-2014), pp. 1431–1439 (2014)
23. Salimans, T., Goodfellow, I., Zaremba, W., Cheung, V., Radford, A., Chen, X.: Improved techniques for training gans. In: Advances in Neural Information Processing Systems, pp. 2226–2234 (2016)
24. Siddharth, N., Paige, B., Van de Meent, J.W., Desmaison, A., Wood, F., Goodman, N.D., Kohli, P., Torr, P.H.S.: Learning disentangled representations with semi-supervised deep generative models. ArXiv e-prints, June 2017
25. Sønderby, C.K., Caballero, J., Theis, L., Shi, W., Huszár, F.: Amortised MAP inference for image super-resolution. CoRR abs/1610.04490 (2016). http://arxiv.org/abs/1610.04490
26. Sønderby, C.K., Raiko, T., Maaløe, L., Sønderby, S.K., Winther, O.: Ladder variational autoencoders. In: Advances in Neural Information Processing Systems, pp. 3738–3746 (2016)
27. Tenenbaum, J.B., Freeman, W.T.: Separating style and content with bilinear models. Neural Comput. **12**(6), 1247–1283 (2000). https://doi.org/10.1162/089976600300015349
28. Zagoruyko, S., Komodakis, N.: Wide residual networks. CoRR abs/1605.07146 (2016). http://arxiv.org/abs/1605.07146
29. Zhu, Z., Luo, P., Wang, X., Tang, X.: Deep learning multi-view representation for face recognition. ArXiv e-prints, June 2014

Scatteract: Automated Extraction of Data from Scatter Plots

Mathieu Cliche[1]([✉]), David Rosenberg[1], Dhruv Madeka[2], and Connie Yee[1]

[1] Bloomberg LP, 731 Lexington Ave., New York, NY, USA
{mcliche,drosenberg44,cyee3}@bloomberg.net
[2] Amazon.com Inc., 7 West 34th Street, New York, NY, USA
maded@amazon.com

Abstract. Charts are an excellent way to convey patterns and trends in data, but they do not facilitate further modeling of the data or close inspection of individual data points. We present a fully automated system for extracting the numerical values of data points from images of scatter plots. We use deep learning techniques to identify the key components of the chart, and optical character recognition together with robust regression to map from pixels to the coordinate system of the chart. We focus on scatter plots with linear scales, which already have several interesting challenges. Previous work has done fully automatic extraction for other types of charts, but to our knowledge this is the first approach that is fully automatic for scatter plots. Our method performs well, achieving successful data extraction on 89% of the plots in our test set.

Keywords: Computer vision · Information retrieval
Data visualization

1 Introduction

Charts are used in many contexts to give a visual representation of data. However, there is often reason to extract the data represented by the chart back into a numeric form, which is easy to use for further analysis or processing. This problem has been studied for many years, and the solution process may be divided into four steps: chart detection [4,17], chart classification [9,15,18], text detection and recognition [5,14], and data extraction [14,18]. Our work focuses on the latter two areas.

We have built Scatteract[1], a system that takes the image of a scatter plot as input and produces a table of the data points represented in the plot, in the coordinate system defined by the plot axes. There are two main steps in this process. First, we localize three key types of objects in the image: (1) *tick marks*, the visual representation of units on an axis (e.g. short lines on the axis),

---

D. Madeka—Work done while the author was at Bloomberg L.P.
[1] https://github.com/bloomberg/scatteract.

© Springer International Publishing AG 2017
M. Ceci et al. (Eds.): ECML PKDD 2017, Part I, LNAI 10534, pp. 135–150, 2017.
https://doi.org/10.1007/978-3-319-71249-9_9

(2) *tick values*, the representation of scale on the axis (e.g. numbers written near the tick marks), and (3) *points*, the visual elements representing data tuples of the form[2] (x, y). In the chart in Fig. 1, we show the bounding boxes for objects of these three types, as produced by our object detection system. The locations of these bounding boxes are initially described in *pixel coordinates*, as is natural for image processing. The second key step in our process is to map the pixel coordinates of the points into *chart coordinates*, as defined by the scale on the chart axes. The end result of our process, a table of data points, is also illustrated in Fig. 1, along with the ground truth (x, y) coordinates of the data.

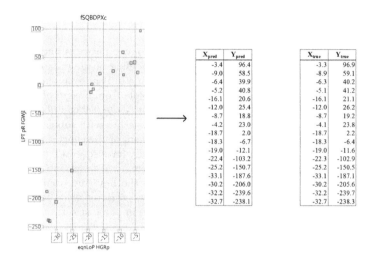

Fig. 1. Illustration of input and output of the end-to-end system. Bounding boxes are shown to illustrate the result of the object detection and are not present on the original image. Red bounding boxes identify the tick values, green bounding boxes identify the tick marks, and blue bounding boxes identify the points. (Color figure online)

While information extraction from charts and other infographics has been widely studied, to our knowledge, our system is the first to extract data from an image of a scatter plot and represent it in the coordinate system of the chart, fully automatically. The closest in aim is [3], an unpublished work that describes an approach to such a system, but only gives test results for the point detection part. Semi-automatic software tools are available, but they require the user to manually define the coordinate system of the chart and click on the data points, or to provide metadata about the axes and data [11,14,20,26].

There exist several systems that attempt to extract key components of charts, but do not attempt to convert from pixel coordinates to chart coordinates. Many use heuristics, such as connected component analysis, edge detection and k-median filtering [8,12,13,15,16]. Recent work has taken a machine learning approach. [2] classifies the role of extracted chart components (e.g. bar or legend).

---

[2] As is traditional, we refer to the horizontal axis as "X" and the vertical axis as "Y".

In [25], a deep learning object detection model is trained to detect sub-figures in compound figures. FigureSeer uses a convolutional neural network to extract features for the localization of lines and heuristics for the localization of tick values [21]. The end result is the localization of all the relevant chart components, but does not include the pixel-to-chart coordinate transformation.

For charts other than scatter plots, there are two systems that aim to convert from pixel coordinates to chart coordinates. The automatic data extraction in [1] is for bar charts. It extracts the text and numerical values using OCR, and then recovers the numerical values for each bar by multiplying its height in pixels by the pixel-per-data ratio. Inaccuracies of the OCR tool resulted in a significant number of the charts having incorrectly calculated Y-axis scale values. The ReVision system proposed in [18] recovers the raw data encoded in bar and pie charts by extracting the labels with OCR and using a scaling factor to map from image space to data space. All these systems are highly dependent on the accuracy of the OCR.

In this paper, we propose Scatteract, an algorithm that leverages deep learning techniques and OCR to retrieve the chart coordinates of the data points in scatter plots. Scatteract has three key advantages over the systems mentioned above. First, to our knowledge, Scatteract is the only system to use deep learning methods to identify all the key components of a scatter plot. The second novel aspect of our work is that we created a system for procedurally generating a large training dataset of scatter plots, for which the ground truth is known, without requiring any manual labeling. This allows our system to be much more extensible than heuristic methods built around a set of assumptions. The plot distribution can be easily modified to accommodate new plot aesthetics, while heuristic methods may need a complete redesign to accommodate these changes. The third key advantage of our system is that our method for determining the mapping from pixel coordinates to chart coordinates is fairly robust to OCR errors.

This paper is organized as follows. In Sect. 2 we describe the datasets used to train and test Scatteract. In Sect. 3 we expand on the methodology and present test results for the building blocks of Scatteract. In Sect. 4 we present a performance analysis of the end-to-end system, and we outline our main conclusions in Sect. 5.

2 Datasets

Scatteract uses an object detection model, which requires a large amount of annotated data for training. One way to achieve this is to collect a large sample of scatter plots from the web and manually label the bounding boxes for the objects of interest (points, tick marks, and tick values). A more efficient approach is to generate the scatter plots procedurally, so that the bounding boxes are known. Using this approach, we generated 25,000 train images and 600 test images. Some examples are shown in Figs. 6 and 8. Besides these artificial charts, we scraped an additional 50 scatter plots from the web (Figs. 7 and 9). More details on how our datasets were obtained are below.

2.1 Procedurally Generated Scatter Plots

To achieve randomness in the scatter plots, we developed a script to randomly select values that affect the aesthetics and the data of the scatter plot. We used the Python library Matplotlib [10] to generate the plots since it allows for easy extraction of the bounding boxes of the points, tick marks, and tick values. The factors we used to randomize the plot aesthetics and data are listed below.

1. Plot aesthetics
   (a) Plot style (default styling values e.g. "classic", "seaborn", "ggplot", etc.)
   (b) Size and resolution of the plot
   (c) Number of style variations on the points
   (d) Point styles (markers, sizes, colors)
   (e) Padding between axis and tick values, axis and axis-labels, and plot area and plot title
   (f) Axis and ticks locations
   (g) Tick sizes if using non-default ticks
   (h) Rotation angle of the tick values, if any
   (i) Presence and style of minor ticks
   (j) Presence and style of grid lines
   (k) Font and size of the tick values
   (l) Fonts, sizes and contents of the axis-labels and plot title
   (m) Colors of the plot area and background area
2. Plot data
   (a) Data points distribution (uniformly random, or random around a linear or quadratic distribution)
   (b) Number of points
   (c) Order of magnitude of the X and Y coordinates
   (d) X and Y coordinates ranges around the selected order of magnitude
   (e) Actual values of the X and Y coordinates given the order of magnitude, ranges and distribution

These parameters allow us to build a wide variety of scatter plots, for which we have full ground-truth labels. Some of the plots generated from this procedure are very difficult and sometimes impossible to read, even for a human. For example, the tick values can overlap with each other or with the data points, and the randomly selected font for the tick values can be unreadable. Although we did not eliminate such unrealistic plots from the training set, we did manually remove them from the test set.

2.2 Scatter Plots from the Web

To see how our system, trained on randomly generated scatter plots, generalizes to real charts, we also collected a small test set of scatter plots that were generated by humans. We downloaded 50 scatter plots from a Google image search for "scatter plot". The only inclusion criteria was the presence of tick values without units.

3   Methodology

We take as input the image of a scatter plot, and the output is a set of the $(X_{\mathrm{pred}}, Y_{\mathrm{pred}})$ chart coordinates of the data points detected. The pipeline is as follows:

1. Use the object detection model to find bounding boxes for the tick marks, tick values, and points.
2. Apply OCR on the images inside the bounding boxes of the tick values.
3. Find the closest tick mark to each tick value.
4. Use clustering to assign each (tick mark, tick value) pair to either the X or Y axis.
5. Apply robust regression to determine the mapping from pixel coordinates to chart coordinates, for each axis.
6. Apply the mapping to the pixel locations of the detected points to build the table of chart coordinates.

Below we expand on the most important steps.

3.1   Object Detection

Object detection is the task of putting bounding boxes around objects that appear in an image. For our task, we use object detection to localize points, tick marks, and tick values. There is a wide variety of object detection models, but all the state-of-the-art methods use deep learning. We chose ReInspect [23] as our object detection method because it is very effective in crowded scenes, and the points in scatter plots are often very close together, even overlapping. ReInspect uses the OverFeat algorithm [19] but adds a long short-term memory network at the end of it, such that predictions are not made independently of each other. We used the Tensorflow implementation of ReInspect, TensorBox[3]. It is standard practice to initialize the core of the model with a pretrained convolutional network, for which we selected Google Inception V1 [24]. We trained three separate models: one for tick marks, one for tick values, and one for points. We resized our scatter plot images to 800×800 pixels for the tick mark and tick value models, and to 1440×1440 for the point model to help resolve a high density of points. Each model was trained for 5 epochs on a Geforce GTX Titan X GPU, and the total training time was about 60 h. Given an image, a model outputs a set of bounding boxes for the object it is detecting, along with a confidence score for each. We only keep detections with a confidence above 0.3, based on validation experiments.

We can evaluate the performance of the object detection on the procedurally generated test set. We define a true positive as a predicted bounding box which has an intersection-over-union ratio with a true bounding box above 50%. We can then vary the threshold on the confidence of the predicted bounding boxes to generate recall-precision curves, see Fig. 2. We get an average precision of 87.1%

---

[3] https://github.com/TensorBox/TensorBox.

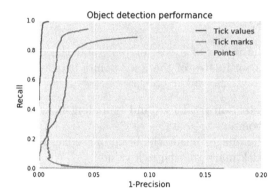

Fig. 2. Recall-Precision curves for the object detection.

for the detection of points, 99.1% for the detection of tick values and 93.0% for the detection of tick marks. The detection of points is the most difficult, particularly for plots with many overlapping points. However, in the end-to-end system, failures on the detection of tick marks and tick values can be much more costly, since they are essential for determining the pixel-to-chart coordinate conversion.

3.2   Optical Character Recognition

To extract the tick values, we apply OCR to the content of the bounding boxes produced by the tick-value model. We use Google's open source OCR engine, Tesseract [22], because of its ease of use. It comes pretrained on several fonts, and while it is possible to retrain it on custom fonts, we used it as is. However, we found that performing several preprocessing steps on a tick-value image before applying Tesseract significantly improves OCR accuracy. We first convert the image to gray scale and then rescale it so that its height is 130 pixels, while maintaining its aspect ratio. The most important transformation, however, is a heuristic procedure to rotate the tick-value images such that they are horizontally aligned, since this is what Tesseract is expecting. This procedure[4] finds the minimum area rectangle that encloses the thresholded tick-value image, and then rotates the image by an angle that takes into account the angle of the rectangle and the number of characters in the tick-value image. The result of this procedure is illustrated in Fig. 3(a). It works well except for tick values that are rotated by precisely ±90° or upside-down.

To test the OCR piece, we used true tick-value images along with their corresponding tick values, for each of the plots in the procedurally generated test set. Then, for each tick-value image, we ran our image preprocessing technique,

---

[4] The procedure is largely inspired by this blog post: http://felix.abecassis.me/2011/ 10/opencv-rotation-deskewing/.

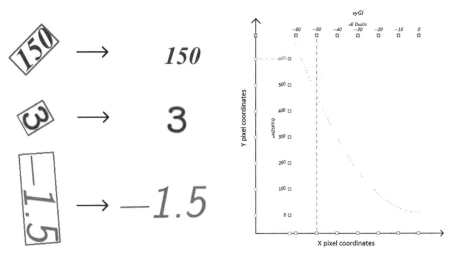

(a) Illustration of the results from the rotation-fixing heuristics. The blue bounding box represents the minimum area rectangle that encloses the thresholded image.

(b) Illustration of the clustering problem used to assign the tick marks to either the X- or Y-axis.

Fig. 3. .

followed by the Tesseract engine, and compared the predicted tick values with the ground truth. This gives an accuracy of 91.2%, but without the image preprocessing the accuracy drops to 63.8%.

3.3 Axis Splitting

Now that we have extracted the tick value associated with each tick-value image, we need to associate each value with its corresponding tick mark and assign it to the X-axis or Y-axis. We make use of the observation that if we project the centers of the tick-mark bounding boxes to the X-axis, the projection of the ticks from the Y-axis are clustered in one area, while the projection of ticks from the X-axis are spread out, as illustrated in Fig. 3(b). This therefore effectively becomes a clustering problem. The algorithm is as follows:

1. Assign each tick value to its closest tick mark.
2. Perform a 1-dimensional DBSCAN [6] clustering algorithm twice: once to obtain a set of clusters for the X pixel coordinates of the ticks, and another one to obtain a set of clusters for the Y pixel coordinates of the ticks. The expected result from the two DBSCANs is one cluster with the X coordinates, these are the ticks that belong on the Y-axis, and one cluster with the Y coordinates, these are the ticks that belong on the X-axis.

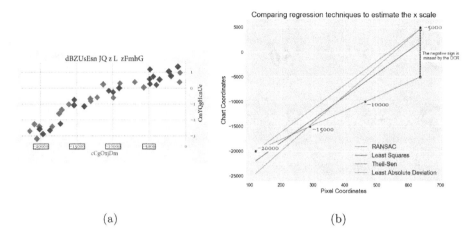

(a)   (b)

Fig. 4. In Fig. 4(a) above, the OCR fails to detect the minus sign in front of the number −5000. While other regression techniques are thrown off by this outlier, Fig. 4(b) shows that RANSAC regression (visualized by the green line) provides a more robust estimation of the transformation from pixel to chart coordinates. (Color figure online)

3.4 RANSAC Regression

At this point, we need to find the mapping from pixel to chart coordinates. We assume the scales are linear, so the conversion is an affine transformation. For the X-axis, it takes the form $X_{\mathrm{chart}} = \alpha_x X_{\mathrm{pixel}} + \beta_x$, and similarly for the Y-axis. We now create a pair of the form $(X_{\mathrm{pixel}}, X_{\mathrm{chart}})$ for each tick mark[5]/tick value pair we find on the X-axis, and similarly for the Y-axis. For the chart in Fig. 4(a), we have extracted these pairs for the X-axis and plotted them as black points in Fig. 4(b). In principle, we can determine our affine transformation just by performing linear regression on these points. However, standard least-squares regression is very sensitive to outliers, which means that a single OCR error can cause a complete failure to estimate the transformation. We therefore explored alternatives that are more robust to outliers, such as least absolute deviations (LAD) regression, Theil-Sen Regression, and RANSAC regression. While all three fared better than linear regression, we obtained the best results with RANSAC regression [7]. RANSAC regression is an iterative algorithm in which linear regression is applied to samples of the data points in order to detect inliers and outliers from the out-of-sample data points. The model with the most inliers and the smallest residuals on the inliers is chosen. For RANSAC we need to choose a loss function, as well as a maximum residual $R_{\max}$ beyond which a point is classified as an outlier. We use a square loss function, and choose $R_{\max}$ as the median absolute deviation of the tick values squared and divided by 50, such that for the X-axis:

$$R_{\mathrm{max},x} = \frac{\mathrm{median}\left(X_{\mathrm{chart},j} - \mathrm{median}(X_{\mathrm{chart}})\right)^2}{50}. \tag{1}$$

---

[5] We use the center of the tick-mark bounding box as the tick mark location.

We apply RANSAC regression for both the X- and Y-axes, such that we end up with a set of 4 parameters $(\alpha_x, \alpha_y, \beta_x, \beta_y)$ for each scatter plot. It is then straightforward to apply these affine transformations to the pixel coordinates of the center of the bounding box of each point detected, and thus obtain their location in chart coordinates. Figure 4 shows how RANSAC improves the estimate of the affine transformation compared to other regression techniques.

4 Results

4.1 Performance Analysis

To evaluate the performance of the end-to-end system, we need to have an evaluation metric. We define true positives as predicted points that are within 2% of true points in chart coordinates:

$$\frac{|X_{\text{pred}} - X_{\text{true}}|}{\Delta X_{\text{true}}} <= 0.02 \quad \text{and} \quad \frac{|Y_{\text{pred}} - Y_{\text{true}}|}{\Delta Y_{\text{true}}} <= 0.02 \tag{2}$$

where $\Delta X_{\text{true}} = \max_j(X_{\text{true},j}) - \min_j(X_{\text{true},j})$ and $\Delta Y_{\text{true}} = \max_j(Y_{\text{true},j}) - \min_j(Y_{\text{true},j})$ are the true ranges of values for the X and Y chart coordinates. We compute all pairs of distances between predicted and true points, and we first select the predicted points closest to each true point in order to find true positives. If a pair of nearby predicted and true points satisfy the true positive criteria, both true and predicted points are removed from the true and predicted set such that we do not overcount true positives. We repeat this process until either no points are remaining or none of the closest points satisfy the true positive criteria. With this definition we can evaluate the precision and recall for each plot. Taking the average across all plots in the procedurally generated test set, we end up with an average precision of 88% and an average recall of 87%. We can also compute the F_1 score for each plot. Figure 5 shows the distribution of F_1 scores across all plots in the procedurally generated test set. Based on Fig. 5, we can see that plots either have a very high F_1 score or a very low one. The latter case indicates that the pixel/chart coordinate transformation was not determined correctly. We can therefore define a successful data extraction for a plot if the F_1 score is above 80% for that plot. By this definition, our method is successful for 89% of the plots in the procedurally generated test. We can also use this definition to evaluate alternative versions of our method. In Table 1, note that the version that uses linear regression on 2 tick values instead of RANSAC is analogous to what [3] used to find the conversion factors between pixel and chart coordinates. RANSAC has a net advantage over the other methods, giving an absolute performance boost of 5.4% over Theil-Sen, the next best method. We note that in its current implementation, Scatteract requires 3.5 s to extract the data from a single plot.

Without the ground truth available for the test set from the web, we cannot evaluate it as systematically as the procedurally generated one. However, a visual inspection of the plots is sufficient to determine if the data extraction is successful. To quickly see if the axes were correctly decoded, we sampled predicted

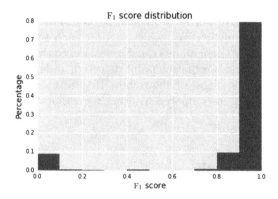

Fig. 5. F_1 score distribution across the plots in the procedurally generated test set.

Table 1. Test results on variations of our system.

System	Success rate
Scatteract	**89.2%**
Scatteract without image preprocessing before OCR	43.3%
Scatteract with Theil-Sen instead of RANSAC	83.8%
Scatteract with LAD instead of RANSAC	82.8%
Scatteract with linear regression on 2 tick values instead of RANSAC	70.7%

points and verified that they correspond to actual points on the plots. Moreover, to see if most points were detected correctly, we inspected the predicted bounding boxes of the points directly on the plot image. Through visual inspection we conclude that our method gave a successful data extraction on 39 plots out of the 50, or 78%. As one might expect, this is lower than on the procedurally generated test set due to the fact that these plots occasionally exhibit features which are not present in the training data. These aspects make identifying all the relevant objects more challenging for the object detection.

4.2 Error Analysis

Let us now comment on results for individual plots. First, Fig. 6 shows six successful data extraction plots from the procedurally generated test set, while Fig. 7 displays six successful data extraction plots from the web test set. Note that these plots cover a wide range of features occasionally seen in scatter plots. Second, Fig. 8 displays two unsuccessful data extraction plots from the procedurally generated test set, while Fig. 9 displays two unsuccessful data extraction plots from the web test set. By examining the picture with the predicted bounding boxes and

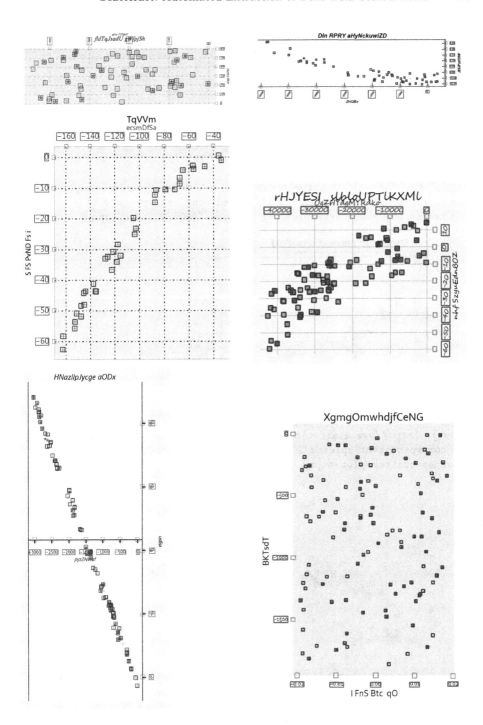

Fig. 6. Successful examples from the procedurally generated test set.

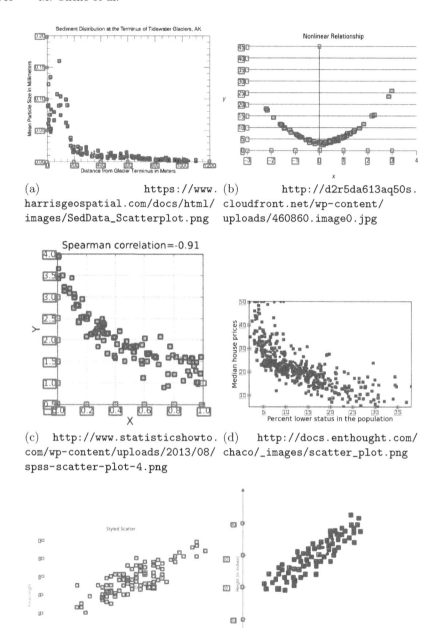

(a)   https://www.
harrisgeospatial.com/docs/html/
images/SedData_Scatterplot.png

(b)   http://d2r5da613aq50s.
cloudfront.net/wp-content/
uploads/460860.image0.jpg

(c) http://www.statisticshowto.
com/wp-content/uploads/2013/08/
spss-scatter-plot-4.png

(d)   http://docs.enthought.com/
chaco/_images/scatter_plot.png

(e)   https://plot.ly/~RPlotBot/
4326/styled-scatter.png

(f)   https://statsmethods.
files.wordpress.com/2013/05/
scatter-plot-2.png

Fig. 7. Successful examples from the web test set.

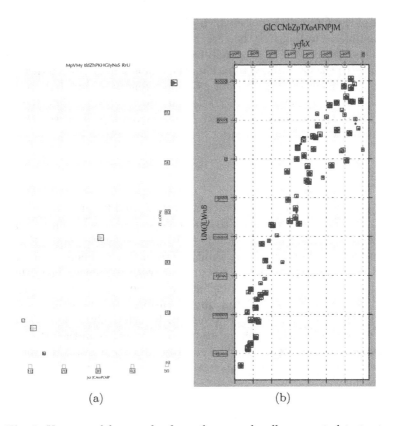

(a)   (b)

Fig. 8. Unsuccessful examples from the procedurally generated test set.

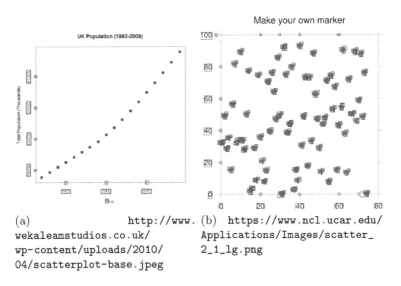

(a)   http://www. (b) https://www.ncl.ucar.edu/
wekaleamstudios.co.uk/   Applications/Images/scatter_
wp-content/uploads/2010/   2_1_lg.png
04/scatterplot-base.jpeg

Fig. 9. Unsuccessful examples from the web test set.

the result of the OCR we can see which part of the system failed on the unsuccessful examples. On Fig. 8(a) the tick detection failed on the Y-axis because it is ambiguous whether the ticks should be on the left or right of the tick values, and on Fig. 8(b) the OCR failed to pick up the minus sign on several tick values of the X-axis. Similarly, on Fig. 9(a) our rotation fixing procedure fails on the 90°-rotated Y tick values, and on Fig. 9(b) the shamrock marker is detected as three separate points by the object detection. Those errors represent well the kind of failure that our method can encounter.

5   Conclusion

In this paper we introduced Scatteract, a method used to automatically extract data from scatter plots. Arguably the most important innovation in Scatteract is the deep learning object detection models used to locate points, tick marks, and tick values, combined with the ability to train these models on procedurally generated scatter plots, for which the ground truth is known. The other main contribution of Scatteract is the pipeline, from the OCR preprocessing to the RANSAC regression, for finding the affine transformation between pixel coordinates and chart coordinates. There seems to be relatively little prior work on this, and we have shown our method to be more robust than the other methods we are aware of in the literature. The end-to-end system is able to successfully extract data from 89% of procedurally generated scatter plots and 78% of scatter plots from the web.

Extending our models to handle new chart aesthetics is straightforward if we can generate training data with the new aesthetics. While the performance on the web test set is promising, it is somewhat worse than performance on the procedurally generated test set. We leave to future work an investigation of how to improve Scatteract's generalization ability. We will also pursue a deeper extraction of information from scatter plots, including point categories, legends, axis titles, nonlinear scales, scales with units, and non-numeric scales, such as dates. The ultimate goal is to extend the method to other chart types as well, so that an end-to-end system would detect charts, classify them (line chart, bar chart, pie chart, scatter plot, etc.), and then use an appropriate pipeline to extract the data from the chart.

References

1. Al-Zaidy, R.A., Giles, C.L.: Automatic extraction of data from bar charts. In: Proceedings of the 8th International Conference on Knowledge Capture, p. 30. ACM (2015)
2. Al-Zaidy, R.A., Giles, C.L.: A machine learning approach for semantic structuring of scientific charts in scholarly documents. In: Twenty-Ninth IAAI Conference (2017)
3. Baucom, A., Echanique, C.: Scatterscanner: data extraction and chart restyling of scatterplots (2013)

4. Browuer, W., Kataria, S., Das, S., Mitra, P., Giles, C.L.: Segregating and extracting overlapping data points in two-dimensional plots. In: Proceedings of the 8th ACM/IEEE-CS Joint Conference on Digital Libraries, pp. 276–279. ACM (2008)
5. Chen, Z., Cafarella, M., Adar, E.: Diagramflyer: a search engine for data-driven diagrams. In: Proceedings of the 24th International Conference on World Wide Web, pp. 183–186. ACM (2015)
6. Ester, M., Kriegel, H.P., Sander, J., Xu, X., et al.: A density-based algorithm for discovering clusters in large spatial databases with noise. In: KDD, vol. 96, pp. 226–231 (1996)
7. Fischler, M.A., Bolles, R.C.: Random sample consensus: a paradigm for model fitting with applications to image analysis and automated cartography. Commun. ACM **24**(6), 381–395 (1981)
8. Huang, W., Tan, C.L.: A system for understanding imaged infographics and its applications. In: Proceedings of the 2007 ACM Symposium on Document Engineering, pp. 9–18. ACM (2007)
9. Huang, W., Tan, C.L., Leow, W.K.: Model-based chart image recognition. In: Lladós, J., Kwon, Y.-B. (eds.) GREC 2003. LNCS, vol. 3088, pp. 87–99. Springer, Heidelberg (2004). https://doi.org/10.1007/978-3-540-25977-0_8
10. Hunter, J.D.: Matplotlib: a 2D graphics environment. Comput. Sci. Eng. **9**(3), 90–95 (2007)
11. Jung, D., Kim, W., Song, H., Hwang, J.i., Lee, B., Kim, B., Seo, J.: ChartSense: interactive data extraction from chart images. ACM (2017)
12. Kataria, S., Browuer, W., Mitra, P., Giles, C.L.: Automatic extraction of data points and text blocks from 2-dimensional plots in digital documents (2008)
13. Lu, X., Kataria, S., Brouwer, W.J., Wang, J.Z., Mitra, P., Giles, C.L.: Automated analysis of images in documents for intelligent document search. Int. J. Document Anal. Recogn. (IJDAR) **12**(2), 65–81 (2009)
14. Mishchenko, A., Vassilieva, N.: Chart image understanding and numerical data extraction. In: 2011 Sixth International Conference on Digital Information Management (ICDIM), pp. 115–120. IEEE (2011)
15. Nair, R.R., Sankaran, N., Nwogu, I., Govindaraju, V.: Automated analysis of line plots in documents. In: 2015 13th International Conference on Document Analysis and Recognition (ICDAR), pp. 796–800. IEEE (2015)
16. Poco, J., Heer, J.: Reverse-engineering visualizations: recovering visual encodings from chart images. In: Computer Graphics Forum, vol. 36, pp. 353–363. Wiley Online Library (2017)
17. Ray Choudhury, S., Giles, C.L.: An architecture for information extraction from figures in digital libraries. In: Proceedings of the 24th International Conference on World Wide Web, pp. 667–672. ACM (2015)
18. Savva, M., Kong, N., Chhajta, A., Fei-Fei, L., Agrawala, M., Heer, J.: Revision: automated classification, analysis and redesign of chart images. In: Proceedings of the 24th Annual ACM Symposium on User Interface Software and Technology, pp. 393–402. ACM (2011)
19. Sermanet, P., Eigen, D., Zhang, X., Mathieu, M., Fergus, R., LeCun, Y.: Overfeat: integrated recognition, localization and detection using convolutional networks. arXiv preprint arXiv:1312.6229 (2013)
20. Shadish, W.R., Brasil, I.C., Illingworth, D.A., White, K.D., Galindo, R., Nagler, E.D., Rindskopf, D.M.: Using ungraph to extract data from image files: verification of reliability and validity. Behav. Res. Methods **41**(1), 177–183 (2009)

21. Siegel, N., Horvitz, Z., Levin, R., Divvala, S., Farhadi, A.: FigureSeer: parsing result-figures in research papers. In: Leibe, B., Matas, J., Sebe, N., Welling, M. (eds.) ECCV 2016. LNCS, vol. 9911, pp. 664–680. Springer, Cham (2016). https://doi.org/10.1007/978-3-319-46478-7_41

22. Smith, R.: An overview of the Tesseract OCR engine. In: Ninth International Conference on Document Analysis and Recognition, ICDAR 2007, vol. 2, pp. 629–633. IEEE (2007)

23. Stewart, R., Andriluka, M., Ng, A.Y.: End-to-end people detection in crowded scenes. In: Proceedings of the IEEE Conference on Computer Vision and Pattern Recognition, pp. 2325–2333 (2016)

24. Szegedy, C., Liu, W., Jia, Y., Sermanet, P., Reed, S., Anguelov, D., Erhan, D., Vanhoucke, V., Rabinovich, A.: Going deeper with convolutions. In: Proceedings of the IEEE Conference on Computer Vision and Pattern Recognition, pp. 1–9 (2015)

25. Tsutsui, S., Crandall, D.: A data driven approach for compound figure separation using convolutional neural networks. arXiv preprint arXiv:1703.05105 (2017)

26. Yang, L., Huang, W., Tan, C.L.: Semi-automatic ground truth generation for chart image recognition. In: Bunke, H., Spitz, A.L. (eds.) DAS 2006. LNCS, vol. 3872, pp. 324–335. Springer, Heidelberg (2006). https://doi.org/10.1007/11669487_29

Unsupervised Diverse Colorization via Generative Adversarial Networks

Yun Cao$^{(\boxtimes)}$, Zhiming Zhou, Weinan Zhang, and Yong Yu

Shanghai Jiao Tong University, Shanghai, China
{yuncao,heyohai,wnzhang,yyu}@apex.sjtu.edu.cn

Abstract. Colorization of grayscale images is a hot topic in computer vision. Previous research mainly focuses on producing a color image to recover the original one in a supervised learning fashion. However, since many colors share the same gray value, an input grayscale image could be diversely colorized while maintaining its reality. In this paper, we design a novel solution for unsupervised diverse colorization. Specifically, we leverage conditional generative adversarial networks to model the distribution of real-world item colors, in which we develop a fully convolutional generator with multi-layer noise to enhance diversity, with multi-layer condition concatenation to maintain reality, and with stride 1 to keep spatial information. With such a novel network architecture, the model yields highly competitive performance on the open LSUN bedroom dataset. The Turing test on 80 humans further indicates our generated color schemes are highly convincible.

1 Introduction

Image colorization assigns a color to each pixel of a target grayscale image. Early colorization methods [15,21] require users to provide considerable scribbles on the grayscale image, which is apparently time-consuming and requires expertise. Later research provides more automatic colorization methods. Those colorization algorithms differ in the ways of how they model the correspondence between grayscale and color.

Given an input grayscale image, non-parametric methods first define one or more color reference images (provided by a human or retrieved automatically) to be used as source data. Then, following the Image Analogies framework [10], the color is transferred onto the input image from analogous regions of the reference image(s) [4,9,17,24]. Parametric methods, on the other hand, learn prediction functions from large datasets of color images in the training stage, posing the colorization problem as either regression in the continuous color space [3,6,26] or classification of quantized color values [2], which is a supervised learning fashion.

Whichever seeking the reference images or learning a color prediction model, all above methods share a common goal, i.e. to provide a color image closer to the original one. But as we know, many colors share the same gray value. Purely from a grayscale image, one cannot tell what color of clothes a girl is wearing or what color a bedroom wall is. Those methods all produce a deterministic

© Springer International Publishing AG 2017
M. Ceci et al. (Eds.): ECML PKDD 2017, Part I, LNAI 10534, pp. 151–166, 2017.
https://doi.org/10.1007/978-3-319-71249-9_10

Fig. 1. Left: the original grayscale image. Middle: image colorized by non-adversarial CNNs. Right: image colorized by human on Reddit. (Figure from [14])

mapping function. Thus when an item could have diverse colors, their models tend to provide a weighted average brownish color as pointed out in [14] (See Fig. 1 as an example).

In this paper, to avoid this sepia-toned colorization, we use conditional generative adversarial networks (GANs) [8] to generate diverse colorizations for a single grayscale image while maintaining their reality. GAN is originally proposed to generate vivid images from some random noise. It is composed of two adversarial parts: a generative model G that captures the data distribution, and a discriminative model D that estimates the probability of whether an image is real or generated by G. The generator part tries to map an input noise to a data distribution closer to the ground truth data distribution, while the discriminator part tries to distinguish the generated "fake" data, which comes to an adversarial situation. By careful designation of both generative and discriminative parts, the generator will eventually produce results, forming a distribution very close to the ground truth distribution, and by controlling the input noise we can get various results of good reality. Thus conditional GAN is a much more suitable framework to handle diverse colorization than other CNNs. Meanwhile, as the discriminator only needs the signal of whether a training instance is real or generated, which is directly provided without any human annotation during the training phase, the task is in an unsupervised learning fashion.

On the aspect of model designation, unlike many other conditional GANs [12] using convolution layers as the encoder and deconvolution layers as the decoder, we build a fully convolutional generator and each convolutional layer is splinted by a concatenate layer to continuously render the conditional grayscale information. Additionally, to maintain the spatial information, we set all convolution stride to 1 to avoid downsizing data. We also concatenate noise channels to the first half convolutional layers of the generator to attain more diversity in the color image generation process. As the generator G would capture the color distribution, we can alter the colorization result by changing the input noise. Thus we no longer need to train an additional independent model for each color scheme like [3].

As our goal alters from producing the original colors to producing realistic diverse colors, we conduct questionnaire surveys as a Turing test instead of calculating the root mean squared error (RMSE) comparing the original image to measure our colorization result. The feedback from 80 subjects indicates that our model successfully produces high-reality color images, yielding more than 62.6% positive feedback while the rate of ground truth images is 70.0%. Furthermore, we perform a significance t-test to compare the percentages of human judges as real color images for each test instance (i.e. a real or generated color image). The resulting p-value is $0.1359 > 0.05$, which indicates that there is no significant difference between our generated color images and the real ones. We share the repeatable experiment code for further research[1].

2   Related Work

2.1   Diverse Colorization

The problem of colorization was proposed in the last century, but the research of diverse colorization was not paid much attention until this decade. In [3], they used additionally trained model to handle diverse colorization of a scene image particularly in day and dawn. [26] posed the colorization problem as a classification task and use class re-balancing at training time to increase the colorfulness of the result. And in the work of [5], a low dimensional embedding of color fields using a variational auto-encoder (VAE) is learned. They constructed loss terms for the VAE decoder that avoid blurry outputs and take into account the uneven distribution of pixel colors and finally developed a conditional model for the multi-modal distribution between gray-level image and the color field embeddings.

Compared with above work, our solution uses conditional generative adversarial networks to achieve unsupervised diverse colorization in a generic way with little domain knowledge of the images.

2.2   Conditional GAN

Generative adversarial networks (GANs) [8] have attained much attention in unsupervised learning research during the recent 3 years. Conditional GANs have been widely used in various computer vision scenarios. [22] used text to generate images by applying adversarial networks. [12] provided a general-purpose image-to-image translation model that handles tasks like label to scene, aerial to map, day to night, edges to photo and also grayscale to color.

Some of the above works may share a similar goal with us, but our conditional GAN structure differs a lot from previous work in several architectural choices mainly for the generator. Unlike other generators which employ an encoder-like front part consisting of multiple convolution layers and a decoder-like end part consisting of multiple deconvolution layers, our generator uses only convolution layers all over the architecture, and does not downsize data shape by applying convolution stride no more than 1 and no pooling operation. Additionally, we add

---

[1] Experiment code is available at https://github.com/ccyyatnet/COLORGAN.

multi-layer noise to generate more diverse colorization, while using multi-layer conditional information to keep the generated image highly realistic.

3   Methods

3.1   Problem Formulation

GANs are generative models that learn a mapping from random noise vector z to an output color image x: $G : z \rightarrow x$. Compared with GANs, conditional GANs learn a mapping from observed grayscale image y and random noise vector z, to x: $G : \{y, z\} \rightarrow x$. The generator G is trained to produce outputs that cannot be distinguished from "real" images by an adversarially trained discriminator D, which is trained with the aim of detecting the "fake" images produced by the generator. This training procedure is illustrated in Fig. 2.

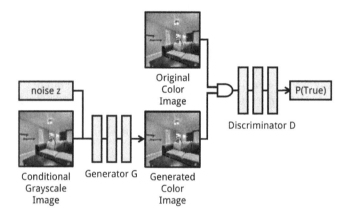

Fig. 2. The illustration of conditional GAN. Generator G is given conditional information (Grayscale image) together with noise z, and produces generated color channels. Discriminator D is trained over the real color image and the generated color image by G. The goal of D is to distinguish real images from the fake ones. Both nets are trained adversarially. (Color figure online)

The objective of a GAN can be expressed as

$$\mathcal{L}_{\mathrm{GAN}}(G, D) = \mathbb{E}_{x \sim P_{\mathrm{data}}(x)}[\log D(x)] \\ + \mathbb{E}_{z \sim P_z(z)}[\log(1 - D(G(z)))], \tag{1}$$

while the objective of a conditional GAN is

$$\mathcal{L}_{\mathrm{cGAN}}(G, D) = \mathbb{E}_{x \sim P_{\mathrm{data}}(x)}[\log D(x)] \\ + \mathbb{E}_{y \sim P_{\mathrm{gray}}(y), z \sim P_z(z)}[\log(1 - D(G(y, z)))], \tag{2}$$

where G tries to minimize this objective against an adversarial D that tries to maximize it, i.e.

$$G^* = \arg \min_G \max_D \mathcal{L}_{cGAN}(G, D). \tag{3}$$

Without z, the generator could still learn a mapping from y to x, but would produce deterministic outputs. That is why GAN is more suitable for diverse colorization tasks than other deterministic neural networks.

3.2   Architecture and Implementation Details

The high-level structure of our conditional GAN is consistent with traditional ones [5,12], while the detailed architecture of our generator G differs a lot.

Convolution or deconvolution. Convolution and deconvolution layers are two basic components of image generators. Convolution layers are mainly used to exact conditional features. And additionally, many researches [5,12,26] use superposition of multiple convolution layers with stride more than 1 to downsize the data shape, which works as a data encoder. Deconvolution layers are then used to upsize the data shape as a decoder of the data representation [5,12,18]. While many other researches share this encoder-decoder structure, we choose to use only convolution layers in our generator G. Firstly, convolution layers are well capable of feature extraction and transmission. Meanwhile, all the convolution stride is set to 1 to prevent data shape from downsizing, thus the important spatial information can be kept along the data flow till the final generation layer. Some other researches [12,23] also takes this spatial information into consideration. They add skip connections between each layer i and layer $n - i$ to form a "U-Net" structure, where n is the total number of layers. Each skip connection simply concatenates all channels at layer i with those at layer $n - i$. Whether adding skip connections or not, the encoder-decoder structure more tends to extract global features and generate images by this overall information which is more suitable for global shape transformation tasks. But in image colorization, we need a very detailed spatial local guidance to make sure item boundaries will be accurately separated by different color parts in generated channels. Let alone our modification is more straightforward and easy to implement. See the structural difference between "U-Net" and our convolution model in Fig. 3.

YUV or RGB. A color image can be represented in different forms. The most common representation is RGB form which splits a color pixel into red, green, blue three channels. Most computer vision tasks use RGB representation [6,12, 19] due to its generality. Other kinds of representations are also included like YUV (or $YCrCb$) [3] and Lab [2,16,26]. In colorization tasks, we have grayscale image as conditional information, thus it is straightforward to use YUV or Lab representation because the Y and L channel or so called Luminance channel represents exactly the grayscale information. So while using YUV representation, we can just predict 2 channels and then concatenate with the grayscale channel to give a full color image. Additionally, if you use RGB as image representation,

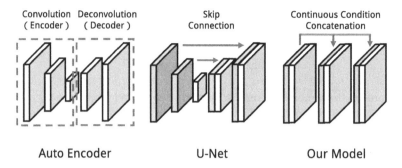

Fig. 3. Structure comparison of auto encoder, U-Net and our generator. Left: auto encoder with convolutional encoder part and deconvolutional decoder part. Middle: U-Net structure [12,23] with skip connections between layer i and $n - i$. Right: our fully convolutional generator with continuous condition concatenation

all result channels are predicted, thus to keep the grayscale of generated color image consistent with the original grayscale image, we need to add an additional $L1$ loss as a controller to make sure $Gray(G(\boldsymbol{y}, \boldsymbol{z})) \simeq \boldsymbol{y}$:

$$\mathcal{L}_{L1}(G) = \mathbb{E}_{\boldsymbol{y} \sim P_{\text{gray}}(\boldsymbol{y}), z \sim P_z(\boldsymbol{z})}[\|\boldsymbol{y} - Gray(G(\boldsymbol{y}, \boldsymbol{z}))\|], \tag{4}$$

where for any color image $\boldsymbol{x} = (\boldsymbol{r}, \boldsymbol{g}, \boldsymbol{b})$, the corresponding grayscale image (or the Luminance channel Y) can be calculated by the well-known psychological formulation:

$$Gray(\boldsymbol{x}) = 0.299\boldsymbol{r} + 0.587\boldsymbol{g} + 0.114\boldsymbol{b}. \tag{5}$$

Note that Eq. (4) can still maintain good colorization diversity, because this $L1$ loss term only lays a constraint on one dimension out of three channels. Then the objective function will be modified to:

$$G^* = \arg \min_G \max_D \mathcal{L}_{\text{cGAN}}(G, D) + \lambda \mathcal{L}_{L1}(G). \tag{6}$$

Since there is no equality constraint between the recovered grayscale $Gray(G(\boldsymbol{y}, \boldsymbol{z}))$ and the original one $\boldsymbol{y}$, the $\mathcal{L}_{L1}(G)$ factor will normally be non-zero, which makes the training unstable due to this additional trade-off. The results will be shown in Sect. 4.2 with experimental comparison on both RGB and YUV representations.

Multi-layer noise. The authors in [12] mentioned noise ignorance while training the generator. To handle this problem, they provide noise only in the form of dropout, applied on several layers of the generator at both training and test time. We also noticed this problem. Traditional GANs and conditional GANs receive noise information at the very start layer, during the continuous data transformation through the network, the noise information is attenuated a lot.

To overcome this problem and make the colorization results more diversified, we concatenate the noise channel onto the first half of the generator layers (the first three layers in our case). We conduct experimental comparison on both one-layer noise and multi-layer noise representations, with results shown in Sect. 4.2.

Multi-layer conditional information. Other conditional GANs usually add conditional information only in the first layer, because the layer shape of previous generators changes along their convolution and deconvolution layers. But due to the consistent layer shape of our generator, we can apply concatenation of conditional grayscale information throughout the whole generator layers which can provide sustained conditional supervision. Though the "U-Net" skip structure of [12] can also help posterior layers receive conditional information, our model modification is still more straightforward and convenient.

Wasserstein GAN. The recent work of Wasserstein GAN [1] has acquired much attention. The authors used Wasserstein distance to help getting rid of problems in original GANs like mode collapse and gradient vanishing and provide a measurable loss to indicate the progress of GAN training. We also try implementing Wasserstein GAN modification into our model, but the results are no better than our model. We make comparison between the results of Wasserstein GAN and our GAN in Sect. 4.2.

The illustration of our model structure is shown in Fig. 4.

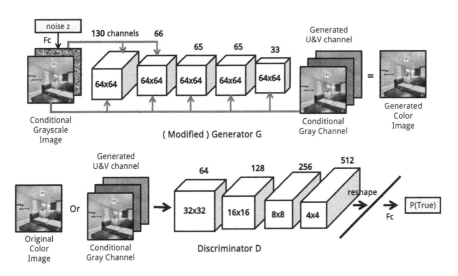

Fig. 4. Detailed structure of our conditional GAN. Top: generator G. Each cubic part represents a Convolution-BatchNorm-ReLU structure. Red connections represent our modifications of the traditional conditional GANs. Bottom: discriminator D (Color figure online)

Algorithm 1. Training phase of our conditional GANs, with the default parameters $k_D = 1, k_G = 1, m = 64, s_z = 100, s = 64$.

1: **for** the number of training iterations **do**
2:     **for** k_D steps **do**
3:         Generate minibatch of m randomly sampled noise $\{z^{(1)}, \ldots, z^{(m)}\}$ each of size $[s_z]$.
4:         Sample minibatch of m grayscale images $\{y^{(1)}, \ldots, y^{(m)}\}$ each of shape $[s, s, 1]$.
5:         Get the corresponding minibatch of m color images $\{x^{(1)}, \ldots, x^{(m)}\}$ from data distribution $p_{\text{data}}(x)$.
6:         Update the discriminator D by:

$$\nabla_{\theta_d} \frac{1}{m} \sum_{i=1}^{m} [\log D(x^{(i)}) + \log(1 - D(G(y^{(i)}, z^{(i)})))]$$

7:     **end for**
8:     **for** k_G steps **do**
9:         Generate minibatch of m randomly sampled noise $\{z^{(1)}, \ldots, z^{(m)}\}$
10:        Sample minibatch of m grayscale images $\{y^{(1)}, \ldots, y^{(m)}\}$
11:        Update the generator by:

$$\nabla_{\theta_g} \frac{1}{m} \sum_{i=1}^{m} \log(1 - D(G(y^{(i)}, z^{(i)})))$$

12:    **end for**
13: **end for**

Algorithm 2. Testing phase of our conditional GANs with the default parameters $m = 64, s_z = 100$.

1: Sample minibatch of m grayscale images $\{y^{(1)}, \ldots, y^{(m)}\}$
2: **for** round i in k_{test} rounds **do**
3:     Generate minibatch of m randomly sampled noise $\{z^{(1,i)}, \ldots, z^{(m,i)}\}$ each of size $[s_z]$.
4:     Generate color image using the trained model G:

$$\{x^{(1,i)}, \ldots, x^{(m,i)}\} \leftarrow G(\{y^{(1)}, \ldots, y^{(m)}\}, \{z^{(1,i)}, \ldots, z^{(m,i)}\})$$

5: **end for**
6: Rearrange generated results x into
$$\{x^{(1,1)}, \ldots, x^{(1,k_{\text{test}})}\}, \ldots, \{x^{(m,1)}, \ldots, x^{(m,k_{\text{test}})}\}$$

3.3   Training and Testing Procedure

The training phase of our conditional GANs is presented in Algorithm 1. To assure the BatchNorm layers to work correctly, one cannot feed an image batch of the same images to test various noise responses. Thus we use multi-round testing with same batch and rearrange them to test different noise responses of each image, which is described in Algorithm 2.

4   Experiments

4.1   Dataset

There are various kinds of color image datasets, and we choose the open LSUN bedroom dataset[2] [25] to conduct our experiment. LSUN is a large color image dataset generated iteratively by human labeling with automatic deep neural classification. It contains around one million labeled images for each of 10 scene categories and 20 object categories. Among them we choose indoor scene bedroom because it has enough samples (more than 3 million) and unlike outdoor scenes in which trees are almost always green and sky is always blue, items in indoor scenes like bedroom have various colors, as shown in Fig. 5. This is exactly what we need to fully explore the capability of our conditional GAN model. In experiments, we use $503,900$ bedroom images randomly picked from the LSUN dataset. We crop a maximum center square out of each image and reshape them into 64×64 pixel as preprocessing.

Fig. 5. Demonstration of LSUN dataset. Top: outdoor scene (church). Always blue sky and green trees. Bottom: indoor scene (bedroom). Various item colors. (Color figure online)

4.2   Comparison Experiments

YUV and RGB. The generated colorization results of a same grayscale image using YUV representation and RGB representation with additional $L1$ loss are shown in Fig. 6. Each group of images consists of 3×4 images generated from a same grayscale image by each model at a same epoch. Focus on the results in red boxes, we can see RGB representation suffers from structural miss due to the additional trade-off between $L1$ loss and the GAN loss. Take the enlarged image on the top right in Fig. 6 as an example, both the wall and the bed on the left are split by unnaturally white and orange colors, while the results of YUV setting acquire more smooth transitions. Moreover, the model using RGB representation

---

[2] LSUN dataset is available at http://lsun.cs.princeton.edu.

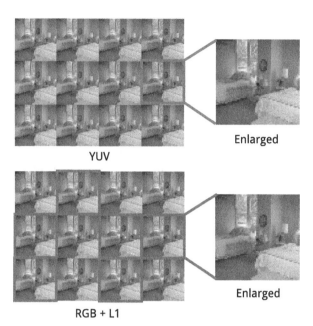

Fig. 6. Comparison of different color space representation. Top: training and testing with YUV representation. Bottom: training with RGB representation and $L1$ loss. Focus on the results in red boxes, RGB representation results lack of item continuity due to the additional trade-off between $L1$ loss and the GAN loss. (Color figure online)

shall predict 3 color channels while YUV representation only predicts 2 channels given the grayscale Y channel as fixed conditional information, which makes the YUV model training much more stable.

Single-layer and multi-layer noise. The generated colorization results of the same grayscale images using single-layer noise model and multi-layer noise model are shown in Fig. 7. Each group consists of 8×8 images generated from a grayscale image by each model at the same epoch. We can see from the results that multi-layer noise possesses our generator G of higher diversity as those results on the right in Fig. 7 are apparently more colorful.

Single-layer and multi-layer condition. The generated colorization results of the same grayscale images using single-layer conditional model and multi-layer conditional model are shown in Fig. 8. We show 2×5 images generated by single-layer condition setting and multi-layer condition setting at the same epoch. We can see from the results that the multi-layer condition model makes the generator more structural information and thus the results of multi-layer condition model are more stable while the single-layer conditional model suffers from colorization derivation like those images in red boxes in Fig. 8.

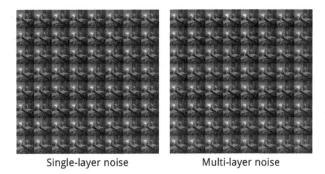

Single-layer noise   Multi-layer noise

Fig. 7. Comparison of single-layer and multi-layer noise model results. Left: results of single-layer noise model. Right: results of multi-layer noise model. Apparently multi-layer noise possesses our generator G of higher diversity.

Single-layer condition

Multi-layer condition

Fig. 8. Comparison of single-layer and multi-layer condition model result. Top: results of single-layer condition model, suffer from colorization derivation (red box). Bottom: results of multi-layer condition model, smooth transition. (Color figure online)

Wasserstein GAN. Three groups of colorization results of the same grayscale images using GAN and Wasserstein GAN are shown in Fig. 9. We can see from the result that Wasserstein GAN can produce comparable results as the first two column of Wasserstein GAN shows, but there are still failed results by Wasserstein GAN like the last column, even to note that the Wasserstein GAN results (40 epoch) come after training twice the number of epochs of the GAN results (20 epoch). This is mainly due to that our training LSUN bedroom dataset is quite large (503,900 image), the discriminator will not overfit easily, which prevents the gradient vanishing problem. And also because of the large dataset, the discriminator needs quite a lot times of optimization to convergence, not to mention Wasserstein GAN shall not use momentum based optimization strategies like Adam due to the non-linear parameter value clipping, Wasserstein GAN needs much longer training to produce comparable results as our model. Since Wasserstain GAN only helps to improve the stability of GAN training at a price of much longer training time and we have achieved results of good reality through our GAN, we did not use Wassserstein GAN structure.

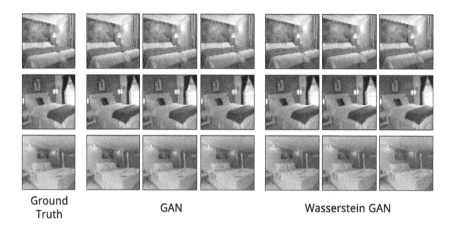

Ground
Truth   GAN   Wasserstein GAN

Fig. 9. Comparison between the results of GAN and Wasserstein GAN. Each line consists of the leftmost ground truth color image and three results by GAN, then three results by Wasserstein GAN. The rightmost images are failed results by Wasserstein GAN. (Color figure online)

More results and discussion of our final model will be shown in the next section.

5   Results and Evaluation

5.1   Colorization Results

A variety of image colorization results by our conditional GANs are provided in Fig. 10. Apparently, our fully convolutional (without stride) generator with multi-layer noise and multi-layer condition concatenation produces various kinds of colorization schemes while maintaining good reality. Almost all color parts are kept within correct components without deviation.

5.2   Evaluation via Human Study

Previous methods share a common goal to provide a color image closer to the original one. That is why many of their models [5, 13, 20] take image distance

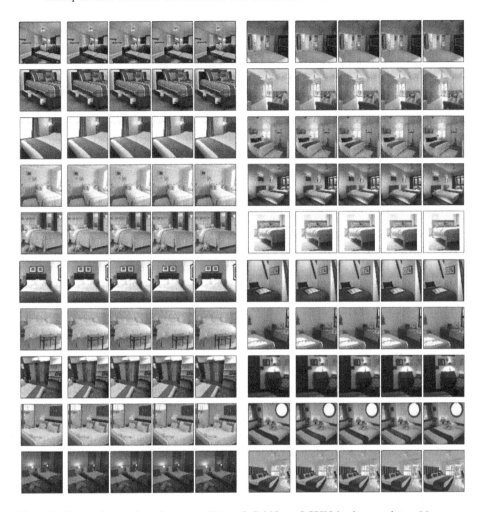

Fig. 10. Example results of our conditional GAN on LSUN bedroom data. 20 groups of images, each consists of the **leftmost** ground truth color image and 4 different colorizations generated by our conditional GANs given the grayscale version of the ground truth image. One can clearly see that our novel structure generator produces various colorization schemes while maintaining good reality. (Color figure online)

like RMSE (Root Mean Square Error) and PSNR (Peak Signal-to-Noise Ratio) as their measurements. And others [11,12] use additional classifiers to predict if colorized image can be detected or still correctly classified. But our goal is to generate diverse colorization schemes, so we cannot take those distance as our measurements as there exist reasonable colorizations that diverge a lot from the original color image. Note that some previous work on image colorization [3,7,18,19] does not provide quantified measurements.

Therefore, just like some previous researches [12, 26], we provide questionnaire surveys as a Turing test to measure our colorization results. We ask each of the total 80 participants 20 questions. In each question, we display 5 color images, one of which is the ground truth image, the others are our generated colorizations of the grayscale image of the ground truth, and ask them if any one of them is of poor reality. We add ground truth image among them as a reference in case of participants do not think any of them is real. And we arrange all images randomly to avoid any position bias for participants. The feedback from 80 participants indicates more than 62.6% of our generated color images are convincible while the rate of ground truth images is 70.0%. Furthermore, we run significance t-test between the ground truth and the generated images on the percentages of humans rating as real image for each question. The p-value is $0.1359 > 0.05$, indicating our generated results have no significant difference with the ground truth images. Also we calculate the credibility rank within each group of the ground truth image and the four corresponding generated images. An image gets higher rank if higher percentage of participants mark it real. And the average credibility rank of the ground truth images is only 2.5 out of 5, which means at least $(2.5-1)/(5-1) = 37.5\%$ of our generated results are even more convincible than true images.

6   Conclusion

In this paper, we proposed a novel solution to automatically generate diverse colorization schemes for a grayscale image while maintaining their reality by exploiting conditional generative adversarial networks which not only solved the sepia-toned problem of other models but also enhanced the colorization diversity. We introduced a novel generator architecture which consists of fully convolutional non-stride structure with multi-layer noise to enhance diversity and multi-layer condition concatenation to maintain reality. With this structure, our model successfully generated diversified high-quality color images for each input grayscale image. We performed a questionnaire survey as a Turing test to evaluate our colorization result. The feedback from 80 participants indicates our generated colorization results are highly convincible.

For future work, as so far we have investigated methods to generate color images by conditional GAN given only corresponding grayscale images, which provides the model maximum freedom to generate all kinds of colors, we can also lay additional constraints on the generator to guide the colorization procedure. Those conditions include but not limited to (i) specified item color, such as blue bed and white wall etc.; and (ii) global color scheme, such as warm tone or cool tone etc. And note that given those constraints, generative adversarial networks shall still produce various vivid colorizations.

References

1. Arjovsky, M., Chintala, S., Bottou, L.: Wasserstein GAN. CoRR abs/1701.07875 (2017)
2. Charpiat, G., Hofmann, M., Schölkopf, B.: Automatic image colorization via multimodal predictions. In: Forsyth, D., Torr, P., Zisserman, A. (eds.) ECCV 2008. LNCS, vol. 5304, pp. 126–139. Springer, Heidelberg (2008). https://doi.org/10.1007/978-3-540-88690-7_10
3. Cheng, Z., Yang, Q., Sheng, B.: Deep colorization. In: ICCV 2015, Santiago, Chile, 7–13 December 2015, pp. 415–423 (2015)
4. Chia, A.Y.S., Zhuo, S., Gupta, R.K., Tai, Y.W., Cho, S.Y., Tan, P., Lin, S.: Semantic colorization with internet images. ACM Trans. Graph. **30**(6), 156:1–156:8 (2011)
5. Deshpande, A., Lu, J., Yeh, M.C., Forsyth, D.A.: Learning diverse image colorization. CoRR abs/1612.01958 (2016)
6. Deshpande, A., Rock, J., Forsyth, D.A.: Learning large-scale automatic image colorization. In: ICCV 2015, Santiago, Chile, 7–13 December 2015, pp. 567–575 (2015)
7. Dong, H., Neekhara, P., Wu, C., Guo, Y.: Unsupervised image-to-image translation with generative adversarial networks. CoRR abs/1701.02676 (2017)
8. Goodfellow, I.J., Pouget-Abadie, J., Mirza, M., Xu, B., Warde-Farley, D., Ozair, S., Courville, A.C., Bengio, Y.: Generative adversarial nets. In: Advances in Neural Information Processing Systems 27: Annual Conference on Neural Information Processing Systems 2014, 8–13 December 2014, Montreal, Quebec, Canada, pp. 2672–2680 (2014). http://papers.nips.cc/paper/5423-generative-adversarial-nets
9. Gupta, R.K., Chia, A.Y.S., Rajan, D., Ng, E.S., Huang, Z.: Image colorization using similar images. In: Proceedings of the 20th ACM Multimedia Conference, MM 2012, Nara, Japan, 29 October–02 November 2012, pp. 369–378 (2012)
10. Hertzmann, A., Jacobs, C.E., Oliver, N., Curless, B., Salesin, D.: Image analogies. In: SIGGRAPH 2001, Los Angeles, California, USA, 12–17 August 2001, pp. 327–340 (2001)
11. Iizuka, S., Simo-Serra, E., Ishikawa, H.: Let there be color!: joint end-to-end learning of global and local image priors for automatic image colorization with simultaneous classification. ACM Trans. Graph. **35**(4), 110:1–110:11 (2016)
12. Isola, P., Zhu, J.Y., Zhou, T., Efros, A.A.: Image-to-image translation with conditional adversarial networks. CoRR abs/1611.07004 (2016)
13. Jung, M., Kang, M.: Variational image colorization models using higher-order mumford-shah regularizers. J. Sci. Comput. **68**(2), 864–888 (2016)
14. Koo, S.: Automatic colorization with deep convolutional generative adversarial networks (2016)
15. Levin, A., Lischinski, D., Weiss, Y.: Colorization using optimization. ACM Trans. Graph. **23**(3), 689–694 (2004)
16. Limmer, M., Lensch, H.P.A.: Infrared colorization using deep convolutional neural networks. In: ICMLA 2016, Anaheim, CA, USA, 18–20 December 2016, pp. 61–68 (2016)
17. Liu, X., Wan, L., Qu, Y., Wong, T.T., Lin, S., Leung, C.S., Heng, P.A.: Intrinsic colorization. ACM Trans. Graph. **27**(5), 152:1–152:9 (2008)
18. Nguyen, T.D., Mori, K., Thawonmas, R.: Image colorization using a deep convolutional neural network. CoRR abs/1604.07904 (2016)
19. Nguyen, V., Sintunata, V., Aoki, T.: Automatic image colorization based on feature lines. In: VISIGRAPP 2016, VISAPP, Rome, Italy, 27–29 February 2016, vol. 4, pp. 126–133 (2016)

20. Perarnau, G., van de Weijer, J., Raducanu, B., lvarez, J.M.A.: Invertible conditional gans for image editing. CoRR abs/1611.06355 (2016)
21. Qu, Y., Wong, T.T., Heng, P.A.: Manga colorization. ACM Trans. Graph. **25**(3), 1214–1220 (2006)
22. Reed, S.E., Akata, Z., Yan, X., Logeswaran, L., Schiele, B., Lee, H.: Generative adversarial text to image synthesis. In: ICML 2016, New York City, NY, USA, 19–24 June 2016, pp. 1060–1069 (2016)
23. Ronneberger, O., Fischer, P., Brox, T.: U-net: convolutional networks for biomedical image segmentation. In: Proceedings of the MICCAI 2015–18th International Conference Munich, Germany, 5–9 October 2015, Part III, pp. 234–241 (2015)
24. Welsh, T., Ashikhmin, M., Mueller, K.: Transferring color to greyscale images. In: SIGGRAPH 2002, San Antonio, Texas, USA, 23–26 July 2002, pp. 277–280 (2002)
25. Yu, F., Zhang, Y., Song, S., Seff, A., Xiao, J.: LSUN: construction of a large-scale image dataset using deep learning with humans in the loop. CoRR abs/1506.03365 (2015)
26. Zhang, R., Isola, P., Efros, A.A.: Colorful image colorization. In: Leibe, B., Matas, J., Sebe, N., Welling, M. (eds.) ECCV 2016. LNCS, vol. 9907, pp. 649–666. Springer, Cham (2016). https://doi.org/10.1007/978-3-319-46487-9_40

Ensembles and Meta Learning

Dynamic Ensemble Selection with Probabilistic Classifier Chains

Anil Narassiguin[1,2(✉)], Haytham Elghazel[1], and Alex Aussem[1]

[1] LIRIS UMR CNRS 5205, Université Lyon 1, 69622 Villeurbanne, France
{haytham.elghazel,aaussem}@univ-lyon1.fr
[2] EASYTRUST, 71 Boulevard National, 92250 La garenne colombes, France
anil.narassiguin@easytrust.com

Abstract. Dynamic ensemble selection (DES) is the problem of finding, given an input **x**, a subset of models among the ensemble that achieves the best possible prediction accuracy. Recent studies have reformulated the DES problem as a multi-label classification problem and promising performance gains have been reported. However, their approaches may converge to an incorrect, and hence suboptimal, solution as they don't optimize the true - but non standard - loss function directly. In this paper, we show that the label dependencies have to be captured explicitly and propose a DES method based on Probabilistic Classifier Chains. Experimental results on 20 benchmark data sets show the effectiveness of the proposed method against competitive alternatives, including the aforementioned multi-label approaches. This study is reproducible and the source code has been made available online (https://github.com/naranil/pcc_des).

Keywords: Dynamic ensemble selection · Multi-label learning
Probabilistic classifier chains

1 Introduction

The ubiquity of ensemble models in several interesting machine learning problems stems primarily from their potential to significantly increase prediction accuracy over individual classifier models [10,18,29]. Ensemble methods can be divided into two categories, depending on how they generate the committee of the classifiers. When the same classification algorithm is used to generate all the models of the ensemble, the ensemble method is called *homogeneous*, otherwise it is called *heterogeneous*. In the last decade, there has been a great deal of research focused on the problem of selecting good subensembles of base classifiers prior to combination in order to improve generalization and prediction efficiency.

The process of selecting a subset of classifiers is called *ensemble selection* or *ensemble pruning*. When the same subset of models is selected for all test instances, the process is referred to as *static selection* [14]. In that case, the simplest idea is to select the ensemble members from a set of individual classifiers

© Springer International Publishing AG 2017
M. Ceci et al. (Eds.): ECML PKDD 2017, Part I, LNAI 10534, pp. 169–186, 2017.
https://doi.org/10.1007/978-3-319-71249-9_11

that are subject to less resource consumption and response time with accuracy that performs at least as good as the original ensemble. A natural follow-up is to determine this subset dynamically, i.e. according to the current input feature **x**. This process is referred to as *dynamic ensemble selection* (DES).

Several DES methods have been recently proposed in the literature. A comprehensive coverage of *individual-based* and *group-based* DES methods is provided in [3]. In individual-based methods, the selection of a subset of models for each test instance is done by estimating the competence level of the base classifiers *individually*, that is, without taking their dependency structure of the model errors into account. *Group-based* methods make one step further by modeling the error co-occurrences.

As noted in [16,17,20], DES may be cast as a distinct special case of multi-label classification (MLC) problem with a specific zero-one error expressing the fact that at least, half of the base classifiers selected for inclusion of the subensemble should be correct for the overall class to be correct (i.e. *precision* > $1/2$, yes or no?). The question raised by these authors was: What should be the properties of the MLC algorithm to minimize this non-standard loss? This question was addressed from an experimental point of view only, pointing out that *precision* was found experimentally a good surrogate loss candidate for the success of DES. Yet, many loss functions has been proposed in the literature and it is now well understood that a MLC method performing optimally for one loss is likely to perform suboptimally for another loss [8]. For simple loss functions, analytic expressions of the Bayes (optimal) classifier can be derived. For example, the Hamming loss minimizer coincides with the marginal modes of the conditional distribution of the class labels given an instance. Conversely, for the subset $0/1$ loss, the risk minimizer is given by the joint mode of the conditional distribution, for which *individual-based* methods might not be good choices. For more complex multi-label loss functions like the one associated with the DES problem, the Bayes (optimal) classifier is unknown and the minimization of such losses requires more involved procedures. In this paper, we show that the minimization of the true loss function necessitates the modeling of dependencies between labels (i.e. co-occurrence of errors) and we use Probabilistic Classifier Chains (PCC), with Monte Carlo sampling, as a "plug-in rule approach" for optimizing this loss directly.

The rest of the paper is organized as follows: Sect. 2 reviews recent studies on DES and introduces our contribution in this context using MLC. Experiments using relevant benchmarks data sets are presented in Sect. 3. Finally, Sect. 4 concludes with a summary of our contributions and raises issues for future work.

2   Problem Statement and Contribution

In this section, we first survey and appraise the recent literature reporting the use of machine learning techniques devoted to the DES problem, giving some prominence to the use of multi-label classification methods. We then present our proposed DES approach based on Probabilistic Classifier Chains.

2.1   Dynamic Ensemble Selection (DES)

The key assumption on which DES methods hinge is that each model in the ensemble has distinct prediction abilities on different subsets of the input space. A criterion to measure the level of competence of a base classifier (*e.g.* accuracy) is needed. The literature reports several of DES methods, considers the classifiers either *individually* or in *groups* [3]. In the first category, the classifiers are selected based on their individual competence on the whole or on a local region of the feature space using a validation set. Most of the methods proposed for this purpose are based either on the nearest neighbors algorithm [12,26] or on clustering techniques [13]. Regarding the number of classifiers they select, these individual-based selection procedures are organized into two groups: *dynamic selection*, for the methods that only select the best classifier; and *dynamic combination*, for the methods that are not restricted in the number of classifiers they select.

In the (*group-based*) DES category, the selection procedures decide for the appropriate subset of the initial ensemble by taking into account the dependencies between the classification errors of the individual models. The most famous methods in this category are *meta-learning* based procedures. Recently, the authors in [6] proposed a DES "meta-learning" framework: instead of using only single criterion to estimate the competence level of the classifiers, several meta-features are used to capture distinct desirable "properties" characterizing the behavior of the base classifiers. These meta-features are extracted from the training data and used by the meta-classifier to decide whether a base classifier is competent on a given input sample **x**.

2.2   DES as a Multi-label Classification Problem

The DES problem has recently been reformulated as a multi-label classification (MLC) problem [16,17,20]. The multi-label training set is constructed on a validation set. The labels are 0–1 indicator random variables indicating whether the corresponding model has made an error on input sample **x**. The transformation process is illustrated in Table 1. This formulation allows us to cast the DES problem as a standard MLC problem, which can be efficiently solved using standard MLC techniques. The IBEP-MLC method in [16,17] was the first framework to use MLC approaches for DES: ML-KNN [27] and Calibrated Label Ranking (CLR) [11]. Significant improvements in accuracy have been reported using a heterogeneous ensemble of 200 classifiers. Another recent proposal, called CHADE (for CHAined Dynamic Ensemble) algorithm [20] is based the classifier chain (CC) technique [22]. This approach was evaluated on a bagging ensemble of 100 decision stumps using a large set of classification data sets.

However, the literature leaves open the question of deciding what MLC algorithm should work best, and more importantly how to exploit the dependencies between the labels, implicitly giving the misleading impression that any MLC method could solve the DES task. The benefit of exploiting label dependence is known to be closely depend on the type of loss to be minimized. Rather than

Table 1. Problem transformation

Validation set		Classifier predictions			Multi-label metabase		
$\mathbf{X}_{val}$	$\mathbf{Y}_{val}$	c_1	c_2	c_3	$\mathbf{X}_{val}$	$\hat{\mathbf{Y}}_{val}$	
$\mathbf{x}_1$	0	1	1	0	$\mathbf{x}_1$	0	0	1
$\mathbf{x}_2$	1	0	1	1	$\mathbf{x}_2$	0	1	1
...		...			...	...	
$\mathbf{x}_n$	0	0	1	0	$\mathbf{x}_n$	1	0	1

proposing yet another MLC algorithm, the aim of this paper is to elaborate more closely on the idea of exploiting label dependence to solve the DES task.

2.3 DES Loss Function

When the multi-label training set is constructed for an ensemble of classifiers $\Psi = \{\psi_1, \ldots ; \psi_n\}$, the goal is to output a subset $\Psi_{\mathbf{x}}$ of classifiers ($\Psi_{\mathbf{x}} \subset \Psi$) using a multi-label classifier for a given test instance $\mathbf{x}$. A natural question is what should be learned from the labels dependency structure to solve the DES task, and what is the appropriate loss function for training the MLC method to obtain a "good" subset of classifiers.

Let's denote the subset of classifiers that correctly classify $\mathbf{x}$ as $\Phi_{\mathbf{x}}$ and suppose that $\mathbf{h}_{\mathbf{x}} = (h_i)_{i=1}^n$ ($h_i \in \{0,1\}$) and $\mathbf{w}_{\mathbf{x}} = (w_i)_{i=1}^n$ ($w_i \in \{0,1\}$) are the binary representations for respectively $\Psi_{\mathbf{x}}$ and $\Phi_{\mathbf{x}}$, an intuitive way of obtaining a correct final prediction in a two-class classification task is to have at least 50% of the classifiers from $\Psi_{\mathbf{x}}$ to be in $\Phi_{\mathbf{x}}$ [16,17]. This condition can be written in different ways:

$$\frac{|\Psi_{\mathbf{x}} \cap \Phi_{\mathbf{x}}|}{|\Psi_{\mathbf{x}}|} > 0.5 \Leftrightarrow \frac{\mathbf{h}_{\mathbf{x}}.\mathbf{w}_{\mathbf{x}}}{\mathbf{h}_{\mathbf{x}}.\mathbf{h}_{\mathbf{x}}} > 0.5 \Leftrightarrow \frac{\sum\limits_{i=1}^{n} h_i.w_i}{\sum\limits_{i=1}^{n} h_i} > 0.5$$

This yields the following actual loss function (also referred to as task loss),

$$Task\_loss(\mathbf{h}_{\mathbf{x}}, \mathbf{w}_{\mathbf{x}}) = \begin{cases} 0, & \text{if } \dfrac{\mathbf{h}_{\mathbf{x}}.\mathbf{w}_{\mathbf{x}}}{\mathbf{h}_{\mathbf{x}}.\mathbf{h}_{\mathbf{x}}} > 0.5 \\ 1, & \text{otherwise.} \end{cases} = 1 - [[\dfrac{\mathbf{h}_{\mathbf{x}}.\mathbf{w}_{\mathbf{x}}}{\mathbf{h}_{\mathbf{x}}.\mathbf{h}_{\mathbf{x}}} > 0.5]] \quad (1)$$

Unfortunately, there is no closed-form of the Bayes optimal multi-label classifier, that is, a mapping $\mathbf{h}^*$ from the input features $\mathcal{X}$ to the labels $\mathcal{Y}$ that minimizes the expected loss (or risk) L of the model h, defined as:

$$R_L(\mathbf{h}) = \mathbb{E}_{\mathbf{XY}} L(\mathbf{Y}, \mathbf{h}(\mathbf{X})) = \sum_{\mathbf{x},\mathbf{y} \in \mathcal{X} \times \mathcal{Y}} P(\mathbf{x}, \mathbf{y}) L(\mathbf{y}, \mathbf{h}(\mathbf{x})) \quad (2)$$

The optimal classifier, $\mathbf{h}^*$, commonly referred to as Bayes classifier, minimizes the risk conditioned on $\mathbf{x}$: $\mathbf{h}^*(\mathbf{x}) = arg\,min_h \sum_{\mathbf{y} \in \mathcal{Y}} P(\mathbf{y}|\mathbf{x})L(\mathbf{y}, \mathbf{h}(\mathbf{x}))$. Finding $\mathbf{h}^*(\mathbf{x})$ directly by brute force search leads to intractable optimization problems and only very few loss functions have a (known) closed-form solution. For simple loss functions, analytic expressions of the Bayes optimal classifier have been derived in [8]. For example, the Hamming loss minimizer was shown to coincide with the marginal modes of the conditional distribution of the labels given an instance $\mathbf{x}$, and methods such as Binary Relevance (BR), perform particularly well in this case. Conversely, for the subset 0/1 loss, the risk minimizer was proven to be the joint mode of the conditional distribution, for which methods such as the Label Powerset classifier (LP) is a good choice. Further results have been established for the ranking loss [8], and more recently for the F-measure loss [7]. However, as far as we know, there is no closed-form expression of the Bayes classifier that minimizes the DES task loss. In such situations, the true loss is usually replaced by a surrogate loss that is easier to cope with.

2.4   MLC Approaches to the DES Problem

With the above difficulty in mind, Markatopoulou et al. [16,17], used the precision loss as surrogate loss:

$$Precision\_loss(\mathbf{h_x}, \mathbf{w_x}) = 1 - \frac{\mathbf{h_x}.\mathbf{w_x}}{\mathbf{h_x}.\mathbf{h_x}} = 1 - Pr(\mathbf{h_x}, \mathbf{w_x}) \qquad (3)$$

To solve the problem, two multi-label learning algorithms (ML-KNN [27] and CLR [11]) were used. Each algorithm outputs a score vector for each label. There were used in tandem with a thresholding strategy as an attempt to optimize the task loss. Despite the performance improvements reported, we shall see next that a method performing optimally for the precision loss may not perform well for the DES task loss, even upon tuning the threshold value. More problematic is the fact that the standard version of ML-KNN does not consider the correlation between labels and, as such, is devoted to minimize the Hamming loss (L_H) [8]:

$$L_H(\Psi_\mathbf{x}, \Phi_\mathbf{x}) = \frac{|(\Psi_\mathbf{x} \cap \Phi_\mathbf{x}) \cup (\overline{\Psi_\mathbf{x}} \cap \overline{\Phi_\mathbf{x}})|}{|\Psi|}, \quad L_H(\mathbf{h_x}, \mathbf{w_x}) = \frac{1}{n}\sum_{i=1}^{n}[[h_i = w_i]] \quad (4)$$

Tuning automatically the threshold via cross-validation was performed to overcome the theoretical shortcomings of the base MLC approaches. Clearly, choosing higher confidence thresholds for inclusion in the final pool tends to reduce the precision loss. Threshold values greater than 0.75 have been considered in their work.

In [20], the Classifier Chains (CC) [22] classifier was used to take the correlation between labels into account. However Dembczynski et al. [8] argued that CC is more appropriate for the subset 0/1 loss as it tends to approximate the joint mode of the conditional distribution of label vectors in a greedy manner. The 0/1 loss is given by:

$$L_{0/1}(\Psi_\mathbf{x}, \Phi_\mathbf{x}) = [[\forall \psi \in \Psi_\mathbf{x}, \psi \in \Phi_\mathbf{x}]], \quad L_{0/1}(\mathbf{h_x}, \mathbf{w_x}) = [[\mathbf{h_x} = \mathbf{w_x}]] \qquad (5)$$

Table 2. A DES example cast as a multi-label problem: different loss functions yield distinct minimizers.

y_1	y_2	y_3	y_4	$P(y_1, \ldots, y_4 \mid \mathbf{x})$
1	1	0	1	3/7
1	1	1	0	2/7
1	0	1	1	1/7
0	0	1	1	1/7

The above methods have several shortcomings. Consider the simple DES example in Table 2. The ensemble consists of 4 models, each having a mean accuracy exceeding 50%. The joint conditional distribution $P(y_1, \ldots, y_4 \mid \mathbf{x})$ is displayed.

It is easy to show that in this toy example, the optimal solution for the Hamming loss, 0/1 loss, DES task loss and Precision loss respectively are given by $\mathbf{h}^*_{hl} = (1, 1, 1, 1)$, $\mathbf{h}^*_{0/1} = (1, 1, 0, 1)$, $\mathbf{h}^*_{DEStaskloss} \in \{(0, 1, 1, 1), (1, 0, 1, 1)\}$ and $\mathbf{h}^*_{Precisionloss} = (1, 0, 0, 0)$. This illuminating toy example is important to caution the hurried researcher against using "off-the-shelf" MLC techniques to solve the DES problem. Indeed, IBEP-MLC which minimizes the Hamming loss implicitly, would select all the classifiers, whereas CHADE, based on CC that attempts to minimize the 0/1 loss, would output $\{c_1, c_2, c_4\}$. As may be observed, both methods fail to recover the optimal solution for the DES actual loss function, $\{c_2, c_3, c_4\}$ or $\{c_1, c_3, c_4\}$. It is also worth noting that the thresholding strategy based on the marginal label probabilities is unable cope with this problem. In fact, some information on the label dependency structure has to be captured to optimize the DES actual loss function. The following result shows that the precision loss tends to favor the best performing model,

Lemma 1. *The mapping* $\mathbf{h}(.) = (h_1(.), \ldots, h_n(.))$ *defined by:*

$$
\begin{cases}
h_k(\mathbf{x}) = 1, & k = \arg\max_{i \in \{1, \ldots, n\}} P(Y_i = 1 \mid \mathbf{x}). \\
h_j(\mathbf{x}) = 0, & j \neq k
\end{cases}
\tag{6}
$$

minimizes the expected precision score loss.

Proof. Minimizing the expected precision loss is equivalent to maximizing the expected precision which can easily be bounded above:

$$
\mathbb{E}_{\mathbf{Y}|\mathbf{X}} Pr(\mathbf{h}, \mathbf{Y}) = \sum_{\mathbf{y} \in \mathcal{Y}} P(\mathbf{y}|\mathbf{x}) \frac{\sum_{i=1}^{n} h_i \cdot y_i}{\sum_{i=1}^{n} h_i} = \frac{\sum_{i=1}^{n} h_i P(y_i = 1|\mathbf{x})}{\sum_{i=1}^{n} h_i} \leq \max_i P(y_i = 1|\mathbf{x})
$$

The mapping $\mathbf{h}(.)$ defined above reaches this bound and is thus Bayes optimal for the expected precision. This concludes the proof.

Therefore, picking the label having the highest confidence is a Bayes optimal solution to the MLC problem under the precision loss. However, we have just seen that on a toy problem that the best performing model is not always a good solution to the DES problem even if it is straightforward to identify. We may conclude that Precision loss is not a valid surrogate loss for this task. In this paper we focus on a general technique capable of minimizing the DES actual loss function based on a combination of Probabilistic Classifier Chains and Monte Carlo sampling. A similar approach was successfully applied to maximize the F-measure in [7]. This constitutes our main contribution.

2.5   Probabilistic Classifier Chains and Monte Carlo Inference

We have seen that some information on the joint conditional distribution $P(\mathbf{Y} \mid \mathbf{x})$ has to be captured to minimize the DES task loss. Brute-force search is however intractable as the number of possible labels permutations grows as $\mathcal{O}(2^n)$. One idea to cope with this issue is to infer a label combination probability in a step-wise manner using the chain rule of probability. Given a test instance $\mathbf{x}$, the joint conditional probability of the labels $\mathbf{y} = (y_1, \ldots, y_n)$ can be expressed by the chain rule of probability:

$$P_{\mathbf{x}}(\mathbf{y}) = P(\mathbf{y}|\mathbf{x}) = P(y_1|\mathbf{x}) \cdot \prod_{i=2}^{n} P(y_i|\mathbf{x}, y_1, \ldots y_{i-1}) \tag{7}$$

The rationale behind *Probabilistic Classifier Chains* [5] (PCC) is to estimate the joint conditional probability using this chain rule. PCC is the probabilistic counterpart of the Classifier Chain [22] algorithm. The method goes as follows: n probabilistic classifiers are used to estimate the probability distributions $P(y_i|\mathbf{x}, y_1, \ldots, y_{i-1})$ for each label $i = 1, \ldots, n$. Therefore, the ith classifier h_i is trained on a training data set composed of the original training data $\mathbf{X}_{tr}$ and $(y_{tr_1}, \ldots, y_{tr_{i-1}})$. While the training stage is rather straightforward, several approaches have been proposed in the literature for performing inference during the testing stage. CC is the simplest approach: each h_i predicts in sequential fashion the label y_i with the highest marginal conditional probability, taking as input the current input $\mathbf{x}$ and the previous predicted labels $(\hat{y}_1, \ldots, \hat{y}_{i-1})$. Therefore, CC may be regarded as a greedy approximation of PCC, focusing on the 0/1 loss minimization as the method estimates the mode of the joint distribution in a greedy fashion. In contrast, inference with PCC amounts to explore exhaustively the probability tree to estimate the Bayes optimal solution for any type of loss. This approach called Exhaustive Search (ES) estimates the true risk minimizer at the cost of extensive computation time since the tree diagram grows exponentially with n. Several methods have been proposed to reduce the computational burden of ES: ϵ-Approximation, Beam Search and Monte Carlo sampling (MC) (see for instance [19] and references therein for further details and experimental comparisons). However, ϵ-Approximation and Beam Search also tend to minimize of the 0/1 loss instead of the DES task loss. In this paper, we use Monte Carlo MC sampling technique [21] due to its ability to minimize arbitrary loss functions. The procedure is rather straightforward: given a new unlabeled instance $\mathbf{x}$, the labels are sampled in sequence, by taking the previously sampled labels $\hat{y}_1, \ldots, \hat{y}_i$ as input to the classifier h_i in order to estimate the marginal conditional probability of the next label y_{i+1}. Finally, the label combination $\hat{\mathbf{y}}_{pcc}$ that exhibits the lowest DES task loss value among the n_{MC} samples is chosen as the final prediction. Note that the DES task loss minimizer is estimated over a subset of n_{MC} samples drawn randomly instead of the whole set of possible labels, in order to keep the computational burden as low as possible. Once the n_{MC} samples are drawn, the search for the DES task loss minimizer requires $O(n_{MC}^2)$ further operations (calls to the loss functon) which can be prohibitive for large values of n_{MC}. Of course, the preference for smaller values of n_{MC} should be traded off against the prediction performance of the selected classifiers. In our experiments, we set $n_{MC} = 1000$. The PCC + Monte Carlo method applied to DES is termed PCC-DES in the sequel.

3   Experiments

In this section, we report on the experiments performed to evaluate the use of the proposed PCC-DES method on several data sets and we compare its predictive performance against other multi-label based DES methods. The following experiments were performed on 20 binary classification data sets primarily selected from the UCI Machine Learning Repository [2] and some other online repositories, covering a wide variety of topics including health, education, business, science etc., and exhibiting various dimensionalities as described in Table 3.

3.1   Ensemble Generation

In order to make fair comparisons, we used two ensemble generation techniques that appeared in the literature and investigated the performance of PCC-DES against other multi-label based DES techniques.

The *First (ensemble) generation* was used in [16,17]. An heterogeneous ensemble of 200 classifiers was constructed consisting of: (1) 40 *multilayer perceptrons* (**MLPs**) with $\{1, 2, 4, 8, 16\}$ hidden units, momentum varying in $\{0, 0.2, 0.5, 0.9\}$ and two learning rates: 0.3 and 0.6, (2) 60 *k nearest neighbors* (**kNNs**) with 20 values for k evenly distributed between 1 and the number of training observations, 3 weighting methods: no weights, inverse-weighting and similarity-weighting, (3) 80 *support vector machines* (**SVMs**) composed of 16 polynomial SVMs with a kernel of degree 2 and 3 and a complexity parameter C varying from 10^{-5} to 10^2 in steps of 10, and 64 radial SVMs with the same values of C and a width γ in $\{0.001, 0.005, 0.01, 0.05, 0.1, 0.5, 1, 2\}$, and (4) 20 *decision trees* (**DTs**), half of which are trained using Gini and half using entropy as split criteria; five values of the maximum depth pruning option 1, 2, 3, 4 and None, 8 decision trees using also Gini and entropy, varying the number of features to consider when looking for the best split (square root, log2, 50%

Table 3. Characteristics of the data sets used in the study

Datasets	# Instances	# Features	# Classes	Ref.
Adult	48842	14	2	[2]
AutoMoto	1980	2159	2	[23]
BaseHock	1993	4862	2	[23,28]
Breast cancer wisconsin (original)	699	9	2	[2]
Colic	368	27	2	[2]
Colon	62	2000	2	[1]
Credit Approval	690	15	2	[2]
EleCrypt	1973	2514	2	[23]
German credit	1000	24	2	[2]
GunMid	1847	2917	2	[23]
Hepatitis	155	19	2	[2]
Ionosphere	351	34	2	[2]
Chess (Krvskp)	3196	36	2	[2]
Madelon	2600	500	2	[2]
Ovarian	54	1536	2	[25]
PcMac	1943	3289	2	[23,28]
RelAthe	1427	4322	2	[23,28]
Connectionist bench (Sonar)	208	60	2	[2]
Spambase	4601	57	2	[2]
Congressional voting records (Vote)	435	16	2	[2]

and 100%) of the total number of features, and 2 decision trees using Gini and 2 values for the minimum number of samples per leaf 2, 3.

The *Second (ensemble) generation* was used in [4]. A pool of 200 heterogeneous models was constructed consisting of: (1) 50 *bagged trees* (**BAG-DTs**) using 25 trees for each splitting criterion (Gini and entropy), (2) 50 *random subspace trees* (**RSM-DTs**) consisting of 25 trees per splitting criterion, (3) 8 *Boosting decision trees* (**BST-DTs**) obtained by boosting a decision tree for each splitting criterion (Gini and entropy) and since Boosting can overfit, boosted DTs were added to the pool after 2, 4, 8, 16 steps of boosting, (4) 14 *Boosting stumps* (**BST-STMP**) obtained by boosting single level decision trees with both splitting criteria, each boosted 2, 4, 8, 16, 32, 64, 128 steps, (5) 24 *multilayer perceptrons* (**MLPs**) with $\{1, 2, 4, 8, 32, 128\}$ hidden units and a momentum varying in $\{0, 0.2, 0.5, 0.9\}$, and (6) 54 *support vector machines* (**SVMs**) composed of 6 linear SVMs with complexity parameter C varying from 10^{-3} to 10^2 in steps of 10, 48 radial SVMs with the same values of C and a width γ in $\{0.001, 0.005, 0.01, 0.05, 0.1, 0.5, 1, 2\}$.

These two strategies have many classifiers (**MLPs** and **SVMs**) in common. Yet, the *second generation* is expected to perform better as more powerful models (**BAG-DTs, RSM-DTs, BST-DTs, BST-STMP**) are generated. The overall mean error rate, averaged over the 20 data sets, is 0.340 with the *first generation* and 0.288 with the *second generation*. This should be kept in mind when analyzing the results.

3.2   Compared Methods and Evaluation Protocol

To gauge the practical relevance of our PCC-DES method, we compared its performance to four multi-label based DES methods in terms of accuracy improvements.

- BR-DES: Binary Relevance based DES method. BR resolves the MLC problem by training a classifier for each label separately. It is tailored for the Hamming loss [8].
- LP-DES: Label Powerset based DES method. LP reduces the MLC problem to multi-class classification, considering each label subset as a distinct meta-class. LP is tailored for the subset 0/1 loss [8].
- PM-DES: *Precision loss* minimizer based DES technique. As discussed in Sect. 2, this approach attempts to select the best classifier in the pool, given **x**.
- CHADE: CHAined Dynamic Ensemble algorithm [20]. It is based on the classifier chain (CC) technique. CC tailored for the subset 0/1 loss [8].
- BEST: the classifier with the highest accuracy in the validation data is selected (static method) [24].
- ENSEMBLE: the complete ensemble is classically used (baseline method).

Following [8], the logistic regression chosen as the base classifier of the MLC methods in our experiments. As noted earlier, a set of $n_{MC} = 1000$ samples was considered during the MC inference stage. The performance of the models was tested using a 5-fold cross-validation experiment. At each step of the cross-validation, 75% of the training data set was used to train the ensemble and the remaining 25% as a validation set to train the meta-learners for DES. This process was repeated 5 times for each DES method. The overall accuracy was computed by averaging over those 25 iterations.

3.3   Results and Discussion

The average accuracies of the compared methods for all 20 data sets using the first and the second generation strategies are reported respectively in Tables 4 and 5. We follow in this study the methodology proposed by [9] for the comparison of several algorithms over multiple data sets. In this study, the non-parametric Friedman test is firstly used to determine if there is a statistically significant difference between the rankings of the compared techniques. The Friedman test reveals here statistically significant differences ($p < 0.05$) for each ensemble generation strategy. Next, as recommended by Demsar [9], we perform the Nemenyi post hoc test with average rank diagrams. These diagrams are given on Fig. 1. The ranks are depicted on the axis, in such a manner that the best ranking algorithms are at the rightmost side of the diagram. The algorithms that do not differ significantly (at $p = 0.05$) are connected with a line. The critical difference (CD) is shown above the graph (CD $= 2.0139$ here). As may be observed from CD plots and the results in Tables 4 and 5 PCC-DES outperform the other models most of the time.

As far as the *first ensemble generation* is concerned (*c.f.* Table 4 and Fig. 1), the performances of PCC-DES are not statistically distinguishable from the performances of the single best classifier in the ensemble (BEST). As mentioned before, the first generation produces a pool containing several weak classifiers. Selecting the best single model from this pool yields remarkably good performance. The nonparametric statistical tests we used are very conservative. To further support these rank comparisons, we compared the 25 accuracy values obtained over each data set split for each pair of algorithms according to the paired t-test (with $p = 0.05$). The results of these pairwise comparisons are depicted in the last row of Table 4 in terms of "win/tie/loss" statuses of all methods against PCC-DES; the three values respectively indicate how times many the corresponding approach is significantly better/not significantly different/significantly worse than PCC-DES. Inspection of this win/tie/loss values reveals that DES using PCC (PCC-DES) is the only MLC-based DES method able to outperform the best single model BEST. The win/tie/loss values triples are statistically better with PCC-DES on 10 data sets, poorer on 1 data set only, and not significant on 8 data sets. Overall, PCC-DES compares more favorably to the other approaches, sometimes by a noticeable margin, in terms of accuracy.

Regarding the *second ensemble generation* strategy, here again PCC-DES outperforms the other algorithms, except BR-DES (*c.f.* Table 5 and Fig. 1). PCC-DES ranks first as well. Yet, it is not statistically better than BR-DES according the post hoc test. On the other hand, the win/tie/loss counts in Table 5 are statistically better for PCC-DES on 4 data sets and not significant on 16 data sets.

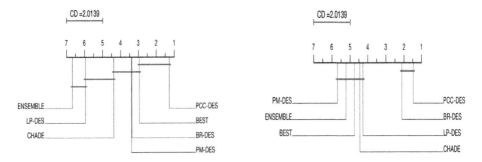

Fig. 1. Average rank diagrams of the compared DES methods using the first (left) and second (right) ensemble generation strategies.

Table 4. Means and standard deviations of accuracy for compared algorithms on the benchmark data sets with the *first ensemble generation* strategy

Data set	ENSEMBLE	PM-DES	BR-DES	LP-DES	CC-DES	PCC-DES	BEST
Adult	$0.752 \pm 0.06^{\bullet}$	$0.781 \pm 0.04^{\bullet}$	$0.798 \pm 0.06^{\bullet}$	$0.755 \pm 0.06^{\bullet}$	$0.790 \pm 0.06^{\bullet}$	0.803 ± 0.04	$0.791 \pm 0.04^{\bullet}$
Auto moto	$0.631 \pm 0.16^{\bullet}$	$0.872 \pm 0.04^{\bullet}$	$0.852 \pm 0.04^{\bullet}$	$0.774 \pm 0.06^{\bullet}$	$0.818 \pm 0.06^{\bullet}$	0.902 ± 0.05	$0.845 \pm 0.04^{\bullet}$
BaseHock	$0.643 \pm 0.19^{\bullet}$	$0.911 \pm 0.02^{\bullet}$	$0.867 \pm 0.07^{\bullet}$	$0.808 \pm 0.06^{\bullet}$	$0.824 \pm 0.11^{\bullet}$	0.933 ± 0.03	$0.912 \pm 0.03^{\bullet}$
Breast-cancer	$0.960 \pm 0.02^{\bullet}$	0.965 ± 0.02	0.970 ± 0.02	$0.961 \pm 0.02^{\bullet}$	0.970 ± 0.02	0.970 ± 0.02	0.968 ± 0.02
Colic	$0.678 \pm 0.03^{\bullet}$	0.812 ± 0.05	$0.737 \pm 0.05^{\bullet}$	$0.709 \pm 0.05^{\bullet}$	$0.735 \pm 0.05^{\bullet}$	0.822 ± 0.04	0.821 ± 0.06
Colon	$0.684 \pm 0.20^{\bullet}$	0.781 ± 0.13	0.794 ± 0.15	$0.774 \pm 0.16^{\bullet}$	0.791 ± 0.17	0.813 ± 0.14	0.779 ± 0.15
Credit approval	$0.828 \pm 0.06^{\bullet}$	$0.852 \pm 0.03^{\bullet}$	0.871 ± 0.03	$0.831 \pm 0.05^{\bullet}$	0.870 ± 0.03	0.872 ± 0.04	0.866 ± 0.03
Elecrypt	$0.774 \pm 0.23^{\bullet}$	$0.909 \pm 0.02^{\bullet}$	$0.882 \pm 0.05^{\bullet}$	$0.818 \pm 0.07^{\bullet}$	$0.833 \pm 0.10^{\bullet}$	0.938 ± 0.02	$0.918 \pm 0.03^{\bullet}$
German credit	$0.700 \pm 0.04^{\bullet}$	$0.727 \pm 0.05^{\bullet}$	0.736 ± 0.05	$0.722 \pm 0.04^{\bullet}$	$0.724 \pm 0.04^{\bullet}$	0.745 ± 0.05	0.733 ± 0.05
Gunmid	$0.582 \pm 0.11^{\bullet}$	$0.768 \pm 0.04^{\bullet}$	$0.756 \pm 0.05^{\bullet}$	$0.715 \pm 0.05^{\bullet}$	$0.738 \pm 0.06^{\bullet}$	0.806 ± 0.04	$0.784 \pm 0.04^{\bullet}$
Hepatitis	$0.794 \pm 0.13^{\bullet}$	0.806 ± 0.11	$0.795 \pm 0.13^{\bullet}$	0.795 ± 0.13	$0.795 \pm 0.13^{\bullet}$	0.808 ± 0.12	0.815 ± 0.12
Ionosphere	$0.641 \pm 0.19^{\bullet}$	0.909 ± 0.05	$0.766 \pm 0.15^{\bullet}$	$0.661 \pm 0.19^{\bullet}$	$0.765 \pm 0.14^{\bullet}$	0.919 ± 0.04	$0.927 \pm 0.05^{\circ}$
Krvskp	$0.662 \pm 0.12^{\bullet}$	$0.946 \pm 0.02^{\bullet}$	$0.916 \pm 0.05^{\bullet}$	$0.801 \pm 0.09^{\bullet}$	$0.912 \pm 0.06^{\bullet}$	0.966 ± 0.02	$0.952 \pm 0.03^{\bullet}$
Madelon	$0.501 \pm 0.05^{\bullet}$	0.584 ± 0.05	0.574 ± 0.04	$0.546 \pm 0.04^{\bullet}$	0.563 ± 0.04	0.590 ± 0.05	0.580 ± 0.06
Ovarian	$0.369 \pm 0.37^{\bullet}$	0.778 ± 0.15	0.833 ± 0.09	$0.745 \pm 0.13^{\bullet}$	0.833 ± 0.10	0.823 ± 0.04	0.771 ± 0.16
PcMac	$0.602 \pm 0.15^{\bullet}$	$0.838 \pm 0.03^{\bullet}$	$0.802 \pm 0.07^{\bullet}$	$0.725 \pm 0.06^{\bullet}$	$0.759 \pm 0.11^{\bullet}$	0.882 ± 0.03	$0.836 \pm 0.03^{\bullet}$
Relathe	$0.562 \pm 0.06^{\bullet}$	$0.849 \pm 0.04^{\bullet}$	$0.801 \pm 0.06^{\bullet}$	$0.789 \pm 0.06^{\bullet}$	$0.674 \pm 0.07^{\bullet}$	0.888 ± 0.03	$0.855 \pm 0.04^{\bullet}$
Sonar	$0.093 \pm 0.18^{\bullet}$	0.438 ± 0.13	$0.298 \pm 0.14^{\bullet}$	$0.220 \pm 0.19^{\bullet}$	$0.303 \pm 0.14^{\bullet}$	0.465 ± 0.12	$0.396 \pm 0.14^{\bullet}$
Spambase	$0.756 \pm 0.06^{\bullet}$	0.890 ± 0.03	$0.854 \pm 0.03^{\bullet}$	$0.754 \pm 0.06^{\bullet}$	$0.812 \pm 0.05^{\bullet}$	0.898 ± 0.03	$0.883 \pm 0.03^{\bullet}$
Vote	0.945 ± 0.04	0.928 ± 0.05	0.940 ± 0.05	$0.930 \pm 0.05^{\bullet}$	0.939 ± 0.05	0.945 ± 0.04	0.938 ± 0.05
(Win/tie/loss)	0/0/19	0/9/10	0/6/13	0/0/20	0/5/14		1/8/10

Table 5. Means and standard deviations of accuracy for compared algorithms on the benchmark data sets with the *second ensemble generation* strategy

Data set	ENSEMBLE	PM-DES	BR-DES	LP-DES	CC-DES	PCC-DES	BEST
Adult	0.950 ± 0.03	0.947 ± 0.04	0.952 ± 0.03	0.948 ± 0.03	0.950 ± 0.03	0.952 ± 0.03	0.952 ± 0.03
Automoto	0.878 ± 0.05•	0.859 ± 0.04•	0.905 ± 0.04	0.879 ± 0.05•	0.893 ± 0.04•	0.908 ± 0.04	0.856 ± 0.04•
BaseHock	0.883 ± 0.06•	0.904 ± 0.03•	0.921 ± 0.04	0.903 ± 0.03•	0.868 ± 0.04•	0.930 ± 0.03	0.898 ± 0.04•
Breast-Cancer	0.963 ± 0.02	0.951 ± 0.03•	0.966 ± 0.02	0.964 ± 0.02	0.963 ± 0.02	0.964 ± 0.02	0.942 ± 0.03•
Colic	0.825 ± 0.04•	0.808 ± 0.05•	0.832 ± 0.03•	0.825 ± 0.05•	0.823 ± 0.04•	0.847 ± 0.04	0.807 ± 0.05•
Colon	0.784 ± 0.15•	0.765 ± 0.14•	0.846 ± 0.13	0.854 ± 0.12	0.842 ± 0.12	0.844 ± 0.11	0.797 ± 0.14•
Credit Approval	0.898 ± 0.03	0.858 ± 0.03•	0.902 ± 0.02	0.877 ± 0.03•	0.902 ± 0.02	0.905 ± 0.02	0.874 ± 0.03•
Elecrypt	0.899 ± 0.03•	0.899 ± 0.03•	0.917 ± 0.03	0.911 ± 0.02•	0.897 ± 0.03•	0.922 ± 0.02	0.912 ± 0.02
German Credit	0.722 ± 0.05•	0.696 ± 0.04•	0.744 ± 0.05	0.735 ± 0.04	0.731 ± 0.05•	0.748 ± 0.04	0.717 ± 0.05•
Gunmid	0.747 ± 0.05•	0.747 ± 0.04•	0.807 ± 0.05	0.772 ± 0.04•	0.780 ± 0.05•	0.806 ± 0.04	0.776 ± 0.05•
Hepatitis	0.815 ± 0.11	0.790 ± 0.11•	0.823 ± 0.10	0.818 ± 0.10	0.812 ± 0.11	0.831 ± 0.09	0.788 ± 0.15•
Ionosphere	0.910 ± 0.06•	0.891 ± 0.05•	0.910 ± 0.06•	0.907 ± 0.06•	0.910 ± 0.06•	0.920 ± 0.05	0.891 ± 0.07•
Krvskp	0.952 ± 0.03•	0.954 ± 0.02	0.960 ± 0.02	0.956 ± 0.02	0.953 ± 0.02	0.959 ± 0.03	0.958 ± 0.02
Madelon	0.548 ± 0.05•	0.540 ± 0.05•	0.592 ± 0.04	0.563 ± 0.05•	0.573 ± 0.05•	0.599 ± 0.04	0.553 ± 0.05•
Ovarian	0.762 ± 0.15•	0.740 ± 0.15•	0.841 ± 0.08	0.738 ± 0.15•	0.820 ± 0.08•	0.845 ± 0.07	0.764 ± 0.14•
PcMac	0.828 ± 0.03•	0.847 ± 0.03•	0.886 ± 0.02	0.836 ± 0.03•	0.859 ± 0.04•	0.894 ± 0.02	0.847 ± 0.04•
Relathe	0.815 ± 0.05•	0.850 ± 0.03•	0.863 ± 0.04•	0.844 ± 0.04•	0.830 ± 0.05•	0.879 ± 0.03	0.867 ± 0.05
Sonar	0.323 ± 0.13•	0.477 ± 0.11°	0.382 ± 0.09•	0.392 ± 0.08	0.340 ± 0.12•	0.415 ± 0.08	0.467 ± 0.13°
Spambase	0.900 ± 0.02	0.886 ± 0.03•	0.903 ± 0.02	0.900 ± 0.02	0.898 ± 0.02	0.906 ± 0.02	0.882 ± 0.03•
Vote	0.950 ± 0.03	0.947 ± 0.04	0.952 ± 0.03	0.948 ± 0.03	0.950 ± 0.03	0.952 ± 0.03	0.952 ± 0.03
(Win/tie/loss)	0/6/14	1/3/16	0/16/4	0/9/11	0/8/12		1/5/14

For a better understanding of the behavior of PCC-DES in comparison with the others DES approaches, we explored in the sequel the relation between the diversity-accuracy of the ensemble and the performance of the dynamic ensemble selection. To measure the diversity within the ensemble, we consider the kappa metric (κ) used in [15]. κ evaluates the level of agreement between two classifier outputs. The plots in Fig. 2 are representative examples of the effects of individual classifier average error and diversity (respectively) on the ability of DES methods for accuracy improvement under the *first generation* and *second generation* strategies.

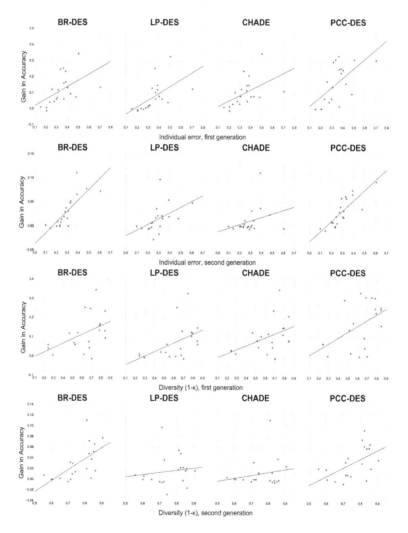

Fig. 2. Gain in accuracy of PCC-DES over the other DES methods vs. individual classifier average error (top plots) and diversity ($1 - \kappa$, lower plots) with the first and second ensemble generation strategies.

Table 6. Average number of classifiers selected by DES methods for the first and second ensemble generation strageies.

Data set	First generation				Second generation			
	BR-DES	LP-DES	CHADE	PCC-DES	BR-DES	LP-DES	CHADE	PCC-DES
Adult	187 ± 40	200 ± 6	185 ± 49	27 ± 20	193 ± 16	189 ± 27	196 ± 21	63 ± 43
Auto moto	125 ± 36	122 ± 40	118 ± 36	106 ± 19	161 ± 23	138 ± 35	165 ± 27	136 ± 30
BaseHock	139 ± 42	128 ± 46	130 ± 44	107 ± 24	164 ± 24	140 ± 38	168 ± 27	137 ± 24
Breast-cancer	176 ± 33	182 ± 30	177 ± 32	159 ± 47	190 ± 10	191 ± 9	191 ± 10	164 ± 45
Colic	160 ± 54	172 ± 47	161 ± 54	84 ± 44	164 ± 31	140 ± 55	183 ± 30	107 ± 42
Colon	172 ± 34	153 ± 49	172 ± 35	161 ± 39	154 ± 38	144 ± 45	155 ± 38	145 ± 32
Credit approval	159 ± 37	166 ± 40	161 ± 38	110 ± 63	173 ± 23	153 ± 39	178 ± 25	111 ± 43
Elecrypt	135 ± 40	128 ± 46	135 ± 41	102 ± 42	167 ± 20	140 ± 45	181 ± 20	122 ± 38
German credit	166 ± 64	187 ± 43	169 ± 68	21 ± 12	171 ± 42	145 ± 56	179 ± 52	51 ± 33
Gunmid	135 ± 36	120 ± 37	129 ± 37	90 ± 33	149 ± 24	124 ± 35	150 ± 36	82 ± 19
Hepatitis	195 ± 20	194 ± 30	196 ± 21	94 ± 61	195 ± 10	159 ± 53	198 ± 1	126 ± 60
Ionosphere	173 ± 44	188 ± 22	172 ± 48	39 ± 17	191 ± 15	187 ± 25	196 ± 8	121 ± 63
krvskp	144 ± 46	151 ± 49	149 ± 45	82 ± 24	167 ± 18	154 ± 29	172 ± 21	151 ± 23
Madelon	115 ± 50	99 ± 59	116 ± 65	47 ± 27	102 ± 27	100 ± 32	112 ± 39	55 ± 9
Ovarian	125 ± 45	118 ± 44	121 ± 43	103 ± 37	161 ± 31	140 ± 27	162 ± 31	123 ± 25
PcMac	144 ± 33	120 ± 40	139 ± 33	100 ± 24	162 ± 23	131 ± 36	163 ± 24	125 ± 12
Relathe	154 ± 54	133 ± 58	170 ± 46	88 ± 17	170 ± 28	139 ± 39	185 ± 27	122 ± 23
Sonar	149 ± 46	149 ± 48	147 ± 52	65 ± 33	163 ± 31	142 ± 44	182 ± 29	86 ± 43
Spambase	172 ± 41	198 ± 13	180 ± 36	63 ± 28	189 ± 14	180 ± 29	198 ± 4	112 ± 36
Vote	168 ± 26	166 ± 31	168 ± 26	160 ± 40	185 ± 13	185 ± 16	186 ± 12	165 ± 32
Mean	152 ± 48	152 ± 52	153 ± 51	85 ± 50	168 ± 32	151 ± 44	175 ± 34	112 ± 49

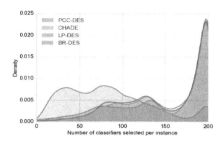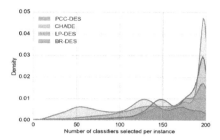

Fig. 3. Histogram of the number of classifiers selected per instance, by each DES method with the first and second ensemble generation strategies.

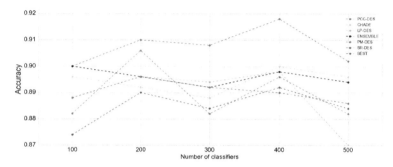

Fig. 4. Accuracy averaged over 20 data sets, as a function of the ensemble size.

A closer inspection of plots in this figure reveals the following: (1) not surprisingly, as the individual classifiers become less accurate (respectively more diverse), the dynamic ensemble selection becomes crucial for ensemble learning, (2) a significant accuracy gain was obtained with large values of errors (respectively diversity) with PCC-DES compared to the others MLC-based DES techniques, especially for ensemble models obtained using the *first generation* strategy.

In Table 6, the average number of models selected by BR-DES, LP-DES, CHADE and PCC-DES across all test instances and for all data sets is displayed. Our prime conclusion is that PCC-DES is a promising approach to DES. Concentrating on the actual DES task loss pays off in terms of performance. Compared to all others DES approaches, it appears that PCC-DES selects a far smaller number of models on average, especially with the first ensemble generation strategy containing weaker models as well (*c.f.* Fig. 3). We also plotted in Fig. 4 the overall accuracy on the 20 data sets as a function of the size of the ensemble, varying from 100 to 500. This confirms that our conclusions are rather insensitive to the size of the original ensemble. The running times are not shown here due to space restrictions. We would like to stress again that the MC inference step with PCC-DES requires $O(n_{MC}^2)$ calls to the loss function. On Madelon, for instance, PCC-DES takes about 90 s to label as single test example. This computational overhead prevents PCC-DES from being used in real-time.

4   Conclusion

In this work, we reformulated the dynamic ensemble selection (DES) problem as a multi-label classification problem and derived the actual multi-label loss associated to the DES problem. Contrary to other approaches that use state-of-art multi-label classification methods, we addressed the problem of optimizing the non-standard actual loss directly, since an analytic expression (or characterization) of the Bayes classifier that minimizes the actual DES loss is missing. We showed that the dependencies of the errors made by each model in the ensemble have to be exploited to optimize this loss. As the problem is intractable for realistic ensemble sizes, we discussed a more sophisticated multi-label procedure based on Probabilistic Classifier Chains and Monte Carlo sampling capable that allows to minimize the actual loss function directly. The experimental results on 20 benchmark data sets demonstrated the effectiveness of the proposed method against competitive alternatives using standard "off-the-shelf" multi-label learning techniques. Our experimental results show that optimizing the actual DES loss pays off in terms of performance. Compared to all others DES approaches, the proposed method was found to select a significantly smaller number of models, especially in the presence of many weak models. Future work should aim to characterize the Bayes classifier for the actual DES loss in order to reduce the computational burden of the training phase and to increase the performance further.

References

1. Alon, U., Barkai, N., et al.: Broad patterns of gene expression revealed by clustering analysis of tumor and normal colon tissues probed by oligonucleotide arrays. Natl. Acad. Sci. **96**, 6745–6750 (1999)
2. Blake, C.L., Merz, C.J.: UCI repository of machine learning databases (1998)
3. Britto Jr., A.S., Sabourin, R., Oliveira, L.E.S.: Dynamic selection of classifiers-a comprehensive review. Pattern Recogn. **47**(11), 3665–3680 (2014)
4. Caruana, R., Niculescu-Mizil, A., Crew, G., Ksikes, A.: Ensemble selection from libraries of models. In: ICML (2004)
5. Cheng, W., Hüllermeier, E., Dembczynski, K.J.: Bayes optimal multilabel classification via probabilistic classifier chains. In: Proceedings of the 27th International Conference on Machine Learning (ICML 2010), pp. 279–286 (2010)
6. Cruz, R.M.O., Sabourin, R., Cavalcanti, G.D.C., Ren, T.I.: META-DES: a dynamic ensemble selection framework using meta-learning. Pattern Recogn. **48**(5), 1925–1935 (2015)
7. Dembczynski, K., Jachnik, A., et al.: Optimizing the F-measure in multi-label classification: plug-in rule approach versus structured loss minimization. In: ICML, vol. 28, pp. 1130–1138 (2013)
8. Dembczyński, K., Waegeman, W., Cheng, W., Hüllermeier, E.: On label dependence and loss minimization in multi-label classification. Mach. Learn. **88**(1), 5–45 (2012)
9. Demšar, J.: Statistical comparisons of classifiers over multiple data sets. J. Mach. Learn. Res. **7**, 1–30 (2006)

10. Dietterich, T.G.: Ensemble methods in machine learning. In: First International Workshop on Multiple Classifier Systems, pp. 1–15 (2000)
11. Fürnkranz, J., Hüllermeier, E., Mencía, E.L., Brinker, K.: Multilabel classification via calibrated label ranking. Mach. Learn. **73**(2), 133–153 (2008)
12. Ko, A.H.R., Sabourin, R., Britto Jr., A.S.: K-nearest oracle for dynamic ensemble selection. In: ICDAR, pp. 422–426 (2007)
13. Kuncheva, L.I.: Clustering-and-selection model for classifier combination. In: KES, pp. 185–188 (2000)
14. Li, N., Yu, Y., Zhou, Z.-H.: Diversity regularized ensemble pruning. In: Flach, P.A., De Bie, T., Cristianini, N. (eds.) ECML PKDD 2012. LNCS (LNAI), vol. 7523, pp. 330–345. Springer, Heidelberg (2012). https://doi.org/10.1007/978-3-642-33460-3_27
15. Margineantu, D.D., Dietterich, T.G.: Pruning adaptive boosting. In: ICML, pp. 211–218 (1997)
16. Markatopoulou, F., Tsoumakas, G., et al.: Instance-based ensemble pruning via multi-label classification. In: ICTAI (2010)
17. Markatopoulou, F., Tsoumakas, G., Vlahavas, I.P.: Dynamic ensemble pruning based on multi-label classification. Neurocomputing **150**, 501–512 (2015)
18. Martínez-Muñoz, G., Hernández-Lobato, D., Suárez, A.: An analysis of ensemble pruning techniques based on ordered aggregation. IEEE Trans. Pattern Anal. Mach. Intell. **31**(2), 245–259 (2009)
19. Mena, D., Quevedo, J.R., Montañés, E., del Coz, J.J.: A heuristic in a* for inference in nonlinear probabilistic classifier chains. Knowl.-Based Syst. **126**, 78–90 (2017)
20. Pinto, F., Soares, C., Mendes-Moreira, J.: CHADE: metalearning with classifier chains for dynamic combination of classifiers. In: Frasconi, P., Landwehr, N., Manco, G., Vreeken, J. (eds.) ECML PKDD 2016. LNCS (LNAI), vol. 9851, pp. 410–425. Springer, Cham (2016). https://doi.org/10.1007/978-3-319-46128-1_26
21. Read, J., Martino, L., Luengo, D.: Efficient monte carlo methods for multidimensional learning with classifier chains. Pattern Recogn. **47**(3), 1535–1546 (2014)
22. Read, J., Pfahringer, B., Holmes, G., Frank, E.: Classifier chains for multi-label classification. Mach. Learn. **85**(3), 333–359 (2011)
23. Rennie, J.: Newsgroups data set, sorted by date (2000)
24. Ruta, D., Gabrys, B.: Classifier selection for majority voting. Inf. Fusion **6**(1), 63–81 (2005)
25. Schummer, M., Ng, W.L.V., Bumgarner, R.E., et al.: Comparative hybridization of an array of 21 500 ovarian cdnas for the discovery of genes overexpressed in ovarian carcinomas. Gene **238**(2), 375–385 (1999)
26. Woods, K., Kegelmeyer, W.P., Bowyer, K.: Combination of multiple classifiers using local accuracy estimates. IEEE Trans. Pattern Anal. Mach. Intell. **19**, 405–410 (1997)
27. Zhang, M.-L., Zhou, Z.-H.: ML-KNN: a lazy learning approach to multi-label learning. Pattern Recogn. **40**(7), 2038–2048 (2007)
28. Zhao, Z., Morstatter, F., et al.: Advancing feature selection research-ASU feature selection repository. School of Computing, Informatics, and Decision Systems Engineering, Arizona State University, Tempe (2010)
29. Zhou, Z.-H., Wu, J., Tang, W.: Ensembling neural networks: many could be better than all. Artif. Intell. **137**(1–2), 239–263 (2002)

Ensemble-Compression: A New Method for Parallel Training of Deep Neural Networks

Shizhao Sun[1(✉)], Wei Chen[2], Jiang Bian[2], Xiaoguang Liu[1], and Tie-Yan Liu[2]

[1] College of Computer and Control Engineering, Nankai University,
Tianjin, People's Republic of China
sunshizhao@mail.nankai.edu.cn, liuxg@nbjl.nankai.edu.cn
[2] Microsoft Research, Beijing, People's Republic of China
wche@microsoft.com, jiabia@microsoft.com, tyliu@microsoft.com

Abstract. Parallelization framework has become a necessity to speed up the training of deep neural networks (DNN) recently. Such framework typically employs the *Model Average* approach, denoted as MA-DNN, in which parallel workers conduct respective training based on their own local data while the parameters of local models are periodically communicated and averaged to obtain a global model which serves as the new start of local models. However, since DNN is a highly non-convex model, averaging parameters cannot ensure that such global model can perform better than those local models. To tackle this problem, we introduce a new parallel training framework called *Ensemble-Compression*, denoted as EC-DNN. In this framework, we propose to aggregate the local models by ensemble, i.e., averaging the outputs of local models instead of the parameters. As most of prevalent loss functions are convex to the output of DNN, the performance of ensemble-based global model is guaranteed to be at least as good as the average performance of local models. However, a big challenge lies in the explosion of model size since each round of ensemble can give rise to multiple times size increment. Thus, we carry out model compression after each ensemble, specialized by a distillation based method in this paper, to reduce the size of the global model to be the same as the local ones. Our experimental results demonstrate the prominent advantage of EC-DNN over MA-DNN in terms of both accuracy and speedup.

Keywords: Parallel machine learning · Distributed machine learning
Deep learning · Ensemble method

1 Introduction

Recent rapid development of deep neural networks (DNN) has demonstrated that its great success mainly comes from big data and big models [13, 25]. However, it is extremely time-consuming to train a large-scale DNN model over big data.

S. Sun—This work was done when the author was visiting Microsoft Research Asia.

M. Ceci et al. (Eds.): ECML PKDD 2017, Part I, LNAI 10534, pp. 187–202, 2017.
https://doi.org/10.1007/978-3-319-71249-9_12

To accelerate the training of DNN, parallelization frameworks like MapReduce [7] and Parameter Server [6,18] have been widely used. A typical parallel training procedure for DNN consists of continuous iterations of the following three steps. First, each worker trains the local model based on its own local data by stochastic gradient decent (SGD) or any of its variants. Second, the parameters of the local DNN models are communicated and aggregated to obtain a global model, e.g., by averaging the identical parameter of each local models [20,27]. Finally, the obtained global model is used as a new starting point of the next round of local training. We refer the method that performs the aggregation by averaging model parameters as *MA*, and the corresponding parallel implementation of DNN as *MA-DNN*.

However, since DNN is a highly non-convex model, the loss of the global model produced by MA cannot be guaranteed to be upper bounded by the average loss of the local models. In other words, the global model obtained by MA-DNN might even perform worse than any local model, especially when the local models fall into different local-convex domains. As the global model will be used as a new starting point of the successive iterations of local training, the poor performance of the global model will drastically slow down the convergence of the training process and further hurt the performance of the final model.

To tackle this problem, we propose a novel framework for parallel DNN training, called *Ensemble-Compression (EC-DNN)*, the key idea of which is to produce the global model by ensemble instead of MA. Specifically, the ensemble method aggregates local models by averaging their outputs rather than their parameters. Equivalently, the global model produced by ensemble is a larger network with one additional layer which takes the outputs of local models as inputs with weights as $1/K$, where K is the number of local models. Since most of widely-used loss functions for DNN (e.g., cross entropy loss, square loss, and hinge loss) are convex with respect to the output vector of the model, the loss of the global model produced by ensemble can be upper bounded by the average loss of the local models. Empirical evidence in [5,25] even show that the ensemble model of DNN, i.e., the global model, is usually better than any base model, i.e., the local model. According to previous theoretical and empirical studies [17,24], ensemble model tend to yield better results when there exists a significant diversity among local models. Therefore, we train the local models for a longer period for EC-DNN to increase the diversity among them. In other words, EC-DNN yields less communication frequency than MA-DNN, which further emphasizes the advantages of EC-DNN by reducing communication cost as well as increasing robustness to limited network bandwidth.

There is, however, no free lunch. In particular, the ensemble method will critically increase the model complexity: the resultant global model with one additional layer will be K times wider than any of the local models. Several ensemble iterations may result in explosion of the size of the global model. To address this problem, we further propose an additional compression step after the ensemble. This approach cannot only restrict the size of the resultant global model to be the same size as local ones, but also preserves the advantage of ensemble over MA. Given that both ensemble and compression steps are dispensable in our new parallelization framework, we name this framework as EC-DNN. As a specialization of the EC-DNN framework, we adopt the distillation based compression [1,14,22], which produces model compression by distilling the predictions of big models. Nevertheless, such distillation method requires extra time for training the compressed model. To tackle this problem, we seek to integrate the model compression into the process of local training by designing a new combination loss, a weighted interpolation between the loss based on the pseudo labels produced by global model and that based on the true labels. By optimizing such combination loss, we can achieve model compression in the meantime of local training.

We conducted comprehensive experiments on CIFAR-10, CIFAR-100, and ImageNet datasets, w.r.t. different numbers of local workers and communication frequencies. The experimental results illustrate a couple of important observations: (1) Ensemble is a better model aggregation method than MA consistently. MA suffers from that the performance of the global model could vary drastically and even be much worse than the local models; meanwhile, the global model obtained by the ensemble method can consistently outperform the local models. (2) In terms of the end-to-end results, EC-DNN stably achieved better test accuracy than MA-DNN in all the settings. (3) EC-DNN can achieve better performance than MA-DNN even when EC-DNN communicates less frequently than MA-DNN, which emphasizes the advantage of EC-DNN in training a large-scale DNN model as it can significantly reduce the communication cost.

2 Preliminary: Parallel Training of DNN

In the following of this paper, we denote a DNN model as $f(\mathbf{w})$ where $\mathbf{w}$ represents the parameters of this DNN model. In addition, we denote the outputs of the model $f(\mathbf{w})$ on the input x as $f(\mathbf{w}; x) = (f(\mathbf{w}; x, 1), \ldots, f(\mathbf{w}; x, C))$, where C is the number of classes and $f(\mathbf{w}; x, c)$ denotes the output (i.e., the score) for the c-th class. DNN is a highly non-convex model due to the non-linear activations and poolings after many layer.

In the parallel training of DNN, suppose that there are K workers and each of them holds a local dataset $D_k = \{(x_{k,1}, y_{k,1}), \ldots, (x_{k,m_k}, y_{k,m_k})\}$ with size m_k, $k \in \{1, \ldots, K\}$. Denote the weights of the DNN model at the iteration t on the worker k as $\mathbf{w}_k^t$. The communication between the workers is invoked after every τ iterations of updates for the weights, and we call τ the communication frequency. A typical parallel training procedure for DNN implements the following three steps in an iterative manner until the training curve converges.

1. Local training: At iteration t, worker k updates its local model by using SGD. Such local model is updated for τ iterations before the cross-machine synchronization.

2. Model aggregation: The parameters of local models are communicated across machines. Then, a global model is produced by aggregating local models according to certain aggregation method.

3. Local model reset: The global model is sent back to the local workers, and set as the starting point for the next round of local training.

We denote the aggregation method in the second step as $G(\mathbf{w}_1^t, \ldots, \mathbf{w}_K^t)$ and the weights of the global model as $\bar{\mathbf{w}}^t$. That is, $f(\bar{\mathbf{w}}^t) = G(\mathbf{w}_1^t, \ldots, \mathbf{w}_K^t)$, where $t = \tau, 2\tau, \cdots$. A widely-used aggregation method is *model average (MA)*, which averages each parameter over all the local models, i.e.,

$$G_{MA}\left(\mathbf{w}_1^t, \ldots, \mathbf{w}_K^t\right) = f\left(\frac{1}{K}\sum_{k=1}^{K}\mathbf{w}_k^t\right), t = \tau, 2\tau, \cdots. \tag{1}$$

We denote the parallel training method of DNN that using MA as MA-DNN for ease of reference.

With the growing efforts in parallel training for DNN, many previous studies [2,3,6,20,26,27] have paid attention to MA-DNN. NG-SGD [20] proposes an approximate and efficient implementation of Natural Gradient for SGD (NG-SGD) to improve the performance of MA-DNN. EASGD [26] improves MA-DNN by adding an elastic force which links the weights of the local models with the weights of the global model. BMUF [3] leverages data parallelism and blockwise model-update filtering to improve the speedup of MA-DNN. All these methods aim at solving different problems with us, and our method can be used with those methods simultaneously.

3 Model Aggregation: MA vs. Ensemble

In this section, we first reveal why the MA method cannot guarantee to produce a global model with better performance than local models. Then, we propose to use ensemble method to perform the model aggregation, which in contrast can ensure to perform better over local models.

MA was originally proposed for convex optimization. If the model is convex w.r.t. the parameters and the loss is convex w.r.t. the model outputs, the performance of the global model produced by MA can be guaranteed to be no worse than the average performance of local models. This is because, when $f(\cdot)$ is a convex model, we have,

$$\mathcal{L}\left(f\left(\bar{\mathbf{w}}^t; x\right), y\right) = \mathcal{L}\left(f\left(\frac{1}{K}\sum_{k=1}^{K}\mathbf{w}_k^t; x\right), y\right) \leq \mathcal{L}\left(\frac{1}{K}\sum_{k=1}^{K}f\left(\mathbf{w}_k^t; x\right), y\right). \tag{2}$$

Moreover, when the loss is also convex w.r.t. the model outputs $f(\cdot; x)$, we have,

$$\mathcal{L}\left(\frac{1}{K}\sum_{k=1}^{K}f\left(\mathbf{w}_k^t; x\right), y\right) \leq \frac{1}{K}\sum_{k=1}^{K}\mathcal{L}\left(f\left(\mathbf{w}_k^t; x\right), y\right). \tag{3}$$

By combining inequalities (2) and (3), we can see that it is quite effective to apply MA in the context of convex optimization, since the loss of the global model by MA is no greater than the average loss of local models in such context.

However, DNN is indeed a highly non-convex model due to the existence of activation functions and pooling functions (for convolutional layers). Therefore, the above properties of MA for convex optimization does not hold for DNN such that the MA method cannot produce any global model with guaranteed better performance than local ones. Especially, when the local models are in the neighborhoods of different local optima, the global model based on MA could be even worse than any of the local models. Furthermore, given that the global model is usually used as the starting point of the next round of local training, the performance of the final model could hardly be good if a global model in any round fails to achieve good performance. Beyond the theoretical analysis above, the experimental results reported in Sect. 5.3 and previous studies [3, 20] also revealed such problem.

While the DNN model itself is non-convex, we notice that most of widely-used loss functions for DNN are convex w.r.t. the model outputs (e.g., cross entropy loss, square loss, and hinge loss). Therefore, Eq. (3) holds, which indicates that averaging the output of the local models instead of their parameters guarantees to yield better performance than local models. To this end, we propose to *ensemble* the local models by averaging their outputs as follows,

$$G_E\left(\mathbf{w}_1^t, \ldots, \mathbf{w}_K^t\right) = \frac{1}{K} \sum_{k=1}^{K} f\left(\mathbf{w}_k^t; x\right), t = \tau, 2\tau, \cdots . \tag{4}$$

4 EC-DNN

In this section, we first introduce the EC-DNN framework, which employs ensemble for model aggregation. Then, we introduce a specific implementation of EC-DNN that adopts distillation for the compression. At last, we take some further discussions for the time complexity of EC-DNN and the comparison with traditional ensemble methods.

4.1 Framework

The details of EC-DNN framework is shown in Algorithm 1. Note that, in this paper, we focus on the synchronous case[1] within the MapReduce framework, but EC-DNN can be generalized into the asynchronous case and parameter server framework as well. Similar to other popular parallel training methods for DNN, EC-DNN iteratively conducts local training, model aggregation, and local model reset.

1. Local training: The local training process of EC-DNN is the same as that of MA-DNN, in which the local model is updated by SGD. Specifically, at iteration t, worker k updates its local model from $\mathbf{w}_k^t$ to $\mathbf{w}_k^{t+1}$ by minimizing the training loss using SGD, i.e., $\mathbf{w}_k^{t+1} = \mathbf{w}_k^t - \eta \Delta(\mathcal{L}(f(\mathbf{w}_k^t; x_k), y_k))$, where η is the learning rate, and $\Delta(\mathcal{L}(f(\mathbf{w}_k^t; x_k), y_k))$ is the gradients of the empirical loss $\mathcal{L}(f(\mathbf{w}_k^t; x_k), y_k)$ of the local model $f(\mathbf{w}_k^t)$ on one mini batch of the local dataset D_k. One local model will be updated for τ iterations before the cross-machine synchronization.

---

[1] As shown in [2], MA-DNN in synchronous case converges faster and achieves better test accuracy than that in asynchronous case.

2. Model aggregation: The goal of model aggregation is to communicate the parameters of local models, i.e., $\mathbf{w}_1^t \ldots \mathbf{w}_K^t$, across machines. To this end, a global model is produced by ensemble according to $G_E(\mathbf{w}_1^t, \ldots, \mathbf{w}_K^t)$ in Eq. (4), i.e., averaging the outputs of the local models. Equivalently, the global model produced by ensemble is a larger network with one additional layer, whose outputs consist of C nodes representing C classes, and whose inputs are those outputs from local models with weights as $1/K$, where K is the number of local models. Therefore, the weights of global model $\bar{\mathbf{w}}^t$ can be denoted as $\bar{\mathbf{w}}^t = \{\mathbf{w}_1^t, \ldots, \mathbf{w}_K^t, \frac{1}{K}\}$, $t = \tau, 2\tau, \ldots$

Note that such ensemble-produced global model (i.e., $f(\bar{\mathbf{w}}_t)$) is one layer deeper and K times wider than the local model (i.e., $f(\mathbf{w}_t^k)$). Therefore, continuous rounds of ensemble process will easily give rise to a global model with exploding size. To avoid this problem, we propose introducing a compression process (i.e., Compression($\mathbf{w}_k^t, \bar{\mathbf{w}}^t, D_k$) in Algorithm 1) after ensemble process, to compress the resultant global model to be the same size as those local models while preserving the advantage of the ensemble over MA. We denote the compressed model for the global model $\bar{\mathbf{w}}_t$ on the worker k as $\tilde{\mathbf{w}}_k^t$.

3. Local model reset: The compressed model will be set as the new starting point of the next round of local training, i.e., $\mathbf{w}_k^t = \tilde{\mathbf{w}}^t$ where $t = \tau, 2\tau, \cdots$.

At the end of the training process, EC-DNN will output K local models and we choose the model with the smallest training loss as the final one. Note that, we can also take the global model (i.e., the ensemble of K local models) as the final model if there are enough computation and storage resources for the test.

Algorithm 1. EC-DNN(D_k)

Randomly initialize $\mathbf{w}_k^0$ and set $t = 0$;
while *stop criteria are not satisfied* **do**
  $\quad \mathbf{w}_k^{t+1} \leftarrow \mathbf{w}_k^t - \eta \Delta(\mathcal{L}(f(\mathbf{w}_k^t; x_k), y_k))$;
  $\quad t \leftarrow t + 1$;
  $\quad$**if** τ *divides* t **then**
  $\quad\quad \bar{\mathbf{w}}^t \leftarrow \{\mathbf{w}_1^t, \ldots, \mathbf{w}_K^t, \frac{1}{K}\}$;
  $\quad\quad \tilde{\mathbf{w}}_k^t \leftarrow$ Compression($\mathbf{w}_k^t, \bar{\mathbf{w}}^t, D_k$);
  $\quad\quad \mathbf{w}_k^t \leftarrow \tilde{\mathbf{w}}_k^t$.

return $\mathbf{w}_k^t$

4.2   Implementations

Algorithm 1 contains two sub-problems that need to be addressed more concretely: (1) how to train local models that can benefit more to the ensemble model; (2) how to compress the global model without costing too much extra time.

Diversity Driven Local Training. In order to improve the performance of ensemble, it is necessary to generate diverse local models other than merely accurate ones [17,24]. Therefore, in the local training phase, i.e., the third line in Algorithm 1, we minimize both the loss on training data and the similarity between the local models, which we call *diversity regularized local training loss*. For the k-th worker, it is defined as follows,

$$\mathcal{L}_{LS}^k(f(\mathbf{w}_k; x_{k,i}), y_{k,i}) = \sum_{i=1}^{m_k} (\mathcal{L}(f(\mathbf{w}_k; x_{k,i}), y_{k,i}) + \alpha \mathcal{L}_{sim}(f(\mathbf{w}_k; x_{k,i}), \bar{z}_{k,i})), \quad (5)$$

where $\bar{z}_{y,i}$ is the average of the outputs of the latest compressed models for input $x_{k,i}$. In our experiments, the local training loss $\mathcal{L}$ takes the form of cross entropy, and the similarity loss $\mathcal{L}_{sim}$ takes the form of $-l_2$ distance. The smaller $\mathcal{L}_{sim}$ is, the farther the outputs of a local model is from the average of outputs of the latest compressed models, and hence the farther (or the more diverse) the local models are from (or with) each other.

Distillation Based Compression. In order to compress the global model to the one with the same size as the local model, we use distillation base compression method[2] [1,14,22], which obtains a compressed model by letting it mimic the predictions of the global model. In order to save the time for compression, in compression process, we minimize the weighted combination of the local training loss and the pure compression loss, which we call *accelerated compression loss*. For the k-th worker, it is defined as follows,

$$\mathcal{L}_{LC}^k(f(\mathbf{w}_k; x_{k,i}), y_{k,i}) = \sum_{i=1}^{m_k} (\mathcal{L}(f(\mathbf{w}_k; x_{k,i}), y_{k,i}) + \beta \mathcal{L}_{comp}(f(\mathbf{w}_k; x_{k,i}), \bar{y}_{k,i})), \quad (6)$$

where $\bar{y}_{k,i}$ is the output of the latest ensemble model for the input $x_{k,i}$. In our experiments, the local training loss $\mathcal{L}$ and the pure compression loss $\mathcal{L}_{comp}$ both take the form of cross entropy loss. By reducing the loss between $f(\mathbf{w}_k; x_{k,i})$ and the pseudo labels $\{\bar{y}_{k,i}; i \in [m_k]\}$, the compressed model can play the similar function as the ensemble model.

We denote the distillation based compression process as Compression$_{distill}$ $(\mathbf{w}_k^t, \bar{\mathbf{w}}^t, D_k)$, and show its details in Algorithm 2. First, on the k-th local worker, we construct a new training data $\hat{D}_k$ by relabeling the original dataset D_k with the pseudo labels produced by the global model, i.e., $\{\bar{y}_{k,i}, i \in [m_k]\}$. Specifically, when producing pseudo labels, we first produce the predictions of each local models respectively, and then average the predictions of all the local models. By this way, we keep using the same amount of GPU memory as MA-DNN throughout the training, because the big global model, which is K times larger than the local model, has never been established in GPU memory. Then, we optimize the accelerated compression loss $\mathcal{L}_{LC}^k$ in Eq. (6) by SGD for p iterations. We initialize the parameters of the compressed model $\tilde{\mathbf{w}}_k^t$ by the parameters of the latest local model $\mathbf{w}_k^t$ instead of random numbers. Finally, the obtained compressed model $\tilde{\mathbf{w}}_k^{t+p}$ is returned, which will be set as the new starting point of next round of the local training.

---

[2] Other algorithms for the compression [4,8–11,21] can also be used for the same purpose, but different techniques may be required in order to plug these compression algorithms into the EC-DNN framework.

We can find that min-
imizing the diversity regu-
larized loss $\mathcal{L}_{\mathrm{LS}}^k$ for local
training (Eq. (5)) and mini-
mizing the accelerated com-
pression loss $\mathcal{L}_{\mathrm{LC}}^k$ for compression (Eq. (6)) are two
opposite but complemen-
tary tasks. They need to
leverage information generated by each other into their
own optimization. Specif-
ically, the local training
phase leverages $\bar{z}_{k,i}$ based

Algorithm 2. Compression$_{\mathrm{distill}}(\mathbf{w}_k^t, \bar{\mathbf{w}}_k^t, D_k)$

for $j \in [m_k]$ **do**
  for $c \in [C]$ **do**
    $\bar{y}_{k,j,c} \leftarrow \frac{1}{K} \sum_{r=1}^K f(\mathbf{w}_r^t; x_{k,j}, c);$
    $\bar{y}_{k,j} = (\bar{y}_{k,j,1}, \ldots, \bar{y}_{k,j,C});$
$\hat{D}_k \leftarrow$
$\{(x_{k,1}, y_{k,1}, \bar{y}_{k,1}), \ldots, (x_{k,m_k}, y_{k,m_k}, \bar{y}_{k,m_k})\};$
Set $\tilde{\mathbf{w}}_k^t = \mathbf{w}_k^t$ and $i = 0;$
while $i \leq p$ **do**
  $\tilde{\mathbf{w}}_k^{t+i+1} \leftarrow \tilde{\mathbf{w}}_k^{t+i} - \eta \Delta(\mathcal{L}_{\mathrm{LC}}^k(f(\tilde{\mathbf{w}}_k^{t+i}; x_k), y_k));$
  $i \leftarrow i + 1;$
return $\tilde{\mathbf{w}}_k^{t+p}.$

---

on compressed model while the compression process uses $\bar{y}_{k,i}$ provided by local
models. Due to such structural duality, we take advantage of a new optimization framework, i.e. dual learning [12], to improve the performance of both tasks
simultaneously.

4.3   Time Complexity

We compare the time complexity of MA-DNN and EC-DNN from two aspects:

1. Communication time: Parallel DNN training process is usually sensitive to the
communication frequency τ. Different parallelization frameworks yield various
optimal τ. In particular, EC-DNN prefers larger τ compared to MA-DNN. Essentially, less frequent communication across workers can give rise to more diverse
local models, which will lead to better ensemble performance for EC-DNN. On
the other hand, much diverse local models may indicate greater probability that
local models are in the neighboring of different local optima such that the global
model in MA-DNN is more likely to perform worse than local ones. The poor
performance of the global model will significantly slow down the convergence
and harm the performance of the final model. Therefore, EC-DNN yields less
communication time than MA-DNN.

2. Computational time: According to the analysis in Sect. 4.2, EC-DNN does not
consume extra computation time for model compression since the compression
process has been integrated into the local training phase, as shown in Eq. (6).
Therefore, compared with MA-DNN, EC-DNN only requires additional time to
relabel the local data using the global model, which approximately equals to the
maximal time of the feed-forward propagation over the local dataset. We call
such extra time *"relabeling time"* for ease of reference. To limit the relabeling
time on large datasets, we choose to relabel a portion of the local data, denoted
as μ. Our experimental results in Sect. 5.3 will demonstrate that the relabeling
time can be controlled within a very small amount compared to the training
time of DNN. Therefore, EC-DNN can cost only a slightly more or roughly
equal computational time over MA-DNN.

Overall, compared to MA-DNN, EC-DNN is essentially more time-efficient as it can reduce the communication cost without significantly increasing computational time.

4.4   Comparison with Traditional Ensemble Methods

Traditional ensemble methods for DNN [5,25] usually first train several DNN models independently without communication and make ensemble of them in the end. We denote such method as E-DNN. E-DNN was proposed to improve the accuracy of DNN models by reducing variance and it has no necessity to train base models with parallelization framework. In contrast, EC-DNN is a parallel algorithm aiming at training DNN models faster without the loss of the accuracy by leveraging a cluster of machines.

Although E-DNN can be viewed as a special case of EC-DNN with only one final communication and no compression process, the intermediate communications in EC-DNN will make it outperform E-DNN. The reasons are as follows: (1) local workers has different local data, the communications during the training will help local models to be consensus towards the whole training data; (2) the local models of EC-DNN can be continuously optimized by compressing the ensemble model after each ensemble process. Then, another round of ensemble will result in more advantage for EC-DNN over E-DNN since the local models of EC-DNN has been much improved.

5   Experiments

5.1   Experimental Setup

Platform. Our experiments are conducted on a GPU cluster interconnected with an InfiniBand network, each machine of which is equipped with two Nvdia's K20 GPU processors. One GPU processor corresponds to one local worker.

Data. We conducted experiments on public datasets CIFAR-10, CIFAR-100 [16] and ImageNet (ILSVRC 2015 Classification Challenge) [23]. For all the datasets, each image is normalized by subtracting the per-pixel mean computed over the whole training set. The training images are horizontally flipped but not cropped, and the test data are neither flipped nor cropped.

Model. On CIFAR-10 and CIFAR-100, we employ NiN [19], a 9-layer convolutional network. On ImageNet, we use GoogLeNet [25], a 22-layer convolutional network. We used the same tricks, including random initialization, l_2-regularization, Dropout, and momentum, as the original paper. All the experiments are implemented using Caffe [15].

Parallel Setting. On experiments on CIFAR-10 and CIFAR-100, we explore the number of workers $K \in \{4, 8\}$ and the communication frequency $\tau \in \{1, 16,$

$2000, 4000\}$ for both MA-DNN and EC-DNN. On experiments on ImageNet, we explore $K \in \{4, 8\}$ and $\tau \in \{1, 1000, 10000\}$. The communication across local workers is implemented using MPI.

Hyperparameters Setting of EC-DNN. There are four hyperparameters in EC-DNN, including (1) the coefficient of the regularization in terms of similarity between local models, i.e., α in Eq. (5), (2) the coefficient of the model compression loss, i.e., β in Eq. (6), (3) the length of the compression process, i.e., p in Algorithm 2, and (4) the portion of the data needed to be relabeled in the compression process μ as mentioned in Sect. 4.3. We tune these hyperparameters by exploring a certain range of values and then choose the one resulting in best performance. In particular, we explored α among $\{0.2, 0.4, \ldots, 1\}$, and finally choose $\alpha = 0.6$ on all the datasets. To decide the value of β, we explored two strategies, one of which uses consistent β at each compression process while the other employs increasing β after a certain percentage of compression process. In the first strategy, we explored β among $\{0.2, 0.4, \ldots, 1\}$, and in the second one, we explored β among $\{0.2, 0.4, \ldots, 1\}$, the incremental step of β among $\{0.1, 0.2\}$, and the percentage of compression process from which β begins to increase among $\{10\%, 20\%, 30\%\}$. On CIFAR datasets, we choose to use $\beta = 0.4$ for the first 20% of compression processes and $\beta = 0.6$ for all the other compression processes. And, on ImageNet, we choose to use consistent $\beta = 1$ throughout the compression. Moreover, we explored p's value among $\{5\%, 10\%, \ldots, 20\%\}$ of the number of the mini-batches that the whole training lasts, and finally pick $p = 10\%$ on all the datasets. Furthermore, we explored μ's value among $\{30\%, 50\%, 70\%\}$. And, we finally select $\mu = 70\%$ on CIFAR datasets, and $\mu = 30\%$ on ImageNet.

5.2   Compared Methods

We conduct performance comparisons on four methods:

- S-DNN denotes the sequential training on one GPU until convergence [19,25].
- E-DNN denotes the method that trains local models independently and makes ensemble of the local models merely at the end of the training [5,25].
- MA-DNN refers the parallel DNN training framework with the aggregation by averaging model parameters [2,3,6,20,26,27].
- EC-DNN refers the parallel DNN training framework with the aggregation by averaging model outputs. EC-DNN applies Compression$_{\text{distill}}$ for the compression for all the experiments in this paper.

Furthermore, we use EC-DNN$_L$, MA-DNN$_L$ and E-DNN$_L$ to denote the corresponding methods that take the local model with the smallest training loss as the final model, and use EC-DNN$_G$, MA-DNN$_G$ and E-DNN$_G$ to represent the respective methods that take the global model (i.e., the ensemble of local models for EC-DNN and E-DNN, and the average parameters of local models for MA-DNN) as the final model.

5.3   Experimental Results

Model Aggregation. We first compare the performance of aggregation methods, i.e. MA and Ensemble. We employ Diff_{LG} as the evaluation metric, which measures the improvement of the test error of the global model compared to that of the local models, i.e.,

$$\text{Diff}_{LG} = \frac{1}{K} \sum_{k=1}^{K} \text{error}_k - \text{error}_{\text{global}}, \tag{7}$$

where error_k denotes the test error of the local model on worker k, and $\text{error}_{\text{global}}$ denotes the test error of the corresponding global model produced by MA (or ensemble) in MA-DNN (or EC-DNN). The positive (or negative) Diff_{LG} means performance improvement (or drop) of global models over local models. On each dataset, we produce a distribution for Diff_{LG} over all the communications and all the parallel settings (including numbers of workers and communication frequencies). We show the distribution for Diff_{LG} of MA and ensemble on CIFAR datasets in Figs. 1 and 2 respectively, in which red bars (or blue bars) stand for that the performance of the global model is worse (or better) than the average performance of local models.

For MA, from Fig. 1, we can observe that, on both datasets, over 10% global models achieve worse performance than the average performance of local models, and the average performance of locals model can be worse than the global model by a large margin, e.g., 30%. On the other hand, for ensemble, we can observe from Fig. 2 that the performance of the global model is consistently better than the average performance of the local models on both datasets. Specifically, the performances of over 20% global models are 5+% better than the average performance of local models on both datasets.

Model Compression. In order to avoid model size explosion, the ensemble model is compressed before the next round of local training. However, such compression may result in a risk of performance loss. To examine if the performance improvement of the compressed global models over those local ones in EC-DNN can still outperform such kind of improvement in MA-DNN, we compare Diff_{LG} in MA-DNN (see Eq. (7)) with Diff_{LC} in EC-DNN,

$$\text{Diff}_{LC} = \frac{1}{K} \sum_{k=1}^{K} \left(\text{error}_k - \text{error}_{\text{compress},k} \right), \tag{8}$$

where error_k denotes the test error of the local model on worker k, and $\text{error}_{\text{compress},k}$ denotes the test error of the corresponding compressed model after compressing the ensemble model of those local models on worker k. The positive (or negative) Diff_{LC} means performance improvement (or drop) of the compressed model over local ones. Figure 3 illustrates the distribution of Diff_{LC} over all the communications and various settings of communication frequency and the number of workers on two CIFAR datasets.

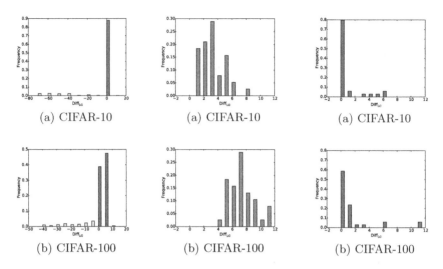

(a) CIFAR-10   (a) CIFAR-10   (a) CIFAR-10

(b) CIFAR-100   (b) CIFAR-100   (b) CIFAR-100

Fig. 1. MA. (Color figure online)   **Fig. 2.** Ensemble. (Color figure online)   **Fig. 3.** Compression.

From Fig. 3, we can observe that the average performance of the compressed models is consistently better than that of the local models on both datasets in EC-DNN, while Fig. 1 indicates that there are over 10% global models do not reaching better performance than the local ones in MA-DNN. In addition, the average improvement of compressed models over local ones in EC-DNN is greater than that in MA-DNN. Specifically, the average of such improvements in EC-DNN are 1.03% and 1.95% on CIFAR-10 and CIFAR-100, respectively, while the average performance difference in MA-DNN are -3.53% and 1.72% on CIFAR-10 and CIFAR-100 respectively. All these results can indicate that EC-DNN is a superior method than MA-DNN.

Accuracy. In the following, we examine the accuracy of compared methods. Figure 4 shows the test error of the global model during the training process w.r.t. the overall time, and Table 1 reports the final performance after the training process converges. For EC-DNN, the relabeling time has been counted in the overall time when plotting the figure and the table. We report EC-DNN and MA-DNN that achieve best test performance among all the communication frequencies.

From Fig. 4, we can observe that EC-DNN$_G$ outperforms MA-DNN$_G$ and S-DNN on both datasets for all the number of workers, which demonstrates that EC-DNN is superior to MA-DNN. Specifically, at the early stage of training, EC-DNN$_G$ may not outperform MA-DNN$_G$. We hypothesize the reason as the very limited number of communications among local works at the early stage of EC-DNN training. Along with increasing rounds of communications, EC-DNN will catch up with and then keep outperforming MA-DNN after the certain time slot. Besides, EC-DNN$_G$ outperforms E-DNN$_G$ consistently for different datasets and number of workers, indicating that technologies in EC-DNN are not trivial improvements of E-DNN but is the key factor of the success of EC-DNN.

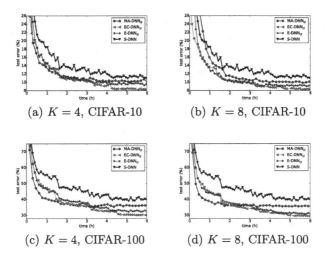

(a) $K = 4$, CIFAR-10   (b) $K = 8$, CIFAR-10

(c) $K = 4$, CIFAR-100   (d) $K = 8$, CIFAR-100

Fig. 4. Test error curves on CIFAR datasets.

Table 1. Test error (%), speed and τ on CIFAR datasets.

	CIFAR-10						CIFAR-100					
	K = 4			K = 8			K = 4			K = 8		
	Error	Speed	τ	Error	Speed	τ	Error	Speed	τ	Error	Speed	τ
MA-DNN$_G$	10.3	1	16	9.99	1	2k	36.18	1	16	35.55	1	16
E-DNN$_G$	9.44	1.58	-	9.05	1.92	-	32.49	1.95	-	30.9	1.97	-
EC-DNN$_G$	**8.43**	**1.92**	4k	**8.19**	**2.05**	4k	**30.26**	**2.52**	4k	**29.31**	**2.48**	2k
MA-DNN$_L$	10.55	0	16	10.54	0	2k	36.39	0	16	35.56	0	16
E-DNN$_L$	11.04	0	-	10.95	0	-	39.57	0	-	39.55	0	-
EC-DNN$_L$	**10.04**	**1.36**	4k	**9.88**	**1.26**	4k	**34.8**	**1.42**	4k	**35.1**	**1.27**	2k
S-DNN	10.41						35.68					

In Table 1, each EC-DNN$_G$ outperforms MA-DNN$_L$ and MA-DNN$_G$. The average improvements of EC-DNN$_G$ over MA-DNN$_L$ and MA-DNN$_G$ are around 1% and 5% for CIFAR-10 and CIFAR-100 respectively. Besides, we also report the final performance of EC-DNN$_L$ considering that it can save test time and still outperform both MA-DNN$_L$ and MA-DNN$_G$ when we do not have enough computational and storage resource. Specifically, the best EC-DNN$_L$ achieved test errors of 10.04% and 9.88% for $K = 4$ and $K = 8$ respectively on CIFAR-10, while it achieved test errors of 34.8% and 35.1% for $K = 4$ and $K = 8$ respectively on CIFAR-100. In addition, E-DNN$_L$ never outperforms MA-DNN$_L$ and MA-DNN$_G$.

Speed. According to our analysis in Sect. 4.3, EC-DNN is more time-efficient than MA-DNN because it communicates less frequently than MA-DNN and thus costs less time on communication. To verify this, we measure the overall time cost by each method to achieve the same accuracy. Table 1 shows the speed of compared methods. In this table, we denote the speed of MA-DNN$_G$ as 1, and normalize the speed of other methods by dividing that of MA-DNN$_G$. If one method never achieves the same performance with MA-DNN$_G$, we denote its speed as 0. Therefore, larger value of speed indicates better speedup.

From Table 1, we can observe that EC-DNN can achieve better speedup than MA-DNN on all the datasets. On average, EC-DNN$_G$ and EC-DNN$_L$ runs about 2.24 and 1.33 times faster than MA-DNN$_G$, respectively. Furthermore, EC-DNN consistently results in better speedup than E-DNN on all the datasets. On average, E-DNN$_G$ only runs about 1.85 times faster than MA-DNN$_G$ while EC-DNN$_G$ can reach about 2.24 times faster speed. From this table, we can also find that E-DNN$_L$ never achieves the same performance with MA-DNN$_G$ while EC-DNN$_L$ can contrarily run much faster than MA-DNN$_G$.

Furthermore, Table 1 demonstrates the communication frequency τ that makes compared methods achieve the corresponding speed. We can observe that EC-DNN tend to communicate less frequently than MA-DNN. Specifically, MA-DNN usually achieves the best performance with a small τ (i.e., 16), while EC-DNN cannot reach its best performance before τ is not as large as 2000.

Large-Scale Experiments. In the following, we will conduct experiments to compare the performance of MA-DNN with that of EC-DNN with the setting of much bigger model and more data, i.e., GoogleNet on ImageNet. Figure 5 shows the test error of the global model w.r.t the overall time. The communication frequencies τ that makes MA-DNN and EC-DNN achieve best performance are 1 and 1000 respectively. We can observe that EC-DNN consistently achieves better test performance than S-DNN, MA-DNN and E-DNN throughout the training. Besiedes, we can observe that EC-DNN outperforms MA-DNN even at the early stage of the training, while EC-DNN cannot achieve this on CIFAR datasets because it communicates less frequently than MA-DNN. The reason is that frequent communication will make the training much slower for very big model, i.e., use less mini-batches of data within the same time. When the improvements introduced by MA cannot compensate the decrease of the number of used data, MA-DNN no longer outperforms EC-DNN at the early stage of the training. In this case, the advantage of EC-DNN becomes even more outstanding.

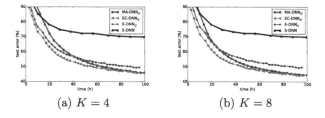

(a) $K = 4$ (b) $K = 8$

Fig. 5. Test error curves on ImageNet.

6   Conclusion and Future Work

In this paper, we propose EC-DNN, a new Ensemble-Compression based parallel training framework for DNN. As compared to the traditional approach, MA-DNN, that averages the parameters of different local models, our proposed method uses the ensemble method to aggregate local models. In this way, we can guarantee that the error of the global model in EC-DNN is upper bounded by the average error of the local models and can consistently achieve better performance than MA-DNN. In the future, we plan to consider other compression methods for EC-DNN. Besides, we will investigate the theoretical properties of the ensemble method, compression method, and the whole EC-DNN framework.

Acknowledgments. This work is partially supported by NSF of China (grant numbers: 61373018, 61602266, 11550110491) and NSF of Tianjin (grant number: 4117JCY-BJC15300).

References

1. Bucilua, C., Caruana, R., Niculescu-Mizil, A.: Model compression. In: Proceedings of the 12th ACM Conference on Knowledge Discovery and Data Mining, pp. 535–541. ACM (2006)
2. Chen, J., Monga, R., Bengio, S., Jozefowicz, R.: Revisiting distributed synchronous SGD. arXiv preprint arXiv:1604.00981 (2016)
3. Chen, K., Huo, Q.: Scalable training of deep learning machines by incremental block training with intra-block parallel optimization and blockwise model-update filtering. In: 2016 IEEE International Conference on Acoustics, Speech and Signal Processing (ICASSP), pp. 5880–5884. IEEE (2016)
4. Chen, W., Wilson, J.T., Tyree, S., Weinberger, K.Q., Chen, Y.: Compressing neural networks with the hashing trick. In: Proceedings of the 32st International Conference on Machine Learning (2015)
5. Ciresan, D., Meier, U., Schmidhuber, J.: Multi-column deep neural networks for image classification. In: IEEE Conference on Computer Vision and Pattern Recognition, pp. 3642–3649. IEEE (2012)
6. Dean, J., Corrado, G., Monga, R., Chen, K., Devin, M., Mao, M., Senior, A., Tucker, P., Yang, K., Le, Q.V., et al.: Large scale distributed deep networks. In: Advances in Neural Information Processing Systems, pp. 1223–1231 (2012)
7. Dean, J., Ghemawat, S.: MapReduce: simplified data processing on large clusters. Commun. ACM **51**(1), 107–113 (2008)
8. Denil, M., Shakibi, B., Dinh, L., de Freitas, N., et al.: Predicting parameters in deep learning. In: Advances in Neural Information Processing Systems, pp. 2148–2156 (2013)
9. Denton, E.L., Zaremba, W., Bruna, J., LeCun, Y., Fergus, R.: Exploiting linear structure within convolutional networks for efficient evaluation. In: Advances in Neural Information Processing Systems, pp. 1269–1277 (2014)
10. Gong, Y., Liu, L., Yang, M., Bourdev, L.: Compressing deep convolutional networks using vector quantization. arXiv preprint arXiv:1412.6115 (2014)
11. Han, S., Pool, J., Tran, J., Dally, W.: Learning both weights and connections for efficient neural network. Adv. Neural Inf. Process. Syst. **28**, 1135–1143 (2015)

12. He, D., Xia, Y., Qin, T., Wang, L., Yu, N., Liu, T., Ma, W.Y.: Dual learning for machine translation. Adv. Neural Inf. Process. Syst. **29**, 820–828 (2016)
13. He, K., Zhang, X., Ren, S., Sun, J.: Delving deep into rectifiers: surpassing human-level performance on ImageNet classification. In: Proceedings of the IEEE International Conference on Computer Vision, pp. 1026–1034 (2015)
14. Hinton, G., Vinyals, O., Dean, J.: Distilling the knowledge in a neural network. arXiv preprint arXiv:1503.02531 (2015)
15. Jia, Y., Shelhamer, E., Donahue, J., Karayev, S., Long, J., Girshick, R., Guadarrama, S., Darrell, T.: Caffe: convolutional architecture for fast feature embedding. arXiv preprint arXiv:1408.5093 (2014)
16. Krizhevsky, A.: Learning multiple layers of features from tiny images. University of Toronto, Technical report (2009)
17. Kuncheva, L., Whitaker, C.: Measures of diversity in classifier ensembles. In: Machine Learning, pp. 181–207 (2003)
18. Li, M., Andersen, D.G., Park, J.W., Smola, A.J., Ahmed, A., Josifovski, V., Long, J., Shekita, E.J., Su, B.Y.: Scaling distributed machine learning with the parameter server. In: 11th USENIX Symposium on Operating Systems Design and Implementation, pp. 583–598 (2014)
19. Min, L., Qiang, C., Yan, S.: Network in network. arXiv preprint arXiv:1312.4400 (2014)
20. Povey, D., Zhang, X., Khudanpur, S.: Parallel training of DNNs with natural gradient and parameter averaging. arXiv preprint arXiv:1410.7455 (2014)
21. Rigamonti, R., Sironi, A., Lepetit, V., Fua, P.: Learning separable filters. In: IEEE Conference on Computer Vision and Pattern Recognition, pp. 2754–2761. IEEE (2013)
22. Romero, A., Ballas, N., Kahou, S.E., Chassang, A., Gatta, C., Bengio, Y.: FitNets: hints for thin deep nets. arXiv preprint arXiv:1412.6550 (2014)
23. Russakovsky, O., Deng, J., Su, H., Krause, J., Satheesh, S., Ma, S., Huang, Z., Karpathy, A., Khosla, A., Bernstein, M., Berg, A.C., Fei-Fei, L.: ImageNet large scale visual recognition challenge. Int. J. Comput. Vis. (IJCV) **115**(3), 211–252 (2015)
24. Sollich, P., Krogh, A.: Learning with ensembles: how overfitting can be useful. Adv. Neural Inf. Process. Syst. **8**, 190–196 (1996)
25. Szegedy, C., Liu, W., Jia, Y., Sermanet, P., Reed, S., Anguelov, D., Erhan, D., Vanhoucke, V., Rabinovich, A.: Going deeper with convolutions. arXiv preprint arXiv:1409.4842 (2014)
26. Zhang, S., Choromanska, A.E., LeCun, Y.: Deep learning with elastic averaging SGD. Adv. Neural Inf. Process. Syst. **28**, 685–693 (2015)
27. Zhang, X., Trmal, J., Povey, D., Khudanpur, S.: Improving deep neural network acoustic models using generalized maxout networks. In: IEEE International Conference on Acoustics, Speech and Signal Processing, pp. 215–219. IEEE (2014)

Fast and Accurate Density Estimation with Extremely Randomized Cutset Networks

Nicola Di Mauro[1(✉)], Antonio Vergari[1], Teresa M. A. Basile[2,3],
and Floriana Esposito[1]

[1] Department of Computer Science, University of Bari "Aldo Moro", Bari, Italy
nicola.dimauro@uniba.it
[2] Department of Physics, University of Bari "Aldo Moro", Bari, Italy
[3] National Institute for Nuclear Physics (INFN), Bari Division, Bari, Italy

Abstract. Cutset Networks (CNets) are density estimators leveraging context-specific independencies recently introduced to provide exact inference in polynomial time. Learning a CNet is done by firstly building a weighted probabilistic OR tree and then estimating tractable distributions as its leaves. Specifically, selecting an optimal OR split node requires cubic time in the number of the data features, and even approximate heuristics still scale in quadratic time. We introduce Extremely Randomized Cutset Networks (XCNets), CNets whose OR tree is learned by performing random conditioning. This simple yet surprisingly effective approach reduces the complexity of OR node selection to constant time. While the likelihood of an XCNet is slightly worse than an optimally learned CNet, ensembles of XCNets outperform state-of-the-art density estimators on a series of standard benchmark datasets, yet employing only a fraction of the time needed to learn the competitors. Code and data related to this chapter are available at: https://github.com/nicoladimauro/cnet.

Keywords: Tractable probabilistic models · Cutset Networks

1 Introduction

Density estimation is the unsupervised task of learning an estimator for the joint probability distribution over a set of random variables (RVs) that generated the observed data. Once such an estimator is learned, it is used to do *inference*, i.e., computing the probability of the queries about certain states of the RVs. Since a perfect estimate of the real distribution would allow to solve many learning tasks exactly when reframed as different kinds of inference[1], density estimation classifies as one of the most general task in machine learning [13].

The main challenge in density estimation is balancing the *representation expressiveness* of the learned model against the *cost of learning* it and *performing*

N. Di Mauro and A. Vergari—Both authors contributed equally.

[1] E.g., classification can be framed as Most Probable Explanation (MPE) inference.

© Springer International Publishing AG 2017
M. Ceci et al. (Eds.): ECML PKDD 2017, Part I, LNAI 10534, pp. 203–219, 2017.
https://doi.org/10.1007/978-3-319-71249-9_13

inference on it. Probabilistic Graphical Models (PGMs), like Bayesian Networks (BNs) and Markov Networks (MNs), are able to model highly complex probability distributions. However, exact inference with them is generally *intractable*, i.e., not solvable in polynomial time, and even some approximate inference routines are intractable in practice [23]. With the aim of performing exact and polynomial inference, a series of *tractable probabilistic models* (TPMs) have been recently proposed: either by restricting the expressiveness of PGMs by bounding their treewidth [24], e.g., tree distributions and their mixtures [18], or by exploiting local structures in a distribution [4]. It is worth noting that inference tractability is not a global property, but it is associated to classes of queries. For instance, computing exact marginals on a TPM may be feasible, while MPE may be not [1]. TPMs like Arithmetic Circuits [6], Sum-Product Networks (SPNs) [19], and Cutset Networks (CNets) [21] promise a good compromise between expressive power and tractable inference by compiling high treewidth distributions in compact and efficient data structures. Even if learning such TPMs may be done in polynomial time, thanks to several recent algorithmic schemes, making these algorithms scale to high dimensional data is still an issue. We focus on CNets since they (i) exactly and tractably compute several inference query types like marginals, conditionals and MPE inference [7], and (ii) promise faster learning times, when compared to other TPMs.

CNets have been introduced in [21] as weighted probabilistic model trees having tree-structured models as the leaves of an OR tree. They exploit context-specific independencies (CSIs) [2] by embedding Pearl's conditioning algorithm. While the learning algorithm originally proposed in [21] provides a heuristic approach, it still requires quadratic time w.r.t. the number RVs to select each tree inner node to condition on. A theoretically principled and more accurate version, presented in [9], overcomes many of the initial version issues, like the tendency to overfit. However, in order to do so, it increases the complexity of performing a single split to cubic time. We tackle the problem of scaling CNet learning to high dimensional data while preserving inference accuracy.

Here we introduce Extremely Randomized CNets (XCNets), as CNets that can be learned in a simple, fast and yet effective approach by performing random conditioning to grow the OR tree. In such a way, selecting a node to split on reduces to constant time w.r.t. the number of features. As we will see, while the likelihood of a single XCNet is not greater than an optimally learned CNet, ensembles of XCNets outperform state-of- the-art density estimators on a series of standard benchmark datasets, yet employing a fraction of the time needed to learn the competitors. To further reduce the learning complexity, we investigate the exploitation of a naive factorization as leaf distribution in XCNets. As a result, we can build an extremely fast mixture of density estimators that is more accurate than several CNets and comparable to a BN exploiting CSI [3].

2   Background

Notation. Let RVs be denoted by upper-case letters, e.g., X, and their values as the corresponding lower-case letters, e.g., $x \sim X$. We denote sets of RVs as

$\mathbf{X}$, and their combined values as $\mathbf{x}$. For a set of RVs $\mathbf{X}$ we denote with $\mathbf{X}_{\backslash i}$ the set $\mathbf{X}$ deprived of X_i, and with $\mathbf{X}_{|\mathbf{Y}}$ the restriction of $\mathbf{X}$ to $\mathbf{Y} \subseteq \mathbf{X}$ (the same applies to assignments $\mathbf{x}$). W.l.o.g., we assume RVs we deal with in the following to be binary valued.

Density Estimation. Let $\mathcal{D} = \{\xi^j\}_{j=1}^m$ be a set of m n-dimensional samples drawn i.i.d. according to an unknown joint probability distribution $\mathsf{p}(\mathbf{X})$, with $\mathbf{X} = \{X_i\}_{i=1}^n$. We refer to $\xi^j[X_i]$ as the value assumed by the sample ξ^j in correspondence of the RV X_i. We are interested in learning a model $\mathcal{M}$ from $\mathcal{D}$ such that its estimate of the underlying distribution, denoted as $\mathsf{p}_{\mathcal{M}}(\mathbf{X})$, is as close as possible to the original one [13]. Generally, measuring this closeness is done via the *log-likelihood* function, or one of its variants, defined as: $\ell_{\mathcal{D}}(\mathcal{M}) = \sum_{j=1}^m \log \mathsf{p}_{\mathcal{M}}(\xi^j)$. In the next sub-sections we review the approaches to density estimation as the building blocks of XCNets we propose in Sect. 4.

2.1 Product of Bernoulli Distributions

The simplest representation assumption for $\mathsf{p}(\mathbf{X})$ over RVs $\mathbf{X}$, allowing tractable inference, involves considering all RVs in $\mathbf{X}$ to be independent: $\mathsf{p}(\mathbf{x}) = \prod_{i=1}^n \mathsf{p}(x_i)$. For binary RVs, this *naive* factorization leads to the product of Bernoulli distributions (PoBs) model, where building $\mathsf{p}_{\mathcal{M}}$ equals to estimate the $\mathsf{p}_{\mathcal{M}}(x_i^0) = \theta_i^0$.

Proposition 1 (LearnPoB time complexity). *Learning a PoB from $\mathcal{D}$ over RVs $\mathbf{X}$ has time complexity $O(nm)$, where $m = |\mathcal{D}|$ and $n = |\mathbf{X}|$.*

Proof. For each Bernoulli RV $X_i \in \mathbf{X}$, estimating θ_i requires a single pass over $\{\xi^j[X_i]\}_{j=1}^m$, hence taking $O(m)$. Consequently, for all RVs in $\mathbf{X}$, it takes $O(mn)$.

Similarly to what Naive Bayes provides for classification, PoBs deliver a cheap and very fast baseline for tractable density estimation, even if the total independence assumption clearly does not hold on real data. Moreover, mixtures of PoBs, sometimes simply referred to mixtures of Bernoulli distributions (MoBs), have proved as an effective way to increase the representation expressiveness of PoBs [16]. However, while inference on MoBs is still tractable, learning them in a principled way requires running the EM algorithm for k iterations and r restarts, thus increasing the complexity up to $O(rkmn)$ [16].

2.2 Probabilistic Tree Models

A *directed tree-structured model* [18] over $\mathbf{X}$ is a BN in which each node $X_i \in \mathbf{X}$ has at most one parent, Pa_{X_i}. It encodes a distribution that factorizes as: $\mathsf{p}(\mathbf{x}) = \prod_{i=1}^n \mathsf{p}(x_i|\mathrm{Pa}_{x_i})$, where Pa_{x_i} denotes the projection of the assignment $\mathbf{x}$ on the parent of X_i. By modeling such dependencies, tree-structured models can be more expressive than PoBs, yet still performing exact complete and marginal inference in $O(n)$ [18]. To learn a model $\mathcal{M} = \langle \mathcal{T}, \{\theta_{i|\mathrm{Pa}_{X_i}}\}_{i=1}^n \rangle$, now one has to estimate *both* a tree structure $\mathcal{T}$ and the conditional probabilities $\theta_{i|\mathrm{Pa}_{X_i}} = \mathsf{p}_{\mathcal{M}}(X_i|\mathrm{Pa}_{X_i})$. Growing an optimal model, according to the KL-divergence, can be done by employing the classical result from Chow and Liu [5]. We will refer to tree-structured models as Chow-Liu trees, or CLtrees, assuming the Chow-Liu algorithm (LearnCLTree) has been employed to learn them.

Proposition 2 (LearnCLTree time complexity [5]**).** *Learning a CLtree from $\mathcal{D}$ over RVs $\mathbf{X}$ has time complexity $O(n^2(m+\log n))$, where $m = |\mathcal{D}|$ and $n = |\mathbf{X}|$.*

Proof. For each pair of RVs in $\mathbf{X}$, their mutual information (MI) can be estimated from $\mathcal{D}$ in $O(mn^2)$ steps. Building a maximum spanning tree on the weighted graph induced by the adjacency matrix MI takes $O(n^2 \log n)$. Lastly, both arbitrarily rooting the tree, traversing it, and estimating the conditional probabilities $\theta_{i|\mathrm{Pa}_{X_i}}$ can be done in $O(n)$.

All in all, the complexity of learning a CLTree is quadratic in n. While this is a huge gain w.r.t. learning a higher order dependency BN, it still poses a practical issue when LearnCLTree is applied as a routine in larger learning schemes and on datasets with thousand features. Nevertheless, CLTrees have been employed as the core components of many tractable probabilistic models ranging from mixtures of them [18], SPNs [26] and CNets [8,9,21]. We will specifically tackle the problem of scaling CNet learning in the following sections.

3   Cutset Networks

Cutset Networks are TPMs introduced in [21] as a hybrid of OR trees and CLTrees as the tree leaves. Here we generalize their definition to comprise generic TPMs as leaf distributions. A CNet $\mathcal{C}$ over a set of RVs $\mathbf{X}$, is a probabilistic weighted model tree defined via a rooted OR tree $\mathcal{G}$ and a set of TPMs $\{\mathcal{M}_i\}_{i=1}^{L}$, in which each $\mathcal{M}_i$ encodes a distribution $\mathsf{p}_{\mathcal{M}_i}$ over a subset of $\mathbf{X}$, called *scope* and denoted as $\mathsf{sc}(\mathcal{M}_i)$. The scope of a CNet $\mathcal{C}$, $\mathsf{sc}(\mathcal{C})$, is the set of RVs appearing in it. A CNet may be defined recursively as follows.

Definition 1 (Cutset network). *Given binary RVs $\mathbf{X}$, a CNet is: (1) a TPM $\mathcal{M}$, with $\mathsf{sc}(\mathcal{M}) = \mathbf{X}$; or (2) a weighted disjunction of two CNets $\mathcal{C}_0$ and $\mathcal{C}_1$ graphically represented as an OR node conditioned on RV $X_i \in \mathbf{X}$, with associated weights w_i^0 and w_i^1 s.t. $w_i^0 + w_i^1 = 1$, where $\mathsf{sc}(\mathcal{C}_0) = \mathsf{sc}(\mathcal{C}_1) = \mathbf{X}_{\backslash i}$.*

A CNet over binary RVs is shown in Fig. 1: each circled node is an OR tree node and labeled by a variable X_i. Each edge emanating from it is weighted by the probability w_i^0, resp.w_i^1, of conditioning X_i to the value 0, resp. 1. The distribution encoded by a CNet $\mathcal{C}$ can be written as:

$$\mathsf{p}(\mathbf{x}) = \mathsf{p}_l(\mathbf{x}_{|\mathsf{sc}(\mathcal{C})\backslash\mathsf{sc}(\mathcal{M}_l)})\mathsf{p}_{\mathcal{M}_l}(\mathbf{x}_{|\mathsf{sc}(\mathcal{M}_l)}), \tag{1}$$

where $\mathsf{p}_l(\mathbf{x}_{|\mathsf{sc}(\mathcal{C})\backslash\mathsf{sc}(\mathcal{M}_l)}) = \prod_i (w_i^0)^{1-x_i}(w_i^1)^{x_i}$ is a factor obtained by multiplying all the weights attached to the edges of the path in the OR tree starting from the root of $\mathcal{C}$ and reaching a unique leaf node l; on the other hand, $\mathsf{p}_{\mathcal{M}_l}(\mathbf{x}_{|\mathsf{sc}(\mathcal{M}_l)})$ is the distribution encoded by the reached leaf l. $\mathsf{p}_{\mathcal{M}_l}$ can be interpreted as a conditional distribution $\mathsf{p}(\mathbf{x}_{|\mathsf{sc}(\mathcal{M}_l)}|\mathbf{x}_{|\mathsf{sc}(\mathcal{C})\backslash\mathsf{sc}(\mathcal{M}_l)})$.

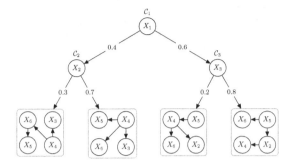

Fig. 1. Example of a CNet over binary RVs. Inner (rounded) nodes on variables X_i are OR nodes, while leaf (squared) nodes represent CLtrees.

3.1 Learning CNets

Learning both the structure and parameters of a CNet from data equals to perform searching in the space of all probabilistic weighted model trees. This would require an exponential time: for a dataset $\mathcal{D}$ over RVs $\mathbf{X}$ learning a full binary OR tree with height k has time complexity $O(n^k 2^k(n^2(m + \log n))) = O(m2^k n^{k+2})$, with $m = |\mathcal{D}|$ and $n = |\mathbf{X}|$. In practice, this problem is tackled in a *two-stage greedy fashion* by: (i) first performing a top-down search in the space of weighted OR trees, and then (ii) learning TPMs as leaf distributions according to a conditioned subset of the data. The first structure learning algorithm for CNets is the one introduced in [21], leveraging a heuristic approach to induce the OR tree and demanding pruning to combat overfitting. A following approach has been introduced in [9], growing the OR tree by a principled Bayesian search maximizing the data likelihood. In the following, we introduce a general scheme to learn CNets, showing how, by properly determining a splitting criterion to grow the OR tree, one can recover both the algorithms from [21] and [9]. This, in turn, highlights how the splitting criterion time complexity determines that of learning the whole OR tree, and hence the whole CNet. In Sect. 4, we propose a variation of the splitting procedure drastically reducing its cost.

General Learning Scheme. Algorithm 1 reports a general approach for CNets structure learning. In particular, the procedure tries to select a variable X_i on the input data slice $\mathcal{D}$ (line 4). If a such a variable exists (line 5), it then recursively (line 8) tries to decompose the two new slices $\mathcal{D}_0$ and $\mathcal{D}_1$ over $\mathbf{X}_{\setminus i}$. When the slice $\mathcal{D}$ has few instances, or it is defined on few variables, then a leaf distribution is learned (line 10). Both, the algorithms reported in [9,21] use CLtrees as leaf distribution, i.e., the learnDistribution procedure on line 10 corresponds to call the LearnCLTree algorithm.

By deriving the time complexity of both growing the OR tree and learning the leaf distributions, one can derive the whole time complexity of LearnCNet. In turn, the time complexity of growing the OR tree clearly depends by the cost of selecting the RV to split on at each step. If we assume the variations of LearnCNet

Algorithm 1. LearnCNet($\mathcal{D}$, $\mathbf{X}$, α, δ, σ)

1: **Input:** a dataset $\mathcal{D}$ over RVs $\mathbf{X}$; α: Laplace smoothing factor; δ min number of samples to split; σ min number of features to split
2: **Output:** a CNet $\mathcal{C}$ encoding $\mathsf{p}_{\mathcal{C}}(\mathbf{X})$ learned from $\mathcal{D}$
3: **if** $|\mathcal{D}| > \delta$ **and** $|\mathbf{X}| > \sigma$ **then**
4:      X_i, success $\leftarrow$ select($\mathcal{D}$, $\mathbf{X}$, α)
5:      **if** success **then**
6:          $\mathcal{D}_0 \leftarrow \{\xi \in \mathcal{D} : \xi[X_i] = 0\}$, $\mathcal{D}_1 \leftarrow \{\xi \in \mathcal{D} : \xi[X_i] = 1\}$
7:          $w_0 \leftarrow |\mathcal{D}_0|/|\mathcal{D}|$, $w_1 \leftarrow |\mathcal{D}_1|/|\mathcal{D}|$
8:          $\mathcal{C} \leftarrow w_0 \cdot$ LearnCNet($\mathcal{D}_0$, $\mathbf{X}_{\setminus i}$, α, δ, σ) $+ w_1 \cdot$ LearnCNet($\mathcal{D}_1$, $\mathbf{X}_{\setminus i}$, α, δ, σ)
9: **else**
10:     $\mathcal{C} \leftarrow$ learnDistribution($\mathcal{D}$, $\mathbf{X}$, α)
11: **return** $\mathcal{C}$

have grown the same sized OR trees, the time complexity of each implementation of select determines the whole OR tree growing phase complexity. Concerning learning leaf distributions, its complexity is determined by the cost of learning a single distribution, that in case of CLTrees is $O(n^2(m + \log(n)))$ (see Proposition 2). As a consequence, assuming to learn L leaves for a tree, then it would take $O(Ln^2(m + \log(n)))$ for all variations to learn such leaves. In the following Sections we revise and analyze the two variations of LearnCNet reported in [9,21].

Proposition 3. *Growing a full binary OR tree with* LearnCNet *on $\mathcal{D}$ over RVs $\mathbf{X}$ has time complexity $O(k(S + m))$, where $m = |\mathcal{D}|$, $n = |\mathbf{X}|$, k is the height of the OR tree, and $S = T(m, n)$, assumed to grow linearly w.r.t. m holding n constant, is the time required to compute the OR split node selection procedure on $\mathcal{D}$ (select function in Algorithm 1, line 4).*

Proof. A set $\mathcal{D}_t^h \subset \mathcal{D}$ of samples falls in each internal node t with height h, such that $\forall i \neq j : \mathcal{D}_i^h \cap \mathcal{D}_j^h = \emptyset$, and $\cup_{i=1}^{2^h}\mathcal{D}_i^h = \mathcal{D}$. Furthermore, for each internal node t with height h, $T(|\mathcal{D}_t^h|, n-h)$ has been the time required to compute the OR split selection, and $|\mathcal{D}_t^h|$ is the time required to split the samples $\mathcal{D}_t^h$. Assuming that $T(m, n)$ grows linearly w.r.t. m holding n constant, then for each height h we have a time complexity equal to $O(\sum_{i=1}^{2^h}(T(|\mathcal{D}_i^h|, n - h) + |\mathcal{D}_i^h|)) = O(T(|\mathcal{D}|, n - h) + m)$. Since the OR tree has height k, then the overall time is $O(\sum_{i=0}^{k-1}(T(|\mathcal{D}|, n - i) + m)) = O(k(T(|\mathcal{D}|, n) + m))$.

Information Gain Splitting Heuristic. The algorithm to learn CNet structures proposed in [21], that here we will call entCNet, performs a greedy top-down search in the OR-trees space that can be reframed in Algorithm 1. It implements the select function as a procedure to determine the RV X_i that maximizes a generative reformulation of the information gain from decision tree theory. Since computing the joint entropy over RVs $\mathbf{X}_{\setminus i}$ would be unfeasible to calculate, it heuristically approximates it by computing the average over marginal entropies.

To cope with the systematic overfitting showed by CNets learned by entCNet, always in [21], a post-pruning method on a validation set is introduced. Leveraging this decision tree technique, on a fully grown CNet, by advancing bottom-up, leaves are pruned and inner nodes without children replaced with a CLtree (that needs to be learned from data), if the network validation data likelihood after this operation is higher than that scored by the not pruned network.

Proposition 4 (select time complexity in entCNet [21]**).** *The time complexity for selecting the best splitting node on a slice $\mathcal{D}$ over RVs $\mathbf{X}$ in entCNet is $O(mn^2)$, where $m = |\mathcal{D}|$ and $n = |\mathbf{X}|$.*

Corollary 1. *Growing a full binary OR tree for entCNet when learning a CNet on $\mathcal{D}$ over RVs $\mathbf{X}$ has time complexity $O(kmn^2)$, where $m = |\mathcal{D}|$, $n = |\mathbf{X}|$, and k is the height of the OR tree.*

Proof. From Propositions 3 and 4, the overall time complexity to grow a full binary OR tree is $O(k(mn^2 + m)) = O(km(n^2 + 1))$.

dCSN: likelihood guided splitting. In [9], the authors proposed the dCSN algorithm that exploits a different approach from that in [21], by avoiding decision tree heuristics while choosing the best variable directly maximizing the data log-likelihood. As already reported in [9], the log-likelihood function of a CNet may be decomposed as follows. Given a CNet $\mathcal{C}$ learned on $\mathcal{D}$ over $\mathbf{X}$, its log-likelihood $\ell_{\mathcal{D}}(\mathcal{C})$ can be computed as follows: $\ell_{\mathcal{D}}(\mathcal{C}) = \sum_{\xi \in \mathcal{D}} \sum_{i=1,\ldots,n} \log \mathsf{p}(\xi[X_i]|\xi[\mathrm{Pa}_{X_i}])$, when $\mathcal{C}$ corresponds to a CLtree. While, in the case of a OR tree rooted on the variable X_i, the log-likelihood is:

$$\ell_{\mathcal{D}}(\mathcal{C}) = \sum_{j=0,1} m_j \log w_i^j + \ell_{\mathcal{D}_j}(\mathcal{C}_j), \tag{2}$$

being $\mathcal{C}_j$ the CNet involved in the OR, $\mathcal{D}_j = \{\xi \in \mathcal{D} : \xi[X_i] = j\}$, $m_j = |\mathcal{D}_j|$, and $\ell_{\mathcal{D}_j}(\mathcal{C}_j)$ is the log-likelihood of the sub-CNet $\mathcal{C}_j$ on the slice $\mathcal{D}_j$, for $j = 0, 1$.

By exploiting this recursive nature of CNets, a CNet is grown top-down, allowing further expansion, i.e., the substitution of a CLtree with an OR node, only if it improves the structure log-likelihood, since it is clear to see that maximizing the second term in Eq. 2, results in maximizing the global score.

As reported in [9], one starts with a single CLtree, learned from $\mathcal{D}$ over $\mathbf{X}$, and then it checks whether there is a decomposition, i.e., an OR node on the best variable X_i applied on two CLtrees, providing a better log-likelihood than that scored by the initial tree. If such a decomposition exists, than the decomposition process is recursively applied to the sub-slices $\mathcal{D}_0$ and $\mathcal{D}_1$ over $\mathbf{X}_{\setminus i}$, testing each leaf for a possible substitution.

Proposition 5 (select time complexity in dCSN). *The time complexity for selecting the best splitting node on a slice $\mathcal{D}$ over RVs $\mathbf{X}$ in dcsn is $O(n^3(m + \log n))$, where $m = |\mathcal{D}|$ and $n = |\mathbf{X}|$.*

Proof. For each variable $X_i \in \mathbf{X}$, two CLTrees have been computed on $\mathcal{D}_0$ and $\mathcal{D}_1$ leading to a splitting complexity $O(n^2(m + \log n))$. Since n splits have to be checked, the overall complexity to select the best split is $O(n^3(m + \log n))$.

Corollary 2. *Growing a full binary OR tree on $\mathcal{D}$ over RVs $\mathbf{X}$ with dCSN has time complexity $O(kmn^3)$, where $m = |\mathcal{D}|$, $n = |\mathbf{X}|$, and k is the height of the OR tree.*

Proof. From Propositions 3 and 5, the overall time complexity to grow a full binary OR tree is $O(k(mn^3 + m)) = O(km(n^3 + 1))$.

3.2   Learning Ensembles of CNets

To mitigate issues like the scarce accuracy of a single model and their tendency to overfit, since [21] CNets have been employed as the components of a mixture of the form: $\mathsf{p}(\mathbf{X}) = \sum_{i=1}^{c} \lambda_i \mathcal{C}_i(\mathbf{X})$, being $\lambda_i \geq 0 : \sum_{i=1}^{c} \lambda_i = 1$ the mixture coefficients. The first approach to learn such a mixture employs EM to alternatively learn both the weights and the mixture components. With this approach, the time complexity of learning CNets grows at least of a factor of ct, where t is the number of iterations of EM. All the classic issues about convergence and instability of EM make this approach less practical then the following ones. A more efficient method to learn Mixtures of CNets, presented in [9], adopts bagging as a cheap and yet more effective way to only increase time complexity by a factor c. For bagged CNets, mixture coefficients are set equally probable and the mixture components can be learned independently on different bootstrapped data samples. An approach adding random subspace projection to bagged CNets learned with dCSN has been introduced in [8]. While its worst case complexity is the same as for bagging, the cost of growing the OR tree reduced by random sub-spacing is effective in practice. Mixtures of CNets have been learned by exploiting three boosting approaches proposed in [20], having time complexity equals to that for bagging or even worst.

4   Extremely Randomized CNets

XCNets (Extremely Randomized CNets) are CNets that are built by LearnCNet where the OR split node procedure (the select function in Algorithm 1, line 4) is simplified in the most straightforward way: selecting a RV uniformly at random. We denote this algorithmic variant of LearnCNet as XCNet. As a consequence, the cost of the new select function in XCNet does not directly depend anymore on the number of features n and can be considered to be constant.

Proposition 6 (select time complexity in XCNet). *The time complexity for selecting the splitting node on a slice $\mathcal{D}$ over $\mathbf{X}$ in XCNet is $O(1)$.*

Proof. The time required to randomly choose a number in $(1, \ldots, |\mathbf{X}|)$.

Corollary 3. *Growing a full binary OR tree on $\mathcal{D}$ over $\mathbf{X}$ with XCNet has time complexity $O(km)$, where k is the height of the OR tree.*

Proof. From Propositions 3 and 6, the overall time complexity to grow a full binary OR tree is $O(k(1 + m))$.

While we introduce this variation with the obvious aim of speeding up a CNet OR tree learning process, we argue that XCNet should still provide accurate density estimators. We support this conjecture with the following motivations.

A CNet can be seen as a sort-of *mixture of experts* in which the gating function role is demanded to the OR tree, the leaf distributions act as the local experts, and the gating function operates by selecting only one expert per input sample. Let $g : \mathbf{X} \to \{\mathcal{M}_i\}_{i=1}^{L}$ be a gating function that associates each configuration $\xi \sim \mathbf{X}$ to only one leaf model, $\mathcal{M}_\xi$. For a CNet $\mathcal{C}$, g can be built by associating to each ξ a path p in the OR tree structure $\mathcal{G}$ of $\mathcal{C}$. A path $p = p_{(1)}p_{(2)} \cdots p_{(k)}$ of length k is grown as a sequence of observed values $v_1 v_2 \cdots v_k$ in the same fashion as one performs inference according to Eq. 1: starting from the root of $\mathcal{C}$, for each OR node i traversed, corresponding to RV $X_{p(i)}$, the branching corresponding to the value $v_i = \xi[X_{p(i)}]$ is followed. At the end of the path p, a leaf model $\mathcal{M}_p = \mathcal{M}_\xi$ is reached. Alternatively, one can express g as a function of all possible combinations one can build over a set of observed RVs $\mathbf{X}$: $g(\xi) = \sum_{p \in \mathcal{G}} \prod_{i=1}^{|p|} \mathbb{1}\{\xi[X_{p(i)}] = v_i\}\mathcal{M}_p$. Now, from this construction of g, one can derive that permuting the order of appearance of the RVs values v_i does not change the value of g. In the same way, from the factorization in Eq. 1, it follows that neither the joint probability mass associated to the configuration ξ changes after such a permutation. This follows from the fact that the portion of the likelihood assigned to ξ that depends on the path p can be exactly recovered by choosing another sequence of conditionings, as different applications of the chain rule of probability still model the same joint distribution. This permutation invariance suggests that given a way to associate a sample to a leaf distribution, the way in which conditionings are performed can be irrelevant. Clearly, while this is true for an already learned CNet, for algorithms inducing the OR tree in a top-down fashion, the order in which conditionings are performed during learning obviously matter. Nevertheless, in practice, it might matter less than expected. From another perspective, building an OR tree, and hence g, is likely to perform a clustering of all possible sample configurations. For all LearnCNet variants, this clustering performs a trivial aggregation of samples based on their equal observed values for the conditioned RVs. This is one of the issues why algorithms like entCNet are very prone to overfit. For XCNets, however, the randomization introduced in this clustering phase behaves as a regularizer and helps to overcome the aforementioned issue. All in all, we argue that it is more demanding to estimate good distributions at the leaves than an overoptimized gating function.

Moreover, an additional motivation to the introduction of XCNets comes from ensemble theory. From the interpretation of CNets as mixture of experts, the leaf distribution of a CNet acts as an ensemble of density estimators.

Employing a randomized selection criterion increases the diversification of the leaf distributions, and, on the other hand, a strong diversification helps ensembles to better generalize [12]. To better understand this aspect consider a run of entCNet in which the select function has chosen a RV X_i instead of RV X_j to condition on as the first reduces the model entropy more than the second. In both branches x_i^0 and x_i^1 of such a conditioning, it is likely that RV X_j would still be considered as one of the top ranked RVs to be split on in the following iterations. By repeating this argument, it might be likely that the leaf distributions appearing in the sub trees generated by conditioning on x_i^0 and x_i^1 would have very similar scopes.

When constructing ensembles of CNets we expect this diversification effect introduced by randomization to be even more prominent and effective. In ensemble methods like bagging one employs bootstrapping as a source of randomness to diversify the ensemble components [12]. This is also the case for mixtures of CNets built by bagging (see Sect. 3.2). Differently from bagged CNets, ensemble of XCNets do not need an additional way to produce strongly different components. Therefore, when learning mixtures of XCNets, we aggregate the components by learning each component independently on the full dataset.

Lastly, we review Extremely Randomized Tree, or simply ExtraTrees [10] as they are similar in spirit and by name to XCNets. An ExtraTree is a decision tree that is learned by considering only a random subset of features for the introduction of an OR node (like for random forests [12]) and by randomly selecting a threshold for the actual split. Among those randomly generated hyperplanes, the best according to an optimization criterion is chosen. XCNets differ from ExtraTrees from several perspectives. First, they are density estimators and therefore each OR node in them has to split over all the possible values the chosen RV is defined on, otherwise the modeled distribution would not be a valid probability density. Consequently, an OR node in an XCNet is totally selected at random, while for ExtraTrees the best of the random selection is actually employed. Lastly, an XCNet only slightly underperforms a corresponding non-random model, while a single ExtraTree is generally a weak learner whose "raison d'etre" is to be a component in an ensemble [10].

It is tempting to further reduce the complexity of XCNet by substituting CLTrees with even simpler models. As stated in Proposition 1, learning PoBs reduces the complexity to be linear w.r.t. n. Clearly, we do not expect a CNet with PoBs as leaves to achieve a better likelihood than one with CLtrees. Nevertheless, we intend to measure how the likelihood degrades with less expressive leaf distributions and, at the same time, how faster this variant can be.

5   Experiments

The research questions we are validating are: (**Q1**) how much does extreme randomization affect the performance of an XCNet when compared to the optimal one learned with dCSN on real data? (**Q2**) how accurate are ensembles of XCNets and how do they compare against all other CNet ensembling techniques

Table 1. Datasets used and their statistics.

Dataset	n	m_{train}	m_{val}	m_{test}	Dataset	n	m_{train}	m_{val}	m_{test}
NLTCS	16	16181	2157	3236	DNA	180	1600	400	1186
MSNBC	17	291326	38843	58265	Kosarek	190	33375	4450	6675
KDDCup2k	65	180092	19907	34955	MSWeb	294	29441	3270	5000
Plants	69	17412	2321	3482	Book	500	8700	1159	1739
Audio	100	15000	2000	3000	EachMovie	500	4525	1002	591
Jester	100	9000	1000	4116	WebKB	839	2803	558	838
Netflix	100	15000	2000	3000	Reuters-52	889	6532	1028	1540
Accidents	111	12758	1700	2551	20NewsG	910	11293	3764	3764
Retail	135	22041	2938	4408	BBC	1058	1670	225	330
Pumsb-star	163	12262	1635	2452	Ad	1556	2461	327	491

and state-of-the-art density estimators? (**Q3**) how scalable are and how much time do actually XCNets save in practice?

We answer all the above questions by performing our experiments[2] on 20 de-facto standard benchmark datasets for density estimation. Introduced by [15] and [11], they are binarized versions of real data from different tasks like frequent itemset mining, recommendation and classification. We adopt their classic splits for training, validation (hyperparameter selection) and testing. Detailed names and statistics are reported in Table 1. Additionally, for the qualitative experiments in Sect. 5.1 we employ the first 10000 training 28×28 pixel images of digits of MNIST, binarized as in [14].

5.1 (Q1) Single Model Performances

Likelihood Performances. Table 2 reports the results, as the average test log-likelihoods, for all the benchmarks for a entropy-based CNet (entCNet) as reported in [9], a CNet learned with dCSN, and a XCNet (XCNet). Furthermore, we learned a CNet (dCSN$_{PoB}$) and a XCnet (XCNet$_{PoB}$) with PoBs as leaf distributions[3]. For the two XCnets variants for each dataset the reported results are the average and the standard deviation over 10 different runs. Clearly, the best scores are achieved by dCSN with entCNet following it soon after. Nevertheless, all the log-likelihoods achieved by XCNet are only slightly worse and always on the same order of magnitude if compared to non random models, while PoB variants perform considerably worse. We plot the training and test log-likelihoods achieved by dCSN and XCNet models, both ran with $\delta = 100$, while adding

---

[2] Source code of dCSN and XCNet in C++11 and the scripts to replicate the experiments are made available at https://github.com/nicoladimauro/cnet. All experiments have been run on a 4-core Intel Xeon E312xx (Sandy Bridge) @2.0 GHz with 8 Gb of RAM and Ubuntu 14.04.1, kernel 3.13.0-39.

[3] The following grid search to learn CNets with dCSN, XCNet, dCSN$_{PoB}$, and XCNet$_{PoB}$ has been performed: $\delta \in \{300, 500, 1000, 2000\}$, $\alpha \in \{0.1, 0.2, 0.5, 1, 2\}$ and $\sigma = 4$.

Table 2. Average test log likelihoods for all (for XCNet models mean and standard deviation over 10 runs are reported).

Dataset	entCNet	dCSN	XCNet	dCSN$_{PoB}$	XCNet$_{PoB}$
NLTCS	−6.06	**−6.03**	6.06 ± 0.01	−6.09	−6.17 ± 0.05
MSNBC	**−6.05**	**−6.05**	−6.09 ± 0.02	−6.05	−6.18 ± 0.03
KDDCup2k	-	**−2.18**	−2.19 ± 0.01	−2.19	−2.21 ± 0.01
Plants	**−13.25**	**−13.25**	−13.43 ± 0.07	−14.89	−15.66 ± 0.22
Audio	**−42.05**	−42.10	−42.66 ± 0.14	−42.95	−44.02 ± 0.22
Jester	−55.56	**−55.40**	−56.10 ± 0.19	−56.23	−57.39 ± 0.15
Netflix	**−58.71**	**−58.71**	−59.21 ± 0.06	−60.20	−61.40 ± 0.25
Accidents	−30.69	**−29.84**	−31.58 ± 0.24	−36.24	−40.22 ± 0.46
Retail	**−10.94**	−11.24	−11.44 ± 0.09	−11.06	−11.19 ± 0.04
Pumsb-star	−24.42	**−23.91**	−25.55 ± 0.34	−32.11	−39.91 ± 2.48
DNA	−87.59	**−87.31**	−87.67 ± 0.00	−98.83	−99.84 ± 0.05
Kosarek	**−11.04**	−11.20	−11.70 ± 0.13	−11.38	−11.80 ± 0.07
MSWeb	**−10.07**	−10.10	−10.47 ± 0.10	−10.19	−10.43 ± 0.07
Book	**−37.35**	−38.93	−42.36 ± 0.28	−38.21	−39.47 ± 0.33
EachMovie	−58.37	**−58.06**	−60.71 ± 0.89	−59.70	−62.58 ± 0.38
WebKB	−162.17	**−161.92**	−167.45 ± 1.59	−168.7	−174.78 ± 0.81
Reuters-52	**−88.55**	−88.65	−99.52 ± 1.93	−90.51	−100.25 ± 0.57
20NewsG	-	**−161.72**	−172.6 ± 1.40	−162.25	−167.39 ± 0.74
BBC	−263.08	**−261.79**	**−261.79** ± 0.00	−264.56	−274.83 ± 1.15
Ad	−16.92	**−16.34**	−18.70 ± 1.44	−36.44	−36.94 ± 1.41

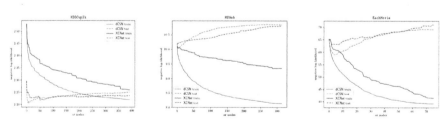

Fig. 2. Negative log-likelihood during learning CNets and XCNets.

nodes during learning in Fig. 2. It is possible to note how, on those datasets, dCSN grows CNets that start overfitting much earlier, while the aleatory nature of XCNet slows the process down and mitigates the effect.

The worst performance is obtained on Ad, with XCNet scoring a relative decrease of 14.46% of the log-likelihood w.r.t. dCSN, while PoB degrade it up to %126.07[4]. These results are very encouraging but not highly surprisingly given

---

[4] The relative decrease is computed as $\frac{\ell_{\mathcal{D}}(\text{XCNet}) - \ell_{\mathcal{D}}(\text{dCSN})}{\ell_{\mathcal{D}}(\text{dCSN})} \cdot 100$.

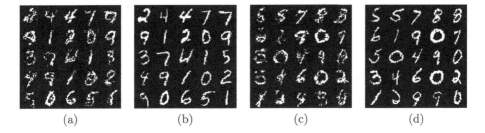

Fig. 3. Samples obtained from a CNet (a), resp. XCNet (c), learned on samples of the binarized MNIST dataset, and their nearest neighbor in training set (b), resp. (d).

our interpretation of CNets as mixture of experts. Moreover this stresses the difference between XCNets with CLTrees and ExtraTrees [10], since a single extremely randomized tree performs much worse than a non-random tree, a behavior we can associate to XCNets with PoBs as leaf distributions.

Generating Samples. It is worth investigating how good is an XCNet at generating samples w.r.t. a CNet learned by dCSN. While results from the previous section can give us a fairly confident estimate according to sample log-likelihoods, these values may not align to the human evaluation of a sample quality [25]. For this reason we perform a qualitative evaluation on samples drawn from XCNets and CNets learned on the first 10000 samples of a binarized version of MNIST with fixed parameters $\delta = 50$, $\alpha = 0.01$, and $\sigma = 4$.

We randomly sampled 25 digits from both models comparing them to the nearest neighbor in the training set, ensuring that the generated samples are not simple memorization, as reported in Fig. 3. It is evident how both models have not memorized the training samples. Since it is not possible to visually spot very relevant differences between the two sample sets, we can confirm that close log-likelihoods correspond to qualitatively similar samples for XCNets and CNets.

5.2 (Q2) Ensemble Performances

To investigate the performance of ensembles of XCNets we build ensembles of 40 components to be comparable with the approaches reported in Sect. 3.2 and introduced in [9,20,21]. We report in the first half of Table 3 the best results for ensembles of bagged ($CNet^{40}$) and boosted ($CNet^{40}_{boost}$) entropy-based CNets taken from [20][5]. Additionally, we learn an ensemble of 40 bagged CNets learned with dCSN as in [9] ($dCSN^{40}$) with a grid search over $\delta \in \{1000, 2000\}$, $\alpha \in \{0.1, 0.2\}$ and $\sigma = 4$. Lastly, we train an ensemble of 40 XCNets ($XCNet^{40}$) and another ensemble of 40 XCNet with PoBs as leaf distributions by running a grid search over $\delta \in \{300, 500, 1000, 2000\}$, $\alpha \in \{0.1, 0.2, 0.5, 1, 2\}$ and $\sigma = 4$. For

---

[5] Note that we report the best log-likelihood across more than one algorithmic variant, hence these results can be considered to be derived from models optimized over more parameters.

Table 3. Average test log likelihoods for all ensembles and other competitors.

Dataset	Ensembles					Competitors			
	CNet^{40}	CNet^{40}_{boost}	dCSN^{40}	XCNet^{40}_{PoB}	XCNet^{40}	XCNet^{500}	ID-SPN	ACMN	WM
NLTCS	**−6.00**	−6.01	**−6.00**	−6.01 ± 0.00	**−6.00** ± 0.00	**−5.99**	−6.02	−6.00	−6.02
MSNBC	−.08	−6.15	**−6.05**	−6.11 ± 0.00	−6.06 ± 0.00	−6.06	**−6.04**	**−6.04**	**−6.04**
KDDCup2k	−2.14	−2.15	−2.15	**−2.13** ± 0.00	**−2.13** ± 0.00	**−2.13**	**−2.13**	−2.17	−2.16
Plants	−12.32	−12.67	−12.59	−13.09 ± 0.01	**−11.99** ± 0.00	**−11.84**	−12.54	−12.80	−12.65
Audio	−40.09	−39.84	−40.19	−40.30 ± 0.02	**−39.77** ± 0.02	**−39.39**	−39.79	−40.32	−40.50
Jester	−52.88	−52.82	−52.99	−53.64 ± 0.03	**−52.65** ± 0.02	**−52.21**	−52.86	−53.31	−53.85
Netflix	−56.55	−56.44	−56.69	−57.64 ± 0.03	**−56.38** ± 0.03	**−55.93**	−56.36	−57.22	−57.03
Accidents	−29.88	−29.45	**−29.27**	−36.92 ± 0.05	−29.31 ± 0.02	−29.10	−26.98	−27.11	**−26.32**
Retail	−10.84	**−10.81**	−11.17	−10.88 ± 0.00	−10.93 ± 0.01	−10.91	−10.85	−10.88	−10.87
Pumsb-star	−23.98	−23.46	−23.78	−32.91 ± 0.02	**−23.44** ± 0.01	−23.31	−22.41	−23.55	**−21.72**
DNA	**−81.07**	−85.67	−85.95	−98.28 ± 0.06	−84.96 ± 0.03	−84.17	−81.21	**−80.03**	−80.65
Kosarek	−10.74	**−10.60**	−10.97	−10.91 ± 0.01	−10.72 ± 0.01	−10.66	**−10.60**	−10.84	−10.83
MSWeb	−9.77	−9.74	−9.93	−9.83 ± 0.00	**−9.66** ± 0.01	**−9.62**	−9.73	−9.77	−9.70
Book	−35.55	**−34.46**	−37.38	−34.77 ± 0.02	−36.35 ± 0.08	−35.45	**−34.14**	−35.56	−36.41
EachMovie	−53.00	**−51.53**	−54.14	−51.66 ± 0.11	−51.72 ± 0.12	**−50.34**	−51.51	−55.80	−54.37
WebKB	−153.12	**−152.53**	−155.47	−155.83 ± 0.30	−153.01 ± 0.28	**−149.20**	−151.84	−159.13	−157.43
Reuters-52	−83.71	**−83.69**	−86.19	−85.16 ± 0.15	−84.05 ± 0.24	**−81.87**	−83.35	−90.23	−87.55
20NewsG	−156.09	−153.12	−156.46	**−152.21** ± 0.19	−153.89 ± 0.15	**−151.02**	−151.47	−161.13	−158.95
BBC	**−237.42**	−247.01	−248.84	−251.31 ± 0.52	−238.47 ± 0.69	**−229.21**	−248.93	−257.10	−257.86
Ad	−15.28	−14.36	−15.55	−26.25 ± 0.08	**−14.20** ± 0.08	**−14.00**	−19.05	−16.53	−18.35
Avrg rank	**2.7**	**2.3**	**3.85**	**3.9**	**1.95**				
	4.7	**4.35**	**6.4**	**6.75**	**3.95**	**2.2**	**3.15**	**6.35**	**6.05**

these two random models Table 3 reports the average and the standard deviation over 10 different runs. Note that we are not performing bagging for our XCNet ensembles, since we do not draw bootstrapped samples of the data. This is motivated by the intuition that randomization is a form of diversification in the ensemble by itself, and it has been confirmed with a preliminary experimentation.

Next we compare CNet ensembles to other state-of-the-art TPMs learned by employing much more sophisticated models as ID-SPN [22], ACMN [17]. The first learns a complex hybrid architecture of SPNs and ACs while the latter learns high treewidth MNs represented as tractable ACs. Lastly, we employ the WinMine toolkit (WM) [3]. WM learns a treewidth *unbounded* BN exploiting context sensitive independencies by modeling its CPTs as trees. These models results are taken from [22]. The 40 component ensemble XCNet^{40} already delivers log-likelihoods comparable to those of the aforementioned models on more than half datasets. Nevertheless, we investigate the effect of building a large ensemble, up to 500 components (XCNet^{500}) by running a grid search over $\delta \in \{300, 500, 1000, 2000\}$, $\alpha = 0.1$ and $\sigma = 4$. On many datasets the log-likelihood scores of such an ensemble are the best achieved in the literature. Compared to XCNet^{40}, results from XCNet^{500} generally improve, however, on datasets like Nltcs and KDDCup2k the improvement saturated, suggesting that adding more components does not diversify the ensemble anymore. It is worth noting that XCNet^{40}_{PoB} is competitive on half datasets against a far more complex model like WM, yet outperforming it in terms of speed of learning and inference.

Table 4. Numbers of victories for the algorithms on the rows w.r.t those on columns.

Model	CNet^{40}	CNet$^{40}_{boost}$	dCSN^{40}	XCNet$^{40}_{PoB}$	XCNet^{40}	XCNet^{500}	ID-SPN	ACMN	WM	**Avrg**
CNET^{40}		7	15	16	6	1	7	14	15	**10.12**
CNET$^{40}_{boost}$	13		15	16	7	3	4	16	14	**11.00**
dCSN^{40}	4	4		12	2	1	3	11	12	**6.12**
XCNet$^{40}_{PoB}$	4	3	8		4	2	1	7	9	**4.75**
XCNet^{40}	13	13	17	15		0	7	14	15	**11.75**
XCNet^{500}	19	17	19	17	18		12	16	15	**16.62**
ID-SPN	13	15	17	18	11	7		16	13	**13.75**
ACMN	4	4	8	12	5	4	3		7	**5.87**
WM	5	6	8	11	5	5	5	12		**7.12**

We summarize comparisons among the algorithms in the first half of the table (resp. all algorithms) through ranking over the twenty datasets. For each dataset, we ranked the performance of the algorithms in the first half of the table (resp. all the algorithms) from 1 to 5 (resp. 9). The average rank of the algorithms is reported in the last two rows of the Table 3, showing that a mixture of XCNets performs the best. Finally, Table 4, reporting the number of victories for each algorithm w.r.t. the others, shows again the performances of mixtures of XCNet against the competitors that obtains 16.62 victories on average.

Table 5. Times (in seconds) taken to learn the best models on each dataset for dCSN, XCNet, dCSN$_{PoB}$, XCNet$_{PoB}$, their ensembles and ID-SPN with default parameters.

Dataset	dCSN	XCNet	dCSN$_{PoB}$	XCNet$_{PoB}$	dCSN^{40}	XCNet$^{40}_{PoB}$	XCNet^{40}	XCNet^{500}	ID-SPN
NLTCS	0	0.2	0.1	0.01	10	0.2	0	3	310
MSNBC	12	0.3	0.7	0.01	499	13.1	13	155	46266
KDDCup2k	112	0.5	12.0	0.32	4126	21.2	16	247	32067
Plants	15	0.3	45.5	0.22	325	1.0	6	77	18833
Audio	58	0.3	74.8	0.48	980	0.8	6	136	21009
Jester	50	0.2	95.6	0.26	989	0.3	4	83	10412
Netflix	75	0.2	2.8	0.02	1546	0.4	9	118	30294
Accidents	54	0.2	153.7	0.04	996	0.7	11	138	15472
Retail	263	0.8	5.8	0.01	3780	3.2	13	164	4041
Pumsb-star	118	0.6	26.2	0.02	2260	0.8	23	290	20952
DNA	30	0.1	4.4	0.01	224	0.06	3	40	3040
Kosarek	588	2.4	41.2	0.01	10033	10.8	43	524	17799
MSWeb	1215	7.2	7.4	0.01	17123	13.2	129	1592	19682
Book	9235	9.7	113.0	0.04	155634	1.9	316	3476	61248
EachMovie	1297	7.1	4.7	0.01	16962	1.1	127	2601	118782
WebKB	4997	11.0	238.0	0.03	18875	0.9	190	2237	45451
Reuters-52	9947	39.3	24.3	0.05	65498	2.7	414	8423	70863
20NewsG	16866	51.3	40.7	0.01	153908	4.4	506	9883	163256
BBC	21381	8.4	7.3	0.02	69572	0.4	256	4251	61471
Ad	5212	116.5	134.0	0.08	75694	4.2	2403	30538	87522

5.3 (Q3) Running Times

We derived the complexity for all considered variants of CNet learning schemes thus proving that XCNets are the ones scaling better w.r.t. the number of the features. Nevertheless, we empirically analyze XCNets learning times since we want (i) to evaluate whether and how much learning the leaf distribution actually impacts on real data, (ii) to compare the learning times of the density estimators employed in the previous sections. While a non-theoretical comparison may fall into the pitfalls of comparing different programming languages and optimization schemes, we provide it as a rule of thumb for practitioners to decide on which off-the-shelf density estimator toolbox to use.

In Table 5 we report the time, in seconds, spent by each algorithm to learn the best model on each dataset. Even increasing the number of components one order of magnitude more than what competitors are able to do in a reasonable time, XCNet still learn a competitive model (see Table 3) in time lesser than that of the competitors (see for instance the comparison w.r.t. ID-SPN).

6   Conclusions

We introduced XCNets, simplifying CNet learning through random conditioning. When learned in ensembles, XCNets achieve the new state-of-the-art results for density estimation on several benchmark datasets. Due to their simplicity to implement, fast learning times, and accurate inference performances, XCNets set the new baseline to compare against for density estimation with TPMs. As future work we plan to exploit their mixture of experts interpretation to devise more expressive gating functions that still perform exact and fast inference.

References

1. Bekker, J., Davis, J., Choi, A., Darwiche, A., Van den Broeck, G.: Tractable learning for complex probability queries. In: NIPS (2015)
2. Boutilier, C., Friedman, N., Goldszmidt, M., Koller, D.: Context-specific independence in Bayesian networks. In: UAI (1996)
3. Chickering, M.: The Winmine Toolkit. Microsoft, Redmond (2002)
4. Choi, A., Van den Broeck, G., Darwiche, A.: Tractable learning for structured probability spaces: a case study in learning preference distributions. In: IJCAI (2015)
5. Chow, C., Liu, C.: Approximating discrete probability distributions with dependence trees. IEEE Trans. Inf. Theory **14**, 462–467 (1968)
6. Darwiche, A.: A differential approach to inference in Bayesian networks. JACM **50**, 280–305 (2003)
7. Di Mauro, N., Vergari, A., Esposito, F.: Multi-label classification with cutset networks. In: PGM (2016)
8. Di Mauro, N., Vergari, A., Basile, T.: Learning Bayesian random cutset forests. In: ISMIS (2015)
9. Di Mauro, N., Vergari, A., Esposito, F.: Learning accurate cutset networks by exploiting decomposability. In: AIXIA (2015)

10. Geurts, P., Ernst, D., Wehenkel, L.: Extremely randomized trees. MLJ **63**, 3–42 (2006)
11. Haaren, J.V., Davis, J.: Markov network structure learning: a randomized feature generation approach. In: AAAI (2012)
12. Hastie, T., Tibshirani, R., Friedman, J.: The Elements of Statistical Learning. Springer, Heidelberg (2009). https://doi.org/10.1007/978-0-387-21606-5
13. Koller, D., Friedman, N.: Probabilistic Graphical Models: Principles and Techniques. MIT Press, Cambridge (2009)
14. Larochelle, H., Murray, I.: The neural autoregressive distribution estimator. In: AISTATS (2011)
15. Lowd, D., Davis, J.: Learning Markov network structure with decision trees. In: ICDM (2010)
16. Lowd, D., Domingos, P.: Naive Bayes models for probability estimation. In: ICML (2005)
17. Lowd, D., Rooshenas, A.: Learning Markov networks with arithmetic circuits. In: AISTATS (2013)
18. Meil, M., Jordan, M.I.: Learning with mixtures of trees. JMLR **1**, 1–48 (2000)
19. Poon, H., Domingos, P.: Sum-product network: a new deep architecture. In: NIPS Workshop on Deep Learning and Unsupervised Feature Learning (2011)
20. Rahman, T., Gogate, V.: Learning ensembles of cutset networks. In: AAAI (2016)
21. Rahman, T., Kothalkar, P., Gogate, V.: Cutset networks: a simple, tractable, and scalable approach for improving the accuracy of Chow-Liu trees. In: ECML/PKDD (2014)
22. Rooshenas, A., Lowd, D.: Learning sum-product networks with direct and indirect variable interactions. In: ICML, pp. 710–718 (2014)
23. Roth, D.: On the hardness of approximate reasoning. AI **82**, 273–302 (1996)
24. Scanagatta, M., Corani, G., de Campos, C.P., Zaffalon, M.: Learning treewidth-bounded Bayesian networks with thousands of variables. In: NIPS (2016)
25. Theis, L., van den Oord, A., Bethge, M.: A note on the evaluation of generative models. In: ICLR (2016)
26. Vergari, A., Di Mauro, N., Esposito, F.: Simplifying, regularizing and strengthening sum-product network structure learning. In: ECML/PKDD (2015)

Feature Selection and Extraction

Deep Discrete Hashing with Self-supervised Pairwise Labels

Jingkuan Song, Tao He, Hangbo Fan, and Lianli Gao[✉]

University of Electronic Science and Technology of China,
2006th Xiyuan Ave, Chengdu, China
jingkuan.song@gmail.com, tao.he@gmail.com, hbfan@gmail.com,
lianli.gao@uestc.edu.cn

Abstract. Hashing methods have been widely used for applications of large-scale image retrieval and classification. Non-deep hashing methods using handcrafted features have been significantly outperformed by deep hashing methods due to their better feature representation and end-to-end learning framework. However, the most striking successes in deep hashing have mostly involved discriminative models, which require labels. In this paper, we propose a novel unsupervised deep hashing method, named Deep Discrete Hashing (DDH), for large-scale image retrieval and classification. In the proposed framework, we address two main problems: (1) how to directly learn discrete binary codes? (2) how to equip the binary representation with the ability of accurate image retrieval and classification in an unsupervised way? We resolve these problems by introducing an intermediate variable and a loss function steering the learning process, which is based on the neighborhood structure in the original space. Experimental results on standard datasets (CIFAR-10, NUS-WIDE, and Oxford-17) demonstrate that our DDH significantly outperforms existing hashing methods by large margin in terms of mAP for image retrieval and object recognition. Code is available at https://github.com/htconquer/ddh.

1 Introduction

Due to the popularity of capturing devices and the high speed of network transformation, we are witnessing the explosive growth of images, which attracts great attention in computer vision to facilitate the development of multimedia search [35,42], object segmentation [22,34], object detection [26], image understanding [4,33] etc. Without a doubt, the ever growing abundance of images brings an urgent need for more advanced large-scale image retrieval technologies. To date, high-dimensional real-valued features descriptors (e.g., deep Convolutional Neural Networks (CNN) [30,37,39] and SIFT descriptors) demonstrate superior discriminability, and bridge the gap between low-level pixels and high-level semantic information. But they are less efficient for large-scale retrieval due to their high dimensionality.

Therefore, it is necessary to transform these high-dimensional features into compact binary codes which enable machines to run retrieval in real-time and

© Springer International Publishing AG 2017
M. Ceci et al. (Eds.): ECML PKDD 2017, Part I, LNAI 10534, pp. 223–238, 2017.
https://doi.org/10.1007/978-3-319-71249-9_14

with low memory. Existing hashing methods can be classified into two categories: *data-independent* and *data-dependent*. For the first category, hash codes are generated by randomly projecting samples into a feature space and then performing binarization, which is independent of any training samples. On the other hand, *data-dependent* hashing methods learn hash functions by exploring the distribution of the training data and therefore, they are also called learning to hashing methods (L2H) [43]. A lot of L2H methods have been proposed, such as Spectral hashing (SpeH) [45], iterative quantization (ITQ) [9], Multiple Feature Hashing (MFH) [36], Quantization-based Hashing (QBH) [32], K-means Hashing (KMH) [12], DH [24], DPSH [20], DeepBit [23], etc. Actually, those methods can be further divided into two categories: supervised methods and unsupervised methods. The difference between them is whether to use supervision information, e.g., classification labels. Some representative unsupervised methods include ITQ, Isotropic hashing [16], and DeepBit which achieves promising results, but are usually outperformed by supervised methods. By contrast, the supervised methods take full advantage of the supervision information. One representative is DPSH [20], which is the first method that can perform simultaneous feature learning and hash codes learning with pairwise labels. However, the information that can be used for supervision is also typically scarce.

To date, hand-craft floating-point descriptors such as SIFT, Speeded-up Robust Features (SURF) [2], DAISY [41], Multisupport Region Order-Based Gradient Histogram (MROGH) [8], the Multisupport Region Rotation and Intensity Monotonic Invariant Descriptor (MRRID) [8] etc., are widely utilized to support image retrieval since they are distinctive and invariant to a range of common image transformations. In [29], they propose a content similarity based fast reference frame selection algorithm for reducing the computational complexity of the multiple reference frames based inter-frame prediction. In [40], they develop a so-called correlation component manifold space learning (CCMSL) to learn a common feature space by capturing the correlations between the heterogeneous databases. Many attempts [21,25] were focusing on compacting such high quality floating-point descriptors for reducing computation time and memory usage as well as improving the matching efficiency. In those methods, the floating-point descriptor construction procedure is independent of the hash codes learning and still costs a massive amounts of time-consuming computation. Moreover, such hand-crafted feature may not be optimally compatible with hash codes learning. Therefore, these existing approaches might not achieve satisfactory performance in practice.

To overcome the limitation of existing hand-crafted feature based methods, some deep feature learning based deep hashing methods [7,10,11,20,46,47] have recently been proposed to perform simultaneous feature learning and hash-code learning with deep neural networks, which have shown better performance than traditional hashing methods with hand-crafted features. Most of these deep hashing methods are supervised whose supervision information is given as triplet or pairwise labels. An example is the deep supervised hashing method by Li *et al.* [20], which can simultaneously learn features and hash codes. Another example is Supervised Recurrent Hashing (SRH) [10] for generating hash codes of videos. Cao *et al.* [3] proposed a continuous method to learn binary codes, which can avoid the relaxation of binary constraints [10] by first learning continuous representations and then thresholding them to get the hash codes. They also added weight to data for balancing similar and dissimilar pairs.

In the real world, however, the vast majority of training data do not have labels, especially for scalable dataset. To the best of our knowledge, DeepBit [23] is the first to propose a deep neural network to learn binary descriptors in an unsupervised manner, by enforcing three criteria on binary codes. It achieves the state-of-art performance for image retrieval, but DeepBit does not consider the data distribution in the original image space. Therefore, DeepBit misses a lot of useful unsupervised information.

So can we obtain the pairwise information by exploring the data distribution, and then use this information to guide the learning of hash codes? Motivated by this, in this paper, we propose a Deep Discrete Hashing (DDH) with pseudo pairwise labels which makes use the self-generated labels of the training data as supervision information to improve the effectiveness of the hash codes. It is worth highlighting the following contributions:

1. We propose a general end-to-end learning framework to learn discrete hashing code in an unsupervised way to improve the effectiveness of hashing methods. The discrete binary codes are directly optimized from the training data. We solve the discrete hash, which is hard to optimize, by introducing an intermediate variable.
2. To explore the data distribution of the training images, we learn on the training dataset and generate the pairwise information. We then train our model by using this pairwise information in a self-supervised way.
3. Experiments on real datasets show that DDH achieves significantly better performance than the state-of-the-art unsupervised deep hashing methods in image retrieval applications and object recognition.

2   Our Method

Given N images, $\mathbf{I} = \{\mathbf{I}_i\}_{i=1}^{N}$ without labels, our goal is to learn their compact binary codes $\mathbf{B}$ such that: (a) the binary codes can preserve the data distribution in the original space, and (b) the discrete binary codes could be computed directly.

As shown in Fig. 1, our DDH consists of two key components: construction of pairwise labels, and hashing loss definition. For training, we first construct the neighborhood structure of images and then train the network. For testing, we obtain the binary codes of an image by taking it as an input. In the remainder of this section, we first describe the process of constructing the neighborhood structure, and then introduce our loss function and the process of learning the parameters.

2.1   Construction of Pairwise Labels

In our unsupervised approach, we propose to exploit the neighborhood structure of the images in a feature space as information source steering the process of hash learning. Specifically, we propose a method based on the K-Nearest Neighbor (KNN) concept to create a neighborhood matrix **S**. Based on [13], we extract 2,048-dimensional features from the pool5-layer, which is last layer of ResNet [13]. This results in the set $\mathbf{X} = \{\mathbf{x}_i\}_{i=1}^N$ where $\mathbf{x}_i$ is the feature vector of image $\mathbf{I}_i$.

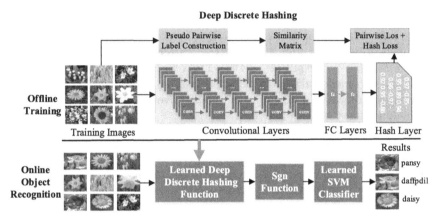

Fig. 1. The structure of our end-to-end framework. It has two components, construction of pairwise labels, and hashing loss definition. We first construct the neighborhood structure of images and then train the network based on the define loss function. We utilize the deep neural network to extract the features of the images.

For the representation of the neighboring structure, our task is to construct a matrix $\boldsymbol{S} = (s_{ij})_{i,j=1}^N$, whose elements indicate the similarity ($s_{ij} = 1$) or dissimilarity ($s_{ij} = -1$) of any two images i and j in terms of their features $\mathbf{x}_i$ and $\mathbf{x}_j$.

We compare images using cosine similarity of the feature vectors. For each image, we select K_1 images with the highest cosine similarity as its neighbors. Then we can construct an initial similarity matrix $\mathbf{S}_1$:

$$(S_1)_{ij} = \begin{cases} 1, & \text{if } \mathbf{x}_j \text{ is } K_1\text{-NN of } \mathbf{x}_i \\ 0, & \text{otherwise} \end{cases} \tag{1}$$

Here we use $\mathbf{L}_1, \mathbf{L}_2, \ldots, \mathbf{L}_N$ to denote the ranking lists of points $\mathbf{I}_1, \mathbf{I}_2, \ldots, \mathbf{I}_N$ by K_1-NN. The precision-recall curve in Fig. 2 indicates the quality of the constructed neighborhood for different values of K_1. Due to the rapidly decreasing precision with the increase of K_1, creating a large-enough neighborhood by simply increasing K_1 is not the best option. In order to find a better approach, we borrow the ideas from the domain of graph modeling. In an undirected graph, if a node v is connected to a node u and if u is connected to a node w, we can infer that v is also connected to w. Inspired by this, if we treat every training image as a node in an undirected graph, we can expand the neighborhood of an image i by exploring the neighbors of its neighbors. Specifically, if $\mathbf{x}_i$ connects to $\mathbf{x}_j$ and $\mathbf{x}_j$ connects to $\mathbf{x}_k$, we can infer that $\mathbf{x}_i$ has the potential to be connected to $\mathbf{x}_k$ as well.

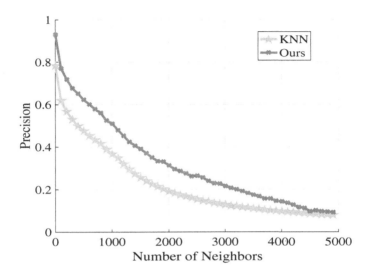

Fig. 2. Precision of constructed labels on cifar-10 dataset with different K, and different methods.

Based on the above observations, we construct $\mathbf{S}_1$ using the deep CNN features. If we only use the constructed labels by $\mathbf{S}_1$, each image has too few positive labels with high precision. So we increase the number of neighbors based on $\mathbf{S}_1$ to obtain more positive labels. Specifically, we calculate the similarity of two images by comparing the two ranking lists of K_1-NN using the expression $\frac{1}{\|\mathbf{L}_i - \mathbf{L}_j\|^2}$. Actually, if two images have the same labels, they should have a lot of intersection points based on K_1-NN, i.e., they have similar K_1-NN ranking list. Then we again construct a ranking list of K_2 neighbors, based on which we generate the second similarity matrix $\mathbf{S}_2$ as:

$$(\mathbf{S}_2)_{ij} = \begin{cases} 1, & \text{if } \mathbf{L}_j \text{ is } K_2\text{-NN of } \mathbf{L}_i \\ 0, & \text{otherwise} \end{cases} \tag{2}$$

Finally, we construct the neighborhood structure by combining the direct and indirect similarities from the two matrices together. This results in the final similarity matrix $\mathbf{S}$:

$$S_{ij} = \begin{cases} 1, & \text{if } (\mathbf{S}_2)_{ik} = 1 \text{ and } j \text{ in } \mathbf{L}_k \\ 0, & \text{otherwise} \end{cases} \tag{3}$$

where the $\mathbf{L}_k$ is the ranking list after K_1-NN. The whole algorithm is shown in Algorithm 1. After the two steps KNN, the constructed label precision is shown in Fig. 2. We note that we could have also omitted this preprocessing step and construct the neighborhood structure directly during the learning of our neural network. We found, however, that the construction of neighborhood structure is time-consuming, and that updating of this structure based on the updating of image features in each epoch does not have significant impact on the performance. Therefore, we chose to obtain this neighborhood structure as described above and fix it for the rest of the process.

Algorithm 1. Construction of neighborhood structure

Input: Images $\mathbf{X} = \{\mathbf{x}_i\}_{i=1}^N$, the number of neighbors K_1, the number of neighbors K_2 for the neighbors expansion;
Output: Neighborhood matrix $\mathbf{S} = \{s_{ij}\}$;
 1: First ranking: Use cosine similarity to generate the index of K_1-NN of each image $L_1, L_2, \ldots, L_N$;
 2: Neighborhood expansion:
 3: **for** $i=1,\ldots,N$ **do**
 4:     Initialize $num \leftarrow \emptyset$;
 5:     **for** $j = 1, \ldots N$ **do**
 6:         $num_j \leftarrow$ the size of $L_i \cap L_j$;
 7:     **end for**
 8:     Sort num by descending order and keep the top K_2 $\{L_j\}$;
 9:     Set new $L'_i \leftarrow$ union of the top K_2 $\{L_j\}$;
10: **end for**
11: **for** $j=1,\ldots,N$ **do**
12:     Construct $\mathbf{S}$ with new L'_j base on Eq. 3;
13: **end for**
14: **return** $\mathbf{S}$;

2.2   Architecture of Our Network

We introduce an unsupervised deep framework, dubbed Deep Discrete Hashing (DDH), to learn compact yet discriminative binary descriptors. The framework includes two main modules, feature learning part and hash codes learning part, as shown in Fig. 1. More specifically, for the feature learning, we use a similar network as in [48], which has seven layers and the details are shown in Table 1. In the experiment, we can easily replace the CNN-F network with other deep networks such as [13,18,38]. Our framework has two branches with the shared weights and both of them have the same weights and same network structure.

We discard the last softmax layer and replace it with a hashing layer, which consists of a fully connected layer and a sgn activation layer to generate compact codes. Specifically, the output of the $full_7$ is firstly mapped to a L-dimensional real-value code, and then a binary hash code is learned directly, by converting the L-dimensional representation to a binary hash code $\mathbf{b}$ taking values of either $+1$ or -1. This binarization process can only be performed by taking the sign function $\mathbf{b} = sgn(\mathbf{u})$ as the activation function on top of the hash layer.

$$\mathbf{b} = \text{sgn}(\mathbf{u}) = \begin{cases} +1, & if \ \mathbf{u} \geq 0 \\ -1, & \text{otherwise} \end{cases} \tag{4}$$

Table 1. The configuration of our framework

Layer	Configure
$conv_1$	filter $64 \times 11 \times 11$, stride 4×4, pad 0, LRN, pool 2×2
$conv_2$	filter $256 \times 5 \times 5$, stride 1×1, pad 2, LRN, pool 2×2
$conv_3$	filter $256 \times 3 \times 3$, stride 1×1, pad 1
$conv_4$	filter $256 \times 3 \times 3$, stride 1×1, pad 1
$conv_5$	filter $256 \times 3 \times 3$, stride 1×1, pad 1, pool 2×2
$full_6$	4096
$full_7$	4096
$hash\ layer$	L

2.3 Objective Function

Suppose we denote the binary codes as $\mathbf{B} = \{\mathbf{b}_i\}_{i=1}^{N}$ for all the images. The neighborhood structure loss models the loss in the similarity structure in data, as revealed in the set of neighbors obtained for an image by applying the hash code of that image. We define the loss function as below:

$$\min J_1 = \frac{1}{2} \sum_{s_{ij} \in S} \left(\frac{1}{L} \mathbf{b}_i^T \mathbf{b}_j - s_{ij} \right)^2 \tag{5}$$

where L is the length of hashing code and $s_{ij} \in \{-1, 1\}$ indicates the similarity of image i and j. The goal of optimizing for this loss function is clearly to bring the binary codes of similar images as close to each other as possible.

We also want to minimize the quantization loss between the learned binary vectors $\mathbf{B}$ and the original real-valued vectors $\mathbf{Z}$. It is defined as:

$$\min J_2 = \frac{1}{2} \sum_{i=1}^{N} \|\mathbf{z}_i - \mathbf{b}_i\|^2 \tag{6}$$

where $\mathbf{z}_i$ and $\mathbf{b}_i$ are the real-valued representation and binary codes of the i-th image in the hashing layer. Then we can obtain our final objective function as:

$$\min J = J_1 + \lambda_1 J_2, \qquad \mathbf{b}_i \in \{-1, 1\}^L, \quad \forall i = 1, 2, 3, \ldots N \tag{7}$$

where λ_1 is the parameter to balance these two terms.

Obviously, the problem in (7) is a discrete optimization problem, which is hard to solve. LFH [48] solves it by directly relaxing $\mathbf{b}_i$ from discrete to continuous, which might not achieve satisfactory performance [15]. In this paper, we design a novel strategy which can solve the problem 5 by introducing an intermediate variable. First, we reformulate the problem in 5 as the following equivalent one:

$$\min J = \frac{1}{2} \sum_{s_{ij} \in S} \left(\frac{1}{L} \mathbf{b}_i{}^T \mathbf{b}_j - s_{ij} \right)^2 + \frac{\lambda_1}{2} \sum_{i=1}^{N} \|\mathbf{z}_i - \mathbf{b}_i\|^2$$

$$s.t \quad \mathbf{u}_i = \mathbf{b}_i, \forall i = 1, 2, 3, \ldots N$$

$$\mathbf{u}_i \in \mathbb{R}^{L \times 1}, \forall i = 1, 2, 3, \ldots N \tag{8}$$

$$\mathbf{b}_i \in \{-1, 1\}^L, \forall i = 1, 2, 3, \ldots N$$

where $\mathbf{u}_i$ is an intermediate variable and $\mathbf{b}_i = sgn(\mathbf{u}_i)$. To optimize the problem in 8, we can optimize the following regularized problem by moving the equality constraints in 8 to the regularization terms:

$$\min J = \frac{1}{2} \sum_{s_{ij} \in \mathbf{S}} \left(\frac{1}{L} \mathbf{u}_i{}^T \mathbf{u}_j - s_{ij} \right)^2 + \frac{\lambda_1}{2} \sum_{i=1}^{N} \|\mathbf{z}_i - \mathbf{b}_i\|^2 + \frac{\lambda_2}{2} \sum_{i=1}^{N} \|\mathbf{b}_i - \mathbf{u}_i\|^2$$

$$s.t \quad \mathbf{u}_i \in \mathbb{R}^{L \times 1}, \forall i = 1, 2, 3, \ldots N$$

$$\mathbf{b}_i \in \{-1, 1\}^L, \forall i = 1, 2, 3, \ldots N$$

$$\tag{9}$$

where λ_2 is the hyper-parameter for the regularization term. Actually, introducing an intermediate variable $\mathbf{u}$ is equivalent to adding another full-connected layer between $\mathbf{z}$ and $\mathbf{b}$ in the hashing layer. To reduce the complexity of our model, we let $\mathbf{z} = \mathbf{u}$, and then we can have a simplified objective function as:

$$\min J = \frac{1}{2} \sum_{s_{ij} \in \mathbf{S}} \left(\frac{1}{K} \mathbf{z}_i{}^T \mathbf{z}_j - s_{ij} \right)^2 + \frac{\lambda_1}{2} \sum_{i=1}^{N} \|\mathbf{z}_i - \mathbf{b}_i\|^2$$

$$s.t \quad \mathbf{b}_i \in \{-1, 1\}^L, \forall i = 1, 2, 3, \ldots N \tag{10}$$

Equation 10 is not discrete and $\mathbf{z}_i$ is derivable, so we can use back-propagation (BP) to optimize it.

2.4   Learning

To learning DDH Model, we need to obtain the parameters of neural networks. We set

$$\mathbf{z}_i = \mathbf{W}^T \phi(x_i; \theta) + \mathbf{c}$$

$$\mathbf{b}_i = sgn(\mathbf{z}_i) = sgn(\mathbf{W}^T \phi(x_i; \theta) + \mathbf{c}) \tag{11}$$

where θ denotes all the parameters of CNN-F network for learning the features. $\phi(x_i; \theta)$ denotes the output of the $full_7$ layer associated with image x_i. $\mathbf{W} \in$

$\mathbb{R}^{4096 \times L}$ denotes hash layer weights matrix, and $\mathbf{C} \in \mathbb{R}^{L \times 1}$ is a bias vector. We add regularization terms on the parameters and change the loss function with 10 constrains as:

$$\min J = \frac{1}{2} \sum_{s_{ij} \in S} \left(\frac{1}{L}\Theta_{ij} - s_{ij}\right)^2$$
$$+ \frac{\lambda_1}{2} \sum_{i=1}^{N} \left\|\mathbf{b}_i - (\mathbf{W}^T \phi(x_i; \theta) + \mathbf{c})\right\|^2 \qquad (12)$$
$$+ \frac{\lambda_2}{2} (\|\mathbf{W}\|_F^2 + \|\mathbf{c}\|_F^2)^2$$

where $\Theta_{ij} = z_i^T z_j$, λ_1 and λ_2 are two parameters to balance the effect of different terms. Stochastic gradient descent (SGD) is used to learn the parameters. We use CNN-F network trained on ImageNet to initialize our network. In particular, in each iteration we sample a mini-batch of points from the whole training set and use back-propagation (BP) to optimize the whole network. Here, we compute the derivatives of the loss function as follows:

$$\frac{\partial J}{\partial \mathbf{z}_i} = \frac{1}{L^2}(\mathbf{z}_i^T \mathbf{z}_j)\mathbf{z}_j - \frac{1}{L}s_{ij}\mathbf{z}_j + \lambda_1(\mathbf{z}_i - \mathbf{b}_i) \qquad (13)$$

2.5 Out-of-Sample Extension

After the network has been trained, we still need to obtain the hashing codes of the images which are not in the training data. For a novel image, we obtain its binary code by inputing it into the DDH network and make a forward propagation as below:

$$\mathbf{b}_i = sgn(\mathbf{z}_i) = sgn(\mathbf{W}^T \phi(x_i; \theta) + \mathbf{c})$$

3 Experiment

Our experiment PC is configured with an Intel(R) Xeon(R) CPU E5–2650 v3 @ 2.30 GHz with 40 cores and the the RAM is 128.0 GB and the GPU is GeForce GTX TITAN X with 12 GB.

3.1 Datasets

We conduct experiments on three challenging datasets, the Oxford 17 Category Flower Dataset, the CIFAR-10 color images, and the NUS-WIDE. We test our binary descriptor on various tasks, including image retrieval and image classification.

1. **CIFAR-10 Dataset** [17] contains 10 object categories and each class consists of 6,000 images, resulting in a total of 60,000 images. The dataset is split into training and test sets, with 50,000 and 10,000 images respectively.
2. **NUS-WIDE dataset** [5] has nearly 270,000 images collected from the web. It is a multi-label dataset in which each image is annotated with one or multiple class labels from 81 classes. Following [19], we only use the images associated with the 21 most frequent classes. For these classes, the number of images of each class is at least 5000. We use 4,000 for training and 1,000 for testing.

3. **The Oxford 17 Category Flower Dataset** [27] contains 17 categories and each class consists of 80 images, resulting in a total of 1,360 images. The dataset is split into the training (40 images per class), validation (20 images per class), and test (20 images per class) sets.

3.2   Results on Image Retrieval

To evaluate the performance of the proposed DDH, we test our method on the task of image retrieval. We compare DDH with other hashing methods, such as LSH [1], ITQ [9], HS [31], Spectral hashing (SpeH) [45], Spherical hashing (SphH) [14], KMH [12], Deep Hashing (DH) [24] and DeepBit [23], Semi-supervised PCAH [44] on the CIFAR-10 dataset and NUS-WIDE. We set the $K_1 = 15$ and $K_2 = 6$ to construct labels, and the learning rate as 0.001, $\lambda_1 = 15$, $\lambda_2 = 0.00001$ and batch-size $= 128$. Table 2 shows the CIFAR-10 retrieval results based on the mean Average Precision (mAP) of the top 1,000 returned images with respect to different bit lengths, while Table 3 shows the mAP value of NUS-WIDE dataset calculated based on the top 5,000 returned neighbors. The precision/recall in CIFAR-10 dataset is shown in Fig. 3.

Table 2. Performance comparison (mAP) of different unsupervised hashing algorithms on the CIFAR-10 dataset. The mean Average Precision (mAP) are calculated based on the top 1,000 returned images with respect to different number of hash bits.

Method	16 bit	32 bit	64 bit
Method	16 bit	32 bit	64 bit
KMH	0.136	0.139	0.145
SphH	0.145	0.146	0.154
SH	0.130	0.141	0.139
PCAH	0.129	0.126	0.121
LSH	0.126	0.138	0.157
PCA-ITQ	0.157	0.162	0.166
DH	0.162	0.166	0.170
DeepBit	0.194	0.249	0.277
DDH	**0.447**	**0.486**	**0.535**

From these results, we have the following observations:

(1) Our method significantly outperforms the other deep or non-deep hashing methods in all datasets. In CIFAR-10, the improvement of DDH over the other methods is more significant, compared with that in NUS-WIDE dataset. Specifically, it outperforms the best counterpart (DeepBit) by 25.3%, 23.7% and 25.8% for 16, 32 and 64-bit hash codes. One possible reason is that CIFAR-10 contains simple images, and the constructed neighborhood

Table 3. Performance comparison (mAP) of different unsupervised hashing algorithms on the NUS-WIDE dataset. The mAP is calculated based on the top 5,000 returned neighbors for NUS-WIDE dataset.

Method	12 bit	24 bit	32 bit	48 bit
CNNH	0.611	0.618	0.625	0.608
FastH	**0.621**	**0.650**	**0.665**	**0.685**
SDH	0.568	0.600	0.608	0.637
KSH	0.556	0.572	0.581	0.588
LFH	0.571	0.568	0.568	0.585
ITQ	0.452	0.468	0.472	0.477
SH	0.454	0.406	0.405	0.400
DDH	**0.675**	**0.680**	**0.701**	**0.712**

structure is more accurate than the other two datasets. DDH improves the state-of-the-arts by 5.4%, 3.0%, 3.6% and 2.7% in NUS-WIDE dataset.

(2) Table 2 shows that DeepBit and FashH are strong competitors in terms of mAP in CIFAR-10 and NUS-WIDE dataset. But the performance gap of DeepBit and our DDH is still very large, which is probably due to that DeepBit uses only 3 fully connected layers to extract the features. Figure 3 shows that most of the hashing methods can achieve a high recall for small number of retrieved samples (or recall). But obviously, our DDH significantly outperforms the others.

(3) With the increase of code length, the performance of most hashing methods is improved accordingly. An exception is PCAH, which has no improvement with the increase of code length.

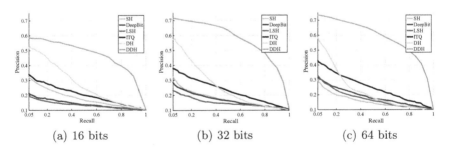

| (a) 16 bits | (b) 32 bits | (c) 64 bits |

Fig. 3. Precision/recall curves of different unsupervised hashing methods on the CIFAR-10 dataset with respect to 16, 32 and 64 bits, respectively

To make fair comparison with the non-deep hashing methods, and validate that our improvement is not only caused by the deep features, we conduct non-deep hashing methods with deep features extracted by the CNN-F pre-trained on ImageNet. The results are reported in Table 4, where "ITQ+CNN" denotes the ITQ method with deep features and other methods have similar notations. When we run the non-deep hashing methods on deep features, the performance is usually improved compared with the hand-crafted features.

Table 4. Performance comparison (mAP) of different hashing algorithms with deep features on the CIFAR-10 dataset.

Method	12 bit	24 bit	32 bit	48 bit
ITQ + CNN	0.237	0.246	0.255	0.261
SH + CNN	0.183	0.164	0.161	0.161
SPLH + CNN	**0.299**	**0.330**	**0.335**	**0.330**
LFH + CNN	0.208	0.242	0.266	0.339
DDH	**0.414**	**0.467**	**0.486**	**0.512**

By constructing the neighborhood structure using the labels, our method can be easily modified as a supervised hashing method. Therefore, we also compared with supervised hashing methods, and show the mAP results on NUS-WIDE dataset in Table 5. It is obvious that our DDH outperforms the state-of-the-art deep and non-deep supervised hashing algorithms by a large margin, which are 5.7%, 5.8%, 7.8% and 8.1% for 12, 24, 32, and 48-bits hash codes. This indicates that the performance improvement of DDH is not only due to the constructed neighborhood structure, but also the other components.

Table 5. Results on NUS-WIDE. The table shows other deep network with supervised pair-wise labels. The mAP value is calculated based on the top 5000 returned neighbors for NUS-WIDE dataset.

Method	16 bit	24 bit	32 bit	48 bit
DRSCH	**0.618**	**0.622**	**0.623**	0.628
DSCH	0.592	0.597	0.611	0.609
DSRH	0.609	0.618	0.621	**0.631**
DDH	**0.675**	**0.680**	**0.701**	**0.712**

3.3 Results on Object Recognition

In the task of object recognition, the algorithm needs to recognize very similar object (daisy, iris and pansy etc.). So it requires more discriminative binary codes to represent images that look very similar. In this paper, we use the Oxford 17 Category Flower Dataset to evaluate our method on object recognition and we compared with several real-valued descriptors such as HOG [6] and SIFT.

Due to the variation of color distributions, pose deformations and shapes, "Flower" recognition becomes more challenging. Besides, we need to consider the computation cost while one wants to recognize the flowers in the wild using mobile devices, which makes us generate very short and efficient binary codes to discriminate flowers. Following the setting in [27], we train a multi-class SVM classifier with our proposed binary descriptor. Table 6 compares the classification accuracy of the 17 categories flowers using different descriptors proposed in [27], [28], including low-level (Color, Shape, Texture), and high-level (SIFT and HOG) features. Our proposed binary descriptor with 256 dimensionality improves previous best recognition accuracy by around 5.01% (80.11% vs. 75.1%). We also test our proposed method with 64 bits, which still outperforms the state-of-art result (76.35% vs. 75.1%). We also test the computational complexity during SVM classifier training with only costing 0.3 s training on 256 bits and 0.17s on 64 bits. Compared with other descriptors, such as Color, Shape, Texture, HOG, HSV and SIFT, DDH demonstrates its efficiency and effectiveness.

Table 6. The categorization accuracy (mean%) and training time for different features on the Oxford 17 Category Flower Dataset

Descriptors	Accuracy	Time
Colour	60.9	3
Shape	70.2	4
Texture	63.7	3
HOG	58.5	4
HSV	61.3	3
SIFT-boundary	59.4	5
SIFT-internal	70.6	4
DeepBit (256bits)	**75.1**	**0.07**
DDH (64bits)	76.4	0.12
DDH(256bits)	**80.5**	**0.30**

4   Conclusion and Future Work

In this work, we address two central problems remaining largely unsolved for image hashing: (1) how to directly generate binary codes without relaxation? (2) how to equip the binary representation with the ability of accurate image retrieval? We resolve these problems by introducing an intermediate variable and a loss function steering the learning process, which is based on the neighborhood structure in the original space. Experiments on real datasets show that our method can outperform other unsupervised and supervised methods to achieve the state-of-the-art performance in image retrieval and object recognition. In the future, it is necessary to improve the classification accuracy by incorporating a classification layer at the end of this architecture.

Acknowledgment. This work was supported in part by the National Natural Science Foundation of China under Project 61502080, Project 61632007, and the Fundamental Research Funds for the Central Universities under Project ZYGX2016J085, Project ZYGX2014Z007.

References

1. Andoni, A., Indyk, P.: Near-optimal hashing algorithms for approximate nearest neighbor in high dimensions. Commun. ACM **51**(1), 117–122 (2008)
2. Bay, H., Ess, A., Tuytelaars, T., Gool, L.V.: Speeded-up robust features (SURF). Comput. Vis. Image Underst. **110**(3), 346–359 (2008). Similarity Matching in Computer Vision and Multimedia
3. Cao, Z., Long, M., Wang, J., Yu, P.S.: HashNet: deep learning to hash by continuation. arXiv preprint arXiv:1702.00758 (2017)
4. Chen, L., Zhang, H., Xiao, J., Nie, L., Shao, J., Chua, T.S.: SCA-CNN: spatial and channel-wise attention in convolutional networks for image captioning. arXiv preprint arXiv:1611.05594 (2016)
5. Chua, T.S., Tang, J., Hong, R., Li, H., Luo, Z., Zheng, Y.: NUS-WIDE: a real-world web image database from national university of Singapore. In: ACM International Conference on Image and Video Retrieval, p. 48 (2009)
6. Dalal, N., Triggs, B.: Histograms of oriented gradients for human detection. In: CVPR, vol. 1, pp. 886–893. IEEE (2005)
7. Do, T.-T., Doan, A.-D., Cheung, N.-M.: Learning to hash with binary deep neural network. In: Leibe, B., Matas, J., Sebe, N., Welling, M. (eds.) ECCV 2016. LNCS, vol. 9909, pp. 219–234. Springer, Cham (2016). https://doi.org/10.1007/978-3-319-46454-1_14
8. Fan, B., Wu, F., Hu, Z.: Rotationally invariant descriptors using intensity order pooling. IEEE Trans. Pattern Anal. Mach. Intell. **34**(10), 2031–2045 (2012)
9. Gong, Y., Lazebnik, S., Gordo, A., Perronnin, F.: Iterative quantization: a procrustean approach to learning binary codes for large-scale image retrieval. IEEE Trans. Pattern Anal. Mach. Intell. **35**(12), 2916–2929 (2013)
10. Gu, Y., Ma, C., Yang, J.: Supervised recurrent hashing for large scale video retrieval. In: ACM Multimedia, pp. 272–276. ACM (2016)
11. Guo, J., Zhang, S., Li, J.: Hash learning with convolutional neural networks for semantic based image retrieval. In: Bailey, J., Khan, L., Washio, T., Dobbie, G., Huang, J.Z., Wang, R. (eds.) PAKDD 2016. LNCS (LNAI), vol. 9651, pp. 227–238. Springer, Cham (2016). https://doi.org/10.1007/978-3-319-31753-3_19
12. He, K., Wen, F., Sun, J.: K-means hashing: an affinity-preserving quantization method for learning binary compact codes. In: NIPS, pp. 2938–2945 (2013)
13. He, K., Zhang, X., Ren, S., Sun, J.: Deep residual learning for image recognition. arXiv preprint arXiv:1512.03385 (2015)
14. Heo, J.P., Lee, Y., He, J., Chang, S.F., Yoon, S.E.: Spherical hashing. In: CVPR, pp. 2957–2964. IEEE (2012)
15. Kang, W.C., Li, W.J., Zhou, Z.H.: Column sampling based discrete supervised hashing. In: AAAI (2016)

16. Kong, W., Li, W.J.: Isotropic hashing. In: Pereira, F., Burges, C.J.C., Bottou, L., Weinberger, K.Q. (eds.) Advances in Neural Information Processing Systems, vol. 25, pp. 1646–1654. Curran Associates, Inc. (2012)
17. Krizhevsky, A.: Learning multiple layers of features from tiny images (2012)
18. Krizhevsky, A., Sutskever, I., Hinton, G.E.: ImageNet classification with deep convolutional neural networks. In: NIPS, pp. 1097–1105 (2012)
19. Lai, H., Pan, Y., Liu, Y., Yan, S.: Simultaneous feature learning and hash coding with deep neural networks. In: CVPR, pp. 3270–3278 (2015)
20. Li, W., Wang, S., Kang, W.: Feature learning based deep supervised hashing with pairwise labels. In: IJCAI, pp. 1711–1717 (2016)
21. Li, X., Shen, C., Dick, A., van den Hengel, A.: Learning compact binary codes for visual tracking. In: CVPR, pp. 2419–2426 (2013)
22. Li, Y., Liu, J., Wang, Y., Lu, H., Ma, S.: Weakly supervised RBM for semantic segmentation. In: IJCAI, pp. 1888–1894 (2015)
23. Lin, K., Lu, J., Chen, C.S., Zhou, J.: Learning compact binary descriptors with unsupervised deep neural networks. In: CVPR, June 2016
24. Liong, V.E., Lu, J., Wang, G., Moulin, P., Zhou, J.: Deep hashing for compact binary codes learning. In: CVPR, pp. 2475–2483, June 2015
25. Brown, M., Hua, G., Winder, S.: Discriminative learning of local image descriptors. IEEE Trans. Pattern Anal. Mach. Intell. (2010)
26. Nguyen, T.V., Sepulveda, J.: Salient object detection via augmented hypotheses. In: IJCAI, pp. 2176–2182 (2015)
27. Nilsback, M.E., Zisserman, A.: A visual vocabulary for flower classification. In: CVPR, vol. 2, pp. 1447–1454. IEEE (2006)
28. Nilsback, M.E., Zisserman, A.: Automated flower classification over a large number of classes. In: Sixth Indian Conference on Computer Vision, Graphics & Image Processing. ICVGIP 2008, pp. 722–729. IEEE (2008)
29. Pan, Z., Jin, P., Lei, J., Zhang, Y., Sun, X., Kwong, S.: Fast reference frame selection based on content similarity for low complexity HEVC encoder. J. Vis. Commun. Image Represent. 40, 516–524 (2016)
30. Russakovsky, O., Deng, J., Su, H., Krause, J., Satheesh, S., Ma, S., Huang, Z., Karpathy, A., Khosla, A., Bernstein, M.S., Berg, A.C., Li, F.: Imagenet large scale visual recognition challenge. Int. J. Comput. Vis. 115(3), 211–252 (2015)
31. Salakhutdinov, R., Hinton, G.: Semantic hashing. Int. J. Approximate Reasoning 50(7), 969–978 (2009)
32. Song, J., Gao, L., Liu, L., Zhu, X., Sebe, N.: Quantization-based hashing: a general framework for scalable image and video retrieval. Pattern Recognition (2017)
33. Song, J., Gao, L., Nie, F., Shen, H.T., Yan, Y., Sebe, N.: Optimized graph learning using partial tags and multiple features for image and video annotation. IEEE Trans. Image Process. 25(11), 4999–5011 (2016)
34. Song, J., Gao, L., Puscas, M.M., Nie, F., Shen, F., Sebe, N.: Joint graph learning and video segmentation via multiple cues and topology calibration. In: ACM Multimedia, pp. 831–840 (2016)
35. Song, J., Yang, Y., Yang, Y., Huang, Z., Shen, H.T.: Inter-media hashing for large-scale retrieval from heterogeneous data sources. In: SIGMOD, pp. 785–796 (2013)
36. Song, J., Yang, Y., Huang, Z., Shen, H.T., Luo, J.: Effective multiple feature hashing for large-scale near-duplicate video retrieval. IEEE Trans. Multimedia 15(8), 1997–2008 (2013)
37. Szegedy, C., Ioffe, S., Vanhoucke, V.: Inception-v4, inception-ResNet and the impact of residual connections on learning. CoRR abs/1602.07261 (2016)

38. Szegedy, C., Liu, W., Jia, Y., Sermanet, P., Reed, S., Anguelov, D., Erhan, D., Vanhoucke, V., Rabinovich, A.: Going deeper with convolutions. In: CVPR, pp. 1–9 (2015)
39. Targ, S., Almeida, D., Lyman, K.: Resnet in Resnet: generalizing residual architectures. CoRR abs/1603.08029 (2016)
40. Tian, Q., Chen, S.: Cross-heterogeneous-database age estimation through correlation representation learning. Neurocomputing **238**, 286–295 (2017)
41. Tola, E., Lepetit, V., Fua, P.: Daisy: an efficient dense descriptor applied to wide-baseline stereo. IEEE Trans. Pattern Anal. Mach. Intell. **32**(5), 815–830 (2010)
42. Wan, J., Wu, P., Hoi, S.C.H., Zhao, P., Gao, X., Wang, D., Zhang, Y., Li, J.: Online learning to rank for content-based image retrieval. In: IJCAI, pp. 2284–2290 (2015)
43. Wang, J., Zhang, T., Song, J., Sebe, N., Shen, H.T., et al.: A survey on learning to hash. IEEE Trans. Pattern Anal. Mach. Intell. (2017)
44. Wang, J., Kumar, S., Chang, S.F.: Semi-supervised hashing for scalable image retrieval. In: CVPR, pp. 3424–3431 (2010)
45. Weiss, Y., Torralba, A., Fergus, R.: Spectral hashing. In: Koller, D., Schuurmans, D., Bengio, Y., Bottou, L. (eds.) NIPS, pp. 1753–1760 (2009)
46. Xia, R., Pan, Y., Lai, H., Liu, C., Yan, S.: Supervised hashing for image retrieval via image representation learning. In: AAAI, pp. 2156–2162 (2014)
47. Yang, H.F., Lin, K., Chen, C.S.: Supervised learning of semantics-preserving hash via deep convolutional neural networks. IEEE Trans. Pattern Anal. Mach. Intell. (2017)
48. Zhang, P., Zhang, W., Li, W.J., Guo, M.: Supervised hashing with latent factor models. In: SIGIR, pp. 173–182 (2014)

Including Multi-feature Interactions and Redundancy for Feature Ranking in Mixed Datasets

Arvind Kumar Shekar[1(✉)], Tom Bocklisch[2], Patricia Iglesias Sánchez[1], Christoph Nikolas Straehle[1], and Emmanuel Müller[2]

[1] Robert Bosch GmbH, Stuttgart, Germany
{arvindkumar.shekar,patricia.iglesiassanchez,
Christoph-Nikolas.Straehle}@de.bosch.com
[2] Hasso Plattner Institute, Potsdam, Germany
{tom.bocklisch,emmanuel.mueller}@hpi.de

Abstract. Feature ranking is beneficial to gain knowledge and to identify the relevant features from a high-dimensional dataset. However, in several datasets, few features by itself might have small correlation with the target classes, but by combining these features with some other features, they can be strongly correlated with the target. This means that multiple features exhibit interactions among themselves. It is necessary to rank the features based on these interactions for better analysis and classifier performance. However, evaluating these interactions on large datasets is computationally challenging. Furthermore, datasets often have features with redundant information. Using such redundant features hinders both efficiency and generalization capability of the classifier. The major challenge is to efficiently rank the features based on relevance and redundance on mixed datasets. In this work, we propose a filter-based framework based on **R**elevance and **R**edundancy (RaR), RaR computes a single score that quantifies the feature relevance by considering interactions between features and redundancy. The top ranked features of RaR are characterized by maximum relevance and non-redundance. The evaluation on synthetic and real world datasets demonstrates that our approach outperforms several state-of-the-art feature selection techniques. Code and data related to this chapter are available at: https://doi.org/10.6084/m9.figshare.5418706.

1 Introduction

In high-dimensional feature spaces, feature ranking is an essential step for feature analysis and elimination of irrelevant features. Such irrelevant features affect the prediction and performance of classifiers [14]. In automotive applications, the data from several sensors (continuous values), status bits, gear-position (categorical values) and calculations forms a mixed dataset with a large number of features. In such a feature space, a set of features interact amongst themselves and these interactions are strongly correlated to the target class. For example,

© Springer International Publishing AG 2017
M. Ceci et al. (Eds.): ECML PKDD 2017, Part I, LNAI 10534, pp. 239–255, 2017.
https://doi.org/10.1007/978-3-319-71249-9_15

engine-temperature and fuel quality are two essential features required to predict engine-performance. On analyzing its individual correlations to the target, each feature is weakly correlated to the engine's performance. However, engine-performance is a combined outcome of engine-temperature and fuel quality. That is, their interactions contribute to the target predictions when used together. In such cases, assigning low relevance scores based on individual correlations is misleading. Hence, to draw conclusions on the relevance of engine temperature, it is necessary to assess its role in multiple subspaces. In addition to the multi-feature interactions, some features may have redundant information. Following our automotive example, certain signals are measured or calculated multiple times in a vehicle for safety reasons. These redundant signals provide similar information, but are not necessarily identical. In such a scenario, two redundant features have the same magnitude of relevance to the target class. However, using both features for a prediction model is unnecessary as they provide similar information. Elimination of redundant features reduces the computational load and enhances the generalization ability of the classifier [23]. All aforementioned problems are motivated with examples from our application, but they exist in several other domains such as Bio-informatics [7] and Media [4].

The first challenge lies in estimating the feature relevance based on interactions between the features and the target. Evaluating all possible feature combinations for these interactions results in an exponential runtime w.r.t. the total number of features. Thus, it is necessary to perform the evaluations in an efficient way. The second major challenge lies in measuring the redundancy of each feature while still acknowledging its relevance w.r.t. the target class. A final challenge is to evaluate relevance and redundancy in mixed feature space. Nevertheless, existing filter-based feature selection methods [14,15,23,27] do not focus on considering all three challenges together: relevance based on multi-feature interactions, redundance and mixed data.

In this work, we propose a feature ranking framework (RaR) to address all three challenges. We begin with computing relevance scores of multiple subspaces. These subspace relevance scores are decomposed to evaluate the individual feature contributions. In order to include the multi-feature interactions, the relevance of a feature is computed based on these individual contributions to multiple subspace relevance scores. The relevance estimation is followed by the redundance calculation. The relevance and redundancy scores are unified such that the relevance of a feature is penalized based on its redundancy. The major contributions of the paper are as follows:

(1) A feature relevance score, that considers the multi-feature interactions.
(2) A measure of redundancy to evaluate the novelty of a feature w.r.t. a subset.
(3) Experimental studies on both synthetic and real world datasets to show that several state-of-the-art approaches underestimate the importance of such interacting features.

Our extensive experiments show that our approach has better ranking quality and lower run times in comparison to several existing approaches.

2 Related Work

Feature selection is an extensively researched topic and can be broadly classified into filter, wrapper, hybrid, embedded and unsupervised approaches [14,16,24,25]. We compare the related work based on the four properties summarized in Table 1.

Wrapper approaches with sequential forward selection (SFS) can handle redundancy, but it is not capable of evaluating feature interactions. Using recursive elimination addresses the problem of multi-feature interactions [28]. However, the major problem of this paradigm is efficiency, as the selection always depends on training the classifier several times.

To overcome this computational challenge, hybrid approaches were introduced. A well-known hybrid approach, Mixed Feature Selection (MFS) [27] is based on the decomposition of continuous feature space along the states of each categorical feature. A hybrid approach presented in [8], addresses the problem of inefficiency by building fewer classifier models. Hybrid paradigms are still inefficient on high-dimensional datasets, as it involves training of classifier multiple times. Hence, this work focuses on the filter-based paradigm which does not require training of a classifier multiple times.

Correlation-based Filter Selection (CFS) is an advanced version of Pearson's correlation, that is capable of handling redundance among features [15]. Similarly, the correlation measure mRmR [23] ranks the features based on relevance and redundancy. Tree-based embedded techniques are also well-known techniques for handling mixed data and redundancy [24]. However, CFS, mRmR and embedded techniques do not address interactions amidst features.

Unlike the aforementioned methods, unsupervised subspace search techniques [16,20] consider multi-feature interactions. However, these approaches focus on providing a score for the entire subspace. In contrast to this, we intend to rank individual features by including their interactions with other features and the target. Moreover, the above discussed subspace methods are incapable of redundancy elimination. CMIM [10] and JMI [2] take relevance and redundancy for feature evaluation. However, CMIM is limited to boolean features and both have limitations for computing feature interactions between more than two features.

We propose a feature ranking framework RaR. RaR is an efficient filter-based feature ranking framework for evaluating relevance based on multi-feature interactions and redundancy on mixed datasets.

Table 1. Overview of the related work on feature selection

Paradigm	Approach	Mixed data	Redundancy	Feature interactions	Efficiency
Wrapper	SFS [28]	✓	✓	✗	✗
	Recursive elimination [28]	✓	✓	✓	✗
Hybrid	MFS [27]	✓	✗	✗	✗
	Doquire [8]	✓	✓	✗	✗
Subspace Ranking	HiCs [16]	✗	✗	✓	✓
Embeddedd	C4.5 [24]	✓	✓	✗	✓
	mRmR [23]	✓	✓	✗	✓
Filter	CFS [15]	✓	✓	✗	✓
	RaR	✓	✓	✓	✓

3   Problem Overview

In this section, we define the problem that we aim to solve. Let $\mathcal{F}$ be a d-dimensional mixed dataset $f_j \in \mathcal{F} \mid j = 1, \cdots, d$ with N instances. As a supervised learning process, the target $\boldsymbol{Y}$ is a collection of discrete classes. The mixed feature space $\mathcal{F}$ is defined by a set $\boldsymbol{X} \subseteq \mathcal{F}$ of continuous and set $\boldsymbol{Z} \subseteq \mathcal{F}$ of categorical features, i.e., $\mathcal{F} = \boldsymbol{X} \cup \boldsymbol{Z}$. In the following, we denote $error(\boldsymbol{S})$ as the error function of the classifier, trained using a subset of features $\boldsymbol{S} \subseteq \mathcal{F}$. For the given mixed dataset, we aim to (1) compute feature relevance by including their interactions with other features, as well as (2) evaluate the redundancy score of each feature.

Evaluation of feature interactions requires a multivariate correlation measure, that quantifies the relevance of $\boldsymbol{S}$ to $\boldsymbol{Y}$. Given such a subspace relevance score $rel : \boldsymbol{S} \mapsto \mathbb{R}$, rel is a function of individual feature relevancies, i.e., $rel(\boldsymbol{S}) = \phi(\{r(f_j) \mid \forall f_j \in \boldsymbol{S}\})$, where ϕ is an unknown function such that $\phi : \mathbb{R}^{|\boldsymbol{S}|} \mapsto \mathbb{R}$. To infer the individual feature relevancies $r : f_j \mapsto \mathbb{R}$, the first challenge is to decompose the subspace scores into individual feature scores. However, individual feature relevance cannot be inferred from a single feature subset because of possible interactions of f_j in other subspaces. To include the multi-feature interactions, it is necessary to evaluate M different subspaces. Thus, we aim to deduce a valid relevance score of a feature $r(f_j)$, based on the contribution of f_j to M different subspace scores.

Additionally, we aim to estimate the redundance of information a feature has, w.r.t a subspace, i.e., $red : (f_j, \boldsymbol{S}) \mapsto \mathbb{R}$. Given a feature $f_i \in \boldsymbol{S}$, non-redundant to $\boldsymbol{S}$ and $f_j \in \boldsymbol{S} \mid i \neq j$, with redundant information to $\boldsymbol{S}$, we intend to quantify a redundancy score such that, $red(f_j, \boldsymbol{S}) > red(f_i, \boldsymbol{S})$. Addition of redundant feature information to a classifier does not contribute to the prediction quality, i.e., $error(\boldsymbol{S}) \approx error(\boldsymbol{S} \setminus f_j)$ [23]. A major challenge for filter-based feature selection approaches is to evaluate this efficiently without training a classifier. Finally, the features are ranked based on the unification of two scores.

4   Relevance and Redundancy Ranking (RaR)

RaR consists of three major steps, computing the feature relevance by including feature interactions, redundancy and finally combining the two scores. To evaluate the feature relevance in a mixed dataset by including the feature interactions, we begin by computing the relevance (to the target class) scores of multiple subsets. We aim to infer the feature relevance based on their contribution to various subspace scores. Thus, the relevance of a feature to the target is decided based on its interaction with other features in multiple subspaces. This requires a multivariate correlation measure that can quantify the relevance of a subspace to the target. Hence, we begin with the introduction of a subspace correlation measure that we employ. This section is followed by the introduction of our heuristic to estimate the feature relevance based on multi-feature interactions. Finally, we elaborate our redundancy estimation and unification of the two scores.

4.1 Subspace Relevance

In the following, we introduce the definition of subspace relevance and a method to calculate it. To estimate the relevance of a subspace to the target, we use the concept of conditional independence. For an uncorrelated subspace, the law of statistical independence is not violated. The degree of violation is quantified by measuring the difference between the conditional and marginal distributions [16,21].

Definition 1. *Subspace Relevance.*
Given a subspace $S \subseteq \mathcal{F}$ and a divergence function D, the subspace relevance score $rel(S)$ to the target Y is defined as:

$$rel(S) = D\Big(p(Y \mid S) \parallel p(Y) \Big).$$

For a set of discrete target classes Y, the marginal of the target is compared to its conditional distribution. This definition enables the measuring of multivariate and non-linear correlations [16] in mixed datasets. For $f_j \in X$, the conditional is estimated based on a slice of continuous instances drawn from f_j. Similarly, for a $f_j \in Z$, the conditional is based on a slice of instances that have a particular categorical state. The magnitude of divergence between these two distributions can be estimated with Kullback–Leibler (KLD) or Jensen-Shannon divergence functions [19]. However, the instantiation of divergence function D will be done in the Sect. 4.5.

4.2 Decomposition for Feature Relevance Estimation

A simple solution to estimate the relevance of f_j using Definition 1 is by computing $rel(\{f_j\})$. Such individual feature relevance scores lacks information about feature interactions. The aim of our approach is to evaluate feature relevance $r(f_j)$ by including its interactions with other features and not to compute subspace scores $rel(S)$. The subspace relevance score represents the contribution of all features present in the subspace. Hence, the subspace score can be seen as a function of individual feature relevancies. We estimate the feature relevance $r(f_j)$ by decomposing the subspace score, which is the result of individual feature relevancies.

Example 1. Assume a dataset $\mathcal{F} = \{f_1, f_2, f_3, f_4\}$, such that there exists multi-feature interactions between $\{f_1, f_2, f_3\}$. Hence, relevance of a subset with all interacting features $(rel(S_1) \mid S_1 = \{f_1, f_2, f_3\})$ is greater than the relevance of a subset $(rel(S_2) \mid S_2 = \{f_1, f_2, f_4\})$ with an incomplete interactions.

A naïve decomposition is to decompose $rel(S)$ as the sum of individual feature relevancies. On applying naïve decomposition to our Example 1, we obtain $rel(S_1) = r(f_1) + r(f_2) + r(f_3)$ and $rel(S_2) = r(f_1) + r(f_2) + r(f_4)$. With an incomplete interaction structure, $rel(S_2)$ will underestimate the values of $r(f_1)$ and $r(f_2)$. Such underestimations are misleading as there exists another subspace where f_1 and f_2 in combination with f_3 forms a complete interaction structure to be more relevant to Y. This necessitates to rewrite the decomposition rule, such that it holds true for both cases. Hence, we define the decomposition as an upper bound of the subspace relevance.

Definition 2. *Feature Constraint.*
Let $r(f_j) \in \mathbb{R}$ be the relevance of individual features within the subspace $S \subseteq \mathcal{F}$, we define the feature constraint as:

$$rel(S) \leq \sum_{f_j \in S} r(f_j)$$

The defined inequality applies for a subspace with a complete or an incomplete interaction structure. The relevance of a feature f_j is to be estimated based on multiple subspaces, i.e., $S \mid S \in 2^{\mathcal{F}}$ and $f_j \in S$. Hence, a single inequality is not sufficient to estimate feature relevance based on multi-feature interactions. Moreover, a single inequality does not enable us to compute the relevance of all features $f_j \in \mathcal{F}$ in the high-dimensional feature space. However, it is computationally not feasible to deduce constraints (c.f. Definition 2) for all possible feature combinations. We address this challenge by running M Monte Carlo iterations. For each iteration, we select a subspace S and define a constraint based on the subspace relevance $rel(S)$ score and the features belonging to S. The constraints provide information on how a feature interacts in multiple subspaces. From these constraints, we aim to estimate the relevance of a feature $r(f_j)$.

Table 2 shows an illustrative example of how our idea of generating constraints works for a dataset (in Example 1) with multi-feature interactions. Our approach draws several random subspaces as shown in Table 2. With the calculated subspace relevancies, we build 3 constraints for estimating the bounds of the individual feature relevance. The constraints of $i = 2$ and 3 underestimate the relevance of the individual features. However, constraint of $i = 1$ increases the boundaries of individual feature relevance. The relevance of a feature $r(f_j)$ is decided by considering multiple subspaces where f_j is a part of. Hence, our approach prevents underestimation of $r(f_1)$ and $r(f_2)$ and enable inclusion of multi-feature interactions.

Table 2. Illustrative example of feature constraints for 3 Monte Carlo iterations

i	S	$rel(S)$	Constraint
1	$\{f_1, f_2, f_3\}$	0.9	$r(f_1) + r(f_2) + r(f_3) \geq 0.9$
2	$\{f_1, f_4\}$	0.12	$r(f_1) + r(f_4) \geq 0.12$
3	$\{f_2, f_1, f_4\}$	0.15	$r(f_1) + r(f_2) + r(f_4) \geq 0.15$

Our approach generates M inequalities for M Monte Carlo iterations. Solving the system of M inequalities does not lead to a unique value of $r(f_j)$.

The inequalities provide only the boundaries for feature relevancies. We aim to deduce a reasonable estimate of the relevancies such that all constraints are satisfied. As these constraints denote the lower bounds of the feature relevancies, we aim to minimize the contributions of individual features. Therefore, we define an objective function that estimates $r(f) \mid f \in \mathcal{F}$ subject to the defined constraints,

$$\min_{r(f)} \left[\sum_{f \in \mathcal{F}} r(f) + \sum_{f \in \mathcal{F}} (r(f) - \mu)^2 \right] \text{ s.t. } rel(\boldsymbol{S}_i) \le \sum_{f \in \boldsymbol{S}_i} r(f) \mid i = 1, \cdots, M, \quad (1)$$

such that, $\mu = (1/|\mathcal{F}|) \sum_{f \in \mathcal{F}} r(f)$. The first term denotes the sum of individual feature relevance. The second part of the optimization function is a standard L2-regularization term to ensure that all relevancies $r(f)$ contribute equally to the boundary. Finally, we apply quadratic programming in order to optimize Eq. 1 subject to the M affine inequalities. The inequalities define a feasible region in which the solution to the problem must be located for the constraints to be satisfied. Thus, we obtain the relevance score for each feature. Computing the subspace relevance (c.f. Definition 1) for each iteration requires the estimation of conditional probability distributions. However, evaluating the empirical conditional probabilities for large $|\boldsymbol{S}|$ is inaccurate. We demonstrate this by empirical evaluation in Sect. 5.3. Hence, it is necessary to restrict the size of the subspace to a maximum of k. That is, each randomly drawn $\boldsymbol{S}_i \mid \boldsymbol{S}_i \subseteq \mathcal{F}$ and $|\boldsymbol{S}_i| \le k$. Algorithm 1 shows the pseudo-code for feature relevance estimation.

Algorithm 1. Estimation of Feature Relevance

Input: $\mathcal{F}, \boldsymbol{Y}, M, k$
1: $C = \emptyset$
2: **for** $i = 1 \rightarrow M$ **do**
3:     Sample $\{\boldsymbol{S}_i \mid \boldsymbol{S}_i \subseteq \mathcal{F} \wedge |S_i| \le k\}$
4:     Compute $rel(\boldsymbol{S}_i)$ using Definition 1
5:     Construct constraint (cf. Definition 2)
6:     Add constraint to set C
7: **end for**
8: Optimize objective function Eq. 1 subject to C
9: **return** $r(f) \mid \forall f \in \mathcal{F}$

4.3   Redundancy Estimation

The feature relevance estimation does not include the effect of redundancy. This means, two identical features are ranked the same based on its relevance scores. A major challenge lies in the detection of redundant features which do not have identical values as explained in Sect. 3. Hence, redundancy is not a binary decision. A pair of redundant features can only have a certain magnitude of information shared among them. Therefore, it is necessary to incorporate this specific information into the final score that exemplifies redundancy and relevance. The principle of redundancy estimation is similar to the relevance measurement. We use the same property of comparing marginal and conditional distributions as in Definition 1 to evaluate redundancy.

Definition 3. *Feature Redundancy.*
Given a set of features $\mathbf{R} \subseteq \mathcal{F}$, a feature $f_j \mid (f_j \in \mathcal{F}$ and $f_j \notin \mathbf{R})$ is non-redundant w.r.t $\mathbf{R}$ iff:

$$P\Big(p(f_j \mid \mathbf{R}) = p(f_j)\Big) = 1.$$

Our feature redundancy estimation is a two step process. Step 1: All features $f_j \in \mathcal{F}$ are ranked based on relevance $r(f_j)$ score. Step 2: For an ordered set $\mathbf{R}_n$ that denotes a set of features until relevance rank n, we compute redundancy score of nth ranked feature based on the redundancy it imposes on features with relevance rank 1 to $n-1$. By following this methodology, if two redundant features have similar relevance scores, the second feature will obtain a higher redundancy score. This redundancy score is used to devalue the redundant contribution of that feature.

$$red(f_j, \mathbf{R}) \equiv D\Big(p(f_j \mid \mathbf{R}) \parallel p(f_j)\Big) \tag{2}$$

If f_j is independent of $\mathbf{R}$, the marginal and the conditional probability distributions will be the same. In other words, if f_j has non-redundant information w.r.t the features $f \in \mathbf{R}$, the deviation between the distributions in Eq. 2 will be 0. We illustrate the steps with an example.

Example 2. Assume a feature space $\mathcal{F} = \{f_1, f_2, \ldots, f_5\}$ in which f_1 and f_3 are redundant features.

For the given feature space in Example 2, the features are sorted based on relevance scores following the step 1, i.e., $\mathbf{R}_n = \{f_5, f_3, f_1, f_2, f_4\} \mid n = |\mathcal{F}|$. The highest relevant feature f_5 is not evaluated for redundancy as, it has no preceding ranked features to be redundant with. The redundancy that f_3 imposes on $\mathbf{R}_1 = \{f_5\}$ is estimated by applying Eq. 2. Therefore, we rank the features based on their relevance and use the top n-relevant features to compute the redundancy of f_{n+1}. The pseudo-code for this estimation is shown in Algorithm 2.

Algorithm 2. Estimation of Redundancy
***
Input: $\mathcal{F}, Y$
1: $\mathbf{R}_n$=Sort $\forall f_j \in \mathcal{F}$ based on $r(f_j)$ from Algorithm 1
2: **for** $n = 2 \rightarrow |\mathcal{F}|$ **do**
3:     Compute $red(\mathbf{R}_n \triangle \mathbf{R}_{n-1}, \mathbf{R}_{n-1})$ c.f. Eq. 2 ▷ $\triangle$ dentotes symmetric difference
4: **end for**
5: **return** Calculate redundancy scores $red \ \forall f_j \in \mathcal{F}$

For estimation of feature relevance, we restricted the subspace size to k (c.f. Sect. 4.2). This avoids inaccurate conditional probability estimates. Algorithm 2 also involves estimation of conditional probabilities. For a large $|\boldsymbol{R}_n|$, the conditional probability estimations using Eq. 2 are not accurate. For example: for estimating the redundancy of the 100th ranked feature, we need to estimate the conditional based on the 99 features ahead in the rank. Thus, for estimation of redundancy score of the nth ranked feature, we sample subspaces $\forall s \subseteq \boldsymbol{R}_{n-1}$. From $\boldsymbol{R}_{n-1}$, various subspaces s of size k are sampled without replacement, i.e., $\binom{n-1}{k}$ number of subsets. The maximal imposed redundancy of the nth ranked feature on the list of subspaces is the redundancy of the nth feature.

4.4 RaR: Relevance and Redundancy Scoring

Having estimated the relevance and redundance of the features in Sects. 4.2 and 4.3, our final goal is to rank features based on a single score that combines both the properties.

Definition 4. *RaR score.*
Given the relevance $r(f_j)$ and redundancy score $red(f_j, \boldsymbol{R})$ of feature f_j, we define RaR(f_j) score as,

$$RaR(f_j) = \left[\frac{2 \cdot r(f_j) \cdot (1 - red(f_j, \boldsymbol{R}))}{r(f_j) + (1 - red(f_j, \boldsymbol{R}))} \right].$$

$RaR(f_j)$ is the harmonic mean of relevance and redundancy scores. The harmonic mean penalizes the relevance score with the information based on redundancy.

Example 3. Assume a feature space $\mathcal{F} = \{f_1, f_2, \ldots, f_5\}$ in which f_1 and f_3 are relevant and exhibit feature interactions. Additionally, f_4 and f_5 are features with redundant information.

In such a case, RaR ranks the feature based on multi-feature interactions and redundancy. Hence, RaR ensures that the non-redundant and the features with interactions, i.e., $\{f_1, f_3\}$ to be present ahead in the feature ranks.

Time complexity analysis of RaR consists of three major phases: subspace sampling for constraint generation (Lines 2–7 of Algorithm 1), quadratic optimization (Line 8 of Algorithm 1) and redundancy estimation (Algorithm 2). In the following, we discuss the time complexity of each part and finally present the overall time complexity of our approach.

For each Monte Carlo iteration, we compute the subspace relevance based on the slicing method presented in [16]. This requires to iterate the instances in the selected slice. In the worst case scenario, all instances are included in the slice with a time complexity of $\mathcal{O}(N)$. The selection of a slice is done for each dimension in subspace S_i. Since $|S_i| \leq k$, it leads to a complexity of $\mathcal{O}(N \cdot k)$ for calculating $rel(S_i)$ (Line 4 of Algorithm 1). The total time complexity for extracting M constraints takes $\mathcal{O}(M \cdot N \cdot k)$. The final step of estimating the relevance of each feature, requires to optimize Eq. 1 subject to M constraints (Line 8).

A quadratic programming algorithm with M constraints and d-dimensional feature space has a time complexity $\mathcal{O}(M + \sqrt{d} \cdot ln\frac{1}{\epsilon})$ [12]. The complexity considers that the optimizer converges to an ϵ-accurate solution. To compute the redundancy of a feature, we group subspaces of size k with all features ahead of it and compute the maximal redundancy using Eq. 2. Thus redundancy takes a total time of $\mathcal{O}\left(d \cdot \frac{d-1}{k} \cdot N\right)$. Finally, ranking the features requires to sort the features based on their relevance and redundancy scores. This procedure requires $\mathcal{O}(d \cdot log(d))$. Considering the complexity of computing the harmonic mean of relevance and redundancy as constant, the total complexity of RaR is represented as,

$$\mathcal{O}\left(M \cdot N \cdot k + \frac{d^2}{k} \cdot N\right).$$

4.5   Instantiations for RaR

In Algorithm 1, a random subspace S is selected with maximum dimensionality k for each iteration. In order to estimate $rel(S)$, we compute the distribution of Y under some conditional slice of S. That is, we aim to obtain a slice of S which satisfies a specific set of conditions, i.e., $D(p(Y \mid S \in [c_1, \cdots, c_{|S|}]), p(Y))$. Defining explicit conditions is a tedious task. Hence, we use adaptive subspace slicing, more details can be found in [16]. After calculating the subspace relevance, we extract an inequality and the set C is updated with this constraint. Finally, we obtain a set of M constraints and optimize the objective function of Eq. 1 subject to these constraints.

RaR requires a divergence function to quantify the difference between distributions. We use KLD for our experiments. As KLD is formulated for both continuous and discrete probability distribution, it is directly applicable for redundancy estimation (c.f. Eq. 2) on mixed feature types. As a non-symmetric measure, we instantiate RaR with $KLD\big(p(Y \mid S) \mid\mid p(Y)\big)$ as it converges to mutual information and $KLD\big(p(Y) \mid\mid p(Y \mid S)\big)$ does not[1].

5   Experiments

5.1   Experimental Setup

In this section we compare the run times and quality of our approach against several existing techniques as competitors. We consider techniques from different paradigms, i.e., filters, wrappers, embedded and hybrid techniques for mixed data as competitors. As wrappers, we test Sequential Forward Selection (SFS) [28] with K-Nearest Neighbors (KNN) [17], capable of handling redundant features. As hybrid technique, we consider the heuristic of Doquire [8]. The scheme requires a correlation measure and a classifier, hence we employ mRmR [23] and KNN with the heuristic of Doquire [8]. As filter approach, we test Maximal Information Coefficient (MIC) [18], mRmR [7,23], ReliefF [25] and

---

[1] https://hpi.de//mueller/rar.html.

Correlation Filter Selection (CFS) [15]. Finally, we test the embedded scheme of decision trees (C4.5 [24]). We provide the implementation of RaR, competitor approaches, synthetic data generator[2] and the parameters[3] of our experiments. The results of our experiments on other classifiers are also made available. Additionally, we employ Gurobi [13] optimizer for the optimization of relevancies in RaR. We evaluate and compare our approach with the above mentioned competitors on synthetic and real world datasets.

Synthetic datasets were generated with varying database sizes and dimensionality. We employ the synthetic data generation program of NIPS [22] to generate continuous feature sets with normal distribution in any proportion of relevant (with multi-feature interactions) and noisy features. For a generated continuous feature f and v number of states, we discretized f to form a categorical feature of v unique values. In our experiments, we generated mixed datasets with equal number of categorical and continuous features. As a measure of feature ranking quality, we use Cumulative Gain (CG) from Information Retrieval [1].

For evaluation of our feature ranking framework, we also use 6 public datasets from the UCI repository with different dimensionalities and database sizes. The datasets contain both continuous and categorical features. The NIPS feature selection challenge [5] (2000 Instances/500 features), Ionosphere [26] (351 Instances/24 features), Musk2 [6] (6598 Instances/166 features), Isolet [9] (2000 Instances/500 features), Semeion [3] (1593 Instances/164 features) and Advertisement [11] (3279 Instances/1558 features) datasets. Experiments that had run times more that one day are denoted as ** in Tables 3 and 4.

5.2   Synthetic Data

We perform scalability analysis by evaluating the run times with increasing dimensionality and database size. Figure 1 shows the efficiency of RaR with

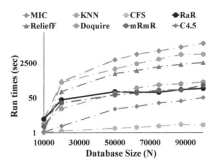 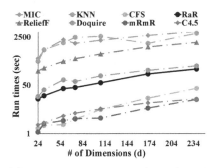

(a) Fixed dimensionality (50) and increasing database size

(b) Fixed database size (20000) and increasing dimensionality

Fig. 1. Run time evaluation: run times of RaR vs. competitor approaches

---

[2] https://github.com/tmbo/rar-mfs.
[3] https://hpi.de//mueller/rar.html.

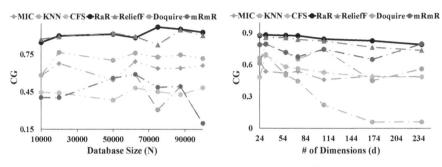

(a) Fixed dimensionality (50) and increasing database size

(b) Fixed database size (20000) and increasing dimensionality

Fig. 2. Quality evaluation: CG of RaR vs. competitor techniques

increasing database size and dimensionality. In general, methods that do not evaluate for feature interactions, i.e., C4.5, mRmR and CFS, have lower run times than RaR. By evaluating these interactions, RaR has better feature ranking quality (c.f. Fig. 2). In comparison to ReliefF, which ranks features based on multi-feature interactions, RaR has lower run times and better feature ranking quality.

5.3 Parameter Analysis

The k parameter of RaR decides the maximum size of the subset drawn for every iteration $i \mid i = 1, \cdots, M$. From our experiments (c.f. Fig. 3(a)) on synthetic data, we observe that the CG decreases with increasing k. The size of the conditional slices is determined by the α parameter [16]. For a dataset of $N = 1000$ and $\mid \mathcal{F} \mid = 100$, setting $\alpha = 0.1$ and a large value of k $(k = 50)$ leads to a conditional slice of size $\alpha^{\frac{1}{k}} \cdot N$ [16]. Hence, the conditional slice has

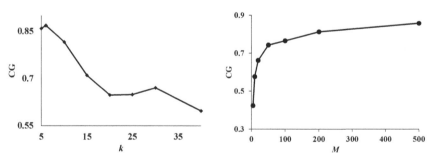

(a) Influence of k on feature ranking $(M=100)$

(b) Influence of M on feature ranking $(k=5)$

Fig. 3. Parameter study, on synthetic dataset of 50 features and 20000 instances

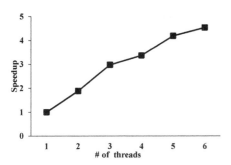

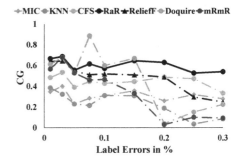

Fig. 4. Speedup of RaR   **Fig. 5.** Robustness of feature ranking

approximately 95% of all the instances. This leads to a very similar conditional and marginal distributions and distorted feature ranking. In Fig. 3(b), we vary M and evaluate its influence on feature ranking. The experiment shows that the ranking quality is stable for a large range of M. Thus, we recommend to restrict k to small values and increase M for better accuracy. Choosing large M affects run times of selection process. However, the task of sampling and building constraints can be distributed over multiple processor threads. Figure 4 shows the efficiency gained by distributed computations of RaR. Speedup denotes the number of folds of decrease in run times (w.r.t single thread) on distributing the Monte Carlo iterations to multiple processor threads.

5.4   Robustness w.r.t. Erroneous Labels

In several application scenarios, the target labels Y are assigned by domain experts. This manual process is prone to errors. With such datasets, it is necessary to ensure that the feature ranking is robust to erroneous target labels. To test this, we manually induced label errors in the synthetic datasets. The hybrid approach from Doquire [8] was able to perform well on a few cases (c.f. Fig. 5). However, as a filter approach, RaR defines the feature relevance score based on constraints defined by multiple subsets. Thus, RaR is more robust to label errors.

5.5   Real World Datasets

Table 3 shows the results w.r.t. the prediction quality of each feature selection technique. Overall, we observe that application of feature selection improves the quality of prediction. By evaluating the feature interactions in the dataset, RaR has the best accuracy in comparison to the competitor approaches. Especially, the existing feature selection techniques do not show improvement of f-score in the case of NIPS challenge dataset. NIPS dataset contains multi-feature interactions, noisy and large number of redundant features. As the competitor approaches do not evaluate feature interactions, they assign lower scores to such interacting features.

Table 3. Average f-score of 3 fold cross-validation using KNN (K = 20) classifier

Selection	NIPS	Ionosphere	Musk2	Isolet	Semeion	Advertisement
Full-dimension	0.57	0.70	0.8	0.58	0.1	0.73
C4.5	0.58	0.87	0.9	0.63	0.79	0.9
MIC	0.78	0.83	0.86	0.78	0.8	0.91
SFS (KNN)	0.84	0.85	0.91	**	**	**
CFS	0.82	0.81	0.86	0.82	0.9	0.91
ReliefF	0.87	0.79	0.84	0.82	0.87	0.87
mRmR	0.55	0.89	0.9	0.57	0.9	0.9
Doquire	0.56	0.88	0.9	0.56	0.93	0.9
RaR	**0.88 ± 0.006**	**0.88 ± 0.00**	**0.91 ± 0.008**	**0.87 ± 0.002**	**0.92 ± 0.005**	**0.92 ± 0.005**

Table 4. Feature ranking run times in *sec* of RaR vs. competitor approaches

Selection	NIPS	Ionosphere	Musk2	Isolet	Semeion	Advertisement
C4.5	1.2	0.5	3.1	3.8	0.21	15.58
MIC	37.7	0.47	40.79	37.25	81.2	49.35
SFS (KNN)	105741.3	6.9	14132.9	**	**	**
CFS	36.7	1.8	8.3	37.5	2.51	417.9
ReliefF	29.3	0.18	98.08	32.7	5.46	95.07
mRmR	42.3	0.5	4.5	59.27	6.1	78.81
Doquire	44.6	4.25	9.19	62.15	9.8	131.42
RaR	**10.35**	**2.05**	**5.3**	**7.9**	**4.37**	**50.26**

Table 4 shows that our approach is several times more efficient in comparison to the competitor filter and wrapper methods. Embedded approach C4.5 has lower run times in comparison to RaR. However, C4.5 is unable to identify feature interactions and has lower prediction quality (c.f. Table 3). Similar to our experiments on synthetic datasets (c.f. Figs. 1 and 2), we observe that methods that have lower run times than RaR have lower f-scores as they no not evaluate feature interactions. For dataset with few features (Ionosphere data), simple bivariate correlation measures (MIC and CFS) was a better choice w.r.t run times.

5.6   Evaluation of the Ranking

To evaluate the quality of feature ranking, i.e., to experimentally show that the top ranked features of RaR are maximally relevant and non-redundant, we follow a 2 step evaluation process on real world datasets. First, we rank the features using each approach. Then, we iteratively add the features ranked by each technique to a classifier (KNN [17]) in the order (best to worst) of their ranks. As shown in Fig. 6, after including each feature, the average f-score of 3 fold cross-validation is calculated. As the top ranked features of RaR are non-redundant, we observe the best quality with the least number of features. However, other approaches do not take into account the effect of redundancy. For example,

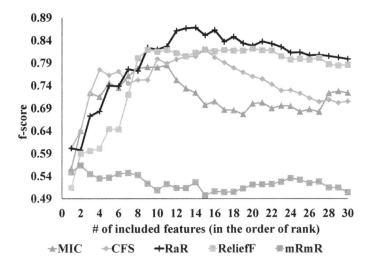

Fig. 6. f-scores of top 30 features on Isolet dataset

Table 5. Number of features required to obtain the quality in Table 3

Selection	NIPS	Ionosphere	Musk2	Isolet	Semeion	Advertisement
MIC	11	2	163	11	82	14
SFS (KNN)	5	2	135	**	**	**
CFS	15	2	155	15	119	7
ReliefF	20	4	136	20	173	54
mRmR	5	5	117	2	151	13
Doquire	2	4	117	2	156	15
RaR	**12**	**2**	**16**	**11**	**17**	**9**

ReliefF has very similar prediction quality (c.f. Table 3) to RaR. By ranking the non-redundant features ahead, RaR achieves better f-score with fewer features (c.f. Fig. 6), i.e., RaR obtains an f-score of 0.87 with 14 features and ReliefF obtains an f-score of 0.82 with 20 features. We performed the experiment on the public datasets and we show the number of features (c.f. Table 5) at which the maximum f-score (c.f. Table 3) was observed. Table 5 shows the number of top ranked features required to obtain the quality in Table 3, and RaR achieves the best f-score with fewer features.

6   Conclusions and Future Works

The results of various state-of-the-art algorithms on the synthetic and real world datasets, show that our feature ranking method is suitable for high-dimensional datasets exhibiting complex feature interactions. By ranking the non-redundant features ahead, RaR achieves better prediction quality with fewer features.

As future works, we intend to address two directions to enhance our approach. In the event where two features are exactly identical to each other and are also maximally relevant, after penalization for redundancy, one of the feature can have a RaR score lower than the noisy features. This calls for a more sophistication in the combining of relevance and redundancy scores. RaR is based on the distribution of target class. Hence, RaR is currently limited to non-sparse datasets.

References

1. Baeza-Yates, R., Ribeiro-Neto, B.: Modern Information Retrieval, vol. 463. ACM Press, New York (1999)
2. Brown, G., Pocock, A., Zhao, M.J., Luján, M.: Conditional likelihood maximisation: a unifying framework for information theoretic feature selection. J. Mach. Learn. Res. **13**(Jan), 27–66 (2012)
3. Buscema, M.: Metanet*: the theory of independent judges. Subst. Use Misuse **33**(2), 439–461 (1998)
4. Chen, L.S., Liu, C.C.: Using feature selection approaches to identify crucial factors of mobile advertisements. In: Proceedings of the International MultiConference of Engineers and Computer Scientists, vol. 1 (2015)
5. Chen, Y.W., Lin, C.J.: Combining SVMs with various feature selection strategies. In: Guyon, I., Nikravesh, M., Gunn, S., Zadeh, L.A. (eds.) Feature Extraction: Foundations and Applications. Studies in Fuzziness and Soft Computing, vol. 207, pp. 315–324. Springer, Heidelberg (2006)
6. Dietterich, T.G., Jain, A.N., Lathrop, R.H., Lozano-Perez, T.: A comparison of dynamic reposing and tangent distance for drug activity prediction. In: Advances in Neural Information Processing Systems, p. 216 (1994)
7. Ding, C., Peng, H.: Minimum redundancy feature selection from microarray gene expression data. J. Bioinform. Comput. Biol. **3**(02), 185–205 (2005)
8. Doquire, G., Verleysen, M.: An hybrid approach to feature selection for mixed categorical and continuous data. In: KDIR, pp. 394–401 (2011)
9. Fanty, M.A., Cole, R.A.: Spoken letter recognition. In: NIPS, pp. 220–226 (1990)
10. Fleuret, F.: Fast binary feature selection with conditional mutual information. J. Mach. Learn. Res. **5**(Nov), 1531–1555 (2004)
11. Fradkin, D., Madigan, D.: Experiments with random projections for machine learning. In: Proceedings of the ninth ACM SIGKDD International Conference on Knowledge Discovery and Data Mining, pp. 517–522. ACM (2003)
12. Gondzio, J.: Interior point methods 25 years later. Eur. J. Oper. Res. **218**(3), 587–601 (2012)
13. Gurobi Optimization, Inc.: Gurobi optimizer reference manual (2015). http://www.gurobi.com
14. Guyon, I., Elisseeff, A.: An introduction to variable and feature selection. J. Mach. Learn. Res. **3**(Mar), 1157–1182 (2003)
15. Hall, M.A.: Correlation-based feature selection of discrete and numeric class machine learning (2000)

16. Keller, F., Müller, E., Bohm, K.: HiCS: high contrast subspaces for density-based outlier ranking. In: 2012 IEEE 28th International Conference on Data Engineering, pp. 1037–1048. IEEE (2012)
17. Keller, J.M., Gray, M.R., Givens, J.A.: A fuzzy K-nearest neighbor algorithm. IEEE Trans. Syst. Man Cybern. (4), 580–585 (1985)
18. Lin, C., Miller, T., Dligach, D., Plenge, R., Karlson, E., Savova, G.: Maximal information coefficient for feature selection for clinical document classification. In: ICML Workshop on Machine Learning for Clinical Data, Edingburgh, UK (2012)
19. Lin, J.: Divergence measures based on the Shannon entropy. IEEE Trans. Inf. Theory **37**(1), 145–151 (1991)
20. Nguyen, H.V., Müller, E., Vreeken, J., Keller, F., Böhm, K.: CMI: an information-theoretic contrast measure for enhancing subspace cluster and outlier detection. In: 13th SIAM International Conference on Data Mining (SDM), Austin, TX, pp. 198–206. SIAM (2013)
21. Nilsson, R., Peña, J.M., Björkegren, J., Tegnér, J.: Consistent feature selection for pattern recognition in polynomial time. J. Mach. Learn. Res. **8**(Mar), 589–612 (2007)
22. NIPS: Workshop on variable and feature selection (2001). http://www.clopinet.com/isabelle/Projects/NIPS2001/
23. Peng, H., Long, F., Ding, C.: Feature selection based on mutual information criteria of max-dependency, max-relevance, and min-redundancy. IEEE Trans. Pattern Anal. Mach. Intell. **27**(8), 1226–1238 (2005)
24. Quinlan, J.R.: C4.5: Programs for Machine Learning. Elsevier, Amsterdam (2014)
25. Robnik-Šikonja, M., Kononenko, I.: Theoretical and empirical analysis of ReliefF and RReliefF. Mach. Learn. **53**(1–2), 23–69 (2003)
26. Sigillito, V.G., Wing, S.P., Hutton, L.V., Baker, K.B.: Classification of radar returns from the ionosphere using neural networks. Johns Hopkins APL Tech. Dig. **10**(3), 262–266 (1989)
27. Tang, W., Mao, K.: Feature selection algorithm for mixed data with both nominal and continuous features. Pattern Recogn. Lett. **28**(5), 563–571 (2007)
28. Theodoridis, S., Pikrakis, A., Koutroumbas, K., Cavouras, D.: Introduction to Pattern Recognition: A Matlab Approach: A Matlab Approach. Academic Press, Cambridge (2010)

Non-redundant Spectral Dimensionality Reduction

Yochai Blau$^{(\boxtimes)}$ and Tomer Michaeli

Technion–Israel Institute of Technology, Haifa, Israel
yochai@campus.technion.ac.il, tomer.m@ee.technion.ac.il

Abstract. Spectral dimensionality reduction algorithms are widely used in numerous domains, including for recognition, segmentation, tracking and visualization. However, despite their popularity, these algorithms suffer from a major limitation known as the "repeated eigen-directions" phenomenon. That is, many of the embedding coordinates they produce typically capture the same direction along the data manifold. This leads to redundant and inefficient representations that do not reveal the true intrinsic dimensionality of the data. In this paper, we propose a general method for avoiding redundancy in spectral algorithms. Our approach relies on replacing the orthogonality constraints underlying those methods by unpredictability constraints. Specifically, we require that each embedding coordinate be unpredictable (in the statistical sense) from all previous ones. We prove that these constraints necessarily prevent redundancy, and provide a simple technique to incorporate them into existing methods. As we illustrate on challenging high-dimensional scenarios, our approach produces significantly more informative and compact representations, which improve visualization and classification tasks.

1 Introduction

The goal in nonlinear dimensionality reduction is to construct compact representations of high dimensional data, which preserve as much of the variability in the data as possible. Such techniques play a key role in diverse applications, including recognition and classification [3,12,18], tracking [24,25,38], image and video segmentation [21,28], pose estimation [11,29], age estimation [15], spatial and temporal super-resolution [7,28], medical image and video analysis [5,34] and data visualization [26,37,40].

Many of the dimensionality reduction methods developed in the last two decades are based on spectral decomposition of some data-dependent (kernel) matrix. These include, e.g., Locally Linear Embedding (LLE) [30], Laplacian Eigenmaps (LEM) [2], Isomap [35], Hessian Eigenmaps (HLLE) [9], Local Tangent Space Alignment (LTSA) [41], Diffusion Maps (DFM) [8], and Kernel

---

Electronic supplementary material The online version of this chapter (https://doi.org/10.1007/978-3-319-71249-9_16) contains supplementary material, which is available to authorized users.

M. Ceci et al. (Eds.): ECML PKDD 2017, Part I, LNAI 10534, pp. 256–271, 2017.
https://doi.org/10.1007/978-3-319-71249-9_16

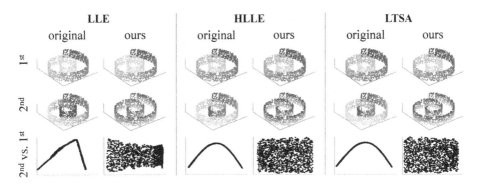

Fig. 1. The first two projections of data points lying on a Swiss roll manifold, as obtained with the original LLE, HLLE and LTSA algorithms and with our non-redundant versions of those algorithms. Top rows: the points colored by the projections. The original algorithms redundantly capture progression along the angular direction twice. In contrast, with our modifications, the second projection captures the vertical direction. Bottom row: scatter plot of the 2nd projection vs. the 1st. In the original algorithms, the 2nd projection is a function of the 1st, while in our algorithms it is not. (Color figure online)

Principal Component Analysis (KPCA) [32]. Methods in this family differ in how they construct the kernel matrix, but in all of them the eigenvectors of the kernel serve as the low-dimensional embedding of the data points [4,17,36].

A significant shortcoming of spectral dimensionality reduction algorithms is the "repeated eigen-directions" phenomenon [10,13,14]. That is, successive eigenvectors tend to represent directions along the data manifold which were already captured by previous ones. This leads to redundant representations that are unnecessarily larger than the intrinsic dimensionality of the data. To illustrate this effect, Fig. 1 visualizes the two dimensional embeddings of a Swiss roll, as obtained by several popular algorithms. In all the examined methods, the second dimension of the embedding carries no additional information with respect to the first. Specifically, although the first dimension already completely characterizes the position along the long axis (angular direction) of the manifold, the second dimension is also a function of this axis. Progression along the short axis (vertical direction) is captured only by the third eigenvector in this case (not shown). Therefore, the two dimensional representation we obtain is 50% redundant: Its second feature is a deterministic function of the first.

In fact, the redundancy of spectral methods can be arbitrarily high. To see this, consider for example the embedding obtained by the LEM method, whose kernel approximates the Laplace-Beltrami operator on the manifold. The Swiss-roll corresponds to a two dimensional strip with edge lengths L_1 and L_2. Thus, the eigenfunctions and eigenvalues (with Neumann boundary conditions) are given in this case by

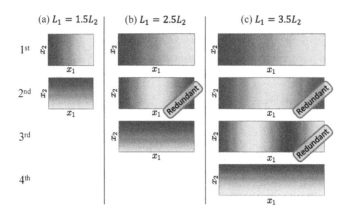

Fig. 2. A 2D strip with edge lengths (a) $L_1 = 1.5L_2$, (b) $L_1 = 2.5L_2$ and (c) $L_1 = 3.5L_2$, colored according to the first few coordinates of the Laplacian Eigenmaps embedding. Coordinates $2, \ldots, \lfloor L_1/L_2 \rfloor$ are redundant as they are all functions of only x_1, which is already fully represented by the first coordinate. (Color figure online)

$$\phi_{k_1 k_2}(x_1, x_2) = \cos\left(\frac{k_1 \pi x_1}{L_1}\right) \cos\left(\frac{k_2 \pi x_2}{L_2}\right), \tag{1}$$

$$\lambda_{k_1 k_2} = \left(\frac{k_1 \pi}{L_1}\right)^2 + \left(\frac{k_2 \pi}{L_2}\right)^2, \tag{2}$$

for $k_1, k_2 = 0, 1, 2, \ldots$, where x_1 and x_2 are the coordinates along the strip. Ignoring the trivial function $\phi_{0,0}(x_1, x_2) = 1$, it can be seen that the first $\lfloor L_1/L_2 \rfloor$ eigenfunctions (corresponding to the smallest eigenvalues) are functions of only x_1 and not x_2 (see Fig. 2). Thus, at least $\lfloor L_1/L_2 \rfloor + 1$ projections are required to capture the two dimensions of the manifold, which leads to a very inefficient representation when L_1 is much larger than L_2. Projections $2, \ldots, \lfloor L_1/L_2 \rfloor$ are all functions of projection 1, and are thus redundant. For example, when $L_1 > 2L_2$, the first two eigenfunctions are $\phi_{1,0}(x_1, x_2) = \cos(\pi x_1/L_1)$ and $\phi_{2,0}(x_1, x_2) = \cos(2\pi x_1/L_1)$, which clearly satisfy $\phi_{2,0}(x_1, x_2) = 2\phi_{1,0}^2(x_1, x_2) - 1$. Notice that this redundancy appears despite the fact that the functions $\{\phi_{k_1 k_2}\}$ are orthogonal. This highlights the fact that *orthogonality does not imply non-redundancy*.

The above analysis is not unique to the LEM method. Indeed, as shown in [14], spectral methods produce redundant representations whenever the variances of the data points along different manifold directions vary significantly. This observation, however, cannot serve to solve the problem as in most cases the underlying manifold is not known a-priori.

In this paper, we propose a general framework for eliminating the redundancy caused by repeated eigen-directions. Our approach applies to all spectral dimensionality reduction algorithms, and is based on replacing the orthogonality constraints underlying those methods, by unpredictability ones. Namely, we restrict subsequent projections to be unpredictable (in the statistical sense) from all previous ones. As we show, these constraints guarantee that the projections

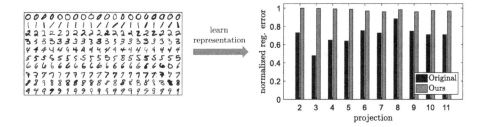

Fig. 3. A 10-dimensional representation of 15K MNIST handwritten digits [23] was learned with LEM and our non-redundant LEM. The bar plots show the normalized errors attained in regressing each projection against all previous ones, indicating to what extent the projection is redundant (higher is less redundant) [10].

be non-redundant. Therefore, once a manifold dimension is fully represented by a set of projections in our method, the following projections must capture a new direction along the manifold. As we demonstrate on several high-dimensional data-sets, the embeddings produced by our algorithm are significantly more informative than those learned by conventional spectral methods.

2   Related Work

Very few works suggested ways to battle the repeated eigen-directions phenomenon. Perhaps the simplest approach is to identify the redundant projections in a post-processing manner [10]. In this method, one begins by computing a large set of projections. Each projection is then regressed against all previous ones (via nonparametric regression). Projections with low regression errors (i.e. which can be accurately predicted from the preceding ones) are discarded. This approach is quite efficient but usually works well only in simple situations. Its key limitation is that it is restricted to choose the projections from a given finite set of functions, which may not necessarily contain a "good" subset. Indeed, as we demonstrate in Fig. 3, in real-world high-dimensional settings all the projections tend to be partially predictable from previous ones. Yet, there usually does not exist any single projection which can be considered fully redundant. Therefore, despite the obvious dependencies, almost no projection is practically discarded in this approach. In contrast, our algorithm produces projections which cannot be predicted from the previous ones (with normalized regression errors ~100%). Therefore, we are able to preserve more information about the data.

Another simple approach is to compute the projections sequentially, by eliminating the variations in the data which can be attributed to the projections that have already been computed. A naive way of doing so, would be to subtract from the data points their reconstructions based on all the previous projections. However, perhaps counter-intuitively, this *sequential regression* process does not necessarily prevent redundancy. This is because the data points may fall off the manifold during the iterations, as demonstrated in Fig. 4(b).

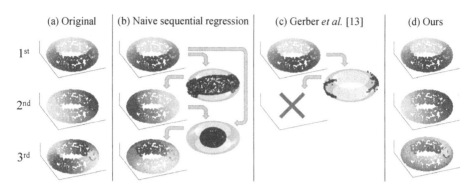

Fig. 4. (a) The first three projections of points lying on a ring manifold, obtained with the original LEM algorithm. The projections correspond to $\cos(\theta)$, $\sin(\theta)$ and $\sin(2\theta + c)$, where θ is the outer angle of the ring. In this case, Projection 2 is not a function of Projection 1 and is thus non-redundant. But Projection 3 is a function of Projections 1 and 2, and is thus redundant. (b) The projections obtained with the naive sequential regression approach (Sect. 2). Here, Projection 3 is still redundant. The right column shows the points after subtracting their prediction from previous projections, which causes them to fall off the manifold. (c) The projections obtained with the algorithm of [13]. Here, the algorithm halts after one projection. The right column shows the points after the advection process along the manifold, which results in two clusters forming an unconnected graph. (d) The projections obtained with our non-redundant version of LEM. Our algorithm extracts a non-redundant third projection, which captures progression along the inner angle of the ring.

A more sophisticated approach, suggested by Gerber et al. [13], is to collapse the data points *along the manifold* in the direction of the gradient of the previous projection. In this approach, the points always remain on the manifold. However, this method fails whenever a projection is a non-monotonic function of some coordinate along the manifold. This happens, for example, in the ring manifold of Fig. 4. In this case, the first projection extracted by LEM corresponds to $\cos(\theta)$, where θ is the outer angle of the ring. Therefore, before computing the second projection, the advection process moves the points along the θ coordinate towards the locations at which $\cos(\theta)$ attains its mean value, which is 0. This causes the points with $\theta \in (0, \pi)$ to collapse to $\theta = \pi/2$, and the points with $\theta \in (\pi, 2\pi)$ to collapse to $\theta = 3\pi/2$. The two resulting clusters form an unconnected graph, so that LEM cannot be applied once more. An additional drawback of this method is that it requires a-priori knowledge of the manifold dimension. Furthermore, it is very computationally intensive and thus impractical for high-dimensional big data applications.

In this paper, we propose a different approach. Similarly to the methods described above, our algorithm is sequential. However, rather than heuristically modifying the data points in each stage, we propose to directly incorporate constraints which guarantee that the projections are not redundant.

3   Eliminating Redundancy

Nonlinear dimensionality reduction algorithms seek a set of *non-linear* projections $f_i : \mathbb{R}^D \to \mathbb{R}$, $i = 1, \cdots, d$ which map D-dimensional data points $\boldsymbol{x}_n \in \mathbb{R}^D$ into a d-dimensional feature space $(d < D)$.

Definition 1. *We call a sequence of projections* $\{f_i\}$ **non-redundant** *if none of them can be expressed as a function of the preceding ones. That is, for every i,*

$$f_i(\boldsymbol{x}) \neq g(f_{i-1}(\boldsymbol{x}), \cdots, f_1(\boldsymbol{x})) \tag{3}$$

for every function $g : \mathbb{R}^{i-1} \to \mathbb{R}$.

Let us see why existing spectral dimensionality reduction algorithms do not necessarily yield non-redundant projections. Spectral algorithms obtain the ith projection of all the data points, denoted by $\boldsymbol{f}_i = (f_i(\boldsymbol{x}_1), \cdots, f_i(\boldsymbol{x}_N))^T$, as the solution to the optimization problem[1]

$$\begin{aligned}
\max_{\boldsymbol{f}_i} \quad & \boldsymbol{f}_i^T \boldsymbol{K} \boldsymbol{f}_i \\
\text{s.t.} \quad & \boldsymbol{1}^T \boldsymbol{f}_i = 0 \\
& \boldsymbol{f}_i^T \boldsymbol{f}_i = 1 \\
& \boldsymbol{f}_i^T \boldsymbol{f}_j = 0, \quad \forall j < i.
\end{aligned} \tag{4}$$

Here, $\boldsymbol{K}$ is an $N \times N$ algorithm-specific positive definite (kernel) matrix constructed from the data points [14,36], and $\boldsymbol{1}$ is an $N \times 1$ vector of ones. The first constraint in (4) ensures that the projections have zero means. The last two constraints restrict the projections to be orthonormal. The solution to Problem (4) is given by the d top eigenvectors of the centered kernel matrix $(\boldsymbol{I} - \frac{1}{N}\boldsymbol{1}\boldsymbol{1}^T)\boldsymbol{K}(\boldsymbol{I} - \frac{1}{N}\boldsymbol{1}\boldsymbol{1}^T)$. When $\boldsymbol{K}$ is a stochastic matrix (e.g. LLE, LEM), the solution is simply eigenvectors $2, \ldots, d+1$ of $\boldsymbol{K}$ (without centering).

The orthogonality constraints in (4) guarantee that the projections be linearly independent. However, they do not guarantee non-redundancy. To see this, it is insightful to interpret them in statistical terms. Assume that the data points $\{\boldsymbol{x}_n\}$ correspond to independent realizations of some random vector X. Then orthogonality corresponds to zero statistical correlation, as

$$\mathbb{E}[f_i(X)f_j(X)] \approx \frac{1}{N} \sum_n f_i(\boldsymbol{x}_n)f_j(\boldsymbol{x}_n) = \frac{1}{N}\boldsymbol{f}_i^T \boldsymbol{f}_j = 0. \tag{5}$$

Therefore, in particular, the constraints in (4) guarantee that each projection be uncorrelated with any linear combination of the preceding projections, so that

---

[1] Note that LEM and DFM rather use *weighted* orthogonality constraints, but they can also be brought into the form of (4) (see supplementary material). Also, note that some methods (e.g. LEM, LLE) rather *minimize* the objective in (4). These problems can be cast as (4) with the kernel $\check{\boldsymbol{K}} = \lambda_{\max}\boldsymbol{I} - \boldsymbol{K}$ [4,17].

$f_i(X)$ *cannot be a linear function of the previous projections* $\{f_j(X)\}_{j<i}$. However, these constraints do not prevent $f_i(X)$ from being a *nonlinear* function of the previous projections, which would lead to redundancy, as in Figs. 1, 2 and 4.

To enforce non-redundancy, i.e. that each projection is not a function of the previous ones, we propose to use the following observation.

Lemma 1. *A sequence of non-trivial zero-mean projections* $\{f_i\}$ *is* **non-redundant** *if each of them is* **unpredictable** *from the preceding ones, namely*

$$\mathbb{E}\left[f_i(X)|f_{i-1}(X),\cdots,f_1(X)\right]=0. \tag{6}$$

Proof. Assume (6) holds and suppose to the contrary that the ith projection is non-trivial and redundant, so that $f_i(X)=h(f_{i-1}(X),\ldots,f_1(X))$ for some function h. From the orthogonality property of the conditional expectation,

$$\mathbb{E}[(f_i(X)-\mathbb{E}[f_i(X)|f_{i-1}(X),\cdots,f_1(X)])\,g(f_{i-1}(X),\cdots,f_1(X))]=0 \tag{7}$$

for every function g. Substituting (6), this property implies that

$$\mathbb{E}\left[f_i(X)\,g(f_{i-1}(X),\cdots,f_1(X))\right]=0,\quad \forall g. \tag{8}$$

Therefore, in particular, for $g\equiv h$ we get that $\mathbb{E}[f_i^2(X)]=0$, contradicting our assumption that $f_i(X)$ is non-trivial. □

Notice that by enforcing unpredictability, we in fact restrict each projection to be uncorrelated with *any function* of the previous projections (see (8)). This constraint is stronger than the original zero correlation constraint (5), yet less restrictive than statistical independence. Specifically, two random variables Y,Z are independent if and only if $\mathbb{E}\left[g(Y)h(Z)\right]=\mathbb{E}\left[g(Y)\right]\mathbb{E}\left[h(Z)\right],\,\forall g,h$, whereas for Y to be unpredictable from Z it is only required that $\mathbb{E}\left[Y\,h(Z)\right]=\mathbb{E}\left[Y\right]\mathbb{E}\left[h(Z)\right],\,\forall h$ (corresponding to (8) in the case of zero-mean variables).

4   Algorithm

The unpredictability condition (6) is in fact an infinite set (a continuum) of constraints, as it restricts the conditional expectation of $f_i(X)$ to be zero, given every possible value that the previous projections $\{f_j(X)\}_{j<i}$ may take. However, in practice, spectral methods compute the projections only at the sample points. Therefore, to obtain a practical method, we propose to enforce these restrictions only at the sample embedding points, leading to a discrete set of N constraints

$$\mathbb{E}\left[f_i(X)|\{f_j(X)=f_j(\boldsymbol{x}_n)\}_{j<i}\right]=0,\quad n=1,\ldots,N. \tag{9}$$

These N conditional expectations can be approximated using a kernel smoother matrix $\boldsymbol{P}_i\in\mathbb{R}^{N\times N}$ for regressing $\boldsymbol{f}_i$ against $\boldsymbol{f}_{i-1},\ldots,\boldsymbol{f}_1$, so that the nth entry of the vector $\boldsymbol{P}_i\boldsymbol{f}_i$ approximates the nth conditional expectation in (9),

$$[\boldsymbol{P}_i\boldsymbol{f}_i]_n\approx\mathbb{E}\left[f_i(X)|\{f_j(X)=f_j(\boldsymbol{x}_n)\}_{j<i}\right]. \tag{10}$$

When using the Nadaraya-Watson estimator [27, 39], the accuracy of this approximation is $\mathcal{O}(N^{-4/(i+3)})$. We therefore propose to replace the zero-correlation constraints $\boldsymbol{f}_i^T \boldsymbol{f}_j = 0$ in (4), by the unpredictability restrictions $\boldsymbol{P}_i \boldsymbol{f}_i = \boldsymbol{0}$. Our proposed redundancy-avoiding version of the spectral dimensionality reduction problem (4) is thus

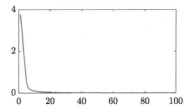

Fig. 5. Top 100 of 15K singular values of the matrix $\boldsymbol{P}_2$ in the MNIST experiment of Fig. 3. The matrix is very close to being low-rank: 0.1% of its singular values account for over 99.9% of its Frobenius norm.

$$\begin{aligned} \max_{\boldsymbol{f}_i} \quad & \boldsymbol{f}_i^T \boldsymbol{K} \boldsymbol{f}_i \\ \text{s.t.} \quad & \boldsymbol{1}^T \boldsymbol{f}_i = 0 \\ & \boldsymbol{f}_i^T \boldsymbol{f}_i = 1 \\ & \boldsymbol{P}_i \boldsymbol{f}_i = \boldsymbol{0}, \qquad \forall i > 1. \end{aligned} \tag{11}$$

In the continuous domain, the conditional expectation operator has a non-empty null space. However, this property is usually not maintained by non-parametric sample approximations, like kernel regressors. As a result, the matrix $\boldsymbol{P}_i$ will typically be only *approximately* low-rank. Figure 5 shows a representative example, where 0.1% of the singular values account for over 99.9% of the Frobenius norm. To ensure that $\boldsymbol{P}_i$ is strictly low-rank (so that $\boldsymbol{P}_i \boldsymbol{f}_i = \boldsymbol{0}$ is not an empty set), we truncate its negligible singular values.

The solution to problem (11) is no longer given by the spectral decomposition of $\boldsymbol{K}$. However, it can be brought into a convenient form by using the following lemma[2] (see proof in Appendix A).

Lemma 2. *Denote the compact SVD of $\boldsymbol{P}_i$ by $\boldsymbol{U}_i \boldsymbol{D}_i \boldsymbol{V}_i^T$. Then the vectors $\boldsymbol{f}_1, \ldots, \boldsymbol{f}_d$ which optimize Problem (11), also optimize*

$$\begin{aligned} \max_{\boldsymbol{f}_i} \quad & \boldsymbol{f}_i^T \tilde{\boldsymbol{K}}_i \boldsymbol{f}_i \\ \text{s.t.} \quad & \boldsymbol{1}^T \boldsymbol{f}_i = 0 \\ & \boldsymbol{f}_i^T \boldsymbol{f}_i = 1, \end{aligned} \tag{12}$$

where $\tilde{\boldsymbol{K}}_i = (\boldsymbol{I} - \boldsymbol{V}_i \boldsymbol{V}_i^T) \boldsymbol{K} (\boldsymbol{I} - \boldsymbol{V}_i \boldsymbol{V}_i^T)$ and $\boldsymbol{V}_1 = \boldsymbol{0}$.

From this lemma, it becomes clear that $\boldsymbol{f}_i$ is precisely the top eigenvector of $\tilde{\boldsymbol{K}}_i$. This implies that we can determine the non-redundant projections sequentially. In the ith step, we first modify the kernel $\boldsymbol{K}$ according to the previous projections $\boldsymbol{f}_{i-1}, \ldots, \boldsymbol{f}_1$ to obtain $\tilde{\boldsymbol{K}}_i$. Then, we compute its top eigenvector to obtain projection $\boldsymbol{f}_i$. This is summarized in Algorithm 1, where for concreteness, we chose $\boldsymbol{P}_i$ to be the Nadaraya-Watson smoother with a Gaussian-kernel.

---

[2] Note that this lemma holds true only for *maximization* problems.

Algorithm 1. Non-redundant dimensionality reduction.

Input: High-dimensional data points $x_n \in \mathbb{R}^D$.
Output: Embeddings $\boldsymbol{f}_i = (f_i(\boldsymbol{x}_1), \cdots, f_i(\boldsymbol{x}_N))^T$.
 1: Construct the kernel matrix $\boldsymbol{K}$ as in the original algorithm (e.g. LLE, LEM, Isomap, etc.).
 2: If the original algorithm *minimizes* the objective of (4) (e.g. LLE, LEM), then set $\boldsymbol{K} \leftarrow \lambda_{\max}\boldsymbol{I} - \boldsymbol{K}$.
 3: Assign the top (non-trivial) eigen-vector of $\boldsymbol{K}$ to $\boldsymbol{f}_1$.
 4: **for** $i = 2, \ldots, d$ **do**
 5:     Construct smoothing matrix

$$[\boldsymbol{P}_i]_{j,k} \leftarrow \exp\left\{-\tfrac{1}{2h^2}\sum_{\ell=1}^{i-1}\left(f_\ell\left(x_j\right) - f_\ell\left(x_k\right)\right)^2\right\}, \quad [\boldsymbol{P}_i]_{j,k} \leftarrow \frac{[\boldsymbol{P}_i]_{j,k}}{\sum_{n=1}^{N}[\boldsymbol{P}_i]_{j,n}}.$$

 6:     Compute $\boldsymbol{V}_i \in \mathbb{R}^{N \times r}$, the top r right singular vectors of $\boldsymbol{P}_i$ accounting for all non-negligible singular values.
 7:     Form the modified kernel matrix

$$\tilde{\boldsymbol{K}}_i \leftarrow \left(\boldsymbol{I} - \boldsymbol{V}_i\boldsymbol{V}_i^T\right)\boldsymbol{K}\left(\boldsymbol{I} - \boldsymbol{V}_i\boldsymbol{V}_i^T\right).$$

 8:     Assign the top eigen-vector of $\tilde{\boldsymbol{K}}_i$ to $\boldsymbol{f}_i$.
 9: **end for**

Efficient Implementation. We use the fast method of [16] to compute the top eigenvector of $\tilde{\boldsymbol{K}}_i$ (step 8). Each iteration of [16] involves multiplication by $\tilde{\boldsymbol{K}}_i$, which can be broken into efficient multiplications by $\boldsymbol{V}_i$ and $\boldsymbol{V}_i^T$ which are $N \times r$ and $r \times N$ with $r \ll N$, and by $\boldsymbol{K}$ which is usually sparse by construction (e.g. in LEM, LLE, LTSA). Thus, we never explicitly form the matrix $\tilde{\boldsymbol{K}}_i$ (step 7).

When memory resources are restrictive, we construct a *sparse* smoothing matrix $\boldsymbol{P}_i$ (step 5) by using only the k nearest neighbors of each sample. To minimize the degradation in the representation quality we use the maximal k such that $\boldsymbol{P}_i$ fits in memory.

4.1   Relation to Independent Component Analysis (ICA)

Our method may seem similar to ICA [19, 22], however, they are quite distinct. First, the ICA *objective* is independence (without preservation of geometrical structure), while in our method the objective is to preserve geometric structure subject to a statistical constraint on the embedding coordinates. Second, *non-linear* ICA is an under-determined problem, making it necessary to impose assumptions or to restrict the class of non-linear functions [20, 33]. Finally, independence is a stronger constraint than unpredictability, and would thus narrow the set of possible solutions. This is while, as we saw, unpredictability is enough for avoiding redundancy.

5   Experiments

We tested our non-redundant algorithm on three high-dimensional data sets. In all our experiments, we report results with the Nadaraya-Watson smoother [27,39], as specified in Algorithm 1. We also experimented with a locally linear smoother and did not observe a significant difference. The kernel smoother bandwidth h was set adaptively: for computing $\boldsymbol{P}_i$, we took $h = \alpha(\sum_{j=1}^{i-1} \frac{1}{N}\|\boldsymbol{f}_j\|^2)^{1/2}$, where the parameter $\alpha \in [0.1, 0.6]$ was chosen using a tune set in the classification task and manually in the visualization tasks. Singular vectors corresponding to singular values smaller than 3% of the largest singular value were truncated. We used the largest number of nearest neighbors such that $\boldsymbol{P}_i$ could still be stored in memory (10K in our case). A hyper-parameter analysis is included in the supplementary material.

5.1   Artificial Head Images

The artificial head image dataset [35] is a popular test bed for manifold learning techniques. It contains 64×64 computer-rendered images of a head, with varying vertical and horizontal camera positions (denoted by θ and ϕ) and lighting directions (denoted by ψ). Linear methods (e.g. PCA, ICA) fail to detect these underlying parameters [35]. However, most (non-linear) spectral methods manage to non-redundantly extract those parameters with the first three projections, since each of the parameters (θ, ϕ, ψ) varies significantly across this data set.

Here, to make the representation learning task more challenging, we chose a 257 subset of the original data set, corresponding to the reduced parameter range $\theta \in [-75°, 75°]$, $\phi \in [-8°, 8°]$, $\psi \in [105°, 175°]$. Figures 6(a), (c) visualize the projections extracted by LEM and LTSA in this case. As can be seen, both algorithms produce redundant representations, as their second projection is a deterministic function of the first. When incorporating our unpredictability constraints, we are able to avoid this repetition and to reveal additional information with the second projection, as evident from Figs. 6(b), (d). We quantify this by reconstructing the images from their two-dimensional embeddings using leave-one-out prediction with a non-parametric regressor (Nadaraya-Watson [27,39]). The average reconstruction peak signal to noise ratio (PSNR) is 18.0/18.2 for the original LEM/LTSA and 19.2/19.9 with our non-redundant LEM/LTSA.

To analyze what the projections capture, we plot in Fig. 6(e)–(h) each of the embedding coordinates vs. the horizontal and vertical camera positions. From Figs. 6(e), (g) we see that in the original algorithms, Projections 1 and 2 are both correlated only with the horizontal angle θ. In our approach, on the other hand, Projection 1 captures the horizontal angle θ while Projection 2 reveals the vertical angle ϕ (see Figs. 6(f), (h)).

5.2   Image Patch Representation

To visualize the effect of non-redundancy in low-level vision tasks, we extracted all 7×7 patches with 3 pixel overlap from an image (taken from [31]), and learned

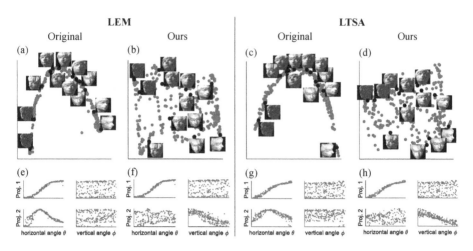

Fig. 6. Two-dimensional embeddings of computer rendered head images with varying pose and lighting directions. (a) The original LEM method. (b) Our non-redundant LEM. (c) The original LTSA method. (d) Our non-redundant LTSA. In the original algorithms, the second coordinate is a function of the first. In our method, the second coordinate clearly carries additional information w.r.t. first, and is thus non-redundant. (e)–(h) The first two projections of the head images vs. the horizontal and vertical angles (θ, ϕ) of the heads. The two projections extracted by the original algorithms are *both* correlated only with the horizontal angle θ. In our non-redundant algorithms, on the other hand, the second projection is correlated with the vertical angle ϕ.

a three dimensional representation using Isomap and using our non-redundant version of Isomap. Figure 7 visualizes the first three projections by coloring each pixel according to the embedding value of its surrounding patch. Observe that in the original algorithm, the first two projections redundantly capture brightness attributes, and the third captures mainly vertical edges with some brightness attributes still remaining (e.g. the sky, the left poolside). In contrast, in our algorithm, the second and third projections capture the vertical *and horizontal* edges (without redundantly capturing brightness multiple times), thus providing additional information. The redundancy of the 2nd Isomap projection can be seen in the scatter plot of the 2nd projection vs. the 1st. With our non-redundant algorithm, the 2nd projection is clearly not a function of the 1st, and thus captures new informative features. To quantify the amount of redundancy, we reconstruct the patches from their three-dimensional embeddings using leave-one-out prediction with the Nadaraya-Watson regressor, and then form an image by averaging overlapping patches (Fig. 8). The reconstruction PSNR is 32.9 using Isomap and 33.2 using our non-redundant Isomap.

Notice that the brightness and gradient features are linear functions of the input patches. Thus, our extracted 3D manifold is in fact linear and would be also correctly revealed by linear methods, such as PCA (not shown). Nevertheless, Isomap which is a nonlinear method, fails to extract this linear manifold due to redundancy (similarly to Fig. 2). In contrast, our non-redundant algorithm can reveal the underlying manifold regardless of its complexity.

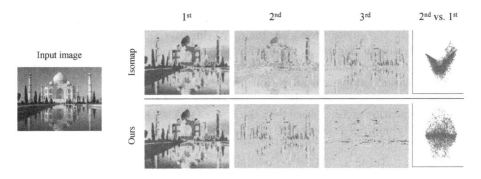

Fig. 7. Three-dimensional embedding of all 7×7 patches with a 3 pixel overlap, obtained with Isomap and with our non-redundant version of Isomap. Each pixel is colored according to the projection of its surrounding patch. In both methods, the first projection captures brightness. However, the original Isomap redundantly captures brightness-related features again with the second projection, and captures vertical edges only with the third projection. In contrast, our non-redundant version captures vertical *and horizontal* edges with the second and third projections. The scatter plot reveals that in the original Isomap, the *2nd* projection is a function of the 1st, while in ours it is not. (Color figure online)

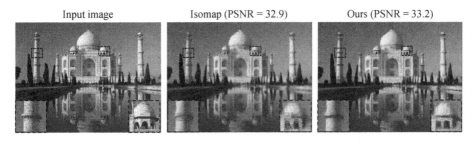

Fig. 8. The image of Fig. 7 reconstructed from the 3-dimensional patch embeddings obtained with Isomap and with our non-redundant Isomap. Note how horizontal edges are not preserved by the Isomap projections, but are preserved by our method.

5.3   MNIST Handwritten Digits

In most applications, the "correct" parametrization of the data manifold is not as obvious as in the head experiment. One example is the MNIST database [23], which contains 28×28 images of handwritten digits. In such settings, determining the quality of a low-dimensional representation can be done by measuring its impact on the performance in downstream tasks, like classification.

In the next experiment, we randomly chose a subset of 15K images from the MNIST data set, based on which we learned low-dimensional representations with LEM and with three modifications of LEM: (i) the sequential regression technique (Sect. 2), (ii) the algorithm of Dsilva et al. [10], and (iii) our non-redundant method. We then split the data into 10K/2.5K/2.5K for training/tuning/testing and trained a third degree polynomial-kernel SVM [6] to classify the digits based on their low-dimensional representations. The SVM's soft margin parameter c and kernel parameter γ were tuned based on performance on the tune set (within the range $c \in [1, 10]$, $\gamma \in [0.1, 0.2]$). Table 1 shows the classification error for various representation sizes. As can be seen, our non-redundant representation leads to the largest and most consistent decrease in the classification error. Notice that the *linear* PCA/ICA[3] baselines are inferior in this highly non-linear scenario.

Table 1. MNIST experiment classification errors [%].

# of proj.	15K examples all labeled						# of proj.	15K examples 300 labeled	
	LEM	Dsilva et. al.	Sequential regression	Ours	PCA	ICA		LEM	Ours
3	17.6	17.6	17.3	**12.0**	47.8	59.0	5	12.6	10.3
5	8.8	8.8	14.4	**7.6**	25.7	34.1	16	8.4	6.6
7	6.9	6.9	14.2	**6.0**	14.2	19.7	24	7.2	7.2
9	6.5	6.5	14.2	**5.6**	9.9	12.9	35	7.8	8.1
11	6.0	5.4	13.8	**5.0**	7.7	6.6	50	8.8	8.8

To demonstrate the importance of *compact* representations, particularly in the semi-supervised scenario, we repeated the experiment where only 300 of the examples are labeled for the SVM training (right pan of Table 1). Notice that the error reaches a minimum at 16/24 projections with our/LEM method, and then begins to rise as the representation dimension increases. This illustrates that unnecessarily large representations result in inferior performance in downstream tasks. Our method, which is designed to construct compact representations, achieves a lower minimal error (6.6% vs. 7.2%).

Run-time. Attaining the non-redundant projections comes at the expense of increased run-time. For example, obtaining 11 projections of 15K MNIST examples takes 14 min on a 4-core Intel i5 desktop with 16 GB RAM, whereas obtaining the original LEM projections takes 13 s.

6   Conclusions

We presented a general approach for overcoming the redundancy phenomenon in spectral dimensionality reduction algorithms. As opposed to prior attempts, which fail in complex high-dimensional situations, our approach provably produces non-redundant representations. This is achieved by replacing the orthogonality constraints underlying spectral methods, by unpredictability constraints.

---

[3] In ICA, the number of independent components is equal to the dimension of the data. To obtain a low-dimensional embedding, we applied ICA on the low-dimensional embedding produced by PCA [1].

Our solution reduces to applying a sequence of spectral decompositions, where in each step, the kernel matrix is modified according to the projections computed so far. Our experiments illustrate the ability of our method to capture more informative compact representations of high-dimensional data.

A A Proof of Lemma 2

We start by proving that any $\boldsymbol{f}_i$ solving (12) necessarily satisfies $\boldsymbol{P}_i \boldsymbol{f}_i = \boldsymbol{0}$. First, note that this constraint is equivalent to $\boldsymbol{V}_i \boldsymbol{V}_i^T \boldsymbol{f}_i = \boldsymbol{0}$, since $\boldsymbol{D}_i$, $\boldsymbol{U}_i$, and $\boldsymbol{V}_i$ have empty null spaces. Now, suppose that $\boldsymbol{f}_i$ maximizes the objective of (12) and satisfies the constraints $\|\boldsymbol{f}_i\| = 1$ and $\boldsymbol{1}^T \boldsymbol{f}_i = \boldsymbol{0}$, but does not satisfy $\boldsymbol{V}_i \boldsymbol{V}_i^T \boldsymbol{f}_i = \boldsymbol{0}$. Then define the alternative solution

$$\tilde{\boldsymbol{f}}_i = \frac{(\boldsymbol{I} - \boldsymbol{V}_i \boldsymbol{V}_i^T)\boldsymbol{f}_i}{\|(\boldsymbol{I} - \boldsymbol{V}_i \boldsymbol{V}_i^T)\boldsymbol{f}_i\|}, \tag{13}$$

which clearly satisfies the constraints $\|\tilde{\boldsymbol{f}}_i\| = 1$ and $\boldsymbol{1}^T \tilde{\boldsymbol{f}}_i = \boldsymbol{0}$, but additionally also satisfies $\boldsymbol{V}_i \boldsymbol{V}_i^T \boldsymbol{f}_i = \boldsymbol{0}$. Notice that $\boldsymbol{I} - \boldsymbol{V}_i \boldsymbol{V}_i^T$ is a projection matrix (as $\boldsymbol{V}_i$ is orthogonal), so that $(\boldsymbol{I} - \boldsymbol{V}_i \boldsymbol{V}_i^T)^2 = \boldsymbol{I} - \boldsymbol{V}_i \boldsymbol{V}_i^T$ and $\|(\boldsymbol{I} - \boldsymbol{V}_i \boldsymbol{V}_i^T)\boldsymbol{f}_i\|^2 \leq \|\boldsymbol{f}_i\|^2 = 1$. Therefore,

$$\tilde{\boldsymbol{f}}_i^T \tilde{\boldsymbol{K}}_i \tilde{\boldsymbol{f}}_i = \frac{\boldsymbol{f}_i^T (\boldsymbol{I} - \boldsymbol{V}_i \boldsymbol{V}_i^T)^2 \boldsymbol{K} (\boldsymbol{I} - \boldsymbol{V}_i \boldsymbol{V}_i^T)^2 \boldsymbol{f}_i}{\|(\boldsymbol{I} - \boldsymbol{V}_i \boldsymbol{V}_i^T)\boldsymbol{f}_i\|^2} \geq \boldsymbol{f}_i^T \tilde{\boldsymbol{K}}_i \boldsymbol{f}_i, \tag{14}$$

with equality only when $\boldsymbol{V}_i \boldsymbol{V}_i^T \boldsymbol{f}_i = \boldsymbol{0}$. In other words, $\tilde{\boldsymbol{f}}_i$ achieves a higher objective value than $\boldsymbol{f}_i$, contradicting our assumption that $\boldsymbol{f}_i$ is a solution to (12). This proves that any $\boldsymbol{f}_i$ that solves problem (12) necessarily also satisfies the constraints of problem (11). Therefore, effectively, the solutions to (11) and (12) satisfy the same constraints.

Next, observe that if $\boldsymbol{f}_i$ satisfies the constraint $\boldsymbol{V}_i \boldsymbol{V}_i^T \boldsymbol{f}_i = \boldsymbol{0}$ then the objectives of (11) and (12) are equivalent, since

$$\boldsymbol{f}_i^T \tilde{\boldsymbol{K}}_i \boldsymbol{f}_i = \boldsymbol{f}_i^T (\boldsymbol{I} - \boldsymbol{V}_i \boldsymbol{V}_i^T) \boldsymbol{K} (\boldsymbol{I} - \boldsymbol{V}_i \boldsymbol{V}_i^T) \boldsymbol{f}_i = \boldsymbol{f}_i^T \boldsymbol{K} \boldsymbol{f}_i. \tag{15}$$

Therefore, $\boldsymbol{f}_i$ solves (12) if and only if it solves (11).

References

1. Bartlett, M.S., Movellan, J.R., Sejnowski, T.J.: Face recognition by independent component analysis. IEEE Trans. Neural Netw. **13**(6), 1450–1464 (2002)
2. Belkin, M., Niyogi, P.: Laplacian eigenmaps for dimensionality reduction and data representation. Neural Comput. **15**(6), 1373–1396 (2003)
3. Belkin, M., Niyogi, P.: Semi-supervised learning on Riemannian manifolds. Mach. Learn. **56**(1–3), 209–239 (2004)

4. Bengio, Y., Paiement, J.F., Vincent, P., Delalleau, O., Le Roux, N., Ouimet, M.: Out-of-sample extensions for LLE, Isomap, MDS, Eigenmaps, and spectral clustering. In: Advances in Neural Information Processing Systems (NIPS) 16, pp. 177–184 (2004)

5. Brun, A., Park, H.-J., Knutsson, H., Westin, C.-F.: Coloring of DT-MRI fiber traces using Laplacian eigenmaps. In: Moreno-Díaz, R., Pichler, F. (eds.) EUROCAST 2003. LNCS, vol. 2809, pp. 518–529. Springer, Heidelberg (2003). https://doi.org/10.1007/978-3-540-45210-2_47

6. Chang, C.C., Lin, C.J.: LIBSVM: a library for support vector machines. ACM Trans. Intell. Syst. Technol. (TIST) 2(3), 27 (2011)

7. Chang, H., Yeung, D.Y., Xiong, Y.: Super-resolution through neighbor embedding. In: IEEE Conference on Computer Vision and Pattern Recognition (CVPR), vol. 1, p. I (2004)

8. Coifman, R.R., Lafon, S.: Diffusion maps. Appl. Comput. Harmon. Anal. 21(1), 5–30 (2006)

9. Donoho, D.L., Grimes, C.: Hessian eigenmaps: locally linear embedding techniques for high-dimensional data. Proc. Natl. Acad. Sci. 100(10), 5591–5596 (2003)

10. Dsilva, C.J., Talmon, R., Coifman, R.R., Kevrekidis, I.G.: Parsimonious representation of nonlinear dynamical systems through manifold learning: a chemotaxis case study. Appl. Comput. Harmon. Anal. (2015, in press)

11. Elgammal, A., Lee, C.S.: Inferring 3D body pose from silhouettes using activity manifold learning. In: IEEE Conference on Computer Vision and Pattern Recognition (CVPR), vol. 2, pp. II–681 (2004)

12. Geng, X., Zhan, D.C., Zhou, Z.H.: Supervised nonlinear dimensionality reduction for visualization and classification. IEEE Trans. Syst. Man Cybern. Part B (Cybern.) 35(6), 1098–1107 (2005)

13. Gerber, S., Tasdizen, T., Whitaker, R.: Robust non-linear dimensionality reduction using successive 1-dimensional Laplacian eigenmaps. In: International Conference on Machine Learning (ICML), pp. 281–288 (2007)

14. Goldberg, Y., Zakai, A., Kushnir, D., Ritov, Y.: Manifold learning: the price of normalization. J. Mach. Learn. Res. 9, 1909–1939 (2008)

15. Guo, G., Fu, Y., Dyer, C.R., Huang, T.S.: Image-based human age estimation by manifold learning and locally adjusted robust regression. IEEE Trans. Image Process. 17(7), 1178–1188 (2008)

16. Halko, N., Martinsson, P.G., Tropp, J.A.: Finding structure with randomness: probabilistic algorithms for constructing approximate matrix decompositions. SIAM Rev. 53(2), 217–288 (2011)

17. Ham, J., Lee, D.D., Mika, S., Schölkopf, B.: A kernel view of the dimensionality reduction of manifolds. In: International Conference on Machine Learning (ICML), p. 47 (2004)

18. He, X., Yan, S., Hu, Y., Niyogi, P., Zhang, H.J.: Face recognition using Laplacianfaces. IEEE Trans. Pattern Anal. Mach. Intell. 27(3), 328–340 (2005)

19. Hyvärinen, A.: Fast and robust fixed-point algorithms for independent component analysis. IEEE Trans. Neural Netw. 10(3), 626–634 (1999)

20. Hyvärinen, A., Pajunen, P.: Nonlinear independent component analysis: existence and uniqueness results. Neural Netw. 12(3), 429–439 (1999)

21. Isola, P., Zoran, D., Krishnan, D., Adelson, E.H.: Crisp boundary detection using pointwise mutual information. In: Fleet, D., Pajdla, T., Schiele, B., Tuytelaars, T. (eds.) ECCV 2014. LNCS, vol. 8691, pp. 799–814. Springer, Cham (2014). https://doi.org/10.1007/978-3-319-10578-9_52

22. Jutten, C., Herault, J.: Blind separation of sources, part I: an adaptive algorithm based on neuromimetic architecture. Sig. Process. **24**(1), 1–10 (1991)
23. LeCun, Y., Bottou, L., Bengio, Y., Haffner, P.: Gradient-based learning applied to document recognition. Proc. IEEE **86**(11), 2278–2324 (1998)
24. Lee, C.S., Elgammal, A.: Modeling view and posture manifolds for tracking. In: International Conference on Computer Vision (ICCV), pp. 1–8 (2007)
25. Lim, H., Camps, O.I., Sznaier, M., Morariu, V.I.: Dynamic appearance modeling for human tracking. In: IEEE Conference on Computer Vision and Pattern Recognition (CVPR), vol. 1, pp. 751–757 (2006)
26. Lim, I.S., de Heras Ciechomski, P., Sarni, S., Thalmann, D.: Planar arrangement of high-dimensional biomedical data sets by Isomap coordinates. In: IEEE Symposium on Computer-Based Medical Systems, pp. 50–55 (2003)
27. Nadaraya, E.A.: On estimating regression. Theory Probab. Appl. **9**(1), 141–142 (1964)
28. Pless, R.: Image spaces and video trajectories: using Isomap to explore video sequences. In: International conference on Computer Vision (ICCV), vol. 3, pp. 1433–1440 (2003)
29. Raytchev, B., Yoda, I., Sakaue, K.: Head pose estimation by nonlinear manifold learning. In: International Conference on Pattern Recognition (ICPR), vol. 4, pp. 462–466 (2004)
30. Roweis, S.T., Saul, L.K.: Nonlinear dimensionality reduction by locally linear embedding. Science **290**(5500), 2323–2326 (2000)
31. Rubinstein, M., Gutierrez, D., Sorkine, O., Shamir, A.: A comparative study of image retargeting. ACM Trans. Graph. (TOG) **29**, 160 (2010)
32. Schölkopf, B., Smola, A., Müller, K.-R.: Kernel principal component analysis. In: Gerstner, W., Germond, A., Hasler, M., Nicoud, J.-D. (eds.) ICANN 1997. LNCS, vol. 1327, pp. 583–588. Springer, Heidelberg (1997). https://doi.org/10.1007/BFb0020217
33. Singer, A., Coifman, R.R.: Non-linear independent component analysis with diffusion maps. Appl. Comput. Harmon. Anal. **25**(2), 226–239 (2008)
34. Souvenir, R., Zhang, Q., Pless, R.: Image manifold interpolation using free-form deformations. In: IEEE International Conference on Image Processing (ICIP), pp. 1437–1440 (2006)
35. Tenenbaum, J.B., De Silva, V., Langford, J.C.: A global geometric framework for nonlinear dimensionality reduction. Science **290**(5500), 2319–2323 (2000)
36. Van Der Maaten, L., Postma, E., Van den Herik, J.: Dimensionality reduction: a comparative review. J. Mach. Learn. Res. **10**, 66–71 (2009)
37. Vlachos, M., Domeniconi, C., Gunopulos, D., Kollios, G., Koudas, N.: Non-linear dimensionality reduction techniques for classification and visualization. In: International Conference on Knowledge Discovery and Data Mining, pp. 645–651 (2002)
38. Wang, Q., Xu, G., Ai, H.: Learning object intrinsic structure for robust visual tracking. In: IEEE Conference on Computer Vision and Pattern Recognition (CVPR), vol. 2, pp. II-227 (2003)
39. Watson, G.S.: Smooth regression analysis. Sankhyā: Indian J. Stat. Ser. A **26**(4), 359–372 (1964)
40. Zhang, Z., Chow, T.W., Zhao, M.: Trace ratio optimization-based semi-supervised nonlinear dimensionality reduction for marginal manifold visualization. IEEE Trans. Knowl. Data Eng. **25**(5), 1148–1161 (2013)
41. Zhang, Z., Zha, H.: Principal manifolds and nonlinear dimensionality reduction via tangent space alignment. J. Shanghai Univ. **8**(4), 406–424 (2004)

Rethinking Unsupervised Feature Selection: From Pseudo Labels to Pseudo Must-Links

Xiaokai Wei[1]([✉]), Sihong Xie[2], Bokai Cao[3], and Philip S. Yu[3]

[1] Facebook Inc., Menlo Park, CA, USA
weixiaokai@gmail.com
[2] CSE Department, Lehigh University, Bethlehem, PA, USA
sxie@cse.lehigh.edu
[3] Department of Computer Science, University of Illinois at Chicago,
Chicago, IL, USA
{caobokai,psyu}@uic.edu

Abstract. High-dimensional data are prevalent in various machine learning applications. Feature selection is a useful technique for alleviating the curse of dimensionality. Unsupervised feature selection problem tends to be more challenging than its supervised counterpart due to the lack of class labels. State-of-the-art approaches usually use the concept of pseudo labels to select discriminative features by their regression coefficients and the pseudo-labels derived from clustering is usually inaccurate. In this paper, we propose a new perspective for unsupervised feature selection by Discriminatively Exploiting Similarity (DES). Through forming similar and dissimilar data pairs, implicit discriminative information can be exploited. The similar/dissimilar relationship of data pairs can be used as guidance for feature selection. Based on this idea, we propose hypothesis testing based and classification based methods as instantiations of the DES framework. We evaluate the proposed approaches extensively using six real-world datasets. Experimental results demonstrate that our approaches achieve significantly outperforms the state-of-the-art unsupervised methods. More surprisingly, our unsupervised method even achieves performance comparable to a supervised feature selection method. Code related to this chapter is available at: http://bdsc.lab.uic.edu/resources.html.

Keyword: Feature selection

1 Introduction

Feature selection [5,7,10,20,22], as a dimension reduction technique, can help improve the performance of machine learning tasks [14], by selecting a subset of features. Besides, it can also enhance the efficiency of subsequent learning process, and provide easier interpretation of the problem.

X. Wei—The work was performed when the first author was a Ph.D. student at University of Illinois at Chicago.

© Springer International Publishing AG 2017
M. Ceci et al. (Eds.): ECML PKDD 2017, Part I, LNAI 10534, pp. 272–287, 2017.
https://doi.org/10.1007/978-3-319-71249-9_17

Depending on the availability of supervision information, feature selection methods can be categorized into two classes: Supervised feature selection and unsupervised feature selection. For supervised feature selection, the criterion on good features is more straightforward: good features should be highly correlated with class labels, such as Fisher Score [3] and HSIC [13]. Without guidance from class labels, it is difficult to evaluate the discriminativeness of features. Different heuristics (e.g., frequency based, variance based) have been proposed to perform unsupervised feature selection. Similarity-preserving approaches [5, 24] have gained much popularity among others. In such similarity preserving framework, a feature is considered to be of good quality if it can preserve the local manifold structure well.

However, such simple heuristics do not necessarily lead to discriminative features. Recently, pseudo label based algorithms [8,22] have been developed. Since class labels are not available, such methods attempt to generate pseudo labels via certain clustering methods. They select features based on their utility to predicting pseudo labels. One major drawback of such approach is that the pseudo labels are usually far from accurate (accuracy is usually about 30%–70% as reported in previous work [8,11]). So such inaccurate pseudo labels can mislead feature selection. Also, in unsupervised explorative analysis, the number of classes is often not known as apriori.

The central issue in unsupervised feature selection is how to effectively uncover the discriminative information embedded in the data. Is the concept of pseudo-label the only and best way to achieving discriminativeness? In this paper, we present a novel perspective, **pseudo must-link**, and show how it can be a better alternative than pseudo-labels. We refer to the proposed framework as DES (Discriminatively Exploiting Similarity), which performs unsupervised feature selection based on similarity in a discriminative manner. Specifically, DES aims to exploit the key intuition that similar data points are more likely to be of the same class than two random data points. We present a pair-wise formulation to effectively utilize the such difference between similar and dissimilar data points.

By regarding similar pair/dissimilar pair as two classes, the rich arsenal of supervised feature selection approaches can be employed. We propose two instantiations of this framework for unsupervised feature selection: *Hypothesis Testing* based HT-DES and *Classification* based CL-DES. The proposed approaches are conceptually simple, easy to implement but highly effective. Experimental results shows that DES can achieve significantly better performance than state-of-the-art unsupervised methods. Besides, the performance of CL-DES is even comparable to that of supervised mutual information-based method on most datasets. To our best knowledge, this is the first time an unsupervised method reportedly achieves comparable performance as a classic supervised feature selection method, which illustrates the strength of the proposed framework.

2   Related Work

In this section, we review related work on feature selection.

Feature selection has attracted considerable amount of attention in the research community. The goal is to alleviate the curse of dimensionality, enabling machine learning model to achieve comparable, if not better, performance. In supervised feature selection, the criterion for good features is more straightforward: good features should be highly correlated with class labels. Many methods have been proposed to capture the correlation between label and feature, such as Mutual Information, Fisher Score [3] and Hilbert-Schmidt Independence Criterion (HSIC) [13] and LASSO [16].

In the unsupervised setting, heuristic-based feature selection algorithms tend to evaluate the importance of features individually [5,23], which has a limitation of neglecting correlation among features. Recent methods [11,12,22] overcome this issue by evaluating feature utility with $L_{2,1}$ norm-based sparse regression, which are typically in the following form:

$$\min_{\mathbf{W},\mathbf{U}} l(\mathbf{XW} - \mathbf{U}) + \sum_{i=1}^{R} \alpha_i \cdot reg_i(\mathbf{U}) + \lambda ||\mathbf{W}||_{2,1}$$

$$s.t. \ Constraint(\mathbf{U})$$

where $\mathbf{X}$ is the feature matrix and $\mathbf{U}$ is cluster indicators/latent factors. The $L_{2,1}$ norm $||\mathbf{W}||_{2,1}$ promotes row sparsity in the coefficient matrix $\mathbf{W}$ and hence achieves the effect the feature selection. Different sparse regression-based methods usually differ in $l(\cdot)$, $reg(\cdot)$ and $Constraint(\mathbf{U})$ they use. A typical choice for $l(\cdot)$ is Frobenius norm, such as in UDFS [22], NDFS [8] and FSASL [2], while RUFS [11] and RSFS [12] employ robust $L_{2,1}$ loss and Huber Loss as $l(\cdot)$, respectively. Concerning $reg(\mathbf{U})$, different choices can also be made. For example, UDFS, NDFS and RSFS use local structure-based regularizations in the objective function. RUFS further adds NMF (Non-negative Matrix Factorization) based regularization and FSASL adds sparse regression [21] based regularization. Most methods also put certain constraints on $\mathbf{U}$, such as non-negative constraint and orthogonal constraint [8,11,12].

However, all these pseudo-label approaches have similar drawbacks: the cluster labels are usually not accurate enough [8,12]. The wrongful information contained in the pseudo labels can further mislead feature selection. Besides, they have 3–4 parameters (e.g., number of pseudo labels and several regularization terms) to be specified in the objective function. In unsupervised setting, it is difficult to know the optimal parameters, which limits their practical utility.

Recently, feature selection for non-traditional data types has drawn increasing attention, such as feature selection for linked data [6,15,17,19] and multiview data [4,18].

3 Formulations

3.1 Notations

Suppose we have n data points $\mathbf{X} = [\mathbf{x}_1, \mathbf{x}_2, \ldots, \mathbf{x}_n]$ and the number of features is D. Let f_p denote the p-th feature ($p = 1, 2, \ldots, D$) and $\mathbf{x}_{ip}$ denotes the value of p-th feature of $\mathbf{x}_i$. Our goal is to select d ($d < D$) discriminative features.

As other similarity based feature selection methods [5], we first construct a kNN similarity graph $\mathcal{G}$ from $\mathbf{X}$ (e.g., by cosine similarity). In this graph $\mathcal{G}$, each data instance is connected with its k nearest neighbors. For each data instance $\mathbf{x}_i$ ($i = 1, \ldots, n$), we denote its neighbors as $\mathcal{N}(\mathbf{x}_i)$ and the set of non-neighbors as $\mathcal{NN}(\mathbf{x}_i)$.

3.2 Discriminatively Exploiting Similarity

In general, we want to select features highly indicative of certain classes/topics. In supervised setting, labels provide clear guidance for feature selection: those features that lead to good separability of different classes should be good features. If a term is usually shared by data points from the same class and rarely shared by data points from different classes, then this term is likely to be a discriminative one. In comparison, generic terms are shared indiscriminatively by data points from different classes. Consider the example shown in Fig. 1. Feature 1 and 8 are discriminative ones but feature 5 is not discriminative.

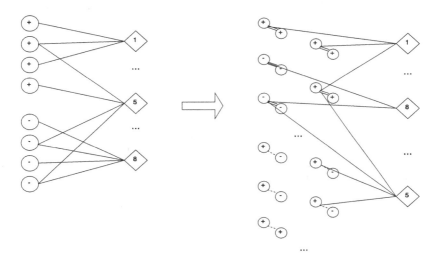

Fig. 1. Supervised Scenario with real labels (left) v.s. Unsupervised Scenario with Sim-Labels (right). Circles are data points and diamonds are terms/features. Link between data instance and terms indicates the occurrence of term. On the left, '+' and '−' denote two classes. Double-edged line denotes similar pair and dashed line denotes dissimilar pair

Our goal is to exploit this intuition in unsupervised case. Based on the kNN similarity graph, we first define *SimLabel* to divide pairs of data points into to classes: must-link (similar pairs) and cannot-link (dissimilar pairs).

Definition 1 SimLabel: *Given the kNN graph $\mathcal{G}$ constructed from pairwise similarity, we label each pair of data points $(\mathbf{x}_i, \mathbf{x}_j)$ $(i, j \in \{1, 2, \ldots, n\}, i \neq j)$, with SimLabel l_{ij}^s defined as below:*

$$l_{ij}^s = \begin{cases} 1 & if \ \mathbf{x}_i \in \mathcal{N}(\mathbf{x}_j) \ or \ \mathbf{x}_j \in \mathcal{N}(\mathbf{x}_i) \\ -1 & otherwise \end{cases} \tag{1}$$

We refer to pairs with $l^s = 1$ as *must-links (similar pairs)* and pairs with $l^s = -1$ as *cannot-links (dissimilar pairs)*. Let us denote the set of similar pairs and set of dissimilar pairs as Ω_s and Ω_d, respectively. We also use $\Omega = \Omega_s \cup \Omega_d$ to denote the set of all pairs. Note that some non-neighbors are not necessarily very different (e.g. data instance i and its $(k + 1)$ closest data instance), but most of them are not as close as neighbors.

To make better use of similarity for selecting discriminative features, we exploit the following intuition: two neighbors are more likely to be of the same class than two random non-neighbors. If a term is shared more often by neighbors than by non-neighbors, it is likely to be discriminative for certain class. In this sense, whether two data points are similar or not can serve the similar functionality of class labels to guide feature selection. We refer to this approach as Discriminatively Exploiting Similarity (DES). Consider feature 1, 5 and 8 in Fig. 1 for example. Discriminative features such as 1 and 8 are shared more often by similar pairs, while generic feature 5 is shared by almost equal amount of similar pairs and dissimilar pairs. For a real-world example, one can consider a collection of research papers on different topics (e.g., Machine Learning, Database and OS). Discriminative terms such as *SVM* and *classification* appear more often in pairs of similar papers. In comparison, generic terms such as *compare* and *propose* appear equally likely in similar papers and dissimilar papers.

In supervised feature selection, each data instance is an instance for learning. In the framework of DES, a pair of data points, rather than a single data instance, becomes the basic instance for learning. Since the instance in DES is a pair of data points, data features cannot be directly used for feature selection. We derive *SimFeatures* from for each pair.

Definition 2 SimFeature: *Given a pair of data points $(\mathbf{x}_i, \mathbf{x}_j)$, the p-th SimFeature is defined as the product of corresponding feature values:*

$$(\mathbf{x}_i, \mathbf{x}_j)_p = \mathbf{x}_{ip} \cdot \mathbf{x}_{jp} \tag{2}$$

For example, for data points with binary features (i.e., term occurrence), the p-th SimFeature for pair $(\mathbf{x}_i, \mathbf{x}_j)$ is whether two data points $\mathbf{x}_i$ and $\mathbf{x}_j$ both have term p.

Our approach uses neighbor/non-neighbor relationship to serve as a proxy of class-belonging information. We aim to exploit the contrast between similar and dissimilar pairs rather than preserving the similarity itself. The contrast can provide useful information for selecting discriminative features. Compared with pseudo label based approaches, we do not explicitly construct labels and therefore the number of classes does not need to be specified explicitly. By discriminatively preserving similarity, DES combines the strength of similarity based approach and pseudo-label based approach.

4 Instantiations of the DES

In previous section, we introduce the general idea of DES. In this section, we adapt ideas from supervised feature selection and combine them with DES.

4.1 Hypothesis Test Based DES (HT-DES)

There are statistical test based supervised feature selection methods, such as Chi-square test [9]. Chi-square tests the null hypothesis that the given feature is independent of the class label. The features with higher test statistics are selected since this indicates the null hypothesis should be rejected and hence such features are highly correlated with the class label.

Inspired by Chi-square test, we propose a hypothesis testing based approach to exploit the difference of similar/dissimilar pairs. We test whether a feature has higher proportion in similar pairs than in dissimilar pairs. If a feature appears more often in similar pairs than in dissimilar pairs, it is likely to be an informative feature.

Specifically, we perform two proportion one-tailed z-test. For a feature, we use p_s to denote the proportion of its presence in similar pairs and p_d the proportion of this feature in dissimilar pairs. The null hypothesis and alternative hypothesis can be formed as follows.

$$\begin{aligned} H_0 &: p_s = p_d \\ H_1 &: p_s > p_d \end{aligned} \tag{3}$$

Pooled sample proportion. Since the null hypothesis assumes $p_s = p_d$, we use a pooled proportion $\hat{p}$ to calculate the standard error.

$$\hat{p} = \frac{p_s \cdot n_s + p_d \cdot n_d}{n_s + n_d} \tag{4}$$

where n_s and n_d are the numbers of sampled similar and dissimilar pairs, respectively.

Standard error. With the pooled proportion p, we can compute the standard error.

$$SE = \sqrt{\hat{p}(1 - \hat{p})(\frac{1}{n_s} + \frac{1}{n_d})} \tag{5}$$

Test statistics. We use the following *z-score* as test statistic for the difference in proportions.

$$z = \frac{p_s - p_d}{SE} \tag{6}$$

Features with high z-scores are shared significantly more often by similar pairs than by dissimilar pairs. Low z-score means that the feature is almost equally possible to appear in both similar and dissimilar pairs and hence less discriminative. For the example in Fig. 1, feature 1 and 8 will have high z-score and feature 5 will have low z-scores. To obtain high-quality features, we can select the top features with high z-scores.

4.2   Classification-Based DES (CL-DES)

HT-DES evaluates features individually and the z-score of one feature is not influenced by the z-scores of other features. So the selected subset of features can be highly redundant. Such redundancy would lead to higher computational cost and potentially degenerated performance.

In this section, we present a classification based approach to evaluate features jointly. The intuition is that discriminative features should be able to distinguish similar pairs from dissimilar pairs. Class labels establish the difference between data instances. In our DES framework, similarity can establish the difference between instance pairs and acts similarly as the class labels do in supervised setting. If a feature is highly indicative of similarity relationship, it is likely to be a useful feature.

To perform feature selection, we first introduce a weight vector $\mathbf{w}$ as features' importance scores since not all features are equally important. By using $\mathbf{w}$, we define *Weighted Similarity* between two instances:

Definition 3 Weighted Similarity: *For two instances $(\mathbf{x}_i, \mathbf{x}_j)$ $(i, j \in \{1, 2, \ldots, n\}, i \neq j)$, their weighted similarity $s_{ij}^{\mathbf{w}}$ w.r.t weight vector $\mathbf{w}$ is defined as:*

$$s_{ij}^{\mathbf{w}} = \mathbf{x}_i^T diag(\mathbf{w}) \mathbf{x}_j \tag{7}$$

where $diag(\mathbf{w})$ is the diagonal matrix with $\mathbf{w}$ as diagonal elements.

It is desirable that the weighted similarity can distinguish similar pairs from dissimilar pairs:

$$s_{ij}^{\mathbf{w}} \cdot l_{ij}^s > 1, \ \forall (i, j) \in \Omega \tag{8}$$

This objective makes our formulation essentially different from the similarity preserving framework, since our goal is to separate similar pairs from dissimilar pairs rather than preserving similarity itself.

In supervised feature selection, L_1-regularization [16] is able to take into consideration the redundancy among features and achieves great success due to its simplicity and effectiveness. To get sparse weight vector $\mathbf{w}$, we add L1 regularization to our DES framework:

$$\min_{\mathbf{w}} \quad \|\mathbf{w}\|_1$$
$$\text{s.t. } s_{ij}^{\mathbf{w}} \cdot l_{ij}^s \geq 1, \forall (i, j) \in \Omega \tag{9}$$

However, similar/dissimilar pairs may not always be separable given the weight vector $\mathbf{w}$, since the original similar/dissimilar pairs constructed from features can be noisy. So, to address this issue, we add an slack variable μ_{ij} to impose soft margin.

$$\min_{\mathbf{w}} \quad \frac{1}{|\Omega|} \sum_{(i,j) \in \Omega} \mu_{ij} + \lambda \|\mathbf{w}\|_1$$

$$\text{s.t. } s_{ij}^{\mathbf{w}} \cdot l_{ij}^s \geq 1 - \mu_{ij}, \forall (i,j) \in \Omega \tag{10}$$

Equation (10) can also be interpreted as L_1 regularized SVM on pair-wise instances with SimLabel and SimFeatures. Discriminative features are more likely to appear in similar pairs and would have relatively larger positive weights. Indiscriminative features have little utility in differentiating similar pairs and dissimilar pairs. As a result, the weights of such features are close to zero or negative. If we rank the features w.r.t their weights, we can select the top ones as high quality features.

5 Optimization

For HT-DES, the optimization is straightforward: one can simply calculate the z-scores of each feature and select the top ones. There are $O(nk)$ similar pairs and $O(n(n-k))$ dissimilar pairs. So the number of dissimilar pairs is much larger than number of similar pairs since $k \ll n$. To avoid imbalanced distribution of SimLabels, we employ a bootstrapping based approach to sample equal amounts of similar and dissimilar pairs for HT-DES and CL-DES.

For CL-DES, the objective function is not differentiable due to hinge loss and L_1 regularization, we calculate subgradient for the objective function and optimize it by stochastic subgradient descent. For a data instance pair $(\mathbf{x}_i, \mathbf{x}_j)$, the subgradient w.r.t. w_p $(p = 1, \ldots, D)$ is:

$$\frac{\partial \mathcal{L}(\mathbf{x}_i, \mathbf{x}_j)}{\partial w_p} = \frac{\partial}{\partial w_p} \mu_{ij} + \lambda \cdot \text{sign}(w_p) \tag{11}$$

where the subdifferential $\frac{\partial}{\partial w_p} \mu_{ij}$ can be calculated as follows.

$$\frac{\partial}{\partial w_p} \mu_{ij} = \begin{cases} x_{ip} \cdot x_{jp} \cdot l_{ij}^s, & \text{if } s_{ij}^{\mathbf{w}} \cdot l_{ij}^s < 1 \\ 0, & \text{otherwise} \end{cases} \tag{12}$$

The Stochastic Sub-gradient Descent method is shown in Algorithm 1 and the time complexity is $O(mT)$, where m is the average number non-zero features in each data instance and T is the total number of iterations.

Algorithm 1. Stochastic Subgradient Descent Algorithm for CL-DES

1: $\mathbf{w}^0 \leftarrow [0, 0, \dots, 0]$
2: **for** (t in $1..T$) **do**
3:     Generate random number $\alpha \in (0, 1)$
4:     **if** $\alpha > 0.5$ **then**
5:         Sample a similar pair $(\mathbf{x}_i, \mathbf{x}_j)$ $(\mathbf{x}_j \in \mathcal{N}(\mathbf{x}_i))$
6:     **else**
7:         Sample a dissimilar pair $(\mathbf{x}_i, \mathbf{x}_j)$ $(\mathbf{x}_j \in \mathcal{N}\mathcal{N}(\mathbf{x}_i))$
8:     **end if**
9:     Update $\mathbf{w}^t$ using the sampled pair with formula (11)
10: **end for**
11: Sort features w.r.t. $w[i]$ and output the top d features

---

6   Experiment

In this section, we conduct experiments on six publicly available datasets. We compare DES with several state-of-the-art approaches.

6.1   Baselines

We compared our approach to four unsupervised feature selection methods and one supervised method. LS is a similarity-preserving approach. UDFS and NDFS are regression based methods which also consider the similarity information.

- All Features: It uses all the features for evaluation.
- Laplacian Score (LS): Laplacian score [5] selects the features which can best preserve the local manifold structure of data points.
- UDFS: Unsupervised Discriminative Feature Selection [22] exploits the local structure with $L_{2,1}$ norm regularized subspace learning.
- RSFS: Robust Spectral Feature Selection [12] selects features by the robust spectral analysis with $L_{2,1}$ norm regularized regression.
- FSASL: A recently proposed approach [2] which performs joint local structure learning and feature selection based on $L_{2,1}$ norm.
- Mutual Information (MI): We also include a widely-used supervised feature selection method which evaluates features by their mutual information with class labels. Since there are multiple classes, for each feature we use its average mutual information with different classes.

6.2   Datasets

We use six publicly available datasets: CNN dataset[1], Handwritten digits Dataset[2], BBCSport dataset[3], Guardian dataset[4], BlogCatalog[5] blog-posts

---
[1] https://sites.google.com/site/qianmingjie/home/datasets/cnn-and-fox-news.
[2] https://archive.ics.uci.edu/ml/datasets/Multiple+Features.
[3] http://mlg.ucd.ie/datasets/bbc.html.
[4] http://mlg.ucd.ie/datasets/3sources.html.
[5] http://dmml.asu.edu/users/xufei/datasets.html.

dataset, Newsgroup[6]. The baseline methods UDFS, RSFS and FSASL are prohibitively slow for large datasets. The original data of the latter two datasets are too large and therefore we sample a subset of them.

- CNN: CNN Web news with 7 classes (the category information contained in the RSS feeds for each news article can be viewed as reliable ground truth). Titles, abstracts, and text body contents are extracted as the text features.
- Handwritten Digits: 2000 images of handwritten digits 0–9 and we use the image pixels as features.
- BBCSport: It consists of 737 documents from the BBC Sport website corresponding to sports news articles in five topical areas from 2004–2005. The dataset has 5 classes: *athletics, cricket, football, rugby, tennis.*
- Guardian: It consists of 302 news stories from Guardian during the period February - April 2009. Each story is annotated with one of the six topical labels based on the dominant topic: *business, entertainment, health, politics, sport, tech.*
- Newsgroup: A subset of Newsgroup dataset on four topics: *comp.graphics, rec.sport.baseball, rec.motorcycles, sci.electronics.*
- BlogCatalog: A subset of users' blogposts from BlogCatalog in the following categories (100 posts are sampled for each category): *cycling, military, architecture, commodities/futures, vacation rentals.*

The statistics of six datasets are summarized in Table 1.

6.3 Experimental Setting

In this section, we evaluate the quality of selected features by their clustering performance. Following the typical setting of evaluation for unsupervised feature selection [8, 20, 22], we use Accuracy and Normalized Mutual Information (NMI) to evaluate the result of clustering. Accuracy is defined as follows.

$$Accuracy = \frac{1}{n} \sum_{i=1}^{n} \mathcal{I}(c_i = map(p_i)) \qquad (13)$$

where p_i is the clustering result of data instance i and c_i is its ground truth label. $map(\cdot)$ is a permutation mapping function that maps p_i to a class label using Kuhn-Munkres Algorithm.

Normalized Mutual Information (NMI) is another popular metric for evaluating clustering performance. Let C be the set of clusters from the ground truth and C' obtained from a clustering algorithm. Their mutual information $MI(C, C')$ is defined as follows:

$$MI(C, C') = \sum_{c_i \in C, c'_j \in C'} p(c_i, c'_j) \log \frac{p(c_i, c'_j)}{p(c_i)p(c'_j)} \qquad (14)$$

---

[6] http://www.cs.umb.edu/~smimarog/textmining/datasets/.

where $p(c_i)$ and $p(c_j')$ are the probabilities that a random data instance from the data set belongs to c_i and c_j', respectively, and $p(c_i, c_j')$ is the joint probability that the data instance belongs to the cluster c_i and c_j' at the same time. In our experiments, we use the normalized mutual information.

$$NMI(C, C') = \frac{MI(C, C')}{max(H(C), H(C'))} \tag{15}$$

where $H(C)$ and $H(C')$ are the entropy of C and C'. Higher value of NMI indicates better quality of clustering.

We set $k = 5$ for the kNN neighbor size in both our approach and the baseline methods following previous evaluation convention [8]. For λ in CL-DES, we set it to be 10^{-4} for all datasets, since in preliminary experiments we found the performance is not sensitive to λ when it is in $(10^{-6}, 10^{-3})$. For HT-DES and CL-DES, we sample 40000 pairs for the optimization, as we observe sampling more pairs usually have similar performance.

For the number of pseudo-classes/latent dimensions in UDFS, RSFS and FSASL, we use the ground-truth number of classes. Note that it actually benefits these pseudo-label based baselines with extra information about the data (and therefore certain advantages) since our approach does not need the number of classes as input. For the parameter to enforce the orthogonal constraint in pseudo-label methods [12], we use 10^8 as in the original papers. However, UDFS, RSFS and FSASL also require specifying the values of several other regularization parameters. In supervised learning, one can perform grid search for the parameters on a validation dataset; but there is no good way to determine the parameter values in unsupervised learning since we assume class labels are not available. In their original papers, all the class labels are used to find the best parameters. However, this violates the assumption of no supervision and favors the methods with best overfitting ability. Nonetheless, we perform grid search in the range of $\{0.1, 1, 10\}$ for the regularization parameters in UDFS, RSFS and FSASL (except for in FSASL, for which we do grid search in $\{0.001, 0.01, 0.1\}$ since γ should be a value in the range of 0–1, as suggested in [2]). Besides the best performance, we also report the median performance for them.

Following the convention in previous work [1, 22], we use KMeans[7] with cosine similarity for clustering evaluation. Since KMeans is affected by the initial seeds, we repeat the experiment for 20 times and report the average performance. We vary the number of features d in the range of $\{100, 200, 300, 400, 600\}$ (except for Handwritten digit dataset, which only has 240 features).

6.4 Clustering Results

The clustering performance on six datasets is shown in Figs. 2 and 3. The performance of baseline methods shown in Figs. 2 and 3 are under their best parameter

---

[7] We use the code at http://www.cad.zju.edu.cn/home/dengcai/Data/Clustering.html.

values. HT-DES and Mutual Information cannot handle continuous feature values and hence are not applied to the handwritten dataset.

The experimental results show that feature selection is a very effective technique for clustering. With much less features, DES can obtain better accuracy and NMI than using all the features. For instance, compared with using all 4612 features, CL-DES with only 100 features improves the clustering accuracy by 28% on BBCSport dataset. Another thing worth noting is that, when the number of selected features is small (such as 100 and 200), the improvement of DES over using all the features is also significant. This means that DES is capable of ranking high-quality features at the top. Besides the improved accuracy and NMI, using selected features rather than all features can also lead to better interpretability for human to analyze.

Among the two DES instantiations, CL-DES tends to have better clustering performance than HT-DES. This demonstrates the importance of evaluating features in a joint manner. HT-DES does not take into consideration correlation between features and there could be more redundancy in selected features.

When comparing DES with other unsupervised baseline methods, we observe that DES methods (especially CL-DES) with fixed λ perform better than baseline methods (with best parameter settings) in terms of both accuracy and NMI on most datasets. For example, on BlogCatalog dataset, CL-DES outperforms the most competitive baseline by 16% with 200 features.

Although HT-DES does not evaluate feature jointly, it still outperforms most unsupervised baseline methods substantially. This illustrates the power of exploiting the implicit class information contained in similar/dissimilar pairs. The baseline methods also utilize similarity in certain ways. For example, LS attempts to preserve the local manifold. RSFS generates pseudo-labels through spectral clustering on the similarity graph. But the inaccurate clustering labels can be viewed as a lossy compression of similarity information and may mislead feature selection. DES directly exploits the implicit class information embedded in similarity pairs without generating intermediate labels. The experimental results show that DES is a much more effective way for utilizing similarity information.

If we compare DES with the supervised method MI, we can see the performance of CL-DES is close to MI. It is usually very difficult for an unsupervised method to achieve performance comparable to a supervised method. This further illustrates strength of Discriminatively Exploiting Similarity (DES).

Table 1. Statistics of datasets

Statistics	CNN	Handwritten Digits	BBC Sport	BlogCatalog	Guardian	Newsgroup
# of instances	2107	2000	737	500	302	1575
# of features	6262	240	4612	4547	3631	2849
# of classes	7	10	5	5	6	4

Table 2. Relative performance (%) of median Accuracy/NMI (percentage) compared to the performance reported in Figs. 2 and 3

Statistics	CNN	Handwritten	BBCSport	BlogCatalog	Guardian	Newsgroup
UDFS	$-3.56/-6.74$	$-3.96/-2.06$	$-10.67/-23.91$	$-8.74/-18.33$	$-11.72/-25.33$	$-10.9/-31.64$
RSFS	$-19.18/-34.07$	$-7.68/-4.80$	$-7.55/-11.07$	$-16.37/-37.49$	$-6.39/-9.50$	$-19.08/-38.97$
FSASL	$-14.19/-17.72$	$-8.06/-12.15$	$-6.41/-4.30$	$-9.49/-16.95$	$-6.84/-12.40$	$-12.17/-24.49$
CL-DES	$-0.53/-0.68$	$-0.55/-0.47$	$-0.32/-0.19$	$-0.06/-0.03$	$-0.75/-0.46$	$+0.09/-0.48$

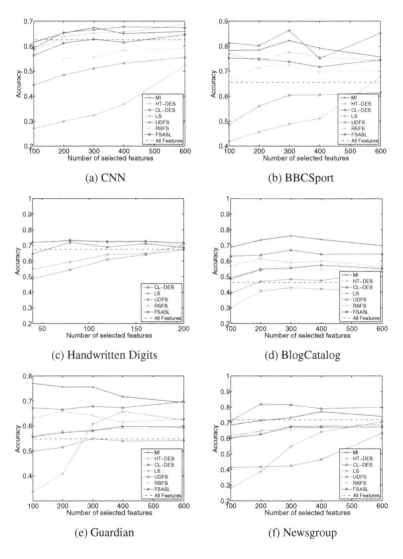

(a) CNN

(b) BBCSport

(c) Handwritten Digits

(d) BlogCatalog

(e) Guardian

(f) Newsgroup

Fig. 2. Accuracy of clustering results

6.5   Sensitivity Analysis

CL-DES has one regularization parameter λ and we study how this parameter affects the quality of selected features. In Fig. 4, we can observe that CL-DES performs consistently well as long as λ is smaller than 10^{-3}.

We also show in Table 2 how the median performance for UDFS, RSFS and FSASL compares with the best performance shown in Fig. 2. It can be observed that these baseline methods are sensitive to the parameter values and the median performance is usually 5%–40% lower than their best performance.

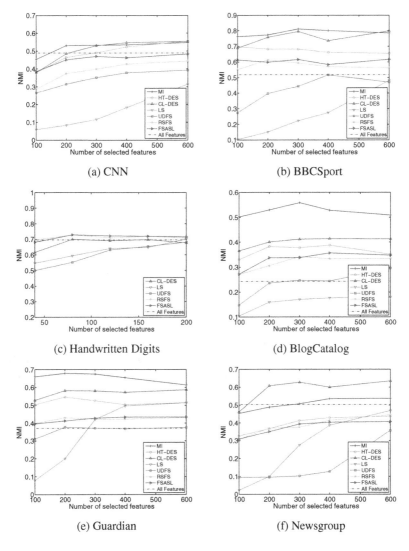

(a) CNN

(b) BBCSport

(c) Handwritten Digits

(d) BlogCatalog

(e) Guardian

(f) Newsgroup

Fig. 3. NMI of clustering results

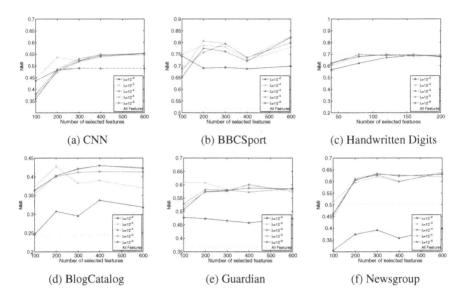

Fig. 4. NMI of clustering results with features selected by CL-DES under different values of λ

Since one cannot know the best parameter combination for these methods in unsupervised setting, the median performance is more realistic to expect in practice. In contrast, we also report the median perform for CL-DES from $\lambda = \{10^{-6}, 10^{-5}, 10^{-4}, 10^{-3}, 10^{-2}\}$ and we observe that the median performance is very close to performance of $\lambda = 10^{-4}$. This makes the proposed method more practical for real-world applications.

7   Conclusion

In this paper, we propose a new perspective for unsupervised feature selection which considers the similarity relationship as pseudo must-link/cannot-links. This new perspective enables us to adapt classic supervised feature selection ideas into our pair-wise formulation. We present hypothesis testing based and classification based approaches as instantiations of our framework. Empirical results show that the proposed method, although frustratingly simple, can select more discriminative features than state-of-the-art unsupervised approaches and even achieve comparable performance as the supervised mutual information approach.

References

1. Cai, D., Zhang, C., He, X.: Unsupervised feature selection for multi-cluster data. In: KDD, pp. 333–342 (2010)
2. Du, L., Shen, Y.-D.: Unsupervised feature selection with adaptive structure learning. In: KDD (2015)

3. Duda, R.O., Hart, P.E., Stork, D.G.: Pattern Classification, 2nd edn. Wiley, Hoboken (2001)
4. Feng, Y., Xiao, J., Zhuang, Y., Liu, X.: Adaptive unsupervised multi-view feature selection for visual concept recognition. In: Lee, K.M., Matsushita, Y., Rehg, J.M., Hu, Z. (eds.) ACCV 2012. LNCS, vol. 7724, pp. 343–357. Springer, Heidelberg (2013). https://doi.org/10.1007/978-3-642-37331-2_26
5. He, X., Cai, D., Niyogi, P.: Laplacian score for feature selection. In: NIPS (2005)
6. Li, J., Hu, X., Jian, L., Liu, H.: Toward time-evolving feature selection on dynamic networks. In: IEEE 16th International Conference on Data Mining (ICDM), 12–15 December 2016, Barcelona, Spain, pp. 1003–1008 (2016)
7. Li, J., Tang, J., Liu, H.: Reconstruction-based unsupervised feature selection: an embedded approach. In: IJCAI (2017)
8. Li, Z., Yang, Y., Liu, J., Zhou, X., Lu, H.: Unsupervised feature selection using nonnegative spectral analysis. In: AAAI (2012)
9. Liu, H., Setiono, R.: Chi2: feature selection and discretization of numeric attributes. In: Proceedings of 7th IEEE International Conference on Tools with Artificial Intelligence (1995)
10. Nie, F., Huang, H., Cai, X., Ding, C.H.Q.: Efficient and robust feature selection via joint l2, 1-norms minimization. In: NIPS, pp. 1813–1821 (2010)
11. Qian, M., Zhai, C.: Robust unsupervised feature selection. In: IJCAI (2013)
12. Shi, L., Du, L., Shen, Y.-D.: Robust spectral learning for unsupervised feature selection. In: ICDM (2014)
13. Song, L., Smola, A.J., Gretton, A., Borgwardt, K.M., Bedo, J.: Supervised feature selection via dependence estimation. In: ICML, vol. 227, pp. 823–830. ACM (2007)
14. Sun, L., Li, Z., Yan, Q., Srisa-an, W., Pan, Y.: SigPID: significant permission identification for android malware detection. In: 2016 11th International Conference on Malicious and Unwanted Software (MALWARE), pp. 1–8. IEEE (2016)
15. Tang, J., Liu, H.: Unsupervised feature selection for linked social media data. In: KDD, pp. 904–912 (2012)
16. Tibshirani, R.: Regression shrinkage and selection via the lasso. J. R. Stat. Soc. (Ser. B) **58**, 267–288 (1996)
17. Wei, X., Cao, B., Yu, P.S.: Unsupervised feature selection on networks: a generative view. In: AAAI, pp. 2215–2221 (2016)
18. Wei, X., Cao, B., Yu, P.S.: Multi-view unsupervised feature selection by cross-diffused matrix alignment. In: International Joint Conference on Neural Networks (IJCNN), pp. 494–501 (2017)
19. Wei, X., Xie, S., Yu, P.S.: Efficient partial order preserving unsupervised feature selection on networks. In: SDM, pp. 82–90 (2015)
20. Wei, X., Yu, P.S.: Unsupervised feature selection by preserving stochastic neighbors. In: AISTATS (2016)
21. Wright, J., Yang, A.Y., Ganesh, A., Sastry, S.S., Ma, Y.: Robust face recognition via sparse representation. IEEE Trans. Pattern Anal. Mach. Intell. **31**, 210–227 (2009)
22. Yang, Y., Shen, H.T., Ma, Z., Huang, Z., Zhou, X.: L2, 1-norm regularized discriminative feature selection for unsupervised learning. In: IJCAI, pp. 1589–1594 (2011)
23. Zhao, Z., Liu, H.: Spectral feature selection for supervised and unsupervised learning. In: ICML, vol. 227, pp. 1151–1157 (2007)
24. Zhao, Z., Wang, L., Liu, H.: Efficient spectral feature selection with minimum redundancy. In: AAAI (2010)

SetExpan: Corpus-Based Set Expansion via Context Feature Selection and Rank Ensemble

Jiaming Shen[✉], Zeqiu Wu, Dongming Lei, Jingbo Shang, Xiang Ren,
and Jiawei Han[✉]

Department of Computer Science, University of Illinois at Urbana-Champaign,
Urbana, IL, USA
{js2,zeqiuwu1,dlei5,shang7,xren7,hanj}@illinois.edu

Abstract. *Corpus-based set expansion* (i.e., finding the "complete" set of entities belonging to the same semantic class, based on a given corpus and a tiny set of seeds) is a critical task in knowledge discovery. It may facilitate numerous downstream applications, such as information extraction, taxonomy induction, question answering, and web search.

To discover new entities in an expanded set, previous approaches either make *one-time entity ranking* based on distributional similarity, or resort to *iterative pattern-based bootstrapping*. The core challenge for these methods is how to deal with noisy context features derived from free-text corpora, which may lead to entity intrusion and semantic drifting. In this study, we propose a novel framework, *SetExpan*, which tackles this problem, with two techniques: (1) a context feature selection method that selects clean context features for calculating entity-entity distributional similarity, and (2) a ranking-based unsupervised ensemble method for expanding entity set based on denoised context features. Experiments on three datasets show that *SetExpan* is robust and outperforms previous state-of-the-art methods in terms of mean average precision.

Code related to this chapter is available at:
https://github.com/mickeystroller/SetExpan

Data related to this chapter are available at: https://goo.gl/1suS3Z

Keywords: Set expansion · Information extraction · Bootstrapping
Unsupervised ranking-based ensemble

1 Introduction

Set expansion refers to the problem of expanding a small set of seed entities into a complete set of entities that belong to the same semantic class [29]. For example, if a given seed set is { *Oregon, Texas, Iowa* }, set expansion should return a hopefully complete set of entities in the same semantic class, "*U.S. states*". Set expansion can benefit various downstream applications, such as knowledge extraction [8], taxonomy induction [27], and web search [2].

J. Shen and Z. Wu—Equal Contribution.

© Springer International Publishing AG 2017
M. Ceci et al. (Eds.): ECML PKDD 2017, Part I, LNAI 10534, pp. 288–304, 2017.
https://doi.org/10.1007/978-3-319-71249-9_18

One line of work for solving this task includes *Google Set* [26], *SEAL* [29], and *Lyretail* [2]. In this approach, a query consisting of seed entities is submitted to a search engine to mine top-ranked webpages. While this approach can achieve relatively good quality, the required seed-oriented online data extraction is costly. Therefore, more studies [10,17,21,23,28] are proposed in a *corpus-based* setting where sets are expanded by offline processing based on a specific corpus.

For *corpus-based* set expansion, there are two general approaches, *one-time entity ranking* and *iterative pattern-based bootstrapping*. Based on the assumption that similar entities appear in similar contexts, the first approach [10,17,23] makes a one-time ranking of candidate entities based on their distributional similarity with seed entities. A variety of "contexts" are used, including Web table, Wikipedia list, or just free-text patterns, and entity-entity distributional similarity is calculated based on *all* context features. However, blindly using *all* such features can introduce undesired entities into the expanded set because many context features are not representative for defining the target semantic class although they do have connections with some of the seed entities. For example, when expanding the seed set { *Oregon, Texas, Iowa* }, "*located in* _" can be a pattern feature (the entity is replaced with a placeholder) strongly connected to all the three seeds. However, it does not clearly convey the semantic meaning of "*U.S. states.*" and can bring in entities like *USA* or *Ontario* when being used to calculate candidate entity's similarity with seeds. This is *entity intrusion* error. Another issue with this approach is that it is hard to obtain the full set at once without back and forth refinement. In some sense, iteratively bootstrapped set expansion is a more conservative way and leads to better precision.

The second approach, iterative pattern-based bootstrapping [8,9,22], starts from seed entities to extract quality patterns, based on a predefined pattern scoring mechanism, and it then applies extracted patterns to obtain even higher quality entities using another entity scoring method. This process iterates and the high-quality patterns from all previous iterations are accumulated into a pattern pool which will be used for the next round of entity extraction. This approach works only when patterns/entities extracted at each iteration are highly accurate, otherwise, it may cause severe *semantic shift* problem. Suppose in the previous example, "*located in* _" is taken as a good pattern from the seed set { *Oregon, Texas, Iowa* }, and this pattern brings in *USA* and *Ontario*. These undesired entities may bring in even lower quality patterns and iteratively cause the set shifting farther away. Thus, the pattern and entity scoring methods are crucial but sensitive in iterative bootstrapping methods. If they are not defined perfectly, the semantic shift can cause big problems. However, it is hard to have a perfect scoring mechanism due to the diversity and noisiness of unstructured text data.

This study proposes a new set expansion framework, SetExpan, which addresses both challenges posed above for corpus-based set expansion on free text. It carefully and conservatively extracts each candidate entity and iteratively improves the results. First, to overcome the entity intrusion problem, instead of using all context features, context features are carefully selected by

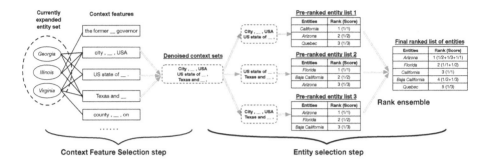

Fig. 1. An example showing two steps in one iteration of SetExpan.

calculating distributional similarity. Second, to overcome the semantic drift problem, different from other bootstrapped approaches, our high-quality feature pool will be reset at the beginning of each iteration. Finally, our carefully designed unsupervised ranking-based ensemble method is used at each iteration to further refine entities and make our system robust to noisy or wrongly extracted pattern features.

Figure 1 shows the pipeline at each iteration. SetExpan iteratively expands an entity set through a context feature selection step and an entity selection step. At the context feature selection, each context feature is scored based on its strength with currently expanded entities and top-ranked context features are selected. At the entity selection step, multiple subsets of the selected representative context features are sampled and each subset is used to obtain a ranked entity list. Finally, all the ranked lists are collected to compute the final ranking list of each candidate entity for expansion.

The major contributions of this paper are: (1) we propose an iterative set expansion framework with a novel context feature selection approach, to handle the issues of entity intrusion and semantic drift; (2) we develop an unsupervised ranking-based ensemble algorithm for entity selection to make our system robust and further reduce the impact of semantic drift. To evaluate the SetExpan method, we use three publicly available datasets and manually label expanded results of 65 queries over 13 semantic classes. Empirical results show that SetExpan outperforms the state-of-the-art baselines in terms of Mean Average Precision. Code[1] and datasets[2] described in this paper are publicly.

2   Related Work

The problem of completing an entity set given several seed entities has attracted extensive research efforts due to its practical importance. Google Sets [26] was among the earliest work dealing with this problem. It used proprietary algorithms and is no longer publicly accessible. Later, Wang and Cohen proposed *SEAL*

---

[1] https://github.com/mickeystroller/SetExpan.
[2] https://tinyurl.com/SetExpan-data.

system [29], which first submits a query consisting of all seed entities into a general search engine and then mines the top-ranked webpages. Recently, Chen et al. [2] improved this approach by leveraging a "page-specific" extractor built in a supervised manner and showed good performance on long-tail (i.e., rare) term expansion. All these methods need an external search engine and require seed-oriented data extraction. In comparison, our approach conducts corpus-based set expansion without resorting to online data extraction from specific webpages.

To tackle the corpus-based set expansion problem, Ghahramani and Heller [6] used a Bayesian method to model the probability that a candidate entity belongs to some unknown cluster that contains the input seeds. Pantel et al. [17] developed a web-scale set expansion pipeline by exploiting distributional similarity on context words for each candidate entity. He et al. proposed the SEISA system [10] that used query logs along with web lists as external evidence besides free text, and designed an iterative similarity aggregation function for set expansion. Recently, Wang et al. [28] leveraged web tables and showed very competitive results when not only seed entities but also intended class name were given. While these semi-structured lists and tables are helpful, they are not always available for some specific domain corpus such as PubMed articles or DBLP papers. Perhaps the most relevant work to ours is by Rong [21]. In that paper, the authors used the skip-gram feature combined with additional user-generated ontologies (i.e., Wikipedia list) for set expansion. However, they targeted the multifaceted expansion and exploited all skip-gram features for calculating the similarity because two entities. In our work, we keep the core idea of distributional similarity but calculate such similarity using only carefully selected *denoised* context features.

In a broader sense, our work is also related to information extraction and named entity recognition. Without given enough training data, bootstrapped entity extraction system [5,7,8] is the most popular and effective choice. At each bootstrap iteration, the system will first create patterns around entities; score patterns based on their ability to extract more positive entities and less negative entities (if provided), and use top-ranked patterns to extract more candidate entities. Multiple pattern scoring and entity scoring functions are proposed. For example, Riloff [20] scored each pattern by calculating the ratio of positive entities among all entities extracted by it, and scored each candidate entity by the number and quality of its matched patterns. Gupta et al. [7] scored patterns using the ratio of scaled frequencies of positive entities among all entities extracted by it. All these methods are heuristic and sensitive to different model parameters.

More generally, our work is also related to class label acquisition [24,30] which aims to propagate class labels to data instances based on labeled training examples, and entity clustering [1,12] where the goal is to find clusters of entities. However, the class label acquisition methods require a much larger number of training examples than the typical size of user input seed set, and the entity clustering algorithms can only find semantically related entities instead of entities strictly in the same semantic class.

3 Our Methodology: The SetExpan Framework

This section introduces first the context features and data model used by SetExpan in Sect. 3.1 and then our context-dependent similarity measure in Sect. 3.2. It then discusses how to select context features in Sect. 3.3 and presents our novel unsupervised ranking-based ensemble method for entity selection in Sect. 3.4.

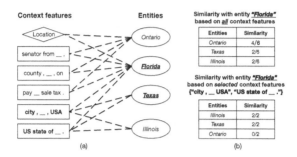

Fig. 2. (a) A simplified bipartite graph data model. (b) Similarity with seed entity conditioned on two different sets of context features.

3.1 Data Model and Context Features

We explore two types of context features obtained from the plain text: (1) skip-grams [21] and (2) coarse-grained types [8]. Data is modeled as a bipartite graph (Fig. 2(a)), with candidate entities on one side and their context features on the other. Each type of context features are described as follows.

Skip-gram: Given a target entity e_i in a sentence, one of its skip-gram is "$w_{-1}\_w_1$" where w_{-1} and w_1 are two context words and e_i is replaced with a placeholder. For example, one skip-gram of entity *"Illinois"* in sentence *"We need to pay Illinois sales tax."* is "pay _ sales". As suggested in [21], we extract up to six skip-grams of different lengths for one target entity e_i in each sentence. One advantage of using skip-grams is that it imposes strong positional constraints.

Coarse-grained type: Besides the unstructured skip-gram features, we use coarse-grained type to filter those obviously-wrong entities. For examples, when we expand the "U.S. states", we will not consider any entity that is typed "Person". After this process, we can obtain a cleaner subset of candidate entities. This mechanism is also adopted in [8].

After obtaining the "nodes" in bipartite graph data model, we need to model the edges in the graph. In this paper, we assign the weight between each pair of entity e and context feature c using the *TF-IDF transformation* [21], which is calculated as follows:

$$f_{e,c} = \log(1 + X_{e,c}) \left[\log |E| - \log \left(\sum_{e'} X_{e',c} \right) \right], \tag{1}$$

where $X_{e,c}$ is the raw co-occurrence count between entity e and context feature c, $|E|$ is the total number of candidate entities. We refer to such scaling as the *TF-IDF transformation* since it resembles the *tf-idf* scoring in information retrieval if we treat each entity e as a "document" and each of its context feature c as a "term". Empirically, we find such weight scaling performs outperforms some other alternatives such as point-wise mutual information (PMI) [10], truncated PMI [15], and BM25 scoring [19].

3.2   Context-Dependent Similarity

With the bipartite graph data model constructed, the task of expanding an entity set at each iteration can be viewed as finding a set of entities that are most "similar" to the currently expanded set. In this study, we use the weighted Jaccard similarity measure. Specifically, given a set of context features F, we calculate the *context-dependent* similarity as follows:

$$Sim(e_1, e_2 | F) = \frac{\sum_{c \in F} \min(f_{e_1,c}, f_{e_2,c})}{\sum_{c \in F} \max(f_{e_1,c}, f_{e_2,c})}. \tag{2}$$

Notice that if we change context feature set F, the similarity between entity pair is likely to change, as demonstrated in the following example.

Example 1. *Figure 2(a) shows a simplified bipartite graph data model where all edge weights are equal to 1 (and thus omitted from the graph for clarity). The entity "Florida" connects with all 6 different context features, while the entity "Ontario" is associated with top 4 context features including 1 type feature and 3 skip-gram features. If we add all the 6 possible context features into the context feature set F, the similarity between "Florida" and "Ontario" is $\frac{1+1+1+1}{1+1+1+1+1+1} = \frac{4}{6}$. On the other hand, if we put only two context features "city, __, USA", "US state of __." into F, the similarity between same pair of entities will change to $\frac{1+1}{1+1} = \frac{2}{2}$. Therefore, we refer such similarity as context-dependent similarity.*

Finally, we want to emphasize that our proposed method is general in the sense that other common similarity metrics such as cosine similarity can also be used. In practice, we find the performance of a set expansion method depends not really on the exact choice of base similarity metrics, but more on which contexts are selected for calculating *context-dependent* similarity. Similar results were also reported in a previous study [10].

3.3   Context Feature Selection

As shown in Example 1, the similarity between two entities really depends on the selected feature set F. The motivation of context feature selection is to find a feature subset F^* of fixed size Q that best "profiles" the target semantic class. In other words, we want to select a feature set F^* based on which entities within target class are most "similar" to each other. Given such F^*, the entity-entity similarity conditioned on it can best reflect their distributional similarity with regard to the target class. In some sense, such F^* best profiles the target semantic class. Unfortunately, to find such F^* of fixed size Q, we need to solve the following optimization problem which turns out to be NP-Hard, as shown in [3].

$$F^* = \arg\max_{|F|=Q} \sum_{i=1}^{|X|} \sum_{j>i}^{|X|} Sim(e_i, e_j|F), \tag{3}$$

where X is the set of currently expanded entities. Initially, we treat the user input seed set S as X. As iterations proceed, more entities will be added into X.

Given the NP-hardness of finding the optimal context feature set, we resort to a heuristic method that first scores each context feature based on its accumulated strength with entities in X and then selects top Q features with maximum scores. This process is illustrated in the following example:

Example 2. *For demonstration purpose, we again assume all edge weights in Fig. 2(a) are equal to 1 and let the currently expanded entity set X be { "Florida", "Texas"}. Suppose we want to select two "denoised" context features, we will first score each context feature based on its associated entities in X. The top 4 contexts will obtain a score 1 since they match only one entity in X with strength 1, and the 2 contexts below will get a score 2 because they match both entities in X. Then, we rank context features based on their scores and select 2 contexts with highest scores: "city, _, USA", "US state of _." into F.*

Finally, we want to emphasize two major differences of our context feature selection method from other heuristic "pattern selection" methods. First, most pattern selection methods require either users to explicitly provide the "negative" examples for the target semantic class [8,11,22], or implicitly expand multiple mutually exclusive classes in which instances in one class serve as negative examples for all the other classes [4,15]. Our method requires only a small number of "positive" examples. In most cases, it is hard for humans to find good discriminative negative examples for one class, or to provide both mutually exclusive and somehow related comparative classes. Second, the bootstrapping method will add its selected "quality patterns" during each iteration into a quality pattern pool, while our method will select high quality context features at each iteration from scratch. If one noisy pattern is selected and added into the pool, it will continue to introduce more irrelevant entities at all the following iterations. Our method can avoid such noise accumulation.

3.4   Entity Selection via Rank Ensemble

Intuitively, the entity selection problem can be viewed as finding those entities that are most similar to the currently expanded set X conditioned on the selected context feature set F. To achieve this, we can rank each candidate entity based on its score in Eq. (4) and then add top-ranked ones into the expanded set:

$$score(e|X, F) = \frac{1}{|X|} \sum_{e' \in X} Sim(e, e'|F). \tag{4}$$

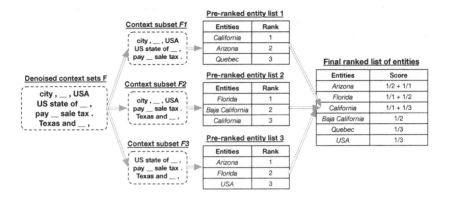

Fig. 3. A toy example to show entity selection via rank ensemble.

However, due to the ambiguity of natural language in free-text corpora, the selected context feature set F may still be noisy in the sense that an irrelevant entity is ranked higher than a relevant one. To further reduce such errors, we propose a novel ranking-based ensemble method for entity selection.

The key insight of our method is that an inferior entity will not appear frequently in multiple pre-ranked entity lists at top positions. Given a selected context set F, we first use sampling without replacement method to generate T subsets of context features $F_t, t = 1, 2, \ldots, T$. Each subset is of size $\alpha|F|$ where α is a model parameter within range $[0, 1]$. For each F_t, we can obtain a pre-ranked list of candidate entities L_t based on $score(e|X, F_t)$ defined in Eq. (4). We use r_t^i to denote the rank of entity e_i in list L_t. If entity e_i does not appear in L_t, we let $r_t^i = \infty$. Finally, we collect T pre-ranked lists and score each entity based on its mean reciprocal rank (mrr). All entities with average rank above r, namely $mrr(e) \leq T/r$, will be added into entity set X.

$$mrr(e_i) = \sum_{t=1}^{T} \frac{1}{r_t^i}, \qquad r_t^i = \sum_{e_j \in E} I\left(score(e_i|X, F_t) \leq score(e_j|X, F_t)\right), \quad (5)$$

where $I(\cdot)$ is the indicator function. Naturally, a relevant entity will rank at top position in multiple pre-ranked lists and thus accumulate a high mrr score, while an irrelevant entity will not consistently appear in multiple lists at high position which leads to low mrr score. Finally, we use the following example to demonstrate the whole process of entity selection.

Example 3. *In Fig. 3, we want to expand the "US states" semantic class given a selected context feature set F with 4 features. We first sample a subset of 3 context features $F_1 = \{$ "city, _, USA", "US state of _,", "pay _ sales tax."$\}$, and then use F_1 to obtain a pre-ranked entity list $L_1 = \langle$ "California", "Arizona", "Quebec"$\rangle$. By repeating this process three times, we get 3 pre-ranked lists and ensemble them into a final ranked list in which entity "Arizona" is scored 1.5 because it is ranked in the 2nd position in L_1 and 1st position in L_3. Finally,*

Algorithm 1. SetExpan

1: **Input:** Candidate entity set E, initial seed set S, entity-context graph G, expected size of output
   set K, model parameters $\{Q, T, \alpha, r\}$.
2: **Output:** The expanded set X.
3: $X = S$.
4: **while** $|X| \leq K$ **do**
5:     Set $F = \emptyset$ // Select denoised contexts from scratch
6:     Score context features based on X and add top Q denoised contexts into F.
7:     // Entity-selection via rank ensemble
8:     **for** $t = 1, 2, \ldots, T$ **do**
9:         Uniformly sample αQ contexts and construct feature subset F_t.
10:         Score entities based on Eq. (4) given F_t and obtain the pre-ranked list L_t.
11:         Update the mrr score of each entity based on Eq. (5).
12:     **end for**
13:     $X = X \cup \{e | mrr(e) \geq \frac{T}{r}\}$ // Add entities into expanded set X.
14: **end while**
15: Return X.

---

we add those entities with mrr score larger than 1, meaning this entity is ranked at 3rd position on average, into the expanded set X. In this simple example, the model parameters $T = 3, \alpha = \frac{|F_1|}{|F|} = 0.75$, and $r = 3$.

Put all together. Algorithm 1 summarizes the whole SetExpan process. The candidate entity set E and bipartite graph data model G are pre-calculated and stored. A user needs only to specify the seed set S and the expected size of output set K. There is a total of 4 model parameters: the number of top quality context features selected in each iteration Q, the number of pre-ranked entity lists T, the relative size of feature subset $0 < \alpha < 1$, and final mrr threshold r. The tuning and sensitivity of these parameters will be discussed in the experiment section.

4 Experiments

4.1 Experimental Setup

Datasets preparation. *SetExpan* is a *corpus-based* entity set expansion system and thus we use three corpora to evaluate its performance. Table 1 lists 3 datasets we used in experiments. (1) **APR** is constructed by crawling all 2015 news articles from AP and Reuters. (2) **Wiki** is a subset of English Wikipedia used in [13]. (3) **PubMed-CVD** is a collection of research paper abstracts about cardiovascular disease retrieved from PubMed.

For APR and PubMed-CVD datasets, we adopt a data-driven phrase mining tool [14] to obtain entity mentions and type them using ClusType [18]. Each entity mention is mapped heuristically to an entity based on its lemmatized surface name. We then extract variable-length skip-grams for all entity mentions as features for their corresponding entities, and construct the bipartite graph data model as introduced in the previous section. For Wiki dataset, the entities have already been extracted and typed using distant supervision. For the type information in each dataset, there are 16 coarse-grained types in APR and 4 coarse-grained types in PubMed-CVD. For Wiki, since it originally has about 50 fine-grained types, which may reveal too much information, we manually mapped them to 11 more coarse-grained types.

Query construction. A query is a set of seed entities of the same semantic class in a dataset, serving as the input for each system to expand the set. The process of query generation is as follows. For each dataset, we first extract 2000 most frequent entities in it and construct an entity list. Then, we ask three volunteers to manually scan the entity lists and propose a few semantic classes for each list. The proposed class should be interesting, relatively unambiguous and has a reasonable coverage in its corresponding corpus. These semantic classes cover a wide variety of topics, including locations, companies as well as political parties, and have different degrees of difficulty for set expansion. After finalizing the semantic classes for each dataset, the students randomly select entities of each semantic class from the frequent entity list to form 5 queries of size 3. To select the queries for PubMed-CVD, we seek help from two additional students with biomedical expertise, following the same previous approach. Due to the large size of PubMed-CVD dataset and runtime limitation, we only select 1 semantic class (hormones) with 5 queries.

With all queries selected, we have humans to label all the classes and instances returned by each of the following 7 compared methods. For APR and Wiki datasets, the inter-rater agreements (kappa-value) over three students are 0.7608 and 0.7746, respectively. For PubMed-CVD dataset, the kappa-value is 0.9236. All entities with conflicting label results are further resolved after discussions among all human labelers. Thus, we have our ground truth datasets.

Table 1. Datasets statistics and query descriptions

Dataset	FileSize	#Sentences	#Entities	#Test queries
APR	775 MB	1.01M	122K	40
Wiki	1.02 GB	1.50M	710K	20
PubMed-CVD	9.3 GB	23M	179K	5

Compared methods. Since the focus on this work is the corpus-based set expansion, we do not compare with other methods that require online data extractions. Also, to further analyze the effectiveness of each module in SetExpan framework. We implement 3 variations of our framework.

- word2vec [16]: We use the "skip-gram" model in word2vec to learn the embedding vector for each entity, and then return k nearest neighbors around seed entities as the expanded set.
- PTE [25]: We first construct a heterogeneous information network including entity, skip-gram features, and type features. PTE model is then applied to learn the entity embedding which is used to determine the k nearest neighbors around seed entities.

– SEISA [10]: An entity set expansion algorithm based on iterative similarity aggregation. It uses the occurrence of entities in web list and query log as entity features. In our experiments, we replace the web list and query log with our skip-gram and coarse-grained context features.

– EgoSet [21]: A multifaceted set expansion system based on skip-gram features, word2vec embeddings and WikiList. The original system is proposed to expand a seed set to multiple entity sets, considering the ambiguities in seed set. To achieve this, we use a community detection method to separate the extracted entities into several communities. However, in order to better compare with EgoSet, we carefully select queries that have little ambiguity or at least the seed set in the query is dominating in one semantic class. Thus, we discard the community detection part in EgoSet and treat all extracted entities as in one semantic class.

– SetExpan^{-cs}: Disable the context feature selection module in SetExpan, and use all context features to calculate distributional similarity.

– SetExpan^{-re}: Disable the rank ensemble module in SetExpan. Instead, we use all selected context feature to rank candidate entities at one time and add top-ranked ones into the expanded set.

– SetExpan^{full}: The full version of our proposed method, with both context feature selection and rank ensemble components enabled.

For fair comparison, we try different combinations of parameters and report the best performance for each baseline method.

Evaluation Metrics. For each test case, the input is a query, which is a set of 3 seed entities of the same semantic class. The output will be a ranked list of entities. For each query, we use the conventional average precision $AP_k(c, r)$ at $k (k = 10, 20, 50)$ for evaluation, given a ranked list of entities c and an unordered ground-truth set r. For all queries under a semantic class, we calculate the mean average precision (MAP) at k as $\frac{1}{N} \sum_i AP_k(c_i, r)$, where N is the number of queries. To evaluate the performance of each approach on a specific dataset, we calculate the mean-MAP (MMAP) at k over all queried semantic classes as $MMAP_k = \frac{1}{T} \sum_{t=1}^{T} [(\frac{1}{N_t}) \sum_i AP_k(c_{ti}, r_t)]$, where T is the number of semantic classes, N_t is the number of queries of t-th semantic class, c_{ti} is the extracted entity list for i-th query for t-th semantic class, and r_t is the ground truth set for t-th semantic class.

4.2   Experimental Results

Comparison with four baseline methods. Table 2 shows the MMAP scores of all methods on 3 datasets[3]. We can see the MMAP scores of SetExpan outperforms all four baselines a lot. We further look at their performances on each concept class, as shown in Fig. 4. We can see that the performance of these baseline methods varies a lot on different semantic classes, while our SetExpan can consistently beat them. One reason is that none of these methods applies context

---

[3] Results of SEISA on PubMed-CVD are omitted due to the scalability issue.

Table 2. Overall end-to-end performance evaluation on 3 datasets over all queries.

Methods	APR			Wiki			PubMed-CVD		
	MAP@10	MAP@20	MAP@50	MAP@10	MAP@20	MAP@50	MAP@10	MAP@20	MAP@50
EgoSet	0.3949	0.3942	0.3706	0.5899	0.5754	0.5622	0.0511	0.0410	0.0441
SEISA	0.7423	0.6090	0.3892	0.7643	0.6606	0.4998	-	-	-
word2vec	0.6054	0.5385	0.4180	0.7193	0.6289	0.4510	0.8427	0.7701	0.6895
PTE	0.3144	0.2777	0.1996	0.6817	0.5596	0.3839	0.9071	0.7654	0.5641
SetExpan^{-cs}	0.8240	0.7997	0.7674	0.9540	0.8955	0.7439	1.000	1.000	0.5991
SetExpan^{-re}	0.8509	0.7792	0.7681	0.9392	0.8680	0.7291	1.000	0.9605	0.7371
SetExpan^{full}	**0.8967**	**0.8621**	**0.7885**	**0.9571**	**0.9010**	**0.7457**	**1.000**	**1.000**	**0.7454**

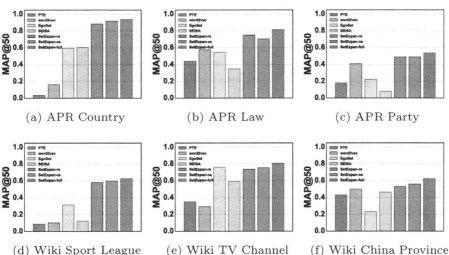

(a) APR Country   (b) APR Law   (c) APR Party

(d) Wiki Sport League   (e) Wiki TV Channel   (f) Wiki China Province

Fig. 4. Evaluation results for each semantic class.

feature selection or rank ensemble, and a single set of unpruned features can lead to various levels of noise in the results. Another reason is the lack of an iterative mechanism in some of those approaches. For example, even if EgoSet includes the results from word2vec to help it boost the performance, it still achieves low MAP scores in some semantic classes. Finding the nearest neighbors in only one iteration can be a key reason. And although SEISA is applying the iterative technique, instead of adding a small number of new entities in each iteration, it expands a full set in each iteration based on the coherence score of each candidate entity with the previously expanded set. It pre-calculates the size of the expanded set with the assumption that the feature similarities follow a certain distribution, which does not always hold to all datasets or semantic classes. Thus, if the size is far different from the actual size or is too big to extract a confident set at once, each iteration will introduce a lot of noise and cause semantic drift.

Comparison with SetExpan^{-re} and SetExpan^{-cs}. At the dataset level, the MMAP scores of SetExpan^{full} outperforms its two variation approaches. In the

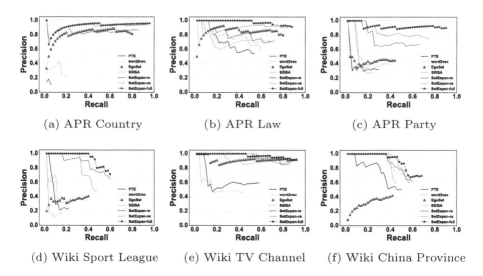

(a) APR Country   (b) APR Law   (c) APR Party

(d) Wiki Sport League   (e) Wiki TV Channel   (f) Wiki China Province

Fig. 5. Evaluation results for each concept class on individual query

semantic class level, we can see that SetExpan^{-re} and SetExpan^{-cs} sometimes have their MAP much lower than SetExpan^{full} while sometimes they almost achieve the same performance with SetExpan^{full}. This means they fail to stably extract entities with good quality. The main reason is still that a single set of features or ensembles over unpruned features can lead to various levels of noise in the results. Only under the circumstances that the single set of features or the unpruned features happen to be nicely selected without too much noise, which tends to happen when the query is relatively "easy", these variation approaches can achieve good results.

Effects of Context Feature Selection. We already see that adding the context feature selection component helps improve the performance. What's also noticeable is that the addition of context selection process becomes more obvious as the size of the corpus increases. The difference between MMAP scores of SetExpan^{-cs} and SetExpan^{full} is much larger in PubMed-CVD compared with APR and Wiki datasets. This is because that as the corpus size increases, we will have more noisy features and more candidate entities while the good features to define the target entity set may be limited. Thus, without context selection, noise can damage the performance much more. The evidence can also be found from the performance of EgoSet across the three datasets. It can achieve reasonably good results in APR and Wiki, however, it performs much worse in PubMed-CVD (Table 2).

Effect of Rank Ensemble. From the above experiments, the effect of rank ensemble has variance across the different semantic classes, however, it seems to be more stable across datasets, compared with the effect of context selection. This is because we apply the default set of parameter values in each test case above. In the parameter analysis part, we will show that the number of ensemble batches and the percentage of features to be randomly sampled can affect the contribution of rank ensemble to the set expansion performance.

Parameter Analysis. There are totally 4 parameters in SetExpan – Q (the number of selected context features), α (the percentage of features to be sampled), T (the number of ensemble batches), and r (the threshold of a candidate entity's average rank). We study the influence of each parameter by fixing all other parameters to default values, and present one graph showing the MMAP scores of SetExpan on APR dataset versus the changes of that parameter (Fig. 6).

- α: From the graph, the performance increases sharply as α increases until it reaches about 0.6. Then, it starts to stay stable and decreases after 0.7.
- Q: In the range of 50–150, the performance increases sharply as Q increases, which means the majority of top 150 context features can provide rich information to identify entities belonging to the target semantic class. The available information gets more and more saturated after Q reaches 150 and start to introduce noises and hamper the performance after around 300.
- r: Our experiments show that the performance is not very sensitive to the threshold of a candidate entity's average rank.
- T: The performance keeps increasing as we increase the ensemble batches, due to the robustness to noise of ensembling. The performance becomes more stable after 60 batches.

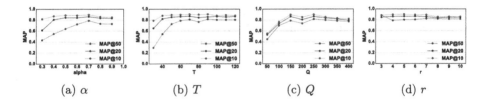

Fig. 6. Parameter sensitivity on two datesets

Case Studies. Figure 7 presents three case studies for SetExpan. We show one query for each dataset. In each case, we show top 3 ranked entities and top/bottom 3 skip-gram features after context feature selection for the first 3 iterations as well as the coarse-grained type. In all cases, our algorithm successfully extracts correct entities in each iteration, and the top-ranked skip-grams are representative in defining the target semantic class. On the other hand, we notice that most of the bottom 3 skip-grams selected are very general or not representative at all. These context features could potentially introduce noisy entities and thus the rank ensemble can play a rival role in improving the results.

Dataset	Query	Top ranked entities in first 3 iterations	Top/Bottom skip-gram features selected in first 3 iterations	Coarse-grained type
APR	{Patriot Act, Obamacare, Clery Act}	Iteration 1: USA Patriot Act, USA Freedom Act, Voting Rights Act, ... Iteration 2: Stock Act, Religious Freedom Restoration Act, Foreign Intelligence Surveillance Act, ... Iteration 3: Americans with Disabilities Act, Healthy Families Act, Goonda Act, ...	Iteration 1: Top 3: "the __ provisions", "provisions of the __", "defund __." Bottom 3: "2010 __ ,", "also known as __ .", "under the __ , and" Iteration 2: Top 3: "under the __ to", "provisions of the __ ,", "the __ into law." Bottom 3: "the __ - which has", "the _The House", "the _ , first" Iteration 3: Top 3: "under the __ to", "Under the __ ,", "the __ into law" Bottom 3: "of the __ passed", "the __ , the most", "replacing __ ."	Event
Wiki	{ESPN, ESPN2, Spike TV}	Iteration 1: ABC, CBS, NBC, ... Iteration 2: BBC, ITV, Channel 4, ... Iteration 3: TBS, ITV1, BBC Two, ...	Iteration 1: Top 3: "telecast on __ .", "televised on __ .", "televised by __." Bottom 3: "on __ , to the", ", and perhaps __ .", "from an __ website" Iteration 2: Top 3: "the __ sitcom", "the __ television network", "ABC , __ ," Bottom 3: "on __ on September", "broadcast on __ on", "the __ soap opera The" Iteration 3: Top 3: "the __ sitcom", "the __ soap", "the __ soap opera", ... Bottom 3: "aired on __ between", "of the __ show", "on the __ crime"	Organization
PubMed -CVD	{FSH, TSH, MSH}	Iteration 1: LH, GH, ACTH, ... Iteration 2: LHRH, AMH, GHRH, ... Iteration 3: Renin, GnRH-I, AVP, ...	Iteration 1: Top 3: "stimulating hormone (__)", "hormone (__) ,", "hormone (__) and", ... Bottom 3: "g/L , __ =", ", __ and prolactin", "hormone (__) -" Iteration 2: Top 3: "hormone (__) ,", "hormone (__) and", "hormone (__) .", ... Bottom 3: ", __ , estradiol ,", ", __ , and PRL", "hormone (__) -" Iteration 3: Top 3: "hormone (__) ,", "hormone (__) and", "hormone (__) .", ... Bottom 3: "(__) and insulin-like", ", TSH , __ ,", "levels of __ , FSH"	Proteins and Genes (PRGE)

Fig. 7. Three case studies on each dataset.

5   Conclusion and Future Work

In this paper, we study the problem of corpus-based set expansion. First, we propose an iterative set expansion framework with a context feature selection method, to deal with the problem of entity intrusion and semantic drift. Second, we develop a novel unsupervised ranking-based ensemble algorithm for entity selection, to further reduce context noise in free-text corpora. Experimental results on three publicly available datasets corroborate the effectiveness and robustness of our proposed SetExpan.

The proposed framework is general and can incorporate other context features besides skip-grams, such as Part-Of-Speech tags or syntactic head tokens. Besides, it would be interesting to study more rank ensemble methods for aggregating multiple pre-ranked lists. In addition, our current framework treats each feature independently, it would be interesting to study how the interaction of context features can influence the expansion result. We leave it for future work.

Acknowledgments. Research was sponsored in part by the U.S. Army Research Lab. under Cooperative Agreement No. W911NF-09-2-0053 (NSCTA), National Science Foundation IIS-1320617, IIS 16-18481, and NSF IIS 17-04532, and grant 1U54GM114838 awarded by NIGMS through funds provided by the trans-NIH Big Data to Knowledge (BD2K) initiative (www.bd2k.nih.gov).

References

1. Balasubramanyan, R., Dalvi, B., Cohen, W.W.: From topic models to semi-supervised learning: biasing mixed-membership models to exploit topic-indicative features in entity clustering. In: Blockeel, H., Kersting, K., Nijssen, S., Železný, F. (eds.) ECML PKDD 2013. LNCS (LNAI), vol. 8189, pp. 628–642. Springer, Heidelberg (2013). https://doi.org/10.1007/978-3-642-40991-2_40

2. Chen, Z., Cafarella, M., Jagadish, H.: Long-tail vocabulary dictionary extraction from the web. In: WSDM, pp. 625–634. ACM (2016)

3. Chierichetti, F., Kumar, R., Pandey, S., Vassilvitskii, S.: Finding the Jaccard median. In: SODA (2010)

4. Curran, J.R., Murphy, T., Scholz, B.: Minimising semantic drift with mutual exclusion bootstrapping (2007)

5. Etzioni, O., Cafarella, M.J., Downey, D., Popescu, A.-M., Shaked, T., Soderland, S., Weld, D.S., Yates, A.: Unsupervised named-entity extraction from the web: an experimental study. Artif. Intell. **165**, 91–134 (2005)

6. Ghahramani, Z., Heller, K.A.: Bayesian sets. In: NIPS (2005)

7. Gupta, S., MacLean, D.L., Heer, J., Manning, C.D.: Research and applications: induced lexico-syntactic patterns improve information extraction from online medical forums. JAMIA **21**, 902–909 (2014)

8. Gupta, S., Manning, C.D.: Improved pattern learning for bootstrapped entity extraction. In: CoNLL, pp. 98–108 (2014)

9. Gupta, S., Manning, C.D.: Distributed representations of words to guide bootstrapped entity classifiers. In: HLT-NAACL (2015)

10. He, Y., Xin, D.: SEISA: set expansion by iterative similarity aggregation. In: WWW (2011)

11. Jindal, P., Roth, D.: Learning from negative examples in set-expansion. In: 2011 IEEE 11th International Conference on Data Mining (2011)

12. Lin, D., Wu, X.: Phrase clustering for discriminative learning. In: ACL/IJCNLP (2009)

13. Ling, X., Weld, D.S.: Fine-grained entity recognition. In: AAAI (2012)

14. Liu, J., Shang, J., Wang, C., Ren, X., Han, J.: Mining quality phrases from massive text corpora. In: SIGMOD, pp. 1729–1744. ACM (2015)

15. McIntosh, T., Curran, J.R.: Weighted mutual exclusion bootstrapping for domain independent lexicon and template acquisition (2008)

16. Mikolov, T., Sutskever, I., Chen, K., Corrado, G.S., Dean, J.: Distributed representations of words and phrases and their compositionality. CoRR, abs/1310.4546 (2013)

17. Pantel, P., Crestan, E., Borkovsky, A., Popescu, A.-M., Vyas, V.: Web-scale distributional similarity and entity set expansion. In: EMNLP (2009)

18. Ren, X., El-Kishky, A., Wang, C., Tao, F., Voss, C.R., Han, J.: Clustype: effective entity recognition and typing by relation phrase-based clustering. In: WWW, pp. 995–1004. ACM (2015)

19. Ren, X., Lv, Y., Wang, K., Han, J.: Comparative document analysis for large text corpora. CoRR, abs/1510.07197 (2017)

20. Riloff, E.: Automatically generating extraction patterns from untagged text. In: AAAI/IAAI, vol. 2 (1996)

21. Rong, X., Chen, Z., Mei, Q., Adar, E.: EgoSet: exploiting word ego-networks and user-generated ontology for multifaceted set expansion. In: WSDM, pp. 645–654. ACM (2016)

22. Shi, B., Zhang, Z., Sun, L., Han, X.: A probabilistic co-bootstrapping method for entity set expansion. In: COLING (2014)
23. Shi, S., Zhang, H., Yuan, X., Wen, J.-R.: Corpus-based semantic class mining: distributional vs. pattern-based approaches. In: COLING (2010)
24. Talukdar, P.P., Reisinger, J., Pasca, M., Ravichandran, D., Bhagat, R., Pereira, F.: Weakly-supervised acquisition of labeled class instances using graph random walks. In: EMNLP (2008)
25. Tang, J., Qu, M., Mei, Q.: PTE: predictive text embedding through large-scale heterogeneous text networks. In: KDD, pp. 1165–1174. ACM (2015)
26. Tong, S., Dean, J.: System and methods for automatically creating lists. US Patent 7,350,187 (2008)
27. Velardi, P., Faralli, S., Navigli, R.: Ontolearn reloaded: a graph-based algorithm for taxonomy induction. Comput. Linguist. **39**(3), 665–707 (2013)
28. Wang, C., Chakrabarti, K., He, Y., Ganjam, K., Chen, Z., Bernstein, P.A.: Concept expansion using web tables. In: WWW (2015)
29. Wang, R.C., Cohen, W.W.: Language-independent set expansion of named entities using the web. In: ICDM (2007)
30. Wang, Y.-Y., Hoffmann, R., Li, X., Szymanski, J.: Semi-supervised learning of semantic classes for query understanding: from the web and for the web. In: CIKM (2009)

Kernel Methods

Bayesian Nonlinear Support Vector Machines for Big Data

Florian Wenzel[1]($\boxtimes$), Théo Galy-Fajou[1], Matthäus Deutsch[2],
and Marius Kloft[1]

[1] Humboldt University of Berlin, Berlin, Germany
{wenzelfl,galy,kloft}@hu-berlin.de
[2] G+J Digital Products Hamburg, Hamburg, Germany
mdeutsch@outlook.com

Abstract. We propose a fast inference method for Bayesian nonlinear support vector machines that leverages stochastic variational inference and inducing points. Our experiments show that the proposed method is faster than competing Bayesian approaches and scales easily to millions of data points. It provides additional features over frequentist competitors such as accurate predictive uncertainty estimates and automatic hyperparameter search.
Code related to this chapter is available at:
https://doi.org/10.6084/m9.figshare.5443627
Data related to this chapter are available at:
https://doi.org/10.6084/m9.figshare.5443624 and
https://doi.org/10.6084/m9.figshare.5443621

Keywords: Bayesian approximative inference
Support vector machines · Kernel methods · Big data

1 Introduction

Statistical machine learning branches into two classic strands of research: Bayesian and frequentist. In the classic supervised learning setting, both paradigms aim to find, based on training data, a function f_β that predicts well on yet unseen test data. The difference in the Bayesian and frequentist approach lies in the treatment of the parameter vector β of this function. In the *frequentist* setting, we select the parameter β that minimizes a certain loss given the training data, from a restricted set $\mathcal{B}$ of limited complexity. In the *Bayesian* school of thinking, we express our prior belief about the parameter, in the form of a probability distribution over the parameter vector. When we observe data, we adapt our belief, resulting in a posterior distribution over β

Advantages of the Bayesian approach include automatic treatment of hyperparameters and direct quantification of the uncertainty[1] of the prediction in the

---

[1] Note that frequentist approaches can also lead to other forms of uncertainty estimates, e.g. in form of confidence intervals. But since the classic SVM does not exhibit a probabilistic formulation these uncertainty estimates cannot be directly computed.

© Springer International Publishing AG 2017
M. Ceci et al. (Eds.): ECML PKDD 2017, Part I, LNAI 10534, pp. 307–322, 2017.
https://doi.org/10.1007/978-3-319-71249-9_19

form of class membership probabilities which can be of tremendous importance in practice. As examples consider the following. (1) We have collected blood samples of cancer patients and controls. The aim is to screen individuals that have increased likelihood of developing cancer. The knowledge of the uncertainty in those predictions is invaluable to clinicians. (2) In the domain of physics it is important to have a sense about the certainty level of predictions since it is mandatory to assert the statistical confidence in any physical variable measurement. (3) In the general context of decision making, it is crucial that the uncertainty of the estimated outcome of an action can be reliably determined.

Recently, it was shown that the support vector machine (SVM) [1]—which is a classic supervised classification algorithm— admits a Bayesian interpretation through the technique of data augmentation [2,3]. This so-called *Bayesian nonlinear SVM* combines the best of both worlds: it inherits the geometric interpretation, its robustness against outliers, state-of-the-art accuracy [4], and theoretical error guarantees [5] from the frequentist formulation of the SVM, but like Bayesian methods it also allows for flexible feature modeling, automatic hyperparameter tuning, and predictive uncertainty quantification.

However, existing inference methods for the Bayesian support vector machine (such as the expectation conditional maximization method introduced in [3]) scale rather poorly with the number of samples and are limited in application to datasets with thousands of data points [3]. Based on stochastic variational inference [6] and inducing points [7], we develop in this paper a *fast* and *scalable* inference method for the nonlinear Bayesian SVM.

Our experiments show superior performance of our method over competing methods for uncertainty quantification of SVMs such as Platt's method [8]. Furthermore, we show that our approach is faster (by one to three orders of magnitude) than the following competitors: expectation conditional maximization (ECM) for nonlinear Bayesian SVM by [3], Gaussian process classification [9], and the recently proposed scalable variational Gaussian process classification method [10]. We apply our method to the domain of particle physics, namely on the SUSY dataset [11] (a standard benchmark in particle physics containing 5 million data points) where our method takes only 10 min to train on a single CPU machine.

Our experiments demonstrate that Bayesian inference techniques are mature enough to compete with corresponding frequentist approaches (such as nonlinear SVMs) in terms of scalability to big data, yet they offer additional benefits such as uncertainty estimation and automated hyperparameter search.

Our paper is structured as follows. In Sect. 2 we discuss related work and review the Bayesian nonlinear SVM model in Sect. 3. In Sect. 4 we propose our novel scalable inference algorithm, show how to optimize hyperparameters and obtain an approximate predictive distribution. We discuss also the special case of the linear SVM, for which we propose a specially tailored fast inference algorithm. Section 5 concludes with experimental results.

2   Related Work

There has recently been significant interest in utilizing max-margin based discriminative Bayesian models for various applications. For example, [12] employs a max-margin based Bayesian classification to discover latent semantic structures for topic models, [13] uses a max-margin approach for efficient Bayesian matrix factorization, and [14] develops a new max-margin approach to Hidden Markov models.

All these approaches apply the Bayesian reformulation of the classic SVM introduced by [2]. This model is extended by [3] to the nonlinear case. The authors show improved accuracy compared to standard methods such as (non-Bayesian) SVMs and Gaussian process (GP) classification.

However, the inference methods proposed in [2,3] have the drawback that they partially rely on point estimates of the latent variables and do not scale well to large datasets. In [15] the authors apply mean field variational inference to the linear case of the model, but their proposed technique does not lead to substantial performance improvements and neglects the nonlinear model.

Uncertainty estimation for SVMs is usually done via Platt's technique [8], which consists of applying a logistic regression on the function scores produced by the SVM. In contrast, our technique directly yields a sound predictive distribution instead of using a heuristically motivated transformation. We make use of the idea of inducing point GPs to develop a scalable inference method for the Bayesian nonlinear SVM. Sparse GPs using pseudo-inputs were first introduced in [16]. Building on this idea Hensman et al. developed a stochastic variational inference scheme for GP regression and GP classification [7,10]. We further extend this ideas to the setting of Bayesian nonlinear SVM.

3   The Bayesian SVM Model

Let $\mathcal{D} = \{x_i, y_i\}_{i=1}^n$ be n observations where $x_i \in \mathbb{R}^d$ is a feature vector with corresponding labels $y_i \in \{-1, 1\}$. The SVM aims to find an optimal score function f by solving the following regularized risk minimization objective:

$$\underset{f}{\arg\min} \ \gamma R\left(f\right) + \sum_{i=1}^n \max\left(0, 1 - y_i f(x_i)\right), \tag{1}$$

where R is a regularizer function controlling the complexity of the decision function f, and γ is a hyperparameter to adjust the trade-off between training error and the complexity of f. The loss $\max\left(0, 1 - yf(x)\right)$ is called hinge loss. The classifier is then defined as $\text{sign}(f(x))$.

For the case of a linear decision function, i.e. $f(x) = x^T \beta$, the SVM optimization problem (1) is equivalent to estimating the mode of a pseudo-posterior

$$p(\beta|\mathcal{D}) \propto \prod_{i=1}^n L(y_i|x_i, \beta)p(\beta).$$

Here $p(\beta)$ denotes a prior such that $\log p(\beta) \propto -2\gamma R(\beta)$. In the following we use the prior $\beta \sim \mathcal{N}(0, \Sigma)$, where $\Sigma \in \mathbb{R}^{d \times d}$ is a positive definite matrix. From a frequentist SVM view, this choice generalizes the usual L^2-regularization to non-isotropic regularizers. Note that our proposed framework can be easily extended to other regularization techniques by adjusting the prior on β (e.g. block $\ell_{(2,p)}$-norm regularization which is known as multiple kernel learning [17]). In order to obtain a Bayesian interpretation of the SVM, we need to define a pseudolikelihood L such that the following holds,

$$L\left(y|x, f(\cdot)\right) \propto \exp\left(-2 \max(1 - y_i f(x_i), 0)\right). \tag{2}$$

By introducing latent variables $\lambda := (\lambda_1, \ldots, \lambda_n)^\top$ (data augmentation) and making use of integral identities stemming from function theory, [2] show that the specification of L in terms of the following marginal distribution satisfies (2):

$$L(y_i|x_i, \beta) = \int_0^\infty \frac{1}{\sqrt{2\pi\lambda_i}} \exp\left(-\frac{1}{2}\frac{\left(1 + \lambda_i - y_i x_i^T \beta\right)^2}{\lambda_i}\right) d\lambda_i. \tag{3}$$

Writing $X \in \mathbb{R}^{d \times n}$ for the matrix of data points and $Y = \text{diag}(y)$, the full conditional distributions of this model are

$$\begin{aligned} \beta|\lambda, \Sigma, \mathcal{D} &\sim \mathcal{N}\left(B(\lambda^{-1} + 1), B\right), \\ \lambda_i|\beta, \mathcal{D}_i &\sim \mathcal{GIG}\left(1/2, 1, (1 - y_i x_i^\top \beta)^2\right), \end{aligned} \tag{4}$$

with $Z = YX$, $B^{-1} = Z\Lambda^{-1}Z^\top + \Sigma^{-1}$, $\Lambda = \text{diag}(\lambda)$ and where $\mathcal{GIG}$ denotes a generalized inverse Gaussian distribution. The n latent variables λ_i of the model scale the variance of the full posteriors locally. The model thus constitutes a special case of a normal variance-mean mixture, where we implicitly impose the improper prior $p(\lambda) = \mathbb{1}_{[0,\infty)}(\lambda)$ on λ. This could be generalized by using a generalized inverse Gaussian prior on λ_i, leading to a conjugate model for λ_i. Henao et al. show that in the case of an exponential prior on λ_i, this leads to a skewed Laplace full conditional for λ_i. Note that this, however, destroys the equivalency to the frequentist linear SVM.

By using the ideas of Gaussian processes [9], Henao et al. develop a nonlinear (kernelized) version of this model [3]. They assume a continuous decision function $f(x)$ to be drawn from a zero-mean Gaussian process $\text{GP}(0, k)$, where k is a kernel function. The random Gaussian vector $f = (f_1, \ldots, f_n)^\top$ corresponds to $f(x)$ evaluated at the data points. They substitute the linear function $x_i^\top \beta$ by f_i in (3) and obtain the conditional posteriors

$$\begin{aligned} f|\lambda, \mathcal{D} &\sim \mathcal{N}\left(CY(\lambda^{-1} + 1), C\right), \\ \lambda_i|f_i, \mathcal{D}_i &\sim \mathcal{GIG}\left(1/2, 1, (1 - y_i f_i)^2\right), \end{aligned} \tag{5}$$

with $C^{-1} = \Lambda^{-1} + K^{-1}$. For a test point x_* the conditional predictive distribution for $f_* = f(x_*)$ under this model is

$$f_*|\lambda, x_*, \mathcal{D} \sim \mathcal{N}\left(k_*^\top (K + \Lambda)^{-1} Y(1 + \lambda), k_{**} - k_*^\top (K + \Lambda)^{-1} k_*\right),$$

where $K := k(X, X)$, $k_{X*} := k(X, x_*)$, $k_{**} := k(x_*, x_*)$. The conditional class membership probability is

$$p(y_* = 1|\lambda, x_*, \mathcal{D}) = \Phi\left(\frac{k_*^T(K + \Lambda)^{-1}Y(1 + \lambda)}{1 + k_{**} - k_*^\top(K + \Lambda)^{-1}k_*}\right),$$

where $\Phi(.)$ is the probit link function.

Note that the conditional posteriors as well as the class membership probability still depend on the local latent variables λ_i. We are interested in the marginal predictive distributions, but unfortunately the latent variables cannot be integrated out analytically. Both [2,3] propose MCMC-algorithms and stepwise inference schemes similar to EM-algorithms to overcome this problem. These methods do not scale well to big data problems and the probability estimation still relies on point estimates of the n-dimensional λ. We overcome these problems proposing a scalable inference method and obtaining approximate marginal predictive distributions (that are not conditioned on λ).

4 Scalable Inference and Automated Hyperparameter Tuning

In the following we develop a fast and reliable inference method for the Bayesian nonlinear SVM. Our method builds on the idea of using inducing points for Gaussian Processes in a stochastic variational inference setting [7] that scales easily to millions of data points. We proceed by first discussing a standard batch variational scheme in Sect. 4.1 and then in Sect. 4.2 we develop our fast and scalable inference method. We show how to automatically tune hyperparameters in Sect. 4.3 and obtain uncertainty estimates for predictions in Sect. 4.4. Finally, we discuss the special case of the Bayesian linear SVM in Sect. 4.5.

4.1 Batch Variational Inference

The idea of variational inference is to approximate the typically intractable posterior of a probabilistic model by a variational (typically factorized) distribution. We find the optimal approximating distribution by maximizing a lower bound on the evidence (the so-called ELBO) with respect to the parameters of the variational distribution, which is equivalent to minimizing the Kullback-Leibler divergence between the variational distribution and the posterior [18,19].

In this section we first develop a batch variational inference scheme [18,19], which uses the full dataset in every iteration. We follow the structured mean field approach and choose the variational distributions within the same families as the full conditional distributions $q(f, \lambda) = q(f) \prod_{i=1}^{n} q(\lambda_i)$, with $q(f) \equiv \mathcal{N}(\mu, \zeta)$ and $q(\lambda_i) \equiv \mathcal{GIG}(1/2, 1, \alpha_i)$. The coordinate ascent updates can be computed by the expected natural parameters of the corresponding full conditionals (5) leading to

$$\alpha_i = \mathbb{E}_{q(f)}[(1 - y_i f_i)^2] = (1 - y_i^\top \mu)^2 + y_i^\top \zeta y_i,$$

$$\zeta = \mathbb{E}_{q(\lambda)}[(\Lambda^{-1} + K^{-1})^{-1}] = \left(A^{-\frac{1}{2}} + K^{-1}\right)^{-1},$$

$$\mu = \zeta \mathbb{E}_{q(\lambda)}[Y(\lambda^{-1} + 1)] = \zeta Y(\alpha^{-\frac{1}{2}} + 1).$$

This concludes the batch variational inference scheme.

The downside of this approach is that it does not scale to big datasets. The covariance matrix of the variational distribution $q(f)$ has dimension $n \times n$ and has to be updated and inverted at every inference step. This operation exhibits the computational complexity $\mathcal{O}(n^3)$, where n is the number of data points. Furthermore, in this setup we cannot apply stochastic gradient descent. We show how to overcome both problems in the next section paving the way to perform inference on big datasets.

4.2 Stochastic Variational Inference Using Inducing Points

We aim to develop a stochastic variational inference (SVI) scheme using only minibatches of the data in each iteration. The Bayesian nonlinear SVM model does not exhibit a set of global variables. Both the number of latent variables λ and the observations of the latent GP f grow with number of data points (c.f. Eq. 5), i.e. they are local variables. This hinders us from directly developing a SVI scheme. We make use of the concept of inducing points [7] imposing a sparse GP acting as global variable. This allows us to apply SVI and reduces the complexity to $\mathcal{O}(m^3)$, where m is the number of inducing points, which is independent of the number of data points.

We augment our original model (5) with $m < n$ inducing points. Let $u \in \mathbb{R}^m$ be pseudo observations at inducing locations $\{\hat{x}_1, \dots, \hat{x}_m\}$. We employ a prior on the inducing points, $p(u) = \mathcal{N}(0, K_{mm})$ and connect f and u setting

$$p(f|u) = \mathcal{N}(K_{nm}K_{mm}^{-1}u, \widetilde{K}) \tag{6}$$

where K_{mm} is the kernel matrix resulting from evaluating the kernel function between all inducing points locations, K_{nm} is the cross-covariance between the data points and the inducing points and $\widetilde{K}$ is given by $\widetilde{K} = K_{nn} - K_{nm}K_{mm}^{-1}K_{mn}$. The augmented model exhibits the joint distribution

$$p(y, u, f, \lambda) = p(y, \lambda|f)p(f|u)p(u).$$

Note that we can recover the original joint distribution by marginalizing over u. We now aim to apply the methodology of variational inference to the marginal joint distribution $p(y, u, \lambda) = \int p(y, u, f, \lambda)\mathrm{d}f$. We impose a variational distribution $q(u) = \mathcal{N}(u|\mu, \zeta)$ on the inducing points u. We follow [7] and apply Jensen's inequality to obtain a lower bond on the following intractable conditional probability,

$$\begin{aligned}
\log p(y, \lambda|u) &= \log \mathbb{E}_{p(f|u)}\left[p(y, \lambda|f)\right] \\
&\geq \mathbb{E}_{p(f|u)}\left[\log p(y, \lambda|f)\right] \\
&= \sum_{i=1}^{n} \mathbb{E}_{p(f_i|u)}\left[\log p(y_i, \lambda_i|f_i)\right] \\
&= \sum_{i=1}^{n} \mathbb{E}_{p(f_i|u)}\left[\log\left((2\pi\lambda_i)^{-\frac{1}{2}}\exp\left(-\frac{1}{2}\frac{(1 + \lambda_i - y_i f_i)^2}{\lambda_i}\right)\right)\right]
\end{aligned}$$

$$\overset{c}{=} -\frac{1}{2}\sum_{i=1}^{n} \mathbb{E}_{p(f_i|u)}\left[\log \lambda_i + \frac{(1+\lambda_i - y_i f_i)^2}{\lambda_i}\right]$$

$$= -\frac{1}{2}\sum_{i=1}^{n}\left(\log \lambda_i + \frac{1}{\lambda_i}\mathbb{E}_{p(f_i|u)}\left[(1+\lambda_i - y_i f_i)^2\right]\right)$$

$$= -\frac{1}{2}\sum_{i=1}^{n}\left(\log \lambda_i + \frac{1}{\lambda_i}\left(\widetilde{K}_{ii} + \left(1+\lambda_i - y_i K_{im}K_{mm}^{-1}u\right)^2\right)\right)$$

$$=: \mathcal{L}_1.$$

Plugging the lower bound $\mathcal{L}_1$ into the standard evidence lower bound (ELBO) [18] leads to the new variational objective

$$\log p(y) \geq \mathbb{E}_q\left[\log p(y,\lambda,u)\right] - \mathbb{E}_q\left[\log q(\lambda,u)\right]$$

$$= \mathbb{E}_q\left[\log p(y,\lambda|u)\right] + \mathbb{E}_q\left[\log p(u)\right] - \mathbb{E}_q\left[\log q(\lambda,u)\right]$$

$$\geq \mathbb{E}_q\left[\mathcal{L}_1\right] + \mathbb{E}_q\left[\log p(u)\right] - \mathbb{E}_q\left[\log q(\lambda,u)\right] \tag{7}$$

$$= -\frac{1}{2}\sum_{i=1}^{n}\mathbb{E}_q\left[\log \lambda_i + \frac{1}{\lambda_i}\left(\widetilde{K}_{ii} + \left(1+\lambda_i - y_i K_{im}K_{mm}^{-1}u\right)^2\right)\right]$$

$$\quad - \mathrm{KL}\left(q(u)||p(u)\right) - \mathbb{E}_{q(\lambda)}\left[\log q(\lambda)\right]$$

$$=: \mathcal{L}.$$

The expectations can be computed analytically (details are given in the appendix) and we obtain $\mathcal{L}$ in closed form,

$$\mathcal{L} \overset{c}{=} \frac{1}{2}\log|\zeta| - \frac{1}{2}\mathrm{tr}(K_{mm}^{-1}\zeta) - \frac{1}{2}\mu^\top K_{mm}^{-1}\mu + y^\top \kappa\mu$$

$$+ \sum_{i=1}^{n}\left\{\log(\mathrm{B}_{\frac{1}{4}}(\sqrt{\alpha_i})) + \frac{1}{2}\log(\alpha_i)\right\} \tag{8}$$

$$- \sum_{i=1}^{n}\frac{1}{2}\alpha_i^{-\frac{1}{2}}\left(1 - \alpha_i - 2y_i\kappa_{i.}\mu + \left(\kappa(\mu\mu^\top + \zeta)\kappa^\top + \widetilde{K}\right)_{ii}\right),$$

where $\kappa = K_{nm}K_{mm}^{-1}$ and $\mathrm{B}_{\frac{1}{2}}(.)$ is the modified Bessel function with parameter $\frac{1}{2}$ [20]. This objective is amenable to stochastic optimization where we subsample from the sum to obtain a noisy gradient estimate. We develop a stochastic variational inference scheme by following noisy natural gradients of the variational objective $\mathcal{L}$. Using the natural gradient over the standard euclidean gradient is often favorable since natural gradients are invariant to reparameterization of the variational family [21,22] and provide effective second-order optimization updates [6,23]. The natural gradients of $\mathcal{L}$ w.r.t. the Gaussian natural parameters $\eta_1 = \zeta^{-1}\mu$, $\eta_2 = -\frac{1}{2}\zeta^{-1}$ are

$$\widetilde{\nabla}_{\eta_1}\mathcal{L} = \kappa^\top Y(\alpha^{-\frac{1}{2}} + 1) - \eta_1 \tag{9}$$

$$\widetilde{\nabla}_{\eta_2}\mathcal{L} = -\frac{1}{2}(K_{mm}^{-1} + \kappa^\top A^{-\frac{1}{2}}\kappa) - \eta_2, \tag{10}$$

with $A = \text{diag}(\alpha)$. Details can be found in the appendix. The natural gradient updates always lead to a positive definite covariance matrix[2] and in our implementation ζ has not to be parametrized in any way to ensure positive-definiteness. The derivative of $\mathcal{L}$ w.r.t. α_i is

$$\nabla_\alpha \mathcal{L} = \frac{(1 - y_i \kappa_i \mu)^2 + y_i(\kappa_i \zeta \kappa_i^\top + \widetilde{K}_{ii})y_i}{4\sqrt{\alpha_i}^3} - \frac{1}{4\sqrt{\alpha_i}}. \tag{11}$$

Setting it to zero gives the coordinate ascent update for α_i,

$$\alpha_i = (1 - y_i \kappa_i \mu)^2 + y_i(\kappa_i \zeta \kappa_i^\top + \widetilde{K}_{ii})y_i.$$

Details can be found in the appendix. The inducing point locations can be either treated as hyperparameters and optimized while training [24] or can be fixed before optimizing the variational objective. We follow the first approach which is often preferred in a stochastic variational inference setup [7,10]. The inducing point locations can be either randomly chosen as subset of the training set or via a density estimator. In our experiments we have observed that the k-means clustering algorithm (kMeans) [25] yields the best results. Combining our results, we obtain a fast stochastic variational inference algorithm for the Bayesian nonlinear SVM which is outlined in Algorithm 1. We apply the adaptive learning rate method described in [26].

Algorithm 1. Inducing point SVI

1: set the learning rate schedule ρ_t appropriately
2: initialize η_1, η_2
3: select m inducing points locations (e.g. via kMeans)
4: compute kernel matrices K_{mm}^{-1} and $\widetilde{K} = K_{nn} - K_{nm}K_{mm}^{-1}K_{mn}$
5: **while** not converged **do**
6:     get $\mathcal{S}$ = minibatch index set of size s
7:     update $\alpha_i = (1 - y_i \kappa_i \mu)^2 + y_i(\kappa_i \zeta \kappa_i^\top + \widetilde{K}_{ii})y_i$
8:     compute $A_\mathcal{S} = \text{diag}(\alpha_i, \ i \in \mathcal{S})$
9:     compute $\hat{\eta}_1 = \kappa^\top Y(\alpha^{-\frac{1}{2}} + 1)$
10:    compute $\hat{\eta}_2 = -\frac{1}{2}(K_{mm}^{-1} + \kappa^\top A^{-\frac{1}{2}}\kappa)$
11:    update $\eta_1 = (1 - \rho_t)\eta_1 + \rho_t\hat{\eta}_1$
12:    update $\eta_2 = (1 - \rho_t)\eta_2 + \rho_t\hat{\eta}_2$
13:    compute $\zeta = -\frac{1}{2}\eta_2^{-1}$
14:    compute $\mu = \zeta\eta_1$
15: **return** $\alpha_1, \ldots, \alpha_n, \mu, \zeta$

4.3   Auto Tuning of Hyperparameters

The probabilistic formulation of the SVM lets us directly learn the hyperparameters while training. To this end we maximize the marginal likelihood $p(y|X, h)$,

---

[2] This follows directly since K_{mm} and $A^{-\frac{1}{2}}$ are positive definite.

where h denotes the set of hyperparameters (this approach is called empirical Bayes [27]). We follow an approximate approach and optimize the fitted variational lower bound $\mathcal{L}(h)$ over h by alternating between optimization steps w.r.t. the variational parameters and the hyperparameters [28]. We include a gradient ascent step w.r.t. h after multiple variational updates in the SVI scheme, this is commonly known as Type II maximum likelihood (ML-II) [9]

$$h^{(t)} = h^{(t-1)} + \tilde{\rho}_t \nabla_h \mathcal{L}(\alpha^{(t-1)}, \mu^{(t-1)}, \zeta^{(t-1)}, h). \tag{12}$$

Since the standard SVM does not exhibit a probabilistic formulation, the hyperparameters have to be tuned via computationally very expensive methods as grid search and cross validation. Our approach allows us to estimate the hyperparameters during training time and lets us follow gradients instead of only evaluating single hyperparameters.

In the appendix we provide the gradient of the variational objective $\mathcal{L}$ w.r.t. to a general kernel and show how to optimize arbitrary differentiable hyperparameters. Our experiments exemplify our automated hyperparameter tuning approach by optimizing the hyper parameter of an RBF kernel.

4.4 Uncertainty Predictions

Besides the advantage of automated hyperparameter tuning, the probabilistic formulation of the SVM leads directly to uncertainty estimates of the predictions. The standard SVM lacks this capability, and only heuristic approaches as e.g. Platt [8] exist. Using the approximate posterior $q(u|\mathcal{D}) = \mathcal{N}(u|\mu, \zeta)$ obtained by our stochastic variational inference method (Algorithm 1) we compute the class membership probability for a test point x^*,

$$
\begin{aligned}
p(f^*|x^*, \mathcal{D}) &= \int p(y^*|u, x^*)p(u|\mathcal{D})\mathrm{d}u \\
&\approx \int p(y^*|u, x^*)q(u|\mathcal{D})\mathrm{d}u \\
&= \mathcal{N}\left(y^*|K_{*m}K_{mm}^{-1}m, \ K_{**} - K_{*m}K_{mm}^{-1}(K_{m*} + \zeta K_{mm}^{-1}K_{m*})\right) \\
&=: q(f^*|x^*, \mathcal{D}),
\end{aligned}
$$

where K_{*m} denotes the kernel matrix between test and inducing points and K_{**} the kernel matrix between test points. This leads to the approximate class membership distribution

$$q(y^*|x^*, \mathcal{D}) = \Phi\left(\frac{K_{*m}K_{mm}^{-1}m}{K_{**} - K_{*m}K_{mm}^{-1}(K_{m*} + \zeta K_{mm}^{-1}K_{m*}) + 1}\right) \tag{13}$$

where $\Phi(.)$ is the probit link function. Note that we already computed inverse K_{mm}^{-1} for the training procedure leading to a computational overhead stemming only from simple matrix multiplication. Our experiments show that (13) leads to reasonable uncertainty estimates.

4.5 Special Case of Linear Bayesian SVM

We now consider the special case of using a linear kernel. If we are interested in this case we may consider the Bayesian model for the linear SVM proposed by Polson et al. (c.f. Eq. 4). This can be favorable over using the nonlinear version since this model is formulated in primal space and, therefore, the computational complexity depends on the dimension d and not on the number of data points n. Furthermore, focusing directly on the linear model allows us to optimize the true ELBO, $\mathbb{E}_q [\log p(y, \lambda, \beta)] - \mathbb{E}_q [\log q(\lambda, \beta)]$, without the need of relying on a lower bound (as in Eq. 7). This typically leads to a better approximate posterior.

We again follow the structured mean field approach and chose our variational distributions to be in the same families as the full conditionals (4),

$$q(\lambda_i) \equiv \mathcal{GIG}(\tfrac{1}{2}, 1, \alpha_i) \text{ and } q(\beta) \equiv \mathcal{N}(\mu, \zeta).$$

We use again the fact that the coordinate updates of the variational parameters can be obtained by computing the expected natural parameters of the corresponding full conditionals (4) and obtain

$$\begin{aligned} \alpha_i &= (1 - z_i^T \mu)^2 + z_i^T \zeta z_i \\ \zeta &= (ZA^{-\frac{1}{2}}Z^T + \Sigma^{-1})^{-1} \\ \mu &= \zeta Z(\alpha^{-\frac{1}{2}} + 1), \end{aligned} \tag{14}$$

where $\alpha = (\alpha_i)_{1 \leq i \leq n}$, $A = \operatorname{diag}(\alpha)$ and $Z = YX$. Since the Bayesian Linear SVM model exhibits global and local variables we can directly employ stochastic variational inference by subsampling the data and only updating minibatches of α. Note that for the linear case the covariance matrices have size $d \times d$, i.e. being independent of the number of data points. Therefore, the SVI Algorithm (14) for the Bayesian Linear SVM exhibits the computational complexity $\mathcal{O}(d^3)$. Luts et al. develop a batch variational inference scheme for the Bayesian linear SVM but do not scale to big datasets.

The hyperparameter can be tuned analogously to (12). The class membership probabilities are

$$p(y_* = 1 | x^*, \mathcal{D}) \approx \int \Phi(f_*) p(f_* | f, x^*) q(f | \mathcal{D}) \mathrm{d}f \mathrm{d}f_* = \Phi \left(\frac{x_*^\top \mu}{x_*^\top \zeta x_* + 1} \right),$$

where x_* are the test points and $q(f | \mathcal{D}) = \mathcal{N}(f | \mu, \zeta)$ the approximate posterior obtained by the above described SVI scheme.

5   Experiments

We compare our approach against the expectation conditional maximization (ECM) method proposed by Henao et al. [3], Gaussian process classification (GPC) [9], its recently proposed scalable stochastic variational inference version (S-GPC) [10], and libSVM with Platt scaling [8,29] (SVM + Platt). For all experiments we use an RBF kernel[3] with length-scale parameter θ. We perform all experiments using only one CPU core with 2.9 GHz and 386 GB RAM. Code is available at github.com/theogf/BayesianSVM.

5.1 Prediction Performance and Uncertainty Estimation

We experiment on seven real-world datasets and compare the prediction performance, the quality of the uncertainty estimates and run time of the methods. The results are presented in Table 1. We show that our method (S-BSVM) is up to 22 times faster than the direct competitor ECM and up to 700 times faster than Gaussian process classification[4] while outperforming the competitors in terms of prediction performance and quality of uncertainty estimates in most cases. The non-probabilistic SVM is naturally the fastest method. Combined with the heuristic Platt scaling approach it leads to class membership probabilities but,

Table 1. Average prediction error and Brier score with one standard deviation.

Dataset	n	dim		S-BSVM	ECM	GPC	SVM + Platt
Breast cancer	263	9	Error	**.26 ± .07**	.27 ± .10	.27 ± .07	.27 ± .09
			Brier score	**.18 ± .03**	.19 ± .05	**.18 ± .03**	.19 ± .04
			Time [s]	0.32	1.4	6.7	**0.04**
Diabetes	768	8	Error	**.22 ± .06**	.25 ± .07	.23 ± .07	.24 ± .07
			Brier score	.16 ± .04	.17 ± .04	**.15 ± .04**	.16 ± .04
			Time [s]	3.9	33	67	**0.11**
Flare	144	9	Error	**.36 ± .12**	**.36 ± .12**	**.36 ± .11**	**.36 ± .12**
			Brier score	**.22 ± .05**	.25 ± .07	.24 ± .03	.24 ± .04
			Time [s]	0.08	0.26	1.8	**0.01**
German	1000	20	Error	**.24 ± .11**	.25 ± .12	.25 ± .13	.27 ± .10
			Brier score	**.17 ± .06**	**.17 ± .05**	**.17 ± .06**	.18 ± .05
			Time [s]	12	80	115	**0.15**
Heart	270	13	Error	**.16 ± .06**	.19 ± .09	**.16 ± .06**	.17 ± .07
			Brier score	.13 ± .04	.14 ± .04	**.12 ± .03**	**.12 ± .04**
			Time [s]	0.34	2.2	6	**0.04**
Splice	2991	60	Error	.13 ± .03	**.11 ± .03**	.32 ± .14	.14 ± .01
			Brier score	.17 ± .01	.18 ± .01	.40 ± .14	**.11 ± .01**
			Time [s]	18	406	419	**1.3**
Waveform	5000	21	Error	**.09 ± .02**	.10 ± .02	.10 ± .02	.10 ± .02
			Brier score	**.06 ± .01**	.15 ± .01	**.06 ± .01**	**.06 ± .01**
			Time [s]	12.5	264	8691	**2.3**

---

[3] The RBF kernel is defined as $k(x_1, x_2, \theta) = \exp\left(-\frac{\|x_1 - x_2\|}{\theta^2}\right)$, where θ is the length scale parameter.

[4] For a comparison with the stochastic variational inference version of GPC, see Sect. 5.3.

however, still lacks the advantages of a probabilistic model (as e.g. uncertainty quantification of the learned parameters and automatic hyperparameter tuning).

To evaluate the quality of the uncertainty estimates we compute the Brier score which is considered as a good performance measure for probabilistic predictions [30] being defined as $BS = \frac{1}{n} \sum_{i=1}^{N} (y_i - q(x_i))^2$, where $y_i \in \{0, 1\}$ is the observed output and $q(x_i) \in [0, 1]$ is the predicted class membership probability. Note that smaller Brier score indicates better performance.

The datasets are all from the Rätsch benchmark datasets [31] commonly used to test the accuracy of binary nonlinear classifiers. We perform a 10-fold cross-validation and use an RBF kernel with fixed parameters for all methods. For S-BSVM we choose the number of inducing points as 20% of the training set size, except for the datasets *Splice*, *German* and *Waveform* where we use 100 inducing points. For each dataset minibatches of 10 samples are used.

5.2   Big Data Experiments

We demonstrate the scalability of our method on the SUSY dataset [11] containing 5 million points with 17 features. This dataset size is very common in particle physics due to the simplicity of artificially generating new events as well as the quantity of data coming from particle detectors. Since it is important to have a sense of the confidence of the predictions for such datasets the Bayesian SVM is an appropriate choice. We use an RBF kernel[5], 64 inducing points and minibatches of 100 points. The training of our model takes only 10 min without any parallelization. We use the area under the receiver operating characteristic (ROC) curve (AUC) as performance measure since it is a standard evaluation measure on this dataset [11].

Our method achieves an AUC of 0.84 and a Brier score of 0.22, whereby the state-of-the-art obtains an AUC of 0.88 using a deep neural network (5 layers, 300 hidden units each) [11]. Note that this approach takes much longer to train and does not include uncertainty estimates.

5.3   Run Time

We examine the run time of our methods and the competitors. We include both the batch variational inference method (B-BSVM) described in Sect. 4.1 and our fast and scalable inference method (S-BSVM) described in Sect. 4.2 in the experiments. For each method we iteratively evaluate the prediction performance on a held-out dataset given a certain training time budget. The prediction error as function of the training time is shown in Fig. 1. We experiment on the *Waveform* dataset from the Rätsch benchmark dataset ($N = 5000$, $d = 21$). We use an RBF kernel with fixed length-scale parameter $\theta = 5.0$ and for the stochastic variational inference methods, S-BSVM and S-GPC, we use a batch size of 10 and 100 inducing points.

---

[5] The length scale parameter tuning is not included in the training time. We found $\theta = 5.0$ by our proposed automatic tuning approach.

Our scalable method (S-BSVM) is around 10 times faster than the direct competitor ECM while having slightly better prediction performance. The batch variational inference version (B-BSVM) is the slowest of the Bayesian SVM inference methods. The related probabilistic model, Gaussian process classification, is around 5000 times slower than S-BSVM. Its stochastic inducing point version (S-GPC) has comparable run time to S-BSVM but is very unstable leading to bad prediction performance. S-GPC showed these instabilities for multiple settings of the hyperparameters. The classic SVM (libSVM) has a similar run time as our method. The speed and prediction performance of S-BSVM depend on the number of inducing points. See Sect. 5.5 for an empirical study. Note that the run time in Table 1 is determined after the methods have converged.

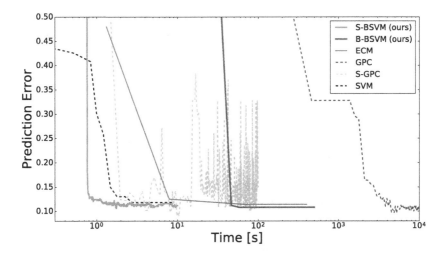

Fig. 1. Prediction error on held-out dataset vs. training time.

5.4 Auto Tuning of Hyperparameters

In Sect. 4.3 we show that our inference method possesses the ability of automatic hyperparameter tunning. In this experiment we demonstrate that our method, indeed, finds the optimal length-scale hyperparameter of the RBF kernel. We use the optimizing scheme (12) and alternate between 10 variational parameter updates and one hyperparameter update. We compute the true validation loss of the length-scale parameter θ by a grid search approach which consists of training our model (S-BSVM) for each θ and measuring the prediction performance using 10-fold cross validation. In Fig. 2 we plot the validation loss and the length-scale parameter found by our method. We find the true optimum by only using 5 hyperparameter optimization steps. Training and hyperparameter optimization takes only 0.3 s for our method, whereas grid search takes 188 s (with a grid size of 1000 points).

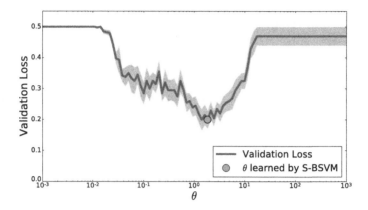

Fig. 2. Average validation loss as function of the RBF kernel length-scale parameter θ, computed by grid search and 10-fold cross validation. The red circle represents the hyperparameter found by our proposed automatic tuning approach. (Color figure online)

5.5 Inducing Points Selection

The sparse GP model used in our inference scheme builds on a set of inducing points where both the number and the locations of the inducing points are free parameters. We investigate three different inducing point selection methods: random subset selection from the training set, the Gaussian Mixture Model (GMM), and the k-means clustering algorithm with an improved k-means++ seeding (kMeans) [32]. Furthermore we show how the number of inducing points affects the prediction accuracy and the run time. We test the three inducing point selection methods on the USPS dataset [33] which we reduced to a binary problem using only the digits 3 and 5 ($N = 1350$ and $d = 256$). For all methods we progressively increase the number of inducing points and compute the prediction error by 10-fold cross validation. We present our results in Fig. 3.

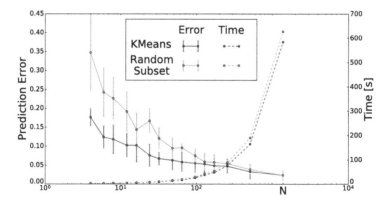

Fig. 3. Average prediction error and training time as functions of the number of inducing points selected by two different methods with one standard deviation (using 10-fold cross validation).

The GMM is unable to fit large numbers of samples and dimensions and fails to converge for almost all datasets tried, therefore, we do not include it in the plot. Using the k-means selection algorithm leads for small numbers of inducing points to much better prediction performance than random subset selection. Furthermore, we show that using only a small fraction of inducing points (around 1% of the original dataset) leads to a nearly optimal prediction performance by simultaneously significantly decreasing the run time. We observe similar results on all datasets we considered.

6   Conclusion

We presented a fast, scalable and reliable approximate inference method for the Bayesian nonlinear SVM. While previous methods were restricted to rather small datasets our method enables the application of the Bayesian nonlinear SVM to large real world datasets containing millions of samples. Our experiments showed that our method is orders of magnitudes faster than the state-of-the-art while still yielding comparable prediction accuracies. We showed how to automatically tune the hyperparameters and obtain prediction uncertainties which is important in many real world scenarios.

In future work we plan to further extend the Bayesian nonlinear SVM model to deal with missing data and account for correlations between data points building on ideas from [34]. Furthermore, we want to develop Bayesian formulations of important variants of the SVM as for instance one-class SVMs [35].

Acknowledgments. We thank Stephan Mandt, Manfred Opper and Patrick Jähnichen for fruitful discussions. This work was partly funded by the German Research Foundation (DFG) award KL 2698/2-1.

References

1. Cortes, C., Vapnik, V.: Support-vector networks. Mach. Learn. **20**(3), 273–297 (1995)
2. Polson, N.G., Scott, S.L.: Data augmentation for support vector machines. Bayesian Anal. **6**(1), 1–24 (2011)
3. Henao, R., Yuan, X., Carin, L.: Bayesian nonlinear support vector machines and discriminative factor modeling. In: NIPS (2014)
4. Fernández-Delgado, M., Cernadas, E., Barro, S., Amorim, D.: Do we need hundreds of classifiers to solve real world classification problems? JMLR **15**(1), 3133–3181 (2014)
5. Mohri, M., Rostamizadeh, A., Talwalkar, A.: Foundations of Machine Learning. MIT press, Cambridge (2012)
6. Hoffman, M.D., Blei, D.M., Wang, C., Paisley, J.: Stochastic variational inference. JMLR **14**, 1303–1347 (2013)
7. Hensman, J., Fusi, N., Lawrence, N.D.: Gaussian processes for big data. In: Conference on Uncertainty in Artificial Intelligence (2013)
8. Platt, P.J.C.: Probabilistic outputs for support vector machines and comparisons to regularized likelihood methods. Adv. Large Margin Classif. **10**(3), 61–74 (1999)

9. Rasmussen, C.E., Williams, C.K.I.: Gaussian Processes for Machine Learning (Adaptive Computation and Machine Learning). MIT Press, Cambridge (2005)
10. Hensman, J., Matthews, A.: Scalable variational Gaussian process classification. In: AISTATS (2015)
11. Baldi, P., Sadowski, P., Whiteson, D.: Searching for exotic particles in high-energy physics with deep learning. Nature Commun. **4** (2014). Article no. 4308
12. Zhu, J., Chen, N., Perkins, H., Zhang, B.: Gibbs max-margin topic models with data augmentation. JMLR **15**(1), 1073–1110 (2014)
13. Xu, M., Zhu, J., Zhang, B.: Fast max-margin matrix factorization with data augmentation. In: ICML, pp. 978–986 (2013)
14. Zhang, A., Zhu, J., Zhang, B.: Max-margin infinite hidden Markov models. In: ICML (2014)
15. Luts, J., Ormerod, J.T.: Mean field variational Bayesian inference for support vector machine classification. Comput. Stat. Data Anal. **73**, 163–176 (2014)
16. Snelson, E., Ghahramani, Z.: Sparse GPs using pseudo-inputs. In: NIPS (2006)
17. Kloft, M., Brefeld, U., Sonnenburg, S., Zien, A.: lp-norm multiple kernel learning. JMLR **12**, 953–997 (2011)
18. Jordan, M.I., Ghahramani, Z., Jaakkola, T.S., Saul, L.K.: An introduction to variational methods for graphical models. Mach. Learn. **37**(2), 183–233 (1999)
19. Wainwright, M.J., Jordan, M.I.: Graphical models, exponential families, and variational inference. Found. Trends Mach. Learn. **1**(1–2), 1–305 (2008)
20. Jørgensen, B.: Statistical Properties of the Generalized Inverse Gaussian Distribution. Springer, New York (2012). https://doi.org/10.1007/978-1-4612-5698-4
21. Amari, S., Nagaoka, H.: Methods of Information Geometry. American Mathematical Society, Providence (2007)
22. Martens, J.: New insights and perspectives on the natural gradient method. Arxiv Preprint (2017)
23. Amari, S.: Natural gradient works efficiently in learning. Neural Comput. **10**, 251–276 (1998)
24. Titsias, M.K.: Variational learning of inducing variables in sparse Gaussian processes. In: Artificial Intelligence and Statistics, vol. 12, pp. 567–574 (2009)
25. Murphy, K.P.: Machine Learning: A Probabilistic Perspective. The MIT Press, Cambridge (2012)
26. Ranganath, R., Wang, C., Blei, D.M., Xing, E.P.: An adaptive learning rate for stochastic variational inference. In: ICML (2013)
27. Maritz, J., Lwin, T.: Empirical Bayes Methods with Applications: Monographs on Statistics and Applied Probability. Chapman & Hall/CRC, Boca Raton (1989)
28. Mandt, S., Hoffman, M., Blei, D.: A variational analysis of stochastic gradient algorithms. In: ICML (2016)
29. Chang, C.C., Lin, C.J.: LIBSVM: a library for support vector machines. ACM Trans. Intell. Syst. Technol. **2**(3), 27:1–27:27 (2011)
30. Brier, G.W.: Verification of forecasts expressed in terms of probability. Mon. Weather Rev. **78**(1), 1–3 (1950)
31. Diethe, T.: 13 benchmark datasets derived from the UCI, DELVE and STATLOG repositories (2015)
32. Bachem, O., Lucic, M., Hassani, H., Krause, A.: Fast and provably good seedings for k-means. In: NIPS (2016)
33. Lichman, M.: UCI machine learning repository (2013)
34. Mandt, S., Wenzel, F., Nakajima, S., Cunningham, J.P., Lippert, C., Kloft, M.: Sparse probit linear mixed model. Mach. Learn. **106**(9–10), 1621–1642 (2017)
35. Perdisci, R., Gu, G., Lee, W.: Using an ensemble of one-class SVM classifiers to H. P.-based anomaly detection systems. In: Data Mining (2006)

Entropic Trace Estimates for Log Determinants

Jack Fitzsimons[1(✉)], Diego Granziol[1], Kurt Cutajar[2], Michael Osborne[1], Maurizio Filippone[2], and Stephen Roberts[1]

[1] Department of Engineering, University of Oxford, Oxford, UK
{jack,diego,mosb,sjrob}@robots.ox.ac.uk
[2] Department of Data Science, EURECOM, Biot, France
{kurt.cutajar,maurizio.filippone}@eurecom.fr

Abstract. The scalable calculation of matrix determinants has been a bottleneck to the widespread application of many machine learning methods such as determinantal point processes, Gaussian processes, generalised Markov random fields, graph models and many others. In this work, we estimate log determinants under the framework of maximum entropy, given information in the form of moment constraints from stochastic trace estimation. The estimates demonstrate a significant improvement on state-of-the-art alternative methods, as shown on a wide variety of matrices from the SparseSuite Matrix Collection. By taking the example of a general Markov random field, we also demonstrate how this approach can significantly accelerate inference in large-scale learning methods involving the log determinant.

1 Introduction

Scalability is a key concern for machine learning in the big data era, whereby inference schemes are expected to yield optimal results within a constrained computational budget. Underlying these algorithms, linear algebraic operations with high computational complexity pose a significant bottleneck to scalability, and the log determinant of a matrix [5] falls firmly within this category of operations. The canonical solution involving Cholesky decomposition [16] for a general $n \times n$ positive definite matrix, A, entails time complexity of $\mathcal{O}(n^3)$ and storage requirements of $\mathcal{O}(n^2)$, which is infeasible for large matrices. Consequently, this term greatly hinders widespread use of the learning models where it appears, which includes determinantal point processes [24], Gaussian processes [31], and graph problems [36].

The application of kernel machines to vector valued input data has gained considerable attention in recent years, enabling fast linear algebra techniques. Examples include Gaussian Markov random fields [32] and Kronecker-based algebra [33], while similar computational speed-ups may also be obtained for sparse matrices. Nonetheless, such structure can only be expected in selected applications, thus limiting the widespread use of such techniques.

In light of this computational constraint, several approximate inference schemes have been developed for estimating the log determinant of a matrix

© Springer International Publishing AG 2017
M. Ceci et al. (Eds.): ECML PKDD 2017, Part I, LNAI 10534, pp. 323–338, 2017.
https://doi.org/10.1007/978-3-319-71249-9_20

more efficiently. Generalised approximation schemes frequently build upon iterative stochastic trace estimation techniques [4]. This includes polynomial approximations such as Taylor and Chebyshev expansions [2,20]. Recent developments shown to outperform the aforementioned approximations include estimating the trace using stochastic Lanczos quadrature [35], and a probabilistic numerics approach based on Gaussian process inference which incorporates bound information [12]. The latter technique is particularly significant as it introduces the possibility of quantifying the numerical uncertainty inherent to the approximation.

In this paper, we present an alternative probabilistic approximation of log determinants rooted in information theory, which exploits the relationship between stochastic trace estimation and the moments of a matrix's eigenspectrum. These estimates are used as moment constraints on the probability distribution of eigenvalues. This is achieved by maximising the entropy of the probability density $p(\lambda)$ given our moment constraints. In our inference scheme, we circumvent the issue inherent to the Gaussian process approach [12], whereby positive probability mass may occur in the region of negative densities. In contrast, our proposed entropic approach implicitly encodes the constraint that densities are necessarily positive. Given equivalent moment information, we achieve competitive results on matrices obtained from the SuiteSparse Matrix Collection [11] which consistently outperform competing approximations to the log-determinant [7,12].

The most significant contributions of this work are listed below.[1]

1. We develop a novel approximation to the log determinant of a matrix which relies on the principle of maximum entropy enhanced with moment constraints derived from stochastic trace estimation.
2. We present the theory motivating the use of maximum entropy for solving this problem, along with insights on why we expect particularly significant improvements over competing techniques for large matrices.
3. We directly compare the performance of our entropic approach to other state-of-the-art approximations to the log-determinant. This evaluation covers real sparse matrices obtained from the SuiteSparse Matrix Collection [11].
4. Finally, to showcase how the proposed approach may be applied in a practical scenario, we incorporate our approximation within the computation of the log-likelihood term of a Gaussian Markov random field, where we obtain a significant increase in speed.

1.1 Related Work

The methodology presented in this work predominantly draws inspiration from the recently introduced probabilistic numerics approach to estimating the log

---

[1] Code for algorithms proposed in this paper are available at https://github.com/OxfordML/EntropicTraceEstimation.

determinant of a matrix [12]. In that work, the computation of the log determinant is reinterpreted as a probabilistic estimation problem, whereby results obtained from budgeted computations are used to infer accurate estimates for the log determinant. In particular, within that proposed framework, the eigenvalues of a matrix A are modelled from noisy observations of $\text{Tr}(A^k)$ obtained from stochastic trace estimation [4] using the Taylor approximation method. By modelling such noisy observations using a Gaussian process [31], Bayesian quadrature [29] can then be invoked for making predictions on the infinite series of the Taylor expansion, and in turn estimating the log determinant. Of particular interest is the uncertainty quantification inherent to this approach, which is a notable step forward in the direction of measuring the complete numerical uncertainty associated with approximating large-scale inference models. The estimates obtained using this Bayesian set-up may be further improved by considering known upper and lower bounds on the value of the log determinant [5]. In this paper, we provide an alternative to this approach by interpreting the observed moments as being constraints on the probability distribution of eigenvalues underlying the computation of the log determinant. As we shall explore, our novel entropic formulation makes better calibrated prior assumptions than the previous work, and consequently yields superior performance.

More traditional approaches to approximating the log determinant build upon iterative algorithms, and exploit the fact that the log determinant may be rewritten as the trace of the logarithm of the matrix. This features in both the Chebyshev expansion approximation [20], as well as the widely-used Taylor series approximation upon which the aforementioned probabilistic inference approaches are built. Recently, an approximation to the log determinant using stochastic Lanczos quadrature [35] has been shown to outperform the aforementioned polynomial approaches, while also providing probabilistic error bounds. Finally, given that the logarithm of a matrix often appears multiplied by a vector (for example the log likelihood term of a Gaussian process [31]), the spline approximation proposed in [10] may be used to accelerate computation.

2   Background

In this section, we shall formally introduce the concepts underlying the proposed maximum entropy approach to approximating the log determinant. We start by describing stochastic trace estimation and demonstrate how this can be applied to estimating the trace term of matrix powers. Subsequently, we illustrate how the latter terms correspond to the raw moments of the matrix's eigenspectrum, and show how the log determinant may be inferred from the distribution of eigenvalues constrained by such moments.

2.1   Stochastic Trace Estimation

Estimating the trace of implicit matrices is a central component of many approaches to approximate the log determinant of a matrix. Stochastic trace estimation [4] builds a Monte Carlo estimate of the trace of a matrix A by repeatedly multiplying it by *probing vectors* $\mathbf{z}$,

$$\text{Tr}(A) \approx \frac{1}{m} \sum_{i=1}^{m} \mathbf{z}_i^T A \mathbf{z}_i,$$

such that the expectation of $\mathbf{z}_i \mathbf{z}_i^T$ is the identity, $\mathbb{E}[\mathbf{z}_i \mathbf{z}_i^T] = I$. This can be readily verified using the expectation of $\text{Tr}(\mathbf{z}_i^T A \mathbf{z}_i)$ by exploiting the cyclical property of the trace operation. As such, many choices of how to sample the probing vectors have emerged. Possibly the most naïve choice involves sampling from the columns of the identity matrix; however, due to poor expected sample variance this is not widely used in the literature. Sampling from vectors on the unit hyper-sphere, and correspondingly sampling normal random vectors (Gaussian estimator), significantly reduces the sample variance, but more random bits are required to generate each sample. A major progression for stochastic trace estimation was the introduction of Hutchinson's method [21], which sampled each element as a Bernoulli random variable requiring only a linear number of random bits, while also reducing the sample variance even further. A more recent approach involves sampling from sets of mutually unbiased bases (MUBs) [13], requiring only a logarithmic number of bits. Table 1 (adapted from [13]) provides a concise overview of the landscape of probing vectors.

Table 1. Comparison of single shot variance V, worst case single shot variance V^{worst} and number of random bits R required for commonly used trace estimators and the MUBs estimator. (* *required for floating point precision*)

	V	V^{worst}	R
Fixed basis estimator	$d \sum_{i=1}^{d} M_{ii}^2 - \text{Tr}(A)^2$	$(d-1)\text{Tr}(A)^2$	$\log_2(d)$
MUBs estimator	$\frac{d}{d+1}\text{Tr}(A^2) - \frac{1}{d+1}\text{Tr}(A)^2$	$\frac{d-1}{d+1}\text{Tr}(A^2)$	$2\log_2(d)$
Hutchinson estimator	$2\left(\text{Tr}(A^2) - \sum_{i=1}^{d} A_{ii}^2\right)$	$\frac{2(d-1)}{d}\text{Tr}(A^2)$	d
Gaussian estimator	$2\text{Tr}(A^2)$	$2\text{Tr}(A^2)$	$\mathcal{O}(d)^*$

A notable application of stochastic trace estimation is the approximation of the trace term for matrix powers, $\text{Tr}(A^k)$. Stochastic trace estimation enables vector-matrix multiplications to be propagated right to left, costing $\mathcal{O}(n^2)$, rather than the $\mathcal{O}(n^3)$ complexity required by matrix multiplication. This simple trick has been applied in several domains such as counting the number of triangles in graphs [3], string pattern matching [1] and of course estimating the log determinant of matrices, as discussed in this work.

2.2   Raw Moments of the Eigenspectrum

The relation between the raw moments of the eigenvalue distribution and the trace of matrix powers allows us to exploit stochastic trace estimation for estimating the log determinant. Raw moments are defined as the mean of the random variables raised to integer powers. Given that the function of a matrix is implicitly applied to its eigenvalues, in the case of matrix powers this corresponds to raising the eigenvalues to a given power. For example, the kth raw moment of the distribution over the eigenvalues (a mixture of Dirac delta functions) is $\sum_{i=1}^{m} \lambda^k p(\lambda)$, where $p(\lambda)$ is the distribution of eigenvalues. The first few raw moments of the eigenvalues are trivial to compute. Denoting the kth raw moment as $\mathbb{E}[\lambda^k]$, we have that $\mathbb{E}[\lambda^0] = 1$, $\mathbb{E}[\lambda^1] = \frac{1}{n}\text{Tr}(A)$ and $\mathbb{E}[\lambda^2] = \frac{1}{n}\sum_{i,j} A_{i,j}^2$. More generally, the kth raw moment can be formulated as $\mathbb{E}[\lambda^k] = \frac{1}{n}\text{Tr}(A^k)$, which can be estimated using stochastic trace estimation. These identities can be easily derived using the definitions and well known identities of the trace term and Frobenius norm.

2.3   Approximating the Log Determinant

In view of the relation presented in the previous subsection, we can reformulate the log determinant of a matrix in terms of its eigenvalues using the following derivation:

$$\log\big(\text{Det}(A)\big) = \sum_{i=1}^{n} \log(\lambda_i) := n\mathbb{E}\left[\log(\lambda)\right] \approx n \int p(\lambda)\log(\lambda)\mathrm{d}\lambda, \qquad (1)$$

where the approximation is introduced due to our estimation of $p(\lambda)$, the probability distribution of eigenvalues. If we knew the true distribution of $p(\lambda)$ it would hold with equality.

Given that we can obtain information about the moments of $p(\lambda)$ through stochastic trace estimation, we can solve this integral by employing the principle of maximum entropy, while treating the estimated moments as constraints. While not explored in this work, it is worth noting that in the event of moment information combined with samples of eigenvalues, we would use the method of maximum relative entropy with data constraints, which is in turn a generalisation of Bayes' rule [9]. This can be applied, for example, in the quantum linear algebraic setting [28].

3   Estimating the Log Determinant Using Maximum Entropy

The maximum entropy method (MaxEnt) [30] is a procedure for generating the most conservatively uncertain estimate of a probability distribution possible with the given information, which is particularly valued for being maximally non-committal with regard to missing information [22]. In particular, to determine a probability density $p(\boldsymbol{x})$, this corresponds to maximising the functional

$$S = -\int p(\boldsymbol{x})\log p(\boldsymbol{x})\mathrm{d}\boldsymbol{x} - \sum_i \alpha_i \left[\int p(\boldsymbol{x})f_i(\boldsymbol{x})\mathrm{d}\boldsymbol{x} - \mu_i\right] \qquad (2)$$

with respect to $p(\boldsymbol{x})$, where $\mathbb{E}[f_i(\boldsymbol{x})] = \mu_i$ are given constraints on the probability density. In our case, each μ_i constraint denotes the stochastic trace estimate of the ith raw moment of the matrix eigenvalues. The first term in the above equation is referred to as the Boltzmann-Shannon-Gibbs (BSG) entropy, which has been applied in multiple fields, ranging from condensed matter physics [15] to finance [8,27]. Along with its path equivalent, maximum caliber [17], it has been successfully used to derive statistical mechanics [18], non-relativistic quantum mechanics, Newton's laws and Bayes' rule [9,17]. Under the axioms of consistency, uniqueness, invariance under coordinate transformations, sub-set and system independence, it can be proved that for constraints in the form of expected values, drawing self-consistent inferences requires maximising the entropy [30,34]. Crucial for our investigation are the functional forms $f_i(\boldsymbol{x})$ of constraints for which the method of maximum entropy is appropriate. The axioms of Johnson and Shore [34] assert that the entropy must have a unique maximum and that the BSG entropy is convex. The entropy hence has a unique maximum provided that the constraints are convex. This is satisfied for any polynomial in $\boldsymbol{x}$ and hence, maximising the entropy given moment constraints constitutes a self-consistent inference scheme [30].

3.1 Implementation

Our implementation of the system follows straight from stochastic trace estimation to estimate the raw moments of the eigenvalues, maximum entropy distribution given these moments and, finally, determining the log of the geometric mean of this distribution. The log geometric mean is an estimate of the log determinant divided by the dimensionality of $A \in \mathbb{R}^{n \times n}$. We explicitly step through the subtleties of the implementation in order to guide the reader through the full procedure.

By taking the partial derivatives of S from Eq. (2), it is possible to show that the maximum entropy distribution given moment information is of the form

$$p(\lambda) = \exp(-1 + \sum_i \alpha_i \mu_i).$$

The goal is to find the set of α_i which match the raw moments of $p(\lambda)$ to the observed moments $\{\mu_i\}$. While this *may* be performed symbolically, this becomes intractable for larger number of moments, and our experience with current symbolic libraries [25,37] is that they are not extendable beyond more than 3 moments. Instead, we turn our attention to numerical optimisation. Early approaches to optimising maximum entropy coefficients worked well for a small number of coefficients but became highly unstable as the number of observed moments grew [26]. However, building on these concepts, more stable approaches emerged [6]. Algorithm 1 outlines a stable approach to this optimisation under the conditions that λ_i is strictly positive and the moments lie between zero and one. We can satisfy these conditions by normalising our positive definite matrix by the maximum of the Gershgorin intervals [14].

Algorithm 1. Optimising the Coefficients of the MaxEnt Distribution

---

Input: Moments $\{\mu_i\}$, Tolerance ϵ
Output: Coefficients $\{\alpha_i\}$
1: $\alpha_i \sim \mathcal{N}(0,1)$
2: $i \leftarrow 0$
3: $p(\lambda) \leftarrow \exp(-1 - \sum_k \alpha_k \lambda^k)$
4: **while** error $< \epsilon$ **do**
5:      $\delta \leftarrow \log\left(\frac{\mu_i}{\int \lambda^i p(\lambda)d\lambda}\right)$
6:      $\alpha_i \leftarrow \alpha_i + \delta$
7:      $p(\lambda) \leftarrow p(\lambda|\alpha)$
8:      error $\leftarrow \max |\int \lambda^i p(\lambda)d\lambda - \mu_i|$
9:      $i \leftarrow \mathrm{mod}(i+1, \mathrm{length}(\mu))$

Given Algorithm 1, the pipeline of our approach can be pieced together. First, the raw moments of the eigenvalues are estimated using stochastic trace estimation. These moments are then passed to the maximum entropy optimisation algorithm to produce an estimate of the distribution of eigenvalues, $p(\lambda)$. Finally, $p(\lambda)$ is used to estimate the log geometric mean of the distribution, $\int \log(\lambda)p(\lambda)d\lambda$. This term is multiplied by the dimensionality of the matrix and if the matrix is normalised, the log of this normalisation term is added again. These steps are laid out more concisely in Algorithm 2.

Algorithm 2. Entropic Trace Estimation for Log Determinants

Input: PD Symmetric Matrix A, Order of stochastic trace estimation k, Tolerance ϵ
Output: Log Determinant Approximation $\log|A|$
1: $B = A/\|A\|_2$
2: μ (moments)$\leftarrow$ StochasticTraceEstimation(B, k)
3: α (coefficients) $\leftarrow$ MaxEntOpt(μ, ϵ)
4: $p(\lambda) \leftarrow p(\lambda|\alpha)$
5: $\log|A| \leftarrow n \int \log(\lambda)p(\lambda)d\lambda + n \log(\|A\|_2)$

---

4   Insights for Large Matrices

The method of entropic trace estimation has the interesting property where we expect the relative error to decrease as the matrix size N increases. Colloquially, we can liken maximum entropy to a maximum likelihood over distributions, where this likelihood functional is raised to the number of eigenvalues in the matrix. This corresponds to the number of particles in the system, in traditional particle physics parlance. Given that there is a global maximum, as the number of eigenvalues increases the functional tends to a delta functional around the $p(x)$ of maximum entropy. This confirms that within the scope of our problem's continuous distribution over eigenvalues, whenever the number of eigenvalues (and correspondingly the dimensionality of the matrix) tends towards infinity, we expect the maximum entropy solution to converge to the true solution. This gives further credence to the suitability of our method when applied to large matrices. We substantiate this claim by delving into the fundamentals of maximum entropy for physical systems and extending the analogy to functionals over the space of densities. We show that in the limit of $N \to \infty$ the maximum entropy distribution dominates the space of solutions satisfying the constraints. We demonstrate the practical significance of this assertion by setting up an experiment using synthetically constructed random matrices, where this is in fact verified.

4.1   Law of Large Numbers for Maximum Entropy in Matrix Methods

In order to demonstrate our result, we consider the quantity W, which represents the number of ways in which our candidate probability distribution can recreate the observed moment information. In order to make this quantity finite we consider the discrete distribution characterised by machine precision ϵ. We show that $W = \exp(NS)$, where S is the entropy. Hence maximising the entropy is equivalent to maximising W, as N is fixed. In the continuous limit, we consider the ratio of two such terms $F_i = W_i / \sum_j W_j$, which is also finite. We consider this quantity F_i to represent the probability of a candidate solution i occurring, given the space of all possible solutions. We further show in the discrete and continuous space that for large N, the candidate distribution maximising S occurs with probability 1.

Consider the analogy of having a physical system made up of particles. The different ways, W, in which we can organise this system of N particles with T distinguishable groups each containing n_t particles, can be expressed as the combinatorial

$$W = \frac{N!}{\prod_{t=1}^{T} n_t!},\tag{3}$$

where $\sum_t n_t = N$. If we consider the logarithm of the above term, we can invoke Stirling's approximation that $\log(n!) \approx n \log(n) - n$, which is exact in the limits $N \to \infty$ and $n_i \to \infty$. Using this relation, we obtain

$$\log W = N \left(- \sum_{t=1}^{T} \frac{n_t}{N} \log \left[\frac{n_t}{N} \right] \right) = NS,\tag{4}$$

where S is the Boltzmann-Shannon-Gibbs entropy and in the continuous case we identify $p(t) = n_t/N$, where $p(t)$ represents the probability of being in group t. Hence, $W = \exp(NS)$.

The number of formulations, W_{maxent}, in which the maximum entropy realisation is more probable than any other realisation can be succinctly expressed as

$$\frac{W_{\text{maxent}}}{W_{\text{other}}} = \exp\big(N(S_{\text{maxent}} - S_{\text{other}})\big), \tag{5}$$

which exposes that in the limit of large N, the maximum entropy solution dominates other solutions satisfying the same constraints. More significantly, we can also show that it dominates the space of *all* solutions. Let $\sum_i W_i$ denote the total number of ways in which the system can be configured for all possible underlying densities satisfying the constraints.

If we consider the ratio between this term and the number of ways the maximum entropy distribution can be configured, we observe that

$$\frac{W_{\text{maxent}}}{\sum_i W_i} = \frac{\exp(NS(P_{\text{maxent}}))}{\sum_i \exp(NS_i)} = \frac{1}{\sum_i \exp\left(N(S(P_i(\boldsymbol{x})) - S(P_{\text{maxent}}(\boldsymbol{x})))\right)} \xrightarrow{N\to\infty} 1,$$

$$\tag{6}$$

with $S(P(\boldsymbol{x}))$ denoting the entropy of the probability of a probability distribution $P(\boldsymbol{x})$ and where we have exploited the fact that one of the S_i is $S_{\max}$ and that $S_{i\neq j} < S_{\max}$. More formally, we consider the probability mass about maxima in the functional describing all possible configurations, which is characterised via their entropy, S:

$$W_{\text{total}} = \int \exp(NS)[\mathcal{D}P] = \int \cdots \int_{-\infty}^{\infty} \exp(NS) \prod_x dP(\boldsymbol{x}). \tag{7}$$

When $N \to \infty$, the maximum value of S accounts for the majority of the integral's mass. To see this consider the ratio of functional integrals,

$$\frac{W_{\text{maxent}}}{W_{\text{total}}} = \frac{\int \exp\left(NS(P_{\text{maxent}}(\boldsymbol{x}))\right) dP_{\text{maxent}}(\boldsymbol{x})}{\int_{-\infty}^{\infty} \exp\left(NS(P(\boldsymbol{x}))\right) \prod_x dP(\boldsymbol{x})}, \tag{8}$$

which tends to 1 as $N \to \infty$. The argument is the continuous version of that which is displayed in Eq. (6), while the convergence to 1 as N approaches infinity follows directly from Laplace's method in the multivariate case, as well as the definition of a probability density and the functional integral.

Laplace's method gives a theoretical basis for the canonical distributions in statistical mechanics. Its equivalent in the complex space, the method of steepest descent, in Feynman's path integral formulation, shows that points in the vicinity of the extrema of the action functional (the classical mechanical solution), contribute maximally to the path integral. This is the corresponding result for the matrix eigenvalue distributions.

4.2   Validation on Synthetic Data

We generate random, diagonally dominant positive semi-definitive matrices, M, which are constructed as

$$M = \frac{A^\top A}{||A^\top A||_2} + I_N, \tag{9}$$

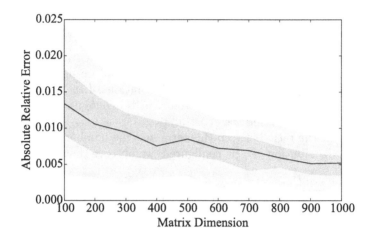

Fig. 1. The absolute relative error with respect to matrix dimensionality. Plotted is the median error, with the 30–70 and 10–90 percentile regions shaded in dark and light blue respectively. (Color figure online)

where $A \in \mathbb{R}^{N \times N}$ is an $N \times N$ matrix filled with Gaussian random variables and I is the identity. In order to test the hypothesis that the maximum entropy solution dominates the space of possible solutions with increasing matrix size, we investigate the relative error of the log determinant L, for $100 \leq N \leq 1000$. As can be seen in Fig. 1, there is a clear decrease in relative error for all plotted percentiles with increasing matrix size N.

5   Experiments

So far, we have supplemented the theoretic foundations of our proposal by devising experiments on synthetically constructed matrices. In this section, we extend our evaluation to include real matrices obtained from a variety of problem domains, and demonstrate how the results obtained using our approach consistently outperform competing state-of-the-art approximations. Moreover, in order to demonstrate the applicability of our method within a practical domain, we highlight the benefits of replacing the exact computation of the log determinant term appearing in the log likelihood of a Gaussian Markov random field with our maximum entropy approximation.

5.1   SparseSuite Matrix Collection

While the ultimate goal of this work is to accelerate inference of large-scale machine learning algorithms burdened by the computation of the log determinant, this is a general approach which can be applied to a wide variety of application domains. The SuiteSparse Matrix Collection [11] (commonly referred to as the set of UFL datasets) is a collection of sparse matrices obtained from

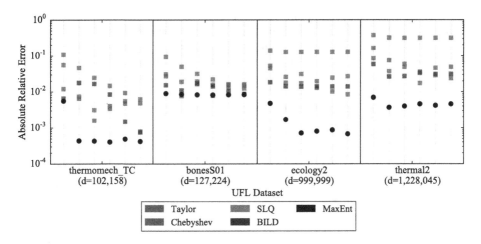

Fig. 2. Comparison of competing approaches over four UFL datasets. Results are also shown for increasing computational budgets, i.e. 5, 10, 15, 20, 25 and 30 moments respectively. Our method obtains substantially lower error rates across 3 out of 4 datasets, and still performs very well on 'bonesS01'.

various real problem domains. In this section, we shall consider a selection of these matrices as 'matrices in the wild' for comparing our proposed algorithm against established approaches. In this experiment we compare against Taylor [2] and Chebyshev [20] approximations, stochastic lanczos quadrature (SLQ) [35] and Bayesian inference of log determinants (BILD) [12]. In Fig. 2, we report the absolute relative error of the approximated log determinant for each of the competing approaches over four different UFL datasets. Following [12], we assess the performance of each method for an increasing computational budget, in terms of matrix vector multiplications, which in this case corresponds to the number of moments considered. It can be immediately observed that our entropic approach vastly outperforms the competing techniques across all datasets, and for any given computational budget. The overall accuracy also appears to consistently improve when more moments are considered.

Complementing the previous experiment, Table 2 provides a further comparison on a range of other sample matrices which are large, yet whose determinants can be computed by standard machines in reasonable time (by virtue of being sparse). For this experiment, we consider 10 estimated moments using 30 probing vectors, and their results are reported for the aforementioned techniques. The results presented in Table 2 are the relative error of the log determinants *after* they have been normalised using Gershgorin intervals [14]. We note, however, that the methods improve at different rates as more raw moments are taken.

5.2   Computation of GMRF Likelihoods

Gaussian Markov random fields (GMRFs) [32] specify spatial dependence between nodes of a graph with Markov properties, where each node denotes a

Table 2. Comparison of competing approximations to the log determinant over additional sparse UFL datasets. The technique yielding the lowest relative error is highlighted in bold, and our approach is consistently superior to the alternatives. Approximations are computed using 10 moments estimated with 30 probing vectors.

Dataset	Dimension	Taylor	Chebyshev	SLQ	BILD	MaxEnt
Shallow_water1	81,920	**0.0023**	0.7255	0.0058	0.0163	0.0030
Shallow_water2	81,920	0.5853	0.9846	0.9385	1.1054	**0.0051**
Apache1	80,800	0.4335	0.0196	0.4200	0.1117	**0.0057**
Finan512	74,752	0.1806	0.1158	0.0142	**0.0005**	0.0171
Obstclae	40,000	0.0503	0.5269	0.0423	0.0733	**0.0026**
Jnlbrng1	40,000	0.1084	0.2079	0.0465	0.0805	**0.0158**

random variable belonging to a multivariate joint Gaussian distribution defined over the graph. These models appear in a wide variety of applications, ranging from interpolation of spatio-temporal data to computer vision and information retrieval. While we refer the reader to [32] for a more comprehensive review of GMRFs, we highlight the fact that the model relies on a positive-definite precision matrix Q_θ parameterised by θ, which defines the relationship between connected nodes; given that not all nodes in the graph are connected, we can generally expect this matrix to be sparse. Nonetheless, parameter optimisation of a GMRF requires maximising the following equation:

$$\log p(\mathbf{x} \mid \theta) = \frac{1}{2} \log\big(\mathrm{Det}(Q_\theta)\big) - \frac{1}{2}\mathbf{x}^\top Q_\theta \mathbf{x} - \frac{n}{2} \log(2\pi),$$

where computing the log determinant poses a computational bottleneck, even where Q_θ is sparse. This arises because it is possible for the Cholesky decomposition of a sparse matrix with zeros outside a band of size k to be nonetheless dense *within* that bound. Thus, the Cholesky decomposition is still expensive to compute.

Following the experimental set-up and code provided in [19], in this experiment we evaluate how incorporating our approximation into the log likelihood term of a GMRF improves scalability when dealing with large matrices, while still maintaining precision. In particular, we construct lattices of increasing dimensionality and in each case measure the time taken to compute the log likelihood term using both approaches. The precision kernel is parameterised by κ and τ [23], and is explicitly linked to the spectral density of the Matérn covariance function for a given smoothness parameter. We repeat this evaluation for the case where a nugget term, which denotes the variance of the non-spatial error, in included in the constructed GMRF model. Note that for the maximum entropy approach we employ 30 sample vectors in the stochastic trace estimation procedure, and consider 10 moments. As illustrated in Fig. 3, the computation of the log likelihood is orders of magnitude faster when computing the log determinant using our proposed maximum entropy approach. In line with our expectations,

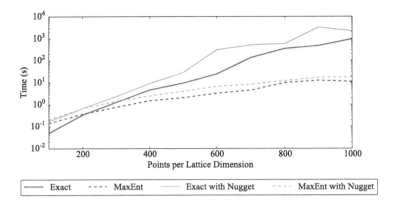

Fig. 3. Time in seconds for computing the log likelihood of a GMRF via Cholesky decomposition or using our proposed MaxEnt approach for estimating the log determinant term. Results are shown for GMRFs constructed on square lattices with increasing dimensionality, with and without a nugget term.

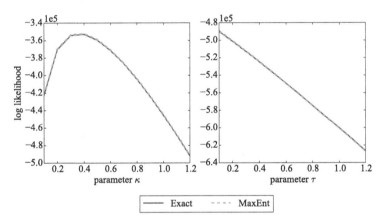

Fig. 4. The above plots indicate the difference of log likelihood between exact computation of the likelihood and the maximum entropy approach for a range of hyperparameters of the model. We note that the extrema of both exact and approximate inference align and it is difficult to distinguish the two lines.

this speed-up is particularly significant for larger matrices. Similar improvements are observed when a nugget term is included. Note that we set $\kappa = 0.1$ and $\tau = 1$ for this experiment.

Needless to say, improvements in computation time mean little if the quality of inference degrades. Figure 4 illustrates the comparable quality of the log likelihood for various settings of κ and τ, and the results confirm that our method enables faster inference without compromising on performance.

6   Conclusion

Inspired by the probabilistic interpretation introduced in [12], in this work we have developed a novel approximation to the log determinant which is rooted in information theory. While lacking the uncertainty quantification inherent to the aforementioned technique, this formulation is appealing because it uses a comparatively less informative prior on the distribution of eigenvalues, and we have also demonstrated that the method is theoretically expected to yield superior approximations for matrices of very large dimensionality. This is especially significant given that the primary scope for undertaking this work was to accelerate the log determinant computation in large-scale inference problems. As illustrated in the experimental section, the proposed approach consistently outperforms all other state-of-the-art approximations by a sizeable margin.

Future work will include incorporating the empirical Monte Carlo variance of the stochastic trace estimates into the inference scheme, extending the method of maximum entropy to include noisy constraints, and explicitly evaluating the ratio of the functional integrals for large matrices to obtain uncertainty estimates similar to those in [12]. We hope that the combination of these advancements will allow for an apt active sampling procedure given pre-specified computational budgets.

References

1. Atallah, M.J., Grigorescu, E., Wu, Y.: A lower-variance randomized algorithm for approximate string matching. Inf. Process. Lett. **113**(18), 690–692 (2013)
2. Aune, E., Simpson, D.P., Eidsvik, J.: Parameter estimation in high dimensional Gaussian distributions. Stat. Comput. **24**(2), 247–263 (2014)
3. Avron, H.: Counting triangles in large graphs using randomized matrix trace estimation. In: Workshop on Large-Scale Data Mining: Theory and Applications, vol. 10, pp. 10–9 (2010)
4. Avron, H., Toledo, S.: Randomized algorithms for estimating the trace of an implicit symmetric positive semi-definite matrix. J. ACM (JACM) **58**(2), 8:1–8:34 (2011)
5. Bai, Z., Golub, G.H.: Bounds for the trace of the inverse and the determinant of symmetric positive definite matrices. Ann. Numer. Math. **4**, 29–38 (1997)
6. Bandyopadhyay, K., Bhattacharya, A.K., Biswas, P., Drabold, D.: Maximum entropy and the problem of moments: a stable algorithm. Phys. Rev. E **71**(5), 057701 (2005)
7. Boutsidis, C., Drineas, P., Kambadur, P., Zouzias, A.: A randomized algorithm for approximating the log determinant of a symmetric positive definite matrix. CoRR abs/1503.00374 (2015)
8. Buchen, P.W., Kelly, M.: The maximum entropy distribution of an asset inferred from option prices. J. Financ. Quant. Anal. **31**(01), 143–159 (1996)
9. Caticha, A.: Entropic Inference and the Foundations of Physics (monograph commissioned by the 11th Brazilian Meeting on Bayesian Statistics-EBEB-2012) (2012)
10. Chen, J., Anitescu, M., Saad, Y.: Computing f(A)b via least squares polynomial approximations. SIAM J. Sci. Comput. **33**(1), 195–222 (2011)

11. Davis, T.A., Hu, Y.: The University of Florida sparse matrix collection. ACM Trans. Math. Softw. (TOMS) **38**(1), 1 (2011)
12. Fitzsimons, J., Cutajar, K., Osborne, M., Roberts, S., Filippone, M.: Bayesian inference of log determinants (2017)
13. Fitzsimons, J., Osborne, M., Roberts, S., Fitzsimons, J.: Improved stochastic trace estimation using mutually unbiased bases. arXiv preprint arXiv:1608.00117 (2016)
14. Gershgorin, S.: Uber die Abgrenzung der Eigenwerte einer matrix. Izvestija Akademii Nauk SSSR Serija Matematika **7**(3), 749–754 (1931)
15. Giffin, A., Cafaro, C., Ali, S.A.: Application of the maximum relative entropy method to the physics of ferromagnetic materials. Phys. A: Stat. Mech. Appl. **455**, 11–26 (2016). http://www.sciencedirect.com/science/article/pii/S03784371 16002478
16. Golub, G.H., Van Loan, C.F.: Matrix Computations, 3rd edn. The Johns Hopkins University Press, Baltimore (1996)
17. González, D., Davis, S., Gutiérrez, G.: Newtonian dynamics from the principle of maximum caliber. Found. Phys. **44**(9), 923–931 (2014)
18. Granziol, D., Roberts, S.: An information and field theoretic approach to the grand canonical ensemble (2017)
19. Guinness, J., Ipsen, I.C.F.: Efficient computation of Gaussian likelihoods for stationary Markov random fields (2015)
20. Han, I., Malioutov, D., Shin, J.: Large-scale log-determinant computation through stochastic Chebyshev expansions. In: Bach, F.R., Blei, D.M. (eds.) Proceedings of the 32nd International Conference on Machine Learning, ICML 2015, Lille, France, 6–11 July 2015
21. Hutchinson, M.: A stochastic estimator of the trace of the influence matrix for Laplacian smoothing splines. Commun. Stat. - Simul. Comput. **19**(2), 433–450 (1990)
22. Jaynes, E.T.: Information theory and statistical mechanics. Phys. Rev. **106**, 620–630 (1957). http://link.aps.org/doi/10.1103/PhysRev.106.620
23. Lindgren, F., Rue, H., Lindström, J.: An explicit link between Gaussian fields and Gaussian Markov random fields: the stochastic partial differential equation approach. J. Roy. Stat. Soc.: Ser. B (Stat. Methodol.) **73**(4), 423–498 (2011)
24. Macchi, O.: The coincidence approach to stochastic point processes. Adv. Appl. Probab. **7**, 83–122 (1975)
25. Meurer, A., Smith, C.P., Paprocki, M., Čertík, O., Kirpichev, S.B., Rocklin, M., Kumar, A., Ivanov, S., Moore, J.K., Singh, S., Rathnayake, T., Vig, S., Granger, B.E., Muller, R.P., Bonazzi, F., Gupta, H., Vats, S., Johansson, F., Pedregosa, F., Curry, M.J., Terrel, A.R., Roučka, Š., Saboo, A., Fernando, I., Kulal, S., Cimrman, R., Scopatz, A.: SymPy: symbolic computing in Python. PeerJ Comput. Sci. **3**, e103 (2017). https://doi.org/10.7717/peerj-cs.103
26. Mohammad-Djafari, A.: A Matlab program to calculate the maximum entropy distributions. In: Smith, C.R., Erickson, G.J., Neudorfer, P.O. (eds.) Maximum Entropy and Bayesian Methods. FTPH, vol. 50, pp. 221–233. Springer, Dordrecht (1992). https://doi.org/10.1007/978-94-017-2219-3_16
27. Neri, C., Schneider, L.: Maximum entropy distributions inferred from option portfolios on an asset. Finan. Stochast. **16**(2), 293–318 (2012)
28. Nielsen, M.A., Chuang, I.: Quantum computation and quantum information (2002)
29. O'Hagan, A.: Bayes-Hermite quadrature. J. Stat. Plan. Infer. **29**, 245–260 (1991)
30. Pressé, S., Ghosh, K., Lee, J., Dill, K.A.: Principles of maximum entropy and maximum caliber in statistical physics. Rev. Mod. Phys. **85**, 1115–1141 (2013). http://link.aps.org/doi/10.1103/RevModPhys.85.1115

31. Rasmussen, C.E., Williams, C.: Gaussian Processes for Machine Learning. MIT Press, Cambridge (2006)
32. Rue, H., Held, L.: Gaussian Markov Random Fields: Theory and Applications, Monographs on Statistics and Applied Probability, vol. 104. Chapman & Hall, London (2005)
33. Saatçi, Y.: Scalable inference for structured Gaussian process models. Ph.D. thesis, University of Cambridge (2011)
34. Shore, J., Johnson, R.: Axiomatic derivation of the principle of maximum entropy and the principle of minimum cross-entropy. IEEE Trans. Inf. Theory **26**(1), 26–37 (1980)
35. Ubaru, S., Chen, J., Saad, Y.: Fast estimation of tr($F(A)$) via stochastic Lanczos quadrature (2016)
36. Wainwright, M.J., Jordan, M.I.: Log-determinant relaxation for approximate inference in discrete Markov random fields. IEEE Trans. Signal Process. **54**(6–1), 2099–2109 (2006)
37. Wolfram Research Inc.: Mathematica. https://www.wolfram.com/mathematica/

Fair Kernel Learning

Adrián Pérez-Suay[✉], Valero Laparra, Gonzalo Mateo-García,
Jordi Muñoz-Marí, Luis Gómez-Chova, and Gustau Camps-Valls

Image Processing Laboratory (IPL), Universitat de València, Spain,
C/Cat. José Beltrán, 2, 46980 Paterna, València, Spain
Adrian.Perez@uv.es
http://isp.uv.es

Abstract. New social and economic activities massively exploit big data
and machine learning algorithms to do inference on people's lives. Applications include automatic curricula evaluation, wage determination, and
risk assessment for credits and loans. Recently, many governments and
institutions have raised concerns about the lack of fairness, equity and
ethics in machine learning to treat these problems. It has been shown
that not including sensitive features that bias fairness, such as gender
or race, is not enough to mitigate the discrimination when other related
features are included. Instead, including fairness in the objective function has been shown to be more efficient.

We present novel fair regression and dimensionality reduction methods built on a previously proposed fair classification framework. Both
methods rely on using the Hilbert Schmidt independence criterion as the
fairness term. Unlike previous approaches, this allows us to simplify the
problem and to use multiple sensitive variables simultaneously. Replacing the linear formulation by kernel functions allows the methods to deal
with nonlinear problems. For both linear and nonlinear formulations the
solution reduces to solving simple matrix inversions or generalized eigenvalue problems. This simplifies the evaluation of the solutions for different trade-off values between the predictive error and fairness terms. We
illustrate the usefulness of the proposed methods in toy examples, and
evaluate their performance on real world datasets to predict income using
gender and/or race discrimination as sensitive variables, and contraceptive method prediction under demographic and socio-economic sensitive
descriptors.

Keywords: Equity · Fairness · Machine learning · Regression
Dimensionality reduction · Kernel methods

1 Introduction

"Perfect objectivity is an unrealistic goal; fairness, however, is not."
–M. Pollan, 2004

The research was funded by the Spanish Ministry of Economy and Competitiveness
(MINECO) through the projects TIN2015-64210-R and TEC2016-77741-R (ERDF).

© Springer International Publishing AG 2017
M. Ceci et al. (Eds.): ECML PKDD 2017, Part I, LNAI 10534, pp. 339–355, 2017.
https://doi.org/10.1007/978-3-319-71249-9_21

Current and upcoming application of machine learning to real-life's problems is overwhelming. Applications have enormous consequences in people's life, and impact decisions on education, economy, health care, and climate policies. The issue is certainly relevant. New social and economic activities massively exploit big data and machine learning algorithms to do inferences, and they decide on the best curriculum to fill in a position [15], to determine wages and in pre-trial risk assessment [4,9], and to evaluate risk of violence [8]. Companies, governments and institutions have raised concerns about the lack of fairness, equity and ethics in machine learning to treat this kind of problems[1]. Machine learning methods are actually far from being fair, just, or equitable in any way. After all, standard pattern analysis is often about model fitting and not the gender issue. Undoubtedly, attaining fair machine learning algorithms is a timely important concern. Fairness is an elusive concept though, so it is the inclusion of such *qualitative* measure in machines that only learn from data.

Several approaches exist in the literature to account for fairness in machine learning. One of the earliest approaches tackled the bias problem through the definition of classification rules [22,24]. Later, some other works focused on (mainly) pre-processing the data [12,16,21,23]: down-weighting sensitive features or directly removing them have been the preferred choices. Perhaps the most naive approach is to simply discard the *sensitive* input features that bias discrimination [31]. Removing gender, disability or race, to predict monthly income is, however, not a good choice because model's accuracy may be largely impacted by the lack of informative features, and because some other correlated features enter the model anyway. This effect is known in statistics as the *omitted variable bias* [6].

Another simple approach consists on including *ad hoc* weights and data normalization to match the prior belief about fairness. Noting that data pre-processing is a quite arbitrary approach, Kamiran and Calders et al. proposed three solutions to learn fair classifiers [17]. Classifiers basically used the sensitive features only during learning and not at the prediction time. A step forward in this direction was presented in [12], where authors proposed pre-processing the data by removing information from all attributes correlated with the objective variable. The intuition behind this approach is that training on discrimination-free data is likely to yield more equitable predictions. A discussion of several more algorithms for binary protected and outcome variables can be found in [18]. Other authors have focused on finding transformations of the input space in order to extract features that do not retain information about the sensitive input variables [30].

All in all, the relevance of fair methods in machine learning is ever increasing, and a wide body of literature and approaches exist. We focus in this paper in a field known as 'disparate impact', in which outcomes should not differ based on individuals' protected class membership. Many definitions for the elusive concept of fairness in machine learning are available (see [3,5,7,11,12,16,22,29]): redlining, negative legacy, underestimation or subset targeting, to name a few. We frame our methods in the 'indirect discrimination' subfield.

---

[1] http://www.fatml.org/.

Recently, an interesting regularization framework for fair classification was proposed in [19]. The framework optimizes a functional that jointly minimizes the classification error and the dependence between predictions and the sensitive variables using mutual-information concepts. We build our proposal upon this framework, and extend it to regression, and to unsupervised dimensionality reduction problems with kernel methods. The proposed kernel machines exploit cross-covariance operators in Hilbert spaces. Both theoretical and empirical advantages are gained. Advantageously, the solutions only involve solving simple matrix inversion or generalized eigenproblems. This allows to check different solutions when the trade-off between prediction and fairness is modified. The methods are able to deal naturally with input variables of several dimensions for the regular as well as for the sensitive variables. Note that this is especially important for the fairness term, where a robust measure of dependence is needed. On top of this, the proposed methods can incorporate prior knowledge about the fairness, invariances and interestingness of the feature relations. We illustrate performance in toy data as well as in two real problems: income prediction subject to gender and/or race discrimination, and contraceptive method prediction under demographic and socio-economic sensitive descriptors.

The remainder of the paper is organized as follows. Section 2 describes the problem statement, introduces notation and presents the fair kernel regression framework in the input and the Hilbert space. Section 3 extends the fair kernel learning framework to dimensionality reduction problems. Toy examples guide the presentation of the two approaches. Experimental evidence of performance is given in Sect. 4. Conclusions finalize the paper in Sect. 5.

2 Fair Regression Framework

This section starts by defining the notation and the concept of fair predictions. Then we introduce the proposed framework for performing fair regression learning based on cross-covariance operators for dependence estimation in Hilbert spaces. We conclude with an illustrative example.

2.1 Notation, Preliminaries, and the Regularization Framework

Let us define the notation and the problem setting. We are given n samples of a response (or target) data matrix $\mathbf{Y} \in \mathbb{R}^{n \times c}$, and $d + q$ prediction variables: d unprotected $\mathbf{X}_u \in \mathbb{R}^{n \times d}$ and q sensitive $\mathbf{S} \in \mathbb{R}^{n \times q}$. The goal is to obtain an accurate prediction function (or model) f for the target variable $\mathbf{Y}$ from the input data, $\mathbf{X} = (\mathbf{X}_u, \mathbf{S})$. This function is said to be *totally fair* if the predictions are statistically independent of the sensitive (protected) features [5,10].

Therefore, two main ingredients are needed to perform fair predictions: we need to ensure independence of the predictions on the sensitive variables, and simultaneously to obtain a good approximation of the target variables. The regularization framework proposed in [19] tackles the problem of finding a fair function f for classification by including a term to enforce fair classification. In our proposal the proposed function f tries to learn the relation between observed input-output data pairs $(\mathbf{x}_1, \mathbf{y}_1), \ldots, (\mathbf{x}_n, \mathbf{y}_n) \in \mathcal{X} \times \mathcal{Y}$ such that generalizes well (good predictions $\hat{\mathbf{y}}_* = f(\mathbf{x}_*) \in \mathcal{Y}$ for the unseen input data point $\mathbf{x}_* \in \mathcal{X}$), and the predictions should be as independent as possible of the sensitive features. Then, the following functional needs to be optimized:

$$\mathcal{L} = \frac{1}{n} \sum_{i=1}^{n} V(f(\mathbf{x}_i), \mathbf{y}_i) + \lambda \, \Omega(\|f\|_{\mathcal{H}}) + \mu \, I(f(\mathbf{x}), \mathbf{s}), \tag{1}$$

where V is the error cost function, $\Omega(\|f\|_{\mathcal{H}})$ acts as a regularizer of the predictive function and controls the smoothness and complexity of the model, and $I(f(\mathbf{x}), \mathbf{s})$ measures the independence between model's predictions and the protected variables. Note that one aims to minimize the amount of information that the model shares with the sensitive variables while controlling the trade-off between fitting and independence through hyperparameters λ and μ. By setting $\mu = 0$ one obtains the ordinary Tikhonov's regularized functional, and by setting $\lambda = 0$ one obtains the unregularized versions of this framework.

The framework admits many variants depending of the cost function V, regularizer Ω and the independence measure, I. For example, in [19], the function f was the logistic regression classifier and I was a simplification of the mutual information estimate. Despite the good results reported in [19], these choices did not allow to solve the problem in closed-form, nor coping with more than one sensitive variable at the same time, since the proposed mutual information is an uni-dimensional dependence measure. In the following section, we elaborate further this framework under the concept of cross-covariance operators in Hilbert spaces, which lead to closed-form solutions and allow to deal with several sensitive variables simultaneously.

2.2   Fair Linear Regression

Let us now provide a straightforward instantiation of the proposed framework for fair linear regression (FLR). We will adopt a linear predictive model for f, i.e. the matrix of predictions for a test data matrix $\mathbf{X}_*$ is given by $\hat{\mathbf{Y}}_* = \mathbf{X}_* \mathbf{W}$, the mean square error for the cost function $V = \|\mathbf{Y} - \mathbf{X}\mathbf{W}\|_2^2$ and the standard ℓ_2 regularization for model weights $\Omega := \|\mathbf{W}\|_2^2$. Other choices could be taken, leading to alternative formulations. In order to measure dependence, we will rely on the cross-covariance operator between the predictions and the sensitive variables in Hilbert space. Let us consider two spaces $\mathcal{Y} \subseteq \mathbb{R}^c$ and $\mathcal{S} \subseteq \mathbb{R}^q$, where random variables $(\hat{\mathbf{y}}, \mathbf{s})$ are sampled from the joint distribution $\mathbb{P}_{\mathbf{ys}}$. Given a set of pairs $\mathcal{D} = \{(\hat{\mathbf{y}}_1, \mathbf{s}_1), \ldots, (\hat{\mathbf{y}}_n, \mathbf{s}_n)\}$ of size n drawn from $\mathbb{P}_{\mathbf{ys}}$, an empirical estimator of HSIC [14] allows us to define

$$I := \mathrm{HSIC}(\mathcal{Y}, \mathcal{S}, \mathbb{P}_{\mathbf{ys}}) = \|\mathbf{C}_{ys}\|_{\mathrm{HS}}^2 = \|\tilde{\mathbf{Y}}^\top \tilde{\mathbf{S}}\|^2 = \frac{1}{n^2} \mathrm{Tr}(\tilde{\mathbf{Y}}^\top \tilde{\mathbf{S}} \tilde{\mathbf{S}}^\top \tilde{\mathbf{Y}}),$$

where $\| \cdot \|_{\mathrm{HS}}$ is the Hilbert-Schmidt norm, $\mathbf{C}_{ys}$ is the empirical cross-covariance matrix between predictions and sensitive variables[2], $\tilde{\mathbf{Y}}$ and $\tilde{\mathbf{S}}$ represent the feature-centered $\mathbf{Y}$ and $\mathbf{S}$ respectively, and $\mathrm{Tr}(\cdot)$ denotes the trace of the matrix. We want to stress that HSIC allows us to estimate dependencies between multi-dimensional variables, and that HSIC is zero if an only if there is no second-order dependence between $\hat{\mathbf{y}}$ and $\mathbf{s}$. In the next section we extend the formulation to higher-order dependencies with the use of kernels [25, 26].

Plugging these definitions of f, V, Ω and I in Eq. (1), one can easily show that the solution has the following closed-form solution for weight estimates

$$\widehat{\mathbf{W}} = (\tilde{\mathbf{X}}^\top \tilde{\mathbf{X}} + \lambda\,\mathbf{I} + \frac{\mu}{n^2}\,\tilde{\mathbf{X}}^\top \tilde{\mathbf{S}}\tilde{\mathbf{S}}^\top \tilde{\mathbf{X}})^{-1}\tilde{\mathbf{X}}^\top \mathbf{Y}, \tag{2}$$

where fairness is trivially controlled with μ, which acts as an additional regularization term. Also note that when $\mu = 0$ the ordinary regularized least squares solution is obtained.

2.3   Fair Kernel Regression

Let us now extend the previous model to the nonlinear case in terms of the prediction function, the regularizer and the dependence measure by means of reproducing kernels [25, 26]. We call this method the fair kernel regression (FKR) model. We proceed in the standard way in kernel machines by mapping data $\mathbf{X}$ and $\mathbf{S}$ to a Hilbert space $\mathcal{H}$ via the mapping functions $\phi(\cdot)$ and $\psi(\cdot)$ respectively. This yields $\boldsymbol{\Phi}, \boldsymbol{\Psi} \in \mathcal{H} \subseteq \mathbb{R}^{d_\mathcal{H}}$, where $d_\mathcal{H}$ is the (unknown and possibly infinite) dimensionality of mapped points in $\mathcal{H}$. The corresponding kernel matrices can be defined as: $\tilde{\mathbf{K}} = \tilde{\boldsymbol{\Phi}}\tilde{\boldsymbol{\Phi}}^\top$ and $\tilde{\mathbf{K}}_S = \tilde{\boldsymbol{\Psi}}\tilde{\boldsymbol{\Psi}}^\top$. Now the prediction function is $\hat{\mathbf{Y}} = \boldsymbol{\Phi}\mathbf{W}_\mathcal{H}$, the regularizer is $\Omega := \|\mathbf{W}_\mathcal{H}\|_2^2$, and the dependence measure I is the HSIC estimate between predictions $\hat{\mathbf{Y}}$ and sensitive variables $\mathbf{S}$, which can now be estimated in Hilbert spaces: $I := \mathrm{HSIC}(\mathcal{Y}, \mathcal{H}, \mathbb{P}_{\mathbf{ys}}) = \|\mathbf{C}_{ys}\|_{\mathrm{HS}}^2$. Now, by plugging all these terms in the cost function, using the representer's theorem $\mathbf{W}_\mathcal{H} = \tilde{\boldsymbol{\Phi}}^\top \boldsymbol{\Lambda}$ and after some simple linear algebra, we obtain the dual weights in closed-form

$$\boldsymbol{\Lambda} = (\tilde{\mathbf{K}} + \lambda\mathbf{I} + \frac{\mu}{n^2}\tilde{\mathbf{K}}\tilde{\mathbf{K}}_S)^{-1}\mathbf{Y}, \tag{3}$$

which can be used for prediction with a new point $\mathbf{x}_*$ by using $\hat{\mathbf{y}}_* = \mathbf{k}_*\boldsymbol{\Lambda}$, where $\mathbf{k}_* = [K(\mathbf{x}_*, \mathbf{x}_1), \ldots, K(\mathbf{x}_*, \mathbf{x}_n)]^\top$. In the case where $\mu = 0$ the method reduces to standard kernel ridge regression (KRR) method [26]. Centering points in feature spaces can be done implicitly with kernels [26]: a kernel matrix $\mathbf{K}$ is centered by doing $\tilde{\mathbf{K}} = \mathbf{H}\mathbf{K}\mathbf{H}$, where $\mathbf{H} = \mathbf{I} - \frac{1}{n}\mathbf{1}\mathbf{1}^\top$.

Lemma 1. *Both KRR and FKR model weights are bounded in norm by the same quantitiy.*

---

[2] The covariance matrix is $\mathcal{C}_{\mathbf{ys}} = \mathbb{E}_{\mathbf{ys}}(\mathbf{ys}^\top) - \mathbb{E}_{\mathbf{y}}(\mathbf{y})\mathbb{E}_{\mathbf{s}}(\mathbf{s}^\top)$, where $\mathbb{E}_{\mathbf{ys}}$ is the expectation with respect to $\mathbb{P}_{\mathbf{ys}}$, and $\mathbb{E}_{\mathbf{y}}$ is the marginal expectation with respect to $\mathbb{P}_{\mathbf{y}}$ (hereafter we assume that all these quantities exist).

Proof. Let us assume the same kernel matrix $\tilde{\mathbf{K}}$ for KRR and FKR, and also suppose $\lambda, \mu \geq 0$, then the following bound is satisfied: $\|(\tilde{\mathbf{K}}+\lambda\mathbf{I}+\frac{\mu}{n^2}\tilde{\mathbf{K}}_S\tilde{\mathbf{K}})^{-1}\| \leq \|(\tilde{\mathbf{K}}+\lambda\mathbf{I})^{-1}\|$. Given $\mu \geq 0$ the following inequality, with $\succeq$ meaning the standard PSD order, holds true: $\tilde{\mathbf{K}} + \lambda\mathbf{I} + \frac{\mu}{n^2}\tilde{\mathbf{K}}_S\tilde{\mathbf{K}} \succeq \tilde{\mathbf{K}} + \lambda\mathbf{I}$. Then also holds $(\tilde{\mathbf{K}} + \lambda\mathbf{I} + \frac{\mu}{n^2}\tilde{\mathbf{K}}_S\tilde{\mathbf{K}})^{-1} \preceq (\tilde{\mathbf{K}} + \lambda\mathbf{I})^{-1}$, and by taking norms we have the following inequality $\|(\tilde{\mathbf{K}} + \lambda\mathbf{I} + \frac{\mu}{n^2}\tilde{\mathbf{K}}_S\tilde{\mathbf{K}})^{-1}\| \leq \|(\tilde{\mathbf{K}} + \lambda\mathbf{I})^{-1}\|$. FKR model weights can be bounded

$$
\begin{aligned}
\|\mathbf{\Lambda}_{\mathrm{FKR}}\| &= \left\|(\tilde{\mathbf{K}} + \lambda\mathbf{I} + \frac{\mu}{n^2}\tilde{\mathbf{K}}_S\tilde{\mathbf{K}})^{-1}\mathbf{Y}\right\| \leq \left\|(\tilde{\mathbf{K}} + \lambda\mathbf{I} + \frac{\mu}{n^2}\tilde{\mathbf{K}}_S\tilde{\mathbf{K}})^{-1}\right\| \|\mathbf{Y}\| \\
&\leq \left\|(\tilde{\mathbf{K}} + \lambda\mathbf{I})^{-1}\right\| \|\mathbf{Y}\|,
\end{aligned}
\tag{4}
$$

which is the same bound for KRR weights:

$$
\|\mathbf{\Lambda}_{\mathrm{KRR}}\| = \left\|(\tilde{\mathbf{K}} + \lambda\mathbf{I})^{-1}\mathbf{Y}\right\| \leq \left\|(\tilde{\mathbf{K}} + \lambda\mathbf{I})^{-1}\right\| \|\mathbf{Y}\|.
$$

Illustrative example. Here we illustrate the performance of the proposed methods in a controlled synthetic experiment. The data considers a sensitive variable drawn from a zero mean Gaussian with standard deviation σ_s, $\mathbf{s} \sim \mathcal{N}(0, \sigma_s)$, and a parametric function $f_s(\mathbf{s})$ that yields an intermediate variable $\mathbf{a}$ buried in additive white Gaussian noise (AWGN), i.e. $\mathbf{a} = f_s(\mathbf{s}) + \mathbf{n}_f$, where $\mathbf{n}_f \sim \mathcal{N}(0, \sigma_f)$. System's output combines both the sensitive as well as its arbitrarily transformed version affected by AWGN $\mathbf{y} = g_s(\mathbf{s}) + g_r(\mathbf{a}) + \mathbf{n}_y$, where $\mathbf{n}_y \sim \mathcal{N}(0, \sigma_y)$. In this example we used $f_s(x) = \log(x)$, and $g_r(x) = g_s(x) = x^2$. This system ensures that, even without using variable $\mathbf{s}$ explicitly as an input factor in the regression model, the information conveyed in $\mathbf{s}$ is embedded in $\mathbf{a}$ indirectly. Two settings are considered, with and without using the sensitive variable $\mathbf{s}$ as an input feature. In both experiments we used the RBF kernel function and fitted hyperparameters (λ, μ and the kernel widths) to be optimal for each μ value. Figure 1 shows the results for the four different configurations (linear and non-linear, with and without considering $\mathbf{s}$), the horizontal axis represents the mean square error (MSE) of the prediction and the vertical axis the HSIC between the prediction and the protected variable. An ideal fair model would obtain zero MSE and zero HSIC. For each configuration, we give the family of solutions that can be obtained by modifying the parameter μ. Classical solutions that do not include the fairness term show that KRR improves the ordinary LR results in MSE terms, but both methods obtain similar HSIC values. On the other hand, the inclusion of the sensitive variable $\mathbf{s}$ as input feature obtains more fair results in HSIC terms but worst results in MSE terms. The fairness paths are obtained for different μ values. The nonlinear regression methods outperform in general the linear counterparts. Including the sensitive variable as input returns better trade-off results. For example, the FKR can be tuned to have the same fairness level as KRR\S but obtaining around 30% lower prediction error. A similar conclusion can be extracted in the linear case, yet the improvement is smaller.

3   Fair Dimensionality Reduction Framework

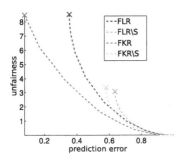

Fig. 1. Regression curves of the prediction error (MSE) versus the unfairness (independence of predictions with sensitive variables) in four different configurations (FLR and FKR, with and without the sensitive variable **s**) and different values of μ (crosses indicate $\mu = 0$ solutions).

Let us now show a different frame for fair machine learning. Rather than optimizing a regression model, we are here concerned about obtaining fair feature representations. We rely on the field of multivariate analysis to develop both linear and nonlinear (kernel) dimensionality reduction methods.

3.1   Fair Dimensionality Reduction

Let us define two training data matrices as before, the full design matrix $\mathbf{X} \in \mathbb{R}^{n \times d}$, and the *sensitive* data matrix $\mathbf{S} \in \mathbb{R}^{n \times q}$, and a labelled data matrix $\mathbf{Y} \in \mathbb{R}^{n \times c}$ (here we use 1-of-c encoding). The goal here is to find a *fair projection matrix* $\mathbf{V} \in \mathbb{R}^{d \times n_p}$ such that the projected data, $\mathbf{XV}$ keeps as much information as possible from the input features yet minimally aligned with the protected, sensitive features. We denoted $\mathbf{V} = [\mathbf{v}_1 | \cdots | \mathbf{v}_{n_p}]$, where $\mathbf{v}_i$ is the i-th projection vector and n_p is the dimension of the projection subspace. Hereafter, the terms alignment and statistical dependency will be used interchangeably. As before, in order to minimize alignment (dependence) between random variables $\mathbf{X}$ and $\mathbf{S}$, we will use the cross-covariance operator, whose empirical estimate reduces to compute the norm of the corresponding empirical cross-covariance given by the HSIC estimator. We will also use HSIC to maximize the dependence between the projected and the original data. The problem can thus be easily formalized as the maximization of the following Rayleigh quotient:

$$\mathbf{V}^* = \arg\max_{\mathbf{V}} \left\{ \frac{\text{HSIC}(\tilde{\mathbf{X}}\mathbf{V}, \tilde{\mathbf{X}})}{\text{HSIC}(\tilde{\mathbf{X}}\mathbf{V}, \tilde{\mathbf{S}})} \right\} = \arg\max_{\mathbf{V}} \left\{ \frac{\text{Tr}(\mathbf{V}^\top (\tilde{\mathbf{X}}^\top \tilde{\mathbf{X}}\tilde{\mathbf{X}}^\top \tilde{\mathbf{X}})\mathbf{V})}{\text{Tr}(\mathbf{V}^\top (\tilde{\mathbf{X}}^\top \tilde{\mathbf{S}}\tilde{\mathbf{S}}^\top \tilde{\mathbf{X}})\mathbf{V})} \right\},$$

where $\tilde{\mathbf{X}}$ represents the feature-centered $\mathbf{X}$. This leads to solving a generalized eigenvalue problem with the empirical covariance $\mathbf{C}_{xx} = \frac{1}{n}\tilde{\mathbf{X}}^\top \tilde{\mathbf{X}}$ and input-sensitive cross-covariance $\mathbf{C}_{xs} = \frac{1}{n}\tilde{\mathbf{X}}^\top \tilde{\mathbf{S}}$:

$$\mathbf{C}_{xx}\mathbf{C}_{xx}^\top \mathbf{v} = \lambda \mathbf{C}_{xs}\mathbf{C}_{xs}^\top \mathbf{v}.$$

The solution resembles that of the standard orthonormalized partial least squares (OPLS) [27]. Note that the generalized eigenproblem involves symmetric matrices. The matrix projection operator $\mathbf{V}$ can then been used to obtain *fair scores* for a new data point $\mathbf{x}_* \in \mathbb{R}^{d \times 1}$ as follows $\tilde{\mathbf{x}}'_* = \mathbf{V}^\top \mathbf{x}_* \in \mathbb{R}^{n_p \times 1}$.

3.2 Kernel Fair Dimensionality Reduction

Let us now derive a nonlinear version of FDR by means of reproducing kernels [25, 26]. We proceed in the standard way by mapping data $\mathbf{X}$ and $\mathbf{S}$ to a Hilbert space $\mathcal{H}$ via mapping functions $\phi(\cdot), \psi(\cdot)$, which yield $\boldsymbol{\Phi}, \boldsymbol{\Psi} \in \mathbb{R}^{n \times d_\mathcal{H}}$ respectively, where $d_\mathcal{H}$ is the dimensionality of $\mathcal{H}$. The FDR ratio now translates into finding a projection matrix $\mathbf{U} = [\mathbf{u}_1 | \cdots | \mathbf{u}_{n_p}] \in \mathbb{R}^{d_\mathcal{H} \times n_p}$ such that:

$$\mathbf{U}^* = \arg\max_{\mathbf{U}} \left\{ \frac{\mathrm{Tr}(\mathbf{U}^\top (\tilde{\boldsymbol{\Phi}}^\top \tilde{\boldsymbol{\Phi}} \tilde{\boldsymbol{\Phi}}^\top \tilde{\boldsymbol{\Phi}}) \mathbf{U})}{\mathrm{Tr}(\mathbf{U}^\top (\tilde{\boldsymbol{\Phi}}^\top \tilde{\boldsymbol{\Psi}} \tilde{\boldsymbol{\Psi}}^\top \tilde{\boldsymbol{\Phi}}) \mathbf{U})} \right\},$$

where $\tilde{\boldsymbol{\Phi}}$, and $\tilde{\boldsymbol{\Psi}}$ contain the *centered* data in Hilbert space. Now, by applying the representer's theorem $\mathbf{U} = \tilde{\boldsymbol{\Phi}}^\top \boldsymbol{\Lambda}$ (where $\boldsymbol{\Lambda} = [\boldsymbol{\alpha}_1 | \cdots | \boldsymbol{\alpha}_{n_p}]^\top$), replacing dot products with kernel functions, $\tilde{k}_x(\mathbf{x}, \mathbf{x}') = \tilde{\phi}(\mathbf{x})^\top \tilde{\phi}(\mathbf{x}')$, $\tilde{k}_s(\mathbf{s}, \mathbf{s}') = \tilde{\psi}(\mathbf{s})^\top \tilde{\psi}(\mathbf{s}')$, and defining kernel matrices, $\tilde{\mathbf{K}}_x = \tilde{\boldsymbol{\Phi}} \tilde{\boldsymbol{\Phi}}^\top$, and $\tilde{\mathbf{K}}_s = \tilde{\boldsymbol{\Psi}} \tilde{\boldsymbol{\Psi}}^\top$, we obtain a dual problem:

$$\boldsymbol{\Lambda}^* = \arg\max_{\boldsymbol{\Lambda}} \left\{ \frac{\mathrm{Tr}(\boldsymbol{\Lambda}^\top (\tilde{\mathbf{K}}_x \tilde{\mathbf{K}}_x \tilde{\mathbf{K}}_x) \boldsymbol{\Lambda})}{\mathrm{Tr}(\boldsymbol{\Lambda}^\top (\tilde{\mathbf{K}}_x \tilde{\mathbf{K}}_s \tilde{\mathbf{K}}_x) \boldsymbol{\Lambda})} \right\},$$

which reduces again to solving a generalized eigenproblem:

$$\tilde{\mathbf{K}}_x \tilde{\mathbf{K}}_x \boldsymbol{\alpha} = \lambda \tilde{\mathbf{K}}_s \tilde{\mathbf{K}}_x \boldsymbol{\alpha}.$$

This problem can be solved iteratively by first computing the leading pair $\{\lambda_i, \boldsymbol{\alpha}_i\}$, and then deflating the matrices. The deflation equation for KFDR can be written as:

$$\tilde{\mathbf{K}}_x \tilde{\mathbf{K}}_x \tilde{\mathbf{K}}_x \leftarrow \tilde{\mathbf{K}}_x \tilde{\mathbf{K}}_x \tilde{\mathbf{K}}_x - \lambda_i \tilde{\mathbf{K}}_x \tilde{\mathbf{K}}_s \tilde{\mathbf{K}}_x \boldsymbol{\alpha}_i \boldsymbol{\alpha}_i^\top \tilde{\mathbf{K}}_x \tilde{\mathbf{K}}_s \tilde{\mathbf{K}}_x.$$

which is equivalent to

$$\tilde{\mathbf{K}}_x \leftarrow \tilde{\mathbf{K}}_x - \sqrt{\lambda_i} \tilde{\mathbf{K}}_x \tilde{\mathbf{K}}_s \boldsymbol{\alpha}_i,$$

i.e., at each step we remove from the kernel matrix the best approximation based on the newly extracted projections of the sensitive data $\tilde{\mathbf{K}}_x \tilde{\mathbf{K}}_s \boldsymbol{\alpha}_i$. The deflation procedure decreases by 1 the rank of the matrix, so the maximum number of features that can be extracted with KFDR is $\mathrm{rank}(\tilde{\mathbf{K}}_x \tilde{\mathbf{K}}_s)$, which for most mapping functions will be $n_p = \min\{n, c\}$.

The KFDR method is again similar to the KOPLS in [1, 2], but here we seek for independent projections from the inequitable variables $\mathbf{S}$ while maximizing the variance. As for any kernel multivariate analysis method, projecting a new test point $\mathbf{x}_* \in \mathbb{R}^{d \times 1}$ is possible, $\tilde{\mathbf{x}}'_* = \mathbf{U}^\top \tilde{\phi}(\mathbf{x}_*) = \boldsymbol{\Lambda}^\top \tilde{\boldsymbol{\Phi}} \tilde{\phi}(\mathbf{x}_*) = \boldsymbol{\Lambda}^\top \tilde{\mathbf{k}}_*$, where $\tilde{\mathbf{k}}_* = [k_x(\mathbf{x}_1, \mathbf{x}_*), \ldots, k_x(\mathbf{x}_n, \mathbf{x}_*)]^\top$.

Invariant feature extraction. This example considers $n = 1000$ points drawn from a sinus function buried in noise, $b_i = \sin(a_i) + n_i$, where $\mathbf{a} \sim \mathcal{U}(0, 1.5\pi)$ and $n_i \sim \mathcal{N}(0, 0.1)$. We compare the maps of PCA and FDR and their kernel counterparts. For illustration purposes we consider two different configurations of the inputs, by switching the *sensitive* variable to be either **a** or **b**. Note that this only changes the results for FDR since PCA and KPCA do not distinguish between *sensitive* and *unprotected* variables. Figure 2 shows the first component projection as a color map in the background for the different methods. Essentially PCA and FDR methods cannot account for the nonlinear feature relations, but FDR allows one to easily force invariance to a pre-specified dimension of interest. Compare for instance the first (PCA) and the second (FDR, **S** = **a**) plots. The first component found by PCA is diagonal, revealing it has information about both components (**a** and **b**). On the other hand, the first component found by FDR is vertical, thus avoiding the information in the horizontal axis, i.e. it is insensitive to the information in **a**, as expected. Similar effects are observed in the kernel versions, yet recovering the nonlinear structure of the manifold.

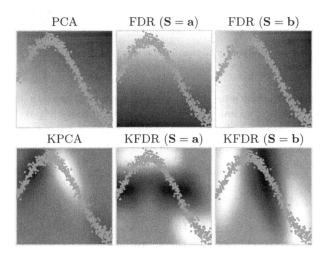

Fig. 2. Linear and kernel feature extraction in a noisy sinusoid distribution. For the sake of simplicity we only show projections onto the first component. The value of the projection is shown as a color map in the background, where dark tones mean small values and brighter tones mean big values. See details in the text. (Color figure online)

Noise-aware feature extraction. We generated a bidimensional banana-shaped distribution corrupted by correlated noise in the $\pi/4$-direction to which we want to be independent, cf. Fig. 3. We compare results of KFDR with those from standard KPCA. Projections #2 and #3 capture the noise distribution while for the KFDR all extracted projections are invariant to variations in the $\pi/4$ direction where the noise is mostly present. The method is intimately related to the kernel signal to noise ratio in [13].

4   Experimental Evidence

The aim of this section is to empirically test the proposed methods on real data. We will see that using regular models and removing the sensitive variables is not

KPCA $n_p = 1$ KPCA $n_p = 2$ KPCA $n_p = 3$

KFDR $n_p = 1$ KFDR $n_p = 2$ KFDR $n_p = 3$

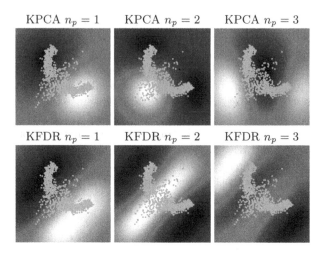

Fig. 3. Dimensionality reduction in a noisy two-dimensional example.

enough to obtain fair solutions. First, we will present the data used and then we will evaluate both proposals, regression and dimensionality reduction, using two different datasets from the UCI Machine Learning repository. Source code and illustrative demos are available in http://isp.uv.es/soft_regression.html for the interested reader.

4.1 Datasets

We consider two datasets from the UCI repository [20]: the Adult dataset and the Contraceptive dataset. Both of them involve sensitive attributes and pose problems of equitable prediction.

Dataset 1: Adult Dataset. This dataset is readily pre-processed and available from the libsvm website[3], and has been used in several studies about fair machine learning methods for classification and feature extraction [12,19,28,30]. The original Adult data set contains 14 features, among which six are continuous and eight are categorical. In this data set, continuous features were discretized into quantiles, and each quantile was represented by a binary feature. Also, a categorical feature with m categories is converted to m binary features. Finally, the original 14 features are pre-processed into 123 features. Details on how each feature is finally converted can be found in Table 1. The dataset is already split into 2 sets, the first one for training the models, which consists of 32561 instances, and the second one was used for testing the results and contains 16281 instances. Both in regression and dimensionality reduction experiments we fit the hyper-parameters using 5000 instances to train and 5000 to validate both randomly

---

[3] https://www.csie.ntu.edu.tw/~cjlin/libsvmtools/datasets/binary.html.

Table 1. Original and processed features for the adult dataset from the UCI repository. The type column distinguishes between a continuous (c) or discrete (d) attribute.

# Feature	Original feature	type (c/d)	# Processed feat
1	Age	c	[1,5]
2	Workclass	d	[6,13]
3	Final weight	c	[14,18]
4	Education	d	[19,34]
5	ed_num	c	[35,39]
6	Marital_status	d	[40,46]
7	Occupation	d	[47,60]
8	Relationship	d	[61,66]
9	Race	d	[67,71]
10	Sex	d	[72,73]
11	Capital_gain	c	[74,75]
12	Capital_loss	c	[76,77]
13	Hours × week	c	[78,82]
14	Country	d	[83,123]

selected from the training set. Afterwards we evaluate those models using the whole test set. All the presented results are the mean of twenty-five realizations of each experiment.

Dataset 2: Contraceptive Method Choice Data Set. In the second problem we study the drivers for adoption of contraceptive types by a women cohort. We used the Contraceptive Method Choice (CMC) Data Set from the UCI repository, which can be downloaded from https://archive.ics.uci.edu/ml/datasets/Contraceptive+Method+Choice. This dataset is a subset of the 1987 National Indonesia Contraceptive Prevalence Survey. The samples are married women who were either not pregnant or do not know if they were at the time of interview. The problem is to predict the current contraceptive method choice leading to three possibilities: 'no use', 'long-term methods', or 'short-term methods' of a woman based on demographic and socio-economic descriptors. We simplified the problem and considered the classes using/not-using a contraceptive method. Table 2 summarizes the total number of features and the class attributes.

The data set consists of 1473 samples with 9 features, and one variable to infer, the contraceptive method. In order to train our algorithms, we split the data into train (500 samples), validation (500 samples) and test sets (the remaining 473 samples). The experiment is performed 25 times, and results are averaged to avoid skewed conclusions.

Table 2. Original and processed features for the contraceptive method choice data set from the UCI repository. The type column distinguishes between a continuous (c) or discrete (d) attributes.

# Feature	Original feature	Type (c/d)
1	Wife's age	c
2	Wife's education	d
3	Husband's education	d
4	Number of children ever born	c
5	Wife's religion	d
6	Wife's now working	d
7	Husband's occupation	d
8	Standard-of-living index	d
9	Media exposure	d
10	Contraceptive method used	Class attribute

4.2   Experimental Setup

In the regression experiment, we optimize the hyperparameters λ (model regularization), σ (kernel width) and σ_S (the kernel parameter for the dependence estimation) using different logarithmically spaced values. Specifically we tried seven values in the interval $[10^{-4}, 10^3]$ for λ, 10 values in $[10^{-4}, 10^4]$ for σ, and 10 values in $[10^{-1}, 10^2]$ for σ_S. We start by seeking the optimal λ and σ parameters that minimize the error in the validation data. Once these two parameters are fixed we explore the kernel parameter for the dependence estimation in order to maximize the dependence between the model and the sensitive data. Finally, we try 25 different logarithmic spaced values in the interval $[10^{-7}, 10^3]$ for the μ fairness hyperparameter (large μ values imply more fair models).

In the FDR experiment the only hyperparameter to tune is σ_S, which is optimized to maximize the dependence between the transformed data and the sensitive variables. We optimized this parameter trying 15 values in the interval $[10^{-5}, 10^3]$. Different number of components n_p were extracted.

4.3   Results for Fair Regression

We analyze the performance of both linear (FLR) and the nonlinear kernel (FKR) formulations. As done in the toy example, we explore the possibility of including or not the sensitive variables **S** in the models. Figure 4 shows the results for different values of μ. Since the original data was collected for a classification problem, we binarized the outputs ($c = 2$), and treat it as a regression problem, afterwards we use max-vote to obtain the predicted class. We analyze two different situations: one where the methods avoid discriminating only by gender and another when the methods avoid discriminating by gender and

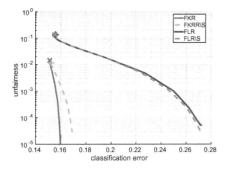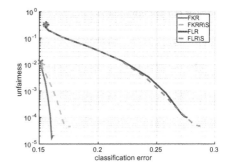

Fig. 4. Error in income classification versus (un)fairness of the sensitive variables for the Adult dataset, avoiding discrimination by gender (left), and by both gender and race (right).

race simultaneously. Note that in the latter case the sensitive variable is bidimensional. While this situation is quite general, using complicated information measures like mutual information (as proposed in [19]) increases dramatically the complexity of the problem. However, in our case, it is straightforward to deal with multidimensional sensitive variables.

In both cases we observe a similar behavior as in the toy example. Both the linear and kernel classical versions (LR and KRR) obtain relatively good classification error rates, but their dependence with the sensitive variables is relatively high. The use of fair versions open the possibility of decreasing this dependence while yielding similar classification errors. Results are better when using the kernel version FKR, which is capable of learning a model with low classification error rate and virtually independent of the sensitive features. Including the sensitive variables when using our proposed method obtains better results in the kernel case. In the linear case, removing the sensitive features has almost no impact on the results.

When it comes to the second dataset, we performed our experimentation over the sensitive variables: wife's education, husband education, number of children ever born and also media exposure (features 2, 3, 4 and 9 respectively). The experiments were done by considering only one sensitive variable at a time. Figure 5 shows the results for all these protected variables. Several conclusions can be derived: (1) kernel fair regression outperforms the linear counterpart in all the hyperparameter space (both on error and fairness); (2) removing the sensitive feature degrades results, as its information is implicitly conveyed by other included features; and (3) one can achieve arbitrary fairness levels tuning the μ hyperparameter, at the cost of a moderately increased prediction error (+2–5% increase in classification error).

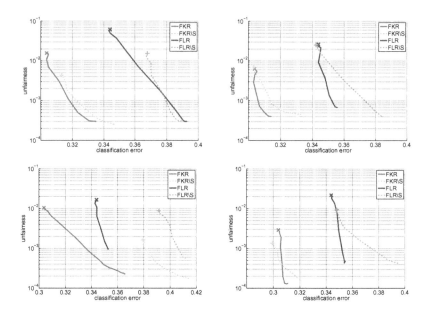

Fig. 5. Error in contraceptive usage classification versus (un)fairness of the sensitive variables for the CMC dataset. Top row (left) wife's education, (right) husband's education, and bottom row (left) number of children ever born and (right) media exposure.

4.4   Results for Fair Dimensionality Reduction

We analyzed the performance of the proposed dimensionality reduction in the income prediction dataset. We present results of using a k-nn classifier ($k = 1$) after reducing the dimensionality of the data set using different methods. In particular, we analyzed the standard Principal Components Analysis (PCA), Kernel PCA (KPCA), and the proposed fair dimensionality Reduction (FDR) and its kernel counterpart (KFDR). As in the previous experiment, we also analyze the solution with and without the sensitive features as inputs.

Figure 6 shows the solutions of different methods. In the case of PCA and KPCA we show results for different numbers of features, which affects the classification error but minimally the fairness score. In both experiments the best fairness-accuracy trade-offs are given by the FDR and KFDR when using all variables as inputs. In particular, when avoiding the gender discrimination, the proposed framework shows better classification error for the KFDR. When we use as sensitive variables gender and race the differences of the proposals with regard the classical methods are more noticeable since the classification errors are similar but the dependence achieved by the proposals are several orders of magnitude lower.

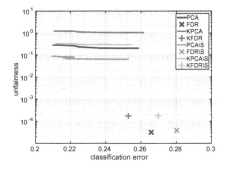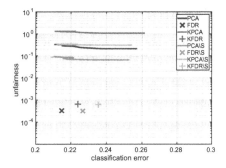

Fig. 6. Error rate in income classification versus independence between predictions and the sensitive variables to avoid discrimination by gender (left), and by both gender and race (right). PCA and KPCA has been evaluated for different number of features, n_p.

5   Conclusions

We have presented novel fair nonlinear regression and dimensionality reduction methods. We included a term to the cost function based on the Hilbert-Schmidt independence criterion which enforces fairness in the solutions and allows to deal with several sensitive variables simultaneously. We presented the methods in linear fashion and extended them to deal with nonlinear problems by using kernel functions. For both the linear and nonlinear cases, the solution for the regression weights and the basis functions in dimensionality reduction are expressed in closed-form, as they only involve solving matrix inversion or generalized eigenproblems respectively.

Tuning the fairness hyperparameter in regression allows us to input sensitive variables to the regression model while keeping the solution fair. This increases the information that can be used by the model during the prediction rather than just ignoring them. Methods performance were successfully illustrated using both synthetic and real data.

We would like to highlight that introducing kernels (and adopting HSIC) for fairness is not incidental: it allows us to achieve closed-form solutions, to trim fairness-fitness with a single hyperparameter, and to encode prior knowledge in a simple way. Interpretability of the models is obviously an issue and will be explored in the near future. While the framework aims to deal with 'population fairness', not with 'individuals' fairness', this refinement can be easily included in our kernel formulations by defining an individual/group diagonal matrix $\mathbf{F}$ and replacing $\mathbf{X}$ with $\mathbf{XF}$ ($\mathbf{I}$ with $\mathbf{F}^{-1}$ for the kernel formulations). As a future work, we also aim to include kernel conditional independence tests. The proposed framework could be easily extended to other machine learning algorithms, from neural networks to Gaussian processes.

References

1. Arenas-García, J., Petersen, K.B., Hansen, L.K.: Sparse kernel orthonormalized PLS for feature extraction in large data sets. In: NIPS, vol. 19. MIT Press (2007)
2. Arenas-García, J., Camps-Valls, G.: Efficient kernel orthonormalized PLS for remote sensing applications. IEEE Trans. Geos. Remote Sens. **46**, 2872–2881 (2008)
3. Barocas, S., Selbst, A.D.: Big Data's Disparate Impact. SSRN eLibrary (2014)
4. Brennan, T., Dieterich, W., Ehret, B.: Evaluating the predictive validity of the compas risk and needs assessment system. Crim. Justice Behav. **36**(1), 21–40 (2009)
5. Chouldechova, A.: Fair prediction with disparate impact: a study of bias in recidivism prediction instruments. CoRR abs/1610.07524 (2016)
6. Clarke, K.A.: The phantom menace: omitted variable bias in econometric research. Conflict Manag. Peace Sci. **22**(4), 341–352 (2005)
7. Corbett-Davies, S., Pierson, E., Feller, A., Goel, S., Huq, A.: Algorithmic decision making and the cost of fairness. CoRR abs/1701.08230 (2017)
8. Cunningham, M.D., Sorensen, J.R.: Actuarial models for assessing prison violence risk. Assessment **13**(3), 253–265 (2006)
9. Dieterich, W., Mendoza, C., Brennan, T.: COMPAS risk scales: demonstrating accuracy equity and predictive parity. Working paper, Northpointe Inc., Res. Dep. (2016)
10. Dimitrakakis, C., Liu, Y., Parkes, D., Radanovic, G.: Subjective fairness: fairness is in the eye of the beholder. Technical report arXiv: 1706.00119 (2017)
11. Dwork, C., Hardt, M., Pitassi, T., Reingold, O., Zemel, R.: Fairness through awareness. In: ITCS 2012, pp. 214–226. ACM, New York (2012)
12. Feldman, M., Friedler, S.A., Moeller, J., Scheidegger, C., Venkatasubramanian, S.: Certifying and removing disparate impact. In: Proceedings of the 21th ACM SIGKDD International Conference on Knowledge Discovery and Data Mining, KDD 2015, pp. 259–268. ACM, New York (2015)
13. Gómez-Chova, L., Nielsen, A.A., Camps-Valls, G.: Explicit signal to noise ratio in reproducing kernel Hilbert spaces. In: IGARSS, pp. 3570–3573, July 2011
14. Gretton, A., Herbrich, R., Hyvärinen, A.: Kernel methods for measuring independence. J. Mach. Learn. Res. **6**, 2075–2129 (2005)
15. Hoffman, M., Kahn, L.B., Li, D.: Discretion in hiring, Working Paper 16–055. Harvard Business School (2015)
16. Kamiran, F., Calders, T.: Classifying without discriminating. In: 2009 2nd International Conference on Computer, Control and Communication, pp. 1–6, February 2009
17. Kamiran, F., Calders, T.: Data preprocessing techniques for classification without discrimination. Knowl. Inf. Syst. **33**(1), 1–33 (2012)
18. Kamishima, T., Akaho, S., Asoh, H., Sakuma, J.: The independence of fairness-aware classifiers. In: 2013 IEEE 13th International Conference on Data Mining Work, pp. 849–858 (2013)
19. Kamishima, T., Akaho, S., Asoh, H., Sakuma, J.: Fairness-Aware Classifier with Prejudice Remover Regularizer. In: Flach, P.A., De Bie, T., Cristianini, N. (eds.) ECML PKDD 2012. LNCS (LNAI), vol. 7524, pp. 35–50. Springer, Heidelberg (2012). https://doi.org/10.1007/978-3-642-33486-3_3
20. Lichman, M.: UCI machine learning repository (2013)
21. Luo, L., Liu, W., Koprinska, I., Chen, F.: Discrimination-aware association rule mining for unbiased data analytics. In: Madria, S., Hara, T. (eds.) DaWaK 2015. LNCS, vol. 9263, pp. 108–120. Springer, Cham (2015). https://doi.org/10.1007/978-3-319-22729-0_9

22. Pedreschi, D., Ruggieri, S., Turini, F.: Discrimination-aware data mining. In: Proceedings of the 14th ACM SIGKDD International Conference on Knowledge Discovery and Data Mining, KDD 2008, pp. 560–568. ACM, New York (2008)

23. Ristanoski, G., Liu, W., Bailey, J.: Discrimination aware classification for imbalanced datasets. In: CIKM 2013, pp. 1529–1532. ACM, New York (2013)

24. Ruggieri, S., Pedreschi, D., Turini, F.: Data mining for discrimination discovery. ACM Trans. Knowl. Discov. Data 4(2), 9:1–9:40 (2010)

25. Schölkopf, B., Smola, A.: Learning with Kernels - Support Vector Machines, Regularization, Optimization and Beyond. MIT Press Series, Cambridge (2002)

26. Shawe-Taylor, J., Cristianini, N.: Kernel Methods for Pattern Analysis. CUP, Cambridge (2004)

27. Worsley, K.J., Poline, J.B., Friston, K.J., Evans, A.C.: Characterizing the response of pet and fMRI data using multivariate linear models. Neuroimage 6, 305–319 (1998)

28. Zafar, M.B., Valera, I., Rodriguez, M.G., Gummadi, K.P.: Learning fair classifiers, May 2016. http://arxiv.org/abs/1507.05259

29. Zafar, M.B., Valera, I., Gomez Rodriguez, M., Gummadi, K.P.: Fairness constraints: mechanisms for fair classification. In: Singh, A., Zhu, J. (eds.) Proceedings of the 20th International Conference on Artificial Intelligence and Statistics Proceedings of Machine Learning Research, vol. 54, pp. 962–970. PMLR, Fort Lauderdale, FL, USA, 20–22 April 2017

30. Zemel, R.S., Wu, Y., Swersky, K., Pitassi, T., Dwork, C.: Learning fair representations. In: ICML (3), vol. 28, pp. 325–333 (2013)

31. Zeng, J., Ustun, B., Rudin, C.: Interpretable classification models for recidivism prediction. J. R. Stat. Soc. Ser. A (Stat. Soc.) 180, 689–722 (2016)

GaKCo: A Fast Gapped k-mer String Kernel Using Counting

Ritambhara Singh, Arshdeep Sekhon, Kamran Kowsari, Jack Lanchantin, Beilun Wang, and Yanjun Qi$^{(\boxtimes)}$

Department of Computer Science, University of Virginia, Charlottesville, USA
`yanjun@virginia.edu`

Abstract. String Kernel (SK) techniques, especially those using gapped k-mers as features (gk), have obtained great success in classifying sequences like DNA, protein, and text. However, the state-of-the-art gk-SK runs extremely slow when we increase the dictionary size (Σ) or allow more mismatches (M). This is because current gk-SK uses a trie-based algorithm to calculate co-occurrence of mismatched substrings resulting in a time cost proportional to $O(\Sigma^M)$. We propose a **fast** algorithm for calculating Gapped k-mer Kernel using Counting (GaKCo). GaKCo uses associative arrays to calculate the co-occurrence of substrings using cumulative counting. This algorithm is fast, scalable to larger Σ and M, and naturally parallelizable. We provide a rigorous asymptotic analysis that compares GaKCo with the state-of-the-art gk-SK. Theoretically, the time cost of GaKCo is independent of the Σ^M term that slows down the trie-based approach. Experimentally, we observe that GaKCo achieves the same accuracy as the state-of-the-art and outperforms its speed by factors of 2, 100, and 4, on classifying sequences of DNA (5 datasets), protein (12 datasets), and character-based English text (2 datasets). (GaKCo is shared as an open source tool at https://github.com/QData/GaKCo-SVM). Code and data related to this chapter are available at: https://doi.org/10.6084/m9.figshare.5434825.

Keywords: Fast learning · String kernels · Sequence classification
Gapped k-mer string kernel · Counting statistics

1   Introduction

Sequence classification is one of the most important machine learning tasks, with widespread uses in fields like biology and natural language processing. Besides accuracy, speed is a critical requirement for modern sequence classification methods. For example, with the advancement of sequencing technologies, a massive amount of protein and DNA sequence data is produced daily [14]. There is an urgent need to analyze these sequences quickly for assisting time-sensitive experiments. Similarly, on-line information retrieval systems need to

Electronic supplementary material   The online version of this chapter (https://doi.org/10.1007/978-3-319-71249-9_22) contains supplementary material, which is available to authorized users.

© Springer International Publishing AG 2017
M. Ceci et al. (Eds.): ECML PKDD 2017, Part I, LNAI 10534, pp. 356–373, 2017.
https://doi.org/10.1007/978-3-319-71249-9_22

classify text sequences, for instance when quickly assessing customer reviews or categorizing documents to different topics.

In this paper, we focus on the String Kernels (SK) in the Support Vector Machine (SVM) framework for supervised sequence classification. SK-SVM methods have been successfully used for classifying sequences like DNA [1,10,12,15], protein [8] or character based natural language text [16]. They have provided state-of-the-art classification accuracy and can guarantee nice asymptotic behavior due to SVM's convex formulation and theoretical property [17]. Through comparing length-k local substrings (k-mers) and incorporating mismatches and gaps, this category of models calculates the similarity (i.e., so-called kernel function) among sequence samples. Then, using such similarity measures, SVM is trained to classify sequences. Recently, Ghandi et al. [6] developed the state-of-the-art SK-SVM tool called gkm-SVM. gkm-SVM uses a gapped k-mer formulation [7] that reduces the feature space considerably compared to other k-mer based SK approaches.

Existing k-mer based SK methods can become very slow or even unfeasible when we increase (1) the number of allowed mismatches (M) or (2) the size of the dictionary (Σ) (detailed asymptotic analysis in Sect. 2). Allowing mismatches during substring comparisons is important since most sequences in biology are prone to mutations, i.e., insertions, deletions or substitution of sequence characters. Also, the size of the dictionary varies from one sequence classification domain to another. While DNA sequence is composed of only four characters ($\Sigma = 4$), most other domains have bigger dictionary sizes like for proteins, $\Sigma = 20$ and for character-based English text, $\Sigma = 36$. The state-of-the-art tool, gkm-SVM, may work well for cases with small values of Σ and M (like for DNA sequences with $\Sigma = 4$ and $M < 4$), however, its kernel calculation is slow for cases like DNA with larger M, protein (dictionary size $= 20$), or character-based English text sequences (dictionary size $= 36$). Its trie-based implementation, in the worst case, scales exponentially with the dictionary size and the number of mismatches ($O(\Sigma^M)$). For example, gkm-SVM takes more than 5 h to calculate the kernel matrix for one protein sequence classification task with only 3312 sequences. This speed limitation hinders the practical applications of SK-SVM.

This paper proposes a **fast** algorithmic strategy, GaKCo: **Ga**pped k-mer **K**ernel using **Co**unting to speed up the gapped k-mer kernel calculation. GaKCo uses a "sort and count" approach to calculate kernel similarity through cumulative k-mer counting [10]. GaKCo groups the counting of co-occurrence of substrings at each fixed number of mismatches ($\{0, \ldots, M\}$) into an independent procedure. Such grouping significantly reduces the number of updates on the kernel matrix (an operation that dominates the time cost). This algorithm is naturally parallelizable; therefore we present a multithread variation as our ultimate tool that improves the calculation speed even further.

We provide a rigorous theoretical analysis showing that GaKCo has a better asymptotic kernel computation time cost than gkm-SVM. Our empirical experiments, on three different real-world sequence classification domains, validate our theoretical analysis. For example, for the protein classification task mentioned

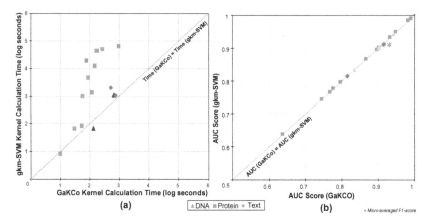

Fig. 1. (a) Kernel calculation times (log(seconds)) of GaKCo (X-axis) versus gkm-SVM (Y-axis) for 19 different datasets - protein (12), DNA (5), and text (2). GaKCo is faster than gkm-SVM for 16/19 datasets. (b) Empirical performance for the same 19 datasets (DNA, protein, and text) of GaKCo (X-axis) versus gkm-SVM (Y-axis). GaKCo achieves the same AUC-scores as gkm-SVM.

above where gkm-SVM took more than 5 h, GaKCo takes only 4 min. Compared to GaKCo, gkm-SVM slows down considerably especially when $M \geq 4$ and for tasks with $\Sigma \geq 4$. Experimentally, GaKCo provides a speedup by factors of 2, 100 and 4 for sequence classification on DNA (5 datasets), protein (12 datasets) and text (2 datasets), respectively, while achieving the same accuracy as gkm-SVM. Figure 1(a) compares the kernel calculation times of GaKCo (X-axis) with gkm-SVM (Y-axis). We plot the kernel calculation times for the best performing (g, k) parameters (see supplementary GitHub) for 19 different datasets. We see that GaKCo is faster than gkm-SVM for 16 out of 19 datasets that we have tested. Similarly, we plot the empirical performance (AUC scores or F1-score) of GaKCo (horizontal axis) versus gkm-SVM (vertical axis) for the best performing (g, k) parameters (see supplementary) for the 19 different datasets in Fig. 1(b). It shows that the empirical performance of GaKCo is same as gkm-SVM with respect to the AUC scores. In summary, the main contributions of this work are:

- **Fast:** GaKCo is a novel combination of two efficient concepts: (1) reduced gapped k-mer feature space and (2) associative array based counting method, making it faster than the state-of-the-art gapped k-mer string kernel, while achieving the same accuracy.
- GaKCo can **scale up** to larger values of m and Σ.
- **Parallelizable:** GaKCo algorithm lends itself to a naturally parallelizable implementation.
- We also present a detailed **theoretical analysis** of the asymptotic time complexity of GaKCo versus state-of-the-art gkm-SVM. This analysis, to our knowledge, has not been reported before.

Table 1. List of symbols and their descriptions that are used.

Notations	Descriptions
D	Dataset under consideration, $D = \{x_1, x_2, \ldots, x_N\}$
N	Number of sequences in a given dataset D
x, x'	Pair of strings in D that are compared for kernel calculation
$K(x, x')$	Kernel Function; Eq. (7) is for the gapped k-mer case
$\phi(x)$	Feature space representation of the string x
l	Average length of sequences in a given dataset D
Σ	Size of the dictionary of a given dataset D
g	Length of the gapped instance or g-mer (specified by the user)
k	Length of k-mer inside a gapped instance (specified by the user)
M	$M = (g - k)$; maximum number of mismatches allowed between two g-mers;
m	Number of mismatches between two g-mers. $m \in \{0, \ldots M\}$
c_{gk}	$c_{gk} = \sum_{m=0}^{M=(g-k)} \binom{g}{m}$.
u	Number of unique g-mers in a given dataset D
z	Number of unique g-mers with > 1 occurrence in a given dataset D
$\mathbf{N}_m(x, x')$	Mismatch profile: number of matching g-mer pairs between x and x' when allowing m mismatches; See Eq. (9)
$\mathbf{C}_m(x, x')$	Cumulative mismatch profile: number of matching $\{g - m\}$-mer pairs between x and x'. Each $\{g - m\}$-mer is generated from a g-mer by removing characters from a total of m different positions; See Eq. (8)
η	Average size of the *nodelist* of leafnodes in gkm-SVM's trie. Each leafnode is a unique g-mer whose *nodelist* includes all g-mers in the trie whose hamming distance to this leaf is up to M; See Eq. (10)

The rest of the paper is organized as follows: Sect. 2 introduces the details of GaKCo and theoretically proves that asymptotically GaKCo runs faster than gkm-SVM for a large dictionary or allowing for more mismatches. Then Sect. 3 provides the experimental results we obtain on three major benchmark applications: TFBS binding prediction (DNA), Remote Protein Homology prediction (Proteins) and Text Classification (categorization and sentiment analysis). Empirically, GaKCo shows consistent improvements over gkm-SVM in computation speed across different types of datasets. When allowing a higher number of mismatches, the disparity in speed between GaKCo and the baseline becomes more apparent. Table 1 summarizes the important notations we use. Due to the space limitation, we discuss the related studies in the supplementary. Recently, Deep Neural Networks (NNs) have provided state-of-the-art performances for various sequence classification tasks. We compare GaKCo's empirical performance with a state-of-the-art deep convolutional neural network (CNN) model [11]. On datasets with few training samples, GaKCo achieves an average accuracy improvement of 20% over the CNN model (details in the supplementary).

2   Method

2.1   Background: Gapped k-mer String Kernels

The key idea of string kernels is to apply a function $\phi(\cdot)$, which maps strings of arbitrary length into a vectorial feature space of fixed dimension. In this space, we apply a standard classifier such as Support Vector Machine (SVM) [17]. Kernel versions of SVMs calculate the decision function for an input x as:

$$f(x) = \sum_{i=1}^{N} \alpha_i y_i K(x_i, x) + b \tag{1}$$

where N is the total number of training samples and $K(\cdot, \cdot)$ is a *kernel function*. String kernels ([6,10,12]), implicitly compute $K(x, x')$ as an inner product in the feature space:

$$K(x, x') = \langle \phi(x), \phi(x') \rangle, \tag{2}$$

where $x = (s_1, \ldots, s_{|x|})$. $x, x' \in \mathcal{S}$. $|x|$ denotes the length of the string x. $\mathcal{S}$ represents the set of all strings composed from a dictionary Σ. The mapping $\phi : \mathcal{S} \to \mathbb{R}^p$ takes a sequence $x \in \mathcal{S}$ to a p-dimensional feature vector.

The feature representation $\phi(\cdot)$ plays a vital role in string analysis since it is hard to describe strings as feature vectors. One classical method is to represent it as an unordered set of k-mers, or combinations of k adjacent characters. A feature vector indexed by all k-mers records the number of occurrences of each k-mer in the current string. The string kernel using this representation is called spectrum kernel [13], where the spectrum representation counts the occurrences of each k-mer in a string. Kernel scores between strings are computed by taking an inner product between corresponding "k-mer-indexed" feature vectors:

$$K(x, x') = \sum_{\gamma \in \Gamma_k} c_x(\gamma) \cdot c_{x'}(\gamma) \tag{3}$$

where γ represents a k-mer, Γ_k is the set of all possible k-mers, and $c_x(\gamma)$ is the number of occurrences (normalized) of k-mer γ in string x. Many variations of spectrum kernels [4,9,10,18] exist in the literature that mostly extend it by including mismatched k-mers when calculating the number of occurrences.

Spectrum kernel and its mismatch variations generate extremely sparse feature vectors for even moderately sized values of k, since the size of Γ_k is Σ^k. To solve this issue, Ghandi et al. [7] introduced a new set of feature representations, called *gapped k-mers*. It is characterized by two parameters: (1) g, the size of a substring with gaps (we call this gapped instance as g-mer hereafter) and (2) k, the size of non-gapped substring in a g-mer (we call it k-mer). The number of gaps is $(g - k)$. The inner product to compute the gapped k-mer kernel function includes sum over all possible k-mer feature counts obtained from the g-mers:

$$K(x, x') = \sum_{\gamma \in \Theta_g} c_x(\gamma) \cdot c_{x'}(\gamma) \tag{4}$$

where γ represents a k-mer, Θ_g is the set of all possible gapped k-mers that can appear in all the g-mers (each with $(g-k)$ gaps) in a given dataset (denoted as D hereafter) of sequence samples.

This formulation's advantage is that it drastically reduces the number of k-mers to consider. If we sum over all k-mers, as in Eq. (3), each of the $\binom{g}{k}$ "non-gap" positions in the g-mer may be filled with any of Σ letters. Thus, the sum has $\binom{g}{k}\Sigma^k$ terms — the number of possible gapped k-mers. This feature space grows rapidly with both Σ and k. In contrast, Eq. (4) (implemented as gkm-SVM [6]) includes only those k-mers whose gapped formulation has appeared in the dataset, D. Θ_g includes all unique g-mers of the dataset D, whose size $|\Theta_g|$ is normally much smaller than $\binom{g}{k}\Sigma^k$ because the new feature space is restricted to only observable gapped k-mers in D. Ghandi et al. [6] use this intuition to reformulate Eq. (4) into:

$$K(x, x') = \sum_{i=0}^{l_1}\sum_{j=0}^{l_2} h_{gk}(g_i^x, g_j^{x'}) \tag{5}$$

For two sequences x and x' of lengths l_1 and l_2 respectively. g_i^x and $g_j^{x'}$ are the ith and j^{th} g-mers of sequences x and x' (i.e., g_i^x is a continuous substring of x starting from the i-th position and ending at the $(i+g-1)^{th}$ position of x). h_{gk} represents the inner product (or similarity) between g_i^x and $g_j^{x'}$ using the co-occurrence of gapped k-mers as features. $h_{gk}(g_i^x, g_j^{x'})$ is non-zero only when g_i^x and $g_j^{x'}$ have common k-mers.

Definition 1. g-pair$_m(x, x')$ denotes a pair of g-mers $(g_1^x, g_2^{x'})$ whose Hamming distance is exactly m. g_1^x is from sequence x and $g_2^{x'}$ is from sequence x'.

Each **g-pair$_m$**(.) has $\binom{g-m}{k}$ common k-mers, therefore its h_{gk} can be directly calculated as $h_{gk}(\textbf{g-pair}_m) = \binom{g-m}{k}$. Ghandi et al. [6] formulate this observation formally into the coefficient h_m:

$$h_m = \begin{cases} \binom{g-m}{k}, & \text{if } g - m \geq k \\ 0, & \text{otherwise.} \end{cases} \tag{6}$$

h_m describes the co-occurrence count of common k-mers for each possible **g-pair$_m$**(.) in D. $h_m > 0$ only for cases of $m \leq (g-k)$ or $(g-m) \geq k$. This is because there will be no common k-mers when the number of mismatches (m) between two g-mers is more than $(g-k)$. Now we can reformulate Eq. 5 by grouping **g-pairs$_m(x, x')$** with respect to different values of m. This is because **g-pairs$_m$**(.) with same m contribute the same number of co-occurrence counts: h_m. Thus, Eq. 5 can be adapted into the following compact form:

$$K(x, x') = \sum_{m=0}^{g-k} N_m(x, x') h_m \tag{7}$$

$N_m(x, x')$ represents the number of **g-pair**$_m(x, x')$ between sequence x and x'. $N_m(x, x')$ is named as *mismatch profile* by [6]. Now, to compute kernel function $K(x, x')$ for gapped k-mer SK, we only need to calculate $N_m(x, x')$ for $m \in \{0, \ldots g - k\}$, since h_m can be precomputed[1]. The state-of-the-art tool gkm-SVM [6] calculates $N_m(x, x')$ using a trie based data structure that is similar to [12] (with some modifications, details in Sect. 2.3).

2.2 Proposed: Gapped k-mer Kernel with Counting (GaKCo)

In this paper, we propose GaKCo, a fast and novel algorithm for calculating gapped k-mer string kernel. GaKCo provides superior time performance over the state-of-the-art gkm-SVM and is different from it in three aspects:

- **Data Structure.** gkm-SVM uses a trie based data structure (plus a separate nodelist at each leafnode) for calculating N_m (see Fig. 2(c)). In contrast, GaKCo uses simple arrays with a "sort-and-count" approach.
- **Algorithm.** GaKCo performs g-mer based cumulative counting of co-occurrence to calculate N_m.
- **Parallelization.** GaKCo groups computations for each value of m into an independent function, making it naturally parallelizable. We, therefore, provide a parallel version that uses multithread implementation.

Intuition: When calculating $\mathbf{N}_m$ between all pairs of sequences in D for each value of m ($m \in \{0, \ldots, M = g - k\}$), we can use counting to process all **g-pairs**$_m(.)$ (details below) from D together. Then we can calculate $\mathbf{N}_m$ from such count statistics of **g-pairs**$_m(.)$. This method is entirely different from gkm-SVM that uses a trie to organize g-mers such that each leafnode's (a unique g-mer's) nodelist points to its mismatched g-mer neighbors in D.

Algorithm. GaKCo calculates $N_m(x, x')$ as follows (pseudo code: Algorithm 1):

1. GaKCo first extracts all possible g-mers from all the sequences in D and puts them in a simple array. Given that there are N number of sequences with average length l[2], the total number of g-mers is $N \times (l - g + 1) \sim Nl$ (see Fig. 2 (a)).
2. $N_{m=0}(x, x')$ represents the number of **g-pair**$_{m=0}(x, x')$ (pairs of g-mers whose Hamming distance is 0) between x and x'. To compute $N_{m=0}(x_i, x_j)$ $\forall i, \forall j = 1, \ldots, N$, GaKCo sorts all the g-mers lexicographically (see Fig. 2(a) [Step 1]) and counts the occurrences (if > 1) of each unique g-mer. Then we use these counts and the associated indexes of sequences to update all the kernel entries for sequences that include the matching g-mers (Fig. 2(a)

---

[1] For convenience, we will occasionally identify the map $N_m(\cdot, \cdot)$ with the $N \times N$ matrix $\mathbf{N}_m$, consisting of the application of N_m to each pair of sequences $x, x' \in D$. This convention is also followed for the kernel function, $K(\cdot, \cdot) \to \mathbf{K}$, and the cumulative mismatch profile (introduced later), $C_m(\cdot, \cdot) \to \mathbf{C}_m$.

[2] A simplification of real world datasets in which sequence length varies across samples.

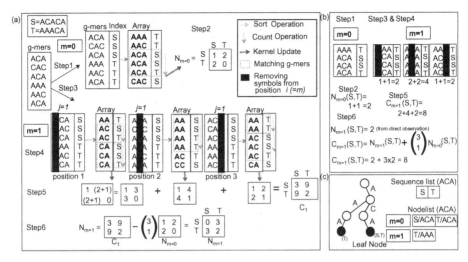

Fig. 2. (a) Overview of GaKCo algorithm for calculating mismatch profile $N_m(S,T)$, where $S = ACACA$ and $T = AAACA$, and $g = 3$ forming g-mers $\{ACA, CAC, ACA\}$ and $\{AAA, AAC, ACA\}$ respectively. [Step 1] For $m = 0$, all g-mers are sorted lexicographically. [Step 2] $N_{m=0}(S,T)$ is calculated directly by sorting and counting. [Step 3] For $m = 1$, we perform over counting of the $g - 1$-mers by picking 1 position at a time (from $\binom{g=3}{1}$ positions) and removing symbols to obtain $(g - 1)$-mers. [Step 4] We sort and count to find the number of matching $(g-1)$-mers for each picked position. [Step 5] Summing up over all $\binom{g=3}{1}$ positions, we get *cumulative mismatch profile* $\mathbf{C}_{m=1}$. [Step 6] Using Eq. 9 we get $N_{m=1}(S,T) = 3$ from $C_{m=1}(S,T) = 9$ and $N_{m=0}(S,T) = 2$. This count is equal to the actual number of pairs of g-mers at Hamming distance $m = 1$ between s and t (i.e. $\{ACA : s/2, AAA : t/1\}$, $\{CAC : s/1, AAC : t/1\}$). A case demonstration of (b) the overcounting when calculating $\mathbf{C}_{m=1}$ (c) two leafnode g-mers and associated nodelist for leaf $\{ACA\}$ in the trie used by gkm-SVM.

[Step 2]). This computation is straight-forward and the sort and count step takes $O(gNl)$ time cost while the kernel update costs $O(zN^2)$ (at the worst case). Here, z is the number of g-mers that occur > 1 times.

3. For cases when $m = 1, \dots (g - k)$, we use a statistics measure $C_m(x, x')$, called *cumulative mismatch profile* between two sequences x and x'. This measure describes the number of matching $(g - m)$-mers between x and x'. Each $(g - m)$-mer is generated from a g-mer by removing a total number of m positions. We can calculate the exact *mismatch profile* $\mathbf{N}_m$ from the cumulative mismatch profile $\mathbf{C}_m$ for $m > 0$ (see Step 4).

By sorting the lists of g-mers with m ignored entries, we compute $\mathbf{C}_m$. First, we first pick m positions and remove the symbols in those positions from all observed g-mers, generating a list of $(g - m)$-mers (Fig. 2(a) [Step 3]). We then sort and count this list to get the number of matching $(g - m)$-mers (Fig. 2(a) [Step 4]). For the sequences that have matching $(g - m)$-mers, we add the counts into their corresponding entries in matrix $\mathbf{C}_m$. This sequence of operations is repeated for all $\binom{g}{m}$ selections of m positions. Then, $\mathbf{C}_m$ is equal to the sum of counts from all $\binom{g}{m}$ runs (Fig. 2(a) [Step 5]).

4. We compute $\mathbf{N}_m$ using $\mathbf{C}_m$ and $\mathbf{N}_j$ for $j = 0, \ldots, m - 1$.

Given two g-mers g_1 and g_2, we remove symbols from the same set of m positions of both g-mers to get two $(g - m)$-mers: g_1' and g_2'. If the Hamming distance between g_1' and g_2' is zero, then we can conclude that the Hamming distance between the original two g-mers is less than or equal to m (formal proof in supplementary). For instance, $C_{m=1}(x, x')$ records the statistic of matching $(g - 1)$-mers between x and x'. It includes the matching statistics of all g-mer pairs with Hamming distance exactly 1, but it also over-counts the matching statistics of all g-mer pairs with Hamming distance 0. This is because the matching g-mers for $m = 0$ also match for $m = 1$ and contribute to the matching statistics $\binom{g}{1}$ times! This over-counting occurs for other values of m as well. Therefore we can calcualte the cumulative mismatch profile $\mathbf{C}_m$ as: $\forall m \in \{0, \ldots, g - k\}$

$$\mathbf{C}_m = \mathbf{N}_m + \sum_{j=0}^{m-1} \binom{g-j}{m-j} \mathbf{N}_j \tag{8}$$

We demonstrate this over-counting in Fig. 2(b). Rearranging Eq. 8, we get the exact mismatch profile N_m as:

$$\mathbf{N}_m = \mathbf{C}_m - \sum_{j=0}^{m-1} \binom{g-j}{m-j} \mathbf{N}_j \tag{9}$$

We subtract $\mathbf{N}_j$ from $\mathbf{C}_m$ to compensate for the over-counting described above.

Parallelization: For each value of m from $\{0, \ldots M = g - k\}$, calculating $\mathbf{C}_m$ is independent from other values of m. Therefore, GaKCo's algorithm can be easily revised into a parallel version. Essentially, we just need to revise Step 9 in Algorithm 1 (pseudo code) — "For each value of m" — into, "For each value of m per machine/per core/per thread". In our current implementation, we create a thread for each value of m from $\{0, \ldots M = g - k\}$ and calculate $\mathbf{C}_m$ in parallel. In the end, we compute the final kernel matrix K using all the resulting $\mathbf{C}_m$ matrices. Figure 4 show the improvement of kernel calculation speed of the multi-thread version over the single-thread implementation of GaKCo.

2.3   Theoretical Comparison of Time Complexity

In this section, we conduct asymptotic analysis to compare the time complexities of GaKCo with the state-of-the-art toolbox gkm-SVM.

Time Complexity of GaKCo: The time cost of GaKCo splits into two groups: (1) Pre-processing: those operations that indirectly update the matching statistics among sequences; (2) Kernel updates: those operations that directly update the matching statistics among sequences.

Pre-processing: For each possible m ($m \in \{0, \ldots M = g - k\}$), GaKCo needs to choose m positions for symbol removing (Fig. 2(a) [Step 3]), and then sort and count the possible $(g - m)$-mers from D (Fig. 2(a) [Step 4]). Therefore the time cost of pre-processing is $O(\Sigma_{m=0}^{M=g-k} \binom{g}{m}(g-m)Nl) \sim O(\Sigma_{m=0}^{M} \binom{g}{m}gNl)$. To simplifying notations, we use c_{gk} to represent $c_{gk} = \Sigma_{m=0}^{M=(g-k)} \binom{g}{m}$ hereafter.

Kernel Updates: These operations update the entries of $\mathbf{C}_m$ or $\mathbf{N}_m$ matrices when GaKCo finishes each round of counting the number of matching $(g - m)$-mers. Assuming z denotes the number of unique $(g - m)$-mers that occur > 1 times, the time cost of kernel update operations is (at the worst case) equivalent to $O(\Sigma_{m=0}^{M} \binom{g}{m}zN^2) \sim O(c_{gk}zN^2)$. Therefore, the overall time complexity of GaKCo is $O(c_{gk}[gNl + zN^2])$.

gkm-SVM Algorithm: Now we introduce the algorithm of gkm-SVM briefly. Given that there are N sequences in a dataset D, gkm-SVM first constructs a trie recording all the unique g-mers in D. Each leafnode in the trie stores a unique g-mer (more precisely by its path to the rootnode) of D. We use u to denote the total number of the unique g-mers in this trie. Next, gkm-SVM traverses the tree in a depth-first ordering. For each leafnode (like ACA in (Fig. 2(c)), it maintains a *nodelist* that includes all those g-mers in D whose Hamming distance to the leafnode g-mer $\leq M$. When accessing a leafnode, all mismatch profile matrices $N_m(x, x')$ for $m \in \{0, \ldots, M = (g - k)\}$ are updated for all possible pairs of sequences x and x'. Here x consists of the g-mer of the current leafnode (like S/ACA in (Fig. 2(c)). x' belongs to the *nodelist*'s sequence list. x' includes a g-mer whose Hamming distance from the leafnode is m (like $T/ACA(m = 0)$ or $T/AAA(m = 1)$ in (Fig. 2 (c)).

Time Complexity of gkm-SVM: We also split operations of gkm-SVM into those indirectly (pre-processing) or directly (kernel-update) updating $\mathbf{N}_m$.

Pre-processing: To construct the trie, gkm-SVM iterates over every possible starting position for a g-mer. Given, there are N sequences each of average length l, then there are approximately Nl starting positions. Furthermore, each g-mer must be inserted into the trie (g steps). Therefore, the time taken to construct the *trie* is $O(gNl)$. Besides, for each node (a unique g-mer) in the trie, the algorithm maintains a list of pointers that point to all other g-mers in the trie whose hamming distance to this node is M. Let the number of such g-mers be η and total number of nodes are ug, then this operation costs $O(\eta ug)$.

Kernel Update: For each leafnode of the trie (total u nodes), for each g-mer in its nodelist (assuming average size of nodelist is η), gkm-SVM uses the matching count among g-mers to update involved sequences' entries in N_m (if Hamming distance between two g-mers is m). Therefore the time cost is $O(\eta uN^2)$ (at the worst case). Essentially η represents on average the number of unique g-mers

Table 2. Comparing time complexity. gvm-SVM's time cost is $O(gNl + \eta ug + \eta u N^2)$. GaKCo's time complexity is $O(c_{gk}[gNl + zN^2])$. In gkm-SVM the $\eta u N^2$ dominates, while the $c_{gk}zN^2$ term dominates for GaKCo.

	GaKCo	gkm-SVM
Pre-processing	$c_{gk}gNl$	$gNl + \eta ug$
Kernel updates	$c_{gk}zN^2$	$\eta u N^2$

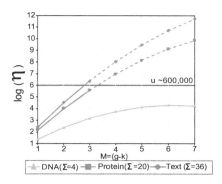

Fig. 3. With a growing $g - k$, the growth curve of η (Eq. (10): the estimated *nodelist* size in gkm-SVM). Both arguments of the min function are plotted; η grows exponentially until $c_{gk}(\Sigma - 1)^M$ exceeds the number of unique g-mers, u.

Algorithm 1. GaKCo
***
Require: L, g, k ▷ L=Array list of g-mers
1: **procedure** CALCULATEKERNEL(L,g,k)
2:     $M \leftarrow g - k$
3:     $\mathbf{N} \leftarrow$ MISMATCHPROFILE(L,g,M)
4:     $K \leftarrow 0$
5:     **for** $m : 0 \rightarrow M$ **do**
6:         $h_m \leftarrow \binom{g-m}{k}$
7:         $K \leftarrow K + \mathbf{N}_m \cdot h_m$
8: **procedure** MISMATCHPROFILE(L,g,M)
9:     **for** $m : 0 \rightarrow M$ **do** ▷ Parallel threads
10:         $\mathbf{C}_m \leftarrow 0$ ▷ Cumulative Profile
11:         $n_{pos} \leftarrow \binom{g}{m}$ ▷ Number of positions
12:         **for** $i : 0 \rightarrow n_{pos}$ **do**
13:             $\mathbf{C}_m^i \leftarrow 0$
14:             $L^i \leftarrow removePosition(L, i)$
15:             $L^i \leftarrow sort(L^i)$
16:             $\mathbf{C}_m^i \leftarrow countAndUpdate(L^i)$
17:             $\mathbf{C}_m \leftarrow \mathbf{C}_m + \mathbf{C}_m^i$
18:     **for** $m : 0 \rightarrow M$ **do**
19:         $\mathbf{N}_m \leftarrow \mathbf{C}_m$
20:         **for** $j : 0 \rightarrow m - 1$ **do**
21:             $\mathbf{N}_m \leftarrow \mathbf{N}_m - \binom{g-j}{m-j}\mathbf{N}_j$
        return N   ▷ $\mathbf{N} = [N_0, \ldots, N_M]$
Ensure: K   ▷ Kernel Matrix
***

(in the trie) that are at a Hamming distance up to M from the current leafnode. η can be formulated as:

$$\eta = \min \left(u, \sum_{m=0}^{M=(g-k)} \binom{g}{m}(\Sigma - 1)^m \right) \sim \min \left(u, c_{gk}(\Sigma - 1)^M \right) \quad (10)$$

Figure 3 shows that η grows exponentially to M until reaching its maximum u. The total complexity of time cost from gkm-SVM is thus $O(gNl + \eta ug + u\eta N^2)$.

Comparing Time Complexity of GaKCo with gkm-SVM: Table 2 compares the asymptotic time cost of GaKCo with gkm-SVM. In gkm-SVM the term $O(\eta u N^2)$ dominates the overall time asymptotically. For GaKCo the term $O(c_{gk}zN^2)$ dominates the time cost asymptotically. For simplicity, we assume that $z = u$ even though $z \leq u$. Upon comparing $O(\eta \times u N^2)$ of gkm-SVM with $O(c_{gk} \times u N^2)$ of GaKCo, clearly the difference lies between the terms η and c_{gk}.

In gkm-SVM, for a given g-mer, the number of all possible g-mers that are at a distance M from it is $c_{gk}(\Sigma - 1)^M$. That is because $\binom{g}{M}$ positions can be substituted with $(\Sigma-1)^M$ possible characters. This means η grows exponentially with the number of allowed mismatches M. We show the trend of function $f = c_{gk}(\Sigma - 1)^M$ in Fig. 3 for three different applications - TF-DNA ($\Sigma = 4$), SCOP-protein ($\Sigma = 20$) and text ($\Sigma = 36$) when varying the values of M for $g = 10$.

However, in real-world datasets, η is upper bounded by the number of unique g-mers in a dataset: u. We show this by thresholding the curves in Fig. 3 at $u = 6 \times 10^4$, which is the average observed value of u across multiple datasets. This means, two possible cases for comparing η with c_{gk}:

- When $\eta \sim c_{gk}(\Sigma - 1)^M$: For cases whose dictionary size Σ is small (e.g. 4), $c_{gk}(\Sigma-1)^M$ is mostly smaller than u. Therefore η will be close to $c_{gk}(\Sigma-1)^M$. This indicates the costs of gkm-SVM grow with a speed proportional to Σ^M. In contrast, the term c_{gk} of GaKCo is independent of the size Σ.
- When $\eta \sim u$: For cases whose Σ is larger than 4, $c_{gk}(\Sigma - 1)^M$ gets larger than u for $M \geq 4$. Therefore η is approximately by u (the number of unique g-mers in the trie built by gkm-SVM). The comparison between u and c_{gk} then depends on the specific application. The size u depends on data, but normally grows fast for $M \geq 4$. For example, for one of the SCOP datasets, when $g = 10$, the count of unique g-mers $u = 6 \times 10^4$ at $M = 4$ (close to u shown in Fig. 3). This means $\eta = 6 \times 10^6$ for gkm-SVM while for the same case $c_{gk} = 210$ for GaKCo. The former is approximately 300 times higher than GaKCo.

Table 3. Details of datasets used for different prediction tasks. All tasks, except WebKB, are binary classification tasks. WebKB is a multi-class (4) classification dataset.

Prediction Task	Repo	Datasets	Training		Testing		Sample properties		
			Pos seq	Neg seq	Pos seq	Neg seq	N	Σ	Max(l)
TF Binding Site (DNA)	ENCODE	CTCF	1000	1000	1000	1000	4000	5	100
		EP300							
		JUND							
		RAD21							
		SIN3A							
Remote protein homology (Protein)	SCOP	1.1	1150	1189	8	1227	3574	20	905
		1.34	866	1209	6	1231	3312		
		2.19	110	1235	9	1206	2560		
		2.31	1063	1235	8	1194	3500		
		2.1	4763	1229	120	950	7062		
		2.34	286	1215	6	1231	2738		
		2.41	192	1235	6	1213	2646		
		2.8	56	1185	8	1231	2480		
		3.19	922	1181	7	1231	3341		
		3.25	1187	1208	11	1231	3637		
		3.33	466	1214	7	1231	2918		
		3.50	105	1231	8	1205	2549		
Text classification	Stanford treebank Dataset from [2]	Sentiment	3883	3579	877	878	9217	36	260
		WebKB	335, 620, 744, 1083		166, 306, 371, 538		4163	36	14218

3   Experiments

3.1   Experimental Setup

19 different sequence datasets: We perform 19 different classification tasks to evaluate the performance of GaKCo. These tasks belong to three categories: (1) Transcription Factor (TF) binding site prediction (DNA dataset), (2) Remote Protein Homology prediction (protein dataset), and (3) Character-based English text classification (text dataset). Table 3 summarizes of data statistics of all datasets we used. Details of these datasets and their associated applications are present in the supplementary.

Baselines: We compare the kernel calculation times and empirical performance of GaKCo with gkm-SVM [6]. We also compare GaKCo to the CNN implementation from [11] for all the datasets (results in supplementary).

Classification: After calculation, we input the $N \times N$ kernel matrix into an SVM classifier as an empirical feature map using a linear kernel in LIBLINEAR [5]. Here N is the number of sequences in each dataset. For the multi-class classification of WebKB data, we use the multi-class version of LIBSVM [3].

Model parameters: We vary the hyperparameters $g \in \{7, 8, 9, 10\}$ and $k \in \{1, 2, \ldots, g-1\}$ of both GaKCo and gkm-SVM. $M = (g-k)$ for all these cases. We also tune the hyperparameter $C \in \{0.01, 0.1, 1, 10, 100, 1000\}$ for the SVM. We present the results for the best g, k, and C values based on the empirical performance metric.

Evaluation Metrics: *Running time:* We compare the kernel calculation times of GaKCo and gkm-SVM in seconds. All run-time experiments were performed on an AMD Opteron^{TM} Processor 6376 @ 2.30 GHz with 250 GB memory.

Empirical performance: We use the Area Under Curve (AUC) score (from the Receiver Operating Characteristic (ROC) curve) as our empirical evaluation metric for 18 binary classification tasks. We report the results of WebKB multi-class classification using micro-averaged F1 score.

3.2   Experimental Results

GaKCo is as accurate as gkm-SVM: Figure 1(b) demonstrated that GaKCo achieves the same empirical performance as gkm-SVM across all 19 tasks (on AUC scores or F1-score). This is because GaKCo's gapped k-mer formulation is the same as gkm-SVM but with an improved (faster) implementation. Besides, in the supplementary, we also compare GaKCo's empirical performance with a state-of-the-art CNN model [11]. For 16/19 tasks, GaKCo outperforms the CNN model with an average of ~20% improvements.

GaKCo scales better than gkm-SVM for larger dictionary size (Σ) and larger number of mismatches (M): Fig. 1(a) shows that GaKCo is faster than gkm-SVM for 16/19 tasks. The three tasks for which GaKCo cost

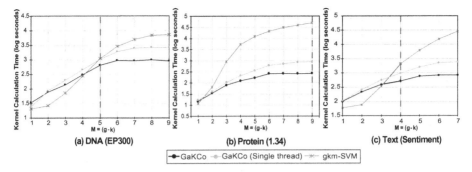

(a) DNA (EP300)   **(b) Protein (1.34)**   **(c) Text (Sentiment)**

—●—GaKCo   —●—GaKCo (Single thread)   —✳—gkm-SVM

Fig. 4. Kernel calculation times (lower is better) for best g and varying k with $M = (g - k) = \{1, 2, \ldots g - 1\}$ hyperparameters for (a) EP300 (DNA, $\Sigma = 5$), (b) 1.34 (protein, $\Sigma = 20$), and (c) Sentiment (text, $\Sigma = 36$) datasets. The best performing hyperparameters (g, k or $M = (g - k)$) are highlighted as red colored dashed lines. GaKCo (single thread) outperforms gkm-SVM for a large dictionary size ($\Sigma > 5$) and a large number of mismatches $M \geq 4$. The final GaKCo (multi-thread) implementation further improves the performance. For protein dataset (b) gkm-SVM takes > 5 h to calculate the kernel, while GaKCo calculates it in 4 min. (Color figure online)

similar time in kernel calculation as gkm-SVM are three DNA sequence prediction tasks. This is as expected since these tasks have a smaller dictionary ($\Sigma = 5$) and thus, for a small number of allowed mismatches (M) gkm-SVM gives comparable speed performance as GaKCo.

Figure 4 shows the kernel calculation times of GaKCo versus gkm-SVM for the best-performing g and varying $k \in \{1, 2, \ldots (g - 1)\}$ for three binary classification datasets: (a) EP300 (DNA), (b) 1.34 (protein), and (c) Sentiment (text) respectively. We select these three datasets as they achieve the best AUC scores out of all 19 tasks (see supplementary). We fix g and vary k to show time performance for any number of allowed mismatches from 1 to $g - 1$. For GaKCo, the results are shown for both single-thread and the multi-thread implementations. We refer to the multi-thread implementation as GaKCo because that is our final code version. Our results show that GaKCo (single-thread) scales better than gkm-SVM for a large dictionary size (Σ) and a large number of mismatches (M). The final version of GaKCo (multi-thread) further improves the performance. Details for each dataset are as follows:

– DNA dataset ($\Sigma = 5$): In Fig. 4(a), we plot the kernel calculation times for best $g = 10$ and varying k with $M \in \{1, 2, \ldots 9\}$ for EP300 dataset. As expected, since the dictionary size of DNA dataset (Σ) is small, gkm-SVM performs fast kernel calculations for $M = (g - k) < 4$. However, for large $M \geq 4$, its kernel calculation time increases considerably compared to GaKCo. This result connects to Fig. 3 in Sect. 2, where our analysis showed that the *nodelist* size becomes closer to u as M increases, thus increasing the time cost.

– Protein dataset ($\Sigma = 20$): Fig. 4(b), shows the kernel calculation times for best $g = 10$ and varying k with $M = (g - k) \in \{1, 2, \ldots 9\}$ for 1.34 dataset. Since the dictionary size of protein dataset (Σ) is larger than DNA, gkm-SVM's kernel calculation time is worse than GaKCo even for smaller values of $M < 4$. This also connects to Fig. 3 where the size of $nodelist \sim u$ even for small M for protein dataset, resulting in higher time cost. For best-performing parameters $g = 10, k = 1 (M = 9)$, gkm-SVM takes 5 h to calculate the kernel, while GaKCo uses less than 4 min.

– Text dataset ($\Sigma = 36$): Fig. 4(c), shows the kernel calculation times for best $g = 8$ and varying k with $M \in \{1, 2, \ldots 7\}$ for Sentiment dataset. For large $M \geq 4$, kernel calculation of gkm-SVM is slower as compared to GaKCo. One would expect that with large dictionary size (Σ) the performance difference will be same as that for protein dataset. However, unlike protein sequences, where the substitution of all 20 characters in a g-mer is roughly equally likely, text dataset has a more skewed underlying distribution. The chance of substituting some characters in a g-mer are higher than other characters for English text. For example, in a given g-mer "my nam", the last position is more likely to be occupied by 'e' than 'z'. Though the dictionary size is large here, the growth of the $nodelist$ is restricted by the underlying distribution. While GaKCo's time performance is consistent across all three datasets, gkm-SVM's time performance varies due to the distribution properties.

According to our asymptotic analysis in Sect. 2, GaKCo should always be faster than gkm-SVM. However, in Fig. 4 we notice that for certain cases (e.g. for DNA when $M < 4$ in Fig. 4) GaKCo's is slower than gkm-SVM. This is because, in our analysis, we theoretically estimate the size of gkm-SVM's $nodelist$. In practice, we see that the actual $nodelist$ size is smaller than our estimated for some cases. Among those cases for some gkm-SVM is faster than GaKCo. However, when with a larger value of $M (\geq 4)$ or a larger dictionary ($\Sigma > 5$), the nodelist size in practice matches our theoretical estimation; therefore, GaKCo always has lower kernel calculation time complexity than gkm-SVM for these cases.

GaKCo is independent of dictionary size (Σ): GaKCo's time complexity analysis (Sect. 2) shows that it is independent of the Σ^M term, which controls the size of gkm-SVM's $nodelist$. In Fig. 5(a), we plot the average kernel calculation times for the best performing (g, k) parameters for DNA ($\Sigma = 5$), protein ($\Sigma = 20$), and text ($\Sigma = 36$) datasets respectively. The results validate our analysis. We find that gkm-SVM takes similar time as GaKCo to calculate the kernel for DNA dataset due to the small dictionary size. However, when the dictionary size increases for protein and text datasets, it slows down considerably. GaKCo, on the other hand, is consistently faster for all three datasets, despite the increase in dictionary size.

GaKCo algorithm benefits from parallelization: As discussed earlier, the calculation of $\mathbf{C}_m$ (with $m \in \{0, 1, \ldots, M = (g - k)\}$) is an independent procedure in GaKCo's algorithm. This property makes GaKCo naturally parallelizable. We implement the final parallelized version of GaKCo by distributing calculation of each $\mathbf{C}_m$ on its thread. In Fig. 4 we see that the multi-threaded version of GaKCo performs faster than its single-threaded counterpart. Next, in Fig. 5(b), we plot the average kernel calculation times across DNA (5), protein (12) and text (2) datasets for both multi-thread and single thread implementations. Hence, we demonstrate that the improvement in speed by parallelization is consistent across all datasets.

GaKCo scales better than gkm-SVM for increasing number of sequences (N): We now compare the kernel calculation times of GaKCo versus gkm-SVM for increasing number of sequences (N). In Fig. 5(c), we plot the kernel calculation times of GaKCo and gkm-SVM for best performing parameters (g, k) for three binary classification datasets: EP300 (DNA), 1.34 (protein), and Sentiment (text). We select these three datasets as they provide the best AUC scores out of all 19 tasks (see supplementary). To show the effect of increasing $N \in \{100, 250, 500, 750\}$ on kernel calculation times, we fix the length of the sequences for all three datasets to $l = 100$. As expected, the time grows for both the algorithms with the increase in the number of sequences. However, this growth in time is more drastic for gkm-SVM than for GaKCo across all three datasets. Therefore, GaKCo is ideal for adaptive training since its kernel calculation time increases more gradually than gkm-SVM as new sequences are added. Besides, GaKCo's time improvement over the baseline is achieved with almost no added memory cost (see supplementary).

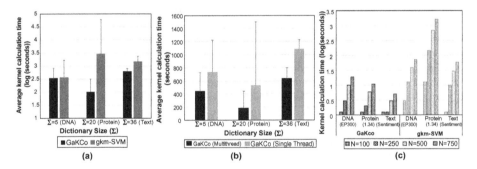

Fig. 5. Average kernel calculation times (lower is better) (a) for the best performing (g, k) parameters for DNA $(\Sigma = 5)$, protein $(\Sigma = 20)$, and text $(\Sigma = 36)$ datasets. gkm-SVM slows down considerably for protein and text datasets but GaKCo is consistently faster for all three datasets. (b) across DNA (5), protein (12) and text (2) datasets. Multi-thread GaKCo implementation improves the kernel calculation speed of the single-thread GaKCo by a factor of 2. (c) Kernel calculation times (lower is better) of GaKCo and gkm-SVM for best performing parameters (g, k) for: EP300 (DNA), 1.34 (protein), and Sentiment (text) datasets. Length of the sequences for all three datasets is fixed to $l = 100$ and number of sequences are varied for $N \in \{100, 250, 500, 750\}$. With increasing number of sequences, the increase in kernel calculation time is more drastic for gkm-SVM than for GaKCo across all three datasets.

4   Conclusion

In this paper, we presented GaKCo, a fast and naturally parallelizable algorithm for gapped k-mer based string kernel calculation. The advantages of this work are:

- **Fast:** GaKCo is a novel combination of two efficient concepts: (1) reduced gapped k-mer feature space and (2) associative array based counting method, making it faster than the state-of-the-art gapped k-mer string kernel, while achieving same accuracy (Fig. 1).
- GaKCo can **scale up** to larger values of m and Σ (Figs. 4 and 5(a)).
- **Parallelizable:** GaKCo algorithm naturally leads to a parallelizable implementation (Figs. 4 and 5(b)).
- We have provided a detailed **theoretical analysis** comparing the asymptotic time complexity of GaKCo with gkm-SVM. This analysis, to the best of the authors' knowledge, has not been reported before (Sect. 2.3).

References

1. Arvey, A., Agius, P., Noble, W.S., Leslie, C.: Sequence and chromatin determinants of cell-type-specific transcription factor binding. Genome Res. **22**(9), 1723–1734 (2012)
2. Cachopo, A.M.D.J.C.: Improving methods for single-label text, categorization, PdD thesis. Instituto Superior Tecnico, Universidade Tecnica de Lisboa (2007)
3. Chang, C.-C., Lin, C.-J.: LIBSVM: a library for support vector machines. ACM Trans. Intell. Syst. Technol. **2**, 1–27 (2011). http://www.csie.ntu.edu.tw/ cjlin/libsvm
4. Chapelle, O., Weston, J., Schölkopf, B.: Cluster kernels for semi-supervised learning. In: Advances in neural information processing systems, pp. 585–592 (2002)
5. Fan, R.-E., Chang, K.-W., Hsieh, C.-J., Wang, X.-R., Lin, C.-J.: Liblinear: a library for large linear classification. J. Mach. Learn. Res. **9**(Aug), 1871–1874 (2008)
6. Ghandi, M., Lee, D., Mohammad-Noori, M., Beer, M.A.: Enhanced regulatory sequence prediction using gapped k-mer features. PLoS Comput. Biol. **10**(7), e1003711 (2014)
7. Ghandi, M., Mohammad-Noori, M., Beer, M.A.: Robust k-mer frequency estimation using gapped k-mers. J. Math. Biol. **69**(2), 469–500 (2014)
8. Jaakkola, T., Diekhans, M., Haussler, D.: A discriminative framework for detecting remote protein homologies. J. Comput. Biol. **7**(1–2), 95–114 (2000)
9. Kuang, R., Ie, E., Wang, K., Wang, K., Siddiqi, M., Freund, Y., Leslie, C.: Profile-based string kernels for remote homology detection and motif extraction. J. Bioinf. Comput. Biol. **3**(03), 527–550 (2005)
10. Kuksa, P.P., Huang, P.-H., Pavlovic, V.: Scalable algorithms for string kernels with inexact matching. In: NIPS 2008, pp. 881–888 (2008)
11. Lanchantin, J., Singh, R., Wang, B., Qi, Y.: Deep motif dashboard: visualizing and understanding genomic sequences using deep neural networks. arXiv preprint arXiv:1608.03644 (2016)
12. Leslie, C., Kuang, R.: Fast string kernels using inexact matching for protein sequences. J. Mach. Learn. Res. **5**, 1435–1455 (2004)

13. Leslie, C.S., Eskin, E., Noble, W.S.: The spectrum kernel: a string kernel for SVM protein classification. In: Pacific Symposium on Biocomputing, pp. 566–575 (2002)
14. Marx, V.: Biology: the big challenges of big data. Nature **498**(7453), 255–260 (2013)
15. Setty, M., Leslie, C.S.: SeqGl identifies context-dependent binding signals in genome-wide regulatory element maps. PLoS Comput. Biol. **11**(5), e1004271 (2015)
16. Singh, R., Qi, Y.: Character based string kernels for bio-entity relation detection. In: ACL 2016, p. 66 (2016)
17. Vapnik, V.N.: Statistical Learning Theory. Wiley-Interscience, Hoboken (1998)
18. Vishwanathan, S.V.N., Smola, A.J., et al.: Fast kernels for string and tree matching. In: Kernel Methods in Computational Biology, pp. 113–130 (2004)

Graph Enhanced Memory Networks
for Sentiment Analysis

Zhao Xu[1](✉), Romain Vial[1,2], and Kristian Kersting[3]

[1] NEC Laboratories Europe, Heidelberg, Germany
zhao.xu@neclab.eu
[2] MINES ParisTech, PSL Research University, Paris, France
romain.vial@mines-paristech.fr
[3] Technical University of Darmstadt, Darmstadt, Germany
kersting@cs.tu-darmstadt.de

Abstract. Memory networks model information and knowledge as memories that can be manipulated for prediction and reasoning about questions of interest. In many cases, there exists complicated relational structure in the data, by which the memories can be linked together into graphs to propagate information. Typical examples include tree structure of a sentence and knowledge graph in a dialogue system. In this paper, we present a novel graph enhanced memory network GEMN to integrate relational information between memories for prediction and reasoning. Our approach introduces graph attentions to model the relations, and couples them with content-based attentions via an additional neural network layer. It thus can better identify and manipulate the memories related to a given question, and provides more accurate prediction about the final response. We demonstrate the effectiveness of the proposed approach with aspect based sentiment classification. The empirical analysis on real data shows the advantages of incorporating relational dependencies into the memory networks.

1   Introduction

Memory network [12,23,39,45] has recently attracted increasing attention due to its success in many applications, such as machine reading and understanding, visual and textual question answering [2,13,15,42,46,47]. In general, a memory network embeds a set of facts and knowledge in vector spaces as memory cells (shorten as memories). Given a question (typically represented with natural language), the model searches the supporting memories, and infers the final answer via manipulating the retrieved memories based on attention mechanism [1,26]. The major advantage of memory networks is that they introduce an external memory component and the associated computational modules in the neural network framework to explicitly store, update, access, and manipulate the knowledge and facts for prediction, inference and reasoning given the questions. The reader and writer functions of memory networks are fully differentiable such that the entire architecture can be learned end-to-end with backpropagation.

© Springer International Publishing AG 2017
M. Ceci et al. (Eds.): ECML PKDD 2017, Part I, LNAI 10534, pp. 374–389, 2017.
https://doi.org/10.1007/978-3-319-71249-9_23

Most of the recent works on memory networks mainly focus on the contents of the facts and knowledge. However the relations between them are not taken into account. In many cases, the facts and knowledge are not independent of each other, but are linked into a relational structure. The information exists not only in the content of the facts, but also in the relations between them. The importance of relational information has been demonstrated in the literature, see e.g., probabilistic models [7,10,29] and neural network models [3,6,21,37].

In this paper, we propose a novel graph enhanced memory network (GEMN) to integrate the relational information into (deep) memory networks. GEMN allows for information propagation between memories and can thus better identify and manipulate the related memories to predict or reason the final response to a question. In particular, we link memories into a graph with their relations, and introduce an extra attention, *graph attention*, which is a weight vector, to capture the relational information. We model the graph attention with a Gaussian random field, i.e., a Gaussian distribution having graph Laplacian as kernels [35,48,49]. The memories with a short distance on the graph show strong correlation and thus likely have similar importance (i.e. weights). The graph attentions are then combined with the content-based attentions with an additional neural network layer. This introduces extra flexibility to automatically learn the combination of the two types of information (content and relations) from the data. The GEMN approach can effectively identify and leverage the important memories for a given question, and thus leads to a better inference and reasoning about the final response. There are few works investigating relational information in memory networks. An recent literature on structured attentions [19] is most relevant to our work. It models probabilistic dependencies with conditional random field that mainly focuses on sequence structures, rather than relations in general. In contrast, our approach incorporates relational information into memory networks and can model both content and relations of memories in an elegant and flexible way. The relational structure modeled in our approach can be any form, such as sequences, trees or graphs. The proposed graph attentions, combined with the content-based attentions, improves the inference and reasoning of memory networks.

We apply the proposed GEMN method to aspect level sentiment classification. With the exponential growth of user-generated content on online social network services, extracting useful insights such as preferences and opinions of users is of growing interest. Sentiment analysis [5,25,31,36] focuses on detecting opinions and emotions of users on products, services and social events from large collections of texts. Typically the sentiment analysis is to estimate the positive or negative polarity of a given sentence. A more important and complicated task is to extract the sentiment polarity towards aspects [25,31,33,34]. For example, in a customer review on a laptop, *"price is ok, but resolution is low!"*, there are positive emotion on the aspect *"price"*, and negative emotion on the aspect *"resolution"*. Simply classifying the sentence as positive or negative may not properly elicit user's opinions, thus a fine-grained analysis on aspect level sentiment is necessary. We consider a supervised case where the aspects are given. There are different approaches explored in the literature, such as SVM [4,22,44], conditional random field [14,43], and neural networks [8,30,41,42]. Our approach exploits graph- and content-attentions to position the related words (i.e. memories) in a sentence w.r.t. a given aspect (i.e. question of interest), and estimates the aspect level sentiment polarity based on the discovered relevant words. The empirical analysis on the real data about customer reviews on laptops and restaurants [33] demonstrates the superiority of the proposed approach.

We start the rest of this paper with a brief review on memory networks, and then introduce the graph enhanced memory network with the application to aspect based sentiment classification in the Sect. 3. Before conclusion, the empirical analysis of the GEMN approach is presented in the Sect. 4.

2 Memory Networks

Memory Networks [12,23,39,45] are a class of learning methods with a memory component that can be read and written for prediction, inference and reasoning. The memory networks typically consist of memories and four computational modules, including I (input), G (generalization), O (output), and R (response) [45]. They are defined as follows:

- Memories are an array of vectorized objects or facts;
- Input module computes the feature representation of the input information;
- Generalization module updates the old memory with the new input;
- Output module produces an output vector given the question of interest and the current memories;
- Response module generates the final response (such as a textual answer to a question) conditioned on the output.

In general, the input and the generalization modules map the facts and the question q (e.g. a question sentence for a question-answering system) into a feature space, and get the vector representation $\mathbf{m}_i$'s and $\mathbf{u}$ for the facts and the question, respectively. The output module manipulates the memories $\mathbf{m}_i$'s and the question vector $\mathbf{u}$ to generate a single output $\mathbf{o}$ for computing the final response. In the recent literature, the output module is typically based on the attention mechanism [1,26]. In particular, the output can be computed as (see e.g. [39]):

$$p_i = softmax(f(\mathbf{m}_i, \mathbf{u})), \quad \mathbf{o} = \sum_i p_i \mathbf{c}_i, \tag{1}$$

Intuitively p_i specifies how important the memory $\mathbf{m}_i$ is w.r.t. the question $\mathbf{u}$, and is scaled with the softmax function, i.e. $softmax(x_i) = \exp(x_i)/\sum_j exp(x_j)$ to ensure the constraints $\sum_i p_i = 1$, $p_i \in [0, 1]$. The weight function $f(\mathbf{m}_i, \mathbf{u})$ quantifies the relevance or similarity between $\mathbf{m}_i$ and $\mathbf{u}$. There are different definitions on the weight function. The typical choices include:

$$f(\mathbf{m}_i, \mathbf{u}) = \begin{cases} \mathbf{u}^T \mathbf{m}_i & \text{dot} \\ \mathbf{u}^T A \mathbf{m}_i & \text{general} \\ \mathbf{v}^T \tanh(A[\mathbf{u}; \mathbf{m}_i]) & \text{concatenation,} \end{cases} \tag{2}$$

where the matrix A and the vector $\mathbf{v}$ are parameters to be learned with back-propagation [26]. The output $\mathbf{o}$ is a weighted sum of $\mathbf{c}_i$. $\mathbf{c}_i$ is known as output memory, i.e. the vector representation of the fact i in the output feature space. In many cases, one can use the same embedding function and get $\mathbf{c}_i \equiv \mathbf{m}_i$ [39]. Given the output $\mathbf{o}$, the final answer to the question $\mathbf{u}$ is modeled as a classification problem. The probability of the label is predicted with softmax

$$p(s) = softmax(g(\mathbf{o}, \mathbf{u})),\tag{3}$$

where the function g can be similarly defined as (2).

3   Graph Enhanced Memory Networks

Memory networks provide a sophisticated neural network architecture to jointly model the facts for answering the questions of interest in an end-to-end fashion. However most existing methods in the literature mainly consider the content of the facts without the relations between them (such as the sentence tree structures and the knowledge graphs). In this paper, we propose a graph enhanced memory network (GEMN), which introduces additional graph attentions to model the relational information for better positioning and manipulating the relevant memories w.r.t. the given questions.

Attentions can be viewed as an additional hidden layer in a neural network framework to estimate a categorical distribution $(p_1, \ldots, p_N)$ for soft selection over the number N of memories. It is obvious that integrating the relational information into the learning process can lead to a more accurate attention distribution. Inspired by [35,48,49], we introduce an auxiliary random variable z_i to each memory. The value of z_i specifies graph attention weight, i.e., to what extent the memory i contributes to the output $\mathbf{o}$ based on its relations to other memories. z_i's are not independent of each other, but are interconnected into a weighted graph $\mathcal{G} = (Z, E, W)$, where $z_i \in Z$ (one for each memory) is the vertex of the graph, and $e \in E$ is the edges between z_i's. The graph $\mathcal{G}$ is represented as an adjacency matrix W of size $N \times N$, where N denotes the number of memories. Each entry $W_{i,j}$ represents the weight of an edge $e_{i,j}$ between the memories i and j. Intuitively, the larger the weight, the stronger the correlation between the two memories, and thus the more likely the memories are assigned similar graph attentions for the output. We formulate the weight as a function of the distance $d_{i,j}$ between i and j on the graph $\mathcal{G}$. The function can be of any form, but non-negative and monotonically decreasing. It can be defined as, e.g.,:

$$\text{Squared exponential: } \exp(-d^2/2\ell^2)\tag{4}$$

$$\text{Rational quadratic: } (1 + d^2/2\alpha\ell^2)^{-\alpha}\tag{5}$$

$$\gamma\text{-exponential: } \exp(-(d/\ell)^\gamma),\ 0 < \gamma \leq 2\tag{6}$$

With the adjacency matrix, we now model the distribution of z_i's for a soft selection over memories. The distribution is modeled as Gaussian random field

[48–50]. In particular, the state of z_i is only conditioned on the connected random variables, and follows a Gaussian distribution. The energy, i.e. sum of clique potentials of the Gaussian random field, is thus defined as [49]:

$$E(\mathbf{z}) = \frac{1}{4} \sum_{i,j} W_{i,j} (z_i - z_j)^2. \tag{7}$$

Therefore, the distribution of z_i's is

$$p(\mathbf{z}) \propto \exp\left(-E(\mathbf{z})\right),$$
$$= \exp\left(-\frac{1}{2}\mathbf{z}^T \Delta \mathbf{z}\right), \tag{8}$$

which is a Gaussian with mean zero and covariance Δ^{-1}. Δ denotes combinatorial graph Laplacian: $\Delta = D - W$, where D is a diagonal degree matrix with $D_{i,i} = \sum_j W_{i,j}$. Putting everything together, we now have the graph based output $\mathbf{o}_g$:

$$\mathbf{o}_g = \sum_i z_i \mathbf{m}_i, \ \mathbf{z} \sim \mathcal{N}(\mathbf{0}, \Delta^{-1}). \tag{9}$$

The content based output $\mathbf{o}_c$ is computed as usual, see (1). To learn from both content and relational information, we mix the two types of outputs with different ways, e.g.:

$$\mathbf{o} = h(\mathbf{o}_c, \mathbf{o}_g) = \begin{cases} \mathbf{o}_c + \mathbf{o}_g & \text{addition} \\ \mathbf{o}_c \otimes \mathbf{o}_g & \text{multiplication} \\ B[\mathbf{o}_c; \mathbf{o}_g] & \text{concatenation} \end{cases} \tag{10}$$

Here multiplication is defined as:

$$\mathbf{o}_c \otimes \mathbf{o}_g = \sum_i a_i \mathbf{m}_i, \ a_i = softmax(z_i \cdot f(\mathbf{m}_i, \mathbf{u})) \tag{11}$$

The parameter matrix B in concatenation makes a linear transformation from the concatenation space to the memory space, and will be learned from the data with backpropagation. Addition is actually a special case of concatenation (i.e. a special weight matrix B). Compared with addition, concatenation can provide more flexibility in learning complex combination of content and graph information from the data (e.g. different weights on dimensions).

We also extend our model to a multiple level version. The structure of the deep network is stacked as follows:

$$\mathbf{u}^{(t)} = A\mathbf{u}^{(t-1)} + \mathbf{o}^{(t-1)}, \ p_i^{(t)} = softmax(f(\mathbf{m}_i, \mathbf{u}^{(t)})), \tag{12}$$
$$\mathbf{o}_c^{(t)} = \sum_i p_i^{(t)} \mathbf{m}_i, \qquad \mathbf{o}^{(t)} = h(\mathbf{o}_c^{(t)}, \mathbf{o}_g), \tag{13}$$

where the stacking strategy of memory embeddings $\{\mathbf{m}_1, \dots, \mathbf{m}_N\}$ is RNN-like, i.e., keeping the memories the same across layers [39]. At the top of the network, the final response is computed with softmax: $p(s) = softmax(g(\mathbf{o}^{(t)}, \mathbf{u}^{(t)}))$.

3.1   The GEMN for Sentiment Analysis

We now illustrate the graph enhanced memory network with aspect based sentiment classification. The network structure is shown as Fig. 1. Assume that there
is a sentence consisting of a sequence of words $\{w_1, \ldots, w_N\}$ and multiple aspects
$\{a_1, \ldots, a_M\}$. For instance, let consider a guest comment on a restaurant, *"food
is ok, but service is bad"*, with two aspect words *"food"* and *"service"*. The task
is to detect aspect level sentiment (i.e., positive emotion on *"food"* and negative
emotion on *"service"*) by exploiting the semantic meanings of words and the tree
structure of the sentence. Here we assume each aspect only involves a single word
in the sentence (e.g., *"food"* and *"service"*). In the case of multi-word aspects,
the computation will be similar.

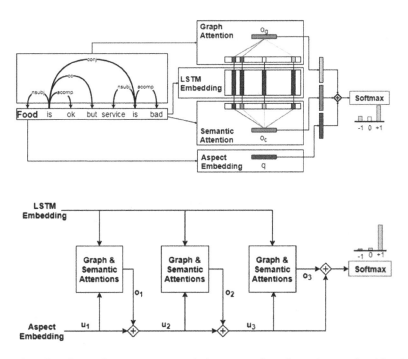

Fig. 1. Graph enhanced memory network for aspect-based sentiment classification: a
single layer version (top) and a multiple layers version (bottom).

Let start with the single level version of our model, shown as the top panel
of Fig. 1. In the memory network framework, the words $\{w_1, \ldots, w_N\}$ of the
sentence are the facts, and the aspect word (e.g. *"food"*) is formulated as the
question q. The final response is the aspect level polarity (positive, negative or
neutral) of the sentence. For the input module, we use word embedding [24,28,32]
and long short-term memory (LSTM) [9,11,16,40], i.e., the LSTM with pre-
trained word vectors as the freezed embedding matrix. The output vector of the

Table 1. Statistics of the datasets

Dataset	Positive	Negative	Neutral
Laptop train	987	866	460
Laptop test	341	128	169
Restaurant train	2164	805	633
Restaurant test	728	196	196

LSTM cell, one for each word w_i, is the memory $\mathbf{m}_i$. The question q (aspect word) is mapped as a vector $\mathbf{u}$ with word embedding. For the output module, the output vector $\mathbf{o}$ consists of two components: content-based $\mathbf{o}_c$ and graph-based $\mathbf{o}_g$. They are computed with (1) and (9), and mixed with (10). Here the activation function for computing p_i can be flexible, e.g., we can replace the softmax with the tanh function, which is theoretically more reasonable (refer to categorical distributions of multi-label classification problem). The graph attention weights z_i's follow a Gaussian distribution (8). Since the aspect word is given, we can compute the maximum likelihood estimations (i.e. mean of the Gaussian conditioned on the aspect word) as the values of z_i's. To characterize the properties of z_i's explicitly in terms of matrix operations, the distribution (8) is expanded as:

$$\begin{bmatrix} z_a \\ \mathbf{z}_m \end{bmatrix} \sim \mathcal{N}\left(\mathbf{0}, \begin{bmatrix} D_{a,a} - W_{a,a} & -W_{a,m} \\ -W_{m,a} & D_{m,m} - W_{m,m} \end{bmatrix}^{-1} \right) \tag{14}$$

where z_a denotes the graph attention weight of the aspect word, which is known as $z_a \equiv 1$, since the word is directly related to the aspect. The vector $\mathbf{z}_m$ denotes the unknown graph attention weights of the other words in the sentence. The Laplacian Δ is split into four corresponding blocks for the aspect word and the other words. Then the maximum likelihood estimation of $\mathbf{z}_m$ conditioned on the attention weight z_a is:

$$\mathbf{z}_m = (D_{m,m} - W_{m,m})^{-1} W_{m,a} z_a. \tag{15}$$

Finally the response module computes the final response with the softmax (3) to predict the probability of the aspect level sentiment polarity. We also model the sentiment classification problem with a multiple level version of the GEMN. The network structure is shown as the bottom panel of Fig. 1.

4   Experiments

To evaluate the performance of the graph enhanced memory network, we apply the approach to address the aspect-based sentiment classification problem. The experimental analysis is performed on real data with comparison against the state-of-the-art methods.

4.1   Datasets

The data is from the Task 4.2 of SemEval2014 [33], which includes two domain-specific English datasets for laptop and restaurant customer reviews. Each dataset has been manually labeled with annotations at the sentence level. The statistics of the datasets are summarized in Table 1. We follow the settings as in [42] that removes the sentences with the label *conflict* due to the small size of the category. The goal of the experiment is to predict the aspect level polarity (three polarities: positive, negative and neutral) of a sentence given the labeled aspect terms. Note that one sentence can include multiple aspects. For example, given the sentence *"Great food but the service was dreadful!"* and the aspect terms { *"food"* and *"service"* }, successful predictions would be { *"food"*: positive and *"service"*: negative}.

4.2   Baselines

The proposed method is compared with multiple recent baselines to demonstrate its performance on aspect based sentiment prediction. The baselines include:

- *Majority*: assigns to each sentence in the test set the majority sentiment label in the training set.
- *SVM* [22]: is ranked at the 1st (Laptops) and 2nd (Restaurants) places in the SemEval2014 contest. The features used in the method are sophisticated hand-crafted, including n-gram, lexicon and parse features.
- Three LSTM based models [41]: the *LSTM* directly uses the output vector of the LSTM cell for the last word of a sentence as input of a softmax to estimate the sentiment polarity. The *TDLSTM* extends the LSTM to consider the content similarity with the aspect words. The *TDLSTM+ATT* further extends the TDLSTM with the attention mechanism.
- *MemNet* [42]: uses several layers of attentions over the word embeddings. MemNet(t) denotes that the model uses t layers of attentions.

Table 2. Classification accuracy of different methods

Baselines	Laptops	Restaurants	GEMN	Laptops	Restaurants
Majority	53.45	65.00	Semantic Attention	70.69	78.84
Feature+SVM	72.10	80.89	Graph Attention	73.51	80.36
LSTM	66.45	74.28	Graph + Semantic (1 hop)	73.82	80.00
TDLSTM	68.13	75.63	Graph + Semantic (2 hops)	**74.29**	80.71
TDLSTM+ATT	66.24	74.31	Graph + Semantic (3 hops)	73.20	80.18
MemNet(1)	67.66	76.10	Graph + Semantic (4 hops)	72.88	80.54
MemNet(3)	71.74	79.06	Graph + Semantic (5 hops)	72.72	80.80
MemNet(5)	71.89	80.14	Graph + Semantic (6 hops)	72.41	**81.43**
MemNet(7)	72.37	80.32	Graph + Semantic (7 hops)	72.72	80.09
MemNet(9)	72.21	80.95	Graph + Semantic (8 hops)	72.26	80.62

4.3 Quantitative Analysis

We first perform quantitative analysis of the proposed method. In the experiments, the GEMN is used to predict aspect level sentiment polarity (positive, negative and neutral) for each test sentence. The performance is measured with classification accuracy.

The graph structure of a sentence, used in the proposed approach, is extracted with Stanford's CoreNLP Toolkit [27]. Here we use the constituency tree of a sentence. The adjacency matrix is computed using squared exponential kernel with $\ell = 0.1$. The distance $d_{i,j}$ between two words is defined as the number of edges of the shortest path connecting them. The distance is normalized by the diameter of the sentence tree. The questions (i.e. the aspects) and the words are mapped as 300-dimensional Glove vectors [32], and the weights of the embedding matrix are freezed during training. The LSTM is then used to map each word in a sentence into a 128-dimensional memory space. We use an aggressive dropout of 0.7 before the final softmax layer to prevent the model from overfitting [38]. Dropout of 0.5 and 0.3 are respectively used at the input nodes and the recurrent connections of the LSTM cells. The optimization is done with Adam method [20]. The learning rate is set to 0.005. The model learns during 10 epochs with a batch size of 32 training sentences.

Table 3. Classification accuracy of the GEMN approach with the constituency and the dependency tree structures of the sentences

	Laptops		Restaurants	
	Constituency	Dependency	Constituency	Dependency
Semantic attention	70.69	73.04	78.84	78.67
Graph attention	73.51	73.51	80.36	79.20
Graph + semantic (1 hops)	73.82	73.51	80.00	80.09
Graph + semantic (2 hops)	74.29	74.76	80.71	80.18
Graph + semantic (5 hops)	72.72	74.45	80.80	81.07
Graph + semantic (10 hops)	71.16	75.39	80.54	80.08

To get detailed performance of the proposed approach, we consider different ways to compute the output vectors for the final softmax layer:

- Variant 1: only models content based output, $\mathbf{o} \equiv \mathbf{o}_c$. In this case we do not use any information extracted from the graph structure of the sentence.
- Variant 2: only models graph based output, $\mathbf{o} \equiv \mathbf{o}_g$. Here the content based information (i.e. the semantic meanings of the words) is ignored.
- Variant 3: combines both outputs with multiplication, $\mathbf{o} = \mathbf{o}_c \otimes \mathbf{o}_g$.
- Variant 4: models the stacked and combined outputs $\mathbf{o}^{(t)}$ with (12) (t hops).

The experimental results are summarized in Table 2. For a fair comparison, we directly use the results of the baselines reported in [42]. Our approach, which models both graph and content attentions and refines over multiple layers, outperforms the baselines. It is interesting to note that our approach with only the graph attentions performs rather well. It reveals that the relational structures are pretty informative in predicting the relevant memories in the given context. The semantic information of the words, which may not be fully contained in the parse trees, can further improve the predictions. Therefore, combining both graph and semantic attentions leads to a notable gain in prediction accuracy. Stacking several layers to get a deep network performs well, e.g., a 0.47% increase in accuracy on the Laptops dataset. In summary, the empirical results demonstrate that, as the relational information reveals additional correlations among memories, the proposed graph attentions help the memory model to focus on the important memories w.r.t. the given aspects, and thus the GEMN achieves better predictions.

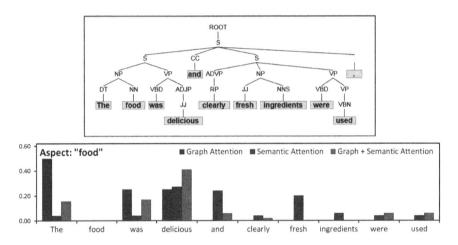

Fig. 2. Example sentence with a single aspect *food*: the constituency tree of the sentence (top) and the learned attention weights (bottom).

We also investigate the influence of the different graph structures of the sentences on the performance of the proposed method. Here we consider two types of tree structures: constituency and dependency. Dependency tree models one-to-one correspondence, and focuses on word grammars, while constituency tree models one-to-one-or-more correspondence, and focuses on phrase structure grammars. The dependency trees are smaller than the constituency ones. The differences between the two types of trees are investigated in details in [18]. Here we do not use the chain structure of the sentences (i.e. distance between the indices of the words), as the chain may meet difficulties in modeling some scenarios, e.g. *"service is bad, but food of the restaurant is good"*. The word *"food"* should be related to *"good"*, rather than *"bad"*, although the index distance is larger. The tree structures of the sentences can better encode such complicated cases. In the experiments, we parse the sentences with Stanford's CoreNLP Toolkit [27] to get the constituency trees, and with the spaCy Toolkit [17] to get the dependency ones. As summarized in Table 3, the GEMN approach with the different tree structures of the sentences achieves similar performance. One can find that both types of tree structures can be well integrated with graph attentions, and provide improvement of classification accuracy. The experimental results further demonstrate the advantages of the proposed approach.

4.4   Qualitative Analysis

To better understand the performance of the proposed approach, we further analyse the computed attentions and reveal interesting insights. Figures 2 and 3 show two example sentences with one and two aspects, respectively. One can see the constituency trees of the sentences and the learned attention weights with respect to the corresponding aspect word. On one hand, graph attention, which only models the relations between memories (i.e. sentence structure), appears to effectively identify the important memories related to the context (aspect), and assigns them high weights. On the other hand, content-based (i.e. semantic) attention only considers the meanings of the words without syntactic clues. As a consequence, it highlights all the sentiment keywords, even if it is not related to the context. For example, *"fresh"* is actually not for the aspect word *"food"* in

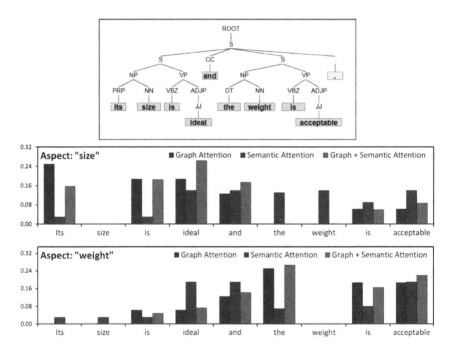

Fig. 3. Example sentence with two aspects: the constituency tree of the sentence (top), and the learned attention weights for the aspect *size* (middle) as well as the aspect *weight* (bottom).

Fig. 2, and *"ideal"* not for the aspect word *"weight"* in Fig. 3. The mixed attention (graph + semantic) takes advantage of both relational and content information to identify in-context memories, and thus better discovers the important words w.r.t. the given aspects. One can see that the words *"delicious"* (Fig. 2 for the aspect *"food"*), *"ideal"* (Fig. 3 for the aspect *"size"*) and *"acceptable"* (Fig. 3 for the aspect *"weight"*) capture larger attentions with scores of 0.41, 0.26 and 0.22 respectively.

We also perform analysis on the effectiveness of a deep structure (i.e., multiple layers attentions). As shown in Fig. 4, one can find that one layer of mixed attention is not always enough to handle complex sentences. With more layers, the attention weights appear to be refined at each pass and gradually focus on the important words. For example, the word *"delicious"* has its score increasing from 0.41 to 0.59 with the number of layers, and simultaneously the noise caused by other words in the sentence is reduced. In addition, we investigate the influence of the different tree structures of the sentences on the graph attentions. Figure 5 illustrates with an example sentence with two aspects. Although the exact values of the attention weights learned from the two tree structures are slightly different, they reveal similar tendency: the words having close syntactic relations with the aspect words get larger attention weights and vice versa.

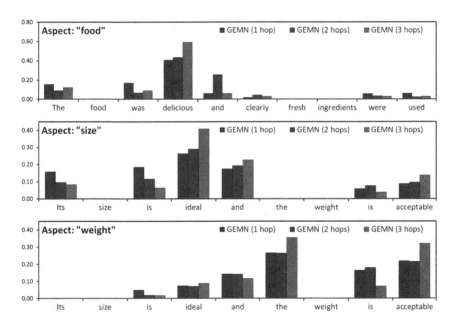

Fig. 4. Examples of the learned attention weights at each hop.

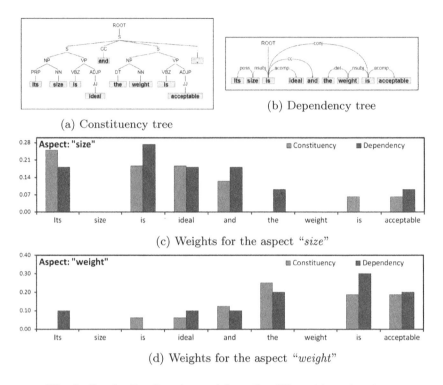

(a) Constituency tree

(b) Dependency tree

(c) Weights for the aspect *"size"*

(d) Weights for the aspect *"weight"*

Fig. 5. Graph attentions learned from the different tree structures

5   Conclusion

In this paper we present a graph enhanced memory network (GEMN) to incorporate relational information for better predicting and reasoning the final response to the question of interest. We introduce a new type of attentions, graph attentions, to model the graph structure of the memories. The graph attentions are mixed with the content based attentions via an additional neural network layer, which flexibly learns the combination of both relational and content information. In turn the GEMN can better identify and manipulate the relevant memories w.r.t. the given question. The GEMN is applied to aspect-based sentiment classification, and the empirical analysis on real data demonstrates superior performance. Our work provides interesting avenues for future work, such as graph enhanced memory networks for question-answering and knowledge graph reasoning.

References

1. Bahdanau, D., Cho, K., Bengio, Y.: Neural machine translation by jointly learning to align and translate. In: Proceedings of the International Conference on Learning Representations (2015)
2. Bordes, A., Usunier, N., Chopra, S., Weston, J.: Large-scale simple question answering with memory networks. arXiv preprint: arXiv:1506.02075 (2015)
3. Bordes, A., Usunier, N., Garcia-Duran, A., Weston, J., Yakhnenko, O.: Translating embeddings for modeling multi-relational data. Adv. Neural Inf. Process. Syst. **26**, 2787–2795 (2013)
4. Brychcin, T., Konkol, M., Steinberger, J.: UWB: machine learning approach to aspect-based sentiment analysis. In: Proceedings of the 8th International Workshop on Semantic Evaluation, pp. 817–822 (2014)
5. Dai, A.M., Le, Q.V.: Semi-supervised sequence learning. In: Advances in Neural Information Processing Systems (2015)
6. Dai, H., Dai, B., Song, L.: Discriminative embeddings of latent variable models for structured data. In: ICML (2016)
7. De Raedt, L., Kersting, K., Natarajan, S., Poole, D.: Statistical Relational Artificial Intelligence: Logic, Probability, and Computation. Synthesis Lectures on Artificial Intelligence and Machine Learning. Morgan & Claypool Publishers, San Rafael (2016)
8. Dong, L., Wei, F., Tan, C., Tang, D., et al.: Adaptive recursive neural network for target-dependent Twitter sentiment classification. In: Proceedings of the 52nd Annual Meeting of the Association for Computational Linguistics (2014)
9. Gers, F.A., Schmidhuber, J., Cummins, F.: Learning to forget: Continual prediction with LSTM. Neural Comput. **12**(10), 2451–2471 (2000)
10. Getoor, L., Taskar, B. (eds.): Introduction to Statistical Relational Learning. MIT Press, Cambridge (2007)
11. Graves, A.: Supervised Sequence Labelling with Recurrent Neural Networks. Studies in Computational Intelligence. Springer, Heidelberg (2012). https://doi.org/10.1007/978-3-642-24797-2
12. Graves, A., Wayne, G., Danihelka, I.: Neural turing machines. arXiv preprint: arXiv:1410.5401 (2014)
13. Graves, A., Wayne, G., Reynolds, M., Harley, T., Danihelka, I., et al.: Hybrid computing using a neural network with dynamic external memory. Nature **538**, 471–476 (2016)
14. Hamdan, H., Bellot, P., Bechet, F.: Lsislif: CRF and logistic regression for opinion target extraction and sentiment polarity analysis. In: Proceedings of the 9th International Workshop on Semantic Evaluation, pp. 719–724 (2015)
15. Hill, F., Bordes, A., Chopra, S., Weston, J.: The goldilocks principle: Reading children's books with explicit memory representations. arXiv preprint: arXiv:1511.02301 (2015)
16. Hochreiter, S., Schmidhuber, J.: Long short-term memory. Neural Comput. **9**(8), 1735–1780 (1997)
17. Honnibal, M., Johnson, M.: An improved non-monotonic transition system for dependency parsing. In: Proceedings of the 2015 Conference on Empirical Methods in Natural Language Processing, pp. 1373–1378 (2015)
18. Hudson, R.: Constituency and depdency. Linguistics **18**, 179–198 (1980)
19. Kim, Y., Denton, C., Hoang, L., Rush, A.M.: Structured attention networks. In: Proceedings of the International Conference on Learning Representations (2017)

20. Kingma, D., Ba, J.: Adam: a method for stochastic optimisation. In: Proceedings of the International Conference on Learning Representations (2015)
21. Kipf, T.N., Welling, M.: Semi-supervised classification with graph convolutional networks. In: Proceedings of ICLR (2017)
22. Kiritchenko, S., Zhu, X., Cherry, C., Mohammad, S.: NRC-Canada-2014: Detecting aspects and sentiment in customer reviews. In: Proceedings of the 8th International Workshop on Semantic Evaluation, pp. 437–442 (2014)
23. Kumar, A., Irsoy, O., Ondruska, P., Iyyer, M., Bradbury, J., Gulrajani, I., Zhong, V., Paulus, R., Socher, R.: Ask me anything: dynamic memory networks for natural language processing. In: Proceedings of the International Conference on Machine Learning (2016)
24. Le, Q.V., Mikolov, T.: Distributed representations of sentences and documents. In: Proceedings of the 31th International Conference on Machine Learning, pp. 1188–1196 (2014)
25. Liu, B.: Sentiment Analysis and Opinion Mining. Morgan & Claypool Publishers, San Rafael (2012)
26. Luong, M.T., Pham, H., Manning, C.D.: Effective approaches to attention based neural machine translation. In: Proceedings of the Conference on Empirical Methods in NLP (2015)
27. Manning, C.D., Surdeanu, M., Bauer, J., Finkel, J., et al.: The Stanford CoreNLP natural language processing toolkit. In: Association for Computational Linguistics (ACL) System Demonstrations, pp. 55–60 (2014)
28. Mikolov, T., Sutskever, I., Chen, K., Corrado, G.S., Dean, J.: Distributed representations of words and phrases and their compositionality. Adv. Neural Inf. Process. Syst. **26**, 3111–3119 (2013)
29. Miller, K.T., Griffiths, T.L., Jordan, M.I.: Nonparametric latent feature models for link prediction. In: Advances in Neural Information Processing Systems (2009)
30. Nguyen, T.H., Shirai, K.: PhraseRNN: phrase recursive neural network for aspect-based sentiment analysis. In: Proceedings of the 2015 Conference on Empirical Methods in Natural Language Processing, pp. 2509–2514 (2015)
31. Pang, B., Lee, L.: Opinion mining and sentiment analysis. Found. Trends Inf. Retrieval **2**(1–2), 1–135 (2008)
32. Pennington, J., Socher, R., Manning, C.D.: Glove: global vectors for word representation. In: Proceedings of the Conference on Empirical Methods in Natural Language Processing, pp. 1532–1543 (2014)
33. Pontiki, M., Galanis, D., Pavlopoulos, J., Papageorgiou, H., et al.: Semeval-2014 task 4: aspect based sentiment analysis. In: Proceedings of the 8th International Workshop on Semantic Evaluation, pp. 27–35 (2014)
34. Schouten, K., Frasincar, F.: Survey on aspect-level sentiment analysis. IEEE Trans. Knowl. Data Eng. **28**(3), 813–830 (2016)
35. Smola, A.J., Kondor, R.: Kernels and regularization on graphs. In: Schölkopf, B., Warmuth, M.K. (eds.) COLT-Kernel 2003. LNCS (LNAI), vol. 2777, pp. 144–158. Springer, Heidelberg (2003). https://doi.org/10.1007/978-3-540-45167-9_12
36. Socher, R., Perelygin, A., Wu, J.Y., Chuang, J., et al.: Recursive deep models for semantic compositionality over a sentiment treebank. In: EMNLP (2013)
37. Socher, R., Chen, D., Manning, C., Ng, A.Y.: Reasoning with neural tensor networks for knowledge base completion. In: Advances in Neural Information Processing Systems, pp. 926–934 (2013)
38. Srivastava, N., Hinton, G., Krizhevsky, A., Sutskever, I., Salakhutdinov, R.: Dropout: a simple way to prevent neural networks from overfitting. J. Mach. Learn. Res. **15**(1), 1929–1958 (2014)

39. Sukhbaatar, S., Szlam, A., Weston, J., Fergus, R.: End-to-end memory networks. In: NIPS, pp. 2431–2439 (2015)
40. Sutskever, I.: Training Recurrent Neural Networks, Ph.D. thesis. University of Toronto (2013)
41. Tang, D., Qin, B., Feng, X., Liu, T.: Target-dependent sentiment classification with long short term memory. arXiv preprint: 1512.01100 (2015)
42. Tang, D., Qin, B., Liu, T.: Aspect level sentiment classification with deep memory network. In: EMNLP, pp. 214–224 (2016)
43. Toh, Z., Wang, W.: Dlirec: aspect term extraction and term polarity classification system. In: Proceedings of the 8th International Workshop on Semantic Evaluation, pp. 235–240 (2014)
44. Wagner, J., Arora, P., Cortes, S., et al.: DCU: Aspect-based polarity classification for semeval task 4. In: Proceedings of the 8th International Workshop on Semantic Evaluation, pp. 223–229 (2014)
45. Weston, J., Chopra, S., Bordes, A.: Memory networks. In: ICLR (2015)
46. Xiong, C., Merity, S., Socher, R.: Dynamic memory networks for visual and textual question answering. In: ICML (2016)
47. Xu, H., Saenko, K.: Ask, attend and answer: exploring question-guided spatial attention for visual question answering. In: Leibe, B., Matas, J., Sebe, N., Welling, M. (eds.) ECCV 2016. LNCS, vol. 9911, pp. 451–466. Springer, Cham (2016). https://doi.org/10.1007/978-3-319-46478-7_28
48. Xu, Z., Kersting, K., Tresp, V.: Multi-relational learning with Gaussian processes. In: IJCAI, pp. 1309–1314 (2009)
49. Zhu, X., Ghahramani, Z., Lafferty, J.: Semi-supervised learning using Gaussian fields and harmonic functions. In: ICML (2003)
50. Zhu, X., Lafferty, J., Ghahramani, Z.: Semi-supervised learning: from Gaussian fields to Gaussian processes. Technical report CMU-CS-03-175. Carnegie Mellon University (2003)

Kernel Sequential Monte Carlo

Ingmar Schuster[1], Heiko Strathmann[2], Brooks Paige[3], and Dino Sejdinovic[4](✉)

[1] FU Berlin, Berlin, Germany
`ingmar.schuster@fu-berlin.de`
[2] Gatsby Unit, University College London, London, UK
`heiko.strathmann@gmail.com`
[3] Alan Turing Institute and University of Cambridge, Cambridge, UK
`bpaige@turing.ac.uk`
[4] University of Oxford and Alan Turing Institute, Oxford, UK
`dino.sejdinovic@stats.ox.ac.uk`

Abstract. We propose kernel sequential Monte Carlo (KSMC), a framework for sampling from static target densities. KSMC is a family of sequential Monte Carlo algorithms that are based on building emulator models of the current particle system in a reproducing kernel Hilbert space. We here focus on modelling nonlinear covariance structure and gradients of the target. The emulator's geometry is adaptively updated and subsequently used to inform local proposals. Unlike in adaptive Markov chain Monte Carlo, continuous adaptation does not compromise convergence of the sampler. KSMC combines the strengths of sequental Monte Carlo and kernel methods: superior performance for multimodal targets and the ability to estimate model evidence as compared to Markov chain Monte Carlo, and the emulator's ability to represent targets that exhibit high degrees of nonlinearity. As KSMC does not require access to target gradients, it is particularly applicable on targets whose gradients are unknown or prohibitively expensive. We describe necessary tuning details and demonstrate the benefits of the the proposed methodology on a series of challenging synthetic and real-world examples.

1 Introduction

Monte Carlo methods for estimating integrals have become one of the main inference tools of statistics and machine learning over the last thirty years. They are used to numerically approximate intractable integrals with respect to Bayesian posterior distributions. Importantly, they also provide means to quantify uncertainty in the form of variance estimates, credible intervals and regions of high posterior density. The most widely adopted Monte Carlo method is Markov Chain Monte Carlo (MCMC), which constructs a Markov chain that admits the desired target as its stationary distribution; MCMC generates approximate samples from the target when the chain is run sufficiently long. Poorly tuned

---

I. Schuster and H. Strathmann—Equal contribution.

© Springer International Publishing AG 2017
M. Ceci et al. (Eds.): ECML PKDD 2017, Part I, LNAI 10534, pp. 390–409, 2017.
https://doi.org/10.1007/978-3-319-71249-9_24

MCMC samplers may need to run 'burn in' for a very long time before reaching its equilibrium distribution, and successive samples may be highly correlated.

In contrast, sequential Monte Carlo (SMC) methods are based on iterative importance sampling, and have traditionally been applied to inference in filtering problems with a sequence of time-varying target distributions [9], e.g. in state-space models, where each intermediate distribution is typically defined on a successively larger latent space. In this paper, we focus on static SMC methods, which recently have generated increasing interest as an alternative to MCMC for Bayesian inference on a single target distribution [3,4,6,11]. Static SMC frames inference over a fixed target distribution as a sequential problem by defining an artificial series of incremental targets. This can be done by tempering the target density [6], by including data points sequentially [4], or by targeting the full density at every iteration. The latter is a special case known as population Monte Carlo (PMC, [2]).

Kernel methods have recently been employed to construct efficient adaptive MCMC algorithms: via modelling a Markov chain trajectory in a reproducing kernel Hilbert space (RKHS) and using geometry therein, it is possible to significantly improve mixing on target distributions with nonlinear interactions between components. Covariance in the RKHS can be used to construct an adaptive random walk scheme, kernel adaptive Metropolis Hastings (KAMH), with proposals that are locally aligned with the target density [20]. Gradients of exponential families in the RKHS can be used to construct kernel Hamiltonian Monte Carlo (KHMC), an algorithm that behaves similar to Hamiltonian Monte Carlo (HMC) but without requiring access to gradient information [22]. Both KAMH and KHMC fall back to a random walk in yet unexplored regions, inheriting convergence properties such as geometric ergodicity on log-concave targets (c.f. Proposition 3 in [22]).

In this paper, we develop a framework for *kernel sequential Monte Carlo (KSMC)* for sampling from static models. Similarly to the previous work in adaptive MCMC [20,22], KSMC represents the (weighted) particle system of SMC algorithms in a RKHS. The learned geometry of the corresponding 'emulator' model is used to construct proposal distributions for both MCMC rejuvenation and importance sampling steps inside SMC.

We apply this framework to two existing SMC algorithms, combining the strengths of SMC with those of kernel adaptive MCMC. Firstly, we introduce *kernel adaptive sequential Monte Carlo (KASMC)*, where the global covariance estimate in the adaptive SMC sampler [ASMC, 11] is replaced by a kernel-informed local covariance [20]. Similar to ASMC, KASMC's proposals start as a standard random walk and then smoothly transition to taking locally aligned steps. As a result, sampling efficiency can be significantly improved over ASMC. Secondly, we use an infinite dimensional exponential family model [21] to estimate target gradients as in Strathmann et al. [22]. This results in *kernel gradient importance sampling (KGRIS)*, a gradient-free version of gradient importance sampling (GRIS) [19]. KGRIS is a novel adaptation of kernel gradient estimation ideas for constructing Langevin diffusions, and inherits their sampling efficiency

---

Algorithm 1. Sequential Monte Carlo for Static Models

Input: Sequence of target densities $\pi_0, \ldots, \pi_T$ (where $\pi_T = \pi$), size of particle system N

Output: sets $\mathbf{X}_1, \ldots, \mathbf{X}_T$ and $\mathbf{W}_1, \ldots, \mathbf{W}_T$ of samples and accompanying weights

Initialise $\mathbf{X}_0$ to N samples from π_0, and $\mathbf{W}_0$ to equal weights $1/N$

for $t = 1$ through $t = T$ **do**

$\quad \widetilde{\mathbf{W}}_t = \{W_{t-1}^i \pi_t(X_{t-1}^i)/\pi_{t-1}(X_{t-1}^i)\}_{i=1}^N$

$\quad$ construct $\widetilde{\mathbf{X}}_t$ by re-sampling $(\mathbf{X}_{t-1}, \widetilde{\mathbf{W}}_t)$, resulting in N copies of samples in $\mathbf{X}_{t-1}$

$\quad$ construct or update proposal q_t

$\quad$ **if** using an MH transition kernel **then**

$\quad\quad$ Set $\mathbf{X}_t$ to

$\quad\quad\quad \{X_t^i \sim$ MH kernel with proposal $q_t(\cdot|\widetilde{X}_t^i)\}_{i=1}^N$

$\quad\quad \mathbf{W}_t = \{1/N\}_{i=1}^N$

$\quad$ **else**

$\quad\quad$ Set $\mathbf{X}_t$ to N samples from

$\quad\quad\quad q_t^{\mathrm{Mixt}}(\cdot) = \frac{1}{N}\sum_{i=1}^N q_t(\cdot|\widetilde{X}_t^i)$

$\quad\quad \mathbf{W}_t = \{\pi_t(X_{t,i})/q_t^{\mathrm{Mixt}}(X_{t,i})\}_{i=1}^N$

$\quad$ **end if**

end for

return $\mathbf{X}_1, \ldots, \mathbf{X}_T$ and $\mathbf{W}_1, \ldots, \mathbf{W}_T$

---

compared to random walks. Our contribution includes crucial implementation details, such as Rao-Blackwelisation, stratification, and tuning of the presented algorithms.

Unlike for Langevin diffusions or Hamiltonian dynamics, our framework does not require gradients or higher-order information of the target. Consequently, the KSMC framework is particularly useful in combination with importance sampling frameworks such as SMC^2 [5] and IS^2 [23] for sampling from doubly intractable targets, where gradient information is unavailable.

We finally argue that (adaptive) SMC is a more natural framework for employing RKHS-based representations. Adaptive MCMC samplers require a vanishing adaptation schedule in order to ensure convergence to the correct target [18], creating a difficult to tune exploration-exploitation trade-off with limited principled guidance on selecting such adaptation schedules. In contrast, SMC proposals can continuously be adapted and the choice of an adaptation schedule is thus entirely circumvented. An easy to use Python package implementing the proposed methods is available under an open source licence.[1]

2   Background

Sequential Monte Carlo algorithms [6,8] approximate a target density π by iteratively targeting a sequence of incremental densities $\pi_0, \ldots, \pi_T$, with $\pi_T = \pi$. These incremental densities are typically defined such that the initial density π_0

---

[1] Source code available at https://github.com/ingmarschuster/kameleon_rks.

is easy to sample from (e.g. the prior in a Bayesian model). Consecutive distributions π_t, π_{t+1} are 'close', in the sense that drawing samples from π_{t+1} given samples from π_t is easier than drawing samples from π_{t+1} directly. At each stage t, we approximate the target density π_t with a set of N samples $\mathbf{X}_t = \{X_t^i\}_{i=1}^N$ with associated importance weights $\mathbf{W}_t = \{W_t^i\}_{i=1}^N$, with

$$\hat{\pi}_t(X) = \sum_{i=1}^N W_t^i \delta_{X_t^i}(X) \tag{1}$$

where $\delta_{X_t^i}$ is a Dirac point mass on X_t^i. In contrast to SMC as applied to state space models, in a static SMC setting each target density π_t is defined on the same space $\mathcal{X}$.

We initialise the algorithm by sampling an initial set of N samples $\mathbf{X}_0$ from the initial density q_0, with equal importance weights $1/N$. For each subsequent $t = 1, \ldots, T$, given a particle set $(\mathbf{X}_{t-1}, \mathbf{W}_{t-1})$ approximating π_{t-1}, we construct a new particle set which approximates π_t. This is a three-step process, summarised in Algorithm 1. First, we re-weight each particle relative to the new target density, setting

$$\widetilde{W}_t^i = W_{t-1}^i \frac{\pi_t(X_{t-1}^i)}{\pi_{t-1}(X_{t-1}^i)}.$$

Weighting the points in $\mathbf{X}_{t-1}$ by $\{\widetilde{W}_t^i\}_{i=1}^N$ yields an approximation to π_t in the same manner as in (1) — the new importance weights correct for the change from π_{t-1} to π_t.

Static SMC then applies re-sampling, constructing an equally-weighted set of particles $\widetilde{\mathbf{X}}_t = \{\widetilde{X}_t^i\}_{i=1}^N$ by sampling with replacement from $\mathbf{X}_{t-1}$ with weights proportional to $\widetilde{W}_t^i$, [7]. Together, these samples form an approximation to π_t, where values from $\mathbf{X}_{t-1}$ with high weight under π_t have been duplicated and those with low weight under π_{t-1} have been discarded. This duplication of values, however, can lead to a sample impoverishment problem: many of the re-sampled values $\widetilde{X}_t^i$ may have identical values. This can be avoided by applying a so-called *rejuvenation* step after re-sampling [4], constructing an overall approximation $(\mathbf{X}_t, \mathbf{W}_t)$ to π_t with a diverse set of values of X_t^i.

The rejuvenation step consists of a proposal $q_t(\mathbf{X}_t|\widetilde{\mathbf{X}}_t)$. We here consider two ways of incorporating such a proposal. One traditional option is to use a Markov density q_t as a proposal in a Metropolis-Hastings (MH) kernel which leaves π_t invariant: For each $\widetilde{X}_t^i$ in $\widetilde{\mathbf{X}}_t$, we propose a new value X_t^i from $q_t(X_t^i|\widetilde{X}_t^i)$ and accept it according to a standard MH acceptance ratio targeting π_t. In this case, each importance weight in $\mathbf{W}_t$ will be identically $1/N$.

An alternative is to consider the mixture proposal of all such Markov densities q_t as an importance sampling proposal over π_t, a common approach in PMC. We can define

$$q_t^{\text{Mixt}}(X_t) = \frac{1}{N} \sum_{i=1}^N q_t(X_t|\widetilde{X}_t^i),$$

and draw N samples X_t from q_t^{Mixt} to generate $\mathbf{X}_t$. Now we set importance weights in $\mathbf{W}_t$ to $W_t^i = \pi_t(X_t^i)/q_t^{\text{Mixt}}(X_t^i)$ for $i = 1, \dots, N$.

2.1   Existing SMC Algorithms

In SMC algorithms, we are free in choosing a proposal q_t. In contrast to MCMC, it may be directly informed by the previous samples $\mathbf{X}_{t-1}$ and their weights $\mathbf{W}_{t-1}$. The following two existing SMC algorithms are examples that we will extend to kernel-based alternatives.

Adaptive SMC. The adaptive SMC sampler (ASMC) studied by Fearnhead and Taylor [11] is based on continuously estimating the global covariance Σ_t of π_t, and updating a scaling parameter ν^2. This is done from the re-weighted particle system, which is subsequently moved through a Markov kernel. The proposal distribution used within the MH kernel at point X in Algorithm 1 is $q_t(\cdot|X) = \mathcal{N}(\cdot|X, \nu^2\Sigma_t + \gamma^2 I)$.

Gradient importance sampling. In addition to using the estimated covariance Σ_t of π as in ASMC, gradient importance sampling [GRIS, 19] incorporates a drift term based on the log target gradient. For target gradient $\nabla \log \pi$ and previous sample X, the proposal distribution in Algorithm 1 is $q_t(\cdot) = \mathcal{N}(\cdot|X + D(\nabla \log \pi(X)), \nu^2\Sigma_t)$, for each individual particle X in the current (unweighted) particle set. A typical choice for the drift function is $D(y) = \delta y$ with $0 < \delta < 1$. Rather than incorporating a MH step, the updated values are importance weighted — GRIS is a population Monte Carlo (PMC) algorithm. In numerical experiments, GRIS compares favourably to its closest MCMC relatives like the adaptive MALTA algorithm and adaptive Metropolis [19].

2.2   Kernel Adaptive MCMC Proposals

The previously described SMC algorithms are based on target covariance and gradients. We now review how these quantities were previously modelled using kernel methods in the context of MCMC. Note that any form of adaptation in MCMC requires care in order to preserve ergodicity of the resulting Markov chain, and some form of vanishing adaptation is needed [1,18]. This can be achieved e.g. by updating the proposal family with vanishing probability [20,22].

Covariance emulator. Sejdinovic et al. [20] introduced a kernel covariance emulator as a method for adapting the proposal distribution in a Metropolis-Hastings MCMC algorithm, based on the history of the Markov chain $\mathbf{X} = \{X_1, X_2, \dots\}$. The idea is to represent covariance of the target as an empirical Gaussian measure with mean $\mu_{\mathbf{X}} := \frac{1}{|\mathbf{X}|}\sum_{X \in \mathbf{X}} \mathbf{k}(X, \cdot)$ and covariance $\frac{1}{|\mathbf{X}|}\sum_{X \in \mathbf{X}} \mathbf{k}(X, \cdot) \otimes \mathbf{k}(X, \cdot) - \mu_{\mathbf{X}} \otimes \mu_{\mathbf{X}}$ in a RKHS with kernel $\mathbf{k}$. This measure can be sampled from exactly, and it is possible to (approximately) map samples back to the original space.

Sejdinovic et al. [20] showed that it is possible to integrate out the RKHS proposal analytically, which elegantly results in a *closed form* Gaussian proposal density in the input space. For a Gaussian kernel, the proposal at particle X_j locally aligns to the structure of the posterior at X_j, and is given by

$$q_{\text{KAMH}}(\cdot|X_j) = \mathcal{N}(\cdot|X_j, \gamma^2 I + \nu^2 M_{\mathbf{X},X_j} C M_{\mathbf{X},X_j}^\top),$$

where $C = I - \frac{1}{n}11^\top$ is a centering matrix and $M_{\mathbf{X},X_j}$ collects kernel gradients with respect to all particles,

$$M_{\mathbf{X},X_j} = 2[\nabla_x \mathbf{k}(x, X_1)|_{x=X_j}, ..., \nabla_x \mathbf{k}(x, X_N)|_{x=X_j}].$$

Additional exploration noise with variance γ^2 avoids that the proposal collapses in unexplored regions of the input space.

Gradient emulator. To overcome random walk behaviour of KAMH, Strathmann et al. [22] constructed an algorithm that adaptively learns the gradient structure of the Markov chain history, and mimics Hamiltonian dynamics using the learned gradients. This is done by fitting an un-normalised infinite dimensional exponential family model with density function $\exp(\langle f, k(x, \cdot)\rangle_\mathcal{H} - A(f))$. Here, $\langle f, k(x, \cdot)\rangle_\mathcal{H} = f(x)$ is the inner product between natural parameters f and sufficient statistics $k(x, \cdot)$ in a RKHS $\mathcal{H}$, and $A(f)$ is the (intractable) log-partition function. Remarkably, it is possible to efficiently estimate f via minimising the expected L^2 error of $\nabla_x f(x)$ without dealing with $A(f)$. Combining this model with a further approximation, based on random basis functions [KMC finite; 22], allows for efficient on-line updates of the emulator. Similar to Hamiltonian Monte Carlo, the resulting KHMC algorithm offers substantial improvements over random walks. Tt does so, however, *without* requiring gradient information of the target. This allows application to intractable likelihood models, where we cannot evaluate the target densities π_t even up to a normalizing constant, and gradients are similarly unavailable.

3   Kernel Sequential Monte Carlo

We now develop a kernel sequential Monte Carlo framework. KSMC is based on combining classical adaptive SMC with the emulator based proposals of kernel adaptive MCMC. In general, once a kernel emulator is fitted to past particle systems, we can use it in either of two ways: as proposals for MH rejuvenation steps inside SMC or as importance densities in PMC.

Key contributions. Our main contribution is to combine several yet unconnected pieces of literature into a novel framework that performs favourably compared to its individual parts: adaptive SMC proposals, SMC for intractable likelihoods, and kernel emulators for efficient proposals. This combination is simple yet very natural: As compared to (kernel) adaptive MCMC, the KSMC framework (i) circumvents the need for vanishing adaptation, (ii) can represent multimodality,

(iii) allows to estimate model evidence in a straight-forward manner. On the other hand, as compared to plain adaptive SMC and PMC, the use of kernel emulators (iv) leads to faster convergence for nonlinear targets.

We present two novel algorithms, KASMC and KGRIS, both of which are weighted and kernelised generalisations of existing kernel MCMC and SMC respectively. These modifications can lead to significant mixing improvements in practice. Our contribution furthermore includes variance reduction techniques that are critical in practice. In particular, Naïve implementations can suffer from high variance induced by simplifications. As this results in lower quality emulators, too high variance would be self-reinforcing and is to be strictly avoided.

3.1   Kernel Adaptive Rejuvenation: KASMC

We can use both kernel emulators for the rejuvenation step of SMC. More specifically, at time-step $t + 1$, we target distribution π_{t+1}, based on a particle system approximating π_t. After re-weighting, the new system $\{(W_{t+1,i}, X_{t+1,i})\}_{i=1}^N$ is a weighted approximation to π_{t+1}. We here focus on the nonlinear covariance emulator which can be either fitted using the equally-weighted re-sampled values $\widetilde{\mathbf{X}}_t$, or the original particle set with weights $\widetilde{\mathbf{W}}_t$. The proposal distribution for Algorithm 1 at X then is exactly q_{KAMH}. As in KAMH, this results in covariance matrices for Gaussian proposals which locally align with the target [20], now taking the SMC particle weights into account. The resulting kernel adaptive SMC sampler (KASMC) inherits KAMH's ability to explore non-linear targets more efficiently than proposals based on estimating global covariance structure such as in Fearnhead and Taylor [11] and Haario et al. [13]. Figure 1 (left) shows a simple illustration of a global (ASMC) and local proposal distribution (KASMC). Compared to previous work on kernel induced local covariance matrices for MCMC [20], we implement a random features approximation in order to enable computationally efficient updates with information gained from new samples [17].

3.2   Kernel Induced Importance Densities: KGRIS

Another way to use kernel-based emulators is for generating proposals which are corrected by importance sampling, i.e. in PMC. In our second approach, a kernel emulator is fitted to weighted particles, which were previously corrected via importance weights. As an example, we here use the kernel gradient emulator by Strathmann et al. [22], in its finite dimensional approximation (KMC finite), c.f. [22, Proposition 2].

The log density of the approximate estimator takes the simple form $f(x) = \theta^\top \phi_x$, where $\phi_x \in \mathbb{R}^m$ is an embedding of x into an m-dimensional feature space, and $\theta \in \mathbb{R}^m$ is estimated by $\hat{\theta} = C^{-1} b$ from samples x. Given a weighted particle system $\{(W_{t,i}, X_{t,i})\}_{i=1}^N$, then b, C are weighted averages of the form

$$b := -\frac{1}{\sum_{i=1}^{N} W_{t,i}} \sum_{i=1}^{N} W_{t,i} \sum_{\ell=1}^{d} \ddot{\phi}_x^\ell,$$

$$C := \frac{1}{\sum_{i=1}^{N} W_{t,i}} \sum_{i=1}^{N} W_{t,i} \sum_{\ell=1}^{d} \dot{\phi}_x^\ell \left(\dot{\phi}_x^\ell \right)^\top,$$

with element-wise derivatives $\dot{\phi}_x^\ell := \frac{\partial}{\partial x_\ell} \phi_x$ and $\ddot{\phi}_x^\ell := \frac{\partial^2}{\partial x_\ell^2} \phi_x$. Note that the estimator can be updated in an online fashion once the particle system changes. Rather than simulating Hamiltonian dynamics to generate a proposal, we here take single gradient steps, i.e. the Markov density at in Algorithm 1 at X is $q_t(\cdot|X) = \mathcal{N}(\cdot|X + \delta \nabla f(X), \nu^2 \Sigma_t)$ for some parameters $\delta > 0, \nu^2 > 0$. This keeps the risk of divergence due to wrongly estimated gradients low. We arrive at kernel GRIS, a gradient-free variant of GRIS [19].

3.3   Controlling Emulator Variance in PMC

PMC is somewhat sensitive to badly scaled proposals, as these are not rejected as in a Metropolis-Hastings step. In particular for gradient emulators used within PMC, variance reduction is important to avoid numerical divergence. The original PMC paper introduces re-sampling in order to deal with un-weighted instead of weighted samples [2], though at the cost of an increased variance. While some approaches avoid re-sampling altogether [3], we consider re-sampling here as a way to obtain a set of locations $\{\widetilde{X}_t^i\}_{i=1}^{N}$ for our Markov proposal components of the mixture $q_t^{\mathrm{Mixt}}(\cdot) = \frac{1}{N} \sum_{i=1}^{N} q_t(\cdot|\widetilde{X}_t^i)$, due to better behaving variance in high dimension. With re-sampling, Monte Carlo variance only grows as $O(D)$ rather than $O(\exp(D))$ without re-sampling, where D is the dimensionality [8].

Given a re-sampled number of N particles and the updated emulator q_t, we simulate from the mixture distribution q_t^{Mixt} with stratification, i.e. we draw exactly one sample from each of the equally weighted mixture components. Another view of this scheme is to draw a single realisation from $q_t(\cdot|\widetilde{X}_t^i)$ for all $i = 1, \ldots, N$ and Rao-Blackwellise. Finally, we can view the scheme as an instance of the deterministic mixture idea [10]. Without this technique, i.e. using weights $\pi(\cdot)/q_t(\cdot|\widetilde{X}_t^i)$, variance might grow catastrophically large, as too high variance can be self-reinforcing by resulting in emulators of low quality.

4   Evaluation

We empirically evaluate performance of KASMC on a simple non-linear target, on a multi-modal sensor network localisation problem, and in estimating Bayesian model evidence in a model with an intractable likelihood on a real-world dataset. The final experiment uses a challenging stochastic volatility model with S&P 500 data from Chopin et al. [5] to evaluate KGRIS.

For the KASMC experiments on static target distributions, a sequence of incremental target densities can be defined using a geometric bridge with $\pi_t \propto \pi_0^{1-\rho_t} \pi^{\rho_t}$

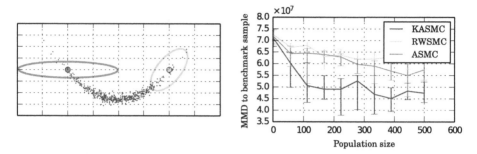

Fig. 1. Left: Proposal distributions around one of many particles (blue) for each KASMC (red) and ASMC (green). KASMC proposals locally align to the target density while ASMC's global covariance estimate might result in poor MH rejuvenation moves. **Right:** Improved convergence of all mixed moments up to order 3 of KASMC compared to using SMC with static or adaptive Metropolis-Hastings steps. (Color figure online)

for some initial distribution π_0, where $(\rho_t)_{t=1}^T$ is an increasing sequence satisfying $\rho_T = 1$. The bandwidth parameter of the kernel emulator models is set to the median distance between particles [12].

We also note these algorithms have a free scaling parameter ν^2, which we would like to adapt online. To accomplish parameter tuning, we use the standard framework of stochastic approximation for tuning MCMC kernels [1], i.e. tuning acceptance rate α_t towards an asymptotically optimal acceptance rate $\alpha_{\text{opt}} = 0.234$ for random walk proposals [18]. After the MCMC rejuvenation step, a Rao-Blackwellised estimate $\hat{\alpha}_t$ of expected acceptance probability is available by simply averaging the acceptance probabilities for all MH proposals. Then, set $\nu_{t+1}^2 = \nu_t^2 + \lambda_t(\hat{\alpha}_t - \alpha_{\text{opt}})$ for some non-increasing sequence $\lambda_1, \ldots, \lambda_T$. This strategy of approximating optimal scaling assumes that consecutive targets are close enough so that the acceptance rate when using ν_t^2 to target π_t provides information about the expected acceptance rate when using ν_t^2 with target π_{t+1}. This is discussed further in the supplemental material.

4.1 KASMC: Improved Convergence on Synthetic Nonlinear Target

We begin by studying convergence of KASMC compared to existing algorithms on a simple benchmark example: the strongly twisted banana-shaped distribution in $D = 8$ dimensions used in Sejdinovic et al. [20]. This distribution is a multivariate Gaussian with a non-linearly transformed second component, defined as

$$\mathcal{B}(y; b, v) = \mathcal{N}(y_1; 0, v)\mathcal{N}(y_2; b(y_1^2 - v), 1) \prod_{j=3}^{D} \mathcal{N}(y_j; 0, 1).$$

We compare SMC algorithms using different rejuvenation MH steps: a static random walk Metropolis move (RWSMC) with fixed scaling $\nu = 2.38/\sqrt{D}$,

ASMC, and KASMC using a Gaussian RBF kernel. For the latter two algorithms, all particles are used to compute the proposal, and a fixed learning rate of $\lambda = 0.1$ is chosen to adapt scale parameters. Starting with particles from a multivariate Gaussian $\mathcal{N}(0, 50^2)$, we use a geometric bridge that reaches the target $\mathcal{B}(y; b = 0.1, v = 100)$ in 20 steps. We repeat the experiment over 30 runs. Figure 1 (right) shows that KASMC achieves faster convergence of the first 3 moments, i.e. in MMD^2 distance to a large benchmark sample.

4.2   A Multi-modal Application: Sensor Network Localisation

We next study performance of KASMC on a multi-modal target arising in a real-world application: inferring the locations of S sensors within a network, as discussed in [14,15]. We here focus on the static case: assume a number of stationary sensors that measure distance to each other in a 2-dimensional space; a distance measurement is successful with a probability that decays exponentially in the squared distance, and the observation is missing otherwise. If distance is measured, it is corrupted by Gaussian noise. The posterior over the unknown sensor locations forms an extremely constrained non-linear and multi-modal distribution induced by the spatial set-up.

Assume S sensors with unknown locations $\{x_i\}_{i=1}^S \subseteq \mathbb{R}^2$. Define an indicator variable $Z_{i,j} \in \{0, 1\}$ for the distance $Y_{ij} \in \mathbb{R}^+$ between a pair of sensors (x_i, x_j) being either observed $(Z_{i,j} = 1)$ or not $(Z_{i,j} = 0)$, according to

$$Z_{i,j} \sim \text{Binom}\left(1, \exp\left(-\frac{\|x_i - x_j\|_2^2}{2R^2}\right)\right).$$

If the distance is observed, then Y_{ij} is corrupted by Gaussian noise, i.e.

$$Y_{i,j}|Z_{i,j} = 1 \sim \mathcal{N}\left(\|x_i - x_j\|, \sigma^2\right),$$

and $Y_{i,j} = 0$ otherwise.

Previously, [14] focussed on MAP estimation of the sensor locations, and [15] focussed on a well-conditioned case ($S = 8$ sensors and $B = 3$ base sensors with known locations) that results in almost no ambiguity in the posterior. We argue that Bayesian quantification of uncertainty is more important for cases where noise and missing measurements *does not* allow to reconstruct the sensor locations exactly. We therefore reuse the dataset from [15] ($R = 0.3$, $\sigma^2 = 0.02$)[3], but only use the first $S = 3$ locations/observations. In order to encourage ambiguities in the localisation task, we only use the first 2 base sensors of [15] with known locations that each do observe distances to the S unknown sensors but not of each other. Unlike [15], we use a Gaussian prior $\mathcal{N}(\mathbf{0.5}, I)$ to avoid the posterior being situated in a bounded domain.

---

[2] The maximum mean discrepancy, here using a polynomial kernel of order 3, quantifies differences of all mixed moments up to order 3 of two independent sets of samples.

[3] Downloaded from http://www.ics.uci.edu/~slan/lanzi/CODES_files/WHMC-code.zip on 8/Oct/2015.

Figure 2 shows the marginalised posterior for one run each of KASMC (SMC) and KAMH (MCMC), where we matched the number of likelihood evaluations (500,000). We run KASMC using 10,000 particles and a bridge length of 50, and MCMC-KAMH for $50 \times 10,000$ iterations of which we discard half as burn-in; both were initialized with samples from the prior. Tuning parameters ν^2 are set using a diminishing adaptation schedule $\lambda_t = 1/\sqrt{t}$ for KAMH and a fixed learning rate $\lambda_t = 1$ for KASMC. MCMC is not able to traverse between the multiple modes and interpretations of the data, in contrast to SMC.

In order to compare ASMC to KASMC, we created a benchmark sample via running 100 standard MCMC chains (randomly initialised to cover all modes) each for 50000 iterations, discarding half the samples as burn-in, and randomly down-sampling to a size of 100. We then compute the empirical MMD distance to the output of the individual algorithms, averaged over 10 runs. For the chosen number of sensors, ASMC and KASMC perform similarly. With less sensors, i.e. more ambiguity, KASMC produces samples with both less MMD distance from a benchmark sample and less variance. For example, for a set-up with $S = 2$ and 1000 particles, we get a MMD distance to a benchmark sample of 0.76 ± 0.4 for KASMC and 0.94 ± 0.7 for ASMC.

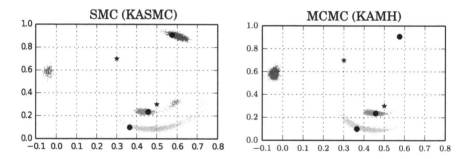

Fig. 2. Posterior samples of unknown sensor locations (in color) by kernel-based SMC and MCMC on the sensors dataset. The set-up of the true sensor locations (black dots) and base sensors (black stars) causes uncertainty in the posterior. SMC recovers all modes while MCMC does not. The posterior has a clear non-linear structure. (Color figure online)

4.3 KASMC: Evidence Estimation in Gaussian Process Classification

Following Sejdinovic et al. [20], we consider Bayesian classification on the UCI Glass dataset, discriminating window glass from non-window glass, using a Gaussian process (GP). It was found that the induced posterior is indeed non-linear [20,22]. In Sejdinovic et al. [20], samples from the marginal posterior over GP hyper-parameters were simulated (the GP latent variables integrated out). We emphasise a different point here: KSMC's ability to estimate the model evidence as compared to KAMH, and its faster convergence compared to ASMC.

Consider the joint distribution of latent variables $\mathbf{f}$, labels $\mathbf{y}$ (with covariate matrix X), and hyper-parameters θ, given by

$$p(\mathbf{f}, \mathbf{y}, \theta) = p(\theta)p(\mathbf{f}|\theta)p(\mathbf{y}|\mathbf{f}),$$

where $\mathbf{f}|\theta \sim \mathcal{N}(0, \mathcal{K}_\theta)$, with $\mathcal{K}_\theta$ modelling the covariance between latent variables evaluated at the input covariates. Consider the binary logistic classifier, i.e. $p(y_i|f_i) = \frac{1}{1 - \exp(-y_i f_i)}$ where $y_i \in \{-1, 1\}$. In order to perform Bayesian model selection (i.e. comparing different covariance functions), we need to estimate the model evidence of the marginal posterior given the hyper-parameters. Here, the marginal likelihood $p(\mathbf{y}|\theta)$ is intractable for non-Gaussian likelihoods $p(\mathbf{y}|\mathbf{f})$.

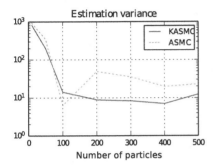

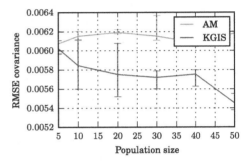

Fig. 3. Estimating model evidence of a GP using the IS^2 framework. The plot shows the MC variance over 50 runs as a function of the size of the particle system.

Fig. 4. Convergence of RMSE for estimating all elements of the posterior covariance matrix of the stochastic volatility model.

We estimate model evidence for the GP classifier equipped with a standard Gaussian Automatic Relevance Determination (ARD) covariance kernel; an unbiased estimate can be obtained using importance sampling

$$\hat{p}(\mathbf{y}|\theta) := \frac{1}{n_{\text{imp}}} \sum_{i=1}^{n_{\text{imp}}} p(\mathbf{y}|\mathbf{f}^{(i)}) \frac{p(\mathbf{f}^{(i)}|\theta)}{q(\mathbf{f}^{(i)}|\theta)}, \qquad (2)$$

where $\{\mathbf{f}^{(i)}\}_{i=1}^{n_{\text{imp}}} \sim q(\mathbf{f}|\theta)$ are n_{imp} importance samples, e.g. from a Laplace approximation of $p(\mathbf{f}|\mathbf{y}, \theta)$. We here do not tune the number of 'inner' importance samples, but follow [20] and use $n_{\text{imp}} = 100$.

Figure 3 shows that evidence estimates of KASMC exhibit less variance than those of ASMC. The ground truth model evidence was established via running 20 SMC instances using $N = 1000$ particles and a bridge length of 30, and averaging their evidence estimates. The experiment is performed 50 times, using $N = 100$ particles and a bridge length of 20, starting from he prior on the log hyper-parameters $\pi_0 = p(\theta_d) \equiv \mathcal{N}(0, 5^2)$. The learning rate is constant $\lambda_t = 1$, and adaptation is towards an acceptance rate of 0.23.

4.4 KGRIS: Stochasitic Volatility Model with Intractable Likelihood

A particularly challenging class of Bayesian inverse problems are stochastic volatility models. As time series models, they often involve high-dimensional nuisance variables, which usually cannot be integrated out analytically. Furthermore, risk management necessitates to account for parameter and model uncertainty, and models have to capture the non-linearities in the data [5]. We here concentrate on the prediction of daily volatility of asset prices, reusing the model and dataset studied by Chopin et al. [5] to evaluate KGRIS. Due to the lack of analytically available gradients for this model, we compare two gradient free PMC versions: KGRIS and a random walk PMC with global covariance adaptation in the style of Haario et al. [13].

Let s_t be the value of some financial asset on day t, then $y_t = 10^{(5/2)} \log(s_t/s_{t-1})$ is called the log-returns (upscaling for numerical reasons). We model the observed log-return y_t as dependent on a latent v_t by the observation equation

$$y_t = \mu + \beta v_t + \sqrt{v_t}\epsilon_t$$

for $t \leq 1$. Here ϵ_t is a sequence of i.i.d. standard Gaussian errors and v_t is assumed to be a stationary stochastic process known as the *actual volatility*. Chopin et al. [5] develop a hierarchical model for v_t based on the idea of analytically integrating a continuous time volatility model over daily intervals [for details see 5]. Using this construction, the (discrete time) v_t is parameterised by stationary mean ξ and variance ω^2 of the so called *spot volatility* and the exponential decay λ of its auto-correlation. This results in the following model for the actual volatility v_t:

$$k \sim \mathrm{Pois}(\lambda\xi^2/\omega^2), c_{1:k} \sim \mathrm{U}(t, t+1), e_{1:k} \sim \mathrm{Exp}(\xi/\omega^2)$$

$$z_{t+1} = z_t \exp(-\lambda) + \sum_{j=1}^{k} e_j \exp(-\lambda(t+1-c_j))$$

$$v_{t+1} = \lambda^{-1}\left(z_t - z_{t+1} + \sum_{j=1}^{k} e_j\right), x_{t+1} = (v_{t+1}, z_{t+1})^\top$$

where z_t is the discretely sampled spot volatility process and $(v_{t+1}, z_{t+1})^\top$ is the Markovian representation of the state process. The variables $k, c_{1:k}$ and $e_{1:k}$ are generated independently for each time period. For $k = 0$, the set $1:k$ is defined to be empty. The dynamics imply $\Gamma(\xi^2/\omega^2, \xi^2/\omega^2)$ to be the stationary distribution for z_t, which is also used as the initial distribution on z_0. The parameters of the model are $\theta = (\mu, \beta, \xi, \omega^2, \lambda)$ and the likelihood is intractable. Chopin et al. [5] developed a sampler for θ based on iterated batch importance sampling using nested SMC with pseudo-marginal MCMC moves for integrating out the x_t and dubbed their approach SMC^2.

In our experiment, we use KGRIS proposals in a population Monte Carlo setting, i.e. without resorting to MCMC moves at all. We re-use the code developed for the original SMC^2 paper in order to integrate out the x_t and thus get likelihood estimates, with the same settings for algorithm parameters. The observed s_t are the 753 observations from consecutive days of the S&P 500 index also used by Chopin et al. [5]. KGRIS uses a particle system of increasing sizes with each particle going through 100 iterations. See Fig. 5 in the Appendix for a plot of the pair-wise marginals of this posterior.

We use the same vague priors as Chopin et al. [5],

$$\mu \sim \mathcal{N}(0, \sigma^2 = 2), \beta \sim \mathcal{N}(0, \sigma^2 = 2), \xi \sim \text{Exp}(0.2)$$
$$\omega^2 \sim \text{Exp}(0.2), \lambda \sim \text{Exp}(1).$$

Figure 4 shows that the incorporated gradients lead to better performance of KGRIS in estimating the target covariance matrix. This is in-line with the finding that GRIS improves over pure random walk methods [19].

5   Discussion

In this paper, we developed a framework for kernel sequential Monte Carlo. KSMC adaptively learns the target geometry via kernel emulators and subsequently uses this information for local proposals. KSMC is especially attractive in the case where likelihoods and gradients are intractable. We instantiated two algorithms within KSMC: estimating nonlinear covariance in combination with MCMC rejuvenation and estimating gradients in combination with importance sampling proposals. Both significantly outperform state-of-the-art gradient-free SMC algorithms in practice. We conclude with some discussion on computational complexity, more general usage of the learned emulators, and on the relative benefits of PMC in the kernel setting.

Computational costs & increasing dimensions. While adaptive schemes for SMC (and MCMC) can increase statistical efficiency of the sampling scheme, they impose additional computational costs. Somewhat surprisingly, however, these relatively large costs do not severely impact the efficiency per runtime ratio in practice. The reason is that in the context of intractable likelihoods, the computational cost of fitting a kernel emulator is typically dominated by the larger cost of evaluating model likelihood. In our real-world experiments on GP classification and a stochastic volatility model in Sects. 4.3 and 4.4, a profiler reveals that less than 5% of the overall wall-clock time is spent in computing kernel informed proposals. This effect increases with dataset size and model complexity, as evaluating likelihood gets more costly. Clearly however, in the case where we need not resort to pseudo-marginal or SMC^2 type samplers, the application of kernel based estimators might result in slower sampling without much gain in Monte Carlo error.

In growing dimensions, the number of data required to sufficiently estimate nonlinear covariance and gradients quickly becomes infeasible. High dimensional sampling problems typically arise in non-parametric models, e.g. Gaussian processes, where each data point comes with additional parameters.

In the intractable likelihood framework that we consider here, however, the marginal posterior over hyper-parameters typically is independent of such latent variables — and therefore usually of moderate dimension. Random walk methods, which are the default choice for intractable likelihoods, scale badly in high dimensions themselves [16]. Our method is an improvement in the intermediate case: closed form gradients are not available, but the dimensionality of the problem allows to estimate the target geometry just accurately enough to improve mixing. Strathmann et al. [22] reported their gradient estimator to scale up from dozens to a hundred dimensions on laptop computers, depending on smoothness properties of the target. It is an active area of research to further scale up these techniques by exploiting structure in the target density.

Emulators as a posterior approximation. The kernel approximation of the target density could be considered itself as an output of our algorithms, representing the posterior directly instead of using the kernel approximation within a sampler. There are a number of problems with this approach though: firstly, we note that our emulator models do not need to be perfect to generate useful proposals, therefore allowing us to exploit posterior structure much earlier (even with non-perfect model fit) during sampling, still resulting in a correct SMC sampler. Also, approximating integrals of test functions with respect to the posterior using the kernel approximation is not possible in closed form, while it is straight forward using a Monte Carlo sum. For example, assume a log density model $f(x) = \sum_i \alpha_i \mathbf{k}(x_i, x)$. For the Gaussian kernel $\mathbf{k}(x, y) = \exp(-||x - y||^2)$, the density is the exponential of a sum of Gaussian centred at the points x_i. Computing an integral as simple as the posterior mean, $\mu = Z^{-1} \int x \exp(\sum_i \alpha_i \exp(-||x_i - x||^2)) \mathrm{d}x$, already is intractable, even if the evidence Z were known. Thirdly, it is not possible to sample from the kernel emulator directly using ordinary Monte Carlo. One could imagine running a second MCMC/SMC targeting the emulator model. Not only would this defeat the purpose of the algorithm (this is the problem we are trying to solve in the first place), it also leads to samples that are not guaranteed to consistently estimate posterior expectations unlike kernel SMC or kernel MCMC.

SMC versus PMC for kernel based proposals. The consensus in the wider SMC community is that using an artificial sequence of proposal distributions for sampling from a static target is preferable to the PMC approach. This is based on the fact that the coverage of the final target is better in these tempering-style algorithms. It however results in a considerable computational investment for those iterations where an intermediate target is considered.

We also note that on-line updates of the kernel emulator are not possible: the target changes in every iteration. The contrary is true in PMC, where the the actual distribution of interest is targeted in every iteration. Here, a popular approximation technique of kernels is a good fit: By expressing the emulator model in terms of finite dimensional random Fourier features, we can perform cheap on-line updates [22]. The emulator therefore can accumulate information from all PMC iterations without the computational efforts of re-computing its solution, providing a relative advantage to SMC in this context.

Acknowledgments. I.S. was supported by a PSL postdoc grant and DFG through grant CRC 1114 "Scaling Cascades in Complex Systems", Project B03 "Multilevel coarse graining of multiscale problems". H.S. was supported by the Gatsby Chaitable foundation. B.P. was supported by The Alan Turing Institute under the EPSRC grant EP/N510129/1.

A   Implementation details

In this section, we cover a number of implementation details for using KASMC in practice, such as optimal scaling, adaptive re-sampling and re-weighting between iterations.

A.1   Scaling Parameters

Similar to other MH proposals, KAMH has a free scaling parameter denoted ν^2 which we would like to adapt after one SMC iteration. To accomplish parameter tuning, we use the standard framework of stochastic approximation for tuning MCMC kernels [1], i.e. tuning acceptance rate α_t towards an asymptotically optimal acceptance rate $\alpha_{\text{opt}} = 0.234$ for random walk proposals [18]. More precisely, after the MCMC rejuvenation step, a Rao-Blackwellised estimate $\hat{\alpha}_t$ of expected acceptance probability is available by simply averaging the acceptance probabilities for all MH proposals. Then, set

$$\nu_{t+1}^2 = \nu_t^2 + \lambda_t(\hat{\alpha}_t - \alpha_{\text{opt}}) \tag{3}$$

for some non-increasing sequence $\lambda_1, \ldots, \lambda_T$. This strategy of approximating optimal scaling assumes that consecutive targets are close enough so that the acceptance rate when using ν_t^2 to target π_t provides information about the expected acceptance rate when using ν_t^2 with target π_{t+1}. As an alternative to this, one could treat ν_t^2 as an auxiliary random variable and define a distribution over it designed to maximise expected utility, an approach taken in the adaptive SMC sampler [11].

A.2   Construction of a Target Sequence

One possibility for constructing a sequence of distributions is the geometric bridge defined by

$$\pi_t \propto \pi_0^{1-\rho_t} \pi^{\rho_t}$$

for some initial distribution π_0, where $(\rho_t)_{t=1}^T$ is an increasing sequence satisfying $\rho_T = 1$. This is the construction used in the experimental section. Another construction is to use a mixture $\pi_t \propto (1 - \rho_t)\pi_0 + \rho_t\pi$. When π is a Bayesian posterior, one can also add more data with increasing t, e.g. by defining the intermediate distributions as $\pi_t(X) = \pi(X|d_1, \ldots, d_{\lfloor \rho_t D \rfloor})$ where d_j is a datapoint and D is the number of data points. This results in an online inference algorithm called Iterated Batch Important Sampling (IBIS) [4]. In IBIS especially, we can apply non-diminishing adaptation, unlike in adaptive MCMC.

When using a distribution sequence that computes the posterior density π using the full dataset (such as the geometric bridge or the mixture sequence), one can reuse the intermediate samples when targeting π_t for posterior estimation. As the value of π is computed for the geometric bridge and the mixture sequence, we re-use the weight $\pi(X)/\pi_{t-1}(X)$ for posterior estimation while employing $\pi_t(X)/\pi_{t-1}(X)$ to inform proposal distributions at iteration t. This way, the evaluation of π (which is typically costly) is put to good use for improving the posterior estimate.

As a simple alternative, leading to the algorithm known as Population Monte Carlo (PMC) [2], we can simply target the final distribution π at each iteration, i.e. with all $\pi_t = \pi$. The original work on PMC exhibited striking resemblance of commonly used MCMC methods such as Random Walk metropolis, often finding that the same proposal kernel with PMC produces better estimates than with MCMC [2].

A.3   Re-weighting and Adaptive Re-sampling

The fact that the weighted approximation to the final target is returned in our algorithm stems from the fact that this approximation has lower variance than the re-sampled particle system [8]. This is why in practice re-sampling might not be performed at every iteration. Rather, re-sampling only when Effective Sample Size (ESS) for the current target falls below a certain threshold will decrease Monte Carlo variance. For details we refer to reviews on SMC [8,9]. Furthermore, care should be taken with respect to implementation of re-weighting: caching values between iterations saves much computation time.

A.4   Intractable Likelihoods and Evidence Estimation

In the the case where likelihoods are intractable, SMC is still a valid algorithm when likelihood values can be estimated unbiasedly. This can be done using e.g. importance sampling or SMC [5,23]. A simple way to think about such nested estimation schemes is in terms of an extended sampling space that spans the actual parameters of interest as well as any nuisance variables. Intractable likelihoods usually result in unavailability of gradients. Consequently, efficient gradient-based sampling schemes based such as GRIS or HMC are unavailable. Current practice there is based on moving particles using random walk schemes solely.

An important issue in Bayesian model selection and averaging is that of estimating the normalizing constant, or *evidence*. The evidence is the marginal probability of the data under a model and can easily be estimated in SMC instantiations [8,11] – as compared to MCMC. This enables routine computation of Bayes factors and posterior model probabilities while also sampling from a posterior over parameters of each model. Under the assumption that the normalizing

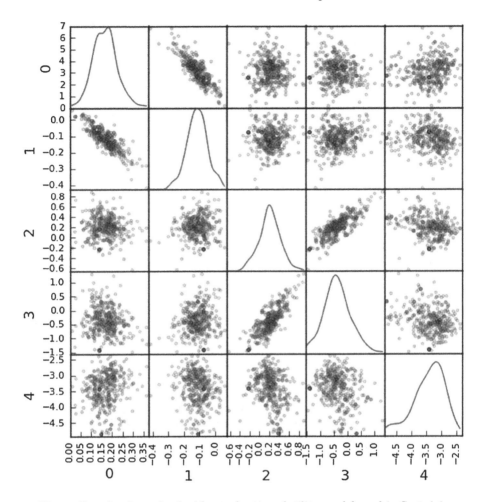

Fig. 5. Samples from the doubly stochastic volatility model used in Sect. 4.4.

constant Z_0 of π_0 (the distribution that is used for initially setting up the particle system) is known, one can estimate the ratio of normalizing constants of any two consecutive targets by

$$\frac{Z_t}{Z_{t-1}} \approx \frac{1}{N} \sum_{j=1}^{N} W_{t,j} \tag{4}$$

for $W_{t,j} = \pi_t(X_{t-1,j})/\pi_{t-1}(X_{t-1,j})$ and thus an estimate for $Z = Z_T$ can be found recursively by

$$Z = Z_T \approx Z_0 \prod_{t=1}^{T} \frac{1}{N} \sum_{j} W_{t,j} \tag{5}$$

starting with known value Z_0. When the likelihood is intractable and importance weights are noisy, evidence estimation is still valid [23, Lemma 3].

References

1. Andrieu, C., Thoms, J.: A tutorial on adaptive MCMC. Stat. Comput. **18**, 343–373 (2008). ISSN 09603174
2. Cappé, O., Guillin, A., Marin, J.M., Robert, C.P.: Population Monte Carlo. J. Comput. Graph. Stat. **13**(4), 907–929 (2004). ISSN 1061–8600
3. Cappé, O., Douc, R., Guillin, A., Marin, J.M., Robert, C.P.: Adaptive importance sampling in general mixture classes. Stat. Comput. **18**(4), 447–459 (2008). ISSN 0960–3174
4. Chopin, N.: A sequential particle filter method for static models. Biometrika **89**(3), 539–552 (2002). ISSN 0006–3444
5. Chopin, N., Jacob, P.E., Papaspiliopoulos, O.: SMC2: an efficient algorithm for sequential analysis of state space models. J. Royal Stat. Soc. Ser. B (Stat. Methodol.) **75**(3), 397–426 (2013)
6. Del Moral, P., Doucet, A., Jasra, A.: Sequential Monte Carlo samplers. J. Royal Stat. Soc. Ser. B (Stat. Methodol.) **68**(3), 411–436 (2006)
7. Douc, R., Cappé, O.: Comparison of resampling schemes for particle filtering. In: Proceedings of the 4th International Symposium on Image and Signal Processing and Analysis, pp. 64–69 (2005). ISBN 953-184-089-X
8. Doucet, A., Johansen, A.: A tutorial on particle filtering and smoothing: fifteen years later. Handb. Nonlinear Filtering, 4–6 (2009)
9. Doucet, A., de Freitas, N., Gordon, N.: An introduction to sequential Monte Carlo methods. In: Doucet, A., de Freitas, N., Gordon, N. (eds.) Sequential Monte Carlo Methods in Practice. Statistics for Engineering and Information Science, pp. 3–14. Springer, Heidelberg (2011). https://doi.org/10.1007/978-1-4757-3437-9_1
10. Elvira, V., Martino, L., Luengo, D., Bugallo, M.F.: Generalized multiple importance sampling. Technical report (2015)
11. Fearnhead, P., Taylor, B.M.: An adaptive sequential Monte Carlo sampler. Bayesian Anal. **2**, 411–438 (2013)
12. Gretton, A., Borgwardt, K.M., Rasch, M.J., Schölkopf, B., Smola, A.: A kernel two-sample test. J. Mach. Learn. Res. **13**(1), 723–773 (2012)
13. Haario, H., Saksman, E., Tamminen, J.: An adaptive metropolis algorithm. Bernoulli **7**(2), 223–242 (2001). ISSN 13507265
14. Ihler, A.T., Fisher, J.W., Moses, R.L., Willsky, A.S.: Nonparametric belief propagation for self-localization of sensor networks. IEEE J. Sel. Areas Commun. **23**(4), 809–819 (2005)
15. Lan, S., Streets, J., Shahbaba, B.: Wormhole Hamiltonian Monte Carlo. In: Twenty-Eighth AAAI Conference on Artificial Intelligence (2014)
16. Neal, R.M.: MCMC using Hamiltonian dynamics. Handb. Markov Chain Monte Carlo. Chapman and Hall/CRC (2011)
17. Rahimi, A., Recht, B.: Random features for large-scale kernel machines. In: Advances in Neural Information Processing Systems, pp. 1177–1184 (2007)
18. Rosenthal, J.S.: Optimal proposal distributions and adaptive MCMC. Handb. Markov Chain Monte Carlo **4**, 93–112 (2011). Chapman and Hall
19. Schuster, I.: Gradient importance sampling. arXiv preprint arXiv:1507.05781 (2015)

20. Sejdinovic, D., Strathmann, H., Garcia, M.L., Andrieu, C., Gretton, A.: Kernel adaptive metropolis-hastings. In: International Conference on Machine Learning (ICML), pp. 1665–1673 (2014)
21. Sriperumbudur, B., Fukumizu, K., Kumar, R., Gretton, A., Hyvärinen, A.: Density estimation in infinite dimensional exponential families. arXiv preprint arXiv:1312.3516 (2014)
22. Strathmann, H., Sejdinovic, D., Livingstone, S., Szabo, Z., Gretton, A.: Gradient-free Hamiltonian Monte Carlo with efficient kernel exponential families. In: NIPS (2015)
23. Tran, M.-N., Scharth, M., Pitt, M.K., Kohn, R.: Importance sampling squared for Bayesian inference in latent variable models. arXiv preprint arXiv:1309.3339 (2013)

Learning Łukasiewicz Logic Fragments by Quadratic Programming

Francesco Giannini[(✉)], Michelangelo Diligenti, Marco Gori,
and Marco Maggini

Department of Information Engineering and Mathematics, University of Siena,
Via Roma 56, Siena, Italy
{fgiannini,diligmic,marco,maggini}@diism.unisi.it

Abstract. In this paper we provide a framework to embed logical constraints into the classical learning scheme of kernel machines, that gives rise to a learning algorithm based on a quadratic programming problem. In particular, we show that, once the constraints are expressed using a specific fragment from the Łukasiewicz logic, the learning objective turns out to be convex. We formulate the *primal* and *dual* forms of a general multi–task learning problem, where the functions to be determined are predicates (of any arity) defined on the feature space. The learning set contains both supervised examples for each predicate and unsupervised examples exploited to enforce the constraints. We give some properties of the solutions constructed by the framework along with promising experimental results.

Keywords: Learning from constraints · Kernel machines
First–order logic · Quadratic programming · Łukasiewicz logic

1 Introduction

Learning from constraints is an extension of classical supervised learning in which the concept of *constraint* is used to refer to a more general class of knowledge than simple labelled examples. In particular a multi–task learning paradigm is assumed in which the functions to be learnt are subject to a set of relational constraints. An extensive investigation of different kinds of constraints can be found in [7], where variational calculus is exploited to generalize the Representation Theorem for kernel machines. In this paper, we focus on logical constraints, which provide a natural, expressive and formally well–defined representation for abstract knowledge as well as being supported by a strong mathematical theory.

In literature, logic is employed in a wide class of learning problems according to different approaches. Diligenti et al. bridge logic and kernel machines considering the logical constraints by means of a functional representation of logical formulas, extending the classical regularization scheme of kernel machines to incorporate logical constraints [4,5]. However, the objective function, depending on the logic clauses, is not guaranteed to be convex in general. A different

M. Ceci et al. (Eds.): ECML PKDD 2017, Part I, LNAI 10534, pp. 410–426, 2017.
https://doi.org/10.1007/978-3-319-71249-9_25

approach with kernel machines can be found in [3] where authors introduce a family of kernels parameterized by a Feature Description Language. Logic rules are also combined with neural networks in different contexts. For instance, Hu et al. [8] propose an iterative procedure that transfers abstract knowledge encoded by the logic rules into the parameters of a deep neural network. Further, it is worth to mention some related works in *inductive logic programming* [11] and in *Markov logic networks* [1,13] that derive probabilistic graphical models from the rule set. However, like other recent papers do [15], we decide to focus on logic and to not consider probabilistic aspects.

In this paper we provide a logical fragment of propositional Łukasiewicz Logic whose corresponding functional constraints turn out to be convex. It is worth to notice that formulas belonging to this fragment are referred as *simple Łukasiewicz clausal forms* in the literature and, among others, the satisfiability problem for these formulas has linear–time complexity [2]. Functions corresponding to formulas in the proposed fragment are concave (or convex for its dual) as also shown in [10], in which such formulas are used to get concave payoff functions. In our framework, formulas built up from the fragment derive convex functional constraints and we also prove that they coincide with the whole class of concave (convex) Łukasiewicz functions. In addition, thanks to McNaughton Theorem, the functions corresponding to Łukasiewicz formulas are continuous piecewise linear with integer coefficients. As a result, the learning task can be formulated as a quadratic programming problem.

The paper is organized as follows: in the next section we introduce and prove the main results about both concave and convex fragments of Łukasiewicz logic. In Sect. 3 we describe the different kinds of constraints we will deal with, paying particular attentions to logical constraints. The primal and dual forms, as well as the Lagrangian associated to the problem, are formulated in Sect. 4. Finally, in Sect. 5 we provide some preliminary experimental results and a toy example to illustrate the fundamental features of the theory.

2   Concave Fragment of Łukasiewicz Logic

Łukasiewicz Logic **Ł** is a suitable framework to express logical constraints for several reasons. First of all, there exist n-valued (for all finite n) as well as infinitely many–valued variants of this logic, thus we can deal with arbitrary degrees of truth. Among the three fundamental fuzzy logics given by a continuous t-norm (Łukasiewicz, Gödel and Product logic), **Ł** is the only one with an involutive negation and such that every formula has an equivalent *prenex normal form* (i.e. quantifiers followed by a quantifier–free part). Furthermore, for the propositional case, the algebra of formulas of **Ł** on n variables is isomorphic to the algebra of McNaughton functions defined on $[0, 1]^n$. As we will see, this is a crucial property and it allows us to get a substantial advantage with respect to the employment of other logics. In particular, we are interested in the fragment of **Ł** whose formulas correspond to concave functions.

We assume to deal with a knowledge base KB made of first order Łukasiewicz formulas. The syntax of formulas is defined in the usual way [12]. We recall some basic notions of Łukasiewicz logic and some fundamental results for its algebraic semantics.

The propositional logic **L** is sound and complete with respect to the variety[1] of **MV**-algebras and in particular with respect to the standard algebra $[0,1]_L$ which generates the whole variety. $[0,1]_L = ([0,1], 0, 1, \neg, \wedge, \vee, \otimes, \oplus, \rightarrow)^2$, with operations defined as follows:

$$\neg x = 1 - x, \quad x \wedge y = \min\{x, y\}, \quad x \vee y = \max\{x, y\},$$

$$x \otimes y = \max\{0, x + y - 1\}, \quad x \oplus y = \min\{1, x + y\}, \quad x \rightarrow y = \min\{1, 1 - x + y\}.$$

The operations $\wedge$ and $\otimes$ are the interpretations of *weak* and *strong* conjunction[3] of **L** respectively, while $\vee$ and $\oplus$ represent *weak* and *strong* disjunction. Further, $\otimes$ and $\rightarrow$ are Łukasiewicz t-norm and its residuum, where $x \rightarrow y$ is definable as $\neg x \oplus y$. In the paper, for short we write $\bigoplus_{i=1}^{n} x_i$ for $x_1 \oplus \ldots \oplus x_n$ and ax, with $a > 0$ for $\bigoplus_{i=1}^{a} x$, where $x, x_1, \ldots, x_n$ denote any literals.

Some fundamental properties of these operations are:

$$\neg\neg x = x, \quad \neg(x \wedge y) = \neg x \vee \neg y, \quad \neg(x \otimes y) = \neg x \oplus \neg y, \quad x \otimes \neg x = 0, \quad x \oplus \neg x = 1,$$

$$x \otimes (y \wedge z) = (x \otimes y) \wedge (x \otimes z), \, x \otimes (y \vee z) = (x \otimes y) \vee (x \otimes z), \, x \oplus (y \wedge z) = (x \oplus y) \wedge (x \oplus z),$$

$$x \oplus (y \vee z) = (x \oplus y) \vee (x \oplus z), \, x \wedge (y \vee z) = (x \wedge y) \vee (x \wedge z), \, x \vee (y \wedge z) = (x \vee y) \wedge (x \vee z).$$

The distributive property does not hold between strong conjunction and strong disjunction.

Since $[0,1]_L$ generates the whole algebraic variety, the algebra of **L**-formulas on n variables is isomorphic to the algebra of functions from $[0,1]^n$ to $[0,1]$ constructible from the projections by means of pointwise defined operations. In particular, the zero of the algebra is the constant function equal to 0 and for every $(t_1, \ldots, t_n) \in [0,1]^n$:

$$\neg f(t_1, \ldots, t_n) = 1 - f(t_1, \ldots, t_n),$$

$$(f \circ g)(t_1, \ldots, t_n) = f(t_1, \ldots, t_n) \circ g(t_1, \ldots, t_n), \quad \text{with } \circ \in \{\wedge, \vee, \otimes, \oplus, \rightarrow\}.$$

In addition, we know exactly which kind of functions corresponds to Łukasiewicz formulas. The result is expressed by the the well–known McNaughton Theorem.

---

[1] A variety is a class of algebras closed under the taking of homomorphic images, subalgebras and direct products.

[2] Through the paper we use the same symbols for connectives and algebraic operations.

[3] The difference can be explained as follows: $\alpha \wedge \beta$ gives us the availability of both α and β but we can pick just one of them; $\alpha \otimes \beta$ forces us to pick both formulas in the pair (α, β).

Definition 1. *Let* $f : [0,1]^n \to [0,1]$ *be a continuous function with* $n \geq 0$, f *is called a McNaughton function if it is piecewise linear with integer coefficients, that is, there exists a finite set of linear functions* $p_1, \ldots, p_m$ *with integer coefficients such that for all* $(t_1, \ldots, t_n) \in [0,1]^n$, *there exists* $i \leq m$ *such that*

$$f(t_1, \ldots, t_n) = p_i(t_1, \ldots, t_n).$$

Theorem 1 (McNaughton Theorem). *For each integer* $n \geq 0$, *the class of* $[0,1]$-*valued functions defined on* $[0,1]^n$ *and corresponding to formulas of propositional Łukasiewicz logic coincides with the class of McNaughton functions* $f : [0,1]^n \to [0,1]$ *equipped with pointwise defined operations.*

As a consequence, for every **Ł**-formula $\varphi(x_1, \ldots, x_n)$ depending on propositional variables $x_1, \ldots, x_n$, we can consider its corresponding function $f_\varphi : (x_1, \ldots, x_n) \in [0,1]^n \mapsto f_\varphi(x_1, \ldots, x_n) \in [0,1]$, whose value on each point is exactly the evaluation of the formula with respect to the same variable assignment. For McNaughton Theorem f_φ is a McNaughton function.

2.1   Concave McNaughton Functions

In this framework the logic constraints will be expressed by McNaughton functions. Hence we are interested in operations (corresponding to logical connectives) that preserve concavity or convexity of McNaughton functions in order to establish a Łukasiewicz fragment in which concavity (or convexity) is guaranteed.

Lemma 1. *Let* f, g *be two* $[0,1]$-*valued functions defined on* $[0,1]^n$, *then*

1. f *is convex if and only if the function* $\neg f$ *is concave;*
2. *if* f, g *are concave then the functions* $f \wedge g$ *and* $f \oplus g$ *are concave;*
3. *if* f, g *are convex then the functions* $f \vee g$ *and* $f \otimes g$ *are convex.*

Proof

1. This point is obvious remembering that for all x, $\neg f(x) := 1 - f(x)$.
2. If f, g are concave then for all $x, y \in [0,1]^n$, $\lambda \in [0,1]$, $(f \wedge g)(\lambda x + (1-\lambda)y) = \min\{f(\lambda x + (1-\lambda)y), g(\lambda x + (1-\lambda)y)\} \geq \min\{\lambda f(x) + (1-\lambda)f(y), \lambda g(x) + (1-\lambda)g(y)\} \geq \{\lambda(f \wedge g)(x) + (1-\lambda)(f \wedge g)(y)\}$.
   Moreover, by definition $(f \oplus g)(x) = \min\{1, f(x) + g(x)\}$ thus if $(f \oplus g)(\lambda x + (1-\lambda)y) = 1$ then obviously it is greater or equal than $\lambda(f \oplus g)(x) + (1-\lambda)(f \oplus g)(y)$. Otherwise $f \oplus g = f + g$ and sum preserves concavity (and it preserves convexity too) so the thesis easily follows.
3. This point follows from 1 and 2 plus recalling that $f \vee g = \neg(\neg f \wedge \neg g)$ and $f \otimes g = \neg(\neg f \oplus \neg g)$.

As a result we get that, if f is a convex function and g is concave, then $f \to g = \neg f \oplus g$ is concave. This implies an immediate consequence.

Corollary 1. *Every Horn clause in Łukasiewicz logic, namely every formula of the form* $(x_1 \otimes \ldots \otimes x_m) \to y$ *with* $x_1, \ldots, x_m, y$ *propositional variables, has a corresponding concave McNaughton function.*

Proof. Projections functions $f_{x_1}, \ldots, f_{x_m}, f_y$ are both convex and concave functions. So from the fact that $(f_{x_1} \otimes \ldots \otimes f_{x_m}) \to f_y = \neg(f_{x_1} \otimes \ldots \otimes f_{x_m}) \oplus f_y = \neg f_{x_1} \oplus \ldots \oplus \neg f_{x_m} \oplus f_y$, for the previous lemma, to this formula corresponds a concave function.

Example 1. Horn clauses can be embedded in the $(\wedge, \oplus)^*$-fragment, but the converse does not hold, as shown by the following formulas:

$$x \wedge y, \qquad x \oplus (y \wedge z), \qquad x \wedge (x \to y), \qquad (x \otimes y) \to (x \wedge z).$$

Lemma 1 allows us to determine two Łukasiewicz fragments whose corresponding McNaughton functions are either concave or convex respectively.

Corollary 2. *Let $(\wedge, \oplus)^*$ and $(\otimes, \vee)^*$ be the smallest sets such that:*

- *if y is a propositional variable, then $y, \neg y \in (\wedge, \oplus)^*$ and $y, \neg y \in (\otimes, \vee)^*$;*
- *if $\varphi_1, \varphi_2 \in (\wedge, \oplus)^*$, then $\varphi_1 \wedge \varphi_2$, $\varphi_1 \oplus \varphi_2 \in (\wedge, \oplus)^*$;*
- *if $\varphi_1, \varphi_2 \in (\otimes, \vee)^*$, then $\varphi_1 \otimes \varphi_2$, $\varphi_1 \vee \varphi_2 \in (\otimes, \vee)^*$;*
- *if $\varphi \in (\wedge, \oplus)^*$ $(\varphi \in (\otimes, \vee)^*)$ and φ is equivalent to ψ, then $\psi \in (\wedge, \oplus)^*$ $(\psi \in (\otimes, \vee)^*)$.*

Every Ł-formula $\varphi \in (\wedge, \oplus)^*$ $(\varphi \in (\otimes, \vee)^*)$ corresponds to a concave (convex) McNaughton function f_φ.

In the paper we will make use of the following result and we refer to Theorem 2.49 in [14] for a proof.

Theorem 2. *Any convex piecewise linear function on $\mathbb{R}^n$ can be expressed as a max of a finite number of affine functions.*

This means that, for every McNaughton function $f : [0,1]^n \to [0,1]$, if f is convex then there exist $A_1, \ldots, A_k \in \mathbb{Z}^n$, $b_1, \ldots, b_k \in \mathbb{Z}$ such that:

$$\text{for all } x \in [0,1]^n \qquad f(x) = \max_{i=1}^{k} A_i' \cdot x + b_i \tag{1}$$

On the other hand, every concave McNaughton function can be expressed as the minimum of a finite number of affine functions.

The coefficients of the linear functions are constructively determined by the shape of the considered formula. For instance, given any convex McNaughton function f_φ, such that $\varphi \in (\otimes, \vee)^*$, we can rewrite φ as a weak disjunction of strong conjunctions of literals according to the distributive laws. Finally, every disjointed term corresponds to a linear function or to the identically 0 function.

Example 2. Let $\varphi = (\neg x \oplus (y \wedge z)) \oplus \neg z$ be, then $\varphi \in (\wedge, \oplus)^*$ and it is equivalent to $(\neg x \oplus y \oplus \neg z) \wedge (\neg x \oplus z \oplus \neg z)$. From this latter expression, we get:

$$\text{for all } x, y, z \in [0,1]: \qquad f_\varphi(x, y, z) = \min\{1, -x + y - z + 2, -x + 2\}.$$

So far, we showed that $(\wedge, \oplus)^*$-formulas bring to concave corresponding McNaughton functions and, as we will see, to convex logical constraints. The next result shows that indeed $(\wedge, \oplus)^*$ contains all the concave ones.

Proposition 1. *Let $f_\varphi : [0,1]^n \to [0,1]$ be a concave McNaughton function, then $\varphi \in (\wedge, \oplus)^*$.*

Proof. By hypothesis f_φ is a concave piecewise linear function hence there exist some elements $a_{i_j}, b_i \in \mathbb{Z}$ for $i = 1, \ldots, m$ and $j = 1, \ldots, n$, such that:

$$f_\varphi(x) = \min_{i=1}^m a_{i_1} x_1 + \ldots + a_{i_n} x_n + b_i, \qquad x \in [0,1]^n.$$

If we set $p_i(x) = a_{i_1} x_1 + \ldots + a_{i_n} x_n + b_i$ for $i = 1, \ldots, m$, our claim follows provided every p_i corresponds to a formula in $(\wedge, \oplus)^*$, indeed the operation of minimum is exactly performed by the connective $\wedge$. Let us fix $i \in \{1, \ldots, m\}$, then we can write

$$p_i(x) = \sum_{j \in P_i} a_{i_j} x_j + \sum_{j \in N_i} a_{i_j} x_j + b_i,$$

where $P_i = \{j \leq n : a_{i_j} > 0\}$ and $N_i = \{j \leq n : a_{i_j} < 0\}$. It is worth noticing that in general p_i will assume values out of the unit interval. However $p_i(x) \geq 0$ for all $x \in [0,1]^n$ and the values greater than 1 obviously do not contribute to f_φ. Thanks to this last remark, we can take into account a formula whose corresponding McNaughton function corresponds to p_i truncated to 1 and restricted in $[0,1]^n$. Let us consider the formula

$$\varphi_i = \bigoplus_{j \in P_i} |a_{i_j}| x_j \oplus \bigoplus_{j \in N_i} |a_{i_j}| \neg x_j \oplus q_i,$$

where $q_i = (\neg x \oplus x) \oplus \ldots \oplus (\neg x \oplus x)$, q_i times. Indeed the first strong disjunction corresponds to all the positive monomials of p_i. The second one corresponds to all the negative monomials of p_i, but it also introduces the quantity $\sum_{j \in N_i} |a_{i_j}|$. Finally $q_i = b_i - \sum_{j \in N_i} |a_{i_j}|$, with $q_i \geq 0$ since $p_i(x) \geq 0$ for all $x \in [0,1]^n$ and in particular $p_i(\bar{x}) = b_i - \sum_{j \in N_i} |a_{i_j}| \geq 0$ where $\bar{x}$ is the vector with 0 in positive and 1 in negative monomial positions respectively. The overall formula can be written as $\varphi = \varphi_1 \wedge \ldots \wedge \varphi_m$.

This result allows us to conclude that we are taking into account the largest fragment of Łukasiewicz Logic whose McNaughton functions are concave.

Corollary 3. $(\otimes, \vee)^*$ *is the fragment coinciding with the class of all convex McNaughton functions.*

3   Constraints

We consider a learning problem in which each learnable task function is a predicate on a given domain. The considered constraints belong to the following three categories.

1. *Pointwise constraints* are given as supervised examples in training set.
2. *Logical constraints* establish some relations that must hold between predicates. They can be expressed as first order formulas in any logic. In this paper we restrict our attention to Łukasiewicz logic. They are enforced on an unsupervised training set.
3. *Consistency Constraints* derive from the need to limit the possible values of predicates into the real unit interval to guarantee the consistency in the definition of the logical constraints.

Let $KB = \{\varphi_h : h \in \mathbb{N}_H{}^4\}$ be a knowledge base of first order Łukasiewicz formulas. We assume KB contains predicates from the set $\mathbf{P} = \{p_j^{(a_j)} : j \in \mathbb{N}_J\}$ (with their own fixed arity) each one defined on a subset of a power of $\mathbb{R}$. More specifically, for every j we have $p_j^{(a_j)} : \mathcal{R}_{j_1} \times \ldots \times \mathcal{R}_{j_{a_j}} = \mathcal{R}_{n_j} \to \mathbb{R}$, where for $k = 1, \ldots, a_j$, $\mathcal{R}_{j_k} \subseteq \mathbb{R}^{j_k}$ and so $\mathcal{R}_{n_j} \subseteq \mathbb{R}^{n_j}$. However, throughout the paper, we will always evaluate predicates in finite, already fixed, training sets. It is worth noting that, different predicates can share some domains of their arguments. For instance, if $\mathcal{R}_{j_k} = \mathcal{R}_{i_l}$ (for some j, k, i, l), this means the k-th argument of predicate p_j is the same as the l-th argument of predicate p_i. For short, we will write $\mathbf{p} = (p_1, \ldots, p_J) : \mathcal{R}^n \to \mathbb{R}^J$, where $n = n_1 + \ldots + n_J$, for the overall task function.

3.1 Pointwise Constraints

We can formulate the learning task as a *classification* or a *regression* problem. However, even if regression would exploit fuzzy predicates in a narrow sense (allowing label in the whole unit interval), for the moment we only consider the classification case.

For $j = 1, \ldots, J$, we consider a supervised training set for the j-th task p_j:

$$\mathscr{L}_j = \{(\mathbf{x}_l^j, y_l^j) : l \in \mathbb{N}_{l_j}, \ \mathbf{x}_l^j \in \mathbb{R}^{n_j}, y_l^j \in \{-1, 1\}\}.$$

In the following, we will omit the superscript j in the case it is clear from the context. For instance, we will write $p_j(\mathbf{x}_l)$ instead of $p_j(\mathbf{x}_l^j)$, because we assume the j-th task applies only to example points from j-th training set. If $y_l = 1$ then $\mathbf{x}_l$ belongs to the true–class, so we would like to have $p_j(\mathbf{x}_l) \geq 1$, whereas if $y_l = -1$ then $\mathbf{x}_l$ belongs to the false–class and we would like to have $p_j(\mathbf{x}_l) \leq 0$. Therefore, supervisions correspond to the following hard–constraints[5]:

$$y_l(2p_j(\mathbf{x}_l) - 1) \geq 1 - 2\xi_{j_l} \text{ with } \xi_{j_l} \geq 0 \qquad \text{for } j = 1, \ldots, J \text{ and } l = 1, \ldots, l_j.$$

where we introduced a slack variable ξ for each point in the training sets to allow soft violations.

---

[4] Where, for every $j \in \mathbb{N}$, $\mathbb{N}_j = \{n \in \mathbb{N} : n \leq j\}$.
[5] If we consider the usual 0–1 logic values as labels, we can consider the condition $(2y_l - 1)p_j(\mathbf{x}_l) \geq y_l$.

3.2   Logical Constraints

Logical constraints arise from the Knowledge Base KB that is any collection of *first order* formulas. For Łukasiewicz logic, we can assume without loss of generality that formulas are in *prenex normal form*. We suppose that each variable occurring in predicates takes its values from a fixed set $\mathcal{U}_{j_k}$. The Cartesian product of the sets $\mathcal{U}_{j_k}$ constitutes the unsupervised training set $\mathcal{U}_j$ for the j-th task,

$$\mathcal{U}_j = \{\mathbf{x}_u^j = (\mathbf{x}_1^j, \ldots, \mathbf{x}_{a_j}^j) \in \mathbb{R}^{n_j} : \mathbf{x}_k^j \in \mathcal{U}_{j_k}, u \in \mathbb{N}_{u_j}\},$$

whose elements will be used to evaluate p_j in logical constraints. In the following we will omit the superscript j when it is clear from the context. Finally we set $U = u_1 + \ldots + u_J$.

Assuming a finite domain for variables, each quantified formula can be replaced with a propositional one applying the following equivalence once per quantifier:

$$\forall x\, \psi(x) \simeq \bigwedge_{\mathbf{x}_k \in \mathcal{U}_{j_k}} \psi[\mathbf{x}_k/x], \qquad \exists x\, \psi(x) \simeq \bigvee_{\mathbf{x}_k \in \mathcal{U}_{j_k}} \psi[\mathbf{x}_k/x], \qquad (2)$$

where x is the k-th argument of a certain predicate p_j occurring in the formula ψ and $[\mathbf{x}_k/x]$ represents the substitution of the element $\mathbf{x}_k$ in place of x in ψ. If two or more predicates share a variable, in possibly different arguments, then that variable will be evaluated on the same set of values for such predicates.

By applying (2), we get a new set of formulas KB', where the propositional variables are exactly all possible groundings of predicates occurring in formulas. Each grounding of a predicate returns a value in the unit interval and since predicate functions have to be determined, these values can be thought of as (propositional) variables.

Example 3. Let us consider $\mathbf{P} = \{p_1^{(1)}, p_2^{(2)}\}$ and $KB = \{\varphi_1, \varphi_2\}$ with

$$\varphi_1 : \forall x \exists y (p_1(x) \rightarrow p_2(x, y)), \qquad \varphi_2 : \forall x (p_1(x)).$$

Now let us suppose $\mathcal{U}_{1_1} = \mathcal{U}_{2_1} = \{a, b\}$ and $\mathcal{U}_{2_2} = \{a, c\}$, then $KB' = \{\varphi_1', \varphi_2'\}$ with φ_1' and φ_2' respectively:

$$[(p_1(a) \rightarrow p_2(a, a)) \vee (p_1(a) \rightarrow p_2(a, c))] \wedge [(p_1(b) \rightarrow p_2(b, a)) \vee (p_1(b) \rightarrow p_2(b, c))],$$

$$p_1(a) \wedge p_1(b).$$

We can simplify the notation introducing new variables, for instance with:

$$y_1 = p_1(a),\ y_2 = p_1(b),\ y_3 = p_2(a, a),\ y_4 = p_2(a, c),\ y_5 = p_2(b, a),\ y_6 = p_2(b, c),$$

$$\varphi_1' : [(y_1 \rightarrow y_3) \vee (y_1 \rightarrow y_4)] \wedge [(y_2 \rightarrow y_5) \vee (y_2 \rightarrow y_6)], \qquad \varphi_2' : y_1 \wedge y_2.$$

Formulas in KB' are expressed in the context of propositional Łukasiewicz Logic **L** and so we can use all the results reported in the previous section. In particular, every n–ary formula φ is isomorphic to a McNaughton function $f_\varphi : [0,1]^n \rightarrow [0,1]$. For the sake of simplicity, we write f_h for the function corresponding to φ'_h formula.

Every formula in KB' depends on all possible groundings of their occurring predicates, so it is useful for $j = 1, \ldots, J$ to indicate with $\bar{\mathbf{p}}_j$ the vector of all groundings of p_j, namely $\bar{\mathbf{p}}_j = (p_j(\mathbf{x}_1), \ldots, p_j(\mathbf{x}_{u_j}))$. To make all uniform, with a little abuse of notation, we will write every formula $\varphi' \in KB'$ as depending on all groundings of all predicates, $f_{\varphi'} = f_{\varphi'}(\bar{\mathbf{p}})$ where $\bar{\mathbf{p}} = (\bar{\mathbf{p}}_1, \ldots, \bar{\mathbf{p}}_J) \in \mathbb{R}^U$.

Formulas in KB' depend on the parameters which determine each predicate function. Let $p_j \in \mathbf{P}$ be, such that $p_j : \mathcal{R}_{n_j} \subseteq \mathbb{R}^{n_j} \rightarrow \mathbb{R}$. We assume that there exists a feature map $\phi_j^6 : \mathbb{R}^{n_j} \rightarrow \mathbb{R}^{N_j}$ (where it may be $n_j << N_j$), such that $p_j(\mathbf{x}) = \omega'_j \cdot \phi_j(\mathbf{x}) + b_j$ with $\omega_j \in \mathbb{R}^{N_j}, b_j \in \mathbb{R}$. If we set $\hat{\omega}_j = (\omega'_j, b_j)', \hat{\phi}_j = (\phi'_j, 1)'$ we have

$$p_j(\mathbf{x}) = \hat{\omega}_j \cdot \hat{\phi}_j(\mathbf{x}). \tag{3}$$

On a fixed training set, the values of predicates are totally determined by the matrices $\hat{\omega}_1, \ldots, \hat{\omega}_J$. This entails that in the weight space formulas will be evaluated by composition as functions on $\mathbb{R}^{N+J}$ where $N = N_1 + \ldots + N_J$. Hence we need to guarantee the convexity of McNaughton functions corresponding to formulas in the weight spaces.

Lemma 2. *Let $f : Y \subseteq \mathbb{R}^m \rightarrow \mathbb{R}$ be a convex or concave function and $g : X \subseteq \mathbb{R}^d \rightarrow Y$ such that $g(x) = Ax + b$ with $A \in \mathbb{R}^{m,d}, b \in \mathbb{R}^m$. Then the function $h : X \rightarrow \mathbb{R}$ defined by $h = f \circ g$ is convex or concave in X.*

Corollary 4. *If a logical constraint is concave in its propositional variables space then it is also concave if expressed in the weight space of the predicates.*

Finally, given $KB' = \{f_1, \ldots, f_H\}$ we enforce the satisfaction of logical constraints by requiring $1 - f_h(\bar{\mathbf{p}}) = 0$, for all $h = 1, \ldots, H$. In order to allow soft violations of these constraints, we introduce new slack variables rewriting the constraints as:

$$1 - f_h(\bar{\mathbf{p}}) \leq \xi_h \text{ with } \xi_h \geq 0 \quad \text{for } h \in \mathbb{N}_H. \tag{4}$$

If f_h is a concave function, then $g_h = 1 - f_h$ is a convex one and we note g_h is the function corresponding to the formula $\neg\varphi'_h$. This means $g_h(\bar{\mathbf{p}}) \leq \xi_h$ is a convex constraint. This formulation brings us to formulate our learning problem as a *convex* mathematical program.

However, in the specific case of Łukasiewicz logic, each formula corresponds to a McNaughton function. Furthermore, we consider only concave functions which

---

[6] For every $j = 1, \ldots, J$, ϕ_j is determined by the j-th kernel function k_j of the RKHS $\mathcal{H}_j$ and it is such that $k_j(\mathbf{x}, \mathbf{y}) = \langle \phi_j(\mathbf{x}), \phi_j(\mathbf{y}) \rangle_{\mathbb{R}^{N_j}}$.

derive convex constraints like (4). Since every convex McNaughton function can be written according to (1) as the maximum of affine functions, we have

$$g_h(\bar{\mathbf{p}}) = 1 - f_h(\bar{\mathbf{p}}) = \max_{i \in \mathbb{N}_{I_h}} M_i^h \cdot \bar{\mathbf{p}} + q_i^h \leq \xi_h \quad \text{for } h \in \mathbb{N}_H,$$

where $M_i^h \in \mathbb{R}^{1,U}$ and $q_i^h \in \mathbb{R}$ are integer coefficients determined by the shape of the formula $\neg\varphi_h'$.

Remark 1. Let f_φ be the convex McNaughton function of a Łukasiewicz formula φ, then $\varphi \in (\otimes, \vee)^*$ and we can write

$$\varphi(y_{1_1}, \ldots, y_{I_{k_I}}) = \bigvee_{i=1}^{I} \bigotimes_{k=1}^{k_i} (\neg) y_{i_k}.$$

Therefore for all i, each term $((\neg)y_{i_1} \otimes \ldots \otimes (\neg)y_{i_{k_i}})$ is related to an affine function of $y_{i_1}, \ldots, y_{i_{k_i}}$ corresponding to f_φ in a certain piece of its domain. For instance,

if $\varphi = (y_1 \otimes y_1 \otimes \neg y_2) \vee (y_1 \otimes \neg y_2 \otimes \neg y_2)$, then $f_\varphi = \max\{2y_1 - y_2 - 1, y_1 - 2y_2, 0\}$.

Now, since for every h, $g_h(\bar{\mathbf{p}})$ is a *max* of terms, we have $g_h(\bar{\mathbf{p}}) \leq \xi_h$ if and only if

$$M_i^h \cdot \bar{\mathbf{p}} + q_i^h \leq \xi_h \quad \text{for all } i \in \mathbb{N}_{I_h}. \tag{5}$$

This means for every $h \in \mathbb{N}_H$, we can consider the I_h logical constraints expressed by (5) in place of (4). Even if the number of constraints is increased, each of them is now an affine function, so we can deal with a quadratic programming problem.

3.3   Consistency Constraints

The function $\mathbf{p}$ represents a tuple of logical predicates. In principle, when we write any p_j according to (3), we have no guarantees p_j is evaluated in $[0, 1]$. This is the reason why we need to add the following hard constraints:

$$0 \leq p_j(\mathbf{x}_s^j) \leq 1, \quad \text{for } j \in \mathbb{N}_J, \mathbf{x}_s^j \in \mathscr{S}_j = \mathscr{U}_j \cup \mathscr{L}_j', \tag{6}$$

with $\mathscr{L}_j' = \{\mathbf{x}_l^j : (\mathbf{x}_l^j, y_l^j) \in \mathscr{L}_j\}$. In the following, we indicate with s_j the cardinality of $\mathscr{S}_j$ and $S = s_1 + \ldots + s_J$.

4   Learning by Quadratic Programming

We can formulate the multi–task learning problem as an optimization problem both in the primal and in the dual space. In the primal space, for the overall predicate $\mathbf{p}$ exploiting (3) for every $j = 1, \ldots, J$, we get:

Primal Problem

$$\min \tfrac{1}{2} \sum_{j\in\mathbb{N}_J} \|\omega_j\|^2 + C_1 \sum_{\substack{j\in\mathbb{N}_J \\ l\in\mathbb{N}_{l_j}}} \xi_{j_l} + C_2 \sum_{h\in\mathbb{N}_H} \xi_h \qquad \text{subject to}$$

$$
\begin{aligned}
&y_l(2p_j(\mathbf{x}_l)-1) \geq 1 - 2\xi_{j_l} && \xi_{j_l} \geq 0 \quad j\in\mathbb{N}_J, l\in\mathbb{N}_{l_j}, (\mathbf{x}_l, y_l)\in\mathscr{L}_j \\
&1 - f_h(\bar{\mathbf{p}}) \leq \xi_h && \xi_h \geq 0 \quad h\in\mathbb{N}_H \\
&0 \leq p_j(\mathbf{x}_s) \leq 1 && s\in\mathbb{N}_{s_j}, \mathbf{x}_s\in\mathscr{S}_j.
\end{aligned}
\tag{7}
$$

Here $C_1, C_2 > 0$ express the degree of satisfaction of the constraints.

The primal problem (7) is a convex optimization problem, since we have to minimize a convex function subject to convex or affine constraints.

We reformulate the above optimization problem as the minimization of a Lagrangian function. In order to get a quadratic optimization problem, we consider the constraints (5) with $\xi_h \geq 0$ for all $h\in\mathbb{N}_H$ in place of (4), obtaining

$$
\begin{aligned}
\mathcal{L}(\hat{\omega}, \xi, \lambda, \mu, \eta) = &\frac{1}{2}\sum_{j=1}^{J}\|\omega_j\|^2 + C_1\sum_{j=1}^{J}\sum_{l=1}^{l_j}\xi_{j_l} + C_2\sum_{h=1}^{H}\xi_h - \sum_{j=1}^{J}\sum_{l=1}^{l_j}\mu_{j_l}\xi_{j_l} + \\
&-\sum_{j=1}^{J}\sum_{l=1}^{l_j}\lambda_{j_l}(y_l(2p_j(\mathbf{x}_l)-1)-1+2\xi_{j_l}) - \sum_{h=1}^{H}\sum_{i=1}^{I_h}\lambda_{h_i}(\xi_h - M_i^h\cdot\bar{\mathbf{p}} - q_i^h) - \sum_{h=1}^{H}\mu_h\xi_h + \\
&-\sum_{j=1}^{J}\sum_{s=1}^{s_j}\eta_{j_s}p_j(\mathbf{x}_s) - \sum_{j=1}^{J}\sum_{s=1}^{s_j}\bar{\eta}_{j_s}(1-p_j(\mathbf{x}_s)).
\end{aligned}
\tag{8}
$$

with the **KKT** conditions, for all $j\in\mathbb{N}_J, l\in\mathbb{N}_{l_j}, h\in\mathbb{N}_H, i\in\mathbb{N}_{I_h}, s\in\mathbb{N}_{s_j}$:

$$\xi_{j_l} \geq 0,\ \xi_h \geq 0,\ \lambda_{j_l} \geq 0,\ \mu_{j_l} \geq 0,\ \lambda_{h_i} \geq 0,\ \mu_h \geq 0,\ \eta_{j_s} \geq 0,\ \bar{\eta}_{j_s} \geq 0,$$
$$\lambda_{j_l}(y_l(2p_j(\mathbf{x}_l)-1)-1+2\xi_{j_l}) = 0,\ \mu_{j_l}\xi_{j_l} = 0,\ \lambda_{h_i}(\xi_h - M_i^h\cdot\bar{\mathbf{p}} - q_i^h) = 0,$$
$$\mu_h\xi_h = 0,\ \eta_{j_s}p_j(\mathbf{x}_s) = 0,\ \bar{\eta}_{j_s}(1-p_j(\mathbf{x}_s)) = 0,\ y_l(2p_j(\mathbf{x}_l)-1)-1+2\xi_{j_l} \geq 0,$$
$$\xi_h - M_i^h\cdot\bar{\mathbf{p}} - q_i^h \geq 0,\ p_j(\mathbf{x}_s) \geq 0,\ 1-p_j(\mathbf{x}_s) \geq 0.$$

To derive the dual space formulation we apply the null gradient condition to the derivatives of $\mathcal{L}$ with respect to every $\omega_j, b_j, \xi_{j_l}, \xi_h$.

$$\nabla_{\omega_j}\mathcal{L} = \omega_j - 2\sum_{l}\lambda_{j_l}y_l\phi_j(\mathbf{x}_l) + \sum_{h,i}\lambda_{h_i}\sum_{u}M_{i,u}^h\phi_j(\mathbf{x}_u) - \sum_{s}(\eta_{j_s}-\bar{\eta}_{j_s})\phi_j(\mathbf{x}_s) = 0;$$

$$\frac{\partial\mathcal{L}}{\partial b_j} = -2\sum_{l}\lambda_{j_l}y_l + \sum_{h,i}\lambda_{h_i}\sum_{u}M_{i,u}^h - \sum_{s}(\eta_{j_s}-\bar{\eta}_{j_s}) = 0;$$

$$\frac{\partial\mathcal{L}}{\partial\xi_{j_l}} = C_1 - 2\lambda_{j_l} - \mu_{j_l} = 0;$$

$$\frac{\partial\mathcal{L}}{\partial\xi_h} = C_2 - \sum_{i}\lambda_{h_i} - \mu_h = 0.$$

Hence if we substitute, we get

$$\theta(\lambda, \eta) = \mathcal{L}(\hat{\omega}^*, \xi^*, \lambda, \mu, \eta) = -\frac{1}{2}\sum_{j}[4\sum_{l,l'}\lambda_{j_l}\lambda_{j_{l'}}y_ly_{l'}k_j(\mathbf{x}_l, \mathbf{x}_{l'}) +$$

$$+ \sum_{\substack{h,i \\ h',i'}} \lambda_{h_i} \lambda_{h'_{i'}} \sum_{u,u'} M^h_{i,u} M^{h'}_{i',u'} k_j(\mathbf{x}_u, \mathbf{x}_{u'}) + \sum_{s,s'} (\eta_{j_s} - \bar{\eta}_{j_s})(\eta_{j_{s'}} - \bar{\eta}_{j_{s'}}) k_j(\mathbf{x}_s, \mathbf{x}_{s'}) +$$

$$-4 \sum_{l,h,i} \lambda_{j_l} \lambda_{h_i} y_l \sum_u M^h_{i,u} k_j(\mathbf{x}_l, \mathbf{x}_u) + 4 \sum_{l,s} \lambda_{j_l} y_l (\eta_{j_s} - \bar{\eta}_{j_s}) k_j(\mathbf{x}_l, \mathbf{x}_s) +$$

$$-2 \sum_{h,i,s} \lambda_{h_i}(\eta_{j_s} - \bar{\eta}_{j_s}) \sum_u M^h_{i,u} k_j(\mathbf{x}_u, \mathbf{x}_s)] + \sum_{j,l} \lambda_{j_l} + \sum_{h,i} \lambda_{h_i}(\frac{1}{2} \sum_u M^h_{i,u} + q^h_i) +$$

$$-\frac{1}{2} \sum_{j,s} (\eta_{j_s} + \bar{\eta}_{j_s}).$$

Finally, we can formulate the dual problem as:

Dual Problem

$$\max \ \theta(\lambda, \eta) \qquad \text{subject to}$$

$$\sum_{h=1}^{H} \sum_{i=1}^{I_h} \lambda_{h_i} \sum_{u=1}^{u_j} M^h_{i,u} = 2 \sum_{l=1}^{l_j} \lambda_{j_l} y_l + \sum_{s=1}^{s_j} (\eta_{j_s} - \bar{\eta}_{j_s}) \qquad j \in \mathbb{N}_J \qquad (9)$$

$$0 \le \lambda_{j_l} \le C_1 \qquad\qquad j \in \mathbb{N}_J, \ l \in \mathbb{N}_{l_j}$$
$$0 \le \lambda_{h_i} \le C_2 \qquad\qquad h \in \mathbb{N}_H, \ i \in \mathbb{N}_{I_h}$$
$$\eta_{j_s} \ge 0, \ \bar{\eta}_{j_s} \ge 0 \qquad\qquad j \in \mathbb{N}_J, \ s \in \mathbb{N}_{s_j}.$$

From (9) we can find the optimal values $\lambda^*_{j_l}, \lambda^*_{h_i}, \eta^*_{j_s}, \bar{\eta}^*_{j_s}$ for all j, l, h, i, s. Then, for all $j = 1, \ldots, J$, we get the formulation of the j-th task in the dual space as:

$$p_j(\mathbf{x}) = \omega^*_j \cdot \phi_j(\mathbf{x}) + b^*_j = 2 \sum_{l=1}^{l_j} \lambda^*_{j_l} y_l k_j(\mathbf{x}_l, \mathbf{x}) +$$

$$-\sum_{h=1}^{H} \sum_{i=1}^{I_h} \lambda^*_{h_i} \sum_{u=1}^{u_j} M^h_{i,u} k_j(\mathbf{x}_u, \mathbf{x}) + \sum_{s=1}^{s_j} (\eta^*_{j_s} - \bar{\eta}^*_{j_s}) k_j(\mathbf{x}_s, \mathbf{x}) + b^*_j. \qquad (10)$$

5   Experimental Results

In this section, we report the experimental evaluation of the proposed method on two datasets. The first one is a toy problem that allows us to enlighten how logical constraints contribute to the solution of a given problem. The second one is based on the Winston benchmark for image classification.

5.1   Toy Problem

Let us consider the following multi–task problem, where we want to determine three predicate functions p_1, p_2, p_3 defined on $\mathbb{R}^2$ and let L_1, L_2, L_3 be such that $L_1 = \{(0.1, 0.5, -1), (0.4, 0.4, -1), (0.3, 0.8, 1), (0.9, 0.7, 1)\}$, $L_2 = \{(0.1, 0.3, -1), (0.6, 0.4, -1), (0.2, 0.8, 1), (0.7, 0.6, 1)\}$ and $L_3 = \{(0.4, 0.2, -1), (0.9, 0.3, -1),$

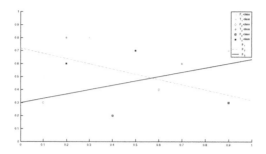

Fig. 1. The dotted, dashed and solid lines represent the task functions p_1, p_2 and p_3 respectively. Data points with different shape relate to different predicates and while empty figures correspond to the false–class, the filled ones correspond to the true–class.

$(0.2, 0.6, 1)$, $(0.5, 0.7, 1)$} the sets of supervised examples. In Fig. 1, we show the (unique) solution for the standard kernel machine scheme in which we only take into account pointwise constraints and we set $C_1 = 15$.

Now let us suppose to know, apart from supervisions, some additional relational informations about task functions, expressed by the logical clauses

$$\varphi_1 = \forall x(p_1(x) \to p_2(x)), \qquad \varphi_2 = \forall x(p_2(x) \to p_3(x)),$$

whose intuitive meaning is: the p_1 true–class is contained in the p_2 true–class which in turn is contained in the p_3 true–class. To evaluate logical constraints we fix a few unsupervised examples. For instance, in Fig. 2 we make a comparison between the effect of a single point $P(0.8, 0.3)$ (in which logical constraints were violated) and the effect of a larger unsupervised training set $U = \{(0.1, 0.5), (0.3, 0.7), (0.5, 0.4), (0.8, 0.3), (0.9, 0.2), (1, 0.5)\}$. The last plot shows that with just few unsupervised points the boundaries of the predicates are correctly placed in the (unique) solution in order to satisfy the given logic constraints in the considered domain.

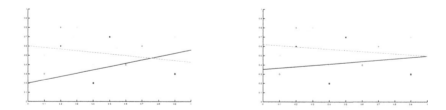

Fig. 2. The pictures show, respectively from left to right, the effect of considering as unsupervised examples the single point P and the set U. In both cases we set $C_1 = C_2 = 2.5$.

5.2   Winston Benchmark

The second experimental analysis is based on the animal identification problem originally proposed by Winston [16]. This benchmark was initially designed to show the ability of logic programming to guess the animal type from some initial clues.

The dataset is composed of 5605 images, taken from the *ImageNet*[7] database, equally divided into 7 classes, each one representing one animal category: albatross, cheetah, giraffe, ostrich, penguin, tiger and zebra. The vector of numbers used to represent each image is composed by two parts: one representing the colors in the image, and one representing its shape. In particular, the feature representation contains a 12-dimension normalized color histogram for each channel in the RGB color space. Furthermore, the SIFT descriptors [9] have been built by sampling a set of images from the dataset and then detecting all the SIFT representations present in at least one of the sampled images. Finally, the SIFT representations have been clustered into 600 *visual words*. The final representation of an image contains 600 values, where the i-th element represents the normalized count of the i-th visual word for the given image (bag-of-descriptors). As previously done in Diligenti et al. [6], the test phase does not get as input a sufficient set of clues to perform classification, but the image representations are used by the learning framework to develop the intermediate clues over which to perform inference[8].

The images have been split into two initial sets: the first one is composed of 2100 images utilized for building the visual vocabulary, while the second set is composed of 3505 images used in the learning process. The experimental analysis has been carried out using by randomly sampling from the overall set of the supervisions the labels to keep as training, validation and test set, randomly sampling 50%, 25%, 25% of the supervisions, respectively.

The knowledge about the classification task is expressed by a set of FOL rules. A total of 33 classes are referenced by the KB, the 7 animal classes plus other intermediate classes each of which is either representing a subset of animals in a taxonomy (like the classes *mammal* or *bird*) or indicating some specific feature of the animals (like *hair* or *darkspots*). Table 1 shows the set of rules used in this task. The first 15 rules are the same as stated in the original problem definition by Winston. The fact that each image shows one and only one animal classification is expressed by another rule stating that each pattern should belong to only one of the classes representing an animal. Another set of rules forces the semantic consistency among the intermediate classes.

All the images are available at training time, but only the training sample of the labels is made available during training. For each reported experiment, one Kernel Machine is trained for each of the 33 predicates in the KB. Gaussian kernels have been selected to be used in the experiments as they were clearly out-

---

[7] http://www.image-net.org.

[8] The dataset and code to reproduce the results can be downloaded from https://github.com/diligmic/ECML2017_1 .

Table 1. The KB used for training the models in the Winston image classification benchmark.

Rule
hair(x) → mammal(x)
milk(x) → mammal(x)
feather(x) → bird(x)
layeggs(x) → bird(x)
mammal(x) ∧ meat(x) → carnivore(x)
mammal(x) ∧ pointedteeth(x) ∧ claws(x) ∧ forwardeyes(x) → carnivore(x)
mammal(x) ∧ hoofs(x) → ungulate(x)
mammal(x) ∧ cud(x) → ungulate(x)
carnivore(x) ∧ tawny(x) ∧ darkspots(x) → cheetah(x)
carnivore(x) ∧ tawny(x) ∧ blackstripes(x) → tiger(x)
ungulate(x) ∧ longlegs(x) ∧ longneck(x) ∧ tawny(x) ∧ darkspots(x) → giraffe(x)
ungulate(x) ∧ white(x) ∧ blackstripes(x) → zebra(x)
bird(x) ∧ longlegs(x) ∧ longneck(x) ∧ black(x) → ostrich(x)
bird(x) ∧ swim(x) ∧ blackwhite(x) → penguin(x)
bird(x) ∧ goodflier(x) → albatross(x)
cheetah(x) ⊕ tiger(x) ⊕ giraffe(x) ⊕ zebra(x) ⊕ ostrich(x) ⊕ penguin(x) ⊕ albatross(x)
mammal(x) ⊕ bird(x)
hair(x) ⊕ feather(x)
(darkspots(x)) → ¬ blackstripes(x)
(blackstripes(x)) → ¬ darkspots(x)
tawny(x) → ¬ black(x)) ∧¬ white(x)
black(x) → ¬ tawny(x)) ∧¬ white(x)
white(x) → ¬ black(x)) ∧¬ tawny(x)
black(x) → ¬ white(x)
black(x) → ¬ tawny(x)
white(x)) → ¬ black(x)
white(x) → ¬ tawny(x)
tawny(x) → ¬ white(x)
tawny(x)) → ¬ black(x)

performing both the linear kernel and all the tested variations of the polynomial kernel.

Table 2 reports the summary of the results. Learning with the logic knowledge is significantly improving the results. The convex Łukasiewicz fragment yields the highest F1 metric on this benchmark.

Table 2. F1 values for the baseline (no logic knowledge) and using the KB integrated into the learning process using different t-norms.

	Baseline	Łukasiewicz	Goedel	Product	Convex Łukasiewicz
F1	0.45	0.53	0.52	0.55	**0.56**

6 Conclusions

In this paper we proposed a theoretical framework that allows the formulation of learning with logical constraints as a quadratic problem for kernel machines. This is guaranteed when logical constraints are expressed by means of $(\wedge, \oplus)^*$-formulas of Łukasiewicz Logic. As a result, the optimal solution is unique. We evaluated the learning scheme on one image classification benchmark. Future developments include the inference of new logical statements from examples. This is straight connected with the development of a *parsimonious agent* able to learn from the environment and to propose new abstract knowledge concerning its tasks.

References

1. Bach, S.H., Broecheler, M., Huang, B., Getoor, L.: Hinge-loss markov random fields and probabilistic soft logic. arXiv preprint arXiv:1505.04406 (2015)
2. Bofill, M., Manya, F., Vidal, A., Villaret, M.: Finding hard instances of satisfiability in lukasiewicz logics. In: 2015 IEEE International Symposium on Multiple-Valued Logic (ISMVL), pp. 30–35. IEEE (2015)
3. Cumby, C.M., Roth, D.: On kernel methods for relational learning. In: Proceedings of the 20th International Conference on Machine Learning (ICML-03), pp. 107–114 (2003)
4. Diligenti, M., Gori, M., Maggini, M., Rigutini, L.: Bridging logic and kernel machines. Mach. Learn. **86**(1), 57–88 (2012)
5. Diligenti, M., Gori, M., Saccà, C.: Semantic-based regularization for learning and inference. Artif. Intell. (2015). https://doi.org/10.1016/j.artint.2015.08.011
6. Diligenti, M., Gori, M., Scoca, V.: Learning efficiently in semantic based regularization. In: Frasconi, P., Landwehr, N., Manco, G., Vreeken, J. (eds.) ECML PKDD 2016. LNCS (LNAI), vol. 9852, pp. 33–46. Springer, Cham (2016). https://doi.org/10.1007/978-3-319-46227-1_3
7. Gnecco, G., Gori, M., Melacci, S., Sanguineti, M.: Foundations of support constraint machines. Neural comput. **27**(2), 388–480 (2015)
8. Hu, Z., Ma, X., Liu, Z., Hovy, E., Xing, E.: Harnessing deep neural networks with logic rules. arXiv preprint arXiv:1603.06318 (2016)
9. Lowe, D.G.: Object recognition from local scale-invariant features. In: The Proceedings of the Seventh IEEE International Conference on Computer Vision, 1999, vol. 2, pp. 1150–1157. IEEE (1999)
10. Marchioni, E., Wooldridge, M.: Łukasiewicz games: a logic-based approach to quantitative strategic interactions. ACM Trans. Comput. Logic (TOCL) **16**(4), 33 (2015)

11. Muggleton, S., Lodhi, H., Amini, A., Sternberg, M.J.E.: Support vector inductive logic programming. In: Hoffmann, A., Motoda, H., Scheffer, T. (eds.) DS 2005. LNCS (LNAI), vol. 3735, pp. 163–175. Springer, Heidelberg (2005). https://doi.org/10.1007/11563983_15

12. Novák, V., Perlieva, I., Močkoř, J.: Mathematical Principles of Fuzzy Logic, vol. 517. Springer Science & Business Media, Heidelberg (2012)

13. Richardson, M., Domingos, P.: Markov logic networks. Mach. Learn. **62**(1), 107–136 (2006)

14. Rockafellar, R.T., Wets, R.J.B.: Variational Analysis, vol. 317. Springer Science and Business Media, Berlin (2009)

15. Serafini, L., Garcez, A.D.: Logic tensor networks: deep learning and logical reasoning from data and knowledge. arXiv preprint arXiv:1606.04422 (2016)

16. Winston, P.H., Horn, B.K.: Lisp. Addison Wesley Pub., Reading (1986)

Nyström Sketches

Daniel J. Perry$^{(\boxtimes)}$, Braxton Osting, and Ross T. Whitaker

University of Utah, Salt Lake City, USA
{dperry,whitaker}@cs.utah.edu, osting@math.utah.edu

Abstract. Despite prolific success, kernel methods become difficult to use in many large scale unsupervised problems because of the evaluation and storage of the full Gram matrix. Here we overcome this difficulty by proposing a novel approach: compute the optimal small, out-of-sample Nyström sketch which allows for fast approximation of the Gram matrix via the Nyström method. We demonstrate and compare several methods for computing the optimal Nyström sketch and show how this approach outperforms previous state-of-the-art Nyström coreset methods of similar size. We further demonstrate how this method can be used in an online setting and explore a simple extension to make the method robust to outliers in the training data.

Keywords: Nyström approximation · Kernel PCA · Kernel preimage

1 Introduction

Kernel-based learning methods have gained popularity in the last few decades due to their simplicity and powerful capability. Specifically, for unsupervised problems kernel PCA [18], spectral clustering [12], and many other methods have been used successfully on a variety of problems, see, *e.g.*, [19,22,25]. However, like other non-parametric learning models, their application to large data sets can be limited by the need to retain all the training data.

One of the most successful approximation techniques, generally referred to as the Nyström approximation, was introduced in [26] and further analyzed in [5] where a small subset of the data, which we will call a Nyström coreset, is used to approximate the full Gram matrix. The primary difficulty with Nyström approximation in an unsupervised setting is to select the right coreset to populate the dictionary—if the coreset is too large unneeded computational cost is incurred and if the coreset doesn't contain a representative set then too much error is incurred. While there has been significant work in the area of coreset selection (see, *e.g.*, [1–3,13,20,21,24]), these methods all select a subset of the training data for use in the Nyström approximation. This is counterintuitive because it is well known [7,14] that the optimal basis on which to project data in an unsupervised setting are the PCA basis vectors, which for clarity we will refer to as a sketch instead of a coreset because the summary basis are a derivation (via SVD) rather than a subset fo the input data. We will show that, for very small element sizes, a Nyström sketch obtained by solving a particular optimization

© Springer International Publishing AG 2017
M. Ceci et al. (Eds.): ECML PKDD 2017, Part I, LNAI 10534, pp. 427–442, 2017.
https://doi.org/10.1007/978-3-319-71249-9_26

problem (similar to the kPCA optimization problem) are better able to describe the data (have smaller projection error) than a Nyström coreset of similar size. In applications, one frequently needs to project a dataset onto the coreset, which is a computationally intensive task, and a smaller coreset/sketch (with similar projection error) reduces these computational costs.

More recently, there has been significant effort to develop techniques that make use of random projections to estimate the lifting function directly [9,11,15,16]. These methods can be thought of as a type of random sketch for kernel approximation, where instead of storing a coreset of the data from the input set, a set of random directions are retained for computing the nonlinear lifting. While theoretically interesting and demonstrably useful, there has also been work exhibiting significant advantages of the Nyström coreset approach [29], leading to the conclusion that, with regards to projection error, sampling from the dataset itself will always yield specific advantages to random projections. Here we demonstrate that with the same size dictionary, the proposed Nyström sketch obtains superior results than a random projection based kernel approximation.

In this paper, we propose a novel approach to kernel approximation that uses a Nyström sketch, similar to a PCA basis, instead of a Nyström coreset [5] or a random projection approximation [9,15]. By formulating this Nyström sketch as an optimization problem, we also incorporate well known optimality ideas from the immensely successful PCA basis selection.

2   Background

While Nyström approximation of the Gram matrix is generally useful in both supervised and unsupervised settings, in order to focus the discussion we will primarily be investigating how different methods compare in estimating the kernel PCA (kPCA), and our primary measure of accuracy will be projection error resulting from the different approximations to kPCA.

2.1   Nyström Coreset and Sketch

Because there is some ambiguity around the terms matrix coreset and matrix sketch and to clarify discussion within this paper, we now introduce a more strict definition on *coreset* and *sketch* with respect to a given data matrix $X \in \mathbb{R}^{d \times n}, X = [x_1, \ldots, x_n]$ for the purposes of Nyström approximation.

Definition 1. *A Nyström coreset is a subset matrix $R \in \mathbb{R}^{d \times m}, R = [r_1, \ldots, r_m]$ where each of the columns r_i has a corresponding column in the input dataset, $r_i = x_j$, so that $R = XQ$, where $Q \in \mathbb{R}^{n \times m}$ is a column sampling matrix chosen for use in a Nyström approximation.*

Definition 2. *A Nyström sketch is a matrix $S \in \mathbb{R}^{d \times c}, S = [s_1, \ldots, s_c]$ where each of the columns is derived from the input dataset for purposes of Nyström approximation, but the columns do not necessarily need to come from X–in this sense the columns are "out of sample".*

A Nyström sketch is a more general matrix, and so while a sketch could be a coreset (if the derived columns happened to correspond to columns in the dataset), a coreset is *not* necessarily a sketch.

2.2 PCA and kPCA

Linear PCA can be computed via a truncated SVD of $X = U\Sigma V^T$ at cost $\mathcal{O}(n^3)$, or the top p components can be computed using the Lanczos method (similar to power iteration) at cost $\mathcal{O}(npz)$ where z is the number of matrix multiplies required to compute one eigenvector, and depends on the singular value gap. For computational efficiency the eigendecomposition is typically performed on the outer product or covariance matrix, $XX^T U = U\Sigma^2$ when $d \ll n$, or on the inner product or Gram matrix, $X^T X V = V\Sigma^2$ situations where $d \gg n$. In either case, a subspace of dimension p is selected based on the eigenvalues, by selecting the subspace spanned by the p eigenvectors corresponding to the largest p eigenvalues. KPCA is a non-linear extension of PCA, where the data elements are lifted in a higher dimensional feature space $\mathbb{H}$ using a non-linear lifting function $\phi \colon \mathbb{R}^d \to \mathbb{H}$ prior to PCA, resulting in a data matrix in $\mathbb{H}$, $\Phi := \phi(X) = [\phi(x_1), \ldots, \phi(x_n)]$, and corresponding decomposition $\Phi = U\Sigma V^T$. The idea is that PCA in $\mathbb{H}$ will reveal nonlinear relationships in the lower dimensional input space $\mathbb{R}^d$. Because the dimension of $\mathbb{H}$ is potentially infinite, we compute the right singular vectors spanning $\mathrm{Im}(\Phi^T)$ using the Gram matrix eigendecomposition, $\Phi^T \Phi V = V\Sigma^2$. We recover the left singular vectors of Φ spanning $\mathrm{Im}(\Phi)$: $U = \Phi V\Sigma^{-1}$. A projection onto the kPCA subspace can be computed using the reproducing kernel, $y_i = \phi(x_i)^T U = \phi(x_i)^T \Phi V\Sigma^{-1} = k(x_i, X)V\Sigma^{-1}$, where the kernel function $k(a,b) = \phi(a)^T \phi(b)$ corresponds to an inner product in feature space. The lifting function $\phi(\cdot)$ and corresponding reproducing kernel $k(\cdot, \cdot)$ are typically chosen so that computing $k(a,b)$ is more efficient than the explicit expansion and evaluation of the inner product in $\mathbb{H}$.

2.3 kPCA Projection Error and Nyström Approximation

For a collection of data points $\{x_i\}_{i=1}^n$, the mean projection error, E_{kpca}^p, for projection onto a p-dimensional PCA subspace is the mean of the lengths of the orthogonal projections,

$$E_{\mathrm{kpca}}^p = \frac{1}{n}\sum_{i=1}^n \|(I - V_p V_p^T)\phi(x_i)\|^2 = \frac{1}{n}\|(I - V_p V_p^T)\Phi\|_F^2. \tag{1}$$

Here $\|\cdot\|_F$ denotes the Frobenius norm. The error can be computed using a kernel trick,

$$E_{\mathrm{kpca}}^p = \frac{1}{n}\sum_{i=1}^n \|(I - V_p V_p^T)\phi(x_i)\|^2 \tag{2}$$

$$= \frac{1}{n}\sum_{i=1}^n k(x_i, x_i) - k(X, x_i)^T k(X, X)_p^{-1} k(X, x_i),$$

where $k(X, X)_p$ is the best rank-p approximation of the matrix $k(X, X)$.

Consider a Nyström coreset specified by the matrix $R \in R^{d \times m}$. Let $\Phi_R = \phi(R)$ have SVD $\Phi_R^T = U_R \Sigma_R V_R^T$ and $k(R, R) = \Phi_R^T \Phi_R \in \mathbb{R}^{m \times m}$ be the Gram matrix of the dictionary elements. We define the mean projection error, $E_{\mathrm{kpca}}(R)$, for projection onto Φ_R analogously to (1) and simplify this expression using the kernel trick from (2),

$$E_{\mathrm{kpca}}(R) := \frac{1}{n} \|(I - V_R V_R^T)\Phi\|^2 \tag{3}$$

$$= \frac{1}{n} \sum_{i=1}^{n} k(x_i, x_i) - k(R, x_i)^T k(R, R)^{-1} k(R, x_i).$$

This approximation to $k(x_i, x_i)$ is known as the Nyström approximation, which we aim to optimize.

3 General Formulation of the Nyström Sketch Learning Problem

We now consider learning an optimal Nyström sketch with $m < n$ elements that describes the collection of n data points $\{x_i\}_{i=1}^{n} \subset \mathbb{R}^d$. Consider a sketch specified by the matrix $R \in R^{d \times m}$ with columns given by m sketch elements $\{r_j\}_{j=1}^{m} \subset \mathbb{R}^d$.

Definition 3. Nyström sketch learning problem. *Compute the Nyström sketch $R^\star$ by solving the optimization problem,*

$$\min_{R \in R^{d \times m}} E_{kpca}(R), \tag{4}$$

where the objective function is defined in (2).

We will show below that the solution to this Nyström sketch learning problem for a linear kernel is the PCA basis of dimension m. For any nonlinear space the Nyström sketch learning problem is essentially tracking the preimage of the kPCA basis as it is computed. We discuss later how this approach differs from precomputing the kPCA basis and then finding a preimage afterwards, as well as some empirical evidence showing that the proposed sketch learning method performs much better in practice.

3.1 Optimal Nyström Sketch is the PCA Basis for the Trivial Lifting Function

As an example, we consider the Nyström sketch learning problem (4) for the trivial lifting function, $\phi(x) = x$, with corresponding reproducing kernel $k(x, y) = x^T y$. In this case, the objective function for the Nyström sketch learning problem can be written

$$E_{\text{kpca}}(R) = \frac{1}{n}\sum_{i=1}^{n} x_i^T x_i - x_i^T R(R^T R)^{-1} R^T x_i. \tag{5a}$$

Assume $m < d$. We observe that minimizing E_{kpca} over all $R \in \mathbb{R}^{d \times m}$ is equivalent to maximizing

$$\sum_{i=1}^{n} x_i^T R(R^T R)^{-1} R^T x_i = \text{tr}\left[X^T R(R^T R)^{-1} R^T X\right]$$

$$= \langle XX^T, R(R^T R)^{-1} R^T \rangle_F.$$

where $\langle \cdot, \cdot, \cdot \rangle_F$ denote the Frobenius inner product. The quantity $R(R^T R)^{-1} R^T$ is simply the projection onto the image of R. Since the optimal Nyström sketch clearly has rank m, we can let $\tilde{U} \in \mathbb{R}^{d \times m}$ have columns which are an orthonormal basis for the image of R. Thus, $R(R^T R)^{-1} R^T = \tilde{U}\tilde{U}^T$. It follows from Von Neumann's trace inequality that $\langle XX^T, \tilde{U}\tilde{U}^T \rangle_F \leq \sum_{i=1}^{m} \lambda_i(XX^T)$ with equality only when span$(\tilde{U})$ is the span of the first m left singular vectors of X. Thus the optimal Nyström sketch is any rank m matrix, R, with image equal to the span of the first m left singular vectors of X, or the rank-m PCA basis.

4   Optimal Nyström Sketch Using Non-linear Least-Squares Methods

In this section we show how the Nyström sketch learning problem (4) can be formulated as a nonlinear least squares problem, for which a variety of optimization algorithms, such as Gauss-Newton or Levenberg-Marquardt, can be applied to find the optimal Nyström sketch, R [27]. To apply such methods, we will require the gradient and Hessian of $E_{\text{kpca}}(R)$ with respect to the sketch, R.

We assume that the sketch R and kernel k have been chosen such that $k(R, R)$ is an invertible matrix. We first compute

$$\frac{\partial}{\partial R_{jk}} k(R, x)^T k(R, R)^{-1} k(R, x) =$$

$$2k(R, x)^T k(R, R)^{-1}\left[\frac{\partial}{\partial R_{jk}} k(R, x)\right]$$

$$+ k(R, x)^T\left[\frac{\partial}{\partial R_{jk}} k(R, R)^{-1}\right] k(R, x)$$

To compute the second term, we use the identity

$$\frac{\partial}{\partial R_{jk}} k(R, R)^{-1} = -k(R, R)^{-1}\left[\frac{\partial}{\partial R_{jk}} k(R, R)\right] k(R, R)^{-1}.$$

Thus we have

$$\frac{\partial}{\partial R_{jk}} k(R,x)^T k(R,R)^{-1} k(R,x) = \tag{6}$$

$$2k(R,x)^T k(R,R)^{-1} \left[\frac{\partial}{\partial R_{jk}} k(R,x) \right]$$

$$- k(R,x)^T k(R,R)^{-1} \left[\frac{\partial}{\partial R_{jk}} k(R,R) \right] k(R,R)^{-1} k(R,x).$$

Now write the objective

$$E_{\text{kpca}}(R) = \frac{1}{n} \sum_{i=1}^{n} \|(I - V_R V_R^T) \phi(x_i)\|^2$$

$$= \frac{1}{n} \sum_{i=1}^{n} \|r_i\|^2 = \frac{1}{n} \|R\|^2,$$

where the i-th residual vector and total residual vector are defined

$$r_i = \phi(x_i) - V_R V_R^T \phi(x_i) \quad \text{and} \quad R = \left(r_1^T \quad \cdots \quad r_n^T \right)^T.$$

Using (6), the gradient of $E_{\text{kpca}}(R)$ is computed

$$\frac{\partial E_{\text{kpca}}}{\partial R_{jk}} = \frac{1}{n} \sum_{i=1}^{n} \frac{\partial \|r_i\|^2}{\partial R_{jk}}$$

$$= -\frac{1}{n} \sum_{i=1}^{n} 2k(R,x_i)^T k(R,R)^{-1} \left[\frac{\partial}{\partial R_{jk}} k(R,x_i) \right]$$

$$- k(R,x_i)^T k(R,R)^{-1} \left[\frac{\partial}{\partial R_{jk}} k(R,R) \right] k(R,R)^{-1} k(R,x_i).$$

In particular, we find that the gradient of $E_{\text{kpca}}(R)$ can be evaluated in terms of the kernel function.

As an example, we compute the gradient for the linear and a general symmetric stationary kernels below.

4.1 Linear Kernel

For the linear kernel, we have $k(x,y) = x^T y$, $k(R,x) = R^T x$, and $k(R,R) = R^T R$.

$$\frac{\partial [R^T R]_{l,m}}{\partial R_{j,k}} = R_{j,m} \delta_{k,l} + R_{j,l} \delta_{k,m} \tag{7}$$

$$\frac{\partial [R^T x]_l}{\partial R_{j,k}} = x_j \delta_{k,l} \tag{8}$$

where $\delta_{a,b} = \begin{cases} 1 & a = b \\ 0 & a \neq b \end{cases}$.

Algorithm 1. A least-squares based method for solving the Nyström sketch learning problem (4).

Input Mercer kernel $k(\cdot, \cdot)$, input data X, and sketch size m
Output Nyström sketch $R \in \mathbb{R}^{d \times m}$
Initialize: Let R be a random subset of X or use, *e.g.*, [13]
Solve (4) using the Levenberg-Marquardt method (or alternative non-linear least squares method) to find the optimal parameter R.

4.2 Symmetric Stationary Kernel

For any symmetric stationary kernel of the form $k(x, y) = f(\|x - y\|^2)$.

$$\frac{\partial [k(R, R)]_{l,m}}{\partial R_{j,k}} = f'(\|R_l - R_m\|^2) 2 \left[(R_{j,k} - R_{j,m}) \delta_{l,k} \right.$$

$$\left. + (R_{j,m} - R_{j,l}) \delta_{m,k} \right]$$

$$\frac{\partial [k(R, x)]_l}{\partial R_{j,k}} = f'(\|R_l - x\|^2) 2 \delta_{l,k} \left[R_{j,k} - x_j \right]$$

4.3 Batch Algorithms

An algorithm for solving the Nyström sketch problem (4) is given in Algorithm 1. In this formulation we explicitly take advantage of the least-squares aspect of the optimal Nyström sketch problem. By using the special structure of a non-linear least-squares problem we are able to perform better than comparable gradient descent based methods. In Sect. 5, we compare the performance of Algorithm 1 with current state-of-the-art Nyström coreset selection and using other optimization strategies such as BFGS on both real and simulated data sets.

A second algorithm for estimating the optimal Nyström sketch is to first solve for the optimal basis in feature space using kPCA and then compute the preimage of the basis. Because there often is not a one-to-one mapping between input space and feature space, the preimage is an estimation problem, however it seems intuitive that if the optimal basis in feature space is the kPCA basis, then finding the corresponding preimage of those bases should provide a reasonable Nyström sketch.

While this approach has been used in other contexts—for example [4] use preimages of their incremental kPCA basis as a sketch—we have not seen it proposed as an explicit solution to the Nyström sketch learning problem. We consequently propose this technique as a "straw hat" approach to solving the sketch learning problem (details Algorithm 2). However, in Sect. 5, we exhibit several instances where a direct solution to the Nyström sketch learning problem via Algorithm 1 results in a more descriptive Nyström sketch. Furthermore the online and robust extensions to Algorithm 1 introduced in Sects. 4.4 and 4.5 are not directly transferable to the preimage approach.

Algorithm 2. A preimage based solution to the Nyström sketch learning problem.

Input Mercer kernel $k(\cdot, \cdot)$, input data X, and sketch size m
Output Approximate optimal Nyström sketch $R \in \mathbb{R}^{d \times m}$
Initialize: Run kPCA to compute the bases $\{V_1, \ldots, V_m\}$ of the kPCA space of dimension m.
Apply a preimage algorithm, such as [17], to each V_i independently, resulting in m sketch elements, $d_i = \text{preimage}(V_i)$.

Algorithm 3. A stochastic gradient descent based solution to the online Nyström sketch learning problem (4).

Input Mercer kernel $k(\cdot, \cdot)$, input data X, sketch size m, and batch size b
Output Optimal Nyström sketch $R \in \mathbb{R}^{m \times d}$
Initialize stage: Let R be the first m elements observed of X.
Solve (4) using stochastic gradient descent over the next b observed data elements, $\{x_i, \ldots, x_{i+b}\}$.

4.4   Online Algorithm

In addition to the batch formulation in Algorithm 1, we also propose two explicit extensions that are useful in online and noisy contexts.

First, by design, a least squares solver will need to revisit each data point repeatedly, which becomes costly as the data set grows. While Algorithm 1 can be used for small- to medium-sized data sets, it does not scale to very large datasets. To address this problem we modify Algorithm 1 to use a stochastic gradient descent (SGD) method, resulting in Algorithm 3. We explore how this algorithm performs on large data sets (large enough that the Gram matrix can't be explicitly formed, making the approach in Algorithm 2 untenable) by comparing to current state-of-the-art kernel approximation techniques such as Random Kitchen Sinks [15,16] and Fastfood [9]. We are able to demonstrate using the same sized dictionaries that the proposed optimal Nyström sketch obtains a smaller projection error on an unseen test set than these state-of-the-art kernel approximation techniques.

4.5   Robust Online Algorithm

A second extension is the use of an alternative loss to least-squares, possibly biasing the solution for certain desirable properties. Because the batch and online algorithms can use gradient descent methods instead of least squares methods, it is possible extend the underlying model to be something other than an ℓ_2-norm loss. One particular loss of interest in data mining is the least-absolute-deviations or ℓ_1-norm loss, which has some nice properties, such as being more robust to outliers than the ℓ_2-norm loss. Other work has adapted batch kPCA to use ℓ_1-norm loss, for example [28] developed an algorithm to solve for the kPCA basis using an ℓ_1-norm loss which yields a subspace basis which characterizes

Algorithm 4. A gradient descent based solution to the robust Nyström sketch learning problem (9).

Input Mercer kernel $k(\cdot, \cdot)$, input data X, sketch size m, batch size b, and regularization ϵ

Output Optimal Nyström sketch $R \in \mathbb{R}^{m \times d}$

Initialize stage: Let R be the first m elements observed of X.

Apply (stochastic) gradient descent to find the optimal parameter R to the least-absolute-deviations problem in (9) after observing b new data elements $\{x_i, \ldots, x_{i+b}\}$.

---

the primary signal in the data more than the outliers. We propose the following modified projection error,

$$E_{\text{rkpca}}(R) = \frac{1}{n}\sum_{i=1}^{n}\|(I - V_p V_p^T)\phi(x_i)\|_1$$

$$\approx \frac{1}{n}\sum_{i=1}^{n}\sqrt{k(x_i, x_i) - k(R, x_i)^T k(R, R)^{-1}k(R, x_i) + \epsilon} \qquad (9)$$

where the approximation is valid as $\epsilon \to 0$. This approximation is (i) differentiable for $\epsilon > 0$ and (ii) allows for evaluation using the kernel trick. The derivative of (9) with respect to the sketch can be calculated as in Sect. 4. To compute solutions of (9), we propose a gradient descent based method in the batch setting or SGD in the online setting, as detailed in Algorithm 4.

5 Experiments

We ran several experiments to explore how well the proposed Nyström sketch learning strategy works compared to current state-of-the-art methods in Nyström coreset selection.

To compare methods in batch mode, we generate a simple simulated dataset of a swirl with outliers, very similar to the swirl data set presented in [28]. This synthesized dataset facilitates exploring and visualizing the performance of the various selection techniques, including the effects of the robust loss function. For a view of real-world data, we also compared the batch algorithms on a subsample of the large forest cover data set from th UCI repository [10].

To compare the methods in an online setting, we use the forest cover, cpu, and two gas datatsets, gas-CO and gas-methane from the UCI repository [6,10]. The datasets are large enough that it becomes difficult to compute kPCA exactly, making Algorithms 1 or 2 unusable. Instead we compare to a random projection based kernel approximation methods. We first ran the proposed algorithm on a training subset, and compare the resulting Nyström approximation to the current state-of-the-art method for the Gaussian kernels on large data sets - random Fourier features (RFF) [15]. We note that the proposed method is learning from the training data while the RFF method is not; this is part of the point - by learning a small sketch for Nyström approximation, the proposed method is able to outperform the current state-of-the-art kernel approximation methods for similar sized sketches.

5.1 Methods

We present results using the least-squares method or *lsq* (Algorithm 1) using the *preimage* method (Algorithm 2), and the stochastic gradient descent method or *sgd* (Algorithm 3). For comparison we also include two non-stochastic gradient based variations on Algorithm 3, the first uses the BFGS method which we refer to as *bfgs*, and the second uses a typical first order gradient descent method *gd*. For details on the Levenberg-Marquadt method used in *lsq*, gradient descent, and the BFGS method we refer the reader to [27]. To compare to the current state-of-the-art kernel approximations, we use the seminal random Fourier features *rff* from [15,16].

The *preimage* method makes use of the preimage approach in [17], which uses a fixed-point iteration solver to compute the preimage. We initially evaluated the preimage approaches in [8] which uses an interpolation of a neighborhood of points and [23] which combines both methods, but found that [17] performed best overall on the problems considered.

To compare against Nyström coreset methods we use the well studied uniform random coreset selection method *random* [5]. Although simple, this method is a common benchmark for Nyström coreset selection; we include it here for reference to other coreset methods.

5.2 Parameter Selection

We specifically present examples using the Gaussian kernel, $k(x_1, x_2) = \frac{\|x_1 - x_2\|_2^2}{2\sigma^2}$, because of it's universal use and to demonstrate that even on problems where the feature space has high curvature and infinite feature size our method performs well. The parameter σ for the synthetic swirl data set was chosen small enough "spread out" the curve as a line in feature space. For the other datasets σ was selected according to the average inter-point distance of the data set, specific parameters are in Table 1.

For each dataset we estimated the number of kPCA components, p, needed to capture 90% of the kPCA spectral energy and set that as the smallest Nyström sketch. We then explore sketches of that size and larger, $m \geq p$.

5.3 Optimal Nyström sketch learning

The first experiment used a synthetic swirl dataset. This type of dataset is interesting in an unsupervised case, because it provides some view into how well the method is learning the (obvious) structure in the data. We generated a swirl with some clear outlier data in a way similar to [28], specifically we sample from the following region uniformly,

$$\{(0.07\gamma + \alpha)(\cos\gamma, \sin\gamma) \colon \alpha \in [0, 0.1], \gamma \in [0, 6\pi]\},$$

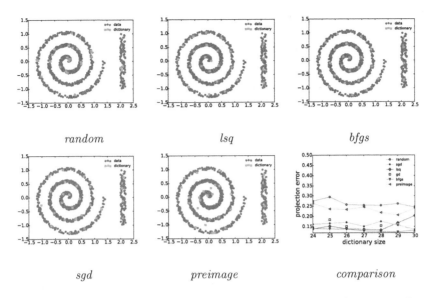

Fig. 1. Results running the various solution methods in batch mode on a synthesized swirl dataset. Top row and bottom left figures are visualizations comparing the sketch and coreset selection for each method. Bottom right figure is a plot of the projection error of each method.

as well as uniformly from a square region off to the side to provide some indication on the effect of obvious outliers on the result. We sample a total of 1000 points with 10% of those sampled from the outlier region. A summary of attributes for all data sets used is given in Table 1.

We are specifically interested in how the results change when the restriction of Nyström coreset selection to a more general Nyström sketch affects results. We have shown that the optimal Nyström sketch for a linear kernel are the first p PCA basis vectors. We hypothesize that relaxing the coreset restriction to a sketch learning problem in the non-linear kernel case also provides an advantage.

We present a visual and quantitative comparison of the Nyström coreset selection and sketch learning using the *random*, *lsq*, and *preimage* methods on the swirl data set in Fig. 1. The plot of projection error shows that the proposed *lsq* approach performs the best for each size of coreset or sketch, and that both *lsq* and *preimage* sketches perform better than the random coreset of the same size. Visually we can see that each method obtains good coverage of the structure of the swirl, but *lsq* places the sketch elements at more strategic locations, for example the center of the swirl with high curvature area has more samples, while areas with less curvature have less sampling density.

Note that the *random* method maintains coreset, i.e. "in-sample" points, while the *preimage* and *lsq* methods can obtain sketch elements that are "out-of-sample". We emphasize that this flexibility in sketch learning provides advantages coreset selection when the sketch/coreset is restricted in size.

Table 1. Data sets and parameters

Data set	d	n	σ
Swirl	2	1000	$\sqrt{0.10}$
Forest (subsampled)	54	1000	2.59×10^3
cpu-train	21	6554	5.88×10^5
cpu-test	21	819	5.88×10^5
Forest-train	54	522910	2.53×10^3
Forest-test	54	58102	2.53×10^3
Gas-CO-train	16	3708261	6.44×10^3
Gas-CO-test	16	500000	6.44×10^3
Gas-methane-train	16	3678504	4.68×10^3
Gas-methane-test	16	500000	4.68×10^3

We are also interested in understanding how various methods perform at solving the optimal Nyström sketch learning problem.

For this we ran a second experiment on a uniformly randomly subsample of the UCI forest cover data set [10]. We solved for the optimal sketch using *lsq, preimage, random*, as well as two gradient descent methods, *bfgs* and *sgd*. The results are plotted in Fig. 2. Note that all three solvers for the optimal formulation, *lsq, bfgs*, and *sgd* perform the best for all sketch sizes. It's interesting that the *preimage* method doesn't perform as well as the sketch size increases.

We also plot the runtime of the various methods. Note that the optimal solvers also all take the most amount of time, and grow in cost as the sketch size increases. This indicates that computing an optimal sketch will primarily be worth the effort for very small sketches. We note that *lsq* runs faster than *sgd*, only because we gave a very loose convergence criteria for *lsq* (1.0×10^{-3}), while *sgd* must run through the entire data set. The *bfgs* method had the same convergence criteria but takes longer than both *lsq* and *sgd*.

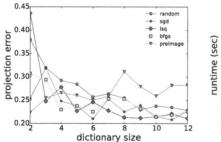 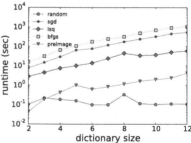

Fig. 2. Projection error (left) and runtime (right) for various solution methods in batch mode on a small subsample of the forest cover dataset.

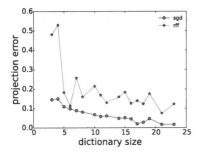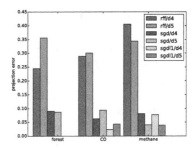

Fig. 3. Results for various solution methods in online mode on the real world datasets. (left) The projection error for the *sgd* and *rff* methods on the cpu dataset. (right) A summary of similar results for the forest, gas-CO, and gas-methane datasets, using sketch of size 4 and 5 (*d4* and *d5*).

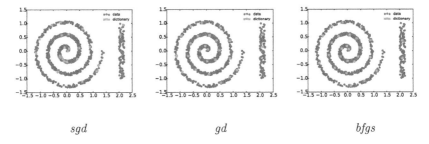

sgd	*gd*	*bfgs*

Fig. 4. Results for the gradient-descent-based methods on the robust sketch learning problem with a synthesized swirl dataset. Compare to Fig. 1.

5.4   Online Extension

As described previously, using a gradient descent method to solve the least-squares problem, opens the possibility of using a method like stochastic gradient descent to extend our model to be used in an online setting, as detailed in Algorithm 3.

The basic idea is that while observing new data we can continually update the sketch so that we alway have a very representative Nyström sketch.

To evaluate this idea, we simulated an online setting by splitting the UCI forest and cpu datasets into a training set and a test set each. Then we run the sketch learning algorithm on the (large) training set to learn the modified sketch as we "observe" new data. Since we are interested in using the kPCA in some kind of dimension reduction or other application, we evaluate the learned sketch on unseen data (the test set).

Because the datasets in an online setting are potentially very large, methods that need to revisit each data point repeatedly, like *lsq* and *bfgs*, are too slow. However the *sgd* method is specifically useful in an online setting. Note that because the data set is too large for direct decomposition, the *preimage* method cannot be used on a dataset that large. Consequently we compare the sgd method with the current state-of-the-art in kernel approximation using *rff*, for the online setting.

The results of this experiment for the cpu dataset are shown in Fig. 3 (left). Note that while the *sgd* was trained over a large training data set (cpu-train), the *rff* method doesn't require any training. Consequently projection error for the *rff* consists of projecting the data point onto the preselected random directions, taking the inner product, and then computing a difference with exact kernel inner product. Because the *sgd* method is trained on the same distribution of data and computes an approximate optimal sketch, it performs better for all sketch sizes evaluated.

The results on the forest, gas-CO, and gas-methane datasets are summarized in Fig. 3 (right). This plot summarizes the projection error for the three datasets using the *rff*, *sgd*, and the ℓ_1 version of *sgd*. Each method was evaluated on a sketch of size 4 or 5, as indicated.

Again these experiments specifically point out the advantage of using a learned sketch and the Nyström approximation over using a random basis of the same size. While random projection methods excel when the basis size is very large, for very small sketches the Nyström approximation obtains a better projection error, as shown here.

5.5   Robust Extension

We ran similar experiments using the robust extension described in Algorithm 4. The results for the swirl dataset are shown in Fig. 4. Note that compared to the least-squares results in Fig. 1 (especially *lsq*), the robust formulation places fewer sketch elements in the outlier section of the swirl data set. This is an important effect caused by the use of the ℓ_1 norm, reducing the effect of outliers on the sketch learning.

We also used the robust extension in an online setting on the gas-CO and gas-methane datasets, with results in Fig. 3 (right). Note how in the gas-CO case the ℓ_1 result improved on the ℓ_2 result, while in the gas-methane dataset the results were similar to the ℓ_2 results. This real world example illustrates one case where a robust result improved generalization for the gas-CO dataset.

6   Conclusion

We introduced a novel out-of-sample dictionary learning method, and shown that it better describes the data than current state-of-the-art methods; in particular for the same sized-dictionaries, the proposed method yields dictionaries with smaller projection error. We proved that for the linear kernel, the proposed method gives a dictionary that is equivalent to the PCA subspace of the given size. We also demonstrated that because of the difficulty in mapping back into the input space, the proposed method for finding the optimal dictionary performs better than the pre-image of the kPCA subspace basis.

Acknowledgements. B. Osting was partially funded by NSF DMS 16-19755.

References

1. Alaoui, A.E., Mahoney, M.W.: Fast randomized kernel methods with statistical guarantees. arXiv:1411.0306, 1–17 (2014)
2. Alzate, C., Suykens, J.A.K.: Kernel component analysis using an epsilon-insensitive robust loss function. IEEE Trans. Neural Netw. **19**(9), 1583–1598 (2008)
3. Calandriello, D., Lazaric, A., Valko, M.: Distributed adaptive sampling for kernel matrix approximation. In: Proceedings of the 20th International Conference on Artificial Intelligence and Statistics, vol. 54, pp. 1421–1429 (2017)
4. Chin, T.-J., Suter, D.: Incremental kernel principal component analysis. IEEE Trans. Image Process. **16**(6), 1662–1674 (2007)
5. Drineas, P., Mahoney, M.W.: On the Nyström method for approximating a gram matrix for improved kernel-based learning. J. Mach. Learn. Res. **6**, 2153–2175 (2005)
6. Fonollosa, J., Sheik, S., Huerta, R., Marco, S.: Reservoir computing compensates slow response of chemosensor arrays exposed to fast varying gas concentrations in continuous monitoring. Sens. Actuators B: Chem. **215**, 618–629 (2015)
7. Jolliffe, I.: Principal component analysis. Wiley Online Library, Hoboken (2002)
8. Kwok, J.T.-Y., Tsang, I.W.: The pre-image problem in kernel methods. IEEE Trans. Neural Netw. **15**(6), 1517–1525 (2004)
9. Le, Q., Sarlós, T., Smola, A.: Fastfood-approximating kernel expansions in loglinear time. In: Proceedings of the International Conference on Machine Learning (2013)
10. Lichman, M.: UCI machine learning repository (2013)
11. Lopez-Paz, D., Sra, S., Smola, A., Ghahramani, Z., Schölkopf, B.: Randomized nonlinear component analysis. arXiv preprint arXiv:1402.0119 (2014)
12. Ng, A.Y., Jordan, M.I., Weiss, Y., et al.: On spectral clustering: analysis and an algorithm. In: Advances in Neural Information Processing Systems, vol. 2, pp. 849–856 (2002)
13. Ouimet, M., Bengio, Y.: Greedy spectral embedding. In: Proceeding of 10th International Workshop on Artificial Intelligence and Statistics, pp. 253–260. Citeseer (2005)
14. Pearson, K.: Principal components analysis. London, Edinb. Dublin Philos. Mag. J. Sci. **6**(2), 559 (1901)
15. Rahimi, A., Recht, B.: Random features for large-scale kernel machines. In: Advances in Neural Information Processing Systems, pp. 1177–1184 (2007)
16. Rahimi, A., Recht, B.: Weighted sums of random kitchen sinks: replacing minimization with randomization in learning. In: Advances in Neural Information Processing Systems, pp. 1313–1320 (2009)
17. Schölkopf, B., Mika, S., Smola, A., Rätsch, G., Müller, K.R.: Kernel PCA pattern reconstruction via approximate pre-Images. In: Niklasson, L., Ziemke, T. (eds.) ICANN 1998. Perspectives in Neural Computing, pp. 147–152. Springer, London (1998). https://doi.org/10.1007/978-1-4471-1599-1_18
18. Schölkopf, B., Smola, A., Müller, K.-R.: Nonlinear component analysis as a kernel eigenvalue problem. Neural Comput. **10**(5), 1299–1319 (1998)
19. Schubert, E., Zimek, A., Kriegel, H.-P.: Generalized outlier detection with flexible kernel density estimates. In: Proceedings of the 14th SIAM International Conference on Data Mining (SDM), Philadelphia, PA, pp. 542–550 (2014)
20. Smola, A.J., Mangasarian, O.L., Schölkopf, B.: Sparse kernel feature analysis. In: Gaul, W., Ritter, G. (eds.) Classification, Automation, and New Media. Studies in Classification, Data Analysis, and Knowledge Organization, pp. 167–178. Springer, Heidelberg (2002). https://doi.org/10.1007/978-3-642-55991-4_18

21. Smola, A.J., Schökopf, B.: Sparse greedy matrix approximation for machine learning. In: Proceedings of the 17th International Conference on Machine Learning, pp. 911–918. Morgan Kaufmann Publishers Inc., (2000)

22. Snape, P., Zafeiriou, S.: Kernel-PCA analysis of surface normals for shape-from-shading. In: 2014 IEEE Conference on Computer Vision and Pattern Recognition (CVPR), pp. 1059–1066. IEEE (2014)

23. Teixeira, A.R., Tomé, A.M., Stadlthanner, K., Lang, E.W.: KPCA denoising and the pre-image problem revisited. Digit. Sig. Process. **18**(4), 568–580 (2008)

24. Tipping, M.E.: Sparse kernel principal component analysis. In: Advances in Neural Information Processing Systems, pp. 633–639 (2001)

25. Wang, J., Lee, J., Zhang, C.: Kernel trick embedded Gaussian mixture model. In: Gavaldá, R., Jantke, K.P., Takimoto, E. (eds.) ALT 2003. LNCS (LNAI), vol. 2842, pp. 159–174. Springer, Heidelberg (2003). https://doi.org/10.1007/978-3-540-39624-6_14

26. Williams, C., Seeger,M.: Using the Nyström method to speed up kernel machines. In: Proceedings of the 14th Annual Conference on Neural Information Processing Systems, number EPFL-CONF-161322, pp. 682–688 (2001)

27. Wright, S.J., Nocedal, J.: Numerical optimization, vol. 2. Springer, New York (1999)

28. Xiao, Y., Wang, H., Wenli, X., Zhou, J.: L1 norm based KPCA for novelty detection. Pattern Recogn. **46**(1), 389–396 (2013)

29. Yang, T., Li, Y.-F., Mahdavi, M., Jin, R., Zhou, Z.-H.: Nyström method vs random fourier features: a theoretical and empirical comparison. In: Advances in Neural Information Processing Systems, pp. 476–484 (2012)

Learning and Optimization

Crossprop: Learning Representations by Stochastic Meta-Gradient Descent in Neural Networks

Vivek Veeriah, Shangtong Zhang$^{(\boxtimes)}$, and Richard S. Sutton

Department of Computing Science, University of Alberta, Edmonton, AB, Canada
{vivekveeriah,shangtong.zhang,rsutton}@ualberta.ca

Abstract. Representations are fundamental to artificial intelligence. The performance of a learning system depends on how the data is represented. Typically, these representations are hand-engineered using domain knowledge. Recently, the trend is to *learn* these representations through stochastic gradient descent in multi-layer neural networks, which is called *backprop*. Learning representations directly from the incoming data stream reduces human labour involved in designing a learning system. More importantly, this allows in scaling up a learning system to difficult tasks. In this paper, we introduce a new incremental learning algorithm called *crossprop*, that learns incoming weights of hidden units based on the meta-gradient descent approach. This meta-gradient descent approach was previously introduced by Sutton (1992) and Schraudolph (1999) for learning step-sizes. The final update equation introduces an additional memory parameter for each of these weights and generalizes the backprop update equation. From our empirical experiments, we show that crossprop learns and reuses its feature representation while tackling new and unseen tasks whereas backprop relearns a new feature representation.

Keywords: Supervised learning · Learning representations
Meta-gradient descent · Continual learning

1   Introduction

The type of representation used for presenting the data to a learning system plays a key role in artificial intelligence and machine learning. Typically, the performance of the system, such as its speed of learning or its error rate, directly depends on how the data is represented internally by the system. Hand-engineering these representations using some special domain knowledge was the norm. More recently, these representations are learned hierarchically and directly from the data through stochastic gradient descent. Learning such representations significantly improves the learning performance and reduces the human effort

---

V. Veeriah and S. Zhang—These authors contributed equally to this work.

© Springer International Publishing AG 2017
M. Ceci et al. (Eds.): ECML PKDD 2017, Part I, LNAI 10534, pp. 445–459, 2017.
https://doi.org/10.1007/978-3-319-71249-9_27

involved in designing the system. Importantly, this allows in scaling up systems to handle bigger and harder problems.

Learning hierarchical representations directly from the data has recently gained a lot of popularity. Designing deep neural networks has allowed the learning systems to tackle incredibly hard problems: classifying or recognizing the objects from natural scene images (Deng et al. 2009; Szegedy et al. 2017), automatically translating text and speeches (Cho et al. 2014; Bahdanau et al. 2014; Wu et al. 2016), achieving and surpassing human-level baseline in Atari (Mnih et al. 2015), achieving super-human performance in Poker (Moravčík et al. 2017) and in improving robot control from learning experiences (Levine et al. 2016). It is important to note that in many of these problems it is difficult to hand-engineer a data representation and an inadequate representation generally limits performance or scalability of the system.

The algorithm behind the training of such deep neural networks is called *backprop* (or *backpropagation*), which was introduced by Rumelhart et al. (1988). It extended stochastic gradient descent, via chain rule, for learning the weights in the hidden layers of a neural network.

Though backprop has produced many successful results, it suffers from some fundamental issues which makes it slow in learning a useful representation that solves many tasks. Specifically, backprop tends to interfere with the previously learned representations because the units that have so far been found to be useful are the ones that are most likely to be changed (Sutton 1986). One of the reasons for this is that the weights of each hidden layer is assumed to be independent with each other, and because of this, the parameters of the neural network race against each other to minimize the error for a given example. In order to overcome this issue, the neural network needs to be trained over multiple sweeps (epochs) with the data so that algorithm can settle down with one representation that encompasses all the data it has seen so far.

In this paper, we introduce a meta-gradient descent approach for learning the weights connecting the hidden units of a neural network. Previously, the meta-gradient descent approach was introduced by Sutton (1992) and Schraudolph (1999) for learning parameter-specific step-sizes, which is adapted here for learning incoming weights of hidden units. Our proposed method is called *crossprop*.

This specifically addresses the racing problem which is observed in backprop. Furthermore, from our continual learning experiments where a learning system experiences a sequence of related tasks, we observed that crossprop tends to find the features that best generalize across these multiple tasks. Backprop, on the other hand, tends to *unlearn* and *relearn* the features with each task that it experiences. From a continual learning perspective, where a learning system experiences a sequence of tasks that are related with each other, it is desirable to have a learning system that can leverage its learning from its past experiences for solving unseen and more difficult tasks that it experiences in its future.

2   Related Methods

There are three fundamental approaches for learning representations, via a neural network, directly from the data.

The first and the most popular approach for learning such representations is through stochastic gradient descent over the supervised learning error function, like the mean squared or the cross-entropy error (Rumelhart et al. 1988). This approach is proved successful in many applications, ranging from difficult problems in computer vision to patient diagnoses. Although this method has a strong track record, it is not perfect yet. Particularly, learning representations by backpropagating the supervised error signal often learns slowly and poorly in many problems (Sutton 1986; Jacobs 1988). In order to address this, many modifications to backprop are introduced, like adding momentum (Jacobs 1988), RMSProp (Tieleman and Hinton 2012), ADADELTA (Zeiler 2012), ADAM (Kingma and Ba 2014) etc. and its not quite clear which variation of backprop will work well for a given task. However, all these variations of backprop still tend to interfere with the previously learned representations, thereby causing the network to unlearn and relearn representation even when the task can be solved by leveraging the learning from previous experiences.

Another promising approach for learning representations is by the generate and test process (Klopf and Gose 1969; Mahmood and Sutton 2013). The underlying principle behind these approaches is to generate many features in a random manner and then test the usability for each of these features. Based on certain heuristics, the features are either preserved or discarded. Furthermore, the generate and test approach can be combined with backprop to achieve a better rate of learning in supervised learning tasks. The primary motivation behind these generate and test approaches is to design a distributed and a computationally inexpensive representation learning method.

Some researchers have also looked at learning representations that fulfil certain unsupervised learning objectives, like clustering, sparsity, statistic independence or reproducibility of data, which takes us to the third fundamental approach towards learning representations (Olshausen and Field 1997; Comon 1994; Vincent et al. 2010; Coates and Ng 2012). Recently, learning such unsupervised representations has allowed in designing an effective clinical decision making system (Miotto et al. 2016). However, its not exactly clear on how to design a learning system for a continual and online learning setting using representations obtained through unsupervised learning, because we do not have access to data prior to the beginning of a learning task.

3   Algorithm

We consider a single-hidden layer neural network with a single output unit for presenting our algorithm. The parameters $U \in \mathbb{R}^{m \times n}$ and $W \in \mathbb{R}^n$ are the incoming and outgoing weights of the neural network where m is the number of input units and n is the number of hidden units. Each element of U is denoted

---

Algorithm 1. Crossprop algorithm

INPUT: α, η, m, n

1: Initialize h_{ij} to 0
2: Initialize u_{ij} and w_{ij} as desired where $i = 1, 2, \cdots, m; j = 1, 2, \cdots, n$
3: **for** each new example (X_t, y_t^*) **do**
4:     $y \leftarrow \sum_{j=1}^{n} \phi_{j,t} w_{j,t}$
5:     $\delta_t \leftarrow y_t^* - y_t$
6:     **for** $j = 1, 2, \cdots, n$ **do**
7:         **for** $i = 1, 2, \cdots, m$ **do**
8:             $u_{ij,t+1} \leftarrow u_{ij,t} + \alpha\delta_t \left[(1 - \eta)\phi_{j,t}h_{ij,t} + \eta w_{j,t}\frac{\partial\phi_{j,t}}{u_{ij,t}} \right]$
9:             $h_{ij,t+1} \leftarrow h_{ij,t}\left(1 - \alpha(1-\eta)\phi_{i,t}^2\right) + \alpha\left(\delta_t - \eta w_{j,t}\phi_{j,t}\right)\frac{\partial\phi_{j,t}}{\partial u_{ij,t}}$
10:        **end for**
11:        $w_{j,t+1} \leftarrow w_{j,t} + \alpha\delta_t\phi_{j,t}$
12:    **end for**
13: **end for**

---

as u_{ij} where i refers to the corresponding input unit and j refers to the hidden unit. Likewise, each element of W is denoted as w_j.

Our proposed method is summarized as a pseudo-code in Algorithm 1 (and the code is available on github[1]). A learning system (for simplicity, consider a single-hidden layer network), at time step t, receives an example $X_t \in \mathbb{R}^m$ where each element of this vector is denoted as $x_{i,t}$. This is mapped onto the hidden units through the incoming weight matrix U and a nonlinearity, like $tanh$, $sigmoid$ or $relu$, is applied over this summed-product. The activations for each hidden unit for a given example at time step t using a tanh activation function is expressed mathematically as, $\phi_{j,t} = \tanh\left(\sum_{i=1}^{m} x_{i,t} u_{ij,t}\right)$. These hidden units are successively mapped onto a scalar output $y_t \in \mathbb{R}$ using the weights W, which is expressed as $y_t = \sum_{j=1}^{n} \phi_{j,t} w_{j,t}$.

Let $\delta_t^2 = \left(y_t^* - y_t\right)^2$ be a noisy objective function where y_t^* is the scalar target and y_t is an estimate made by an algorithm for an example at time step t. The incoming and outgoing weights (U and W) are incrementally learned after processing an example one after the other.

The outgoing weights W are updated using the least mean squares (LMS) learning rule after processing an example at time step t as follows:

$$
\begin{aligned}
w_{j,t+1} &= w_{j,t} - \frac{1}{2}\alpha\frac{\partial\delta_t^2}{\partial w_{j,t}} \\
&= w_{j,t} - \alpha\delta_t\frac{\partial\delta_t}{\partial w_{j,t}} \\
&= w_{j,t} - \alpha\delta_t\frac{\partial[y_t^* - y_t]}{\partial w_{j,t}}
\end{aligned}
$$

---
[1] https://github.com/ShangtongZhang/Crossprop.

$$= w_{j,t} + \alpha\delta_t \frac{\partial y_t}{\partial w_{j,t}}$$

$$w_{j,t+1} = w_{j,t} + \alpha\delta_t \frac{\partial}{\partial w_{j,t}} \left[\sum_{i=1}^{n} \phi_{i,t} w_{i,t} \right]$$

$$w_{j,t+1} = w_{j,t} + \alpha\delta_t \phi_{j,t} \tag{1}$$

We diverge from the conventional way (i.e., through *backprop*) for learning the incoming weights U. Specifically, for learning the weights U, we consider the influence of all the past values of $U_1, U_2, \cdots U_t$ on the current error δ_t^2. We would like to learn the values of $u_{ij,t+1}$ by making an update using the partial derivative term $\frac{\partial \delta_t^2}{\partial u_{ij}}$ where u_{ij} refers to all its past values.

This is interesting because most of the current research on representation learning usually consider only the influence of the weight at the current time step $u_{ij,t}$ on the squared error δ_t^2: $\frac{\partial \delta_t^2}{\partial u_{ij,t}}$. This ignores the effects of the previous possible values of these weights on the squared error at the current time step.

We now derive the update rule for the incoming weights as follows:

$$u_{ij,t+1} = u_{ij,t} - \frac{1}{2}\alpha \frac{\partial \delta_t^2}{\partial u_{ij}}$$

$$= u_{ij,t} - \alpha\delta_t \frac{\partial [y_t^* - y_t]}{\partial u_{ij}}$$

$$u_{ij,t+1} = u_{ij,t} + \alpha\delta_t \frac{\partial y_t}{\partial u_{ij}} \tag{2}$$

Adapting the meta-gradient descent approach, that was introduced by Sutton (1992) and Schraudolph (1999), we derive the update rule for the incoming weights U as follows:

$$\frac{\partial y_t}{\partial u_{ij}} = \sum_k \frac{\partial y_t}{\partial w_{k,t}} \frac{\partial w_{k,t}}{\partial u_{ij}} + \sum_k \frac{\partial y_t}{\partial \phi_{k,t}} \frac{\partial \phi_{k,t}}{\partial u_{ij}}$$

$$= \sum_k \frac{\partial y_t}{\partial w_{k,t}} \frac{\partial w_{k,t}}{\partial u_{ij}} + \sum_k \frac{\partial y_t}{\partial \phi_{k,t}} \frac{\partial \phi_{k,t}}{\partial u_{ij,t}}$$

$$\frac{\partial y_t}{\partial u_{ij}} \approx \frac{\partial y_t}{\partial w_{j,t}} \frac{\partial w_{j,t}}{\partial u_{ij}} + \frac{\partial y_t}{\partial \phi_{j,t}} \frac{\partial \phi_{j,t}}{\partial u_{ij,t}} \tag{3}$$

Any error made during estimation of y_t by the learning system is attributed to both the outgoing weights of the features and to the activations of the hidden units. The approximations of $\sum_k \frac{\partial y_t}{\partial w_{k,t}} \frac{\partial w_{k,t}}{\partial u_{ij}} \approx \frac{\partial y_t}{\partial w_{j,t}} \frac{\partial w_{j,t}}{\partial u_{ij}}$ and $\sum_k \frac{\partial y_t}{\partial \phi_{k,t}} \frac{\partial \phi_{k,t}}{\partial u_{ij,t}} \approx \frac{\partial y_t}{\partial \phi_{j,t}} \frac{\partial \phi_{j,t}}{\partial u_{ij,t}}$ are reasonable because the primary effect on the input weight u_{ij} will be through the corresponding output weight $w_{j,t}$ and feature $\phi_{j,t}$.

By defining $h_{ij,t} = \frac{\partial w_{j,t}}{\partial u_{ij}}$, we can obtain a simple form for Eq. (3):

$$\frac{\partial y_t}{\partial u_{ij}} \approx \frac{\partial y_t}{\partial w_{j,t}}\frac{\partial w_{j,t}}{\partial u_{ij}} + \frac{\partial y_t}{\partial \phi_{j,t}}\frac{\partial \phi_{j,t}}{\partial u_{ij,t}}$$

$$= \left(\frac{\partial}{\partial w_{j,t}}\sum_k \phi_{k,t}w_{k,t}\right)h_{ij,t} + \frac{\partial y_t}{\partial \phi_{j,t}}\frac{\partial \phi_{j,t}}{\partial u_{ij,t}}$$

$$\frac{\partial y_t}{\partial u_{ij}} \approx \phi_{j,t}h_{ij,t} + \frac{\partial y_t}{\partial \phi_{j,t}}\frac{\partial \phi_{j,t}}{\partial u_{ij,t}} \tag{4}$$

The partial derivative $\frac{\partial y_t}{\partial \phi_{j,t}}\frac{\partial \phi_{j,t}}{\partial u_{ij,t}}$ is the conventional backprop update. However, in our proposed algorithm, we have an additional update term $\phi_{j,t}h_{ij,t}$ that captures the dependencies of all the previous values of u_{ij} on the current estimate y_t and on the current squared error δ_t^2.

$h_{ij,t}$ is an additional memory parameter corresponding to the input weight $u_{ij,t}$ and can be written as a recursive update equation as follows:

$$h_{ij,t+1} \approx \frac{\partial w_{j,t+1}}{\partial u_{ij}}$$

$$= \frac{\partial}{\partial u_{ij}}\Big[w_{j,t} + \alpha\delta_t\phi_{j,t}\Big]$$

$$= \frac{\partial w_{j,t}}{\partial u_{ij}} + \alpha\frac{\partial}{\partial u_{ij}}\Big[\delta_t\phi_{j,t}\Big]$$

$$= h_{ij,t} + \alpha\delta_t\frac{\partial \phi_{j,t}}{\partial u_{ij,t}} + \alpha\frac{\partial \delta_t}{\partial u_{ij}}\phi_{j,t}$$

$$\approx h_{ij,t} + \alpha\delta_t\frac{\partial \phi_{j,t}}{\partial u_{ij,t}} + \alpha\frac{\partial \delta_t}{\partial y_t}\frac{\partial y_t}{\partial u_{ij,t}}\phi_{j,t}$$

$$\approx h_{ij,t} + \alpha\delta_t\frac{\partial \phi_{j,t}}{\partial u_{ij,t}} + \alpha\frac{\partial \delta_t}{\partial y_t}\frac{\partial y_t}{\partial w_{j,t}}\frac{\partial w_{j,t}}{\partial u_{ij}}\phi_{j,t} + \alpha\frac{\partial \delta_t}{\partial y_t}\frac{\partial y_t}{\partial \phi_{j,t}}\frac{\partial \phi_{j,t}}{\partial u_{ij,t}}\phi_{j,t}$$

$$= h_{ij,t} + \alpha\delta_t\frac{\partial \phi_{j,t}}{\partial u_{ij,t}} - \alpha\frac{\partial y_t}{\partial w_{j,t}}\frac{\partial w_{j,t}}{\partial u_{ij}}\phi_{j,t} - \alpha\frac{\partial y_t}{\partial \phi_{j,t}}\frac{\partial \phi_{j,t}}{\partial u_{ij,t}}\phi_{j,t}$$

$$= h_{ij,t} + \alpha\delta_t\frac{\partial \phi_{j,t}}{\partial u_{ij,t}} - \alpha\phi_{j,t}^2 h_{ij,t} - \alpha w_{j,t}\phi_{j,t}\frac{\partial \phi_{j,t}}{\partial u_{ij,t}}$$

$$h_{ij,t} \approx h_{ij,t}\left(1 - \alpha\phi_{j,t}^2\right) + \alpha\left(\delta_t - w_{j,t}\phi_{j,t}\right)\frac{\partial \phi_{j,t}}{\partial u_{ij,t}} \tag{5}$$

By substituting Eq. (4) in Eq. (2), we define a recursive update equation for the weights $u_{ij,t}$ and thereby summarize the complete algorithm as follows:

$$u_{ij,t+1} = u_{ij,t} + \alpha\delta_t\Big[\phi_{j,t}h_{ij,t} + w_{j,t}\frac{\partial \phi_{j,t}}{\partial u_{ij,t}}\Big]$$

$$h_{ij,t+1} = h_{ij,t}\left(1 - \alpha\phi_{j,t}^2\right) + \alpha\left(\delta_t - w_{j,t}\phi_{j,t}\right)\frac{\partial \phi_{j,t}}{\partial u_{ij,t}} \tag{6}$$

$$w_{i,t+1} = w_{i,t} + \alpha\delta_t\phi_{i,t}$$

Depending on the nonlinearity used for the hidden units, $\frac{\partial \phi_{j,t}}{\partial u_{ij,t}}$ can be reduced to a closed-form equation.

For instance, if a logistic function is used, then $\phi_{j,t} = \sigma\left(\sum_{i=1}^{m} x_{i,t} u_{ij,t}\right)$,

$$\frac{\partial \phi_{j,t}}{\partial u_{ij,t}} = \frac{\partial \phi_{j,t}}{\partial u_{ij,t}}$$

$$= \frac{\partial}{\partial u_{ij,t}} \sigma\left(\sum_{i=1}^{m} x_{i,t} u_{ij,t}\right)$$

$$\frac{\partial \phi_{j,t}}{\partial u_{ij,t}} = \phi_{j,t}\left(1 - \phi_{j,t}\right) x_{i,t}$$

Another frequently used activation function is tanh, which implies that $\phi_{j,t} = \tanh\left(\sum_{i=1}^{m} x_{i,t} u_{ij,t}\right)$,

$$\frac{\partial \phi_{j,t}}{\partial u_{ij,t}} = \frac{\partial \phi_{j,t}}{\partial u_{ij,t}}$$

$$= \frac{\partial}{\partial u_{ij,t}} \tanh\left(\sum_{i=1}^{m} x_{i,t} u_{ij,t}\right)$$

$$\frac{\partial \phi_{j,t}}{\partial u_{ij,t}} = \left(1 - \phi_{j,t}^2\right) x_{i,t}$$

We could also introduce a weighting factor $\eta \in [0, 1]$ in Eq. (4), which allows in smoothly mixing backprop and meta-gradient updates,

$$\frac{\partial y_t}{\partial u_{ij}} \approx (1 - \eta)\phi_{j,t} h_{ij,t} + \eta \frac{\partial y_t}{\partial \phi_{j,t}} \frac{\partial \phi_{j,t}}{\partial u_{ij,t}}$$

which results in the following update equations for learning the weights U and W of the neural network:

$$u_{ij,t+1} = u_{ij,t} + \alpha \delta_t \left[(1 - \eta)\phi_{j,t} h_{ij,t} + \eta w_{j,t} \frac{\partial \phi_{j,t}}{\partial u_{ij,t}}\right]$$

$$h_{ij,t+1} = h_{ij,t}\left(1 - \alpha(1 - \eta)\phi_{j,t}^2\right) + \alpha\left(\delta_t - \eta w_{j,t}\phi_{j,t}\right) \frac{\partial \phi_{j,t}}{\partial u_{ij,t}} \tag{7}$$

$$w_{i,t+1} = w_{i,t} + \alpha \delta_t \phi_{i,t}$$

The algorithm that was derived and presented in Eqs. (7) and (6) are computationally expensive when there are more number of outgoing weights per hidden unit. Specifically, when there are k output units, then δ_t becomes a k-dimensional vector with dimensions equal to that of the output units. This leads to a large computational cost involved in computing $h_{ij,t}$, which can be avoided by approximating the $h_{ij,t}$ parameter. The approximation involves in accumulating the error assigned to each of the hidden units through its outgoing weights and using this to compute the update term. This approximated algorithm is referred to as crossprop-approx. in our experiments and has the following update equations:

$$u_{ij,t+1} = u_{ij,t} + \alpha \sum_k \delta_{k,t} \Big[(1-\eta)\phi_{j,t}h_{jk,t} + \eta w_{jk,t}\Big] \frac{\partial \phi_{j,t}}{\partial u_{ij,t}}$$

$$h_{jk,t+1} = h_{jk,t}\Big(1 - \alpha(1-\eta)\phi_{j,t}^2\Big) + \alpha\Big(\delta_{k,t} - \eta w_{jk,t}\phi_{j,t}\Big) \qquad (8)$$

$$w_{jk,t+1} = w_{jk,t} + \alpha \delta_{k,t}\phi_{j,t}$$

4 Experiments and Results

Here we empirically investigate whether crossprop is effective in finding useful representations for continual learning tasks and compare them with backprop and its many (such as adding momentum, RMSProp and ADAM). By continual learning tasks, we refer to an experiment setting where supervised training examples are generated and presented to a learning system from a sequence of related tasks. The learning system does not know when the task is switched.

4.1 GEOFF Tasks

The GEneric Online Feature Finding (GEOFF) problem was first introduced by Sutton (2014) as a generic, synthetic, feature-finding test bed for evaluating different representation learning algorithms. The primary advantage of this test bed is that infinitely many supervised-learning tasks can be generated without any experimenter bias.

The test bed consists of a single hidden layer neural network, called the *target network*, with a real-valued scalar output. Each input example, $X_t \in \{0,1\}^m$, is a m-dimensional binary input vector where each element in the vector can take a value of 0 or 1. The hidden layer consists of n Linear Threshold Units (LTUs), $\phi_t^* \in \{0,1\}^n$, with a threshold parameter of β. The β parameter controls the sparsity in the hidden layer. The weights $U^* \in \{-1,+1\}^{m \times n}$ maps the input vector to the hidden units and the weights $W^* \in \{-1,0,+1\}^n$ linearly combine the LTUs (features) to produce a scalar target output y^*. The weights U^* and W^* are generated using a uniform probability distribution and remain fixed throughout a task, representing a stationary function mapping a given input vector X_t to a scalar target output y_t^*. The input vector X_t is generated randomly using a uniform probability distribution. For each input vector, this target network is used to produce a scalar target output $y_t^* = \sum_{i=1}^n \phi_{i,t}^* w_i^* + \mathcal{N}(0,1)$. For our experiments, we fix $m = 20$, $n = 1000$ and $\beta = 0.6$ (the parameters of the target network).

Experiment setup. For our experiments, we create an instance of the GEOFF task. This is called Task A and use this to generate a set of 5000 examples. These examples are then used for training the learning systems in an online manner, where each example is processed once and then discarded. After processing the examples from Task A, we generate a Task B by randomly choosing and regenerating 50% of the outgoing target weights W^*. Another set of 5000 training examples is generated for training using this modified task. Similarly, after processing the examples from Task B, Task C is produced which is used for generating another 5000 training examples. The learning systems learn online from training examples produced by a sequence of related tasks (Tasks A, B & C) where the representations learned from one can be used for solving the other tasks. It is important to point out here that all these tasks share the same feature representation (i.e. the weights U^* remain fixed throughout) and the system can leverage from its previous learning experiences.

This experiment was designed from a continual learning perspective where a learning system will experience examples generated from a sequence of related tasks and learning from one task can help in learning other similar tasks. The step-size for all the algorithms was fixed at a constant value ($\alpha = 0.0005$) as the objective here is to show how the features are learned by different algorithms for a sequence of related learning tasks. The learning network consisted of a single hidden layer neural network with a single output unit. It had 20 input units and 500 hidden units using tanh activation function. The squared error function was used for learning the parameters of this network. These were the parameters of the learning network used for evaluating multiple algorithms.

Results. We compare the behavior of crossprop with backprop and its variations on a sequence of related tasks generated using the GEOFF testbed. Fig. 1 *Left* shows the learning curve for different algorithms. After every 5000 examples, the task switches to a new and related task as previously described. It is important to note here that the learning system does not know that the task has changed.

The learning curves show that crossprop reaches a similar asymptotic value to that of backprop, implying that the introduced algorithm produces a similar solution as backprop. In terms of asymptotic values, backprop achieves a significantly better asymptotic value compared to crossprop and the other variations of backprop. However, it is interesting to note that these learning algorithms approach the solution differently.

Figure 2 *Left* shows the euclidean norm (l^2 norm) between the weights U after processing the nth training example and the initialized value of the same weights. Though all the algorithms reach similar asymptotic values, the way backprop achieves this is clearly different from that of crossprop. Backprop tends to frequently modify the features even though it has seen examples that are generated using a previously learned function. Specifically, backprop fails to leverage from its previous learning experiences in solving new tasks even when it is possible. Because of this, backprop tends to take a lot of time in finding a feature representation which can sufficiently solve this continual problem. This is clearly not the case with crossprop. Our proposed algorithm tends to find a feature representation much quicker than backprop that can sufficiently solve the sequence of continual problems and reuses this for solving new tasks that it encounters in the future.

Figure 3 *Left* shows the euclidean norm between the weights W after processing the nth example and the initialized value of the same weights. Because

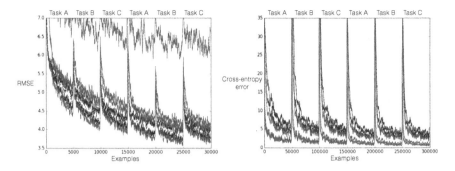

Fig. 1. The learning curves of crossprop (i.e., crossprop, crossprop-approx with $\eta = 0, 0.5$) are colored blue and backprop (backprop, momentum, RMSProp, ADAM) are colored red. The learning systems do not know that the task has changed. *Left:* The learning curves on a series of GEOFF tasks, averaged from 30 independent runs where each run used a different target network. *Right:* The learning curves on a series of MNIST tasks, obtained from a single run where the MNIST training set was used. Also, in the MNIST experiments, only crossprop-approx. was evaluated. (Color figure online)

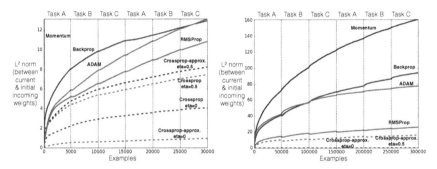

Fig. 2. *Left:* generated from GEOFF tasks. *Right:* generated from MNIST tasks. The plots shows the change in the incoming weights U after processing each example. Specifically, the plots shows l^2 norm between the incoming weights U after processing the nth example and its initialized value for different learning algorithms. From the plots, it can be observed that crossprop tends to change the incoming weights the least even when the task is significantly changed, implying that crossprop tends to find a reusable feature representation that can sufficiently solve the sequence of tasks that it experiences. On the other hand, backprop tends to significantly *relearn* the feature representation throughout the experiment even when the task can be solved by leveraging from previous learning experiences.

crossprop tends to find the set of features much quicker than backprop and reuses these features while solving a new task, it reduces the error by moving the outgoing weights rather than modifying its feature representation. Furthermore, all the tasks presented to the learning system can be solved by using a single feature representation and from our plots, it can be clearly seen that crossprop recognizes this.

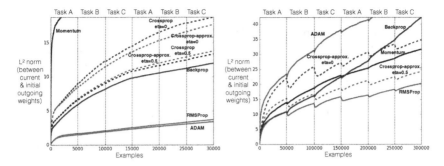

Fig. 3. *Left:* generated from GEOFF tasks. *Right:* generated from MNIST tasks. The plots shows the change in the outgoing weights W after processing each example. It shows the l^2 norm between the outgoing weights W after processing the nth example and its initialized value for different learning algorithms. From the plots, it can be observed that crossprop tends to change the outgoing weights the most as it is needed to map the feature representation to an estimate y_t. In backprop, each parameter independently minimizes the error and because of this, each parameter race with each other in reducing the error without coordinating their efforts. So, it tends to change its feature representation for accommodating a new example.

4.2 MNIST Tasks

The MNIST dataset of handwritten digits was introduced by LeCun et al. (1988). Though the MNIST dataset is old, it is still viewed as a standard supervised learning benchmark task for testing out new learning algorithms (Sironi et al. 2015; Papernot et al. 2016).

The dataset consists of grayscale images each with 28×28 dimensions. These images are obtained from handwritten digits and their corresponding labels denote the supervised learning target for a given image. The objective of a learning system in a MNIST task would be to learn a mapping function that maps each of these images to a label.

Experiment setup. We adapt the MNIST dataset to a continual learning setting, where in each task the label for the training images is shifted by one. For example, Task A uses the standard MNIST training images and their labels, Task B uses the same training examples as Task A, but now the labels get shifted by one. Similarly, for Task C the label for the training examples get further shifted by one. As in our previous experiment, we fix the step-size ($\alpha = 0.0005$) for the different algorithms as our objective here is to study how the representations are learned between these algorithms for a continual learning setting, where the learning system experiences examples from a sequence of related tasks. The learning system consisted of a single hidden layer neural network with 784 input units, 1024 hidden units and 10 output units. The hidden units used a tanh activation function and the output units used a softmax activation function. The cross-entropy error function was used for training the network.

Results. Figure 1 *Right* shows the learning curves for all the methods evaluated on the MNIST tasks. As observed in the GEOFF tasks, the learning curves for the different algorithms converge to almost similar points which means that all the methods reach similar solutions. However, ADAM and RMSProp achieves a significantly better asymptotic error value compared to the other learning algorithms.

Figures 2 *Right* and 3 *Right* show the euclidean norm of the change in weights U and W respectively. As seen in our previous experiments, crossprop tends to find the features much quicker than backprop and its variations. Also, crossprop tends to reuse these features in solving the new tasks that it faces. It is interesting to observe that backprop does not seem to settle down on a good feature representation for solving a sequence of continual learning problems. It tends to naïvely *unlearn* and *relearn* its feature representation even when the tasks are similar to each other and can be solved by using the feature representation learned from the first task. Specifically, backprop does not seem to leverage its previous learning experiences while encountering a new task.

5   Visualizing the Learned Features

We visualize the features that are obtained while training the learning systems using crossprop and backprop. These visualization are obtained using the t-SNE approach, which was developed by Maaten and Hinton (2008) for visualizing high-dimensional data by giving each datapoint a location in a two or three-dimensional map. Here, we show only the two-dimensional map generated using the features learned by the different learning algorithms.

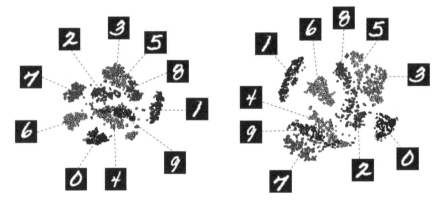

Fig. 4. Backprop (*Left*) and crossprop (*Right*) seem to learn similar feature representations, by clustering together similar examples. The plot shows the visualizations of the features (i.e. activations of the hidden units) learned by backprop and crossprop on the standard MNIST task. Both the learning algorithms were trained online on the MNIST dataset and the parameters learned by these algorithms were used for generating these visualizations. The η was set to 0 for crossprop in order to draw out differences between the conventional and the meta-gradient descent approach for learning the features. The plot was generated by using 2500 training examples, uniformly sampled from the MNIST dataset.

The features learned by backprop and crossprop (with η set to 0) on a standard MNIST task are plotted in Fig. 4. From the visualizations, it can be observed that both these algorithms produce similar feature representations on the task. Both these algorithms learn a feature representation that clusters examples according to their labels. There does not seem to be much of a difference between them by looking at their features.

6 Discussions

Neural networks and backprop form a powerful, hierarchical feature learning paradigm. Designing deep neural network architectures has allowed many learning systems to achieve levels of performance comparable to that of humans in many domains. Many of the recent research works, however, fail to notice or ignore the fundamental issues that are present with backprop, even though it is important to address them.

Some research works even tend to provide ad-hoc solutions to overcome these fundamental problems posed by backprop, but these are usually not scalable to general domains. Over time, many modifications were introduced to backprop, but they still fail to address the fundamental issue with backprop, which is that backprop tends to interfere with its previously learned representations in order to accommodate a new example. This prevents in directly applying backprop to continual learning domains, which is critical for achieving Artificial Intelligence (Ring 1997; Kirkpatrick et al. 2017).

In a continual learning setting, a learning system needs to progressively learn and hierarchically accumulate knowledge from its experiences, using them to solve many difficult, unseen tasks. In such a setting, it is not desirable to have a learning system that naïvely *unlearns* and *relearns* even when it sees a task that can be solved by reusing its learning from its past experiences. Particularly, for a continual learning setting, it is necessary to have a learning system that can hierarchically build knowledge from its previous experiences and use them in solving a completely new and unseen task.

In this paper, we present two continual learning tasks that were adapted from standard supervised learning domains: the GEOFF testbed and MNIST dataset. On these tasks, we evaluate backprop and its variations (momentum, RMSProp and ADAM). We also evaluate our proposed meta-gradient descent approach for learning the features in a neural network, called crossprop. We show that backprop (and its variations) tends to relearn its feature representations for every task, even when these tasks can be solved by reusing the feature representation learned from previous experiences. Crossprop, on the other hand, tends to reuse its previously learned representations in tackling new and unseen tasks. The process of consistently failing to leverage from previous learning experiences is not particularly desirable in a continual learning setting which prevents in directly applying backprop to such settings. Addressing this particular issue is the primary motivation for our work.

As an immediate future work, we would like to study the performances of this meta-gradient descent approach on deep neural networks and comprehensively evaluate them on more difficult benchmarks, like IMAGENET (Deng et al. 2009) and the Arcade Learning Environment (Bellemare et al. 2013).

7   Conclusions

In this paper, we introduced a meta-gradient descent approach, called crossprop, for learning the incoming weights of hidden units in a neural network and showed that such approaches are complementary to backprop, which is the popular algorithm for training neural networks. We also show that by using crossprop, a learning system can learn to reuse the learned features for solving new and unseen tasks. However, we see this as the first general work towards comprehensively addressing and overcoming the fundamental issues posed by backprop, particularly for continual learning domains.

References

Bahdanau, D., Cho, K., Bengio, Y.: Neural machine translation by jointly learning to align and translate (2014). arXiv preprint arXiv:1409.0473

Bellemare, M.G., Naddaf, Y., Veness, J., Bowling, M.: The arcade learning environment: an evaluation platform for general agents. J. Artif. Intell. Res. (JAIR) **47**, 253–279 (2013)

Cho, K., Van Merrinboer, B., Gulcehre, C., Bahdanau, D., Bougares, F., Schwenk, H., Bengio, Y.: Learning phrase representations using RNN encoder-decoder for statistical machine translation (2014). arXiv preprint arXiv:1406.1078

Coates, A., Ng, A.Y.: Learning feature representations with K-means. In: Montavon, G., Orr, G.B., Müller, K.-R. (eds.) Neural Networks: Tricks of the Trade. LNCS, vol. 7700, pp. 561–580. Springer, Heidelberg (2012). https://doi.org/10.1007/978-3-642-35289-8_30

Comon, P.: Independent component analysis, a new concept? Sig. process. **36**(3), 287–314 (1994)

Deng, J., Dong, W., Socher, R., Li, L.J., Li, K., Fei-Fei, L.: Imagenet: a large-scale hierarchical image database. In: 2009 IEEE Conference on Computer Vision and Pattern Recognition, CVPR 2009, pp. 248–255. IEEE, June 2009

Jacobs, R.A.: Increased rates of convergence through learning rate adaptation. Neural Netw. **1**(4), 295–307 (1988)

Kingma, D., Ba, J.: Adam: a method for stochastic optimization (2014). arXiv preprint arXiv:1412.6980

Kirkpatrick, J., Pascanu, R., Rabinowitz, N., Veness, J., Desjardins, G., Rusu, A.A., Milan, K., Quan, J., Ramalho, T., Grabska-Barwinska, A., Hassabis, D.: Overcoming catastrophic forgetting in neural networks. Proc. Nat. Acad. Sci. 201611835 (2017)

Klopf, A., Gose, E.: An evolutionary pattern recognition network. IEEE Trans. Syst. Sci. Cybern. **5**(3), 247–250 (1969)

LeCun, Y., Cortes, C., Burges., C.: The MNIST database of handwritten digits (1988)

Levine, S., Finn, C., Darrell, T., Abbeel, P.: End-to-end training of deep visuomotor policies. J. Mach. Learn. Res. **17**(39), 1–40 (2016)

Maaten, L.V.D., Hinton, G.: Visualizing data using t-SNE. J. Mach. Learn. Res. **9**(Nov), 2579–2605 (2008)

Mahmood, A.R., Sutton, R.S.: Representation search through generate and test. In: AAAI Workshop, Learning Rich Representations from Low-Level Sensors, June 2013

Miotto, R., Li, L., Kidd, B.A., Dudley, J.T.: Deep patient: an unsupervised representation to predict the future of patients from the electronic health records. Scientific reports 6, 26094 (2016)

Mnih, V., Kavukcuoglu, K., Silver, D., Rusu, A.A., Veness, J., Bellemare, M.G., Graves, A., Riedmiller, M., Fidjeland, A.K., Ostrovski, G., Petersen, S.: Human-level control through deep reinforcement learning. Nature **518**(7540), 529–533 (2015)

Moravčík, M., Schmid, M., Burch, N., Lis, V., Morrill, D., Bard, N., Davis, T., Waugh, K., Johanson, M., Bowling, M.: Deepstack: Expert-level artificial intelligence in no-limit poker (2017). arXiv preprint arXiv:1701.01724

Olshausen, B.A., Field, D.J.: Sparse coding with an overcomplete basis set: a strategy employed by V1? Vis. Res. **37**(23), 3311–3325 (1997)

Papernot, N., McDaniel, P., Jha, S., Fredrikson, M., Celik, Z.B., Swami, A.: The limitations of deep learning in adversarial settings. In: 2016 IEEE European Symposium on Security and Privacy (EuroS&P), pp. 372–387. IEEE, March 2016

Ring, M.B.: CHILD: a first step towards continual learning. Mach. Learn. **28**(1), 77–104 (1997)

Rumelhart, D.E., Hinton, G.E., Williams, R.J.: Learning representations by back-propagating errors. Cogn. Model. **5**(3), 1 (1988)

Schraudolph, N.N.: Local gain adaptation in stochastic gradient descent (1999)

Sironi, A., Tekin, B., Rigamonti, R., Lepetit, V., Fua, P.: Learning separable filters. IEEE Trans. Pattern Anal. Mach. Intell. **37**(1), 94–106 (2015)

Sutton, R.S.: Two problems with backpropagation and other steepest-descent learning procedures for networks. In: Proceeding of 8th Annual Conference on Cognitive Science Society, pp. 823–831. Erlbaum, May 1986

Sutton, R.S.: Adapting bias by gradient descent: an incremental version of delta-bar-delta. In: AAAI, pp. 171–176, July 1992

Sutton, R.S.: Myths of representation learning. In: ICLR (2014). Lecture

Szegedy, C., Ioffe, S., Vanhoucke, V., Alemi, A.A.: Inception-v4, inception-ResNet and the impact of residual connections on learning. In: AAAI, pp. 4278–4284 (2017)

Tieleman, T., Hinton, G.: Lecture 6.5-rmsprop: divide the gradient by a running average of its recent magnitude. COURSERA: Neural Netw. Mach. Learn. **4**(2), 26–31 (2012)

Vincent, P., Larochelle, H., Lajoie, I., Bengio, Y., Manzagol, P.A.: Stacked denoising autoencoders: learning useful representations in a deep network with a local denoising criterion. J. Mach. Learn. Res. **11**(Dec), 3371–3408 (2010)

Wu, Y., Schuster, M., Chen, Z., Le, Q.V., Norouzi, M., Macherey, W., ... Klingner, J.: Google's neural machine translation system: bridging the gap between human and machine translation (2016). arXiv preprint arXiv:1609.08144

Zeiler, M.D.: ADADELTA: an adaptive learning rate method (2012). arXiv preprint arXiv:1212.5701

Distributed Stochastic Optimization of Regularized Risk via Saddle-Point Problem

Shin Matsushima[1]([✉]), Hyokun Yun[2], Xinhua Zhang[3],
and S. V. N. Vishwanathan[2,4]

[1] The University of Tokyo, Tokyo, Japan
shin_matsushima@mist.i.u-tokyo.ac.jp
[2] Amazon.com, Seattle, WA 98170, USA
yunhyoku@amazon.com
[3] University of Illinois at Chicago, Chicago, IL 60607, USA
xzhang@uic.edu
[4] University of California, Santa Cruz, CA 95064, USA
vishy@ucsc.edu

Abstract. Many machine learning algorithms minimize a regularized risk, and stochastic optimization is widely used for this task. When working with massive data, it is desirable to perform stochastic optimization in parallel. Unfortunately, many existing stochastic optimization algorithms cannot be parallelized efficiently. In this paper we show that one can rewrite the regularized risk minimization problem as an equivalent saddle-point problem, and propose an efficient distributed stochastic optimization (DSO) algorithm. We prove the algorithm's rate of convergence; remarkably, our analysis shows that the algorithm scales almost linearly with the number of processors. We also verify with empirical evaluations that the proposed algorithm is competitive with other parallel, general purpose stochastic and batch optimization algorithms for regularized risk minimization.

1 Introduction

Regularized risk minimization is a well-known paradigm in machine learning:

$$\min_{\mathbf{w}} P(\mathbf{w}) := \lambda \sum_j \phi_j(w_j) + \frac{1}{m} \sum_{i=1}^{m} \ell(\langle \mathbf{w}, \mathbf{x}_i \rangle, y_i). \tag{1}$$

Here, we are given m training data points $\mathbf{x}_i \in \mathbb{R}^d$ and their corresponding labels y_i, while $\mathbf{w} \in \mathbb{R}^d$ is the parameter of the model. Furthermore, w_j denotes the j-th component of $\mathbf{w}$, while $\phi_j(\cdot)$ is a convex function which penalizes complex models. $\ell(\cdot, \cdot)$ is a loss function, which is convex in $\mathbf{w}$. Moreover, $\langle \cdot, \cdot \rangle$ denotes the Euclidean inner product, and $\lambda > 0$ is a scalar which trades-off between the average loss and the regularizer. For brevity, we will use $\ell_i(\langle \mathbf{w}, \mathbf{x}_i \rangle)$ to denote $\ell(\langle \mathbf{w}, \mathbf{x}_i \rangle, y_i)$.

© Springer International Publishing AG 2017
M. Ceci et al. (Eds.): ECML PKDD 2017, Part I, LNAI 10534, pp. 460–476, 2017.
https://doi.org/10.1007/978-3-319-71249-9_28

Many well-known models can be derived by specializing (1). For instance, if $y_i \in \{\pm 1\}$, then setting $\phi_j(w_j) = \frac{1}{2}w_j^2$ and $\ell_i(\langle \mathbf{w}, \mathbf{x}_i \rangle) = \max(0, 1 - y_i \langle \mathbf{w}, \mathbf{x}_i \rangle)$ recovers binary linear support vector machines (SVMs) [23]. On the other hand, using the same regularizer but changing the loss function to $\ell_i(\langle \mathbf{w}, \mathbf{x}_i \rangle) = \log(1 + \exp(-y_i \langle \mathbf{w}, \mathbf{x}_i \rangle))$ yields regularized logistic regression [11]. Similarly, setting $\phi_j(w_j) = |w_j|$ leads to sparse learning such as LASSO [11] with $\ell_i(\langle \mathbf{w}, \mathbf{x}_i \rangle) = \frac{1}{2}(y_i - \langle \mathbf{w}, \mathbf{x}_i \rangle)^2$.

A number of specialized as well as general purpose algorithms have been proposed for minimizing the regularized risk. For instance, if both the loss and the regularizer are smooth, as is the case with logistic regression, then quasi-Newton algorithms such as L-BFGS [17] have been found to be very successful. On the other hand, for smooth regularizers but non-smooth loss functions, Teo et al. [27] proposed a bundle method for regularized risk minimization (BMRM). Another popular first-order solver is alternating direction method of multipliers (ADMM) [4]. These optimizers belong to the broad class of batch minimization algorithms; that is, in order to perform a parameter update, at every iteration they compute the regularized risk $P(\mathbf{w})$ as well as its gradient

$$\nabla P(\mathbf{w}) = \lambda \sum_{j=1}^{d} \nabla \phi_j(w_j) \mathbf{e}_j + \frac{1}{m} \sum_{i=1}^{m} \nabla \ell_i(\langle \mathbf{w}, \mathbf{x}_i \rangle) \mathbf{x}_i, \tag{2}$$

where $\mathbf{e}_j$ denotes the j-th standard basis vector. Both $P(\mathbf{w})$ as well as the gradient $\nabla P(\mathbf{w})$ take $O(md)$ time to compute, which is computationally expensive when m, the number of data points, is large. Batch algorithms can be efficiently parallelized, however, by exploiting the fact that the empirical risk $\frac{1}{m} \sum_{i=1}^{m} \ell_i(\langle \mathbf{w}, \mathbf{x}_i \rangle)$ as well as its gradient $\frac{1}{m} \sum_{i=1}^{m} \nabla \ell_i(\langle \mathbf{w}, \mathbf{x}_i \rangle) \mathbf{x}_i$ decompose over the data points, and therefore one can compute $P(\mathbf{w})$ and $\nabla P(\mathbf{w})$ in a distributed fashion [7].

Batch algorithms, unfortunately, are known to be unfavorable for large scale machine learning both empirically and theoretically [3]. It is now widely accepted that stochastic algorithms which process one data point at a time are more effective for regularized risk minimization. In a nutshell, the idea here is that (2) can be stochastically approximated by

$$\mathbf{g}_i = \lambda \sum_{j=1}^{d} \nabla \phi_j(w_j) \mathbf{e}_j + \nabla \ell_i(\langle \mathbf{w}, \mathbf{x}_i \rangle) \mathbf{x}_i, \tag{3}$$

when i is chosen uniformly random in $\{1, \ldots, m\}$. Note that $\mathbf{g}_i$ is an unbiased estimator of the true gradient $\nabla P(\mathbf{w})$; that is, $\mathbb{E}_{i \in \{1, \ldots, m\}}[\mathbf{g}_i] = \nabla P(\mathbf{w})$. Now we can replace the true gradient by this *stochastic* gradient to approximate a gradient descent update as

$$\mathbf{w} \leftarrow \mathbf{w} - \eta \mathbf{g}_i, \tag{4}$$

where η is a step size parameter. Computing $\mathbf{g}_i$ only takes $O(d)$ effort, which is independent of m, the number of data points. Bottou and Bousquet [3] show that stochastic optimization is asymptotically faster than gradient descent and other second-order batch methods such as L-BFGS for regularized risk minimization.

However, a drawback of update (4) is that it is not easy to parallelize anymore. Usually, the computation of $\mathbf{g}_i$ in (3) is a very lightweight operation for which parallel speed-up can rarely be expected. On the other hand, one cannot execute multiple updates of (4) simultaneously, since computing $\mathbf{g}_i$ requires *reading* the latest value of $\mathbf{w}$, while updating (4) requires *writing* to the components of $\mathbf{w}$. The problem is even more severe in distributed memory systems, where the cost of communication between processors is significant.

Existing parallel stochastic optimization algorithms try to work around these difficulties in a somewhat ad-hoc manner (see Sect. 4). In this paper, we take a fundamentally different approach and propose a reformulation of the regularized risk (1), for which one can *naturally* derive a parallel stochastic optimization algorithm. Our technical contributions are:

- We reformulate regularized risk minimization as an equivalent saddle-point problem, and show that it can be solved via a new distributed stochastic optimization (DSO) algorithm.
- We prove $O\left(1/\sqrt{T}\right)$ rates of convergence for DSO which is independent of number of processors and theoretically described almost linear dependence of total required time with the number of processors.
- We verify with empirical evaluations that when used for training linear support vector machines (SVMs) or binary logistic regression models, DSO is comparable to general-purpose stochastic (*e.g.*, Zinkevich et al. [33]) or batch (*e.g.*, Teo et al. [27]) optimizers.

2   Reformulating Regularized Risk Minimization

We begin by reformulating the regularized risk minimization problem as an equivalent saddle-point problem. Towards this end, we first rewrite (1) by introducing an auxiliary variable u_i for each $\mathbf{x}_i$:

$$\min_{\mathbf{w},\mathbf{u}} \lambda \sum_{j=1}^{d} \phi_j\left(w_j\right) + \frac{1}{m} \sum_{i=1}^{m} \ell_i\left(u_i\right) \quad \text{s.t. } u_i = \langle \mathbf{w}, \mathbf{x}_i \rangle \quad \forall i = 1,\ldots,m. \quad (5)$$

By introducing Lagrange multipliers α_i to eliminate the constraints, we obtain

$$\min_{\mathbf{w},\mathbf{u}} \max_{\alpha} \lambda \sum_{j=1}^{d} \phi_j\left(w_j\right) + \frac{1}{m} \sum_{i=1}^{m} \ell_i\left(u_i\right) + \alpha_i(u_i - \langle \mathbf{w}, \mathbf{x}_i \rangle).$$

Here $\mathbf{u}$ denotes a vector whose components are u_i. Likewise, α is a vector whose components are α_i. Since the objective function (5) is convex and the constraints are linear, strong duality applies [5]. Therefore, we can switch the maximization over α and the minimization over $\mathbf{w}, \mathbf{u}$. Note that $\min_{u_i} \alpha_i u_i + \ell_i(u_i)$ can be written $-\ell_i^\star(-\alpha_i)$, where $\ell_i^\star(\cdot)$ is the Fenchel-Legendre conjugate of $\ell_i(\cdot)$ [5]. The above transformations yield to our formulation:

$$\max_{\alpha} \min_{\mathbf{w}} f(\mathbf{w}, \boldsymbol{\alpha}) := \lambda \sum_{j=1}^{d} \phi_j(w_j) - \frac{1}{m} \sum_{i=1}^{m} \alpha_i \langle \mathbf{w}, \mathbf{x}_i \rangle - \frac{1}{m} \sum_{i=1}^{m} \ell_i^\star(-\alpha_i). \quad (6)$$

If we analytically minimize $f(\mathbf{w}, \boldsymbol{\alpha})$ in terms of $\mathbf{w}$ to eliminate it, then we obtain so-called *dual* objective which is only a function of $\boldsymbol{\alpha}$. Moreover, any combination of $\mathbf{w}^*$ which is a solution of the primal problem (1), and $\boldsymbol{\alpha}^*$ which is a solution of the dual problem, forms a saddle point of $f(\mathbf{w}, \boldsymbol{\alpha})$ [5]. In other words, minimizing the primal, maximizing the dual, and finding a saddle point of $f(\mathbf{w}, \boldsymbol{\alpha})$ are all equivalent problems.

2.1 Stochastic Optimization

Let x_{ij} denote the j-th coordinate of $\mathbf{x}_i$, and $\Omega_i := \{j : x_{ij} \neq 0\}$ denote the non-zero coordinates of $\mathbf{x}_i$. Similarly, let $\bar{\Omega}_j := \{i : x_{ij} \neq 0\}$ denote the set of data points where the j-th coordinate is non-zero and $\Omega := \{(i, j) : x_{ij} \neq 0\}$ denotes the set of all non-zero coordinates in the training dataset $\mathbf{x}_1, \ldots, \mathbf{x}_m$. Then, $f(\mathbf{w}, \boldsymbol{\alpha})$ can be rewritten as

$$f(\mathbf{w}, \boldsymbol{\alpha}) = \sum_{(i,j) \in \Omega} \frac{\lambda \phi_j(w_j)}{|\bar{\Omega}_j|} - \frac{\ell_i^\star(-\alpha_i)}{m |\Omega_i|} - \frac{\alpha_i w_j x_{ij}}{m}$$

$$=: \sum_{(i,j) \in \Omega} f_{i,j}(w_j, \alpha_i),$$

where $|\cdot|$ denotes the cardinality of a set. Remarkably, each component $f_{i,j}$ in the above summation depends only on one component w_j of $\mathbf{w}$ and one component α_i of $\boldsymbol{\alpha}$. This allows us to derive an optimization algorithm which is stochastic in terms of both i and j. Let us define

$$\mathbf{g}_{i,j} := \left(|\Omega| \left(\frac{\lambda \nabla \phi_j(w_j)}{|\bar{\Omega}_j|} - \frac{\alpha_i x_{ij}}{m} \right) \mathbf{e}_j, |\Omega| \left(\frac{\nabla \ell_i^\star(-\alpha_i)}{m |\Omega_i|} - \frac{w_j x_{ij}}{m} \right) \mathbf{e}_i \right). \quad (7)$$

Under the uniform distribution over $(i, j) \in \Omega$, one can easily see that $\mathbf{g}_{i,j}$ is an unbiased estimate of the gradient of $f(\mathbf{w}, \boldsymbol{\alpha})$, that is, $\mathbb{E}_{\{(i, j) \in \Omega\}}[\mathbf{g}_{i,j}] = (\nabla_{\mathbf{w}} f(\mathbf{w}, \boldsymbol{\alpha}), -\nabla_{\boldsymbol{\alpha}} - f(\mathbf{w}, \boldsymbol{\alpha}))$. Since we are interested in finding a saddle point of $f(\mathbf{w}, \boldsymbol{\alpha})$, our stochastic optimization algorithm uses the stochastic gradient $\mathbf{g}_{i,j}$ to take a *descent* step in $\mathbf{w}$ and an *ascent* step in $\boldsymbol{\alpha}$ [19]:

$$w_j \leftarrow w_j - \eta \left(\frac{\lambda \nabla \phi_j(w_j)}{|\bar{\Omega}_j|} - \frac{\alpha_i x_{ij}}{m} \right), \quad \alpha_i \leftarrow \alpha_i + \eta \left(\frac{\nabla \ell_i^\star(-\alpha_i)}{m |\Omega_i|} - \frac{w_j x_{ij}}{m} \right).$$

$$(8)$$

Surprisingly, the time complexity of update (8) is independent of the size of data; it is $O(1)$. Compare this with the $O(md)$ complexity of batch update and $O(d)$ complexity of regular stochastic gradient descent.

Note that in the above discussion, we implicitly assumed that $\phi_j(\cdot)$ and $\ell_i^\star(\cdot)$ are differentiable. If that is not the case, then their derivatives can be replaced by sub-gradients [5]. Therefore this approach can deal with wide range of regularized risk minimization problem.

3   Parallelization

The minimax formulation (6) not only admits an efficient stochastic optimization algorithm, but also allows us to derive a distributed stochastic optimization (DSO) algorithm. The key observation underlying DSO is the following: Given (i, j) and (i', j') both in Ω, if $i \neq i'$ and $j \neq j'$ then one can simultaneously perform update (8) on (w_j, α_i) and $(w_{j'}, \alpha_{i'})$. In other words, the updates to w_j and α_i are independent of the updates to $w_{j'}$ and $\alpha_{i'}$, as long as $i \neq i'$ and $j \neq j'$.

Before we formally describe DSO we would like to present some intuition using Fig. 1. Here we assume that we have 4 processors. The data matrix X is an $m \times d$ matrix formed by stacking $\mathbf{x}_i^\top$ for $i = 1, \ldots, m$, while $\mathbf{w}$ and $\boldsymbol{\alpha}$ denote the parameters to be optimized. The non-zero entries of X are marked by an x in the figure. Initially, both parameters as well as rows of the data matrix are partitioned across processors as depicted in Fig. 1 (left); colors in the figure denote ownership $e.g.$, the first processor owns a fraction of the data matrix and a fraction of the parameters $\boldsymbol{\alpha}$ and $\mathbf{w}$ (denoted as $\mathbf{w}^{(1)}$ and $\boldsymbol{\alpha}^{(1)}$) shaded with red. Each processor samples a non-zero entry x_{ij} of X within the dark shaded rectangular region (active area) depicted in the figure, and updates the corresponding w_j and α_i. After performing updates, the processors stop and exchange coordinates of $\mathbf{w}$. This defines an *inner iteration*. After each inner iteration, ownership of the $\mathbf{w}$ variables and hence the active area change, as shown in Fig. 1 (right). If there are p processors, then p inner iterations define an *epoch*. Each coordinate of $\mathbf{w}$ is updated by each processor at least once in an epoch. The algorithm iterates over epochs until convergence.

Four points are worth noting. First, since the active area of each processor does not share either row or column coordinates with the active area of other processors, as per our key observation above, the updates can be carried out by each processor in parallel without any need for intermediate communication with other processors. Second, we partition and distribute the data only once. The coordinates of $\boldsymbol{\alpha}$ are partitioned at the beginning and are not exchanged by the processors; only coordinates of $\mathbf{w}$ are exchanged. This means that the cost of communication is independent of m, the number of data points. Third, our algorithm can work in both shared memory, distributed memory, and hybrid (multiple threads on multiple machines) architectures. Fourth, the $\mathbf{w}$ parameter is distributed across multiple machines and there is no redundant storage, which makes the algorithm scale linearly in terms of space complexity. Compare

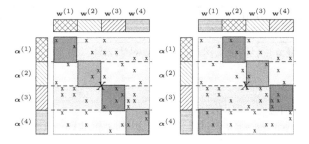

Fig. 1. Illustration of DSO with 4 processors. The rows of the data matrix X as well as the parameters $\mathbf{w}$ and $\boldsymbol{\alpha}$ are partitioned as shown. Colors denote ownerships. The active area of each processor is in dark colors. Left: the initial state. Right: the state after one bulk synchronization. (Color figure online)

this with the fact that most parallel optimization algorithms require each local machine to hold a copy of $\mathbf{w}$.

To formally describe DSO, suppose p processors are available, and let $I_1, \ldots, I_p$ denote a fixed partition of the set $\{1, \ldots, m\}$ and $J_1, \ldots, J_p$ denote a fixed partition of the set $\{1, \ldots, d\}$ such that $|I_q| \approx |I_{q'}|$ and $|J_r| \approx |J_{r'}|$ for any $1 \leq q, q', r, r' \leq p$. We partition the data $\{\mathbf{x}_1, \ldots, \mathbf{x}_m\}$ and the labels $\{y_1, \ldots, y_m\}$ into p disjoint subsets according to $I_1, \ldots, I_p$ and distribute them to p processors. The parameters $\{\alpha_1, \ldots, \alpha_m\}$ are partitioned into p disjoint subsets $\boldsymbol{\alpha}^{(1)}, \ldots, \boldsymbol{\alpha}^{(p)}$ according to $I_1, \ldots, I_p$ while $\{w_1, \ldots, w_d\}$ are partitioned into p disjoint subsets $\mathbf{w}^{(1)}, \ldots, \mathbf{w}^{(p)}$ according to $J_1, \ldots, J_p$ and distributed to p processors, respectively. The partitioning of $\{1, \ldots, m\}$ and $\{1, \ldots, d\}$ induces a $p \times p$ partition of Ω:

$$\Omega^{(q,r)} := \{(i,j) \in \Omega \ : \ i \in I_q, j \in J_r\}, \quad \forall q, r \in \{1, \ldots, p\}.$$

The execution of DSO proceeds in epochs, and each epoch consists of p inner iterations; at the beginning of the r-th inner iteration $(r \geq 1)$, processor q owns $\mathbf{w}^{(\sigma_r(q))}$ where $\sigma_r(q) := \{(q + r - 2) \bmod p\} + 1$, and executes stochastic updates (8) on coordinates in $\Omega^{(q,\sigma_r(q))}$. Since these updates only involve variables in $\boldsymbol{\alpha}^{(q)}$ and $\mathbf{w}^{(\sigma(q))}$, no communication between processors is required to execute them. After every processor has finished its updates, $\mathbf{w}^{(\sigma_r(q))}$ is sent to machine $\sigma_{r+1}^{-1}(\sigma_r(q))$ and the algorithm moves on to the $(r+1)$-st inner iteration. Detailed pseudo-code for the DSO algorithm can be found in Algorithm 1.

3.1 Convergence Analysis

It is known that the stochastic procedure in Sect. 2.1 is guaranteed to converge to a saddle point of $f(\mathbf{w}, \boldsymbol{\alpha})$ if (i, j) is randomly accessed [19]. The main technical difficulty in proving convergence in our case is due to the fact that DSO does not sample (i, j) coordinates uniformly at random due to its distributed nature. Therefore, first we prove that DSO is serializable in a certain sense, that is, there exists an ordering of the updates such that *replaying* them on a single machine

Algorithm 1. Distributed stochastic optimization (DSO) for finding saddle point of (6)

1: Each processor $q \in \{1, 2, \ldots, p\}$ initializes $\mathbf{w}^{(q)}, \boldsymbol{\alpha}^{(q)}$
2: $t \leftarrow 1$
3: **repeat**
4:    $\eta_t \leftarrow \eta_0 / \sqrt{t}$
5:    **for all** $r \in \{1, 2, \ldots, p\}$ **do**
6:      **for all processors** $q \in \{1, 2, \ldots, p\}$ **in parallel do**
7:        **for** $(i, j) \in \Omega^{(q, \sigma_r(q))}$ **do**
8:         $w_j \leftarrow w_j - \eta_t \left(\frac{\lambda \nabla \phi_j(w_j)}{|\bar{\Omega}_j|} - \frac{\alpha_i x_{ij}}{m} \right), \; \alpha_i \leftarrow \alpha_i + \eta_t \left(\frac{\nabla \ell_i^\star(-\alpha_i)}{m |\Omega_i|} - \frac{w_j x_{ij}}{m} \right)$
9:        **end for**
10:       send $\mathbf{w}^{(\sigma_r(q))}$ to machine $\sigma_{r+1}^{-1}(\sigma_r(q))$ and receive $\mathbf{w}^{(\sigma_{r+1}(q))}$
11:      **end for**
12:    **end for**
13:    $t \leftarrow t + 1$
14: **until** convergence

---

would recover the same solution produced by DSO. We then analyze this serial algorithm to establish convergence. We believe that this proof technique is of independent interest, and differs significantly from convergence analysis for other parallel stochastic algorithms which typically assume correlation between data points e.g. [6,15]. We first formally state the main theorem and then prove 3 lemmas. Finally we give a proof of the main theorem in the last part of this section.

Theorem 1. *Let* $(\mathbf{w}^t, \boldsymbol{\alpha}^t)$ *and* $(\tilde{\mathbf{w}}^t, \tilde{\boldsymbol{\alpha}}^t) := \left(\frac{1}{t} \sum_{s=1}^{t} \mathbf{w}^s, \frac{1}{t} \sum_{s=1}^{t} \boldsymbol{\alpha}^s \right)$ *denote the parameter values, and the averaged parameter values respectively after the t-th epoch of Algorithm 1. Moreover, assume that* $\|\mathbf{w}\|, \|\boldsymbol{\alpha}\|, |\nabla \phi_j(w_j)|, |\nabla \ell_i^\star(-\alpha_i)|,$ *and* λ *are upper bounded by a constant* $c > 1$*. Then, there exists a constant* C*, which is dependent only on* c*, such that after* T *epochs the duality gap is*

$$\max_{\boldsymbol{\alpha}} f\left(\tilde{\mathbf{w}}^T, \boldsymbol{\alpha}\right) - \min_{\mathbf{w}} f\left(\mathbf{w}, \tilde{\boldsymbol{\alpha}}^T\right) \leq C \frac{\sqrt{d}}{\sqrt{T}}. \tag{9}$$

On the other hand, if $\phi_j(s) = \frac{1}{2} s^2$*,* $\sqrt{\max_i |\Omega_i|} < m$ *and* $\eta_t < \frac{1}{\lambda}$ *hold, then there exists a different constant* C' *dependent only on* c *and satisfying*

$$\max_{\boldsymbol{\alpha}'} f\left(\tilde{\mathbf{w}}^T, \boldsymbol{\alpha}'\right) - \min_{\mathbf{w}'} f\left(\mathbf{w}', \tilde{\boldsymbol{\alpha}}^T\right) \leq \frac{C'}{\sqrt{T}}. \tag{10}$$

The first lemma states that there exists an ordering of the pairs of coordinates (i, j)s that recovers the solution produced by DSO.

Lemma 1. *On the inner iterations of the t-th epoch of Algorithm 1, let us index all* $(i, j) \in \Omega$ *as* (i_k, j_k) *by* $k = 1, \ldots, |\Omega|$ *as follows:* $a < b$ *if updates to* (w_{j_a}, α_{i_a}) *were performed before updating* (w_{j_b}, α_{i_b})*. On the other hand, if* (w_{j_a}, α_{i_a}) *and*

(w_{j_b}, α_{i_b}) were updated at the same time because we have p processors simultaneously updating the parameters, then the updates are ordered according to the rank of the processor performing the update[1]. Then, denote the parameter values after k updates by $(\mathbf{w}_k^t, \boldsymbol{\alpha}_k^t)$. For all k and t we have

$$\mathbf{w}_k^t = \mathbf{w}_{k-1}^t - \eta_t \nabla_{\mathbf{w}} f_k \left(\mathbf{w}_{k-1}^t, \boldsymbol{\alpha}_{k-1}^t \right) \tag{11}$$

$$\boldsymbol{\alpha}_k^t = \boldsymbol{\alpha}_{k-1}^t - \eta_t \nabla_{\boldsymbol{\alpha}} - f_k \left(\mathbf{w}_{k-1}^t, \boldsymbol{\alpha}_{k-1}^t \right), \tag{12}$$

where

$$f_k(\mathbf{w}, \boldsymbol{\alpha}) := f_{i_k, j_k} \left(w_{j_k}, \alpha_{i_k} \right).$$

Proof. Let q be the processor which performed the k-th update in the $(t + 1)$-st epoch. Moreover, let $(k - \delta)$ be the most recent previous update done by processor q. There exists $1 \leq \delta', \delta'' \leq \delta$ such that $\left(\mathbf{w}_{k-\delta'}^t, \boldsymbol{\alpha}_{k-\delta''}^t \right)$ be the parameter values read by the q-th processor to the perform k-th update. Because of our data partitioning scheme, only q can change the value of the i_k-th component of $\boldsymbol{\alpha}$ and the j_k-th component of $\mathbf{w}$. Therefore, we have

$$\alpha_{k-1, i_k}^t = \alpha_{\kappa, i_k}^t, \text{ and } w_{k-1, j_k} = w_{\kappa, j_k}^t, \quad k - \delta \leq \forall \kappa \leq k - 1.$$

Since f_k is invariant to changes in any coordinate other than (i_k, j_k), we have

$$f_k \left(\mathbf{w}_{k-\delta'}^t, \boldsymbol{\alpha}_{k-\delta''}^t \right) = f_k \left(\mathbf{w}_{k-1}^t, \boldsymbol{\alpha}_{k-1}^t \right).$$

The claim holds because we can write the k-th update formula as

$$\mathbf{w}_k^t = \mathbf{w}_{k-1}^t - \eta_t \nabla_{\mathbf{w}} f_k \left(\mathbf{w}_{k-\delta'}^t, \boldsymbol{\alpha}_{k-\delta''}^t \right) \text{ and} \tag{13}$$

$$\boldsymbol{\alpha}_k^t = \boldsymbol{\alpha}_{k-1}^t - \eta_t \nabla_{\boldsymbol{\alpha}} - f_k \left(\mathbf{w}_{k-\delta'}^t, \boldsymbol{\alpha}_{k-\delta''}^t \right). \tag{14}$$

$\square$

Next, we prove the following technical lemma that shows a sufficient condition to establish a global convergence of general iterative algorithms on general convex-concave functions. Note that it is closely related to well-known results on convex functions (e.g., Theorem 3.2.2 in [20], Lemma 14.1. in [24]).

Lemma 2. *Suppose there exists $D > 0$ and $C > 0$ such that for all $(\mathbf{w}, \boldsymbol{\alpha})$ and $(\mathbf{w}', \boldsymbol{\alpha}')$ we have $\|\mathbf{w} - \mathbf{w}'\|^2 + \|\boldsymbol{\alpha} - \boldsymbol{\alpha}'\|^2 \leq D$, and for all $t = 1, \ldots, T$ and all $(\mathbf{w}, \boldsymbol{\alpha})$ we have*

$$\left\| \mathbf{w}^{t+1} - \mathbf{w} \right\|^2 + \left\| \boldsymbol{\alpha}^{t+1} - \boldsymbol{\alpha} \right\|^2 \leq \left\| \mathbf{w}^t - \mathbf{w} \right\|^2 + \left\| \boldsymbol{\alpha}^t - \boldsymbol{\alpha} \right\|^2$$
$$- 2\eta_t \left(f \left(\mathbf{w}^t, \boldsymbol{\alpha} \right) - f \left(\mathbf{w}, \boldsymbol{\alpha}^t \right) \right) + C\eta_t^2, \tag{15}$$

then setting $\eta_t = \sqrt{\dfrac{D}{2Ct}}$ ensures that

$$\max_{\boldsymbol{\alpha}'} f \left(\tilde{\mathbf{w}}^T, \boldsymbol{\alpha}' \right) - \min_{\mathbf{w}'} f \left(\mathbf{w}', \tilde{\boldsymbol{\alpha}}^T \right) \leq \sqrt{\frac{2DC}{T}}. \tag{16}$$

---

[1] Any other tie-breaking rule would also suffice.

Proof. Rearrange (15) and divide by η_t to obtain

$$2\left(f\left(\mathbf{w}^t,\boldsymbol{\alpha}\right)-f\left(\mathbf{w},\boldsymbol{\alpha}^t\right)\right)\leq\eta_t C+\frac{1}{\eta_t}\left(\left\|\mathbf{w}^t-\mathbf{w}\right\|^2+\left\|\boldsymbol{\alpha}^t-\boldsymbol{\alpha}\right\|^2\right.$$
$$\left.-\left\|\mathbf{w}^{t+1}-\mathbf{w}\right\|^2-\left\|\boldsymbol{\alpha}^{t+1}-\boldsymbol{\alpha}\right\|^2\right).$$

Summing the above for $t=1,\ldots,T$ yields

$$2\sum_{t=1}^{T}f\left(\mathbf{w}^t,\boldsymbol{\alpha}\right)-2\sum_{t=1}^{T}f\left(\mathbf{w},\boldsymbol{\alpha}^t\right)\leq\sum_{t=1}^{T}\eta_t C+\frac{1}{\eta_1}\left(\left\|\mathbf{w}^1-\mathbf{w}\right\|^2+\left\|\boldsymbol{\alpha}^1-\boldsymbol{\alpha}\right\|^2\right)$$
$$+\sum_{t=2}^{T-1}\left(\frac{1}{\eta_{t+1}}-\frac{1}{\eta_t}\right)\left(\left\|\mathbf{w}^t-\mathbf{w}\right\|^2+\left\|\boldsymbol{\alpha}^t-\boldsymbol{\alpha}\right\|^2\right)$$
$$-\frac{1}{\eta_T}\left(\left\|\mathbf{w}^{T+1}-\mathbf{w}\right\|^2+\left\|\boldsymbol{\alpha}^{T+1}-\boldsymbol{\alpha}\right\|^2\right)$$
$$\leq\sum_{t=1}^{T}\eta_t C+\frac{1}{\eta_1}D+\sum_{t=2}^{T-1}\left(\frac{1}{\eta_{t+1}}-\frac{1}{\eta_t}\right)D$$
$$\leq\sum_{t=1}^{T}\eta_t C+\frac{1}{\eta_T}D. \tag{17}$$

On the other hand, thanks to convexity in $\mathbf{w}$ and concavity in $\boldsymbol{\alpha}$, we see

$$f\left(\tilde{\mathbf{w}}^T,\boldsymbol{\alpha}\right)\leq\frac{1}{T}\sum_{t=1}^{T}f\left(\mathbf{w}^t,\boldsymbol{\alpha}\right),\quad\text{and}\quad-f\left(\mathbf{w},\tilde{\boldsymbol{\alpha}}^T\right)\leq\frac{1}{T}\sum_{t=1}^{T}-f\left(\mathbf{w},\boldsymbol{\alpha}^t\right).$$

Using them for (17) and letting $\eta_t=\sqrt{\frac{D}{2Ct}}$ leads to the following inequalities

$$f\left(\tilde{\mathbf{w}}^T,\boldsymbol{\alpha}\right)-f\left(\mathbf{w},\tilde{\boldsymbol{\alpha}}^T\right)\leq\frac{\sum_{t=1}^{T}\eta_t C+\frac{1}{\eta_T}D}{2T}\leq\frac{\sqrt{DC}}{2T}\sum_{t=1}^{T}\frac{1}{\sqrt{2t}}+\sqrt{\frac{DC}{2T}}.$$

The claim in (16) follows by using $\sum_{t=1}^{T}\frac{1}{\sqrt{2t}}\leq\sqrt{2T}$.   □

In order to derive (9), C of (15) has to be the order of d. In case of L_2-regularizer, it has to be dependent only on c to obtain (10). The last lemma validates them. The proof is technical, and related to techniques outlined in Nedić and Bertsekas [18]. See Appendix for the proof.

Lemma 3. *Under the assumptions outlined in Theorem 1, (13) and (14), (15) is satisfied with C of the form of $C=C_1 d$. It does with $C=C_2$ in case of L_2-regularizer. Here C_1 and C_2 are dependent only on c.*

The proof of Theorem 1 can be shown in a very simple form given those 3 lemmas.

Proof. Because the parameter produced by Algorithm 1 is the same as one defined by (13) and (14), it is sufficient to show (13) and (14) lead to the statements in the theorem. From Lemma 3 and the fact that $\|\mathbf{w} - \mathbf{w}'\|^2 + \|\boldsymbol{\alpha} - \boldsymbol{\alpha}'\|^2 \leq 8c^2$, (16) of Lemma 2 holds with $\sqrt{CD} = 2c\sqrt{2C_1 d}$ for general case and $\sqrt{CD} = 2c\sqrt{2C_2}$ in case of L_2-regularizer, where C_1 and C_2 are dependent only on c. This immediately implies (9) and (10). □

To understand the implications of the above theorem, let us assume that Algorithm 1 is run with $p \leq \min(m, d)$ processors with a partitioning of Ω such that $\left|\Omega^{(q,\sigma_r(q))}\right| \approx \frac{|\Omega|}{p^2}$ and $|J_q| \approx \frac{d}{p}$ for all q. Let us denote time amount taken in performing updates in one epoch by T_u, which is of $\mathcal{O}(|\Omega|)$. Moreover, let us assume that communicating $\mathbf{w}$ across the network takes time amount denoted by T_c, which is of $\mathcal{O}(d)$, and communicating a subset of $\mathbf{w}$ takes time proportional to its cardinality. Under these assumptions, the time for each inner iteration of Algorithm 1 can be written as

$$\frac{\left|\Omega^{(q,\sigma_r(q))}\right|}{|\Omega|} T_u + \frac{\left|J_{\sigma_r(q)}\right|}{d} T_c \approx \frac{T_u}{p^2} + \frac{T_c}{p}.$$

Since there are p inner iterations per epoch, the time required to finish an epoch is $T_u/p + T_c$. As per Theorem 1 the number of epochs to obtain an ϵ accurate solution is independent of p. Therefore, one can conclude that DSO scales linearly in p as long as $T_u/T_c \gg p$ holds. As is to be expected, for large enough p the cost of communication T_c will eventually dominate.

4 Related Work

Effective parallelization of stochastic optimization for regularized risk minimization has received significant research attention in recent years. Because of space limitations, our review of related work will unfortunately only be partial.

The key difficulty in parallelizing update (4) is that gradient calculation requires us to *read*, while updating the parameter requires us to *write* to the coordinates of $\mathbf{w}$. Consequently, updates have to be executed in serial. Existing work has focused on working around the limitation of stochastic optimization by either (a) introducing strategies for computing the stochastic gradient in parallel (*e.g.*, Langford et al. [15]), (b) updating the parameter in parallel (*e.g.*, Bradley et al. [6], Recht et al. [21]), (c) performing independent updates and combining the resulting parameter vectors (*e.g.*, Zinkevich et al. [33]), or (d) periodically exchanging information between processors (*e.g.*, Bertsekas and Tsitsiklis [2]). While the former two strategies are popular in the shared memory setting, the latter two are popular in the distributed memory setting. In many cases the convergence bounds depend on the amount of correlation between data points and are limited to the case of strongly convex regularizer (Hsieh et al. [12], Yang [30], Zhang and Xiao [32]). In contrast our bounds in Theorem 1 do not depend on such properties of data and more general.

Table 1. Summary of the datasets used in our experiments. m is the total # of examples, d is the # of features, s is the feature density (% of features that are non-zero). K/M/G denotes a thousand/million/billion.

Name	m	d	$|\Omega|$	$s(\%)$
real-sim	57.76K	20.95K	2.97M	0.245
webspam-t	350.00K	16.61M	1.28G	0.022

Algorithms that use so-called parameter server to synchronize variable updates across processors have recently become popular (*e.g.*, Li et al. [16]). The main drawback of these methods is that it is not easy to "serialize" the updates, that is, to replay the updates on a single machine. This makes proving convergence guarantees, and debugging such frameworks rather difficult, although some recent progress has been made [16].

The observation that updates on individual coordinates of the parameters can be carried out in parallel has been used for other models. In the context of Latent Dirichlet Allocation, Yan et al. [29] used a similar observation to derive an efficient GPU based collapsed Gibbs sampler. On the other hand, for matrix factorization Gemulla et al. [10] and Recht and Ré [22] independently proposed parallel algorithms based on a similar idea. However, to the best of our knowledge, rewriting (1) as a saddle point problem in order to discover parallelism is our novel contribution.

5   Experimental Results

5.1   Dataset and Implementation Details

We implemented DSO, SGD, and PSGD ourselves, while for BMRM we used the optimized implementation that is available from the toolkit for advanced optimization (TAO, https://bitbucket.org/sarich/tao-2.2). All algorithms are implemented in C++ and use MPI for communication. In our multi-machine experiments, each algorithm was run on four machines with eight cores per machine. DSO, SGD, and PSGD used AdaGrad [8] step size adaptation. We also used stochastic variance reduced gradient (SVRG) of Johnson and Zhang [14] to accelerate updates of DSO. In the multi-machine setting DSO initializes parameters of each MPI process by locally executing twenty iterations of dual coordinate descent [9] on its local data to locally initialize w_j and α_i parameters; then w_j values were averaged across machines. We chose binary logistic regression and SVM as test problems, i.e., $\phi_j(s) = \frac{1}{2}s^2$ and $\ell_i(u) = \log(1 + \exp(-u))$, $[1 - u]_+$. To prevent degeneracy in logistic regression, values of α_i's are restricted to $(10^{-14}, 1 - 10^{-14})$, while in the case of linear SVM they are restricted to $[0, 1]$. Similarly, the w_j's are restricted to lie in the interval $[-1/\sqrt{\lambda}, 1/\sqrt{\lambda}]$ for linear SVM and $[-\sqrt{\log(2)/\lambda}, \sqrt{\log(2)/\lambda}]$ for logistic regression, following the idea of Shalev-Shwartz et al. [25].

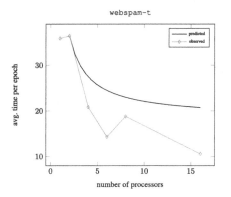

Fig. 2. The average time per epoch using p machines on the `webspam-t` dataset.

5.2 Scalability of DSO

We first verify, that the per epoch complexity of DSO scales as $T_{\mathrm{u}}/p + T_{\mathrm{c}}$, as predicted by our analysis in Sect. 3.1. Towards this end, we took the `webspam-t` dataset of Webb et al. [28], which is one of the largest datasets we could comfortably fit on a single machine. We let $p = \{1, 2, 4, 8, 16\}$ while fixing the number of cores on each machine to be 4.

Using the average time per epoch on one and two machines, one can estimate T_{u} and T_{c}. Given these values, one can then predict the time per iteration for other values of p. Figure 2 shows the predicted time and the measured time averaged over 40 epochs. As can be seen, the time per epoch indeed goes down as $\approx 1/p$ as predicted by the theory. The test error and objective function values on multiple machines was very close to the test error and objective function values observed on a single machine, thus confirming Theorem 1.

5.3 Comparison with Other Solvers

In our single machine experiments we compare DSO with stochastic gradient descent (SGD) and bundle methods for regularized risk minimization (BMRM) of Teo et al. [27]. In our multi-machine experiments we compare with parallel stochastic gradient descent (PSGD) of Zinkevich et al. [33] and BMRM. We chose these competitors because, just like DSO, they are general purpose solvers for regularized risk minimization (1), and hence can solve non-smooth problems such as SVMs as well as smooth problems such as logistic regression. Moreover, BMRM is a specialized solver for regularized risk minimization, which has similar performance to other first-order solvers such as ADMM.

We selected two representative datasets and two values of the regularization parameter $\lambda = \{10^{-5}, 10^{-6}\}$ to present our results. For the single machine experiments we used the `real-sim` dataset from Hsieh et al. [13], while for the multi-machine experiments we used `webspam-t`. Details of the datasets can be

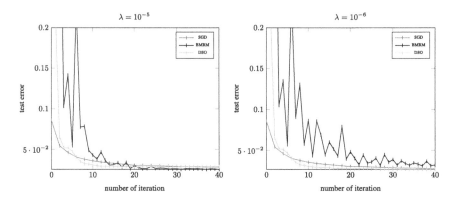

Fig. 3. The test error of different optimization algorithms on linear SVM with `real-sim` dataset, as a function of the number of iteration.

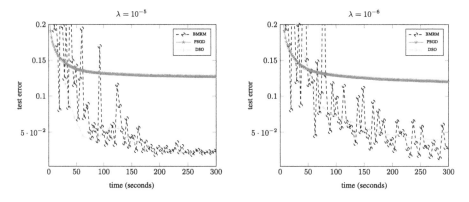

Fig. 4. The test error of different optimization algorithms on logistic regression with `webspam-t` dataset. Test error as a function of elapsed time.

found in Table 1 in the appendix. We use test error rate as comparison metric, since stochastic optimization algorithms are efficient in terms of minimizing generalization error, not training error [3]. The results for single machine experiments on linear SVM training can be found in Fig. 3. As can be seen, DSO shows comparable efficiency to that of SGD, and outperforms BMRM. This demonstrates that saddle-point optimization is a viable strategy even in serial setting.

Our multi-machine experimental results for linear SVM training can be found in Fig. 5. As can be seen, PSGD converges very quickly, but the quality of the final solution is poor; this is probably because PSGD only solves processor-local problems and does not have a guarantee to converge to the global optimum. On the other hand, both BMRM and DSO converges to similar quality solutions, and at fairly comparable rates. Similar trends we observed on logistic regression. Therefore we only show the results with 10^{-5} in Fig. 4.

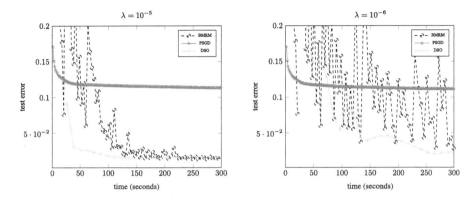

Fig. 5. Test errors of different parallel optimization algorithms on linear SVM with webspam-t dataset, as a function of elapsed time.

5.4 Terascale Learning with DSO

Next, we demonstrate the scalability of DSO on one of the largest publicly available datasets. Following the same experimental setup as Agarwal et al. [1], we work with the splice site recognition dataset [26] which contains 50 million training data points, each of which has around 11.7 million dimensions. Each datapoint has approximately 2000 non-zero coordinates and the entire dataset requires around **3 TB** of storage. Previously [26], it has been shown that sub-sampling reduces performance, and therefore we need to use the entire dataset for training.

Similar to Agarwal et al. [1], our goal is not to show the best classification accuracy on this data (this is best left to domain experts and feature designers). Instead, we wish to demonstrate the scalability of DSO and establish that (a) it can scale to such massive datasets, and (b) the empirical performance as measured by AUPRC (Area Under Precision-Recall Curve) improves as a function of time.

We used 14 machines with 8 cores per machine to train a linear SVM, and plot AUPRC as a function of time in Fig. 6. Since PSGD did not perform well in earlier experiments, here we restrict our comparison to BMRM. This experiment demonstrates one of the advantages of stochastic optimization, namely that the test performance increases steadily as a function of the number of iterations. On the other hand, for a batch solver like BMRM the AUPRC fluctuates as a function of the iteration number. The practical consequence of this observation is that, one usually needs to wait for a batch optimizer to converge before using the resulting solution. On the other hand, even the partial solutions produced by a stochastic optimizer such as DSO usually exhibit good generalization properties.

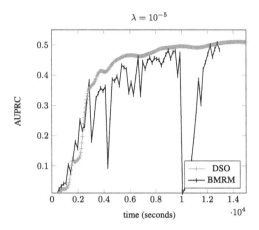

Fig. 6. AUPRC (Area Under Precision-Recall Curve) as a function of elapsed time on linear SVM with splice site recognition dataset.

6   Discussion and Conclusion

We presented a new reformulation of regularized risk minimization as a saddle point problem and showed that one can derive an efficient distributed stochastic optimizer (DSO). We also proved rates of convergence of DSO. Unlike other solvers, our algorithm does not require strong convexity and thus has wider applicability. Our experimental results show that DSO is competitive with state-of-the-art optimizers such as BMRM and SGD, and outperforms simple parallel stochastic optimization algorithms such as PSGD.

A natural next step is to derive an asynchronous version of DSO algorithm along the lines of the NOMAD algorithm proposed by Yun et al. [31]. We can see that our convergence proof which only relies on having an equivalent serial sequence of updates will still apply. Of course, there is also more room to further improve the performance of DSO by deriving better step size adaptation schedules, and exploiting memory caching to speed up random access.

Acknowledgments. This work is partially supported by MEXT KAKENHI Grant Number 26730114 and JST-CREST JPMJCR1304.

References

1. Agarwal, A., Chapelle, O., Dudík, M., Langford, J.: A reliable effective terascale linear learning system. JMLR **15**, 1111–1133 (2014)
2. Bertsekas, D., Tsitsiklis, J.: Parallel and Distributed Computation: Numerical Methods (1997)
3. Bottou, L., Bousquet, O.: The tradeoffs of large-scale learning. In: Optimization for Machine Learning (2011)

4. Boyd, S., Parikh, N., Chu, E., Peleato, B., Eckstein, J.: Distributed optimization and statistical learning via the alternating direction method of multipliers. Found. Trends ML **3**(1), 1–123 (2010)
5. Boyd, S., Vandenberghe, L.: Convex Optimization. Cambridge University Press, Cambridge (2004)
6. Bradley, J., Kyrola, A., Bickson, D., Guestrin, C.: Parallel coordinate descent for L1-regularized loss minimization. In: ICML, pp. 321–328 (2011)
7. Chu, C.T., Kim, S.K., Lin, Y.A., Yu, Y., Bradski, G., Ng, A.Y., Olukotun, K.: Map-reduce for machine learning on multicore. In: NIPS, pp. 281–288 (2006)
8. Duchi, J., Hazan, E., Singer, Y.: Adaptive subgradient methods for online learning and stochastic optimization. JMLR **12**, 2121–2159 (2010)
9. Fan, R.E., Chang, J.W., Hsieh, C.J., Wang, X.R., Lin, C.J.: LIBLINEAR: a library for large linear classification. JMLR **9**, 1871–1874 (2008)
10. Gemulla, R., Nijkamp, E., Haas, P.J., Sismanis, Y.: Large-scale matrix factorization with distributed stochastic gradient descent. In: KDD, pp. 69–77 (2011)
11. Hastie, T., Tibshirani, R., Friedman, J.: The Elements of Statistical Learning. (2009)
12. Hsieh, C.J., Yu, H.F., Dhillon, I.S.: PASSCoDe: parallel asynchronous stochastic dual coordinate descent. In: ICML (2015)
13. Hsieh, C.J., Chang, K.W., Lin, C.J., Keerthi, S.S., Sundararajan, S.: A dual coordinate descent method for large-scale linear SVM. In: ICML, pp. 408–415 (2008)
14. Johnson, R., Zhang, T.: Accelerating stochastic gradient descent using predictive variance reduction. In: NIPS, pp. 315–323 (2013)
15. Langford, J., Smola, A.J., Zinkevich, M.: Slow learners are fast. In: NIPS (2009)
16. Li, M., Andersen, D.G., Smola, A.J., Yu, K.: Communication efficient distributed machine learning with the parameter server. In: Neural Information Processing Systems (2014)
17. Liu, D.C., Nocedal, J.: On the limited memory BFGS method for large scale optimization. Math. Program. **45**(3), 503–528 (1989)
18. Nedić, A., Bertsekas, D.P.: Incremental subgradient methods for nondifferentiable optimization. SIAM J. Optim. **12**(1), 109–138 (2001)
19. Nemirovski, A., Juditsky, A., Lan, G., Shapiro, A.: Robust stochastic approximation approach to stochastic programming. SIAM J. Optim. **19**(4), 1574–1609 (2009)
20. Nesterov, Y.: Introductory Lectures on Convex Optimization: A Basic Course. Springer, Heidelberg (2004). https://doi.org/10.1007/978-1-4419-8853-9
21. Recht, B., Re, C., Wright, S., Niu, F.: Hogwild: a lock-free approach to parallelizing stochastic gradient descent. In: NIPS, pp. 693–701 (2011)
22. Recht, B., Ré, C.: Parallel stochastic gradient algorithms for large-scale matrix completion. Math. Program. Comput. **5**(2), 201–226 (2013)
23. Schölkopf, B., Smola, A.J.: Learning with Kernels (2002)
24. Shalev-Shwartz, S., Ben-David, S.: Understanding Machine Learning. Cambridge University Press, Cambridge (2014)
25. Shalev-Shwartz, S., Singer, Y., Srebro, N.: Pegasos: Primal estimated sub-gradient solver for SVM. In: ICML (2007)
26. Sonnenburg, S., Franc, V.: COFFIN: a computational framework for linear SVMs. In: ICML (2010)
27. Teo, C.H., Vishwanthan, S.V.N., Smola, A.J., Le, Q.V.: Bundle methods for regularized risk minimization. JMLR **11**, 311–365 (2010)
28. Webb, S., Caverlee, J., Pu, C.: Introducing the webb spam corpus: using email spam to identify web spam automatically. In: CEAS (2006)

29. Yan, F., Xu, N., Qi, Y.: Parallel inference for latent Dirichlet allocation on graphics processing units. In: NIPS, pp. 2134–2142 (2009)
30. Yang, T.: Trading computation for communication: distributed stochastic dual coordinate ascent. In: NIPS (2013)
31. Yun, H., Yu, H.F., Hsieh, C.J., Vishwanathan, S.V.N., Dhillon, I.S.: NOMAD: non-locking, stOchastic multi-machine algorithm for asynchronous and decentralized matrix completion. VLDB **7**, 975–986 (2014)
32. Zhang, Y., Xiao, L.: DiSCO: distributed optimization for self-concordant empirical loss. In: ICML (2015)
33. Zinkevich, M., Smola, A.J., Weimer, M., Li, L.: Parallelized stochastic gradient descent. In: NIPS, pp. 2595–2603 (2010)

Speeding up Hyper-parameter Optimization by Extrapolation of Learning Curves Using Previous Builds

Akshay Chandrashekaran[(⊠)] and Ian R. Lane

Department of Electrical and Computer Engineering,
Carnegie Mellon University, Pittsburgh, PA, USA
{akshayc,lane}@cmu.edu

Abstract. Recent work has shown that the usage of extrapolation of learning curves to determine when to terminate a training build has been shown to be effective in reducing the number of epochs of training required for finding a good performing hyper-parameter configuration. However, the current technique uses the information only from the current build to make the prediction. We propose the usage of a simple regression based extrapolation model that uses the trajectories from previous builds to make predictions of new builds. This can be used to terminate poorly performing builds and thus, speed up hyper-parameter search with performance comparable to non-augmented hyper-parameter optimization techniques. We compare the predictions made by our model against that of the existing extrapolation technique in different tasks. We incorporate our approach into a pre-existing termination criterion. We incorporate this termination criterion into an existing hyper-parameter optimization toolkit. We analyze the performance of our approach and contrast it against a baseline in terms of quality of prediction in three different tasks. We show that our approach yields builds with performance comparable to the non-augmented version with fewer epochs, and outperforms an existing parametric extrapolation technique for two out of three tasks in terms of number of required epochs.

Keywords: Hyper-parameter optimization
Extrapolation of learning curves · Bayesian optimization
Deep neural networks

1 Introduction

Deep neural networks have become ubiquitous in various fields in machine learning to solve complex problems like speech recognition, machine translation, image recognition, etc. Training the neural network involves specifying a model architecture, and passing training data in the form of randomized mini-batches to find the gradient of a selected loss function. This gradient is then propagated back through the network, whose parameters are changed to minimize the loss.

© Springer International Publishing AG 2017
M. Ceci et al. (Eds.): ECML PKDD 2017, Part I, LNAI 10534, pp. 477–492, 2017.
https://doi.org/10.1007/978-3-319-71249-9_29

Typically, training a neural network this way requires iterating over multiple epochs of training. The surface of the loss distributed over the model architecture is not convex, and its properties are generally unknown, thus making structure specification a non-trivial problem. Automated black-box hyper-parameter optimization techniques have shown to outperform manual and grid search for hyper-parameter optimization. However, these techniques rely on the complete training of the model for a given hyper-parameter setting.

In the presence of large amounts of data, training neural networks up to completion, which entails either a termination condition based on the performance over a validation data-set or a bound a maximum number of epochs can take multiple days to finish. Given this, training multiple neural network architectures is computationally very expensive, and can quickly become infeasible.

Humans can use information gained from previous model builds and current model trajectory to determine how a model is likely to perform at the end of training. Some work has been done to automate the above intuition using a parametric model in [5]. Empirically, learning curve trajectories across different hyper-parameter configurations look remarkably similar. Exploration on the usage of the above observation to build better extrapolation strategies have only recently gained interest [9].

However, the aforementioned work requires a significantly large number of previous builds before which it gives comparable performance in terms of mean square error of the loss function.

We propose an alternative extension to a probabilistic model that extrapolates the performance of a model at the early stages of training using the trajectories of previous model builds to identify and terminate poorly performing builds quickly.

Our contributions in this paper are as follows:

– We present a simple, yet effective approach to leverage the learning curves of previously completed builds to predict what the learning curve of the current build is likely to terminate in Sect. 3.1.
– We incorporate this approach with an existing termination criterion algorithm in Sect. 3.2.
– We analyze the quality of predictions obtained from this approach in terms of overall fit to the actual learning curve, as well as the performance at the final epoch in Sect. 4.2 for three different tasks.
– We incorporate this new termination criterion within an existing hyper-parameter optimization toolkit.
– We analyze the performance of the proposed termination criterion and compare it against the baseline technique in Sect. 4.3.

2   Related Work

We first review hyper-parameter optimization techniques popular in modern literature and previous attempts to model learning curves.

2.1   Hyper-parameter Optimization Techniques

Other than manual tuning, one of the most utilized hyper-parameter optimization techniques in machine learning has been the grid search. Recently, simple random sampling [1] has been shown to outperform grid search, particularly in high dimensional problems. Sophisticated sequential model based global optimization techniques employing Bayesian Optimization [12], tree of Parzen estimators [2], SMAC [8] have been shown to perform even better on several data-sets and tasks. For our task, we have used relatively few hyper-parameters to optimize. Hence, based on [6], we integrate our proposed approach with Spearmint [12] for our experiments. However, this approach can be integrated with any other hyper-parameter optimization technique with relative ease.

2.2   Modelling Learning Curves

In this work, we are trying to model the performance of an iterative ML algorithm as a function of iterations. We consider the performance of the ML algorithm in terms of validation accuracy over epochs to be the learning curve. The goal is to predict the validation performance and terminate a build that is unlikely to beat the best one so far. Freeze-thaw Bayesian Optimization [15] is a GP based Bayesian Optimization technique that includes a parametric exponential decay model for modelling the learning curve. They use this to perform exploration of several configurations by temporarily pausing the training of models, and then resuming the training. This technique has been shown to work on matrix factorization, online latent Dirichlet allocation (LDA) and logistic regression, but has not performed well on deep neural networks [5]. In [5], the authors model the learning curve for each selected hyper-parameter configuration using a weighted combination of parametrized exponential decay models. The following is an explanation of their termination criterion as understood from the code, which varies a bit compared to the explanation given in their paper. Every m epochs, the performance values are gathered, and Markov chain Monte Carlo (MCMC) is run to estimate the distribution of the parameters of the extrapolation model. Using this, the predicted termination value and uncertainty of the prediction is computed. If the standard deviation of the prediction falls below a specified threshold, the probability of the current model outperforming the best model so far is computed. If this probability falls below another specified threshold, the training is terminated, with the expected value returned in lieu of the real loss. Otherwise, the training is allowed to proceed. This method is shown to give builds with performance comparable to un-augmented optimization techniques while speeding up optimization by a factor of 2. We will be using this approach as the baseline for comparison.

In [9], the authors have built Bayesian neural networks based on trajectories obtained from random samples of hyper-parameter configurations, which have shown to have better fit compared to [5]. However, they require the presence of a large number of prior trajectories to effectively train their network, which is problematic if the task is new, and very computationally expensive.

In [16], the authors propose a Euclidean distance based distance measure similar to the one we will propose below to rank classifiers and their performance on previously evaluated data-sets to determine which combination of data-set and classifier matches the current condition correctly. However, their mapping function uses only the result after the prior classifiers have converged, and does not take iterative classifiers into consideration.

To our knowledge, we are not aware of any prior research into prediction of learning curve trajectories that uses only a few previous builds to make accurate and fast predictions of the performance in the termination stage.

3   Proposed Approach

Given a machine learning algorithm $\mathcal{A}$ having hyper-parameters $\lambda_1, \ldots, \lambda_n$ with respective domains $\Lambda_1, \ldots, \Lambda_n$, we define the hyper-parameter space as the hyper-cube $\Lambda = \Lambda_1 \times \ldots \Lambda_n$. For each hyper-parameter setting $\lambda \in \Lambda$, we use $\mathcal{A}_\lambda$ to denote the learning algorithm $\mathcal{A}$ using this setting. We further use $l_k(\lambda) = \mathcal{L}_k(\mathcal{A}_\lambda, \mathcal{D}_{train}, \mathcal{D}_{valid})$ to denote the best validation accuracy till epoch k that $\mathcal{A}_\lambda$ achieves on data $\mathcal{D}_{valid}$ when trained on $\mathcal{D}_{train}$. We further denote the hyper-parameter optimization $l_*(\lambda) = \max_k l_k(\lambda)$ to be the maximum validation accuracy obtained for the algorithm using λ over all epochs. The hyper-parameter optimization problem is to find $\lambda^* = \arg \max_{\lambda \in \Lambda} l_*(\lambda)$.

3.1   Ensemble of Learning Curves Model

In this section, we describe a novel approach to use a short initial portion of the learning curve for a given model and information from previous builds to give a probabilistic estimate of the validation accuracy at a later point. For simplicity, let $y_{r,k}$ indicate the best validation accuracy observed till epoch k for the rth build, i.e. $y_{r,k} = l_k(\lambda_r)$.

$\mathbf{y}_{r,1:n} = [y_{r,1}, y_{r,2}, \ldots, y_{r,n}]$ denotes the observed performance values for the rth build till epoch n. m is the final epoch after which a build will be terminated $(m > n)$.

$\mathbf{Y}_{1:r-1,1:m} = [\mathbf{y}_{1,m}, \mathbf{y}_{2,m}, \ldots, \mathbf{y}_{r-1,m}]$ is the observed performance values from previous $r - 1$ builds.

Our objective is, given $R - 1$ completed previous builds $(\mathbf{Y}_{1:R-1,1:m})$, and n epochs from the current build $(\mathbf{y}_{R,1:n})$, to predict the performance of the current build at termination: $y_{R,m}$. We solve this problem using an ensemble of learning curve models as described below.

Our approach is to transform the data from previous builds to approximately match the current build. We propose that the learning curve for the current build is a simple affine transformation of a previous build with an additive noise component. For each previous build $r = 1 : R - 1$ for a given epoch k,

$$\hat{y}_{Rr,k} = a_r y_{r,k} + b_r + \epsilon$$
$$\epsilon \sim \mathcal{N}(0, \sigma^2) \tag{1}$$

We assume the noise ϵ to be normally distributed with variance σ^2.

Using this definition, we construct a loss function comprised of all available information from the current build. This takes the form of a linear regression problem.

$$L_r = \frac{1}{n}||\mathbf{y}_{R,1:n} - \hat{\mathbf{y}}_{Rr,1:n}||_2^2 + \frac{\theta_1}{2}\frac{(1 - a_r)^2}{e^{\theta_2 n}} \tag{2}$$

where $\hat{\mathbf{y}}_{Rr,1:n} = [\hat{y}_{Rr,1}, \hat{y}_{Rr,2}, \cdots \hat{y}_{Rr,n}]$

The second term in the loss function is a hand-crafted regularization term developed to avoid pathological conditions of overfitting when only the initial few epochs of the current build have completed. θ_1 indicates how much with this regularization dominate in the loss function. θ_2 controls for how many epochs of the current build will the loss function dominate. Larger values of θ_2 will nullify the second term with fewer number of epochs available in the current build. For our experiments, we set their values to be 1.

The derivatives of the loss function with respect to the parameters a_r and b_r are given by:

$$\frac{\partial L_r}{\partial a_r} = -\frac{1}{n}(\mathbf{y}_{R,1:n} - \hat{\mathbf{y}}_{Rr,1:n})^T\mathbf{y}_{r,1:n} - \theta_1\frac{(1 - a_r)}{e^{\theta_2 n}}$$

$$\frac{\partial L_r}{\partial b_r} = -\frac{1}{n}(\mathbf{y}_{R,1:n} - \hat{\mathbf{y}}_{Rr,1:n})^T\mathbf{1}_{1:n} \tag{3}$$

where $.^T$ indicates the transpose of the vector, and $\mathbf{1}_{1:n}$ is an identity vector of length n.

Using the above loss and gradients, we perform simple gradient descent with multiple random start points for all the available previous builds. Once this is done, we select the top S parameter combinations and corresponding builds. We project the selected builds with their respective parameters. From this, we can get the expected value of the prediction and it's standard deviation for all subsequent epochs.

$$\mu_{\hat{y}_{R,k}} = \frac{1}{S}\sum_{i=1}^{S}\hat{y}_{Rr_i,k}$$

$$\sigma_{\hat{y}_{R,k}}^2 = \frac{1}{S-1}\sum_{i=1}^{S}(\hat{y}_{Rr_i,k} - \mu_{\hat{y}_{R,k}})^2 \tag{4}$$

Using the above mean and standard deviation definitions, we can estimate the probability that y_m exceeds a certain value $\tilde{y}$ as

$$P(y_m \geq \tilde{y}|\mathbf{y}_{R,1:n}, \mathbf{Y}_{1:R-1,1:m}) = 1 - \Phi(\tilde{y}; \mu_{\hat{y}_{R,m}}, \sigma_{\hat{y}_{R,m}}^2) \tag{5}$$

where $\Phi(.; \mu, \sigma^2)$ is the cumulative distribution function of a Gaussian with mean μ and variance σ^2.

The rationale behind the above method is inspired by the theory of wisdom of crowds [14,17]. This is an empirical concept based on aggregation of information in groups, which empirically have been shown to yield better results and any individual member of the group under the right circumstances. Our method satisfies the criteria required to form a wise crowd as described therein.

- *Diversity of Opinion-* Each individual should have private information: Each estimator uses the validation accuracy trajectory of only it's own previous build.
- *Independence-* Opinions of the individuals must be autonomously generated: Each estimator trains its own parameters independent of any previous builds or concurrently built estimator.
- *Decentralization-* An individual should be able to specialize and draw on local knowledge: Each estimator's regression depends only on the validation accuracy trajectories of the current build and the one previous build it is based on.
- *Aggregation-* A methodology should be available to arrive at a common answer: Eqs. 4 and 5 are the mechanisms to convert private judgments (estimate of prediction) to a collective decision.

3.2   Termination Criterion

We use a modified version of the termination criterion as described in [5] using the predictive model described above to speed up hyper-parameter search. We keep track of the best performance $\hat{y}$ found so far. Each time the optimizer queries the performance $l_*(\boldsymbol{\lambda})$ for a hyper-parameter setting $\boldsymbol{\lambda}$, we begin the training of the DNN using λ as usual. At the end of each epoch n, we record the performance over the validation set $y_{R,1:n}$. We run the algorithm described in Sect. 3.1 to determine $m_{\hat{y}_{R,m}}, \sigma^2_{\hat{y}_{R,k}}$ and $P(y_m \geq \hat{y}|y_{R,1:n}, y_{1:R-1,1:m})$. We use this information in Algorithm 1 to determine whether or not to terminate the current build. If the criterion dictates that the build be terminated, we replace it's final value $l_*(\boldsymbol{\lambda})$ to be the expected value derived by the ensemble method.

Practical Considerations. Empirically, we have observed that, on integrating the above with our hyper-parameter optimization toolkit, for some of the tasks, if either of the initial build points performed really well, then the termination criterion became very aggressive in pruning off paths, which could result in builds that would have become viable in later epochs getting pruned very quickly. Hence, we have enforced a condition that a few initial builds be trained till completion. This empirically results in a better spread of validation accuracy trajectories that could be used for comparison, resulting in lower prediction loss. Another reason for this enforcement is to satisfy the independence condition for the wisdom of crowds. In the initial builds, Bayesian optimization is much more explorative in nature. Hence, the first few builds can, in some sense be considered to be independent of each other.

Another empirical observation showed that the optimization technique usually under-predicted the performance of the current build, despite the enforced

ConservativeTerminationCriterion *Procedure*

   Data: current learning curve: $\mathbf{y}_{R,1:n}$,
     previous learning curves: $\mathbf{Y}_{1:R-1,1:m}$ $(n < m)$,
     predictive mean: $\mu_{\hat{y}_{R,m}}$,
     predictive standard deviation: $\sigma_{\hat{y}_{R,k}}$,
     probability of exceeding best observed value: $P(y_m \geq \tilde{y}|\mathbf{y}_{R,1:n}, \mathbf{Y}_{1:R-1,1:m})$,
     probability threshold: δ,
     standard deviation threshold: σ_{thresh}
   if $y_n > \hat{y}$ **then**
     | Continue Training till completion
   else
     if $P(y_m \geq \tilde{y}|\mathbf{y}_{R,1:n}, \mathbf{Y}_{1:R-1,1:m}) \geq \delta$ **then**
       | Continue Training for next epoch
     else
       if $\sigma_{\hat{y}_{R,k}} \geq \sigma_{thresh}$ **then**
         | Continue Training for next epoch
       else
         Terminate build
         return $\mu_{\hat{y}_{R,m}}$
       end
     end
   end

Algorithm 1. Procedure for the Conservative Termination Criterion

condition that the validation accuracy is non-decreasing. Hence, we have employed a recency weighting schema during training, which has a geometrically decreasing weight-age to the residual at earlier epochs as opposed to the residual of later epochs.

The weights are given by the formula for each available epoch till epoch k:

$$w(i) = (i * 10^{(\frac{1}{i})})^i \text{ for } i = 1 : k \tag{6}$$

Finally, we have also incorporated a monotonicity check during the gradient computation: The predicted accuracy of the estimator cannot be lesser than the best observed value so far. This is a logical check based on our definition of the validation accuracy at a given epoch.

4   Experimental Setup

4.1   Tasks

To test the validity of our proposed approach, we performed empirical tests in three different tasks. We describe the training and evaluation procedure for each of the tasks below.

For our experiments, we had a starting point for hyper-parameters given from the existing scripts in the MXNet toolkit for the MNIST and CIFAR-10 tasks, and from a pre-existing manually optimized script for the WSJ task. We specified a much wider range around those values, and used that to formulate the search space. We feel this is what a naive user would do given a baseline script.

MNIST Image Classification. The MNIST image classification task deals with classifying hand-written digit images of size 28×28 pixels. We use a simple fully connected DNN network here to take in a single image and return a distribution over the 10 labels. The hyper-parameters and ranges are given in Table 1. The data-set consists of 60000 training images and 10000 test images. Each network is trained up to 20 epochs, with the epoch giving the best test accuracy being selected as the final model for the given hyper-parameter configuration. The training was done using MXNet training toolkit using a single GPU. The performance metric is the accuracy of classification over the validation data-set, which is to be maximized.

Table 1. Hyper-parameters for MNIST classification task

Hyper-parameter	Type	Minimum	Maximum	Step
Number of layers	Int	1	8	1
Number of neurons per layer	Int	32	3200	32
Learning rate	Float	0.001	0.1	-
Learning rate factor	Float	0.5	0.9	0.1
Learning rate step epochs	List	$\{[4, 8, 12, 16], [5, 10, 15], [10]\}$		

CIFAR-10 Image Classification. The CIFAR-10 data-set [10] consists of 60000 32×32 images belonging to 10 different classes, with 6000 images per class as the training corpus, and 10000 images in the test corpus. For this task, we implement a variation of the ResNet [7] architecture for classifying an image into the 10 classes. The hyper-parameters and ranges for this model are given in Table 2. For the number of conv units in each segment, we had a geometric progression based on the segment number. The first segment would have 16 conv units per layer, the second 32, the third 64, and so on. The training was done using MXNet training toolkit on a single GPU. The performance metric is the accuracy of classification over the validation data-set, which is to be maximized.

Large Vocabulary Continuous Speech Recognition. For this task, we selected the optimization of a simple fully-connected deep neural network (DNN) hidden Markov model (HMM) hybrid speech recognition system trained on the Wall Street Journal (WSJ) corpus. This consists of 286 h of training data, with a validation set of 503 utterances. The following is a brief description of the training pipeline. An initial Gaussian Mixture Model based system is trained using the Kaldi recipes [11] up to the tri4b stage. Once this is done, a training

Table 2. Hyper-parameters for CIFAR-10 classification task using the ResNet architecture

Hyper-parameter	Type	Minimum	Maximum	Step
Number of residual segments	Int	1	8	1
Number of conv layers per segment	Int	1	5	1
Learning rate	Float	0.001	0.1	-
Learning rate factor	Float	0.5	0.9	0.1
Learning rate step epochs	List	$\{[30, 60, 90...., 270], [60, 120, 180, 240],$ $[90, 180, 270], [120, 240], [150], [200, 250]\}$		

corpus of extracted log-mel filter-bank features [4], with labels obtained by the alignment of the trained text to the audio is generated. We use 40 mel-filterbanks based on [13]. 10 iterations of DNN training using stochastic gradient descent is performed using the training framework used in [3]. The iteration giving the best validation set word error rate is selected as the final model trained for the given hyper-parameter configuration. The structure of the DNN model depends on the model hyper-parameters of number of input frames spliced, the number of hidden layers, and the number of neurons per hidden layer. The hyper-parameters and ranges for this task are given in Table 3. The performance metric for this task is word error rate (WER), which is the ratio of the total number of substitutions, insertions, and deletions to be performed on a hypothesis to match the reference. This needs to be minimized for this task. To fit the description of the proposed approach, we take 1 - WER as the metric to maximize.

Table 3. Hyper-parameters for the WSJ LVCSR Task

Hyper-parameter	Type	Minimum	Maximum	Step
Frame context splicing	Int	0	9	1
Number of hidden layers	Int	1	6	1
Number of neurons per hidden layer	Int	512	2944	128

4.2 Experiments on Quality of Predictions

The first set of experiments involve finding the quality of the predictions made by the proposed method. We look at two different metrics for this. The first, is the average root mean square error (RMSE) of the prediction across all the epochs. This gives us an idea of how well the overall prediction fits with the actual observations. The second is the average RMSE of the prediction over only the final epochs. This gives an indication of the performance of the estimators for the termination criterion that will be integrated to any hyper-parameter optimization technique.

Experimental Setup. We compute these numbers while progressively increasing the number of previous builds, and progressively increasing the number of available epochs from the current build. For the MNIST and CIFAR-10 tasks, we conducted 100 builds with randomly varied hyper-parameter setting to construct the initial data-set. For the WSJ task, we conducted 30 builds with random hyper-parameters. For each build, we select the specified number previous builds from the remaining builds randomly. We use our proposed approach to compute the predictions for all the remaining epochs for the current build. We conduct the above 50 times, and average the results over all the resultant combinations.

For the ensemble approach, for each previous build, we perform 100 iterations of gradient descent with random initialization of the affine parameters a_r and b_r. We then select the top 100 estimators with the least training loss to construct the ensemble estimator.

Results. The results for assessing the quality of predictions are given in Figs. 1 and 2. Figures (1a, c and e) show the variation of the RMSE of the predictions over all epochs as a functions of number of epochs completed in the current build, while Figs. (1b, d and f) show the corresponding average of the predictive standard deviations computed using Eq. (4) for all the runs for a given number of previous builds and given number of epochs of a current build, across all epochs.

Figures (2a, c and e) show the variation of the RMSE of the predictions of the final epoch as a functions of number of epochs completed in the current build, while Figs. (2b, d and f) show the corresponding average predictive standard deviations over only the final epoch.

We show RMSE here instead of the actual prediction since we want to observe how much they deviate from the actual value at all the operating points of interest.

We observe the for all the cases, in the proposed method, using successively more number of previous builds to predict the result for the current build improves the prediction loss. Looking at the predictions, we observe the average RMSE of predictions across epochs and average RMSE of the prediction for the final epoch goes down as the number of previously completed builds increases. In the initial epochs for all three, our proposed approach performs significantly better than the baseline prediction method in both metrics. For all the tasks, we observe that, if given 5 initial builds, we match or outperform the baseline method for at least the initial 20% of the epochs. For the remaining epochs, the results are more of a mixed bag, but the proposed and baseline method perform comparably across all the tasks.

We observe that for the CIFAR-10 task, we observe a higher loss in terms of prediction over all the different scenarios, however the baseline is dramatically more confident in it's prediction. This may result in potentially terminating builds in the initial epochs, even if those builds can potentially get better at later stages. In contrast, the standard deviations of our proposed ensemble approach is more in line with the average RMSE across all tasks.

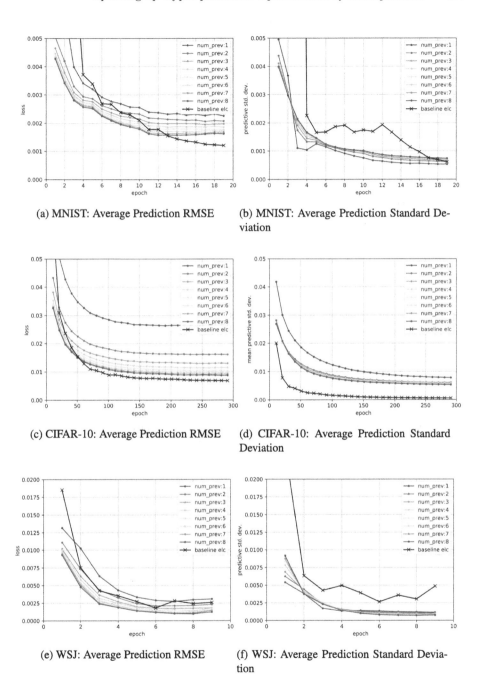

(a) MNIST: Average Prediction RMSE

(b) MNIST: Average Prediction Standard Deviation

(c) CIFAR-10: Average Prediction RMSE

(d) CIFAR-10: Average Prediction Standard Deviation

(e) WSJ: Average Prediction RMSE

(f) WSJ: Average Prediction Standard Deviation

Fig. 1. Quality of predictions in three tasks over all epochs. The left plots show the average RMSE, while the right shows the predictive standard deviation over all epochs. *num_prev* indicates the number of previous builds used to build the estimator.

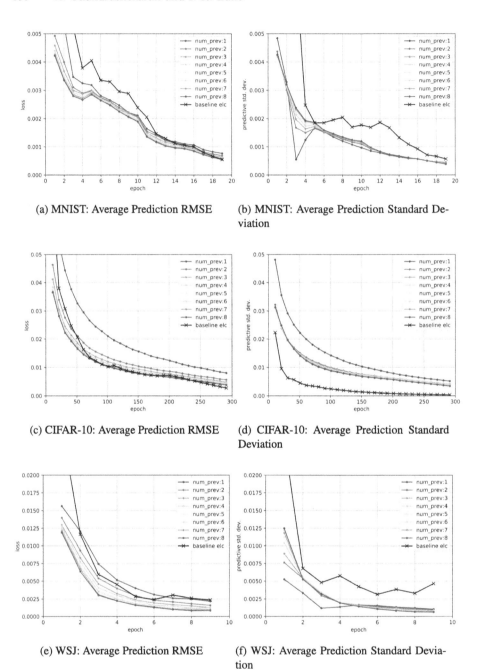

(a) MNIST: Average Prediction RMSE

(b) MNIST: Average Prediction Standard Deviation

(c) CIFAR-10: Average Prediction RMSE

(d) CIFAR-10: Average Prediction Standard Deviation

(e) WSJ: Average Prediction RMSE

(f) WSJ: Average Prediction Standard Deviation

Fig. 2. Quality of predictions in three tasks at the final epoch. The left side shows the average RMSE at the final epoch, while the right shows the average standard deviation over the final epoch. *num_prev* indicates the number of previous builds used to build the estimator.

4.3   Experiments on Integration with Bayesian Optimization

The second set of experiments evaluate the performance of integrating the proposed termination criterion with an existing hyper-parameter optimization toolkit. The experiments performed here use a combination of integral and floating point hyper-parameters which do not have any hierarchy. Also, the number of hyper-parameters to be optimized is low for our current test cases. Hence, based on the findings of [6], we have selected Bayesian Optimization as the sequential model based optimization method for our overarching automated hyper-parameter optimization technique. We have modified the Spearmint toolkit [12] to incorporate the termination criterion.

Experimental Setup. For each task, we perform three sub-experiments. The first performs normal Bayesian Optimization without any early stopping (Non-augmented). So, each build will run till completion. The second experiment incorporates the termination criterion using the baseline extrapolation of learning curves as described in [5] into Bayesian optimization (Baseline ELC). The third incorporates the termination criterion using the proposed extrapolation of learning curves technique as described in Sect. 3 into the Bayesian optimization (Proposed ELC).

For each sub-experiment for MNIST and WSJ, we perform 30 builds in a single iteration of optimization. For CIFAR-10, we perform 20 builds for a single optimization iteration. For all the sub-experiments, we analyze the best validation performance metric obtained at the end of the optimization run, and the total number of epochs of learning required to reach the end of optimization. We are aware that each epoch within each model build can have dramatically different time requirements. However, because we are trying to reduce the number of epochs required for training, we do not report the actual wall time required to finish the optimization run.

Results. Table 4 shows a comparison of the performance of the two extrapolation approaches. Each sub-table performs the comparison for each individual task. We observe that in the MNIST and WSJ tasks, the proposed ensemble approach requires fewer number of epochs to find the best performing build compared to the baseline extrapolation technique, with little to no loss in performance. In the case of the CIFAR-10 task, though the baseline performs fewer epochs, there is a substantial decrease in performance in terms of accuracy. We presume this is due to the learning rate annealing feature (the last row in Table 2). Empirically, we saw that at the epochs where the learning rate was annealed, there was a significant improvement in the validation accuracy. This information of probable future improvement is not captured by the baseline extrapolation technique. However, since similar behavior was probably observed in the previously completed builds used by the proposed extrapolation method, the termination criterion does not kill the job immediately. This can also be tied in to the observations of the loss and standard deviation trajectories in Figs. (2c and d).

Table 4. Performance comparison on integration with Bayesian optimization

Extrapolation type	Total Epochs (Relative improvement)	Best Accuracy(%) (Relative Improvement)
Non-augmented	600	98.76%
Baseline ELC	370 (38.3)%	98.76% (0.0%)
Proposed ELC	257 (57.16%)	98.63% (-0.1%)

(a) Task: MNIST

Extrapolation type	Total Epochs (Relative improvement)	Best Accuracy(%) (Relative Improvement)
Non-augmented	6000	91.359%
Baseline ELC	4292 (28.9%)	90.5% (-0.9%)
Proposed ELC	5019 (16.3%)	90.92% (-0.4%)

(b) Task: CIFAR-10

Extrapolation type	Total Epochs (Relative improvement)	Best WER(%) (Relative Improvement)
Non-augmented	300	7.85%
Baseline ELC	201 (33%)	7.85% (0%)
Proposed ELC	171 (43%)	7.85% (0%)

(c) Task: WSJ

Although we have shown only one run of hyper-parameter optimization for each, we do not expect to see a large variance between multiple runs.

5   Limitations and Future Directions

The proposed approach currently uses only the information from completed builds to make the prediction for new builds. Due to this, it can make predictions only up to the number of epochs in the current build. The baseline method [5], although not explored, does not suffer from this limitation. One possible work-around could be to use parametric models like the baseline approach on the previously completed builds, and train an ensemble predictor using those.

In it's present form, it relies on the presence diversity in the initial builds to come up with good extrapolation models for new builds. Though we have used a single probability threshold, standard deviation threshold, and minimum number of required initial builds across the tasks, there is no guarantee that this value will produce the optimal result, i.e. best build in least number of epochs. Other tasks could potentially have differing thresholds.

Our technique relies on the random initialization of the affine transformation parameters to result in the presence of a variety in the training loss values which will guide which weak estimators are selected. This can be made more formal using Markov Chain Monte Carlo (MCMC), which we plan to explore in the future. Using this could also incorporate some of the practical considerations mentioned in Sect. 3.2 in a more formal manner.

Finally, the proposed approach does not use the build hyper-parameters directly to guide the ensemble construction. This is another future direction we plan to explore.

6   Conclusion

We have studied an ensemble approach for modelling learning curves of machine learning algorithms. We have analyzed the performance of this technique across three different tasks. We have demonstrated its performance when integrated with an automated optimization technique over these three tasks. This approach requires relatively small overhead for the estimation, and performs remarkably well in a majority of the tasks. It lends itself well to sequential optimization techniques since it relies on the presence of few builds that have run to completion to make predictions for subsequent builds, and shows improvement in prediction accuracy as more builds get completed. We have discussed the strengths and limitations of this new approach compared to previous approaches, and have discussed some potential ideas for further improvement of this technique.

Note

The source code is available at https://github.com/akshayc11/Spearmint under the branch `elc`. Please refer to this for more details.

References

1. Bergstra, J., Bengio, Y.: Random search for hyper-parameter optimization. J. Mach. Learn. Res. **13**(Feb), 281–305 (2012)
2. Bergstra, J.S., Bardenet, R., Bengio, Y., Kégl, B.: Algorithms for hyper-parameter optimization. In: Advances in Neural Information Processing Systems, pp. 2546–2554 (2011)
3. Chan, W., Lane, I.: Deep recurrent neural networks for acoustic modelling. arXiv preprint arXiv:1504.01482 (2015)
4. Davis, S., Mermelstein, P.: Comparison of parametric representations for mono-syllabic word recognition in continuously spoken sentences. IEEE Trans. Acoust. Speech Signal Process. **28**(4), 357–366 (1980)
5. Domhan, T., Springenberg, J.T., Hutter, F.: Speeding up automatic hyperparameter optimization of deep neural networks by extrapolation of learning curves. In: International Joint Conference on Artificial Intelligence (IJCAI), 2015, pp. 3460–3468. AAAI Press (2015)
6. Eggensperger, K., Feurer, M., Hutter, F., Bergstra, J., Snoek, J., Hoos, H., Leyton-Brown, K.: Towards an empirical foundation for assessing Bayesian optimization of hyperparameters. In: NIPS workshop on Bayesian Optimization in Theory and Practice, pp. 1–5 (2013)

7. He, K., Zhang, X., Ren, S., Sun, J.: Deep residual learning for image recognition. In: Proceedings of the IEEE Conference on Computer Vision and Pattern Recognition, pp. 770–778 (2016)
8. Hutter, F., Hoos, H.H., Leyton-Brown, K.: Sequential model-based optimization for general algorithm configuration (extended version). Technical report TR-2010-10, University of British Columbia, Computer Science (2010)
9. Klein, A., Falkner, S., Springenberg, J.T., Hutter, F.: Learning curve prediction with Bayesian neural networks. In: Proceedings of ICLR 2017 (2017)
10. Krizhevsky, A.: Learning multiple layers of features from tiny images. Citeseer (2009)
11. Povey, D., Ghoshal, A., Boulianne, G., Burget, L., Glembek, O., Goel, N., Hanne-mann, M., Motlicek, P., Qian, Y., Schwarz, P., et al.: The kaldi speech recognition toolkit. In: IEEE 2011 Workshop on Automatic Speech Recognition and Under-standing, IEEE Signal Processing Society (2011)
12. Snoek, J., Larochelle, H., Adams, R.P.: Practical Bayesian optimization of machine learning algorithms. In: Advances in Neural Information Processing Systems, pp. 2951–2959 (2012)
13. Soltau, H., Saon, G., Kingsbury, B.: The IBM Attila speech recognition toolkit. In: Spoken Language Technology Workshop (SLT), 2010 IEEE, pp. 97–102. IEEE (2010)
14. Surowiecki, J.: The wisdom of crowds. Anchor (2005)
15. Swersky, K., Snoek, J., Adams, R.P.: Freeze-thaw Bayesian optimization. arXiv preprint arXiv:1406.3896 (2014)
16. van Rijn, J.N., Abdulrahman, S.M., Brazdil, P., Vanschoren, J.: Fast algorithm selection using learning curves. In: Fromont, E., De Bie, T., van Leeuwen, M. (eds.) IDA 2015. LNCS, vol. 9385, pp. 298–309. Springer, Cham (2015). https://doi.org/10.1007/978-3-319-24465-5_26
17. Yampolskiy, R.V., Ashby, L., Hassan, L.: Wisdom of artificial crowds-a metaheuris-tic algorithm for optimization. J. Intell. Learn. Syst. Appl. 4(2), 98 (2012)

Thompson Sampling for Optimizing Stochastic Local Search

Tong Yu[1(✉)], Branislav Kveton[2], and Ole J. Mengshoel[1]

[1] Electrical and Computer Engineering, Carnegie Mellon University,
Pittsburgh, USA
tongy1@andrew.cmu.edu, ole.mengshoel@sv.cmu.edu
[2] Adobe Research, San Jose, USA
kveton@adobe.com

Abstract. Stochastic local search (SLS), like many other stochastic optimization algorithms, has several parameters that need to be optimized in order for the algorithm to find high quality solutions within a short amount of time. In this paper, we formulate a stochastic local search bandit (SLSB), which is a novel learning variant of SLS based on multi-armed bandits. SLSB optimizes SLS over a sequence of stochastic optimization problems and achieves high average cumulative reward. In SLSB, we study how SLS can be optimized via low degree polynomials in its noise and restart parameters. To determine the coefficients of the polynomials, we present polynomial Thompson Sampling (PolyTS). We derive a regret bound for PolyTS and validate its performance on synthetic problems of varying difficulty as well as on feature selection problems. Compared to bandits with no assumptions of the reward function and other parameter optimization approaches, our PolyTS assuming polynomial structure can provide substantially better parameter optimization for SLS. In addition, due to its simple model update, PolyTS has low computational cost compared to other SLS parameter optimization methods.

1 Introduction

Stochastic optimization algorithms are among the most popular approaches to solving NP-hard and black box computational problems [15,33]. Stochastic optimization algorithms include stochastic local search (SLS). Even in their simplest forms, SLS algorithms typically have several parameters that dramatically impact performance.

The problem of choosing parameters for a stochastic optimization algorithm is essential, since the optimal values for these parameters are problem-specific and have dramatic impact on the performance of the algorithm. As an example, consider the feature selection problem for sensor-based human activity recognition (HAR). The mobile sensor data for a given activity, for example jogging, can vary substantially from person to person and depending on where the mobile devices are placed on the body. Compared to training models on a large-scale

© Springer International Publishing AG 2017
M. Ceci et al. (Eds.): ECML PKDD 2017, Part I, LNAI 10534, pp. 493–510, 2017.
https://doi.org/10.1007/978-3-319-71249-9_30

general dataset, training models on a small personal dataset can yield better accuracy [22,39]. Thus, it is better for the HAR model to be trained periodically when new labeled personal data is available. In such a sequence of problems, using fixed parameters for feature selection methods may not always provide the optimal accuracy. In this paper, we study how to optimize the parameters for stochastic optimization algorithms in such sequences of problems, in order to achieve the highest cumulative reward over time.

Several previous works could potentially be applied to optimize the parameters of SLS. UCB1 [6] is used in adaptive operator selection [11,12,21]. Reinforcement learning is used in single parameter tuning for local search [7,30]. To optimize the parameters of a general algorithm, there also exists a wide range of methods, such as racing algorithms (e.g., irace) [8,23], ParamILS [18], sequential model-based optimization (SMAC) [16,17], gender-based genetic algorithm (GGA) [3], and Bayesian optimization (e.g., GPUCB and GPEIMCMC) [35,36]. However, most previous works have one or more of these limitations, when being applied in our problem:

1. **Not optimizing cumulative reward for a sequence of problems:** Most parameter optimization methods [3,8,14,16–18,23,26,32] are for best-arm identification, where they explore aggressively in fixed steps and then exploit. However, in our problem, we aim to maximize a long-term reward over a sequence of problems.[1]
2. **No regret bound or suboptimal regret bound:** Previous bandit methods in Adaptive Operator Selection [11,12,21] have no regret bound analysis. In fact, we show that directly applying a bandit method without parametric approximation leads to suboptimal regret in Sect. 4.2.
3. **Intensive computational cost:** Most methods have high computational cost (see Table 1). For example, inference of Gaussian process is computationally expensive in Bayesian optimization [35,36].
4. **Optimizing only a single parameter for local search:** Previous work on reinforcement learning only tunes a single parameter for local search [7,30]. However, it has been shown that multiple parameters should be tuned to optimize SLS [25].

This paper seeks to jointly address all four limitations above through our *stochastic local search bandit* (SLSB). In SLSB, each set of parameters for SLS corresponds to a bandit arm. We assume that the learning problem at each time of training is drawn i.i.d. from some distribution over learning problems, which is unknown to the learning agent. When SLSB is applied to feature selection, there is a bandit, SLS, and a classifier. At each time t, they interact as follows: (1) The bandit chooses SLS's parameters $A \in [0,1]^2$. (2) SLS find κ feature subsets using the chosen parameters A. (3) With each feature subset, a classifier

---

[1] As a motivation for our focus on maximizing cumulative reward, consider parameter tuning for feature selection on a sequence of HAR datasets. In this application, it is important to provide usable predictions even in the initial steps of tuning feature selection, else users may stop using the HAR system.

is trained and evaluated by the accuracy on the data available at time t. (4) The maximum among the accuracies on κ feature subsets is the bandit reward.

SLSB is a framework for learning from adaptively chosen parameter sets with the objective of finding the best SLS parameter set. It differs from classical approaches [3,8,14,16–18,23,26,32] to parameter optimization in two aspects. First, the objective is to identify the best set of parameters without perfectly fitting a model, which may not be necessary. Second, the training parameter sets are chosen adaptively. We would like to stress that our goal of maximizing cumulative reward[1] is different from that of best arm identification [5], where the learning can explore aggressively in initial steps and its performance is measured only at some fixed later step.

The contributions of this paper are as follows:

- We solve the problem of optimizing SLS parameters to achieve the highest cumulative reward over a sequence of problems by means of a multi-armed bandit, which we call a stochastic local search bandit (SLSB).
- Inspired by Markov chain analyses of SLS, we study polynomial Thompson Sampling (PolyTS), which learns low-degree polynomials in the noise and restart parameters of SLS. In addition, we show that the parametric approximation setting of PolyTS can effectively improve the regret bound of SLSB, compared to the setting without parametric approximation.
- Experimentally, we compare PolyTS to several state-of-the-art parameter optimization methods. We demonstrate that PolyTS can effectively find optimal SLS parameters leading to high reward (*i.e.*, high accuracy) in a sequence of feature selection problems. We show that PolyTS induces substantially lower computational cost compared to other prominent parameter tunning methods.

2   Related Work

2.1   Stochastic Local Search

Stochastic local search (SLS) algorithms are simple but broadly applicable [15], and among the best-performing algorithms for NP-hard problems including satisfiability [15,33] as well as computing the most probable explanation [26,27] and the maximum a posteriori hypothesis in Bayesian networks.

SLS algorithms are optimization methods that combine greedy, noisy, and restart heuristics among others. Noise and restart operations, combined with powerful initialization heuristics, have proven to be valuable additions to greedy search in many SLS algorithms [4,26,32,33]. This raises the question of how to optimally balance between the greedy, noisy, and restart operations, ideally in a problem-specific manner.

One method to optimize SLS parameters is adaption. Hoos explores adaptive noise [14]. Adaptive noise and restart were recently integrated [25]. Ryvchin and Strichman advocate a related concept, namely localized restarts [32]. A second approach to optimizing SLS parameters is to perform a Markov chain analysis.

Markov chain models of SLS have been used to compute expected hitting times [25,26], which can be used for parameter optimization.

2.2 Parameter Optimization Methods

Reinforcement learning was used to optimize the performance of reactive search optimization [7] and local search [30]. Markov decision processes were used to optimize the prohibition value in reactive tabu search, the noise parameter in Adaptive Walksat, and the smoothing probability in reactive scaling and probabilistic smoothing algorithms [7]. R-learning was used to dynamically optimize noise in standard SAT local search algorithms [30]. These methods optimize algorithm parameters during execution.

Other algorithms for parameter optimization exist. Racing [8,23] evaluates a set of candidate parameter configurations by discarding bad ones when statistically sufficient evidence is gathered against them. ParamILS [18] optimizes parameters based on Iterated Local Search. Sequential model-based optimization [16,17] iterates between constructing a regression model and obtaining additional data for it. Previous works also model an algorithm's performance as samples from a Gaussian process [35,36].

2.3 Bandits

A multi-armed bandit [6,20] is a popular framework for learning to act under uncertainty. The framework has been successfully applied to many complex structured problems, ranging from stochastic combinatorial optimization to reinforcement learning. UCB1 [6] is used in adaptive operator selection [11,12,21]. The proposed SLSB in this paper is, to our knowledge, the first work that formulates the problem of parameter learning in SLS as a multi-armed bandit problem. There are potentially other bandit algorithms suitable for SLSB [9,19,24]. Besides, Thompson Sampling [10] is applied to automatically choose the best technique and its associated hyper-parameters in a machine learning task [13]. Our novelty compared to previous bandit research is in the formulation of SLSB to optimize SLS, the parametric approximation adaptation of the existing bandit algorithms to our problem, the regret bound analysis, and their empirical applications in feature selection for HAR.

3   Background on Stochastic Local Search

We consider *stochastic local search (SLS)* [15] as search in the space of *bit-strings* $B = \{0,1\}^N$. The goal is find the maximum of the *fitness function* $V : B \to \mathbb{R}$:

$$b^* = \arg \max_{b \in B} V(b).$$

The function V characterizes the problem being solved.

SLS is, in our case, parameterized by search parameters $A = (p_r, p_n)$ and the maximum number of optimization operations κ. SLS is initialized at a bit-string $b \in B$. The next, and any subsequent, SLS operation is either a restart operation, a noise operation, or a greedy operation. These operations are chosen randomly according to the following scheme [25]. SLS chooses a restart operation with probability p_r.

Definition 1 *(Restart Operation). The restart operation restarts SLS to a new random bit-string b.*

If SLS does not restart, then a noise operation is chosen with probability p_n.

Definition 2 *(Noise Operation). The noise operation randomly flips a bit in b.*

Finally, if SLS does not choose a noise operation, SLS performs a greedy operation.

Definition 3 *(Greedy Operation). The greedy operation chooses the highest-fitness neighbor of b. The neighbor of b is any bit-string that differs from b in one bit.*

The parameters $A = (p_n, p_r)$ have a dramatic impact on the performance of SLS. When $p_n = p_r = 0$, SLS behaves greedily. When $p_n = 1$ and $p_r = 0$, SLS is a random walk. The challenge is to set p_n and p_r, such that SLS finds a bit-string $\hat{b}$ such that $V(\hat{b})$ is close to $V(b^*)$. Let $b_1, \ldots, b_\kappa$ be the sequence of bit-strings in the κ operations of SLS with parameters $A = (p_n, p_r)$. Then we define the *reward* of this step of the SLSB (see Sect. 4) as

$$f(A) = \max_{i \in [\kappa]} V(b_i). \tag{1}$$

Note that f depends on both A and V, and also on the randomness in choosing the operations in SLS. To simplify notation, we do not make this dependence explicit.

We propose a variant of SLS, a learning agent, in the next section. The goal of the learning agent is to find $A = (p_n, p_r)$ that maximizes $\mathbb{E}[f(A)]$, where the expectation is taken with respect to both (i) randomly choosing V from some class of problems and (ii) the randomness of the operations in SLS.

4   Stochastic Local Search Bandit

We formulate the problem of learning optimal SLS parameters as a *stochastic local search bandit* (SLSB). Formally, a stochastic local search bandit is a tuple $B = (\Theta, P)$, where $\Theta = [0, 1]^2$ is the *space of arms* (all pairs of the restart and noise-operation probabilities in SLS) and P is a probability distribution over *reward functions* $f : \Theta \to \mathbb{R}$. For any arm $A \in \Theta$, we denote by $f(A)$ the *reward* of A under f, which is the same as in (1); and by $\bar{f}(A) = \mathbb{E}[f(A)]$ the *expected reward* of using SLS with parameters A.[2] The expectation is with respect to

---

[2] For example, in HAR, $f(A)$ is the activity recognition accuracy of a particular machine learning model trained on a HAR dataset, when using SLS with parameters A for feature selection.

a potentially random fitness function, which represents different i.i.d. chosen problems from some set, and also the randomness in choosing the operations in SLS. We assume that $\bar{f}(A)$ is Lipschitz in A.

Let $(f_t)_{t=1}^n$ be an i.i.d. sequence of n reward functions drawn from P, where f_t is the reward function at time t. The learning agent interacts with our SLS algorithm as follows. At time t, it selects arm $A_t \in \Theta$, which depends on the observations of the agent up to time t. Then it observes the stochastic reward of arm A_t, $f_t(A_t)$, and receives it as a reward.

The goal of the learning agent is to maximize its cumulative reward over time. The quality of the agent's policy is measured by its *expected cumulative regret*. Let

$$A^* = \arg \max_{A \in \Theta} \bar{f}(A) \tag{2}$$

be the *optimal solution* in hindsight. Then the expected cumulative regret is

$$R(n) = \mathbb{E}\left[\sum_{t=1}^n R(A_t, f_t)\right], \tag{3}$$

where

$$R(A_t, f_t) = f(A^*, f_t) - f(A_t, f_t) \tag{4}$$

is the *instantaneous stochastic regret* of the agent at time t. For simplicity of exposition, we assume that A^* is unique.

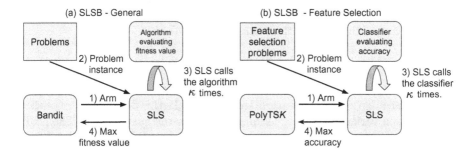

Fig. 1. In (a), SLSB solves a sequence of stochastic optimization problems. At each step t, there are four actions: 1) Based on the fitness values up to steps $t-1$, the bandit algorithm chooses SLS's parameters, $A = (p_r, p_n)$. 2) We receive a problem instance. 3) Run SLS for κ operations on the problem with the chosen parameters. 4) Feed the maximum of the fitness value in the κ operations as the reward. The goal of SLSB is to maximize the cumulative accuracy over T steps, where T is the total number of steps. In (b), SLSB solves a sequence of feature selection problems. At each step t, 1) PolyTSK suggests the parameters to SLS. 2) We receive a dataset. 3) SLS searches κ feature subsets using the suggested parameters. Under the evaluation of a classifier on this dataset, the feature subset with the highest validation accuracy is selected. 4) This feature subset's corresponding validation accuracy is the reward $f(A)$ returned to PolyTSK.

4.1 SLSB in General and SLSB's Application in Feature Selection

The different algorithms of SLSB and how those algorithms interact are illustrated in Fig. 1. Figure 1(a) shows a general example of SLSB solving a sequence of stochastic optimization problems. One key research problem that we discuss in Sects. 4.2, 4.3, and 4.4 is how to design an efficient algorithm for the bandit in SLSB in Fig. 1(a), such that the cumulative reward is maximized.

The SLSB can be applied in feature selection, as shown in Fig. 1(b). We reduce the feature selection problems to optimization problems solved by SLS [25]. For training data with N dimension, if the ith bit is 1, we select the ith feature, else we ignore it. In total we have 2^N possible feature subsets (*i.e.*, bit-strings). At time t, we select arm A_t, the parameters of the SLS; and randomly divide the dataset at time t into the training, validation, and test sets. Then we run SLS for κ operations. The i-th operation of SLS is associated with bit-string b_i, an indicator vector of features. At operation i, we learn a classifier on the training set with features b_i and evaluate it by its accuracy $V(b_i)$ on the validation set. The SLS solution is the best performing b_i, as in (1). Our method can also be applied to other problems, not just feature selection. SLS is used for sparse signal recovery [29] and finding the most probable explanation in Bayesian networks [26,27], where the SLS can also be optimized by SLSB.

4.2 Thompson Sampling (TS)

The key idea in our first design is to discretize the space of arms Θ into a set of points $\mathcal{A}$ and estimate $\bar{f}(A)$ separately in each $A \in \mathcal{A}$. We refer to each A as an *arm* and denote the number of arms by $|\mathcal{A}|$. We assume that $\mathcal{A}$ is an ε-grid with granularity ε, which satisfies $\|A - A'\|_\infty \le \varepsilon$ for any $A, A' \in \mathcal{A}$. Since $\mathcal{A}$ is an ε grid on a two-dimensional space, $|\mathcal{A}| = 1/\varepsilon^2$.

Our first learning algorithm is a variant of *Thompson sampling (TS)* [37] for normally distributed rewards with σ^2 variance. We assume that the conjugate prior on the expected reward of each arm is $\mathcal{N}(0, \sigma_0^2)$, a normal distribution with a zero mean and σ_0^2 variance. At each step t, our SLSB algorithm operates in three stages. First, we sample the expected reward $\bar{f}_t(A)$ of each arm $A \in \mathcal{A}$ from its posterior. By our assumptions and Bayes rule, the expected reward of arm A at time t is distributed as

$$\bar{f}_t(A) \sim \mathcal{N}\left(\frac{\sigma_{T_{t-1}(A)}^2}{\sigma^2} T_{t-1}(A)\hat{f}_{T_{t-1}(A)}(A),\ \sigma_{T_{t-1}(A)}^2 \right),$$

where $T_t(A)$ is the number of times arm A is chosen in t steps, $\hat{f}_s(A)$ is the average of s observed rewards of arm A, and $\sigma_s^2 = (\sigma_0^{-2} + s\sigma^{-2})^{-1}$ is the variance of the posterior after s observations. Second, we choose the arm with the highest posterior reward:

$$A_t = \arg\max_{A \in \mathcal{A}} \bar{f}_t(A).$$

Finally, we observe $f_t(A_t)$ and receive it as a reward.

Proposition 1. *Let c be the Lipschitz factor of f and $\mathcal{A}$ be an ε-grid over Θ. Then for $\varepsilon = c^{-\frac{1}{2}} n^{-\frac{1}{4}}$, the regret of* TS *is $O(c^{\frac{1}{2}} n^{\frac{3}{4}})$.*

Proof. Since $\mathcal{A}$ is an ε grid on a two-dimensional space, $|\mathcal{A}| = 1/\varepsilon^2$. Thus, the regret of TS with respect to the best arm on the ε-grid is $O(\sqrt{n|\mathcal{A}|})$ [1]. Considering $\sqrt{n|\mathcal{A}|} = \sqrt{n/\varepsilon^2} = \sqrt{n}/\varepsilon$, the regret of TS with respect to the best arm is $O(\sqrt{n}/\varepsilon)$. Because the Lipschitz factor of f is c and $\mathcal{A}$ is an ε-grid, the expected rewards of the best arms in Θ and $\mathcal{A}$ differ by at most $c\varepsilon$. Therefore, the regret of PolyTS with respect to the best arm in Θ is $O(\sqrt{n}/\varepsilon) + c\varepsilon n$. Now substitute our suggest value of ε to minimize $O(\sqrt{n}/\varepsilon) + c\varepsilon n$, and our claim follows.

The regret of TS is suboptimal in n and we will show PolyTS with an improved regret bound in Sect. 4.3.

4.3 Polynomial Thompson Sampling (PolyTS)

The shortcoming of TS is that its regret is suboptimal in n. We now study how to learn a parametric approximation of $\bar{f}$. Based on existing work on SLS [25,26] and our discussion in Sect. 4.4, we assume that $\bar{f}$ is a low-order polynomial over Θ.

Our learning algorithm is a variant of *linear Thompson sampling* (LinTS) [1] for learning polynomials of degree K. Let i-th entry of A be $A[i]$. We assume that the reward of arm $A = (A[1], A[2]) \in \Theta$ at time t is normally distributed as

$$f_t(A) \sim \mathcal{N}(\mathbf{x}_A \theta^*, \sigma^2), \tag{5}$$

where $\mathbf{x}_A \in [0,1]^{1 \times L}$ is the feature vector of all possible $L = (K+1)(K+2)/2$ terms $A[1]^i A[2]^j$, and $i \geq 0$ and $j \geq 0$ are any integers such that $i + j \leq K$. When $K = 2$, we get a quadratic polynomial and:

$$\mathbf{x}_A = (A[1]^2, A[1]A[2], A[2]^2, A[1], A[2], 1).$$

The vector $\theta^* \in \mathbb{R}^{L \times 1}$ represents L unknown parameters of $\bar{f}$, which are shared by all arms. We assume that the conjugate prior on θ^* is

$$\mathcal{N}(\mathbf{0}, \lambda^{-1} I_L), \tag{6}$$

a normal distribution with a zero mean and isotropic λ^{-1} variance; where I_L denotes an $L \times L$ identity matrix. For degree K, we refer to our learning algorithm as PolyTSK.

At each step t, our algorithm operates in three stages. First, we sample the parameters of $\bar{f}$ from its posterior. Denote $\bar{\theta}_{t-1}$ as the mean of the parameters of $\bar{f}$ after $t-1$ steps. By our assumptions and Bayes rule, the posterior at time t is normally distributed as $\theta_t \sim \mathcal{N}(\bar{\theta}_{t-1}, S_{t-1})$, where the *covariance matrix* is $S_t = (\sigma^{-2} X_t^\mathsf{T} X_t + \lambda I_L)^{-1}$, the *mean* is $\bar{\theta}_t = \sigma^{-2} S_t X_t^\mathsf{T} F_t$, $X_t = (\mathbf{x}_{A_1}^\mathsf{T}, \dots, \mathbf{x}_{A_t}^\mathsf{T})^\mathsf{T}$ is a $t \times L$ matrix of the features of all chosen arms in

Algorithm 1. SLS Algorithm (see Fig. 1(a) and (b)).

Function SLS(A)

$p_n \leftarrow A[1]$; $p_r \leftarrow A[2]$;
Randomly initialize $b \in B = \{0,1\}^N$ {See Sect. 3.}
$V_{\max} \leftarrow V(b)$
for *SLS operation* $i \leftarrow 2$ **to** κ **do**
  if *rand(0,1)* $< p_r$ **then** $b \leftarrow$ RestartOperation()   {See Definition 1. The *rand(0,1)* generates a uniformly distributed random variable in $[0, 1]$.};
  else
    if *rand(0,1)* $< p_n$ **then** $b \leftarrow$ NoiseOperation(b)   {See Definition 2.};
    else $b \leftarrow$ GreedyOperation(b, D)   {See Definition 3.} ;
  end
  if $V_{max} < V(b)$ **then** $V_{\max} \leftarrow V(b)$ {$V_{\max}$ is the $f(A)$ in Eq. 1.} ;
end
Return $V_{\max}$

Algorithm 2. PolyTSK, which calls Algorithm 1 (see Fig. 1(b)) .

Input : Parameters σ and λ^{-1} in (5) and (6), total number of steps T.
for *Step* $t \leftarrow 1$ **to** T **do**
  $X_{t-1} \leftarrow (\mathbf{x}_{A_1}^\mathsf{T}, \ldots, \mathbf{x}_{A_{t-1}}^\mathsf{T})^\mathsf{T}$
  $F_{t-1} \leftarrow (f_1(A_1), \ldots, f_{t-1}(A_{t-1}))^\mathsf{T} \in \mathbb{R}^{(t-1)\times 1}$
  $S_{t-1} \leftarrow (\sigma^{-2}X_{t-1}^\mathsf{T}X_{t-1} + \lambda I_L)^{-1}$
  $\bar{\theta}_{t-1} \leftarrow \sigma^{-2}S_{t-1}X_{t-1}^\mathsf{T}F_{t-1}$
  Sample the parameters of $\bar{f}$ from its posterior by $\theta_t \sim \mathcal{N}(\bar{\theta}_{t-1}, S_{t-1})$
  Choose the arm with the highest expected reward $A_t \leftarrow \text{argmax}_{A\in\mathcal{A}}\mathbf{x}_A\theta_t$
  Observe and record $f_t(A_t) \leftarrow$ SLS(A_t)
end

---

t steps, and $F_t = (f_1(A_1), \ldots, f_t(A_t))^\mathsf{T} \in \mathbb{R}^{t\times 1}$ are all observed rewards in t steps. After we sample θ_t, we choose the arm with the highest expected reward with respect to θ_t:

$$A_t = \arg\max_{A\in\mathcal{A}} \mathbf{x}_A\theta_t. \tag{7}$$

Finally, we observe $f_t(A_t)$ as a reward. We bound the regret of PolyTS below.

Proposition 2. *Let c be the Lipschitz factor of f and $\mathcal{A}$ be an ε-grid over Θ. Then for $\varepsilon = L/(c\sqrt{n})$, the regret of PolyTS is $\tilde{O}(L\sqrt{n})$, where $\tilde{O}(\cdot)$ is the big-O notation up to logarithmic factors.*

Proof. The regret of PolyTS with respect to the best arm on the grid is $\tilde{O}(L\sqrt{n})$ [1]. Because the Lipschitz factor of f is c and $\mathcal{A}$ is an ε-grid, the expected rewards of the best arms in Θ and $\mathcal{A}$ differ by at most $c\varepsilon$. Therefore, the regret of PolyTS with respect to the best arm in Θ is $\tilde{O}(L\sqrt{n})+c\varepsilon n$. Now substitute our suggested value of ε and our claim follows.

When the degree K is small, PolyTSK gives us a reasonable tradeoff between the bias and variance of our model. Asymptotically, we expect that TS can learn better approximations than PolyTS when the number of discretization points $|\mathcal{A}|$ is large, because TS does not make any assumptions on $\bar{f}$. However, in a finite time, we hypothesize that PolyTS can learn good-enough approximations at a much lower regret than TS because it learns fewer parameters. We study this issue empirically, using synthetic problems in Sect. 5 and real world problems in Sect. 6.

4.4   Discussion of the Polynomial Approximation

Markov chains were used to analytically optimize SLS [25,26]. The expected hitting times for SLS is shown to be rational functions $\frac{P(p_n)}{Q(p_n)}$ of noise p_n, where $P(p_n)$ and $Q(p_n)$ are polynomials [26]. The analysis was extended to the analysis of both noise p_n and restart p_r parameters [25]. Although it was not mentioned explicitly, the hitting time can be written as a rational function $H(p_n, p_r) = \frac{P(p_n, p_r)}{Q(p_n, p_r)}$, where $P(p_n, p_r)$ and $Q(p_n, p_r)$ are polynomials of p_n and p_r. A possible way to optimize $H(p_n, p_r)$ is to compute p_n and p_r for $H'(p_n, p_r) = 0$. Unfortunately, obtaining p_n and p_r via $H'(p_n, p_r) = 0$ is difficult practically.

Minimizing the hitting time of SLS in step t of SLSB is closely related to minimizing the instantaneous stochastic regret in step t, $R(A_t, f_t)$, which is equivalent to maximizing reward $f(A_t, f_t)$ in step t in (4).

Based on the Stone-Weierstrass theorem, we have: if f is a continuous real-valued function defined on the range $[a, b] \times [c, d]$ and $\psi > 0$, then we have a polynomial function P in two variables x and y such that $|f(x, y) - P(x, y)| < \psi$ for all $x \in [a, b]$ and $y \in [c, d]$. So we can approximate the expected instantaneous reward $f(A_t, f_t)$ by polynomials in the noise and restart parameters, where $a = c = 0$, $b = d = 1$, $x = p_r$, and $y = p_n$. What degree polynomial P do we need to get a good approximation? We study this question experimentally, see Sects. 5 and 6.

5   Synthetic Experiments

5.1   Problems

All synthetic problems are defined on 20-bit bit-strings. Therefore, the cardinality of the search space in our synthetic problems is 2^{20}. These synthetic problems are designed to exercise the greedy, noise, and restart operations of SLS to varying degrees. All fitness functions are symmetric and therefore we can visualize them in a single dimension (Fig. 2). The maximum of all our functions is at the all-ones bit-string, $\|b\|_1 = 20$.

The fitness function in the first problem is $V^1(b) = 3\|b\|_1 + 2 + \epsilon$, where $\epsilon \sim \mathcal{N}(0, 1)$ is a random scalar which is the same for all b, in any random realization of V^1. The maximum of $V^1(b)$ can be be found greedily from any b and for any ϵ. The fitness function in the second problem is $V^2(b) = \max\{16 - \|b\|_1, 3\|b\|_1 - 40\} + \epsilon$. The maximum of $V^2(x)$ can be found greedily

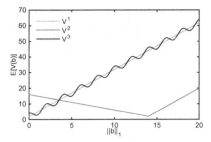

Fig. 2. 1D projections of the fitness functions in our synthetic problems, V^1, V^2 and V^3.

if SLS starts at $\|b\|_1 \geq 14$ for any ϵ. Restarts are very beneficial in $V^2(x)$. The fitness function for the third problem is $V^3(b) = 3\|b\|_1 + 2 + 2\cos(\pi\|b\|_1) + \epsilon$. The maximum of $V^3(b)$ can be found by mixing greedy and noise operations from any b and for any ϵ.

The expected reward functions for all three problems are shown in Fig. 3. Clearly, the optimal probability values p_r^* and p_n^* (in white regions) for these problems are at quite different values of the probabilities p_r and p_n.

5.2   Setting of Baseline Methods and Our Methods

For the synthetic problems, we use the following baseline methods. In UCB1 [6], the UCB of arm A at time t is computed as $U_t(A) = \hat{f}_{T_{t-1}(A)}(A) + c_{t-1,T_{t-1}(A)}$, where $T_t(A)$ is the number of times that arm A is chosen in t steps, $\hat{f}_s(A)$ is the average of s observed rewards of arm A, $c_{t,s} = \gamma\sqrt{(1.5\log t)/s}$ is the radius of the confidence interval around $\hat{f}_s(A)$ after t steps, and γ is the difference between the maximum and minimum rewards. irace [23] evaluates a set of candidate parameters by discarding bad ones when statistically sufficient evidence has been gathered. SMAC [16] iterates between constructing a regression model and obtaining additional data based on this model. GPEIMCMC [35] uses Bayesian optimization to model SLS performance as samples from a Gaussian process. GPUCB [36] formulates parameter optimization as a bandit problem, where the reward function is sampled from a Gaussian process or has low RKHS norm. The selected baselines irace, SMAC, and GPEIMCMC have competitive performance [16] compared to other approaches, such as GGA [3] and ParamILS [18].

For those methods not optimizing cumulative regret (irace, SMAC, and GPEIMCMC), they explore N_{budget} steps and then exploit. We fine-tune their performance on each problem using different N_{budget}. This gives them a significant competitive advantage over our SLSB approach, which is not fine-tuned on each problem. For irace and SMAC, we try budget $N_{\text{budget}} \in \{200, 500, 1000, 2000, 5000, 10000, 20000\}$. For GPEIMCMC, we try $N_{\text{budget}} \in \{200, 500\}$.[3] For GPUCB, we set $N_{\text{budget}} = 2000$.[3] Finally, we report these methods' results at

---

[3] We did not try other N_{budget} for GPEIMCMC and GPUCB, because their computation time is prohibitive.

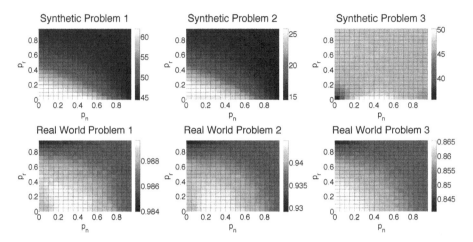

Fig. 3. The expected reward $\bar{f}$ of the three synthetic problems in Fig. 2 (top row) and the three real world problems (bottom row): (1) Breast Cancer, (2) Wine, and (3) Australian. The six problems are diverse, with different optimal p_r and p_n (lighter, meaning higher reward, is better). These are used as ground truth for evaluation purposes only, and not known to the optimization algorithms.

N_{budget} with the lowest cumulative regret in 2×10^4 steps. After `irace`, `SMAC`, and `GPEIMCMC` run out of budget, the slopes of the curves show the quality of the parameters they find (smaller slope is better here).

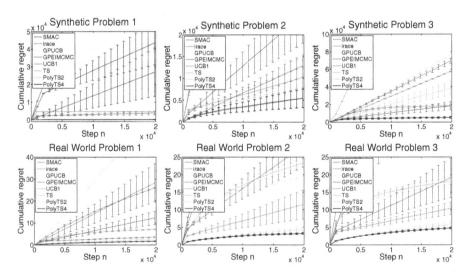

Fig. 4. The cumulative regret of different approaches in up to $n = 2 \times 10^4$ steps on the three synthetic problems V^1, V^2 and V^3 in Fig. 2 and the three real world problems: (1) Breast Cancer, (2) Wine, and (3) Australian.

For UCB1 and TS, we discretize the space of arms Θ on a 20×20 uniform grid. Let γ be the difference between the maximum and minimum rewards, which can be estimated from random samples. For TS, we set $\sigma_0 = \gamma$, because we expect that the learned parameters in TS are on the same order as the rewards. We put $\sigma = \frac{\gamma}{10}$, which means that the noise in observations of any arm $A \in \Theta$ is much lower than the range of rewards. For the same reasons, in PolyTS, $\sigma = \frac{\gamma}{10}$ and $\lambda^{-1} = \gamma^2$. This setting of γ is also used in UCB1 to adjust for the range of rewards. We show results of PolyTS2 and PolyTS4, which are variants of PolyTS for learning degree-2 (quadratic) and degree-4 polynomials respectively. We run all bandit methods for $n = 2 \times 10^4$ steps and compute their regret as in (3). We find A^* in (2) by discretizing the space of arms Θ on a 20×20 uniform grid. Results are averaged over 10 runs. We also show error bars indicating standard error, a measure of confidence in the estimate of the mean. For SLS in all experiments, we initialize b uniformly at random, and set $\kappa = 200$.

5.3   Results

We compare the regret of all methods in Fig. 4. We observe several major trends. First, in most cases our PolyTS2 and PolyTS4 learn as t increases in Fig. 4. This is apparent from the concave shape of their regret plots. Second, we observe that the regrets of PolyTS2 and PolyTS4 are consistently and substantially lower than those of UCB1 and TS. Third, PolyTS4 performs better than irace and SMAC, while PolyTS2 only outperforms irace and SMAC in synthetic problems V^1 and V^2. Fourth, PolyTS generally performs better than GPEIMCMC and GPUCB. GPEIMCMC performs better than GPUCB, which is consistent with previous results [35]. Finally, the error bars of PolyTS4 are small, showing that PolyTS4 is stable.

The regret of PolyTS is lower than the regrets of UCB1 and TS, indicating that the polynomial approximation of the expected reward is reasonable. Compared to PolyTS4, PolyTS2's higher regret in problem V^3 could be caused by the lightest and abruptly changing regions as showed in Fig. 3. It may be difficult to approximate well those regions by quadratic polynomials. The reason why GPEIMCMC is performing worse here may be because we only try $N_{\text{budget}} \in \{200, 500\}$, due to high computational cost.

The parameter tuning times of irace, SMAC, GPEIMCMC, GPUCB and PolyTS4 on problem V^3 are shown in Table 1. We use V^3 as an example and the tuning time comparison results are similar on V^1, V^2 and V^3. We have several observations. First, PolyTS4 has the shortest running time. Second, the running times of irace, SMAC, and PolyTS4 grow approximately linearly in N_{budget}. Third, the running times of methods based on Gaussian process grow in a super-linear fashion. When N_{budget} is large, these Gaussian process methods may not be suitable. These results suggest that PolyTS4 has very low computational cost, compared to other parameter optimization methods.

Table 1. The parameter tuning time (in seconds) of different methods with different N_{budget}. We use a Linux server with two Intel Xeon E5530 2.40 GHz CPUs and 24 GB memory. The running time of SLS is excluded and we only compare parameter tuning time. Results are averaged over 10 runs. Marker '-' means the experiment did not finish within 40,000 s (about 11 h).

	N_{budget}						
	200	500	1000	2000	5000	10000	20000
irace	2.07	3.32	5.00	8.14	17.95	35.62	74.86
SMAC	59.72	98.90	273.98	504.04	1,443.60	3,249.51	6,583.36
GPEIMCMC	1,975.73	10,309.59	39,716.12	-	-	-	-
GPUCB	14.87	84.19	577.38	4,284.76	-	-	-
PolyTS4	**0.06**	**0.13**	**0.21**	**0.39**	**0.90**	**1.73**	**3.51**

6   Real World Experiments

6.1   Problems: Feature Selection and Its Applications in HAR

In this section, we evaluate our methods on five real world feature selection problems: Breast Cancer (9 features, 700 samples), Wine (13 features, 178 samples), Australian (14 features, 690 samples), HAR data 1 (561 features, 3, 433 samples) [2] and HAR data 2 (416 features, 12, 590 samples) [34].[4]

For HAR, limiting the number of features reduces power consumption of mobile devices [38]. Thus, here we select $N_{feature} \leq 20$ features from engineered features.[5] We initialize SLS with $b = 0 \ldots 0$ and set $\kappa = 20$.[6] We simulate two scenarios when generating a sequence of feature selection problems. In each step, the training, validation and test data are sampled from (i) 10 of 30 users on HAR data 1 and (ii) data collected from one body position (left pocket, right pocket, wrist, upper arm, and belt) on HAR data 2.

6.2   Setting of Baseline Methods and Our Methods

For the small scale datasets, the settings of baselines and our methods are the same as in Sect. 5. For the large scale HAR experiments, we use the following baseline methods. The irace method is used as baseline, as it outperforms SMAC in the small scale problems. The GPEIMCMC method is also evaluated, as it outperforms GPUCB in the small scale problems. Besides, GPEIMCMC is more

---

[4] Breast Cancer, Wine and Australian can be found at https://archive.ics.uci.edu/ml/datasets.html.

[5] Feature descriptions for HAR data 1 are in [2]. The features for HAR data 2 include mean, variance and FFT values of sensors signals for each 1 s sliding window.

[6] If $b = 0 \ldots 0$ and $\kappa = 20$, we have $N_{feature} \leq 20$. The reason is that after $\kappa = 20$ SLS operations, the bit-string found by SLS can differ from the initial b in at most 20 bit.

tunable with regard to different N_{budget}. The budget N_{budget} is smaller here than in the small scale problems. Two traditional feature selection methods are also compared to PolyTS4: a sparse learning based method RFS [28] and a similarity based method reliefF [31]. For the traditional methods, we try budget $N_{budget} \in \{150, 200, 250, 300\}$ and report the result for N_{budget} with the highest average cumulate reward over 300 steps. We control RFS and reliefF such that they select the same number of features as PolyTS4. We use decision tree as the classifier. We evaluate different methods based on test accuracy, since this is a better measure of generalization than the validation accuracy. In practice, A^* in (2) for larger dataset is computationally expensive to obtain, so it is difficult to calculate regret as evaluation criterion. On the HAR datasets, we thus report the average test accuracy up to step n. All results are averaged over 10 runs.

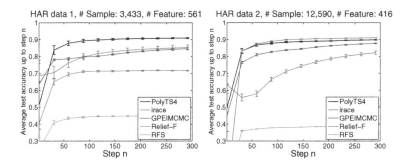

Fig. 5. The average test accuracy up to step n on HAR problems.

6.3   Results

Figure 4 shows the regret on the small scale datasets. We have similar observations as for the synthetic problems. The regrets of PolyTS2 and PolyTS4 are better than the regrets of irace, SMAC, GPEIMCMC, and other bandit methods.

The results on the HAR datasets are in Fig. 5. In the very early steps, PolyTS4 has slightly lower average cumulative accuracy than irace and GPEIMCMC. As learning goes on, after about 10 steps, PolyTS4 is better than irace and GPEIMCMC. PolyTS is not better initially because it explores the parameter space. Later on, PolyTS exploits more and is better. Besides, PolyTS4 outperforms RFS and reliefF in most cases, except that it has slightly lower accuracy (1.2%) than that of reliefF on HAR data 2. The results suggest that PolyTS4 can find useful feature subsets that have high cumulative accuracy on average. Further, PolyTS4 also provides reasonably good accuracy in initial steps, which is very useful in real world HAR scenarios.

7 Conclusion

This paper is motivated by the strong performance of SLS in a multitude of applications [15], along with the need to carefully optimize SLS search parameters in order to achieve such performance. To optimize SLS parameters to achieve the highest cumulative reward over a sequence of stochastic optimization problems, we propose SLSB. We study different bandit methods to implement SLSB, specifically: UCB1, TS, and PolyTS. Both UCB1 and TS work, but their regret bound is suboptimal. Motivated by Markov chain analyses of SLS and the observation that our reward function is smooth in the coordinates of arms, we approximate the reward by polynomial functions, and present PolyTS with an improved regret bound. Compared to other parameter optimization methods, we show superior performance of PolyTS on synthetic and real-world problems. In addition, we show that the model update of PolyTS is very simple, such that PolyTS requires lower computational cost than other parameter optimization methods.

Acknowledgment. This material is based, in part, upon work supported by the National Science Foundation under Grant No. 1344768.

References

1. Agrawal, S., Goyal, N.: Thompson sampling for contextual bandits with linear payoffs. In: ICML-2013, pp. 127–135 (2013)
2. Anguita, D., Ghio, A., Oneto, L., Parra Perez, X., Reyes Ortiz, J.L.: A public domain dataset for human activity recognition using smartphones. In: ESANN-2013, pp. 437–442 (2013)
3. Ansótegui, C., Sellmann, M., Tierney, K.: A gender-based genetic algorithm for the automatic configuration of algorithms. In: Gent, I.P. (ed.) CP 2009. LNCS, vol. 5732, pp. 142–157. Springer, Heidelberg (2009). https://doi.org/10.1007/978-3-642-04244-7_14
4. Audemard, G., Simon, L.: Refining restarts strategies for SAT and UNSAT. In: Milano, M. (ed.) CP 2012. LNCS, pp. 118–126. Springer, Heidelberg (2012). https://doi.org/10.1007/978-3-642-33558-7_11
5. Audibert, J.Y., Bubeck, S., Munos, R.: Best arm identification in multi-armed bandits. In: COLT 2010, pp. 41–53 (2010)
6. Auer, P., Cesa-Bianchi, N., Fischer, P.: Finite-time analysis of the multiarmed bandit problem. Mach. Learn. **47**, 235–256 (2002)
7. Battiti, R., Campigotto, P.: An investigation of reinforcement learning for reactive search optimization. In: Hamadi, Y., Monfroy, E., Saubion, F. (eds.) Autonomous Search, pp. 131–160. Springer, Heidelberg (2011). https://doi.org/10.1007/978-3-642-21434-9_6
8. Birattari, M., Stützle, T., Paquete, L., Varrentrapp, K.: A racing algorithm for configuring metaheuristics. In: GECCO-2002, pp. 11–18 (2002)
9. Cappé, O., Garivier, A., Maillard, O.A., Munos, R., Stoltz, G.: Kullback-Leibler upper confidence bounds for optimal sequential allocation. Ann. Stat. **41**(3), 1516–1541 (2013)
10. Chapelle, O., Li, L.: An empirical evaluation of Thompson sampling. In: NIPS, vol. 24, pp. 2249–2257 (2011)

11. DaCosta, L., Fialho, A., Schoenauer, M., Sebag, M.: Adaptive operator selection with dynamic multi-armed bandits. In: GECCO-2008, pp. 913–920 (2008)

12. Fialho, Á., Da Costa, L., Schoenauer, M., Sebag, M.: Analyzing bandit-based adaptive operator selection mechanisms. Ann. Math. Artif. Intell. **60**(1–2), 25–64 (2010)

13. Hoffman, M., Shahriari, B., Freitas, N.: On correlation and budget constraints in model-based bandit optimization with application to automatic machine learning. In: Artificial Intelligence and Statistics, pp. 365–374 (2014)

14. Hoos, H.H.: An adaptive noise mechanism for WalkSAT. In: AAAI-2002, pp. 655–660 (2002)

15. Hoos, H.H., Stützle, T.: Stochastic Local Search: Foundations and Applications. Morgan Kaufmann, San Francisco (2005)

16. Hutter, F., Hoos, H.H., Leyton-Brown, K.: Sequential model-based optimization for general algorithm configuration. In: Coello, C.A.C. (ed.) LION 2011. LNCS, vol. 6683, pp. 507–523. Springer, Heidelberg (2011). https://doi.org/10.1007/978-3-642-25566-3_40

17. Hutter, F., Hoos, H.H., Leyton-Brown, K., Murphy, K.: Time-bounded sequential parameter optimization. In: Blum, C., Battiti, R. (eds.) LION 2010. LNCS, vol. 6073, pp. 281–298. Springer, Heidelberg (2010). https://doi.org/10.1007/978-3-642-13800-3_30

18. Hutter, F., Hoos, H.H., Leyton-Brown, K., Stützle, T.: ParamILS: an automatic algorithm configuration framework. JAIR **36**, 267–306 (2009)

19. Kleinberg, R.D.: Nearly tight bounds for the continuum-armed bandit problem. NIPS, vol. 17, pp. 697–704 (2004)

20. Lai, T.L., Robbins, H.: Asymptotically efficient adaptive allocation rules. Adv. Appl. Math. **6**(1), 4–22 (1985)

21. Li, K., Fialho, A., Kwong, S., Zhang, Q.: Adaptive operator selection with bandits for a multiobjective evolutionary algorithm based on decomposition. IEEE Trans. Evol. Comput. **18**(1), 114–130 (2014)

22. Lockhart, J.W., Weiss, G.M.: The benefits of personalized smartphone-based activity recognition models. In: SDM-2014, pp. 614–622 (2014)

23. López-Ibáñez, M., Dubois-Lacoste, J., Stützle, T., Birattari, M.: The irace package: iterated racing for automatic algorithm configuration (2011)

24. Maillard, O.-A., Munos, R.: Online learning in adversarial Lipschitz environments. In: Balcázar, J.L., Bonchi, F., Gionis, A., Sebag, M. (eds.) ECML PKDD 2010. LNCS (LNAI), vol. 6322, pp. 305–320. Springer, Heidelberg (2010). https://doi.org/10.1007/978-3-642-15883-4_20

25. Mengshoel, O.J., Ahres, Y., Yu, T.: Markov chain analysis of noise and restart in stochastic local search. In: IJCAI, pp. 639–646 (2016)

26. Mengshoel, O.J.: Understanding the role of noise in stochastic local search: analysis and experiments. Artif. Intell. **172**(8), 955–990 (2008)

27. Mengshoel, O.J., Wilkins, D.C., Roth, D.: Initialization and restart in stochastic local search: computing a most probable explanation in Bayesian networks. IEEE Trans. Knowl. Data Eng. **23**(2), 235–247 (2011)

28. Nie, F., Huang, H., Cai, X., Ding, C.H.: Efficient and robust feature selection via joint l2, 1-norms minimization. In: NIPS, vol. 23, pp. 1813–1821 (2010)

29. Pal, D.K., Mengshoel, O.J.: Stochastic CoSaMP: randomizing greedy pursuit for sparse signal recovery. In: Frasconi, P., Landwehr, N., Manco, G., Vreeken, J. (eds.) ECML PKDD 2016. LNCS (LNAI), vol. 9851, pp. 761–776. Springer, Cham (2016). https://doi.org/10.1007/978-3-319-46128-1_48

30. Prestwich, S.: Tuning local search by average-reward reinforcement learning. In: Maniezzo, V., Battiti, R., Watson, J.-P. (eds.) LION 2007. LNCS, vol. 5313, pp. 192–205. Springer, Heidelberg (2008). https://doi.org/10.1007/978-3-540-92695-5_15

31. Robnik-Šikonja, M., Kononenko, I.: Theoretical and empirical analysis of ReliefF and RReliefF. Mach. Learn. **53**(1–2), 23–69 (2003)

32. Ryvchin, V., Strichman, O.: Local restarts in SAT. Constraint Program. Lett. (CPL) **4**, 3–13 (2008)

33. Selman, B., Levesque, H., Mitchell, D.: A new method for solving hard satisfiability problems. In: AAAI, pp. 440–446 (1992)

34. Shoaib, M., Scholten, H., Havinga, P.J.: Towards physical activity recognition using smartphone sensors. In: UIC/ATC-2013, pp. 80–87. IEEE (2013)

35. Snoek, J., Larochelle, H., Adams, R.P.: Practical Bayesian optimization of machine learning algorithms. In: NIPS, vol. 25, pp. 2951–2959 (2012)

36. Srinivas, N., Krause, A., Seeger, M., Kakade, S.M.: Gaussian process optimization in the bandit setting: no regret and experimental design. In: ICML-2010, pp. 1015–1022 (2010)

37. Thompson, W.R.: On the likelihood that one unknown probability exceeds another in view of the evidence of two samples. Biometrika **25**(3–4), 285–294 (1933)

38. Yu, M.C., Yu, T., Wang, S.C., Lin, C.J., Chang, E.Y.: Big data small footprint: the design of a low-power classifier for detecting transportation modes. Proc. VLDB Endow. **7**(13), 1429–1440 (2014)

39. Yu, T., Zhuang, Y., Mengshoel, O.J., Yagan, O.: Hybridizing personal and impersonal machine learning models for activity recognition on mobile devices. In: MobiCASE-2016, pp. 117–126 (2016)

Matrix and Tensor Factorization

Comparative Study of Inference Methods for Bayesian Nonnegative Matrix Factorisation

Thomas Brouwer[1(✉)], Jes Frellsen[2], and Pietro Lió[1]

[1] Computer Laboratory, University of Cambridge, Cambridge, UK
tab43@cam.ac.uk
[2] Department of Computer Science, IT University of Copenhagen,
Copenhagen, Denmark

Abstract. In this paper, we study the trade-offs of different inference approaches for Bayesian matrix factorisation methods, which are commonly used for predicting missing values, and for finding patterns in the data. In particular, we consider Bayesian nonnegative variants of matrix factorisation and tri-factorisation, and compare non-probabilistic inference, Gibbs sampling, variational Bayesian inference, and a maximum-a-posteriori approach. The variational approach is new for the Bayesian nonnegative models. We compare their convergence, and robustness to noise and sparsity of the data, on both synthetic and real-world datasets. Furthermore, we extend the models with the Bayesian automatic relevance determination prior, allowing the models to perform automatic model selection, and demonstrate its efficiency. Code and data related to this chapter are availabe at: https://github.com/ThomasBrouwer/BNMTF_ARD.

1 Introduction

Matrix factorisation methods have been used extensively in recent years to decompose matrices into latent factors, helping us reveal hidden structure and predict missing values. In particular we decompose a given matrix into two smaller matrices so that their product approximates the original one (see Fig. 1). Nonnegative matrix factorisation models [9] have been particularly popular, as the nonnegativity constraint makes the resulting matrices easier to interpret, and is often inherent to the problem—such as in image processing or bioinformatics [9,20]. A related problem is that of matrix tri-factorisation, first introduced by Ding et al. [6], where the observed dataset is decomposed into three smaller matrices, which again are constrained to be non-negative.

Both matrix factorisation and tri-factorisation methods have found many applications in recent years, such as for collaborative filtering [5,13], sentiment classification [11], predicting drug-target interaction [8] and gene functions [12], and image analysis [23]. Methods can be categorised as either non-probabilistic

Electronic supplementary material The online version of this chapter (https://doi.org/10.1007/978-3-319-71249-9_31) contains supplementary material, which is available to authorized users.

© Springer International Publishing AG 2017
M. Ceci et al. (Eds.): ECML PKDD 2017, Part I, LNAI 10534, pp. 513–529, 2017.
https://doi.org/10.1007/978-3-319-71249-9_31

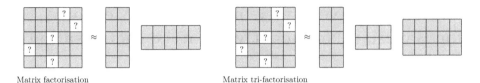

Matrix factorisation   Matrix tri-factorisation

Fig. 1. Overview of matrix factorisation and matrix tri-factorisation methods, with missing values (?-entries).

or Bayesian. For the former, finding the factorisation (*inference*) is commonly done using multiplicative updates, whereas for the latter we use approximate Bayesian inference methods. Non-probabilistic or maximum a posteriori (MAP) solutions give a single point estimate, which can lead to overfitting more easily and neglects uncertainty. Bayesian approaches address this issue, by instead finding a full distribution over the matrices, where we define prior distributions over the matrices and then compute their posterior after observing the actual data. This can greatly reduce overfitting. A key question that arises is: what exactly are the trade-offs between different matrix factorisation inference approaches? In particular, which perform better in terms of speed of convergence, predictive performance, and robustness to noise and sparsity?

In this paper we answer these questions by performing a thorough empirical study to explore these trade-offs between non-probabilistic and Bayesian inference approaches, which to our knowledge had not been done before. We consider the popular non-probabilistic matrix factorisation model from Lee and Seung [10], and a Bayesian nonnegative matrix factorisation and tri-factorisation model from Schmidt et al. [15] and Brouwer and Lió [4], respectively. These models use exponential priors to enforce nonnegativity, giving Gibbs sampling algorithms for inference. The former paper also introduced a MAP algorithm, called iterated conditional modes (ICM). Both of these approaches rely on a sampling procedure to eventually converge to draws of the desired distribution—in this case the posterior of the matrices. This means that we need to inspect the values of the draws to determine when our method has converged (burn-in), and then take additional draws to estimate the posteriors.

We introduce a fourth inference technique for the Bayesian nonnegative models, based on variational Bayesian inference (VB), where instead of relying on random draws we obtain deterministic convergence to a solution. We do this by introducing a new distribution that is easier to compute, and optimise it to be as similar to the true posterior as possible. Some papers (for instance [14]) assert that variational inference gives faster but less accurate inference than sampling methods like Gibbs. One study investigating this for latent dirichlet allocation can be found in [1], but ours is the first paper giving a thorough empirical study of the trade-offs for matrix factorisation. We furthermore extend the Bayesian models with automatic relevance determination (ARD), to eliminate the need for model selection.

We perform extensive experiments on both artificial and real-world data to explore the trade-offs between speed of inference, and robustness to sparsity and noise for predicting missing values. We show that Gibbs sampling is the most robust, while VB and ICM give significant run-time speedups but sacrifice some robustness, and that non-probabilistic inference tends to be fast but not robust. Finally, we show that ARD is an effective way of performing automatic model selection, and increases the robustness of matrix factorisation models if they are given the wrong dimensionality.

Although we study a specific Bayesian nonnegative matrix factorisation and tri-factorisation model, we believe that many of our findings and insights apply to the broad range of other matrix factorisation and tri-factorisation methods, as well as tensor and Tucker decomposition methods—their three-dimensional extensions.

2   Models

2.1   Nonnegative Matrix Factorisation

We follow the notation used by Schmidt et al. [15] for nonnegative matrix factorisation (NMF), which can be formulated as decomposing a matrix $R \in \mathbb{R}^{I \times J}$ into two latent (unobserved) matrices $U \in \mathbb{R}_+^{I \times K}$ and $V \in \mathbb{R}_+^{J \times K}$, whose values are constrained to be positive. In other words, solving $R = UV^T + E$, where noise is captured by matrix $E \in \mathbb{R}^{I \times J}$. The dataset R need not be complete—the indices of observed entries can be represented by the set $\Omega = \{(i,j)|R_{ij} \text{is observed}\}$. These entries can then be predicted by UV^T.

We take a probabilistic approach to this problem. We express a likelihood function for the observed data, and treat the latent matrices as random variables. As the likelihood we assume each value of R comes from the product of U and V, with some Gaussian noise added,

$$R_{ij} \sim \mathcal{N}(R_{ij}|U_i \cdot V_j, \tau^{-1})$$

where U_i, V_j denote the ith and jth rows of U and V, and $\mathcal{N}(x|\mu,\tau) = \tau^{\frac{1}{2}}(2\pi)^{-\frac{1}{2}} \exp\left\{-\frac{\tau}{2}(x-\mu)^2\right\}$ is the density of the Gaussian distribution with precision τ. The set of parameters for our model is denoted $\theta = \{U, V, \tau\}$. In the Bayesian approach to inference, we want to find the distributions over parameters θ after observing the data $D = \{R_{ij}\}_{i,j\in\Omega}$. We can use Bayes' theorem,

$$p(\theta|D) \propto p(D|\theta)p(\theta).$$

We need priors over the parameters, allowing us to express beliefs for their values—such as constraining U, V to be nonnegative. We can normally not compute the posterior $p(\theta|D)$ exactly, but some choices of priors allow us to obtain a good approximation. Schmidt et al. choose an exponential prior over

U and V, so that each element in U and V is assumed to be independently exponentially distributed with rate parameters $\lambda_{ik}^U, \lambda_{jk}^V > 0$.

$$U_{ik} \sim \mathcal{E}(U_{ik}|\lambda_{ik}^U) \qquad V_{jk} \sim \mathcal{E}(V_{jk}|\lambda_{jk}^V)$$

where $\mathcal{E}(x|\lambda) = \lambda \exp\{-\lambda x\} u(x)$ is the density of the exponential distribution, and $u(x)$ is the unit step function. For the precision τ we use a Gamma distribution with shape $\alpha_\tau > 0$ and rate $\beta_\tau > 0$,

$$p(\tau) \sim \mathcal{G}(\tau|\alpha_\tau, \beta_\tau) = \frac{\beta_\tau{}^{\alpha_\tau}}{\Gamma(\alpha_\tau)} x^{\alpha_\tau - 1} e^{-\beta_\tau x}$$

where $\Gamma(x) = \int_0^\infty x^{t-1} e^{-x} dt$ is the gamma function.

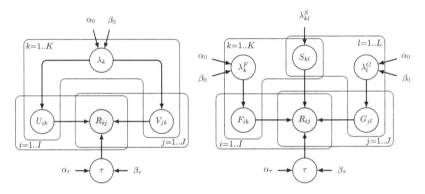

Fig. 2. Graphical model representation of Bayesian nonnegative matrix factorisation (left) and tri-factorisation (right), with ARD.

2.2   Nonnegative Matrix Tri-Factorisation

The problem of nonnegative matrix tri-factorisation (NMTF) can be formulated similarly to that of nonnegative matrix factorisation, and was introduced by Brouwer and Lió [4]. We now decompose R into three matrices $F \in \mathbb{R}_+^{I \times K}$, $S \in \mathbb{R}_+^{K \times L}$, $G \in \mathbb{R}_+^{J \times L}$, so that $R = FSG^T + E$. This decomposition has the advantage of extracting row and column factor values separately (through F and G), allowing us to identify both row and column clusters. We again use a Gaussian likelihood and Exponential priors for the latent matrices.

$$R_{ij} \sim \mathcal{N}(R_{ij}|F_i \cdot S \cdot G_j, \tau^{-1}) \qquad\qquad \tau \sim \mathcal{G}(\tau|\alpha_\tau, \beta_\tau)$$
$$F_{ik} \sim \mathcal{E}(F_{ik}|\lambda_{ik}^F) \quad S_{kl} \sim \mathcal{E}(S_{kl}|\lambda_{kl}^S) \qquad G_{jl} \sim \mathcal{E}(G_{jl}|\lambda_{jl}^G)$$

2.3   Automatic Relevance Determination

Automatic relevance determination (ARD) is a Bayesian prior which helps perform automatic model selection. It works by replacing the individual λ parameters in the factor matrix priors by one that is shared by all entries in the same column (in other words, shared for each factor). We then place a further Gamma prior over all these λ_k parameters. For the NMF model, the priors become

$$U_{ik} \sim \mathcal{E}(U_{ik}|\lambda_k) \qquad V_{jk} \sim \mathcal{E}(V_{jk}|\lambda_k) \qquad \lambda_k \sim \mathcal{G}(\lambda_k|\alpha_0, \beta_0).$$

Since this parameter is shared by all entries in the same column, the entire factor k is either activated (if λ_k^t has a low value) or "turned off" (if λ_k^t has a high value), pushing factors that are active for only a few entities further to zero. This prior has been used for both real-valued [18,19] and nonnegative matrix factorisation [17]. Instead of having to choose the correct K, we give an upper bound and the model will automatically determine the number of factors to use. A similar approach can be found in [7], which incorporates the elimination of unused factors into their expectation-maximisation inference algorithm. ARD is implemented on a model level, and therefore works with all inference approaches.

For NMTF we use two ARD's, one for $\boldsymbol{F}(\lambda_k^F)$ and another for $\boldsymbol{G}(\lambda_l^G)$,

$$F_{ik} \sim \mathcal{E}(F_{ik}|\lambda_k^F)\lambda_k^F \sim \mathcal{G}(\lambda_k^F|\alpha_0, \beta_0) \qquad G_{jl} \sim \mathcal{E}(G_{jl}|\lambda_l^G) \quad \lambda_l^G \sim \mathcal{G}(\lambda_l^G|\alpha_0, \beta_0).$$

The graphical models for Bayesian NMF and NMTF are given in Fig. 2.

3   Inference

In this section we give details for four different types of inference for nonnegative matrix factorisation (NMF) and tri-factorisation (NMTF) models. Non-probabilistic inference gives a point estimate solution. Gibbs sampling and variational Bayesian inference both give a full posterior estimate, whereas iterated conditional modes gives a maximum a posteriori (MAP) point estimate.

3.1   Non-probabilistic Inference

A non-probabilistic (NP) approach for NMF can be found in Lee and Seung [10]. Their algorithm relies on multiplicative updates, where at each iteration the values in the $\boldsymbol{U}$ and $\boldsymbol{V}$ matrices are updated using the following values:

$$U_{ik} = U_{ik}\frac{\sum_{j\in\Omega_i} R_{ij}V_{jk}/(\boldsymbol{U}_i\boldsymbol{V}_j)}{\sum_{j\in\Omega_i} V_{jk}} \qquad V_{jk} = V_{jk}\frac{\sum_{i\in\Omega_j} R_{ij}U_{ik}/(\boldsymbol{U}_i\boldsymbol{V}_j)}{\sum_{i\in\Omega_j} U_{ik}}$$

where $\Omega_i = \{j|(i,j) \in \Omega\}$ and $\Omega_j = \{i|(i,j) \in \Omega\}$. These updates can be shown to minimise the I-divergence (generalised KL-divergence),

$$D(\boldsymbol{R}||\boldsymbol{U}\boldsymbol{V}^T) = \sum_{(i,j)\in\Omega} \left(R_{ij} \log \frac{R_{ij}}{(\boldsymbol{U}\boldsymbol{V}^T)_{ij}} - R_{ij} + (\boldsymbol{U}\boldsymbol{V}^T)_{ij} \right).$$

Yoo and Choi [22] extended this approach to NMTF, giving the following multiplicative updates, with $\boldsymbol{S}_{\cdot l}$ denoting the lth column of $\boldsymbol{S}$:

$$F_{ik} = F_{ik} \frac{\sum_{j \in \Omega_i} R_{ij}(\boldsymbol{S}_k \boldsymbol{G}_j)/(\boldsymbol{F}_i \boldsymbol{S} \boldsymbol{G}_j)}{\sum_{j \in \Omega_i}(\boldsymbol{S}_k \boldsymbol{G}_j)} \qquad G_{jl} = G_{jl} \frac{\sum_{i \in \Omega_j} R_{ij}(\boldsymbol{F}_i \boldsymbol{S}_{\cdot l})/(\boldsymbol{F}_i \boldsymbol{S} \boldsymbol{G}_j)}{\sum_{i \in \Omega_j}(\boldsymbol{F}_i \boldsymbol{S}_{\cdot l})}$$

$$S_{kl} = S_{kl} \frac{\sum_{(i,j) \in \Omega} R_{ij} F_{ik} G_{jl}/(\boldsymbol{F}_i \boldsymbol{S} \boldsymbol{G}_j)}{\sum_{(i,j) \in \Omega} F_{ik} G_{jl}}.$$

3.2   Gibbs Sampling

Schmidt et al. [15] introduced a Gibbs sampling algorithm for approximating the posterior distribution—a similar NMF model that uses Gibbs sampling can be found in [24,25]. Gibbs sampling works by sampling new values for each parameter θ_i from its marginal distribution given the current values of the other parameters $\boldsymbol{\theta}_{-i}$, and the observed data D. If we sample new values in turn for each parameter θ_i from $p(\theta_i|\boldsymbol{\theta}_{-i}, D)$, we will eventually converge to draws from the posterior, which can be used to approximate the posterior $p(\boldsymbol{\theta}|D)$. We have to discard the first n draws because it takes a while to converge (*burn-in*), and since consecutive draws are correlated we only use every ith value (*thinning*).

For NMF this means that we need to be able to draw from distributions

$$p(\tau|\boldsymbol{U}, \boldsymbol{V}, \boldsymbol{\lambda}, D) \qquad\qquad p(U_{ik}|\tau, \boldsymbol{U}_{-ik}, \boldsymbol{V}, \boldsymbol{\lambda}, D)$$
$$p(\lambda_k|\tau, \boldsymbol{U}, \boldsymbol{V}, D) \qquad\qquad p(V_{jk}|\tau, \boldsymbol{U}, \boldsymbol{V}_{-jk}, \boldsymbol{\lambda}, D).$$

where $\boldsymbol{U}_{-ik}$ denotes all elements in $\boldsymbol{U}$ except U_{ik}, and similarly for $\boldsymbol{V}_{-jk}$. $\boldsymbol{\lambda}$ is a vector including all λ_k values. Using Bayes theorem we obtain the following posterior distributions:

$$p(\tau|\boldsymbol{U}, \boldsymbol{V}, \boldsymbol{\lambda}, D) = \mathcal{G}(\tau|\alpha_\tau^*, \beta_\tau^*) \qquad p(U_{ik}|\tau, \boldsymbol{U}_{-ik}, \boldsymbol{V}, \boldsymbol{\lambda}, D) = \mathcal{TN}(U_{ik}|\mu_{ik}^U, \tau_{ik}^U)$$
$$p(\lambda_k|\tau, \boldsymbol{U}, \boldsymbol{V}, D) = \mathcal{G}(\lambda_k|\alpha_k^*, \beta_k^*) \qquad p(V_{jk}|\tau, \boldsymbol{U}, \boldsymbol{V}_{-jk}, \boldsymbol{\lambda}, D) = \mathcal{TN}(V_{jk}|\mu_{jk}^V, \tau_{jk}^V)$$

where

$$\mathcal{TN}(x|\mu, \tau) = \begin{cases} \dfrac{\sqrt{\frac{\tau}{2\pi}} \exp\left\{-\frac{\tau}{2}(x - \mu)^2\right\}}{1 - \Phi(-\mu\sqrt{\tau})} & \text{if } x \geq 0 \\ 0 & \text{if } x < 0 \end{cases}$$

is a truncated normal: a normal distribution with zero density below $x = 0$ and renormalised to integrate to one. $\Phi(\cdot)$ is the cumulative distribution function of $\mathcal{N}(0, 1)$.

For NMTF we can derive a Gibbs sampling algorithm similarly, as done by Brouwer and Lió [4]. The posteriors, together with the parameter values for both Gibbs samplers, are given in the supplementary materials.

3.3   Iterated Conditional Modes

The iterated conditional models (ICM) algorithm for inference in the NMF model was given in Schmidt et al. [15]. It works very similarly to the Gibbs sampler, but instead of randomly drawing a value from the conditional posteriors, we take the mode at each iteration. This gives a maximum a posteriori (MAP) point estimate $\theta_{\text{MAP}} = \max_\theta p(\theta|D)$, rather than a full posterior distribution. We furthermore still need to use thinning and burn-in. For random variables $X \sim \mathcal{G}(a, b)$, $Y \sim \mathcal{TN}(\mu, \tau)$, the modes are $\frac{a-1}{b}$ and $\max(0, \mu)$, respectively.

In practice ICM often converges to solutions where multiple columns in the matrices are all zeros, leading to poor approximations. We have addressed this issue by resetting zeros to a small positive value like 0.1 at each iteration.

3.4   Variational Bayesian Inference

Variational Bayesian inference (VB) has been used for other matrix factorisation models before [8], but not for the nonnegative model in this paper. We therefore now introduce a new VB algorithm for our model. Like Gibbs sampling, VB is a way to approximate the true posterior $p(\theta|D)$. The idea behind VB is to introduce an approximation $q(\theta)$ to the true posterior that is easier to compute, and to make our variational distribution $q(\theta)$ as similar to $p(\theta|D)$ as possible (as measured by the KL-divergence). We assume the variational distribution $q(\theta)$ factorises completely, so all variables are independent in the posterior,

$$q(\theta) = \prod_{\theta_i \in \theta} q(\theta_i).$$

This is called the mean-field assumption. We use the same forms of $q(\theta_i)$ as we used in Gibbs sampling,

$$q(\tau) = \mathcal{G}(\tau|\alpha_\tau^*, \beta_\tau^*) \qquad\qquad q(\lambda_k) = \mathcal{G}(\lambda_k|\alpha_k^*, \beta_k^*)$$
$$q(U_{ik}) = \mathcal{TN}(U_{ik}|\mu_{ik}^U, \tau_{ik}^U) \qquad\qquad q(V_{jk}) = \mathcal{TN}(V_{jk}|\mu_{jk}^V, \tau_{jk}^V).$$

It can be shown [3] that the optimal distribution for the ith parameter, $q^*(\theta_i)$, can be expressed as follows (for some constant C), allowing us to find the optimal updates for the variational parameters.

$$\log q^*(\theta_i) = \mathbb{E}_{q(\theta_{-i})}\left[\log p(\theta, D)\right] + C.$$

We now take the expectation with respect to the distribution $q(\theta_{-i})$ over the parameters but excluding the ith one. This gives rise to an iterative algorithm: for each parameter θ_i we update its distribution to that of its optimal variational distribution, and then update the expectation and variance with respect to q. We therefore need updates for the variational parameters, and to be able to compute the expectations and variances of the random variables. This algorithm is guaranteed to maximise the Evidence Lower Bound (ELBO)

$$\mathcal{L} = \mathbb{E}_q\left[\log p(\theta, D) - \log q(\theta)\right],$$

which is equivalent to minimising the KL-divergence.

We use $\widetilde{f(X)}$ as a shorthand for $\mathbb{E}_q[f(X)]$, where X is a random variable and f is a function over X. For random variables $X \sim \mathcal{G}(a,b)$ and $Y \sim \mathcal{TN}(\mu,\tau)$ the variance and expectation are

$$\widetilde{X} = \frac{a}{b} \qquad \widetilde{Y} = \mu + \frac{1}{\sqrt{\tau}}\lambda\left(-\mu\sqrt{\tau}\right) \qquad \mathrm{Var}\left[Y\right] = \frac{1}{\tau}\left[1 - \delta\left(-\mu\sqrt{\tau}\right)\right],$$

where $\psi(x) = \frac{d}{dx}\log\Gamma(x)$ is the digamma function, $\lambda(x) = \phi(x)/[1 - \Phi(x)]$, and $\delta(x) = \lambda(x)[\lambda(x) - x]$. $\phi(x) = \frac{1}{\sqrt{2\pi}}\exp\{-\frac{1}{2}x^2\}$ is the density function of $\mathcal{N}(0,1)$.

The updates for NMF are given in the supplementary materials. Our VB algorithm for NMTF follows the same steps as before, but now has an added complexity due to the term $\mathbb{E}_q\left[(R_{ij} - \boldsymbol{F}_i \cdot \boldsymbol{S} \cdot \boldsymbol{G}_j)^2\right]$. Before, all covariance terms for $k' \neq k$ were zero due to the factorisation in q, but we now obtain some additional non-zero covariance terms:

$$\mathbb{E}_q\left[(R_{ij} - \boldsymbol{F}_i \cdot \boldsymbol{S} \cdot \boldsymbol{G}_j)^2\right] = \left(R_{ij} - \sum_{k=1}^{K}\sum_{l=1}^{L}\widetilde{F_{ik}}\widetilde{S_{kl}}\widetilde{G_{jl}}\right)^2$$

$$+ \sum_{k=1}^{K}\sum_{l=1}^{L}\mathrm{Var}_q\left[F_{ik}S_{kl}G_{jl}\right] \tag{1}$$

$$+ \sum_{k=1}^{K}\sum_{l=1}^{L}\sum_{k'\neq k}\mathrm{Cov}\left[F_{ik}S_{kl}G_{jl}, F_{ik'}S_{k'l}G_{jl}\right] \tag{2}$$

$$+ \sum_{k=1}^{K}\sum_{l=1}^{L}\sum_{l'\neq l}\mathrm{Cov}\left[F_{ik}S_{kl}G_{jl}, F_{ik}S_{kl'}G_{jl'}\right]. \tag{3}$$

The above variance and covariance terms are equal to the following, respectively, leading to the variational updates given in the supplementary materials.

$$\widetilde{F_{ik}^2}\widetilde{S_{kl}^2}\widetilde{G_{jl}^2} - \widetilde{F_{ik}}^2\widetilde{S_{kl}}^2\widetilde{G_{jl}}^2, \ \mathrm{Var}_q\left[F_{ik}\right]\widetilde{S_{kl}}\widetilde{G_{jl}}\widetilde{S_{kl'}}\widetilde{G_{jl'}}, \ \widetilde{F_{ik}}\widetilde{S_{kl}}\mathrm{Var}_q\left[G_{jl}\right]\widetilde{F_{ik'}}\widetilde{S_{k'l}}.$$

3.5 Complexity

Each of the four approaches have the same time complexities, but vary in how efficiently the updates can be computed, and how quickly they converge. The time complexity per iteration for NMF is $\mathcal{O}(IJK^2)$, and $\mathcal{O}(IJ(K^2L + KL^2))$ for NMTF. However, the updates in each column of $\boldsymbol{U}, \boldsymbol{V}, \boldsymbol{F}, \boldsymbol{G}$ are independent of each other and can therefore be updated in parallel. For Gibbs and ICM this means we can draw these values in parallel, but for VB and NP we can jointly update the columns using a single matrix operation. Modern computer architectures can exploit this using vector processors, leading to a great speedup.

Furthermore, after the VB algorithm converges we have our approximation to the posterior distributions immediately, whereas with Gibbs and ICM we need to obtain further draws after convergence and use a thinning rate to obtain an accurate MAP (ICM) or posterior (Gibbs) estimate. This deterministic behaviour of VB and NP makes them easier to use. Although additional variables need to be stored to represent the posteriors, this does not result in a worse space complexity, as the Gibbs sampler needs to store draws over time.

3.6 Initialisation

Initialising the parameters of the models can vastly influence the quality of convergence. This can be done by using the hyperparameters λ_{ik}^U, λ_{jk}^V, λ_{ik}^F, λ_{kl}^S, λ_{jl}^G, α, β, α_0, β_0, α_0^F, β_0^F, α_0^G, β_0^G to set the initial values to the mean of the priors of the model, or using random draws. We found that random draws tend to give faster and better convergence than the expectation, as it provides a better initial guess of the right patterns in the matrices. For matrix tri-factorisation we can initialise F by running the K-means clustering algorithm on the rows as datapoints, and similarly G for the columns, as suggested by Ding et al. [6]. For the VB and NP algorithms we then set the μ parameters to the cluster indicators, and for Gibbs and ICM we add 0.2 for smoothing. We found that this improved the convergence as well, with S initialised using random draws.

Table 1. Overview of the four drug sensitivity datasets, giving the number of cell lines (rows), drugs (columns), and the fraction of entries that are observed.

Dataset	Cell lines	Drugs	Fraction observed
GDSC IC_{50}	707	139	0.806
CTRP EC_{50}	887	545	0.801
CCLE IC_{50}	504	24	0.965
CCLE EC_{50}	504	24	0.630

3.7 Software

Implementations of all methods, the datasets, and experiments described in the next section, are available at https://github.com/ThomasBrouwer/BNMTF_ARD.

4 Experiments

To demonstrate the trade-offs between the four inference methods presented, we conducted experiments on synthetic data and four real-world drug sensitivity datasets. We compare the convergence speed, robustness to noise, and robustness to sparsity.

4.1   Datasets

For the synthetic datasets we generated the latent matrices using unit mean exponential distributions, and adding zero mean unit variance Gaussian noise to the resulting product. For the matrix factorisation model we used $I = 100, J = 80, K = 10$, and for the matrix tri-factorisation $I = 100, J = 80, K = 5, L = 5$.

We considered four drug sensitivity datasets, each detailing the effectiveness (IC_{50} or EC_{50} values) of a range of drugs on different cell lines for cancer and tissue types, where some of the entries are missing. We consider the Genomics of Drug Sensitivity in Cancer (GDSC v5.0 [21], IC_{50}), Cancer Therapeutics Response Portal (CTRP v2 [16], EC_{50}), and Cancer Cell Line Encyclopedia (CCLE [2], IC_{50} and EC_{50}). The four datasets are summarised in Table 1, giving the number of cell lines, drugs, and the fraction of entries that are observed.

In some experiments we focused on a selection of the datasets, but results for all can be found in the supplementary materials, together with preprocessing details. For all models we used weak priors ($\lambda = 0.1, \alpha_\tau = \beta_\tau = \alpha_0 = \beta_0 = 1$).

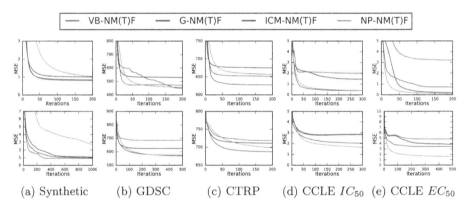

Fig. 3. Convergence of algorithms on the synthetic and drug sensitivity datasets, measuring the training data fit (mean square error) across iterations, for each of the inference approaches for NMF (top row) and NMTF (bottom row).

4.2   Convergence Speed

We firstly measured the convergence speeds of the different inference methods on the datasets, using the versions of NMF and NMTF without ARD. Convergence plots on all datasets are given in Fig. 3, plotting the mean squared error on the training data against the number of iterations, for NMF (top row) and NMTF (bottom row). For the synthetic data we used the correct number of factors, and for the drug sensitivity datasets we used $K = 20$ for NMF and $K = L = 10$ for NMTF. We ran each method 20 times, taking the average training errors.

Although the results are empirical, they show that the inference approaches have different convergence speeds and depths (final training error reached). On the synthetic data VB is the fastest, followed by ICM and Gibbs, and finally NP. All methods reach the optimal MSE of 1 (which is the level of noise added). On the real-world drug sensitivity datasets, all methods reach their lowest depth at roughly the same number of iterations. However, ICM and NP generally converge much deeper than VB and Gibbs. Although this initially seems good, this is a sign of overfitting to the training data, and can lead to poor predictions for unseen data. We will see this later in the noise and sparsity experiments (Sects. 4.4 and 4.5), where VB and Gibbs are more robust than ICM and NP.

In the supplementary materials we also give the convergence speed against time taken, which shows that the NP approach takes the least amount of time per iteration, followed by ICM, VB, and then Gibbs. In summary, ICM and NP give the fastest convergence, followed by VB, and then Gibbs.

4.3   Cross-Validation

Next we measured the cross-validation performances of the methods on the four drug sensitivity datasets. For each method we performed 10-fold nested cross-validation (nested to pick the dimensionality K—for simplicity we used $L = K$ for the NMTF models), giving the average performance in Fig. 4. For the ARD models we did not need to pick the dimensionality, instead using $K = 20$ for NMF, and $K = 10, L = 10$ for NMTF.

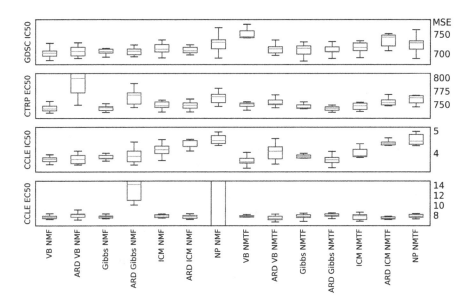

Fig. 4. 10-fold cross-validation results (mean squared error) for drug sensitivity predictions on each of the four datasets. Each boxplot gives the median (red line), standard deviation (blue box), and upper quartiles (black lines). (Color figure online)

We can see that most models perform very similarly, with little to no difference between the matrix factorisation and tri-factorisation versions. Using the ARD models often works equally well as without ARD, but with the added benefit of not having to run nested cross-validation to choose the dimensionality, reducing the running time from hours to minutes. However, sometimes ARD fails to prevent overfitting, such as for VB NMF on CTRP EC_{50}, and Gibbs NMF on CCLE EC_{50}). This is unsurprising as the ARD models are given dimensionalities that are way too high. We will see in Sect. 4.6 that the ARD is actually very efficient at turning off unnecessary factors and reducing overfitting.

We can also see that the VB and Gibbs models often do a bit better than the NP and ICM versions. This is especially obvious on the CCLE IC_{50} dataset, and also on GDSC IC_{50}. On the CCLE EC_{50} dataset the NP NMF model completely overfits on one of the folds, leading to extremely high predictive errors.

4.4   Noise Test

We conducted a noise test on the synthetic data to measure the robustness of the methods. We add different levels of Gaussian noise to the data, with the noise-to-signal ratio being given by the ratio of the standard deviation of the Gaussian noise we add, to the standard deviation of the generated data. For each noise level we split the datapoints randomly into ten folds, and measure the predictive performance of the models on one held-out set. The results are given in Figs. 5a (NMF) and 5b (NMTF), where we can see that the non-probabilistic approach starts overfitting heavily at low levels of noise, whereas the Bayesian approaches achieve the best possible predictive powers even at high levels of noise. In the supplementary materials we also show that adding ARD did not make a difference for the robustness of the Bayesian models.

4.5   Sparsity Test

We furthermore measured the robustness of each inference technique to sparsity of the data. For different fractions of missing values we randomly split the data ten times into train and test sets using those proportions, and measured the average predictive error. We conducted this experiment on the synthetic data, using the true dimensionality K (and L) for each model. We also performed it on the GDSC and CTRP datasets, using the most common dimensionalities in the cross-validation from Sect. 4.3 (given in supplementary materials).

The results are given in Figs. 6a (NMF) and 6d (NMTF) for the synthetic data, Fig. 6b and e for GDSC, and Fig. 6c and f for CTRP. We can see that the non-probabilistic models start overfitting even on very low sparsity levels (with the exception of Fig. 6d)—in Fig. 6a we cannot even see the line. The ICM models are also less robust when the sparsity is high. In contrast, the Gibbs sampling model achieves very good predictive performance even under extreme sparsity. The VB models are similar, but for sparser data it can sometimes not find the best solution, as can be seen in Fig. 6d. We conducted this experiment for the models with ARD as well (results given in supplementary materials), where we show that ARD makes no difference to the robustness of Gibbs and VB (which are already very robust), but for ICM it can sometimes improve results.

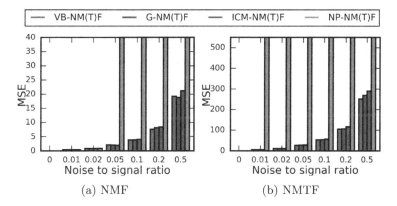

Fig. 5. Noise test performances, measured by average predictive performance on test set (mean square error) for different noise-to-signal ratios.

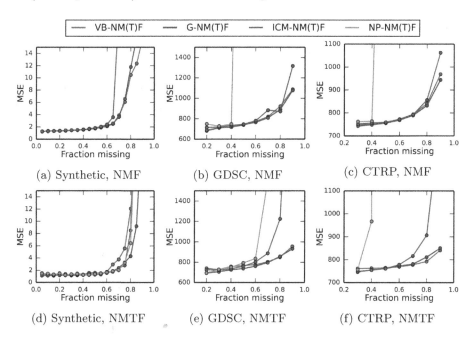

Fig. 6. Sparsity test performances, measured by average predictive performance on test set (mean square error) for different sparsity levels. The top row gives the performances for NMF, and the bottom for NMTF, for the synthetic data (left), GDSC dataset (middle), and CTRP dataset (right).

4.6   Model Selection

Finally, we conducted an experiment to see the extent of overfitting if the model is given a high dimensionality K, and whether this is remedied through the use of ARD. If we give a model a higher dimensionality, it can fit more to the data, but this can lead to overfitting and a higher predictive error. ARD can remedy this by turning off scarsely used factors, hopefully leading to less overfitting.

On the GDSC dataset, we performed 10-fold cross-validation for different values of K (and L for NMTF, using $K = L$) for Gibbs, VB, and ICM. We show these results in Figs. 7a–f, where the results for models without ARD are given by crosses (x) and with ARD by circles (o). We can see that in most graphs, the models with ARD have a much flatter line as the dimensionality increases, hence reducing overfitting. This effect is more apparent for the NMF models than for the NTMF ones. The only exception is NMTF ICM, where the ARD is preventing the model from fitting as much to the data, hence leading to poor predictive results. Results for this experiment on the other three drug sensitivity datasets is given in the supplementary materials, which show that this problem only occurred for NMTF ICM on the GDSC dataset.

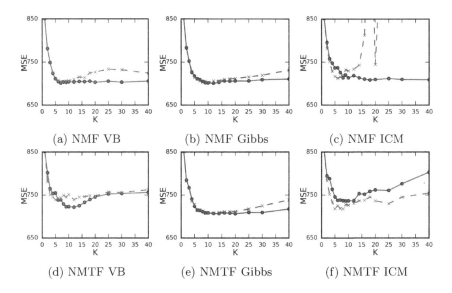

Fig. 7. 10-fold cross-validation performances of the Bayesian models on the GDSC dataset, where we vary the dimensionality K (using $L = K$ for NMTF). The top row gives the performances for NMF, the bottom row for NMTF. Performances for models without ARD are given by dotted lines and crosses (x), with ARD by circles (o).

Table 2. Qualitative comparison of inference methods.

Method	Estimate	Requires sampling	Speed of convergence	Robustness
Non-probabilistic	Point	No	High	Low
Iterated conditional modes	Point (MAP)	Yes	High	Medium
Gibbs sampling	Full posterior	Yes	Low	High
Variational Bayes	Full posterior	No	Medium	Fairly high

5   Conclusion

We have studied the trade-offs between different inference approaches for Bayesian nonnegative matrix factorisation and tri-factorisation models. We considered three methods, namely Gibbs sampling, iterated conditional modes, and non-probabilistic inference, and introduced a fourth one based on variational Bayesian inference. We furthermore extended these models with the Bayesian automatic relevance determination prior, to perform automatic model selection. Through experiments on both synthetic data, and real-world drug sensitivity datasets, we explored the trade-offs in convergence, robustness to noise, and robustness to sparsity.

A qualitative summary based on our quantitative findings can be found in Table 2. We found that the non-probabilistic methods are not very robust to noise and sparsity. Gibbs sampling is the most robust of the methods, especially for sparse datasets, and gives a full Bayesian posterior estimate. However, it converges slowly, and requires additional samples to estimate the posterior. Iterated conditional modes offers a much faster convergence and run-time speed, but sacrifices some robustness, still requires sampling, and no longer returns a full posterior (giving a MAP estimate instead). Our variational Bayesian inference gives good convergence speeds while maintaining more robustness properties.

Finally, we have shown that ARD is an effective way of reducing overfitting when using the wrong dimensionality in matrix factorisation models. This can eliminate the use for performing model selection, or nested cross-validation— although it is not perfect. We also discovered that adding ARD has little impact on performance, or on the robustness of the models to sparsity and noise (except for iterated conditional modes, where ARD increases its robustness to sparsity).

Our experiments were conducted for a specific version of Bayesian matrix factorisation and tri-factorisation, but we believe they offer insights into the trade-offs between different inference techniques in other matrix factorisation models, as well as tensor and Tucker decomposition methods.

Acknowledgements. This work was supported by the UK Engineering and Physical Sciences Research Council (EPSRC), grant reference EP/M506485/1. JF acknowledge funding from the Danish Council for Independent Research 0602-02909B.

References

1. Asuncion, A., Welling, M., Smyth, P., Teh, Y.W.: On smoothing and inference for topic models. In: Proceedings of the Twenty-Fifth Conference on Uncertainty in Artificial Intelligence (2009)
2. Barretina, J., Caponigro, G., Stransky, N., Venkatesan, K., Margolin, A.A., Kim, S., Wilson, C.J., et al.: The Cancer Cell Line Encyclopedia enables predictive modelling of anticancer drug sensitivity. Nature **483**(7391), 603–607 (2012)
3. Beal, M., Ghahramani, Z.: The variational Bayesian EM algorithm for incomplete data: with application to scoring graphical model structures. In: Bayesian Statistics, vol. 7. Oxford University Press (2003)
4. Brouwer, T., Lió, P.: Bayesian hybrid matrix factorisation for data integration. In: Proceedings of the 20th International Conference on Artificial Intelligence and Statistics (AISTATS) (2017)
5. Chen, G., Wang, F., Zhang, C.: Collaborative filtering using orthogonal nonnegative matrix tri-factorization. Inf. Process. Manag. **45**(3), 368–379 (2009)
6. Ding, C., Li, T., Peng, W., Park, H.: Orthogonal nonnegative matrix t-factorizations for clustering. In: Proceedings of the 12th ACM SIGKDD (2006)
7. Figueiredo, M., Jain, A.: Unsupervised learning of finite mixture models. IEEE Trans. Pattern Anal. Mach. Intell. **24**(3), 381–396 (2002)
8. Gönen, M.: Predicting drug-target interactions from chemical and genomic kernels using Bayesian matrix factorization. Bioinformatics **28**(18), 2304–2310 (2012)
9. Lee, D.D., Seung, H.S.: Learning the parts of objects by non-negative matrix factorization. Nature **401**(6755), 788–791 (1999)
10. Lee, D.D., Seung, H.S.: Algorithms for non-negative matrix factorization. In: NIPS, pp. 556–562. MIT Press (2000)
11. Li, T., Zhang, Y., Sindhwani, V.: A non-negative matrix tri-factorization approach to sentiment classification with lexical prior knowledge. In: Proceeding of the 47th Annual Meeting of the Association for Computational Linguistics (2009)
12. Lippert, C., Weber, S., Huang, Y.: Relation prediction in multi-relational domains using matrix factorization. In: NIPS Workshop on Structured Input, Structured Output (2008)
13. Salakhutdinov, R., Mnih, A.: Probabilistic matrix factorization. In: Advances in Neural Information Processing Systems (NIPS), pp. 1257–1264 (2008)
14. Salimans, T., Kingma, D.P., Welling, M.: Markov chain Monte Carlo and variational inference: bridging the gap. In: Proceedings of the 32nd International Conference on Machine Learning (2015)
15. Schmidt, M.N., Winther, O., Hansen, L.K.: Bayesian non-negative matrix factorization. In: Adali, T., Jutten, C., Romano, J.M.T., Barros, A.K. (eds.) ICA 2009. LNCS, vol. 5441, pp. 540–547. Springer, Heidelberg (2009). https://doi.org/10.1007/978-3-642-00599-2_68
16. Seashore-Ludlow, B., Rees, M.G., Cheah, J.H., Cokol, M., Price, E.V., Coletti, M.E., Jones, V., et al.: Harnessing connectivity in a large-scale small-molecule sensitivity dataset. Cancer Discov. **5**(11), 1210–23 (2015)
17. Tan, V.Y.F., Févotte, C.: Automatic relevance determination in nonnegative matrix factorization with the (β)-divergence. IEEE Trans. Pattern Anal. Mach. Intell. **35**(7), 1592–1605 (2013)
18. Virtanen, S., Klami, A., Khan, S., Kaski, S.: Bayesian group factor analysis. In: Proceedings of the 15th International Conference on Artificial Intelligence and Statistics (AISTATS) (2012)

19. Virtanen, S., Klami, A., Kaski, S.: Bayesian CCA via group sparsity. In: Proceedings of the 28th International Conference on Machine Learning (2011)
20. Wang, J.J.Y., Wang, X., Gao, X.: Non-negative matrix factorization by maximizing correntropy for cancer clustering. BMC Bioinform. **14**(1), 107 (2013)
21. Yang, W., Soares, J., Greninger, P., Edelman, E.J., Lightfoot, H., Forbes, S., Bindal, N., et al.: Genomics of drug sensitivity in cancer (GDSC): a resource for therapeutic biomarker discovery in cancer cells. Nucleic Acids Res. **41**(Database issue), D955–D961 (2013)
22. Yoo, J., Choi, S.: Probabilistic matrix tri-factorization. In: IEEE International Conference on Acoustics, Speech, and Signal Processing (2009)
23. Zhang, D., Chen, S., Zhou, Z.-H.: Two-dimensional non-negative matrix factorization for face representation and recognition. In: Zhao, W., Gong, S., Tang, X. (eds.) AMFG 2005. LNCS, vol. 3723, pp. 350–363. Springer, Heidelberg (2005). https://doi.org/10.1007/11564386_27
24. Zhong, M., Girolami, M.: Reversible jump MCMC for non-negative matrix factorization. In: Proceedings of the Twelfth International Conference on Artificial Intelligence and Statistics (AISTATS-2009), pp. 663–670 (2009)
25. Zhong, M., Girolami, M., Faulds, K., Graham, D.: Bayesian methods to detect dye-labelled DNA oligonucleotides in multiplexed Raman spectra. J. Roy. Stat. Soc.: Ser. C (Appl. Stat.) **60**(2), 187–206 (2011)

Content-Based Social Recommendation with Poisson Matrix Factorization

Eliezer de Souza da Silva$^{(\boxtimes)}$, Helge Langseth, and Heri Ramampiaro

Department of Computer Science, Norwegian University of Science and Technology (NTNU), 7491 Trondheim, Norway
{eliezer.souza.silva,helge.langseth,heri}@ntnu.no

Abstract. We introduce Poisson Matrix Factorization with Content and Social trust information (PoissonMF-CS), a latent variable probabilistic model for recommender systems with the objective of jointly modeling social trust, item content and user's preference using Poisson matrix factorization framework. This probabilistic model is equivalent to collectively factorizing a non-negative user–item interaction matrix and a non-negative item–content matrix. The user–item matrix consists of sparse implicit (or explicit) interactions counts between user and item, and the item–content matrix consists of words or tags counts per item. The model imposes additional constraints given by the social ties between users, and the homophily effect on social networks – the tendency of people with similar preferences to be socially connected. Using this model we can account for and fine-tune the weight of content-based and social-based factors in the user preference. We develop approximate *variational* inference algorithm and perform experiments comparing PoissonMF-CS with competing models. The experimental evaluation indicates that PoissonMF-CS achieves superior predictive performance on held-out data for the top-M recommendations task. Also, we observe that PoissonMF-CS generates compact latent representations when compared with alternative models while maintaining superior predictive performance.
Code related to this chapter is available at:
https://github.com/zehsilva/poissonmf_cs
Data related to this chapter are available at: http://files.grouplens.org/datasets/hetrec2011/hetrec2011-lastfm-readme.txt

Keywords: Probabilistic matrix factorization
Non-negative matrix factorization · Hybrid recommender systems
Poisson matrix factorization

1 Introduction

Recommender systems have proven to be a valuable component in many applications of personalization and Internet economy. Traditional recommender systems try to estimate a score function mapping each pair of user and item to a scalar value using the information of previous items already rated or interacted by the

© Springer International Publishing AG 2017
M. Ceci et al. (Eds.): ECML PKDD 2017, Part I, LNAI 10534, pp. 530–546, 2017.
https://doi.org/10.1007/978-3-319-71249-9_32

user [1]. Recent methods have been successful in integrating side information as content of the item, user context, social network, item topics, etc. For this purpose a variety of features should be taken into consideration, such as the routine, the geolocation, spatial correlation of certain preferences, mood and sentiment analysis, as well as social relationships such as "friendship" to others users or "belonging" to a community in a social network [2]. In particular, a rich area of research has explored the integration of topic models and collaborative filtering approaches using principled probabilistic models [3–5]. Another group of models has been developed to integrate social network information into recommender systems using user–item ratings with extra dependencies [6] or constraining and regularizing directly the user latent factors with social features [7,8]. Finally, some models have focused on the collective learning of both social features and content features, constructing hybrid recommender systems [5,9,10].

Our contribution is situated within all these three groups of efforts: we propose a probabilistic model that generalizes both previous models by jointly modeling content and social factors in the preference model applying Poisson-Gamma latent variable models to model the non-negativeness of the user–item ratings and induce sparse non-negative latent representation. Using this joint model we can generate recommendations based on the estimated score of non-observed items. In this article, we formulate the problem (Sect. 1.1), describe the proposed model (Sect. 3), present the variational inference algorithm (Sect. 4) and discuss the empirical results (Sect. 5). Our results indicate improved performance when compared to state-of-the-art methods including Correlated Topic Regression with Social Matrix Factorization (CTR-SMF) [5].

1.1 Problem formulation

Consider that given a set of observations of user–item interactions $R_{\text{train}} = \{(u, d, R_{ud})\}$, with $|R_{\text{train}}| = N_{\text{obs}} \ll U \times D$ (U is the number of users and D the number of documents), using additional item content information and user social network, we aim to learn a function f that estimates the value of each user–item interactions for all pairs of user and items $R_{\text{complete}} = \{(u, d, f(u, d))\}$. In general to solve this problem we assume that users have a set of preferences, and (using matrix factorization) we model these preferences using latent vectors.

Therefore, we have the documents (or items) set $\mathcal{D}$ of size $|\mathcal{D}| = D$, vocabulary set $\mathcal{V}$ of size $|\mathcal{V}| = V$, users set $\mathcal{U}$ of size $|\mathcal{U}| = U$, the social network given by the set of neighbors for each user $\{N(u)\}_{u \in \mathcal{U}}$. So, given the partially observed user–item matrix with integer ratings or implicit counts $\boldsymbol{R} = (R_{ud}) \in \mathbb{N}^{U \times D}$, the observed document–word count matrix $\boldsymbol{W} = (W_{dv}) \in \mathbb{N}^{D \times V}$, and the user social network $\{N(u)\}_{u \in \mathcal{U}}$, we need to estimate a matrix $\widetilde{\boldsymbol{R}} \in \mathbb{N}^{U \times D}$ to complete the user–item matrix $\boldsymbol{R}$. Finally, with the estimated matrix we can rank the unseen items for each user and make recommendations.

2 Related Work

Collaborative Topic Regression (CTR): CTR [3] is a probabilistic model combining topic modeling (using Latent Dirichlet Allocation) and probabilistic matrix factorization (using Gaussian likelihood). Collaborative Topic Regression with Social Matrix Factorization (CTR-SMF) [5] builds upon CTR adding social matrix factorization, creating a joint model Gaussian factorization model with content and social side information. Limited Attention Collaborative Topic Regression (LA-CTR) [9], is another approach with which the authors propose a joint model based on CTR integrating behavioral mechanism of attention. In this case, the amount of attention the user has invested in the social network is limited, and there is a measure of influence implying that the user may favor some friends more than others. In [10], the authors propose a CTR model seamlessly integrated item–tags, item content and social network information. All the models mentioned above combine in some degree LDA with Gaussian based matrix factorization for recommendations. Thus the time complexity for training those models is dominated by LDA complexity, making them difficult to scale. Also, the combination of LDA and Gaussian matrix factorization in CTR is a non-conjugate model that is hard to fit and difficult to work with sparse data.

Poisson Factorization: The basic Poisson factorization is a probabilistic model for non-negative matrix factorization based on the assumption that each user–item interaction R_{ui} can be modelled as a inner product of a user K dimensional latent vector U_u and item latent vector V_i representing the unobserved user preferences and item attributes [11], so that $R_{ui} \sim \text{Poisson}(U_u^T V_i)$. Poisson factorization models for recommender systems have the advantage of principled modeling of implicit feedback, generating sparse latent representations, fast approximate inference with sparse matrix (the likelihood depends only on the consumed items) and improved empirical results compared with the Gaussian-based models [11,12]. Nonparametric Poisson factorization model (BNPPF) [12] extends basic Poisson factorization by drawing user weights from a *Gamma process*. The latent dimensionality in this model is estimated from the data, effectively avoiding the *ad hoc* process of choosing the latent space dimensionality K. Social Poisson factorization (SPF) [6] extends basic Poisson factorization to accommodate preference and social based recommendations, adding a degree of trust variable and making all user–item interaction conditionally dependent on the user friends. With collaborative topic Poisson factorization (CTPF) [4], shared latent factors are utilized to fuse recommendation with topic model using Poisson likelihood and Gamma variables for both.

Non-negative matrix and tensor factorization using Poisson models: Poisson models are also successfully utilized in more general models such as tensor factorization and relational learning, particularly where it can use count data and non-negative factors. In [13], the authors propose a generic Bayesian non-negative tensor factorization model for count data and binary data. In [14], the authors explore the idea of adding constraints between the model variables using side information with hierarchical information, while the approach in [15] uses graph side information jointly modeled with topic modeling with Gamma process – a joint non-parametric model of network and documents.

3 Poisson Matrix Factorization with Content and Social Trust Information (PoissonMF-CS)

The proposed model PoissonMF-CS (see Fig. 1) is an extension and generalization of previous Poisson models, combining social factorization model (social Poisson factorization – SPF) [6], and topic based factorization (collaborative topic Poisson factorization – CTPF) [4].

The main idea is to employ shared latent Gamma factors for topical preference and trust weight variables in the user social network, combining all factors in the rate of a Poisson likelihood of the user–item interaction. We model both sources of information having an additive effect on the observed user–item interactions and add two global multiplicative weights for each group of latent factors. The intuition behind the additive effect of social trust is that users tend to interact with items presented by their peers, so we can imagine a mechanism of "peer pressure" operating, where items offered through the social network have a positive (or neutral) influence on the user. In other words, we believe there is a positive social bias more than an anti-social bias, and we factor this in PoissonMF-CS model.

In the case of Poisson models, this non-negative constraint results in sparseness in the latent factors and can help avoid over-fitting (in comparison the Gaussian-based models [11,12]). Gamma priors on the latent factors, and the fact that the latent factors can only have a positive or a zero effect on the final prediction, induce sparse latent representations in the model. Hence, in the inference process we adjust a factor that decreases the model likelihood by making its value closer to zero.

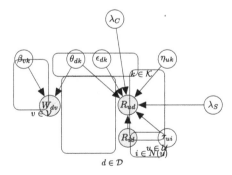

Fig. 1. Plate diagram for PoissonMF-CS model

3.1 Generative Model

In this model, W_{dv} is a counting variable for the number of times word v appears in document d, $\boldsymbol{\beta}_v$ is a latent vector capturing topic distribution of word v and $\boldsymbol{\theta}_d$

is the document–topic intensity vector, both with dimensionality K. Count variable W_{dv} is parametrized by the linear combination of these two latent factors $\boldsymbol{\theta}_d^T \boldsymbol{\beta}_v$. The document–topic latent factor $\boldsymbol{\theta}_d$ influences also the user–document rating variable R_{ud}. Each user has a latent vector $\boldsymbol{\eta}_u$ representing the user–topic propensity, which interacts with the document topic intensity factor $\boldsymbol{\theta}_d$ and document topic offset factor $\boldsymbol{\epsilon}_d$, resulting in the term $\boldsymbol{\eta}_u^T \boldsymbol{\theta}_d + \boldsymbol{\eta}_u^T \boldsymbol{\epsilon}_d$. Here, $\boldsymbol{\eta}_u^T \boldsymbol{\epsilon}_d$ captures the baseline matrix factorization, while $\boldsymbol{\eta}_u^T \boldsymbol{\theta}_d$ connects the rating variable with the content-based part of the model (word–document variable W_{dv}). The trust factor τ_{ui} between user u to user i is equal to zero for all users that are not connected in the social network ($\tau_{ui} > 0 \Leftrightarrow i \in N(u)$). This trust factor adds dependency between social connected users: the user–document rating R_{ud} is influenced by the average rating to item d given by friends of user u in the social network, weighted by the trust user u assigns to his friends ($\sum_{i \in N(u)} \tau_{ui} R_{id}$). We model this social dependency using a conditional specified model, as in [6]. The latent variables λ_C and λ_S are weight variables added in the model to capture and control the general weight of the content and social factors. These variables allow us to infer the importance of content and social factors according to the dataset or domain of usage. Also, instead of estimating these weights from the observed data, we may set λ_C and λ_S to constant values, thus controlling the importance of content and social parts of the model. Specifically if we set $\lambda_C = 0$ and $\lambda_S = 1$ we obtain the SPF model, while setting $\lambda_C = 1$ and $\lambda_S = 0$ result in CTPF, and $\lambda_C = 0$ and $\lambda_S = 0$ is equivalent to the simple Poisson matrix factorization without any side information [11].

Now we present the complete generative model assuming documents (or items) set $\mathcal{D}$ of size $|\mathcal{D}| = D$, vocabulary set $\mathcal{V}$ of size $|\mathcal{V}| = V$, users set $\mathcal{U}$ of size $|\mathcal{U}| = U$, the user social network given by the set of neighbors for each user $\{N(u)\}_{u \in \mathcal{U}}$ D documents, and K latent factors (topics) (with an index set $\mathcal{K}$).

1. Latent parameter distributions:
   (a) for all topics $k \in \mathcal{K}$:
      – for all words $v \in \mathcal{V}$: $\beta_{vk} \sim \text{Gamma}(a_\beta^0, b_\beta^0)$
      – for all documents $d \in \mathcal{D}$: $\theta_{dk} \sim \text{Gamma}(a_\theta^0, b_\theta^0)$ and $\epsilon_{dk} \sim \text{Gamma}(a_\epsilon^0, b_\epsilon^0)$
      – for all users $u \in \mathcal{U}$: $\eta_{uk} \sim \text{Gamma}(a_\eta^0, b_\eta^0)$
      • for all user $i \in N(u)$: $\tau_{ui} \sim \text{Gamma}(a_\tau^0, b_\tau^0)$
   (b) Content weight: $\lambda_C \sim \text{Gamma}(a_C^0, b_C^0)$
   (c) Social weight: $\lambda_S \sim \text{Gamma}(a_S^0, b_S^0)$
2. Observations probability distribution:
   (a) for all observed document–word pairs dv :

$$W_{dv} | \boldsymbol{\beta}_v, \boldsymbol{\theta}_d \sim \text{Poisson}(\boldsymbol{\beta}_v^T \boldsymbol{\theta}_d)$$

   (b) for all observed user–document pairs ud :

$$R_{ud} | \boldsymbol{R}_{N(u),d}, \boldsymbol{\eta}_u, \boldsymbol{\epsilon}_d, \boldsymbol{\theta}_d \sim \text{Poisson}(\lambda_C \boldsymbol{\eta}_u^T \boldsymbol{\theta}_d + \boldsymbol{\eta}_u^T \boldsymbol{\epsilon}_d + \lambda_S \sum_{i \in N(u)} \tau_{ui} R_{id})$$

4   Inference

First, we add a set of auxiliary latent Poisson variables to facilitate the posterior inference of the model. By doing so, the extended model will be complete conjugate, and consequently have analytical equations for the complete conditionals and variational updates [16]. In Appendix A we show that those auxiliary variables can be seen as by-product of a lower bound on the expected value of the log sum of the latent random variables. Variable $Y_{dv,k}$ represent a topic specific latent count for a word–document pair, so that the observed word–document counts is a sum of the latent counts (a property of the Poisson distribution)[1]. We can perform a similar modification for the user–item counts, splitting the latent terms of R_{ud} rate into two groups of topic specific latent count allocation variables: $Z^M_{ud,k}$ for the item content part, $Z^N_{ud,k}$ for the collaborative filtering part and $Z^S_{ud,i}$ for the social trust part (for this part, the intuitive explanation for the latent dimension is the idea of friend specific allocation of trust). The sum of all those latent counts is the observed user–item interaction count variable R_{ud}.

$$
\begin{aligned}
Y_{dv,k}|\beta_{vk},\theta_{dk} &\sim \mathrm{Poisson}(\beta_{vk}\theta_{dk}) \\
Z^M_{ud,k}|\lambda_C,\eta_{uk},\theta_{dk} &\sim \mathrm{Poisson}(\lambda_C\eta_{uk}\theta_{dk}) \\
Z^N_{ud,k}|\eta_{uk},\epsilon_{dk} &\sim \mathrm{Poisson}(\eta_{uk}\epsilon_{dk}) \\
Z^S_{ud,i}|\lambda_S,\tau_{ui},R_{id} &\sim \mathrm{Poisson}(\lambda_S\tau_{ui}R_{id})
\end{aligned}
\tag{1}
$$

with $\sum_k Y_{dv,k} = W_{dv}$, and $\sum_k Z^M_{ud,k} + Z^N_{ud,k} + \sum_{i\in N(U)} Z^S_{ud,i} = R_{ud}$.

The inference problem consists on the estimation of the posterior distribution of the latent variables given the observed rating R, the observed document–word counts W, and the user social network $\{N(u)\}_{u\in\mathcal{U}}$, in other words, computing

$$p(\Theta|R,W,\{N(u)\}_{u\in\mathcal{U}}),$$

where $\Theta = \{\beta,\theta,\eta,\epsilon,\tau,y,z,\lambda_C,\lambda_S\}$ is the set of all latent variables. The exact computation of this posterior probability is intractable for any practical scenario, so we need approximation techniques for efficient parameter learning. In our case, we apply variational techniques to derive the learning algorithm. As an intermediate step towards the variational inference algorithm, we also derive the full conditional distribution for each latent variable. The full conditional distribution of each latent variable is also useful as update equations for Gibbs sampling, meaning that we could use the resulting equations to implement a sampling-based approximation. However, sampling methods are hard to scale and usually requires more memory, so as a design choice for the implementation

---

[1] The change consist in assigning a new Poisson variable to each sum-term in the latent rate of the Poisson likelihood, so if $S \sim \mathrm{Poisson}(\sum_i X_i)$, we add variables $S_i \sim \mathrm{Poisson}(X_i)$, and by the sum property of Poisson random variable $S = \sum_i S_i \sim \mathrm{Poisson}(\sum_i X_i)$.

of the learning algorithm, we refrained from applying the Gibbs sampling method and focus on the variational inference.

In the next sections, we present the full conditional distribution of each of the latent variables in Sect. 4.1, and show the resulting update equation for the variational parameters in Sect. 4.2.

4.1 Full Conditional Distribution

The full conditional distribution of each of the latent variables is the distribution of a variable given all the other variables in the model, except the variable that we are considering. Given a set of indexed random variables X_k, we use the notation $p(X_k|X_{-k})$ (where X_{-k} means all the variables X_i such that $i \neq k$) to represent the full conditional distribution. Given the factorized structure of the model we can simplify the conditional set to the Markov blanket of the node we are considering (children nodes and co-parents nodes)[2] [16]. For conciseness, we show the derivations only for one Gamma latent variables and one Poisson latent count variable.

- **Gamma distributed variables:** We demonstrate how to obtain the full conditional distribution for Gamma distributed variable θ_{dk}, for the remaining Gamma distributed variables we only present the end result without the intermediate steps.

$$
\begin{aligned}
p(\theta_{dk}|*) &= p(\theta_{dk}|\mathrm{MarkovBlanket}(\theta_{dk})) \\
&\propto p(\theta_{dk}) \prod_{v=1}^{V} p(Y_{dv,k}|\beta_{vk}, \theta_{dk}) \prod_{u=1}^{U} p(Z_{ud,k}^M|\lambda_C, \eta_{uk}, \theta_{dk}) \\
&\propto \theta_{dk}^{a_\theta^0 - 1} e^{-b_\theta^0 \theta_{dk}} \prod_v \theta_{dk}^{Y_{dv,k}} e^{-\beta_{vk}\theta_{dk}} \prod_u \theta_{dk}^{Z_{ud,k}^M} e^{-\lambda_C \eta_{uk}\theta_{dk}} \\
&\propto \theta_{dk}^{a_\theta^0 + \sum_v Y_{dv,k} + \sum_u Z_{ud,k}^M - 1} e^{-\theta_{dk}(b_\theta^0 + \sum_v \beta_{vk} + \lambda_C \sum_u \eta_{uk})}
\end{aligned}
\tag{2}
$$

Normalizing equation Eq. 2 over θ_{dk} we obtain the pdf of a Gamma variable with shape $a_\theta^0 + \sum_v Y_{dv,k} + \sum_u Z_{ud,k}^M$ and rate $b_\theta^0 + \sum_v \beta_{vk} + \lambda_C \sum_u \eta_{uk}$. The final solution is written in Eq. 3. Also, notice that because of the way the model is structured all other Gamma latent variable have similar equations, the difference being the set of variables in the Markov blanket.

$$
\begin{aligned}
\theta_{dk}|* &\sim \mathrm{Gamma}(a_\theta^0 + \textstyle\sum_v Y_{dv,k} + \sum_u Z_{ud,k}^M, b_\theta^0 + \sum_v \beta_{vk} + \lambda_C \sum_u \eta_{uk}) \\
\beta_{vk}|* &\sim \mathrm{Gamma}(a_\beta^0 + \textstyle\sum_d Y_{dv,k}, b_\beta^0 + \sum_d \theta_{dk}) \\
\eta_{uk}|* &\sim \mathrm{Gamma}(a_\eta^0 + \textstyle\sum_d Z_{ud,k}^M + Z_{ud,k}^N, b_\eta^0 + \lambda_C \sum_d \theta_{dk} + \sum_d \epsilon_{dk}) \\
\epsilon_{dk}|* &\sim \mathrm{Gamma}(a_\epsilon^0 + \textstyle\sum_u -Z_{ud,k}^N, b_\epsilon^0 + \sum_u \eta_{uk}) \\
\tau_{ui}|* &\sim \mathrm{Gamma}(a_\tau^0 + \textstyle\sum_d Z_{ud,i}^S, b_\tau^0 + \lambda_S \sum_d R_{id}) \\
\lambda_C|* &\sim \mathrm{Gamma}(a_C + \textstyle\sum_{u,d,k} Z_{ud,k}^M, b_C + \sum_{u,d,k} \eta_{uk}\theta_{dk}) \\
\lambda_S|* &\sim \mathrm{Gamma}(a_S + \textstyle\sum_{u,d,i} Z_{ud,i}^S, b_S + \sum_{u,d,i} \tau_{ui}R_{id})
\end{aligned}
\tag{3}
$$

---

[2] We use the notation $\mathrm{MarkovBlanket}(X)$ to denote the Markov blanket of a variable X – the set of children and co-parents nodes of variable X in the graphical model.

– **Multinomial distributed (auxiliary) variables:** looking at the Markov blanket of $\boldsymbol{Y}_{dv}$ we obtain:

$$
\begin{aligned}
p(\boldsymbol{Y}_{dv}|*) &\propto \textstyle\prod_{k=1}^{K} p(Y_{dv,k}|\beta_{vk},\theta_{dk}) = \prod_{k=1}^{K} \text{Poisson}(Y_{dv,k}|\beta_{vk}\theta_{dk}) \\
&\propto \textstyle\prod_{k=1}^{K} \frac{(\beta_{vk}\theta_{dk})^{Y_{dv,k}}}{Y_{dv,k}!}
\end{aligned}
\tag{4}
$$

Given that we know that $\sum_k Y_{dv,k} = W_{dv}$, this functional form is equivalent to the pdf of a Multinomial distribution with parameter probabilities proportional to $\beta_{vk}\theta_{dk}$.

$$
\boldsymbol{Y}_{dv}|* \sim \text{Mult}(W_{dv}; \boldsymbol{\phi_{dv}}) \qquad \text{with } \phi_{dv,k} = \frac{\beta_{vk}\theta_{dk}}{\sum_k \beta_{vk}\theta_{dk}}
\tag{5}
$$

Similarly, $\boldsymbol{Z}_{ud}$ is a Multinomial with parameters proportional to the parent nodes of $\boldsymbol{Z}_{ud}$. For convenience in the previous section, we split $\boldsymbol{Z}_{ud}$ in three blocks of variables and parameters $\boldsymbol{Z}_{ud} = [\boldsymbol{Z}_{ud}^M, \boldsymbol{Z}_{ud}^N, \boldsymbol{Z}_{ud}^S]$ representing the different high-level parts of our model. The dimensionality of the first two blocks is the K, while for the last block is U, resulting that $\boldsymbol{Z}_{ud}$ has dimensionality $2K + U$. Similarly the parameters of the $\boldsymbol{Z}_{ud}$ full conditional Multinomial have a block structure $\boldsymbol{\xi}_{ud} = [\boldsymbol{\xi}_{ud}^M, \boldsymbol{\xi}_{ud}^N, \boldsymbol{\xi}_{ud}^S]$.

$$
Z_{ud}|* \sim \text{Mult}(R_{ud}; \boldsymbol{\xi}_{ud})
$$

$$
\text{with } \xi_{ud,k} = \begin{cases}
\xi_{ud,k}^M = \dfrac{\lambda_C \eta_{uk}\theta_{dk}}{\sum_k \eta_{uk}(\lambda_C\theta_{dk}+\epsilon_{dk})+\lambda_S \sum_{i\in N(u)} \tau_{ui}R_{id}} \\[2ex]
\xi_{ud,k}^N = \dfrac{\eta_{uk}\epsilon_{dk}}{\sum_k \eta_{uk}(\lambda_C\theta_{dk}+\epsilon_{dk})+\lambda_S \sum_{i\in N(u)} \tau_{ui}R_{id}} \\[2ex]
\xi_{ud,i}^S = \dfrac{\lambda_S \tau_{ui}R_{id}}{\sum_k \eta_{uk}(\lambda_C\theta_{dk}+\epsilon_{dk})+\lambda_S \sum_{i\in N(u)} \tau_{ui}R_{id}}
\end{cases}
$$

We present in next section how to use these equations to derive a deterministic optimization algorithm for approximate inference using the *variational* method.

4.2 Variational Inference

Given a family of surrogate distributions $q(\Theta|\Psi)$ for the unobserved variables (latent terms) parametrized by variational parameters Ψ, we want to find an assignment of the variational parameters that minimize the KL-divergence between $q(\Theta|\Psi)$ and $p(\Theta|\boldsymbol{R}, \boldsymbol{W})^3$,

$$
\arg \min_{\Psi} \text{KL}\{q(\Theta|\Psi), p(\Theta|\boldsymbol{R}, \boldsymbol{W})\}.
$$

Then, the optimal surrogate distribution can be used as an approximation the true posterior. However, the optimization problem using directly the KL divergence is not tractable, since it depends on the computation of the evidence

---

[3] To simplify the notation, we use the short-handed $p(\Theta|\boldsymbol{R}, \boldsymbol{W})$ to denote the posterior distribution $p(\Theta|\boldsymbol{R}, \boldsymbol{W}, \{N(u)\}_{u\in\mathcal{U}})$. Also, we drop the explicitly notation indicating the dependency on the social network.

$\log p(\mathbf{R}, \mathbf{W})$. This can be accomplished using Jensen inequality to get lower bounds on the evidence and changing the optimization objective to this lower bound – the Evidence Lower BOund (ELBO):

$$\arg \min_{\Psi} L(\Psi) = \mathrm{E}_q[\log p(\mathbf{R}, \mathbf{W}, \Theta) - \log q(\Theta|\Psi)]$$

Another ingredient in this approximation is the mean field assumption. It consists in assuming that all variables in the variational distribution $q(\Theta|\Psi)$ are mutually independent. As a result the variational surrogate distribution can be expressed as a factorized distribution of each latent factor (Eq. 7). Another implication is that we can compute the updates for each variational X_i factor using the complete conditional of the latent factor [16]. Finally, the inference algorithm consists in iterative updating of variational parameters of each factorized distribution until convergence is reached, resulting in the *coordinate ascent variational inference* algorithm based on the following equation:

$$q(X_i) \propto \exp\{\mathrm{E}_q[\log p(X_i|*)]\} \tag{6}$$

Using Eq. 6, we can take each complete conditional variable that we described in the previous section and create a respective proposal distribution for the variational inference. This proposal distribution is in the same family as the full conditional distribution of the latent variables, meaning that we have a group of Gamma and Multinomial variables. As long as we update the parameters of the variational distribution using Eq. 6, it is guaranteed to minimize the KL divergence between the surrogate variational distribution (Eq. 7) over the latent variables and the posterior distribution of the model.

$$\begin{aligned}
q(\Theta|\Psi) = {} & q(\lambda_C|a_{\lambda_C}, b_{\lambda_C})q(\lambda_S|a_{\lambda_S}, b_{\lambda_S}) \prod_{u,k,i} q(\tau_{ui}|a_{\tau_{ui}}, b_{\tau_{ui}})q(\eta_{uk}|a_{\eta_{uk}}, b_{\eta_{uk}}) \\
& \times \prod_{d,v,k} q(\epsilon_{dk}|a_{\epsilon_{dk}}, b_{\epsilon_{dk}})q(\theta_{dk}|a_{\theta_{dk}}, b_{\theta_{dk}})q(\beta_{vk}|a_{\beta_{vk}}, b_{\beta_{vk}}) \\
& \times \prod_{d,v,u} q(\mathbf{Z}_{dv}|\boldsymbol{\phi}^*_{dv})q(\mathbf{Y}_{ud}|\boldsymbol{\xi}^{M*}_{ud}, \boldsymbol{\xi}^{N*}_{ud}, \boldsymbol{\xi}^{S*}_{ud})
\end{aligned} \tag{7}$$

After applying Eq. 6 together with the expected value properties for each latent variable[4], we obtain the following update equations for the variational parameters.

– **Content and social weights:**

$$a_{\lambda_C} = a_C + \sum_{u,d,k} R_{ud}\xi^{M*}_{ud,k}, \qquad\qquad b_{\lambda_C} = b_C + \sum_{u,d,k} \frac{a_{\eta_{uk}}}{b_{\eta_{uk}}} \frac{a_{\theta_{dk}}}{b_{\theta_{dk}}}$$
$$a_{\lambda_S} = a_S + \sum_u R_{ud}\xi^{M*}_{ud,k} + \sum_v W_{dv}\phi^*_{dv,k}, \; b_{\lambda_S} = b_S + \sum_{u,d,i} R_{id}\frac{a_{\tau_{ui}}}{b_{\tau_{ui}}}$$

– **Content v (topic/tags/etc) parameters:**

$$a_{\beta_{vk}} = a^0_\beta + \sum_d W_{dv}\phi^*_{dv,k}, \; b_{\beta_{vk}} = b^0_\beta + \sum_d \frac{a_{\theta_{dk}}}{b_{\theta_{dk}}}$$

---

[4] Notice that, if $q(X) = \mathrm{Gamma}(X|a_X, b_X)$ (parameterized by shape and rate), then $\mathrm{E}_q[X] = \frac{a_X}{b_X}$ and $\mathrm{E}_q[\log X] = \Psi(a_X) - \log(b_X)$, where $\Psi(.)$ is the Digamma function. If $q(\mathbf{X}) = \mathrm{Mult}(R|\mathbf{p})$, then $\mathrm{E}_q[X_i] = Rp_i$.

- **Item d parameters**:

$$a_{\epsilon_{dk}} = a_\epsilon^0 + \sum_u R_{ud}\xi_{ud,k}^{N*}, \qquad\qquad b_{\epsilon_{dk}} = b_\epsilon^0 + \sum_u \frac{a_{\eta_{uk}}}{b_{\eta_{uk}}}$$

$$a_{\theta_{dk}} = a_\theta^0 + \sum_u R_{ud}\xi_{ud,k}^{M*} + \sum_v W_{dv}\phi_{dv,k}^*, \; b_{\theta_{dk}} = b_\theta^0 + \mathrm{E}_q[\lambda_C]\sum_u \frac{a_{\eta_{uk}}}{b_{\eta_{uk}}} + \sum_v \frac{a_{\beta_{vk}}}{b_{\beta_{vk}}}$$

- **User u parameters**:

$$a_{\eta_{uk}} = a_\eta^0 + \sum_d R_{ud}(\xi_{ud,k}^{M*} + \xi_{ud,k}^{N*}), \; b_{\eta_{uk}} = b_\eta^0 + \sum_d \mathrm{E}_q[\lambda_C]\frac{a_{\theta_{dk}}}{b_{\theta_{dk}}} + \frac{a_{\epsilon_{dk}}}{b_{\epsilon_{dk}}}$$

$$a_{\tau_{ui}} = a_\tau^0 + \sum_d R_{ud}\xi_{ud,i}^{S*}, \qquad\qquad b_{\tau_{ui}} = b_\tau^0 + \mathrm{E}_q[\lambda_S]\sum_d R_{id}$$

- **item–content dv parameters**:

$$\phi_{dv,k}^* \propto \frac{e^{\Psi\left(a_{\beta_{vk}}\right)}}{b_{\beta_{vk}}}\frac{e^{\Psi\left(a_{\theta_{dk}}\right)}}{b_{\theta_{dk}}} \text{ with } \sum_k \phi_{dv,k} = 1$$

- **user–item ud parameters**:

$$\xi_{ud,k}^{M*} \propto e^{E_q[\log\lambda_C]}\frac{e^{\Psi\left(a_{\eta_{uk}}\right)}}{b_{\eta_{uk}}}\frac{e^{\Psi\left(a_{\theta_{dk}}\right)}}{b_{\theta_{dk}}} \quad \xi_{ud,k}^{N*} \propto \frac{e^{\Psi\left(a_{\eta_{uk}}\right)}}{b_{\eta_{uk}}}\frac{e^{\Psi\left(a_{\epsilon_{dk}}\right)}}{b_{\epsilon_{dk}}}$$

$$\xi_{ud,i}^{S*} \propto e^{E_q[\log\lambda_S]}\frac{e^{\Psi\left(a_{\tau_{ui}}\right)}}{b_{\tau_{ui}}}R_{id} \qquad \text{with } \sum_k \xi_{ud,k}^{M*} + \xi_{ud,k}^{N*} + \sum_i \xi_{ud,i}^{S*} = 1$$

Computing the ELBO: The variational updates calculated in the previous sections are guaranteed to non-decrease the ELBO. However, we still need to calculate this lower bound after each iteration to evaluate a stopping condition for the optimization algorithm. We briefly describe a particular lower-bounding for the ELBO involving the log-sum present in the Poisson rate.

Note also that the surrogate distribution is factorized using the mean field assumptions (Eq. 7), so we have a sum of terms corresponding to the expected log probability over the surrogate distribution. The terms comprising the log-probabilities of the Poisson likelihood display a expected value over a sum of logarithms of latent variables (for example $\mathrm{E}_q[\log(\sum_k \beta_{vk}\theta_{dk})]$), this is a challenging computation, but we can apply another lower-bound[5] and simplify it to Eq. 8.

$$\begin{aligned} \mathrm{E}_q[\log(\textstyle\sum_k \beta_{vk}\theta_{dk})] &\geq \sum_k \phi_{dv,k}^* \left(\mathrm{E}_q[\log\beta_{vk}] + \mathrm{E}_q[\log\theta_{dk}]\right) \\ &- \sum_k \phi_{dv,k}^* \log\phi_{dv,k}^* \end{aligned} \tag{8}$$

This same simplification can be done to all Poisson terms independently because of the mean field assumptions. It is equivalent to using the auxiliary latent counts. So, for example, using the latent variable $Z_{dv,k}$, β_{vk} and θ_{dk}, the Poisson term in the ELBO results in Eq. 9.

$$\begin{aligned} \mathrm{E}_q\left[\log\frac{p(Z_{dv})}{q(Z_{dv})}\right] &= \sum_k W_{dv}\phi_{dv,k}^* \mathrm{E}_q[\log(\beta_{vk}\theta_{dk})] \\ &- \mathrm{E}_q[\beta_{vk}\theta_{dk}] - W_{dv}\phi_{dv,k}^* \log(\phi_{dv,k}^*) - \log(W_{dv}!) \end{aligned} \tag{9}$$

---

[5] This lower bound is valid for any $\phi_{dv,k}^*$, with $\sum_k \phi_{dv,k}^* = 1$, check Eq. 13 in Appendix A for details.

For the Gamma terms, the calculations are a direct application of ELBO formula for the appropriate variable. For example, Eq. 10 describes the resulting terms for β_{vk}.

$$
\begin{aligned}
E_q\left[\log \frac{p(\beta_{vk})}{q(\beta_{vk})}\right] = & \log \frac{\Gamma(a_{\beta_{vk}})}{\Gamma(a)} + a \log b + a_{\beta_{vk}}(1 - \log b_{\beta_{vk}}) \\
& + (a - a_{\beta_{vk}})E_q[\log \beta_{vk}] - bE_q[\beta_{vk}]
\end{aligned}
\tag{10}
$$

Recommendations: Once we learn the latent factors of the model from the observations we can infer the user preference over the set of items using the expected value of the user–item rating $E[R_{ud}]$. The recommendation algorithm ranks the unobserved items for each user according to $E[R_{ud}]$ and recommend to top-M items. We utilize the variational distribution to efficiently compute $E[R_{ud}]$ as defined in Eq. 11. This value can be broken down into three non-negative scores: $E_q[\eta_u]^T E_q[\epsilon_d]$, representing the "classic" collaborative filtering matching of users preferences and items features, $E_q[\lambda_C]E_q[\eta_u]^T E_q[\theta_d]$ representing the content factors contribution and $E_q[\lambda_S]\sum_{i \in N(u)} E_q[\tau_{ui}]R_{id}$ the social influence contribution.

$$
E[R_{ud}] \approx E_q[\eta_u]^T (E_q[\lambda_C]E_q[\theta_d] + E_q[\epsilon_d]) + E_q[\lambda_S] \sum_{i \in N(u)} E_q[\tau_{ui}]R_{id}
\tag{11}
$$

Complexity and Convergence: The complexity of each iteration of the variational inference algorithm is linear on the number of latent factors K, non-zero ratings nR, non-zero word-document counts nW, users U, items D, vocabulary set W and neighbors for each user nS, in other words $O(K(nW + nR + nS + U + D + W))$. We have shown that we can obtain closed-form updates for the inference algorithm, which stems from the fact that the model is fully conjugate and in the exponential family of distributions. In this setting variational inference is guaranteed to converge, and we observed in the experiments the algorithm converging after 20 to 40 iterations.

5   Evaluation

In this section, we analyze the predictive power of the proposed model with a real world dataset and compare it with state of the art methods.[6]

Datasets. to be able to compare with the state-of-art method Correlated Topic Regression with Social Matrix Factorization [5], we conducted experiments using the *hetrec2011-lastfm-2k* (Last.fm) dataset [17]. This dataset consists of a set of user–artists weighted interactions ("artists" is item set), a set of user–artists-tags and a set of user–user relations[7]. We process the dataset to create an artist–tags

---

[6] Our C++ implementation of PoissonMF-CS with some of the experiments will be available this repository https://github.com/zehsilva/poissonmf_cs.

[7] The statistics for the dataset are: 1892 users, 17632 artists, 11946 tags, 25434 user–user connections, 92834 user–items interactions, and 186479 user–tag–items entries.

matrix by summing up all the tags given by all users to a given artist, this matrix is the item–content matrix in our model. Also, we discard the user–artists weight, considering a "1" for all observed cases. After the preprocessing, we sample 85% of the user–artists observation for training, and kept 15% held-out for predictive evaluation, selecting only users with more than 5 item ratings for the training part of the split.

Metric: Given the random splits of training and test, we train our model and use the estimated latent factors to predict the entries in the testing datasets. In this setting zero ratings can not be necessarily interpreted as negative, making it problematic to use the precision metric. Instead, we focus on recall metric to be comparable with previous work [5] and because the relevant items are available. Specifically, we calculate the recall at the top M items (recall@M) for a user, defined as:

$$\text{recall@}M = \frac{\text{number of items the user likes in Top}M}{\text{total number of items the user likes}} \quad (12)$$

Recall@M from Eq. 12 is calculated for each user, to obtain a single measure for the whole dataset we average it over all the users obtaining the Avg. Recall@M.

5.1   Experiments

Initially we set all the Gamma hyperparameters to the same values a_{all}[8] and b_{all}[9] equal to 0.1, while varying the latent dimensionality K. For each value of K we ran the experiments on 30 multiple random splits of the dataset in order to be able to generate boxplots of the final recommendation recall. We compare our results with the reported results in [5] for the same dataset and with optimal parameters. In this first experiment we let the algorithm estimate the optimal content weight λ_C and social weight λ_S. It is possible to see in Fig. 2 that PoissonMF-CS is consistently outperforming by large margin CTR-SMF and CTR (Fig. 2a), while outperforming other Poisson factorization methods (Fig. 2b) by a significant margin ($p \leq 1 \cdot 10^{-6}$ in Wilcoxon paired test for each M). This may be indicative that both the choice of Poisson likelihood with non-negative latent factors and the modelling of content and social weights have positive impact in the predictive power of the model.

Model selection. Figure 3 shows the resulting predictive performance of PoissonMF-CS with different values of number of latent factors K in *Hetrec2011-lastfm* dataset. We concluded that the optimal choice for K is 15. This result is important, indicating that the model is generating compact latent representations, given that the optimal choice of K reported for CTR-SMF in the same dataset is 200. In Fig. 5 we show the results for the latent variable hyperparameters. We ran one experiment varying the hyperparameters a_{all} and b_{all} to understand the impact of these hyperparameters in the final recommendation.

---

[8] $a_{all} = a_\beta^0 = a_\eta^0 = a_\theta^0 = a_\epsilon^0 = a_\tau^0 = a_C = a_S = 0.1$.
[9] $b_{all} = b_\beta^0 = b_\eta^0 = b_\theta^0 = b_\epsilon^0 = b_\tau^0 = b_C = b_S = 0.1$.

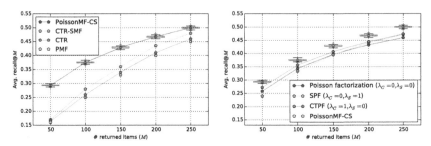

(a) PoissonMF-CS (K=10) and Gaus-(b) PoissonMF-CS (K=10) and other
sian based models   PF models

Fig. 2. Comparison of PoissonMF-CS with alternative models. Each subplot is the
result of running the PoissonMF-CS recommendation algorithm over 30 random splits
of the *Hetrec2011-lastfm* dataset for a fixed number of latent features K (in this case,
$K = \{10, 15\}$). The values for CTR-SMF, CTR, and PMF was taken from [5], and
according to the reported results, they are the best values found after a grid search.

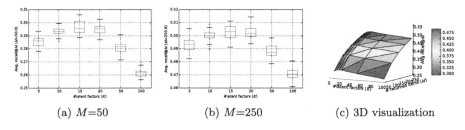

(a) M=50   (b) M=250   (c) 3D visualization

Fig. 3. Impact of the number of latent variables (K) parameter on the Av. Recall@M
metric for different number of returned items (M). Each subplot is the result of running
the PoissonMF-CS recommendation algorithm over 30 random splits of the dataset with
K varying in (5, 10, 15, 20, 50, 100)

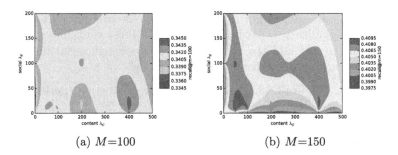

(a) M=100   (b) M=150

Fig. 4. Evaluation of the impact of content and social weight parameters (in all experiments in this figure $K = 10$)

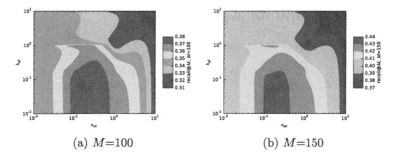

(a) M=100   (b) M=150

Fig. 5. Evaluation of the impact of latent Gamma hyperpriors on the recall (in all experiments in this figure $K = 10$)

We noticed that the optimal values for different values of M for both hyperparameters are between 0.1 and 1, a result consistent with the recommendations in the literature [4,6,12] and with the statistical intuition that Poisson likelihood with Gamma prior with shape parameter $a < 1$ favour sparse latent representation.

The next experiment was to set the content weight and social weight to fixed values and evaluate the impact of these weights on the result. In Fig. 4 we can see that the resulting pattern for different values of M is not evident, but indicates that the resulting recall is less sensitive to change in the content and social weights parameters than on the hyperparameters a_{all} and b_{all}. This is also indicative that the importance of social and content factors is not the same at different points of the ranked list of recommendations.

6 Conclusion

This article describes PoissonMF-CS, a joint Bayesian model for recommendations integrating three sources of information: item textual content, user social network, and user–item interactions. It generalizes existent Poisson factorization models for recommendations by adding both content and social features. Our experiment shows that the proposed model consistently outperforms previous Poisson models (SPF and CTPF) and alternative joint models based on Gaussian probabilistic factorization and LDA (CTR-SMF and CTR) on a dataset containing both content and social side information. These results demonstrate that joint modeling of social and content features using Poisson models improves the recommendations, can have scalable inference and generates more compact latent features. Although the batch variational inference algorithm is already efficient[10], one future improvement will be the design of Stochastic Variational Inference algorithm for very large scale inference.

---

[10] For example, it takes 12 min to train the best performing model in a desktop machine with the *Hetrec2011-lastfm* dataset in a single core without any parallelism.

A   A Lower Bound for $E_q[\log \sum_k X_k]$

The function $\log(\cdot)$ is a concave, meaning that:

$$\log(px_1 + (1-p)x_2) \geq p \log x_1 + (1-p) \log x_2$$
$$\forall p : p \geq 0$$

By induction this property can be generalized to any convex combination of x_k ($\sum_k p_k x_k$ with $\sum_k p_k = 1$ and $p_k \geq 0$): $\log \sum_k p_k x_k \geq \sum_k p_k \log x_k$ Now using random variables we can create a similar convex combination by multiplying and dividing each random variable X_k by $p_k > 0$ and apply the sum of of expectation property:

$$E_q[\log \sum_k X_k] = E_q[\sum_k \log p_k \frac{X_k}{p_k}]$$
$$\log \sum_k p_k \frac{X_k}{p_k} \geq \sum_k p_k \log \frac{X_k}{p_k}$$
$$\Rightarrow E_q[\log \sum_k p_k \frac{X_k}{p_k}] \geq \sum_k p_k E_q[\log \frac{X_k}{p_k}] \tag{13}$$
$$\Rightarrow E_q[\log \sum_k X_k] \geq \sum_k p_k E_q[\log X_k] - p_k \log p_k$$

The lower bound of Eq. 13 is applied in Eq. 8 and it is a general lower bound useful for the log–sum terms in the ELBO computation. If we want a tight lower bound, we should use Lagrange multipliers to choose the set of p_k that maximize the lower-bound given that they sum to 1.

$$L(p_1, \ldots, p_K) = (\sum_k p_k E_q[\log X_k] - p_k \log p_k) + \lambda(1 - \sum_k p_k)$$
$$\frac{\partial L}{\partial p_k} = E_q[\log X_k] - \log p_k - 1 - \lambda = 0$$
$$\frac{\partial L}{\partial \lambda} = 1 - \sum_k p_k = 0$$
$$\Rightarrow E_q[\log X_k] = \log p_k + 1 + \lambda$$
$$\Rightarrow \exp E_q[\log X_k] = p_k \exp(1 + \lambda) \tag{14}$$
$$\Rightarrow \underbrace{\sum_k \exp E_q[\log X_k]}_{=1} = \sum_k p_k \exp(1 + \lambda)$$
$$\Rightarrow p_k = \frac{\exp\{E_q[\log X_k]\}}{\sum_k \exp\{E_q[\log X_k]\}}$$

The final formula for p_k in Eq. 14 is exactly the same that we can find for the parameters of the Multinomial distribution of the auxiliary variables in the Poisson model with sum of Gamma distributed latent variables, which demonstrates that the choice of distribution for the auxiliary variables is optimal for this lower-bound.

References

1. Adomavicius, G., Tuzhilin, A.: Toward the next generation of recommender systems: a survey of the state-of-the-art and possible extensions. IEEE Trans. Knowl. Data Eng. **17**(6), 734–749 (2005)

2. Tang, J., Hu, X., Liu, H.: Social recommendation: a review. Soc. Netw. Anal. Min. **3**(4), 1113–1133 (2013)
3. Wang, C., Blei, D.M.: Collaborative topic modeling for recommending scientific articles. In: Proceedings of the 17th ACM SIGKDD International Conference on Knowledge Discovery and Data Mining, pp. 448–456, 21–24 August 2011, San Diego, CA, USA (2011)
4. Gopalan, P., Charlin, L., Blei, D.M.: Content-based recommendations with Poisson factorization. In: Advances in Neural Information Processing Systems 27: Annual Conference on Neural Information Processing Systems 2014, pp. 3176–3184, 8–13 December 2014, Montreal, Quebec, Canada (2014)
5. Purushotham, S., Liu, Y.: Collaborative topic regression with social matrix factorization for recommendation systems. In: Proceedings of the 29th International Conference on Machine Learning, ICML 2012, Edinburgh, Scotland, UK, 26 June–1 July 2012. icml.cc/Omnipress (2012)
6. Chaney, A.J., Blei, D.M., Eliassi-Rad, T.: A probabilistic model for using social networks in personalized item recommendation. In: Proceedings of the 9th ACM Conference on Recommender Systems, RecSys 2015, pp. 43–50, 16–20 September 2015, Vienna, Austria (2015)
7. Ma, H., Zhou, D., Liu, C., Lyu, M.R., King, I.: Recommender systems with social regularization. In: Proceedings of the Forth International Conference on Web Search and Web Data Mining, WSDM 2011, pp. 287–296, 9–12 February 2011, Hong Kong, China (2011)
8. Yuan, Q., Chen, L., Zhao, S.: Factorization vs. regularization: fusing heterogeneous social relationships in top-n recommendation. In: Proceedings of the Fifth ACM Conference on Recommender Systems, RecSys 2011, pp. 245–252. ACM, New York, NY, USA (2011)
9. Kang, J., Lerman, K.: LA-CTR: a limited attention collaborative topic regression for social media. In: Proceedings of the Twenty-Seventh AAAI Conference on Artificial Intelligence, 14–18 July 2013, Bellevue, Washington, USA (2013)
10. Wang, H., Chen, B., Li, W.: Collaborative topic regression with social regularization for tag recommendation. In: Proceedings of the 23rd International Joint Conference on Artificial Intelligence, IJCAI 2013, pp. 2719–2725, 3–9 August 2013, Beijing, China (2013)
11. Gopalan, P., Hofman, J.M., Blei, D.M.: Scalable recommendation with hierarchical Poisson factorization. In: Meila, M., Heskes, T., (eds.) Proceedings of the Thirty-First Conference on Uncertainty in Artificial Intelligence, UAI 2015, pp. 326–335, 12–16 July 2015. AUAI Press, Amsterdam, The Netherlands (2015)
12. Gopalan, P., Ruiz, F.J.R., Ranganath, R., Blei, D.M.: Bayesian nonparametric Poisson factorization for recommendation systems. In: Proceedings of the Seventeenth International Conference on Artificial Intelligence and Statistics, AISTATS 2014. JMLR Workshop and Conference Proceedings, vol. 33, pp. 275–283, 22–25 April 2014, Reykjavik, Iceland JMLR.org (2014)
13. Hu, C., Rai, P., Chen, C., Harding, M., Carin, L.: Scalable Bayesian non-negative tensor factorization for massive count data. In: Appice, A., Rodrigues, P.P., Santos Costa, V., Gama, J., Jorge, A., Soares, C. (eds.) ECML PKDD 2015. LNCS (LNAI), vol. 9285, pp. 53–70. Springer, Cham (2015). https://doi.org/10.1007/978-3-319-23525-7_4
14. Hu, C., Rai, P., Carin, L.: Non-negative matrix factorization for discrete data with hierarchical side-information. In: Proceedings of the 19th International Conference on Artificial Intelligence and Statistics, AISTATS 2016, pp. 1124–1132, 9–11 May 2016, Cadiz, Spain (2016)

15. Acharya, A., Teffer, D., Henderson, J., Tyler, M., Zhou, M., Ghosh, J.: Gamma process Poisson factorization for joint modeling of network and documents. In: Appice, A., Rodrigues, P.P., Santos Costa, V., Soares, C., Gama, J., Jorge, A. (eds.) ECML PKDD 2015. LNCS (LNAI), vol. 9284, pp. 283–299. Springer, Cham (2015). https://doi.org/10.1007/978-3-319-23528-8_18
16. Bishop, C.M.: Pattern Recognition and Machine Learning. Information Science and Statistics. Springer, New York (2006)
17. Cantador, I., Brusilovsky, P., Kuflik, T.: 2nd workshop on information heterogeneity and fusion in recommender systems (HetRec 2011). In: Proceedings of the 5th ACM conference on Recommender systems, RecSys 2011, New York, NY, USA, ACM (2011)

C-SALT: Mining Class-Specific ALTerations in Boolean Matrix Factorization

Sibylle Hess[(✉)] and Katharina Morik

Computer Science, TU Dortmund University, Dortmund, Germany
{sibylle.hess,katharina.morik}@tu-dortmund.de
http://www-ai.cs.uni-dortmund.de/PERSONAL/hess.html
http://www-ai.cs.uni-dortmund.de/PERSONAL/morik.html

Abstract. Given labeled data represented by a binary matrix, we consider the task to derive a Boolean matrix factorization which identifies commonalities and specifications among the classes. While existing works focus on rank-one factorizations which are either specific or common to the classes, we derive class-specific alterations from common factorizations as well. Therewith, we broaden the applicability of our new method to datasets whose class-dependencies have a more complex structure. On the basis of synthetic and real-world datasets, we show on the one hand that our method is able to filter structure which corresponds to our model assumption, and on the other hand that our model assumption is justified in real-world application. Our method is parameter-free. Code and data related to this chapter are available at: https://doi.org/10.6084/m9.figshare.5441365.

Keywords: Boolean matrix factorization · Shared subspace learning
Nonconvex optimization · Proximal alternating linearized optimization

1 Introduction

When given labeled data, a natural instinct for a data miner is to build a discriminative model that predicts the correct class. Yet in this paper we put the focus on the characterization of the data with respect to the label, i.e., finding similarities and differences between chunks of data belonging to miscellaneous classes. Consider a binary matrix where each row is assigned to one class. Such data emerge from fields such as gene expression analysis, e.g., a row reflects the genetic information of a cell, assigned to one tissue type (primary/relapse/no tumor), market basket analysis, e.g., a row indicates purchased items at the assigned store, or from text analyses, e.g., a row corresponds to a document/article and the class denotes the publishing platform. For various applications a characterization of the data with respect to classes is of particular interest. In genetics, filtering the genes which are responsible for the re-occurrence of a tumor may introduce new possibilities for personalized medicine [14]. In market basket analysis it might be of interest which items sell better in some shops than others and in text analysis one might ask about variations in the vocabulary used when reporting from diverse viewpoints.

© Springer International Publishing AG 2017
M. Ceci et al. (Eds.): ECML PKDD 2017, Part I, LNAI 10534, pp. 547–563, 2017.
https://doi.org/10.1007/978-3-319-71249-9_33

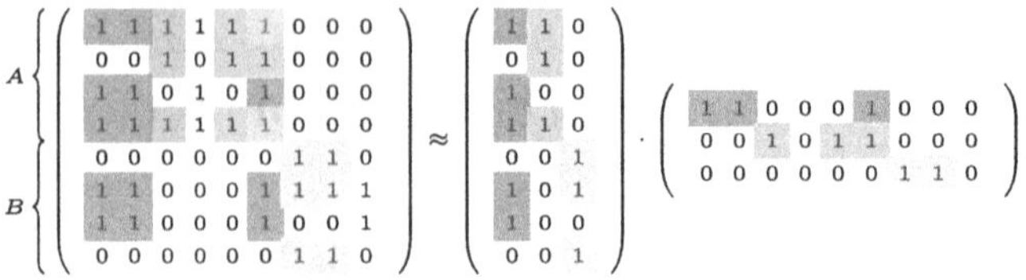

$$
\begin{aligned}
A &\left\{ \begin{array}{c}
\left(\begin{array}{ccccccccc}
1 & 1 & 1 & 1 & 1 & 1 & 0 & 0 & 0 \\
0 & 0 & 1 & 0 & 1 & 1 & 0 & 0 & 0 \\
1 & 1 & 0 & 1 & 0 & 1 & 0 & 0 & 0 \\
1 & 1 & 1 & 1 & 1 & 1 & 0 & 0 & 0 \\
0 & 0 & 0 & 0 & 0 & 0 & 1 & 1 & 0
\end{array} \right. \\
B &\left. \begin{array}{ccccccccc}
1 & 1 & 0 & 0 & 0 & 1 & 1 & 1 & 1 \\
1 & 1 & 0 & 0 & 0 & 1 & 0 & 0 & 1 \\
0 & 0 & 0 & 0 & 0 & 0 & 1 & 1 & 0
\end{array} \right)
\end{array} \right.
\end{aligned}
\approx
\left(\begin{array}{ccc}
1 & 1 & 0 \\
0 & 1 & 0 \\
1 & 0 & 0 \\
1 & 1 & 0 \\
0 & 0 & 1 \\
1 & 0 & 1 \\
1 & 0 & 0 \\
0 & 0 & 1
\end{array} \right)
\cdot
\left(\begin{array}{ccccccccc}
1 & 1 & 0 & 0 & 0 & 1 & 0 & 0 & 0 \\
0 & 0 & 1 & 0 & 1 & 1 & 0 & 0 & 0 \\
0 & 0 & 0 & 0 & 0 & 0 & 1 & 1 & 0
\end{array} \right)
$$

Fig. 1. A Boolean factorization of rank three. The data matrix on the left is composed by transactions belonging to two classes A and B. Each outer product is highlighted. Best viewed in color.

These questions are approached as pattern mining [17] and Boolean matrix factorization [8] problems. Both approaches search for factors or patterns which occur in both or only one of the classes. This is illustrated in Fig. 1; a data matrix is indicated on the left, whose rows are assigned to one class, A or B. While the pink outer product spreads over both classes, the blue and green products concentrate in only one of the classes. We refer to the factorizations of the first kind as common and to those of the second kind as class-specific.

The identification of class specific and common factorizations is key to a characterization of similarities and differences among the classes. Yet, what if meaningful deviations between the classes are slightly hidden underneath an overarching structure? The factorization in Fig. 1 is not exact, we can see that the red colored ones in the data matrix are not taken into account by the model. This is partially desired as the data is expected to contain noise which is supposedly filtered by the model. On the other hand, we can observe concurrence of the red ones and the pink factors – in each class.

1.1 Main Contributions

In this paper we propose a novel Boolean Matrix Factorization (BMF) method which is suitable to compare horizontally concatenated binary data matrices originating from diverse sources or belonging to various classes. To the best of the authors knowledge, this is the first method in the field of matrix factorizations of any kind, combining the properties listed below in one framework:

1. the method can be applied to compare any number of classes or sources,
2. the factorization rank is automatically determined; this includes the number of outer products, which are common among multiple classes, but also the number of discriminative outer products occurring in only one class,
3. in addition to discriminative rank-one factorizations, more subtle characteristics of classes can be derived, pointing out how common outer products deviate among the classes.

While works exist which approach one of the points 1 or 2 (see Sect. 2.2), the focus on subtle deviations among the classes as addressed in point 3 is entirely new. This expands the applicability of the new method to datasets where deviations among the classes have a more complex structure.

2 Preliminaries

We identify items $\mathcal{I} = \{1, \ldots, n\}$ and transactions $\mathcal{T} = \{1, \ldots, m\}$ by a set of indices of a binary matrix $D \in \{0,1\}^{m \times n}$. This matrix represents the data, having $D_{ji} = 1$ iff transaction j contains item i. A set of items is called a *pattern*.

We assume that the data matrix is composed of various sources, identified by an assignment from transactions to classes. Denoting by $[A^{(a)}]_a$ the matrix horizontally concatenating the matrices $A^{(a)}$ for $a \in \{1, \ldots, c\}$, we write

$$D = \left[D^{(a)}\right]_a , \; Y = \left[Y^{(a)}\right]_a \text{ and } V^T = \left[V^{(a)T}\right]_a . \tag{1}$$

The $(m_a \times n)$-matrix $D^{(a)}$ comprises the $m_a < m$ transactions belonging to class a. Likewise, we explicitly notate the class-related $(m_a \times r)$- and $(n \times r)$-dimensional parts of the $m \times r$ and $n \times rc$ factor matrices Y and V as $Y^{(a)}$ and $V^{(a)}$. These factor matrices are properly introduced in Sect. 2.3.

We often employ the function θ_t which rounds a real value $x \geq t$ to one and $x < t$ to zero. We abbreviate $\theta_{0.5}$ to θ and denote with $\theta(X)$ the entry-wise application of θ to a matrix X. We denote matrix norms as $\| \cdot \|$ for the Frobenius norm and $| \cdot |$ for the entry-wise 1-norm. We express with $x^{m \times n}$ the $(m \times n)$-dimensional matrix having all entries equal to x. The operator $\circ$ denotes the Hadamard product. Finally, we denote with log the natural logarithm.

2.1 Boolean Matrix Factorization in Brief

Boolean Matrix Factorization (BMF) assumes that the data $D \in \{0,1\}^{m \times n}$ originates from a matrix product with some noise, i.e.,

$$D = \theta(YX^T) + N, \tag{2}$$

where $X \in \{0,1\}^{n \times r}$ and $Y \in \{0,1\}^{m \times r}$ are the factor matrices of rank r and $N \in \{-1, 0, 1\}^{m \times n}$ is the noise matrix. The Boolean product disjuncts r matrices; the outer products $Y_{.s}X_{.s}^T$ for $1 \leq s \leq r$. We use θ to denote the Boolean disjunction in terms of Boolean algebra. Each outer product is defined by a pattern, indicated by $X_{.s}$, and a set of transactions using the pattern, indicated by $Y_{.s}$. Correspondingly, X is called the pattern and Y the usage matrix.

Unfortunately, solving X and Y from Eq. (2), if only the data matrix D is known, is generally not possible. Hence, surrogate tasks are formulated in which the data is approximated by a matrix product according to specific criteria. The most basic approach is to find the factorization of given rank which minimizes the residual sum of absolute values $|D - \theta(YX^T)|$. This problem, however, cannot be approximated within any factor in polynomial time (unless **NP = P**) [9].

BMF has a very popular relative, called Nonnegative Matrix Factorization (NMF). Here, a nonnegative data matrix $D \in \mathbb{R}_+^{m \times n}$ is approximated by the product of nonnegative matrices $X \in \mathbb{R}_+^{n \times r}$ and $Y \in \mathbb{R}_+^{m \times r}$. NMF tasks often involve minimizing the Residual Sum of Squares (RSS) $\frac{1}{2}\|D - YX^T\|^2$ [18]. Minimizing the RSS subject to binary matrices X and Y introduces the task of binary matrix factorization [19].

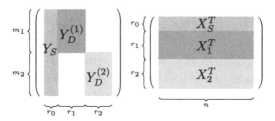

Fig. 2. A Boolean product identifying common (pink) and class-specific outer products (blue and green). Best viewed in color.

2.2   Related Work

If the given data matrix is class-wise concatenated (cf. Eq. (1)), a first approach for finding class-defining characteristics is to separately derive factorizations for each class. However, simple approximation measurements as discussed in Sect. 2.1 are already nonconvex and have multiple local optima. Due to this vagueness of computed models, class-wise factorizations are not easy to interpret; they lack a view on the global structure. Puzzling together the (parts of) patterns defining (dis-)similarities of classes afterwards, is non-trivial.

In the case of nonnegative, labeled data matrices, measures such as Fisher's linear discriminant criterion are minimized to derive weighted feature vectors, i.e., patterns in the binary case, which discriminate most between classes. This variant of NMF is successfully implemented for classification problems such as face recognition [11] and identification of cancer-associated genes [12].

For social media retrieval, Gupta et al. introduce Joint Subspace Matrix Factorization (JSMF) [2]. Focusing on the two-class setting, they assume that data points (rows of the data matrix) emerge not only from discriminative but also from common subspaces. JSMF infers for a given nonnegative data matrix and ranks r_0, r_1 and r_2 a factorization as displayed in Fig. 2. Multiplicative updates minimize the weighted sum of class-wise computed RSS. In Regularized JSNMF (RJSNMF), a regularization term is used to prevent that shared feature vectors swap into discriminative subspaces and vice versa [3]. The arising optimization problem is solved by the method of Lagrange multipliers. Furthermore, a provisional method to determine the rank automatically is evaluated. However, this involves multiple runs of the algorithm with increasing rank of shared and discriminative subspaces, until the approximation error barely decreases. A pioneering extension to the multi-class case is provided in [4].

Miettinen [8] transfers the objective of JSMF into Boolean algebra, solving

$$\min_{X,Y} \sum_{a \in \{1,2\}} \frac{\mu_a}{2} \left| D^{(a)} - \theta \left(\begin{bmatrix} Y_S^{(a)} & Y_D^{(a)} \end{bmatrix} \begin{bmatrix} X_S^T \\ X_a^T \end{bmatrix} \right) \right|$$

for binary matrices D, X and Y, and normalizing constants $\mu_{1/2}^{-1} = |D^{(2/1)}|$. A variant of the BMF algorithm ASSO [9] governs the minimization. A provisional determination of ranks based on the Minimum Description Length (MDL) principle is proposed, computing which of the candidate rank constellations yields the lowest description length. The description length captures model complexity and data fit, and is hence suitable for model order selection [5, 10].

Budhatoki and Vreeken [17] pursue the idea of MDL to derive a set of pattern sets, which characterizes similarities and differences of groups of classes. Identifying the usage of each pattern with its support in the data, the number of derived patterns equates the rank in BMF. In this respect, their proposed algorithm DIFFNORM automatically determines the ranks in the multi-class case. However, the posed constraint on the usage often results in vast amount of returned patterns.

In the two-class nonnegative input matrices case, Kim et al. improve over RJSNMF by allowing small deviations from shared patterns in each class [6]. They found that shared patterns are often marginally altered according to the class. In this paper, we aim at finding these overlooked variations of shared patterns together with strident differences among multiple classes, combining the strengths of MDL for rank detection and the latest results in NMF.

2.3   (Informal) Problem Definition

Given a binary data matrix composed from multiple classes, we assume that the data has an underlying model similar to the one in Fig. 1. There are common or shared patterns (pink) and class-specific patterns (blue and green). Furthermore, there are class-specific patterns, which align within a subset of the classes where a pattern is used (the red ones). We call such aligning patterns class-specific alterations and introduce the matrix V to reflect these.

Definition 1. *Let $X \in \{0,1\}^{n \times r}$ and $V \in \{0,1\}^{n \times cr}$. We say the matrix V models class-specific alterations of X if $\|X \circ V^{(a)}\| = 0$ for all $1 \leq a \leq c$, and $\|V^{(1)} \circ \ldots \circ V^{(c)}\| = 0$.*

Similar to the data decomposition denoted in Eq. (2), we assume that data emerges from a Boolean matrix product; yet, we now consider multiple products, one for each class, which are defined by the class-wise alteration matrix V, its pattern matrix, usage and the noise matrix $N = [N^{(a)}]_a$, such that for $1 \leq a \leq c$

$$D^{(a)} = \theta \left(Y^{(a)} (X + V^{(a)})^T \right) + N^{(a)}. \tag{3}$$

Given a class-wise composed binary data matrix, we consider the task to filter the factorization, defined by X, Y and V, from the noise.

3   The Proposed Method

We build upon the BMF algorithm PRIMP, which combines recent results from numerical optimization with MDL in order to return interpretable factorizations of a suitably estimated rank [5]. The employed description length f reflects the size of the data encoded by a code table as known from algorithms SLIM and

KRIMP [15,16]. Determining a smooth function F, bounding the description length from above, and a function ϕ to penalize non-binary values, locally minimizing matrices of the relaxed objective $F(X,Y) + \phi(X) + \phi(Y)$ are derived. Rounding the local minimizers to binary matrices according to the description length, yields the final result and decides over the rank of the factorization.

The numerical optimization is performed by *Proximal Alternating Linearized Minimization* (PALM) [1]. That are alternatingly invoked *proximal mappings* with respect to ϕ from the gradient descent update with respect to F (cf. lines 6, 8 and 10 in Algorithm 1). The proximal mapping of ϕ returns a matrix satisfying the following minimization criterion:

$$\text{prox}_\phi(X) \in \arg\min_{\hat{X}} \left\{ \frac{1}{2}\|X - \hat{X}\|^2 + \phi(\hat{X}) \right\}.$$

Loosely speaking, X is given a little push into a direction minimizing ϕ. We choose $\phi(X) = \sum_{i,j} \Lambda(X_{ij})$ to penalize non-binary matrix-entries by an entry-wise application of the function Λ. Correspondingly, the prox-operator is computed entry-wise $\text{prox}_{\alpha\phi}(X) = (\text{prox}_{\alpha\Lambda}(X_{ji}))_{ji}$, where

$$\Lambda(x) = \begin{cases} -|1 - 2x| + 1 & x \in [0,1] \\ \infty & x \notin [0,1]. \end{cases}, \quad \text{prox}_{\alpha\Lambda}(x) = \begin{cases} \max\{0, x - 2\alpha\} & x \le 0.5 \\ \min\{1, x + 2\alpha\} & x > 0.5. \end{cases}$$

Notice, the proximal mapping ensures that factor matrices always attain values between zero and one. For further information on prox-operators, see, e.g., [13].

The step sizes of the gradient descent updates are computed by the Lipschitz moduli of partial gradients (cf. lines 5, 7 and 9 in Algorithm 1). Assuming that the infimum of F and ϕ exists and ϕ is proper and lower continuous, PALM generates a nonincreasing sequence of function values which converges to a critical point of the relaxed objective.

3.1   C-Salt

In order to capture class-defining characteristics in the framework of PRIMP, few extensions have to be made. We pose two requirements on the interplay between usage and class-specific alterations of patterns: class-specific alterations ought to fit very well to the corresponding class but as little as possible to other classes. We introduce a regularizing function to penalize nonconformity to this request.

$$S(Y,V) = \sum_{s=1}^{r}\sum_{a=1}^{c}\left(\left|Y_{\cdot s}^{(a)}\right|\left|V_{\cdot s}^{(a)}\right| - Y^{(a)^T}D^{(a)}V_{\cdot s}^{(a)} \right) + \sum_{b\neq a} Y_{\cdot s}^{(b)^T}D^{(b)}V_{\cdot s}^{(a)}$$

$$= \sum_{a=1}^{c} \text{tr}\left(\left(Y^{(a)^T}(1^{m_a \times n} - 2D^{(a)}) + Y^T D\right) V^{(a)} \right).$$

We extend the description length of PRIMP such that class-specific alterations are encoded in the same way as patterns; by standard codes, assigning item

Algorithm 1. C-SALT$(D = [D^{(a)}]_a; \Delta_r = 10, \gamma = 1.00001)$

1: $(X_K, V_K, Y_K) \leftarrow (\emptyset, \emptyset, \emptyset)$
2: **for** $r \in \{\Delta_r, 2\Delta_r, 3\Delta_r, \ldots\}$ **do**
3:      $(X_0, V_0, Y_0) \leftarrow$ INCREASERANK$(X_K, V_K, Y_K, \Delta_r)$ ▷ Append random columns
4:      **for** $k = 0, 1, \ldots$ **do**   ▷ Select stop criterion
5:          $1/\alpha_k \leftarrow \gamma M_{\nabla_X F}(V_k, Y_k)$
6:          $X_{k+1} \leftarrow \text{prox}_{\alpha_k \phi}(X_k - \alpha_k \nabla_X F(X_k, V_k, Y_k))$
7:          $1/\nu_k^{(a)} \leftarrow \gamma M_{\nabla_V^{(a)} F}(X_{k+1}, Y_k)$ ▷ $1 \leq a \leq c$
8:          $V_{k+1}^{(a)} \leftarrow \text{prox}_{\nu_k^{(a)} \phi}\left(V_k^{(a)} - \nu_k^{(a)} \nabla_V^{(a)} F(X_{k+1}, V_k^{(a)}, Y_k)\right)$ ▷ $1 \leq a \leq c$
9:          $1/\beta_k \leftarrow \gamma M_{\nabla_Y F}(X_{k+1}, V_{k+1})$
10:         $Y_{k+1} \leftarrow \text{prox}_{\beta_k \phi}(Y_k - \beta_k \nabla_Y F(X_{k+1}, V_{k+1}, Y_k))$
11:     $(X, V, Y) \leftarrow$ ROUND(f, X_k, V_k, Y_k) ▷ Try thresholds from finite set
12:     **if** $r - r(X, V, Y) > 1$ **then return** (X, V, Y) **end if**

$i \in \mathcal{I}$ a code of length $u_i = -\log(|D_{\cdot i}|/|D|)$. The objective function f adds the description length to the specificity-regularizer

$$f(X, V, Y) = -\sum_{s:|Y_{\cdot s}|>0}\left((|Y_{\cdot s}|+1) \cdot \log\left(\frac{|Y_{\cdot s}|}{|Y|+|N|}\right) + X_{\cdot s}^T u + \sum_a V_{\cdot s}^{(a)^T} u\right)$$
$$- \sum_{i:|N_{\cdot i}|>0}\left((|N_{\cdot i}|+1) \cdot \log\left(\frac{|N_{\cdot i}|}{|Y|+|N|}\right) + u_i\right) + S(Y, V).$$

This determines the relaxed objective $F(X, V, Y) + \phi(X) + \phi(V) + \phi(Y)$, where

$$F(X, V, Y) = \frac{1}{2}\left(\mu \sum_{a=1}^c \|D^{(a)} - Y^{(a)}(X + V^{(a)})^T\|^2 + G(X, V, Y) + S(Y, V)\right),$$

$\mu = 1 + \log(n)$ and G is defined as stated in Appendix A. F has Lipschitz continuous gradients and is suitable for PALM.

Algorithm 1 details C-SALT, which largely follows the framework of PRIMP [5]. C-SALT has as input the data D and two parameters, for which default values are given, which rarely need to be adjusted in practice. Further information about the robustness and significance of these parameters is provided in Algorithm 1. For step-wise increased ranks, PALM optimizes the relaxed objective (lines 4–10). Note that the alternating minimization of more than two matrices corresponds to the extension of PALM for multiple blocks, discussed in [1]. The required gradients and Lipschitz moduli are stated in Appendix A. Subsequently, a rounding procedure returns the binary matrices $X_{t_1} = \theta_{t_1}(X_K)$, $V_{t_1} = \theta_{t_1}(V_K)$ and $Y_{t_2} = \theta_{t_2}(Y_K)$ for thresholds $t_1, t_2 \in \{0.05k \mid k \in \{0, 1, \ldots, 20\}\}$ minimizing f. Thereby, the validity of Definition 1 is ensured by setting unsuitable values in V to zero. Furthermore, *trivial* outer products covering fewer than two transactions or items are removed. The number of remaining outer products defines the rank $r(X, V, Y)$. If the gap between the number of possibly and actually modeled outer products is larger than one, the current factorization is returned (line 12).

4 Experiments

The experimental evaluations concern the following research questions:

1. Given that the data matrix is generated as stated by the informal problem definition in Sect. 2.3, does C-SALT find the original data structure?
2. Is the assumption that real-world data emerge as stated in Eq. (3) reasonable, and what effect has the modeling of class-specific alterations on the results?

We compare against the algorithms DBSSL, the dominated approach proposed in [8], and PRIMP[1]. The first question is approached by a series of synthetic datasets, generated according to Eq. (3). To address the second question, we compare on real-world datasets the RSS, computed factorization ranks and visually inspect derived patterns. Furthermore, we discuss an application in genome analysis where none of the existing methods provides the crucial information.

For C-SALT and PRIMP we use as stop criterion a minimum average function decrease (of last 500 iterations) of 0.005 and maximal $k \leq 10,000$ iterations. We use the Matlab/C implementation of DBSSL which has been kindly provided by the authors upon request. Setting the minimum support parameter of the employed FP-Growth algorithm proved tricky. Choosing the minimum support too low results in a vast memory consumption (we provided 100 GiB RAM); setting it too high yields too few candidate patterns. Hence, this parameter varies between experiments within the range $\{2,\dots,8\}$.

C-SALT is implemented for GPU, as is PRIMP. We provide the source code of our algorithms together with the data generating script[2].

4.1 Measuring the Quality of Factorizations

For synthetic datasets, we compare the computed models against the planted structure by an adaptation of the micro-averaged F-measure. We assume that generated matrices $X^\star, V^\star, Y^\star$ and computed models X, V, Y have the same rank r. Otherwise, we attach columns of zeros to make them match. We compute one-to-one matchings $\sigma_1 : \{1,\dots,r\} \to \{1,\dots,r\}$ between outer products of computed and generated matrices by the Hungarian algorithm [7]. The matching maximizes $\sum_{s=1}^{r} F_{s,\sigma_1(s)}^{(a)}$, where

$$F_{S,T}^{(a)} = 2 \frac{\text{pre}_{S,T}^{(a)} \cdot \text{rec}_{S,T}^{(a)}}{\text{pre}_{S,T}^{(a)} + \text{rec}_{S,T}^{(a)}},$$

---

[1] http://sfb876.tu-dortmund.de/primp.
[2] http://sfb876.tu-dortmund.de/csalt.

for selections of columns S and T. $\text{pre}_{S,T}^{(a)}$ and $\text{rec}_{S,T}^{(a)}$ denote precision and recall w.r.t. the denoted column selection. Writing $X^{(a)} = X + V^{(a)}$, we compute

$$\text{pre}_{S,T}^{(a)} = \frac{\left|(Y_{\cdot S}^{\star} \circ Y_{\cdot T})^{(a)}(X_{\cdot S}^{\star} \circ X_{\cdot T})^{(a)^T}\right|}{\left|Y_{\cdot T}^{(a)} X_{\cdot T}^{(a)^T}\right|}, \quad \text{rec}_{S,T}^{(a)} = \frac{\left|(Y_{\cdot S}^{\star} \circ Y_{\cdot T})^{(a)}(X_{\cdot S}^{\star} \circ X_{\cdot T})^{(a)^T}\right|}{\left|Y_{\cdot S}^{\star\,(a)} X_{\cdot S}^{\star\,(a)^T}\right|}.$$

We calculate then precision and recall such that planted outer products with indices $R = (1, \ldots, r)$ are compared to outer products of the computed factorization with indices $\sigma_1(R) = (\sigma_1(1), \ldots, \sigma_1(r))$. The corresponding F-measure is the micro F-measure, which is identified by $F_{R,\sigma_1(R)}^{(a)}$.

Since class-specific alterations of patterns, reflected by the matrix V, are particularly interesting in the scope of this paper, we additionally state the recall of $V^{\star}$, denoted by rec_V. Therefore, we compute a maximum matching σ_2 between generated class alterations $V^{\star}$ with usage $Y^{\star}$ and computed patterns $X_V = [X \; V]$ (setting V to the $(n \times cr)$ zero matrix for other algorithms than C-SALT) with usage $Y_V = [Y \ldots Y]$ (concatenating c times). The recall $\text{rec}_{R,\sigma_2(R)}^{(a)}$ is then computed with respect to the matrices $V^{\star}, Y^{\star}, X_V$ and Y_V. Furthermore, we compute the class-wise factorization rank $r^{(a)}$ as the number of nontrivial outer products, involving more than only one column or row. Outer products where solely one item or one transaction is involved yield no insight for the user and are therefore always discarded. In following plots, we indicate averaged measures over all classes

$$F = \frac{1}{c}\sum_a F_{R,\sigma_1(R)}^{(a)}, \quad \text{rec}_V = \frac{1}{c}\sum_a \text{rec}_{R,\sigma_2(R)}^{(a)} \text{ and } r = \frac{1}{c}\sum_a r^{(a)}.$$

Therewith, the size of the class is not taken into account; the discovery of planted structure is considered equally important for every class. F-measure and recall have values between zero and one. The closer both approach one, the more similar are the obtained and planted factorizations.

4.2   Synthetic Data Generation

We state the synthetic data generation as a procedure which receives the matrix dimensions $(m_a)_a$ $(m = \sum_a m_a)$ and n, the factorization rank $r^{\star}$, matrix $C \in \{0, 1\}^{c \times r}$ and noise probability p as input. The matrix C indicates for each pattern in which classes it is used.

GenerateData$(n, (m_a)_a, r^{\star}, C, p)$

1. Draw the $(n \times r^{\star})$ and $(m \times r^{\star})$ matrices $X^{\star}$, $V^{(a)^{\star}}$ and $Y^{\star}$ uniformly random from the set of all binary matrices subject to
   - each column $X_{\cdot s}^{\star}(Y_{\cdot s}^{\star})$ has at least $n/100(m/100)$ uniquely assigned bits,
   - the density is bounded by $|X_{\cdot s}^{\star}| \leq n/10$ and $|Y_{\cdot s}^{(a)^{\star}}| \leq C_{sa}m_a/10$
   - $V^{(a)^{\star}}$ models class-specific alterations of $X^{\star}$ and $\left|\sum_{a=1}^c V_s^{(a)^{\star}}\right| \leq 2/3\,|X_{\cdot s}^{\star}|$

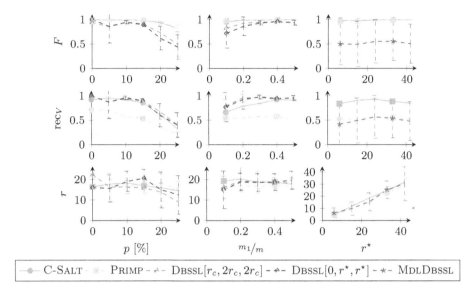

Fig. 3. Variation of noise (left column), class distribution m_1/m (middle column) and the rank (right column). The F-measure, recall of the matrix V (both the higher the better) and the class-wise estimated rank of the calculated factorization is plotted against the varied parameter. Best viewed in color.

2. Set $D^{(a)}$, flipping every bit of $\theta \left(Y^{(a)\star} (X^\star + V^{(a)\star})^T \right)$ with probability p.

By default, the parameters $r^\star = 24$, $m_a = m/2$, where m and n are varied as described in Sect. 4.3, $p = 0.1$, and depending on the number of classes we set

$$C_2 = \left[\begin{pmatrix} 1\,0\,1 \\ 1\,1\,0 \end{pmatrix} \right]_{\frac{r^\star}{3}}, \quad C_3 = \left[\begin{pmatrix} 1\,1\,0\,0 \\ 1\,0\,1\,0 \\ 1\,0\,1\,1 \end{pmatrix} \right]_{\frac{r^\star}{4}}, \quad C_4 = \left[\begin{pmatrix} 1\,1\,0\,0\,0 \\ 1\,0\,1\,0\,0 \\ 1\,0\,1\,1\,0 \\ 1\,0\,1\,1\,1 \end{pmatrix} \right]_{\frac{r^\star}{5}}.$$

4.3   Synthetic Data Experiments

We plot for the following series of experiments the averaged F-measure, recall rec_V, and the rank (cf. Sect. 4.1), against the parameter varied when generating the synthetic data (see Sect. 4.2). Error bars have length 2σ. For every experiment, we generate eight matrices: two for each combination of dimensions $(n, m) \in \{(500, 1600), (1600, 500), (800, 1000), (1000, 800)\}$.

Figure 3 contrasts the results of C-Salt, Primp and Dbssl in the two-class setting. For Dbssl, we consider two instantiations if the rank $r^\star$ is fixed. Both correctly reflect the number of planted specific and common patterns, yet the one rates class-specific alterations as separate patterns and the other counts every pattern with its class-specific alteration as a class-specific pattern. In the

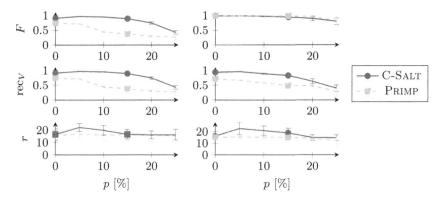

Fig. 4. Variation of noise for generated data matrices with three (left) and four classes (right). The F-measure, recall of the matrix V (both the higher the better) and the class-wise estimated rank of the calculated factorization (between 16 and 24 can be considered correct) is plotted against the varied parameter. Best viewed in color.

experiments varying the rank, we employ the MDL-based selection of the rank proposed for DBSSL. The input candidate constellations of class-specific and common patterns are determined according to the number of planted patterns, i.e., candidate rank constellations are a combination of $r_0 \in r^*/3 \pm \{5, 0\}$ and $r_1 = r_2 \in \{kr^*/3 \mid k \in \{1, 2, 4\}\}$.

Figure 3 shows the performance measures of the competing algorithms when varying three parameters: noise p (left column), ratio of transactions in each class m_1/m (middle column) and rank r^* (right column). We observe an overall high F-measure of C-SALT and PRIMP. Both DBSSL instantiations also obtain high F-values, but only at lower noise levels and if one class is not very dominant over the other. C-SALT and PRIMP differ most notably in the discovery of class specific alterations measured by rec_V. C-SALT shows a similar recall as DBSSL if the noise is varied but a lower recall if classes are imbalanced. The ranks of returned factorizations by all algorithms lie in a reasonable interval, considering that class-specific alterations can also be interpreted as unattached patterns. Hence, a class-wise averaged rank between 16 and 24 is legitimate. When varying the number of planted patterns, the MDL selection procedure of the rank also yields correct estimations for DBSSL. However, the F-measure and recall of V^* decrease to 0.5 if the rank is not set to the correct parameters for DBSSL.

Figure 4 displays the results of PRIMP and C-SALT when varying the noise for generated class-common and class-specific factorizations for three and four classes. The plots are similar to Fig. 3. The more complex constellations of class-overarching outer products, which occur when more than two classes are involved, do not notably affect the ability to discover class-specific alterations by C-SALT and the planted factorization by PRIMP and C-SALT.

Table 1. Comparison of the amount of derived class-specific (r_1, r_2) and class-common patterns (r_0), the overall rank $r = r_0 + r_1 + r_2$ and the RSS of the BMF (scaled by 10^4) for real-world datasets. Values in parentheses correspond to factorizations where outer products with less than four items or transactions are discarded. The last two columns summarize characteristics of the datasets: number of rows belonging to the first and second class (m_1, m_2), number of columns (n) and density $d = |D|/(nm)$ in percent.

	Space-Rel				Politics				Movie			
	C-Salt	Primp	Dbssl1	Dbssl2	C-Salt	Primp	Dbssl1	Dbssl2	C-Salt	Primp	Dbssl1	Dbssl2
r	29(28)	30(30)	40(7)	18(6)	41(40)	30(30)	57(20)	42(15)	26(25)	30(27)	27(4)	12(4)
r_0	4(3)	8(8)	19(1)	7(1)	10(10)	8(8)	16(2)	5(0)	25(25)	29(27)	21(1)	6(0)
r_1	9(9)	8(8)	13(4)	8(4)	19(18)	15(15)	27(14)	18(11)	1(0)	1(0)	3(1)	3(1)
r_2	16(16)	14(14)	8(2)	3(1)	12(12)	7(7)	14(4)	19(4)	0(0)	0(0)	3(2)	3(3)
RSS	76(77)	76(76)	73(79)	76(79)	119(119)	122(122)	110(122)	116(123)	320(320)	319(319)	315(318)	316(318)
	m_1	m_2	n	$d[\%]$	m_1	m_2	n	$d[\%]$	m_1	m_2	n	$d[\%]$
	622	980	2244	2.27	936	775	2985	2.64	998	997	4442	3.68

4.4 Real-World Data Experiments

We explore the algorithms' behavior by three interpretable text-datasets depicted in Table 1. The datasets are composed by two classes to allow a comparison to Dbssl. The dimensions m_1 and m_2 describe how many documents belong to the first, respectively second class. Each document is represented by its occurring lemmatized words, excluding stop words. The dimension n reflects the number of words which occur in 20 documents at least. From the 20 Newsgroup corpus[3], we compose the *Space-Rel* dataset by posts from sci.space and talk.religion.misc, and the *Politics* dataset from talk.politics.mideast and talk.politics.misc. The *Movie* dataset is prepared from a collection of 1000 negative and 1000 positive movie reviews[4].

We consider two instantiations of Dbssl: Dbssl1 is specified by $r_0 = r_1 = r_2 = 30$ and Dbssl2 by $r_0 = r_1 = r_2 = 15$. For a fair comparison, we set a maximum rank of 30 for C-Salt and Primp. Therewith, the returned factorizations have a maximum rank of 90 for Dbssl1, 45 for Dbssl2, 30 for Primp and 60 for C-Salt. Note that C-Salt has the possibility to neglect X and use mainly V to reflect $cr = 60$ class-specific outer products. In practice, we consider patterns $V_{\cdot s}^{(a)} + X_{\cdot s}$ as individual class-specific patterns if $|V_{\cdot s}^{(a)}| > |X_{\cdot s}|$.

Table 1 shows the number of class-specific and common patterns, and the resulting RSS. Since outer products involving only a few items or transactions either provide little insight or are difficult to interpret, we also state in parentheses the values concerning *truncated factorizations*, i.e., outer products reflecting less than four items or transactions are discarded (glossing over the truncating of singletons, which is performed in both cases).

The untruncated factorizations obtained from Dbssl generally obtain a low RSS. However, when we move to the more interesting truncated factorizations,

---

[3] http://qwone.com/~jason/20Newsgroups/.
[4] http://www.cs.cornell.edu/People/pabo/movie-review-data/.

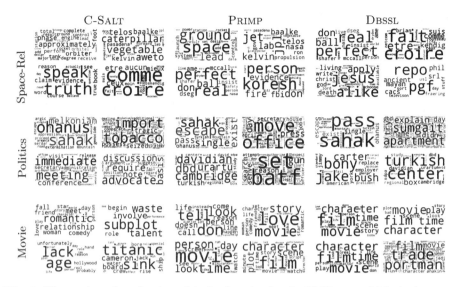

Fig. 5. Illustration of a selection of derived topics for the 20 News and Movie datasets. The size of a word reflects its frequency in the topic ($\sim Y_{.s}^T D_{.i}$) and the color its class affiliation: pink words are class-common, blue words belong to the first and green words to the second class. Best viewed in color.

DBSSL suffers (the rank shrinks to less than a third for factorizations of DBSSL2). On the 20 News datasets this leads to a substantial RSS increase; C-SALT and PRIMP provide the lowest RSS in this case. We also observe, that the integration of the matrix V by C-SALT empowers the derivation of more class-specific factorizations than PRIMP. Nevertheless, both algorithms describe the Movie dataset only by class-common patterns. We inspect these results more closely in the next section, showing that mining class-specific alterations points at exclusively derived class characteristics, especially for the Movie dataset.

4.5   Illustration of Factorizations

Let us inspect the derived most prevalent topics in the form of word clouds. Figure 5 displays for every algorithm the top four topics, whose outer product spans the largest area. Class-common patterns are colored pink whereas class-specific patterns are blue or green. Class-specific alterations within topics become apparent by differently colored words in one word cloud. We observe that the topics displayed for the 20-News data are mostly attributed to one of the classes. The topics are generally interpretable and even comparable among the algorithms (cf. the first topic in the Politics dataset). Here, class-specific alterations of C-SALT point at the context in which a topic is discussed, e.g., the press release from the white house after a conference or meeting took place, whereby the latter may be discussed in both threads (cf. the third topic for the Politics dataset).

Table 2. Average size and empirical standard deviation of patterns ($\cdot 10^3$) and class-specific alterations ($\cdot 10^3$).

Fig. 6. Transposed usage matrix returned by C-SALT on the genome dataset. Class-memberships are signalized by colors. (Color figure online)

$|X|$	$|V^{(N)}|$	$|V^{(T)}|$	$|V^{(R)}|$
10.7 ± 96	2.1 ± 2.5	3.6 ± 4.8	3.8 ± 6.6

The most remarkable contribution of class-specific alterations is given for the movie dataset. Generally, movie reviews addressing a particular genre, actors, etc., are not exclusively bad or good. PRIMP and C-SALT derive accordingly only common patterns. Here, C-SALT can derive the decisive hint which additional words indicate the class membership. We recall from Table 1 that DBSSL returns in total four truncated topics for the Movie dataset. Thus, the displayed topics for the Movie dataset represent all the information we obtain from DBSSL. In addition, the topics display a high overlap in words, which underlines the reasonability of our assumption that minor deviations of major and common patterns can denote the sole class-distinctions.

4.6 Genome Data Analysis

The results depicted in the previous section are qualitatively easy to assess. We easily identify overlapping words and filter the important class characteristics from the topics at hand. In this experiment, the importance or meaning of features is unclear and researchers benefit from any summarizing information which is provided by the method, e.g., the common and class-specific parts of a pattern. We regard the dataset introduced in [14] representing the genomic profile of 18 Neuroblastoma patients. For each patient, samples are taken from three classes: *normal* (N), *primary tumor* (T) and *relapse tumor cell* (R). The data denotes loci and alterations taking place with respect to a reference genome. Alterations denote nucleotide variations such as $A \rightarrow C$, insertions ($C \rightarrow AC$) and deletions ($AC \rightarrow A$). One sample from each of the classes N and T is given for every patient ($m_N = m_T = 18$), one patient lacks one and another has three additional relapse samples ($m_R = 20$), resulting in $m = 56$ samples. We convert the alterations into binary features, each representing one alteration at one locus (position on a chromosome). The resulting matrix has $n \approx 3.7$ million columns.

C-SALT returns on the genome data a factorization of rank 28, of which we omit sixteen patterns solely occurring in one patient. Figure 6 depicts the usage of the remaining twelve outer products, being almost identical for each class. Most notably, all derived patterns are class-common and describe the genetic background of patients instead of class characteristics. Table 2 summarizes the average length of patterns and corresponding class-specific alterations. We see that the average pattern reflects ten thousands of genomic alterations and that among the class-specific alterations, the ones which are attributed to relapse samples are highest in average. These results correspond to the evaluation in [14].

The information provided by C-SALT can not be extracted by existing methods. PRIMP yields only class-common patterns whose usage aligns with patients, regardless of classes. Running PRIMP separately on each class-related part $D^{(a)}$ yields factorizations of rank zero – the genomic alignments between patients can not be differentiated from noise for such few samples. However, using the framework of PRIMP to minimize the RSS without any regularization, yields about 15 patterns for each part $D^{(a)}$. The separately mined patterns overlap over the classes in an intertwined fashion. The specific class characteristics are not easily perceived for such complex dependencies and would require further applications of algorithms which structure the information from the sets of vast amounts of features.

5   Conclusion

We propose C-SALT, an explorative method to simultaneously derive similarities and differences among sets of transactions, originating from diverse classes. C-SALT solves a Boolean Matrix Factorization (BMF) by means of numerical optimization, extending the method PRIMP [5] to incorporate classes. We integrate a factor matrix reflecting class-specific alterations of outer products from a BMF (cf. Definition 1). Therewith, we capture class characteristics, which are lost by unsupervised factorization methods such as PRIMP. Synthetic experiments show that a planted structure corresponding to our model assumption is filtered by C-SALT (cf. Fig. 3). Even in the case of more than two classes, C-SALT filters complex dependencies among them (cf. Fig. 4). These experiments also show that the rank is correctly estimated. On interpretable text data, C-SALT derives meaningful factorizations which provide valuable insight into prevalent topics and their class specific characteristics (cf. Table 1 and Fig. 5). An analysis of genomic data underlines the usefulness of our new factorization method, yielding information which none if the existing algorithms can provide (cf. Sect. 4.6).

Acknowledgments. Part of the work on this paper has been supported by Deutsche Forschungsgemeinschaft (DFG) within the Collaborative Research Center SFB 876 "Providing Information by Resource-Constrained Analysis", project C1 http://sfb876. tu-dortmund.de.

A   Functions, Gradients and Lipschitz-Moduli

The functions, required by Algorithm 1, are stated in relation to $N = [N^{(a)}]_a$, as defined in Eq. (3). F, stated in Sect. 3.1, and its gradients are defined by

$$G(X,V,Y) = -\sum_{s=1}^{r}(|Y_{.s}| + 1)\log\left(\frac{|Y_{.s}| + 1}{|Y| + r}\right) + |X^T u| + \sum_{a=1}^{c}|V^{(a)^T} u| + |Y|,$$

$$\nabla_X F(X,V,Y) = -\mu\sum_{a=1}^{c} N^{(a)^T} Y^{(a)} + u(0.5)^{1\times n},$$

$$\nabla_V^{(a)} F(X, V, Y) = -\mu N^{(a)^T} Y^{(a)} + u(0.5)^{1 \times n} + \nabla_V^{(a)} S(Y, V),$$

$$\nabla_Y^{(a)} F(X, V, Y) = -\mu N^{(a)} X - \frac{1}{2} \left(\log \left(\frac{|Y_{.s}|+1}{|Y|+r} \right) - 1 \right)_{js} + \nabla_Y^{(a)} S(Y, V),$$

$$\nabla_V^{(a)} S(Y, V) = D^T Y + (1^{m_a \times n} - 2D^{(a)})^T Y^{(a)}$$

$$\nabla_Y^{(a)} S(Y, V) = D^{(a)} \left(\sum_{b \neq a} V^{(b)} \right) + (1^{m_a \times n} - D^{(a)}) V^{(a)}.$$

The Lipschitz moduli are $M_{\nabla_X F}(Y, V) = \mu \|YY^T\|$, $M_{\nabla_V^{(a)} F}(X, Y) = \mu \|Y^{(a)} Y^{(a)^T}\|$ and $M_{\nabla_Y^{(a)} F}(X, V) = \mu \|(X + V^{(a)})(X + V^{(a)})^T\| + m_a$, $M_{\nabla_Y F} = \|[(M_{\nabla_Y^{(a)} F}(X, Y)]_a\|$.

References

1. Bolte, J., Sabach, S., Teboulle, M.: Proximal alternating linearized minimization for nonconvex and nonsmooth problems. Math. Program. **146**(1–2), 459–494 (2014)
2. Gupta, S.K., Phung, D., Adams, B., Tran, T., Venkatesh, S.: Nonnegative shared subspace learning and its application to social media retrieval. In: KDD, pp. 1169–1178 (2010)
3. Gupta, S.K., Phung, D., Adams, B., Venkatesh, S.: DAMI. Regularized nonnegative shared subspace learning **26**(1), 57–97 (2013)
4. Gupta, S.K., Phung, D., Adams, B., Venkatesh, S.: A matrix factorization framework for jointly analyzing multiple nonnegative data sources. In: Yada, K. (ed.) Data Mining for Service. SBD, vol. 3, pp. 151–170. Springer, Heidelberg (2014). https://doi.org/10.1007/978-3-642-45252-9_10
5. Hess, S., Morik, K., Piatkowski, N.: The primping routine–tiling through proximal alternating linearized minimization. DAMI **31**(4), 1090–1131 (2017)
6. Kim, H., Choo, J., Kim, J., Reddy, C.K., Park, H.: Simultaneous discovery of common and discriminative topics via joint nonnegative matrix factorization. In: KDD, pp. 567–576 (2015)
7. Kuhn, H.W.: The hungarian method for the assignment problem. Naval Res. Logistics Q. **2**(1–2), 83–97 (1955)
8. Miettinen, P.: On finding joint subspace Boolean matrix factorizations. In: SDM, pp. 954–965 (2012)
9. Miettinen, P., Mielikäinen, T., Gionis, A., Das, G., Mannila, H.: TKDE. The discrete basis problem **20**(10), 1348–1362 (2008)
10. Miettinen, P., Vreeken, J.: Model order selection for Boolean matrix factorization. In: KDD, pp. 51–59 (2011)
11. Nikitidis, S., Tefas, A., Pitas, I.: Projected gradients for subclass discriminant nonnegative subspace learning. IEEE Trans. Cybern. **44**(12), 2806–2819 (2014)
12. Odibat, O., Reddy, C.K.: Efficient mining of discriminative co-clusters from gene expression data. Knowl. Inf. Syst. **41**(3), 667–696 (2014)
13. Parikh, N., Boyd, S.: Proximal algorithms. Found. Trends Optim. **1**(3), 127–239 (2014)

14. Schramm, A., Köster, J., Assenov, Y., Althoff, K., Peifer, M., Mahlow, E., Odersky, A., Beisser, D., Ernst, C., Henssen, A., Stephan, H., Schröder, C., Heukamp, L., Engesser, A., Kahlert, Y., Theissen, J., Hero, B., Roels, F., Altmüller, J., Nürnberg, P., Astrahantseff, K., Gloeckner, C., De Preter, K., Plass, C., Lee, S., Lode, H., Henrich, K., Gartlgruber, M., Speleman, F., Schmezer, P., Westermann, F., Rahmann, S., Fischer, M., Eggert, A., Schulte, J.: Mutational dynamics between primary and relapse neuroblastomas. Nat. Genet. **47**(8), 872–877 (2015)
15. Siebes, A., Vreeken, J., van Leeuwen, M.: Item sets that compress. In: SDM, vol. 6, pp. 393–404 (2006)
16. Smets, K., Vreeken, J.: Slim: directly mining descriptive patterns. In: SDM, pp. 236–247 (2012)
17. Vreeken, J., van Leeuwen, M., Siebes, A.: Characterising the difference. In: KDD, pp. 765–774 (2007)
18. Wang, Y.-X., Zhang, Y.-J.: Nonnegative matrix factorization: a comprehensive review. TKDE **25**(6), 1336–1353 (2013)
19. Zhang, Z., Ding, C., Li, T., Zhang, X.: Binary matrix factorization with applications. In: ICDM, pp. 391–400 (2007)

Feature Extraction for Incomplete Data via Low-rank Tucker Decomposition

Qiquan Shi[1], Yiu-ming Cheung[1($\boxtimes$)], and Qibin Zhao[2,3]

[1] Department of Computer Science, Hong Kong Baptist University,
Hong Kong, China
{csqqshi,ymc}@comp.hkbu.edu.hk
[2] Tensor Learning Unit, Center for Advanced Intelligence Project (AIP),
RIKEN, Wako, Japan
qibin.zhao@riken.jp
[3] School of Automation, Guangdong University of Technology, Guangzhou, China

Abstract. Extracting features from incomplete tensors is a challenging task which is not well explored. Due to the data with missing entries, existing feature extraction methods are not applicable. Although tensor completion techniques can estimate the missing entries well, they focus on data recovery and do not consider the relationships among tensor samples for effective feature extraction. To solve this problem of feature extraction for incomplete data, we propose an unsupervised method, **TDVM**, which incorporates *low-rank Tucker Decomposition* with *feature Variance Maximization* in a unified framework. Based on Tucker decomposition, we impose nuclear norm regularization on the core tensors while minimizing reconstruction errors, and meanwhile maximize the variance of core tensors (i.e., extracted features). Here, the relationships among tensor samples are explored via variance maximization while estimating the missing entries. We thus can simultaneously obtain lower-dimensional core tensors and informative features directly from observed entries. The alternating direction method of multipliers approach is utilized to solve the optimization objective. We evaluate the features extracted from two real data with different missing entries for face recognition tasks. Experimental results illustrate the superior performance of our method with a significant improvement over the state-of-the-art methods.

Keywords: Missing data · Feature extraction
Low-rank tucker decomposition · Variance maximization

1 Introduction

This paper aims to extract features directly from data with missing entries. Many real-world data are multi-dimensional, in the form of *tensors*, which are ubiquitous such as multichannel images and have become increasingly popular [1]. Tucker decomposition is widely used to solve tensor learning problems,

© Springer International Publishing AG 2017
M. Ceci et al. (Eds.): ECML PKDD 2017, Part I, LNAI 10534, pp. 564–581, 2017.
https://doi.org/10.1007/978-3-319-71249-9_34

which decomposes a tensor into a core tensor with factor matrices [2]. Based on Tucker decomposition, many tensor methods are proposed for feature extraction (dimension reduction) [3–7]. For example, multilinear principal component analysis (MPCA) [3] extracts features directly from tensors, which is a popular extension of classical Principal Component Analysis (PCA) [8]. Furthermore, some robust methods such as robust tensor PCA (TRPCA) [9] are well studied for data with corruptions (e.g., noise and outliers) [9–11].

In practice, some entries of tensors are often missing due to the problems in the acquisition process or costly experiments etc. [12]. This missing data problem appears in a wide range of fields such as social sciences, computer vision and medical systems [13]. For example, partial responses in surveys are common in the social sciences, leading to incomplete datasets with arbitrary patterns [14]. Moreover, some images are corrupted during the image acquisition and partial entries are missing [15]. In these scenarios, the above existing feature learning methods cannot work well. How to correctly handle missing data is a fundamental yet challenging problem in machine learning [16], and the problem of *extracting features* from *incomplete tensors* is not well explored.

One natural solution to solving this problem is to recover the missing data and then view the recovered tensors as the extracted features. Tensor completion techniques are widely used for missing data problems and has drawn much attention in many applications such as image recovery [17] and video completion [18]. For example, a high accuracy low-rank tensor completion algorithm (HaLRTC) [17] is proposed to estimate missing values in tensors of visual data, and a generalized higher-order orthogonal iteration (gHOI) [19] achieves simultaneous low-rank Tucker decomposition and completion efficiently. Although these tensor completion methods can recover the missing entries well under certain conditions, they only focus on data recovery without exploring the relationships among samples for effective feature extraction. Besides, taking recovered data as features, the dimension of features cannot be reduced.

Another straightforward solution is a "two-step" strategy, i.e., "tensor completion methods + feature extraction methods": the missing entries are first recovered by the former and then the features are extracted from the completed data by the latter. For example, LRANTD [20] performs nonnegative Tucker decomposition (NTD) for incomplete tensors by realizing "low-rank representation (LRA) + nonnegative feature extraction". It needs a tensor completion method to estimate the missing values in the preceding LRA step. However, this "two-step" strategy probably amplifies the reconstruction errors as the missing entries and features are not learned in one stage, and the errors from tensor completion methods can deteriorate the performance of feature extraction in the succeeding step. Moreover, this approach is generally not computationally efficient.

Recently, a few works apply tensor completion methods to feature classification by incorporating completion model with discriminant analysis [21,22]. These methods are supervised and require labels which are expensive and difficult to obtain. To the best of our knowledge, there is no an unsupervised method to extract features directly from tensors with missing entries.

To solve the problem of extracting features from incomplete tensors, we propose an unsupervised method, i.e., incorporating *Low-rank Tucker Decomposition* with *feature Variance Maximization* in a unified framework, namely **TDVM**. In this framework, based on Tucker decomposition with orthonormal factor matrices (a.k.a., higher-order singular value decomposition (HOSVD) [23]), we impose nuclear norm regularization on the core tensors while minimizing the reconstruction error, and meanwhile maximize the variance of core tensors. In this paper, the learned *core tensors* (analogous to the *singular values* of a matrix) are viewed as the *extracted features*. Compared with tensor completion methods and "two-step" strategies:

– Although Tucker decomposition-based tensor completion methods can also obtain core tensors, these core tensors are learned with aiming to recover the tensor samples and without exploring the relationships among samples for effective feature extraction. Unlike these tensor completion methods, here we focus on low-dimensional feature extraction rather than missing data recovery. Besides, we incorporate a specific term (*feature variance maximization*) to enhance the discriminative properties of learned core tensors.
– Different from the "two-step" strategies, we simultaneously learn the missing entries and features directly from observed entries in the unified framework. Besides, TDVM directly learns low-dimensional features in one step, which saves computational cost.

We optimize our model using alternating direction method of multipliers (ADMM) [24]. After feature extraction, we evaluate the extracted features for face recognition, which empirically demonstrates that TDVM outperforms the competing methods consistently. In a nutshell, the contributions of this paper are twofold:

– We propose an efficient unsupervised feature extraction method, TDVM, based on low-rank Tucker decomposition. TDVM can simultaneously obtain low-dimensional core tensors and features for incomplete data.
– We incorporate nuclear norm regularization with variance maximization on core tensors (features) to explore the relationships among tensor samples while estimating missing entries, leading to informative features extracted directly from observed entries.

2   Preliminaries and Backgrounds

2.1   Notations and Operations

The number of dimensions of a tensor is the *order* and each dimension is a *mode* of it. A vector (i.e. first-order tensor) is denoted by a bold lower-case letter $\mathbf{x} \in \mathbb{R}^{I}$. A matrix (i.e. second-order tensor) is denoted by a bold capital letter $\mathbf{X} \in \mathbb{R}^{I_1 \times I_2}$. A higher-order ($N \geq 3$) tensor is denoted by a calligraphic letter

$\mathcal{X} \in \mathbb{R}^{I_1 \times \cdots \times I_N}$. The ith entry of a vector $\mathbf{a} \in \mathbb{R}^I$ is denoted by $\mathbf{a}_i$, and the (i,j)th entry of a matrix $\mathbf{X} \in \mathbb{R}^{I_1 \times I_2}$ is denoted by $\mathbf{X}_{i,j}$. The $(i_1, \cdots, i_N)$th entry of an Nth-order tensor $\mathcal{X}$ is denoted by $\mathcal{X}_{i_1, \cdots, i_N}$, where $i_n \in \{1, \cdots, I_n\}$ and $n \in \{1, \cdots, N\}$. The Frobenius norm of a tensor $\mathcal{X}$ is defined by $\|\mathcal{X}\|_F = \langle \mathcal{X}, \mathcal{X} \rangle^{1/2}$ [25]. $\boldsymbol{\Omega} \in \mathbb{R}^{I_1 \times I_2}$ is a binary index set: $\boldsymbol{\Omega}_{i_1, \cdots, i_N} = 1$ if $\mathcal{X}_{i_1, \cdots, i_N}$ is observed, and $\boldsymbol{\Omega}_{i_1, \cdots, i_N} = 0$ otherwise. $\mathcal{P}_{\boldsymbol{\Omega}}$ is the associated sampling operator which acquires only the entries indexed by $\boldsymbol{\Omega}$, defined as:

$$(\mathcal{P}_{\boldsymbol{\Omega}}(\mathcal{X}))_{i_1, \cdots, i_N} = \begin{cases} \mathcal{X}_{i_1, \cdots, i_N}, & \text{if}(i_1, \cdots, i_N) \in \boldsymbol{\Omega} \\ 0, & \text{if}(i_1, \cdots, i_N) \in \boldsymbol{\Omega}^c \end{cases}, \tag{1}$$

where $\boldsymbol{\Omega}^c$ is the complement of $\boldsymbol{\Omega}$, and $\mathcal{P}_{\boldsymbol{\Omega}}(\mathcal{X}) + \mathcal{P}_{\boldsymbol{\Omega}^c}(\mathcal{X}) = \mathcal{X}$.

Definition 1 Mode-n Product. *A mode-n product between a tensor $\mathcal{X} \in \mathbb{R}^{I_1 \times \cdots \times I_N}$ and a matrix/vector $\mathbf{U} \in \mathbb{R}^{I_n \times J_n}$ is denoted by $\mathcal{Y} = \mathcal{X} \times_n \mathbf{U}^T$. The size of $\mathcal{Y}$ is $I_1 \times \cdots \times I_{n-1} \times J_n \times I_{n+1} \times \cdots \times I_N$, with entries given by $\mathcal{Y}_{i_1 \cdots i_{n-1} j_n i_{n+1} \cdots i_N} = \sum_{i_n} \mathcal{X}_{i_1 \cdots i_{n-1} i_n i_{n+1} \cdots i_N} \mathbf{U}_{i_n, j_n}$, and we have $\mathbf{Y}_{(n)} = \mathbf{U}^T \mathbf{X}_{(n)}$ [25].*

Definition 2 Mode-n Unfolding. *Unfolding, a.k.a., matricization or flattening, is the process of reordering the elements of a tensor into matrices along each mode [1]. A mode-n unfolding matrix of a tensor $\mathcal{X} \in \mathbb{R}^{I_1 \times \cdots \times I_N}$ is denoted as $\mathbf{X}_{(n)} \in \mathbb{R}^{I_n \times \Pi_{n^* \neq n} I_{n^*}}$.*

2.2 Tucker Decomposition

A tensor $\mathcal{X} \in \mathbb{R}^{I_1 \times I_2 \times \cdots \times I_N}$ is represented as a core tensor with factor matrices in Tucker decomposition model [1]:

$$\mathcal{X} = \mathcal{G} \times_1 \mathbf{U}^{(1)} \times_2 \mathbf{U}^{(2)} \cdots \times_N \mathbf{U}^{(N)}, \tag{2}$$

where $\{\mathbf{U}^{(n)} \in \mathbb{R}^{I_n \times R_n}, n = 1, 2 \cdots N$, and $R_n < I_n\}$ are factor matrices with orthogonal columns and $\mathcal{G} \in \mathbb{R}^{R_1 \times R_2 \times \cdots \times R_N}$ is the core tensor with smaller dimension. Tucker-rank of an Nth-order tensor $\mathcal{X}$ is an N-dimensional vector, denoted as $(R_1, \cdots, R_N)$, where R_N is the rank of the mode-n unfolded matrix $\mathbf{X}_{(n)}$ of $\mathcal{X}$. Figure 1 illustrates this decomposition. In this paper, we regard the *core tensor* consists of the *extracted features* of a tensor.

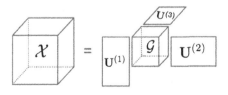

Fig. 1. The Tucker decomposition of tensors (a third-order tensor $\mathcal{X}$ shown for illustration).

3 Feature Extraction for Incomplete Data

3.1 Problem Definition

Given M tensor samples $\{\mathcal{T}_1, \cdots, \mathcal{T}_m, \cdots, \mathcal{T}_M\}$ with *missing entries* in each sample $\mathcal{T}_m \in \mathbb{R}^{I_1 \times \cdots \times I_N}$. I_n is the mode-n dimension. We denote $\mathcal{T} = [\mathcal{T}_1, \cdots, \mathcal{T}_m, \cdots \mathcal{T}_M] \in \mathbb{R}^{I_1 \times \cdots \times I_N \times M}$, where the M are the number of tensor samples concatenated along the mode-$(N+1)$ of $\mathcal{T}$. For feature extraction (dimension reduction), we aim to directly extract low-dimensional features $\mathcal{G} = [\mathcal{G}_1, \cdots, \mathcal{G}_m, \cdots \mathcal{G}_M] \in \mathbb{R}^{R_1 \times \cdots \times R_N \times M} (R_n < I_n, n = 1, \cdots, N)$ from the given high-dimensional incomplete tensors $\mathcal{T}$.

Remark: This problem is different from the case of data with corruptions (e.g., noise and outliers) widely studied in [9, 26–28]: Only if the corruptions are arbitrary, missing data could be regarded as a special case of corruptions (with the location of corruption being known). However, the magnitudes of corruptions in reality are not arbitrarily large. In other words, here we study a new problem and existing feature extraction methods are not applicable.

3.2 Formulation of the Proposed Method: TDVM

To solve this problem, we propose an unsupervised feature extraction method. Based on Tucker decomposition, we impose the nuclear norm on the core tensors of observed tensors while minimizing reconstruction errors, and meanwhile maximize the variance of core tensors (features), i.e., incorporating low-rank **T**ucker **D**ecomposition with feature **V**ariance **M**aximization, namely *TDVM*. Thus, the objective function of TDVM is:

$$\min_{\mathcal{X}_m, \mathcal{G}_m, \mathcal{S}_m, \mathbf{U}^{(n)}} \sum_{m=1}^{M} \frac{1}{2} \|\mathcal{X}_m - \mathcal{G}_m \times_1 \mathbf{U}^{(1)} \cdots \times_N \mathbf{U}^{(N)}\|_F^2$$

$$+ \sum_{m=1}^{M} \|\mathcal{G}_m\|_* - \frac{1}{2} \sum_{m=1}^{M} \|\mathcal{G}_m - \bar{\mathcal{G}}\|_F^2, \quad (3)$$

$$\text{s.t.} \ \ \mathcal{P}_\Omega(\mathcal{X}_m) = \mathcal{P}_\Omega(\mathcal{T}_m), \mathbf{U}^{(n)\top}\mathbf{U}^{(n)} = \mathbf{I}.$$

where $\{\mathbf{U}^{(n)} \in \mathbb{R}^{I_n \times R_n}\}_{n=1}^{N}$ are common factor matrices with orthonormal columns. $\mathbf{I} \in \mathbb{R}^{R_n \times R_n}$ is an identity matrix. $\mathcal{G}_m$ is the core tensor which consists of the *extracted features* (analogous to the *singular values* of a matrix) of an incomplete tensor $\mathcal{T}_m$ with observed entries in Ω. $\|\mathcal{G}_m\|_*$ is the nuclear norm of $\mathcal{G}_m$ (i.e., the summation of the singular values of the unfolded matrices along modes of $\mathcal{G}_m$ [17]). $\bar{\mathcal{G}} = \frac{1}{M}\sum_{m=1}^{M} \mathcal{G}_m$ is the mean of core tensors (extracted features).

Remark: Our objective function (3) integrates three terms into a unified framework:

- The first term: minimizing $\sum_{m=1}^{M} \frac{1}{2}\|\mathcal{X}_m - \mathcal{G}_m \times_1 \mathbf{U}^{(1)} \cdots \times_N \mathbf{U}^{(N)}\|_F^2$, aims to minimize the reconstruction error based on given observed entries.

– The second term: minimizing $\sum_{m=1}^{M} \|\mathcal{G}_m\|_*$, aims to obtain low-dimensional features. It is proved that imposing the nuclear norm on a core tensor $\mathcal{G}_m$ is essentially equivalent to that on its original tensor $\mathcal{X}_m$ [19]. We thus obtain a low-rank solution, i.e. R_n can be small ($R_n < I_n$). Thus, the learned feature subspace is naturally low-dimensional. Besides, imposing nuclear norm on core tensors $\mathcal{G}_m$ instead of original $\mathcal{X}_m$ saves computational cost.

– The third term: minimizing $-\sum_{m=1}^{M} \frac{1}{2}\|\mathcal{G}_m - \bar{\mathcal{G}}\|_F^2$, is equivalent to maximize the variance of extracted features (core tensors) following PCA. We thus explore the relationships of incomplete tensors via *variance maximization* while estimating the missing entries via the first and second term (*low-rank Tucker decomposition*).

By this unified framework, we can efficiently extract low-dimensional informative features directly from observed entries, which is different from *tensor completion methods* (only focusing on data recovery without considering the relationships among samples for effective feature extraction) and *"two-step" strategies* (the reconstruction error from tensor completion step probably deteriorates the performance of feature extraction in the succeeding step, and combining two methods is generally time consuming).

3.3   Optimization by ADMM

To optimize (3) using ADMM, we apply the variable splitting technique and introduce a set of *auxiliary variables* $\{\mathcal{S}_m \in \mathbb{R}^{R_1 \times \cdots \times R_N}, m = 1 \cdots M, n = 1, \cdots, N\}$, and then reformulate (3) as:

$$
\min_{\mathcal{X}_m, \mathcal{G}_m, \mathcal{S}_m, \mathbf{U}^{(n)}} \sum_{m=1}^{M} \frac{1}{2}\|\mathcal{X}_m - \mathcal{G}_m \times_1 \mathbf{U}^{(1)} \cdots \times_N \mathbf{U}^{(N)}\|_F^2
$$

$$
+ \sum_{m=1}^{M} \|\mathcal{S}_m\|_* - \frac{1}{2}\sum_{m=1}^{M}\|\mathcal{G}_m - \bar{\mathcal{G}}\|_F^2, \tag{4}
$$

$$
\text{s.t. } \mathcal{P}_\Omega(\mathcal{X}_m) = \mathcal{P}_\Omega(\mathcal{T}_m), \mathcal{S}_m = \mathcal{G}_m, {\mathbf{U}^{(n)}}^\top \mathbf{U}^{(n)} = \mathbf{I}.
$$

For easy derivation of (4), we reformulate it by unfolding each tensor variable along mode-n and absorbing the constraints. Thus, we get the Lagrange function as follows:

$$
\mathcal{L} = \sum_{m=1}^{M}\sum_{n=1}^{N}\Big(\frac{1}{2}\|\mathbf{X}_m^{(n)} - \mathbf{U}^{(n)}\mathbf{G}_m^{(n)}{\mathbf{P}^{(n)}}^\top\|_F^2 + \|\mathbf{S}_m^{(n)}\|_*
$$

$$
+ \langle \mathbf{Y}_{mn}, \mathbf{G}_m^{(n)} - \mathbf{S}_m^{(n)}\rangle + \frac{\mu}{2}\|\mathbf{G}_m^{(n)} - \mathbf{S}_m^{(n)}\|_F^2 - \frac{1}{2}\|\mathbf{G}_m^{(n)} - \bar{\mathbf{G}}^{(n)}\|_F^2\Big) \tag{5}
$$

where $\mathbf{P}^{(n)} = \mathbf{U}^{(N)} \otimes \cdots \otimes \mathbf{U}^{(n+1)} \otimes \mathbf{U}^{(n-1)} \cdots \otimes \mathbf{U}^{(1)} \in \mathbb{R}^{\Pi_{j\neq n} I_j \times \Pi_{j\neq n} R_j}$ and $\{\mathbf{Y}_{mn} \in \mathbb{R}^{R_n \times \Pi_{j\neq n} R_j}, n = 1, \cdots, N, m = 1, \cdots, M\}$ are the matrices of Lagrange multipliers. $\mu > 0$ is a penalty parameter. $\mathbf{X}_m^{(n)} \in \mathbb{R}^{I_n \times \Pi_{j\neq n} I_j}$ and

$\{\mathbf{G}_m^{(n)}, \mathbf{S}_m^{(n)}, \bar{\mathbf{G}}^{(n)}\} \in \mathbb{R}^{R_n \times \Pi_{j \neq n} R_j}$ are the mode-n unfolded matrices of tensor $\mathcal{X}_m$ and {core tensor $\mathcal{G}_m$, auxiliary variable $\mathcal{S}_m$, mean of features $\bar{\mathcal{G}}$}, respectively.

ADMM solves the problem (5) by successively minimizing $\mathcal{L}$ over $\{\mathbf{X}_m^{(n)}, \mathbf{G}_m^{(n)}, \mathbf{S}_m^{(n)}, \mathbf{U}^{(n)}\}$, and then updating $\mathbf{Y}_{mn}$.

Update $\mathbf{S}_m^{(n)}$. The Lagrange function (5) with respect to $\mathbf{S}_m^{(n)}$ is,

$$\mathcal{L}_{\mathbf{S}_m^{(n)}} = \sum_{m=1}^{M} \sum_{n=1}^{N} \left(\|\mathbf{S}_m^{(n)}\|_* + \frac{\mu}{2} \|(\mathbf{G}_m^{(n)} + \mathbf{Y}_{mn}/\mu) - \mathbf{S}_m^{(n)}\|_F^2 \right). \tag{6}$$

To solve (6), we use the spectral soft-thresholding operation [29] to update $\mathbf{S}_m^{(n)}$:

$$\mathbf{S}_m^{(n)} = prox_{1/\mu}(\mathbf{G}_m^{(n)} + \mathbf{Y}_{mn}/\mu) = \mathbf{U}\text{diag}(\max \sigma - \frac{1}{\mu}, 0)\mathbf{V}^\top, \tag{7}$$

where $prox$ is the soft-thresholding operation and $\mathbf{U}\text{diag}(\max \sigma - \frac{1}{\mu}, 0)\mathbf{V}^\top$ is the Singular Value Decomposition (SVD) of $(\mathbf{G}_m^{(n)} + \mathbf{Y}_{mn}/\mu)$.

Update $\mathbf{U}^{(n)}$. The Lagrange function (5) with respect to $\mathbf{U}^{(n)}$ is:

$$\mathcal{L}_{\mathbf{U}^{(n)}} = \sum_{m=1}^{M} \sum_{n=1}^{N} \frac{1}{2} \|\mathbf{X}_m^{(n)} - \mathbf{U}^{(n)} \mathbf{G}_m^{(n)} \mathbf{P}^{(n)\top}\|_F^2, \quad \text{s.t.} \quad \mathbf{U}^{(n)\top}\mathbf{U}^{(n)} = \mathbf{I}, \tag{8}$$

According to the Theorem 4 in [30], the minimization of the problem (8) over the matrices $\{\mathbf{U}^{(1)}, \cdots, \mathbf{U}^{(N)}\}$ having orthonormal columns is equivalent to the maximization of the following problem:

$$\mathbf{U}^{(n)} = \arg\max \ \text{trace}\left(\mathbf{U}^{(n)\top} \mathbf{X}_m^{(n)} (\mathbf{G}_m^{(n)} \mathbf{P}^{(n)\top})^\top\right) \tag{9}$$

where trace() is the trace of a matrix, and we denote $\mathbf{W}^{(n)} = \mathbf{G}_m^{(n)} \mathbf{P}^{(n)\top}$.

The problem (9) is actually the well-known orthogonal procrustes problem [31], whose global optimal solution is given by the SVD of $\mathbf{X}_m^{(n)} \mathbf{W}^{(n)\top}$, i.e.,

$$\mathbf{U}^{(n)} = \hat{\mathbf{U}}^{(n)} (\hat{\mathbf{V}}^{(n)})^\top, \tag{10}$$

where $\hat{\mathbf{U}}^{(n)}$ and $\hat{\mathbf{V}}^{(n)}$ are the left and right singular vectors of SVD of $\mathbf{X}_m^{(n)} \mathbf{W}^{(n)\top}$, respectively.

Update $\mathbf{G}_m^{(n)}$. The Lagrange function (5) with respect to $\mathbf{G}_m^{(n)}$ is:

$$\mathcal{L}_{\mathbf{G}_m^{(n)}} = \|\mathbf{X}_m^{(n)} - \mathbf{U}^{(n)} \mathbf{G}_m^{(n)} \mathbf{P}^{(n)\top}\|_F^2 + \frac{\mu}{2} \|\mathbf{G}_m^{(n)} \ \mathbf{S}_m^{(n)} + \mathbf{Y}_{mn}/\mu\|_F^2$$
$$-\frac{1}{2} \|(1 - \frac{1}{M})\mathbf{G}_m^{(n)} - \frac{1}{M} \sum_{j \neq m}^{M} \mathbf{G}_j^{(n)}\|_F^2 \Big). \tag{11}$$

Algorithm 1. Low-rank Tucker Decomposition with Feature Variance Maximization (**TDVM**)

1: **Input:** Incomplete tensors $\mathcal{T} = [\mathcal{T}_1, \cdots, \mathcal{T}_m, \cdots \mathcal{T}_M]$, $\boldsymbol{\Omega}$, μ, and the maximum iterations K, the dimension of core tensors (features) $D = [R_1, \cdots, R_N]$, and stopping tolerance tol.
2: **Initialization:** Set $\mathcal{P}_\Omega(\mathcal{X}_m) = \mathcal{P}_\Omega(\mathcal{T}_m), \mathcal{P}_{\Omega^c}(\mathcal{X}_m) = \mathbf{0}, m = 1, \cdots, M$; initialize $\{\mathcal{G}_m\}_{m=1}^M$ and $\{\mathbf{U}^{(n)}\}_{n=1}^N$ randomly; $\mu_0 = 5, \rho = 10, \mu_{max} = 1e10$.
3: **for** $m = 1$ **to** M **do**
4:     **for** $k = 1$ **to** K **do**
5:         **for** $n = 1$ **to** N **do**
6:             Update $\mathbf{S}_m^{(n)}$, $\mathbf{U}^{(n)}$ and $\mathbf{G}_m^{(n)}$ by (7), (10) and (12) respectively.
7:             Update $\mathbf{Y}_m^{(n)}$ by $\mathbf{Y}_m^{(n)} = \mathbf{Y}_m^{(n)} + \mu(\mathbf{G}_m^{(n)} - \mathbf{S}_m^{(n)})$.
8:         **end for**
9:         Update $\mathcal{X}_m$ by (13).
10:         Update $\mu_{k+1} = \min(\rho\mu_k, \mu_{max})$.
11:     **end for**
12:     If $\|\mathcal{G}_m - \mathcal{S}_m\|_F^2 / \|\mathcal{G}_m\|_F^2 < tol$, break; otherwise, continue.
13: **end for**
14: **Output:** Extracted features (core tensors): $\mathcal{G} = [\mathcal{G}_1, \cdots, \mathcal{G}_m, \cdots \mathcal{G}_M]$.

Then we set the partial derivative $\dfrac{\partial \mathcal{L}_{\mathbf{G}_m^{(n)}}}{\partial \mathbf{G}_m^{(n)}}$ to zero, and get:

$$
\mathbf{G}_m^{(n)} = \frac{M^2}{M^2\mu + 2M - 1}\left(\mu\mathbf{S}_m^{(n)} - \mathbf{Y}_{mn} + \mathbf{U}^{(n)^\top}\mathbf{X}_m^{(n)}\mathbf{P}^{(n)}\right)
$$
$$
- \left((\frac{1}{M} - \frac{1}{M^2})\sum_{j \neq m}^M \mathbf{G}_j^{(n)}\right). \tag{12}
$$

Update $\mathcal{X}_m$. The Lagrange function (5) with respect to $\mathcal{X}$ is:

$$
\mathcal{L}_{\mathcal{X}_m} = \frac{1}{2}\sum_{m=1}^M \|\mathcal{X}_m - \mathcal{G}_m \times_1 \mathbf{U}^{(1)} \cdots \times_N \mathbf{U}^{(N)}\|_F^2, \tag{13}
$$
$$
\text{s.t. } \mathcal{P}_\Omega(\mathcal{X}_m) = \mathcal{P}_\Omega(\mathcal{T}_m),
$$

By deriving the Karush-Kuhn-Tucker (KKT) conditions for function (13), we can update $\mathcal{X}_m$ by $\mathcal{X}_m = \mathcal{P}_\Omega(\mathcal{X}_m) + \mathcal{P}_{\Omega^c}(\mathcal{Z}_m)$, where $\mathcal{Z}_m = \mathcal{G}_m \times_1 \mathbf{U}^{(1)} \cdots \times_N \mathbf{U}^{(N)}$.

We summarize the proposed method, TDVM, in Algorithm 1.

3.4 Complexity Analysis

We analyze the complexity of TDVM following [32]. For simplicity, we assume the size of tensor is $I_1 = \cdots = I_N = I$, and the feature dimensions are $R_1 = \cdots = R_N = R$. At each iteration, the time complexity of performing the soft-thresholding operator (7) is $O(MNR^{N+1})$. The time complexities of some multiplication operators in (10)/(12) and (13) are $O(MNRI^N)$ and $O(MRI^N)$, respectively. Hence, the total time complexity of TDVM is $O(M(N+1)RI^N)$ per iteration.

4   Experimental Results

We implemented TDVM[1] in MATLAB and all experiments were performed on a PC (Intel Xeon(R) 4.0 GHz, 64 GB).

4.1   Experimental Setup

Compared Methods: We compare our TDVM with *nine* methods in three categories:

– *Two* tensor completion methods based on Tucker-decomposition: **HaLRTC** [17] and **gHOI** [19]. The recovered tensor are regarded as the features.
– *Six* {tensor completion methods + feature extraction methods} (i.e., "two-step" strategies): HaLRTC + **PCA** [8], gHOI + **PCA**, HaLRTC + **MPCA** [3], gHOI + **MPCA**, HaLRTC + **LRANTD** [20] and gHOI + **LRANTD**.
– *One* robust tensor feature learning method: **TRPCA** [9].

After feature extraction stage, we use the Nearest Neighbors Classifier (**NNC**) to evaluate the extracted features on two real data for face recognition. We had also evaluated TDVM on MNIST handwritten digits [33] for object classification, and TDVM obtains the best results in all cases. We do not report here due to limited space.

Data: We evaluate the proposed TDVM on two real data for face recognition tasks. One is a subset of Facial Recognition Technology database (FERET)[2] [34], which has 721 face samples from 70 subjects. Each subject has 8–31 faces with at most $15°$ of pose variation and each face image is normalized to a 80×60 gray image. The other is a subset of extended Yale Face Database B (YaleB)[3] [35], which has 2414 face samples from 38 subjects. Each subject has 59–64 near frontal images under different illuminations and each face image is normalized to a 32×32 gray image.

Missing Data Settings: We set the tensors with two types of missing data:

– **Pixel-based missing:** we uniformly select $10\% - 90\%$ pixels (entries) of tensors as missing at random. Pixel-based missing setting is widely used in tensor completion domain. One example (e.g., missing 50% entries) is shown in Fig. 2(b).
– **Block-based missing:** we randomly select $B_1 \times B_2$ block pixels of each tensor sample as *missing*. The missing block is random in each sample. One example (e.g., $\{B_1 = 40, B_2 = 30\}$ for FERET and $\{B_1 = 16, B_2 = 16\}$ for YaleB) is shown in Fig. 2(c). In practice, some parts of a face can be covered by some objects such as a sunglass, which can be regarded as the block-based missing case.

---

[1] Codes and data: https://www.dropbox.com/sh/h4k07sstdmthd80/AABMPFEqD Dz-NzKWXIhDnLL0a/Qiquan_TDVM(ECML_198)?dl=0.

[2] http://www.dsp.utoronto.ca/~haiping/MSL.html.

[3] http://www.cad.zju.edu.cn/home/dengcai/Data/FaceData.html.

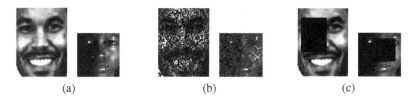

(a)	(b)	(c)

Fig. 2. One example of (a) original images of FERET and YaleB with (b) 50% pixel-based and with (c) 40×30 and 16×16 block-based missing entries, respectively.

Intuitively, handling data with *block-based missing* is more difficult than that with *pixel-based missing* if same number of entries are missing.

Parameter Settings: We set the maximum iterations $K = 200$, $tol = 1e-5$ for all methods and set the feature dimension $D = R_1 \times R_2 = \{40 \times 30, 16 \times 16\}$ for TDVM, gHOI and LRANTD on {FERET, YaleB} respectively. In other words, we directly learn $40 \times 30 \times 721$ features from FERET ($80 \times 60 \times 721$) and extract $16 \times 16 \times 2414$ features from YaleB ($32 \times 32 \times 2414$). Other parameters of the compared methods have followed the original papers.

Applying extracted features for face recognition using NNC, we randomly select $L = \{1, 2, \cdots, 7\}$ extracted feature samples from each subject (with 8–31 samples) of FERET for training in NNC. On YaleB, we randomly select $L = \{5, 10, \cdots, 50\}$ extracted feature samples from each subject (with 59–64 samples) for training.

4.2   Parameter Sensitivity and Convergence Study

Effect of Feature Dimension D: We study the effect of TDVM with different feature dimensions (size of each core tensor) for face recognition on FERET. We set the feature dimension D of each face sample as $R_1 \times R_2$ in TDVM and show the corresponding face recognition results. Figure 3 shows that TDVM with

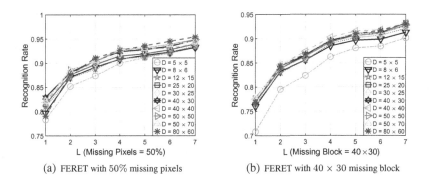

(a) FERET with 50% missing pixels	(b) FERET with 40×30 missing block

Fig. 3. Recognition results on FERET via TDVM with different feature dimension Ds.

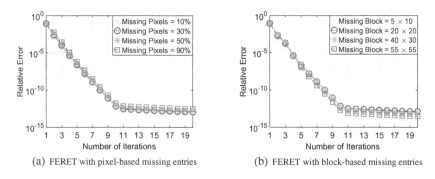

Fig. 4. Convergence curves of TDVM in terms of *Relative Error*: $\|\mathcal{G}_m - \mathcal{S}_m\|_F^2 / \|\mathcal{G}_m\|_F^2$ on FERET with (a) pixel-based/(b) block-based missing entries.

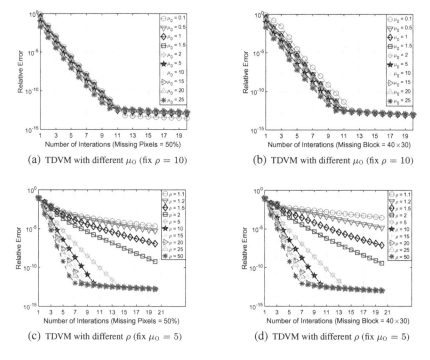

Fig. 5. Convergence curves of feature extraction on FERET with pixel-based (50%)/block-based (40 × 30) missing entries via TDVM with ten different values of μ_0 and ρ, respectively.

different feature dimensions stably yields similar recognition results on FERET in both pixel-based and block-based missing cases, excepting $D = 5 \times 5$ (i.e., only 25 features are extracted from each 80×60 face image) where the number of features are too limited to achieve good results. Since a larger D costs more time and we aim to learn low-dimensional features, here we set $D = R_1 \times R_2 = 40 \times 30$ and 16×16 for TDVM to extract features from FERET and YaleB, respectively.

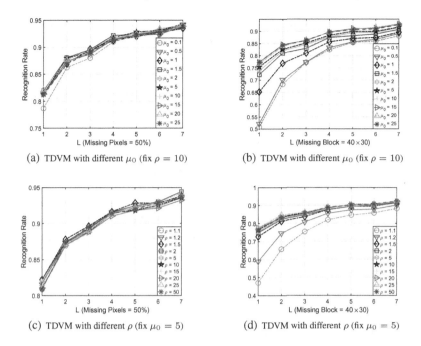

(a) TDVM with different μ_0 (fix $\rho = 10$)   (b) TDVM with different μ_0 (fix $\rho = 10$)

(c) TDVM with different ρ (fix $\mu_0 = 5$)   (d) TDVM with different ρ (fix $\mu_0 = 5$)

Fig. 6. Recognition results on FERET with pixel-based (50%)/block-based (40×30) missing entries via TDVM with ten different values of μ_0 and ρ, respectively.

Convergence: We study the convergence of TDVM in terms of *Relative Error*: $\|\mathcal{G}_m - \mathcal{S}_m\|_F^2 / \|\mathcal{G}_m\|_F^2$ on FERET with pixel/block-based missing entries. Here, we set $\mu_0 = 5$ and $\rho = 10$ for TDVM. Figure 4 shows that the relative error dramatically decreases to a very small value (around 10^{-13} order) with about 10 iterations. In other words, the proposed TDVM converges fast within 5 iterations if we set $tol = 1e - 5$.

Sensitivity Analysis of Parameter μ_0 and ρ : In line 10 of Algorithm 1, we iteratively update the penalty parameter μ with a step size ρ from an initial μ_0, which has been widely used in many methods such as [9] and makes the algorithm converges faster. Figures 5 and 6 show the convergence curves and corresponding recognition results on FERET with 50% missing pixels and 40×30 missing block via TDVM with different μ_0 and ρ respectively. As seen from Figs. 5(a) and (b), with different μ_0, TDVM stably converges to a small value (around 10^{-13} order) with around 10 iterations. In terms of ρ, the relative errors converge to a small value faster if TDVM with a larger ρ (e.g., $\rho = 10$), as shown in Figs. 5(c) and (d). Figures 6(a) and (c) show that the feature extraction performance of our TDVM is stable and not sensitive to the values of μ_0 and ρ on FERET with 50% missing pixels. Besides, as seen from Figs. 6(b) and (d), with a larger μ_0 ($\mu_0 > 1$) and ρ ($\rho > 1.5$) for TVDM on FERET with 40×30 block-based missing entries, the corresponding face recognition results are similar and stable.

In general, we do not need to carefully tune the parameter μ_0 and ρ for the proposed TDVM. In this paper, we set $\mu_0 = 5$ and $\rho = 10$ in Algorithm 1 for all tests.

4.3   Evaluate Features Extracted from Data with Pixel/Block-Based Missing

To save space, for pixel-based missing case, we report the results of FERET and YaleB with $\{10\%, 30\%, 50\%, 90\%\}$ missing pixels in Table 1. For block-based missing case, we report the results of FERET with $\{5 \times 10, 20 \times 20, 40 \times 30, 55 \times 55\}$ missing block and YaleB with $\{5 \times 5, 8 \times 10, 16 \times 16, 30 \times 25\}$ missing block in Table 2 respectively. In each pixel/block-based missing case, we report the recognition rates of randomly selecting $L = \{1, 7\}$ and $L = \{5, 50\}$ extracted feature samples from each subject of FERET and YaleB for training in NNC, respectively. We highlight the **best** results in **bold** fonts and second best in underline respectively. We repeat the runs 10 times of *feature extraction* and of *recognition* separately, and report the average results.

Face Recognition Results on FERET/YaleB with Pixel-Based Missing. TDVM outperforms the other nine methods by $\{34.69\%, 35.72\%, 35.19\%, 46.65\%\}$ in all cases of **FERET** with $\{10\%, 30\%, 50\%, 90\%\}$ missing pixels on average respectively, as shown in the left half of Table 1. Besides, TRPCA achieves the second best results in six cases given more than 50% observations

Table 1. Face recognition results (average recognition rates %) on the FERET and YaleB with $\{10\%, 30\%, 50\%, 90\%\}$ pixel-based missing entries.

Data	FERET (Image size 80 × 60)								YaleB (Image size 32 × 32)							
Missing Pixels	10%		30%		50%		90%		10%		30%		50%		90%	
L	1	7	1	7	1	7	1	7	5	50	5	50	5	50	5	50
HaLRTC	36.44	73.59	36.50	73.07	35.75	72.42	21.75	44.03	35.85	76.71	34.78	75.08	31.55	71.77	12.87	31.83
gHOI	36.41	73.94	36.52	73.94	36.53	74.07	29.65	63.72	35.42	76.38	33.40	73.74	28.68	66.91	10.51	19.82
HaLRTC + PCA	32.23	69.74	26.37	63.38	25.56	59.91	6.84	9.05	22.22	51.26	20.70	48.00	16.47	36.63	13.75	32.59
gHOI + PCA	32.24	68.96	28.74	63.51	26.41	61.26	13.90	24.68	29.71	51.79	15.63	36.69	18.42	41.96	4.95	6.75
HaLRTC + MPCA	40.49	75.37	40.08	73.68	38.94	72.77	4.79	7.62	30.23	73.07	29.62	71.36	28.57	69.34	14.45	38.62
gHOI + MPCA	41.18	75.97	41.01	75.89	40.14	74.68	8.76	14.16	30.02	72.59	27.37	68.07	13.71	43.25	4.79	6.17
HaLRTC + LRANTD	34.85	73.64	34.64	72.60	33.89	71.86	15.73	27.88	23.13	55.93	22.03	52.53	20.52	48.79	9.53	22.55
gHOI + LRANTD	36.44	74.11	36.44	74.07	36.68	74.33	30.06	64.42	22.66	54.30	21.64	52.20	21.28	50.33	10.06	20.53
TRPCA	51.27	84.33	46.24	82.08	41.71	79.09	4.61	10.39	32.66	71.83	29.51	68.72	22.56	53.99	2.65	2.72
TDVM	**85.50**	**96.23**	**84.62**	**95.58**	**82.01**	**94.59**	**58.39**	**79.57**	**93.21**	**98.40**	**92.28**	**98.91**	**91.93**	**98.13**	**87.15**	**97.20**

Table 2. Face recognition results (average recognition rates %) on the FERET and YaleB with block-based missing entries.

Data	FERET (Image size 80 × 60)								YaleB (Image size 32 × 32)							
Missing Block	5 × 10		20 × 20		40 × 30		55 × 55		5 × 5		8 × 10		16 × 16		30 × 25	
L	1	7	1	7	1	7	1	7	5	50	5	50	5	50	5	50
HaLRTC	36.37	73.64	35.75	72.77	33.96	71.21	27.24	59.31	36.02	76.73	35.27	76.30	33.08	74.38	27.43	67.06
gHOI	36.44	73.55	35.53	72.86	31.55	68.10	22.70	52.12	32.86	74.07	31.24	72.86	30.67	70.27	20.13	49.14
HaLRTC + PCA	31.32	66.67	31.27	65.41	20.14	52.68	17.22	40.22	43.49	83.54	33.61	57.57	19.94	46.61	15.06	35.21
gHOI + PCA	26.82	60.13	20.49	47.92	10.86	25.19	5.38	10.48	21.69	58.72	19.52	52.43	13.27	31.89	6.88	11.95
HaLRTC + MPCA	41.09	76.06	41.38	76.32	21.11	53.64	22.52	53.20	30.31	73.56	29.53	71.54	27.46	68.17	17.52	46.01
gHOI + MPCA	41.43	76.58	22.89	56.80	10.81	21.47	18.08	36.32	24.14	64.81	21.37	60.88	9.32	21.01	7.58	16.23
HaLRTC + LRANTD	36.41	73.98	35.73	73.55	33.72	70.39	26.90	58.53	22.27	53.66	21.47	51.63	21.20	50.00	11.98	30.43
gHOI + LRANTD	36.53	74.16	35.12	74.29	31.31	68.27	21.83	50.69	21.45	52.12	20.69	49.92	19.98	47.68	11.26	24.61
TRPCA	39.63	77.40	36.67	74.98	30.55	63.64	21.89	45.97	33.21	72.20	32.09	70.88	30.21	68.35	25.90	58.85
TDVM	**84.21**	**96.45**	**81.67**	**94.81**	**76.82**	**91.99**	**75.04**	**91.95**	**82.51**	**95.76**	**75.54**	**95.14**	**72.79**	**95.29**	**59.91**	**94.63**

while its performance drops dramatically when missing 90% pixels, where the gHOI + LRANTD takes the second place. Moreover, with less training features (e.g. $L = 1$) in NNC, our TDVM has more advantage as it aims to extract low-dimensional informative features.

The right half of Table 1 shows that TDVM outperforms the best performing existing algorithm (HaLRTC) in all cases of **YaleB** with $\{10\%, 30\%, 50\%\}$ missing pixels by $\{39.52\%, 40.67\%, 43.37\%\}$ on average, respectively. When the missing rate achieves 90%, the performance of compared methods drop sharply, excepting HaLRTC + MPCA which wins other existing methods in this scenario, where our TDVM keeps the best performance with 77.45% over all the existing methods.

Face Recognition Results on FERET/YaleB with Block-Based Missing. The left half of Table 2 shows that TDVM outperforms all competing methods by $\{35.99\%, 37.70\%, 44.48\%, 50.68\%\}$ in all cases of **FERET** with $\{5 \times 10, 20 \times 20, 40 \times 30, 55 \times 55\}$ missing blocks on average, respectively. Furthermore, gHOI/HaLRTC + MPCA and HaLRTC share the second place in these cases.

As shown in the right half of Table 2: TDVM outperforms the nine state-of-the-art methods by $\{40.53\%, 40.40\%, 46.07\%, 50.42\%\}$ in all cases of **YaleB** with $\{5 \times 5, 8 \times 10, 16 \times 16, 30 \times 25\}$ missing blocks on average, respectively. Specifically, HaLRTC is the best performing existing algorithm in the cases of missing $\{8 \times 10, 16 \times 16, 30 \times 25\}$ block, but our TDVM outperforms it by $\{29.55\%, 30.31\%, 30.02\%\}$ respectively there.

4.4 Computational Cost

We report the average time cost of feature extraction in Table 3. As shown in Table 3, TDVM is much more efficient than all the compared methods in all cases, as we impose nuclear norm on core tensors instead of original tensors to learn low-dimensional features. Specifically, HaLRTC is the second fastest methods on FERET with block-based missing entries while slower than gHOI and HaLRTC + PCA in two pixel-based missing cases. Besides, HaLRTC is also the second efficient method on YaleB with pixel-based missing entries excepting the case of missing 90% pixels. In the block-based missing cases of YaleB, TRPCA is faster than TDVM, but it yields worse results. Moreover, the "two-step" strategies such as gHOI + MPCA/LRANTD are the most time consuming (more than 10 times slower than TDVM on average).

Table 3. Time cost (seconds) of feature extraction on the FERET and YaleB with pixel/block-based missing entries.

Data	Missing Pixels/Block	HaLRTC [17]	gHOI [30]	HaLRTC + PCA [8]	gHOI + PCA	HaLRTC + MPCA [3]	gHOI + MPCA	HaLRTC + LRANTD [20]	gHOI + LRANTD	TRPCA [9]	TDVM
FERET	10%	101.6	_67.2_	112.7	99.4	270.4	289.9	237.1	383.1	231.0	**32.6**
	30%	_114.2_	149.8	120.8	310.5	313.3	516.8	289.6	413.9	269.2	**33.3**
	50%	_123.7_	230.4	129.9	533.6	332.6	687.5	325.0	612.9	212.7	**23.7**
	90%	120.0	175.3	_104.1_	442.1	132.1	654.8	111.2	563.6	118.2	**20.1**
FERET	5 × 10	_103.1_	521.3	114.8	166.5	191.5	254.5	445.5	521.3	170.0	**29.3**
	20 × 20	_110.5_	599.7	120.4	220.2	178.7	299.1	459.9	599.7	165.9	**24.0**
	40 × 30	_119.1_	555.1	129.9	221.7	209.5	220.2	479.0	555.1	139.8	**21.7**
	55 × 55	166.6	465.9	_117.3_	208.2	245.0	213.1	377.5	465.9	144.5	**30.3**
YaleB	10%	_127.8_	318.2	183.6	721.8	619.6	1095.0	494.6	674.3	156.5	**45.9**
	30%	_150.8_	631.3	203.5	1423.9	650.6	2202.0	575.2	1645.2	160.0	**45.5**
	50%	_160.9_	624.6	216.1	1304.9	684.5	1783.6	565.0	2218.2	160.2	**45.3**
	90%	241.0	1052.8	223.4	1030.3	425.5	1152.7	590.0	997.6	_118.7_	**49.6**
YaleB	5 × 5	177.0	516.3	179.1	653.6	610.7	788.1	499.7	2422.6	_160.4_	**49.5**
	8 × 10	169.3	598.9	206.6	601.9	573.2	1483.3	502.1	1576.4	_163.6_	**47.0**
	16 × 16	210.9	611.2	226.8	611.0	682.8	1600.7	571.7	880.8	_161.8_	**52.5**
	30 × 25	175.4	568.0	212.1	569.0	426.3	1453.0	513.5	548.7	_133.6_	**45.2**

4.5 Summary of Experimental Results

- TDVM outperforms the nine competing methods in all cases of face recognition on two real data, especially on data with more missing entries. Besides, our method is much more efficient than all compared methods. Moreover, with less training features (e.g. $L = 1$ for FERET and $L = 5$ for YaleB) in NNC, TDVM shows more advantage as it extracts low-dimensional informative features. These results verifies the superiority of incorporating low-rank Tucker decomposition with feature variance maximization.

- The tensor learning method (TRPCA) is the best performing existing algorithm in six cases of FERET with pixel-based missing entries. However, it works much worse than TDVM on data with increasing missing entries. For example, on YaleB with 90% missing pixels, TRPCA loses up to 94.48% than TDVM on average.
- Tensor completion methods (HaLRTC and gHOI) obtain similar results in most cases and HaLRTC achieves the second best results in about half of all cases, while TDVM outperforms these two methods by 34.92% and 41.71% on average on FERET and YaleB respectively. These results echo our claim: tensor completion methods focus on recovering missing data and do not explore the relationships among samples for effective feature extraction.
- The "two-step" strategies (e.g., gHOI + PCA/MPCA) do not have much improvement and even perform worse than using only tensor completion methods (e.g., gHOI), as we claimed that reconstruction errors from completion step can deteriorate the performance in feature extraction step. Although gHOI + LRANTD/MPCA and HaLRTC + PCA/MPCA achieve the second best results in a few cases, TDVM outperforms the "two-step" strategies in all cases as we extracts informative features directly from observed entries.

5   Conclusion

In this paper, we have proposed an unsupervised feature extraction method, i.e. TDVM, which solves the problem of feature extraction for tensors with missing data. TDVM incorporates low-rank Tucker decomposition with feature variance maximization in a unified framework, which results in low-dimensional informative features extracted directly from observed entries. We have evaluated the proposed method on two real datasets with different pixel/block-based missing entries and applied the extracted features for face recognition. Experimental results have shown the superiority of TDVM in both pixel-based and block-based missing cases, where the proposed method consistently outperforms the nine competing methods in all cases, especially on data with more missing entries. Moreover, TDVM is not sensitive to parameters and more efficient than the compared methods.

Acknowledgment. This work is supported by the NSFC Grant: 61672444, HKBU Faculty Research Grant: FRG2/16-17/051, the SZSTI Grant: JCYJ2016053119400 6833, Hong Kong PhD Fellowship Scheme, and JSPS KAKENHI Grant: 17K00326. We thank Prof. Canyi Lu, Dr. Guoxu Zhou, Prof. Guangcan Liu, and Dr. Johann Bengua, for their helpful discussion.

References

1. Kolda, T.G., Bader, B.W.: Tensor decompositions and applications. SIAM Rev. **51**(3), 455–500 (2009)
2. Tucker, L.R.: Implications of factor analysis of three-way matrices for measurement of change. Prob. Meas. Change **15**, 122–137 (1963)

3. Lu, H., Plataniotis, K.N., Venetsanopoulos, A.N.: MPCA: multilinear principal component analysis of tensor objects. IEEE Trans. Neural Netw. **19**(1), 18–39 (2008)

4. Lu, J., Tan, Y.-P., Wang, G.: Discriminative multimanifold analysis for face recognition from a single training sample per person. IEEE Trans. Pattern Anal. Mach. Intell. **35**(1), 39–51 (2013)

5. Shi, Q., Lu, H.: Semi-orthogonal multilinear PCA with relaxed start. In: International Conference on Joint Conference on Artificial Intelligence (2015)

6. Cao, B., Lu, C.-T., Wei, X., Yu, P.S., Leow, A.D.: Semi-supervised tensor factorization for brain network analysis. In: Frasconi, P., Landwehr, N., Manco, G., Vreeken, J. (eds.) ECML PKDD 2016. LNCS (LNAI), vol. 9851, pp. 17–32. Springer, Cham (2016). https://doi.org/10.1007/978-3-319-46128-1_2

7. Li, X., Ng, M.K., Cong, G., Ye, Y., Wu, Q.: MR-NTD: manifold regularization nonnegative tucker decomposition for tensor data dimension reduction and representation. IEEE Trans. Neural Netw. Learn. Syst. **28**(8), 1787–1800 (2017). IEEE

8. Jolliffe, I.T.: Principal Component Analysis (2nd edn), In: Springer Serires in Statistics (2002)

9. Lu, C., Feng, J., Chen, Y., Liu, W., Lin, Z., Yan, S.: Tensor robust principal component analysis: exact recovery of corrupted low-rank tensors via convex optimization. In: Proceedings of the IEEE Conference on Computer Vision and Pattern Recognition, pp. 5249–5257 (2016)

10. Peng, Y., Ganesh, A., Wright, J., Wenli, X., Ma, Y.: Rasl: robust alignment by sparse and low-rank decomposition for linearly correlated images. IEEE Trans. Pattern Anal. Mach. Intell. **34**(11), 2233–2246 (2012)

11. Liu, G., Lin, Z., Yan, S., Sun, J., Yu, Y., Ma, Y.: Robust recovery of subspace structures by low-rank representation. IEEE Trans. Pattern Anal. Mach. Intell. **35**(1), 171–184 (2013)

12. Acar, E., Dunlavy, D.M., Kolda, T.G., Mørup, M.: Scalable tensor factorizations for incomplete data. Chemometr. Intell. Lab. Syst. **106**(1), 41–56 (2011)

13. Williams, D., Liao, X., Xue, Y., Carin, L., Krishnapuram, B.: On classification with incomplete data. IEEE Trans. Pattern Anal. Mach. Intell. **29**(3), 427–436 (2007). IEEE

14. Williams, D., Liao, X., Xue, Y., Carin, L.: Incomplete-data classification using logistic regression. In: Proceedings of the 22nd International Conference on Machine Learning, pp. 972–979. ACM (2005)

15. Guleryuz, O.G.: Nonlinear approximation based image recovery using adaptive sparse reconstructions and iterated denoising-part I: theory. IEEE Trans. Image Process. **15**(3), 539–554 (2006)

16. Hazan, E., Livni, R., Mansour, Y.: Classification with low rank and missing data. In: Proceedings of the 32nd International Conference on Machine Learning, pp. 257–266 (2015)

17. Liu, J., Musialski, P., Wonka, P., Ye, J.: Tensor completion for estimating missing values in visual data. IEEE Trans. Pattern Anal. Mach. Intell. **35**(1), 208–220 (2013)

18. Hu, W., Tao, D., Zhang, W., Xie, Y., Yang, Y.: The twist tensor nuclear norm for video completion. IEEE Trans. Neural Netw. Learn. Syst. **28**(12), 2961–2973 (2017). IEEE

19. Liu, Y., Shang, F., Fan, W., Cheng, J., Cheng, H.: Generalized higher-order orthogonal iteration for tensor decomposition and completion. In: Advances in Neural Information Processing Systems, pp. 1763–1771 (2014)

20. Zhou, G., Cichocki, A., Zhao, Q., Xie, S.: Efficient nonnegative tucker decompositions: algorithms and uniqueness. IEEE Trans. Image Process. **24**(12), 4990–5003 (2015)
21. Jia, C., Zhong, G., Fu, Y.: Low-rank tensor learning with discriminant analysis for action classification and image recovery. In: Proceedings of the International Conference Artificial Intelligence, pp. 1228–1234. AAAI Press (2014)
22. Wu, T.T., Lange, K.: Matrix completion discriminant analysis. Comput. Stat. Data Anal. **92**, 115–125 (2015)
23. De Lathauwer, L., De Moor, B., Vandewalle, J.: A multilinear singular value decomposition. SIAM J. Matrix Anal. Appl. **21**(4), 1253–1278 (2000)
24. Boyd, S.: Alternating direction method of multipliers. In: NIPS Workshop on Optimization and Machine Learning (2011)
25. Lu, H., Plataniotis, K.N., Venetsanopoulos, A.: Multilinear Subspace Learning: Dimensionality Reduction of Multidimensional Data. CRC Press, Boca Raton (2013)
26. Wright, J., Yang, A.Y., Ganesh, A., Sastry, S.S., Ma, Y.: Robust face recognition via sparse representation. IEEE Trans. Pattern Anal. Mach. Intell. **31**(2), 210–227 (2009)
27. Liu, G., Yan, S.: Latent low-rank representation for subspace segmentation and feature extraction. In: Proceedings of the IEEE Conference on Computer Vision, pp. 1615–1622. IEEE (2011)
28. Elhamifar, E., Vidal, R.: Sparse subspace clustering: algorithm, theory, and applications. IEEE Trans. Pattern Anal. Mach. Intell. **35**(11), 2765–2781 (2013)
29. Cai, J.-F., Candès, E.J., Shen, Z.: A singular value thresholding algorithm for matrix completion. SIAM J. Optim. **20**(4), 1956–1982 (2010)
30. Shang, F., Liu, Y., Cheng, J.: Generalized higher-order tensor decomposition via parallel ADMM. In: Proceedings AAAI Conference on Artificial Intelligence, pp. 1279–1285. AAAI Press (2014)
31. Higham, N., Papadimitriou, P.: Matrix procrustes problems. Rapport Technique, University of Manchester (1995)
32. Liu, Y., Shang, F., Fan, W., Cheng, J., Cheng, H.: Generalized higher order orthogonal iteration for tensor learning and decomposition. IEEE Trans. Neural Netw. Learn. Syst. **27**(12), 2551–2563 (2016)
33. LeCun, Y., Cortes, C., Burges, C.J.C.: Gradient-based learning applied to document recognition. Proc. IEEE **86**(11), 2278–2324 (1998). IEEE
34. Phillips, P.J., Moon, H., Rizvi, S.A., Rauss, P.J.: The FERET evaluation methodology for face-recognition algorithms. IEEE Trans. Pattern Anal. Mach. Intell. **22**(10), 1090–1104 (2000)
35. Lee, K.C., Ho, J., Kriegman, D.: Acquiring linear subspaces for face recognition under variable lighting. IEEE Trans. Pattern Anal. Mach. Intell. **27**(5), 684–698 (2005)

Structurally Regularized Non-negative Tensor Factorization for Spatio-Temporal Pattern Discoveries

Koh Takeuchi[1,3(✉)], Yoshinobu Kawahara[2,4], and Tomoharu Iwata[1]

[1] NTT Communication Science Laboratories, Kyoto, Japan
{takeuchi.koh,iwata.tomoharu}@lab.ntt.co.jp
[2] The Institute of Scientific and Industrial Research (ISIR), Osaka University,
Osaka, Japan
ykawahara@sanken.osaka-u.ac.jp
[3] Department of Intelligence Science and Technology, Kyoto University,
Kyoto, Japan
[4] Center for Advanced Intelligence Project, RIKEN, Tokyo, Japan

Abstract. Understanding spatio-temporal activities in a city is a typical problem of spatio-temporal data analysis. For this analysis, tensor factorization methods have been widely applied for extracting a few essential patterns into latent factors. Non-negative Tensor Factorization (NTF) is popular because of its capability of learning interpretable factors from non-negative data, simple computation procedures, and dealing with missing observation. However, since existing NTF methods are not fully aware of spatial and temporal dependencies, they often fall short of learning latent factors where a large portion of missing observation exist in data. In this paper, we present a novel NTF method for extracting smooth and flat latent factors by leveraging various kinds of spatial and temporal structures. Our method incorporates a unified structured regularizer into NTF that can represent various kinds of auxiliary information, such as an order of timestamps, a daily and weekly periodicity, distances between sensor locations, and areas of locations. For the estimation of the factors for our model, we present a simple and efficient optimization procedure based on the alternating direction method of multipliers. In missing value interpolation experiments of traffic flow data and bike-sharing system data, we demonstrate that our proposed method improved interpolation performances from existing NTF, especially when a large portion of missing values exists.

1 Introduction

Spatio-temporal data covering a wide area of a city have become available due to the commoditization of sensor-monitoring systems and mobile-phone networks. These monitoring systems observe various types of data, such as vehicle transportation counts on a road network, bike-renting counts of a bike-sharing system, and the purchasing records of shops around a city, where missing values often

© Springer International Publishing AG 2017
M. Ceci et al. (Eds.): ECML PKDD 2017, Part I, LNAI 10534, pp. 582–598, 2017.
https://doi.org/10.1007/978-3-319-71249-9_35

appear due to the failure of sensor nodes, data transmission errors, and trouble with data recording systems. We can find rich and bounteous information in such spatio-temporal data. However, it becomes difficult to grasp what spatio-temporal activities appeared in the data at a glance. Therefore, understanding of such activities via pattern extractions is a typical problem of spatio-temporal data analysis, in which the interpretability of the extracted patterns is regarded as one of the most important property for analysis methods.

Tensor factorization methods have been widely applied to discover spatial and temporal patterns from various kinds of spatio-temporal data [17]. These methods represent spatio-temporal data as a higher-order dimensional array, called a tensor that is a generalization of a matrix. For example, we can represent spatio-temporal data as a three-way tensor whose first, second, and third modes correspond to sensor locations, timestamps for 24 h, and the observed days. We illustrated an example of a tensor for spatio-temporal data analysis in Fig. 1. With this formulation, we can naturally incorporate an assumption that daily or weekly periodicity can be found in data and similar spatial patterns appear on different days. We can extract a few numbers of spatial, temporal, and daily patterns into latent factors by decomposing the tensor. However, since most existing tensor factorization methods do not consider the non-negativity of data where observations only contain non-negative values, they often result in messy and hard to interpret factors.

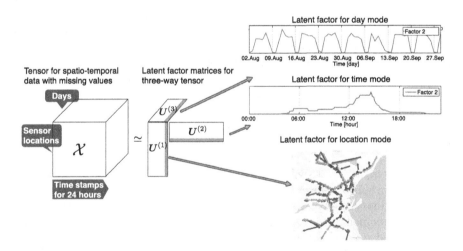

Fig. 1. Example for a non-negative tensor factorization method on analysis of a traffic-flow data set, where latent patterns for location, time, and day modes are extracted in the latent factor.

Unlike those tensor factorization methods, Non-negative Tensor Factorization (NTF) [8], which leverages non-negativity, is effective for extracting interpretable patterns from the non-negative data [13,18]. This method has successively yielded interpretable factors from various kinds of spatio-temporal data, such as location-based social network services [14,25], mobile phone GPS logs [10], log messages of network equipment [16], and traffic records of road networks [32]. However, NTF was not applicable to the existence of missing values. To deal with missing values, NTF was recently extended to learn the latent factors from a subset of elements in a tensor, called the non-negative tensor completion [15,31]. With this NTF, we can interpolate missing values in data by learned latent factors. However, NTF methods for the missing value completion problem suffer from overfitting when just a few observations are available. Because they ignore spatial and temporal contextual information such as the order of time stamps, weekly periodicity, the distances between sensor locations and treats each feature of the tensor independently.

To incorporate such contextual information, most matrix/tensor factorization methods have employed a graph Laplacian based regularizer for encouraging the latent factors to be smooth with spatio-temporal dependencies [21]. The graph regularized non-negative matrix factorization [6] is a variant of such schemes and has been widely utilized in many applications, however, it does not consider scenarios where missing values exist and analyzing higher-order dimensional arrays.

Another choice for representing such auxiliary information is structured regularizers [2] that have become popular in the fields of machine learning, signal processing, and data mining [7,28]. For example, the fused lasso [27], which is also known as the total variation, approximates parameters by piecewise-constant values with the order of parameters. Since its estimated parameters have the same estimated value, this is beneficial for finding segments of parameters. In a pioneering work [29], the penalized matrix decomposition was proposed to utilize the fused lasso as a regularizer on latent factors and was applied to a gene data analysis problem. They presented latent factors easy to find gene segments rather than existing matrix factorization methods incorporated the lasso regularizer. However, this method and its subsequent works have only considered the fused lasso without incorporating more general structured regularizer such as spatial dependencies of sensors, and also ignored the non-negative properties and the existence of missing values.

In this paper, we attempt to solve a problem of extracting latent factors from spatio-temporal data where a lot of missing values exists. To tackle this problem, we propose a novel NTF that learns factors by employing spatial and temporal auxiliary information as regularizers. We utilize this information to represent phenomena often appear in spatio-temporal data, such as counts of vehicles passed roads smoothly grow or decrease or take the same value along with space and time. To exploit such information, we introduce a regularizer that consists of both a graph-based Laplacian regularizer and structured regularizers that incorporate not only the order of features but also more general graph and group based structures [3,24]. With our regularizer, we can utilize various kinds of auxiliary information into NTF including a daily and weekly periodicity, distances between sensor locations, and areas of locations. Our proposed method is highly robust to the presence of a large portion of missing values because it encourages latent factors to be smooth and flat with spatial and temporal structures, where we regard segments of parameters that take the same value as flat. To estimate the latent factors for our proposed method, we present an efficient optimization procedure of the alternating direction method of multipliers [4] that utilizes simple proximity operators of the conjugate gradient method [21] and a parametric network flow algorithm [12].

We conducted missing value interpolation experiments with real-world traffic flow data and compared the performance of our proposed method with existing NTF methods. We demonstrate that our proposed method improved the interpolation performances from existing NTF methods. We also show that our extracted factors were interpretable to detect change points. Because our factors have segments, we can easily find a boundary of segments as a change point.

2   Non-negative Tensor Factorization

We denote a N-th way non-negative tensor as $\mathcal{X} \in \mathbb{R}_{\geq 0}^{I_1 \times \cdots \times I_N}$, where I_n is the number of features in the n-th mode. The n-th mode unfolding of a tensor $\mathcal{X}$ is denoted as $\mathcal{X}_n$. We use $i = (i_1, \ldots, i_N)$ and D to represent an element and the whole set of the elements in the tensor, respectively. A subset of the observed elements in the tensor is denoted by $\Omega = \{i \mid x_i \text{ is observed}, \forall i \in D\}$.

NTF decomposes the observed values of tensor $\mathcal{X}$ into K latent non-negative factors, where $K \ll \min(I_1, \ldots, I_N)$. The n-th mode factor matrix is denoted as $\boldsymbol{A}^{(n)} \in \mathbb{R}_{\geq 0}^{I_n \times K}$ whose k-th column is factor vector $\boldsymbol{a}_k^{(n)} \in \mathbb{R}_{\geq 0}^{I_n}$. We denote a whole set of factor vectors as $A = \{\boldsymbol{a}_k^{(n)} \mid \forall (n, k)\}$. An estimation for element x_i is given by a sum of latent factor vectors $\hat{x}_i = \sum_{k=1}^{K} a_{i_1,k}^{(1)} a_{i_2,k}^{(2)} \cdots a_{i_N,k}^{(n)} \in \hat{\mathcal{X}}$. We denote the transpose operator as $\top$, the Khatri-Rao product as $\odot$, and its series as $\odot_{n=1}^{N} \boldsymbol{A}^{(n)} = \boldsymbol{A}^{(1)} \odot \cdots \odot \boldsymbol{A}^{(N)}$.

The empirical loss function for NTF can be defined as a sum of divergences that indicates a discrepancy between x_i and its estimation $\hat{x}_i$:

$$f(A) = D_\Omega(\mathcal{X}\|\hat{\mathcal{X}}) + \sum_{n=1}^{N}\sum_{k=1}^{K} g^{(n)}(\boldsymbol{a}_k^{(n)}), \tag{1}$$

where $D_\Omega(\mathcal{X}\|\hat{\mathcal{X}}) = \sum_{i \in \Omega} d(x_i\|\hat{x}_i)$. We $d(p\|q)$ to denote a divergence between scalars p and q, and $g^{(n)}$ to denote a penalty function for the n-th mode factor vector. Because loss function f is non-convex with respect to A, an NTF problem is to obtain a local minimizer A^* of the loss under a non-negative constraint:

$$A^* = \arg\min_A f(A) \text{ subject to } \boldsymbol{a}_k^{(n)} \geq 0 \ \ \forall (n, k). \tag{2}$$

The graph regularized non-negative matrix factorization method [6] employs a graph Laplacian regularizer [22] to represent the smoothness in latent factors. An adjacency matrix for the n-th mode features is denoted as $\boldsymbol{W}^{(n)} \in \mathbb{R}^{I_n \times I_n}$ that represents a graph whose nodes and capacities of edges correspond to the features of the n-th mode and the similarity measures between the two features, respectively. The Laplacian matrix can be denoted as $\boldsymbol{L}^{(n)} = \boldsymbol{D}^{(n)} - \boldsymbol{W}^{(n)}$, where $\boldsymbol{D}^{(n)}$ is a diagonal matrix whose elements are the sums of each row of $\boldsymbol{W}^{(n)}$. Then a graph Laplacian regularizer can be defined:

$$g^{(n)}(\boldsymbol{a}_k^{(n)}) = \boldsymbol{a}_k^{(n)^\top} \boldsymbol{L}^{(n)} \boldsymbol{a}_k^{(n)}. \tag{3}$$

This regularizer penalty function encourages smoothness because its formulation equals putting a weighted quadratic term on the difference between the adjacency elements.

3   Proposed Model

We introduce a unified structured regularizer to employ both smooth and piecewise-constant properties with auxiliary structures:

$$g^{(n)}(\boldsymbol{a}_k^{(n)}) = \sum_{m=1}^{3} \lambda_m g_m^{(n)}(\boldsymbol{a}_k^{(n)}) + g_{\geq 0}^{(n)}(\boldsymbol{a}_k^{(n)}), \tag{4}$$

where λ_1, λ_2 and λ_3 are the hyperparameters for each regularizer. We employ a Generalized Fused Lasso (GFL) [5,30] and a Higher-Order Fused Lasso (HOFL) [24] as $g_1^{(n)}$ and $g_2^{(n)}$, respectively. $g_3^{(n)}$ corresponds to the Laplacian regularizer for extracting smooth patterns. We use an indicator function for the non-negative region:

$$g_{\geq 0}^{(n)}(\boldsymbol{a}_k^{(n)}) = \begin{cases} 0 & (\text{if } a_{i,k} \geq 0, \ \forall i) \\ +\infty & (\text{otherwise}) \end{cases}. \tag{5}$$

The GFL penalty is defined:

$$g_1(\boldsymbol{a}_k^{(n)}) = \sum_{j=1}^{I_n} \sum_{j'=1}^{I_n} w_{j,j'}^{(n)} \left| a_{j,k}^{(n)} - a_{j',k}^{(n)} \right|. \tag{6}$$

The GFL prefers parameters with the same value if they are adjacent on the given graph, such as distances between sensor locations and temporal lags between time stamps. The HOFL encourages parameters in a given group to take identical values [24]. With this regularizer, we can utilize auxiliary information, such as sensors placed in a specific area that may output similar values and a group of time stamps when a specific train leaves from a station. We denote the r-th group of features in the n-th mode as $g_r^{(n)} \subseteq D_n$ and a set of groups by $\mathcal{G}^{(n)} = \{g_1^{(n)}, \cdots, g_{R_n}^{(n)}\}$, where D_n and R_n are a set of elements in the n-th mode and the number of groups, respectively. The weights of each element for the r-th group on the n-th mode are denoted by $c_{r,m}^{(n)} = \bar{c}_{r,m}^{(n)}$ if $m \in g_r^{(n)}$, and 0 otherwise, where $\bar{c}_{r,m}^{(n)} > 0$. Then a simplified HOFL penalty $g_2(\boldsymbol{a}_k^{(n)})$ is given:

$$\sum_{r=1}^{R} \sum_{m=1}^{I_n} c_{r,j_m}^{(n)} |a_{j_m,k}^{(n)} - \bar{a}_{r,j_m,k}^{(n)}| + \theta_r^{(n)}(a_{s_r,k}^{(n)} - a_{t_r,k}^{(n)}), \tag{7}$$

where $\theta_r^{(n)} > 0$ is a hyperparameter that controls the consistency of the parameters in a group. $\bar{a}_{r,k}^{(n)}$ is defined as $\bar{a}_{r,m,k}^{(n)} = a_{s_k,k}^{(n)}$ (if $m \geq s_k$), $a_{t_k,k}^{(n)}$ (if $m \leq t_k$) and $a_{j_m,k}^{(n)}$ (otherwise) for distinct indices $j_1, j_2, \ldots, j_{I_n} \in D_n$ that correspond to a permutation that arranges the entries of $\boldsymbol{a}_k^{(n)}$ in a non-increasing order. Thresholding indices s_r and t_r are given as $s_k = \min\{m' \mid \sum_{m=1}^{m'} c_{r,j_m}^{(n)} \geq \theta_r^{(n)}\}$ and $t_k = \min\{m' \mid \sum_{m=m'}^{I_n} c_{r,j_m}^{(n)} < \theta_r^{(n)}\}$.

For convenience, we denote $\bar{g}^{(n)}(\boldsymbol{a}_k^{(n)}) = \sum_{m=1}^{2} \lambda_m g_m^{(n)}(\boldsymbol{a}_k^{(n)}) + g_{\geq 0}^{(n)}(\boldsymbol{a}_k^{(n)})$. By adopting our structured regularizers to the loss of NTF, we define the following minimization problem for our purpose:

$$A^* = \arg\min_{A} D_\Omega(\mathcal{X} \| \hat{\mathcal{X}}) + \sum_{n=1}^{N} \sum_{k=1}^{K} \bar{g}^{(n)}(\boldsymbol{a}_k^{(n)}) + \lambda_3 g_3^{(n)}(\boldsymbol{a}_k^{(n)}). \qquad (8)$$

Note that when $\lambda_1 = \lambda_2 = \lambda_3 = 0$, our method is reduced to an original NTF. When $\lambda_2 = \lambda_3 = 0$, our method can be regarded as a tensor extension of the graph-regularized non-negative matrix factorization. Our method includes those methods as special cases.

4   Parameter Estimation

We present an efficient parameter estimation procedure for obtaining a local minimizer of our proposed method. We employ a scaled formulation of the Alternating Direction Method of Multipliers (ADMM) for NTF [15]. The minimization problem for our proposed method can be rewritten:

$$\min_{A, \mathcal{Z}} D_\Omega(\mathcal{X} \| \mathcal{Z}) + \sum_{n=1}^{N} \sum_{k=1}^{K} \bar{g}^{(n)}(\boldsymbol{b}_k^{(n)}) + g_3^{(n)}(\boldsymbol{a}_k^{(n)})$$

$$\text{subject to } \mathcal{Z} = \hat{\mathcal{X}}, \boldsymbol{a}_k^{(n)} = \boldsymbol{b}_k^{(n)} \; (\forall n, k), \qquad (9)$$

where $\mathcal{Z}$ and $\boldsymbol{b}_k^{(n)}$ are auxiliary variables, and P_Ω is a projection function that only retains the divergence of the observed elements. To solve our problem efficiently with keeping both constraints and separability, we define an augmented Lagrangian for our problem:

$$L_\rho(A, B, \mathcal{Z}) = D_\Omega(\mathcal{X} \| \mathcal{Z}) + \frac{\rho}{2} \| \mathcal{Z} - \hat{\mathcal{X}} + \mathcal{U} \|_{\mathcal{F}}^2$$

$$\sum_{n=1}^{N} \sum_{k=1}^{K} \bar{g}^{(n)}(\boldsymbol{b}_k^{(n)}) + g_3^{(n)}(\boldsymbol{a}_k^{(n)}) + \frac{\rho}{2} \| \boldsymbol{a}_k^{(n)} - \boldsymbol{b}_k^{(n)} + \boldsymbol{u}_k^{(n)} \|_2^2, \qquad (10)$$

where $\mathcal{U}$ and $\boldsymbol{u}_k^{(n)}$ are Lagrangian multipliers, and ρ is a step-size parameter, respectively. We summarize the minimization procedure for our proposed method in Algorithm 1. The minimization for ADMM can be efficiently calculated if a simple minimization operator for each of each $\boldsymbol{a}_k^{(n)}$ and $\boldsymbol{b}_k^{(n)}$ exists.

Algorithm 1. Alternative direction method of multiplier for our proposed non-negative tensor factorization

---

Input   : $\mathcal{X}, \Omega, \lambda_1, \lambda_2, \lambda_3, K, \boldsymbol{W}^{(n)}$
Output: set of factor matrices A

1 Initialize parameters
2 Sample A, B, and $\mathcal{Z}$ from random distributions
3 **repeat**
4 | Alternatively update parameters;
5 | $\mathcal{Z} \leftarrow \arg\min_{\mathcal{Z}} D_{\Omega}(\mathcal{X}\|\mathcal{Z}) + (\rho/2)\|\mathcal{Z} - \hat{\mathcal{X}} + \mathcal{U}\|_F^2$
6 | **for** $n = 1$ **to** N **do**
7 | | Update $\boldsymbol{A}^{(n)}$ by solving Eq. (11)
8 | | **for** $k = 1$ **to** K **do**
9 | | | Update $\boldsymbol{b}_k^{(n)}$ by solving Eq. (12)
10 | | **end**
11 | | $\boldsymbol{U}^{(n)} \leftarrow \boldsymbol{U}^{(n)} + (\boldsymbol{A}^{(n)} - \boldsymbol{B}^{(n)})$
12 | **end**
13 | $\mathcal{U} \leftarrow \mathcal{U} + (\mathcal{X} - \hat{\mathcal{X}})$
14 **until** *convergence*;

The loss function with respect to $\boldsymbol{A}^{(n)}$ and $\boldsymbol{b}_k^{(n)}$ contains the graph Laplacian regularizer and the non-separable graph-based and group-based penalties, respectively. Thus the main difficulty with our proposed method lies in the minimization of $\boldsymbol{A}^{(n)}$ and $\boldsymbol{b}_k^{(n)}$, whose minimization problems can be rewritten:

$$\boldsymbol{A}^{(n)} = \arg\min_{\boldsymbol{A}^{(n)}} \frac{\rho}{2}\|\bar{\mathcal{Z}}_n - \boldsymbol{A}^{(n)}\boldsymbol{V}_n^{\top}\|_2^2 + \frac{\rho}{2}\|\boldsymbol{A}^{(n)} - \bar{\boldsymbol{V}}_n\|_2^2 + \lambda_3\sum_{k=1}^{K} g_3^{(n)}(\boldsymbol{a}_k^{(n)}) \quad (11)$$

$$\boldsymbol{b}_k^{(n)} = \arg\min_{\boldsymbol{b}_k^{(n)}} \bar{g}^{(n)}(\boldsymbol{b}_k^{(n)}) + \frac{\rho}{2}\|\bar{\boldsymbol{v}}_k^{(n)} - \boldsymbol{b}_k^{(n)}\|_2^2, \quad (12)$$

where $\bar{\mathcal{Z}} = \mathcal{Z} + \mathcal{U}$, $\boldsymbol{V}_n = \odot_{n=n'}^{N}\boldsymbol{A}^{(n)}$, $\bar{\boldsymbol{V}}_n = \boldsymbol{B}^{(n)} - \boldsymbol{U}^{(n)}$, and $\bar{\boldsymbol{v}}_k^{(n)} = \boldsymbol{a}_k^{(n)} + \boldsymbol{u}_k^{(n)}$. We efficiently solve the minimization of Eq. (11) by using the fact that it corresponds to the loss function of the graph regularized alternating least squares [21], which approximately runs in $\mathcal{O}(\text{nnz}(\boldsymbol{L}^{(n)})K)$ ($\text{nnz}(\cdot)$ is the number of non-zero elements).

The minimization problem in Eq. (12) corresponds to the calculation of the proximity operator, which is defined as: $\text{prox}_{\gamma h}(\theta) = \arg\min_{\theta} h(\theta) + \frac{1}{2\gamma}\|\hat{\theta} - \theta\|_2^2$. We present a minimization procedure for Eq. (12) by leveraging the properties of the proximity operator, and obtaining a minimizer for the sum of the non-negative indication function and other convex functions by the following property [26]: $\text{prox}_{\bar{g}^{(n)}} = \text{prox}_{g_{\geq 0}^{(n)}} \circ \text{prox}_{\lambda_1 g_1^{(n)} + \lambda_2 g_2^{(n)}}$. Thus, if we have a minimizer for $\lambda_1 g_1^{(n)} + \lambda_2 g_2^{(n)}$, we can attain the exact minimizer for $\bar{g}^{(n)}$ by setting negative parameters to zeros. A minimizer for $\lambda_1 g_1^{(n)} + \lambda_2 g_2^{(n)}$ can be simply calculated by employing a submodular function minimization procedure. Because the penalty functions of GFL and HOFL are the Lovász extensions [19] of the graph-representable submodular functions [11], we can attain a minimizer for the sum of functions $\lambda_1 g_1^{(n)} + \lambda_2 g_2^{(n)}$ by an efficient parametric network flow algorithm [7,24,30]. We show the details of our minimization procedure for this function in the appendix.

5   Related Works

There has been a lot of articles in which NTF was applied to analyze spatio-temporal data. Kimura et al. proposed a special NTF that decomposes a three-way tensor into two-factor matrices and a three-mode tensor for extracting log messages related to network failures [16]. Yang et al. proposed a combination of NTF without regularizers and post-processing for modeling user activities [32]. Takeuchi et al. proposed an NTF that simultaneously decomposes multiple tensors to extract patterns appeared among different tensors [25]. NTF was used to extract spatio-temporal patterns from human-flow data [10]. However, all of those methods did not employ regularizers into NTF and their methods were not applicable to missing values. One exception is a paper of Sun and Axhausen [23], in which they proposed a probabilistic non-negative Tucker decomposition for discovering interactions among factors. However, they did not incorporate the spatial and temporal structures into regularizers. Han and Moutarde proposed an extension of NTF for predicting future observations [14]. However, they did not consider spatial structures. Our method can be applied to their framework to utilizing spatial and temporal regularizers. The estimation procedures of them were based on the multiplicative update rule and EM algorithm. Our proposed method can utilize graphs and groups of spatial and temporal features to regularize parameters and also employ ADMM as an estimation procedure.

6   Experiments

We conducted missing value completion problems with a traffic flow data set provided by City Pulse [1] and two bike-sharing system data sets recorded in Washington D.C.[1] and New York[2] [1].

The traffic flow data consist of the numbers of cars that passed at 419 locations every thirty minutes in Arhus City, Denmark. We picked 30 days from August 2nd to 31st 2014, and constructed three-way tensor $\mathcal{X} \in \mathbb{R}^{48 \times 30 \times 441}$ whose modes corresponded to 48 daily time points, 30 days, and 441 observation locations, respectively. From the bike-sharing system data in Washington D.C. and New York, we employed 15 days from April 1st to the 15th with 351 and 344 bike stations. We constructed three-way tensors $\mathcal{X} \in \mathbb{R}^{24 \times 15 \times 351}$ and $\mathcal{X} \in \mathbb{R}^{24 \times 15 \times 44}$ whose values were the numbers of bikes returned to the station

---

[1] https://www.capitalbikeshare.com.
[2] http://www.citibikenyc.com/.

in an hour. For the time mode, we utilized the adjacency of the time points as a graph. For the day mode, we employed the adjacency of days and the days of the week as a graph and groups. For the location mode, we used the inverse of the Euclid distance of GPS locations and clusters attained by k-nearest neighbors ($k = 5, 10$) for a graph and groups.

We exploited the Euclid distance as the divergence in experiments. We compared our proposed method (Proposed 1) and our proposed method with only the graph Laplacian regularizer (Proposed 2, $\lambda_1 = \lambda_2 = 0$) with NTF estimated by ADMM [15] (ADMM), NTF with the graph Laplacian regularizer [6] estimated by a multiplicative update rule considering missing values (Multi+Lap) [9], and NTF estimated by the multiplicative update rule (Multi). We set the proportion of observations to $p = \{0.1, 0.01, 0.005, 0.001\}$. By five-fold cross validation, we selected K and other hyperparameters from $K = \{3, 5, 10\}$ and $\{0.1, 1, 10\}$. We utilized the normalized RMSE (NRMSE) and the normalized deviation (ND) as error measurements:

$$\text{NRMSE} = \sqrt{(1/|\Omega|) \sum_{(p,t) \in \Omega} (x_{p,t} - \hat{x}_{p,t})^2 / Q}, \tag{13}$$

$$\text{ND} = (1/|\Omega|) \sum_{(p,t) \in \Omega} |x_{p,t} - \hat{x}_{p,t}| / Q, \tag{14}$$

here $Q = (1/|\Omega|) \sum_{(p,t) \in \Omega} |x_{p,t}|$. We ran our experiments five times with randomly selected different missing values.

The results are shown in Tables 1, 2, 3, 4, 5, and 6, where the left and right values in a cell correspond to the average and the standard deviation of those values. We confirmed that our proposed methods showed the best performance in every setting. Our proposed method was robust to the appearance of a large portion of missing values for every data set $p = \{0.01, 0.005, 0.001\}$. Our proposed method with both the graph-based Laplacian and structured regularizer (Proposed 1) showed better or competitive performance with our proposed method with the graph-based Laplacian regularizer (Proposed 2). Furthermore, our proposed method with the graph-based Laplacian regularizer (Proposed 2)

Table 1. NRAME for the traffic flow data of our proposed method (Proposed 1), our proposed method with the graph Laplacian regularizer (Proposed 2), NTF estimated by ADMM (ADMM), NTF with the graph Laplacian regularizer estimated by the multiplicative update rule (Multi+Lap), and NTF (Multi)

	$p = 0.1$	$p = 0.01$	$p = 0.005$	$p = 0.001$
Proposed 1	**0.50 (0.00)**	**0.99 (0.03)**	**1.49 (0.03)**	**1.87 (0.01)**
Proposed 2	0.51 (0.00)	1.12 (0.05)	**1.49 (0.03)**	1.89 (0.01)
ADMM	0.51 (0.00)	1.15 (0.03)	1.50 (0.02)	1.91 (0.00)
Multi+Lap	0.52 (0.00)	2.98 (1.86)	2.92 (1.00)	11.9 (11.7)
Multi	0.52 (0.00)	2.89 (2.22)	3.27 (1.80)	5.98 (6.24)

Table 2. NRAME for the bike-sharing record data of Washington D.C.

Method	$p = 0.1$	$p = 0.01$	$p = 0.005$	$p = 0.001$
Proposed 1	**1.67 (0.02)**	**2.14 (0.02)**	**2.21 (0.04)**	**2.43 (0.02)**
Proposed 2	1.68 (0.01)	**2.14 (0.05)**	2.22 (0.05)	2.62 (0.08)
ADMM	1.68 (0.02)	2.21 (0.05)	2.32 (0.03)	2.47 (0.01)
Multi+Lap	1.69 (0.01)	2.72 (0.22)	2.76 (0.24)	11.2 (4.62)
Multi	1.70 (0.01)	299.1 (405.7)	8.25 (4.87)	16.3 (3.13)

Table 3. NRAME for the bike-sharing record data of New York

Method	$p = 0.1$	$p = 0.01$	$p = 0.005$	$p = 0.001$
Proposed 1	**0.98 (0.00)**	**1.28 (0.02)**	**1.42 (0.01)**	**1.62 (0.01)**
Proposed 2	**0.98 (0.00)**	1.30 (0.03)	1.44 (0.01)	1.63 (0.01)
ADMM	**0.98 (0.00)**	1.34 (0.02)	1.49 (0.01)	1.65 (0.01)
Multi+Lap	1.00 (0.02)	27.2 (41.3)	5.68 (3.22)	1.86 (0.17)
Multi	1.00 (0.02)	53.7 (22.0)	26.6 (29.4)	3.04 (1.03)

Table 4. ND for the traffic flow data of our proposed method (Proposed 1), our proposed method with the graph Laplacian regularizer (Proposed 2), NTF estimated by ADMM (ADMM), NTF with the graph Laplacian regularizer estimated by the multiplicative update rule (Multi+Lap), and NTF (Multi)

Method	$p = 0.1$	$p = 0.01$	$p = 0.005$	$p = 0.001$
Proposed 1	**0.27 (0.00)**	**0.46 (0.01)**	**0.70 (0.01)**	**0.92 (0.01)**
Proposed 2	0.28 (0.00)	0.51 (0.02)	0.71 (0.02)	0.94 (0.00)
ADMM	0.28 (0.00)	0.53 (0.01)	0.73 (0.01)	0.94 (0.00)
Multi+Lap	0.28 (0.00)	0.61 (0.08)	0.78 (0.02)	1.19 (0.16)
Multi	0.28 (0.00)	0.60 (0.08)	0.81 (0.04)	1.18 (0.24)

Table 5. ND for the bike-sharing record data of Washington D.C.

Method	$p = 0.1$	$p = 0.01$	$p = 0.005$	$p = 0.001$
Proposed 1	**0.81 (0.00)**	**0.91 (0.01)**	**1.04 (0.02)**	**1.04 (0.08)**
Proposed 2	**0.81 (0.01)**	**0.91 (0.00)**	**1.04 (0.02)**	1.05 (0.09)
ADMM	**0.81 (0.01)**	**0.91 (0.01)**	1.10 (0.01)	1.20 (0.01)
Multi+Lap	**0.81 (0.00)**	1.19 (0.05)	1.51 (0.15)	1.99 (0.54)
Multi	**0.81 (0.00)**	9.70 (6.91)	1.44 (0.10)	2.56 (0.31)

Table 6. ND for the bike-sharing record data of New York

Method	$p = 0.1$	$p = 0.01$	$p = 0.005$	$p = 0.001$
Proposed 1	**0.60 (0.00)**	**0.72 (0.01)**	**0.81 (0.01)**	**0.93 (0.01)**
Proposed 2	**0.60 (0.00)**	0.73 (0.01)	0.82 (0.01)	**0.93 (0.01)**
ADMM	**0.60 (0.00)**	0.74 (0.01)	0.84 (0.01)	0.94 (0.01)
Multi+Lap	**0.60 (0.00)**	2.13 (1.48)	1.53 (0.21)	1.05 (0.05)
Multi	**0.60 (0.00)**	4.48 (0.79)	2.51 (0.95)	1.21 (0.07)

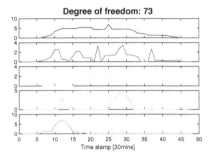

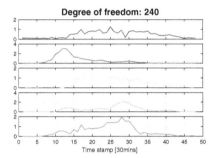

Fig. 2. Time factors of Proposed 1 on the traffic flow data (Color figure online)

Fig. 3. Time factors of Multi+Lap on the traffic flow data (Color figure online)

always outperformed the same model estimated by the multiplicative update rule (Multi+Lap). This result was caused by the benefits of simultaneously combining graph-based and structured regularizers with graph and group structures. Thus our proposed model and parameter estimation procedure both contributed to the improvements on missing value interpolations. The existing methods resulted in poor performances with settings where a large portions of tensor elements were missing.

To check the qualitative performances of the interpretability, we showed the extracted factors of proposed method (Proposed 1) and existing NTF with the Laplacian regularizer (Multi+Lap) from traffic flow data in Figs. 2, 3, 4, 5, 6, and 7, where $p = 0.1$. The degree of freedom (DoF) in Figures corresponded the number of segments in a factor matrix. Thanks to the Laplacian and structured regularizers, proposed method extracted the interpretable latent factors in which both smooth and flat properties appeared, whose DoF of parameters in factor matrices were extremely less than that of NTF with the Laplacian regularizer. Our factors with low DoF were easy to find change points. For example, the blue factor had a change at 3 am and gradually grew until 6 am. Then it took the constant values until 3 pm in Fig. 2. This factor also has the same value from day 2 to day 6 and from day 8 to day 13. Thus, we can easily understand that the blue factor in Figs. 2 and 4 corresponded to activity that occurred in weekday during daylight with a spatial pattern in Fig. 6. However, NMF with the Laplacian

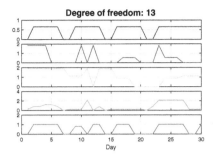

Fig. 4. Day factors of Proposed 1 on the traffic flow data (Color figure online)

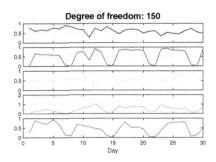

Fig. 5. Day factors of Multi+Lap on the traffic flow data (Color figure online)

Fig. 6. A spatial pattern of the blue factor of Proposed 1 on the traffic flow data (Color figure online)

Fig. 7. A spatial pattern of the blue factor of Multi+Lap on the traffic flow data (Color figure online)

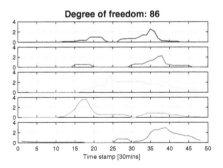

Fig. 8. Time factors of Proposed 1 on the bike-sharing data of Washington D.C. (Color figure online)

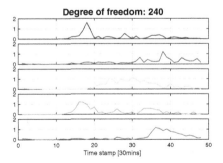

Fig. 9. Time factors of Multi+Lap on the bike-sharing data of Washington D.C. (Color figure online)

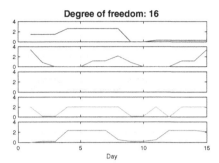

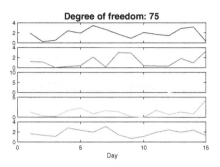

Fig. 10. Day factors of Proposed 1 on the bike-sharing data of Washington D.C. (Color figure online)

Fig. 11. Day factors of Multi+Lap on the bike-sharing data of Washington D.C. (Color figure online)

Fig. 12. A spatial pattern of the yellow factor of Proposed 1 on the bike-sharing data of Washington D.C. (Color figure online)

Fig. 13. A spatial pattern of the yellow factor of Multi+Lap on the bike-sharing data of Washington D.C. (Color figure online)

regularizer resulted in messy factors. We also showed that of bike-sharing data in Washington D.C. in Figs. 8, 9, 10, 11, 12, and 13. Our proposed method also extracted more interpretable patterns than existing NTF. For example, the yellow factor of ours in Fig. 8 had a change point at 8 am. After it had taken a peak at 12 am, it kept the same value from 1 pm to 5 pm. Then its value gradually decreased to zero. The yellow factor in Fig. 10 had the same high value on day 2, 3, 9, and 10. Thus, we confirmed that this factor indicated a weekend afternoon activity with a spatial pattern in Fig. 12. Similar interpretations can be obtained from other factors of ours.

7   Conclusion

In this paper, we proposed a structurally regularized non-negative tensor factorization that incorporated both the graph Laplacian and the structured regularizers on latent factors. For the structured regularizer, we employed the generalized fused lasso and the higher-order fused lasso to represent both graph-based and group-based information in time and space. We introduced a flexible and efficient parameter estimation method based on the alternating direction method of multipliers and showed a proximity operator for our unified structured regularizer. With experiments on a missing value imputation problem of three data sets, we confirmed that our proposed method showed the best quantitative performance and successfully extracted more interpretable latent factors than the existing non-negative tensor factorization methods.

Acknowledgements. The part of this work was supported by JSPS KAKENHI Grant Numbers JP16H01548 and JP26280086, and NICT "Research and Development on Fundamental and Utilization Technologies for Social Big Data".

A   Appendix

Although the issue in Eq. (12) is a general problem containing the previous problems [5,24,30] as special cases, we can solve it in a similar manner as these works. We first briefly introduce the parametric optimization method for a non-decreasing set function. Let $\alpha \in \mathbb{R}_{\geq 0}$, and define set function $l_\alpha(S) = l(S) - \alpha \mathbf{1}(S)$ $(\forall S \subset V)$, where $\mathbf{1}(S) = \sum_{i \in S} 1$. Then if l is a non-decreasing submodular function, then there exists a set of $r + 1$ $(\leq |V|)$ subsets: $S^* = \{S_0 \subset S_1 \subset \cdots \subset S_r\}$, where $S_j \subset V$, $S_0 = \emptyset$, and $S_r = V$, and $r + 1$ subintervals Q_r of α: $Q_0 = [0, \alpha_0), Q_1 = [\alpha_1, \alpha_2), \cdots, Q_r = [\alpha_r, \infty)$, such that, for each $j \in \{0, 1, \cdots, r\}$, S_j is the unique maximal minimizer of $h_\alpha(S), \forall \alpha \in Q_j$. A minimizer of Eq. (12) $\boldsymbol{t}^* = (t_1^*, t_2^*, \cdots, t_{|V|}^*)$ is then determined: $t_i^* = \frac{f(S_{j+1}) - f(S_j)}{\mathbf{1}(S_{j+1} \setminus S_j)}$, $\forall i \in (S_{j+1} \setminus S_j)$, $j = (1, \cdots, r)$. We introduce two lemmas [20] to see that l is a non-decreasing submodular function.

Lemma 1 (Lemma). *For any $\eta \in \mathbb{R}$ and submodular function h, $\boldsymbol{t}^*$ is an optimal solution to $\min_{t \in \mathcal{B}(l)} \|\boldsymbol{t}\|_2^2$ if and only if $\boldsymbol{t}^* - \eta \mathbf{1}$ is an optimal solution to $\min_{t \in \mathcal{B}(l) + \eta \mathbf{1}} \|\boldsymbol{t}\|_2^2$.*

Lemma 2 (Lemma). *Set $\eta = \max_{i=1,\cdots,|V|} \{0, l(V \setminus \{i\}) - l(V)\}$, and then $l + \eta \mathbf{1}$ is a non-decreasing submodular function.*

With Lemma 2, we solve

$$\min_{S \subset V} f(S) - \hat{z}(S) + (\eta - \alpha)\mathbf{1}(S), \tag{15}$$

and apply Lemma 1 to obtain a solution to the original problem. With fixed α, we can efficiently attain the optimal of Eq. (15) because this is a minimum cut problem.

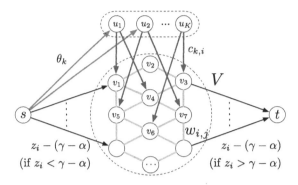

Fig. 14. Minimum s/t-cut problem of Problem (15). Given graph $G = (V', E')$ for our proposed method, capacities of edges $c(v_i', v_j')$ are defined as: $c(s, u_k) = \theta_k$, $c(v_i, v_j) = w_{i,j}$, $c(u_k, v_i) = c_{k,i}$, $c(s, v_i) = z_i - (\gamma - \alpha)$ if $z_i > \gamma - \alpha$, and $c(v_i, t) = (\gamma - \alpha) - z_i$ if $z_i < \gamma - \alpha$. Nodes u_k $k = (1, \cdots, K)$ are hyper nodes that correspond to the groups g_k. And s, t, and v_i are source, sink, and parameters nodes, respectively.

Proposition 1. *The problem in Eq.*(15) *is equivalent to a minimum s/t-cut problem.*

Proof. Each component in f is graph-representable. The graph is obtained due to the additive property of the graph-representative submodular functions, where the groups of parameters are represented with hyper nodes u_1^k, u_0^k that corresponds to each group, and the capacities of the edges between hyper and ordinal nodes $v_i \in V$.

The attained graph includes both of the GFL and HOFL graphs as spacial cases. As a consequence, we can attain a sequence of solutions for all α of the parametric s/t minimun-cut problem (15) using an efficient parametric-flow algorithm, such as [12], that runs in $O(|V'||E'| \log(|V'|^2/|E'|))$ as the worst case and $|V'|$ and $|E'|$ are the number of nodes and edges of the graph (Fig. 14).

References

1. Ali, M.I., Gao, F., Mileo, A.: CityBench: a configurable benchmark to evaluate RSP engines using smart city datasets. In: Arenas, M., Corcho, O., Simperl, E., Strohmaier, M., d'Aquin, M., Srinivas, K., Groth, P., Dumontier, M., Heflin, J., Thirunarayan, K., Staab, S. (eds.) ISWC 2015. LNCS, vol. 9367, pp. 374–389. Springer, Cham (2015). https://doi.org/10.1007/978-3-319-25010-6_25
2. Bach, F.R.: Structured sparsity-inducing norms through submodular functions. In: Proceedings of NIPS, pp. 118–126 (2010)
3. Barbero, A., Sra, S.: Fast Newton-type methods for total variation regularization. In: Proceedings of ICML, pp. 313–320 (2011)
4. Boyd, S., Parikh, N., Chu, E., Peleato, B., Eckstein, J.: Distributed optimization and statistical learning via the alternating direction method of multipliers. Found. Trends Mach. Learn. **3**(1), 1–122 (2011)

5. Boykov, Y., Kolmogorov, V.: An experimental comparison of min-cut/max-flow algorithms for energy minimization in vision. IEEE Trans. Pattern Anal. Mach. Intell. **26**(9), 1124–1137 (2004)
6. Cai, D., He, X., Han, J., Huang, T.S.: Graph regularized nonnegative matrix factorization for data representation. IEEE Trans. Pattern Anal. Mach. Intell. **33**(8), 1548–1560 (2011)
7. Chambolle, A., Darbon, J.: On total variation minimization and surface evolution using parametric maximum flows. Int. J. Comput. Vis. **84**(3), 288 (2009)
8. Cichocki, A., Zdunek, R., Phan, A.H., Amari, S.-I.: Nonnegative Matrix and Tensor Factorizations: Applications to Exploratory Multi-way Data Analysis and Blind Source Separation. Wiley, Hoboken (2009)
9. Dhillon, I.S., Sra, S.: Generalized nonnegative matrix approximations with Bregman divergences. In: Proceedings of NIPS, vol. 18 (2005)
10. Fan, Z., Song, X., Shibasaki, R.: CitySpectrum: a non-negative tensor factorization approach. In: Proceedings of UbiComp, pp. 213–223 (2014)
11. Fujishige, S.: Submodular Functions and Optimization, vol. 58. Elsevier, Amsterdam (2005)
12. Gallo, G., Grigoriadis, M.D., Tarjan, R.E.: A fast parametric maximum flow algorithm and applications. SIAM J. Comput. **18**(1), 30–55 (1989)
13. Gillis, N.: The why and how of nonnegative matrix factorization. In: Regularization, Optimization, Kernels, and Support Vector Machines, pp. 257–291. Chapman and Hall/CRC (2014)
14. Han, Y., Moutarde, F.: Analysis of large-scale traffic dynamics in an urban transportation network using non-negative tensor factorization. Int. J. Intell. Transp. Syst. Res. **14**(1), 36–49 (2016)
15. Huang, K., Sidiropoulos, N.D., Liavas, A.P.: A flexible and efficient algorithmic framework for constrained matrix and tensor factorization. IEEE Trans. Sig. Process. **64**(19), 5052–5065 (2016)
16. Kimura, T., Ishibashi, K., Mori, T., Sawada, H., Toyono, T., Nishimatsu, K., Watanabe, A., Shimoda, A., Shiomoto, K.: Spatio-temporal factorization of log data for understanding network events. In: Proceedings of INFOCOM, pp. 610–618 (2014)
17. Kolda, T.G., Bader, B.W.: Tensor decompositions and applications. SIAM Rev. **51**(3), 455–500 (2009)
18. Lee, D.D., Seung, H.S.: Learning the parts of objects by non-negative matrix factorization. Nature **401**(6755), 788–791 (1999)
19. Lovász, L.: Submodular functions and convexity. In: Bachem, A., Korte, B., Grötschel, M. (eds.) Mathematical Programming The State of the Art, pp. 235–257. Springer, Heidelberg (1983). https://doi.org/10.1007/978-3-642-68874-4_10
20. Nagano, K., Kawahara, Y., Aihara, K.: Size-constrained submodular minimization through minimum norm base. In: Proceedings of ICML, pp. 977–984 (2011)
21. Rao, N., Yu, H.-F., Ravikumar, P.K., Dhillon, I.S.: Collaborative filtering with graph information: consistency and scalable methods. In: Proceedings of NIPS, pp. 2107–2115 (2015)
22. Smola, A.J., Kondor, R.: Kernels and regularization on graphs. In: Schölkopf, B., Warmuth, M.K. (eds.) COLT-Kernel 2003. LNCS (LNAI), vol. 2777, pp. 144–158. Springer, Heidelberg (2003). https://doi.org/10.1007/978-3-540-45167-9_12
23. Sun, L., Axhausen, K.W.: Understanding urban mobility patterns with a probabilistic tensor factorization framework. Transp. Res. Part B: Methodol. **91**, 511–524 (2016)

24. Takeuchi, K., Kawahara, Y., Iwata, T.: Higher order fused regularization for supervised learning with grouped parameters. In: Appice, A., Rodrigues, P.P., Santos Costa, V., Soares, C., Gama, J., Jorge, A. (eds.) ECML PKDD 2015. LNCS (LNAI), vol. 9284, pp. 577–593. Springer, Cham (2015). https://doi.org/10.1007/978-3-319-23528-8_36
25. Takeuchi, K., Tomioka, R., Ishiguro, K., Kimura, A., Sawada, H.: Non-negative multiple tensor factorization. In: Proceedings of ICDM, pp. 1199–1204 (2013)
26. Tandon, R., Sra, S.: Sparse nonnegative matrix approximation: new formulations and algorithms. Rapp. Tech. **193**, 38–42 (2010)
27. Tibshirani, R., Saunders, M., Rosset, S., Zhu, J., Knight, K.: Sparsity and smoothness via the fused lasso. J. R. Stat. Soc. Ser. B (Stat. Methodol.) **67**(1), 91–108 (2005)
28. Wang, Y.-X., Sharpnack, J., Smola, A., Tibshirani, R.J.: Trend filtering on graphs. J. Mach. Learn. Res. **17**(105), 1–41 (2016)
29. Witten, D.M., Tibshirani, R., Hastie, T.: A penalized matrix decomposition with applications to sparse principal components and canonical correlation analysis. Biostatistics **10**, 515–534 (2009)
30. Xin, B., Kawahara, Y., Wang, Y., Gao, W.: Efficient generalized fused lasso with its application to the diagnosis of Alzheimer's disease. In: Proceedings of AAAI, pp. 2163–2169 (2014)
31. Yangyang, X., Yin, W.: A block coordinate descent method for regularized multiconvex optimization with applications to nonnegative tensor factorization and completion. SIAM J. Imaging Sci. **6**(3), 1758–1789 (2013)
32. Yang, D., Zhang, D., Zheng, V.W., Yu, Z.: Modeling user activity preference by leveraging user spatial temporal characteristics in LBSNs. IEEE Trans. Syst. Man Cybern.: Syst. **45**(1), 129–142 (2015)

Networks and Graphs

Attributed Graph Clustering with Unimodal Normalized Cut

Wei Ye[1(✉)], Linfei Zhou[1], Xin Sun[1,2], Claudia Plant[3], and Christian Böhm[1]

[1] Ludwig-Maximilians-Universität München, Munich, Germany
{ye,zhou,boehm}@dbs.ifi.lmu.de
[2] Ocean University of China, Qingdao, China
sunxin@ouc.edu.cn
[3] University of Vienna, Vienna, Austria
claudia.plant@univie.ac.at

Abstract. Graph vertices are often associated with attributes. For example, in addition to their connection relations, people in friendship networks have personal attributes, such as interests, age, and residence. Such graphs (networks) are called attributed graphs. The detection of clusters in attributed graphs is of great practical relevance, e.g., targeting ads. Attributes and edges often provide complementary information. The effective use of both types of information promises meaningful results. In this work, we propose a method called UNCut (for Unimodal Normalized Cut) to detect cohesive clusters in attributed graphs. A cohesive cluster is a subgraph that has densely connected edges and has as many homogeneous (unimodal) attributes as possible. We adopt the *normalized cut* to assess the density of edges in a graph cluster. To evaluate the unimodality of attributes, we propose a measure called *unimodality compactness* which exploits Hartigans' dip test. Our method UNCut integrates the *normalized cut* and *unimodality compactness* in one framework such that the detected clusters have low *normalized cut* and *unimodality compactness* values. Extensive experiments on various synthetic and real-world data verify the effectiveness and efficiency of our method UNCut compared with state-of-the-art approaches. Code and data related to this chapter are available at: https://www.dropbox.com/sh/xz2ndx65jai6num/AAC9RJ5PqQoYoxreItW83PrLa?dl=0.

1 Introduction

Real-world graphs (networks) tend to have attributes associated with vertices. For example, in social networks such as Facebook, Google+ and Twitter, users have their personal information, e.g., interests, ages, living places, and etc., in addition to their friendship relationships. Proteins in a protein-protein internation network may be associated with gene expressions in addition to their interaction relations. Such graphs are referred to as *attributed graphs* in which vertices represent entities, edges represent their relations and attributes describe their own characteristics. Often the attributes and edges provide complementary information [11]. Neither can we infer vertex relationships from their attributes

© Springer International Publishing AG 2017
M. Ceci et al. (Eds.): ECML PKDD 2017, Part I, LNAI 10534, pp. 601–616, 2017.
https://doi.org/10.1007/978-3-319-71249-9_36

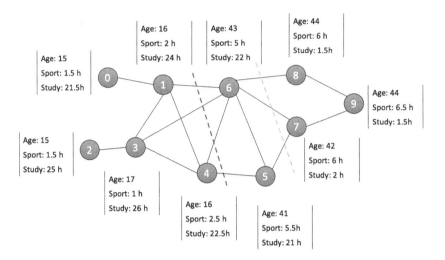

Fig. 1. An example social network. (Color figure online)

nor vice versa. Nevertheless, both types of information can be valuable for the detection of clusters in attributed graphs. Traditional methods for attributed graph clustering consider all attributes to compute the similarity. However, some attributes may be irrelevant to the edge structure and thus clusters only exist in the subsets (subspaces) of attributes. Currently, several methods have been proposed to detect subspace clusters in attributed graphs, such as CoPaM [11] and SSCG [3]. CoPaM uses various pruning strategies to find maximal cohesive patterns in the subspaces of attributes. One major problem with CoPaM is that it outputs a large number of clusters which have few vertices or attributes and which overwhelm data analysts. As for SSCG, it needs to eigen-decompose the graph Laplacian matrix and to update the subspace dependent weight matrix in every iteration, which is not scalable for large-scale graphs. How to effectively find clusters in attributed graphs remains a big challenge.

In this work, we develop an effective and efficient method to find cohesive clusters in attributed graphs. A cohesive cluster is a subgraph that has densely connected edges and has as many homogeneous (unimodal) attributes as possible. Why do we prefer to find cohesive clusters? One proper answer is that the more cohesive a graph cluster is, the more information it can reveal. For example, in social networks, if social networking advertisers know more characteristics of the people, they can do targeting ads more precisely. Figure 1 demonstrates an example social network with three attributes (age, sport time per week, and studying time per week) associated to each vertex. The task is to divide the network into two distinct parts which have as many homogenerous (unimodal) attributes as possible. In this example social network, we have two candidate partitions, i.e., by the orange dashed line and by the blue dashed line. The orange dashed line divides the network into two cohesive clusters $C_1 = \{0, 1, 2, 3, 4, 5, 6\}$ that is cohesive on the attribute studying time and $C_2 = \{7, 8, 9\}$ that is cohesive on all the attributes. The blue dashed line divides the network into another two cohesive clusters $\mathcal{C}_3 = \{0, 1, 2, 3, 4\}$ which is cohesive on all the attributes and $\mathcal{C}_4 = \{5, 6, 7, 8, 9\}$ which is cohesive on the attributes age and sport time. Compared with clusters $\mathcal{C}_1$ and $\mathcal{C}_2$, clusters $\mathcal{C}_3$ and $\mathcal{C}_4$ are more cohesive. Although the normalized cut value increases a little bit from 0.536 to 0.559, the *unimodality compactness* (see Sect. 3) value of attributes dramatically decreases from 3.289 to 1.230. The *unimodal normalized cut* (see Sect. 3) value of the partition by the blue dashed line is 0.895 and that of the partition by the orange dashed line is 1.913. Thus, we prefer clusters $\mathcal{C}_3$ and $\mathcal{C}_4$ to clusters $\mathcal{C}_1$ and $\mathcal{C}_2$.

Our contributions can be summarized as follows,

- **We introduce the univariate statistic hypothesis test called Hartigans' dip test [4] to the problem of attributed graph clustering.**
- **We achieve the cohesive cluster detection by developing an objective function which integrates the proposed measure *unimodality compactness* with the *normalized cut*.** The *unimodality compactness* takes advantage of Hartigans' dip test to measure the degree of the unimodality of attributes in a graph cluster.
- **We show the effectiveness and efficiency of our method UNCut by conducting extensive experiments on synthetic and real-world graphs.**

The paper is organized as follows: We continue in Sect. 2 with a review of preliminaries. Section 3 covers the core ideas and theory behind our approach UNCut, including the *unimodality compactness* and algorithmic details. Using synthetic and real-world data, Sect. 4 compares UNCut to related techniques. Section 5 discusses the related work and Sect. 6 gives concluding remarks.

2 Preliminaries

2.1 Notation

In this work, we use lower-case Roman letters (e.g. a, b) to denote scalars. We denote vectors (column) by boldface lower case letters (e.g. $\mathbf{x}$). Matrices are denoted by boldface upper case letters (e.g. $\mathbf{X}$). We denote entries in a matrix by non-bold lower case letters, such as x_{ij}. Row i of matrix $\mathbf{X}$ is denoted by the vector $\mathbf{x}_{i\cdot}$, column j by the vector $\mathbf{x}_{\cdot j}$. A set is denoted by calligraphic capital letters (e.g. $\mathcal{S}$). An undirected attributed graph is denoted by $\mathsf{G} = (\mathcal{V}, \mathcal{E}, \mathbf{F})$, where $\mathcal{V}$ is a set of graph vertices with number $n = |\mathcal{V}|$ of vertices, $\mathcal{E}$ is a set of graph edges with number $m = |\mathcal{E}|$ of edges and $\mathbf{F} \in \mathbb{R}^{n \times d}$ is a data matrix of attributes associated to vertices, where d is the number of attributes. An adjacency matrix of vertices is denoted by $\mathbf{A} \in \mathbb{R}^{n \times n}$ with $a_{ij} = 1$ if the vertices v_i and v_j are connected, and $a_{ij} = 0$ otherwise. The degree matrix $\mathbf{D}$ is a diagonal matrix associated with $\mathbf{A}$ with $d_{ii} = \sum_j a_{ij}$. The random walk transition matrix $\mathbf{W}$ is defined as $\mathbf{D}^{-1}\mathbf{A}$. The Laplacian matrix is denoted as $\mathbf{L} = \mathbf{I} - \mathbf{W}$, where $\mathbf{I}$ is an identity matrix. A graph cluster is a subset of vertices $\mathcal{S} \in \mathcal{V}$. The indicator function is denoted by $\mathbb{1}(x)$.

2.2 Normalized Cut

The definition of the widely used *normalized cut* [16] objective function is:

$$\text{NCut}(\mathcal{S}) = \frac{\text{cut}(\mathcal{S}, \overline{\mathcal{S}})}{\text{vol}(\mathcal{S})}. \tag{1}$$

where $\text{cut}(\mathcal{S}, \overline{\mathcal{S}}) = \sum_{v_i \in \mathcal{S}, v_j \in \overline{\mathcal{S}}} a_{ij}$ and $\text{vol}(\mathcal{S}) = \sum_{v_i \in \mathcal{S}, v_j \in \mathcal{V}} a_{ij}$.

Equation 1 can be equivalently rewritten as (for a more detailed explanation, please refer to [18]):

$$\text{NCut}(\mathcal{S}) = \mathbf{u}^\mathsf{T} \mathbf{L} \mathbf{u}, \, s.t. \, \mathbf{u}^\mathsf{T} \mathbf{D} \mathbf{u} = \text{vol}(\mathsf{G}), \, \mathbf{D} \mathbf{u} \perp \mathbf{1}. \tag{2}$$

where $\mathbf{u}$ is the cluster indicator vector and $\mathbf{u}^\mathsf{T}\mathbf{L}\mathbf{u}$ is the cost of the cut and $\mathbf{1}$ is a constant vector whose entries are all 1. Note that finding the optimal solution is known to be NP-hard [19] when the values of $\mathbf{u}$ are constrained to $\{1, -1\}$. But if we relax the objective function to allow it take values in $\mathbb{R}$, a near optimal partition of the graph G can be derived from the second smallest eigenvector of $\mathbf{L}$. More generally, k eigenvectors with the k smallest eigenvalues partition the graph into k subgraphs with near optimal normalized cut value.

2.3 The Dip Test

In this paper, we apply a univariate statistic hypothesis test for unimodality called Hartigans' dip test [4] on the vertex attributes to measure the degree of the unimodality of a graph cluster. The dip test has been successfully used in detecting clusters in a sea of noise [10]. The dip measures the departure of a distribution from unimodality. Before introducing the concept of the dip test, let us first introduce the concepts of the greatest convex minorant (g.c.m) and the least concave majorant (l.c.m.). The g.c.m of $F(x)$ in $(-\infty, x_l]$ is sup $G(x)$ for $x \leq x_l$, where the sup is taken over all functions G that are convex in $(-\infty, x_l]$ and nowhere greater than $F(x)$. The l.c.m. of $F(x)$ in $[x_u, \infty)$ is inf $L(x)$ for $x \geq x_u$, where the inf is taken over all functions L that are concave in $[x_u, \infty)$ and nowhere less than $F(x)$. Let $\mathcal{U}$ be the set of all unimodal distributions, the dip test of the distribution function $F(x)$ is computed as follows,

$$D(F) = \inf_{H \in \mathcal{U}} \sup_x |F(x) - H(x)| \tag{3}$$

The dip test is the infimum among the supremum computed between the cumulative distribution function (CDF) of F and the CDF of H from the set of unimodal distributions. The computation of the dip test is: Let $F(x)$ be an empirical distribution function for the sorted samples $x_1, \ldots, x_n$. There are $n \cdot (n-1)/2$ candidate modal intervals. Compute for each candidate $[x_i, x_j], i \leq j \leq n$ the g.c.m. of $F(x)$ in $(-\infty, x_i]$ and the l.c.m. of $F(x)$ in $[x_j, \infty)$ and let d_{ij} be the maximum distance of F to these computed curves (g.c.m. and l.c.m.). Finally, it selects the modal interval with the maximum distance which is the twice of the dip test. For more details, please refer to [4,6].

As pointed out in [4], the class of uniform distributions U is the most suitable for the null hypothesis, because their dip test values are stochastically larger than those of other unimodal distributions. The p-value for the unimodality test is then computed by comparing $D(F)$ with $D(U^r)$ b times, each time with a different n observations from U, and the proportion $\sum_{1 \leq r \leq b} \mathbb{1}(D(F) \leq D(U^r))/b$ is the p-value. If the p-value is greater than a significance level α, say 0.05, the null hypothesis that F is unimodal is accepted.

3   Unimodal Normalized Cut

Our objective is to detect cohesive graph clusters which have densely connected edges (low *normalized cut* value) and have as many homogeneous (unimodal) attributes as possible (low *unimodality compactness* value). To achieve the goal, we need to take both the edge structure and attribute information into account. If we eigen-decompose the Laplacian matrix associated with the edge structure to generate n eigenvectors, the k eigenvectors associated with the k smallest eigenvalues near optimally partition the graph into k subgraphs. However, the procedure does not consider the attribute information. Since each eigenvector bisects the graph into two clusters, our idea is to develop a measure to simultaneously evaluate the density of the edge structure and the homogeneity of vertex attributes of a graph cluster derived from the eigenvector. To this end, we first propose a measure called *unimodality compactness* to assess the homogeneity of attributes of a graph cluster. Then we integrate it with the *normalized cut* and call the combination *unimodal normalized cut*. We select k eigenvectors associated with the k smallest *unimodal normalized cut* values to partition the graph. In the following, we describe our idea in detail. But first let us give the definitions as follows,

Definition 1. *A **unimodal graph cluster** is defined as a set of vertices with at least one attribute following unimodal distributions.*

To compute the degree of the unimodality of a graph cluster, we devise a measure called *unimodality compactness* using the dip test on each attribute of the cluster.

Definition 2. *Given a cluster of vertices $\mathcal{S}$ with number $c > 0$ of unimodal attributes, the unimodality compactness is defined as,*

$$UC(\mathcal{S}) = \log_2 \frac{d}{c} + \frac{1}{c} \sum_{i=1}^{c} D(F_i). \tag{4}$$

where d is the number of attributes, F_i is the empirical distribution function of the i-th unimodal attribute of $\mathcal{S}$ and $D(F_i)$ is the dip test of F_i.

The first summand measures the number of unimodal attributes of a cluster. The second summand measures the average dip test of these unimodal attributes. This measure prefers the cluster that has more unimodal attributes with lower average dip test. Note that the multimodal (irrelevant) attributes are not considered in the computation. If a graph cluster only has one unimodal attribute, its *unimodality compactness* is close to $\log_2 d$ because the second summand in Eq. 4 is very low. If there is no unimodal attribute in a cluster, we simply set its *unimodality compactness* to $2\log_2 d$. When d is large and $c = 1$, the value of $\frac{d}{c}$ is also large. To reduce the effect of $\frac{d}{c}$, we introduce $\log_2$ in the definition. We do not use the sigmoid function $S(x) = \frac{1}{1+\exp(-x)}$ here because its resolution is not good, for example $S(\frac{8}{1}) = 0.9997$ and $S(\frac{8}{2}) = 0.9820$. Also note that a graph cluster will be more cohesive if it has more unimodal attributes.

A cohesive graph cluster is defined as follows,

Definition 3. *A **cohesive graph cluster** is a subgraph that has densely connected edges and has as many homogeneous (unimodal) attributes as possible. The density of edges is measured by the normalized cut, and the homogeneity of attributes is measured by the unimodality compactness.*

To detect cohesive graph clusters, our objective function integrates the *normalized cut* and *unimodality compactness* in one framework which is given as follows,

$$\text{UNCut}(\mathcal{S}) = (1 - \omega) \cdot \text{NCut}(\mathcal{S}) + \omega \cdot \text{UC}(\mathcal{S}). \tag{5}$$

where $\omega(0 \leq \omega \leq 1)$ is a weight parameter to adjust the importance between the *unimodality compactness* value and the *normalized cut* value of a graph cluster.

As said above, we can first eigen-decompose $\mathbf{L}$ to get some eigenvectors. Then, for each eigenvector, we apply 2-means (k-means with the input number of clusters two) to bisect the graph into two clusters and compute our objective function (Eq. 5). Finally, we select the k eigenvectors associated with the k smallest *unimodal normalized cut* values. However, the time complexity to eigen-decompose $\mathbf{L}$ is $\mathcal{O}(n^3)$ which is impractical for large-scale attributed graphs. Instead, in this work, we use the power iteration method [8] to compute a number, say $10 \cdot k$, of pseudo-eigenvectors (approximate eigenvectors) and then choose k pseudo-eigenvectors associated with the k smallest *unimodal normalized cut* values.

The power iteration is a fast method to compute the dominant eigenvector of a matrix. Note that the k largest eigenvectors of $\mathbf{W}$ are also the k smallest eigenvectors of $\mathbf{L}$. The power iteration method starts with a randomly generated vector $\mathbf{v}^0$ and iteratively updates as follows,

$$\mathbf{v}^t = \frac{\mathbf{W}\mathbf{v}^{t-1}}{\|\mathbf{W}\mathbf{v}^{t-1}\|_1}. \tag{6}$$

Suppose $\mathbf{W}$ has eigenvectors $\mathbf{U} = [\mathbf{u}_1; \mathbf{u}_2; \cdots ; \mathbf{u}_n]$ with eigenvalues $\mathbf{\Lambda} = [\lambda_1, \lambda_2, \cdots, \lambda_n]$, where $\lambda_1 = 1$ and $\mathbf{u}_1$ is constant. We have $\mathbf{W}\mathbf{U} = \mathbf{\Lambda}\mathbf{U}$ and in general $\mathbf{W}^t\mathbf{U} = \mathbf{\Lambda}^t\mathbf{U}$. When ignoring renormalization, Eq. (6) can be written as

$$\begin{aligned}
\mathbf{v}^t = \mathbf{W}\mathbf{v}^{t-1} = \mathbf{W}^2\mathbf{v}^{t-2} = \cdots &= \mathbf{W}^t\mathbf{v}^0 \\
&= \mathbf{W}^t (c_1\mathbf{u}_1 + c_2\mathbf{u}_2 + \cdots + c_n\mathbf{u}_n) \\
&= c_1\mathbf{W}^t\mathbf{u}_1 + c_2\mathbf{W}^t\mathbf{u}_2 + \cdots + c_n\mathbf{W}^t\mathbf{u}_n \\
&= c_1\lambda_1^t\mathbf{u}_1 + c_2\lambda_2^t\mathbf{u}_2 + \cdots + c_n\lambda_n^t\mathbf{u}_n.
\end{aligned} \tag{7}$$

where $\mathbf{v}^0$ can be denoted by $c_1\mathbf{u}_1+c_2\mathbf{u}_2+\cdots+c_n\mathbf{u}_n$ which is a linear combination of all the original eigenvectors. By generating different starting vectors, we can get diverse linear combinations. If we let the power iteration method run enough time, it will converge to the dominant eigenvector $\mathbf{u}_1$ which is of little use in clustering. We define the velocity at t to be the vector $\boldsymbol{\delta}^t = \mathbf{v}^t - \mathbf{v}^{t-1}$ and define the acceleration at t to be the vector $\boldsymbol{\epsilon}^t = \boldsymbol{\delta}^t - \boldsymbol{\delta}^{t-1}$ and stop the power iteration when $\|\boldsymbol{\epsilon}^t\|_{max}$ is below a threshold $\hat{\epsilon}$.

Algorithm 1 gives the pseudo-code to find k clusters with the smallest k *unimodal normalized cut* values.

Algorithm 1. UNCut

Input: Adjacency matrix $\mathbf{A}$, data matrix $\mathbf{F}$ and the cluster number k
Output: Cluster indicator $\mathbf{c}$

1 $\omega \leftarrow 0.5, \hat{\epsilon} \leftarrow 0.001$;
2 compute the random walk transition matrix $\mathbf{W}$;
3 $iter \leftarrow 100, K \leftarrow 10 \cdot k$;
4 **for** $i \leftarrow 1$ **to** K **do**
5 $\quad$ $t \leftarrow 0, \mathbf{v}_i^0 \leftarrow \mathbf{randn}\,(1,n)$;   /* $\mathbf{v}_i \in \mathbb{R}^{1 \times n}$ */
$\quad$ /* power iteration   */
6 $\quad$ **repeat**
7 $\quad\quad$ $\mathbf{v}_i^{t+1} \leftarrow \dfrac{\mathbf{W}\mathbf{v}_i^t}{\|\mathbf{W}\mathbf{v}_i^t\|_1}$;
8 $\quad\quad$ $\boldsymbol{\delta}^{t+1} \leftarrow |\mathbf{v}_i^{t+1} - \mathbf{v}_i^t|$;
9 $\quad\quad$ $t \leftarrow t + 1$;
10 $\quad$ **until** $\|\boldsymbol{\delta}_i^{t+1} - \boldsymbol{\delta}_i^t\|_{\max} \leq \hat{\epsilon}$ **or** $t \geq iter$;
11 $\quad$ $\mathcal{S}_i \leftarrow$ **2-means** $(\mathbf{v}_i^t)$;
12 $\quad$ $\mathrm{UNCut}(\mathcal{S}_i) \leftarrow (1 - \omega) \cdot \mathrm{NCut}(\mathcal{S}_i) + \omega \cdot \mathrm{UC}(\mathcal{S}_i)$;
13 select k pseudo-eigenvectors associated with the k smallest *unimodal normalized cut* values;
14 use k-means on the selected k pseudo-eigenvectors to get the cluster indicator $\mathbf{c}$;
15 **return** $\mathbf{c}$;

Complexity analysis. Lines 5–10 in Algorithm 1 use the power iteration method to compute one pseudo-eigenvector, whose time complexity is $\mathcal{O}(m)$ [9], where m is the number of graph edges. Line 11 uses 2-means on each pseudo-eigenvector, whose time complexity is $\mathcal{O}(n)$. At line 12, we compute the *unimodal normalized cut* which is dominated by the complexity of computing the *unimodality compactness* of clusters. We first need to sort each attribute before computing the dip test, which costs $\mathcal{O}\,(n \cdot \log(n))$. The computation of dip test on each attribute costs $\mathcal{O}(n)$ [4]. Thus, the time complexity of lines 4–12 is $\mathcal{O}\,((m + n \cdot \log(n) \cdot d) \cdot k)$. Line 13 uses k-means on the selected k pseudo-eigenvector, whose time complexity is $\mathcal{O}(n \cdot k^2)$. The total time complexity of Algorithm 1 is $\mathcal{O}\,(m \cdot k + n \cdot \log(n) \cdot d \cdot k + n \cdot k^2)$, which is superlinear in the number of vertices n, linear in the numbers of edges m and attributes d, and quadratic in the number of clusters k.

4   Experimental Evaluation

In this section, we compare our method UNCut with state-of-the-art methods from the attributed graph clustering field. As pointed out in [3], the comparison with the overlapping clustering approaches [2,11] would always be biased to one of the paradigms due to their completely different objective from those of paititioning clustering approaches. Thus, following [3] we compare UNCut with the partitioning clustering methods SA-cluster [21], SSCG [3] and NNM [17]. We use the synthetic and real-world data to evaluate the clustering performance. All the experiments are run on the same machine with an Intel Core Quad i7-3770 with 3.4 GHz and 32 GB RAM. We set $\omega = 0.5$ for our method UNCut on all the synthetic and real-world data. The parameters for the competitors are set according to their original papers. For every method, we use the same number of cluster on each dataset. For the evaluation of clustering on synthetic data, we use the Normalized Mutual Information (NMI) and Adjusted Rand Index (ARI) [5] as clustering quality measures. The higher these clustering measures are, the better the clustering is. Because we do not have the ground truth for the real-world data, we use the *normalized cut* and our *unimodality compactness* to evaluate the clustering performance and interpret the results. The code and all the synthetic and real-world data are publicly available at the website[1].

4.1   Synthetic Data

Cluster Quality. We generate synthetic graphs with varying number of vertices n and attributes d. For the case of varying n, we fix the attribute dimension $d = 20$. For the case of varying d, we fix the number of vertices $n = 2000$. All the graphs are generated based on a benchmark graph generator [7], which makes the degree and cluster size follow power law distributions that reflect the real properties of vertices and clusters found in real networks. To add vertex attributes, for each graph cluster, we choose 20% attributes as relevant attributes and generate their values according to a Gaussian distribution with mean value of each attribute randomly sampled from the range $[0, 100]$ and variance value of each attribute randomly sampled from the range $(0, 0.1)$. To render the other attributes of clusters irrelevant to the edge structure, we randomly permute the cluster labels and generate each cluster's irrelevant attribute values according to a Gaussian distribution with mean 0 and variance 1. For each experiment, we test all the methods on the generated ten attributed graphs differing in the edge structure and attribute values and report the average performance of each method.

Figures 2(a) and 3(a) show the performance of all the methods when varying the number of attributes, where we can see that UNCut is superior to its competitors. Compared with SA-cluster and NNM, both UNCut and SSCG exceed them with large margins. UNCut and SSCG are subspace clustering methods, while SA-cluster and NNM are full-space clustering methods which are easily

---

[1] https://www.dropbox.com/sh/xz2ndx65jai6num/AAC9RJ5PqQoYoxreItW83PrLa?
dl=0.

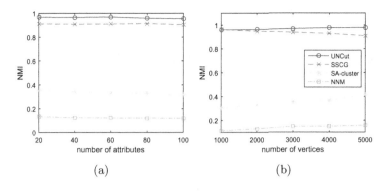

Fig. 2. Quality evaluation (NMI).

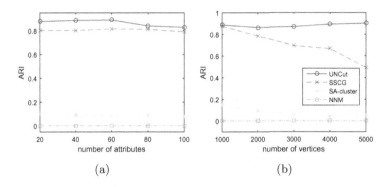

Fig. 3. Quality evaluation (ARI).

deceived by *"the curse of dimensionality"*. Figures 2(b) and 3(b) present the performance of all the methods when varying the number of graph vertices. SSCG has a comparable performance when the number of vertices is 1000. However, our method UNCut beats SSCG when increasing the vertex number. Note that subspace clustering methods UNCut and SSCG are still better than the full-space clustering methods SA-cluster and NNM.

Scalability. We still use the above attributed graph generation method to generate synthetic graphs for the evaluation of the runtime of each method. Figure 4(a) shows the runtime when varying the number of attributes (the number of vertices is fixed to 2000). We can see that NNM is the fastest method and SSCG is the slowest method. SSCG needs to update its subspace dependent weight matrix in every iteration, which is very time consuming. Figure 4(b) demonstrates the runtime when varying the number of vertices (the number of attributes is fixed to 20). NNM still performs the best and SSCG performs the worst. Our method UNCut is the second. Because UNCut is linear in the number of edges, a drop in the runtime when increasing the number of vertices from 4000 to 6000 can be interpreted as caused by the drop in the number of edges.

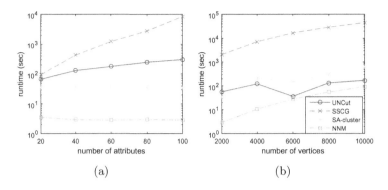

Fig. 4. Runtime evaluation.

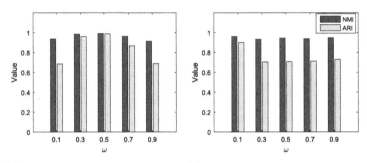

(a) the graph with 100 attributes(b) the graph with 20 attributes
and 2000 vertices   and 1000 vertices

Fig. 5. Varying the parameter ω.

Stability. In this section, we study how the parameter ω affects the clustering performance. Figure 5(a) gives the clustering performance of UNCut on the synthetic graph with 100 attributes and 2000 vertices when varying ω. And Fig. 5(b) gives the clustering performance of UNCut on the synthetic graph with 20 attributes and 1000 vertices when varying ω. From Fig. 5(a), we can see that UNCut achieves the best result when the value of ω is 0.5. From Fig. 5(b), we can see that UNCut achieves the best result when the value of ω is 0.1. For different graphs with different edge structure and attribute values, the values of the best ω are different.

4.2   Real-World Data

In this section, we evaluate UNCut and its competitors on six real-world datasets DISNEY [12], DFB [3], ARXIV [3], POLBLOGS [13], 4AREA [13] and PATENTS [3]. The statistics of the real-world data are given in Table 1. The *normalized cut* and *unimodality compactness* values achieved by each algorithm are listed in Table 2.

Table 1. Statistics of datasets.

Datasets	#vertices	#edges	#attributes	#clusters
DISNEY	124	333	28	9
DFB	100	1,106	5	14
ARXIV	856	2,660	30	19
POLBLOGS	358	1,288	44,839	10
4AREA	26,144	108,550	4	50
PATENTS	100,000	188,631	5	150

Table 2. Normalized cut and unimodality compactness values. (N/A means the results are not available due to the runout of memory.)

Datasets	Normalized cut				Unimodality compatness			
	UNCut	SSCG	SA-cluster	NNM	UNCut	SSCG	SA-cluster	NNM
DISNEY	**2.702**	2.646	3.959	8.058	**1.807**	20.459	10.709	77.266
DFB	**10.596**	13.161	13.116	13.026	**11.541**	20.507	60.692	43.082
ARXIV	**1.889**	17.940	10.606	18.017	**26.621**	176.911	45.940	148.378
POLBLOGS	7.429	**5.436**	8.181	9.071	**1.568**	155.377	217.068	124.404
4AREA	30.120	41.314	**10.813**	N/A	184.000	152.83	**37.075**	N/A
PATENTS	**31.980**	N/A	N/A	N/A	**415.941**	N/A	N/A	N/A

We can see from Table 2 that our method UNCut achieves the best results on the datasets DISNEY, DFB and ARXIV in terms of both the *normalized cut* and *unimodality compactness* values. On the dataset POLBLOGS, SSCG achieves the best *normalized cut* value. However, the *unimodality compactness* value achieved by UNCut is much lower than those of its competitors. On the dataset 4AREA, SA-cluster achieves the best results. Although SSCG is a method detecting subspace clusters, it is defeated by SA-cluster on the datasets DISNEY, ARXIV and 4AREA in terms of the *unimodality compactness* values. For the dataset PATENTS, all the competitors fail due to their much consumption of the memory. Our method UNCut is scalable for large-scale networks. To examine whether UNCut can achieve differing results to those of its competitors, as did in [3], we compute NMI between the results of UNCut and its competitors. A low NMI value indicates that UNCut is able to detect novel cluster insights, without implying that the results of the competitors are worse or meaningless. The NMI values are given in Table 3. From Table 3, we can see that UNCut can find novel cluster insights different from the competitors, especially on the 4AREA dataset. The NMI values between the results of UNCut and its competitors are near 0, which means totally different insights. For case studies, we interpret the detected clusters of all the methods on the datasets DISNEY and POLBLOGS. The results are plotted in Figs. 6 and 7 by the Python toolbox *Networkx*.

Table 3. NMI between the results of UNCut and its competitors. (N/A means the results are not available due to the runout of memory.)

Datasets	UNCut	SSCG	SA-cluster	NNM
DISNEY	1.000	0.724	0.597	0.164
DFB	1.000	0.298	0.246	0.272
ARXIV	1.000	0.096	0.387	0.131
POLBLOGS	1.000	0.488	0.297	0.060
4AREA	1.000	0.027	0.043	N/A
PATENTS	1.000	N/A	N/A	N/A

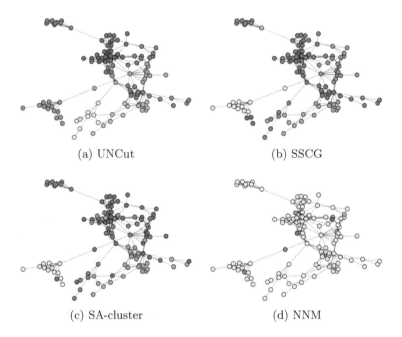

(a) UNCut   (b) SSCG

(c) SA-cluster   (d) NNM

Fig. 6. Clustering results on DISNEY. (Color figure online)

Disney. DISNEY is a subgraph of the Amazon copurchase network. Each movie (vertex) is described by 28 attributes, such as "average vote", "product group", "price" and etc. The green cluster has 14 movies, which is rated as PG (Parental Guidance Suggested) and attributed as "Action & Adventure". It contains movies such as "Spy Kids", "Inspector Gadget" and "Mighty Joe Young". The purple cluster includes 9 read-along movies, which is rated as G (General Audience) and attributed as "Kids & Family". It has movies such as "Beauty and the Beast", "Lilo and Stitch", "Toy Story 2", "The Little Mermaid", and "Monsters, Inc.". The purple cluster has three multimodal attributes "review frequency", "rating of review with most votes", and "rating of most helpful rating". In other

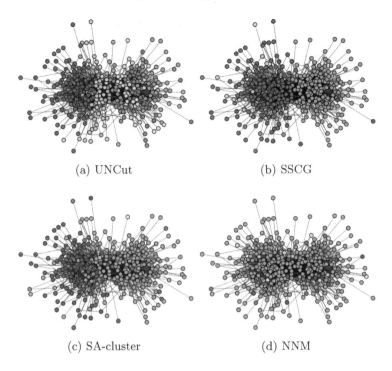

(a) UNCut   (b) SSCG

(c) SA-cluster   (d) NNM

Fig. 7. Clustering results on POLBLOGS. (Color figure online)

words, the movies in the purple cluster are similar in the subspace spanned by the other attributes. The clusters found by our method UNCut are subspace clusters which are cohesive on as many attributes as possible. SSCG splits our purple cluster into two clusters and our green clusters into two clusters. SA-cluster splits our green cluster into two clusters. NNM groups the most of the movies together (yellow cluster), which leads to the highest *unimodality compactness* value as shown in Table 2.

PolBlogs. POLBLOGS is the citation network among a collection of online blogs that discuss political issues. Attributes are the keywords in their text. If a keyword appears in the text, the attribute value is set to 1, otherwise 0. Thus, each attribute only has binary values. The red cluster contains 70 blogs. The top five frequent keywords of the red cluster are "London", "Iraq", "government", "work", and "American". The orange cluster contains 23 blogs. The top six frequent keywords of the orange cluster are "act", "bush", "conservative", "court", "justice", and "law". The blue cluster includes 53 blogs. The top eight frequent keywords of the blue cluster are "people", "post", "right", "political", "issue", "media", "president", and "public". For SSCG and SA-cluster, the sizes of the two main clusters are very big, i.e., the red and green clusters found by SSCG totally have 312 vertices and the blue and green clusters found by SA-cluster totally have 335 vertices. For NNM, the most of the blogs belong to the green cluster which has 306 vertices. Thus, the sizes of the most clusters detected by the competitors are small, which leads to the high probability of having multi-modal attributes as proved by the much higher *unimodality compactness* values in Table 2.

5   Related Work and Discussion

Compared with massive works on the plain graph clustering, there are relatively less work on the attributed graph clustering. Differing from the plain graph clustering that groups vertices only considering the edge structure, the attributed graph clustering achieves grouping vertices with dense edge connectivity and homogeneous attribute values into clusters. NNM [17] first develops a measure called *normalized network modularity* and then proposes a spectral method that combines the costs of clustering numerical vectors and *normalized network modularity* into an eigen-decomposition problem. BAGC (Bayesian Attributed Graph Clustering) [20] develops a Bayesian probabilistic model for attributed graphs, which captures both structure and attribute aspects of a graph. Clustering is accomplished by an efficient variational inference method. BAGC is only capable of categorical attributes. PICS [1] groups vertices into disjoint clusters satisfying that vertices in the same cluster exhibit similar connectivity and feature coherence. It exploits the Minimum Description Length (MDL) principle to automatically select the parameters such as the cluster number. PICS is only capable of graphs with binary feature vectors. SA-cluster [21] designs a unified neighborhood random walk distance to measure the vertex similarity on an augmented graph. It uses k-medoids to partition the graph into clusters with cohesive intracluster structures and homogeneous attribute values.

However, the above methods which take all attributes into consideration may fail because there may be attributes irrelevant to the edge structure. Now more researches focus on detecting subspace clusters to which only subsets of attributes are assigned. CoPaM [11] exploits various pruning strategies to efficiently find maximal cohesive patterns in the subspace of feature vectors. GAMer [2] determines sets of vertices which have high similarity in the subsets of attributes and are densely connected as well by combining the paradigms of subspace clustering and dense subgraph mining together. The twofold clusters are optimized by exploiting various pruning strategies considering the density, size and number of relevant attributes. CoPaM and GAMer exploit the notion of quasi-cliques which poses strong restrictions on the feature range and diameter of the clusters. CoPaM generates a huge number of redundant overlapping clusters. To reduce the redundancy, GAMer introduces additional parameters which are difficult to set for the real-world data. Differing from CoPaM and GAMer, our partitioning method UNCut does not suffer from redundancy. SSCG [3] presents a solution for an objective function called *Minimum Normalized Subspace Cut*, which integrates spectral clustering to the problem of subspace clustering for attributed graphs. It detects an individual set of relevant features for each cluster. Our method UNCut only considers the relevant attributes to the edge structure, i.e., irrelevant attributes are excluded from the computation of the *unimodality compactness*. In other words, UNCut detects subspace clusters with as many unimodal attributes as possible.

Recently, a new research trend is to detect community outliers in attributed graphs. MAM (maximization of attribute-aware modularity) [14] develops attribute compactness to quantify the relevance of the attributes, which is then combined with the conventional modularity for the robust graph clustering with respect to irrelevant attributes and outliers. ConSub (congruent subspace selection) [15] defines a measure to assess the degree of congruence between a set of attributes and the edge structure, which is then used for the statistical selection of the congruent subspaces. FocusCO [13] defines a new graph clustering problem which incorporates the user's preference into graph mining. Given a set of examplar nodes of user's interest, FocusCO infers user's preference by applying a distance metric learning method. New nodes are carefully added to the set of examplar nodes by checking the weighted conductance. Differing from the conventional attributed graph clustering methods, FocusCO performs a local clustering of interest to the user rather than the global partitioning of the entire graph.

6 Conclusion

In this paper, we have proposed UNCut to detect cohesive clusters in attributed graphs. To this end, we develop a measure called *unimodality compactness*, which is then combined with the *normalized cut* to elegantly search for cohesive clusters. Since the complexity of the eigen-decomposition of the graph Laplacian matrix is high, we adopt the power iteration method to approximately compute the eigenvectors. We have tested our method UNCut on various synthetic and real-world data, which verifies that UNCut achieves better results than its competitors. Since in social networks people may belong to multiple groups, an interesting challenge for the future work is to develop a method to detect overlapping cohesive clusters in attributed graphs.

References

1. Akoglu, L., Tong, H., Meeder, B., Faloutsos, C.: PICS: parameter-free identification of cohesive subgroups in large attributed graphs. In: SDM, pp. 439–450. SIAM (2012)
2. Günnemann, S., Färber, I., Boden, B., Seidl, T.: Subspace clustering meets dense subgraph mining: a synthesis of two paradigms. In: ICDM, pp. 845–850 (2010)
3. Günnemann, S., Färber, I., Raubach, S., Seidl, T.: Spectral subspace clustering for graphs with feature vectors. In: ICDM, pp. 231–240 (2013)
4. Hartigan, J.A., Hartigan, P.: The dip test of unimodality. Ann. Stat. **13**, 70–84 (1985)
5. Hubert, L., Arabie, P.: Comparing partitions. J. Classif. **2**(1), 193–218 (1985)
6. Krause, A., Liebscher, V.: Multimodal projection pursuit using the dip statistic. Preprint-Reihe Mathematik, vol. 13 (2005)

7. Lancichinetti, A., Fortunato, S., Radicchi, F.: Benchmark graphs for testing community detection algorithms. Phys. Rev. E **78**(4), 046110 (2008)
8. Lin, F., Cohen, W.W.: Power iteration clustering. In: ICML, pp. 655–662 (2010)
9. Lin, F., Cohen, W.W.: A very fast method for clustering big text datasets. In: ECAI, pp. 303–308 (2010)
10. Maurus, S., Plant, C.: Skinny-dip: clustering in a sea of noise. In: SIGKDD, pp. 1055–1064. ACM (2016)
11. Moser, F., Colak, R., Rafiey, A., Ester, M.: Mining cohesive patterns from graphs with feature vectors. In: SDM, pp. 593–604 (2009)
12. Müller, E., Sánchez, P.I., Mülle, Y., Böhm, K.: Ranking outlier nodes in subspaces of attributed graphs. In: ICDEW, pp. 216–222. IEEE (2013)
13. Perozzi, B., Akoglu, L., Sánchez, P.I., Müller, E.: Focused clustering and outlier detection in large attributed graphs. In: SIGKDD, pp. 1346–1355 (2014)
14. Sánchez, P.I., Müller, E., Böhm, K., Kappes, A., Hartmann, T., Wagner, D.: Efficient algorithms for a robust modularity-driven clustering of attributed graphs. In: SDM, vol. 15. SIAM (2015)
15. Sánchez, P.I., Müller, E., Laforet, F., Keller, F., Böhm, K.: Statistical selection of congruent subspaces for mining attributed graphs. In: ICDM, pp. 647–656 (2013)
16. Shi, J., Malik, J.: Normalized cuts and image segmentation. IEEE Trans. Pattern Anal. Mach. Intell. **22**(8), 888–905 (2000)
17. Shiga, M., Takigawa, I., Mamitsuka, H.: A spectral clustering approach to optimally combining numericalvectors with a modular network. In: SIGKDD, pp. 647–656. ACM (2007)
18. Von Luxburg, U.: A tutorial on spectral clustering. Stat. Comput. **17**(4), 395–416 (2007)
19. Wagner, D., Wagner, F.: Between min cut and graph bisection. In: Borzyszkowski, A.M., Sokołowski, S. (eds.) MFCS 1993. LNCS, vol. 711, pp. 744–750. Springer, Heidelberg (1993). https://doi.org/10.1007/3-540-57182-5_65
20. Xu, Z., Ke, Y., Wang, Y., Cheng, H., Cheng, J.: A model-based approach to attributed graph clustering. In: SIGMOD, pp. 505–516 (2012)
21. Zhou, Y., Cheng, H., Yu, J.X.: Graph clustering based on structural/attribute similarities. PVLDB **2**(1), 718–729 (2009)

K-Clique-Graphs for Dense Subgraph Discovery

Giannis Nikolentzos[1,2]([✉]), Polykarpos Meladianos[1,2], Yannis Stavrakas[3],
and Michalis Vazirgiannis[1,2]

[1] LIX, École Polytechnique, Palaiseau, France
[2] Athens University of Economics and Business, Athens, Greece
{nikolentzos,pmeladianos,mvazirg}@aueb.gr
[3] Institute for the Management of Information Systems RC "Athena", Athens, Greece
yannis@imis.athena-innovation.gr

Abstract. Finding dense subgraphs in a graph is a fundamental graph
mining task, with applications in several fields. Algorithms for identifying dense subgraphs are used in biology, in finance, in spam detection,
etc. Standard formulations of this problem such as the problem of finding the maximum clique of a graph are hard to solve. However, some
tractable formulations of the problem have also been proposed, focusing
mainly on optimizing some density function, such as the degree density
and the triangle density. However, maximization of degree density usually leads to large subgraphs with small density, while maximization of
triangle density does not necessarily lead to subgraphs that are close to
being cliques.

In this paper, we introduce the k-clique-graph densest subgraph problem, $k \geq 3$, a novel formulation for the discovery of dense subgraphs.
Given an input graph, its k-clique-graph is a new graph created from
the input graph where each vertex of the new graph corresponds to a
k-clique of the input graph and two vertices are connected with an edge
if they share a common $k-1$-clique. We define a simple density function, the k-clique-graph density, which gives compact and at the same
time dense subgraphs, and we project its resulting subgraphs back to
the input graph. In this paper, we focus on the triangle-graph densest
subgraph problem obtained for $k = 3$. To optimize the proposed function, we provide an exact algorithm. Furthermore, we present an efficient
greedy approximation algorithm that scales well to larger graphs.

We evaluate the proposed algorithms on real datasets and compare
them with other algorithms in terms of the size and the density of
the extracted subgraphs. The results verify the ability of the proposed
algorithms in finding high-quality subgraphs in terms of size and density. Finally, we apply the proposed method to the important problem
of keyword extraction from textual documents. Code related to this
chapter is available at: https://github.com/giannisnik/k-clique-graphs-
dense-subgraphs.

Electronic supplementary material The online version of this chapter (https://
doi.org/10.1007/978-3-319-71249-9_37) contains supplementary material, which is
available to authorized users.

M. Ceci et al. (Eds.): ECML PKDD 2017, Part I, LNAI 10534, pp. 617–633, 2017.
https://doi.org/10.1007/978-3-319-71249-9_37

1   Introduction

In recent years, graph-based representations have become extremely popular for modelling real-world data. Some examples of data represented as graphs include social networks, protein or gene regulation networks and textual documents. The problem of extracting dense subgraphs from such graphs has received a lot of attention due to its potential applications in many fields. Specifically, in the web graph, dense subgraphs may correspond to link spam [18] and hence, they can be used for spam detection. In bioinformatics, they are used for finding molecular complexes in protein-protein interaction networks [6] and for discovering motifs in genomic DNA [17]. In the field of finance, they are used for discovering migration motifs in financial markets [14]. Other applications include graph compression [10], graph visualization [1], real-time identification of important stories in Twitter [3] and community detection [12].

Given an undirected, unweighted graph $G = (V, E)$, we will denote $|V| = n$ the number of vertices and $|E| = m$ the number of edges. Given a subset of vertices $S \subseteq V$, let $E(S)$ be the set of edges that have both end-points in S. Hence, $G(S) = (S, E(S))$ is the subgraph induced by S. The *density* of the set S is $\delta(S) = |E(S)|/\binom{|S|}{2}$, the number of edges in S over the total possible edges. Finding the set S that maximizes δ is not a meaningful problem, as density δ does not take into account the size of the subgraph. For example, a subgraph consisting of two vertices connected with an edge has higher density δ than a subgraph consisting of 100 vertices and all but one edge between them. However, clearly, we would prefer the latter subgraph from the former even if it achieves a lower value of density δ. Typically, the problem of dense subgraph discovery asks for a set of vertices S which is large and which has high density. Several different functions have been proposed in the literature that aim to solve this problem. Some of these functions can be optimized in polynomial time, however, most of these formulations of extracting dense subgraphs are NP-hard and also hard to approximate.

Recently, there was a growing interest in the extraction of subgraphs whose vertices are highly connected to each other [7,34,35]. However, existing methods do not always find subgraphs with high density δ. Instead, they prefer subgraphs with many vertices even if their density δ is not very high. In many cases, we are interested in discovering sets of vertices where there is an edge between almost all their pairs. In this paper, we introduce a new formulation for extracting dense subgraphs. We define a new family of functions for measuring the density of a subgraph and we provide exact and approximate algorithms that allow the extraction of large subgraphs with high density δ by maximizing these functions. Our contributions are fourfold:

(i) **New formulation:** We introduce the *k-clique-graph densest subgraph* (*k*-clique-GDS) problem, a new formulation for finding large subgraphs with high density δ. Given a value for k, we create a graph whose vertices correspond to k-cliques of the original graph and we draw edges between two k-cliques if they share a common $(k - 1)$-clique. We then extract a dense subgraph from the new graph and we project the result back to the original graph. We focus on the special case obtained for $k = 3$ which we call the *triangle-graph densest subgraph* (TGDS) problem. We define a new density function which is suited to the needs of our problem.

(ii) **Exact algorithm:** We present an algorithm that solves exactly the TGDS problem. The algorithm finds the optimal subgraph by solving a series of supermodular maximization problems.

(iii) **Approximation algorithm:** We propose an efficient greedy approximation algorithm for the TGDS problem which removes one vertex at each iteration. The algorithm achieves nearly-optimal results on real-world networks.

(iv) **Experimental evaluation:** We evaluate our exact and approximation algorithms on several real-world networks. We compare the obtained subgraphs with those outputted by state-of-the-art algorithms and we observe that the proposed algorithms extract subgraphs of high quality. We also present an application of our problem to the task of keyword extraction from textual documents.

2   Related Work

In this section, we review the related work published in the areas of *Clique Finding*, *Dense Subgraph Discovery* and *Triangle Listing*.

Clique Finding. A clique is a graph whose vertices are all connected to each other. Hence, all cliques have density $\delta = 1$. A *maximum* clique of a graph is a clique, such that there is no clique with more vertices. Finding the maximum clique of a graph is an NP-complete problem [22]. The maximum clique problem is also hard to approximate. More specifically, Håstad showed in [20] that for any $\epsilon > 0$, there is no polynomial algorithm that approximates the maximum clique within a factor better than $\mathcal{O}(n^{1-\epsilon})$, unless NP has expected polynomial time algorithms. Feige presented in [15] a polynomial-time algorithm that approximates the maximum clique within a ratio of $\mathcal{O}(n(\log \log n)^2/(\log n)^3)$. A *maximal* clique is a clique that is not included in a larger clique. The Bron–Kerbosch algorithm is a recursive backtracking procedure [9] that lists all maximal cliques in a graph in $\mathcal{O}(3^{n/3})$ time.

Dense Subgraph Discovery. The problem of finding a dense subgraph given an input graph has been widely studied in the literature [24]. As mentioned above, such a problem aims at finding a subset of vertices $S \subseteq V$ of an input graph G that maximizes some notion of density. Among all the functions for evaluating dense subgraphs, degree density has gained increased popularity. The degree density of a set of vertices S is defined as $d(S) = 2|E(S)|/|S|$. The problem of finding the set of vertices that maximizes the degree density is known as the *densest subgraph* (DS) problem. The set of vertices $S \subseteq V$ that maximizes the degree density can be identified in polynomial time by solving a series of minimum-cut problems [19]. Charikar showed in [11] that the DS problem can also be formulated as a linear programming (LP) problem. In the same paper, the author proved that the greedy algorithm proposed by Asahiro et al. [5] provides a $\frac{1}{2}$-approximation to the DS problem in linear time.

Some variations of the DS problem include the *densest k-subgraph* (DkS), the *densest at-least-k-subgraph* (DalkS) and the *densest at-most-k-subgraph* (DamkS) problems. These variations put restrictions on the size of the extracted subgraph. The DkS identifies the subgraph with exactly k vertices that maximizes the degree density and is known to be NP-complete [4]. Feige et al. provided in [16] an approximation algorithm with approximation ratio $\mathcal{O}(n^\delta)$, where $\delta < 1/3$. The DalkS and DamkS problems were introduced by Andersen and Chellapilla [2]. The first problem asks for the subgraph of highest degree density among all subgraphs with at least k vertices and is known to be NP-hard [23], while the second problem asks for the subgraph of highest density among all subgraphs with at most k vertices and is known to be NP-complete [2].

Tsourakakis introduced in [34] the *k-clique densest subgraph* (k-clique-DS) problem which generalizes the DS problem. The k-clique-DS problem maximizes the average number of k-cliques induced by a set $S \subseteq V$ over all possible vertex subsets. For $k = 3$, we obtain the so-called *triangle densest subgraph* (TDS) problem which maximizes the triangle density defined as $d_{tr}(S) = t(S)/|S|$ where $t(S)$ is the number of triangles in S. The author provides two polynomial-time algorithms that identify the exact set of vertices that maximizes the triangle density and a $\frac{1}{3}$-approximation algorithm which runs asymptotically faster than any of the exact algorithms.

There are several other recent algorithms that extract dense subgraphs by maximizing other notions of density [32,35,36]. It is worthwhile mentioning Tsourakakis et al.'s work [35]. The authors defined the *optimal quasi-clique* (OQC) problem which finds the subset of vertices $S \subseteq V$ that maximizes the function $f_\alpha(S) = |E(S)| - \alpha\binom{|S|}{2}$ where $\alpha \in (0,1)$ is a constant. The OQC problem is not polynomial-time solvable and the authors provided a greedy approximation algorithm that runs in linear time and a local-search heuristic.

Triangle Listing. Given a graph G, the *triangle listing* problem reports all the triangles in G. The triangle listing problem has been extensively studied and a large number of algorithms has been proposed [13,21,30]. A listing algorithm requires at least one operation per triangle. In the worst case, there are n^3 triangles in terms of the number of vertices and $m^{3/2}$ in terms of the number of edges. Hence, in the worst case, it takes $m^{3/2}$ time just to report the triangles. The above algorithms require $\mathcal{O}(m^{3/2})$ time to list the triangles and they are thus optimal in the worst case. Recently, Björklund et al. proposed output sensitive algorithms which run asymptotically faster when the number of triangles in the graph is small [8].

3   Problem Definition

In this section, we will introduce the *k-clique-graph densest subgraph* (k-clique-GDS) problem, a novel formulation for finding dense subgraphs. In the following, we will restrict ourselves to the case where $k = 3$, that is to triangles. At the end

---

Algorithm 1. Construct triangle-graph

Input: graph $G = (V, E)$
Output: graph $G' = (V', E')$
  1: Assign a unique label to each edge of the input graph G.
  2: Extract all triangles in G by running a triangle listing algorithm. Let $T(S)$ be the set of the extracted triangles.
  3: Create a new empty graph G'.
  4: For each triangle $t \in T(G)$ create a vertex in the G'.
  5: Connect two vertices in G' with an edge if the corresponding triangles in G share a common edge.
  6: Assign to the new edge the label of the edge that is shared between the two triangles.
  7: Return G'.

---

of the section, we will describe how the proposed approach can be generalized to the case of k-cliques, $k > 3$.

The cornerstone of the proposed method is the transformation of the input graph $G = (V, E)$ into another graph $G' = (V', E')$. The transformed graph G' is a more abstract representation of G. Specifically, it encodes information regarding the triangles of the input graph G and the relationships between them.

As a preprocessing step before applying the transformation, we assign labels to the edges of the input graph G. Given a set of labels $L, \ell : E \to L$ is a function that assigns labels to the edges of the graph. Each edge is assigned a unique label. Hence, the cardinality of the set L is equal to that of set $E, |L| = |E|$. We next proceed with the transformation of G into G'. The first step of the transformation procedure is to run a triangle listing algorithm. There are several available triangle listing algorithms as described in Sect. 2. Let $T(S)$ be the set of triangles extracted from G. For each triangle $t \in T(G)$, we create a vertex in the new graph G'. Therefore, each vertex represents one of the triangles extracted from G. Pairs of triangles that share a common edge in G are considered neighbors and are connected with an edge in G'. In other words, each edge in G' corresponds to a pair of triangles sharing the same edge. The edges of G' are also assigned labels. Each edge in G' is given the label of the edge that is shared between the two corresponding triangles in G. For example, given a pair of triangles $t_1 = (v_1, v_2, v_3)$ and $t_2 = (v_1, v_2, v_4)$ where $t_1, t_2 \in T(G)$, these triangles have a common edge $e = (v_1, v_2)$ and the edge e' that links them in G' gets the same label as e, that is $\ell(e') = \ell(e)$. A triangle has three edges, hence, although it can have any number of adjacent edges in G', its labels come from a limited alphabet consisting of only three items (the labels of the three edges of the triangle in G). We call the transformed graph G' the *triangle-graph* of G. Algorithm 1 describes the steps required to create G' from G and Fig. 1 illustrates how a graph containing 4 triangles is transformed into its triangle-graph.

After creating the triangle-graph G', we can find a subset of vertices $S' \subset V'$ that correponds to a dense subgraph. As mentioned earlier, each vertex $v \in S'$ represents a triangle t of the input graph G. Each triangle t is a set of three vertices. Intuitively, the union of the vertices of all the triangles that belong to the set S' will form a dense subgraph of G. To extract the set of vertices S', we can define a density measure and optimize it. A simple measure we can employ is the well-known degree density defined as $d(S') = 2|E(S')|/|S'|$. However, the above function will not necessarily lead to subgraphs with high density. Consider the two graphs shown in Fig. 2. As can be seen from the Figure, the triangle-graphs emerging from the two input graphs are structurally equivalent, and hence, they have the same degree density. As a result, if the two graphs are components of a larger graph and there are no other subgraphs with higher value, they are equally likely solutions to the DS problem. However, it is obvious that the upper graph suits better our purpose, and we would like our algorithm to prefer this compared to the lower graph.

To account for this problem, we define a new density measure which we call the *triangle-graph density*.

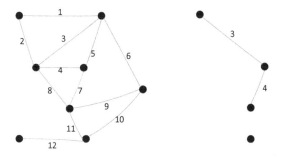

Fig. 1. Example of an input graph (left) and the triangle-graph (right) created from it. There are 4 triangles in the input graph defined by the following triads of edges: $(1, 2, 3)$, $(3, 4, 5)$, $(4, 7, 8)$ and $(9, 10, 11)$. The first two as well as the second and third triangles have a common edge (edge 3 and edge 4 respectively). Hence, these pairs of triangles are connected with an edge in the triangle-graph. The fourth triangle does not share any edges with the other triangles, therefore, it has no adjacent edges in the triangle-graph.

Definition 1 (Triangle-Graph Density). *Given an undirected, unweighted graph $G = (V, E)$, first construct its triangle graph $G' = (V', E')$. For any $S' \subseteq V'$, we define its triangle-graph density as $f(S') = \frac{d(S')}{|S'|}$ where $d(S') = \sum_{v \in S'} \min_{l \in L(v)} \left(deg_{S'}(v, l) \right)$, $L(v)$ the set of labels of the edges adjacent to v (three labels at most), and $deg_{S'}(v, l)$ the number of edges that are adjacent to v in the subgraph induced by S' and are assigned the label l.*

The triangle-graph density will allow the discovery of subgraphs with high values of density δ. This is due to the fact that for each triangle t in G, the function takes into account the number of neighbors from all three edges of t. If a triangle t corresponding to the vertex v in G' shares one of its edges with many other triangles,

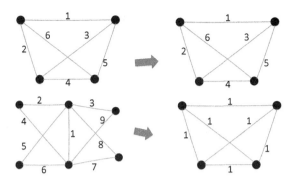

Fig. 2. Two input graphs (left) and their triangle-graphs (right). The two triangle-graphs are structurally equivalent although the input graphs are not.

but the other two edges with no triangles, then $\min_{l \in L(v)} \left(deg_{S'}(v, l) \right) = 0$. Therefore, even if t has many neighbors, it contributes nothing to the triangle-graph density. Triangle-graph density seeks for subgraphs whose vertices belong to edges which all consist of large sets of vertices. Cliques are natural candidates for maximizing the function since all their edges are shared between several triangles.

We next introduce the *triangle-graph densest subgraph* problem, the optimization problem we address in this paper.

Problem 1 (TGDS problem). Given an undirected, unweighted graph $G = (V, E)$, create its triangle-graph $G' = (V', E')$, and find a subset of vertices $S^* \subseteq V'$ such that $f(S^*) = \arg \max_{S' \subseteq V'} f(S')$.

After optimizing the triangle-graph density, we end up with a set of vertices $S' \subseteq V'$ and from these we obtain the set of vertices $S \subseteq V$ that corresponds to the resulting subgraph. The set S consists of all the vertices that form the triangles in S'. It is clear that the TGDS problem can result in subgraphs with high values of density δ.

What needs to be investigated next is what are the properties of the extracted subgraphs and how they differ from the ones extracted from existing methods. The proposed *triangle-graph densest subgraph* (TGDS) problem seems to be very related to the *triangle densest subgraph* (TDS) problem introduced by Tsourakakis in [34]. However, as we will show next, the two problems can result in different solutions, and the subgraphs returned by TGDS are closer to being near-cliques compared to the ones returned by TDS. Consider the graph G and its triangle-graph G' both shown in Fig. 3. The optimal solution of TDS is the whole graph. Conversely, the optimal solution of TGDS is the subgraph induced by the vertices that form the 4-clique. Hence, the optimal solution of the proposed problem is a clique, while the optimal solution of TDS is a larger graph with lower density δ. The above example demonstrates that the optimal solutions of TGDS correspond to subgraphs that exhibit a stronger near-clique structure compared to TDS.

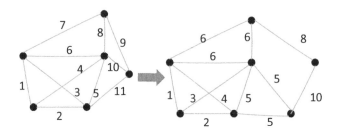

Fig. 3. Example of an input graph (left) and the triangle-graph (right) created from it. There are 7 triangles in the input graph defined by the following triads of edges: $(1, 2, 3)$, $(1, 4, 6)$, $(2, 4, 5)$, $(3, 5, 6)$, $(6, 7, 8)$, $(8, 9, 10)$ and $(5, 10, 11)$.

The process of creating the k-clique graph for $k > 3$ is similar to the one described above for $k = 3$. Specifically, to construct the k-clique graph $G' = (V', E')$, we first extract all the k-cliques from G. Then for each k-clique in G, we create a vertex v in G'. Two vertices $v_1, v_2 \in V'$ are connected with an edge if the corresponding cliques share a common $(k - 1)$-clique in G. For example, for $k = 4$, if two 4-cliques in G share a common triangle, an edge is drawn between them in G'. Each $(k - 1)$-clique in G is assigned a unique label and the edges of the k-clique graph are assigned the labels of the $(k - 1)$-cliques that are shared between their two endpoints. Then, the k-*clique-graph density* and the k-*clique-graph densest subgraph* (k-clique-GDS) problem are defined in a similar way as in the case of triangles. The algorithms presented in the next section for maximizing triangle-graph density can be generalized to maximizing the k-clique-graph density. However, extracting k-cliques for $k > 3$ is a computationally demanding task, and hence, we restrict ourselves to the case where $k = 3$.

4   Proposed Methods

In this section, we present some algorithms for solving the TGDS problem. These algorithms are inspired by previously-introduced algorithms in the field of dense subgraph discovery. More specifically, we provide an algorithm that solves the TGDS problem exactly as well as a greedy approximation algorithm. In what follows, we assume that we have extracted all triangles from the input graph and we have created the triangle-graph. Note that, for simplicity of notation, from now on, we denote by $G = (V, E)$ the triangle-graph and not the input graph. We also denote by $q_S(v)$ the minimum degree of vertex v with respect to the three labels of its adjacent edges in the subgraph induced by S, that is $q_S(v) = \min_{l \in L(v)} \big(deg_S(v, l) \big)$.

4.1   A Supermodular Maximization Approach

In this section, we provide an exact algorithm for finding the set of vertices S^* that maximizes the triangle-graph density. The algorithm is based on the *supermodular maximization* approach proposed by Tsourakakis in [34]. More specifically, maximizing the triangle-graph density can be cast as a supermodular maximization problem. We next introduce a brief background on submodularity and supermodularity.

Submodular and supermodular functions are classes of functions with many useful properties which have found application in several real world problems. The main property of supermodular functions is that given two sets A and B, where $A \subseteq B \subseteq V \backslash v$, the difference in the incremental value of the function that a single element v makes when added to an input set increases as the size of the input set increases. Hence, the incremental value of adding v to sets A and B is larger for B compared to A. Let V be a finite ground set. A function $h : 2^V \rightarrow \mathcal{R}$ that maps subsets $S \subseteq V$ to a real value $h(S)$ is called *supermodular* if the following equation holds for any $A, B \subseteq V$

$$h(A \cup B) + h(A \cap B) \geq h(A) + h(B)$$

We next give a second equivalent definition of supermodularity. Let A, B be two sets such that $A \subseteq B \subseteq V$ and $v \in V \backslash B$. Then, function h is supermodular if

$$h(B \cup \{v\}) - h(B) \geq h(A \cup \{v\}) - h(A)$$

This second form of supermodularity shows that the value added by a new element v never decreases when the context gets larger. Submodular and supermodular functions have gained increased popularity due to the fact that they can be minimized and maximized respectively in strongly polynomial time [31]. More specifically, Orlin proposed in [28] an algorithm for maximizing an integer valued supermodular function f which runs in $\mathcal{O}(n^5 EO + n^6)$ where EO is the time to evaluate $h(S)$ for some $S \subseteq V$.

We next show that for any $\alpha \in \mathcal{R}^+$, the function $h : 2^V \rightarrow \mathcal{R}$ defined by $h(S) = d(S) - \alpha|S|$ is supermodular.

Theorem 1. *Function $h : 2^V \rightarrow \mathcal{R}$ where $h(S) = d(S) - \alpha|S|$ is supermodular.*

Proof. The proof is left to the supplementary material [27].

To find the set of vertices S^* that maximizes the triangle-graph density, we can use Algorithm 2. The algorithm terminates in logarithmic number of rounds. In each iteration, we run the Orlin-Supermodular-Opt procedure in order to find the set of vertices that maximize function h given the current value of parameter α.

Theorem 2. *There exists an algorithm that solves the TGDS problem and runs in $\mathcal{O}\left(m^{3/2} + \left(t^5(t + y) + t^6\right) \log t\right)$ time where m is the number of edges of the input graph, t is the number of triangles in the input graph and y is the number of edges of the triangle-graph.*

Algorithm 2. Supermodular maximization algorithm

Input: triangle-graph $G = (V, E)$
Output: Subset of vertices $S^* \subseteq V$
   $l \leftarrow \frac{d(V)}{n}$
   $u \leftarrow \frac{n(n-1)}{3n}$
   $S^* \leftarrow V$
   while $u - l \geq \frac{1}{n(n-1)}$ **do**
      $\alpha \leftarrow \frac{l+u}{2}$
      $(val, S) \leftarrow$ Orlin-Supermodular-Opt(G, α)
      if $val < 0$ **then**
         $u \leftarrow \alpha$
      else
         $l \leftarrow \alpha$
         $S^* \leftarrow S$
      end if
   end while
   Return S^*

To create the triangle-graph from the input graph, we first need to run a triangle listing algorithm. The one proposed by Itai and Rodeh runs in $\mathcal{O}(m^{3/2})$ time [21]. As mentioned above, the algorithm will run in a logarithmic number of rounds. Furthermore, Orlin's algorithm runs in $\mathcal{O}(n^5 EO + n^6)$ time where n is the size of the ground set and EO is the time to evaluate $h(S)$ for some $S \subseteq V$ [28]. In our case, the ground set corresponds to the vertices of the triangle-graph. Hence, it is equal to the number of triangles t in the input graph. As regards the complexity of computing $h(S)$, it is linear to the number of vertices and number of edges of the triangle-graph. Let y denote the number of edges of the triangle-graph. The overall running time of Algorithm 2 is thus $\mathcal{O}\left(m^{3/2} + \left(t^5(t + y) + t^6\right)\log t\right)$.

Lemma 1. *Algorithm 2 solves the TGDS problem and runs in* $\mathcal{O}\left(m^{3/2} + \left(t^5(t + y) + t^6\right)\log t\right)$ *time.*

Proof. The proof is left to the supplementary material [27].

4.2 A Greedy Approximation Algorithm

In this section, we provide an efficient algorithm for extracting a set of vertices $S \subseteq V$ with high value of triangle-graph density $f(S)$. The proposed algorithm is an adaptation of the greedy algorithm of Asahiro et al. [5]. The algorithm is illustrated as Algorithm 3. The algorithm iteratively removes the vertex v whose value $d(v)$ is the smallest among all vertices. Subsequently, it computes the triangle-graph density of the subgraph induced by the remaining vertices. The output is the subgraph over all the produced subgraphs that maximizes triangle-graph density. The algorithm is linear to the number of vertices and the

---

Algorithm 3. Greedy algorithm

Input: graph $G = (V, E)$
Output: Subset of vertices $S \subseteq V$
$\quad S_{|V|} \leftarrow V$
$\quad$**for** $i \leftarrow |V|$ to 1 **do**
$\quad\quad$ Let v be the vertex whose minimum value of the three degrees is the smallest in
$\quad\quad$ the subgraph induced by S_i
$\quad\quad\quad S_{i-1} \leftarrow S_i \setminus \{v\}$
$\quad$**end for**
$\quad S \leftarrow \arg\max_{i=1,\ldots,|V|} f(S_i)$

---

number of edges of the triangle-graph, hence its complexity is $\mathcal{O}(t + y)$ where t is the number of triangles in the input graph and y is the number of edges of the triangle-graph.

Theorem 3. *Let S be the set of vertices returned after the execution of Algorithm 3 and let S^* be the set of vertices of the optimal subgraph. Consider the iteration of the greedy algorithm just before the first vertex u that belongs in the optimal set S^* is removed, and let S_I denote the vertex set currently kept in that iteration. Let also $q_{S_I}(u)$ be the minimum degree of vertex u in S_I with respect to the three labels of its adjacent edges. Then, it holds that*

$$f(S) \geq \frac{|S^*|}{|S_I|} f_G^* + \left(1 - \frac{|S^*|}{|S_I|}\right) q_{S_I}(u)$$

Proof. The proof is left to the supplementary material [27].

From the above result, we can see that the bound provided by the approximation algorithm highly depends on the relationship between $|S_I|$, the size of the vertex set just before the first vertex of S^* is removed, and $|S^*|$, the size of the optimal set. It also depends on the relationship between the optimal value of the triangle-graph density $f(S^*)$ and the minimum degree $q_{S_I}(u)$ of the first vertex of the optimal set S^* to be removed from S_I with respect to its three labels. The difference between $|S_I|$ and $|S^*|$, and between $f(S^*)$ and $q_{S_I}(u)$ is not very large in practice, and the algorithm leads to subgraphs with quality almost equal to that of the optimal subgraphs.

5 Experiments and Evaluation

In this section, we present the evaluation of the proposed approach for extracting dense subgraphs. We first give details about the datasets that we used for our experiments. We then present the employed experimental settings. And we last report on the results obtained by our approach and some other methods.

5.1 Experimental Setup

For the evaluation of the proposed algorithms, we employed several publicly available graphs. The algorithms are applicable to simple unweighted, undirected graphs. Hence, we made all graphs simple by ignoring the edge direction in the case of directed graphs and by removing self-loops and edge weights, if any. Table 1 shows statistics of these graphs. The first ten datasets were obtained from UCIrvine Network Data Repository[1], while the remaining datasets were obtained from Stanford SNAP Repository[2]. We compared the proposed algorithms with algorithms that solve the *densest subgraph* (DS), the *triangle densest subgraph* (TDS) and the *optimal quasi-clique* (OQC) problems. For the first two (DS and TDS problems) as well as for the proposed problem, there are algorithms that solve these problems exactly in polynomial time. Hence, for small-sized datasets, we present the results obtained from both the exact and greedy approximation algorithms for each problem. For larger datasets, we report only on the results achieved by the greedy approximation algorithms. With regards to the objective function of the OQC problem, we set the value of parameter α equal to $1/3$ as suggested in [35]. All algorithms were implemented in Python[3] and all experiments were conducted on a single machine with a 3.4 GHz Intel Core i7 processor and 32 GB of RAM. To assess the quality of the extracted subgraphs, we employed the following measures: the density of the extracted subgraph $\delta(S) = |E(S)|/\binom{|S|}{2}$, the density with respect to the number of triangles $\tau(S) = t(S)/\binom{|S|}{3}$, that is the number of triangles in S over the total possible triangles, and the size of the subgraph $|S|$. The δ and τ measures take values between 0 and 1. The larger their value, the closer the subgraph to being a clique. Therefore, we are interested in finding large subgraphs (large value of $|S|$) with δ and τ values close to 1.

5.2 Results and Discussion

Table 2 summarizes the results obtained on small-sized graphs. We observe that on the small-sized graphs, the proposed algorithms (Exact TGDS and Greedy TGDS) return in general subgraphs that are closer to being a clique compared to the competing algorithms. As we can see from the Table, the densities δ and τ of the subgraphs extracted by our algorithms are relatively high. Our initial intention was to design an algorithm for finding a set of vertices with many edges between them. The obtained results verify our intuition that the proposed approach is capable of finding near-cliques. Furthermore, we show in Table 3 the triangle-graph density of the subgraphs extracted by the exact and the greedy approximation algorithm. We notice that on four out of the six graphs, the two densities are equal to each other, while on the other two, they are very close

---

[1] https://networkdata.ics.uci.edu/index.php.
[2] http://snap.stanford.edu/data/index.html.
[3] Code is available at https://github.com/giannisnik/k-clique-graphs-dense-subgr aphs.

Table 1. Graphs used for evaluating the algorithms.

Graph	$|V|$	$|E|$
Karate	34	78
Dolphins	62	159
Lesmis	77	254
Adjnoun	112	425
Football	115	613
Polbooks	105	441
Celegansneural	297	2,148
Polblogs	1,224	16,715
Power	4,941	6,594
Wiki-Vote	7,115	100,762
ca-CondMat	23,133	93,439
p2p-Gnutella31	62,586	147,892
Slashdot0902	82,168	504,230
email-EuAll	265,009	364,481
web-NotreDame	325,729	1,497,134
Amazon	334,863	925,872
Youtube	1,134,890	2,987,624
roadNet-CA	1,965,206	2,766,607

Table 2. Comparison of the extracted subgraphs by Goldberg's exact algorithm for the DS problem (Exact DS), Charikar's $\frac{1}{2}$ approximation algorithm for the DS problem (Greedy DS), Tsourakakis's algorithm for the TDS problem (Exact TDS), Tsourakakis's $\frac{1}{3}$ approximation algorithm for the TDS problem (Greedy TDS), Tsourakakis et al.'s greedy approximation algorithm for the OQC problem (Greedy OQC), our exact algorithm for the TGDS problem (Exact TGDS), and our greedy approximation algorithm for the TGDS problem (Greedy TGDS).

Dataset	Exact DS			Greedy DS			Exact TDS			Greedy TDS			Greedy OQC			Exact TGDS			Greedy TGDS		
	$|S|$	δ	τ	$|S|$	δ	τ	$|S|$	δ	τ	$|S|$	δ	τ	$|S|$	δ	τ	$|S|$	δ	τ	$|S|$	δ	τ
Karate	16	0.35	0.05	16	0.35	0.05	6	0.93	0.80	6	0.93	0.80	10	0.55	0.18	6	0.93	0.80	6	0.93	0.80
Dolphins	20	0.32	0.04	36	0.17	0.01	7	0.80	0.54	6	0.93	0.80	13	0.47	0.11	6	0.93	0.80	6	0.93	0.80
Lesmis	23	0.49	0.18	23	0.49	0.18	13	0.88	0.71	13	0.88	0.71	22	0.50	0.19	12	0.93	0.83	12	0.93	0.83
Adjnoun	48	0.20	0.01	44	0.22	0.01	41	0.23	0.01	41	0.23	0.01	16	0.48	0.11	8	0.82	0.51	7	0.85	0.62
Football	115	0.09	0.00	115	0.09	0.00	18	0.48	0.20	18	0.48	0.20	10	0.88	0.66	18	0.48	0.20	18	0.48	0.20
Polbooks	24	0.41	0.09	48	0.19	0.02	20	0.49	0.15	36	0.26	0.04	14	0.67	0.30	16	0.59	0.23	13	0.69	0.34

to each other. The obtained results indicate that the greedy algorithm achieves approximation ratios close to 1 on real-world networks. Hence, the approximation algorithm is nearly-optimal in practice.

Next, we present results obtained on larger graphs. Specifically, Table 4 compares the four approaches on 12 graphs. In general, the proposed algorithm still manages to extract subgraphs with high values of δ and τ. However, on two

Table 3. Triangle-graph densities of the subgraphs extracted by the exact and the greedy approximation algorithms.

Dataset	Exact TGDS	Greedy TGDS
Karate	2.25	2.25
Dolphins	2.25	2.25
Lesmis	7.60	7.60
Adjnoun	2.39	2.36
Football	6.0	6.0
Polbooks	4.02	3.89

Table 4. Comparison of the extracted subgraphs by Charikar's $\frac{1}{2}$ approximation algorithm for the DS problem (Greedy DS), Tsourakakis's $\frac{1}{3}$ approximation algorithm for the TDS problem (Greedy TDS), Tsourakakis et al.'s greedy approximation algorithm for the OQC problem (Greedy OQC), and our greedy approximation algorithm for the TGDS problem (Greedy TGDS).

Dataset	Greedy DS			Greedy TDS			Greedy OQC			Greedy TGDS		
	$|S|$	δ	τ	$|S|$	δ	τ	$|S|$	δ	τ	$|S|$	δ	τ
Celegansneural	127	0.13	0.005	30	0.47	0.13	22	0.61	0.25	24	0.55	0.21
Polblogs	278	0.20	0.020	102	0.54	0.195	100	0.55	0.202	74	0.67	0.343
Power	31	0.20	0.021	12	0.54	0.195	12	0.54	0.195	12	0.54	0.195
Wiki-Vote	828	0.11	0.004	464	0.19	0.014	133	0.47	0.131	152	0.42	0.104
ca-CondMat	26	1.0	1.0	26	1.0	1.0	26	1.0	1.0	26	1.0	1.0
p2p-Gnutella31	1,549	0.005	0.0	10	0.40	0.11	14	0.48	0.0	22	0.15	0.016
soc-Slashdot0902	219	0.39	0.097	171	0.50	0.165	155	0.54	0.200	145	0.56	0.225
email-EuAll	505	0.13	0.005	200	0.29	0.041	97	0.51	0.164	91	0.52	0.179
web-NotreDame	1,367	0.11	0.012	457	0.34	0.114	305	0.51	0.255	155	1.0	1.0
Amazon	9	0.91	0.761	16	0.45	0.178	9	0.91	0.761	170	0.03	0.001
Youtube	1,860	0.049	0.0006	729	0.11	0.005	125	0.46	0.115	442	0.17	0.012
roadNet-CA	19,899	0.0001	0.0	168	0.017	0.0002	5	0.80	0.40	168	0.017	0.0002

graphs (Amazon, roadNet-CA), it fails to discover high-quality subgraphs in terms of density. Overall, the Greedy DS algorithm returns the largest subgraphs, followed by the Greedy TDS algorithm, while the Greedy OQC algorithm and the proposed algorithm return smaller subgraphs with higher values of density. We notice that the subgraphs extracted by the proposed greedy approximation algorithm resemble most those extracted by the Greedy TDS algorithm. On the ca-CondMat dataset, all the algorithms extract the same subgraph.

6   Application

In this section, we apply the proposed algorithm to a central problem in Natural Language Processing: extracting keywords from a textual document. Keyword extraction finds applications in several fields from information retrieval to text classification and summarization. Given a document d, we can represent it as a statistical *graph-of-words*, following earlier approaches in keyword extraction [26,29,33] and in summarization [25]. The construction of the graph is preceded by a preprocessing phase where standard text processing tasks are performed. The processed document is then transformed into an unweighted, undirected graph G whose vertices represent unique terms and whose edges represent co-occurrences between the connected terms within a fixed-size window. We then employ the proposed algorithm to extract a dense subgraph from G. The vertices of the subgraph act as representative keywords of the document.

To demonstrate the ability of the proposed approach to identify meaningful keywords, we extracted the text of this paper and we transformed it into a graph G using a window of size 3 (each word is connected with an edge with each one of its two preceding and two following words, if any). We then extracted a dense subgraph from G using the proposed greedy approximation algorithm. The output subgraph consists of the following 25 vertices:

> set, labels, number, subgraphs, triangles, maximizes, given, density, graph, input, function, triangle, subgraph, cliques, vertex, edges, clique, algorithm, k, vertices, value, edge, supermodular, hence, problem

As we can observe, the extracted keywords capture the main concepts of the paper.

7   Conclusion

In this paper, we propose a novel approach for extracting dense subgraphs. Given a graph, our algorithm first transforms it to a k-clique-graph. We then introduce a simple density measure to extract high-quality subgraphs. We propose an algorithm for exactly maximizing the density function. We also present a greedy approximation algorithm. We evaluate our proposed approach for the case where $k = 3$ on real graphs and we compare it with other popular measures. Overall, our algorithms show good performance in finding large near-cliques, and can serve as useful additions to the list of dense subgraph discovery algorithms.

References

1. Alvarez-Hamelin, J.I., Dall'Asta, L., Barrat, A., Vespignani, A.: Large scale networks fingerprinting and visualization using the k-core decomposition. In: NIPS 2005, pp. 41–50 (2005)
2. Andersen, R., Chellapilla, K.: Finding dense subgraphs with size bounds. In: Avrachenkov, K., Donato, D., Litvak, N. (eds.) WAW 2009. LNCS, vol. 5427, pp. 25–37. Springer, Heidelberg (2009). https://doi.org/10.1007/978-3-540-95995-3_3
3. Angel, A., Koudas, N., Sarkas, N., Srivastava, D., Svendsen, M., Tirthapura, S.: Dense subgraph maintenance under streaming edge weight updates for real-time story identification. VLDB J. **23**(2), 175–199 (2014)

4. Asahiro, Y., Hassin, R., Iwama, K.: Complexity of finding dense subgraphs. Discret. Appl. Math. **121**(1), 15–26 (2002)
5. Asahiro, Y., Iwama, K., Tamaki, H., Tokuyama, T.: Greedily finding a dense subgraph. J. Algorithms **34**(2), 203–221 (2000)
6. Bader, G.D., Hogue, C.W.: An automated method for finding molecular complexes in large protein interaction networks. BMC Bioinform. **4**(1), 1 (2003)
7. Balalau, O.D., Bonchi, F., Chan, T., Gullo, F., Sozio, M.: Finding subgraphs with maximum total density and limited overlap. In: WSDM 2015, pp. 379–388 (2015)
8. Björklund, A., Pagh, R., Williams, V.V., Zwick, U.: Listing triangles. In: Esparza, J., Fraigniaud, P., Husfeldt, T., Koutsoupias, E. (eds.) ICALP 2014. LNCS, vol. 8572, pp. 223–234. Springer, Heidelberg (2014). https://doi.org/10.1007/978-3-662-43948-7_19
9. Bron, C., Kerbosch, J.: Algorithm 457: finding all cliques of an undirected graph. Commun. ACM **16**(9), 575–577 (1973)
10. Buehrer, G., Chellapilla, K.: A scalable pattern mining approach to web graph compression with communities. In: WSDM 2008, pp. 95–106 (2008)
11. Charikar, M.: Greedy approximation algorithms for finding dense components in a graph. In: Jansen, K., Khuller, S. (eds.) APPROX 2000. LNCS, vol. 1913, pp. 84–95. Springer, Heidelberg (2000). https://doi.org/10.1007/3-540-44436-X_10
12. Chen, J., Saad, Y.: Dense subgraph extraction with application to community detection. TKDE **24**(7), 1216–1230 (2012)
13. Chiba, N., Nishizeki, T.: Arboricity and subgraph listing algorithms. In: SICOMP 1985, vol. 14, no. 1, pp. 210–223 (1985)
14. Du, X., Jin, R., Ding, L., Lee, V.E., Thornton Jr., J.H.: Migration motif: a spatial-temporal pattern mining approach for financial markets. In: KDD 2009, pp. 1135–1144 (2009)
15. Feige, U.: Approximating maximum clique by removing subgraphs. In: SIDMA 2004, vol. 18, no. 2, pp. 219–225 (2004)
16. Feige, U., Peleg, D., Kortsarz, G.: The dense k-subgraph problem. Algorithmica **29**(3), 410–421 (2001)
17. Fratkin, E., Naughton, B.T., Brutlag, D.L., Batzoglou, S.: MotifCut: regulatory motifs finding with maximum density subgraphs. Bioinformatics **22**(14), e150–e157 (2006)
18. Gibson, D., Kumar, R., Tomkins, A.: Discovering large dense subgraphs in massive graphs. In: VLDB 2005, pp. 721–732 (2005)
19. Goldberg, A.V.: Finding a maximum density subgraph. Technical report, University of California Berkeley (1984)
20. Håstad, J.: Clique is hard to approximate within $n^{1-\epsilon}$. In: FOCS 1996, pp. 627–636 (1996)
21. Itai, A., Rodeh, M.: Finding a minimum circuit in a graph. In: SICOMP 1978, vol. 7, no. 4, pp. 413–423 (1978)
22. Karp, R.M.: Reducibility Among Combinatorial Problems. Springer, Boston (1972). https://doi.org/10.1007/978-1-4684-2001-2_9
23. Khuller, S., Saha, B.: On finding dense subgraphs. In: Albers, S., Marchetti-Spaccamela, A., Matias, Y., Nikoletseas, S., Thomas, W. (eds.) ICALP 2009. LNCS, vol. 5555, pp. 597–608. Springer, Heidelberg (2009). https://doi.org/10.1007/978-3-642-02927-1_50
24. Lee, V.E., Ruan, N., Jin, R., Aggarwal, C.: A survey of algorithms for dense subgraph discovery. In: Managing and Mining Graph Data, pp. 303–336 (2010)

25. Meladianos, P., Nikolentzos, G., Rousseau, F., Stavrakas, Y., Vazirgiannis, M.: Degeneracy-based real-time sub-event detection in Twitter stream. In: ICWSM 2015, pp. 248–257 (2015)
26. Mihalcea, R., Tarau, P.: TextRank: bringing order into texts. In: EMNLP 2004, pp. 404–411 (2004)
27. Nikolentzos, G., Meladianos, P., Stavrakas, Y., Vazirgiannis, M.: Supplementary material for k-clique-graphs for dense subgraph discovery (2017). http://www.db-net.aueb.gr/nikolentzos/files/ecml_pkdd17_suppl.pdf
28. Orlin, J.B.: A faster strongly polynomial time algorithm for submodular function minimization. Math. Program. **118**(2), 237–251 (2009)
29. Rousseau, F., Vazirgiannis, M.: Main core retention on graph-of-words for single-document keyword extraction. In: Hanbury, A., Kazai, G., Rauber, A., Fuhr, N. (eds.) ECIR 2015. LNCS, vol. 9022, pp. 382–393. Springer, Cham (2015). https://doi.org/10.1007/978-3-319-16354-3_42
30. Schank, T., Wagner, D.: Finding, counting and listing all triangles in large graphs, an experimental study. In: Nikoletseas, S.E. (ed.) WEA 2005. LNCS, vol. 3503, pp. 606–609. Springer, Heidelberg (2005). https://doi.org/10.1007/11427186_54
31. Schrijver, A.: A combinatorial algorithm minimizing submodular functions in strongly polynomial time. JCT **80**(2), 346–355 (2000)
32. Sozio, M., Gionis, A.: The community-search problem and how to plan a successful cocktail party. In: KDD 2010, pp. 939–948 (2010)
33. Tixier, A.J.P., Malliaros, F.D., Vazirgiannis, M.: A graph degeneracy-based approach to keyword extraction. In: EMNLP 2016 (2016)
34. Tsourakakis, C.: The k-clique densest subgraph problem. In: WWW 2015, pp. 1122–1132 (2015)
35. Tsourakakis, C., Bonchi, F., Gionis, A., Gullo, F., Tsiarli, M.: Denser than the densest subgraph: extracting optimal quasi-cliques with quality guarantees. In: KDD 2013, pp. 104–112 (2013)
36. Wang, N., Zhang, J., Tan, K.L., Tung, A.K.: On triangulation-based dense neighborhood graph discovery. VLDB Endow. **4**(2), 58–68 (2010)

Learning and Scaling Directed Networks via Graph Embedding

Mikhail Drobyshevskiy$^{(\boxtimes)}$, Anton Korshunov, and Denis Turdakov

Institute for System Programming of Russian Academy of Sciences, Moscow, Russia
{drobyshevsky,korshunov,turdakov}@ispras.ru

Abstract. Reliable evaluation of network mining tools implies significance and scalability testing. This is usually achieved by picking several graphs of various size from different domains. However, graph properties and thus evaluation results could be dramatically different from one domain to another. Hence the necessity of aggregating results over a multitude of graphs within each domain.

The paper introduces an approach to automatically learn features of a directed graph from any domain and generate similar graphs while scaling input graph size with a real-valued factor. Generating multiple graphs with similar size allows significance testing, while scaling graph size makes scalability evaluation possible. The proposed method relies on embedding an input graph into low-dimensional space, thus encoding graph features in a set of node vectors. Edge weights and node communities could be imitated as well in optional steps.

We demonstrate that embedding-based approach ensures variability of synthetic graphs while keeping degree and subgraphs distributions close to the original graphs. Therefore, the method could make significance and scalability testing of network algorithms more reliable without the need to collect additional data. We also show that embedding-based approach preserves various features in generated graphs which can't be achieved by other generators imitating a given graph.

Keywords: Random graph generating · Graph embedding
Representation learning

1 Introduction

Modeling and generating random graphs is an actively evolving research area. Theoretical aspects include studying features and processes defining connectivity structure in real graphs with mathematical methods. There are also important practical use-cases where random graphs are traditionally employed. A set of random graphs with similar properties could be used for testing significance of results of network mining tools, e.g. community detection. If size of generated graphs is adjustable, scalability of graph algorithms could be tested as well.

© Springer International Publishing AG 2017
M. Ceci et al. (Eds.): ECML PKDD 2017, Part I, LNAI 10534, pp. 634–650, 2017.
https://doi.org/10.1007/978-3-319-71249-9_38

According to our experience, the pipeline of modeling and generating random graphs resembling properties of real data includes the following steps:

1. learn statistical features of real graphs: distributions, dependencies, parameter ranges, etc.;
2. select features to be modeled: degree distribution, clustering, diameter, subgraphs distribution, etc.;
3. define a probability space over all possible graphs with selected features (usually achieved by defining parametrized generative process for graphs);
4. sample random graphs from the defined space.

The fundamental issue here is that each graph domain (social, mobile [19], biological [22], etc.) has its own specific features and many of them may be unknown. Consequently, random graph models created for one domain could be invalid for others. And generally one can't be sure that all essential features of real graph are modeled.

Uncertainty about properties of an arbitrary graph leads to the idea of automatic extraction of features from real data. Suchwise, steps 1–3 of the aforementioned pipeline are replaced with automatic model learning for the given graph. The model could be complicated to be studied analytically and could hardly provide new insights about the data. However, practical use-cases of random graphs could be improved and extended: e.g., data anonymization for publishing a synthetic version of real network preserving its information privacy.

The most popular technique of automatic feature extraction from graphs is representation learning. In recent years this area has attracted much attention in view of recent success in word embedding [17] and adaptation of these ideas to graph domain [23, 25].

Our approach incorporates graph embedding into random graph generation. Embedding result is a set of node vectors which altogether encode some statistical features of the input graph. Experiments suggest that all (or almost all) edges of the input graph could be recovered from the node vectors given proper embedding scheme.

Another advantage of using embedding in context of graph generation is possibility to approximate node vectors distribution. This allows to utilize a unified sampling scheme for random graphs, regardless of input graph domain and features. Finally, arbitrary number of node vectors could be sampled for each synthetic graph, allowing to precisely control resulting graph size.

Our main contributions are as follows:

– We present a pipeline of generating controllable size random graphs similar to a given one based on graph embedding (Embedding-based Random Graph Generator, ERGG).
– We develop an embedding method such that an arbitrary directed graph can be recovered from its embedding with small distortion.

- Within ERGG pipeline we develop ERGG-dwc – a concrete algorithm based on this embedding method which handles $\underline{d}$irected $\underline{w}$eighted graphs with $\underline{c}$ommunity structure[1].
- We show that ERGG-dwc preserves subgraph distribution, clustering, shortest path length distribution, and other properties in generated graphs which can't be achieved by other RGGs imitating real graphs.

Many graph domains are directed naturally (e.g. mobile call graphs), while edge weights contain additional information about the object modelled by the graph (e.g. duration of call). Furthermore, many graphs have community structure which determines high-level organization of a network into groups of nodes sharing similar function, property, role, etc. That's why ERGG-dwc targets edge direction and weight along with community structure of nodes.

Important aspect of task definition is how we define similarity between graphs, especially for the case of different graph sizes. We believe that any fixed set of features could be incomplete. Still, we selected two distributional features which seem to cover all levels of network organisation. We require these distributions in the generated graphs to be close to those of a given graph.

Node degree distribution is an important global characteristic of graph. For instance, many complex networks demonstrate power-law degree distribution. The form of degree distribution is known to determine some other graph characteristics: for instance, power-law distribution determines small **diameter** [5], **clustering coefficient** as a function of node degree [6], etc.

Subgraph distribution is another informative topological characteristic of a graph which defines node clustering behavior and other features. For instance, it allows to categorize graphs over their domains with better precision than other features [2].

The rest of the paper is organized as follows. In Sect. 2 we survey works in related areas, then in Sect. 3 we introduce our ERGG pipeline together with ERGG-dwc algorithm. In Sect. 4 we provide a detailed research of ERGG-dwc steps, and then in Sect. 5 we experimentally evaluate them and perform other tests. Finally, we discuss main results and give a conclusion in Sect. 6.

Notation

$G = (N, E)$ — graph with nodes N and edges $E \subseteq N \times N$, $n = |N|$ — number of nodes, $e = |E|$ — number of edges;
$H = (M, F)$ — generated graph with nodes M and edges F;
i, j, k, l — nodes, (i, j), $i \rightarrow j$ — edges;
$deg(i)$ – degree of node i, $deg_+(i)$ – out-degree of node i;
w_{ij} — weight of edge $i \rightarrow j$;
n-GP — graph profile, L^2-normalized vector of counts of all possible (up to isomorphism) connected subgraphs with n nodes in a graph (see [2] for details). For example, 3-GP is a vector of 13 numbers which reflects distribution of subgraphs with 3 nodes in a graph (see Fig. 1).

---

[1] Online demo: http://ergg.at.ispras.ru/.

Fig. 1. 13 possible connected directed subgraphs with 3 nodes, up to isomorphism.

2   Related Work

2.1   Random Graph Generation

A lot of random graph models were suggested in order to reflect particular properties discovered in real networks: power law degree distribution, small diameter, high clustering coefficient and others. Such models include Erdös-Rènyi model [8], Watts-Strogatz model [26], Barabàsi-Albert model [1], R-MAT [3], Kronecker graphs [14], MFNG [21], dot-product graphs [30], hyperbolic geometric graph [11], etc.

Some models are designed to generate graphs with community structure: stochastic block model [18], LFR [12], CKB [4], ReCoN [24]. Several models aim at producing specified subgraph distribution: triplet model [27], STS [28]. A few models allow specifying of graph features to be produced: Feature constraints method [29], MFNG [21].

To the best of our knowledge, there is no method for generating controlled size directed weighted graphs with communities, capable of automatically reproducing important statistical properties of a given graph, namely degree and subgraph distribution (see Table 1).

2.2   Directed Graph Embedding

The task of mapping nodes of a particular graph into real-valued vectors of some low-dimensional space with preserving useful features of this graph is referred to as graph *embedding* or *representation learning*. In view of recent success of graph embeddings based on machine learning techniques [10, 23, 25] we concentrate on them.

BLM. Bilinear link model (BLM) [10] uses the following model for directed graphs:

$$p(j|i) = \frac{\exp\left(\boldsymbol{u}_i \cdot \boldsymbol{v}_j\right)}{\sum_k \exp\left(\boldsymbol{u}_i \cdot \boldsymbol{v}_k\right)} \tag{1}$$

Here each node $i \in N$ is associated with input and output node vectors: $\boldsymbol{u}_i, \boldsymbol{v}_i$. With joint link probability $p(i, j) = p(i)p(j|i)$, the objective is a log-likelihood of the whole graph:

$$J_\Theta = \sum_{(i,j) \in E} \log p(i, j) \to \max_\Theta \tag{2}$$

For softmax approximating authors implement Noise contrastive estimation (NCE) [9], which was developed for estimation of unnormalized probabilistic models, treating the normalizing constant Z_i as an additional parameter.

LINE. A similar approach for embedding of directed weighted graph was suggested in [25]. Along with (1) which is called 2-nd order proximity, authors also maximize 1-st order proximity of node pairs $p_1(i,j) = \dfrac{1}{1 + \exp\left(-\boldsymbol{u}_i \cdot \boldsymbol{u}_j\right)}$. To overcome a large summation in denominator negative sampling (NEG) [17] is used, which is a simplification of NCE.

Table 1. Comparison of supported properties for several RGG algorithms. Properties notation: **learn** – how features are extracted from a graph; **dir, wei, com** – support of directed, weighted graphs and communities; **DD** – degree distribution; **size** – support of controllable size of generated graphs; **complexity** – complexity of graph generation. Values notation: '' – not supported, '+' – supported (specified to a model); '+−' – partially supported; '=' – exactly reproduced; '≈' – approximately reproduced; '*' – supported in theory, but unchecked in practice.

Algorithm	Learn	Dir	Wei	Com	DD	3-GP	Size	Complexity
Feature constraints	manual	+			=	*	=	$O(|E|)$
SKG	auto	+			≈		+−	$O(|E|\log|N|)$
MFNG	manual	+			≈	*	+	$O(|E|\log|N|)$
ReCoN	auto	+		=	=		+−	$O(|E|)$
Dot product	no	+			+−		+	$O(|N|^2)$

Finally we note that although graph embedding is widely used for learning graph features no one has yet applied it to graph generation. Of our interest are such embedding methods of a directed graph that would (1) encode most of its properties in the representation, and (2) allow to reconstruct from it graphs of various size with these properties preserved.

3   Embedding Based Graph Generating

Now we present the general pipeline of our Embedding-based Random Graph Generator (ERGG) for generation of controllable size random graphs similar to a given one. Then we suggest ERGG-dwc algorithm tailored for directed weighted graphs with community structure.

3.1   General Pipeline of ERGG

The nodes of an original graph are first embedded into vectors in low dimensional space $\mathbb{R}^d$. Then new vectors corresponding to new nodes are sampled from some probability distribution which approximates the distribution of node vectors. Finally, new nodes are connected with edges giving a new graph.

More formally, ERGG input is a graph $G = (N, E)$ and scaling factor $x > 0$ $(x \in \mathbb{R})$. The output is a new random graph $H = (M, F)$ with $|M| \approx \lfloor x|N| \rfloor$ nodes. The steps are as follows:

1. Embed graph $G = (N, E)$ into low-dimensional space, such that its nodes $i \in N$ are mapped into real value vectors $\{r_i\}_{i=1}^{|N|}$.
2. Approximate the distribution of vectors $\{r_i\}_{i=1}^{|N|}$ and sample a new set of $\lfloor x|N| \rfloor$ random vectors $\{q_i\}_{i=1}^{|M|}$ from this distribution. These vectors will correspond to nodes of a new graph $(M, \cdot)$.
3. Connect nodes of graph $(M, \cdot)$ with edges using the embedding model from step 1, obtaining a result graph $H = (M, F)$.

We assume that at step 1 one may use any embedding method providing for a pair of nodes i, j a score function $s_{ij} = s(r_i, r_j)$, characterizing a link $i \rightarrow j$. Since an input graph G is embedded and the distribution of its node vectors $\{r_i\}_{i=1}^{|N|}$ is approximated by some distribution model $\mathcal{R}$, we get a generative model which defines a probability distribution over graphs similar to G. To sample such a random graph one should sample M vectors from $\mathcal{R}$ and use s_{ij} function to connect corresponding nodes. Note that this pipeline covers both directed and undirected graphs and allows for extending it to graphs with weights and/or communities.

3.2 ERGG-dwc — Algorithm for Scaling a Directed Weighted Graph with Communities

Now we suggest ERGG-dwc, a concrete algorithm capable of handling graphs with weights and communities. Weights are treated as edge labels, while communities — as node labels. Suppose an input is a directed weighted graph $G = (N, E)$ with community structure given as node labelling $\{C_i\}_{i=1}^{|N|}$ and a scaling factor x. Additional parameters are noise magnitude ϵ and default edge weight w_0. Steps to be performed are the following:

1. **Embedding.** Embed graph $G = (N, E)$ with a modified embedding method (see Sect. 4.1) into node vectors $\{r_i\}_{i=1}^{|N|}$; and find a threshold t_G which discriminates top E node pairs (i, j) with relatively higher score $s(r_i, r_j)$ from the rest of all node pairs $i, j \in N$. Threshold t_G is utilized in edge generation process during the connecting step.
2. **Approximating + sampling.** Randomly sample (with repetitions) $m = \lfloor x|N| \rfloor$ vectors $\{q_i\}_{i=1}^{m}$ from $\{r_i\}_{i=1}^{|N|}$ adding small gaussian noise $g \sim \mathcal{N}(0, diag(\epsilon, \ldots, \epsilon))$. A mapping φ from nodes of original graph to a new graph is thus defined: $N \overset{\varphi}{\mapsto} M$.
3. **Connecting.** Connect with edges those pairs of nodes k, l from M that have $s(q_k, q_l) > t_G$. Dangling nodes if present are removed, obtaining graph $H = (M', F)$.
4. **Attributing**
   (a) For each edge $(k, l) \in F$, assign weight $w'_{kl} = w_{ij}$, where $i = \varphi^{-1}(k)$, $j = \varphi^{-1}(l)$. If $(i, j) \notin E$, pick a default edge weight w_0.
   (b) For each node $k \in M$, assign community labels $C'_k = C_i$, where $i = \varphi^{-1}(k)$.

4   Detailed Algorithm Description

Here we elaborate on ERGG-dwc steps described in Sect. 3.1. Instead of solving the whole ERGG-dwc task we have splitted it in a sequence of simpler sub-tasks and conducted a research of their possible solutions, optimizing a separate objective for each one. Namely, these subtasks are: embedding + connecting in Sect. 4.1, approximating + sampling in Sect. 4.2 and attributing in Sect. 4.3.

4.1   Embedding + Connecting

The first task is to embed nodes of a given graph into vectors preserving maximal information. In order to achieve this we try to recover edges of the same graph back from the vectors and maximize F_1 score of restored edges versus existed edges. We will refer to it as edge-recovery F_1 or simply F_1 further. Thereby Embedding step is coupled with Connecting step.

Graph Reconstructing. We assume that embedding method provides a score function $s_{ij} = s(\boldsymbol{r}_i, \boldsymbol{r}_j)$ on pairs of nodes, which is trained to score edges of the graph higher than non-edges. Then we just score all node pairs and choose a threshold t_G equal to s_{ij} with rank $E + 1$ and therefore all pairs with $s_{ij} > t_G$ become edges.

Modified Embedding Method. We have elaborated BLM and LINE algorithms according to our goal of separating edges from non-edges, and have suggested a new COMBO algorithm. After some experiments (omitted due to limited space) we found an optimal combination in terms of F_1 for all tested graphs. Namely, negative sampling is used to optimize the objective as done in LINE:

$$J_\theta = \frac{1}{E} \sum_{(i,j) \in E} \left(\log \sigma(s_{ij}) + \sum_{j' \sim p_n(j')}^{\nu} \log \sigma(-s_{ij'}) \right), \qquad (3)$$

where bilinear model from BLM is used as score function:

$$s_{ij} = \boldsymbol{u}_i \cdot \boldsymbol{v}_j - Z_i. \qquad (4)$$

Embedding vectors are initialized as $\boldsymbol{u}_i, \boldsymbol{v}_i \sim \mathcal{U}[-\frac{1}{2\sqrt{d}}, \frac{1}{2\sqrt{d}}]$ (where d is dimensionality of embedding space) and $Z_i = \log |N|$; noise edges are filtered such that $(i, j') \notin E$ only are sampled as negative examples; noise distribution $p_n(j) \propto deg(j)^{3/4}$ [25]. Regularization is eliminated. Number of noise samples for each edge $\nu = 25$, learning rate $\eta = 0.025$, training is performed during 200 epochs.

Thereby, the set of parameters of the algorithm is: $\theta = \{\boldsymbol{u}_i, \boldsymbol{v}_i, Z_i\}_{i=1}^{|N|}$, representation vector learnt for node $i \in N$ is: $\boldsymbol{r}_i = \begin{bmatrix} \boldsymbol{u}_i & \boldsymbol{v}_i & Z_i \end{bmatrix}^T$.

For optimization we used asynchronous stochastic gradient descent like in LINE and BLM. At each step gradient is computed by one edge:

$$\frac{\partial J_\theta^{(i,j)}}{\partial \theta} = \frac{\partial}{\partial \theta} \log \sigma(s_{ij}) + \sum_{j' \sim p_n(j')}^{\nu} \frac{\partial}{\partial \theta} \log \sigma(-s_{ij'})$$

$$= \sigma(-s_{ij}) \frac{\partial s_{ij}}{\partial \theta} + \sum_{j' \sim p_n(j')}^{\nu} \sigma(s_{ij'}) \frac{\partial s_{ij'}}{\partial \theta}. \tag{5}$$

We considered embedding to be successful if graph edges could be restored with $F_1 > 0.99$. This means that obtained representation explains more than 99% of graph edges under the model, while rest 1% may be outliers.

We also discovered that dimensionality of embedding space d is crucial parameter of embedding. Minimal d such that F_1 reaches 0.99 for a particular graph could be viewed as a "complexity" of this graph under the embedding model. We found that it varies for different graphs (see Fig. 2).

4.2   Distribution Approximating + Sampling

At this stage we have a vector representation of nodes of an input graph $\{r_i\}_{i=1}^{|N|}$ such that it can be restored from them with $F_1 > 0.99$. The next task is to model the distribution of node vectors $r_i \sim \mathcal{R}$, such that new node vectors sampled from $\mathcal{R}$ would produce (using reconstruction procedure from Sect. 4.1) graphs with similar properties. This step handles creation of a graph generative model and provides randomization and variability of size of generated graphs. In order to experimentally compare graphs of different size, we use cosine similarity of their 3-GP vectors and "eyeball" similarity of the form of their degree distributions. Since the input graph is embedded, node vectors encode information about graph structure. We assume that key statistical graph properties are reflected in *distribution* of $\{r_i\}_{i=1}^{|N|}$, rather than in individual vectors. Therefore a model captured the distribution $\mathcal{R}$ such that $r_i \sim \mathcal{R}$ would also contain these properties. Furthermore if we sample a new set of node vectors from $\mathcal{R}$ and construct a new graph by the same procedure, it will also demonstrate these properties. This idea is justified well by an example of random dot product graphs, where node vectors distribution analytically determines graph properties [20]. Another benefit of this approach is that number of vectors sampled and therefore size of generated graph may be varied.

To model a distribution $\mathcal{R}$ we suggest the following method called GN based on gaussian noise. GN just memorizes the whole set of vectors $\{r_i\}_{i=1}^{|N|}$ and adds to each one gaussian noise with magnitude ϵ. To draw a sample from $\mathcal{R}_\epsilon$ it randomly samples $i \in \{1..|N|\}$ and returns $r_i + g$, $g \sim \mathcal{N}(0, diag(\epsilon, \ldots, \epsilon))$. This is in fact kernel density estimation with a normal kernel. In case of large graphs storing all N vectors may be excessive, so one could randomly sample a smaller subset from them.

Size of scaled graphs. If we scale a graph with n_0 nodes and e_0 edges by a factor of x, it should have $n_x \approx x n_0$ nodes. What number of edges e_x should it have, or in other words what is $e(n)$ law? In our approach the law $e \sim n^2$ can be proved theoretically for graphs generated by one model despite of probability distribution model:

Theorem 1. *Let $\mathcal{R}$ be a probability distribution, $s_{ij} = s(\mathbf{r}_i, \mathbf{r}_j)$ be a real-valued function. If $\mathbf{r}_i \sim \mathcal{R}$ and edge $i \to j$ is defined by condition $s_{ij} > t_G$, then if sample n vectors the number of edges $e \propto n^2$.*

Proof. Since $\mathbf{r}_i$ and $\mathbf{r}_j$ come from the same distribution edge probability between two randomly chosen nodes $P(s_{ij} > t_G) = p$ is a constant and is determined only by $\mathcal{R}$. Fraction of pairs connected with edges doesn't depend on the number of node-vectors sampled n, i.e. $\mathbb{E}(e/n^2) = p = const$. Hence $e \propto n^2$.   □

As long as real graphs may exhibit different laws of growth, we may treat an ERGG-dwc imitation of a real graph not as its future state, but as its scaled version.

4.3   Attributing

Now we are able to generate controllable size directed graphs structurally similar to a given one. The final step is to handle edge weights and community labels assigning in a generated graph. Their coherence with graph topology should be preserved. How do we make communities in new graphs and endow them with weights in a proper way?

Community structure. We view a community structure of the graph as its nodes labelling: each node i has a (possibly empty) set of community labels $\mathcal{C}_i$ it belongs to. We suggest to inherit these labels in a generated graph using GN sampling method: if a node k of a new graph was sampled from node i, it has the same labels $\mathcal{C}'_k = \mathcal{C}_i$. In this way, given uniform sampling of nodes in GN, communities in a new graph become proportionally scaled images of the original ones.

Weights. In order to assign edge weights in a generated graph we also inherit original weights using GN sampling method. For edge (k, l) of a new graph, if adjacent nodes k, l were sampled from nodes i, j of the original graph, we assign corresponding weight $w'_{kl} = w_{ij}$.

The question left is what if the original graph $G(N, E)$ doesn't have an edge (i, j)? One reason is incorrect embedding of (i, j). Since the fraction of edges $(i, j) \notin E$ after successful embedding is less than 1%, their weights wouldn't affect the results much. In this case a random weight may be chosen: $w_0 \sim \mathcal{U}(\{w_{ij}\}_{(i,j)\in E})$.

Another reason is the noise added to the sampled node vectors. In this case we can consider such an edge a weak connection, which motivates choice of default weight as minimal possible weight: $w_0 = \min_{(i,j)\in E} w_{ij}$. In order to check correctness of weights assignment, we scaled weighted graphs with communities and used modularity measure designed for directed weighted graphs with communities [7]. High value of this metric is a kind of evidence that communities are more densely (accounting also for edge weights) connected within than between each other. Obtaining modularity values as high as in the original graph in experiments supports the hypothesis about weights assigning (see Sect. 5.1).

Complexity of ERGG-dwc is $O((|E|\nu \log \frac{|E|}{|N|} + x^2|N|^2)d)$, details are omitted due to space limit.

5 Experiments

When developing our ERGG-dwc algorithm we used a set of small and medium size directed graphs from different domains and also artificial graphs (see Table 2 for their parameters). Graph embedding is implemented in C++ using pthreads library for thread parallelization.

5.1 ERGG-dwc Steps Elaborating

Embedding Method Parameters. We investigated how F_1 depends on feature space dimensionality d (see Fig. 2). All graphs successfully reach $F_1 = 0.99$ at some d (see Table 2), which we called their "complexity". The value d corresponding to $F_1 = 0.99$ for a particular graph can be determined via binary search: graph is iteratively embedded with different d and F_1 is approximately estimated for each. This usually takes 5–7 iterations.

Table 2. Left: directed graphs; right: directed weighted graphs with detected communities. Graphs parameters: name on plots, number of nodes N, number of edges E, modularity [7] Q, embedding space dimensionality d such that $F_1 > 0.99$.

graph	N	E	d
Karate[2]	34	78	3
Yeast[3]	688	1079	9
VAST[4]	400	1562	12
Foods[5]	128	2106	8
TW [15]	146	1309	12
Kron[6] [16]	2187	11675	24
Words[3]	2704	8300	17
ER[7] [16]	800	8000	26
G+ [15]	1243	106485	62

graph	N	E	Q	d
Protein[3]	95	213	0.6630	7
Resid[8]	217	2672	0.5106	14
VAST[4]	400	1562	0.5743	12
Airport[9]	1574	28236	0.1247	31
LFR[10] [12]	1000	14396	0.7209	26

[2] http://support.sas.com/documentation/cdl/en/procgralg/68145/HTML/default/viewer.htm#procgralg_optgraph_examples07.htm Edges were considered directed.

[3] http://www.weizmann.ac.il/mcb/UriAlon/download/collection-complex-networks.

[4] http://hcil2.cs.umd.edu/newvarepository/VAST%20Challenge%202008/challenges/MC3%20-%20Cell%20Phone%20Calls/.

[5] http://vlado.fmf.uni-lj.si/pub/networks/data/bio/foodweb/foodweb.htm.

[6] Used SNAP generator with parameters "krongen -m:'0 0.783, 0.003, 0.733; 0.147, 0.636, 0.772; 0.028, 0.700, 0.009' -i:9".

[7] Used SNAP generator with parameters "graphgen -g:e -n:800 -m:8000".

[8] http://moreno.ss.uci.edu/data.html#oz.

[9] https://toreopsahl.com/datasets/#usairports.

[10] Run with parameters "-N 1000 -om 3 -on 0.5 -maxk 150 -t1 2.4 -t2 1.6"

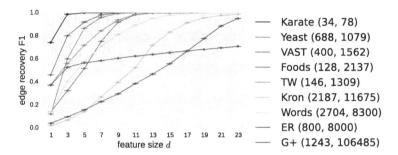

Fig. 2. Edge-recovery F_1 measure of restored graphs versus embedding vectors dimensionality d. Averaged over 5 runs.

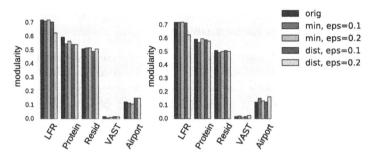

Fig. 3. Modularity of communities in ERGG-dwc generated graphs for two default edge weightings 'min' and 'dist' depending on GN noise magnitude ϵ. Scaling factors $x = 1$ (left) and $x = 4$ (right). Original communities are detected by OSLOM (except for LFR). Values are averaged over 5 runs.

Distribution Approximating and Sampling. Here we fixed an embedding for each test graph such that $F_1 > 0.99$ and experimented with approximating of embedding vectors distribution.

An input here is a set of representation vectors $r_i = \begin{bmatrix} u_i & v_i & Z_i \end{bmatrix}^T$ (of length $2d + 1$) for $i = 1..|N|$ and threshold t_G. Distribution approximating algorithm is applied to fit a given set $\{r_i\}_{i=1}^{|N|}$ and then used to sample new $m = \lfloor x|N| \rfloor$ vectors. Corresponding nodes are connected with edges according to $s_{ij} > t_G$ condition, and finally dangling nodes are removed, giving an output graph $H = (M, F)$.

We measured cosine similarity between 3-GPs of a generated graph and an original one; number of nodes M/xN, and number of edges F/x^2E in relation to their expected values, with different scaling factors x. We found that GN with $\epsilon \in [0.1; 0.2]$ performs best in terms of these metrics.

Community Labels and Edge Weights. For experiments here we used several directed weighted graphs from various domains and applied a community detection method (OSLOM [13]), and also one synthetic graph generated by LFR

[12] (see Table 2, right). We also fixed embeddings with $F_1 > 0.99$ and applied GN inheriting community labels and edge weights as described in Sect. 4.3. We compared modularity of generated communities in the generated graphs for two methods of default weight choice, minimal weight ('min') and random weight from the distribution ('dist'). We varied noise magnitude ϵ and scaling factor x.

We found that when $\epsilon \in [0; 0.2]$ modularity remains approximately the same in average both for $x = 1$ and $x = 4$, while higher $\epsilon = 0.3$ makes it lower especially significant at $x = 4$ (not plotted). This again proved $\epsilon \in [0.1; 0.2]$ to be optimal values, therefore we used them further.

Comparison of default weighting schemes showed that both of them perform almost identically in terms of modularity on different graphs (see Fig. 3) at $\epsilon = 0.1$, while at $\epsilon = 0.2$ on large graphs 'dist' scheme leads to larger modularity loss than 'min' scheme.

5.2   Variability Evaluation

In order to apply ERGG-dwc for significance testing of various network mining tools, generated graphs should be not only similar to the original one but also differ from each other. Variability of graph imitations produced by ERGG-dwc should be wide enough to model the natural variability across real networks. For assessing the variability of ERGG-dwc we considered how different graph statistics vary across graphs from one domain and compared to corresponding variances of different ERGG-dwc imitations of one graph of them. For that we chose a set of twitter-ego nets and picked 15 graphs close in number of nodes ($|N| \in [170; 180]$) and number of edges ($|E| \in [2000; 3000]$) as a dataset. We analyzed in-degree, 3-GP, and clustering coefficient distributions for these graphs and also for 15 ERGG-dwc imitations of one of them: Fig. 4 demonstrates similar variabilities for both sets.

5.3   Significance Testing

Here we demonstrate how to perform significance testing of several network mining tools using ERGG-dwc. For that we chose a set of Google+ ego-nets and picked 8 graphs close in number of nodes ($|N| \in [450; 500]$). For each graph we generated 5 imitations and ran an algorithm on the graph (black triangles at Fig. 5) and on the imitations (blue triangles at Fig. 5). We measured modularity for community detection methods OSLOM and Infomap[2] and running time for diameter computing algorithm, see Fig. 5 (green triangles correspond to modularity of communities generated by ERGG-dwc).

Looking at the plots one can conclude that OSLOM produces insignificant results in terms of modularity, while Infomap's result are much more significant.

---

[2] http://www.mapequation.org/code.html.

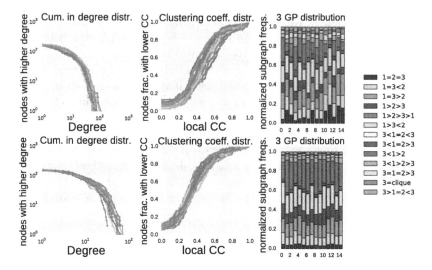

Fig. 4. Variability evaluation for cumulative in-degree distribution, cumulative clustering, 3-GP. Top row: set of 15 twitter-ego graphs of size $|N| \in [170; 180]$, $|E| \in [2000; 3000]$, bottom row: 15 ERGG-dwc imitations of one of them.

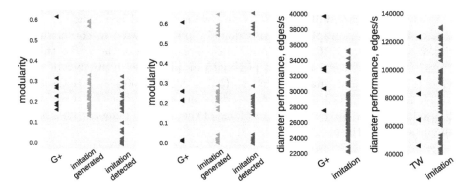

Fig. 5. Significance tests. From left to right: OSLOM CD, Infomap CD, performance of diameter computing Google+ ego-nets and Twitter ego-nets. 8 ego-nets of Google+ ($N \in [450; 500]$) and 8 ego-nets of Twitter ($N \in [170; 180]$) are used. (Color figure online)

We also tested performance of diameter computing algorithm (Fig. 5, 2 plots on the right). We also used 5 Google+ ego-nets and 5 ego-nets from Twitter ($|N| \in [170; 180]$), generating 15 ERGG-dwc imitations for each graph. Everything is as expected for Twitter, but more time is needed to figure out the reasons of Google+ results.

5.4   Comparison to SKG

Now we compare the work of ERGG-dwc with SKG [14] in terms of various graph features. We fitted US Airports graph[3] (Airport) by our algorithm and SKG. Fitting was done by Kronfit with default parameters.

Degree distribution and spectral properties are approximated slightly better via ERGG-dwc than SKG. 3-GP and clustering properties are not captured by SKG at all, while ERGG-dwc matches them well. SKG generated graph doesn't match number of nodes and hence average degree, but has lower diameter 5 that is significantly smaller than the original 8. Reciprocity metric (proportion of reciprocal edges) is 0.02 versus original 0.78 and 0.65 for ERGG-dwc. This may be explained by the fact that edges $i \rightarrow j$ and $j \rightarrow i$ are sampled independently even having same probability and thus reciprocal edge $i \leftrightarrows j$ is much less probable.

5.5   Discussion

In the experiments we showed that directed graphs from various domains can be successfully embedded into low-dimensional space by means of our modified embedding method COMBO. Embedding quality is evaluated in a special sense specific for RGG task: edges of graph can be restored with high F_1 (> 0.99). Optimal dimensionality d of the space depends on graph and is currently found by repetitive trials. Faster determination of optimal d for a given graph could be a future work.

Unfortunately, in our experiments we found that at high scaling factors x the form of degree distribution is not preserved. As it was shown theoretically the number of edges $|E|$ is proportional to $x^2|N|^2$.

Furthermore, GN method provides a simple way to inherit weights and community labels from the original graph. Experiments suggest that this labelling method preserves high modularity of generated communities in scaled graphs. These graphs may be used as benchmarks for testing community detection algorithms. Another future direction is advancing this naive method of labelling: current approach may cause staircase effects in edge weight and community size distributions at high scaling factors x. Besides showing the closeness of generated graphs to the original one in terms of degree distributions and 3-GP, we found their variability wide enough to model a graph domain. This means that ERGG-dwc can be used to generate datasets for significance testing.

We also compared quality of fitting a real graph for ERGG-dwc and SKG in terms of several graph statistics and found that ERGG-dwc reproduces most of them much closer to the original. Although WCC distribution reveals many small connected components in ERGG-dwc graph, it may be not critical since the largest WCC is large enough.

Finally, we evaluated performance of ERGG-dwc and confirmed $O(|N|^2)$ generation time. One way to speed-up edge generation is to optimize finding all pairs

---

[3] https://toreopsahl.com/datasets/#usairports.

(i, j) with $s_{ij} > t_G$. Some techniques used in nearest neighbor search related problem could be employed to overcome an exhaustive search. Another idea is to reduce the search space to pairs corresponding to existing graph edges, which makes finding edges F in $O(x^2|E|)$ steps.

6   Conclusion

We introduced and thoroughly evaluated an approach to modeling real directed graphs without a priori knowledge about their domain and properties. To the best of our knowledge, this is the first successful attempt to employ graph embedding technique in random graph generation. The resulting graphs are statistically similar to the input one, proving that representing graphs in low-dimensional space is the right way to obtain their scaled imitations.

Scalability of our method could be improved by adjusting embedding scheme and/or edge generation process. Also, edge weights and node communities are currently treated as attributes and in fact are cloned from the input graph. Additional research is required towards embedding weights and communities along with edges. The generating scheme should be adjusted accordingly to recover weights and communities from the embedding and preserve their statistical properties.

Acknowledgements. This research was collaborated with and supported by Huawei Technologies Co.,Ltd. under contract YB2015110136.

We are also thankful to Ilya Kozlov and Sergey Bartunov for their ideas and valuable contributions.

References

1. Albert, R., Barabási, A.-L.: Statistical mechanics of complex networks. Rev. Mod. Phys. **74**(1), 47 (2002)
2. Bordino, I., Donato, D., Gionis, A., Leonardi, S.: Mining large networks with subgraph counting. In: 2008 Eighth IEEE International Conference on Data Mining, pp. 737–742. IEEE (2008)
3. Chakrabarti, D., Zhan, Y., Faloutsos, C.: R-mat: a recursive model for graph mining. In: SDM, vol. 4, pp. 442–446. SIAM (2004)
4. Chykhradze, K., Korshunov, A., Buzun, N., Pastukhov, R., Kuzyurin, N., Turdakov, D., Kim, H.: Distributed generation of billion-node social graphs with overlapping community structure. In: Contucci, P., Menezes, R., Omicini, A., Poncela-Casasnovas, J. (eds.) Complex Networks V. SCI, vol. 549, pp. 199–208. Springer, Cham (2014). https://doi.org/10.1007/978-3-319-05401-8_19
5. Cohen, R., Havlin, S.: Scale-free networks are ultrasmall. Phys. Rev. Lett. **90**, 058701 (2003)
6. Dorogovtsev, S.N., Goltsev, A.V., Mendes, J.F.F.: Pseudofractal scale-free web. Phys. Rev. E **65**, 066122 (2002)
7. Drobyshevskiy, M., Korshunov, A., Turdakov, D.: Parallel modularity computation for directed weighted graphs with overlapping communities. Proc. Inst. Syst. Program. **28**(6), 153–170 (2016)

8. Erdos, P., Rényi, A.: On the evolution of random graphs. Publ. Math. Inst. Hungar. Acad. Sci. **5**, 17–61 (1960)
9. Gutmann, M., Hyvärinen, A.: Noise-contrastive estimation: a new estimation principle for unnormalized statistical models. In: AISTATS, vol. 1, p. 6 (2010)
10. Ivanov, O.U., Bartunov, S.O.: Learning representations in directed networks. In: Khachay, M.Y., Konstantinova, N., Panchenko, A., Ignatov, D.I., Labunets, V.G. (eds.) AIST 2015. CCIS, vol. 542, pp. 196–207. Springer, Cham (2015). https:// doi.org/10.1007/978-3-319-26123-2_19
11. Krioukov, D., Papadopoulos, F., Kitsak, M., Vahdat, A., Boguná, M.: Hyperbolic geometry of complex networks. Phys. Rev. E **82**(3), 036106 (2010)
12. Lancichinetti, A., Fortunato, S.: Benchmarks for testing community detection algorithms on directed and weighted graphs with overlapping communities. Phys. Rev. E **80**(1), 016118 (2009)
13. Lancichinetti, A., Radicchi, F., Ramasco, J.J., Fortunato, S.: Finding statistically significant communities in networks. PLoS ONE **6**(4), e18961 (2011)
14. Leskovec, J., Chakrabarti, D., Kleinberg, J., Faloutsos, C., Ghahramani, Z.: Kronecker graphs: an approach to modeling networks. J. Mach. Learn. Res. **11**(Feb), 985–1042 (2010)
15. Leskovec, J., Krevl, A.: SNAP datasets: stanford large network dataset collection, June 2014. http://snap.stanford.edu/data
16. Leskovec, J., Sosič, R.: Snap: a general-purpose network analysis and graph-mining library. ACM Trans. Intell. Syst. Technol. (TIST) **8**(1), 1 (2016)
17. Mikolov, T., Sutskever, I., Chen, K., Corrado, G.S., Dean, J.: Distributed representations of words and phrases and their compositionality. In: Advances in Neural Information Processing Systems, pp. 3111–3119 (2013)
18. Mossel, E., Neeman, J., Sly, A.: Stochastic block models and reconstruction. arXiv preprint arXiv:1202.1499 (2012)
19. Nanavati, A.A., Gurumurthy, S., Das, G., Chakraborty, D., Dasgupta, K., Mukherjea, S., Joshi, A.: On the structural properties of massive telecom call graphs: findings and implications. In: Proceedings of the 15th ACM International Conference on Information and Knowledge Management, CIKM 2006, pp. 435–444, New York, NY, USA. ACM (2006)
20. Nickel, C.L.M.: Random dot product graphs: a model for social networks, vol. 68 (2007)
21. Palla, G., Lovász, L., Vicsek, T.: Multifractal network generator. Proc. Nat. Acad. Sci. **107**(17), 7640–7645 (2010)
22. Pavlopoulos, G.A., Secrier, M., Moschopoulos, C.N., Soldatos, T.G., Kossida, S., Aerts, J., Schneider, R., Bagos, P.G.: Using graph theory to analyze biological networks. BioData Min. **4**(1), 10 (2011)
23. Perozzi, B., Al-Rfou, R., Skiena, S.: Deepwalk: online learning of social representations. In: Proceedings of the 20th ACM SIGKDD International Conference on Knowledge Discovery and Data Mining, pp. 701–710. ACM (2014)
24. Staudt, C.L., Hamann, M., Safro, I., Gutfraind, A., Meyerhenke, H.: Generating scaled replicas of real-world complex networks. arXiv preprint arXiv:1609.02121 (2016)
25. Tang, J., Qu, M., Wang, M., Zhang, M., Yan, J., Mei, Q.: Line: large-scale information network embedding. In: Proceedings of the 24th International Conference on World Wide Web, pp. 1067–1077. ACM (2015)
26. Watts, D.J., Strogatz, S.H.: Collective dynamics of small-worldnetworks. Nature **393**(6684), 440–442 (1998)

27. Wegner, A., et al.: Random graphs with motifs (2011)
28. Winkler, M., Reichardt, J.: Motifs in triadic random graphs based on steiner triple systems. Phys. Rev. E **88**(2), 022805 (2013)
29. Ying, X., Wu, X.: Graph generation with prescribed feature constraints. In: SDM, vol. 9, pp. 966–977. SIAM (2009)
30. Young, S.J., Scheinerman, E.R.: Random dot product graph models for social networks. In: Bonato, A., Chung, F.R.K. (eds.) WAW 2007. LNCS, vol. 4863, pp. 138–149. Springer, Heidelberg (2007). https://doi.org/10.1007/978-3-540-77004-6_11

Local Lanczos Spectral Approximation
for Community Detection

Pan Shi[1], Kun He[1,2(⊠)], David Bindel[2], and John E. Hopcroft[2]

[1] Huazhong University of Science and Technology, Wuhan, China
{panshi,brooklet60}@hust.edu.cn
[2] Cornell University, Ithaca, NY, USA
kh555@cornell.edu, {bindel,jeh}@cs.cornell.edu

Abstract. We propose a novel approach called the Local Lanczos Spectral Approximation (LLSA) for identifying all latent members of a local community from very few seed members. To reduce the computation complexity, we first apply a fast heat kernel diffusing to sample a comparatively small subgraph covering almost all possible community members around the seeds. Then starting from a normalized indicator vector of the seeds and by a few steps of Lanczos iteration on the sampled subgraph, a local eigenvector is gained for approximating the eigenvector of the transition matrix with the largest eigenvalue. Elements of this local eigenvector is a relaxed indicator for the affiliation probability of the corresponding nodes to the target community. We conduct extensive experiments on real-world datasets in various domains as well as synthetic datasets. Results show that the proposed method outperforms state-of-the-art local community detection algorithms. To the best of our knowledge, this is the first work to adapt the Lanczos method for local community detection, which is natural and potentially effective. Also, we did the first attempt of using heat kernel as a sampling method instead of detecting communities directly, which is proved empirically to be very efficient and effective.

Keywords: Community detection · Heat kernel
Local lanczos method

1 Introduction

Community detection aims to find a set of nodes in a network that are internally cohesive but comparatively separated from the remainder of the network. In social networks, community detection is a classical and challenging problem which is very useful for analyzing the topology structure and extracting information from the network, and numerous algorithms and techniques have been proposed [12,34].

Most of the researchers have focused on uncovering the global community structure [1,13,28]. With the rapid growth of the network scale, global community detection becomes very costly or even impossible for very large networks.

© Springer International Publishing AG 2017
M. Ceci et al. (Eds.): ECML PKDD 2017, Part I, LNAI 10534, pp. 651–667, 2017.
https://doi.org/10.1007/978-3-319-71249-9_39

The big data drives researchers to shift their attention from the global structure to the local structure [15,16]. How to adapt the existing effective methods initially designed for the global community detection in order to uncover the local community structure is a natural and important approach for the accurate membership identification from a few exemplary members. Several probabilty diffusion methods, PageRank [16], heat kernel [15] and spectral subspace approximation [14,22] are three main techniques for local community detection.

The Lanczos method [21] is a classic method proposed for calculating the eigenvalues, aka the spectra of a matrix. Through there exists some work using the Lanczos method for the spectral bisection [6], unlike other spectra calculation methods, the Lanczos method is seldom used for community detection and to the best of our knowledge, it has never been used for the local community detection. In this paper, we propose a novel approach called the Local Lanczos Spectral Approximation (LLSA) for local community detection. Specifically, we execute a few steps of Lanczos iteration to attain a local eigenvector that approximates the eigenvector of the transition matrix with the largest eigenvalue. Elements of this local eigenvector is a relaxed indicator for the affiliation probability of the corresponding nodes to the target community. As compared with other spectral approxiamtion methods, the Lanczos iterative method is efficient for computing the top eigen-pairs of large sparse matrices and it is space efficient, which is very helpful for large social networks which are usually sparse.

Our contributions include: (1) To the best of our knowledge, this is the first work to address local community detection by the Lanczos approximation. Also, we adapt the standard Lanczos method which is on a symmetric matrix to the unsymmetrical transition matrix directly. (2) Instead of using the heat kernel method to directly extract a community, we did the first attempt to leverage its very fast diffusion property to sample a localized subgraph to largely reduce the subsequent calculation. (3) Based on the Rayleigh quotient related to the conductance, we provide a theoretical base for the proposed LLSA method. (4) Experiments on five real-world networks as well as seven synthetic networks show that the proposed method considerably outperforms existing local community detection algorithms.

2   Related Work

2.1   Local Community Detection

Techniques for local community detection can be classified into three categories, namely the PageRank, heat kernel and local spectral methods. Other techniques like finding minimum cut [4,5,26] can also be used for local community detection.

PageRank. The PageRank method is widely used for local community detection. Spielman and Teng [31] use degree-normalized, personalized PageRank (DN PageRank) with respect to the initial seeds and do truncation on small probability values. DN PageRank is adopted by several competitive PageRank based clustering algorithms [3,35], including the popular PageRank Nibble method [2].

Kloumann and Kleinberg [16] evaluate different variations of PageRank method and find that the standard PageRank yields higher performance than the DN PageRank.

Heat Kernel. The heat kernel method involves the Taylor series expansion of the exponential of the transition matrix. Chung [8,10] provides a theoretical analysis and a local graph partitioning algorithm based on heat kernel diffusion. Chung and Simpson [9] propose a randomized Monte Carlo method to estimate the diffusion speed, and Kloster and Gleich [15] propose a deterministic method that uses coordinate relaxation on an implicit linear system to estimate the heat kernel diffusion, and the heat value of each node represents the likelihood of affiliation.

Local Spectral. A third branch is to adapt the classic spectral method to locate the target community. Mahoney *et al.* [25] introduce a locally-biased analogue of the second eigenvector, the Fiedler vector associated with the algebraic connectivity, to extract local properties of data graphs, and apply the method for a semi-supervised image segmentation and a local community extraction by finding a sparse-cut around the seeds in small social networks. He *et al.* [14] and Li *et al.* [22] extract the local community by seeking a sparse vector from the local spectral subspaces using ℓ_1 norm optimization.

2.2 Lanczos Method

Many real world problems can be modeled as sparse graphs and be represented as matrices, and the eigenvalue calculation of the matrices is usually a crucial step for the problem solving. All the eigenpairs can be calculated by power method [29], SVD [32], or QR factorization [17]. However, these methods are intractable for large matrices due to the high complexity and memory consumption. As the Lanczos method can significantly reduce the time and space complexity, it is usually applied to large sparse matrices [27].

As a classic eigenvalue calculation method, the original Lanczos method [21] cannot hold the orthogonality of the calculated Krylov subspace and it is not widely used in practice. Paige [27] computes the eigenpairs for very large sparse matrices by an improved Lanczos method, as only a few iterations are typically required to get a good approximation on the extremal eigenvalues. After that, the Lanczos method becomes very attractive for large sparse matrix approximation. For example in the application of graph partitioning and image reconstruction, Barnes [6] illustrates that Lanczos method is an efficient implementation of the spectral bisection method; Wu *et al.* [33] propose an incremental bilinear Lanczos algorithm for high dimensionality reduction and image reconstruction; Bentbib *et al.* [7] illustrate that efficient image restoration can be achieved by Tikhonov regularization based on the global Lanczos method.

To the best of our knowledge, there is no Lanczos based algorithms for local community detection in the literature.

3   Local Lanczos Method

The local community detection problem can be formalized as follows. Given a connected, undirected graph $G = (V, E)$ with n nodes and m edges. Let $\mathbf{A} \in \{0,1\}^{n \times n}$ be the associated adjacency matrix, $\mathbf{I}$ the identity matrix, and $\mathbf{e}$ the vector of all ones. Let $\mathbf{d} = \mathbf{A}\mathbf{e}$ be the vector of node degrees, and $\mathbf{D} = \mathrm{diag}(\mathbf{d})$ the diagonal matrix of node degrees. Let S be the set of a few exemplary members in the target community $T = (V_t, E_t)$ ($S \subseteq V_t \subseteq V$, $|V_t| \ll |V|$). Let $\mathbf{s} \in \{0,1\}^n$ be a binary indicator vector representing the exemplary members in S. We are asked to identify the remaining latent members in the target community T.

There are three key steps in the proposed algorithm: heat kernel sampling, local Lanczos spectral approximation and community boundary truncation.

3.1   Local Heat Kernel Sampling

The heat kernel method [15] runs in linear time and is very fast for community detection on large networks. However, the detection accuracy is not high enough as compared with the local spectral method [14]. In this paper, we use the advantage of heat kernel's fast diffusion speed to do the sampling, using parameter settings such that the resulting subgraph is large enough to cover almost all members in the target community.

The Heat Kernel Diffusion. The heat kernel diffusion model spread the heat across a graph regarding the seed set as the persistent heat source.

The heat kernel diffusion vector is defined by

$$\mathbf{h} = e^{-t} \left[\sum_{k=0}^{\infty} \frac{t^k}{k!} (\mathbf{A}\mathbf{D}^{-1})^k \right] \mathbf{p_0}, \tag{1}$$

where $\mathbf{p_0} = \mathbf{s}/|S|$ is the initial heat values on the source seeds. For simplicity of notation, let

$$\mathbf{h}_N = e^{-t} \left[\sum_{k=0}^{N} \frac{t^k}{k!} (\mathbf{A}\mathbf{D}^{-1})^k \right] \mathbf{p_0} \tag{2}$$

indicate the sum of the first N terms.

In practice, we usually seek a vector $\mathbf{x}$ to approximate $\mathbf{h}$:

$$\|\mathbf{D}^{-1}\mathbf{h} - \mathbf{D}^{-1}\mathbf{x}\|_{\infty} < \epsilon. \tag{3}$$

Premultiplying e^t on both sides, we have

$$\|\mathbf{D}^{-1}e^t\mathbf{h} - \mathbf{D}^{-1}e^t\mathbf{x}\|_{\infty} < e^t\epsilon. \tag{4}$$

If for an integer N,

$$\|\mathbf{D}^{-1}e^t\mathbf{h} - \mathbf{D}^{-1}e^t\mathbf{h}_N\|_{\infty} < e^t\epsilon/2, \tag{5}$$

and $\mathbf{z} = e^t \mathbf{x} \approx e^t \mathbf{h}_N$ satisfies

$$\|\mathbf{D}^{-1} e^t \mathbf{h}_N - \mathbf{D}^{-1} \mathbf{z}\|_\infty < e^t \epsilon / 2, \tag{6}$$

then by the triangle inequality, (4) holds, and then (3) holds.

Kloster and Gleich [15] propose a hk-relax algorithm to guarantee (5) by letting N be no greater than $2t \log(\frac{1}{\epsilon})$ and computing a vector $\mathbf{z}$ that satisfies (6), then use the heat values in $\mathbf{x}$ to identify memberships in the local community. We adapt their method to do the heat kernel sampling, shown in Algorithm 1.

Algorithm 1. The heat kernel sampling

Input: Graph $G = (V, E)$, seed set $S \subset V$, upper bound of the subgraph size N_1, heat kernel diffusion parameters t and ϵ

1: Start from S, calculate heat value vector $\mathbf{x}$ to approximate the heat kernel diffusion vector $\mathbf{h}$

2: Sort elements in $\mathbf{x}$ in decreasing order to get a vector $\tilde{\mathbf{x}}$

3: $G_s \leftarrow$ nodes corresponding to all the nonzero elements in $\tilde{\mathbf{x}}$

4: **if** $|G_s| > N_1$ **then**

5:     $G_s \leftarrow$ top N_1 nodes in G_s according to the heat value

Output: Sampled subgraph G_s

Denote the sampled subgraph as $G_s = (V_s, E_s)$ with n_s nodes and m_s edges in the following discussion. We then extract the local community from this comparatively small subgraph instead of the original large network. This pre-processing procedure runs in milliseconds in large networks with millions of nodes, and significantly reduces the computation cost for the follow-up fine tuning of the community detection.

3.2 Local Lanczos Spectral Approximation

In this subsection, we first provide the necessary theoretical base that finding a low-conductance community corresponds to finding the eigenvector of the transition matrix with the largest eigenvalue. Then we briefly introduce a variant of the Lanczos process on the Laplician matrix to calculate this eigenvector. Finally we propose a local Lanczos spectral approximation method to get a "local" eigenvector indicating the implicit topology structure of the network around the seeds, and provide an convergence analysis on the Lanczos iteration process.

Theoretical Base. Let $\mathbf{L} = \mathbf{D_s} - \mathbf{A_s}$ be the Laplacian matrix of G_s where $\mathbf{A_s}$ and $\mathbf{D_s}$ denotes the adjacency matrix and the diagonal degree matrix of G_s. We define two normalized graph Laplacian matrices:

$$\mathbf{L_{rw}} = \mathbf{I} - \mathbf{N_{rw}} = \mathbf{D_s^{-1} L},$$

$$\mathbf{L_{sym}} = \mathbf{I} - \mathbf{N_{sym}} = \mathbf{D_s}^{-\frac{1}{2}}\mathbf{LD_s}^{-\frac{1}{2}},$$

where $\mathbf{N_{rw}} = \mathbf{D_s}^{-1}\mathbf{A_s}$ is the transition matrix, and $\mathbf{N_{sym}} = \mathbf{D_s}^{-\frac{1}{2}}\mathbf{A_sD_s}^{-\frac{1}{2}}$ is the normalized adjacency matrix.

For a community C, the conductance [30] of C is defined as

$$\Phi(C) = \frac{\mathrm{cut}(C, \overline{C})}{\min\{\mathrm{vol}(C), \mathrm{vol}(\overline{C})\}},$$

where $\overline{C}$ consists of all nodes outside C, $\mathrm{cut}(C, \overline{C})$ denotes the number of edges between, and $\mathrm{vol}(\cdot)$ calculates the "edge volume", i.e. for the subset of nodes, we count their total node degrees in graph G_s. Low conductance gives priority to a community with dense internal links and sparse external links.

Let $\mathbf{y} \in \{0, 1\}^{n_s}$ be a binary indicator vector representing a small community C in the sampled graph G_s. Here for "small community", we mean $\mathrm{vol}(C) \leq \frac{1}{2}\mathrm{vol}(V_s)$. As $\mathbf{y^TD_sy}$ equals the total node degrees of C, and $\mathbf{y^TA_sy}$ equals two times the number of internal edges of C, the conductance $\Phi(C)$ could be written as a generalized Rayleigh quotient:

$$\Phi(C) = \frac{\mathbf{y^TLy}}{\mathbf{y^TD_sy}} = \frac{(\mathbf{D_s^{\frac{1}{2}}y})^T\mathbf{L_{sym}}(\mathbf{D_s^{\frac{1}{2}}y})}{(\mathbf{D_s^{\frac{1}{2}}y})^T(\mathbf{D_s^{\frac{1}{2}}y})}. \tag{7}$$

Theorem 1 *(Cheeger Inequality). Let λ_2 be the second smallest eigenvalue of* $\mathbf{L_{sym}}$ *for a graph G_s, then $\phi(G_s) \geq \frac{\lambda_2}{2}$, where $\phi(G_s) = \min_{\tilde{V} \subset V_s} \Phi(\tilde{V})$.*

The proof refers to [11], and we omit the details here. According to this theorem and the definition of $\Phi(C)$, we have $\frac{\lambda_2}{2} \leq \Phi(C) \leq 1$.

According to the Rayleigh-Ritz theorem [23], if we want to minimize the conductance $\Phi(C)$ by relaxing the indicator vector $\mathbf{y}$ to take arbitrary real values, then the scaled relaxed indicator vector $\mathbf{D_s^{\frac{1}{2}}y}$ should be the eigenvector of $\mathbf{L_{sym}}$ with the smallest eigenvalue 0, which is $\mathbf{D_s^{\frac{1}{2}}e}$.

We know that:

$$\mathbf{L_{rw}v} = \lambda\mathbf{v} \quad \Leftrightarrow \quad \mathbf{L_{sym}}(\mathbf{D^{\frac{1}{2}}v}) = \lambda(\mathbf{D^{\frac{1}{2}}v}),$$

the relaxed indicator vector $\mathbf{y}$ should be the eigenvector of $\mathbf{L_{rw}}$ with the smallest eigenvalue. As $\mathbf{L_{rw}} = \mathbf{I} - \mathbf{N_{rw}}$, the eigenvalue decomposition of the Laplacian matrix is also closely related to the expansion of rapid mixing of random walks. As

$$\mathbf{L_{rw}v} = (\mathbf{I} - \mathbf{N_{rw}})\mathbf{v} = \lambda\mathbf{v} \quad \Leftrightarrow \quad \mathbf{N_{rw}v} = (1 - \lambda)\mathbf{v},$$

it follows that $\mathbf{L_{rw}}$ and $\mathbf{N_{rw}}$ share the same set of eigenvectors and the corresponding eigenvalue of $\mathbf{N_{rw}}$ is $1 - \lambda$ where λ is the eigenvalue of $\mathbf{L_{rw}}$. Equivalently, the relaxed indicator vector $\mathbf{y}$ should be the eigenvector of $\mathbf{N_{rw}}$ with the largest eigenvalue.

The largest eigenvalue of $\mathbf{N_{rw}}$ is 1 and the corresponding eigenvector is $\mathbf{e}$ [24], so the relaxed indicator vector $\mathbf{y} = \mathbf{e}$, corresponding to the whole graph with zero conductance. This relaxed indicator vector $\mathbf{y}$ contains global information while the real solution of the indicator vector $\mathbf{y}$ reveals local property for a small community whose total degree is no greater than half of the total degree of the whole graph. As the Lanczos method is efficient for computing the top eigenpairs of large sparse matrices and it is space efficient, we propose a variant of Lanczos method on $\mathbf{N_{rw}}$ to get a "local" eigenvector indicating the latent local structure around the seeds.

Lanczos Process. Based on a theoretical guarantee [18], there exists an orthogonal matrix $\mathbf{Q}$ and a tridiagonal matrix $\mathbf{T}$ such that

$$\mathbf{Q^T}(\mathbf{D_s}^{-\frac{1}{2}}\mathbf{A_s}\mathbf{D_s}^{-\frac{1}{2}})\mathbf{Q} = \mathbf{T}, \tag{8}$$

$$\mathbf{T} = \begin{bmatrix} \alpha_1 & \beta_1 & & & \\ \beta_1 & \alpha_2 & \beta_2 & & \\ & \beta_2 & \alpha_3 & \ddots & \\ & & \ddots & \ddots & \beta_{n_s-1} \\ & & & \beta_{n_s-1} & \alpha_{n_s} \end{bmatrix}. \tag{9}$$

Designate the columns of $\mathbf{Q}$ by

$$\mathbf{Q} = \begin{bmatrix} \mathbf{q_1} & | & \cdots & | & \mathbf{q_{n_s}} \end{bmatrix}.$$

Let $\tilde{\mathbf{Q}} = \mathbf{D_s}^{-\frac{1}{2}}\mathbf{Q}$, so

$$\begin{aligned}
\tilde{\mathbf{Q}} &= \begin{bmatrix} \mathbf{D_s}^{-\frac{1}{2}}\mathbf{q_1} & | & \cdots & | & \mathbf{D_s}^{-\frac{1}{2}}\mathbf{q_{n_s}} \end{bmatrix} \\
&\triangleq \begin{bmatrix} \tilde{\mathbf{q}}_1 & | & \cdots & | & \tilde{\mathbf{q}}_{n_s} \end{bmatrix}.
\end{aligned}$$

Equation (8) can be rewritten as

$$\tilde{\mathbf{Q}}^\mathbf{T}\mathbf{A_s}\tilde{\mathbf{Q}} = \mathbf{T}. \tag{10}$$

As $\mathbf{Q}$ is an orthogonal matrix,

$$\tilde{\mathbf{Q}}\tilde{\mathbf{Q}}^\mathbf{T} = \mathbf{D_s}^{-\frac{1}{2}}\mathbf{Q}\mathbf{Q}^\mathbf{T}\mathbf{D_s}^{-\frac{1}{2}} = \mathbf{D_s}^{-1}. \tag{11}$$

Premultiplying $\tilde{\mathbf{Q}}$ on both sides of Eq. (10), we have $\mathbf{D_s}^{-1}\mathbf{A_s}\tilde{\mathbf{Q}} = \tilde{\mathbf{Q}}\mathbf{T}$. Equating the columns in this equation, we conclude that for $k \in \{1, ..., n_s\}$,

$$\mathbf{D_s}^{-1}\mathbf{A_s}\tilde{\mathbf{q}}_\mathbf{k} = \beta_{k-1}\tilde{\mathbf{q}}_{\mathbf{k-1}} + \alpha_k\tilde{\mathbf{q}}_\mathbf{k} + \beta_k\tilde{\mathbf{q}}_{\mathbf{k+1}}, \tag{12}$$

by setting $\beta_0\tilde{\mathbf{q}}_\mathbf{0} \triangleq \mathbf{0}$, and $\beta_n\tilde{\mathbf{q}}_{\mathbf{n_s+1}} \triangleq \mathbf{0}$.

By the orthogonality of $\mathbf{Q}$, we have $\tilde{\mathbf{Q}}^T \mathbf{D_s} \tilde{\mathbf{Q}} = \mathbf{I}$. Premultiplying $\tilde{\mathbf{q}}_k^T \mathbf{D_s}$ on both sides of Eq. (12), the $\mathbf{D_s}$-inner product orthonormality of the $\tilde{\mathbf{q}}$-vectors implies

$$\alpha_k = \tilde{\mathbf{q}}_k^T \mathbf{A_s} \tilde{\mathbf{q}}_k. \tag{13}$$

Let the vector $\tilde{\mathbf{r}}_k$ be

$$\tilde{\mathbf{r}}_k = \mathbf{D_s}^{-1} \mathbf{A_s} \tilde{\mathbf{q}}_k - \alpha_k \tilde{\mathbf{q}}_k - \beta_{k-1} \tilde{\mathbf{q}}_{k-1}. \tag{14}$$

If $\tilde{\mathbf{r}}_k$ is nonzero, then by Eq. (12) we have

$$\tilde{\mathbf{q}}_{k+1} = \tilde{\mathbf{r}}_k / \beta_k. \tag{15}$$

With the "canonical" choice such that $\tilde{\mathbf{Q}}^T \mathbf{D_s} \tilde{\mathbf{Q}} = \mathbf{I}$,

$$\beta_k = \|\mathbf{D_s}^{\frac{1}{2}} \tilde{\mathbf{r}}_k\|_2. \tag{16}$$

For any unit vector $\mathbf{q}_1$, let $\beta_0 = 1$, $\tilde{\mathbf{q}}_0 = 0$, and $\tilde{\mathbf{r}}_0 = \mathbf{D_s}^{-\frac{1}{2}} \mathbf{q}_1$. Start from $k = 1$, we could iteratively calculate the entries of α_k, β_k in $\mathbf{T}$ until $k = n_s$. Meanwhile, $\tilde{\mathbf{Q}}$ is also obtained during the iteration.

Spectral Calculation via Lanczos Process. Let $\mathbf{v}$ be the eigenvector of $\mathbf{N_{rw}}$ with the largest eigenvalue λ, we know

$$\mathbf{N_{rw}} \mathbf{v} = \mathbf{D_s}^{-1} \mathbf{A_s} \mathbf{v} = \lambda \mathbf{v}.$$

Premultiplying $\tilde{\mathbf{Q}}^T \mathbf{D_s}$ on both sides, we get

$$\tilde{\mathbf{Q}}^T \mathbf{A_s} \mathbf{v} = \lambda \tilde{\mathbf{Q}}^T \mathbf{D_s} \mathbf{v}. \tag{17}$$

According to Eq. (11), $\tilde{\mathbf{Q}} \tilde{\mathbf{Q}}^T \mathbf{D_s} = \mathbf{I}$. Then by Eq. (10), the left hand side of Eq. (17) equals

$$\tilde{\mathbf{Q}}^T \mathbf{A_s} \tilde{\mathbf{Q}} \tilde{\mathbf{Q}}^T \mathbf{D_s} \mathbf{v} = \mathbf{T} \tilde{\mathbf{Q}}^T \mathbf{D_s} \mathbf{v}.$$

Let $\mathbf{u} = \tilde{\mathbf{Q}}^T \mathbf{D_s} \mathbf{v}$, we get

$$\mathbf{T} \mathbf{u} = \lambda \mathbf{u}. \tag{18}$$

On the other hand, premultiplying $\tilde{\mathbf{Q}}$ and postmultiplying $\mathbf{u}$ on both sides of Eq. (8), we have

$$\mathbf{N_{rw}} \tilde{\mathbf{Q}} \mathbf{u} = \lambda \tilde{\mathbf{Q}} \mathbf{u},$$

so λ is also the largest eigenvalue of $\mathbf{T}$. As

$$\mathbf{v} = \tilde{\mathbf{Q}} \tilde{\mathbf{Q}}^T \mathbf{D_s} \mathbf{v} = \tilde{\mathbf{Q}} \mathbf{u}, \tag{19}$$

we can calculate $\mathbf{v}$ by calculating the eigenvector $\mathbf{u}$ of $\mathbf{T}$ with the largest eigenvalue λ.

Local Lanczos Spectral Approximation. Instead of using the eigenvalue decomposition to get the "global spectra", He *et al.* [14] use short random walks starting from the seed set to get a local proxy for the eigenvectors of $\mathbf{N_{rw}}$, which they call the "local spectra". Here we consider a novel way based on the Lanczos method [27] to approximate the eigenvector of $\mathbf{N_{rw}}$ with the largest eigenvalue. A few steps of the Lanczos iteration lead to the local approximation of this eigenvector.

Let $\mathbf{q_1}$ be the normalized indicator vector for the seed set, set $\beta_0 = 1$, $\tilde{\mathbf{q}}_0 = 0$, and $\tilde{\mathbf{r}}_0 = \mathbf{D_s}^{-\frac{1}{2}} \mathbf{q_1}$. By k steps of Lanczos iteration, we could get the first k by k submatrix of $\mathbf{T}$, denoted by $\mathbf{T_k}$. Correspondingly, let the first k columns of $\tilde{\mathbf{Q}}$ be a matrix $\tilde{\mathbf{Q}}_\mathbf{k}$. Let the eigenvectors of $\mathbf{T_k}$ with larger eigenvalues be a matrix $\mathbf{U_k}$. According to Eq. (19), the columns of $\mathbf{V_k} = \tilde{\mathbf{Q}}_\mathbf{k} \mathbf{U_k}$ approximate the eigenvectors of $\mathbf{N_{rw}}$ with larger eigenvalues. The first column of $\mathbf{V_k}$ approximates the eigenvector of $\mathbf{N_{rw}}$ with the largest eigenvalue, which is the indicator vector $\mathbf{y}$ we want to find.

The Local Lanczos Spectral Approximation (LLSA) procedure on the sampled graph is summarized in Algorithm 2. The slowest step is Step 4 for calculating $\tilde{\mathbf{q}}_\mathbf{k}, \alpha_k, \tilde{\mathbf{r}}_\mathbf{k}$, and β_k. It requires $O(Kn_s^2)$ time to implement the Lanczos iteration, where K is the steps of Lanczos iteration and n_s is number of nodes in graph G_s. Also, note that the Lanczos iteration requires only a few vectors of intermediate storage.

Algorithm 2. Local Lanczos Spectral Approximation

Input: $G_s = (V_s, E_s)$, maximum iteration steps K, initial vector $\mathbf{q_1}$

1: Initialize $k = 0, \beta_0 = 1, \mathbf{q_0} = \mathbf{0}, \tilde{\mathbf{r}}_0 = \mathbf{D_s}^{-\frac{1}{2}} \mathbf{q_1}$
2: **while** $(k < K)$ **do**
3:     $k = k + 1$
4:     Calculate $\tilde{\mathbf{q}}_\mathbf{k}, \alpha_k, \tilde{\mathbf{r}}_\mathbf{k}, \beta_k$ by Eqs. (13), (14), (15) and (16)
5: Let $\mathbf{T_k}$ be the first $k \times k$ entries of $\mathbf{T}$ in Eq. (9)
6: Get the eigenvector $\mathbf{u}$ of $\mathbf{T_k}$ with the largest eigenvalue λ
Output: $\mathbf{y} = \tilde{\mathbf{Q}}_\mathbf{k} \mathbf{u}$

Convergence Analysis. Here we provide an analysis on the convergence of the Lanczos process, i.e. the approximation gap between the local eigenvector which indicates the local structure around the seeds and the global eigenvector of the graph.

By Eqs. (12) and (14), we conclude that for $1 \leq k < n_s$,

$$\mathbf{N_{rw}}\tilde{\mathbf{Q}}_\mathbf{k} = \mathbf{D_s}^{-1}\mathbf{A_s}\tilde{\mathbf{Q}}_\mathbf{k} = \tilde{\mathbf{Q}}_\mathbf{k}\mathbf{T_k} + \tilde{\mathbf{r}}_\mathbf{k}\mathbf{e_k^T}, \tag{20}$$

where $\mathbf{e_k}$ is the kth unit vector with unity in the kth element and zero otherwise.

Let $\mathbf{u}$ be the eigenvector of $\mathbf{T_k}$ with the largest eigenvalue λ, postmultiplying $\mathbf{u}$ on both sides of Eq. (20), we have

$$\mathbf{N_{rw}}\tilde{\mathbf{Q}}_\mathbf{k}\mathbf{u} = \tilde{\mathbf{Q}}_\mathbf{k}\mathbf{T_k}\mathbf{u} + \tilde{\mathbf{r}}_\mathbf{k}\mathbf{e_k^T}\mathbf{u}. \tag{21}$$

Let $\mathbf{y} = \tilde{\mathbf{Q}}_\mathbf{k}\mathbf{u}$, the approximated residual value can be calculated as

$$\|\mathbf{r}\|_2 = \|\mathbf{N_{rw}}\mathbf{y} - \lambda\mathbf{y}\|_2 = \|\mathbf{N_{rw}}\tilde{\mathbf{Q}}_\mathbf{k}\mathbf{u} - \lambda\tilde{\mathbf{Q}}_\mathbf{k}\mathbf{u}\|_2. \tag{22}$$

By Eqs. (21), (22) can be modified as

$$\|\mathbf{r}\|_2 = \|\mathbf{N_{rw}}\tilde{\mathbf{Q}}_\mathbf{k}\mathbf{u} - \tilde{\mathbf{Q}}_\mathbf{k}\mathbf{T_k}\mathbf{u}\|_2 = \|\tilde{r}_\mathbf{k}\mathbf{e_k^T}\mathbf{u}\|_2. \tag{23}$$

Furthermore, by Eq. (15),

$$\|\mathbf{r}\|_2 = \|\beta_k\tilde{\mathbf{q}}_{\mathbf{k+1}}\mathbf{e_k^T}\mathbf{u}\|_2 = \|\beta_k\mathbf{D_s^{-\frac{1}{2}}}\mathbf{q_{k+1}}\mathbf{e_k^T}\mathbf{u}\|_2 = \beta_k|u_k| \cdot \|\mathbf{D_s^{-\frac{1}{2}}}\mathbf{q_{k+1}}\|_2, \tag{24}$$

where u_k is the kth (last) term of eigenvector $\mathbf{u}$.

As $\mathbf{q_{k+1}}$ is a unit vector, according to the Rayleigh-Ritz theorem [23],

$$\|\mathbf{D_s^{-\frac{1}{2}}}\mathbf{q_{k+1}}\|_2 \leq \max_{\|\mathbf{x}\|_2=1} \|\mathbf{D_s^{-\frac{1}{2}}}\mathbf{x}\|_2 = d_{min}^{-\frac{1}{2}}, \tag{25}$$

where d_{min} denotes the minimum degree of the nodes in graph G_s. $d_{min}^{-\frac{1}{2}}$ is also the largest eigenvalue of the diagonal matrix $\mathbf{D_s^{-\frac{1}{2}}}$.

By Eqs. (24) and (25), we have

$$\|\mathbf{r}\|_2 \leq \beta_k d_{min}^{-\frac{1}{2}}|u_k|. \tag{26}$$

Generally, the higher the value of k is, the smaller the residual value $\|\mathbf{r}\|_2$ is. And we need to use a small iteration step to find the "local" eigenvector. Experimental analysis in Sect. 4 shows a suitable value for the iteration step is around 4 or 5.

3.3   Community Boundary Truncation

The value of the kth element of $\mathbf{y}$ indicates how likely node k belongs to the target community. We use a heuristic similar to [35] to determine the community boundary.

We sort the nodes based on the element values of $\mathbf{y}$ in the decreasing order, and find a set S_{k^*} with the first k^* nodes having a comparatively low conductance. Specifically, we start from an index k_0 where set S_{k_0} contains all the seeds. We then generate a sweep curve $\Phi(S_k)$ by increasing index k. Let k^* be the value of k where $\Phi(S_k)$ achieves a first local minimum. The set S_{k^*} is regarded as the detected community.

We determine a local minima as follows. If at some point k^* when we are increasing k, $\Phi(S_k)$ stops decreasing, then this k^* is a candidate point for the local minimum. If $\Phi(S_k)$ keeps increasing after k^* and eventually becomes higher than $\alpha\Phi(S_{k^*})$, then we take k^* as a valid local minimum. We experimented with several values of α on a small trial of data and found that $\alpha = 1.03$ gives good performance across all the datasets.

The overall Local Lanczos Spectral Approximation (LLSA) algorithm is shown in Algorithm 3.

Algorithm 3. The overall LLSA algorithm

Input: $G = (V, E)$, seed set $S \subseteq V$
1: Get sampled subgraph $G_s = (V_s, E_s)$ by Algorithm 1
2: Calculate vector $\mathbf{y}$ by Algorithm 2
3: Sort nodes by the decreasing value of elements in $\mathbf{y}$
4: Find k_0 where S_k contains all the seeds
5: For $k = k_0 : n_s$, compute the conductance $\Phi(S_k)$: $\Phi_k = \Phi(S_k = \{v_i | i \leq k \text{ in the sorted list}\})$
6: Find k^* with the first local minimum $\Phi(S_{k^*})$
Output: Community $C = S_{k^*}$

---

4 Experiments

In this section, we compare LLSA with several state-of-the-art local community detection algorithms, and evaluate the performance by a popular F_1 metric.

4.1 Data Description

Seven synthetic datasets (parameters in Table 1) and five real-world datasets (Table 2) are considered for a comprehensive evaluation.

LFR Benchmark Graphs. Lancichinetti *et al.* [19, 20] proposed a method for generating LFR[1] benchmark graphs with a built-in binary community structure, which simulates properties of real-world networks on heterogeneity of node degree and community size distributions. The LFR benchmark graphs are widely used for evaluating community detection algorithms, and Xie *et al.* [34] performed a thorough performance comparison of different community detection algorithms on the LFR benchmark datasets.

We adopt the same set of parameter settings used in [34] and generate seven LFR benchmark graphs. Table 1 summarizes the parameter settings, among which the mixing parameter μ has a big impact on the network topology. μ controls the average fraction of neighboring nodes that do not belong to any community for each node. μ is usually set to be 0.1 or 0.3 and the detection accuracy usually decays for a larger μ. Each node belongs to either one community or *om* overlapping communities, and the number of nodes in overlapping communities is specified by *on*. A larger *om* or *on* indicates more overlaps on the communities, leading to a harder community detection task.

Real-World Networks. We choose five real-world network datasets with labeled ground truth from the SNAP[2], namely Amazon, DBLP, LiveJ, YouTube

---

[1] http://santo.fortunato.googlepages.com/inthepress2.
[2] http://snap.stanford.edu.

Table 1. Parameters for the LFR benchmarks.

Parameter	Description
$n = 5000$	Number of nodes in the graph
$\mu = 0.3$	Mixing parameter
$\bar{d} = 10$	Average degree of the nodes
$d_{max} = 50$	Maximum degree of the nodes
$[20, 100]$	Range of the community size
$\tau_1 = 2$	Node degree distribution exponent
$\tau_2 = 1$	Community size distribution exponent
$om \in \{2, 3..., 8\}$	Overlapping membership
$on = 500$	Number of overlapping nodes

Table 2. Statistics for real-world networks and their ground truth communities.

	Network			Ground truth communities	
Domain	Name	# Nodes	# Edges	Avg. ± Std. Size	Avg. Cond.
Product	Amazon	334,863	925,872	13 ± 18	0.073
Collaboration	DBLP	317,080	1,049,866	22 ± 201	0.414
Social	LiveJ	3,997,962	34,681,189	28 ± 58	0.388
Social	YouTube	1,134,890	2,987,624	21 ± 73	0.839
Social	Orkut	3,072,441	117,185,083	216 ± 321	0.731

and Orkut in the domains of product, collaboration and social contact [35]. Table 2 summarizes the statistics of the networks and the ground truth communities. We calculate the average and standard deviation of the community size, and the average conductance, where low conductance gives priority to communities with dense internal links and sparse external links.

4.2 Experimental Setup

We implement the proposed LLSA method in Matlab[3] through a C mex interface and conduct experiments on a computer with 2 Intel Xeon processors at 2.30 GHz and 128 GB memory. For the five SNAP datasets, we randomly locate 500 ground truth communities on each dataset, and randomly pick three exemplary seeds from each target community. For the seven LFR datasets, we deal with every ground truth community and randomly pick three exemplary seeds from each ground truth community. To make a fair comparison, we run all baseline algorithms using the same set of random seeds.

For the parameters, we fix $(t, \epsilon, N_1) = (3, 10^{-6}, 5000)$ for Algorithm 1 such that the resulting subgraph is large enough to cover almost all the members in

---

[3] https://github.com/PanShi2016/LLSA.

the target community. We set $K = 4$ for Algorithm 2 to have a good trade-off on real-world datasets as well as the synthetic data.

Comparison Baselines. We select three representative local community detection algorithms as the baselines. All algorithms accept as inputs an adjacency matrix $\mathbf{A}$ and a seed set S, and run on their default parameter settings. They apply different techniques to compute diffusion ranks starting from the seed set, then perform a sweep cut on the resulting ranks.

- **pprpush (PR)** [2]: the popular PageRank Nibble method.
- **hk-relax (HK)** [15]: the current best-performing heat kernel diffusion method.
- **LOSP** [14]: the current best-performing local spectral subspace based method.

Evaluation Metric. We adopt F_1 score to quantify the similarity between the detected local community C and the target ground truth community T. The F_1 score for each pair of (C, T) is defined by:

$$F_1(C,T) = \frac{2 \cdot P(C,T) \cdot R(C,T)}{P(C,T) + R(C,T)},$$

where the precision P and recall R are defined as:

$$P(C,T) = \frac{|C \cap T|}{|C|}, R(C,T) = \frac{|C \cap T|}{|T|}.$$

4.3 Experimental Results

Sampling. Table 3 shows the statistics for the heat kernel sampling on the real datasets. The sampled subgraphs are relatively small with 3200 nodes on average, only sampled about 0.3% of the nodes from the original graph. Nevertheless, there is a very high coverage ratio (ratio of ground truth nodes covered by the subgraph) of 96%, and the sampling procedure is very fast in less than 0.3 seconds. As for the LFR datasets, the sampling almost covers all the 5000 nodes as the synthetic networks are denser and much smaller.

Table 3. Statistics of the average values for the sampling.

Datasets	Coverage	n_s	n_s/n	Time (s)
Amazon	0.999	449	0.0013	0.016
DBLP	0.991	3034	0.0096	0.039
LiveJ	0.998	2639	0.0007	0.258
YouTube	0.919	4949	0.0044	0.437
Orkut	0.900	4990	0.0016	0.620
Average	0.961	3212	0.0035	0.274

Convergence Results. As LLSA involves the local Lanczos iteration, we experimentally investigate the convergence property of Algorithm 2 on two datasets: a synthetic network LFR for $om = 5$ and a real network YouTube. For the output of each iteration, we calculate the residual value $\|\mathbf{r}\|_2 = \|\mathbf{N}_{rw}\mathbf{y} - \lambda\mathbf{y}\|_2$, as shown in Fig. 1. One can see that the output $\mathbf{y}$ of Algorithm 2 converges very quickly, indicating that the spectra becomes global for more than 10 iterations. To gain a "local" eigenvector indicating the implicit topology structure of the local region around the seeds, we set the iteration step $K = 4$ for Algorithm 2.

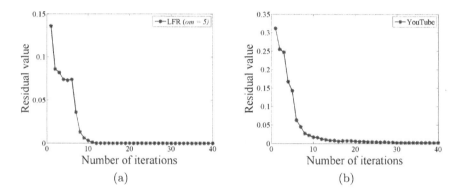

Fig. 1. Convergence analysis on LFR ($om = 5$) network and YouTube network.

Accuracy Comparison. Figure 2(a) illustrates the average detection accuracy of LLSA and the baselines on the LFR networks. LLSA significantly outperforms all baseline methods on all the seven synthetic networks. As $on = 500$ overlapping nodes are assigned to $om = 2$, 3, or 8 communities, a larger om makes the detection more difficult, leading to a lower accuracy.

Figure 2(b) illustrates the average detection accuracy on real-world networks. LLSA outperforms all baseline methods on Amazon, DBLP and LiveJ. LOSP performs the best on Youtube but is in the last place on Orkut; HK and PR show better performance on Orkut but behave poorly on Youtube. Though LLSA does not outperform all other methods on YouTube and Orkut, it is the most robust method and very competitive on average. As a whole, LLSA performs the best on the five real-world datasets.

Table 4 shows more comparisons on real-world datasets. Compared with LLSA and LOSP that finds a local minimal conductance, HK and PR seek for a global minimum conductance, and often find larger communities with lower conductance. On the other hand, as shown in Table 2, the ground truth communities are small with lower conductance for the first three datasets. This may explain why the four algorithms provide favorable results for the first three datasets but are adverse to YouTube and Orkut which have higher conductance, indicating many links to external nodes, hence lower conductance alone is not suited in finding the local, small communities. This may explain why HK and PR show better

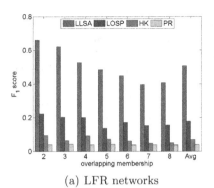

(a) LFR networks

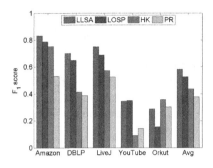

(b) Real-world networks

Fig. 2. Accuracy comparison on all datasets.

Table 4. Average conductance and size of the identified communities and average running time of the algorithms on real-world networks.

	Conductance				Size				Time (s)			
	LLSA	LOSP	HK	PR	LLSA	LOSP	HK	PR	LLSA	LOSP	HK	PR
Amazon	0.227	0.297	0.042	0.030	9	8	48	4485	0.045	0.040	0.008	0.015
DBLP	0.309	0.414	0.110	0.114	12	22	87	9077	1.038	0.546	0.025	0.075
LiveJ	0.243	0.419	0.083	0.086	43	29	119	512	1.191	1.132	0.029	0.264
YouTube	0.618	0.800	0.175	0.302	120	10	122	13840	1.572	2.896	0.038	0.955
Orkut	0.659	0.930	0.513	0.546	920	17	341	1648	4.352	2.662	0.027	1.392
Average	0.411	0.572	0.185	0.216	221	17	143	5912	1.640	1.455	0.025	0.540

performance on Orkut which contains communities with hundreds of nodes but behave poorly on Youtube with small community size. Table 4 shows that LOSP finds small communities with higher conductance, this may explain why LOSP performs the best on Youtube but is in the last place on Orkut.

For the running time, all algorithms are very fast and run in seconds. On average, HK is the fastest in 0.025 s, and the other three are similar in 0.5 to 1.6 s. LLSA and LOSP costs one more second as compared with PR, as they involve finding a community with the local minimal conductance. Also, different methods are implemented in different languages (LLSA and LOSP use Matlab, HK and PR use C++), so the running times could give an indication of the overall trend, and it can not be compared directly.

5   Conclusion

In this paper, we propose a novel Local Lanczos Spectral Approximation (LLSA) approach for local community detection, which is, to the best of our knowledge, the first time to apply Lanczos for local community detection. The favorable results on the synthetic LFR datasets and the real-world SNAP datasets suggest that the Lanczos method could be a new and effective way to detect local communities in large graphs. Based on Rayleigh quotient and conductance, we provide theoretical base for the proposed method. In addition, we also utilize the very fast heat kernel diffusion to get a local sampled subgraph that largely reduces the complexity of the subsequent computation. We wish our work inspire more researches based on the Lanczos method for network analysis and community detection.

Acknowledgments. The work is supported by NSFC (61772219, 61472147), US Army Research Office (W911NF-14-1-0477), and MSRA Collaborative Research (97354136).

References

1. Ahn, Y.Y., Bagrow, J.P., Lehmann, S.: Link communities reveal multiscale complexity in networks. Nature **466**(7307), 761–764 (2010)
2. Andersen, R., Chung, F., Lang, K.: Local graph partitioning using pagerank vectors. In: FOCS, pp. 475–486 (2006)
3. Andersen, R., Lang, K.J.: Communities from seed sets. In: WWW, pp. 223–232 (2006)
4. Andersen, R., Lang, K.J.: An algorithm for improving graph partitions. In: SODA, pp. 651–660 (2008)
5. Andersen, R., Peres, Y.: Finding sparse cuts locally using evolving sets. In: STOC, pp. 235–244 (2009)
6. Barnes, E.R.: An algorithm for partitioning the nodes of a graph. SIAM J. Algebraic Discrete Methods **3**(4), 303–304 (1982)
7. Bentbib, A.H., El Guide, M., Jbilou, K., Reichel, L.: A global Lanczos method for image restoration. J. Comput. Appl. Math. **300**, 233–244 (2016)
8. Chung, F.: The heat kernel as the pagerank of a graph. PNAS **104**(50), 19735–19740 (2007)
9. Chung, F., Simpson, O.: Solving linear systems with boundary conditions using heat kernel pagerank. In: Algorithms and Models for the Web Graph (WAW), pp. 203–219 (2013)
10. Chung, F.: A local graph partitioning algorithm using heat kernel pagerank. Internet Math. **6**(3), 315–330 (2009)
11. Chung, F.: Spectral Graph Theory, vol. 92. American Mathematical Society, Providence (1997)
12. Coscia, M., Giannotti, F., Pedreschi, D.: A classification for community discovery methods in complex networks. Stastical Anal. Data Min. **4**(5), 512–546 (2011)
13. Coscia, M., Rossetti, G., Giannotti, F., Pedreschi, D.: DEMON: a local-first discovery method for overlapping communities. In: KDD, pp. 615–623 (2012)
14. He, K., Sun, Y., Bindel, D., Hopcroft, J., Li, Y.: Detecting overlapping communities from local spectral subspaces. In: ICDM, pp. 769–774 (2015)
15. Kloster, K., Gleich, D.F.: Heat kernel based community detection. In: KDD, pp. 1386–1395 (2014)
16. Kloumann, I.M., Kleinberg, J.M.: Community membership identification from small seed sets. In: KDD, pp. 1366–1375 (2014)

17. Knight, P.A.: Fast rectangular matrix multiplication and QR decomposition. Linear Algebra Appl. **221**, 69–81 (1995)
18. Komzsik, L.: The Lanczos Method: Evolution and Application, vol. 15. SIAM, Philadelphia (2003)
19. Lancichinetti, A., Fortunato, S.: Benchmarks for testing community detection algorithms on directed and weighted graphs with overlapping communities. Phys. Rev. E **80**(1), 016118 (2009)
20. Lancichinetti, A., Fortunato, S., Radicchi, F.: Benchmark graphs for testing community detection algorithms. Phys. Rev. E **78**(4), 046110 (2008)
21. Lanczos, C.: An iteration method for the solution of the eigenvalue problem of linear differential and integral operators. J. Res. Nat. Bur. Stan. **45**, 255–282 (1950)
22. Li, Y., He, K., Bindel, D., Hopcroft, J.: Uncovering the small community structure in large networks. In: WWW, pp. 658–668 (2015)
23. Lütkepohl, H.: Handbook of Matrices, vol. 2. Wiley, Hoboken (1997)
24. von Luxburg, U.: A tutorial on spectral clustering. Stat. Comput. **17**(4), 395–416 (2007)
25. Mahoney, M.W., Orecchia, L., Vishnoi, N.K.: A local spectral method for graphs: with applications to improving graph partitions and exploring data graphs locally. J. Mach. Learn. Res. **13**(1), 2339–2365 (2012)
26. Orecchia, L., Zhu, Z.A.: Flow-based algorithms for local graph clustering. In: SODA, pp. 1267–1286 (2014)
27. Paige, C.: The computation of eigenvalues and eigenvectors of very large sparse matrices. Ph.D. thesis, University of London (1971)
28. Palla, G., Derenyi, I., Farkas, I., Vicsek, T.: Uncovering the overlapping community structure of complex networks in nature and society. Nature **435**(7043), 814–818 (2005)
29. Parlett, B.N., Poole Jr., W.G.: A geometric theory for the QR, LU and power iterations. SIAM J. Numer. Anal. **10**(2), 389–412 (1973)
30. Shi, J., Malik, J.: Normalized cuts and image segmentation. IEEE Trans. Pattern Anal. Mach. Intell. **22**(8), 888–905 (2000)
31. Spielman, D.A., Teng, S.: Nearly-linear time algorithms for graph partitioning, graph sparsification, and solving linear systems. In: STOC, pp. 81–90 (2004)
32. Stewart, G.W.: On the early history of the singular value decomposition. SIAM Rev. **35**(4), 551–566 (1993)
33. Wu, G., Xu, W., Leng, H.: Inexact and incremental bilinear Lanczos components algorithms for high dimensionality reduction and image reconstruction. Pattern Recogn. **48**(1), 244–263 (2015)
34. Xie, J., Kelley, S., Szymanski, B.K.: Overlapping community detection in networks: the state-of-the-art and comparative study. ACM Comput. Surv. (CSUR) **45**(4), 43 (2013)
35. Yang, J., Leskovec, J.: Defining and evaluating network communities based on ground-truth. In: ICDM, pp. 745–754 (2012)

Regularizing Knowledge Graph Embeddings via Equivalence and Inversion Axioms

Pasquale Minervini[1(✉)], Luca Costabello[2], Emir Muñoz[1,2], Vít Nováček[1], and Pierre-Yves Vandenbussche[2]

[1] Insight Centre for Data Analytics, National University of Ireland, Galway, Ireland
{pasquale.minervini,emir.munoz,vit.novacek}@insight-centre.org
[2] Fujitsu Ireland Ltd., Galway, Ireland
{luca.costabello,emir.munoz,pierre-yves.vandenbussche}@ie.fujitsu.com

Abstract. Learning embeddings of entities and relations using neural architectures is an effective method of performing statistical learning on large-scale relational data, such as knowledge graphs. In this paper, we consider the problem of regularizing the training of neural knowledge graph embeddings by leveraging external background knowledge. We propose a principled and scalable method for leveraging equivalence and inversion axioms during the learning process, by imposing a set of model-dependent soft constraints on the predicate embeddings. The method has several advantages: *(i)* the number of introduced constraints does not depend on the number of entities in the knowledge base; *(ii)* regularities in the embedding space effectively reflect available background knowledge; *(iii)* it yields more accurate results in link prediction tasks over non-regularized methods; and *(iv)* it can be adapted to a variety of models, without affecting their scalability properties. We demonstrate the effectiveness of the proposed method on several large knowledge graphs. Our evaluation shows that it consistently improves the predictive accuracy of several neural knowledge graph embedding models (for instance, the MRR of TRANSE on WORDNET increases by 11%) without compromising their scalability properties.

1 Introduction

Knowledge graphs are graph-structured knowledge bases, where factual knowledge is represented in the form of relationships between entities: they are powerful instruments in search, analytics, recommendations, and data integration. This justified a broad line of research both from academia and industry, resulting in projects such as DBPEDIA (Auer et al. 2007), FREEBASE (Bollacker et al. 2007), YAGO (Suchanek et al. 2012), NELL (Carlson et al. 2010), and Google's Knowledge Graph and Knowledge Vault projects (Dong et al. 2014).

However, despite their size, knowledge graphs are often very far from being complete. For instance, 71% of the people described in FREEBASE have no known place of birth, 75% have no known nationality, and the coverage for less used relations can be even lower (Dong et al. 2014). Similarly, in DBPEDIA, 66% of

© Springer International Publishing AG 2017
M. Ceci et al. (Eds.): ECML PKDD 2017, Part I, LNAI 10534, pp. 668–683, 2017.
https://doi.org/10.1007/978-3-319-71249-9_40

the persons are also missing a place of birth, while 58% of the scientists are missing a fact stating what they are known for (Krompaß et al. 2015).

In this work, we focus on the problem of *predicting missing links* in large knowledge graphs, so to discover new facts about the world. In the literature, this problem is referred to as *link prediction* or *knowledge base population*: we refer to Nickel et al. (2016) for a recent survey on machine learning-driven solutions to this problem.

Recently, *neural knowledge graph embedding models* (Nickel et al. 2016) – neural architectures for embedding entities and relations in continuous vector spaces – have received a growing interest: they achieve state-of-the-art link prediction results, while being able to scale to very large and highly-relational knowledge graphs. Furthermore, they can be used in a wide range of applications, including entity disambiguation and resolution (Bordes et al. 2014), taxonomy extraction (Nickel et al. 2016), and query answering on probabilistic databases (Krompaß et al. 2014). However, a limitation in such models is that they only rely on existing facts, without making use of any form of background knowledge. At the time of this writing, how to efficiently leverage preexisting knowledge for learning more accurate neural knowledge graph embeddings is still an open problem (Wang et al. 2015).

Contribution – In this work, we propose a principled and scalable method for leveraging external background knowledge for regularising neural knowledge graph embeddings. In particular, we leverage background axioms in the form $p \equiv q$ and $p \equiv q^-$, where the former denotes that relations p and q are *equivalent*, such as in the case of relations PARTOF and COMPONENTOF, while the latter denotes that the relation p is the *inverse* of the relation q, such as in the case of relations PARTOF and HASPART. Such axioms are used for defining and imposing a set of model-dependent soft constraints on the relation embeddings during the learning process. Such constraints can be considered as regularizers, reflecting available prior knowledge on the distribution of embedding representations of relations.

The proposed method has several advantages: *(i)* the number of introduced constraints is independent on the number of entities, allowing it to scale to large and Web-scale knowledge graphs with millions of entities; *(ii)* relationships between relation types in the embedding space effectively reflect available background schema knowledge; *(iii)* it yields more accurate results in link prediction tasks than state-of-the-art methods; and *(iv)* it is a general framework, applicable to a variety of embedding models. We demonstrate the effectiveness of the proposed method in several link prediction tasks: we show that it consistently improves the predictive accuracy of the models it is applied to, without negative impact on their scalability properties.

2   Preliminaries

Knowledge Graphs – A knowledge graph is a graph-structured knowledge base, where factual information is stored in the form of relationships between entities. Formally, a knowledge graph $\mathcal{G} \triangleq \{\langle s, p, o \rangle\} \subseteq \mathcal{E} \times \mathcal{R} \times \mathcal{E}$ is a set of $\langle s, p, o \rangle$ triples, each consisting of a *subject* s, a *predicate* p and an *object* o, and encoding the statement "s has a relationship p with o". The subject and object $s, o \in \mathcal{E}$ are entities, $p \in \mathcal{R}$ is a relation type, and $\mathcal{E}, \mathcal{R}$ respectively denote the sets of all entities and relation types in the knowledge graph.

Example 1. Consider the following statement: *"Ireland is located in Northern Europe, and shares a border with the United Kingdom."* It can be expressed by the following triples:

Subject	Predicate	Object
IRELAND	LOCATEDIN	NORTHERN EUROPE
IRELAND	NEIGHBOROF	UNITED KINGDOM

A knowledge graph can be represented as a labelled directed multigraph, in which each triple is represented as an edge connecting two nodes: the source and target nodes represent the subject and object of the triple, and the edge label represents the predicate.

Knowledge graph adhere to the *Open World Assumption* (Hayes and Patel-Schneider 2014): a missing triple does not necessarily imply that the corresponding statement holds false, but rather that its truth value is *unknown*, *i.e.* it cannot be observed in the graph. For instance, the fact that the triple ⟨UNITED KINGDOM, NEIGHBOROF, IRELAND⟩ is missing from the graph in Example 1 does not imply that the United Kingdom does not share a border with Ireland, but rather that we do not know whether this statement is true or not.

Equivalence and Inversion Axioms – Knowledge graphs are usually endowed with additional background knowledge, describing classes of entities and their properties and characteristics, such as equivalence and symmetry. In this work, we focus on two types of logical axioms in the form $p \equiv q$ and $p \equiv q^-$, where $p, q \in \mathcal{R}$ are predicates.

A largely popular knowledge representation formalism for expressing schema axioms is the OWL 2 Web Ontology language (Schneider 2012). According to the OWL 2 RDF-based semantics, the axiom $p \equiv q$ implies that predicates p and q share the same property extension, *i.e.* if $\langle s, p, o \rangle$ is true then $\langle s, q, o \rangle$ is also true (and vice-versa). Similarly, the axiom $p \equiv q^-$ implies that the predicate q is the inverse of the predicate p, *i.e.* if $\langle s, p, o \rangle$ is true then $\langle o, q, s \rangle$ is also true (and vice-versa). It is possible to express that a predicate $p \in \mathcal{R}$ is *symmetric* by using the axiom $p \equiv p^-$. Such axioms can be expressed by the OWL 2 `owl:equivalentProperty` and `owl:inverseOf` constructs.

Example 2. Consider the following statement: *"The relation* LOCATEDIN *is the inverse of the relation* LOCATIONOF, *and the relation* NEIGHBOROF *is symmetric."* It can be encoded by the axioms LOCATEDIN $\equiv$ LOCATIONOF$^-$ and NEIGHBOROF $\equiv$ NEIGHBOROF$^-$.

Link Prediction – As mentioned earlier, real world knowledge graphs are often largely incomplete. *Link prediction* in knowledge graphs consists in identifying missing triples (facts) in order to discover new facts about a domain of interest. This task is also referred to as *knowledge base population* in literature. We refer to Nickel et al. (2016) for a recent survey on link prediction methods.

The link prediction task can be cast as a *learning to rank* problem, where we associate a *prediction score* ϕ_{spo} to each triple $\langle s, p, o \rangle$ as follows:

$$\phi_{spo} \triangleq \phi(\langle s, p, o \rangle; \Theta),$$

where the score ϕ_{spo} represents the confidence of the model that the statement encoded by the triple $\langle s, p, o \rangle$ holds true, $\phi(\cdot; \Theta)$ denotes a *triple scoring function*, with $\phi : \mathcal{E} \times \mathcal{R} \times \mathcal{E} \to \mathbb{R}$, and Θ represents the parameters of the scoring function and thus of the link prediction model. Triples associated with a higher score by the link prediction model have a higher probability of encoding a true statement, and are thus considered for a completion of the knowledge graph $\mathcal{G}$.

3   Neural Knowledge Graph Embedding Models

Recently, *neural* link prediction models received a growing interest (Nickel et al. 2016). They can be interpreted as simple multi-layer neural networks, where given a triple $\langle s, p, o \rangle$, its score $\phi(\langle s, p, o \rangle; \Theta)$ is given by a two-layer neural network architecture, composed by an *encoding layer* and a *scoring layer*.

Encoding Layer – in the encoding layer, the subject and object entities s and o are mapped to distributed vector representations $\mathbf{e}_s$ and $\mathbf{e}_o$, referred to as *embeddings*, by an encoder $\psi : \mathcal{E} \mapsto \mathbb{R}^k$ such that $\mathbf{e}_s \triangleq \psi(s)$ and $\mathbf{e}_o \triangleq \psi(o)$. Given an entity $s \in \mathcal{E}$, the encoder ψ is usually implemented as a simple embedding layer $\psi(s) \triangleq [\boldsymbol{\Psi}]_s \in \mathbb{R}^k$, where $\boldsymbol{\Psi} \in \mathbb{R}^{|\mathcal{E}| \times k}$ is an embedding matrix (Nickel et al. 2016).
The distributed representations in this layer can be either pre-trained (Baroni et al. 2012) or, more commonly, learnt from data by back-propagating the link prediction error to the embeddings (Bordes et al. 2013; Yang et al. 2015; Trouillon et al. 2016; Nickel et al. 2016).
Scoring Layer – in the scoring layer, the subject and object representations $\mathbf{e}_s$ and $\mathbf{e}_o$ are scored by a predicate-dependent function $\phi_p^\theta(\mathbf{e}_s, \mathbf{e}_o) \in \mathbb{R}$, parametrised by θ.

The architecture of neural link prediction models can be summarized as follows:

$$\begin{aligned} \phi(\langle s, p, o \rangle; \Theta) &\triangleq \phi_p^\theta(\mathbf{e}_s, \mathbf{e}_o) \\ \mathbf{e}_s, \mathbf{e}_o &\triangleq \psi(s), \psi(o), \end{aligned} \tag{1}$$

and the set of parameters Θ corresponds to $\Theta \triangleq \{\theta, \boldsymbol{\Psi}\}$. Neural link prediction model generate distributed embedding representations for all entities in a knowledge graph, as well as a model of determining whether a triple is more likely than others, by means of a neural network architecture. For such a reason, they are also referred to as *neural knowledge graph embedding models* (Yang et al. 2015; Nickel et al. 2016).

Several neural link prediction models have been proposed in the literature. For brevity, we overview a small subset of these, namely the Translating Embeddings model TRANSE (Bordes et al. 2013); the Bilinear-Diagonal model DISTMULT (Yang et al. 2015); and its extension in the complex domain COM-PLEX (Trouillon et al. 2016). Unlike previous models, such models can scale to very large knowledge graphs, thanks to: (i) a space complexity that grows *linearly* with the number of entities $|\mathcal{E}|$ and relations $|\mathcal{R}|$; and (ii) efficient and scalable scoring functions and parameters learning procedures. In the following, we provide a brief and self-contained overview of such neural knowledge graph embedding models.

TRANSE – The scoring layer in TRANSE is defined as follows:

$$\phi_p(\mathbf{e}_s, \mathbf{e}_o) \triangleq -\|\mathbf{e}_s + \mathbf{r}_p - \mathbf{e}_o\| \in \mathbb{R},$$

where $\mathbf{e}_s, \mathbf{e}_o \in \mathbb{R}^k$ represent the subject and object embeddings, $\mathbf{r}_p \in \mathbb{R}^k$ is a predicate-dependent translation vector, $\| \cdot \|$ denotes either the L_1 or the L_2 norm, and $\|\mathbf{x} - \mathbf{y}\|$ denotes the distance between vectors $\mathbf{x}$ and $\mathbf{y}$. In TRANSE, the score $\phi_p(\mathbf{e}_s, \mathbf{e}_o)$ is then given by the *similarity* between the translated subject embedding $\mathbf{e}_s + \mathbf{r}_p$ and the object embedding $\mathbf{e}_o$.

DISTMULT – The scoring layer in DISTMULT is defined as follows:

$$\phi_p(\mathbf{e}_s, \mathbf{e}_o) \triangleq \langle \mathbf{r}_p, \mathbf{e}_s, \mathbf{e}_o \rangle \in \mathbb{R},$$

where, given $\mathbf{x}, \mathbf{y}, \mathbf{z} \in \mathbb{R}^k$, $\langle \mathbf{x}, \mathbf{y}, \mathbf{z} \rangle \triangleq \sum_{i=1}^{k} \mathbf{x}_i \mathbf{y}_i \mathbf{z}_i$ denotes the standard component-wise multi-linear dot product, and $\mathbf{r}_p \in \mathbb{R}^k$ is a predicate-dependent vector.

COMPLEX – The recently proposed COMPLEX is related to DISTMULT, but uses complex-valued embeddings while retaining the mathematical definition of the dot product. The scoring layer in COMPLEX is defined as follows:

$$
\begin{aligned}
\phi_p(\mathbf{e}_s, \mathbf{e}_o) &\triangleq \mathrm{Re}\left(\langle \mathbf{r}_p, \mathbf{e}_s, \overline{\mathbf{e}_o} \rangle\right) \\
&= \langle \mathrm{Re}\left(\mathbf{r}_p\right), \mathrm{Re}\left(\mathbf{e}_s\right), \mathrm{Re}\left(\mathbf{e}_o\right) \rangle + \langle \mathrm{Re}\left(\mathbf{r}_p\right), \mathrm{Im}\left(\mathbf{e}_s\right), \mathrm{Im}\left(\mathbf{e}_o\right) \rangle \\
&\quad + \langle \mathrm{Im}\left(\mathbf{r}_p\right), \mathrm{Re}\left(\mathbf{e}_s\right), \mathrm{Im}\left(\mathbf{e}_o\right) \rangle - \langle \mathrm{Im}\left(\mathbf{r}_p\right), \mathrm{Im}\left(\mathbf{e}_s\right), \mathrm{Re}\left(\mathbf{e}_o\right) \rangle \in \mathbb{R},
\end{aligned}
$$

where given $\mathbf{x} \in \mathbb{C}^k$, $\overline{\mathbf{x}}$ denotes the complex conjugate of $\mathbf{x}$[1], while $\mathrm{Re}\left(\mathbf{x}\right) \in \mathbb{R}^k$ and $\mathrm{Im}\left(\mathbf{x}\right) \in \mathbb{R}^k$ denote the real part and the imaginary part of $\mathbf{x}$, respectively.

4   Training Neural Knowledge Graph Embedding Models

In neural knowledge graph embedding models, the parameters Θ of the embedding and scoring layers are learnt from data. A widely popular strategy for

---

[1] Given $x \in \mathbb{C}$, its complex conjugate is $\overline{x} \triangleq \mathrm{Re}\left(x\right) - i\mathrm{Im}\left(x\right)$.

Algorithm 1. Learning the model parameters Θ via Projected SGD

Require: Batch size n, epochs τ, learning rates $\boldsymbol{\eta} \in \mathbb{R}^{\tau}$
Ensure: Optimal model parameters $\hat{\Theta}$
1: **for** $i = 1, \ldots, \tau$ **do**
2:     $\mathbf{e}_e \leftarrow \mathbf{e}_e / \|\mathbf{e}_e\|, \forall e \in \mathcal{E}$
3:     {Sample a batch of positive and negative examples $\mathcal{B} = \{(t, \tilde{t})\}$}
4:     $\mathcal{B} \leftarrow \text{SAMPLEBATCH}(\mathcal{G}, n)$
5:     {Compute the gradient of the loss function $\mathcal{J}$ on examples $\mathcal{B}$}
6:     $g_i \leftarrow \nabla \sum_{(t, \tilde{t}) \in \mathcal{B}} \left[\gamma - \phi(t; \Theta_{i-1}) + \phi(\tilde{t}; \Theta_{i-1}) \right]_+$
7:     {Update the model parameters via gradient descent}
8:     $\Theta_i \leftarrow \Theta_{i-1} - \eta_i g_i$
9: **end for**
10: **return** Θ_τ

---

learning the model parameters is described in Bordes et al. (2013); Yang et al. (2015); Nickel et al. (2016). In such works, authors estimate the optimal parameters by minimizing the following pairwise margin-based ranking loss function $\mathcal{J}$ defined on parameters Θ:

$$\mathcal{J}(\Theta) \triangleq \sum_{t^+ \in \mathcal{G}} \sum_{t^- \in \mathcal{C}(t^+)} \left[\gamma - \phi(t^+; \Theta) + \phi(t^-; \Theta) \right]_+ \qquad (2)$$

where $[x]_+ = \max\{0, x\}$, and $\gamma \geq 0$ specifies the width of the margin. Positive examples t^+ are composed by all triples in $\mathcal{G}$, and negative examples t^- are generated by using the following *corruption process*:

$$\mathcal{C}(\langle s, p, o \rangle) \triangleq \{\langle \tilde{s}, p, o \rangle \mid \tilde{s} \in \mathcal{E}\} \cup \{\langle s, p, \tilde{o} \rangle \mid \tilde{o} \in \mathcal{E}\},$$

which, given a triple, generates a set of corrupt triples by replacing its subject and object with all other entities in $\mathcal{G}$. This method of sampling negative examples is motivated by the *Local Closed World Assumption* (LCWA) (Dong et al. 2014). According to the LCWA, if a triple $\langle s, p, o \rangle$ exists in the graph, other triples obtained by corrupting either the subject or the object of the triples not appearing in the graph can be considered as negative examples. The optimal parameters can be learnt by solving the following minimization problem:

$$\begin{aligned} \underset{\Theta}{\text{minimize}} \quad & \mathcal{J}(\Theta) \\ \text{subject to} \quad & \forall e \in \mathcal{E} : \|\mathbf{e}_e\| = 1, \end{aligned} \qquad (3)$$

where Θ denotes the parameters of the model. The norm constraints on the entity embeddings prevent to trivially solve the optimization problem by increasing the norm of the embedding vectors (Bordes et al. 2014). The loss function in Eq. (2) will reach its global minimum 0 iff, for each pair of positive and negative examples t^+ and t^-, the score of the (true) triple t^+ is higher with a margin of at least γ than the score of the (missing) triple t^-. Following Yang et al. (2015), we use the Projected Stochastic Gradient Descent (SGD) algorithm (outlined in

Algorithm 1) for solving the loss minimization problem in Eq. (3), and AdaGrad (Duchi et al. 2011) for automatically selecting the optimal learning rate η at each iteration.

5   Regularizing via Background Knowledge

We now propose a method for incorporating background schema knowledge, provided in the form of equivalence and inversion axioms between predicates, in neural knowledge graph embedding models. Formally, let $\mathcal{A}_1$ and $\mathcal{A}_2$ denote the following two sets of equivalence and inversion axioms between predicates:

$$\mathcal{A}_1 \triangleq \{p_1 \equiv q_1, \ldots, p_m \equiv q_m\} \qquad \mathcal{A}_2 \triangleq \{p_{m+1} \equiv q_{m+1}^-, \ldots, p_n \equiv q_n^-\} \qquad (4)$$

where $1 \leq m \leq n$, and $\forall i \in \{1, \ldots, n\} : p_i, q_i \in \mathcal{R}$. Recall that each axiom $p \equiv q$ encodes prior knowledge that predicates p and q are equivalent, *i.e.* they share the same extension. Similarly, each axiom $p \equiv q^-$ encodes prior knowledge that the predicate p and the *inverse* of the predicate q are equivalent.

Equivalence Axioms – Consider the case in which predicates $p \in \mathcal{R}$ and $q \in \mathcal{R}$ are equivalent, as encoded by the axiom $p \equiv q$. This implies that a model with scoring function $\phi(\,\cdot\,; \Theta)$ and parameters Θ should assign the same scores to the triples $\langle s, p, o \rangle$ and $\langle s, q, o \rangle$, for all entities $s, o \in \mathcal{E}$:

$$\phi(\langle s, p, o \rangle; \Theta) = \phi(\langle s, q, o \rangle; \Theta) \quad \forall s, o \in \mathcal{E}. \qquad (5)$$

A simple method for enforcing the constraint in Eq. (5) during the parameter learning process consists in solving the loss minimization problem in Eq. (3) under the additional equality constraints in Eq. (5). However, this solution results in introducing $\mathcal{O}(|\mathcal{E}|^2)$ constraints in the optimization problem in Eq. (3), a quantity that grows *quadratically* with the number of entities $|\mathcal{E}|$. This solution may not be feasible for very large knowledge graphs, which typically contain millions of entities or more, while $|\mathcal{R}|$ is usually several orders of magnitude lower. A more efficient method consists in enforcing the model to associate *similar embedding representations* to both p and q, *i.e.* $\mathbf{r}_p = \mathbf{r}_q$. This solution can be encoded by *a single constraint*, satisfying all identities in Eq. (5).

Inversion Axioms – Consider the case in which the predicate p (*e.g.* PARTOF) and the inverse of the predicate q (*e.g.* HASPART) are equivalent, as encoded by the axiom $p \equiv q^-$. This implies that a model with scoring function $\phi(\,\cdot\,; \Theta)$ and parameters Θ should assign the same scores to the triples $\langle s, p, o \rangle$ and $\langle o, q, s \rangle$, for all entities $s, o \in \mathcal{E}$:

$$\phi(\langle s, p, o \rangle; \Theta) = \phi(\langle o, q, s \rangle; \Theta) \quad \forall s, o \in \mathcal{E}. \qquad (6)$$

Also in this case we can enforce the identity in Eq. (6) through a single constraint on the embeddings of predicates p and q. In the following, we derive the constraints for the models TRANSE, DISTMULT and COMPLEX. The constraints rely on a function $\Phi(\,\cdot\,)$ that applies a model-dependent transformation to the predicate embedding $\mathbf{r}_q$.

TransE: We want to enforce that, for any pair of s and o embedding vectors $\mathbf{e}_s, \mathbf{e}_o \in \mathbb{R}^k$, the score associated to the triples $\langle s, p, o \rangle$ and $\langle o, q, s \rangle$ are the same. Formally:

$$\| \mathbf{e}_s + \mathbf{r}_p - \mathbf{e}_o \| = \| \mathbf{e}_o + \mathbf{r}_q - \mathbf{e}_s \|, \quad \forall \mathbf{e}_s, \mathbf{e}_o \in \mathbb{R}^k \tag{7}$$

where $\| \cdot \|$ denotes either the L_1 or the L_2 norm.

Theorem 1. *The identity in Eq. (7) is satisfied by imposing:*

$$\mathbf{r}_p = \Phi(\mathbf{r}_q) \quad such\ that \quad \Phi(\mathbf{r}_q) \triangleq -\mathbf{r}_q.$$

Proof. *For any $\mathbf{e}_s, \mathbf{e}_o \in \mathbb{R}^k$, the following result holds:*

$$\| \mathbf{e}_s + \mathbf{r}_p - \mathbf{e}_o \| = \| \mathbf{e}_o - \mathbf{r}_p - \mathbf{e}_s \|,$$

where $\| \cdot \|$ is a norm on $\mathbb{R}^k$. Because of the absolute homogeneity property *of norms we have that, for any $\alpha \in \mathbb{R}$ and $\mathbf{x} \in \mathbb{R}^k$:*

$$\| \alpha \mathbf{x} \| = |\alpha| \| \mathbf{x} \|.$$

It follows that:

$$\begin{aligned}
\| \mathbf{e}_s + \mathbf{r}_p - \mathbf{e}_o \| &= \| -1 \,(\mathbf{e}_o - \mathbf{r}_p - \mathbf{e}_s) \| \\
&= |-1| \| \mathbf{e}_o - \mathbf{r}_p - \mathbf{e}_s \| \quad \textit{(absolute homogeneity property)} \\
&= \| \mathbf{e}_o - \mathbf{r}_p - \mathbf{e}_s \|.
\end{aligned}$$

DistMult: We want to enforce that:

$$\langle \mathbf{r}_p, \mathbf{e}_s, \mathbf{e}_o \rangle = \langle \mathbf{r}_q, \mathbf{e}_o, \mathbf{e}_s \rangle, \quad \forall \mathbf{e}_s, \mathbf{e}_o \in \mathbb{R}^k \tag{8}$$

A limitation in DistMult, addressed by ComplEx, is that its scoring function is *symmetric*, *i.e.* it assigns the same score to $\langle s, p, o \rangle$ and $\langle o, p, s \rangle$, due to the commutativity of the element-wise product.

The identity in Eq. (8) is thus satisfied by imposing $\mathbf{r}_p = \Phi(\mathbf{r}_q)$ such that $\Phi(\mathbf{r}_q) \triangleq \mathbf{r}_q$.

ComplEx: We want to enforce that:

$$\mathrm{Re}\left(\langle \mathbf{r}_p, \mathbf{e}_s, \overline{\mathbf{e}_o} \rangle \right) = \mathrm{Re}\left(\langle \mathbf{r}_q, \mathbf{e}_o, \overline{\mathbf{e}_s} \rangle \right), \quad \forall \mathbf{e}_s, \mathbf{e}_o \in \mathbb{C}^k. \tag{9}$$

The identity in Eq. (9) can be satisfied as follows:

Theorem 2. *The identity in Eq. (9) is satisfied by imposing:*

$$\mathbf{r}_p = \Phi(\mathbf{r}_q) \quad such\ that \quad \Phi(\mathbf{r}_q) \triangleq \overline{\mathbf{r}_q}.$$

Proof. *For any* $\mathbf{e}_s, \mathbf{e}_o \in \mathbb{C}^k$, *the following result holds:*

$$Re\left(\langle \mathbf{r}_p, \mathbf{e}_s, \overline{\mathbf{e}_o}\rangle\right) = Re\left(\langle \overline{\mathbf{r}_p}, \mathbf{e}_o, \overline{\mathbf{e}_s}\rangle\right).$$

Consider the following steps:

$$
\begin{aligned}
Re\left(\langle \mathbf{r}_p, \mathbf{e}_s, \overline{\mathbf{e}_o}\rangle\right) &= Re\left(\overline{\langle \overline{\mathbf{r}_p}, \overline{\mathbf{e}_s}, \mathbf{e}_o\rangle}\right) && (since \overline{\overline{\mathbf{x}}} = \mathbf{x}) \\
&= Re\left(\overline{\langle \overline{\mathbf{r}_p}, \mathbf{e}_o, \overline{\mathbf{e}_s}\rangle}\right) && (commutative\ property) \\
&= Re\left(\langle \overline{\mathbf{r}_p}, \mathbf{e}_o, \overline{\mathbf{e}_s}\rangle\right) && (since Re\left(\overline{\mathbf{x}}\right) = Re\left(\mathbf{x}\right)).
\end{aligned}
$$

Similar procedures for deriving the function $\Phi(\cdot)$ can be used in the context of other knowledge graph embedding models.

5.1   Regularizing via Soft Constraints

One solution for integrating background schema knowledge consists in solving the loss minimization problem in Eq. (3) under additional hard equality constraints on the predicate embeddings, for instance by enforcing $\mathbf{r}_p = \mathbf{r}_q$ for all $p \equiv q \in \mathcal{A}_1$, and $\mathbf{r}_p = \Phi(\mathbf{r}_q)$ for all $p \equiv q^- \in \mathcal{A}_2$. However, this solution does not cover cases in which two predicates are not strictly equivalent but still share very similar semantics, such as in the case of predicates MARRIEDWITH and PARTNEROF.

A more flexible solution consists in relying on *soft constraints* (Meseguer et al. 2006), which are used to formalize *desired properties* of the model rather than requirements that cannot be violated: we propose relying on weighted soft constraints for encoding our background knowledge on latent predicate representations.

Formally, we extend the loss function $\mathcal{J}$ described in Eq. (2) with an additional penalty term $\mathcal{R}_\mathcal{S}$ for enforcing a set of desired relationships between the predicate embeddings. This process leads to the following novel loss function $\mathcal{J}_\mathcal{S}$:

$$
\begin{aligned}
\mathcal{R}_\mathcal{S}(\Theta) &\triangleq \sum_{p \equiv q \in \mathcal{A}_1} D\left[\mathbf{r}_p \| \mathbf{r}_q\right] + \sum_{p \equiv q^- \in \mathcal{A}_2} D\left[\mathbf{r}_p \| \Phi(\mathbf{r}_q)\right] \\
\mathcal{J}_\mathcal{S}(\Theta) &\triangleq \mathcal{J}(\Theta) + \lambda \mathcal{R}_\mathcal{S}(\Theta),
\end{aligned}
\tag{10}
$$

where $\lambda \geq 0$ is the weight associated with the soft constraints, and $D\left[\mathbf{x} \| \mathbf{y}\right]$ is a divergence measure between two vectors $\mathbf{x}$ and $\mathbf{y}$. In our experiments, we use the Euclidean distance as divergence measure, *i.e.* $D\left[\mathbf{x} \| \mathbf{y}\right] \triangleq \|\mathbf{x} - \mathbf{y}\|_2^2$.

In particular, $\mathcal{R}_\mathcal{S}$ in Eq. (10) can be thought of as a schema-aware *regularization term*, which encodes our prior knowledge on the distribution of predicate embeddings. Note that the formulation in Eq. (10) allows us to freely interpolate between *hard constraints* ($\lambda = \infty$) and the original models represented by the loss function $\mathcal{J}$ ($\lambda = 0$), permitting to adaptively specify the relevance of each logical axiom in the embedding model.

6   Related Works

How to effectively improve neural knowledge graph embeddings by making use of background knowledge is a largely unexplored field. Chang et al. (2014); Krompass et al. (2014); Krompaß et al. (2015) make use of type information about entities for only considering interactions between entities belonging to the domain and range of each predicate, assuming that type information about entities is complete. In Minervini et al. (2016), authors assume that type information can be incomplete, and propose to adaptively decrease the score of each missing triple depending on the available type information. These works focus on type information about entities, while we propose a method for leveraging background knowledge about relation types which can be used jointly with the aforementioned methods.

Dong et al. (2014); Nickel et al. (2014); Wang et al. (2015) propose combining observable patterns in the form of rules and latent features for link prediction tasks. However, rules are not used *during* the parameters learning process, but rather *after*, in an ensemble fashion. Wang et al. (2015) suggest investigating how to incorporate logical schema knowledge during the parameters learning process as a future research direction. Rocktäschel et al. (2015) regularize relation and entity representations by grounding first-order logic rules. However, as they state in their paper, adding a very large number of ground constraints does not scale to domains with a large number of entities and predicates.

In this work we focus on *2-way* models rather than *3-way* models (García-Durán et al. 2014), since the former received an increasing attention during the last years, mainly thanks to their scalability properties (Nickel et al. 2016). According to García-Durán et al. (2014), 3-way models such as RESCAL (Nickel et al. 2011; 2012) are more prone to overfitting, since they typically have a larger number of parameters. It is possible to extend the proposed model to RESCAL, whose score for a $\langle s, p, o \rangle$ triple is $\mathbf{e}_s^T \mathbf{W}_p \mathbf{e}_o$. For instance, it is easy to show that $\mathbf{e}_s^T \mathbf{W}_p \mathbf{e}_o = \mathbf{e}_o^T \mathbf{W}_p^T \mathbf{e}_s$. However, extending the proposed method to more complex 3-way models, such as the latent factor model proposed by Jenatton et al. (2012) or the ER-MLP model (Dong et al. 2014) can be less trivial.

7   Evaluation

We evaluate the proposed schema-based soft constraints on three datasets: WORDNET, DBPEDIA and YAGO3. Each dataset is composed by a *training*, a *validation* and a *test* set of triples, as summarized in Table 1. All material needed for reproducing the experiments in this paper is available online[2].

WORDNET (Miller 1995) is a lexical knowledge base for the English language, where entities correspond to word senses, and relationships define lexical relations between them: we use the version made available by Bordes et al. (2013).

---

[2] At https://github.com/pminervini/neural-schema-regularization.

Table 1. Statistics for the datasets used in experiments

Dataset	$|\mathcal{E}|$	$|\mathcal{R}|$	#training	#validation	#test
WORDNET	40,943	18	141,442	5,000	5,000
DBPEDIA	32,510	7	289,825	5,000	5,000
YAGO3	123,182	37	1,079,040	5,000	5,000

YAGO3 (Mahdisoltani et al. 2015) is a large knowledge graph automatically extracted from several sources: our dataset is composed by facts stored in the YAGO3 CORE FACTS component of YAGO3.

DBPEDIA (Auer et al. 2007) is a knowledge base created extracting structured, multilingual knowledge from Wikipedia, and made available using Semantic Web and Linked Data standards. We consider a fragment extracted following the indications from Krompaß et al. (2014), by considering relations in the music domain[3].

The axioms we used in experiments are simple common-sense rules, and are listed in Table 1.

Evaluation Metrics – For evaluation, for each test triple $\langle s, p, o \rangle$, we measure the quality of the ranking of each test triple among all possible subject and object substitutions $\langle \tilde{s}, p, o \rangle$ and $\langle s, p, \tilde{o} \rangle$, with $\tilde{s}, \tilde{o} \in \mathcal{E}$. Mean Reciprocal Rank (MRR) and Hits@k as described by Bordes et al. (2013); Nickel et al. (2016); Trouillon et al. (2016) are widely adopted evaluation measures for evaluating knowledge graph completion algorithms. The measures are reported in the *raw* and *filtered* settings (Bordes et al. 2013). In the *filtered* setting, metrics are computed after removing all the other positive (true) triples that appear in either training, validation or test set from the ranking, whereas in the *raw* setting these are not removed. The filtered setting is motivated by observing that ranking a positive test triple after another true triple should not be considered a mistake (Bordes et al. 2013).

Evaluation Setting – In our experiments we consider three knowledge graph embedding models – TRANSE, COMPLEX and DISTMULT, as described in Sect. 3. For evaluating the effectiveness of the proposed method, we train them using both the standard loss function $\mathcal{J}$, defined in Eq. (2), and the proposed schema-aware loss function $\mathcal{J}_S$, defined in Eq. (10). Models trained by using the proposed method are denoted by the R superscript.

For each model and dataset, hyper-parameters were selected on the validation set by grid search. Specifically, we selected the embedding size $k \in \{20, 50, 100, 150\}$, the regularization weight $\lambda \in \{0, 10^{-4}, 10^{-2}, \ldots, 10^6\}$ and, in TRANSE, the norm $\| \cdot \|$ is selected across the L_1 and the L_2 norm.

---

[3] Following Krompass et al. (2014), such relations are ALBUM, ASSOCIATED BAND, ASSOCIATED MUSICAL ARTIST, GENRE, MUSICAL ARTIST, MUSICAL BAND, and RECORD-LABEL.

Table 2. Link prediction results (Hits@k and Mean Reciprocal Rank, filtered setting) on WORDNET, DBPEDIA and YAGO3.

	WordNet				DBpedia				YAGO3			
	Hits@N (%)			MRR	Hits@N (%)			MRR	Hits@N (%)			MRR
	3	5	10		3	5	10		3	5	10	
TRANSE	79.9	87.3	91.1	0.452	44.3	52.6	59.0	0.245	32.4	40.7	50.5	0.214
TRANSE^R	**86.9**	**91.6**	**93.3**	**0.566**	**47.8**	**54.0**	**60.0**	**0.256**	**33.4**	**42.5**	**52.0**	**0.248**
DISTMULT	91.7	93.2	94.2	0.840	44.6	**50.6**	55.7	0.371	**29.9**	**37.2**	46.3	0.260
DISTMULT^R	**92.4**	**93.8**	**94.9**	**0.851**	**44.9**	50.6	**55.8**	**0.381**	29.9	37.2	**46.4**	0.260
COMPLEX	94.2	94.4	94.6	0.939	52.7	54.2	55.8	0.486	**34.8**	41.5	49.9	0.304
COMPLEX^R	**94.3**	**94.5**	**94.7**	**0.940**	**53.1**	**54.3**	**55.9**	**0.503**	34.7	**41.6**	**50.0**	0.304

Similarly to Yang et al. (2015) we set the margin $\gamma = 1$ and, for each combination of hyper-parameters, we train each model for 1000 epochs. The learning rate in Stochastic Gradient Descent was initially set to 0.1, and then adapted during training by AdaGrad.

Results – We report test results in terms of raw and filtered Mean Reciprocal Rank (MRR), and filtered Hits@k in Table 2. For both the MRR and Hits@k metrics, the higher the results on the test set, the better.

We can see that, in every case, the proposed method – which relies on regularizing relation embeddings by leveraging background knowledge – improves the generalization abilities for each of the models. Results are especially evident for TRANSE, which largely benefits from the novel regularizer. For instance we can see that, in the WORDNET case, the Hits@10 improves from 91.1 to 93.3, while the Mean Reciprocal Rank improves from 0.452 to 0.566. For the remaining models we can only notice marginal improvements, probably because they already able to capture the patterns encoded by the background knowledge.

In Fig. 2 we can see a set of trained WORDNET predicate embeddings (using the model TRANSE), where relationships predicates are described in the axioms in Fig. 1. We can immediately see that, if $p \equiv q^-$, *i.e.* p is the inverse of q, then $\mathbf{r}_p \approx -\mathbf{r}_q$, which means that their embeddings $\mathbf{r}_p$ and $\mathbf{r}_q$ will be similar but will have opposite sign. On the left we set $\lambda = 0$, *i.e.* we do not enforce any soft constraint: we can see that the model is naturally inclined to assign opposite sign embeddings to relations such as PART OF and HAS PART, and HYPONYM and HYPERNYM; however, there is still some error margin in such an assignment, possibly due to the incompleteness of the knowledge graph. On the right we set $\lambda = 10^6$, *i.e.* we enforce the relationships between predicate embeddings via soft constraints: we can see that the aforementioned error margin in modeling the relationships between predicate embeddings is greatly reduced, improving the generalization properties of the model and establishing new state-of-the-art link prediction results on several datasets.

Axioms		
HAS PART	≡	PART OF⁻
HYPERNYM	≡	HYPONYM⁻
INSTANCE HYPERNYM	≡	INSTANCE HYPONYM⁻
M. HOLONYM	≡	S. MERONYM⁻
M. OF DOMAIN REGION	≡	S DOMAIN REGION OF⁻
M. OF DOMAIN TOPIC	≡	S. DOMAIN TOPIC OF⁻
M. OF DOMAIN USAGE	≡	S. DOMAIN USAGE OF⁻
DER. RELATED FORM	≡	DER. RELATED FORM⁻
VERB GROUP	≡	VERB GROUP⁻
ASSOC. BAND	≡	ASSOC. MUSICAL ARTIST
MUSICAL BAND	≡	MUSICALARTIST
ISMARRIEDTO	≡	ISMARRIEDTO⁻
HASNEIGHBOR	≡	HASNEIGHBOR⁻

Predicates embeddings (Real Part | Imaginary Part):

Predicate	Real Part					Imaginary Part				
hypernym	1.0	3.0	-3.1	2.5	-2.7	3.2	2.9	1.7	-3.0	-3.0
hyponym	1.0	3.1	-3.1	2.6	-2.7	-3.4	-2.8	-1.7	2.9	3.0
synset domain topic of	-3.1	-2.7	2.2	3.2	-2.4	-3.0	-1.7	-2.9	-2.8	2.6
member of domain topic	-3.1	-2.7	2.2	3.2	-2.5	2.8	1.7	2.9	2.9	-2.6
member of domain usage	-1.4	-0.1	-2.5	-3.4	2.7	-3.0	1.7	2.6	-0.6	-1.3
synset domain usage of	-1.2	-0.1	-2.3	-3.3	2.6	3.1	-1.8	-2.8	0.7	1.4
instance hypernym	-1.1	-2.8	1.6	2.7	-2.5	3.0	-2.6	2.7	-1.1	-2.8
instance hyponym	-1.0	3.0	1.5	2.9	-2.4	-2.9	2.8	-2.6	1.2	2.8
part of	-2.4	3.2	2.7	-1.5	3.0	-2.4	-0.6	-2.6	2.9	-1.9
has part	-2.5	3.2	2.9	-1.5	3.0	2.4	0.6	2.8	-3.0	1.9
member holonym	2.4	2.8	2.4	1.9	-2.4	2.9	-2.3	2.6	2.7	-2.4
member meronym	2.4	2.9	2.4	1.9	-2.3	-2.9	2.3	-2.5	-2.8	2.5
synset domain region of	-3.1	-0.3	3.1	-3.0	1.9	-0.9	2.0	-2.1	-1.2	1.0
member of domain region	-3.1	-0.3	3.2	-3.4	2.0	1.0	-2.1	2.2	1.3	-1.2
verb group	3.5	3.4	3.3	-1.8	-2.8	0.0	-0.1	0.0	0.0	0.0
derivationally related form	3.5	3.4	-3.2	3.4	3.2	0.0	0.0	-0.0	0.0	0.0

Fig. 1. Axioms used with WORDNET, DBPEDIA and YAGO3 (left) and WORDNET predicate embeddings learned by COMPLEX (right). Note that if $p \equiv q^-$ (e.g. PART OF and HAS PART) then $\mathbf{r}_p \approx \overline{\mathbf{r}_q}$, i.e. $\mathbf{r}_p$ and $\mathbf{r}_q$ have similar real parts and similar but opposite sign imaginary parts.

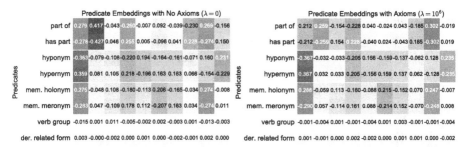

Fig. 2. WORDNET predicate embeddings learned using the TRANSE model, with $k = 10$ and regularization weight $\lambda = 0$ (left) and $\lambda = 10^6$ (right) – embeddings are represented as a heatmap, with values ranging from larger (red) to smaller (blue). Note that, assuming the axiom $p \equiv q^-$ holds, using the proposed method leads to predicate embeddings such that $\mathbf{r}_p \approx -\mathbf{r}_q$. (Color figure online)

A similar phenomenon in Fig. 1 (right), where predicated embeddings have been trained using COMPLEX: we can see that the model is naturally inclined to assign complex conjugate embeddings to inverse relations and, as a consequence, nearly-zero imaginary parts to the embeddings of symmetric predicates – since it is the only way of ensuring $\mathbf{r}_p \approx \overline{\mathbf{r}_p}$. However, we can

Table 3. Average number of seconds required for training.

	Plain	Regularized
WORDNET	31.7 s	32.0 s
DBPEDIA	57.9 s	58.5 s
YAGO3	220.7 s	221.3 s

enforce such relationships explicitly by means of model-specific regularizers, for increasing the predictive accuracy and generalization abilities of the models.

We also benchmarked the computational overhead introduced by the novel regularizers by timing the training time for unregularized (plain) models and for regularized ones – results are available in Table 3. We can see that the proposed method for leveraging background schema knowledge during the learning process adds a negligible overhead to the optimization algorithm – less than 10^{-1} s per epoch.

8   Conclusions and Future Works

In this work we introduced a novel and scalable approach for leveraging background knowledge into neural knowledge graph embeddings. Specifically, we proposed a set of background knowledge-driven regularizers on the relation embeddings, which effectively enforce a set of desirable algebraic relationships among the distributed representations of relation types. We showed that the proposed method improves the generalization abilities of all considered models, yielding more accurate link prediction results without impacting on the scalability properties of neural link prediction models.

Future Works

A promising research direction consists in leveraging more sophisticated background knowledge – *e.g.* in the form of First-Order Logic rules – in neural knowledge graph embedding models. This can be possible by extending the model in this paper to regularize over subgraph pattern embeddings (such as *paths*), so to leverage relationships between such patterns, rather than only between predicates. Models for embedding subgraph patterns have been proposed in the literature – for instance, see (Niepert 2016; Guu et al. 2015). For instance, it can be possible to enforce an equivalence between the path PARENTOF ∘ PARENTOF and GRANDPARENTOF, effectively incorporating a First-Order rule in the model, by regularizing over their embeddings.

Furthermore, a future challenge is also extending the proposed method to more complex models, such as ER-MLP (Dong et al. 2014), and investigating how to mine rules by extracting regularities from the latent representations of knowledge graphs.

Acknowledgements. This work was supported by the TOMOE project funded by Fujitsu Laboratories Ltd., Japan and Insight Centre for Data Analytics at National University of Ireland Galway (supported by the Science Foundation Ireland grant 12/RC/2289).

References

Auer, S., Bizer, C., Kobilarov, G., Lehmann, J., Cyganiak, R., Ives, Z.: DBpedia: a nucleus for a web of open data. In: Aberer, K., et al. (eds.) ASWC/ISWC -2007. LNCS, vol. 4825, pp. 722–735. Springer, Heidelberg (2007). https://doi.org/10.1007/978-3-540-76298-0_52

Baroni, M., Bernardi, R., Do, N-Q., Shan, C.: Entailment above the word level in distributional semantics. In: EACL, pp. 23–32. The Association for Computer Linguistics (2012)

Bollacker, K.D., Cook, R.P., Tufts, P.: Freebase: a shared database of structured general human knowledge. In: AAAI, pp. 1962–1963. AAAI Press (2007)

Bordes, A., Usunier, N., García-Durán, A., Weston, J., Yakhnenko, O.: Translating embeddings for modeling multi-relational data. In: NIPS, pp. 2787–2795 (2013)

Bordes, A., Glorot, X., Weston, J., Bengio, Y.: A semantic matching energy function for learning with multi-relational data - application to word-sense disambiguation. Mach. Learn. **94**(2), 233–259 (2014)

Carlson, A., Betteridge, J., Kisiel, B., Settles, B., Hruschka Jr., E.R., Mitchell, T.M.: Toward an architecture for never-ending language learning. In: AAAI. AAAI Press (2010)

Chang, K.-W., Yih, W., Yang, B., Meek, C.: Typed tensor decomposition of knowledge bases for relation extraction. In: EMNLP, pp. 1568–1579. ACL (2014)

Dong, X., Gabrilovich, E., Heitz, G., Horn, W., Lao, N., Murphy, K., Strohmann, T., Sun, S., Zhang, W.: Knowledge vault: a web-scale approach to probabilistic knowledge fusion. In: KDD, pp. 601–610. ACM (2014)

Duchi, J.C., Hazan, E., Singer, Y.: Adaptive subgradient methods for online learning and stochastic optimization. J. Mach. Learn. Res. **12**, 2121–2159 (2011)

García-Durán, A., Bordes, A., Usunier, N.: Effective blending of two and three-way interactions for modeling multi-relational data. In: ECML-PKDD, pp. 434–449 (2014)

Guu, K., Miller, J., Liang, P.: Traversing knowledge graphs in vector space. In: EMNLP, pp. 318–327. The Association for Computational Linguistics (2015)

Hayes, P., Patel-Schneider, P.: RDF 1.1 semantics. W3C recommendation, W3C, February 2014. http://www.w3.org/TR/2014/REC-rdf11-mt-20140225/

Jenatton, R., Roux, N.L., Bordes, A., Obozinski, G.: A latent factor model for highly multi-relational data. In: NIPS, pp. 3176–3184 (2012)

Krompass, D., Nickel, M., Tresp, V.: Large-scale factorization of type-constrained multi-relational data. In: DSAA, pp. 18–24. IEEE (2014)

Krompaß, D., Nickel, M., Tresp, V.: Querying factorized probabilistic triple databases. In: Mika, P., et al. (eds.) ISWC 2014. LNCS, vol. 8797, pp. 114–129. Springer, Cham (2014). https://doi.org/10.1007/978-3-319-11915-1_8

Krompaß, D., Baier, S., Tresp, V.: Type-constrained representation learning in knowledge graphs. In: Arenas, M., et al. (eds.) ISWC 2015. LNCS, vol. 9366, pp. 640–655. Springer, Cham (2015). https://doi.org/10.1007/978-3-319-25007-6_37

Mahdisoltani, F., Biega, J., Suchanek, F.M.: YAGO3: a knowledge base from multilingual wikipedias. In: CIDR (2015). www.cidrdb.org

Meseguer, P., Rossi, F., Schiex, T.: Soft constraints. In: Handbook of Constraint Programming, of Foundations of Artificial Intelligence, vol. 2, pp. 281–328. Elsevier (2006)

Miller, G.A.: WordNet: a lexical database for english. Commun. ACM **38**(11), 39–41 (1995)

Minervini, P., d'Amato, C., Fanizzi, N., Esposito, F.: Leveraging the schema in latent factor models for knowledge graph completion. In: SAC, pp. 327–332. ACM (2016)

Nickel, M., Tresp, V., Kriegel, H.-P.: A three-way model for collective learning on multi-relational data. In: ICML, pp. 809–816 (2011)

Nickel, M., Tresp, V., Kriegel, H.-P.: Factorizing YAGO: scalable machine learning for linked data. In: WWW, pp. 271–280. ACM (2012)

Nickel, M., Jiang, X., Tresp, V.: Reducing the rank in relational factorization models by including observable patterns. In: NIPS, pp. 1179–1187 (2014)

Nickel, M., Murphy, K., Tresp, V., Gabrilovich, E.: A review of relational machine learning for knowledge graphs. Proc. IEEE **104**(1), 11–33 (2016)

Niepert, M.: Discriminative Gaifman models. In: NIPS, pp. 3405–3413 (2016)

Rocktäschel, T., Singh, S., Riedel, S.: Injecting logical background knowledge into embeddings for relation extraction. In: HLT-NAACL, pp. 1119–1129. The Association for Computational Linguistics (2015)

Schneider, M.: OWL 2 web ontology language RDF-based semantics, 2nd edn. W3C recommendation, W3C, December 2012. http://www.w3.org/TR/2012/REC-owl2-rdf-based-semantics-20121211/

Suchanek, F.M., Kasneci, G., Weikum, G.: YAGO: a core of semantic knowledge. In: WWW, pp. 697–706. ACM (2007)

Trouillon, T., Welbl, J., Riedel, S., Gaussier, É., Bouchard, G.: Complex embeddings for simple link prediction. In: ICML, of JMLR Workshop and Conference Proceedings, vol. 48, pp. 2071–2080. JMLR. org (2016)

Wang, Q., Wang, B., Guo, L.: Knowledge base completion using embeddings and rules. In: IJCAI, pp. 1859–1866. AAAI Press (2015)

Yang, B., Yih, W-t., He, X., Gao, J., Deng, L.: Embedding entities and relations for learning and inference in knowledge bases. In: Proceedings of the International Conference on Learning Representations (ICLR) 2015, May 2015

Survival Factorization on Diffusion Networks

Nicola Barbieri[1], Giuseppe Manco[2], and Ettore Ritacco[2(✉)]

[1] Tumblr, 35 E 21st St., New York 10010, USA
nicola@tumblr.com
[2] ICAR - CNR, via Pietro Bucci 7/11C, 87036 Arcavacata di Rende, CS, Italy
{giuseppe.manco,ettore.ritacco}@icar.cnr.it

Abstract. In this paper we propose a survival factorization framework that models information cascades by tying together social influence patterns, topical structure and temporal dynamics. This is achieved through the introduction of a latent space which encodes: (a) the relevance of a information cascade on a topic; (b) the topical authoritativeness and the susceptibility of each individual involved in the information cascade, and (c) temporal topical patterns. By exploiting the cumulative properties of the survival function and of the likelihood of the model on a given adoption log, which records the observed activation times of users and side-information for each cascade, we show that the inference phase is linear in the number of users and in the number of adoptions. The evaluation on both synthetic and real-world data shows the effectiveness of the model in detecting the interplay between topics and social influence patterns, which ultimately provides high accuracy in predicting users activation times. Code and data related to this chapter are available at: https://doi.org/10.6084/m9.figshare.5411341.

Keywords: Social network analysis · Survival analysis
Information diffusion · Influence propagation · Adoption prediction

1 Introduction

An information cascade is a social process for adoptions, where the decision of each individual depends on the decision of people who have adopted the same content earlier. Such cascades have been identified in settings such as blogging, e-mail, product recommendation, and social Web platforms. The availability of large-scale, time-resolved cascade data on the social Web allows the study of interesting questions, such as: (i) How does information spread on networks? (ii) How far and fast does information flow? (iii) What is the network structure upon that allows the diffusion of information? (iv) How does the network structure affect information flow (and viceversa)? (v) How does the content being propagated affect the structure and shape of information cascades?

Electronic supplementary material The online version of this chapter (https://doi.org/10.1007/978-3-319-71249-9_41) contains supplementary material, which is available to authorized users.

© Springer International Publishing AG 2017
M. Ceci et al. (Eds.): ECML PKDD 2017, Part I, LNAI 10534, pp. 684–700, 2017.
https://doi.org/10.1007/978-3-319-71249-9_41

Understanding the structural, topical and temporal dynamics of information cascades can provide insights on the complex patterns that govern the information propagation process and it can be used to forecast future events. The problem of inferring the topical, temporal and network properties that characterize an observed set of information cascades is complicated by the fact that the diffusion network, transmission rates and the topical structure are hidden. Moreover, in many scenarios of interest for this paper, we are able to only observe cascades, having no information about the network structure (users' interconnections).

In this setting, to infer the diffusion network and the topical structure jointly, a natural approach is to model user's activation times as continuous random variables. Then, we can assume that those variables are generated by a stochastic process that depends on topical pairwise transmission rates $\lambda_{u,v}^k$, explaining the influence exerted by user v on u according to the topic k (see e.g. [18]). This approach has three main drawbacks: a large number of parameters (i.e. it's prone to overfitting); the parameter inference does not scale well; poor estimates when the episodes of information propagation from v to u are limited.

To address these issues, in this paper we introduce a stochastic model that factorizes pairwise transmission rates in terms of general user authoritativeness and susceptibility on a set of topics of interest. According to such a principle, both the side-information and temporal dynamics observed on a given information cascade are explained by 3 low-dimensional latent factors that encode: (i) the topical authority of each user $A_{v,k}$, (ii) the topical susceptibility $S_{u,k}$ and (iii) the relevance of side information w (e.g. hashtag) on topic k, $\varphi_{w,k}$.

The main contributions of this work can be summarized as follows.

- We review previous studies on information diffusion (Sect. 2) and briefly introduce a survival framework for modeling information diffusion (Sect. 3).
- Next, we introduce a factorization model (Sect. 3.1) that expresses topical pairwise transmission rates in terms of user authority and susceptibility, by coupling the topical content of a cascade and the observed activation times.
- We devise a highly scalable expectation maximization algorithms (Sect. 4) for the model parameter learning.
- We run an extensive evaluation (Sect. 5) on both synthetic and real-world data. We assess the capability of the model in detecting the interplay between the topical structure and temporal dynamics.

2   Related Work

Starting from seminal studies [9,13,21], the research on information diffusion has been mainly focused on determining how information spreads across pairs of users, observing the social network structure and the adoption log. A recent line of research [7,8] studies a different perspective, where the social network is not given in input, and the problem is how to uncover the hidden network structure starting from the log of users activity. This problem is addressed by assuming that infections follow a continuous-time independent cascade model. For example, in NetRate [7], if node u succeeds in activating v, then the contagion of the

Table 1. Comparison of the proposed method to the state of the art.

	Time	Req. Network	Inference	Side Info	Clustering
NetRate [7]	Contin.	No	$O(N^2)$	No	No
MONET [17]	Contin.	Yes	$O(N^2)$	Nodes	No
MMRate [18]	Contin.	No	$O(N^2)$	No	Cascades
CSDK [4]	Contin.	No	$O(NM)$	Cascades	No
LIS [19]	Discrete	No	$O(N^2)$	No	No
AIR [1]	Discrete	Yes	$O(N)$	Cascades	Nodes
CCN [2]	Discrete	Yes	$O(N)$	No	Nodes
CWN [3]	Both	No	$O(N)$	No	Nodes
Our method	Contin.	No	$O(N)$	Cascades	Cascades

latter happens after an incubation period sampled from a chosen distribution. According to this propagation model, the likelihood of a propagation cascade can be formulated by applying standard survival analysis [14]. Recent extensions of the survival diffusion process exploit Poisson [12] or Hawkes processes [5,22].

A different research line extends the diffusion process by considering enhancements based on features [17], or topics which characterize cascades [3,6,10,11, 18]. These models assume that the diffusion speed depends on node connections, features characterizing users and cascades, and node topical affinity [6,10,18].

Recent works have also focused on alternative ways of representing interactions between nodes, using latent-dimensional embedding techniques. In [4] authors propose a framework based on a *heat diffusion process* which projects each node into a latent space where the proximity between a pair of nodes reflects the proximity of their activations times in the observed cascades.

The approaches described so far do not explicitly consider the diffusion process as a result of the interaction between influence and susceptibility. In [1,3], the probability of activation is modeled as the effect of the influence of neighbor nodes within the cascades and/or the network. Further, the approaches [2,19] propose factorization techniques which associate two low-dimensional vectors to each node, representing influence and susceptibility. The propagation probability that one user forwards information depends on the product of her activated neighbors' influence vectors and her own susceptibility vector. The drawback of these approaches is that they only model cascades in a discrete-time scenario.

Table 1 compares the approach proposed in this work and some paradigmatic approaches mentioned above, by considering the following dimensions: modeling of time (continuous vs. discrete), whether they require as input the underlying network, complexity of the inference phase, modeling of side information, whether they are able to detect clustering structure. By denoting with N, M the number of nodes and cascades, we can see that all methods based on pairwise transmission rates suffer from the drawback of quadratic complexity in the learning phase. Thus, they do not scale to a large number of users and cascades.

By contrast, linear methods only model discrete time, and they do not necessarily model side information. To the best of our knowledge, our method is the only capable of combining the advantages of linear complexity and comprehensive modeling of temporal dynamics.

3   Modeling Information Diffusion

A cascade represents the propagation of a piece of information (news, post, meme, etc.) over a set of nodes (e.g., users of the system). We can specify each cascade as the activation times of a set of nodes $\mathcal{V}$ with cardinality N (i.e., $|\mathcal{V}| = N$). Formally, $\mathbf{t}^c$ can be represented as a N-dimensional vector $\mathbf{t}^c = (t_1(c), \cdots, t_N(c))$, where $t_u(c) \in [0, T^c] \cup \{\infty\}$ represents the timestamp when the node u becomes active on the cascade $\mathbf{t}^c$. For instance, if each cascade refers to the propagation of a meme, $t_u(c)$ will represent the timestamp at which user u reposted meme c. Without loss of generality, we can assume that each cascade starts at timestamp 0; moreover, $t_u(c) = \infty$ encodes the fact that the node u has not been infected during the observation window $[0, T^c]$. Let $\mathcal{V}^+(c)$ denote the set of active nodes on the cascade c (i.e., $t_u(c) \neq \infty$), while $\mathcal{V}^-(c) = \mathcal{V} \setminus \mathcal{V}^+(c)$ denotes the set of inactive nodes. The term N_c denotes the size of $\mathcal{V}^+(c)$.

Let $\mathbf{w}^c$ denote side information on the cascade c. We represent it as a bag-of-words $\mathbf{w}^c = \{w_1, \cdots, w_{len(c)}\}$, where each w_i is a word from a dictionary $\mathcal{W}$ and $len(c)$ is the number of words associated with the cascade $\mathbf{c}$. Finally, let $\mathcal{C} = \{(\mathbf{t}^1, \mathbf{w}^1) \cdots (\mathbf{t}^M, \mathbf{w}^M)\}$ denote a collection of M cascades over $\mathcal{V}$.

Propagation model. In our setting, we assume that *(i)* an event can trigger further events in the future, within the same cascade; *(ii)* events in different cascades are independent from each other. That is, a node v can trigger the activation of a node u on cascade c if and only if $t_v(c) < t_u(c)$. Hence, each cascade $\mathbf{t}^c$ defines a directed-acyclic graph, where $par_u(c) = \{v \in \mathcal{V} : t_v(c) < t_u(c)\}$. In the following we will use the notation $v \prec_c u$ to represent that v is a potential influencer for the activation of u within the cascade c, i.e. $v \in par_u(c)$.

Similar to the Independent Cascade model [13], we assume that node activations are *binary* (either active or inactive), *progressive* (an active node cannot turn inactive in the future) and all the parents try to infect their child nodes independently. Based on such assumptions, we can model each cascade by expressing the likelihood of activation times for active nodes and the likelihood that the adoption did not happen by time T^c for inactive nodes, according to a chosen propagation model.

Survival analysis for diffusion cascades. Let T denote a non-negative random variable representing the time of occurrence on an event. We can assume that for each pair of nodes (v, u) such that v triggered u's activation within the considered cascade c, there is a dependency between the respective activation times. Following [7], we formalize such dependency by introducing a conditional pairwise transmission likelihood $f(t_u(c)|t_v(c), \lambda_{v,u})$ which depends on the delay $\Delta_{u,v}^c = t_u(c) - t_v(c)$ between activation times and on the transmission rate $\lambda_{v,u}$.

Then, the likelihood of observing the activation times within a cascade can be formulated by applying a survival analysis framework [7]:

$$\Pr(\mathbf{t}^c|\Theta) = \prod_{u \in \mathcal{V}^-(c)} \prod_{v \in \mathcal{V}^+(c)} S(T^c - t_v(c); \lambda_{v,u}) \cdot$$

$$\prod_{u \in \mathcal{V}^+(c)} \prod_{v \prec_c u} S(\Delta_{u,v}^c; \lambda_{v,u}) \cdot \sum_{v' \prec_c u} h(\Delta_{u,v'}^c; \lambda_{v',u}), \tag{3.1}$$

where the *survival function* $S(t - t'; \lambda) = \Pr(T \geq t|t', \lambda) = 1 - \int_{t'}^{t} f(x|t', \lambda)dx$ encodes the probability that an event does not occur by time t and the *hazard function* $h(t - t'|\lambda) = \frac{f(t|t', \lambda)}{S(t - t'|\lambda)}$ is the rate of instantaneous infection at time t. Similarly, let W denote a random variable over words in $\mathcal{W}$; we can consider $\mathbf{w}^c$ as a collection of $len(c)$ i.i.d draws from a distribution Φ over $\mathcal{W}$:

$$\Pr(\mathbf{w}^c|\Phi) = \prod_{w \in \mathbf{w}^c} \Pr(w|\Phi).$$

3.1   Factorization Model

We start from the idea that the temporal dynamics, governing the activations of each node within observed cascades, depends on a set of *hidden* topics. The propagation of a piece of information depends inherently on its content and on pairwise transmission that are topic-dependent. The goal of our framework is to jointly factorize activation times and side information about each cascade to discover a finite set of K topics (where K is given as input), representing both a diffusion pattern and thematic information about the content.

This setting presents two challenges. First, in many practical scenarios we observe only node activations within a cascade, with no knowledge about what (or who) triggered them. Secondly, we observe side information and activation times of nodes within a set of cascades, but both the topical-structure and the relationships between topics and pairwise transmission likelihood are hidden.

To infer hidden topics and diffusion patterns we will introduce a generative process. As aforesaid, $\mathcal{C}$ is governed by a mixture of K underlying topics. Such a mixture is specified by introducing binary random variables $z_{c,k}$ which denote the membership of the cascade within each topic, with the constraint $\sum_{k=1}^{K} z_{c,k} = 1$. Let $\mathbf{Z}$ denote the overall $M \times K$ hidden topic assignments matrix. We characterize each topic k with the following 3 *non-negative* components:

- $A_{u,k}$, the authority degree of node u (i.e. tendency of triggering the activation of other nodes);
- $S_{u,k}$, the susceptibility degree of node u (i.e., tendency of being influenced by other nodes);
- $\varphi_{w,k}$, the relevance of word w.

Our factorization model is based on the assumption that the pairwise transmission rates within topic k can be factorized as a linear combination of users' authority and susceptibility components:

$$\lambda_{v,u,k} = A_{v,k} \cdot S_{u,k}. \tag{3.2}$$

The generation of a cascade unfolds as follows. First, we pick a topic z_c which specifies a topical-diffusion pattern, by drawing upon a multinomial distribution over topics $\Theta = \{\pi_1, \ldots, \pi_k\}$. Then, we adopt a *Poisson language model* [16] to generate the side-information by drawing the number of occurrences of each term w in the cascade c, shorted as $n_{w,c}$ from a *Poisson* distribution governed by the parameter set $\boldsymbol{\Phi}_k = \{\varphi_{w,k}\}_{w \in \mathcal{W}}$. Finally, the observed activation times within a cascade are generated according to a survival model. A summary of the conditional dependencies between latent and observed variables in our model is given in Fig. 1 and discussed below.

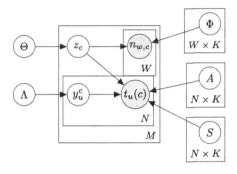

Fig. 1. Graphical model of survival factorization.

The modeling of activation times for each node in the cascade assumes that the delay between the influencer v and the influenced u ($t_v(c) < t_u(c)$) is generated accordingly to a *Weibull* distribution, whose scale parameter is the transmission rate, while the shape ρ is fixed:

$$f(t_u(c)|t_v(c), \lambda_{v,u,k}) = \mathcal{W}eib(\Delta^c_{u,v}; \lambda_{v,u,k}, \rho). \tag{3.3}$$

Here, $\mathcal{W}eib(t; \rho, \lambda) = \rho\lambda t^{\rho-1}e^{-\lambda t^{\rho}}$. Different choices of ρ correspond to different assumptions about the hazard: the hazard is rising if $\rho > 1$, constant if $\rho = 1$ (exponential model), and declining if $\rho < 1$. The corresponding survival and hazard functions are:

$$h(t; \lambda, \rho) = \rho\lambda t^{\rho-1}, \tag{3.4}$$

$$S(t; \lambda, \rho) = e^{-\lambda t^{\rho}}. \tag{3.5}$$

As stated above, we only observe activation times but not who triggered the activation. To model the hidden influencer for the activation of each node u within a cascade, we introduce latent binary variables $y^c_{u,v}$, with the constraint $\sum_{v \in \mathcal{V}} y^c_{u,v} = 1$. Let $\mathbf{Y}$ denote a $M \times N \times N$ binary matrix, where $y^c_{u,v} = 1$ represents the fact that node v triggered the activation of node u in the cascade c.

For each pair of users u and v, the prior probability that $y_{u,v}^c = 1$ is governed by a multinomial distribution Λ^1.

Given the status of the hidden variables $\mathbf{Z}$ and $\mathbf{Y}$, we can finally formalize the likelihood of observing the activation times within a cascade c:

$$\Pr(\mathbf{t}^c|\mathbf{Z}, \mathbf{Y}, \mathbf{A}_k, \mathbf{S}_k) = \prod_k \left(\prod_{u \in \mathcal{V}^-(c)} \prod_{v \in \mathcal{V}^+(c)} S(T^c - t_v(c); \lambda_{v,u,k}, \rho) \cdot \right.$$

$$\left. \prod_{u \in \mathcal{V}^+(c)} \prod_{v \prec_c u} h(\Delta_{u,v}^c; \lambda_{v,u,k}, \rho)^{y_{u,v}^c} \cdot S(\Delta_{u,v}^c; \lambda_{v,u,k}, \rho) \right)^{z_{c,k}} . \tag{3.6}$$

Finally, the overall likelihood of all cascades is:

$$\Pr(\{\mathbf{t}^1, \cdots, \mathbf{t}^M\}|\mathbf{Z}, \mathbf{Y}, \mathbf{A}, \mathbf{S}) = \prod_{c=1}^{M} \Pr(\mathbf{t}^c|\mathbf{Z}, \mathbf{Y}, \mathbf{A}, \mathbf{S}) .$$

Compared to the modeling in Eq. 3.1, the above model exhibits two main differences. First, cascade are characterized by a topic which also governs the propagation speed. Second, we explicitly model influencers by introducing the $\mathbf{Y}$ matrix. In fact, Eq. 3.6 is a refined extension of Eq. 3.1, since the latter can be obtained from the former by assuming $K = 1$ and marginalizing over $\mathbf{Y}$.

Likelihood of side-information. The probability of observing content $\mathbf{w}^c$ under topic k is given by the probability of observing the frequency count $n_{w,c}$ of each word. Within the *homogeneous Poisson model* [16], this frequency under topic k follows a Poisson distribution with parameter $\varphi_{w,k}$. The latter is the expected number of occurrences of w in a unit of time, and the time associated to the generation of side-information $\mathbf{w}^c$ is assumed to be $|\mathbf{w}^c| = len(c)$. Thus, according to this model, the likelihood of observing a bag-of-words $\mathbf{w}^c$ when the topic is k can be expressed as:

$$\Pr(\mathbf{w}^c|\mathbf{\Phi}_k) = \prod_w \frac{(|\mathbf{w}^c| \cdot \varphi_{w,k})^{n_{w,c}} \exp\{-|\mathbf{w}^c| \cdot \varphi_{w,k}\}}{n_{w,c}!} . \tag{3.7}$$

Since each cascade is generated independently from each other, the overall likelihood of side information over all cascades, given hidden topic-assignment $\mathbf{Z}$, can be expressed as:

$$\Pr(\{\mathbf{w}^1, \cdots, \mathbf{w}^M\}|\mathbf{\Phi}, \mathbf{Z}) = \prod_{c=1}^{M} \prod_k \Pr(\mathbf{w}^c|\mathbf{\Phi}_k)^{z_{c,k}} .$$

---

[1] In the next we shall assume that this distribution is uniform, i.e., each v has equal chances of activating u.

4   Inference and Parameter Estimation

Let $\varXi = \{\mathbf{A}, \mathbf{S}, \boldsymbol{\Phi}, \varLambda, \boldsymbol{\Theta}\}$ denote the status of parameters of the model. Given latent assignments $\mathbf{Z}$ and $\mathbf{Y}$, the conditional data likelihood is

$$\Pr(\mathcal{C}|\mathbf{Z}, \mathbf{Y}, \varXi) = \Pr(\{\mathbf{t}^1, \cdots, \mathbf{t}^M\}|\mathbf{Z}, \mathbf{Y}, \varXi) \cdot \Pr(\{\mathbf{w}^1, \cdots, \mathbf{w}^M\}|\mathbf{Z}, \varXi).$$

Thus, the optimal values for $\varXi$ can be obtained by optimizing the likelihood

$$\Pr(\mathcal{C}, \varXi) = \sum_{\mathbf{Z}, \mathbf{Y}} \Pr(\mathcal{C}|\mathbf{Z}, \mathbf{Y}, \varXi) \Pr(\mathbf{Z}, \mathbf{Y}, \varXi). \tag{4.1}$$

Exact inference is intractable, and we have to resort to heuristic optimization strategies. It turns out that the Expectation Maximization algorithm can be easily adapted for estimating the optimal parameters. That is, it is easy to devise an iterative alternating strategy consisting of the following two steps:

E step: estimate the posterior $\Pr(\mathbf{Z}, \mathbf{Y}|\mathcal{C}, \boldsymbol{\varXi}^{(n-1)})$
M step: exploit the posterior to solve

$$\boldsymbol{\varXi}^{(n)} = \underset{\varXi}{\mathrm{argmax}} \sum_{\mathbf{Z}, \mathbf{Y}} \Pr(\mathbf{Z}, \mathbf{Y}|\mathcal{C}, \boldsymbol{\varXi}^{(n-1)}) \cdot \log \Pr(\mathcal{C}, \mathbf{Z}, \mathbf{Y}, \varXi)$$

Both steps are tractable and the estimation produces closed formulas. The details of the derivations can be found in the appendix submitted as supplemental material.

In particular, for the E step the estimation of $\Pr(\mathbf{Z}, \mathbf{Y}|\mathcal{C}, \boldsymbol{\varXi}^{(n)})$ can be decomposed into the specific components, thus yielding

$$\Pr(z_{c,k}, y_{u,v}^c | \mathbf{t}^c, \mathbf{w}^c, \varXi) = \eta_{c,u,v}^k \cdot \gamma_{c,k},$$

where

$$\eta_{c,u,v}^k = \frac{h(\Delta_{u,v}^c; \lambda_{v,u,k}, \rho)}{\sum_{v' \prec_c u} h(\Delta_{u,v'}^c; \lambda_{v',u,k}, \rho)}, \tag{4.2}$$

$$\gamma_{c,k} = \frac{\Pr(\mathbf{t}^c|\mathbf{A}_k, \mathbf{S}_k) \Pr(\mathbf{w}^c|\boldsymbol{\Phi}_k)\pi_k}{\sum_k \Pr(\mathbf{t}^c|\mathbf{A}_k, \mathbf{S}_k) \Pr(\mathbf{w}^c|\boldsymbol{\Phi}_k)\pi_k}. \tag{4.3}$$

Here, $\gamma_{c,k}$ represents the posterior probability that cascade c is relative to topic k, and $\eta_{c,u,v}^k$ the posterior probability that the activation of u was triggered by v within topic k. The component $\Pr(\mathbf{w}^c|\boldsymbol{\Phi}_k)$ is specified by Eq. 3.7, and $\Pr(\mathbf{t}^c|\mathbf{A}_k, \mathbf{S}_k)$ is obtained by marginalizing $\Pr(\mathbf{t}^c|z_c, \mathbf{Y}^c, \mathbf{A}, \mathbf{S})$ in 3.6 with respect to $\mathbf{Y}$.

For the M step, by plugging η and γ into the expected log-posterior we can solve the optimization step with regards to all the available parameters. In particular, optimal values for $\boldsymbol{\Theta}$ and $\boldsymbol{\Phi}$ can be obtained directly:

$$\pi_k = \frac{1}{M} \sum_c \gamma_{c,k} \tag{4.4}$$

$$\varphi_{w,k} = \frac{\sum_c \gamma_{c,k} n_{w,c}}{\sum_c \gamma_{c,k} |\mathbf{w}^c|} \tag{4.5}$$

Concerning $\mathbf{A}$ and $\mathbf{S}$, the expected likelihood expresses an interdependency which can be resolved by block coordinate ascent optimization:

$$S_{u,k} = \frac{\sum_{c:u\in\mathcal{V}+(c)} \gamma_{c,k}}{\sum_{c=1}^M \sum_{v\prec_c u} \gamma_{c,k} \cdot (\Delta_{u,v}^c)^\rho \cdot A_{v,k}} \tag{4.6}$$

$$A_{v,k} = \frac{\sum_{c:v\in\mathcal{V}+(c)} \sum_{\substack{u\in\mathcal{V}+(c)\\v\prec_c u}} \eta_{c;u,v}^k \cdot \gamma_{c,k}}{\sum_{c:v\in\mathcal{V}+(c)} \sum_u \gamma_{c,k} \cdot (\Delta_{u,v}^c)^\rho \cdot S_{u,k}} \tag{4.7}$$

We deliberately choose not to optimize the ρ parameter, and to investigate the case $\rho = 1$.

Table 2. Counters on the cascades.

Term	Definition	Term	Definition
$A_{c,u,k}$	$\sum_{v\prec_c u} A_{v,k}$	$S_{c,u,k}$	$\sum_{v\preceq_c u} S_{v,k}$
$\tilde{A}_{c,u,k}$	$\sum_{v\prec_c u} t_v(c) A_{v,k}$	$\tilde{S}_{c,u,k}$	$\sum_{v\preceq_c u} t_v(c) S_{v,k}$
$A_{c,k}$	$\sum_{v\in\mathcal{V}+(c)} A_{v,k}$	$S_{c,k}$	$\sum_{v\in\mathcal{V}+(c)} S_{v,k}$
$\tilde{A}_{c,k}$	$\sum_{v\in\mathcal{V}+(c)} t_v(c) A_{v,k}$	$\tilde{S}_{c,k}$	$\sum_{v\in\mathcal{V}+(c)} t_v(c) S_{v,k}$
$R_{c,u,k}$	$\sum_{\substack{v\in\mathcal{V}+(c)\\u\prec_c v}} (A_{c,v,k})^{-1}$	S_k	$\sum_v S_{v,k}$
$\tilde{R}_{c,v,k}$	$\sum_{\substack{u\in\mathcal{V}+(c)\\v\prec_c u}} t_u(c) \left(t_u(c) A_{c,u,k} - \tilde{A}_{c,u,k}\right)^{-1}$	$L_{c,k}$	$\sum_{v\in\mathcal{V}+(c)} \log S_{v,k}$

Scaling up the estimation

When $\rho = 1$, the Weibull distribution simplifies to an exponential distribution. In such a case, we can introduce the counters described in Table 2 and rewrite the update equations for $\mathbf{A}$ and $\mathbf{S}$ as shown in Fig. 2 (see appendix (see footnote 2) for details). Algorithm 1 describes the overall procedure for estimating the parameters.

$$S_{u,k} = \frac{\sum_{c:u\in\mathcal{V}+(c)} \gamma_{c,k}}{\sum_{c:u\in\mathcal{V}+(c)} \gamma_{c,k} \left(t_u(c) A_{c,u,k} - \tilde{A}_{c,u,k}\right) + \sum_{c:u\in\mathcal{V}-(c)} \gamma_{c,k} \left(T^c A_{c,k} - \tilde{A}_{c,k}\right)} \tag{4.8}$$

$$A_{v,k} = \frac{A_{v,k}^{(n-1)} \sum_{c:v\in\mathcal{V}+(c)} \gamma_{c,k} R_{c,v,k}}{\sum_{c:v\in\mathcal{V}+(c)} \gamma_{c,k} \left\{ \begin{array}{l} \tilde{S}_{c,k} - \tilde{S}_{c,v,k} \\ + T^c(S_k - S_{c,k}) \\ - t_v(c)(S_k - S_{c,v,k}) \end{array} \right\}} \tag{4.9}$$

$$\log \Pr(\mathbf{t}^c|\mathbf{A}_k, \mathbf{S}_k) = L_{c,k} - (S_k - S_{c,k})(T^c A_{c,k} - \tilde{A}_{c,k})$$
$$+ \sum_{u\in\mathcal{V}+(c)} \left\{ \log A_{c,u,k} - S_{u,k} \left(t_u(c) A_{c,u,k} - \tilde{A}_{c,u,k}\right) \right\} \tag{4.10}$$

Fig. 2. Optimized estimations for the exponential distribution. All equations rely on counters defined in Table 2.

Algorithm 1. Optimized Survival Factorization EM
___
Require: $\mathcal{C}$, the number of latent features K
Ensure: matrices $\mathbf{A}$, $\mathbf{S}$ and $\boldsymbol{\Phi}$
 1: Randomly initialization for $\mathbf{A}$, $\mathbf{S}$, $\boldsymbol{\Phi}$;
 2: Compute all counters of Table 2;
 3: $n \leftarrow 0$
 4: **while** Increment in Likelihood is negligible **do**
 5:    **for all** cascades c and topic k **do**
 6:        Compute $\gamma_{c,k}$ exploiting $\log \Pr(\mathbf{t}^c | \mathbf{A}_k, \mathbf{S}_k)$ as defined in Eq. 4.10;
 7:    **end for**
 8:    **for all** topic k **do**
 9:        Update π_k according to Eq. 4.4;
10:        **for all** users u **do**
11:            Compute $S_{u,k}$ according to Eq. 4.8;
12:        **end for**
13:        Update all counters relative to $\mathbf{S}$ as defined in Table 2;
14:        **for all** users u **do**
15:            Compute $A_{u,k}$ according to Eq. 4.9;
16:        **end for**
17:        Update counters relative to $\mathbf{A}$ as defined in Table 2;
18:        **for all** words w **do**
19:            Compute $\phi_{w,k}$ according to Eq. 4.5;
20:        **end for**
21:    **end for**
22:    $n \leftarrow n + 1$
23: **end while**
___

Theorem 1. *Algorithm 1 has complexity* $O(\sum_c N_c \log N_c + nK(N + W + \sum_c N_c))$ *time (where n is the total number of iterations) and $O(KN)$ space.*

Proof. See appendix (see footnote 2). □

5 Evaluation

The following experimental evaluation is aimed at exploring the following aspects: (1) Investigate the conditions upon which the proposed method can correctly detect authoritativeness and susceptibility from propagation logs; (2) Evaluate the proposed models under two different prediction scenarios: (i) given a partially observed cascade, predict which nodes are more likely to become active within a fixed time window and (ii) inferring the underlying propagation network among nodes; (3) Assess the adequacy of the model at fitting real-world data and at identifying topical diffusion patterns.

To perform such analyses we rely on both synthetic and real data, as reported below. The implementation we we used in the experiments can be found at http://github.com/gmanco/SurvivalFactorization.

5.1 Synthetic Data

The first set of experiments is conducted in a controlled environment. We artificially generate the cascades by hypothesizing a diffusion process and measure the goodness-of-fit of the algorithm to the underlying process.

We base the generation on the assumption (studied, e.g., in [20]) that vertices are connected and the diffusion of information happens through the links of the underlying network. Thus, to generate synthesized data, we, firstly, build networks with a known community structure by varying connectivity structure of the network. To this aim, we borrow the synthetic networks studied in [3].

Given a network $G = (V, E)$, we next generate synthetic propagation cascades by simulating a propagation process which spreads over E. The process generates $|\mathcal{I}|$ propagation traces according to the following protocol. The degree of authoritativeness and susceptibility of each node in each community depend on its connectivity pattern. If the node u belongs to community k the values $A_{u,k}$ and $S_{u,k}$ are sampled from lognormal distributions with means $p \cdot \frac{indegree(u)}{\max_v indegree(v)} + (1-p) \cdot rand(0.1, 1)$ and $p \cdot (1 - \frac{outdegree(u)}{\max_v outdegree(v)}) + (1-p) \cdot rand(0.1, 1)$ respectively. For all the remaning communities $h \neq k$, the values for $A_{u,h}$ and $S_{u,h}$ are randomly sampled within a uniform range lower than $A_{u,k}$ $(S_{u,k})$ by an order of magnitude. The propagation cascades are generated exploiting $\mathbf{A}$ and $\mathbf{S}$: for each cascade to generate, we randomly sample a topic k and a maximal propagation horizon T_{max}. Then, we sample an initial node v with probability proportional to $A_{v,k}$. From this node we start the subsequent diffusion process. Given an active node u and a neighbor v, we sample a hypothetical infection time $t_{u,v}$ using t_v and the rate $A_{u,k} \cdot S_{v,k}$. Node v then becomes active if there exist an influencer u such that $t_{u,v} < T_{max}$. Finally, for each cascade we generate the content. For each topic k, we generate $\varphi_{w,k}$ randomly and then draw word-frequencies according to the Poisson model and to the topic of the cascade.

In the following experiments, we set $p = 0.9$, $|\mathcal{I}| = 2{,}048$ and run the generation of cascades on 4 networks, with different degrees of overlapping. The main properties of the synthesized data are summarized in Table 3.

Table 3. Statistics for the synthesized cascades.

	S1	S2	S3	S4
Communities	9	7	11	6
Activations	215,608	275,633	171,501	313,972
Median activations/cascade	86	139	73	127
Median activations/user	220	276	173	314
Min activations/user	192	250	145	231

Predicting activation times. The first experiment is meant to evaluate the accuracy in estimating the activation times. Given a training and test sets $\mathcal{C}_{train}$ and $\mathcal{C}_{test}$ of cascades, we train the model on $\mathcal{C}_{train}$ and measure the accuracy of the predictions on $\mathcal{C}_{test}$[2]. We chronologically split each cascade $c \in \mathcal{C}_{test}$ into c_1 and c_2 (for each $u \in c_1$ and $v \in c_2$, $t_u(c) < t_v(c)$) and pick a random subset c_3 of vertices that did not participate to corresponding cascade. We use c_1 to predict

---

[2] The two sets are obtained by randomly splitting the original dataset by ensuring that there is no overlap among the cascades of the two sets, but there is no vertex in the test that has not been observed in the training.

the most likely topic k by exploiting Eq. 4.3. Then, for each user in $c_2 \cup c_3$ we compute $\delta_u = \min_{v \in c_1} (A_{v,k} S_{u,k})^{-1}$.

We set a 90:10 training/test proportion and a chronological split proportion of 80%. Given a target delay horizon H, the prediction on u is considered as: true positive (TP) if $\delta_u < H$ and $u \in c_2$; true negative (TN) if $\delta_u > H$ and $u \in c_3$; false positive (FP) if $\delta_u < H$ and $u \in c_3$; and false negative (FN) if $\delta_u > H$ and $u \in c_2$. By varying H, we can plot ROC and F curves.

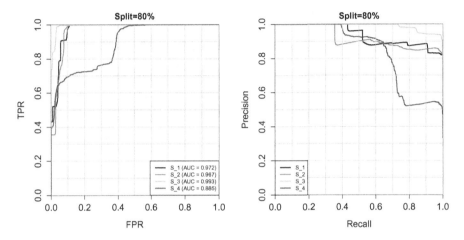

Fig. 3. AUC and precision/recall on predicting the activation time over synthetic data.

The results of the experiments, reported in Fig. 3, show that the proposed method is effective in predicting activation behaviour even when the propagation happens on networks with an overlapping community structure. The best performances are achieved on the network $S3$, despite the fact that some communities are strongly interconnected. A possible explanation is the higher number of communities in the dataset, which also makes cascades shorter and the co-occurrence of nodes less likely in cascades where they are not susceptible/authoritative.

5.2   Real Data

In this section, we assess the performances of the proposed method on real data, from a quantitative and qualitative perspective. First, we evaluate the accuracy of the model at predicting when a user will retweet a post. Secondly, we analyze and discuss topical and diffusion patterns inferred on the Memetracker dataset.

Twitter. The following analysis is based on a sample of real-world propagation cascades crawled from the public timeline of *Twitter* and studied in [2]. The propagation of information on Twitter happens by retweet and in this dataset tracks the propagation of URLs over the Twitter network during a period of one month (July 2012) Each activation/adoption corresponds to the instance when a user tweets a certain URL. Note that this dataset does not provide side-information (e.g. hashtags associated to each tweet, or the actual URL being shared). We also select a subset of the dataset by considering users who participated in at least 15 cascades and retweet cascades that involved at least 5 users. We refer to this dataset as *Twitter-Small*. A summary of the properties of both datasets is shown in Table 4.

Table 4. Summary of the Twitter data used for evaluation.

	Twitter-Large	Twitter-Small
Nodes	28,585	6,030
Edges	1,636,451	259,568
Activations	516,412	187,941
Cascades	8,541	3,983
Max Delay	2,380,651	2,141,136
Avg Delay	36,775	50,117
Median activations/cascade	18	17
Median activations/user	15	26
Min activations/cascade	1	7
Min activations/user	11	15

Predicting activation times. We apply the testing protocol detailed in Sect. 5.1 on the Twitter datasets for predicting users retweet times, by considering two training/test chronological split (80%) and measuring prediction accuracy by ROC analysis. Results, reported in Fig. 4, show that the model achieves high accuracy in predicting which are the users more likely to become active on each cascade within the prediction window. The prediction accuracy is higher on *Twitter-Small*. This result is compatible with the intuition that the inference works better when the focus is on users who actively participate into cascades. Finally, like in the case of synthesized data, the accuracy is not affected by the size of the cascade used for inferring the optimal topic.

Memetracker. The evaluation on the Memetracker dataset [15] is aimed at assessing the alignment between the topical and social influence structure. This dataset tracks phrases and quotes over online-news providers and blogs; textual variants of the same phrase are clustered together and the dataset specifies each timestamp at which a particular blog mentioned a phrase belonging to a cluster. We consider each cluster as a separate cascade, the root-phrase as the content being diffused and the hostname extracted from the url of the blog as vertex identifier. In this case, an activation within a information cascade represent the first timestamp at which a given blog mentioned a phrase belonging to the considered cluster. The raw dataset was cleaned from cascades with less than 10

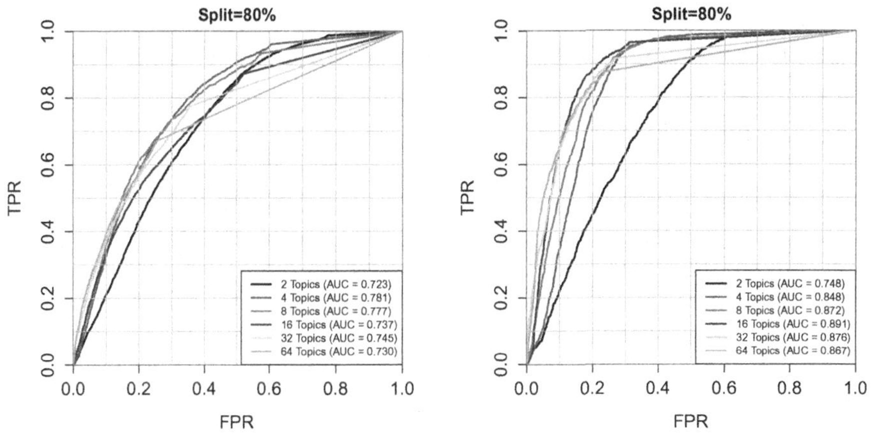

Fig. 4. Accuracy on predicting user's retweet time on *Twitter-Large* (on the left) and *Twitter-Small* (on the right).

Table 5. Most relevant terms for each topic.

Topic 1 (economy)	Topic 2 (France/Germany)
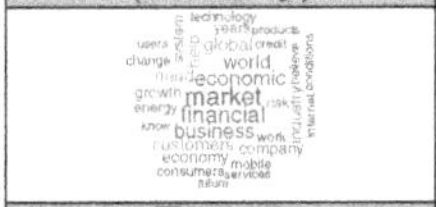	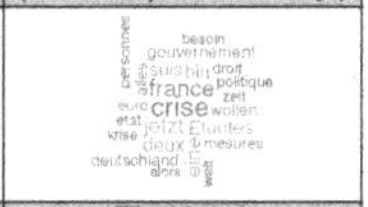
Topic 3 (presidential elections)	**Topic 4 (family)**
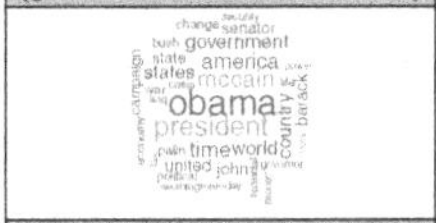	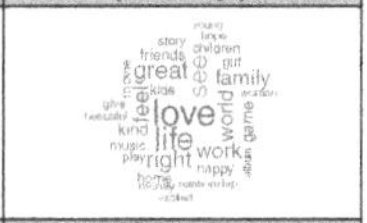
Topic 5 (international crisis)	**Topic 6 (news in spanish)**
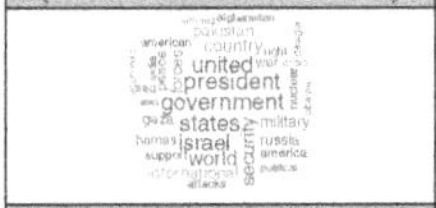	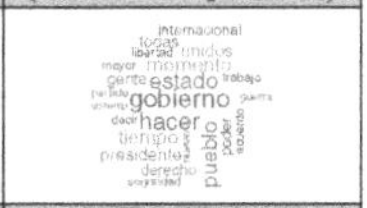
Topic 7 (religion)	**Topic 8 (sport)**
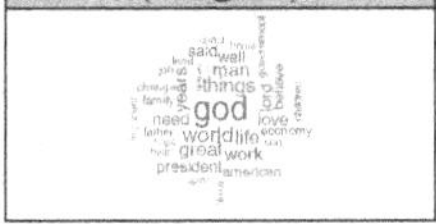	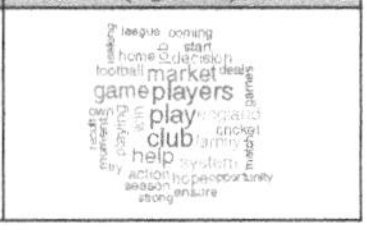

Table 6. Most influential hosts for each topic.

Topic 1 (economy)	Topic 2 (France/Germany)
priceofoil.org finanzas.com invertia.com blogs.wsj.com markets.chron.com forexfactory.com ft.com businessweek.com pr-inside.com prnewswire.co.uk telegram.com money.canoe.ca	de.eonline.com tagblatt.ch stjude.com.ve focus-fen.net de.reuters.com fr.biz.yahoo.com finanzas.com wellingtonhive.blogspot.com presseportal.ch tembam.wordpress.com tagesspiegel.de
Topic 3 (presidential elections)	**Topic 4 (family)**
washingtonindependent.com politicalsalsa.blogspot.com sj-r.com electionfraudblog.com freepressinternational.com cnjonline.com washingtonmonthly.com blogrevolution.com democraticpartyofgeorgia.org mattjohnston.blogspot.com	celebgitz.com koreanpower999.wordpress.com 610cktb.com pentictonherald.ca theusreport.com retroskank.com
Topic 5 (international crisis)	**Topic 6 (news in spanish)**
swampland.blogs.time.com breaktheterror.wordpress.com alertatotal.blogspot.com us.oneworld.net timesofoman.com news.balita.ph earthside.com	ellitoral.com laopiniondegranada.es laurennroth.com abc.com.py elnuevodiario.com.ni elmtreeforge.blogspot.com howrah.org cooperativa.cl eldigital.net
Topic 7 (religion)	**Topic 8 (sport)**
breaktheterror.wordpress.com inspiremagazine.org.uk canadian-catholic.blogspot.com thepaganophina.blogspot.com	cricket.timesofindia.indiatimes.com sportal.com.au sportgate.de uk.eurosport.yahoo.com news.cincinnati.com ecanadanow.com

activations and less than 10 words as content, and from vertices that belong to less than 10 cascades. The final dataset contains 7k vertices and 28k cascades, the word dictionary contains 3.5k tokens, with of 16 words for cascade on average.

For the sake of presentation, we run the survival factorization learning algorithm setting $K = 8$. Table 5 reports the most relevant words for each topic, i.e. the words w which exhibit the highest value of $\varphi_{w,k}$ for each k, and our interpretation of the topic is reported in the headings of the table.

Next, we analyze each cascade and compute:

- The most-likely topic as $\tilde{k}_c = \arg\max_k \gamma_{c,k}$;
- The most-likely cascade tree for each cascade $\tilde{T}_c$ by computing the parent of each active node (excluding the root) as $par(u)_c = \arg\max_v \eta_{c,u,v}^{\tilde{k}_c}$;
- For each cascade c the delay $\Delta_{u,v}^c$ for each pair u, v such that $par(u)_c = v$, and compute the average delay over cascades in each topic;
- The Wiener index for each cascade tree, and use this information to compute the average Wiener index for a topic k as $\bar{w}_k = avg_{c:\ \tilde{k}_c=k}\ W(\tilde{T}_c)$;
- The depth of each cascade tree, which is averaged across cascades in the same topic to compute the average cascade topical depth.

The outcome of this analysis is summarized in Table 7. The topic labeled as "sports" exhibits the shortest average transmission delay, followed by "international crisis" and "news in spanish language". In general, cascade trees are shallow, which suggests that the propagation of information is due to few influencers. The highest average Wiener index is observed on the topic "religion".

Table 7. Characterization of the cascade trees for each topic.

Topic	Average delay	Avg Wiener index	Avg depth
1	20.6 h	1.73	1.77
2	22.8 h	1.69	1.74
3	43.5 h	1.82	1.93
4	21 h	1.76	1.83
5	12.7 h	1.80	1.92
6	12 h	1.85	2.13
7	23.8 h	1.89	2.20
8	7.8 h	1.83	2.08

Finally, Table 6 shows the top influencers for each topic, computed by counting the number of children of each node in each cascade and aggregating this info at the topic level. The top influential blogs are well aligned with the topical structure shown in Table 5.

6   Conclusions

In this work we proposed a model for information diffusion where adoptions can be explained in terms of susceptibility and authoritativeness. The latter concepts can be expressed as latent factors over a low-dimensional space representing topical interests. We showed the adequacy of the resulting probabilistic model both from a mathematical and an experimental point of view. There are different points worth further investigation. For example, we showed that the instantiation based on the Exponential distribution admit an efficient implementation. In future work we will study if this property holds on other models, e.g. Rayleigh. Also, the robustness of the model can be improved by relying on a full bayesian framework.

References

1. Barbieri, N., Bonchi, F., Manco, G.: Topic-aware social influence propagation models. In: ICDM, pp. 81–90 (2012)
2. Barbieri, N., Bonchi, F., Manco, G.: Cascade-based community detection. In: WSDM, pp. 33–42 (2013)
3. Barbieri, N., Bonchi, F., Manco, G.: Influence-based network-oblivious community detection. In: ICDM, pp. 955–960 (2013)
4. Bourigault, S., et al.: Learning social network embeddings for predicting information diffusion. In: WSDM, pp. 393–402 (2014)
5. Du, N., Farajtabar, M., Ahmed, A., Smola, A.J., Song, L.: Dirichlet-Hawkes processes with applications to clustering continuous-time document streams. In: KDD, pp. 219–228 (2015)
6. Du, N., Song, L., Woo, H., Zha, H.: Uncover topic-sensitive information diffusion networks. In: Proceedings of the Sixteenth International Conference on Artificial Intelligence and Statistics (AISTATS), pp. 229–237 (2013)
7. Gomez-Rodriguez, M., Balduzzi, D., Schölkopf, B.: Uncovering the temporal dynamics of diffusion networks. In: ICML, pp. 561–568 (2011)
8. Gomez-Rodriguez, M., Leskovec, J., Krause, A.: Inferring networks of diffusion and influence. In: KDD (2010)
9. Goyal, A., Bonchi, F., Lakshmanan, L.: Learning influence probabilities in social networks. In: WSDM (2010)
10. He, X., Rekatsinas, T., Foulds, J.R., Getoor, L., Liu, Y.: HawkesTopic: a joint model for network inference and topic modeling from text-based cascades. In: ICML, pp. 871–880 (2015)
11. Hu, Q., Xie, S., Lin, S., Fan, W., Yu, P.S.: Frameworks to encode user preferences for inferring topic-sensitive information networks. In: SDM, pp. 442–450 (2015)
12. Iwata, T., Shah, A., Ghahramani, Z.: Discovering latent influence in online social activities via shared cascade poisson processes. In: KDD, pp. 266–274 (2013)
13. Kempe, D., Kleinberg, J., Tardos, É.: Maximizing the spread of influence through a social network. In: KDD, pp. 137–146 (2003)
14. Lee, E.T., Wang, J.: Statistical Methods for Survival Data Analysis. Wiley, Hoboken (2003)
15. Leskovec, J., Backstrom, L., Kleinberg, J.: Meme-tracking and the dynamics of the news cycle. In: KDD, pp. 497–506 (2009)
16. Mei, Q., Fang, H., Zhai, C.X.: A study of Poisson query generation model for information retrieval. In: SIGIR, pp. 319–326 (2007)
17. Wang, L., Ermon, S., Hopcroft, J.E.: Feature-enhanced probabilistic models for diffusion network inference. In: ECMLPKDD, pp. 499–514 (2012)
18. Wang, S., Hu, X., Yu, P.S., Li, Z.: MMRate: inferring multi-aspect diffusion networks with multi-pattern cascades. In: KDD, pp. 1246–1255 (2014)

19. Wang, Y., Shen, H., Liu, S., Cheng, X.: Learning user-specific latent influence and susceptibility from information cascades. In: AAAI (2015)
20. Weng, L., Menczer, F., Ahn, Y.: Virality prediction and community structure in social networks. Sci. Rep. **3**, 2522 (2013)
21. Xiang, R., Neville, J., Rogati, M.: Modeling relationship strength in online social networks. In: WWW, pp. 981–990 (2010)
22. Yang, S., Zha, H.: Mixture of mutually exciting processes for viral diffusion. In: ICML, pp. 1–9 (2013)

The Network-Untangling Problem: From Interactions to Activity Timelines

Polina Rozenshtein$^{(\boxtimes)}$, Nikolaj Tatti, and Aristides Gionis

HIIT, Aalto University, Espoo, Finland
{Polina.Rozenshtein,Nikolaj.Tatti,Aristides.Gionis}@aalto.fi

Abstract. In this paper we study a problem of determining when entities are active based on their interactions with each other. More formally, we consider a set of entities V and a sequence of time-stamped edges E among the entities. Each edge $(u, v, t) \in E$ denotes an interaction between entities u and v that takes place at time t. We view this input as a *temporal network*. We then assume a simple *activity model* in which each entity is *active* during a short time interval. An interaction (u, v, t) can be explained if at least one of u or v are active at time t. Our goal is to reconstruct the activity intervals, for all entities in the network, so as to explain the observed interactions. This problem, which we refer to as the *network-untangling problem*, can be applied to discover timelines of events from complex interactions among entities.

We provide two formulations for the network-untangling problem: (*i*) minimizing the total interval length over all entities, and (*ii*) minimizing the maximum interval length. We show that the sum problem is **NP**-hard, while, surprisingly, the max problem can be solved optimally in linear time, using a mapping to 2-SAT. For the sum problem we provide efficient and effective algorithms based on realistic assumptions. Furthermore, we complement our study with an evaluation on synthetic and real-world datasets, which demonstrates the validity of our concepts and the good performance of our algorithms.

Keywords: Temporal networks · Complex networks
Timeline reconstruction · Vertex cover · Linear programming · 2-SAT

1 Introduction

Data increase in volume and complexity. A major challenge that arises in many applications is to process efficiently large amounts of data in order to synthesize the available bits of information into a concise but meaningful picture.

New data abstractions, emerging from modern applications, require new definitions for data-summarization and synthesis tasks. In particular, for many data that are typically modeled as networks, temporal information is nowadays readily available, leading to *temporal networks* [9,19]. In temporal networks $G = (V, E)$, edges describe interactions over a set of entities V. For each edge $(u, v, t) \in E$, the time of interaction t, between entities $u, v \in V$ is also available.

© Springer International Publishing AG 2017
M. Ceci et al. (Eds.): ECML PKDD 2017, Part I, LNAI 10534, pp. 701–716, 2017.
https://doi.org/10.1007/978-3-319-71249-9_42

In this paper we introduce a new problem for summarizing temporal networks. The main idea is to consider that the entities of the network are *active* over presumably short time intervals. Edges (interactions) of the temporal network between two entities can be explained by *at least one* of the two entities being active at the time of the interaction. Our summarization task is to process the available temporal edges (interactions) and infer the latent activity intervals for all entities. In this way, we can infer an *activity timeline* for the whole network. To motivate the summarization task studied in this paper, consider the following application scenario.

Example. Consider a news story unfolding over the period of several months, or years, such as **Brexit**. There is a sequence of intertwined events (e.g., UK referendum, prime minister resigns, appointment of new prime minister, supreme court decision, invoking article 50, etc.) as well as a roster of key characters who participate in the events (e.g., Cameron, Johnson, May, Tusk, etc.). Consider now a stream of Brexit-related tweets, as events unfold, and hashtags mentioned in those tweets (e.g., #brexit, #remain, #ukip, #indyref2, etc.). For our purposes, we view the twitter stream as a temporal network: a tweet mentioning two hashtags h_1 and h_2 and posted at time t is seen as a temporal edge (h_1, h_2, t). A typical situation is that a hashtag *bursts* during a time interval that is associated with a *main event*, while it may also appear outside the time interval in a connection with other *secondary events*. For instance, the peak activity for #remain may have been during the weeks leading to the referendum, but the same hashtag may also appear later, say, in reference to invoking article 50, by a user who wished that EU had not voted for Brexit. The question that we ask in this paper is whether it is possible to process the temporal network of entity interactions and reconstruct the latent activity intervals for each entity (hashtags, in this example), and thus, infer the complete timeline of the news story.

Motivated by the previous example, and similar application scenarios, we introduce the *network-untangling problem*, where the goal is to reconstruct an activity timeline from a temporal network. Our formulation uses a simple model in which we assume that each network entity is *active* during a time interval. An temporal edge (u, v, t) is *covered* if at least one of u or v are active at time t. The algorithmic objective is to find a set of activity intervals, one for each entity, so that all temporal edges are covered, and the length of the activity intervals is minimized. We consider two definitions for interval length: total length and maximum length.

We show that the problem of minimizing the maximum length over all activity intervals can be mapped to 2-SAT, and be solved optimally and in linear time On the other hand, minimizing the total interval length is an **NP**-hard problem. To confront this challenge we offer two iterative algorithms that rely on the fact that certain subproblems can be solved approximately or optimally. In both cases the subproblems can be solved by linear-time algorithms, yielding overall very practical and efficient methods.

We complement our theoretical results with an experimental evaluation, where we demonstrate that our methods are capable on finding ground-truth activity intervals planted on synthetic datasets. Additionally we conduct a case study where it is shown that the discovered intervals match the timeline of real-world events and related sub-events.

2 Preliminaries and Problem Definition

Our input is a temporal network $G = (V, E)$, where V is a set of vertices and E is a set of time-stamped edges. The edges of the temporal network are triples of the form (u, v, t), where $u, v \in V$ and t is a time stamp indicating the time that an interaction between vertices u and v takes place. In our setting we do not preclude the case that two vertices u and v interact multiple times. As it is customary, we denote by n the number of vertices in the graph, and by m the number of edges. For our algorithms we assume that the edges are given in chronological order, if not, they can be sorted in additional $\mathcal{O}(m \log m)$ time.

Given a vertex $u \in V$, we will write $E(u)$ to be the set of edges adjacent to vertex u, i.e., $E(u) = \{(u, v, t) \in E\}$. We will also write $N(u) = \{v \mid (u, v, t) \in E\}$ to represent the set of vertices adjacent to u, and $T(u) = \{t \mid (u, v, t) \in E\}$ to represent the set of time stamps of the edges containing u. Finally, we write $t(e)$ to denote the time stamp of an edge $e \in E$.

Given a vertex $u \in V$ and two real numbers s_u and e_u, we consider the interval $I_u = [s_u, e_u]$, where s_u is a start time and e_u is an end time. We refer to I_u as the *activity interval* of vertex u. Intuitively, we think of I_u as the time interval in which the vertex u has been active. A set of activity intervals $\mathcal{T} = \{I_u\}_{u \in V}$, one interval for each vertex $u \in V$, is an *activity timeline* for the temporal network G.

Given a temporal network $G = (V, E)$ and an activity timeline $\mathcal{T} = \{I_u\}_{u \in V}$, we say that the timeline $\mathcal{T}$ *covers* the network G if for each edge $(u, v, t) \in E$, we have $t \in I_u$ or $t \in I_v$, that is, when each network edge occurs at least one of its endpoints is active.

Note that each temporal network has a trivial timeline that provides a cover. Such a timeline, defined by $I_u = [\min T(u), \max T(u)]$, may have unnecessarily long intervals. Instead, we aim finding an activity timeline that have as compact intervals as possible. We measure the quality of a timeline by the total duration of all activity intervals in it. More formally, we define the *total span*, or *sum-span*, of a timeline $\mathcal{T} = \{I_u\}_{u \in V}$ by

$$S(\mathcal{T}) = \sum_{u \in V} \sigma(I_u),$$

where $\sigma(I_u) = e_u - s_u$ is the duration of a single interval. An alternative way to measure the compactness of a timeline is by the duration of its longest interval,

$$\Delta(\mathcal{T}) = \max_{u \in V} \sigma(I_u).$$

We refer to $\Delta(\mathcal{T})$ as the *max-span* of the timeline $\mathcal{T}$.

Associated with the above compactness measures we define the following two problems that we consider in this paper.

Problem 1 (MinTimeline). Given a temporal network $G = (V, E)$, find a time-line $\mathcal{T} = \{I_u\}_{u \in V}$ that covers G and minimizes the sum-span $S(\mathcal{T})$.

Problem 2 (MinTimeline$_\infty$). Given a temporal network $G = (V, E)$ find a time-line $\mathcal{T} = \{I_u\}_{u \in V}$ that covers G and minimizes the max-span $\Delta(\mathcal{T})$.

3   Computational Complexity and Algorithms

Surprisingly, while MinTimeline is an **NP**-hard problem, MinTimeline$_\infty$ can be solved optimally efficiently. The optimality of MinTimeline$_\infty$ is a result of the algorithm presented in Sect. 5. In this section we establish the complexity of MinTimeline, and we present two efficient algorithms for MinTimeline and MinTimeline$_\infty$.

Proposition 1. *The decision version of the* MinTimeline *problem is* **NP***-complete. Namely, given a temporal network* $G = (V, E)$ *and a budget* ℓ, *it is* **NP***-complete to decide whether there is timeline* $\mathcal{T}^* = \{I_u\}_{u \in V}$ *that covers* G *and has* $S(\mathcal{T}^*) \leq \ell$.

Proof. We will prove the hardness by reducing VertexCover to MinTime-line. Assume that we are given a (static) network $H = (W, A)$ with n vertices $W = \{w_1, \ldots, w_n\}$ and a budget ℓ. In the VertexCover problem we are asked to decide whether there exists a subset $U \subseteq W$ of at most ℓ vertices ($|U| \leq \ell$) covering all edges in A.

We map an instance of VertexCover to an instance of MinTimeline by creating a temporal network $G = (V, E)$, as follows. The vertices V consists of $2n$ vertices: for each $w_i \in W$, we add vertex v_i and u_i. The edges are as follows: For each edge $(w_i, w_j) \in A$, we add a temporal edge $(v_i, v_j, 0)$ to E. For each vertex $w_i \in W$, we add two temporal edges $(v_i, u_i, 1)$ and $(v_i, u_i, 2n + 1)$ to E.

Let $\mathcal{T}^*$ be an optimal timeline covering G. We claim that $S(\mathcal{T}^*) \leq \ell$ if and only if there is a vertex cover of H with ℓ vertices. To prove the *if* direction, consider a vertex cover of H, say U, with ℓ vertices. Consider the following coverage: cover each u_i at $2n + 1$, and each v_i at 1. For each $w_i \in U$, cover v_i at 0. The resulting intervals are indeed forming a timeline with a total span of ℓ.

To prove the other direction, first note that if we cover each v_i by an interval $[0, 1]$ and each u_i by an interval $[2n + 1, 2n + 1]$, then this yields a timeline $\mathcal{T}^*$ covering G. The total span intervals $\mathcal{T}^*$ is n. Thus, $S(\mathcal{T}^*) \leq n$. This guarantees that if $0 \in I_{v_i}$, then $2n+1 \notin I_{v_i}$, so $2n+1 \in I_{u_i}$. This implies that $1 \notin I_{u_i}$ and so $1 \in I_{v_i}$. In summary, if $0 \in I_{v_i}$, then $\sigma(I_{v_i}) = 1$. This implies that if $S(\mathcal{T}^*) \leq \ell$, then we have at most ℓ vertices covered at 0. Let U be the set of those vertices. Since $\mathcal{T}^*$ is timeline covering G, then U is a vertex cover for H.   □

3.1   Iterative Method Based on Inner Points

As we saw, MinTimeline is an **NP**-hard problem. The next logical question is whether we can approximate this problem. Unfortunately, there is evidence that such an algorithm would be highly non-trivial: we can show that if we extend our problem definition to hyper-edges—the coverage then means that one vertex needs to be covered per edge—then such a problem is inapproximable. This suggests that an approximation algorithm would have to rely on the fact that we are dealing with edges and not hyper-edges.

Luckily, we can consider meaningful subproblems. Assume that we are given a temporal network $G = (V, E)$ and we also given a set of time point $\{m_v\}_{v \in V}$, i.e., one time point m_v for each vertex $v \in V$, and we are asked whether we can find an optimal activity timeline $\mathcal{T} = \{I_u\}_{u \in V}$ so that the interval I_v of vertex v contains the corresponding time point m_v, i.e., $m_v \in I_v$, for each $v \in V$. Note that these inner points can be located *anywhere* within the interval (not just, say, in the center of the interval). This problem definition is useful when we know one time point that each vertex was active, and we want to extend this to an optimal timeline. We refer to this problem as $\mathrm{MINTIMELINE}_m$.

Problem 3 ($\mathrm{MINTIMELINE}_m$). Given a temporal network $G = (V, E)$ and a set of inner time points $\{m_v\}_{v \in V}$, find a timeline $\mathcal{T} = \{I_u\}_{u \in V}$ that covers G, satisfies $m_v \in I_v$ for each $v \in V$, and minimizes the sum-span $S(\mathcal{T})$.

Interestingly, we can show that the $\mathrm{MINTIMELINE}_m$ problem can be solved approximately, in *linear time*, within a factor of 2 of the optimal solution. The 2-approximation algorithm is presented in Sect. 4.

Being able to solve $\mathrm{MINTIMELINE}_m$, motivates the following algorithm for $\mathrm{MINTIMELINE}$, which uses $\mathrm{MINTIMELINE}_m$ as a subroutine: initialize $m_v = (\min T(v) + \max T(v))/2$ to be an inner time point for vertex v; recall that $T(v)$ are the time stamps of the edges containing v. We then use our approximation algorithm for $\mathrm{MINTIMELINE}_m$ to obtain a set of intervals $\{I_v\} = \{[s_v, e_v]\}_{v \in V}$. We use these intervals to set the new inner points, $m_v = (s_v + e_v)/2$, and repeat until the score no longer improves. We call this algorithm `Inner`.

3.2   Iterative Method Based on Budgets

Our algorithm for $\mathrm{MINTIMELINE}_\infty$ also relies on the idea of using a subproblem that is easier to solve.

In this case, we consider as subproblem an instance in which, in addition to the temporal network G, we are also given a set of budgets $\{b_v\}$ of interval durations; one budget b_v for each vertex v. The goal is to find a timeline $\mathcal{T} = \{I_u\}_{u \in V}$ that covers the temporal network G and the length of each activity interval I_v is at most b_v. We refer to this problem as $\mathrm{MINTIMELINE}_b$.

Problem 4 ($\mathrm{MINTIMELINE}_b$). Given a temporal network $G = (V, E)$ and a set of budgets $\{b_v\}_{v \in V}$, find a timeline $\mathcal{T} = \{I_u\}_{u \in V}$ that covers G and satisfies $\sigma(I_v) \leq b_v$ for each $v \in V$.

Surprisingly, the $\mathrm{MINTIMELINE}_b$ problem can be solved *optimally* in *linear time*. The algorithm is presented in Sect. 5. Note that this result is compatible with the **NP**-hardness of $\mathrm{MINTIMELINE}$, since here we know the budgets for *individual* intervals, and thus, there are an exponential number of ways that we can distribute the total budget among the individual intervals.

We can now use binary search to find the optimal value $\Delta(\mathcal{T})$. We call this algorithm Budget. To guarantee a small number of binary steps, some attention is required: Let $T = t_1, \ldots, t_m$ be all the time stamps, sorted. Assume that we have L, the largest known infeasible budget and U, the smallest known feasible budget. To define a new candidate budget, we first define $W(i) = \{t_j - t_i \mid L < t_j - t_i < U\}$. The optimal budget is either U or one of the numbers in $W(i)$. If every $W(i)$ is empty, then the answer is U. Otherwise, we compute $m(i)$ to be the median of $W(i)$, ignore any empty $W(i)$. Finally, we test the weighted median of all $m(i)$, weighted by $|W(i)|$, as a new budget. We can show that at each iteration $\sum |W(i)|$ is reduced by $1/4$, that is, only $\mathcal{O}(\log m)$ iterations is needed. We can determine the medians $m(i)$ and the sizes $|W(i)|$ in linear time since T is sorted, and we can determine the weighted median in linear time by using a modified median-of-medians algorithm. This leads to a $\mathcal{O}(m \log m)$ running time. However, in our experimental evaluation, we use a straightforward binary search by testing $(U + L)/2$ as a budget.

4   Approximation Algorithm for MINTIMELINE$_m$

In this section we design a 2-approximation linear-time algorithm for the MIN-TIMELINE$_m$ problem. As defined in Problem 3, our input is a temporal network $G = (V, E)$ and a set of interior time points $\{m_v\}_{v \in V}$. As before, $T(v)$ denotes the set of time stamps of the edges containing vertex v.

Consider a vertex v and the corresponding interior point m_v. For a time point t we define the *peripheral time stamps* $p(t; v)$ to be the time stamps that are on the other side of t than m_v,

$$p(t; v) = \begin{cases} \{s \in T(v) \mid s \geq t\} & \text{if } t > m_v, \\ \{s \in T(v) \mid s \leq t\} & \text{if } t < m_v, \\ T(v) & \text{if } t = m_v. \end{cases}$$

Our next step is to express MINTIMELINE$_m$ as an integer linear program. To do that we will define a variable x_{vt} for each vertex $v \in V$ and time stamp $t \in T(v)$. Instead of going for the obvious construction, where $x_{vt} = 1$ indicates that v is active at t, we will do a different formulation: in our program $x_{vt} = 1$ indicates that t is either the *beginning* or the *end* of the active region of v. It follows that the integer program

$$\min \sum_{v,t} |t - m_v| x_{vt},$$

$$\text{such that } \sum_{s \in p(v;t)} x_{vs} + \sum_{s \in p(u;t)} x_{us} \geq 1, \text{ for all } (u, v, t) \in E$$

solves MINTIMELINE$_m$. Naturally, here we also require that $x_{vt} \in \{0, 1\}$. Minimizing the first sum corresponds to minimizing the sum-span of the timeline, while the constraint on the second sum ensures that the resulting timeline covers the temporal network. Note that we do not require that each vertex should have exactly one beginning and one end, however, the minimality of the optimal solution ensures that this constraint will be satisfied, too.

Relaxing the integrality constraint and considering the program as linear program, allows us to write the dual. The variables in the dual can be viewed as positive weights α_e on the edges, with the goal of maximizing the total sum of these weights.

To express the constraints on the dual, let us define an auxiliary function $h(v, t, s)$ as the sum of the weights of adjacent edges between t and s,

$$h(v, t, s) = \sum \{\alpha_e \mid e \in E(v),\ t(e)\text{ is between } s\text{ and }t\},$$

where, recall that, $E(v)$ denotes the edges adjacent to v and $t(e)$ denotes the time stamp of edge $e \in E$. The dual can now be formulated as

$$\max \sum_{e \in E} \alpha_e, \quad \text{such that} \quad h(v, t, m_v) \le |t - m_v|, \text{ for all } v \in V,\ t \in T(v),$$

that is, we maximize the total weight of edges such that for each vertex v and for each time stamp t, the sum of adjacent edges is bounded by $|t - m_v|$.

We say that the solution to dual is *maximal* if we cannot increase any edge weight α_e without violating the constraints. An optimal solution is maximal but a maximal solution is not necessarily optimal.

Our next result shows that a maximal solution can be used to obtain a 2-approximation dynamic cover.

Proposition 2. *Consider a maximal solution α_e to the dual program. Define a set of intervals $\mathcal{T} = \{I_v\}$ by $I_v = [\min X_v, \max X_v]$, where*

$$X_v = \{m_v\} \cup \{t \in T(v) \mid h(v, t, m_v) = |t - m_v|\}.$$

Then $\mathcal{T}$ is a 2-approximation solution for the problem $\mathrm{MINTIMELINE}_m$.

Proof. We first show that a maximal dual solution is a feasible timeline. Let $e = (u, v, t)$ be a temporal edge. If $p(t; v) \cap X_v = \emptyset$ and $p(t; u) \cap X_u = \emptyset$, then we can increase the value of α_e without violating the constraints, so the solution is not maximal. Thus $t \in I_v \cup I_u$, making $\mathcal{T}$ a feasible timeline.

Next we show that the resulting solution $\mathcal{T}$ is a 2-approximation to MIN-$\mathrm{TIMELINE}_m$. Write $x_v = \min\{X_v\}$ and $y_v = \max\{X_v\}$. Let $\mathcal{T}^*$ be the optimal solution. Then

$$S(\mathcal{T}) = \sum_{v \in V} |x_v - m_v| + |y_v - m_v| = \sum_{v \in V} h(v, x_v, m_v) + h(v, y_v, m_v)$$

$$\le \sum_{v \in V} \sum_{e \in E(v)} \alpha_e = 2 \sum_{e \in E} \alpha_e \le 2S(\mathcal{T}^*),$$

where the second equality follows from the definition of X_v, the first inequality follows from the fact that $\alpha_e \ge 0$, and the last inequality follows from primal-dual theory. This proves the claim. □

We have established that as long as we can obtain a maximal solution for the dual, we can extract a timeline that is 2-approximation. We will now introduce a linear-time algorithm that computes a maximal dual solution. The algorithm visits each edge $e = (u, v, t)$ in chronological order and increases α_e as much as possible without violating the dual constraints. To obtain a linear-time complexity we need to determine in *constant* time by how much we can increase α_e. The pseudo-code is given in Algorithm 1, and the remaining section is used to prove the correctness of the algorithm.

Algorithm 1. Maximal, yields 2-approximation to $\mathrm{MINTIMELINE}_m$.

$b[v] \leftarrow \infty$ for $v \in V$;
$a[v] \leftarrow 0$ for $v \in V$;
foreach $e = (u, v, t) \in E$ in chronological order **do**
$\quad \alpha_e \leftarrow \min\{z(u), z(v)\}$;   {see Eq. (2)}
$\quad$ **if** $t < m_v$ **then** $b[v] \leftarrow \min\{b[v] - \alpha_e, m_v - t - \alpha_e\}$;
$\quad$ **else** $a[v] \leftarrow a[v] + \alpha_e$;
$\quad$ **if** $t < m_u$ **then** $b[u] \leftarrow \min\{b[u] - \alpha_e, m_u - t - \alpha_e\}$;
$\quad$ **else** $a[u] \leftarrow a[u] + \alpha_e$;

Let us enumerate the edges chronologically by writing e_i for the i-th edge, and let us write α_i to mean α_{e_i}. We will also write t_i for the time stamp of e_i. Finally, let us define k_v to be the smallest index of an edge (u, v, t) with $t \geq m_v$, and o_v to be the largest index of an edge (u, v, t) with $t \leq m_v$.[1]

For simplicity, we rewrite the dual constrains using indices instead of time stamps. Given two indices $i \leq j$, we slightly overload the notation and we write

$$h(v, i, j) = \sum \{\alpha_\ell \mid e_\ell \in E(v), \ \ell \text{ is between } i \text{ and } j\} .$$

The dual constraints can be written as

$$h(v, i, o_v) \leq |t_i - m_v|, \text{ if } i < k_v, \quad \text{and} \quad h(v, i, k_v) \leq |t_i - m_v|, \text{ if } i \geq k_v. \quad (1)$$

Each dual constraint is included in these constraints. Equation (1) may also contain some additional constraints but they are redundant, so the dual constraints hold if and only if constraints in Eq. (1) hold.

As the algorithm goes over the edges, we maintain two counters per each vertex, $a[v]$ and $b[v]$. Let $e_j = (u, v, t)$ be the current edge. The counter $a[v]$ is maintained only if $t \geq m_v$, and the counter $b[v]$ is maintained if $t < m_v$. Our invariant for maintaining the counters $a[v]$ and $b[v]$ is that at the beginning of j-th round they are equal to

$$a[v] = h(v, k_v, j) \quad \text{and} \quad b[v] = \min_{\ell < j}\{t_\ell - m_v - h(v, \ell, j - 1)\}.$$

The following lemma tells us how to update α_j using $a[v]$ and $b[v]$.

---

[1] If there is an edge exactly at m_v, then $k_v = o_v$.

Lemma 1. *Assume that we are processing edge* $e_j = (u, v, t)$. *We can increase* α_j *by at most*

$$
\min\{z(u), z(v)\}, \quad \text{where} \quad z(w) = \begin{cases} t - m_w - a[w] & \text{if } j \geq k_v, \\ \min\{m_w - t, b[w]\} & \text{if } j < k_v. \end{cases} \tag{2}
$$

Proof. We will prove this result by showing that $\alpha_e \leq z(v)$ if and only if all constraints in Eq. (1) related to v are valid. Since the same holds also for u the lemma follows. We consider two cases.

First case: $j < k_v$. In this case we have $z(v) = \min\{m_w - t, b[w]\} = \min_{\ell \leq j}\{t_\ell - m_v - h(v, \ell, o_v)\}$, before increasing α_j. This guarantees that if $\alpha_j \leq z(v)$, then $h(v, \ell, o_v) \leq |t_\ell - m_v|$, for every $\ell \leq j$. Moreover, when $\alpha_j = z(v)$ one of these constraints becomes tight. Since these are the only constraints containing α_j, we have proven the first case.

Second case: $j \geq k_v$. If $\ell < j$, the sum $h(v, \ell, k_v)$ does not contain α_j, so the corresponding constraint remains valid. If $\ell \geq j$, then the corresponding constraint is valid if and only if $h(v, j, k_v) \leq |t_j - m_v|$. This is because $\alpha_\ell = 0$ for all $\ell > j$. But $z(v)$ corresponds exactly to the amount we can increase α_i so that $h(v, j, k_v) = |t_j - m_v|$. This proves the second case. □

Our final step is to how to maintain $a[v]$ and $b[v]$. Maintaining $a[v]$ is trivial: we simply add α_j to $a[v]$. The new $b[v]$ is equal to

$$
\min_{\ell \leq j}\{t_\ell - m_v - h(v, \ell, j)\} = \min\{b[v] - \alpha_j, m_v - t - \alpha_j\}.
$$

Clearly the counters $a[v]$ and $b[v]$ and the dual variables α_e can be maintained in constant time per edge processed, making `Maximal` a linear-time algorithm.

5   Exact Algorithm for MINTIMELINE_b

In this section we develop a linear-time algorithm for the problem MINTIMELINE_b. Here we are given a temporal network G, and a set of budgets $\{b_v\}$ of interval durations, and all activity intervals should satisfy $\sigma(I_v) \leq b_v$.

The idea for this optimal algorithm is to map MINTIMELINE_b into 2-SAT. To do that we introduce a boolean variable x_{vt} for each vertex v and for each timestamp $t \in T(v)$. To guarantee the solution will cover each edge (u, v, t) we add a clause $(x_{vt} \vee x_{ut})$. To make sure that we do not exceed the budget we require that for each vertex v and each pair of time stamps $s, t \in T(v)$ such that $|s - t| > b_v$ either x_{vs} is false or x_{vt} is false, that is, we add a clause $(\neg x_{vs} \vee \neg x_{vt})$. It follows immediately, that MINTIMELINE_b has a solution if and only if 2-SAT has a solution. The solution for MINTIMELINE_b can be obtained from the 2-SAT solution by taking the time intervals that contain all boolean variables set to true. Since 2-SAT is a polynomially-time solvable problem [1], we have the following.

Proposition 3. *MINTIMELINE_b can be solved in a polynomial time.*

Solving 2-SAT can be done in linear-time with respect to the number of clauses [1]. However, in our case we may have $\mathcal{O}(m^2)$ clauses. Fortunately, the 2-SAT instances created with our mapping have enough structure to be solvable in $\mathcal{O}(m)$ time. This speed-up is described in the remainder of the section.

Let us first review the algorithm by Aspvall et al. [1] for solving 2-SAT. The algorithm starts with constructing an *implication graph* $H = (W, A)$. The graph H is directed and its vertex set $W = P \cup Q$ has a vertex p_i in P and a vertex q_i in Q for each boolean variable x_i. Then, for each clause $(x_i \vee x_j)$, there are two edges in A: $(q_i \to p_j)$ and $(q_j \to p_i)$; The negations are handled similarly.

In our case, the edges A are divided to two groups A_1 and A_2. The set A_1 contains two directed edges $(q_{vt} \to p_{ut})$ and $(q_{ut} \to p_{vt})$ for each edge $e = (u, v, t) \in E$. The set A_2 contains two directed edges $(p_{vt} \to q_{vs})$ and $(p_{vs} \to q_{vt})$ for each vertex v and each pair of time stamps $s, t \in T(v)$ such that $|s - t| > b_v$. Note that A_1 goes from Q to P and A_2 goes from P to Q. Moreover, $|A_1| \in \mathcal{O}(m)$ and $|A_2| \in \mathcal{O}(m^2)$.

Next, we decompose H in strongly connected components (SCC), and order them topologically. If any strongly connected component contains both p_{vt} and q_{vt}, then we know that 2-SAT is not solvable. Otherwise, to obtain the solution, we start enumerate over the components, children first: if the boolean variables corresponding to the vertices in the component do not have truth assignment,[2] then we set x_{vt} to be true if p_{vt} is in the component, and x_{vt} to be false if q_{vt} is in the component

The bottleneck of this method is the SCC decomposition, which requires $\mathcal{O}(|W| + |A|)$ time, and the remaining steps can be done in $\mathcal{O}(|W|)$ time. Since $|W| \in \mathcal{O}(m)$, we need to optimize the SCC decomposition to perform in $\mathcal{O}(m)$ time. We will use the algorithm by Kosajaru (see [10]) for the SCC decomposition. This algorithm consists of two depth-first searches, performing constant-time operations on each visited node. Thus, we need to only optimize the DFS.

To speed-up the DFS, we need to design an oracle such that given a vertex $p \in P$ it will return an *unvisited* neighboring vertex $q \in Q$ in *constant* time. Since $|Q| \in \mathcal{O}(m)$, this guarantees that DFS spends at most $\mathcal{O}(m)$ time processing vertices $p \in P$. On the other hand, if we are at $q \in Q$, then we can use the standard DFS to find the neighboring vertex $p \in P$. Since $|A_1| \in \mathcal{O}(m)$, this guarantees that DFS spends at most $\mathcal{O}(m)$ time processing vertices $q \in Q$.

Next, we describe the oracle: first we keep the unvisited vertices Q in lists $\ell[v] = (q_{vt} \in Q; q_{vt}$ is not visited) sorted chronologically. Assume that we are at $p_{vt} \in P$. We retrieve the first vertex in $\ell[v]$, say q_{vs}, and compare if $|s - t| > b_v$. If true, then q_{vs} is a neighbor of p_{vt}, so we return q_{vs}. Naturally, we delete q_{vs} from $\ell[v]$ the moment we visit q_{vs}. If $|s - t| \leq b_v$, then test similarly the *last* vertex in $\ell[v]$, say $q_{vs'}$. If both $q_{vs'}$ and q_{vs} are non-neighbors of p_{vt}, then, since $\ell[v]$ is sorted chronologically, we can conclude that $\ell[v]$ does not have unvisited neighbors of p_{vt}. Since p_{vt} does not have any neighbors outside $\ell[v]$, we conclude that p_{vt} does not have any unvisited neighbors.

---

[2] Due to the property of implication graph, either all or none variables will be set in the component.

Using this oracle we can now perform DFS in $\mathcal{O}(m)$ time, which in turns allows us to do the SCC decomposition in $\mathcal{O}(m)$ time, which then allows us to solve MINTIMELINE$_b$ in $\mathcal{O}(m)$ time.

6 Related Work

To the best of our knowledge, the problem we consider in this paper has not been studied before in the literature. In this section we review briefly the lines of work that are most closely related to our setting.

Vertex cover. Our problem definition can also be considered a temporal version of the classic vertex-cover problem, one of 21 original **NP**-complete problems in Karp's seminal paper [12]. A factor-2 approximation is available for vertex cover, by taking all vertices of a maximal matching [6]. Slightly improved approximations exist for special cases of the problem, while assuming that the unique games conjecture is true, the minimum vertex cover cannot be approximated within any constant factor better than 2 [13]. Nevertheless, our formulation cannot be mapped directly to the static vertex-cover problem, thus, the proposed solutions need to be tailor-made for the temporal setting.

Modeling and discovering burstiness on sequential data. Modeling and discovering bursts in time sequences is a very well-studied topic in data mining. In a seminal work, Kleinberg [14] discovered burstiness using an exponential model over the delays between the events. Alternative techniques are based on modeling event counts in a sliding window: Ihler et al. [11] modeled such a statistic with Poisson process, while Fung et al. [5] used Binomial distribution. Additionally, Zhu and Shasha [26] used wavelet analysis, Vlachos et al. [23] applied Fourier analysis, and He and Parker [8] adopted concepts from Mechanics to discover burst events. Finally, Lappas et al. [15] propose discovering maximal bursts with large discrepancy.

A highly related problem for discovering bursty events is segmentation. Here the goal is to segment the sequence in k coherent pieces. One should expect that time periods of high activity will occur in its own segment. If the overall score is additive with respect to the segments, then this problem can be solved in $\mathcal{O}(n^2 k)$ time [3]. Moreover, under some mild assumptions we can obtain a $(1 + \epsilon)$ approximation in linear time [7].

The difference of all these works with our setting is that we consider networked data, i.e., sequences of interactions among pairs of entities. By assuming that for each interaction only one entity needs to be active, our problem becomes highly combinatorial. In order to counter-balance this increased combinatorial complexity, we consider a simpler burstiness model than previous works: in particular, we assume that each entity has only one activity interval. Extending our definition to more complex activity models (multiple intervals per entity, or multiple activity levels) is left for future work.

Event detection in temporal data. As the input to our problem is a sequence of temporal edges, our work falls in the broad area of mining *temporal networks* [9,19]. More precisely, the network-untangling problem can be considered an event-detection problem, where the goal is to find time intervals and/or sets of nodes with high activity. Typical event-detection methods use text or other meta-data, as they reveal event semantics. One line of work is based of constructing different types of word graphs [4,18,24]. The events are detected as clusters or connected components in such graphs and temporal information is not considered directly.

Another family of methods uses statistical modeling for identify events as trends [2,17]. Leskovec et al. [16] and Yang et al. [25] consider spreading of short quotes in the citation network of social media. These methods rely on clustering "bursty" keywords. Our setting is considerably different as we focus on interactions between entities and explicitly model entity activity by continuous time intervals.

Information maps. From an application point-of-view, our work is loosely related with papers that aim to process large amounts of data and create maps that present the available information in a succinct and easy-to-understand manner. Shahaf and co-authors have considered this problem in the context of news articles [21,22] and scientific publications [20]. However, their approach is not directly comparable to ours, as their input is a set of documents and not a temporal network, and their output is a "metro map" and not an activity timeline.

7   Experimental Evaluation

In this section we empirically evaluate the performance of our methods.[3]

Setup. We first test the algorithms on synthetic datasets and then present a case study on a real-world social-media dataset.

For the *Synthetic* dataset, we start by generating a static background network of $n = 100$ vertices with a power law degree distribution (we use the configuration model with power law exponent set to 2.0). Then for every vertex we generate a ground-truth activity interval and we add 100 interactions with random neighbors. These interactions are placed consequently with unit time distance, and thus each activity interval has length of $\ell = 99$ time units. We place the ground-truth activity intervals on a timeline in an overlapping manner, and we control their temporal overlap using a parameter $p \in [0, 1]$. When $p = 0$, all intervals are disjoint and every timestamp has only one interaction, thus, it should be easy to find the correct activity intervals. When $p = 1$, all intervals are merged into one, and every time stamp has 100 of different interactions, so there is a large number of solutions whose score is even better than the ground-truth solution. In all cases *Synthetic* has 10 000 interactions in total.

For the case study we use a dataset collected from *Twitter*. The dataset records activity of Twitter users in Helsinki during 12.2008–05.2014. We consider only tweets with more than one hashtag (666 487 tweets) and build the

---

[3] The implementation of all algorithms and scripts used for the experimental evaluation is available at https://github.com/polinapolina/the-network-untangling-problem.

co-occurrence network of these hashtags: vertices corresponding to hashtags and time-stamped edges corresponding to a tweet in which two hashtags are mentioned. The temporal network contains 304 573 vertices and 3 292 699 edges.

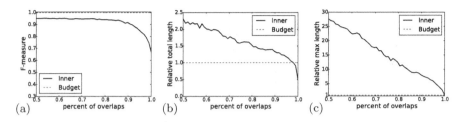

Fig. 1. Output of both algorithms for different overlaps p in the ground truth activity intervals. All values are averaged over 100 runs. (a) F-measure of correctly identifies active time-stamped vertices, (b) L, total activity interval length divided by true total activity interval length, (c) M, maximum activity interval length divided by true maximum activity interval length.

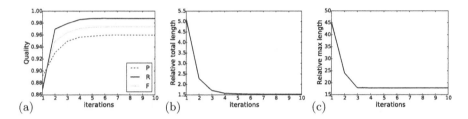

Fig. 2. Convergence of Maximal algorithm. Overlap p is set to 0.5, values are averaged over 100 runs. (a) Precision, recall and F-measure, (b) L, relative total length, (c) M, relative length of the maximum interval.

Results from synthetic datasets. To evaluate the quality of the discovered activity intervals we compare the set of discovered intervals with the ground-truth intervals. For every vertex u we define precision $P_u = \frac{|TP_u|}{|F_u|}$, where TP_u is the set of correctly identified moments of activity of u, and F_u is the set of all discovered moments of activity of u. Similarly, we define the recall for vertex u as $R_u = \frac{|TP_u|}{|A_u|}$, where A_u is the set of true moments of activity of u. We calculate the average precision and recall: $P = \frac{1}{|V|} \sum_{u \in V} P_u$ and $R = \frac{1}{|V|} \sum_{u \in V} R_u$; and report the F-measure $F = \frac{2PR}{P+R}$.

In addition to F-measure, we calculate the relative total length L and the relative maximum length M. Here, L is the total length of the discovered intervals divided by the ground-truth total length of the activity intervals. Similarly, M is the maximum length of the discovered intervals divided by the true maximum length of activity intervals.

We test both algorithms on the *Synthetic* dataset with varying overlap parameter p. The results are shown in Fig. 1. All measures are averaged over

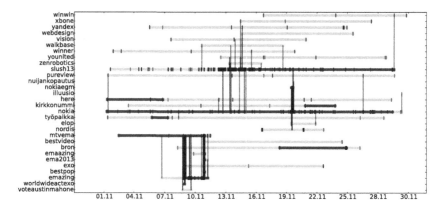

Fig. 3. Part of the output of `Maximal` algorithm on Twitter dataset for November'13. Intervals of activity of co-occurring tags, seeded from hashtags `#slush13`, `#mtvema` and `#nokiaemg`.

100 runs. Note that in the *Synthetic* dataset all activity intervals have the same length, thus, if during binary search the correct value of budget is found, then automatically all vertices receive the correct budget.

Figure 1a demonstrates that for algorithm `Maximal` the F-measure is typically high for all values of the overlap parameter, but drops, when p increases. On the other hand, Fig. 1b shows that algorithm `Maximal` takes advantage of the overlaps and for large values of p it finds solutions that have better score than the ground truth. This however, leads to decrease in accuracy. As for the maximum interval length, shown in Fig. 1c, algorithm `Maximal` is not designed to optimize it and it typically finds few large intervals, while keeping the total length low. `Budget` finds solutions of correct total and maximum lengths on the *Synthetic* dataset for all values of overlap parameter p.

In Fig. 2 we show how the solution of `Maximal` evolves during iterations with re-initialization. After a couple of iterations the value and quality (F-measure, precision and recall) of the solution are improved significantly. During the next iterations the value of the solution does not change, but the quality keeps increasing. The method converges in less than 10 iterations.

Scalability. Both `Budget` and `Inner` use linear-time algorithms in their inner loops and the number of needed outer loop iterations is small. This means that our methods are scalable. To demonstrate this, we were able to run `Maximal` with a network of 1 million vertices and 1 billion interactions in 15 min, despite the large constant factor due to the Python implementation.

Case study. Next we present our results on the *Twitter* dataset. In Fig. 3 we show a subset of hashtags from tweets posted in November 2013. We also depict the activity intervals for those hashtags, as discovered by algorithm `Maximal`. Note that for not cluttering the image, we depict only a subset of all relevant hashtags. In particular, we pick 3 "seed" hashtags: `#slush13`, `#mtvema` and #nokiaemg and the set of hashtags that co-occur with the "seeds." Each of the seeds corresponds to a known event: #slush13 corresponds to Slush'13 – the world's leading startup and tech event, organized in Helsinki in November 13–14, 2013. #mtvema is dedicated to MTV Europe Music Awards, held on 10 November, 2013. #nokiaemg is Extraordinary General Meeting (EGM) of Nokia Corporation, held in Helsinki in November 19, 2013.

For each hashtag we plot its entire interval with a light color, and the discovered activity interval with a dark color. For each selected hashtag, we draw interactions (co-occurrence) with other selected hashtags using black vertical lines, while we mark interactions with non-selected hashtags by ticks.

Figure 3 shows that the tag #slush13 becomes active exactly at the starting date of the event. During its activity this tag covers many technical tags, e.g. #zenrobotics (Helsinki-based automation company), #younited (personal cloud service by local company) and #walkbase (local software company). Then on 19 November, the tag #nokiaemg becomes active: this event is very narrow and covers mentions of Microsoft executive Stephen Elop. Another large event is occurring around 10 November with active tags #emazing, #ema2013 and #mtvema. They cover #bestpop, #bestvideo and other related tags.

8   Conclusions

In this paper we introduced and studied a new problem, which we called network untangling. Given a set of temporal undirected interactions, our goal is to discover activity time intervals for the network entities, so as to explain the observed interactions. We consider two settings: MINTIMELINE, where we aim to minimize the total sum of activity-interval lengths, and MINTIMELINE$_\infty$, where we aim to minimize the maximum interval length. We show that the former problem is **NP**-hard and we develop efficient iterative algorithms, while the latter problem is solvable in polynomial time.

There are several natural open questions: it is not known whether there is an approximation algorithm for MINTIMELINE or whether the problem is inapproximable. Second, our model uses one activity interval for each entity. A natural extension of the problem is to consider k intervals per entity, and/or different activity levels.

Acknowledgements. This work was supported by the Tekes project "Re:Know," the Academy of Finland project "Nestor" (286211), and the EC H2020 RIA project "SoBigData" (654024).

References

1. Aspvall, B., Plass, M.F., Tarjan, R.E.: A linear-time algorithm for testing the truth of certain quantified boolean formulas. IPL **14**(4), 195 (1982)
2. Becker, H., Naaman, M., Gravano, L.: Beyond trending topics: Real-world event identification on Twitter. In: ICWSM (2011)

3. Bellman, R.: On the approximation of curves by line segments using dynamic programming. CACM **4**(6), 284 (1961)
4. Cataldi, M., Di Caro, L., Schifanella, C.: Emerging topic detection on twitter based on temporal and social terms evaluation. In: MDMKDD (2010)
5. Fung, G.P.C., Yu, J.X., Yu, P.S., Lu, H.: Parameter free bursty events detection in text streams. In: VLDB (2005)
6. Gary, M.R., Johnson, D.S.: Computers and intractability: a guide to the theory of NP-completeness (1979)
7. Guha, S., Koudas, N., Shim, K.: Approximation and streaming algorithms for histogram construction problems. TODS **31**(1), 396–438 (2006)
8. He, D., Parker, D.S.: Topic dynamics: an alternative model of bursts in streams of topics. In: KDD (2010)
9. Holme, P., Saramäki, J.: Temporal networks. Phys. Rep. **519**(3), 97–125 (2012)
10. Hopcroft, J.E., Ullman, J.D.: Data Structures and Algorithms, vol. 175. Addison-Wesley, Boston (1983)
11. Ihler, A., Hutchins, J., Smyth, P.: Adaptive event detection with time-varying Poisson processes. In: KDD (2006)
12. Karp, R.M.: Reducibility among combinatorial problems. In: Complexity of computer computations (1972)
13. Khot, S., Regev, O.: Vertex cover might be hard to approximate to within $2 - \varepsilon$. JCSS **74**(3), 335–349 (2008)
14. Kleinberg, J.: Bursty and hierarchical structure in streams. DMKD **7**(4), 373–397 (2003)
15. Lappas, T., Arai, B., Platakis, M., Kotsakos, D., Gunopulos, D.: On burstiness-aware search for document sequences. In: KDD (2009)
16. Leskovec, J., Backstrom, L., Kleinberg, J.: Meme-tracking and the dynamics of the news cycle. In: KDD (2009)
17. Mathioudakis, M., Koudas, N.: TwitterMonitor: trend detection over the Twitter stream. In: KDD (2010)
18. Meladianos, P., Nikolentzos, G., Rousseau, F., Stavrakas, Y., Vazirgiannis, M.: Degeneracy-based real-time sub-event detection in Twitter stream. In: ICWSM (2015)
19. Michail, O.: An introduction to temporal graphs: an algorithmic perspective. Internet Math. **12**(4), 239–280 (2016)
20. Shahaf, D., Guestrin, C., Horvitz, E.: Metro maps of science. In: KDD (2012)
21. Shahaf, D., Guestrin, C., Horvitz, E.: Trains of thought: generating information maps. In: WWW (2012)
22. Shahaf, D., Yang, J., Suen, C., Jacobs, J., Wang, H., Leskovec, J.: Information cartography: creating zoomable, large-scale maps of information. In: KDD (2013)
23. Vlachos, M., Meek, C., Vagena, Z., Gunopulos, D.: Identifying similarities, periodicities and bursts for online search queries. In: SIGMOD (2004)
24. Weng, J., Lee, B.S.: Event detection in Twitter. In: ICWSM (2011)
25. Yang, J., Leskovec, J.: Patterns of temporal variation in online media. In: WSDM (2011)
26. Zhu, Y., Shasha, D.: Efficient elastic burst detection in data streams. In: KDD (2003)

TransT: Type-Based Multiple Embedding Representations for Knowledge Graph Completion

Shiheng Ma[1], Jianhui Ding[1], Weijia Jia[1(✉)], Kun Wang[1,2], and Minyi Guo[1]

[1] Shanghai Jiao Tong University, Shanghai 200240, China
{ma-shh,ding-jh}@sjtu.edu.cn, {jia-wj,guo-my}@cs.sjtu.edu.cn
[2] Nanjing University of Posts and Telecommunications, Nanjing 210042, China
kwang@njupt.edu.cn

Abstract. Knowledge graph completion with representation learning predicts new entity-relation triples from the existing knowledge graphs by embedding entities and relations into a vector space. Most existing methods focus on the structured information of triples and maximize the likelihood of them. However, they neglect semantic information contained in most knowledge graphs and the prior knowledge indicated by the semantic information. To overcome this drawback, we propose an approach that integrates the structured information and entity types which describe the categories of entities. Our approach constructs relation types from entity types and utilizes type-based semantic similarity of the related entities and relations to capture prior distributions of entities and relations. With the type-based prior distributions, our approach generates multiple embedding representations of each entity in different contexts and estimates the posterior probability of entity and relation prediction. Extensive experiments show that our approach outperforms previous semantics-based methods. The source code of this paper can be obtained from https://github.com/shh/transt.

Keywords: Knowledge graph · Representation learning
Multiple embedding

1 Introduction

Knowledge graphs (KGs) have become a key resource for artificial intelligence applications including question answering, recommendation system, knowledge inference, etc. Recently years, several large-scale KGs, such as Freebase [2], DBpedia [1], NELL [4], and Wikidata [25], have been built by automatically extracting structured information from text and manually adding structured information according to human experiences. Although large-scale KGs have contained billions of triples, the extracted knowledge is still a small part of the real-world knowledge and probably contains errors and contradictions. For example, 71% of people in Freebase have no known place of birth, and 75% have

© Springer International Publishing AG 2017
M. Ceci et al. (Eds.): ECML PKDD 2017, Part I, LNAI 10534, pp. 717–733, 2017.
https://doi.org/10.1007/978-3-319-71249-9_43

no known nationality [5]. Therefore, knowledge graph completion (KGC) is a crucial issue of KGs to complete or predict the missing structured information based on existing KGs.

A typical KG transforms real-world and abstract information into triples denoted as (head entity, relation, tail entity), (h, r, t) for short. To complete or predict the missing element of triples, such as $(h, r, ?)$, $(h, ?, t)$, $(?, r, t)$, representation learning (RL) is widely deployed. RL embeds entities and relations into a vector space, and has produced many successful translation models including TransE [3], TransH [26], TransR [16], TransG [29], etc. These models aim to generate precise vectors of entities and relations following the principle $h + r \approx t$, which means t is translated from h by r.

Most RL-based models concentrate on structured information in triples and neglect the rich semantic information of entities and relations, which is contained in most KGs. Semantic information includes types, descriptions, lexical categories and other textual information. Although these models have significantly improved the embedding representations and increased the prediction accuracy, there is still room for improvement by exploiting semantic information in the following two aspects.

Representation of entities. One of the main obstacles of KGC is the polysemy of entities or relations, i.e., each entity or relation may have different semantics in different triples. For example, in the triple (Isaac_Newton, birthplace, Lincolnshire), Newton is a person, while in (Isaac_Newton, author_of, Opticks), Newton is a writer or physicist. This is a very common phenomenon in KGs and it causes difficulty in vector representations. Most works focus on entity polysemy and utilize linear transformations to model different semantics of an entity in different triples to attain the high accuracy. However, they represent each entity as a single vector which cannot capture the uncertain semantics of entities. This is a critical limitation for modeling the rich semantics.

Estimation of posterior probability. Another problem of most previous works is the neglect of prior probability of known triples. Most previous works optimize the maximum likelihood (ML) estimation of vector representations. Few models discuss the posterior probability, which incorporates a prior distribution to augment optimization objectives. Specifically, previous ML models essentially maximize the probability $p(h, r, t)$ that h, r, t form a triple $p(h, r, t)$. When predicting the missing tail of $(h, r, ?)$, however, h and r are already known and they may influence the possible choices of t. Thus, the posterior probability $p(t \mid h, r)$ of predicting t is a more accurate expression of optimization goals than $p(h, r, t)$. In another word, we could prune the possible choices based on the prior probability of the missing element in a triple.

To address the two issues above, we propose a type-based multiple embedding model (TransT). TransT fully utilizes the entity type information which represents the categories of entities in most KGs. Compared with descriptions and other semantic information, types are simpler and more specific because types of an entity are unordered and contain less noise. Moreover, we can construct or extend entity types from other semantic information, if there is no explicit type information in a KG. For example, in Wordnet, we can construct types from the lexical categories of entities. Other semantic information does not have this advantage. In addition to entity types, we construct multiple types of relations from common types of related entities. We measure the semantic similarity of entities and relations based on entity types and relation types. With this type-based semantic similarity, we integrate type information into entity representations and prior estimation which are detailed below.

We model each entity as multiple semantic vectors with type information to represent entities more accurately. Different from using semantics-based linear transformations to separate the mixed representation [16,26,27,31], TransT models the multiple semantics separately and utilizes the semantic similarity to distinguish entity semantics. In order to capture entity semantics accurately, we dynamically generate new semantic vectors for different contexts.

We utilize the type-based semantic similarity to incorporate prior probability in the optimization objective. It is inspired by the observation that the missing element of a triple semantically correlates to the other two elements. Specifically, all entities appearing in the head (or tail) with the same relation have some common types, or these entities have some common type owned by the entities appearing in the tail (or head). In the "Newton" example mentioned above, if the head of (Isaac_Newton, author_of, Opticks) is missing, we can predict the head is an entity with "author" or "physicist" since we know the relation is "author_of" and the tail is "Opticks", a physics book. Therefore, we design a type-based semantic similarity based on the similarity of type sets. With this similarity, TransT captures the prior probability of missing elements in triples for the accurate posterior estimation.

Our contributions are summarized as follows:

- We propose a new approach for fusing structured information and type information. We construct multiple types of relations from entity types and design the type-based semantic similarity for multiple embedding representations and prior knowledge discovering.
- We propose a multiple embedding model that represents each entity as multiple vectors with specific semantics.
- We estimate prior probabilities for entity and relation predictions based on the semantic similarity between elements of triples in KGs.

The rest of this paper is organized as follows. Section 2 shows the recent studies of KGC. Section 3 introduces our approach including multiple embedding model, prior probability estimation, and objective function optimization. Section 4 displays the evaluation of our approach on FB15K and WN18. Section 5 concludes the paper.

2   Related Work

TransE [3] proposes the principle $h + r \approx t$ to assign a single vector for each entity and relation by minimizing the energy function $\|h + r - t\|$ of every triple.

It is a simple and efficient model but unable to capture the rich semantics of entities and relations.

Some models revise function $\| \cdot \|$ in the energy functions for the complex structures in KGs. TransA [13] adaptively finds the optimal loss function without changing the norm function. Tatec [7] utilizes canonical dot products to design different energy functions for different relations. HolE [20] designs the energy function based on a tensor product which captures the interaction in features of entities and relations. ComplEx [24] represents entities and relations as complex-number vectors and calculates Hermitian dot product in the energy function. ManifoldE [28] expands the position of triples from one point to a hyperplane or sphere and calculates energy function for the two manifolds. KG2E [9] models the uncertainty of entities and relations by Gaussian embedding and defines KL divergence of entity and relation distributions as the energy function. ProjE [22] proposes a neural network model to calculate the difference between $h + r$ and t.

Some models design $\| \boldsymbol{h}_r + \boldsymbol{r} - \boldsymbol{t}_r \|$ to make an entity vector adaptive to different relations. They aim to find appropriate representations of $\boldsymbol{h}_r$ and $\boldsymbol{t}_r$. TransH [26] projects entity vector into hyperplanes of different relations. It represents $\boldsymbol{h}_r$ as the projection vector of $\boldsymbol{h}_r$ on the relation hyperplanes. TransR [16] adjusts entity vectors by transform matrices instead of projections. It represents $\boldsymbol{h}_r$ as the result of linear transformation of $\boldsymbol{h}$. TranSparse [12] considers the transform matrix should reflect the heterogeneous and imbalance of entity pairs and improves the transform matrix into two sparse matrices corresponding to the head entity and the tail entity respectively. TransG [29] considers relations also have multiple semantics like entities. It generates multiple vectors for each relation.

Semantic information, such as types, descriptions, and other textual information, is an important supplement to structured information in KGs. DKRL [30] represents entity descriptions as vectors for tuning the entity and relation vectors. SSP [27] modifies TransH by using the topic distribution of entity descriptions to construct semantic hyperplanes. Entity descriptions are also used to derive a better initialization for training models [17]. With type information, type-constraint model [14] selects negative samples according to entity and relation types. TKRL [31] encodes type information into multiple representations in KGs with the help of hierarchical structures. It is a variant of TransR with semantic information and it is the first model introducing type information. However, TKRL also neglects the two issues mentioned above.

There are several other approaches to modeling KGs as graphs. PRA [15] and SFE [8] predict missing relations from existing paths in KGs. These approaches consider that sequences of relations in paths between two entities can comprise the relation between the two entities. RESCAL [21], PITF [6] and ARE [19] complete KGs through retrieving their adjacent matrices. These approaches need to process large adjacent matrices of entities.

3 Methodology

3.1 Overview

The goal of our model is to obtain the vector representations of entities and relations, which maximize the prediction probability over all existing triples. The prediction probability is a conditional probability because except the missing element, the rest two elements in a triple are known. Specifically, when predicting the tail entity for a triple (h, r, t), we expect to maximize the probability of t under the condition that the given triple satisfies the principle $h + r \approx t$ and the head entity and relation are h and r. We denote this conditional probability entity as $p(t \mid h, r, true)$ which means triple $(h, r, *)$ is "true". "true" represents the triple satisfies $h + r \approx t$ principle. "true" triples are also called correct triples in this paper. Maximizing this probability is the aim of the tail prediction. According to Bayes' theorem [10], $p(t \mid h, r, true)$ can be seen as a posterior probability and its correlation with the prior probability is derived as

$$p(t \mid h, r, true) = \begin{cases} \frac{p(true \mid h,r,t)\, p(t \mid h,r)}{p(true \mid h,r)} & p(t \mid h, r) \neq 0 \\ 0 & p(t \mid h, r) = 0, \end{cases} \tag{1}$$

where $p(true \mid h, r, t)$ is the probability that (h, r, t) is "true", $p(t \mid h, r)$ is the prior probability of t. To obtain the most possible entity, we can only compare probabilities of triples $(h, r, *)$. All these probabilities have the same $p(t \mid h, r)$. Thus, we can omit $p(true \mid h, r)$ in (1):

$$p(t \mid h, r, true) \propto p(true \mid h, r, t)\, p(t \mid h, r). \tag{2}$$

Similarly, the objective of the head prediction is

$$p(h \mid r, t, true) \propto p(true \mid h, r, t)\, p(h \mid r, t), \tag{3}$$

and the objective of the relation prediction is

$$p(r \mid h, t, true) \propto p(true \mid h, r, t)\, p(r \mid h, t). \tag{4}$$

All the three formulas have two components: likelihood and prior probability. $p(true \mid h, r, t)$ is the likelihood estimated by the multiple embedding representations. The other component is the prior probability estimated by the semantic similarity. TransT introduces a type-based semantic similarity to estimate the two components and optimizes the vector representations to maximize these posterior probabilities over the training set.

3.2 Type-Based Semantic Similarity

In order to estimate the likelihood and prior probability, we introduce the semantic similarity to measure the distinction of entity semantics with the type information.

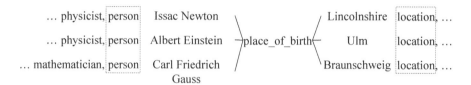

Fig. 1. The entities in the head or tail of a relation have some common types. In this example, all the head entities have "person" type and all the tail entities have "location" type. Therefore, "person" and "location" are the head and tail type of this relation, respectively. Moreover, if we relax this constraint, "physicist" type is also the head type of the relation since most head entities contain this type.

All entities appearing in the head (or tail) with the same relation have some common types. These common types determine this relation as shown in Fig. 1. There are head and tail positions for each relation. Thus, each relation r has two type sets $T_{r,head}$ for entities in the head and $T_{r,tail}$ for entities in the tail. We construct type sets of relations from these common types:

$$T_{r,head} = \bigcap_{\substack{e \in Head_r \\ \rho}} T_e \qquad T_{r,tail} = \bigcap_{\substack{e \in Tail_r \\ \rho}} T_e, \tag{5}$$

where T_e is the type sets of entity e, $Head_r$ and $Tail_r$ are the set of entities appearing respectively in the head and tail with relation r. $\bigcap_\rho$ is a special intersection which contains elements belonging to most of the type sets. This intersection can capture more type information of entities than the normal intersection. However, more information may include more noises. Thus, we balance the influence by the parameter ρ, which is the lowest frequency of types in all T_e.

With the type information of entities and relations, we denote the asymmetric semantic similarity of relations and entities as the following similarity of two sets inspired by Jaccard Index [11]:

$$s(r_{head}, h) = \frac{|T_{r,head} \cap T_h|}{|T_{r,head}|} \quad s(r_{tail}, t) = \frac{|T_{r,tail} \cap T_t|}{|T_{r,tail}|} \quad s(h, t) = \frac{|T_h \cap T_t|}{|T_h|}, \tag{6}$$

where $s(r_{head}, h)$ is the semantic similarity between the relation and the head, $s(r_{tail}, t)$ is the semantic similarity between the relation and the tail, $s(h, t)$ is the semantic similarity between the head and tail.

The type-based semantic similarity plays an important role in the following estimations especially in the prior probability estimation.

3.3   Multiple Embedding Representations

Entities with rich semantics are difficult to be accurately represented in KGC. Thus it is difficult to measure the likelihood $p(true \,|\, h, r, t)$ accurately. In this section, we introduce the multiple embedding representations to capture the entity semantics for the accurate likelihood.

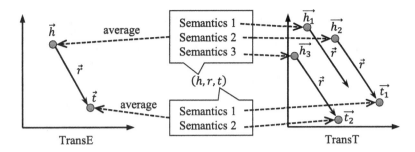

Fig. 2. TransE represents each entity as a single vector which tries to describe all semantics of the entity. Thus the vector representation is not accurate for any entity semantics. In TransT, separate representations of entity semantics describe the relationship among a triple more accurately.

As shown in Fig. 2, there is only one vector representation for one entity in previous work, e.g., TransE. To overcome this drawback, TransT represents each entity semantics as a vector and denotes each entity as a set of semantic vectors. In our approach, we embed each semantics into a vector space. We assume relations have single semantics and entities have multiple semantics. Thus, each relation is represented as a single vector. To adapt the rich entity semantics, we represent each entity as a set of semantic vectors instead of a single vector. Therefore, an entity can be viewed as a random variable of its multiple semantic vectors. Furthermore, the likelihood $p(true \mid h, r, t)$ depends on the expected probability of all possible semantic combinations of random variables h and t. This can define the likelihood of the vector representations for the triple as below

$$p(true \mid h, r, t) = \sum_{i=1}^{n_h} \sum_{j=1}^{n_t} w_{h,i} w_{t,j} p_{true}(v_{h,i}, v_r, v_{t,j}), \tag{7}$$

where n_h and n_t are the number of entity semantics of h and t; $w_h = (w_{h,1}, \ldots, w_{h,n_h})$ and $w_t = (w_{t,1}, \ldots, w_{t,n_t})$ are the distributions of random variables h and t; $v_{h,i}, v_r, v_{t,j}$ are the vectors of h, r, t; $p_{true}(v_{h,i}, v_r, v_{t,j})$ is the likelihood of the component with i-th semantic vector $v_{h,i}$ of h and j-th semantic vector $v_{t,j}$ of t. According to the principle $h + r \approx t$, this likelihood is determined by the difference between $h + r$ and t:

$$p_{true}(v_{h,i}, v_r, v_{t,j}) = \sigma(d(v_{h,i} + v_r, v_{t,j})), \tag{8}$$

where the distance function d measures this difference; the squashing function σ transforms values of d from 0 to $+\infty$ into probability values from 1 to 0 since the probability of a semantic combination is larger if the distance between their corresponding vectors is smaller. To satisfy the property, we set $d(x, y) = \|x - y\|_1$ (1-norm) and $\sigma(x) = e^{-x}$.

In order to capture entity semantics more accurately, we do not assign the specific semantics of entities and the size of their vector sets in advance. We model the generating process of semantic vectors as a random process revised from Chinese restaurant process (CRP), a widely employed form for the Dirichlet process [10]. This avoids the man-made subjectivity for setting n_h and n_t.

In training process, the tail (or head) entity in each triple generates a new semantic vector with the following probability

$$p_{new,tail}(h,r,t) = \left(1 - \max_{t_i \in Semantics_t} s(t_i, r_{tail})\right) \frac{\beta e^{-\|r\|_1}}{\beta e^{-\|r\|_1} + p(true \,|\, h, r, t)}, \quad (9)$$

where β is the scaling parameter in CRP which controls the generation probability. The bracketed formula means t more possibly generates a new semantics when the existing semantics are more different from r; the fraction part is similar to the CRP in TransG [29], which indicates that t possibly generates a new semantics if its current semantic set cannot represent t accurately. Similarly, the new semantic vector of h can be generated with the probability $p_{new,head}(h, r, t)$.

3.4   Prior Probability Estimation

In our model, the prior probability reflects features of a KG from the perspective of semantics. We estimate the prior probabilities (2), (3) and (4) by the type-based semantic similarity.

Note that the type sets of three elements in a triple have obvious relationships. We can estimate the prior distribution of the missing element from the semantic similarity between the missing element and the others.

When we predict t in a triple (h, r, t), the entities with more common types belonging to r and h have higher probability. Therefore, we use the semantic similarity between t and its context $(*, h, r)$ to estimate t's prior probability:

$$p(t \,|\, h, r) \propto s(r_{tail}, t)^{\lambda_{tail}} s(h, t)^{\lambda_{relation}}, \quad (10)$$

where $\lambda_{relation}, \lambda_{head}, \lambda_{tail} \in \{0, 1\}$ are the similarity weights, because h and r have different impacts on the prior probability of t. We use these weights to select different similarity for different situation. Similarly, the prior estimation of head entity h is

$$p(h \,|\, r, t) \propto s(r_{head}, h)^{\lambda_{head}} s(t, h)^{\lambda_{relation}}. \quad (11)$$

By the similar derivation, the prior estimation of relation r is

$$p(r \,|\, h, t) \propto s(r_{head}, h)^{\lambda_{head}} s(r_{tail}, t)^{\lambda_{tail}}. \quad (12)$$

To adapt different datasets, the parameters, $\lambda_{relation}$, λ_{head} and λ_{tail}, should be adjusted.

3.5   Objective Function with Negative Sampling

To achieve the goal of maximizing posterior probabilities, we define the objective function as the sum of prediction errors with negative sampling [18].

For a triple (h, r, t) in the training set Δ, we sample its negative triple $(h', r', t') \notin \Delta$ by replacing one element with another entity or relation. When predicting different elements of a triple, we replace the corresponding elements to obtain the negative triples. Therefore, the prediction error is denoted as a piecewise function:

$$l(h, r, t, h', r', t') = \begin{cases} -\ln p(h \mid r, t, true) + \ln p(h' \mid r, t, true) & h' \neq h \\ -\ln p(t \mid h, r, true) + \ln p(t' \mid h, r, true) & t' \neq t \\ -\ln p(r \mid h, t, true) + \ln p(r' \mid h, t, true) & r' \neq r, \end{cases} \quad (13)$$

where we measure the performance of the probability estimation by the probability difference of the training triple and its negative sample. We define the objective function as the total of prediction errors:

$$\sum_{(h,r,t) \in \Delta} \sum_{(h',r',t') \in \Delta'_{(h,r,t)}} \max\left\{0, \gamma + l(h, r, t, h', r', t')\right\}, \quad (14)$$

where $\Delta'_{(h,r,t)}$ is the negative triple set of (h, r, t).

The total posterior probabilities of predictions are maximized through the minimization of the objective function. Moreover, stochastic gradient descent is applied to optimize the objective function, and we normalize the semantic vectors of entities to avoid overfitting.

4   Experiments

In this paper, we adopt two public benchmark datasets that are the subsets of Freebase and Wordnet, FB15K [3] and WN18 [3], to evaluate our models on knowledge graph completion and triple classification [23]. As for knowledge graph completion, we divide the task into two sub-tasks: entity prediction and relation prediction. Following [3], we split datasets into train, validation and test set. The statistics of datasets are listed in Table 1.

Type information of entities in FB15K has been collected in [31]. There are 4,064 types in FB15K and the average number of types for entities is approximately 12. There is no explicit type information in WN18. Thus we construct type sets of entities from lexical categories. For example, the name of "_trade_name_NN_1" contains its lexical category "NN" (noun), we define the type of "_trade_name_NN_1" as "NN". Because each entity in Wordnet represents the exact semantics, the number of types for entities is 1. There are 4 types in WN18.

The baselines include three semantics-based models: TKRL [31] utilizes entity types; DKRL [30] and SSP [27] take advantage of entity descriptions.

4.1   Entity Prediction

Entity prediction aims at predicting the missing entity when given an entity and a relation, i.e. we predict t given $(h, r, *)$, or predict h given $(*, r, t)$. FB15K and WN18 are the benchmark dataset for this task.

Table 1. Statistics of datasets

Dataset	#Ent	#Rel	#Train	#Valid	#Test
FB15k	14,951	1,345	483,142	50,000	59,071
WN18	40,943	18	141,442	5,000	5,000

Evaluation Protocol. We adopt the same protocol used in previous studies. For each triple (h, r, t) in the test set, we replace the tail t (or the head h) with every entity in the dataset. We calculate the probabilities of all replacement triples and rank these probabilities in descending order. Two measures are considered as evaluation metrics: Mean Rank, the mean rank of original triples in the corresponding probability ranks; HITS@N, the proportion of original triples whose rank is not larger than N. In this task, we use HITS@10. This setting is called "Raw". Some of these replacement triples exist in the training, validation, or test sets, thus ranking them ahead of the original triple is acceptable. Therefore, we filter out these triple to eliminate this case. This filtering setting is called "Filter". In both settings, a higher HITS@10 and a lower Mean Rank mean better performance.

Experiment Settings. As the datasets are the same, we directly reuse the best results of several baselines from the literature [16,26,31]. We have attempted several settings on the validation dataset to get the best configuration. Under the "unif." sampling strategy [26], the optimal configurations are: learning rate $\alpha = 0.001$, vector dimension $k = 50$, margin $\gamma = 3$, CRP factor $\beta = 0.0001$, similarity weights $\lambda_{head} = \lambda_{head} = 0$, $\lambda_{relation}$ is set to 0 or 1 for different relations depending on statistical results of the training set, on WN18; $\alpha = 0.00025$, $k = 300$, $\gamma = 3.5$, $\beta = 0.0001$, $\lambda_{head} = \lambda_{tail} = 1$, $\lambda_{relation} = 0$ on FB15K. We train the model until convergence.

Results. Evaluation results on FB15K and WN18 are shown in Table 2. On FB15K, we compare impacts of multiple vectors and type information. Single or Multiple means entities are represented as single vectors or multiple vectors. Type or no type means type information is used or not. From the result, we observe that:

1. TransT significantly outperforms all baselines on WN18. On FB15K, TransT significantly outperforms all baselines with the filter setting. This demonstrates that our approach successfully utilizes the type information and multiple entity vectors can capture the different semantics of every entity more accurately than linear transformations of single entity vector.
2. Compared with baselines, TransT has the largest difference between the results of Raw and Filter settings on FB15K. This indicates that TransT ranks more correct triples ahead of the original triple. This is caused by the prior estimation of TransT. Specifically, if the predicted element is the head of the original triple, these correct triples have the same relation and tail. Thus, when we learn the prior knowledge from the training set, the head entities

Table 2. Evaluation results on entity prediction

FB15K	Mean rank		HITS@10 (%)	
	Raw	Filter	Raw	Filter
TransE	238	143	46.4	62.1
TransH	212	87	45.7	64.4
TransR	199	77	47.2	67.2
DKRL (CBOW)	236	151	38.3	51.8
DKRL (CNN)	200	113	44.3	57.6
DKRL (CNN)+TransE	181	91	49.6	67.4
TKRL (RHE)	184	68	49.2	69.4
TKRL (WHE+STC)	202	87	50.3	73.4
SSP (Std.)	**154**	77	57.1	78.6
SSP (Joint)	163	82	**57.2**	79.0
TransT (type information)	181	72	54.0	82.3
TransT (multiple vectors)	215	62	50.6	83.6
TransT (multiple+type)	199	**46**	53.3	**85.4**
WN18	Mean Rank		HITS@10 (%)	
	Raw	Filter	Raw	Filter
TransE	263	251	75.4	89.2
TransH	401	338	73.0	82.3
TransR	238	225	79.8	92.0
SSP (Std.)	204	193	81.3	91.4
SSP (Joint)	168	156	81.2	93.2
TransT	**137**	**130**	**92.7**	**97.4**

of these correct triples have higher semantic similarities to the head entity of the original triple than other triples. TransT utilizes these similarities to estimate the prior probability resulting in ranking similar entities higher. In fact, this phenomenon shows that the prior probability improves the prediction performance.

3. There is less difference between the results of Raw and Filter settings on WN18 than FB15K. The reason is that the type-based prior knowledge in WN18 is more accurate than that in FB15K. Specifically, WN18 includes 4 types with simple meanings: noun, verb, adjective and adverb. In addition, an entity in WN18 can only have one type. Thus, types in WN18 have stronger ability to distinguish different entities.

4. Both the two approaches, multiple-vector representation and type information, have their own advantages. Type information performs better in raw setting, while multiple-vector representation performs better in filter setting.

4.2 Relation Prediction

Relation prediction aims at predicting the missing relation when given two entities, i.e., we predict r given $(h, *, t)$. FB15K is the benchmark dataset for this task.

Evaluation Protocol. We adopt the same protocol used in entity prediction. For each triple (h, r, t) in the test set, we replace the relation r with every relation in the dataset. Mean Rank and HITS@1 are considered as evaluation metrics for this task.

Experiment Settings. As the datasets are the same, we directly reuse the experimental results of several baselines from the literature. We have attempted several settings on the validation dataset to get the best configuration. Under the "unif." sampling strategy, the optimal configurations are: learning rate $\alpha = 0.0001$, vector dimension $k = 300$, margin $\gamma = 3.0$, CRP factor $\beta = 0.001$, similarity weights $\lambda_{head} = \lambda_{tail} = 1$, $\lambda_{relation} = 0$.

Table 3. Evaluation results on relation prediction

Method	Mean rank		HITS@1 (%)	
	Raw	Filter	Raw	Filter
TransE	2.91	2.53	69.5	90.2
TransH	8.25	7.91	60.3	72.5
TransR	2.49	2.09	70.2	91.6
DKRL (CBOW)	2.85	2.51	65.3	82.7
DKRL (CNN)+TransE	2.41	2.03	69.8	90.8
TKRL (RHE)	2.12	1.73	71.1	92.8
TKRL (WHE+STC)	2.47	2.07	68.3	90.6
SSP (Std.)	**1.58**	1.22	69.9	89.2
SSP (Joint)	1.87	1.47	70.9	90.9
TransT	1.59	**1.19**	**72.0**	**94.1**

Results. Evaluation results on FB15K are shown in Table 3. From the result, we observe that:

1. TransT significantly outperforms all baselines. Compared with TKRL, which also utilized type information, TransT improves HITS@1 by 3.5% and Mean Rank by 0.88.
2. In the Raw setting, TransT also achieves the best performance. This result is different from the entity prediction task. The reason is more prior knowledge of relation predictions. In the entity prediction task, the prior knowledge is derived from relations. In the relation prediction task, the prior knowledge is

derived from the head and tail entities. The latter has more sources for prior estimation. Thus, TransT ranks more incorrect triples behind the original triple. This further supports the necessity of the prior probability.

4.3   Triple Classification

Triple classification aims at predicting whether a given triple is correct or incorrect, i.e., we predict the correctness of (h, r, t). FB15K is the benchmark dataset of this task.

Evaluation Protocol. We adopt the same protocol used in entity prediction. Since FB15K has no explicit negative samples, we construct negative triples following the same protocol used in [23]. For each triple (h, r, t) in the test set, if the probability of its correctness is below a threshold σ_r, the triple is incorrect; otherwise, it is correct. The thresholds $\{\sigma_r\}$ are determined on the validation dataset with negative samples.

Experiment Settings. As the datasets are the same, we directly reuse the experimental results of several baselines from the literature. We have attempted several settings on the validation dataset to get the best configuration. Under the "unif." sampling strategy, the optimal configurations are: learning rate $\alpha = 0.001$, vector dimension $k = 300$, margin $\gamma = 3.0$, CRP factor $\beta = 0.01$, similarity weights $\lambda_{head} = \lambda_{tail} = \lambda_{relation} = 0$.

Table 4. Evaluation results on triple classification

Method	Accuracy (%)
TransE	85.7
TransH	87.7
TransR	86.4
TKRL (RHE)	86.9
TKRL (WHE+STC)	88.5
TransT	91.0

Results. Evaluation results on FB15K are shown in Table 4. TransT outperforms all baselines significantly. Compared with the best result, TransT improves the accuracy by 2.5% and it is the only model whose accuracy is over 90%. This task shows the ability to discern which triples are correct.

4.4   Semantic Vector Analysis

We analyze the correlations between the semantic vector number and several statistical properties of different entities. We adopt the vector representations obtained by TransT and TransE during the entity prediction task on FB15K.

Figure 3 shows that the correlations between the number of semantic numbers and the average number of relations/types/triples for different entities. For an entity represented by more semantic vectors, it has more types and appears with more different relations and triples in the training set. Thus the entities with more semantic vectors have more complex semantics. Therefore, the result of TransT conforms to our understanding of the entity semantics.

Figure 4 shows that the prediction probabilities of several selected entities. Our approach generates at most 11 semantic vectors for entities. The entities with more semantic vectors have broader concepts. Thus, popular places and people including "Paris", "Alan Turing", have more semantic vectors than awards like "Film Award" and events like "2007 NBA draft". Compared with TransE, multiple semantic vectors improve prediction probability of most entities.

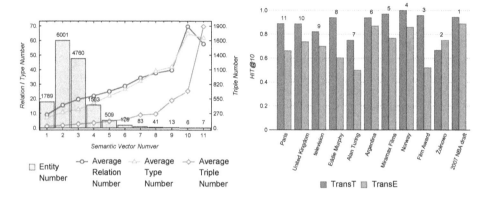

Fig. 3. The bar chart is the number of entities with different semantic numbers. The left y-axis is the number of relations or types. The right y-axis is the number of triples. The x-axis is the number of semantic vectors.

Fig. 4. HIT@10 of 11 entities with different numbers of semantic vectors. The number of semantic vectors are placed above the bars.

5   Conclusion

This paper proposes TransT, a new approach for KGC, which combines structured information and type information. With the type-based prior knowledge, TransT generates semantic vectors for entities in different contexts based on CRP and optimizes the posterior probability estimation. This approach makes full use of type information and accurately captures semantic features of entities. Extensive experiments show that TransT achieves markable improvements against the baselines.

Acknowledgments. This work is supported by Chinese National Research Fund (NSFC) Key Project No. 61532013; National China 973 Project No. 2015CB352401; NSFC No. 61572262; Shanghai Scientific Innovation Act of STCSM No. 15JC1402400; 985 Project of Shanghai Jiao Tong University No. WF220103001; China Postdoctoral Science Foundation No. 2017M610252 and China Postdoctoral Science Special Foundation No. 2017T100297.

References

1. Bizer, C., Lehmann, J., Kobilarov, G., Auer, S., Becker, C., Cyganiak, R., Hellmann, S.: DBpedia - a crystallization point for the web of data. J. Web Semant. **7**(3), 154–165 (2009)
2. Bollacker, K.D., Evans, C., Paritosh, P., Sturge, T., Taylor, J.: Freebase: a collaboratively created graph database for structuring human knowledge. In: Proceedings of the ACM SIGMOD International Conference on Management of Data (SIGMOD), pp. 1247–1250 (2008)
3. Bordes, A., Usunier, N., García-Durán, A., Weston, J., Yakhnenko, O.: Translating embeddings for modeling multi-relational data. In: Advances in Neural Information Processing Systems 26 (NIPS), pp. 2787–2795 (2013)
4. Carlson, A., Betteridge, J., Kisiel, B., Settles, B., Hruschka Jr., E.R., Mitchell, T.M.: Toward an architecture for never-ending language learning. In: Proceedings of the Twenty-Fourth AAAI Conference on Artificial Intelligence (AAAI) (2010)
5. Dong, X., Gabrilovich, E., Heitz, G., Horn, W., Lao, N., Murphy, K., Strohmann, T., Sun, S., Zhang, W.: Knowledge vault: a web-scale approach to probabilistic knowledge fusion. In: The 20th ACM SIGKDD International Conference on Knowledge Discovery and Data Mining (KDD), pp. 601–610 (2014)
6. Drumond, L., Rendle, S., Schmidt-Thieme, L.: Predicting RDF triples in incomplete knowledge bases with tensor factorization. In: Proceedings of the ACM Symposium on Applied Computing (SAC), pp. 326–331 (2012)
7. García-Durán, A., Bordes, A., Usunier, N.: Effective blending of two and three-way interactions for modeling multi-relational data. In: Calders, T., Esposito, F., Hüllermeier, E., Meo, R. (eds.) ECML PKDD 2014. LNCS (LNAI), vol. 8724, pp. 434–449. Springer, Heidelberg (2014). https://doi.org/10.1007/978-3-662-44848-9_28
8. Gardner, M., Mitchell, T.M.: Efficient and expressive knowledge base completion using subgraph feature extraction. In: Proceedings of the 2015 Conference on Empirical Methods in Natural Language Processing (EMNLP), pp. 1488–1498 (2015)
9. He, S., Liu, K., Ji, G., Zhao, J.: Learning to represent knowledge graphs with Gaussian embedding. In: Proceedings of the 24th ACM International Conference on Information and Knowledge Management (CIKM), pp. 623–632 (2015)
10. Hjort, N.L., Holmes, C., Müller, P., Walker, S.G.: Bayesian Nonparametrics, vol. 28. Cambridge University Press, Cambridge (2010)
11. Jaccard, P.: Distribution de la Flore Alpine: dans le Bassin des dranses et dans quelques régions voisines. Rouge (1901)
12. Ji, G., Liu, K., He, S., Zhao, J.: Knowledge graph completion with adaptive sparse transfer matrix. In: Proceedings of the Thirtieth AAAI Conference on Artificial Intelligence (AAAI), pp. 985–991 (2016)

13. Jia, Y., Wang, Y., Lin, H., Jin, X., Cheng, X.: Locally adaptive translation for knowledge graph embedding. In: Proceedings of the Thirtieth AAAI Conference on Artificial Intelligence (AAAI), pp. 992–998 (2016)

14. Krompaß, D., Baier, S., Tresp, V.: Type-constrained representation learning in knowledge graphs. In: Arenas, M., et al. (eds.) ISWC 2015. LNCS, vol. 9366, pp. 640–655. Springer, Cham (2015). https://doi.org/10.1007/978-3-319-25007-6_37

15. Lao, N., Mitchell, T.M., Cohen, W.W.: Random walk inference and learning in a large scale knowledge base. In: Proceedings of the 2011 Conference on Empirical Methods in Natural Language Processing (EMNLP), pp. 529–539 (2011)

16. Lin, Y., Liu, Z., Sun, M., Liu, Y., Zhu, X.: Learning entity and relation embeddings for knowledge graph completion. In: Proceedings of the Twenty-Ninth AAAI Conference on Artificial Intelligence, pp. 2181–2187 (2015)

17. Long, T., Lowe, R., Cheung, J.C.K., Precup, D.: Leveraging lexical resources for learning entity embeddings in multi-relational data. In: Proceedings of the 54th Annual Meeting of the Association for Computational Linguistics (ACL), Short Papers, vol. 2 (2016)

18. Mikolov, T., Sutskever, I., Chen, K., Corrado, G.S., Dean, J.: Distributed representations of words and phrases and their compositionality. In: Advances in Neural Information Processing Systems 26 (NIPS), pp. 3111–3119 (2013)

19. Nickel, M., Jiang, X., Tresp, V.: Reducing the rank in relational factorization models by including observable patterns. In: Advances in Neural Information Processing Systems 27 (NIPS), pp. 1179–1187 (2014)

20. Nickel, M., Rosasco, L., Poggio, T.A.: Holographic embeddings of knowledge graphs. In: Proceedings of the Thirtieth AAAI Conference on Artificial Intelligence (AAAI), pp. 1955–1961 (2016)

21. Nickel, M., Tresp, V., Kriegel, H.: A three-way model for collective learning on multi-relational data. In: Proceedings of the 28th International Conference on Machine Learning (ICML), pp. 809–816 (2011)

22. Shi, B., Weninger, T.: ProjE: Embedding projection for knowledge graph completion. In: Proceedings of the Thirty-First AAAI Conference on Artificial Intelligence, pp. 1236–1242 (2017)

23. Socher, R., Chen, D., Manning, C.D., Ng, A.Y.: Reasoning with neural tensor networks for knowledge base completion. In: Advances in Neural Information Processing Systems 26 (NIPS), pp. 926–934 (2013)

24. Trouillon, T., Welbl, J., Riedel, S., Gaussier, É., Bouchard, G.: Complex embeddings for simple link prediction. In: Proceedings of the 33nd International Conference on Machine Learning (ICML), pp. 2071–2080 (2016)

25. Vrandecic, D.: Wikidata: a new platform for collaborative data collection. In: Proceedings of the 21st World Wide Web Conference (WWW), Companion Volume, pp. 1063–1064 (2012)

26. Wang, Z., Zhang, J., Feng, J., Chen, Z.: Knowledge graph embedding by translating on hyperplanes. In: Proceedings of the Twenty-Eighth AAAI Conference on Artificial Intelligence, pp. 1112–1119 (2014)

27. Xiao, H., Huang, M., Meng, L., Zhu, X.: SSP: semantic space projection for knowledge graph embedding with text descriptions. In: Proceedings of the Thirty-First AAAI Conference on Artificial Intelligence, pp. 3104–3110 (2017)

28. Xiao, H., Huang, M., Zhu, X.: From one point to a manifold: knowledge graph embedding for precise link prediction. In: Proceedings of the Twenty-Fifth International Joint Conference on Artificial Intelligence (IJCAI), pp. 1315–1321 (2016)

29. Xiao, H., Huang, M., Zhu, X.: TransG: a generative model for knowledge graph embedding. In: Proceedings of the 54th Annual Meeting of the Association for Computational Linguistics (ACL), Long Papers, vol. 1 (2016)
30. Xie, R., Liu, Z., Jia, J., Luan, H., Sun, M.: Representation learning of knowledge graphs with entity descriptions. In: Proceedings of the Thirtieth AAAI Conference on Artificial Intelligence (AAAI), pp. 2659–2665 (2016)
31. Xie, R., Liu, Z., Sun, M.: Representation learning of knowledge graphs with hierarchical types. In: Proceedings of the Twenty-Fifth International Joint Conference on Artificial Intelligence (IJCAI), pp. 2965–2971 (2016)

Neural Networks and Deep Learning

A Network Architecture
for Multi-Multi-Instance Learning

Alessandro Tibo[1(✉)], Paolo Frasconi[1], and Manfred Jaeger[2]

[1] Department of Information Engineering, University of Florence,
Via di Santa Marta 3, 50139 Firenze, Italy
{alessandro.tibo,paolo.frasconi}@unifi.it
[2] Department of Computer Science, Aalborg University, Aalborg, Denmark
jaeger@cs.aau.dk

Abstract. We study an extension of the multi-instance learning problem where examples are organized as nested bags of instances (e.g., a document could be represented as a bag of sentences, which in turn are bags of words). This framework can be useful in various scenarios, such as graph classification, image classification and translation-invariant pooling in convolutional neural network. In order to learn multi-multi instance data, we introduce a special neural network layer, called bag-layer, whose units aggregate sets of inputs of arbitrary size. We prove that the associated class of functions contains all Boolean functions over sets of sets of instances. We present empirical results on semi-synthetic data showing that such class of functions can be actually learned from data. We also present experiments on citation graphs datasets where our model obtains competitive results. Code and data related to this chapter are available at: https://doi.org/10.6084/m9.figshare.5442451.

Keywords: Classification · Deep learning · Neural networks
Relational and structured data

1 Introduction

In the multi-instance (MI) setting, data is organized in bags of instances rather than single instances. Only labels of bags are observed, while labels of individual instances are assumed to be unknown. Since the original introduction of this setting in the seminal paper of Dieterich [4], several algorithmic solutions have been proposed in the literature, including diverse density [13] and its expectation-maximization extension [23], and adaptations of various learning algorithms such as kNN [21], neural networks [17,24], and support vector machines [1]. Most methods are based on the classic assumptions that instances are either positive or negative, while bags are positive if and only if they contain at least one positive instance. Several other kinds of underlying assumptions have been formulated (see [6] for a review). In this paper, we extend the MI learning setting by considering data consisting of several nested levels of bags

© Springer International Publishing AG 2017
M. Ceci et al. (Eds.): ECML PKDD 2017, Part I, LNAI 10534, pp. 737–752, 2017.
https://doi.org/10.1007/978-3-319-71249-9_44

of instances. We call this framework multi-multi-instance (MMI) learning, referring specifically to the case of bags-of-bags (the generalization to deeper levels of nesting is however immediate). We focus on the setting where the task is to classify whole bags-of-bags (a setting where instances or sub-bags are to be classified is also conceivable but not addressed in this paper). In our approach, we relax the classic MI assumption of binary instance labels, allowing categorical labels lying in an unspecified alphabet. For MMI learning we propose a method based on neural networks with a special layer called bag-layer. Unlike previous neural network approaches to MI learning, where predicted instance labels are aggregated by the maximum operator, bag-layers aggregate internal representations of instances (or bags of instances) and can be naturally intermixed with other layers commonly used in deep learning. Bag-layers can be in fact interpreted as a generalization of pooling layers commonly used in convolutional neural networks.

The paper is organized as follows. In Sect. 2 we formally introduce the MMI learning setting. In Sect. 3 we present the proposed neural network architecture. In Sect. 4 we discuss some related works. In Sect. 5 we offer a theoretical analysis of the expressiveness of the proposed bag-layer architecture. In Sect. 6 we report an experimental evaluation of our method on some semi-synthetic datasets and on real citation network data. Finally, in Sect. 7 we summarize our findings and discuss possible directions of future research. The code we used for the experiments can be downloaded from https://github.com/alessandro-t/mmi.

2   The Multi-Multi-Instance Learning Setting

For the sake of simplicity, we introduce our setting with data consisting of bags-of-bags of instances. The extension to deeper levels of nesting (i.e., multi^K-instance learning) is immediate. We will also informally use the expression "bag-of-bags" to describe sets with two or more levels of nesting.

Our setting is supervised and the dataset consists of pairs $\{(X^{(i)}, y^{(i)}), i = 1, \ldots, n\}$ where $y^{(i)}$ is a discrete category label and $X^{(i)}$ is a set of sets: $X^{(i)} = \{S_j^{(i)}, j = 1, \ldots, |X^{(i)}|\}$ where $|X|$ denotes the cardinality of X. In turn, $S_j^{(i)} = \{x_{j,\ell}^{(i)}, \ell = 1, \ldots, |S_j^{(i)}|\}$ where $x_{j,\ell}^{(i)} \in \mathcal{X}$ is an *instance* and $\mathcal{X}$ the instance space. We call $X^{(i)}$ *top-bags* and $S_j^{(i)}$ *sub-bags*. We assume that examples are uniformly and independently drawn from a distribution $p(X, y) = p(y|X)p(X)$. Since we take a discriminative approach to supervised learning, we do not make any specific assumptions on $p(X)$. We further assume that *latent* labels are attached to both instances and sub-bags (i.e., only labels attached to top-bags are observed). To simplify notation, we drop the example index (superscript i) when not necessary. We further assume that the label of instance $x_{j,\ell}$, denoted $y_{j,\ell}$, is drawn from a conditional distribution $p(y_{j,\ell}|x_{j,\ell})$, that the label of sub-bag S_j, denoted y_j, is drawn from a conditional distribution $p(y_j|y_{j,1}, \ldots, y_{j,|S_j|})$, and that the label of top-bag X, denoted y, is drawn from a conditional distribution $p(y|y_1, \ldots, y_{|X|})$.

In most cases, both top-bags and sub-bags will actually be sets, i.e., two instances $x_{j,\ell}, x_{j,\ell'}$ contained in a sub-bag S_j will typically be distinct objects, and similarly for the sub-bags contained in a top-bag. However, when considering only the labels of the instances, we obtain proper bags (multi-sets) $\{y_{j,1}, \ldots, y_{j,|S_j|}\}$. Our bag-of-bag terminology is motivated by the fact that we view the data mostly as bags of latent labels, rather than sets of raw data points. By a slight abuse of notation, we also use $\{\}$ to denote multisets. In a context where we speak about multisets, then, for instance, $\{0, 0, 1\} \neq \{0, 1\}$.

Example 1. In this example we consider bags-of-bags of handwritten digits (as in the MNIST dataset). Each instance (a digit) has attached its own category in $\{0, \ldots, 9\}$, which is not observed. We consider binary sub-bag and top-bag labels. Only the top-bag labels are observed. In particular, a sub-bag is positive if it contains an instance of category 7 and does not contain an instance of category 3. A top-bag is positive if it contains at least one positive sub-bag. Figure 1 shows a positive top-bag and a negative top-bag.

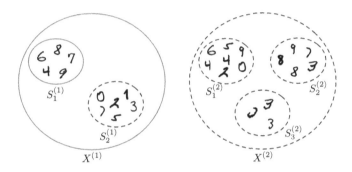

Fig. 1. Two top-bags: on the left a positive top-bag $X^{(1)}$, while on the right a negative top-bag $X^{(2)}$. Solid green lines represent positive bags while dashed red lines represent negative bags. (Color figure online)

3   Model

We model the conditional distribution $p(y|X)$ with a neural network architecture that handles sets of sets of variable sizes by aggregating intermediate internal representations. For this purpose, we introduce a new layer called *bag-layer*. A bag-layer takes as input a bag of vectors $\{\phi_1, \ldots, \phi_n\}$, $\phi_i \in \mathbb{R}^m$ and outputs a $k-$dimensional representation of the bag computed as follows:

$$g(\{\phi_1, \ldots, \phi_n\}; w, b) = \mathop{\Xi}_{i=1}^{n} \alpha \left(w\phi_i + b\right) \qquad (1)$$

where Ξ is component-wise aggregation operator (such as max or average), $w \in \mathbb{R}^{k \times m}$ is the weight matrix, $b \in \mathbb{R}^k$ is the bias, and α is any component-wise activation function (such as ReLU, tanh, or linear). Both w and b are tunable parameters. Note that Eq. 1 works with bags of arbitrary cardinality. A bag-layer is illustrated in Fig. 2. Networks with a single bag-layer can process sets of

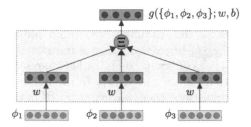

Fig. 2. A bag-layer receiving a bag of cardinality $n = 3$. In this example $k = 4$ and $m = 5$.

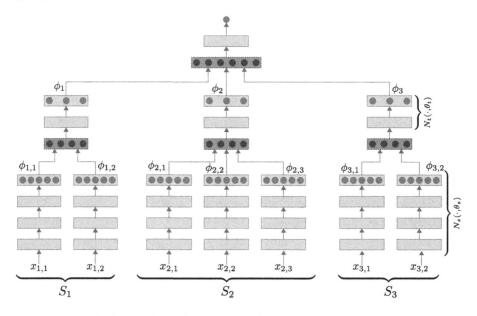

Fig. 3. Network for multi-multi instance learning applied to the bag-of-bags $\{\{x_{1,1}, x_{1,2}\}, \{x_{2,1}, x_{2,2}, x_{2,2}\}, \{x_{3,1}, x_{3,2}\}\}$. Bag-layers are depicted in red with dashed borders. Blue boxes are standard (e.g., dense) neural network layers. Note that parameters θ_s in each of the seven bottom vertical columns are shared, as well parameters θ_t in the middle three columns. (Color figure online)

instances (as in standard multi-instance learning). In order to work in the multi-multi instance setting, two bag-layers are required. The bottom bag-layer aggregates over internal representations of instances, and the top sub-layer aggregates over internal representations of sub-bags, yielding a representation for the entire top-bag. In this case, the representation of each sub-bag $S_j = \{x_{j,1}, \ldots, x_{j,|S_j|}\}$ would be obtained as

$$\phi_j = g(x_{j,1}, \ldots, x_{j,|S_j|}; w_s, b_s) \quad j = 1, \ldots, |X| \tag{2}$$

and the representation of a top-bag $X = \{S_1, \ldots, S_{|X|}\}$ would be obtained as

$$\phi = g(\phi_1, \ldots, \phi_{|X|}; w_t, b_t) \tag{3}$$

where (w_s, b_s) and (w_t, b_t) denote the parameters used to construct sub-bag and top-bag representations.

Note that nothing prevents us from intermixing bag-layers with standard neural network layers, thereby forming networks of arbitrary depth. In this case, each $x_{j,\ell}$ in Eq. (2) would be simply replaced by the last layer activation of a deep network taking $x_{j,\ell}$ as input. Denoting by θ_s the parameters of such network and by $N_s(x; \theta_s)$ its last layer activation when fed with instance x, Eq. (2) becomes

$$\phi_j = g(N_s(x_{j,1}; \theta_s), \ldots, N(x_{j,|S_j|}; \theta_s); w_s, b_s) \quad j = 1, \ldots, |X|. \tag{4}$$

Similarly, we may use a network N_t, with parameters θ_t, to transform sub-bag representations. As a result, Eq. (3) becomes

$$\phi = g(N_t(\phi_1; \theta_t), \ldots, N_t(\phi_{|X|}; \theta_t); w_t, b_t). \tag{5}$$

Of course the top-bag representation can be itself further processed by other layers. An example of the overall architecture is shown in Fig. 3.

4   Related Works

4.1   Multi-instance Neural Networks

The first algorithm for MI learning was based on axis-parallel rectangles [4]. Shortly after, Ramon and De Raedt [17] proposed a neural network solution where each instance x_j in a bag $X = \{x_1, \ldots, x_{|X|}\}$ is first processed by a replica of a neural network f with weights w. In this way, a bag of output values $\{f(x_1; w), \ldots, f(x_{|X|}; w)\}$ computed for each bag of instances. These values are then aggregated by a smooth version of the max function:

$$F(X) = \frac{1}{M} \log \left(\sum_j e^{Mf(x_j; w)} \right)$$

where M is a constant controlling the smoothness of the aggregation. For large enough M, the exact maximum is computed. Note that a single bag-layer (as defined in Sect. 3) can be easily used to learn in the MI setting. Still, a major difference compared to the work of [17] is that bag-layers perform aggregation at the representation level rather than at the output level. In this way, more layers can be added on the top of the aggregated representation, allowing for more expressiveness. In the classic MI setting (where a bag is positive iff at least one instance is positive) this additional expressiveness is not required. However, it allows to work in slightly more complicated MI settings. For example, suppose each instance has a latent variable $y_j \in 0, 1, 2$, and suppose that a bag is positive iff it contains at least one instance with label 0 and no instance with label 2.

In this case, a bag-layer with two units can distinguish positive and negative bags, provided that instance representations can separate instances belonging to the classes $0, 1$ and 2. The network proposed in [17] would not be able to separate positive from negative bags. Indeed, as proved in Sect. 5, networks with bag-layers can represent any Boolean function over sets of instances.

4.2 Convolutional Neural Networks

Convolutional neural networks (CNN) [7,12] are the state-of-the-art method for image classification (see, e.g., [20]). It is easy to see that the representation computed by one convolutional layer followed by max-pooling can be emulated with one bag-layer by just creating bags of adjacent image patches. The representation size k corresponds to the number of convolutional filters. The major difference is that the outputs of a convolutional layer are spatially ordered, whereas a bag-layer outputs a set of vectors (without any ordering). This difference may become significant when two or more layers are sequentially stacked. Figure 4 illustrates the relationship between a convolutional layer and a bag-layer, for simplicity assuming a one-dimensional signal (i.e., a sequence). When applied to signals, a bag-layer essentially correspond to a disordered convolutional layer and its output needs further aggregation before it can be fed into a classifier. The simplest option would be to stack one additional bag-layer before the classification layer. Interestingly, a network of this kind would be able to detect the presence of a short subsequence regardless of its position within the whole sequence, achieving invariance to arbitrarily large translations (an experiment related to translational invariance is presented in Sect. 6.2).

We finally note that it is possible to emulate a CNN with two layers by properly defining the structure of bags-of-bags. For example, a second layer with filter size 3 on the top of the CNN shown in Fig. 4 could be emulated with two bag-layers fed by the bag-of-bags

$$\{\{\{x_{1,1}, x_{1,2}\}, \{x_{2,1}, x_{2,2}\}, \{x_{3,1}, x_{3,2}\}\}, \{\{x_{2,1}, x_{2,2}\}, \{x_{3,1}, x_{3,2}\}, \{x_{4,1}, x_{4,2}\}\}\}.$$

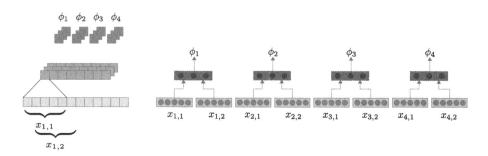

Fig. 4. One convolutional layer with subsampling (left) and the corresponding bag-layer (right). Note that the convolutional layer outputs $[\phi_1, \phi_2, \phi_3, \phi_4]$ whereas the bag-layer outputs $\{\phi_1, \phi_2, \phi_3, \phi_4\}$.

A bag-layer, however, is not limited to pooling adjacent elements in a feature map. One could for example segment the image first (e.g., using a hierarchical strategy [2]) and then create bags-of-bags by following the segmented regions.

The convolutional approach has been also recently employed for learning with graph data. For example the technique proposed in [15] casts graphs into a sequential format suitable for representation learning via CNNs. In [5], CNNs are applied to molecular fingerprint vectors. In [3] a diffusion process across general graph structures generalizes the CNN strategy of scanning a regular grid of pixels or voxels. Finally, [11] propose a neural network model based on spectral graph convolutions. In some settings, the architecture presented in this paper can solve graph learning problems where it is natural to organize graph data as bags-of-bags. Indeed, we present in Sect. 6.3 an application of MMI learning showing that our approach is suitable for citation networks.

4.3 Nested SRL Models

In Statistical Relational Learning (SRL) a great number of approaches have been proposed for constructing probabilistic models for relational data. Relational data has an inherent bag-of-bag structure: each object o in a relational domain can be interpreted as a bag whose elements are all the other objects linked to o via a specific relation. These linked objects, in turn, also are bags containing the objects linked via some relation. A key component of SRL models are the tools employed for aggregating (or combining) information from the bag of linked objects. In many types of SRL models, such an aggregation only is defined for a single level. However, a few proposals have included models for nested combination [9,14]. Like most SRL approaches, these models employ concepts from first-order predicate logic for syntax and semantics, and [9] contains an expressivity result similar in spirit to the one we present in the following Sect. 5.

A key difference between SRL models with nested combination constructs and our MMI network models is that the former build models based on rules for conditional dependencies which are expressed in first-order logic and typically only contain a very small number of numerical parameters (such as a single parameter quantifying a noisy-or combination function for modeling multiple causal influences). MMI network models, in contrast, make use of the high-dimensional parameter spaces of (deep) neural network architectures. Roughly speaking, MMI network models combine the flexibility of SRL models to recursively aggregate over sets of arbitrary cardinalities with the power derived from high-dimensional parameterisations of neural networks.

5 Model Expressiveness

In this section, we focus on a restricted (noiseless) version of the MMI setting described in Sect. 2 where labels are deterministically assigned and no form of counting is involved. We show that under these assumptions, the model of

Sect. 3 has enough expressivity to represent the solution to the MMI learning problem. Our approach relies on classic universal interpolation results for neural networks [8]. Note that existing results hold for vector data, and this section shows that they can be leveraged to bag-of-bag data when using the model of Sect. 3.

Definition 1. *We say that data is generated under the restricted MMI setting if the following conditions hold true:*

1. *instance labels are generated by an unknown function* $\hat{f} : \mathcal{X} \mapsto C = \{c_1, \ldots, c_M\}$, *i.e.,* $y_{j,\ell} = \hat{f}(x_{j,\ell})$, *for* $j = 1, \ldots, |X|$, $\ell = 1, \ldots |S_j|$;
2. *sub-bag labels are generated by an unknown function* $\hat{g} : \mathcal{M}(C) \mapsto D$, *i.e.,* $y_j = \hat{g}(\{y_{j,1}, \ldots, y_{j,|S_j|}\})$, *where* $\mathcal{M}(C)$ *is the set of all multisets of* C;
3. *the top-bag label is generated by an unknown function* $\hat{h} : \mathcal{M}(D) \mapsto \{0,1\}$, *i.e.,* $y = \hat{h}(\{y_1, \ldots, y_{|X|}\})$.

When data is generated in the restricted setting, then the label of a bag-of-bags

$$X = \{\{x_{1,1}, \ldots, x_{1,S_1}\}, \ldots, \{x_{|X|,1}, \ldots, x_{|X|,|S_{|X|}|}\}\},$$

would be produced as

$$y = \hat{h}\left(\left\{\hat{g}\left(\{\hat{f}(x_{1,1}), \ldots, \hat{f}(x_{1,|S_1|})\}\right), \ldots, \hat{g}\left(\{\hat{f}(x_{|X|,1}), \ldots, \hat{f}(x_{|X|,|S_{|X|}|})\}\right)\right\}\right)$$

Note that the classic MI learning formulation [13] is recovered when examples are sub-bags, $C = \{0,1\}$, and $\hat{g}(\{y_1, \ldots, y_{|S|}\}) = \bigvee_{\ell=1}^{|S|} y_\ell$. Other generalized MI learning formulations [6,18,22] can be similarly captured in this restricted setting.

For a multiset s let $set(s)$ denote the set of elements occurring in s. E.g. $set(\{0,0,1\}) = \{0,1\}$.

Definition 2. *We say that data is generated under the non-counting restricted MMI setting if, in addition to the conditions of Definition 1, both* $\hat{g}(s)$ *and* $\hat{h}(s)$ *only depend on* $set(s)$.

The following result indicates that a network containing a bag-layer with max aggregation is sufficient to compute the functions that label both sub-bags and top-bags.

Lemma 1. *Let* $C = \{c_1, \ldots, c_M\}, D = \{d_1, \ldots, d_L\}$ *be sets of labels, and let* $\hat{g} : \mathcal{M}(C) \mapsto D$ *be a labeling function for which* $\hat{g}(s) = \hat{g}(s')$ *whenever* $set(s) = set(s')$. *Then there exist a network with one bag-layer that computes* $\hat{g}$.

Proof. We construct a network N where first a bag-layer maps the multiset input s to a bit-vector representation of $set(s)$, on top of which we can then compute $\hat{g}(s)$ using a standard architecture for Boolean functions.

In detail, N is constructed as follows: the input $s = \{y_1, \ldots, y_{|s|}\}$ is encoded by $|s|$ M-dimensional vectors ϕ_i containing the one-hot representations of the y_i.

We construct a bag-layer with $k = m = M$, w is the $M \times M$ identity matrix, b is zero, α is the identity function, and Ξ is *max*. The output of the bag-layer then is an M-dimensional vector ψ whose i'th component is the indicator function $\mathbb{I}[c_i \in s]$.

For each $j \in \{1, \ldots, L\}$ we can write the indicator function $\mathbb{I}[\hat{g}(set(s)) = d_j]$ as a Boolean function of the indicator functions $\mathbb{I}[c_i \in s]$. Using standard universal approximation results (see, e.g., [8], Corollary 2.5) we can construct a network that on input ψ computes $\mathbb{I}[\hat{g}(set(s)) = d_j]$. L such networks in parallel then produce an L-dimensional output vector containing the one-hot representation of $\hat{g}(s)$.

Theorem 1. *Given a dataset of examples generated under the non-counting restricted MMI setting, there exist a network with two bag-layers that can correctly label all examples in the dataset.*

Proof. We first note that the universal interpolation result of [8] can be applied to a network taking as input an instance x and generating the desired label $\hat{f}(x)$. We then use Lemma 1 twice, first to form a network that computes the sub-bag labeling function $\hat{g}$, and then to form a network that computes the top-bag labeling function $\hat{h}$.

6   Experiments

We evaluated our model on three experimental setups:

1. we constructed a multi-multi instance semi-synthetic dataset from MNIST, in which digits were organized in bags-of-bags of arbitrary cardinality. This setup is a complex version of the example shown in Sect. 3. The aim of this experiment is to show the ability of the network to learn functions that have generated the data according to Theorem 1 in Sect. 5;
2. we constructed semi-synthetic dataset from MNIST, placing digits randomly into a background images of black pixels. The aim of this experiment is to show the disordered convolution property discussed in Sect. 4.2;
3. we focused on real citation network datasets where data can be naturally decomposed into bags-of-bags (MMI data) or bags (MI data). The goal is to understand whether MMI and MI decompositions are reasonable representations for citation networks and whether MMI representation is more suitable than MI representation. Finally we compared our approach with the state-of-art architecture.

6.1   Learning in the Restricted Setting

The results of Sect. 5 show that MMI networks can represent any labelling function in the non-counting restricted case. We show here that MMI networks trained by gradient descent can actually *learn* such functions from MMI data. The setup is similar to Example 1 in Sect. 2, but the classification rule is more complicated. Starting from digit images extracted from the MNIST dataset, we constructed an MMI datasets where each top-bag X is a set of sets of images. For each sub-bag S_j, we denote by $Y_j = \{y_{j,1} \ldots, y_{j,|S_j|}\}$ the set of instance labels in sub-bag S_j, where $y_{j,\ell}$ are derived from the MNIST dataset labels. We then labeled each sub-bag according to the following rules:

- sub-bag S_j has label 0 iff one of the following conditions is satisfied: $\{1,3,5,7\} \subset Y_j$, $\{2,1,3\} \subset Y_j$, $\{3,2,7,9\} \subset Y_j$;
- sub-bag S_j has label 1 iff S_j has not label 0 and one of the following conditions is satisfied: $\{8,9\} \subset Y_j$, $\{4,5,6\} \subset Y_j$, $\{7,2,1\} \subset Y_j$;
- sub-bag S_j has label 2 iff S_j has not label 0 and S_j has not label 1.

Finally each top-bags is positive iff it contains at least a sub-bag of class 1. Observe that data generated according to those rules satisfied the conditions of Definitions 1 and 2. Using these rules, we generated a balanced training set of 5000 top-bags and a balanced test set of 5000 top-bags. Sub-bag and top-bag cardinalities were uniformly sampled in the interval $[2, 6]$. Instances in the training set and in the test set were randomly sampled with replacement from the 60,000 MNIST training images the 10,000 MNIST test images, respectively.

The structure of neural network we used in this experiment is summarized in Table 1. The model was trained by minimizing the L2-regularized binary cross-entropy loss (with L2 penalty $5 \cdot 10^{-4}$). We stress the fact that instance and sub-bag labels were not used to form the training objective. We ran 200 epochs of the Adam optimizer [10] with learning rate 0.001. Results in Table 2 confirm

Table 1. Neural network structure for multi-multi instance MNIST dataset

Layer	Parameters
Convolutional layer Batch normalization ReLU	Kernel size 5×5 with 32 channels
Max pooling	Kernel size 2×2
Convolutional layer Batch normalization ReLU	Kernel size 5×5 with 64 channels
Max pooling	Kernel size 2×2
Dense ReLU	1024 units
Dropout	Probability 0.5
BagLayer (linear activation) ReLU	100 units
BagLayer (linear activation) ReLU	100 units
Dense	1 unit

Table 2. Accuracies on multi-multi instance MNIST dataset

Set	Loss	Accuracy
Training	0.063	98.60%
Test	0.108	96.44%

that the network is able to recover the latent logic function that was used in the data generation process with a reasonably high accuracy.

6.2   Multi-Multi Instance as Disordered Pooling

As discussed in Sect. 4.2, bag-layers can be used as disordered convolutional layers. The aim of the following experiment is to show this property on a simple scenario, and to compare results with a standard convolutional approach. We first constructed a dataset $MNIST^{EXT}$ by placing the 28×28 pixels MNIST digit images on a black background of size $w \times w$ (we generated datasets for several values of w, i.e. $w \in \{105, 135, 165, 195\}$). $MNIST^{EXT}$ images are labelled with the digit they contain. The goal is to classify digits regardless of their location within the images. Our purpose here is to show how the location-invariance of the bag-layer function can be exploited to obtain robust classification results for raw image data without centering or cropping preprocessing. As MNIST, the $MNIST^{EXT}$ dataset, is split in training and test sets containing respectively 60,000 images and 10,000 images. Digits of training and test images are placed in the top and bottom half of the background, respectively. On the same datasets we trained a MMI network and a standard convolutional network.

Concerning the MMI network we constructed MMI data as follows: each top-bag X represents an image in $MNIST^{EXT}$. For each we extracted *macro-patches* of size 15×15 and consecutive macro-patches are overlapped by 5 pixels. Each macro-patch represents a sub-bag $S_j \in X$. For each macro-patch we extracted *micro-patches* of size 5×5 and consecutive micro-patches are overlapped by 3 pixels. Each micro-patch represents an instance $x_{j,\ell} \in S_j$. Note that the choice of the micro-patches, macro-patches and w sizes is in general arbitrary. Here we chose those specific values for avoiding, as far as possibile, the zero padding for adapting the sizes and then waste of computational resources. Furthermore choosing w fixed for each experiment is crucial in order to compare the MMI network with the convolutional neural network. Indeed while our approach can handle bags of different size the convolutional model requires that input images have fixed size. The structure of the MMI network we used in this experiment is summarized in Table 3. The model was trained by minimizing the categorical cross-entropy loss. We ran 20 epochs of the Adam optimizer with learning rate 0.0001.

Concerning the convolutional network we use a model composed of 4 convolutional blocks, a dense layer of 1024 units with ReLU activation, a Dropout layer with probability 0.5 and a Softmax Layer of 10 units. The first convolutional block contains a convolutional layer with a kernel of size 7×7 and 32

Table 3. Structure of MMI network for discoveringspatial-independent features

Layer	Parameters
BagLayer (linear activation) ReLU	100 units
BagLayer (linear activation) ReLU	200 units
Dense ReLU	1024 units
Dropout	Probability 0.5
Dense	10 units

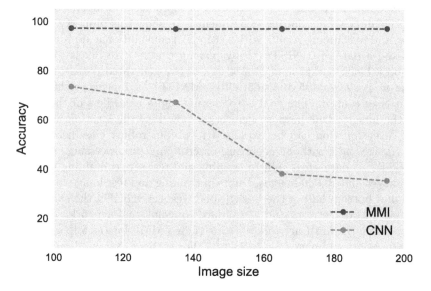

Fig. 5. Accuracies of MMI network (blue) and convolutional neural network (green) in function of window size w. (Color figure online)

channels while the other blocks contain convolutional layers of size 5×5 and 64 channels. Each convolutional layer is followed by a ReLU activation, MaxPooling of size 2×2 and a Dropout layer with probability 0.5. The model was trained by minimizing the categorical cross-entropy loss. We ran 20 epochs of the Adam optimizer with learning rate 0.001.

In Fig. 5 we report the accuracy of MMI network and CNNs as a function of w. We note that the accuracy of the CNN decreases as w grows large, while accuracy the MMI network remains stable. Those results confirm that MMI network is able to learn location-invariant features.

6.3   Citation Network Datasets

In Sect. 4.2 we discussed the possible benefits that graphs can obtain through bag-layers and then MMI networks. We show here a natural way for decomposing graphs into MMI data. We also show a way for decomposing graphs into multi-instance data (which we abbreviate with MI data) which is still a reasonable representation for graphs. We considered three graph citation datasets: Citeseer, Cora and PubMed [19]. Nodes represent papers described by titles and abstracts and edges are citation links. We treat the citation links as undirected edges. The goal is to classify each node of the graph. In Table 4 are reported the statistics for each dataset.

MMI data was constructed from citation networks in which a top-bag X represents a bag of nodes: the neighborhood of a node (including the node itself). A sub-bag $S_j \in X$ represents the bag of words corresponding to text attached to the node. An instance $x_{j,\ell} \in S_j$ is a word represented with GloVe word vectors [16]. MI data was constructed from citation networks in which bags are the set of bag of words of the neighborhood of a node (including the node itself). Note that in general both the cardinality of bags for MMI data and MI data could differ.

The structure of neural networks we used for those experiments is summarized in Table 5 (all bag-layers have linear activation). All models were trained by minimizing the softmax cross-entropy loss. We ran 400 epochs of the Adam optimizer with learning rate 0.0001 on 10 randomly drawn splits: the training set is composed of 20 top-bags for each class and the test set is composed of 1000 top-bags. Results in Table 6 report the average accuracy evaluated on the same training/test splits for MMI network, MI network and the state-of-art architec-

Table 4. Citations dataset

Dataset	#Classes	#Nodes	#Edges
Citeseer	6	3,327	4,732
Cora	7	2,708	5,429
PubMed	3	19,717	44,338

Table 5. Structure of MI networks (left) and MMI networks (right); BL stands for BagLayer

MI			MMI		
Citeseer	Cora	Pubmed	Citeseer	Cora	Pubmed
BL(2000)	BL(2000)	BL(3000)	BL(2000)	BL(2000)	BL(3000)
Sigmoid	ReLU	Sigmoid	Sigmoid	ReLU	Sigmoid
Dense(6)	Dense(7)	Dense(3)	BL(2000)	BL(2000)	BL(2000)
			Sigmoid	ReLU	Sigmoid
			Dense(6)	Dense(7)	Dense(3)

Table 6. Averaged accuracies (in percent)

Method	Citeseer	Cora	PubMed
Multi instance	68.5 ± 2.1	77.6 ± 1.3	76.0 ± 2.0
Multi-multi instance	**71.2 ± 1.7**	**79.2 ± 1.5**	77.1 ± 2.0
GCN [11]	69.6 ± 1.9	78.8 ± 1.1	**78.2 ± 2.1**

ture [11]. In Table 2 of [11], GCN is compared against several methods, outperforming all of them. Hence we only compare against GCN. The results confirm that it is reasonable to represent graphs as MI data, the performance being competitive with the current state-of-art approach. Nevertheless MMI decomposition allows to obtain better results than MI decomposition. This fact suggests that in this scenario graphs benefit from the more nested structure of MMI data. Finally MMI networks trained on such MMI data can slightly improve the state-of-art in two out of three datasets.

7   Conclusions

We have introduced the MMI framework for handling data organized in nested bags. The MMI setting allows for a natural hierarchical organization of data, where components at different levels of the hierarchy are unconstrained in their cardinality. We have identified several learning problems that can be naturally expressed as MMI problems. For instance, image or graph classification are promising application areas, because here the examples can be objects of varying structure and size, for which a bag-of-bag data representation is quite suitable, and can provide a natural alternative to graph kernels or convolutional network for graphs. The fact that bags do not impose an order on their elements directly leads to useful spatial invariance properties when applied to image data.

We proposed a neural network architecture involving the new construct of bag layers for learning in the MMI setting. Theoretical results show the expressivity of this type of model. In the empirical results we have shown that learning MMI models from data is feasible, and the theoretical capabilities of MMI networks can be exploited in practice, e.g., to learn accurate models for noiseless data, or location invariant models for image classification.

In this paper, we have focused on the setting where whole bags-of-bags are to be classified. In conventional MI learning, it is also possible to define a task where individual instances are to be classified. Such a task is however less clearly defined in our setup since we do not assume to know the label spaces at the instance and sub-bag level, nor the functional relationship between the labels at the different levels. It is an interesting subject of future work to extend our approach also to also identify the unknown labels associated with instances and sub-bags. Interestingly, this direction of research might also connect to the interpretation of the behavior of the network, which at the moment, like all related deep learning techniques, is totally opaque.

References

1. Andrews, S., Tsochantaridis, I., Hofmann, T.: Support vector machines for multiple-instance learning. In: Advances in Neural Information Processing Systems, pp. 561–568 (2002). http://machinelearning.wustl.edu/mlpapers/paper_files/AA10.pdf. 00828
2. Arbeláiez, P., Maire, M., Fowlkes, C., Malik, J.: Contour detection and hierarchical image segmentation. IEEE Trans. Pattern Anal. Mach. Intell. **33**(5), 898–916 (2011). http://ieeexplore.ieee.org/document/5557884/. 00000
3. Atwood, J., Towsley, D.: Diffusion-convolutional neural networks. In: Advances in Neural Information Processing Systems, pp. 1993–2001 (2016). http://papers.nips.cc/paper/6212-diffusion-convolutional-neural-networks
4. Dietterich, T.G., Lathrop, R.H., Lozano-Pérez, T.: Solving the multiple instance problem with axis-parallel rectangles. Artif. Intell. **89**(1–2), 31–71 (1997). http://www.sciencedirect.com/science/article/pii/S0004370296000343. 01439
5. Duvenaud, D., Maclaurin, D., Aguilera-Iparraguirre, J., Gómez-Bombarelli, R., Hirzel, T., Aspuru-Guzik, A., Adams, R.P.: Convolutional networks on graphs for learning molecular fingerprints. arXiv:1509.09292 [cs, stat], September 2015
6. Foulds, J., Frank, E.: A review of multi-instance learning assumptions. Knowl. Eng. Rev. **25**(01), 1 (2010). http://www.journals.cambridge.org/abstract_S026988 890999035X. 00081
7. Fukushima, K.: Neocognitron: a self-organizing neural network model for a mechanism of pattern recognition unaffected by shift in position. Biol. Cybern. **36**(4), 193–202 (1980). https://doi.org/10.1007/BF00344251. 01681
8. Hornik, K., Stinchcombe, M., White, H.: Multilayer feedforward networks are universal approximators. Neural Netw. **2**(5), 359–366 (1989). http://www.science direct.com/science/article/pii/0893608089900208. 12001
9. Jaeger, M.: Relational Bayesian networks. arXiv:1302.1550 [cs] (1997). 00252
10. Kingma, D., Ba, J.: Adam: a method for stochastic optimization. arXiv:1412.6980 [cs], December 2014. 00204
11. Kipf, T.N., Welling, M.: Semi-supervised classification with graph convolutional networks. arXiv preprint arXiv:1609.02907 (2016). 00020
12. LeCun, Y., Boser, B., Denker, J.S., Henderson, D., Howard, R.E., Hubbard, W., Jackel, L.D.: Backpropagation applied to handwritten zip code recognition. Neural Comput. **1**(4), 541–551 (1989). http://www.mitpressjournals.org/doi/abs/10.1162/neco.1989.1.4.541. 01543
13. Maron, O., Lozano-Pérez, T.: A framework for multiple-instance learning. In: Advances in Neural Information Processing Systems, pp. 570–576 (1998). http://lamda.nju.edu.cn/zhangml/files/NIPS97.pdf. 00870
14. Natarajan, S., Tadepalli, P., Dietterich, T.G., Fern, A.: Learning first-order probabilistic models with combining rules. Ann. Math. Artif. Intell. **54**(1–3), 223–256 (2008). https://doi.org/10.1007/s10472-009-9138-5. 00069
15. Niepert, M., Ahmed, M., Kutzkov, K.: Learning convolutional neural networks for graphs. New York, NY, USA, May 2016. arXiv:1605.05273. 00001
16. Pennington, J., Socher, R., Manning, C.D.: Glove: global vectors for word representation. In: EMNLP, vol. 14, pp. 1532–1543 (2014). http://llcao.net/cu-deeplearning15/presentation/nn-pres.pdf. 00365
17. Ramon, J., De Raedt, L.: Multi instance neural networks (2000). http://citeseerx.ist.psu.edu/viewdoc/summary?doi=10.1.1.43.682. 00115

18. Scott, S., Zhang, J., Brown, J.: On generalized multiple-instance learning. Int. J. Comput. Intell. Appl. **5**(01), 21–35 (2005). http://www.worldscientific.com/doi/abs/10.1142/S1469026805001453. 00059

19. Sen, P., Namata, G., Bilgic, M., Getoor, L., Galligher, B., Eliassi-Rad, T.: Collective classification in network data. AI Mag. **29**(3), 93 (2008). https://vvvvw.aaai.org/ojs/index.php/aimagazine/article/view/2157. 00567

20. Szegedy, C., Ioffe, S., Vanhoucke, V., Alemi, A.: Inception-v4, Inception-ResNet and the impact of residual connections on learning. arXiv:1602.07261 [cs], February 2016. 00127

21. Wang, J., Zucker, J.D.: Solving multiple-instance problem: a lazy learning approach (2000). http://cogprints.org/2124. 00444

22. Weidmann, N., Frank, E., Pfahringer, B.: A two-level learning method for generalized multi-instance problems. In: Lavrač, N., Gamberger, D., Blockeel, H., Todorovski, L. (eds.) ECML 2003. LNCS (LNAI), vol. 2837, pp. 468–479. Springer, Heidelberg (2003). https://doi.org/10.1007/978-3-540-39857-8_42

23. Zhang, Q., Goldman, S.A., et al.: EM-DD: an improved multiple-instance learning technique. In: NIPS, vol. 1, pp. 1073–1080 (2001). https://papers.nips.cc/paper/1959-em-dd-an-improved-multiple-instance-learning-technique.pdf

24. Zhou, Z.H., Zhang, M.L.: Neural networks for multi-instance learning. In: Proceedings of the International Conference on Intelligent Information Technology, Beijing, China, pp. 455–459 (2002). http://cs.nju.edu.cn/zhouzh/zhouzh.files/publication/techrep02.pdf. 00066

Con-S2V: A Generic Framework for Incorporating Extra-Sentential Context into Sen2Vec

Tanay Kumar Saha[1](✉)[iD], Shafiq Joty[2][iD], and Mohammad Al Hasan[1][iD]

[1] Indiana University – Purdue University Indianapolis, Indianapolis, IN 46202, USA
{tksaha,alhasan}@iupui.edu
[2] Nanyang Technological University, Singapore, Singapore
srjoty@ntu.edu.sg

Abstract. We present a novel approach to learn distributed representation of sentences from unlabeled data by modeling both content and context of a sentence. The content model learns sentence representation by predicting its words. On the other hand, the context model comprises a neighbor prediction component and a regularizer to model distributional and proximity hypotheses, respectively. We propose an online algorithm to train the model components jointly. We evaluate the models in a setup, where contextual information is available. The experimental results on tasks involving classification, clustering, and ranking of sentences show that our model outperforms the best existing models by a wide margin across multiple datasets.
Code related to this chapter is available at:
https://github.com/tksaha/con-s2v/tree/jointlearning
Data related to this chapter are available at: https://www.dropbox.com/sh/ruhsi3c0unn0nko/AAAgVnZpojvXx9loQ21WP_MYa?dl=0

Keywords: Sen2Vec · Extra-sentential context
Embedding of sentences

1 Introduction

For many text processing tasks that involve classification, clustering, or ranking of sentences, vector representation of sentences is a prerequisite. Bag-of-words (BOW) based vector representation has been used traditionally in these tasks, but in recent years, it has been shown that *distributed representation*, in the form of condensed real-valued vectors, learned from *unlabeled* data outperforms BOW based representations [1]. It is now well established that distributed representation captures semantic properties of linguistic units and yields better generalization [2,3].

However, most of the existing methods to devise distributed representation for sentences consider only the content of a sentence or its grammatical structure [1,4] disregarding its context. But, sentences rarely stand on their own in a

© Springer International Publishing AG 2017
M. Ceci et al. (Eds.): ECML PKDD 2017, Part I, LNAI 10534, pp. 753–769, 2017.
https://doi.org/10.1007/978-3-319-71249-9_45

text, rather the meaning of one sentence depends on the meaning of others within its context. For example, sentences in a text segment address a common topic [5]. At a finer level, sentences are connected by certain coherence relations (e.g., *elaboration, contrast*) and acts together to express a coherent message holistically [6].

Our work is built on the following hypothesis: *since the meaning of a sentence can be best interpreted within its context, its representation should also be inferred from its context.* Several recent works attempt to learn sentence representations which support the above hypothesis by utilizing words or word sequences of neighboring sentences [7,8]. However, by learning representations to predict content of neighboring sentences, existing methods may learn semantic and syntactic properties that are more specific to the neighbors rather than the sentence under consideration. Furthermore, these methods either make a simple BOW assumption or disregard context when extracting a sentence vector.

In contrast to the existing works, we consider neighboring sentences as *atomic* linguistic units, and propose novel methods to learn the representations of a given sentence by jointly modeling content and context of a sentence. Our work considers two types of context: *discourse* and *similarity*. The discourse context of a given sentence $\mathbf{v}$ comprises with its previous and the following sentence in the text. On the other hand, the similarity context is based on a user defined similarity function; thus it allows any sentences in the text to be in the context of $\mathbf{v}$ depending on how similar that sentence is with $\mathbf{v}$ based on the chosen function.

Our proposed computational model for learning the vector representation of a sentence comprises three components. The first component models the content by asking the sentence vector to predict its constituent words. The second component models the *distributional* hypotheses [9] of a context. The distributional hypothesis conveys that the sentences occurring in similar contexts should have similar representations. Our computation model captures this preference by using a context prediction component. Finally, the third component models the *proximity* hypotheses of a context, which also suggests that sentences that are proximal should have similar representations. Our method achieves this preference by using a Laplacian regularizer. To this end, we consider the sentence representation learning problem as an optimization problem whose objective function is built with expressions from the above three components and we solve this optimization problem by using an efficient online algorithm.

1.1   Summary of Results

We evaluate our sentence representation for learning models on multiple information retrieval tasks: topic classification and clustering, and single-document summarization. Our evaluation on these tasks across multiple datasets shows impressive results for our model, which outperforms the best existing models by up to 7.7 F_1-score in classification, 15.1 V-score in clustering, 3.2 ROUGE-1 score in summarization. We found that the discourse context performs better on topic classification and clustering tasks, while similarity context performs better on summarization. We make our code[1] and pre-processed dataset[2] publicly available.

2 Related Work

Extensive research has been conducted on learning distributed representation of linguistic units both in supervised (task-specific) and in unsupervised (task-agnostic) settings. In this paper, we focus on learning *sentence* representations from *unlabeled* data.

Two log-linear models are proposed in [10] for learning representations of words: continuous bag-of-words (CBOW) and continuous skip-gram. CBOW learns word representations by predicting a word given its (intra-sentential) context. The skip-gram model on the other hand learns representation of a word by predicting the words in a context. [11] proposed C-PHRASE, an extension of CBOW, where the context is extracted from a syntactic parse of the sentence. Simple averaging or addition of word vectors to construct sentence vectors often works well [12], and serves as baselines in our experiments.

CBOW and skip-gram models are extended in [1] to sentences and documents by proposing distributed memory (DM) and distributed bag-of-words (DBOW) models. In these models, similar to words, a sentence is mapped to an unique id and its representation is learned using contexts of words in the sentence. DM predicts a word given a context and the sentence id, where DBOW predicts all words in a context independently given the sentence id. Since these models are agnostic to sentence structure, they are quite fast to train. However, they disregard extra-sentential context of a sentence.

Sequential denoising autoencoder (SDAE) and FastSent are proposed in [8] for modeling sentences. SDAE employs an encoder-decoder framework, similar to neural machine translation (NMT) [13], to denoise an original sentence (target) from its corrupted version (source). FastSent is an additive model to learn sentence representation from word vectors. Given a sentence as BOW, it predicts the words of its adjacent sentences. The auto-encode version of FastSent also predicts the words of the current sentence. SDAE composes sentence vectors sequentially, but it disregards context of the sentence. FastSent, on the other hand, is a BOW model that considers neighboring sentences.

Another context-sensitive model is Skip-Thought [7], which uses the NMT framework to predict adjacent sentences (target) given a sentence (source). Since the encoder and the decoder use recurrent layers to compose vectors sequentially, SDAE and Skip-Thought are very slow to train. Furthermore, by learning representations to predict content of neighboring sentences, these methods (FastSent and Skip-Thought) may learn linguistic properties that are more specific to the neighbors rather than the sentence under consideration.

---

[1] https://github.com/tksaha/con-s2v/tree/jointlearning.

[2] https://www.dropbox.com/sh/ruhsi3c0unn0nko/AAAgVnZpojvXx9loQ21WP_MYa?dl=0.

By contrast, we encode a sentence by treating it as an atomic unit like word, and similar to DBOW, we predict the words to model its content. Similarly, context is considered in our model by treating neighboring sentences as atomic units. This abstraction makes our model quite fast to train.

3 The Model

We hypothesize that the representation of a sentence depends not only on its content words, but also on other sentences in its context. It will be convenient to present our learning model using graph.

Let $G = (V, E)$ be a graph, where $V = \{\mathbf{v}_1, \mathbf{v}_2, \cdots, \mathbf{v}_{|V|}\}$ represents the set of sentences in our corpus, and edge $(\mathbf{v}_i, \mathbf{v}_j) \in E$ reflects some relation between sentences $\mathbf{v}_i$ and $\mathbf{v}_j$. A sentence $\mathbf{v}_i \in V$ is a sequence of words $(v_i^1, v_i^2, \cdots, v_i^M)$, each coming from a dictionary $\mathcal{D}$. We define $\mathcal{N}(\mathbf{v}_i)$ as the set of neighboring sentences of $\mathbf{v}_i$, which constitutes extra-sentential context for sentence $\mathbf{v}_i$. We formalize relation between sentences and context later in Sect. 3.3.

Let $\phi : V \rightarrow \mathbb{R}^d$ be the mapping function from sentences to their distributed representations, i.e., real-valued vectors of d dimensions. Equivalently, ϕ can be thought of as a look-up matrix of size $|V| \times d$, where $|V|$ is the total number of sentences. Our aim is to learn $\phi(\mathbf{v}_i)$ by incorporating information from two different sources: (i) the content of the sentence, $\mathbf{v}_i = (v_i^1, v_i^2, \cdots, v_i^M)$; and (ii) the context of the sentence in the graph, i.e., $\mathcal{N}(\mathbf{v}_i)$. Let $\langle v_i \rangle_t^l = (v_i^{t-l}, \ldots, v_i^t, \ldots, v_i^{t+l})$ denote a window of $2l + 1$ words around the word v_i^t in sentence $\mathbf{v}_i$, and $C_i = |\mathcal{N}(\mathbf{v}_i)|$ denote the context size for sentence $\mathbf{v}_i$. We define our model as a combination of three different loss functions:

$$J(\phi) = \sum_{\mathbf{v}_i \in V} \sum_{\substack{v \in \langle v_i \rangle_t^l \\ j \sim \mathcal{U}(1, C_i)}} \left[\mathcal{L}_c(\mathbf{v}_i, v) + \mathcal{L}_g(\mathbf{v}_i, \mathbf{v}_j) + \right.$$

$$\left. \mathcal{L}_r(\mathbf{v}_i, \mathcal{N}(\mathbf{v}_i)) \right] \quad (1)$$

where loss $\mathcal{L}_c(\mathbf{v}_i, v)$ is used to model the content of a sentence $\mathbf{v}_i$, and other two loss functions are for modeling the context of the sentence. We define $\mathcal{L}_c(\mathbf{v}_i, v)$ as the cost for predicting the content word v using the sentence vector $\phi(\mathbf{v}_i)$ as input features. Similarly, $\mathcal{L}_g(\mathbf{v}_i, \mathbf{v}_j)$ is defined as the cost for predicting a neighboring node $\mathbf{v}_j \in \mathcal{N}(\mathbf{v}_i)$, again using the sentence vector $\phi(\mathbf{v}_i)$ as input. The third loss $\mathcal{L}_r(\mathbf{v}_i, \mathcal{N}(\mathbf{v}_i))$ is a graph smoothing regularizer defined over the context of $\mathbf{v}_i$, which encourages two proximal sentences to have similar representations.

To learn the representation of a sentence $\mathbf{v}_i$ using Eq. 1, for each content word v in a window $\langle v_i \rangle_t^l$, we sample a neighboring node $\mathbf{v}_j$ from $\mathcal{N}(\mathbf{v}_i)$, uniformly at random, with replacement. We use the sentence vector $\phi(\mathbf{v}_i)$ (under estimation) to predict v and $\mathbf{v}_j$, respectively. A regularization is performed to smooth the estimated vector with respect to the neighboring vectors. Figure 1 shows instances of our model for learning the representation of sentence $\mathbf{v}_2$ within a context of two other sentences: $\mathbf{v}_1$ and $\mathbf{v}_3$.

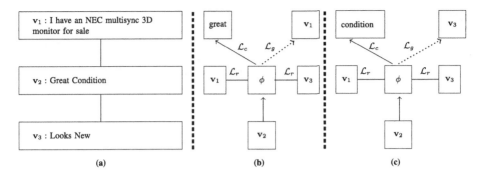

Fig. 1. Two instances (see (b) and (c)) of our model for learning representation of sentence $\mathbf{v}_2$ within a context of two other sentences: $\mathbf{v}_1$ and $\mathbf{v}_3$ (see (a)). Directed and undirected edges indicate prediction loss and regularization loss, respectively, and dashed edges indicate that the node being predicted is randomly sampled. (Collected from: 20news-bydate-train/misc.forsale/74732. The central topic is "forsale".)

We can use the standard *softmax* function for the prediction tasks. Formally, the negative log probability of an item o (can be a content word or a neighboring node) given the sentence vector $\phi(\mathbf{v}_i)$ is

$$- \log p(o|\mathbf{v}_i) = -\mathbf{w}_o^T \phi(\mathbf{v}_i) + \log \sum_{o' \in \mathcal{O}} \exp \left(\mathbf{w}_{o'}^T \phi(\mathbf{v}_i) \right) \qquad (2)$$

where $\mathcal{O}$ is the set of all possible items (i.e., vocabulary of words or set of all nodes), and $\mathbf{w}$'s are the weight parameters. Optimization is typically performed using gradient-based online methods, such as stochastic gradient descend (SGD), where gradients are obtained via backpropagation.

Unfortunately, training could be impractically slow on large corpora due to summation over all items in $\mathcal{O}$ (Eq. 2), which needs to be performed for every training instance $(\mathbf{v}_i, o)$. Several methods have been proposed to address this issue including *hierarchical softmax* [14], *noise contrastive estimation* [15], and *negative sampling* [16]. We use negative sampling, which samples negative examples to approximate the summation term. Specifically, for each training instance $(\mathbf{v}_i, o)$, we add S negative examples $\{(\mathbf{v}_i, o^s)\}_{s=1}^S$ by sampling o^s from a known noise distribution ψ (e.g., *unigram*, *uniform*). The negative log probability in Eq. 2 is then formulated as such to discriminate a *positive* instance o from a *negative* one o^s:

$$- \log \sigma \left(\mathbf{w}_o^T \phi(\mathbf{v}_i) \right) - \log \sum_{s=1}^S \mathbb{E}_{o^s \sim \psi} \, \sigma \left(-\mathbf{w}_{o^s}^T \phi(\mathbf{v}_i) \right) \qquad (3)$$

where σ is the sigmoid function defined as $\sigma(x) = 1/(1+e^{-x})$, and $\mathbf{w}$'s and $\phi(\mathbf{v}_i)$ are similarly defined as before. Negative sampling thus reduces the number of computations needed from $|\mathcal{O}|$ to $S + 1$, where S is a small number $(5 - 10)$ compared to the vocabulary size $|\mathcal{O}|$ (26K – 139K).

In the following, we elaborate on our methods for modeling content and context of a sentence.

3.1   Modeling Content

Our approach for modeling content of a sentence is similar to the distributed bag-of-words (DBOW) model of [1]. Given an input sentence $\mathbf{v}_i$, we first map it to a unique vector $\phi(\mathbf{v}_i)$ by looking up the corresponding vector in the sentence embedding matrix ϕ. We then use $\phi(\mathbf{v}_i)$ to predict each word v sampled from a window of words in $\mathbf{v}_i$. Formally, the loss for modeling content using negative sampling is

$$\mathcal{L}_c(\mathbf{v}_i, v) = - \log \sigma \left(\mathbf{w}_v^T \phi(\mathbf{v}_i) \right)$$

$$- \log \sum_{s=1}^{S} \mathbb{E}_{v^s \sim \psi_c}\, \sigma \left(-\mathbf{w}_{v^s}^T \phi(\mathbf{v}_i) \right) \tag{4}$$

where σ is the sigmoid function as defined before, $\mathbf{w}_v$ and $\mathbf{w}_{v^s}$ are the weight vectors associated with words v and v^s, respectively, and ψ_c is the noise distribution from which v^s is sampled. In our experiments, we use unigram distribution of words raised to the 3/4 power as our noise distribution, in accordance to [16].

By asking the same sentence vector (under estimation) to predict its words, the content model captures the overall semantics of the sentence. The model has $O(d \times (|V| + |\mathcal{D}|))$ parameters.

3.2   Modeling Context

Our content model above attempts to capture the overall meaning of a sentence by looking at its words. However, sentences in a text are not independent, rather the meaning of a sentence depends on its neighboring sentences. For instance, consider the second and the third sentences in Fig. 1(a). When the sentences are considered in isolation, one cannot understand what they are talking about (i.e., *monitor for sale*). This suggests, since meaning of a sentence can be best interpreted within its context, the representation of the sentence should also be inferred from its context. We distinguish between two types of contextual relations between sentences: (*i*) distributional similarity, and (*ii*) proximity. Each of these corresponds to a loss in our model (Eq. 1), as we describe them below.

Modeling Distributional Similarity: Our sentence-level distributional hypothesis [9] is that if two sentences share many neighbors in the graph, their representations should be similar. We formulate this in our model by asking the sentence vector to predict its neighboring nodes. More formally, the loss for predicting a neighboring node $\mathbf{v}_j \in \mathcal{N}(\mathbf{v}_i)$ using the sentence vector $\phi(\mathbf{v}_i)$ is

$$\mathcal{L}_g(\mathbf{v}_i, \mathbf{v}_j) = - \log \sigma \left(\mathbf{w}_j^T \phi(\mathbf{v}_i) \right)$$

$$- \log \sum_{s=1}^{S} \mathbb{E}_{j^s \sim \psi_g}\, \sigma \left(-\mathbf{w}_{j^s}^T \phi(\mathbf{v}_i) \right) \tag{5}$$

where $\mathbf{w}_j$ and $\mathbf{w}_j^s$ are the weight vectors associated with nodes $\mathbf{v}_j$ and $\mathbf{v}_j^s$, respectively, and ψ_g is the noise distribution over nodes from which $\mathbf{v}_j^s$ is sampled. Similar to our content model, ψ_g is defined as unigram distribution of nodes raised to the 3/4 power. The unigram distribution is computed based on the occurrences of the nodes in the neighborhood sets, $\{\mathcal{N}(\mathbf{v}_i)\}_{i=1}^{|V|}$. This model has $O(d \times (|V| + |V|))$ parameters.

Modeling Proximity: According to our proximity hypothesis, sentences that are proximal in their contexts, should have similar representations. We use a Laplacian regularizer to model this. Formally, the regularization loss for modeling proximity for a sentence $\mathbf{v}_i$ in its context $\mathcal{N}(\mathbf{v}_i)$ is

$$\mathcal{L}_r(\mathbf{v}_i, \mathcal{N}(\mathbf{v}_i)) = \frac{\lambda}{C_i} \sum_{\mathbf{v}_k \in \mathcal{N}(\mathbf{v}_i)} ||\phi(\mathbf{v}_i) - \phi(\mathbf{v}_k)||^2 \tag{6}$$

where $C_i = |\mathcal{N}(\mathbf{v}_i)|$ as defined before, and λ is a hyper-parameter to control regularization strength.

Rather than including the Laplacian as a regularizer in the objective function, another option is to first learn sentence embeddings using other components of the model (e.g., first two loss functions in Eq. 1), and then *retrofit* them using the Laplacian as a post-processing step. [17] adopted this approach to incorporate lexical semantics (e.g., synonymy, hypernymy) into word representations. We compare our approach with retrofitting in Sect. 5.

3.3   Context Types

In this section we characterize context of a sentence. We distinguish between two types of context: discourse context and similarity context.

Discourse Context: The discourse context of a sentence is formed by the previous and the following sentences in the text. As explained before, the order of the sentences carries important information. For example, adjacent sentences in a text are logically connected by certain coherence relations (e.g., *elaboration*, *contrast*) to express the meaning [6]. On a coarser level, sentences in a text segment (e.g., paragraph) address a common (sub)topic [5]. The discourse context thus captures both coherence and topic structures of a text.

Similarity Context: While the discourse context covers important discourse phenomena like *coherence* and *cohesion* [18], some applications might require a context type that is based on more direct measures of similarity, and considers relations between all possible sentences in a document and possibly across multiple documents. For example, graph-based methods for topic segmentation [19] and summarization [20] rely on complete graphs of sentences, where edge weights represent cosine similarity between sentences. In an empirical evaluation of data structures for representing discourse coherence, [21] advocate for a graph representation of discourse allowing non-adjacent connections.

Our similarity context allows any other sentence in the corpus to be in the context of a sentence depending on how similar they are. To measure the similarity, we first represent the sentences with vectors learned by Sen2Vec [1], then we measure the cosine between the vectors. We restrict the context size of a sentence for computational efficiency, while still ensuring that it is informative enough. We achieve this by imposing two kinds of constraints. First, we set thresholds for intra- and across-document connections: sentences in a document are connected only if their similarity value is above 0.5, and sentences across documents are connected only if their similarity is above 0.8. Second, we allow up to 20 most similar neighbors.

3.4 Training

Algorithm 1 illustrates the SGD-based algorithm to train our model. We first initialize the model parameters; the sentence vectors ϕ are initialized with small random numbers sampled from uniform distribution $\mathcal{U}(-0.5/d, 0.5/d)$, and the weight parameters $\mathbf{w}$'s are initialized with zero. We then compute the noise distributions ψ_c and ψ_g for $\mathcal{L}_c(\mathbf{v}_i, v)$ and $\mathcal{L}_g(\mathbf{v}_i, \mathbf{v}_j)$ losses, respectively.

We iterate over the sentences in our corpus in each epoch of SGD, as we learn their representations. Specifically, to estimate the representation of a sentence, for each word token in the sentence, we take three gradient steps to account for the three loss functions in Eq. 1. By making the same number of gradient updates, the algorithm weights equally the contributions of content and context.

Algorithm 1. Training Con-S2V with SGD

Input : set of sentences V, graph $G = (V, E)$
Output: learned sentence vectors ϕ
1. Initialize model parameters: ϕ and $\mathbf{w}$'s;
2. Compute noise distributions: ψ_c and ψ_g
3. **repeat**
    for *each sentence* $\mathbf{v}_i \in V$ **do**
        for *each content word* $v \in \mathbf{v}_i$ **do**
            a) Generate a positive pair $(\mathbf{v}_i, v)$ and S negative pairs $\{(\mathbf{v}_i, v^s)\}_{s=1}^S$ using ψ_c;
            b) Take a gradient step for $\mathcal{L}_c(\mathbf{v}_i, v)$;
            c) Sample a neighboring node $\mathbf{v}_j$ from $\mathcal{N}(\mathbf{v}_i)$;
            d) Generate a positive pair $(\mathbf{v}_i, \mathbf{v}_j)$ and S negative pairs $\{(\mathbf{v}_i, \mathbf{v}_j^s)\}_{s=1}^S$ using ψ_g;
            e) Take a gradient step for $\mathcal{L}_g(\mathbf{v}_i, \mathbf{v}_j)$;
            f) Take a gradient step for $\mathcal{L}_r(\mathbf{v}_i, \mathcal{N}(\mathbf{v}_i))$;
        end
    end
until *convergence*;

---

4   Evaluation Tasks

Different methods have been proposed to evaluate sentence representation models [8]. However, unlike most existing methods, our model learns representation of a sentence by exploiting contextual information in addition to the content.[3] To be able to evaluate our models, we thus require corpora of annotated sentences with ordering and document boundaries preserved, i.e., documents with sentence-level annotations. To the best of our knowledge, no previous work has used or released such corpora for learning sentence representation. In this work, we automatically create large corpora of documents with sentence-level topic annotations, which are then used to evaluate our models on topic *classification* and *clustering* tasks. Additionally, we evaluate our models on a *ranking* task of generating *extractive* single-document summaries. In the interest of coherence, we present the summarization task, followed by topic classification and clustering.

4.1   Extractive Summarization

Extractive summarization is often considered as a ranking problem, where the goal is to select the most important sentences to form an abridged version of the source document(s) [22]. Unsupervised methods are the predominant paradigm for determining sentence importance. We use the popular graph-based algorithm LexRank [20]. The input to LexRank is a graph, where nodes represent sentences and edges represent cosine similarity between *vector representations* (learned by models) of the two corresponding sentences. We run the PageRank [23] on the graph to compute importance of each sentence in the graph.[4] The top-ranked sentences are extracted as the summary sentences.

Data: We use the benchmark datasets from DUC-2001 and DUC-2002, and evaluate our representation models on the official task of generating a 100-words summary for each document in the datasets.[5] The sentence representations are learned independently a priori from the same source documents. Table 1 shows some basic statistics about the datasets. For each document, 2–3 short ($\approx$100 words) human authored reference summaries are available, which we use as gold summaries for automatic evaluation.

Metric: We use the widely used automatic evaluation metric ROUGE [24] to evaluate the system-generated summaries. ROUGE computes n-gram recall between a system-generated summary and a set of human-authored reference summaries. Among the vari-

Table 1. Basic statistics about the DUC datasets

Dataset	#Doc.	#Avg. sen.	#Avg. sum.
DUC 2001	486	40	2.17
DUC 2002	471	28	2.04

ants, ROUGE-1 (i.e., $n = 1$) has been shown to correlate well with human judgments for short summaries [24]. Therefore, we only report ROUGE-1 in this paper.

---

[3] For this reason, we did not evaluate our models on tasks previously used to evaluate sentence representation models.

[4] The dumping factor in the PageRank was set to 0.85.

[5] http://www-nlpir.nist.gov/projects/duc/guidelines.

4.2   Topic Classification and Clustering

We evaluate our models by measuring how effective the learned vectors are when they are used as features for classifying or clustering the sentences into topics. Text categorization has now become a standard in evaluating cross-lingual word embeddings [25]. We use a MaxEnt classifier and a K-means++ [26] clustering algorithm for classification and clustering tasks, respectively.

Data: We use the standard text categorization corpora: *Reuters-21578* and *20-Newsgroups*. Reuters-21578 (henceforth Reuters) is a collection of 21,578 news documents covering 672 topics.[6] 20-Newsgroups (hence-

Table 2. Statistics about Reuters and Newsgroups.

Dataset	#Doc.	Total #sen.	Annot. #sen.	Train #sen.	Test #sen.	#Class
Reuters	9,001	42,192	13,305	7,738	3,618	8
Newsgroups	7,781	95,809	22,374	10,594	9,075	8

forth Newsgroups) is a collection of about 20,000 news articles organized into 20 different topics.[7] We used the standard train-test splits (*ModApte* split for Reuters) split, and selected documents only from the 8 most frequent topics in both datasets. The selected topics for Reuters dataset are: *acq, crude, earn, grain, interest, money-fx, ship,* and *trade*. The topics selected for Newsgroups dataset are: *sci.space, sci.med, talk.politics.guns, talk.politics.mideast, rec.autos, rec.sport.baseball, comp.graphics,* and *soc.religion.christian*.

Generating Sentence-level Topic Annotations: As mentioned above, both *Newsgroups* and *Reuters* datasets come with document-level topic annotations. However, we need sentence-level annotations for our evaluation. One option is to assume that all the sentences of a document share the same topic label as the document. However, this naive assumption induces a lot of noise. Although sentences in a document collectively address a common topic, not all sentences are directly linked to that topic, rather they play supporting roles. To minimize this noise, we employ our extractive summarizer introduced in Sect. 4.1 to select the top 20% sentences of each document as representatives of the document, and assign them the same topic label as the topic of the document. We used Sen2Vec [1] representation to compute cosine similarity between two sentences in LexRank. Table 2 shows statistics of the resulting datasets. Note that the sentence vectors are learned independently from an entire dataset (#Total Sen. column in Table 2).

Metrics: We report raw **acc**uracy, macro-averaged **F₁**-score, and Cohen's **κ** for comparing classification performance. For clustering, we report **V**-measure [27] and adjusted mutual information or **AMI** [28]. We use all the annotated sentences (train+test in Table 2) for comparing clustering performance.

---

[6] http://kdd.ics.uci.edu/databases/reuters21578/.
[7] http://qwone.com/~jason/20Newsgroups/.

5   Experiments

In this section, we present our experiments — the models we compare, their settings, and the results.

5.1   Models Compared

We compare our representation learning model against several baselines and existing models. We also experiment with a number of variations of our proposed model considering which components of the model are active, types of context, and how we incorporate the context. For clarity, in our tables we show results divided into five evaluation groups:

(I) Existing Distributed Models: This group includes Sen2Vec [1], W2V-avg, C-PHRASE [11], FastSent [8], and Skip-Thought [7].

   We used Mikolov's implementation[8] of **Sen2Vec**, which gave better results than gensim's version when validated on the sentiment treebank [29]. Following the recommendation by [1], we concatenate the vectors learned by DM and DBOW models. The concatenated vectors also performed better on our tasks.

   For **W2V-avg**, we obtain a sentence vector by averaging the word vectors learned by training a skip-gram Word2Vec [16] on our training set. Since code for **C-Phrase** is not publicly available, we use pre-trained word vectors (of 300 dimensions) available from author's webpage.[9] We first add the word vectors to obtain a sentence vector, then we normalize the vector with l_2 normalization. Normalized vectors performed better on our tasks than the ones obtained by simple addition.

   We use the auto-encode version of **FastSent** (FastSent+AE) since it considers both content and context of a sentence. For **Skip-Thought**, we use the pre-trained combine-skip model that concatenates the vectors encoded by uni- and bi-skip models.[10] The resultant vectors are of 4800 dimensions. The model was originally trained on the BookCorpus[11] with a vocabulary size of $20K$ words, however, it uses publicly available CBOW Word2Vec vectors to expand the vocabulary size to $930,911$ words.

(II) Non-distributed Model: We use **Tf-Idf** as our non-distributed baseline, where a sentence is represented by tf*idf weighting of its words.

(III) Retrofitted Models: We compare our approach of modeling context with the retrofitting method of [17]. We first learn sentence vectors using the content model only (i.e., by turning off contextual components in Eq. 1). Then we retrofit these vectors with the graph Laplacian $\mathcal{L}_r(\mathbf{v}_i, \mathcal{N}(\mathbf{v}_i))$ to encourage the revised

---

[8]   https://code.google.com/archive/p/word2vec/.
[9]   http://clic.cimec.unitn.it/composes/cphrase-vectors.html.
[10]   https://github.com/ryankiros/skip-thoughts.
[11]   http://yknzhu.wixsite.com/mbweb.

vectors to be similar to the vectors of neighboring sentences and also similar to their prior representations. We consider two types of graph contexts: discourse (RET-dis) and similarity (RET-sim).

(IV) Regularized Models: We compare with a variant of our model, where the loss to capture distributional similarity $\mathcal{L}_g(\mathbf{v}_i, \mathbf{v}_j)$ is turned off. This model considers the same information as the retrofitting model (i.e., content and proximity), but trains the vectors in a single step. Its comparison with our complete model will tell us how much distributional similarity contributes to the overall performance. We define regularizers on two types of contexts: discourse (REG-dis) and similarity (REG-sim).

(V) Our Models: We experiment with two variants of our combined model, CON-S2V: one with discourse context (CON-S2V-dis), and the other with similarity context (CON-S2V-sim).

Table 3. Optimal values of the hyper-parameters for different models on different tasks.

Dataset	Task	Sen2Vec	FastSent	W2V-avg	REG-sim	REG-dis	CON-S2V-sim	CON-S2V-dis
		(win. size)			(win. size, reg. str.)		(win. size, reg. str.)	
Reuters	Clas.	8	10	10	(8, 1.0)	(8, 1.0)	(8, 0.8)	(8, 1.0)
	Clus.	12	8	12	(12, 0.3)	(12, 1.0)	(12,0.8)	(12, 0.8)
Newsgroups	Clas.	10	8	10	(10, 1.0)	(10, 1.0)	(10, 1.0)	(10, 1.0)
	Clus.	12	12	12	(12, 1.0)	(12, 1.0)	(12, 0.8)	(10, 1.0)
DUC 2001	Sum.	10	12	12	(10, 0.8)	(10, 0.5)	(10, 0.3)	(10, 0.3)
DUC 2002	Sum.	8	8	10	(8, 0.8)	(8, 0.3)	(8, 0.3)	(8, 0.3)

5.2   Model Settings

The representation dimensions were set to 300 in DM and DBOW models. The concatenation of the two vectors yields 600 dimensions for Sen2Vec. For a fair comparison, the dimensions in all other models that we train (except pre-trained C-PHRASE and Skip-Thought) were fixed to 600. All the prediction-based models were trained with SGD. Retrofitting was done using iterative method [17] with 20 iterations. The number of noise samples (S) in negative sampling was set to 5. We also used subsampling of frequent words [16], which together with negative sampling give significant speed-ups in training.

For each dataset described in Sect. 4, we randomly selected 20% documents from the training set to form a held-out validation set on which we tune the hyper-parameters. *Window size* (k) is a hyper-parameter that is common to all models. The regularized models have an additional hyper-parameter, *regularization strength*(λ). We tuned with $k \in \{8, 10, 12\}$ and $\lambda \in \{0.3, 0.6, 0.8, 1\}$, and we optimized F_1 for classification, AMI for clustering, and ROUGE-1 for summarization. Table 3 shows the hyper-parameters and their optimal values for each task. We evaluated our models on the test sets with these optimal values. We ran each experiment five times and take the average of the evaluation measures to avoid any randomness in results.

5.3 Classification and Clustering Results

Table 4 shows the results of the models on topic classification and clustering tasks, respectively. The scores are shown in comparison to Sen2Vec.

Table 4. Performance of our models on topic classification and clustering tasks in comparison to Sen2Vec.

	Topic classification results						Topic clustering results			
	Reuters			Newsgroups			Reuters		Newsgroups	
	F_1	Acc	κ	F_1	Acc	κ	V	AMI	V	AMI
Sen2Vec	83.25	83.91	79.37	79.38	79.47	76.16	42.74	40.00	35.30	34.74
W2V-avg	(+) 2.06	(+) 1.91	(+) 2.51	(−) 0.42	(−) 0.44	(−) 0.50	(−) 11.96	(−) 10.18	(−) 17.90	(−) 18.50
C-Phrase	(−) 2.33	(−) 2.01	(−) 2.78	(−) 2.49	(−) 2.38	(−) 2.86	(−) 11.94	(−) 10.80	(−) 1.70	(−) 1.44
FastSent	(−) 0.37	(−) 0.29	(−) 0.41	(−) 12.23	(−) 12.17	(−) 14.21	(−) 15.54	(−) 13.06	(−) 34.40	(−) 34.16
Skip-Thought	(−) 19.13	(−) 15.61	(−) 21.8	(−) 13.79	(−) 13.47	(−)15.76	(−) 29.94	(−) 28.00	(−) 27.50	(−) 27.04
Tf-Idf	(−) 3.51	(−) 2.68	(−) 3.85	(−) 9.95	(−) 9.72	(−) 11.55	(−) 21.34	(−) 20.14	(−) 29.20	(−) 30.60
Ret-sim	(+) 0.92	(+) 1.28	(+) 1.65	(+) 2.00	(+) 1.97	(+) 2.27	(+) 3.72	(+) 3.34	(+) 5.22	(+) 5.70
Ret-dis	(+) 1.66	(+) 1.79	(+) 2.30	(+) 5.00	(+) 4.91	(+) 5.71	(+) 4.56	(+) 4.12	(+) 6.28	(+) 6.76
Reg-sim	(+) 2.53	(+) 2.53	(+) 3.28	(+) 3.31	(+) 3.29	(+) 3.81	(+) 4.76	(+) 4.40	(+) 12.78	(+) 12.18
Reg-dis	(+) 2.52	(+) 2.43	(+) 3.17	(+) 5.41	(+) 5.34	(+) 6.20	(+) 7.40	(+) 6.82	(+) 12.54	(+) 12.44
Con-S2V-sim	(+) 3.83	(+) 3.55	(+) 4.62	(+) 4.52	(+) 4.50	(+) 5.21	(+) **14.98**	(+) **14.38**	(+) 13.68	(+) 13.56
Con-S2V-dis	(+) **4.29**	(+) **4.04**	(+) **5.22**	(+) **7.68**	(+) **7.56**	(+) **8.80**	(+) 9.30	(+) 8.36	(+) **15.10**	(+) **15.20**

Unsurprisingly, Sen2Vec outperforms Tf-Idf representation (row 6) by a good margin on both tasks. It gets improvements of up to 11.6 points on classification, and up to 30.6 points on clustering. This is inline with the finding of [1], and demonstrates the benefits of using distributed representation over sparse BOW representations.

Simple averaging of Word2Vec vectors performs quite well for classification, especially, on Reuters, where it outperforms Sen2Vec by 1.9 to 2.5 points. [8] also reported similar findings on five out of six datasets. However, averaging does not perform well on clustering, where the scores are 10.2 to 18.5 points below than Sen2Vec.

Simple addition-based composition of C-Phrase word vectors performs poorly on both tasks – lower than Sen2Vec by 2 to 3 points on classification and by 1.4 to 11.9 points on clustering.

Unexpectedly, FastSent and Skip-Thought perform quite poorly on both tasks. Skip-Thought, in particular, has the worst performance on both tasks. These results contradict the claim made by [7] that skip-thought vectors are generic. To investigate if the poor results are due to shift of domains (*book* vs. *news*), we also trained Skip-Thought on our training corpora with vector size 600 and vocabulary size $30K$. The performance was even worse. We hypothesize, this is due to our training set size, which may not be enough for the heavy model. Also, Skip-Thought does not perform any inference to extract the vector using a context – although the model was trained to generate neighboring sentences, context was ignored when the encoder was used to extract the sentence vector.

Regarding FastSent, although its classification performance on Reuters is comparable to Sen2Vec, it performs poorly on Newsgroups, where the measures are 12.2 to 14.3 points lower than Sen2Vec. The differences get bigger in clustering. The reason could be that FastSent does not learn sentence representations directly, rather it simply adds the word vectors. Note that FastSent was outperformed by Tf-Idf in all classification tasks in [8]. Since both Skip-Thought and FastSent learn representations by predicting contents of adjacent sentences, the learned vectors might capture linguistic properties that are more specific to the neighbors.

We also experimented with SAE and SDAE auto-encoders proposed in [8]. However, they performed poorly on our tasks (thus not shown in the table). For example, SAE gave accuracies of around 40% on reuters and 18% on newsgroups. This is similar to what [8] observed. They propose to use pretrained word embeddings to improve the results. We did not achieve significant gains by using pretrained embeddings on our tasks.

Interestingly, the retrofitting and regularized models improve over Sen2Vec on both tasks, showing gains of up to 6.2 points on classification and up to 12.8 points on clustering. The improvements in most cases are significant. This demonstrates that proximity hypothesis is beneficial for these tasks.

When we compare regularized models with retrofitted ones, we observe that regularized models consistently outperform the retrofitted counterparts on both tasks with improvements of up to 1.6 points on classification and up to 7.6 points on clustering. This demonstrates that incorporating contextual information by means of regularization is more effective than retrofitting. This could be due to the fact that regularization approach induces contextual information while learning the vectors from scratch as opposed to revising them in a post-processing step.

Finally, we observe further improvements for our complete models (CON-S2V variants) on both tasks. Compared to the best regularized models, our models deliver improvements of up to 2.6 points on classification and up to 7.6 points on clustering. This demonstrates that by including the neighbor prediction component to model distributional similarity, our model captures complementary contextual information to what is captured by the regularized models. A comparison between the context types reveals that discourse context is more beneficial than similarity context in most cases, especially for classification. For clustering, similarity context gives better results in a few cases (e.g., on Reuters). Overall, our best model outperforms the best existing model by up to 8.8 and 15.20 points on classification and clustering tasks, respectively.

5.4   Summarization Results

Table 5 shows ROUGE-1 scores of our models on DUC datasets for the summary length of 100 words. W2V-avg performs well achieving comparable score to Sen2Vec on DUC'01 and 1.4 points improvement on DUC'02. C-Phrase outperforms Sen2Vec by 2.5 and 1.7 points on DUC'01 and DUC'02, respectively. FastSent and Skip-Thought again perform disappointingly. Sen2Vec outperforms FastSent by 4.15 and 7.53 points on DUC'01 and DUC'02, respectively. Skip-Thought performs comparably to Sen2Vec on DUC'01, but gets worse on DUC'02.

Interestingly, Tf-Idf performs quite well on this task. It gives the top score on DUC'01 (i.e., 48.7 ROUGE-1), and an improvement of 1.5 points over Sen2Vec on DUC'02. These results suggest that existing distributed representation methods are inferior to traditional methods in modeling aspects that are necessary for measuring sentence importance.

Retrofitted models give mixed results and fail to get significant improvement over Sen2Vec. On the other hand, with similarity context, regularized model improves over Sen2Vec by 2 to 3 points. This again suggests that regularization is a better method to incorporate context proximity. By including the neighbor prediction component to incorporate distributional similarity, our combined model improves the scores further; it achieves the second best result on DUC'01, and becomes top-performer on DUC'02.

Table 5. ROUGE-1 scores of the models on DUC datasets in comparison with Sen2Vec.

	DUC'01	DUC'02
Sen2Vec	43.88	54.01
W2V-avg	(−) 0.62	(+) 1.44
C-Phrase	(+) 2.52	(+) 1.68
FastSent	(−) 4.15	(−) 7.53
Skip-Thought	(+) 0.88	(−) 2.65
Tf-Idf	(+) **4.83**	(+) 1.51
Ret-sim	(−) 0.62	(+) 0.42
Ret-dis	(+) 0.45	(−) 0.37
Reg-sim	(+) 2.90	(+) 2.02
Reg-dis	(−) 1.92	(−) 8.77
Con-S2V-sim	(+) 3.16	(+) **2.71**
Con-S2V-dis	(+) 1.15	(−) 4.46

It is not surprising that similarity context is more suitable than discourse context for this task. From a context of topically similar sentences, our model learns representations that capture linguistic aspects related to information centrality. Given that the existing models fail to beat the Tf-Idf baseline on this task, our results are rather encouraging.

6   Discussion and Future Directions

We have presented a novel model to learn distributed representation of sentences by considering content as well as context of a sentence. Our results on tasks involving classifying, clustering and ranking sentences confirm that extra-sentential contextual information is crucial for modeling sentences, and this information is best captured by our model that comprises a neighbor-based prediction component and a regularization component to capture distributional similarity and contextual proximity, respectively.

One important property of our model is that it encodes a sentence directly, and it considers neighboring sentences as atomic units. Apart from the improvements that we achieve in various tasks, this property makes our model quite efficient to train compared to *compositional* methods like encoder-decoder models (e.g., SDAE, Skip-Thought) that compose a sentence vector from the word vectors. Encoder-decoder approaches attempt to capture the structure of a sentence, which could be beneficial to model long distance relations between words (e.g., negation in sentiment classification). It would be interesting to see how our model compares with compositional models on sentiment classification task. However, this would require creating a new dataset of comments with sentence-level sentiment annotations. We intend to create such datasets and evaluate the models in the future.

Acknowledgments. This research is partially supported by Mohammad Hasan's NSF CAREER Award (IIS-1149851).

References

1. Le, Q.V., Mikolov, T.: Distributed representations of sentences and documents. In: ICML, vol. 14, pp. 1188–1196 (2014)
2. Mikolov, T., Yih, W., Zweig, G.: Linguistic regularities in continuous space word representations. In: Proceedings of the 2013 Conference of the North American Chapter of the Association for Computational Linguistics: Human Language Technologies, Atlanta, Georgia, pp. 746–751. Association for Computational Linguistics, June 2013
3. Bengio, Y., Ducharme, R., Vincent, P., Janvin, C.: A neural probabilistic language model. J. Mach. Learn. Res. **3**, 1137–1155 (2003)
4. Socher, R., Lin, C.C.Y., Ng, A.Y., Manning, C.D.: Parsing natural scenes and natural language with recursive neural networks. In: Proceedings of the 28th International Conference on Machine Learning (ICML), pp. 129–136 (2011)
5. Stede, M.: Discourse Processing. Morgan & Claypool Publishers, San Rafael (2011)
6. Hobbs, J.R.: Coherence and coreference. Cogn. Sci. **3**(1), 67–90 (1979)
7. Kiros, R., Zhu, Y., Salakhutdinov, R., Zemel, R.S., Torralba, A., Urtasun, R., Fidler, S.: Skip-thought vectors. In: Proceedings of the 28th International Conference on Neural Information Processing Systems, NIPS 2015, Montreal, Canada, pp. 3294–3302. MIT Press (2015)
8. Hill, F., Cho, K., Korhonen, A.: Learning distributed representations of sentences from unlabelled data. In: Proceedings of the 2016 Conference of the North American Chapter of the Association for Computational Linguistics: Human Language Technologies, San Diego, California, pp. 1367–1377. Association for Computational Linguistics, June 2016
9. Harris, Z.: Distributional structure. Word **10**, 146–162 (1954)
10. Mikolov, T., Chen, K., Corrado, G., Dean, J.: Efficient estimation of word representations in vector space. arXiv preprint arXiv:1301.3781 (2013)
11. Pham, N.T., Kruszewski, G., Lazaridou, A., Baroni, M.: Jointly optimizing word representations for lexical and sentential tasks with the C-PHRASE model. In: ACL 2015, 26–31 July 2015, Beijing, China, vol. 1: Long Papers, pp. 971–981 (2015)
12. Mitchell, J., Lapata, M.: Composition in distributional models of semantics. Cogn. Sci. **34**(8), 1388–1439 (2010)
13. Cho, K., van Merriënboer, B., Gülçehre, Ç., Bahdanau, D., Bougares, F., Schwenk, H., Bengio, Y.: Learning phrase representations using RNN encoder-decoder for statistical machine translation. In: Proceedings of the 2014 Conference on Empirical Methods in Natural Language Processing (EMNLP), pp. 1724–1734. Association for Computational Linguistics, October 2014

14. Morin, F., Bengio, Y.: Hierarchical probabilistic neural network language model. In: Cowell, R.G., Ghahramani, Z. (eds.) Proceedings of the Tenth International Workshop on Artificial Intelligence and Statistics, Society for Artificial Intelligence and Statistics, pp. 246–252 (2005)

15. Gutmann, M., Hyvärinen, A.: Noise-contrastive estimation: a new estimation principle for unnormalized statistical models. In: Teh, Y., Titterington, M. (eds.) Proceedings of the International Conference on Artificial Intelligence and Statistics (AISTATS). JMLR W&CP, vol. 9, pp. 297–304 (2010)

16. Mikolov, T., Sutskever, I., Chen, K., Corrado, G., Dean, J.: Distributed representations of words and phrases and their compositionality. In: Proceedings of the 26th International Conference on Neural Information Processing Systems, NIPS 2013, pp. 3111–3119 (2013)

17. Faruqui, M., Dodge, J., Jauhar, S.K., Dyer, C., Hovy, E., Smith, N.A.: Retrofitting word vectors to semantic lexicons. In: Proceedings of NAACL (2015)

18. Halliday, M., Hasan, R.: Cohesion in English. Longman, London (1976)

19. Malioutov, I., Barzilay, R.: Minimum cut model for spoken lecture segmentation. In: Proceedings of the 21st International Conference on Computational Linguistics and the 44th Annual Meeting of the Association for Computational Linguistics, ACL-44, Sydney, Australia, pp. 25–32. Association for Computational Linguistics (2006)

20. Erkan, G., Radev, D.R.: LexRank: graph-based lexical centrality as salience in text summarization. J. Artif. Int. Res. 22(1), 457–479 (2004)

21. Wolf, F., Gibson, E.: Representing discourse coherence: a corpus-based study. Comput. Linguist. 31, 249–288 (2005)

22. Nenkova, A., McKeown, K.: Automatic summarization. Found. Trends Inf. Retrieval 5(23), 103–233 (2011)

23. Page, L., Brin, S., Motwani, R., Winograd, T.: The pagerank citation ranking: bringing order to the web (1999)

24. Lin, C.Y.: Rouge: a package for automatic evaluation of summaries. In: Text Summarization Branches Out: Proceedings of the ACL-04 Workshop, Barcelona, Spain, vol. 8 (2004)

25. Hermann, K.M., Blunsom, P.: Multilingual models for compositional distributed semantics. In: Proceedings of the 52nd Annual Meeting of the Association for Computational Linguistics, Baltimore, Maryland, vol. 1: Long Papers, pp. 58–68. Association for Computational Linguistics, June 2014

26. Arthur, D., Vassilvitskii, S.: K-means++: the advantages of careful seeding. In: Proceedings of the Eighteenth Annual ACM-SIAM Symposium on Discrete Algorithms, pp. 1027–1035 (2007)

27. Rosenberg, A., Hirschberg, J.: V-measure: a conditional entropy-based external cluster evaluation measure. In: EMNLP-CoNLL, vol. 7, pp. 410–420 (2007)

28. Vinh, N.X., Epps, J., Bailey, J.: Information theoretic measures for clusterings comparison: variants, properties, normalization and correction for chance. J. Mach. Learn. Res. 11, 2837–2854 (2010)

29. Socher, R., Perelygin, A., Wu, J., Chuang, J., Manning, C.D., Ng, A.Y., Potts, C.: Recursive deep models for semantic compositionality over a sentiment treebank. In: Proceedings of the 2013 Conference on Empirical Methods in Natural Language Processing, Stroudsburg, PA, pp. 1631–1642. Association for Computational Linguistics, October 2013

Deep Over-sampling Framework for Classifying Imbalanced Data

Shin Ando$^{(\boxtimes)}$ and Chun Yuan Huang

School of Management, Tokyo University of Science,
1-11-2 Fujimi, Chiyoda-ku, Tokyo, Japan
ando@rs.tus.ac.jp, 8613095@ed.tus.ac.jp

Abstract. Class imbalance is a challenging issue in practical classification problems for deep learning models as well as traditional models. Traditionally successful countermeasures such as synthetic over-sampling have had limited success with complex, structured data handled by deep learning models. In this paper, we propose *Deep Over-sampling* (DOS), a framework for extending the synthetic over-sampling method to the deep feature space acquired by a convolutional neural network (CNN). Its key feature is an explicit, supervised representation learning, for which the training data presents each raw input sample with a synthetic embedding target in the deep feature space, which is sampled from the linear subspace of in-class neighbors. We implement an iterative process of training the CNN and updating the targets, which induces smaller in-class variance among the embeddings, to increase the discriminative power of the deep representation. We present an empirical study using public benchmarks, which shows that the DOS framework not only counteracts class imbalance better than the existing method, but also improves the performance of the CNN in the standard, balanced settings.

Keywords: Class imbalance · Convolutional neural network
Deep learning · Representation learning · Synthetic over-sampling

1 Introduction

In recent years, deep learning models have contributed to significant advances in supervised and unsupervised learning tasks on complex data, such as speech and imagery. Convolutional neural networks (CNNs), for example, have achieved *state-of-the-art* performances on various image classification benchmarks [8,27]. One of the key features of CNN is representation learning, i.e., the hidden layers of convolutional neural network generates an expressive, non-linear mapping of complex data in a *deep feature* space [12,19]. Such features are shown to be useful for other classification models or similar classification tasks [2,13], enabling further means of enhancements such as multi-task learning [7].

The applications of deep learning, meanwhile, encounter many practical challenges, such as the cost of preparing a sufficient amount of labeled samples. The problem of class imbalance arises when the number of samples significantly differ

© Springer International Publishing AG 2017
M. Ceci et al. (Eds.): ECML PKDD 2017, Part I, LNAI 10534, pp. 770–785, 2017.
https://doi.org/10.1007/978-3-319-71249-9_46

between two or more classes. Such imbalance can affect the traditional classification models [1,11,17] as well as deep learning models [14,15,28], commonly resulting in poor performances over the classes in the minority. For deep learning models, its influence on representation learning can deteriorate the performances on majority classes as well.

There is a rich literature on countermeasures against class imbalance for traditional classification models [4]. A popular and intuitive approach among them is re-sampling, which directly adjusts the sample sizes of respective classes. For example, SMOTE [3] generates synthetic samples, which are interpolations of *in-class* neighboring samples, to augment the minority class. Its underlying assumption is that the interpolations do not deviate from the original class distribution, as in a locally linear feature space. Similar approaches for deep learning models, e.g., re-sampling, cost-sensitive learning, and their combinations [14,28], have also been explored, but in a more limited number of studies. Overall, they introduce complex architectures or sampling schemes, which require significant amount of data- and model-specific configurations and lower the applicability to new problems. It also should be noted that generating synthetic samples of structured, complex data, in order to conduct synthetic over-sampling on such data, is not straightforward.   .

In this work, we extend the synthetic over-sampling method to the convolutional neural network (CNN) using its deep representation. To our knowledge, over-sampling in the acquired, deep feature space have not been explored prior to this work. Integrating synthetic instances, which are not direct mappings of any raw input sample, effectively into a supervised learning framework is a non-trivial challenge. Our main idea is to use synthetic instances as the supervising targets in the deep feature space, to implement a representation learning which induce better class distinction in the acquired space.

The proposed framework, *Deep Over-sampling* (DOS), employs a basic CNN architecture in which the lower layers acquire the embedding function and the top layers acquire the classification function. We implement the training of the CNN with explicit supervising information for both functions, i.e., the network parameters are updated by propagation from the output of the lower layers as well as the top layers. Accordingly, the training data presents with each raw input sample, a class label and a target in the deep feature space. The targets are sampled from a linear subspace of the in-class neighbors around the embedded input. As such targets naturally distribute closer to the class mean, our aim is to induce smaller in-class variance among the embeddings.

DOS provides the framework to address the effect of class imbalance on both classifier and representation learning. First, the training data is augmented by assigning multiple synthetic targets to one input sample. Secondly, an iterative process of learning the CNN and updating the targets with the acquired representation enhances the discriminative power of the deep features.

The main contribution of this work is a general re-sampling framework, which enables the deep neural net to learn the deep representation and the classifier jointly in a class-imbalanced setting without substantial modification on its architecture, thus are applicable to a wide range of deep learning models. We validate the effectiveness of the proposed framework in present an empirical study using public image classification benchmarks. Furthermore, we investigate the effect of the proposed framework outside the class imbalance setting. The rest of this paper is organized as follows. Section 2 introduces the related work and the preliminaries, respectively. Section 3 describes the details of the proposed framework. The empirical results are shown in Sect. 4, and we present our conclusion in Sect. 5.

2   Background

2.1   Class Imbalance

Class imbalance is a common issue in practical classification problems, where a large imbalance in the number of training samples between classes causes the learning algorithms to over-generalize for the classes in the majority. Its effect is critical as retrieving the minority classes is usually the primary interest in practice [11,17]. The countermeasures against class imbalance can be generally categorized into three major approaches. The re-sampling approach attempts to directly adjust the sample sizes by over- or under-sampling on the training set. The instance weighting approach exploits a similar intuition by increasing the importance of the minority class samples. Finally, the cost-sensitive learning approach modifies the loss function and/or the learning algorithm to penalize the errors on the minority class predictions.

For traditional classification models, the synthetic over-sampling methods such as SMOTE [3] have been generally successful in countering the effect of imbalance. Typically, synthetic samples are generated by randomly selecting a minority class sample and taking an interpolation between its neighbors. The re-sampling approach has also been attempted on neural network models, e.g., [28] has combined over-sampling with cost-sensitive learning and [15] has combined under-sampling with synthetic over-sampling.

One limitation of the synthetic over-sampling method is the need for the vector form input data, i.e., it is not applicable to non-vector input domain, e.g., pair-wise distances [1,16]. Moreover, it implicitly assumes that the interpolations do not deviate from the original distribution, as in the locally linear feature space. Such assumption is usually not problematic for the traditional classification models, many of which are developed with similar assumptions. Synthetic over-sampling is generally successful when the features are pre-selected for such models. Meanwhile, the assumption does not hold for complex, structured data often handled by deep neural nets. Generating samples of complex data is substantially difficult, and simple interpolations can easily deviate from the original distribution. While acquiring locally-linear representation is a key advantage of deep neural nets, a recent study has reported that class imbalance can affect their representation learning capability as well [14].

In [14], a sophisticated under-sampling scheme called Large Margin Local Embedding (LMLE) was implemented to generate an abridged training set for representation learning. These samples were selected considering the class and cluster structures such as in-/out-of-class and, in-/out-of-cluster neighbors. It also introduced a new loss function based on class-separating margins, inspired by the Large Margin Nearest Neighbor (LMNN) [26].

The potential demerit of under-sampling is the loss of information from discarding the subset of the training data. As such, computationally-intensive analysis to retain important samples, in this case the analyses of class and cluster structure, is inevitable in the re-sampling process. Another drawback in the above work was the specificity of the loss function and the network architecture. The effect of class imbalance, as we demonstrate in the next section, differ based on the classification model. It is thus not clear whether the experimental results of a modified kNN extends generally to other classifiers, especially given that the proposed margin-based loss function is oriented toward the kNN-based classification model. Additionally, its architecture does not support simultaneous learning of the representation and the classifier, which is a key feature of CNN. Overall, the above implementation is likely to require much task- and model-specific configurations when applying to a new problem.

In this paper, we explore an over-sampling scheme in the deep representation space to avoid computationally intensive analyses. We also attempt to utilize a basic CNN architecture in order to maintain its wide applicability and its advantage of cohesive representation and classifier learning.

2.2 Preliminary Results

To motivate our study, we first present a preliminary result on the effect of class imbalance on deep learning. An artificial imbalanced setting was created with the MNIST-back-rotation images [20] by selecting four digits randomly, and removing 90 % of their samples. We trained two instances of a basic CNN architecture [22] by back-propagation respectively with the original and the imbalanced data.

The training of the two CNNs from initial parameters were repeated ten times and the averages of their class-wise retrieval performances, i.e., Precision, Recall, and F1-score, are reported here. Although the overall accuracy has been reported in prior studies, we preferred the class-wise retrieval measures in order to obtain separate insights for the minority and majority classes.

Tables 1a and b show the class-wise precision, recall, and F1-score of one trial from each experiment. In Table 1b, the minority class digits are indicated by the asterisks. Additionally, significant declines (0.1 or more) in precision or recall, compared to Table 1a, are indicated by double underline and smaller (0.05 or more) drops are indicated by single underlines. Tables 2a and b show the same performance measures by the kNN classifier using the deep representation acquired by the two CNNs. The performance of the kNN, which is a non-inductive lazy learning algorithm, is said to substantially reflect the effect of class imbalance on representation learning [14]. The minority classes and reductions in precision or recall are indicated in the same manner as in Table 2a.

Table 1. Class-wise performance comparison (CNN)

(a) Balanced Data

Digit	Precision	Recall	F1-score
0	0.88	0.92	0.90
1	0.93	0.87	0.90
2	0.63	0.68	0.65
3	0.80	0.81	0.80
4	0.64	0.83	0.72
5	0.71	0.69	0.70
6	0.77	0.67	0.72
7	0.75	0.72	0.74
8	0.78	0.73	0.75
9	0.71	0.70	0.70

(b) Imbalanced Data

Digit	Precision	Recall	F1-score
0*	0.60	0.98	0.75
1	0.92	0.78	0.85
2	0.69	0.42	0.52
3	0.81	0.58	0.68
4*	0.090	0.92	0.16
5*	0.12	0.90	0.22
6*	0.075	0.84	0.14
7	0.77	0.53	0.63
8	0.70	0.62	0.65
9	0.80	0.38	0.51

Table 2. Class-wise performance comparison (deep representation + kNN)

(a) Balanced Data

Digit	Precision	Recall	F1-score
0	0.91	0.86	0.89
1	0.93	0.82	0.87
2	0.60	0.74	0.66
3	0.81	0.74	0.77
4	0.72	0.72	0.72
5	0.64	0.73	0.68
6	0.67	0.78	0.72
7	0.74	0.71	0.72
8	0.74	0.75	0.74
9	0.70	0.64	0.67

(b) Imbalanced Data

Digit	Precision	Recall	F1-score
0*	0.86	0.83	0.85
1	0.91	0.83	0.87
2	0.57	0.62	0.59
3	0.75	0.74	0.74
4*	0.61	0.61	0.61
5*	0.51	0.57	0.54
6*	0.48	0.56	0.52
7	0.69	0.67	0.68
8	0.71	0.67	0.69
9	0.63	0.59	0.61

In Table 1b, there is a clear trend of decrease in precision for the minority classes. Over the majority classes, there are reduction in recall which are smaller but still substantial. In Table 2b, both the precision and the recall decline for most of the minority classes. There are declines in precision or recall of many majority classes as well.

Table 3 shows the average measures of minority and majority classes over ten trials. The digits of the minority classes were chosen randomly in each trial. The trends in Table 3 regarding the precision and the recall are consistent with those of Tables 1 and 2. These preliminary results support our insight that the class imbalance has a negative impact on the representation learning of CNN as well as its classifier training, and the influence differs depending on the classifier.

Table 3. Summary of average retrieval measures

(a) CNN

Setting	Precision	Recall	F1-score
Balanced	0.76	0.76	0.76
Minority	0.32	0.86	0.43
Majority	0.79	0.57	0.66

(b) Deep Representation + kNN

Setting	Precision	Recall	F1-score
Balanced	0.75	0.75	0.75
Minority	0.65	0.67	0.65
Majority	0.71	0.70	0.71

3   Deep Over-sampling Framework

This section describes the details of the proposed framework, *Deep Over-sampling* (DOS). The main idea of DOS is to re-sample the training data in an expressive, nonlinear feature space acquired by the convolutional neural network. While previous over-sampling based methods such as SMOTE have achieved general success for traditional models, their approach of sampling from the linear subspace of the original data has clear limitations for complex, structured data, such as imagery. In contrast, DOS implements re-sampling in the linear subspace of deep feature instances and exploit the re-sampled instances for explicitly supervised representation learning as well as complementing the minority classes.

3.1   Notations

We employ a basic CNN whose architecture can be divided into two groups of layers: the lower layers embedding the raw input into the deep feature space, and the top layers predicting the class label from the deep features. We denote the embedding function of the CNN by $f : \Phi \to \mathbb{R}^d$, where Φ is the raw input domain with a complex data structure. We also denote the discriminative function of the CNN by $g : \mathbb{R}^d \to [0 : 1]^n$, whose output represents a vector of posterior class probabilities $P(C|x)$ over n classes.

Let $\mathcal{X} = \{(x_i, y_i)\}_{i=1}^m$ denote a set of training data, where $x_i \in \Phi$ and y_i takes a class value from $\mathcal{C} = \{c_j\}_{j=1}^n$. The network parameters, denoted by $\mathbf{W}_f$ and $\mathbf{W}_g$ for the embedding layers and the classification layers, respectively, are learned with back-propagation. A class imbalance, such that $\#\{(x, y) : y = c_i\} \gg \#\{(x, y) : y = c_j\}$ for some $\{c_i, c_j\} \subset \mathcal{C}$, may arise from practical issues such as the cost of data collection. A significant imbalance can hinder the performance of the acquired model. The architecture is illustrated in Fig. 1. We further elaborate on its details in Sect. 3.3.

3.2   Deep Feature Overloading

We employ re-sampling in the deep feature space to assign each raw input sample with multiple deep feature instances. As a result, the supervising targets are provided for both the embedding function f and the classification function g.

Let $\mathcal{V}(c_j) = \{f(x_i) : y_i = c_j\}$ denote the set of embeddings whose raw input has the label c_j. A training instance is defined as a triplet $z_i = (x_i, y_i, \mathcal{N}(x_i))$, consisting of an input sample x_i, its class label y_i, and a subset of embeddings $\mathcal{N}(x_i)$. $\mathcal{N}(x_i)$ is a subset of $\mathcal{V}(y_i)$ that includes $f(x_i)$ and its k in-class neighbors,

$$\mathcal{N}(x_i; k) = \underset{\substack{\mathcal{N} \subset \mathcal{V}(y_i) \wedge \\ \#(\mathcal{N}) = k+1}}{\arg\min} \sum_{v \in \mathcal{N}} \|f(x_i) - v\|^2 \tag{1}$$

We refer to the process of pairing each $(x_i, y_i) \in \mathcal{X}$ with its deep feature neighbors as deep feature *overloading*. The process is illustrated in Fig. 2.

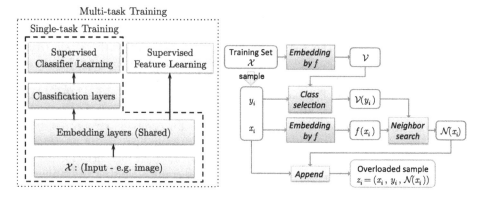

Fig. 1. CNN architecture   **Fig. 2.** Deep feature overloading

We refer to k as the *overloading* parameter and $\mathcal{Z} = \{z_i\}_{i=1}^m$ as the *overloaded* training set.

As we describe in the following section, the deep feature neighbors are used to generate synthetic targets for the minority class samples. The value of k, thus may be varied between the minority and the majority classes as the latter does not need to be synthetically complemented. Note that the minimum value for k is 0, in which case $\mathcal{N}(x_i) = \{f(x_i)\}$.

3.3   Micro-cluster Loss Function

Our CNN architecture, illustrated in Fig. 1, features two outputs, one for the classification and one for the embedding functions. The initial parameters of the network are learned in a single-task training using only the substructure indicated by the dotted box on the left side of the figure with the original imbalanced training set.

The classifier output is given by the softmax layer and the cross-entropy loss [9] $\mathcal{H}$ based on the predicted class probabilities $g(f(x))$ of a given input (x, y)

$$\ell(x, y) = \mathcal{H}\left(g(f(x)), y\right) \tag{2}$$

is used for single-task learning.

After the initialization, the training is expanded to the multi-task learning architecture in Fig. 1 to use propagation from both f and g. The loss function given an overloaded sample z_i is defined with regards to $\mathcal{N}(x_i)$, which can be considered an in-class cluster in the deep feature space. We thus refer to the functions as the *micro-cluster* loss.

The micro-cluster loss for the embedding function f is defined as a sum of squared errors

$$\ell_f(x) = \sum_{v \in \mathcal{N}(x)} \|f(x) - v\|^2 \tag{3}$$

The minimum of (3) is obtained when $f(x)$ is mapped to the mean of $\mathcal{N}(x_i)$. Note that the mean is a synthetic point in the deep feature space, to which no particular original sample is projected.

There are two key intuitions for setting the target representation to local means. First, the summation of the squared errors can add emphases to the minority class samples, by *overloading* them with a larger number of embeddings. Secondly, as the local means distribute closer toward the mean of the original distribution, it induces smaller in-class variance in the learned representation. Smaller in-class variance yields better class distinction, which can be induced further by iterating the procedure with the updated embeddings.

The micro-cluster loss for g is defined as the weighted sum of the cross-entropy losses, i.e.,

$$\ell_g(x, y) = \sum_{v \in \mathcal{N}(x)} \rho(v) \mathcal{H}(g(v), y) \tag{4}$$

where $\rho(v)$ is the normalized exponential weight given the squared errors in (3),

$$\rho(v) = \frac{1}{Z} \exp\left(-\|f(x) - v\|^2\right) \tag{5}$$

and Z denotes a normalizer such that

$$Z = \sum_{\mathcal{N}(x)} \exp\left(-\|f(x) - v\|^2\right)$$

In (5), the largest weight among $\mathcal{N}(x)$, 1, is assigned to the original loss from $f(x)$, and a larger weight is assigned to neighbors in a closer range.

3.4 Deep Over-sampling

Deep Over-sampling uses the overloaded instances to supplement the minority classes. Multiple overloaded instances are generated from each training sample by pairing it with different targets sampled from the linear subspace of its in-class neighbors.

Let $\mathcal{W}$ denote a domain of positive, $\ell 1$-normalized vectors of k-dimensions, i.e., for all $\mathbf{w} \in \mathcal{W}$, $\|\mathbf{w}\|^1 = 1$ and $w_i \geq 0$ for $i = 1, \ldots, k$. Note that k is the overloading parameter. For each overloaded instance $z_i \in \mathcal{Z}$, we sample a set of vectors $\{\mathbf{w}^{(i,j)}\}_{j=1}^r$ from $\mathcal{W}$.

We define the weighted overloading instance as a quadruplet

$$z_i^{(j)} = (x_i, y_i, \mathcal{N}(x_i), \mathbf{w}^{(i,j)}) \tag{6}$$

Note that each element of the weight vector correspond with an element of $\mathcal{N}(x_i)$.

Sampling r vectors for each z_i, we obtain a weighted training set

$$\mathcal{Z}' = \bigcup_{j=1}^r \left\{ z_i^{(j)} \right\}_{i=1}^m \tag{7}$$

We define the following micro-cluster loss functions for the weighted instances. The loss function for f given a quadruplet of x, y, $\mathcal{N}(x) = \{v_i\}_{i=1}^k)$, and $\mathbf{w} = (w_1, \ldots, w_k)$ is written as

$$\ell'_f(x, y, \mathbf{w}, \mathcal{N}(x)) = \sum_{i=1}^k w_i \|f(x) - v_i\|^2 \tag{8}$$

The minimum of (8) is attained when $f(x)$ is at the weighted mean, $\sum_i w_i v_i$.

The weighted micro-cluster loss for g is defined, similarly to (4), as

$$\ell'_g(x, y, \mathbf{w}) = \sum_{i=1}^k \rho'(v_i, w_i) \mathcal{H}(g(v_i), y) \tag{9}$$

where ρ' is the normalized weight

$$\rho'(v_i, w_i) = \frac{1}{Z} \exp(-w_i \|f(x) - v_i\|^2) \tag{10}$$

To summarize, the augmentative samples for the minority classes are generated by pairing each raw input sample with multiple targets for representation learning. The rationale for learning to map one input onto multiple targets can be explained as promoting robustness under the imbalanced setting. Since there are less than sufficient number of samples for the minority classes, strong supervised learning may induce the risk of overfitting. Generating multiple targets within the range of local neighbors is similar in effect as adding noise to the target and can prevent the convergence of the gradient descent to undesired local solutions.

After training the CNN, the targets are recomputed with the updated representation. The iterative process of training the CNN and updating the targets incrementally shifts the targets toward the class mean and improve the class distinction among the embeddings. The pseudo code of the iterative procedure is shown in Algorithm 1.

3.5   Parameter Selection

As mentioned in Sect. 3.2, different values of overloading parameter k may be selected for the minority and the majority classes to placing additional emphases on the former. Let k_{mnr} and k_{mjr} denote the overloading values for the minority and the majority classes, respectively. If the latter is set to the minimum, i.e., $k_{\mathrm{mjr}} = 0$, then the loss for the minority class samples, as given by (3), accounts for $(k_{\mathrm{mnr}} + 1)$ times more squared errors.

In essence, however, k should be chosen to reflect the extent of the neighborhood, as the size of the neighbors, $\mathcal{N}(x)$, can influence the the efficiency of the back-propagation learning. As one increase k, the target shifts closer to the global class mean and, in turn, farther away from the tentative embedding $f(x)$. For better convergence of the gradient descent, the target should be maintained within a moderate range of proximity from the tentative embedding.

Algorithm 1. CNN Training with Deep Over-sampling

1: **Input** $\mathcal{X} = \{(x_i, y_i)\}_{i=1}^m$, class values $\mathcal{C} = \{c_j\}_{j=1}^n$, class-wise overloading $\{k_j\}_{j=1}^n$, class-wise over-sampling size $\{r_j\}_{j=1}^n$, CNN with outputs for functions f and g, deep feature dimensions d, number of iterations T
2: **Output** A trained CNN with functions f and g
3: **function** STL($\mathcal{T}$): Single-task training of CNN with training set $\mathcal{T}$
4: **function** MTL($\mathcal{T}$): Multi-task training of CNN with training set $\mathcal{T}$
5: Initialize CNN parameters: STL($\mathcal{X}$)
6: **for** $t = 1, \ldots, T$ **do**
7:     **for** $i = 1, \ldots, m$ **do**
8:         Compute $v_i = f(x_i)$
9:     **end for**
10:     **for** $j = 1, \ldots, n$ **do**
11:         **if** $k_j > 0$ **then**
12:             Compute the mutual distance matrix $D(c_j)$ over $\mathcal{V}(c_j)$ for neighbor search
13:         **end if**
14:         $\mathcal{Z}_j = \emptyset$
15:         **for** $\{(x_i, y_i) : y_i = c_j\}$ **do**
16:             Select $\mathcal{N}(x_i)$ from $\mathcal{V}(c_j)$
17:             Sample a set of r_j normalized positive vectors $\mathcal{W}$
18:             $\mathcal{Z}_j = \mathcal{Z}_j \cup \{(x_i, y_i, \mathcal{N}(x_i), \mathbf{w})\}_{\mathbf{w} \in \mathcal{W}}$
19:         **end for**
20:     **end for**
21:     $\mathcal{Z} = \bigcup_{j=1}^n \mathcal{Z}_c$
22:     Update CNN parameters: MTL($\mathcal{Z}$)
23: **end for**

Our general guideline for the parameter selection is therefore to choose a value of k_{mnr} from [3:10] by empirical validation and set $k_{\mathrm{mjr}} = 0$ provided that there are sufficient number of samples for the majority classes. Furthermore, we suggest to choose the over-sampling rate r from $[\frac{1}{R} : \frac{k_{\mathrm{mnr}}}{R}]$ where R denotes the average ratio of the minority and the majority class samples, i.e.,

$$R = \frac{\#\{(x, y) : (x, y) \in \mathcal{X} \wedge y = c_{\mathrm{mnr}}\}}{\#\{(x, y) : (x, y) \in \mathcal{X} \wedge y = c_{\mathrm{mjr}}\}} \tag{11}$$

For the number of iterations T, we suggest it to be the same as the number of training rounds, i.e., to re-compute the targets after every training round.

4   Empirical Results

We conducted an empirical study[1] to evaluate the DOS framework in three experimental settings. The first experiment is a baseline comparison, for which we replicated a setting used in the most recently proposed model to address the

---

[1] The source codes for reproducing the datasets and the results are made available at http://www.rs.tus.ac.jp/ando/exp/DOS.html.

class imbalance. Secondly, we evaluated the sensitivity of DOS with different levels of imbalance and parameter choices. Finally, we investigated the effect of deep over-sampling in the standard, balanced settings. The imbalanced settings were created with standard benchmarks by deleting the samples from selected classes.

4.1   Datasets and CNN Settings

We present the results on five public datasets: MNIST [21], MNIST-*back-rotation* images [20], SVHN [23], CIFAR-10 [18], and STL-10 [6]. We have set up the experiment in the image domain, because it is one of the most popular domains in which CNNs have been used extensively, and also it is difficult to apply the SMOTE algorithm directly. Note that we omit the result of preprocessing the imbalanced image set using SMOTE, as it achieved no improvement in the classifier performances.

The MNIST digit dataset consists of a training set of 60000 images and a test set of 10000 images, which includes 6000 and 1000 images for each digit, respectively. The MNIST-*back-rotation*-image (MNISTrb) is an extension of the MNIST dataset contains 28×28 images of rotated digits over randomly inserted backgrounds. The default training and test set consist of 12000 and 50000 images, respectively. The Street View House Numbers (SVHN) dataset consists of 73,257 digits for training and 26,032 digits for testing in 32×32 RGB images. The CIFAR-10 dataset consists of 32×32 RGB images. A total of 60,000 images in 10 categories are split into 50,000 training and 10,000 testing images. The STL-10 dataset contains 96×96 RGB images in 10 categories. All results are reported on the default test sets.

For MNIST, MNISTrb, and SVHN, we employ a CNN architecture consisting of two convolution layers with 6 and 16 filters, respectively, and two fully-connected layers with 400 and 120 hidden units. ReLU is adopted between the convolutional layers and the fully connected layers. For CIFAR-10 and STL-10, we use convolutional layers with 20 and 50 filters and fully-connected layers with 500 and 120 hidden units. The summary of the datasets and the architectures are shown in Table 4.

4.2   Experimental Settings and Evaluation Metrics

Our first experiment follows that of [14] using the MNIST-*back-rot* images. First, the original dataset was augmented 10 times with mirrored and rotated images. Then, class imbalance was created by deleting samples selected with Gaussian distribution until a designated overall reduction rate is reached. We compare the average per-class accuracy (average class-wise recall) with those of Triplet re-sampling with cost-sensitive learning and Large Margin Local Embedding (LMLE) reported in [14]. Triplet loss re-sampling with cost-sensitive learning is a hybrid method that implements the triplet loss function used in [5,24,25] with re-sampling and cost-sensitive learning.

Table 4. Datasets and CNN architectures

Dataset	#$\mathcal{C}$	#TRN	#TST	Image dim	CNN layers	Batch size/ Trn.Rnds
MNIST [21]	10	50,000	10,000	$1 \times 28 \times 28$	C6-C16-F400-F120	60/3
MNISTrb [20]	10	12,000	50,000	$1 \times 28 \times 28$	C6-C16-F400-F120	40/5
SVHN [23]	10	73,257	26,032	$3 \times 32 \times 32$	C6-C16-F400-F120	60/3
CIFAR-10 [18]	10	50,000	10,000	$3 \times 32 \times 32$	C20-C50-F500-F120	50/4
STL-10 [6]	10	5,000	8,000	$3 \times 96 \times 96$	C20-C50-F500-F120	50/5

The second experiment analyzes the sensitivity of DOS over the levels of imbalance and the choices of k, using MNIST, MNIST-back-rotation, and SVHN datasets. The value of k is altered over 3, 5, and 10. In [3], 5 was given as the default value of k and other values have been tested in ensuing studies. The imbalance is created by randomly selecting four classes and removing p portion of their samples. We report the class-wise retrieval measures: precision, recall, F1-score, and the Area Under the Precision-Recall Curve (AUPRC), for the minority and the majority classes, respectively. The precision-recall curve is used in similar scope as the receiver operating characteristic curve (ROC). [10] has suggested the use of AUPRC over AUROC, which provides an overly optimistic estimate of the retrieval performance in some cases. Note that the precision, recall, and F1-score are computed from a multi-class confusion matrix, while the precision-recall curve is computed from the class-wise posterior probabilities.

The third experiment is conducted using the original SVHN, CIFAR-10, and STL-10 datasets. Since the classes are not imbalanced, the overloading value k is set uniformly for all classes, and the over-sampling rate r, from (11), is set to 1. The result of this experiment thus reflect the effect of deep feature overloading in a well-balanced setting. The evaluation metrics are the same as the second experiment, but averaged over all classes.

4.3   Results

Comparison with Existing Work. The result from the first experiment is summarized in Table 5. The overall reduction rate is shown on the first column. The performances of the baseline methods (TL-RS-CSL, LMLE) are shown in the second and the third columns. In the last three columns, the performances of DOS and two classifiers: logistic regression (LR) and k-nearest neighbors (kNN) using its deep representation are shown. While all methods show declining trends of accuracy, DOS showed the slowest decline against the reduction rate.

Sensitivity Analysis on Imbalanced Data. Tables 6 and 7 summarize the results of the second experiment. In Table 6, we compare the performances of the basic CNN, traditional classifiers using the deep representation of the basic CNN (CNN-CL), and DOS, over the reduction rates $p = 0.90, 0.95, 0.99$. On each row, four evaluation measures on MNIST, MNISTbr, and SVHN are shown. Note

Table 5. Baseline comparison (class-wise recall)

Reduction rate	TL-RS-CSL	LMLE	DOS	DOS-LR	DOS-kNN
0	76.12	77.64	77.35	77.62	76.00
20	67.18	75.58	77.13	74.60	73.53
40	56.49	70.13	75.43	73.53	72.98

that the AUPRC of CNN-CL is computed from the predicted class probabilities of the logistic regression classifier and other performances are those of the kNN classifier. The performances of the minority (mnr) and the majority (mjr) classes are indicated on the third column, respectively. We indicate the significant increases (0.1 or more) over CNN and CNN-CL by DOS with double underlines and smaller increases (0.05 or more) with single underlines. In Table 6, DOS exhibit more significant advantage with increasing level of imbalance, over the basic CNN and the classifiers using its deep representation.

Table 7 summarizes the sensitivity analysis on the overloading parameter values $k = 3, 5, 10$ with reduction rate set to $p = 0.01$. Each row shows the four evaluation measures on SVHN, CIFAR-10, and STL-10, respectively. For reference, The performances of the basic CNN and the deep feature classifiers are shown on the top rows. We indicate the significant increases over the baselines in the similar manner as Table 6. The minority and majority classes are indicated on the third column as well. Additionally, we indicate the unique largest values among DOS settings by bold letters. These results show that DOS is generally not sensitive to the choice of k. However, there is a marginal trend that the performances on the minority classes are slightly higher with $k = 3, 5$ and those of the majority classes are slightly higher with $k = 10$. It suggests possible decline in performance with overly large k.

Run-Time Analysis. The deep learning in the above experiment was conducted on NVIDIA GTX 980 graphics card with 704 cores and 6GB of global memory. The average increases in run-time for DOS compared to the basic, single-task learning architecture CNN were 11%, 12%, and 32% for MNIST, MNIST-bak-rot, and SVHN datasets, respectively.

Evaluation on Balanced Data. Table 8 summarizes the performances on the balanced settings. On the top rows, the performances of the basic CNN and the classifiers using its representations are shown for reference. On the bottom rows, the performances of DOS at $k = 3, 5, 10$ are shown. The uniquely best values among the three settings are indicated by bold fonts. While the improvements from the basic CNN were smaller (between 0.01 and 0.03) than in previous experiments, DOS showed consistent improvements across all datasets. This result supports our view that deep feature overloading can improve the discriminative power of the deep representation. We note that the performance of DOS were not sensitive to the values chosen for k.

Table 6. Performance comparison on imbalanced data over reduction rate

Model	p	Class	MNIST				MNISTbr				SVHN			
			Pr	Re	F1	AUC	Pr	Re	F1	AUC	Pr	Re	F1	AUC
CNN	0.90	mnr	0.98	0.93	0.96	0.99	0.31	0.76	0.43	0.68	0.62	0.77	0.55	0.78
		mjr	0.96	0.99	0.97	1.0	0.77	0.56	0.65	0.77	0.80	0.76	0.75	0.89
	0.95	mnr	0.99	0.89	0.94	0.99	0.27	0.76	0.23	0.57	0.13	0.86	0.21	0.61
		mjr	0.93	0.98	0.95	0.99	0.52	0.69	0.58	0.67	0.89	0.61	0.71	0.87
	0.99	mnr	0.65	0.98	0.77	0.96	0.31	0.71	0.43	0.68	0.50	0.72	0.42	0.74
		mjr	0.99	0.82	0.89	0.99	0.78	0.56	0.65	0.77	0.73	0.67	0.60	0.81
CNN-CL	0.90	mnr	0.99	0.90	0.94	1.0	0.22	0.77	0.31	0.69	0.59	0.60	0.42	0.78
		mjr	0.94	0.99	0.96	0.98	0.78	0.53	0.63	0.78	0.77	0.75	0.73	0.86
	0.95	mnr	0.99	0.83	0.90	0.97	0.28	0.77	0.24	0.60	0.039	0.68	0.069	0.61
		mjr	0.89	0.99	0.94	1.0	0.52	0.68	0.57	0.69	0.89	0.60	0.70	0.84
	0.99	mnr	0.75	0.98	0.85	0.95	0.22	0.72	0.31	0.69	0.47	0.71	0.37	0.72
		mjr	0.99	0.86	0.92	0.99	0.78	0.53	0.63	0.78	0.71	0.57	0.56	0.78
DOS ($k=5$)	0.90	mnr	0.99	0.97	0.98	1.0	0.66	0.77	0.71	0.79S	0.71	0.82	0.72	0.84
		mjr	0.98	0.99	0.98	1.0	0.75	0.68	0.71	0.81	0.85	0.79	0.81	0.92
	0.95	mnr	0.98	0.96	0.97	1.0	0.56	0.75	0.63	0.72S	0.40	0.89	0.55	0.73
		mjr	0.97	0.99	0.98	1.0	0.64	0.74	0.69	0.78	0.90	0.69	0.78	0.91
	0.99	mnr	0.91	0.99	0.95	0.99	0.61	0.73	0.66	0.75	0.51	0.91	0.64	0.80
		mjr	0.98	0.94	0.96	1.0	0.77	0.70	0.73	0.82	0.89	0.68	0.77	0.90

Table 7. Performance comparison on imbalanced data over k

Classifier	k		MNIST				MNIST-back-rot				SVHN			
			Pr	Re	F1	AUC	Pr	Re	F1	AUC	Pr	Re	F1	AUC
CNN		mnr	0.65	0.98	0.77	0.96	0.31	0.76	0.43	0.68	0.50	0.72	0.42	0.74
		mjr	0.99	0.82	0.89	0.99	0.78	0.56	0.65	0.77	0.73	0.67	0.60	0.81
CNN-CL		mnr	0.75	0.98	0.85	0.95	0.22	0.77	0.31	0.69	0.47	0.71	0.37	0.72
		mjr	0.99	0.86	0.92	0.99	0.78	0.53	0.63	0.78	0.71	0.57	0.56	0.78
DOS	3	mnr	0.91	0.98	0.95	0.99	0.65	0.77	0.70	0.78	0.67	0.77	0.66	0.83
		mjr	0.99	0.94	0.96	1.0	0.75	0.68	0.71	0.80	0.80	0.74	0.74	0.86
	5	mnr	0.91	0.99	0.95	0.99	0.66	0.77	0.71	0.79	0.51	0.91	0.64	0.80
		mjr	0.98	0.94	0.96	1.0	0.75	0.68	0.71	0.81	0.89	0.68	0.77	0.90
	10	mnr	0.91	0.99	0.95	0.99	0.61	0.73	0.66	0.75	0.40	0.89	0.55	0.73
		mjr	0.99	0.94	0.96	1.0	**0.77**	0.70	**0.73**	0.82	0.90	0.69	0.78	0.91

Table 8. Performance comparison on balanced data over k

Classifier	k	SVHN		CIFAR-10		STL-10	
		F1	AUC	F1	AUC	F1	AUC
CNN		0.85	0.92	0.62	0.68	0.38	0.38
CNN-CL		0.85	0.87	0.61	0.68	0.39	0.40
DOS	3	0.87	0.94	0.64	0.70	0.42	0.41
	5	0.88	**0.95**	0.64	**0.71**	0.42	0.42
	10	0.88	0.94	0.64	0.70	0.42	0.42

5   Conclusion

We proposed the Deep Over-sampling framework for imbalanced classification problem of complex, structured data that allows the CNN to learn the deep representation and the classifier jointly without substantial modification to its architecture. The framework extends the synthetic over-sampling technique by using the synthetic instances not only to complement the minority classes for classifier learning, but also to supervise representation learning and enhance its robustness and class distinction. The empirical results showed that the proposed framework can address the class imbalance more effectively than the existing countermeasures for deep learning, and the improvements were more significant under stronger levels of imbalance. Furthermore, its merit on representation learning were verified from the improved performances in the balanced setting.

References

1. Ando, S.: Classifying imbalanced data in distance-based feature space. Knowl. Inf. Syst. **46**(3), 707–730 (2016)
2. Bengio, Y., Courville, A., Vincent, P.: Representation learning: a review and new perspectives. IEEE Trans. Pattern Anal. Mach. Intell. **35**(8), 1798–1828 (2013)
3. Chawla, N.V., Bowyer, K.W., Hall, L.O., Kegelmeyer, W.P.: SMOTE: synthetic minority over-sampling technique. J. Artif. Int. Res. **16**(1), 321–357 (2002)
4. Chawla, N.V., Cieslak, D.A., Hall, L.O., Joshi, A.: Automatically countering imbalance and its empirical relationship to costs. Data Min. Knowl. Discov. **17**(2), 225–252 (2008)
5. Chechik, G., Shalit, U., Sharma, V., Bengio, S.: An online algorithm for large scale image similarity learning. In: Advances in Neural Information Processing Systems, vol. 22, pp. 306–314 (2009)
6. Coates, A., Lee, H., Ng, A.: An analysis of single-layer networks in unsupervised feature learning. In: Proceedings of the 14th International Conference on Artificial Intelligence and Statistics, vol. 15, pp. 215–223 (2011)
7. Collobert, R., Weston, J.: A unified architecture for natural language processing: deep neural networks with multitask learning. In: Proceedings of the 25th International Conference on Machine Learning, pp. 160–167 (2008)
8. Dong, C., Loy, C.C., He, K., Tang, X.: Learning a deep convolutional network for image super-resolution. In: Fleet, D., Pajdla, T., Schiele, B., Tuytelaars, T. (eds.) ECCV 2014. LNCS, vol. 8692, pp. 184–199. Springer, Cham (2014). https://doi.org/10.1007/978-3-319-10593-2_13
9. Dunne, R.A.: A Statistical Approach to Neural Networks for Pattern Recognition. Wiley Series in Computational Statistics. Wiley-Interscience, Hoboken (2007)
10. Flach, P.A., Hernández-Orallo, J., Ramirez, C.F.: A coherent interpretation of AUC as a measure of aggregated classification performance. In: Proceedings of the 28th International Conference on Machine Learning, pp. 657–664 (2011)
11. He, H., Garcia, E.A.: Learning from imbalanced data. IEEE Trans. Knowl. Data Eng. **21**(9), 1263–1284 (2009)
12. Hinton, G., Deng, L., Yu, D., Dahl, G.E., Mohamed, A., Jaitly, N., Senior, A., Vanhoucke, V., Nguyen, P., Sainath, T.N., Kingsbury, B.: Deep neural networks for acoustic modeling in speech recognition the shared views of four research groups. IEEE Sig. Process. Mag. **29**(6), 82–97 (2012)

13. Hinton, G.E., Salakhutdinov, R.R.: Reducing the dimensionality of data with neural networks. Science **313**(5786), 504–507 (2006)
14. Huang, C., Li, Y., Loy, C.C., Tang, X.: Learning deep representation for imbalanced classification. In: 2016 IEEE Conference on Computer Vision and Pattern Recognition, pp. 5375–5384 (2016)
15. Jeatrakul, P., Wong, K.W., Fung, C.C.: Classification of imbalanced data by combining the complementary neural network and SMOTE algorithm. In: Wong, K.W., Mendis, B.S.U., Bouzerdoum, A. (eds.) ICONIP 2010 Part II. LNCS, vol. 6444, pp. 152–159. Springer, Heidelberg (2010). https://doi.org/10.1007/978-3-642-17534-3_19
16. Köknar-Tezel, S., Latecki, L.J.: Improving SVM classification on imbalanced time series data sets with ghost points. Knowl. Inf. Syst. **28**(1), 1–23 (2011)
17. Krawczyk, B.: Learning from imbalanced data open challenges and future directions. Prog. Artif. Intell. **5**(4), 221–232 (2016)
18. Krizhevsky, A.: Learning multiple layers of features from tiny images. Master's thesis (2009)
19. Krizhevsky, A., Sutskever, I., Hinton, G.E.: Imagenet classification with deep convolutional neural networks. In: Proceedings of the 25th International Conference on Neural Information Processing Systems, pp. 1097–1105 (2012)
20. Larochelle, H., Erhan, D., Courville, A., Bergstra, J., Bengio, Y.: An empirical evaluation of deep architectures on problems with many factors of variation. In: Proceedings of the 24th International Conference on Machine Learning, pp. 473–480 (2007)
21. Lecun, Y., Bottou, L., Bengio, Y., Haffner, P.: Gradient-based learning applied to document recognition. Proc. IEEE **86**(11), 2278–2324 (1998)
22. LeCun, Y., Kavukcuoglu, K., Farabet, C.: Convolutional networks and applications in vision. In: Proceedings of 2010 IEEE International Symposium on Circuits and Systems, pp. 253–256 (2010)
23. Netzer, Y., Wang, T., Coates, A., Bissacco, A., Wu, B., Ng, A.Y.: Reading digits in natural images with unsupervised feature learning. In: NIPS Workshop on Deep Learning and Unsupervised Feature Learning (2011)
24. Schroff, F., Kalenichenko, D., Philbin, J.: Facenet: a unified embedding for face recognition and clustering. In: 2015 IEEE Conference on Computer Vision and Pattern Recognition, pp. 815–823 (2015)
25. Wang, J., Song, Y., Leung, T., Rosenberg, C., Wang, J., Philbin, J., Chen, B., Wu, Y.: Learning fine-grained image similarity with deep ranking. In: Proceedings of the 2014 IEEE Conference on Computer Vision and Pattern Recognition, pp. 1386–1393 (2014)
26. Weinberger, K.Q., Saul, L.K.: Distance metric learning for large margin nearest neighbor classification. J. Mach. Learn. Res. **10**, 207–244 (2009)
27. Zeiler, M.D., Fergus, R.: Visualizing and understanding convolutional networks. In: Fleet, D., Pajdla, T., Schiele, B., Tuytelaars, T. (eds.) ECCV 2014. LNCS, vol. 8689, pp. 818–833. Springer, Cham (2014). https://doi.org/10.1007/978-3-319-10590-1_53
28. Zhou, Z.H., Liu, X.Y.: Training cost-sensitive neural networks with methods addressing the class imbalance problem. IEEE Trans. Knowl. Data Eng. **18**(1), 63–77 (2006)

FCNN: Fourier Convolutional Neural Networks

Harry Pratt$^{(\boxtimes)}$, Bryan Williams, Frans Coenen, and Yalin Zheng

University of Liverpool, Liverpool L69 3BX, UK
{sghpratt,bryan,coenen,yzheng}@liverpool.ac.uk

Abstract. The Fourier domain is used in computer vision and machine learning as image analysis tasks in the Fourier domain are analogous to spatial domain methods but are achieved using different operations. Convolutional Neural Networks (CNNs) use machine learning to achieve state-of-the-art results with respect to many computer vision tasks. One of the main limiting aspects of CNNs is the computational cost of updating a large number of convolution parameters. Further, in the spatial domain, larger images take exponentially longer than smaller image to train on CNNs due to the operations involved in convolution methods. Consequently, CNNs are often not a viable solution for large image computer vision tasks. In this paper a Fourier Convolution Neural Network (FCNN) is proposed whereby training is conducted entirely within the Fourier domain. The advantage offered is that there is a significant speed up in training time without loss of effectiveness. Using the proposed approach larger images can therefore be processed within viable computation time. The FCNN is fully described and evaluated. The evaluation was conducted using the benchmark Cifar10 and MNIST datasets, and a bespoke fundus retina image dataset. The results demonstrate that convolution in the Fourier domain gives a significant speed up without adversely affecting accuracy. For simplicity the proposed FCNN concept is presented in the context of a basic CNN architecture, however, the FCNN concept has the potential to improve the speed of any neural network system involving convolution.

1 Introduction

Convolutional Neural Networks (CNNs) [1] are a popular, state-of-the-art, deep learning approach to computer vision with a wide range of application in domains where data can be represented in terms of three dimensional matrices. For example, in the case of image and video analysis. Historically, CNNs were first applied to image data in the context of handwriting recognition [2]. Since then the viability of CNNs, and deep learning in general, has been facilitated, alongside theoretical improvements, by significant recent advancements in the availability of processing power. For example, Graphics Processing Units (GPUs) allow us to deal with the heavy computation required by convolution.

Electronic supplementary material The online version of this chapter (https://doi.org/10.1007/978-3-319-71249-9_47) contains supplementary material, which is available to authorized users.

© Springer International Publishing AG 2017
M. Ceci et al. (Eds.): ECML PKDD 2017, Part I, LNAI 10534, pp. 786–798, 2017.
https://doi.org/10.1007/978-3-319-71249-9_47

However, there are increasingly larger datasets to which we wish to apply deep learning to [3] and, in the case of deep learning, a growing desire to increase the depth of the networks used in order to achieve better results [4,5]. This not only increases memory utilisation requirements, but also computational complexity. In the case of CNNs, the most computationally expensive element is the calculation of the spatial convolutions. The convolution is typically conducted using a traditional sliding window approach across the data matrix together with the application of a kernel function of some kind [6]. However, this convolution is computationally expensive, which in turn means that CNNs are often not viable for large image computer vision tasks. To address this issue, this paper proposes the idea of a using the Fourier domain. More specifically this paper proposes the Fourier Convolution Neural Network (FCNN) whereby training is conducted entirely in the Fourier domain. The advantage offered is that there is a significant speed up in training time without loss of effectiveness. Using FCNN images are processed and represented using the Fourier domain to which a convolution mechanism is applied in a manner similar to that used in the context of more traditional CNN techniques. The proposed approach offers the advantage that it reduces the complexity, especially in the context of larger images, and consequently provides for significant increase in network efficiency.

The underlying intuition given by the Convolution Theorem which states that for two functions κ and u, we have

$$\mathcal{F}(\kappa * u) = \mathcal{F}(\kappa) \odot \mathcal{F}(u) \tag{1}$$

where $\mathcal{F}$ denotes the Fourier transform, $*$ denotes convolution and $\odot$ denotes the Hadamard Pointwise Product. This allows for convolution to be calculated more efficiently using Fast Fourier Transforms (FFTs). Since convolution corresponds to the Hadamard product in the Fourier domain and given the efficiency of the Fourier transform, this method involves significantly fewer computational operations than when using the sliding kernel spatial method, and is therefore much faster [7]. Working in the Fourier domain is less intuitive as we cannot visualise the filters learned by our Fourier convolution; this is a common problem with CNN techniques and is beyond the scope of this paper. While the Fourier domain is frequently used in the context of image processing and analysis [8–10], there has been little work directed at adopting the Fourier domain with respect to CNNs. Although FFTs, such as the Cooley-Tukey algorithm [11], have been applied in the context of neural networks for image [12] and time series [13] analysis. These applications date from the embryonic stage of CNNs and, at that time, the improvement was minimal.

The concept of using the Fourier domain for CNN operations has been previously proposed [7,14,15]. In both [7,14] the speed-up of convolution in the Fourier domain was demonstrated. Down-sampling within the Fourier domain was used in [15] where the ability to retain more spatial information and obtain faster convergence was demonstrated. However, the process proposed in [7,14,15] involved interchanges between the Fourier and spatial domains at both the training and testing stages which added significant complexity. The FFT required is the computationally intensive part of the process. FFTs, and inverse FFTs, needed to be applied for each convolution; thus giving rise to an undesired computational overhead. In the case of the proposed FCNN the data is converted to the Fourier domain before the process starts, and remains in the Fourier domain; no inverse FFTs are required at any point.

Instead of defining spatial kernel functions, which must then be transformed to the Fourier domain, as in the case of [7], using the proposed FCNN, a bespoke Fourier convolution mechanism is also proposed whereby convolution kernels are initialised in the Fourier domain. This method saves computation time during both the training and utilisation. Pooling in the Fourier domain is implemented in a similar fashion to that presented in [15] with truncation in the Fourier domain. This is not only more efficient than max-pooling, but can achieve better results [15]. The other layers implemented within the FCNN are dense layers and dropout. These Fourier layers are analogous to the equivalent spatial layers. Dropout randomly drops nodes within our network at a probability of p to stop over-fitting. This applies in the Fourier domain as it does in the spatial domain. Likewise, dense layers for learning abstract links within convolved image data operates with respect to Fourier data in the same manner as for spatial data.

The layout of the rest of the paper is as follows. In Sect. 2, we present our method of implementation of the specific layers that constitute our FCNNs, in Sect. 3 we present our experimental results. In Sects. 4 and 5 we present a discussion together with conclusions concerning abilities of the FCNN.

2   The Fourier Convolution Neural Network (FCNN) Approach

The FCNN was implemented using the deep learning frameworks *Keras* [16] and *Theano* [17]. *Theano* is the machine learning backend of *Keras*. This backend was used to code the Fourier layers. The Theano FFT function *Theano* was used to convert our training and test data. The *Theano* FFT function is a tensor representation of the multi-dimensional Cooley-Tukey algorithm. This function is the n-dimensional discrete Fourier transform over any number of axes in an m-dimensional array by using FFT. The multi-dimensional discrete Fourier transform used is defined as:

$$A_{kl} = \sum_{\ell_1=0}^{m-1} \sum_{\ell_2=0}^{n-1} a_{\ell_1 \ell_2} e^{-2\pi i \left(\frac{\ell_1}{m} + \frac{\ell_2}{n} \right)} \tag{2}$$

where the image is of size $m \times n$. The comparative methods of spatial convolution and max-pooling used throughout this paper relate to *Keras* and *Theano*'s implementations. To demonstrate the ability of the FCNNs implementation of all the core CNN layers in the Fourier domain we use the network architectures shown in supplementary.

The well used network architecture from AlexNet [1] was adopted because it provides a simple baseline network structure to compare the results of our equivalent Fourier and spatial CNNs on the MNIST [18] and Cifar10 datasets [19].

The MNIST dataset contains 60,000 grey scale images, 50,000 for training and 10,000 for testing, of hand written numeric digits in the form of 28×28 pixel images, giving a 10 class classification problem. The Cifar10 [19] dataset contains 60,000, 32×32 pixel, colour images containing 10 classes. These datasets are regularly used for standard CNN baseline comparison [4, 20]. Experiments were also conducted using a large fundus image Kaggle data set [3]. This dataset comprised 80,000 RGB fundus images, of around 3M pixels per image, taken from the US diabetic screening process. The images are labelled using five classes describing level of diabetic retinopathy. These images are currently down-sampled during training using established CNN techniques because of the size of the images; this seems undesirable.

2.1   Fourier Convolution Layer

In traditional CNNs discrete convolutions between the images $\mathbf{u}^j$ and kernel functions κ^i are carried out using the sliding window approach. That is, a window the size of the kernel matrix is moved across the image. The convolution is computed as the sum of the Hadamard product $\odot$ of the image patch with the kernel:

$$\mathbf{z}_{k_1,k_2}^{i,j} = \sum_{\ell_1=\lfloor -m_\kappa/2 \rfloor}^{\lfloor m_\kappa/2 \rfloor} \sum_{\ell_2=\lfloor -n_\kappa/2 \rfloor}^{\lfloor n_\kappa/2 \rfloor} \kappa_{\ell_1,\ell_2}^i \odot \mathbf{u}_{k_1-\ell_1,k_2-\ell_2}^j \tag{3}$$

which results in an $(m_\mathbf{u} - m_\kappa) \times (n_\mathbf{u} - n_\kappa)$ image $\mathbf{z}$ since the image is usually re-sized to avoid including boundary artefacts in calculations. At each point (k_1, k_2), there are $m_k n_k$ operations required and so $(m_\mathbf{u} - m_{\kappa+1})(n_\mathbf{u} - n_{\kappa+1})m_k n_k$ operations are needed for a single convolution.

We intend to replace, in the first instance, the sliding window approach with the Fourier transform using the discrete analogue of the convolution theorem:

$$\mathcal{F}(\kappa * \mathbf{u}) = \mathcal{F}(\kappa) \odot \mathcal{F}(\mathbf{u}) \tag{4}$$

where $\mathcal{F}$ denotes the two dimensional discrete Fourier transform:

$$\tilde{\mathbf{u}}_{i_1,i_2} = \sum_{j_1=1}^{m_\mathbf{u}} \sum_{j_2=1}^{n_\mathbf{u}} e^{-2\imath\pi \left(\frac{i_1 j_1 n_\mathbf{u} + i_2 j_2 m_\mathbf{u}}{m_\mathbf{u} n_\mathbf{u}} \right)} \mathbf{u}_{j_1,j_2} \tag{5}$$

The computation of the discrete Fourier transform for an $n \times n$ image $\mathbf{u}$ involves n^2 multiplications and $n(n-1)$ additions, but this can be reduced considerably using an FFT algorithm, such as Cooley-Tukey [11] which can compute the Direct Fourier Transform (DFT) with $n/2 \log_2 n$ multiplications and $n \log_2 n$ additions. This gives an overall improvement from the $O(n^2)$ operations required to calculate the DFT directly to $O(n \log n)$ for the FFT.

Thus, for a convolutional layer which has N^κ kernels κ^i in a network training $N^\mathbf{u}$ images $\mathbf{u}^j$, the output is the set $\mathbf{z}^{i,j} = \kappa^i * \mathbf{u}^j$ where $*$ denotes convolution. The algorithm is then:

1. $\tilde{\kappa}^i = \mathcal{F}\left(\kappa^i\right), i = 1, \ldots, N^\kappa$
2. $\tilde{\mathbf{u}}^i = \mathcal{F}\left(\mathbf{u}^i\right), i = 1, \ldots, N^{\mathbf{u}}$
3. $\tilde{\mathbf{z}}^{i,j} = \tilde{\kappa}^i \odot \tilde{\mathbf{u}}^j, i = 1, \ldots, N^\kappa, j = 1, \ldots, m^{\mathbf{u}}$
4. $\mathbf{z}^{i,j} = \mathcal{F}^{-1}\left(\tilde{\mathbf{z}}^{i,j}\right), i = 1, \ldots, N^\kappa, j = 1, \ldots, N^{\mathbf{u}}$

This decrease in the number of operations gives an increasing relative speed-up for larger images. This is of particular relevance given that larger computer vision (image) datasets are increasingly becoming available [3].

With respect to the proposed FCNN the N^k complex Fourier kernels are initialised using glorot initialisation [21]. The parameter n is equivalent to the number of kernel filters in the spatial network. Glorot initialisation was adopted because it is more efficient than doing FFT transformations of spatial kernels as this would require lots of FFTs during training to update the numerous convolution kernels. The weights for our Fourier convolution layer are defined as our initialised Fourier kernels. Hence, the Fourier kernels are trainable parameters optimised during learning, using back propagation, to find the best Fourier filters for the classification task with no FFT transformations relating to the convolution kernels required. Another benefit of Fourier convolutions is not only the speed of the convolutions, but that we can perform pooling during the convolution phase in order to save more computation cost.

A novel element of our convolution kernels is that, because they remain in the Fourier domain throughout, they have the ability to learn the equivalent of arbitrarily large spatial kernels limited only by initial image size. The image size is significantly larger than the size selected by spatial kernels. That is, our Fourier kernels which match the image size can learn a good representation of a 3×3 spatial kernel or a 5×5 spatial kernel depending on what aids learning the most. This is a general enhancement of kernel learning in neural networks as most networks typically learn kernels of a fixed size, reducing the ability of the network to learn the spatial kernel of the optimal size. In the Fourier domain, we can train to find not only the optimal spatial kernel of a given size but the optimal spatial kernel size and the optimal spatial kernel itself.

2.2 Fourier Pooling Layer

In the Fourier domain, the image data is distributed in a differ manner to the spatial. This allows us to reduce the data size by the same amount that it would be reduced by in the spatial domain but retain more information. High frequency data is found towards the centre of a Fourier matrix and low frequency towards the boundaries. Therefore, we truncate the boundaries of the matrices as the high frequency Fourier data contains more of the spatial information that we wish to retain. Our Fourier pooling layer shown in Fig. 1, operates as follows. Given a complex 3 dimensional tensors of $X \times Y \times Z$ dimensions, and AN arbitrary pool_size variable relating to the amount of data we wish to retain. For $x \in X$,:

$$x_{y\,\min} = (0.5 - \frac{\text{pool\_size}}{2}) \times Y, \quad x_{y\,\max} = (0.5 + \frac{\text{pool\_size}}{2}) \times Y \qquad (6)$$

Fourier Pooling

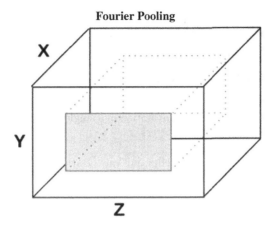

Fig. 1. Our layer initially contains an $X \times Y \times Z$ voxel. The truncation runs through the x-axis of the Fourier data (thus truncating the Y and Z axis).

$$x_{z\,\min} = (0.5 - \frac{\text{pool\_size}}{2}) \times Z, \quad x_{z\,\max} = (0.5 + \frac{\text{pool\_size}}{2}) \times Z \qquad (7)$$

This method provides a straightforward Fourier pooling layer for our FCNN. It has a minimal number of computation operations for the GPU to carry out during training.

The equivalent method in the spatial context is max-pooling, which takes the maximum value in a $k \times k$ window where k is a chosen parameter. For example if $k = 2$, max-pooling reduces the data size by a quarter by taking the maximum value in the 2×2 matrices across the whole data. Similarly, in our Fourier pooling we would take pool_size $= 0.25$ which, using Eqs. 6 and 7, gives us:

$$x_{y\ \min} = 0.375 \times Y, x_{y\ \max} = 0.625 \times Y \qquad (8)$$

$$x_{z\ \min} = 0.375 \times Z, x_{z\ \max} = 0.625 \times Z \qquad (9)$$

which also reduces our data by a quarter.

3   Evaluation

The evaluation was conducted using an Nvidia K40c GPU that contains 2880 CUDA cores and comes with the Nvidia CUDA Deep Neural Network library (cuDNN) for GPU learning. For the evaluation both the computation time and the accuracy of the layers in the spatial and Fourier domains was compared. The FCNN and its spatial counterpart were trained using the 3 datasets introduced above: MNIST, Cifar10 and Kaggle fundus images. Each dataset was used to evaluate different aspects of the proposed FCNN. The MNIST dataset allows us to compare high-level accuracy while demonstrating the speed up of doing convolutions in the Fourier domain. The Cifar10 dataset was used to show that the FCNN can learn a more complicated classification task to the same degree as a spatial CNN with the same number of filters. The results are presented below in terms of speed, accuracy and propagation loss. Finally, the large fundus Kaggle dataset was used to show that the FCNN is better suited to dealing with larger images, than spatial CNNs, because of the nature of the Fourier convolutions.

Table 1. Computation time for the convolution of a single images of varying size, using both Fourier and spatial convolution layers.

Size	FourierConv	SpatialConv	Ratio increase
2^{10}	5×10^{-2}	N/A	N/A
2^9	1×10^{-2}	N/A	N/A
2^8	2.67×10^{-3}	1.48×10^{-1}	55.43
2^7	7.74×10^{-4}	8.4×10^{-2}	10.85
2^6	2.85×10^{-4}	1.74×10^{-3}	6.10
2^5	1.78×10^{-4}	2.51×10^{-4}	1.41
2^4	1.36×10^{-4}	1.56×10^{-4}	1.14

3.1 Fourier Convolution

The small kernels used in neural networks mean that when training on larger images the amount of memory required to store all the convolution kernels on the GPU for parallel training is no longer viable. Using the Nvidia K40c GPU and a spatial convolution with 3×3 kernels the feed forward process of our network architecture cannot run a batch of images once image size approaches 2^9. The proposed Fourier convolution mechanism requires less computational memory when running in parallel. The memory capacity is not reached using the Fourier convolution mechanism until images of a size four times greater to the maximum size using the spatial domain are arrived at. This is due to the operational memory required for spatial convolution compared to the Fourier convolution.

The FCNN is able to train much larger images of the same batch size because the kernels are initialised in the Fourier domain, we initialise a complex matrix with the size matching the image size. Our convolutions are matrix multiplications and we are not required to pass across the image in a sliding window fashion, where extra storage is needed. The only storage we require is for the Fourier kernels, which are the same size as the images.

Table 1 presents a comparison of computation times, using Fourier and spatial convolution, for a sequence of single images of increasing size. From the table it can been seen that the computation time for a small images ($2^4 \times 2^4$ pixels) is similar for spatial and Fourier data in both cases. However, as the image size increases, the spatial convolution starts to become exponentially more time-consuming whereas the Fourier convolution scales at a much slower rate and allows convolution with respect to a much larger image size.

3.2   Fourier Pooling

Table 2 gives a comparison of the computation time, required to process a sequences of images of increasing size using, using the proposed Fourier pooling method in comparison with Max-pooling and Down-sampling. Fourier pooling is similar in terms of computational time to the max-pooling method which is the most basic down-sampling technique. This speed increase is for the same reason as the increase in convolution speed. Max-pooling requires access to smaller matrices within the data and takes the maximum value. On the other hand, in the Fourier domain, we can simply truncate in manner such that spatial information throughout the whole image is retained.

Table 2. Computation time for pooling an image of the given size using: (i) Down-sampling, (ii) Max pooling and (iii) Fourier pooling.

Size	Down-sampling	Max-pooling	Fourier pooling
2^{12}	2.77e-2	9.01	9.42e-2
2^{11}	7.93e-3	2.07	2.44e-2
2^{10}	2.19e-3	4.96e-1	5.30e-3
2^{9}	2.33e-4	1.26e-1	5.27e-4
2^{8}	2.70e-5	3.14e-2	1.01e-5
2^{7}	1.73e-5	6.80e-3	3.20e-6
2^{6}	3.67e-6	1.65e-3	5.29e-6
2^{5}	2.71e-6	3.82e-4	6.03e-6
2^{4}	2.46e-6	8.55e-5	5.35e-6

Figure 2 shows a comparison of pooling using down sampling, max pooling and Fourier pooling. In the figure the images in each image subsequent to the top row were reduced to half the size of the previous row and then up-scaled to the original image size for down-sampling and max-pooling. For Fourier pooling, the Fourier signal was embedded into a zero matrix of the same size as the original image and the Fourier transform is presented. Figure 3 shows how the Fourier pooling retains more spatial information as the best result in terms of visual acuity retained during pooling using mean squared error is the Fourier pooled image. All output images are the same size, but the Fourier retains more information. From the figures it can be seen that the Fourier pooling retains more spatial information than on the case of max-pooling when down-sampling the data by the same factor. This is because of the nature of the Fourier domain, the spatial information of the data is not contained in one specific point.

Pooling Methods

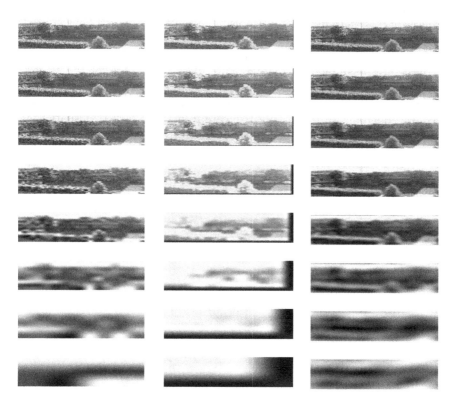

Fig. 2. Comparison of pooling using: (i) down-sampling (col. 1), (ii) max-pooling (col. 2) and (iii) Fourier pooling (col. 3).

3.3 Network Training

The baseline network is trained on both the MNIST and Cifar10 datasets to compare networks. Training was done using the categorical cross-entropy loss function and optimised using the rmsprop algorithm. The results are presented in Figs. 4 and 5 using network one. The fundus training was carried out on network two and epoch speeds were recorded see Table 3. The accuracy achieved on the MNIST and Cifar10 test sets using the FCNN is only marginally below the spatial CNN but the results are achieved with a significant speed up. The MNIST training was twice as fast on the FCNN in comparison the spatial CNN and the Cifar10 dataset was trained in 6 times the speed. This is due to the Cifar dataset containing slightly larger images than MNIST and demonstrates how our FCNN scales better to large images.

Fourier Pooling of fundus image

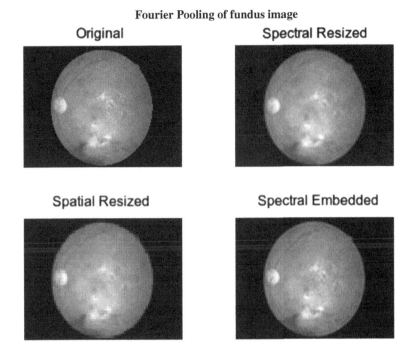

Original · Spectral Resized

Spatial Resized · Spectral Embedded

Fig. 3. (Top-left) Original fundus image, (Bottom-left) normal max-pooling and then resizing to original size; (Top-right) Fourier pooling, back to spatial domain and resize to original size; (Bottom-right) Fourier pooling, embed in a zero matrix and convert back to spatial

Training on the MNIST dataset

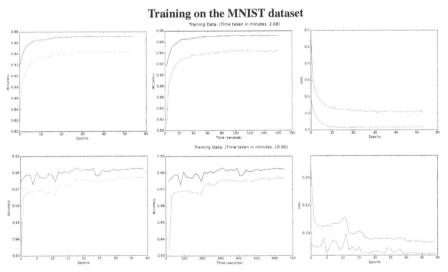

Fig. 4. (Top) FCNN (Bottom) Spatial CNN. Dark blue, black and red are validation values, lighter colours are training values. (Color figure online)

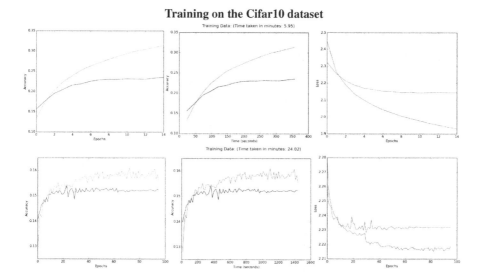

Fig. 5. Training on the Cifar10 dataset: (top) FCNN (bottom) Spatial CNN. Dark blue, black and red are validation values, lighter colours are training values. (Color figure online)

Table 3. Computation time in seconds for an epoch of re-sized fundus images. One epoch is 60,000 training images.

Image size	FCNN epoch	Spatial epoch
2^9	65.56	2435.93
2^8	30.42	1839.12
2^7	14.47	358.90
2^6	8.38	124.63
2^5	3.92	36.91
2^4	0.76	3.72

4   Discussion

The proposed FCNN technique allows training to be conducted entirely in the Fourier domain, in other words only one FFT is required throughout the whole process. The increase in computation time required for the FFT is recovered because of the resulting speed up of the convolution. Compared to spatial approach the evaluation results obtained evidence an exponential increase in efficiency for larger images. Given a more complex network, or a dataset of larger images, the benefit would be even more pronounced.

The results presented demonstrated that using the Fourier representation training time, using the same layer structure, was considerably less than when a spatial representation was used. The analogous Fourier domain convolutions and more spatially accurate pooling method allowed for a retention in accuracy on both datasets introduced. It was conjectured that the higher accuracy achieved using the proposed FCNN on the Cifar10 dataset was due to the larger Fourier domain kernels within the Fourier convolution layer. Due to the Fourier kernel size, more parameters within the network were obtained than in the case of spatial window kernels. This allowed for more degrees of freedom when learning features of the images.

The reason for lower accuracy of the FCNN using the MNIST dataset is likely due to the network being trained on very small images. This creates boundary issues and information loss in the Fourier domain when converting from the spatial. This is particularly relevant with respect to smaller images; it is much less of an issue in larger images. Hence, when dealing with larger images we would expect no reduction in accuracy in the Fourier domain while achieving the speed-ups shown. To combat this, we could consider boundary conditions with respect to all of our Fourier layers, which is what is done in the spatial case.

5   Conclusion

This paper has proposed the idea of a Fourier Convolution Neural Network (FCNNs) which offers run-time advantages, especially during training. The reported performance results were comparable with standard CNNs but with the added advantage of a significant speed increase. As a consequence the FCNN approach can be used to classify image sets featuring large images; not possible using the spatial CNNs. The FCNN layers are not specific to any architecture and therefore can be extended to any network using convolution, pooling and dense layers. This is the case for the vast majority of neural network architectures. For future work the authors intend to investigate how the Fourier layers can be optimised and implemented with respect to other network architectures that have achieved state-of-the-art accuracies [4,5]. The authors speculate that, given the efficiency advantage offered by FCNNs, they would be used to address classification tasks directed at larger images, and in a much shorter time frames, than would be possible using standard CNNs.

Acknowledgement. The authors would like to acknowledge everyone in the Centre for Research in Image Analysis (CRiA) imaging team at the Institute of Ageing and Chronic Disease at the University of Liverpool and the Fight for Sight charity who have supported this work through funding.

References

1. Krizhevsky, A., Sutskever, I., Hinton, G.E.: ImageNet classification with deep convolutional neural networks. In: Pereira, F., Burges, C.J.C., Bottou, L., Weinberger, K.Q. (eds.) Advances in Neural Information Processing Systems, vol. 25, pp. 1097–1105. Curran Associates Inc. (2012)

2. LeCun, Y., Boser, B., Denker, J.S., Howard, R.E., Habbard, W., Jackel, L.D., Henderson, D.: Advances in neural information processing systems, vol. 2, pp. 396–404. Citeseer (1990)
3. Kaggle: Kaggle datasets. https://www.kaggle.com/datasets
4. He, K., Zhang, X., Ren, S., Sun, J.: Deep residual learning for image recognition. CoRR, abs/1512.03385 (2015)
5. Szegedy, C., Liu, W., Jia, Y., Sermanet, P., Reed, S., Anguelov, D., Erhan, D., Vanhoucke, V., Rabinovich, A.: Going deeper with convolutions. In: Computer Vision and Pattern Recognition (CVPR) (2015)
6. Sermanet, P., Eigen, D., Zhang, X., Mathieu, M., Fergus, R., LeCun, Y.: Overfeat: integrated recognition, localization and detection using convolutional networks. CoRR, abs/1312.6229 (2013)
7. Vasilache, N., Johnson, J., Mathieu, M., Chintala, S., Piantino, S., LeCun, Y.: Fast convolutional nets when fbfft: a GPU performance evaluation (2015)
8. Chan, T.F., Wong, C.K.: Total variation blind deconvolution. IEEE Trans. Image Process. **7**(3), 370–375 (1998)
9. Persch, N., Elhayek, A., Welk, M., Bruhn, A., Grewenig, S., Böse, K., Kraegeloh, A., Weickert, J.: Enhancing 3-D cell structures in confocal and STED microscopy: a joint model for interpolation, deblurring and anisotropic smoothing. Measur. Sci. Technol. **24**(12), 125703 (2013)
10. Williams, B.M., Chen, K., Harding, S.P.: A new constrained total variational deblurring model and its fast algorithm. Numer. Algorithms **69**(2), 415–441 (2015)
11. Cooley, J.W., Tukey, J.W.: An algorithm for the machine calculation of complex fourier series. Math. comput. **19**(90), 297–301 (1965)
12. Campisi, P., Egiazarian, K.: Blind Image Deconvolution. CRC Press, Boca Raton (2007)
13. Kumar, R., Gothwal, H., Kedawat, S.: Cardiac arrhythmias detection in an ECG beat signal using fast fourier transform and artificial neural network. J. Biomed. Sci. Eng. **4**, 289–296 (2011)
14. LeCun, Y., Mathieu, M., Henaff, M.: Fast training of convolutional networks through FFTs (2014)
15. Adams, R.P., Rippel, O., Snoek, J.: Spectral representations for convolutional neural networks (2015)
16. Chollet, F.: Keras. https://github.com/fchollet/keras (2015)
17. Theano Development Team. Theano: a python framework for fast computation of mathematical expressions. arXiv e-prints abs/1605.02688, May 2016
18. LeCun, Y., Cortes, C.: MNIST handwritten digit database (2010)
19. Krizhevsky, A.: Learning multiple layers of features from tiny images. https://www.cs.toronto.edu/~kriz/learning-features-2009-TR.pdf
20. Goodfellow, I., Pouget-Abadie, J., Mirza, M., Xu, B., Warde-Farley, D., Ozair, S., Courville, A., Bengio, Y.: Generative adversarial nets. In: Ghahramani, Z., Welling, M., Cortes, C., Lawrence, N.D., Weinberger, K.Q. (eds.) Advances in Neural Information Processing Systems, vol. 27, pp. 2672–2680. Curran Associates Inc. (2014)
21. Glorot, X., Bengio, Y.: Understanding the difficulty of training deep feedforward neural networks. In: Proceedings of the International Conference on Artificial Intelligence and Statistics (AISTATS 2010). Society for Artificial Intelligence and Statistics (2010)

Joint User Modeling Across Aligned Heterogeneous Sites Using Neural Networks

Xuezhi Cao$^{(\boxtimes)}$ and Yong Yu

Apex Data and Knowledge Management Lab,
Shanghai Jiao Tong University, Shanghai, China
{cxz,yyu}@apex.sjtu.edu.cn

Abstract. The quality of user modeling is crucial for personalized recommender systems. Traditional within-site recommender systems aim at modeling user preferences using only actions within target site, thus suffer from cold-start problem. To alleviate such problem, researchers propose cross-domain models to leverage user actions from other domains within same site. Joint user modeling is later proposed to further integrate user actions from aligned sites for data enrichment. However, there are still limitations in existing works regarding the modeling of heterogeneous actions, the requirement of full alignment and the design of preferences coupling. To tackle these, we propose JUN: a joint user modeling framework using neural network. We take advantage of neural network's capability of capturing different data types and its ability for mining high-level non-linear correlations. Specifically, in additional to site-specific preferences models, we further introduce an auxiliary neural network to transfer knowledge between sites by fine-tuning the user embeddings using alignment information. We adopt JUN for item-based and text-based site to demonstrate its performance. Experimental results indicate that JUN outperforms both within-site and cross-site models. Specifically, JUN achieves relative improvement of 2.96% and 2.37% for item-based and text-based sites (5.77% and 13.54% for cold-start scenarios). Besides performance gain, JUN also achieves great generality and significantly extends the use scenarios.

Keywords: Joint user modeling · Personalized recommender system
Cross-domain recommendation · Neural networks

1 Introduction

Personalized recommender system has now become an indispensable tool for the information overload in online environment. A great portion of online services include a native recommender system. Such systems benefit both users and service providers by improving user's online experiences and stimulating potential actions. Driven by its importance, plenty research works emerge in this area. Both research works and online applications prove the success of such systems.

© Springer International Publishing AG 2017
M. Ceci et al. (Eds.): ECML PKDD 2017, Part I, LNAI 10534, pp. 799–815, 2017.
https://doi.org/10.1007/978-3-319-71249-9_48

An accurate and comprehensive user modeling is crucial for the quality of personalized recommender systems. Traditional recommender systems capture user preferences by modeling one's previous actions within the target site. For example, recommending movies based on one's movie rating logs. Although the performances are satisfying, problems and limitations still exist.

A well-known and widely exist problem is cold-start problem: recommender systems may fail when dealing with newly arrived users due to lack of historical data [27]. This problem severely jeopardizes user's first impressions. Furthermore, the problem widely exists in almost all recommender systems, making it one of the most urgent problems in this research direction.

Another limitation is the lack of comprehensiveness. When participating in specific online services, user normally reveals only a portion of his preferences. As traditional user modelings focus only on user actions in target site, they can only capture the heavily revealed parts of his preferences. Although these are the most important parts for future recommendations within same site, we cannot claim the other aspects to be totally irrelevant. For example, it is hard to mine one's political preferences using only movie rating histories, but such preferences can help when recommending political movies or documentaries.

To address these problems, researchers propose cross-domain models to leverage user actions in other domains as supplementary data. The intuition behind is that user's underlying preferences is consistent across domains. However, most existing researches focus on domains within same site and only transfer between homogeneous actions (e.g. between different types of movies [28]). There also exist works target at synthetic multi-domain data generated by subdividing single-domain dataset [1,26]. Although there are works aim at heterogeneous domains, they require aligned actions or additional semantic knowledges [6,21].

More recently, researchers propose joint user modeling which aims at integrating multiple aligned heterogeneous sites for data enrichment [4,31]. Given the reality that most people engage in multiple online sites for various needs (Facebook, Twitter, IMDb, etc.), we can integrate user actions in all these sites for a more comprehensive user modeling. Analogous to cross-domain modeling that integrates actions in different domains at site-level, joint user modeling aims at integrating different sites at Internet-level. The advantage is its comprehensiveness and users only face cold-start scenario once when first exposed to the Internet, while the disadvantage is the challenge of heterogeneousness. Due to the task's recency, existing works only focus on fully aligned sites, while a more realistic scenario would be partial alignment or pairwise score-based alignment generated using network aligners [20,29].

Another common disadvantage of cross-domain recommendation and joint user modeling is that most works employ hard constraints for preferences transfer. Specifically, they follow the framework that first captures user preferences on target and auxiliary domains respectively, and then integrates them using non-personalized hard constraints (e.g. uses user preferences in auxiliary domains as features, or forces the preferences representations to be close). However, these do not match with reality. Although on average users show consistency across domains (sites), the coupling strength should be identical for each user. For example, there are users who always keep their post the same for Facebook and Twitter, while there also exist users who maintain totally different characters.

In this paper, we employ neural network for joint user modeling to tackle the aforementioned limitations. We favor neural network due to its capability of capturing different types of data, including discrete, sequential, image, etc., which suits the heterogeneous actions in this task. Specifically, we design individual neural networks for each site, capturing the site-specific user preferences. Besides, we design auxiliary neural networks for preferences transfer between sites using the alignment information. All these neural networks share same user embedding layer, thus the auxiliary ones serve as fine-tunning networks using the alignment information. In this paper, we focus on text-based and item-based actions because they are fundamentally heterogeneous and cover most online applications. Therefore, it serves as a good representative for the heterogeneous scenarios. The main contributions of this paper are as follows:

- We propose a neural network based framework JUN for joint user modeling over heterogeneous sites, providing a general solution for this task.
- Within JUN, we propose a novel approach for modeling user preferences consistency across sites using fine-tunning networks, which better transfers knowledge between sites comparing to hard couplings as in existing methods.
- We further employ JUN to model item-based and text-based actions as a representative setting.
- We discuss the integration of JUN with existing online recommender systems to show its utility values besides research contributions.
- Extensive experiments indicate that we achieve a relative improvement of 2.96% and 2.37% for item-based and text-based sites. For cold-start scenarios, we further achieve relative improvements of 5.77% and 13.54%.

The rest of the paper is organized as follows. We first discuss related works in Sect. 2. Then, we present the dataset and conduct preliminary analysis in Sect. 3. We propose JUN in Sect. 4, and report experimental results in Sect. 5. Finally, we summarize and discuss future works in Sect. 6.

2   Related Work

2.1   User Modeling in Recommender System

Personalized recommender system is now an indispensable tool for finding wanted contents from the overwhelming online data, thus draws plenty research attentions. As user actions vary, researchers propose plenty user modeling techniques accordingly. Matrix factorization [15,23] is the most widely used method for item-based sites such as e-commercial and movie/music rating sites [13]. Topic models [2] and word embeddings [22] are employed for modeling text-based sites such as Twitter and Tumblr [9]. There are also works targeting at social relationships [18], location-based informations [17], etc.

Recently, several neural network-based recommender systems are proposed. Google proposes Wide & Deep learning which jointly train wide linear models and deep neural networks to combine the benefits of memorization and generalization [5]. He et. al. integrate neural network with matrix factorization, leading to neural collaborative filtering [11]. Neural network's advantage on modeling textual data is also employed for capturing metadata and side-informations [30].

2.2   Cold-Start Problem in Recommender System

The cold-start problem is that recommender system may fail when dealing with newly arrived users with no or only few historical actions. Such problem widely exists in almost all recommender systems, hence is of great importance.

Researchers propose various solutions to alleviate this problem [16, 27]. One direction is to compensate by additional information such as social relations [8], social tags [32], etc. They require specific types of side information, which are not always available in general. Another direction is to introduce an interview process immediately after registration. This is widely adopted in real applications. Decision trees [10] and functional matrix factorization [34] are employed for generating such questions. Nevertheless, this approach requires additional manual works from users thus still have negative impact on user experiences.

2.3   Cross-Domain Recommendation

Following the intuition that user's preferences is consistent across domains, researchers propose cross-domain recommendation to leverage user actions from other domains as auxiliary data [7].

Existing cross-domain approaches mainly focus on homogeneous data. For example, transferring between different types of movies [28]. There is also a good fraction of papers target at synthetic multi-domain data generated by subdividing a single-domain dataset [1, 26]. For those scenarios, different 'domains' actually share a lot in common (action type, behavior pattern, etc.).

There do exist works aiming at heterogeneous data. McAuley et al. aim at understanding product ratings with review text [21]. Bayesian hierarchical approach based on LDA is also proposed [28]. Fernandez et. al. further employ semantic information to link items across domains [6]. However, these approaches require aligned actions, or additional semantic informations to help the knowledge transfer, which are not always available in general settings.

2.4   Joint User Modeling

Nowadays, as Internet infiltrates into various aspects of our daily life, people always participate in multiple online sites. Following similar intuition of cross-domain recommendation, we can further align multiple sites and integrate the user actions for joint modeling. It extents the site-level integration in cross-domain modeling to an Internet-level integration. Existing works indicate that it is a promising direction for data enrichment [4, 31].

Joint user modeling requires the sites to be aligned. Although a large portion of online applications now enable users to login with cross-site accounts (Facebook, Twitter, Google+, etc.), there still exist plenty unaligned users and isolated networks. There are works aiming at automatically alignment by mining personal identifiable information [19], social tags [12], user behaviors [20], etc. An accuracy of over 80% is achieved. Due to the recency of joint user modeling, existing approaches focus only on perfectly aligned sites (with bijection between the user sets). However, when applied to realistic scenarios, we are facing partial alignment or pairwise-score-based alignment recovered by network aligners.

3   Data and Preliminary Analysis

3.1   Data Set

The dataset we use is from previous joint user modeling work [4], collected from Douban[1] and Weibo[2]. We exclude users with no actions in one site as well as users marked as inactive, leaving 27,814 users as the final dataset. For item-based site Douban, on average each user has 171.96 movie rating logs. For text-based site Weibo, we have 623.38 microblog actions per user on average, including both original and re-tweets.

3.2   Preliminary Analysis

User Consistency. The underlying assumption of joint user modeling is that user preferences are to some extent consistent across sites. As user preferences are not explicitly revealed, we can not measure its consistency directly. Because user actions serve as explicit indicators of user preferences, we give evidence that user action similarity is consistent instead.

Specifically, we analyze whether users with similar actions in one site are still tend to be similar in the other. For each pair of users, we evaluate their similarity according to actions in each site. For text-based site, we first employ Latent Dirichlet Allocation (LDA [2]) to model their topic distributions and then apply KL divergence for similarity. For item-based site, we use Jaccard similarity coefficient upon the sets of rated movies. We show the correlation in Fig. 1(a), where the users are grouped by similarities in source site. Results indicate that user pairs with similar actions in source site also tend to be similar in target site for both directions, which support the assumption.

Action Count Distribution. Now we show that leveraging cross-site actions can actually help to enrich the data for cold-start users. Specifically, we analyze whether user's action count in different sites are highly correlated. If the answer is yes, then cold users would still be cold in parallel sites hence joint user modeling may not lead to a valid data enrichment.

---

[1] http://www.douban.com/.

[2] http://www.weibo.com/.

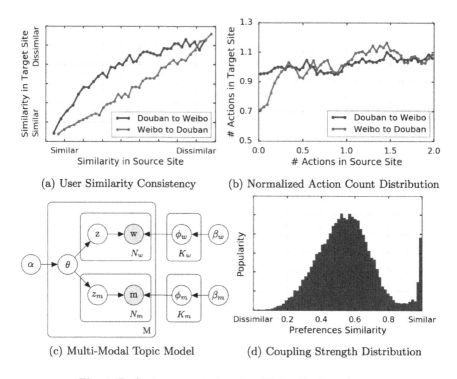

(a) User Similarity Consistency

(b) Normalized Action Count Distribution

(c) Multi-Modal Topic Model

(d) Coupling Strength Distribution

Fig. 1. Preliminary analysis using Weibo-Douban dataset

We group users by normalized action count in source site and report their average normalized action count in target site. Here the normalization is conducted by dividing action count by average action count in corresponding site. We plot the results in Fig. 1(b). Surprisingly, results indicate that user's action counts in different sites are rather independent. For numeric analysis, the Pearson correlation coefficient is only 0.0549. Therefore, the data enrichment is valid.

Coupling Strength Distribution. As stated previously, non-personalized hard coupling methods used in existing works do not match with reality. Now we give evidence by estimating user's coupling strength distribution between sites.

We first train a multi-modal topic model to represent user preferences in both site using common topic space. We consider movies as another set of 'words' and plug them into traditional topic model (showed in Fig. 1(c)). Then, for each user we generate two topic distributions using only microblog actions and movie rating logs respectively. Finally, we estimate coupling strength by cosine similarity between these two topic distributions. We show the distribution in Fig. 1(d). Although the overall trend is towards high similarity, the popularity spreads out among different levels of coupling strength. Therefore, non-personalized hard constraints used in traditional work do not well-represent the reality.

4   Our Approach

Our framework JUN is consist of two parts: the models for site-specific preferences and the model for cross-site preferences consistency. Specifically, we have one individual model for each site to capture the user preferences revealed in it. The design of these models depend on site-specific action types and settings, details are given in Sect. 4.1. The cross-site preferences consistency model is designed to transfer knowledge between site-specific models on user representation level using the alignment information, discussed in Sect. 4.2.

User representation is the foundation of user modeling. Most user modeling techniques fall into the framework that model user actions based on user representation U_i and site-specific information A (item representation, word embedding, etc.). Formally, these models can be formulated as $\hat{y}_{ic} = f(U_i, c, A|\Theta)$ where $\hat{y}_{ic}$ is the estimated action for user i towards content c. Take matrix factorization as an example, latent vectors are used to represent both users and items. The prediction can be modeled as $\hat{y}_{ui} = U_i \cdot A_c$ for the basic setting. User representation provides a good abstraction for user actions. It reveals the required user preferences information while hiding the heterogeneous action details. Therefore, we conduct the integration at user representation level.

In our approach, we employ neural networks for both site-specific preferences model as well as preferences consistency model. Substantial research works successfully employed neural network for varies tasks, including image processing, word embedding, recommender systems, etc. Evidences indicate that neural network has great capability in handling heterogeneous data. Thus, it provides great opportunity for modeling heterogeneous actions. Besides, its ability to capture high-level non-linear correlations is also preferred for preferences consistency modeling. Therefore, we adapt neural network as the underlying technique.

4.1   Modeling the Site-Specific Preferences

There exist plenty of heterogeneous action types in different online services. Item-based and text-based actions are typically considered as the most popular ones, covering a wide range of online applications including e-commerce, rating sites, news, blogging, etc. In this work, we focus on modeling movie rating site (item-based) and microblogging site (text-based) as a representative setting.

To integrate site S into JUN, we need to define the followings: the user representation U_i^S, the site-specific information A^S, and the function f^S parametrized by Θ^S for modeling user actions: $\hat{y}_{ic}^S = f^S(U_i^S, c, A^S|\Theta^S)$. For generality purpose, we focus on modeling implicit feedback actions, i.e. using only the action itself with no additional information such as rating scores, like/dislike tags, etc. Hence, $\hat{y}_{ic}^S$ estimates the probability of user i interacts with content c.

Item-Based Site ($\mathcal{I}$): In item-based sites, actions can be represented as a matrix $Y \in [0, 1]^{n \times m}$, where n, m are the number of users and items respectively, Y_{ij} indicates whether there is an action from user i towards item j.

The most widely used technique for item-based sites is matrix factorization, which models the interaction between user and item by inner product of the

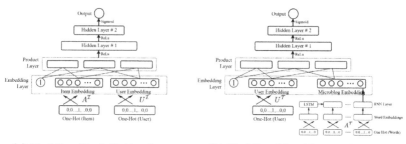

(a) Modeling Movie Rating Logs   (b) Modeling Microblogging Histories

Fig. 2. Neural networks for modeling site-specific preferences

user's latent vector and the item's. For probability estimation instead of rating scores, sigmoid function $\sigma(x) = 1/(1 + e^{-x})$ is often used as the activation function. Formally, we have $f^{\mathcal{I}}_{mf}(U^{\mathcal{I}}_i, c, A^{\mathcal{I}}) = \sigma(U^{\mathcal{I}}_i \cdot A^{\mathcal{I}}_c)$, where $U^{\mathcal{I}} \in \mathbb{R}^{n \times k}$, $A^{\mathcal{I}} \in \mathbb{R}^{m \times k}$ and k is the latent vector dimension.

The underlying methodology of matrix factorization that represents users and items using latent vectors actually matches with embedding layer in neural networks. The difference is that matrix factorization uses simple inner product for action estimation while neural network employs rather complex multi-layer designs. Despite neural network's deep structure, traditional designs can not directly model the inner product as in matrix factorization. To combine their advantages, we employ the Product-based Neural Network (PNN, [25]). Compared to traditional neural networks, PNN further introduces a product layer to capture strong interactions such as inner product between embeddings.

The detailed design for item-based site is depicted in Fig. 2(a). We first use one-hot encoding for both user and item as input, then multiply them by user embedding and item embedding matrices $U^{\mathcal{I}} \in \mathbb{R}^{n \times k}$ and $A^{\mathcal{I}} \in \mathbb{R}^{m \times k}$ respectively for the embedding. A constant 1 is added here as an additional field. For product layer, we calculate outer product between each pair of embeddings. After that, we append two fully connected hidden layers, with rectified linear unit (relu, $relu(x) = \max(0, x)$) as activation function. For final output layer, we use sigmoid function to model the probability of this action. Formally, we have:

$$
\begin{aligned}
e^u_p, e^i_q &= \text{embeddings for user p and item q} \\
h_0 &= \text{concat}(e^u_p, e^i_q, e^u_p \odot e^u_p) \\
h_k &= \text{relu}(W_k h_{k-1} + b_k) \\
\text{output} &= \sigma(h_n W_o + b_o)
\end{aligned}
\tag{1}
$$

where $x \odot y$ indicates the flattened outter product of x and y, W_k and b_k are the parameters for the fully connected hidden layers.

Text-Based Site ($\mathcal{T}$): In microblogging sites, the user actions can be represented by $(i, \{w_k\})$ tuples indicating that user i posted or retweeted a microblog with content $\{w_k\}$, where w_k is the kth word. Traditionally, topic models such as Latent Dirichlet Allocation [2] are widely used for textual inputs. As neural network develops, word embedding [22] has became the new tool.

We illustrate our model for text-based site in Fig. 2(b). For users, we use one-hot input followed by an embedding layer. For microblogs, we use recurrent neural network with long short-term memory cells to capture the microblog embedding after word embedding layer [24]. Then, we append similar PNN layers as in item-based site for product layer and hidden layers. Finally, sigmoid activation is used for probability estimation.

4.2   Modeling the Cross-Site Preferences Consistencies

Most traditional approaches transfer knowledge between sites based on heuristic assumptions and put hard constraints on preferences in different domains. As previously stated, hard transfers do not match with reality.

Instead, we propose JUN based on the assumption that user preferences contain information to reveal the alignment (judging whether two accounts are held by same natural person). Normally, this assumption indicates that we can reveal the alignment by mining user preferences. Because in this task the alignment is given and the goal is to improve user modelings, we reverse the learning direction and further fine-tune the user embeddings using the given alignment.

Specifically, we design a neural network to classify whether two accounts in different sites are held by same natural person, with corresponding user embeddings and side informations as input evidences (showed in Fig. 3). Similar as previous, we use PNN network structure. For side informations, we include username similarity and social ties. Instead of using user embeddings as constant input features, we let them be the same embedding layers as in site-specific action models (Fig. 2), using shared parameters. Then, with the given alignment as supervise training, we can 'push' these information back into user embeddings using back propagation, forming a fine-tunning process. Similar methodology has also been widely used for the fine-tuning of word embeddings.

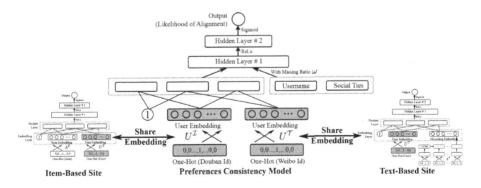

Fig. 3. Fine tuning the user embeddings using alignment information

4.3 Learning Techniques

As there are multiple neural networks in JUN, how to conduct the training to achieve expected outcome is also an interesting question.

We train the neural networks by the followings. We first conduct pre-training over site-specific models to initialize user embeddings to represent user preferences. Then, we update the network parameters for preferences consistency model by using user embeddings as constant features. Finally, we update all these networks simultaneously (round-robin, one batch for each network) to conduct the fine-tuning. With this training process, we force the user embeddings to represent user preferences instead of capturing only alignment information.

In order to successfully conduct the fine-tunning using the alignment information, we need to make sure the classifier for alignment does not fully rely on side informations. This could happen because side informations provide strong signals, while user preferences is rather noisy. To prevent this, we simulate information missing on side informations. Specifically, we randomly hide side informations from the model during training with probability ω.

4.4 Discussion

Comparing to traditional cross-domain or joint user modeling techniques, the major advantage of JUN is its generality in the following aspects:

- **Heterogeneous Actions.** Although in this work we only apply JUN for item-based and text-based sites, JUN's capability is not limited to these two. It can be easily adopted for any type of action by plugging in corresponding user-embedding based site-specific model as in Sect. 4.1.
- **Employ Other Site-Specific Models.** Similarly, JUN can easily adopt other user embedding-based models to capture the site-specific preferences besides the models proposed in Sect. 4.1.
- **Requirement on Alignment.** Existing cross-site works only work with perfect alignment (bijection) between sites. However, in reality this never happens. With JUN, we only have minimum requirements on alignment. Besides full alignment, JUN can also leverage partial alignment, multi-to-multi alignment, and pairwise score-based alignment as auxiliary data.
- **Extend to Multiple Sites.** JUN can be easily extended for more than two sites by having multiple preferences consistency models, one for each alignment among them. We only evaluate for two sites due to dataset limitation.

Besides generality, JUN is also of high utility value. Most existing recommender systems employ user embedding, or latent vector representation, which both can be easily integrated with JUN. Therefore, JUN can benefit real world applications instead of only for research purpose.

The additional computational cost of JUN is also acceptable. For existing recommender systems, the cost is proportional to the number of user actions. When integrated with JUN, these costs remain (for site-specific preferences models).

The extra computational cost required is for the preferences consistency model. The cost is proportional to the size of alignment, which is same scale for number of users. Comparing to number of user actions, such cost is actually ignorable. Therefore, JUN does not raise a computational cost concern when integrating.

5   Experiments

5.1   Experimental Settings

Dataset. We carry out the experiments using data from Weibo and Douban. The details are previously discussed in Sect. 3.1.

Methodology. As there is no ground truth for user preferences, we can only evaluate user modeling through recommender systems. Specifically, we implement recommender system using JUN or existing techniques as user modeling, and then compare the performances of these recommender systems. For Douban and Weibo, the recommender systems rank potential movies and microblogs respectively, based on the estimated likelihoods of action. Area Under the Curve (AUC) is used as the metric to evaluate the ranking.

Comparing Algorithms. For comprehensive evaluation, we compare existing approaches for both within-site and cross-site(domain) models. For traditional within-site models, we compare the followings (where $\mathcal{I}$ and $\mathcal{T}$ indicate item-based and text-based site respectively):

- ($\mathcal{I}$) **PMF:** Probabilistic Matrix Factorization [23].
- ($\mathcal{I}$) **SVD++:** Matrix factorization with implicit feedback [14].
- ($\mathcal{I}$) **PNN-I:** The isolated site-specific model for item site (Fig. 2(a)).
- ($\mathcal{T}$) **LDA:** Latent Dirichlet Allocation, a widely used topic model [2].
- ($\mathcal{T}$) **PNN-T:** The isolated site-specific model for text site (Fig. 2(b)).

For cross-domain and joint user modeling techniques, we compare:

- ($\mathcal{I}, \mathcal{T}$) **mmTM:** Multi-modal topic model (Fig. 1(c)).
- ($\mathcal{I}$) **TMF:** Topic-Based Matrix Factorization, explicitly adding user's topic distribution in Weibo as features into feature-based matrix factorization.
- ($\mathcal{I}, \mathcal{T}$) **JUMA:** A graphical model-based joint user modeling approach [4].
- ($\mathcal{I}, \mathcal{T}$) **JUN:** Our approach proposed in this paper.

Although there are plenty cross-domain recommendation techniques, most of them can not be applied under this experiment setting. As we discussed in Sect. 2, most works focus on homogeneous actions, thus can not transfer between microblogging site and movie rating site. There are works modeling between text-based and item-based actions [21], however they require aligned actions for the model learning. For ComSoc in [33], it focuses on borrowing social relations instead of user actions thus is also not suitable for comparison.

Implementation and Parameters. We implement JUN using TensorFlow[3], and use AdamOptimizer for the learning. Unless indicated otherwise, the embedding dimension is 32 for all users, items and words. For hidden layers #1 and #2, we have 100 and 50 hidden units respectively. ω is set to 0.5. 80% user actions are used for training. For more details, please refer to our project website[4].

5.2   Performance Comparison

We evaluate aforementioned methods given full alignment as auxiliary information, with training ratio from 40% to 80%. We show the results in Table 1.

For item-based site Douban, JUN achieves better performance comparing to both within-site approaches (PMF, SVD++ and PNN-I) and cross-site approaches (TMF, mmTM, JUMA). Specifically, our approach JUN achieves an AUC of 0.9108 and 0.8984 when training ratio is 80% and 40% respectively, while the best within-site recommender system (PNN-T) achieves only 0.8869 and 0.8695. The relative improvement is 2.69% and 3.32% respectively. Comparing to best cross-site approach, JUN also achieves an additional 2.15% improvement.

For text-based site Weibo, JUN also outperforms all comparing approaches. It achieves an AUC of 0.7443 and 0.7312 under training ratio of 80% and 40%, and the corresponding relative improvement comparing to best existing approach is 2.37% and 3.94% respectively.

Note that for both sites, the performance drop of JUN when reducing training ratio from 80% to 40% is smaller comparing to existing models, indicating that JUN is rather robust to training ratio thanks to the successful data enrichment.

These results indicate that JUN out-performs existing user modeling techniques including within-site and cross-site ones. Also, the improvement occurs on both item-based site and text-based site, indicating that JUN can successfully transfer knowledge between site in both directions. By these experiments, we show that both PNN-I (Fig. 2(a)) and PNN-T (Fig. 2(b)) can be successfully integrated with JUN for a further improvement. For other unevaluated embedding-based techniques, we believe that similar improvement can also be achieved.

5.3   Cold-Start Scenario

We simulate cold-start scenarios by limiting the number of user actions used for training. We depict the relative AUC improvements over users with different number of training actions in Fig. 4(a). Results in both sites indicate that performance improvement is much higher when dealing with cold users comparing to non-cold users. We achieve a relative improvement of 13.54% for users with no historical actions in Weibo and 5.77% in Douban, indicating that JUN succeeded in leveraging cross-site actions for cold-start problem. Note that the improvement is more significant in text-based site comparing to item-based site. This

---

[3] https://www.tensorflow.org/.

[4] Implementation and dataset are available at: http://www.apexlab.org/projects/32.

Table 1. Experimental results, varying training ratio

Target	Type	ALGS	AUC score, varying training ratio		
			0.4	0.6	0.8
Item-based (Douban)	Within site	PMF	0.7969 ± 0.0017	0.8087 ± 0.0016	0.8196 ± 0.0015
		SVD++	0.8598 ± 0.0013	0.8738 ± 0.0016	0.8849 ± 0.0014
		PNN-I	0.8695 ± 0.0016	0.8787 ± 0.0013	0.8869 ± 0.0012
	Cross site	mmTM	0.7494 ± 0.0019	0.7510 ± 0.0014	0.7608 ± 0.0012
		TMF	0.8514 ± 0.0016	0.8654 ± 0.0013	0.8763 ± 0.0015
		JUMA	0.8790 ± 0.0018	0.8889 ± 0.0014	0.8916 ± 0.0015
		JUN	$\mathbf{0.8984 \pm 0.0015}$	$\mathbf{0.9052 \pm 0.0013}$	$\mathbf{0.9108 \pm 0.0013}$
Text-Based (Weibo)	Within site	LDA	0.6514 ± 0.0017	0.6694 ± 0.0016	0.6839 ± 0.0015
		PNN-T	0.7035 ± 0.0018	0.7129 ± 0.0015	0.7271 ± 0.0013
	Cross site	mmTM	0.6652 ± 0.0018	0.6776 ± 0.0015	0.6828 ± 0.0016
		JUMA	0.6824 ± 0.0014	0.6976 ± 0.0014	0.7120 ± 0.0012
		JUN	$\mathbf{0.7312 \pm 0.0016}$	$\mathbf{0.7384 \pm 0.0013}$	$\mathbf{0.7443 \pm 0.0014}$

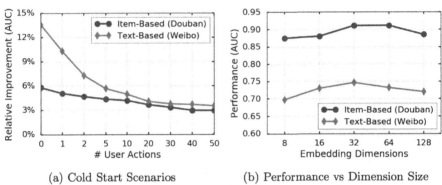

(a) Cold Start Scenarios   (b) Performance vs Dimension Size

Fig. 4. Detailed analysis of JUN.

is because non-personalized recommendation still works well in item-based scenario by directly ranking the items by popularity, but not in text-based scenario (Fig. 4).

5.4 Parameter Tunning

We vary the embedding dimensions of JUN for detailed analysis. We evaluate using dimension size from 8 to 128 and report the results in Fig. 4(b). According to the results, embedding of 32 dimensions gives the best overall performance. Large dimension size leads to over fitting and higher computational cost. Also note that when reducing embedding dimensions, the performance drop for item-based site is smaller than for text-based site. This may also due to the quality of non-personalized recommender system in item-based site is rather good.

Experiments also indicate that performance is not very sensitive to other parameters besides embedding dimension.

5.5   Partial Alignment and Score-Based Alignment

A major limitation of existing joint user modeling or cross-domain user modeling technique is that they can only be applied to fully aligned sites. However, such situation only exist in ideal research environment but not the reality. To demonstrate that JUN can also be applied to partial alignment and score-based alignment generated using network aligners, we conduct experiments for these scenarios. For score-based alignment, we use BASS as the network aligner [3].

We show the results in Table 2. For this set of experiments, we report not only the overall AUC for all users, but also the average AUC for aligned users and unaligned users respectively to show the effect of JUN for different user groups according to the alignment. The results match with our expectation that full alignment results in the most significant improvement according to overall AUC, and the improvement of partial alignment lies between full alignment and no alignment depending on the alignment ratio. For aligned users, the improvements are mostly consistent for all alignment methods. Note that for unaligned users, JUN also achieves a slightly improvement. This may due to the knowledge transfer between aligned users leads to better embedding spaces for both user and items (or words, etc.), thus also improve the quality for unaligned users.

Table 2. Apply JUN for partial alignment and score-based alignment

Alignment	Aligned users		Unaligned users		All users	
	$\mathcal{I}$	$\mathcal{T}$	$\mathcal{I}$	$\mathcal{T}$	$\mathcal{I}$	$\mathcal{T}$
No alignment	n/a	n/a	0.8869	0.7271	0.8869	0.7271
40% alignment	0.9086	0.7425	0.8931	0.7342	0.8993	0.7375
80% alignment	0.9113	0.7438	0.8964	0.7393	0.9083	0.7429
Full alignment	0.9108	0.7443	n/a	n/a	0.9108	0.7443
Score-based alignment	n/a	n/a	n/a	n/a	0.9086	0.7419

6   Conclusion and Future Works

In this paper, we aim at improving the quality of user modeling by conducting joint user modeling across aligned heterogeneous sites using neural networks. To overcome the limitations of existing cross-domain and joint user modeling works regarding the modeling of heterogeneous actions, the requirement of full alignment and the design of coupling strength, we propose a novel neural network-based framework JUN to tackle this task. JUN takes advantage of neural network's capability for capturing heterogeneous data and its ability for mining high-level non-linear correlations. Comparing to existing works, JUN can be further applied for scenarios where only partial alignment or pairwise score-based alignment are available. Also, JUN models preferences consistency using fine-tunning neural network instead of hard constraints, leading to better knowledge transfer across sites. We conduct extensive experiments using real data from

Douban and Weibo to evaluate JUN's performance. Results indicate that JUN outperforms existing works in both sites, and successfully alleviates the cold-start problem. We achieve relative improvement of 2.96% and 2.37% for item-based and text-based sites respectively. For cold-start scenarios, we achieve relative improvement of 5.77% and 13.54% respectively.

For future works, we may consider integrating social relations from the aligned sites into JUN. Also, we are interested in integrating JUN with network aligners to conduct the aligning and user modeling simultaneous as these tasks are high related with each other.

References

1. Berkovsky, S., Kuflik, T., Ricci, F.: Cross-domain mediation in collaborative filtering. In: Conati, C., McCoy, K., Paliouras, G. (eds.) UM 2007. LNCS (LNAI), vol. 4511, pp. 355–359. Springer, Heidelberg (2007). https://doi.org/10.1007/978-3-540-73078-1_44
2. Blei, D.M., Ng, A.Y., Jordan, M.I.: Latent dirichlet allocation. JML **3**, 993–1022 (2003)
3. Cao, X., Yu, Y.: BASS: a bootstrapping approach for aligning heterogenous social networks. In: Frasconi, P., Landwehr, N., Manco, G., Vreeken, J. (eds.) ECML PKDD 2016. LNCS (LNAI), vol. 9851, pp. 459–475. Springer, Cham (2016). https://doi.org/10.1007/978-3-319-46128-1_29
4. Cao, X., Yu, Y.: Joint user modeling across aligned heterogeneous sites. In: Proceedings of the 10th ACM Conference on Recommender Systems, pp. 83–90. ACM (2016)
5. Cheng, H.T., Koc, L., Harmsen, J., Shaked, T., Chandra, T., Aradhye, H., Anderson, G., Corrado, G., Chai, W., Ispir, M., et al.: Wide & deep learning for recommender systems. In: Proceedings of the 1st Workshop on Deep Learning for Recommender Systems, pp. 7–10. ACM (2016)
6. Fernández-Tobías, I., Cantador, I., Kaminskas, M., Ricci, F.: A generic semantic-based framework for cross-domain recommendation. In: Proceedings of the 2nd International Workshop on Information Heterogeneity and Fusion in Recommender Systems, pp. 25–32. ACM (2011)
7. Fernández-Tobías, I., Cantador, I., Kaminskas, M., Ricci, F.: Cross-domain recommender systems: a survey of the state of the art. In: SCIR (2012)
8. Gao, H., Tang, J., Liu, H.: Addressing the cold-start problem in location recommendation using geo-social correlations. Data Min. Knowl. Discov. **29**, 1–25 (2014)
9. Godin, F., Slavkovikj, V., De Neve, W., Schrauwen, B., Van de Walle, R.: Using topic models for Twitter hashtag recommendation. In: WWW (2013)
10. Golbandi, N., Koren, Y., Lempel, R.: Adaptive bootstrapping of recommender systems using decision trees. In: Proceedings of the Fourth ACM International Conference on Web Search and Data Mining, pp. 595–604. ACM (2011)
11. He, X., Liao, L., Zhang, H., Nie, L., Hu, X., Chua, T.S.: Neural collaborative filtering. In: Proceedings of the 26th International World Wide Web Conference (2017)
12. Iofciu, T., Fankhauser, P., Abel, F., Bischoff, K.: Identifying users across social tagging systems. In: ICWSM (2011)

13. Koenigstein, N., Dror, G., Koren, Y.: Yahoo! music recommendations: modeling music ratings with temporal dynamics and item taxonomy. In: Proceedings of the Fifth ACM Conference on Recommender Systems, pp. 165–172. ACM (2011)
14. Koren, Y.: Factorization meets the neighborhood: a multifaceted collaborative filtering model. In: Proceedings of the 14th ACM SIGKDD International Conference on Knowledge Discovery and Data Mining, pp. 426–434. ACM (2008)
15. Koren, Y., Bell, R., Volinsky, C.: Matrix factorization techniques for recommender systems. Computer **42**, 30–37 (2009)
16. Lam, X.N., Vu, T., Le, T.D., Duong, A.D.: Addressing cold-start problem in recommendation systems. In: Proceedings of the 2nd International Conference on Ubiquitous Information Management and Communication, pp. 208–211. ACM (2008)
17. Levandoski, J.J., Sarwat, M., Eldawy, A., Mokbel, M.F.: Lars: a location-aware recommender system. In: 2012 IEEE 28th International Conference on Data Engineering (ICDE), pp. 450–461. IEEE (2012)
18. Liu, H., Maes, P.: Interestmap: harvesting social network profiles for recommendations. Beyond Personalization-IUI, p. 56 (2005)
19. Liu, J., Zhang, F., Song, X., Song, Y.I., Lin, C.Y., Hon, H.W.: What's in a name? An unsupervised approach to link users across communities. In: Proceedings of the 6th ACM International Conference on Web Search and Data Mining, pp. 495–504. ACM (2013)
20. Liu, S., Wang, S., Zhu, F., Zhang, J., Krishnan, R.: Hydra: large-scale social identity linkage via heterogeneous behavior modeling. In: SIGMOD (2014)
21. McAuley, J., Leskovec, J.: Hidden factors and hidden topics: understanding rating dimensions with review text. In: RecSys, pp. 165–172. ACM (2013)
22. Mikolov, T., Chen, K., Corrado, G., Dean, J.: Efficient estimation of word representations in vector space. arXiv preprint arXiv:1301.3781 (2013)
23. Mnih, A., Salakhutdinov, R.: Probabilistic matrix factorization. In: NIPS, pp. 1257–1264 (2007)
24. Palangi, H., Deng, L., Shen, Y., Gao, J., He, X., Chen, J., Song, X., Ward, R.: Deep sentence embedding using long short-term memory networks: analysis and application to information retrieval. IEEE/ACM Trans. Audio Speech Lang. Process. (TASLP) **24**(4), 694–707 (2016)
25. Qu, Y., Cai, H., Ren, K., Zhang, W., Yu, Y., Wen, Y., Wang, J.: Product-based neural networks for user response prediction. arXiv preprint arXiv:1611.00144 (2016)
26. Sahebi, S., Brusilovsky, P.: Cross-domain collaborative recommendation in a cold-start context: the impact of user profile size on the quality of recommendation. In: Carberry, S., Weibelzahl, S., Micarelli, A., Semeraro, G. (eds.) UMAP 2013. LNCS, vol. 7899, pp. 289–295. Springer, Heidelberg (2013). https://doi.org/10.1007/978-3-642-38844-6_25
27. Schein, A.I., Popescul, A., Ungar, L.H., Pennock, D.M.: Methods and metrics for cold-start recommendations. In: SIGIR, pp. 253–260 (2002)
28. Tan, S., Bu, J., Qin, X., Chen, C., Cai, D.: Cross domain recommendation based on multi-type media fusion. Neurocomputing **127**, 124–134 (2014)
29. Tan, S., Guan, Z., Cai, D., Qin, X., Bu, J., Chen, C.: Mapping users across networks by manifold alignment on hypergraph. In: AAAI (2014)
30. Vasile, F., Smirnova, E., Conneau, A.: Meta-Prod2Vec: product embeddings using side-information for recommendation. In: Proceedings of the 10th ACM Conference on Recommender Systems, pp. 225–232. ACM (2016)
31. Zhang, J., Kong, X., Yu, P.S.: Predicting social links for new users across aligned heterogeneous social networks. In: 2013 IEEE 13th International Conference on Data Mining, pp. 1289–1294. IEEE (2013)

32. Zhang, Z.K., Liu, C., Zhang, Y.C., Zhou, T.: Solving the cold-start problem in recommender systems with social tags. EPL **92**(2), Article no. 28002 (2010)
33. Zhong, E., Fan, W., Wang, J., Xiao, L., Li, Y.: ComSoc: adaptive transfer of user behaviors over composite social network. In: Proceedings of the 18th ACM SIGKDD, pp. 696–704. ACM (2012)
34. Zhou, K., Yang, S.H., Zha, H.: Functional matrix factorizations for cold-start recommendation. In: Proceedings of the 34th International ACM SIGIR Conference on Research and Development in Information Retrieval, pp. 315–324. ACM (2011)

Sequence Generation with Target Attention

Yingce Xia[1], Fei Tian[2], Tao Qin[2(✉)], Nenghai Yu[1], and Tie-Yan Liu[2]

[1] University of Science and Technology of China, Hefei,
Anhui, People's Republic of China
`yingce.xia@gmail.com, ynh@ustc.edu.cn`
[2] Microsoft Research, Beijing, People's Republic of China
`fetia@microsoft.com, taoqin@microsoft.com, Tie-Yan.Liu@microsoft.com`

Abstract. Source-target attention mechanism (briefly, source attention) has become one of the key components in a wide range of sequence generation tasks, such as neural machine translation, image caption, and open-domain dialogue generation. In these tasks, the attention mechanism, typically in control of information flow from the encoder to the decoder, enables to generate every component in the target sequence relying on different source components. While source attention mechanism has attracted many research interests, few of them turn eyes to if the generation of target sequence can additionally benefit from attending back to itself, which however is intuitively motivated by the nature of attention. To investigate the question, in this paper, we propose a new *target-target* attention mechanism (briefly, target attention). Along the progress of generating target sequence, target attention mechanism takes into account the relationship between the component to generate and its preceding context within the target sequence, such that it can better keep the coherent consistency and improve the readability of the generated sequence. Furthermore, it complements the information from source attention so as to further enhance semantic adequacy. After designing an effective approach to incorporate target attention in encoder-decoder framework, we conduct extensive experiments on both neural machine translation and image caption. Experimental results clearly demonstrate the effectiveness of our design of integrating both source and target attention for sequence generation tasks.

Keywords: Sequence generation · Target-target attention model
Neural machine translation · Image captioning

1 Introduction

Recurrent Neural Network (RNN) based sequence generation, which aims to decode a target sequence y given source input x, has been widely adopted in real-world applications such as machine translation [1,3], image caption [28,30], and document summarization [19]. Although some RNN variants which include

---

This work was done when Yingce Xia was an intern at Microsoft Research Asia.

© Springer International Publishing AG 2017
M. Ceci et al. (Eds.): ECML PKDD 2017, Part I, LNAI 10534, pp. 816–831, 2017.
https://doi.org/10.1007/978-3-319-71249-9_49

multiplicative gating mechanisms, such as LSTM [7] and GRU [1], can help smooth the flow of historical information, it is not guaranteed to be sufficient, especially when faced with long sequences. To overcome this difficulty, the attention mechanism [1,15,30] is introduced to RNN based neural models. Inspired by human cognitive process, attention mechanism assumes that the generation of each target component (i.e., a word) can rely on different contexts, either in a "soft" form that depends on a weighted combination of all contextual components [1,30], or in a "hard" way that assumes only one contextual component can affect the target one [15]. Accordingly, attention mechanism can enable to discover semantic dependencies between the source and the target sequences in an end-to-end way.

Typical attention mechanisms model the dependency of target sequence on the source input, which implies that the context of the attention only comes from the source side input. Taking the "soft" attention mechanism as an example, to generate the j-th component y_j in y, an attentive weight associated with each source side component x_i is employed to describe how important each x_i is for y_j. The attentive weights in fact follow the multinomial distribution, derived from the matching degree between decoder's hidden state s_{j-1} and every hidden state of the source encoder, represented as $\{h_1, \cdots, h_{T_x}\}$ with T_x as the sequence length of x. In other words, such attention is a source-target mechanism, in which attentive information used in decoding only depends on the source-side encoder.

While the source-target attention (briefly, *source* attention) enables to pass important semantics from source input to target sequence, it overlooks the important coherent information hidden in the target sequence itself. Intuitively, beyond the selected source side parts, the generation of each target side component can be affected by certain preceding components in target side as well, especially when the target sequence is comparatively long. Apparently, such attentive dependency cannot be captured by source attention alone.

Therefore, to build a more comprehensive mechanism for sequence generation, in this paper, we propose the *target-target* attention (briefly, *target* attention), as a powerful complement to the source attention. In the proposed approach, the generation of each target side component depends not only on certain components in source side, but also on its prefix (i.e., the preceding components) in the target sequence. Acting in this way, more accurate probabilistic modelling for target sequence is achieved based on the better characterization for dependency within target side. Furthermore, we observe that in the decoding phase, even the semantics contained in source side could be enhanced due to the stimulation brought by attending to target-side prefix. As a result, compared with the source attention, our new approach can generate the target sequence with a couple of advantages:

1. The coherent consistency gets improved;
2. Eliminated repeated and redundant textual fragments [26];
3. Adequate semantics reflecting source-side information.

Examples in Table 2 clearly demonstrate all these improvements.

We conduct extensive experiments on both neural machine translation (English-to-French, English-to-Germany and Chinese-to-English translations) and image caption. The results show that incorporating the target attention mechanism effectively improves the quality of generated sequences.

The rest of the paper is organized as follows: the mathematical details of target attention are introduced in Sect. 2. We report and analyze the experiments on neural machine translation in Sect. 3, and image caption in Sect. 4. Background related works are summarized in Sect. 5. The paper is concluded in Sect. 6 together with perspectives on future works.

2 Target Attention Framework

In this section, we introduce the proposed target attention mechanism for sequence generation. The overall framework, together with several mathematical notations used in this section, is illustrated in Fig. 1. As a preliminary, we incorporate the target-target attention mechanism into the RNN based sequence-to-sequence network with source-target attention, which is briefly introduced in the next subsection.

2.1 Source Attention Based Sequence Generation

The sequence-to-sequence networks[1] typically include two components: the encoder network which encodes the source-side sequence and the decoder network which decodes and generates the target-side sequence. The attention mechanism acts as a bridge effectively transmitting information between the encoder and decoder.

Concretely speaking, the encoder network reads the source input x and processes it into a source-side memory buffer $M_{src} = \{h_1, h_2, \cdots, h_{T_x}\}$ with size T_x. For each $i \in [T_x]$,[2] the vector h_i acts as a representation for a particular part of x. Here are several examples.

- In neural machine translation [1] and neural dialogue generation [21], the encoder networks are RNNs with LSTM/GRU units, which sequentially process each word in x and generate a sequence of hidden states. The source-side memory M_{src} is composed of RNN hidden states at each time-step.
- In image captioning [28,30], the encoder network is a convolution neural network (CNN) working on an image. In this task, M_{src} contains low level local feature map vectors extracted by the CNN, representing different parts of the input image.

---

[1] In some scenarios such as image caption, the source-side input is not in a typical sequential form. For the ease of statement but with a little inaccuracy, we still use "sequence-to-sequence" as a general name even for these scenarios.

[2] For ease of reference, $[T_x]$ denotes the set $\{1, 2, \cdots, T_x\}$.

The decoder network is typically implemented using LSTM/GRU RNN together with a softmax layer. Specifically the decoder consumes every component (i.e., word) $y_j, j \in [T_y]$ in the target sequence y, meanwhile selectively reads from the source-side memory M_{src} to form attentive contextual vectors c_j^e (Eq. (1)), and finally generates each RNN hidden state s_j for any $j \in [T_y]$ (Eq. (2)). All these signals are then fed into the softmax-layer to generate the next-step component, i.e., y_j (Eq. (3)):

$$c_j^e = q(s_{j-1}, M_{\text{src}}), \tag{1}$$

$$s_j = g(s_{j-1}, y_{j-1}, c_j^e), \tag{2}$$

$$P(y_j | y_{<j}, x) \propto \exp(y_j; s_j, c_j^e). \tag{3}$$

The source attention plays an important role in the generation of c_j^e, i.e., the function $q(\cdot, \cdot)$. Intuitively, the attention mechanism grants different weights on source-side memory vectors $h_i \in M_{\text{src}}$ in generating each c_j^e:

$$\alpha_{ij} = \frac{\exp(A_e(s_{j-1}, h_i))}{\sum_{k=1}^{T_x} \exp(A_e(s_{j-1}, h_k))}, \tag{4}$$

$$c_j^e = \sum_{i=1}^{T_x} \alpha_{ij} h_i. \tag{5}$$

In Eq. (4), $A_e(\cdot, \cdot)$ acts as the key component in source attention, typically implemented as a feed-forward neural network.

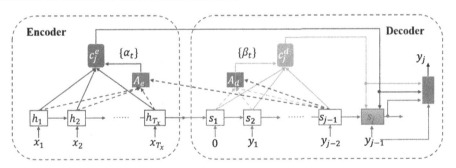

Fig. 1. The structure of sequence-to-sequence learning with target attention.

2.2   Target Attention Based Sequence Generation

From Eqs. (1) and (4), it is not difficult to observe that the attention weights are associated with the source-side memory M_{src}. Apart from that, as we have argued before, in sequence generation, better control over target-side contexts, i.e., the components that have been generated so far in the target sequence, is important as well. To add such target attention, we augment the memory space read by decoder RNN by adding an extra target-side memory M_{tgt} to original source-side one M_{src}. The j-th step slice of such a target-side memory is defined as $M_{\text{tgt}}^j = \{s_1, \cdots, s_{j-1}\}$, i.e., the hidden states before time step[3] j.

---

[3] For $j < 2$, let $M_{\text{tgt}}^1 = \emptyset$. Target attention starts to work when $j \geq 2$.

Afterwards, to decode the word at j-th timestep, M_{tgt}^j is read and weighted averaged to form an extra contextual representation c_j^d (Eq. (7)), where the weights are computed using a new attentive function $A_d(\cdot, \cdot)$ (Eq. (6)). Intuitively speaking, β_{tj} represents how important the t-th (already) generated word is for current decoding at time step j. Such attentive signals rising form target-side memory M_{tgt}^j are integrated into c_j^d:

$$\beta_{tj} = \frac{\exp(A_d(s_{j-1}, s_t))}{\sum_{k=1}^{j-1} \exp(A_d(s_{j-1}, s_k))}, \tag{6}$$

$$c_j^d = \sum_{t=1}^{j-1} \beta_{tj} s_t. \tag{7}$$

Finally, c_j^d is provided as an addition input to derive the hidden state s_j (Eq. (8)) and softmax distribution $P(y_j)$ (Eq. (9)), from which the j-th component is chosen. The target-side memory also consequently gets updated as $M_{tgt}^{j+1} = M_{tgt}^j \cup \{s_j\}$.

$$s_j = g(s_{j-1}, y_{j-1}, c_j^e, \boldsymbol{c_j^d}) \tag{8}$$

$$P(y_j|y_{<j}, x) \propto \exp(y_j; s_j, c_j^e, \boldsymbol{c_j^d}). \tag{9}$$

To make it more clear, we further give the mathematical details of Eqs. (8) and (9) in Eqs. (10) and (11) respectively. (We take GRU as an example here and the mathematical formulation of LSTM can be similarly defined.)

$$\begin{aligned} s_j &= (1 - z_j) \circ s_{j-1} + z_j \circ \tilde{s}_j; \\ \tilde{s}_j &= \tanh(W y_{j-1} + U[r_j \circ s_{j-1}] + C^e c_j^e + \boldsymbol{C^d} \boldsymbol{c_j^d}); \\ z_j &= \sigma(W_z y_{j-1} + U_z s_{j-1} + C_z^e c_j^e + \boldsymbol{C_z^d} \boldsymbol{c_j^d}); \\ r_j &= \sigma(W_r y_{j-1} + U_r s_{j-1} + C_r^e c_j^e + \boldsymbol{C_r^d} \boldsymbol{c_j^d}). \end{aligned} \tag{10}$$

$$P(y_j|y_{<j}, x) \propto \exp(W_y y_{j-1} + W s_j + W^e c_j^e + \boldsymbol{W^d} \boldsymbol{c_j^d}). \tag{11}$$

In Eqs. (8) $\sim$ (11), bold symbols are what make our target attention enhanced model different from conventional sequence-to-sequence model. The W's, U's and C's are the parameters of the network unit, $\sigma(\cdot)$ denotes the sigmoid activation function, and $\circ$ indicates element-wise product.

3   Application to Neural Machine Translation

We evaluate our proposed algorithm on three translation tasks: English→French (En→Fr), English→Germany (En→De) and Chinese→English (Zh→En).

3.1   Settings

For En→Fr and En→De, we conduct experiments on the same datasets as those used in [8]. To be more specific, part of data in WMT'14 is used as the bilingual training data, which consists of 12M and 4.5M sentence pairs for En→Fr and En→De respectively. We remove the sentences with more than 50 words. *newstest2012* and *newstest2013* are concatenated as the validation set and *newstest2014* acts as test set. For En→Fr, we limit the source and target vocabularies as the most frequent 30k words, while for En→De, such vocabulary size is set as 50k. The out-of-vocabulary words will be replaced with a special token "UNK"[4]. For Zh→En, we use 1.25M bilingual sentence pairs from LDC dataset as training set and NIST2003 as validation set. Furthermore, NIST2004, NIST2005 and NIST2006 all act as the test sets. Both source and target vocabulary sizes are set as 30k for Zh→En.

The basic NMT model is the sequence-to-sequence GRU model widely adopted in previous works [1,8]. In such a model, the word embedding size and GRU hidden layer size are respectively set as 620 and 1000. For Zh→En, dropout with a ratio 0.5 is applied to the last layer before softmax.

We use *beam search* algorithm [10] with beam width 12 to generate translations. The most common metric to evaluate translation quality is the BLEU score [16], which is defined as follows:

$$\text{BLEU} = \text{BP} \cdot \exp\left(\sum_{n=1}^{4} \frac{1}{4} \log p_n\right), \quad \text{BP} = \begin{cases} 1, & c > r \\ e^{1-r/c}, & c \le r \end{cases}, \quad (12)$$

where c is the total number of candidate sentence pairs in the corpus, r is the sum of the lengths for maximum perfect aligned translation subsequence in every translation pair, and p_n is a measure for n-gram translation precision.

Following the common practice in NMT, for En→Fr and En→De, the translation quality is evaluated by tokenized case-sensitive BLEU score[5]. For Zh→En, the BLEU scores are case-insensitive. Furthermore, to demonstrate better language modelling brought by target attention, we apply the perplexity [13] of target sentences conditioned on source sentences as another evaluation metric, which measures not only the translative correspondence of (source, target) pair, but also the naturalness of the target sentences. The perplexity is defined as follows:

$$\text{ppl}(D) = \exp\left\{-\frac{\sum_{i=1}^{M} \log P(y(i)|x(i))}{\sum_{i=1}^{M} N_i}\right\}, \quad (13)$$

where $D = \{(x(i), y(i))\}_{i=1}^{M}$ is the test set containing M bilingual language sentence pairs and N_i is the number of words in target sentence $y(i)$.

To reduce training time and stabilize the training process, following the common practice [23,25], for all the three translation tasks, we initialize the target attention accompanied NMT models with the basic NMT models without target attention, i.e., the RNNSearch models [1]. (Note that training from scratch would lead to similar results as those obtained by training from warm-start models.)

---

[4] We focus on the word-level translations, instead of subword-level ones like BPE [20]. The reason is that BPE cannot be applied to some languages like Chinese, although it works well for other languages like German and Czech.

[5] The script is from https://github.com/moses-smt/mosesdecoder/blob/master/scripts/generic/multi-bleu.perl.

The three RNNSearch models used for warm start are all trained by Adadelta [33] with mini-batch size as 80, and these initialized models are able to reproduce the public reported BLEU scores in previous works.

After model initialization, we adopt vanilla SGD with minibatch size 80 to continue model training. According to the validation performance, the initial learning rates are set as 0.4, 0.4 and 0.1 for En→Fr, En→De and Zh→En respectively. During the training process, we halve the learning rates once the BLEU on validation set drops. We clip the gradients [17] with clipping threshold 1.0, 5.0 and 1.0 for En→Fr, En→De and Zh→En respectively. When halving learning rates cannot improve validation BLEU scores, we freeze the word embedding and continue model training for additional several days [8], leading to total training time of roughly one week for all the three translation tasks.

3.2   Quantitative Results

The experimental results are shown in Table 1. In this table, "RNNsearch" refers to the warm-start model, i.e., the sequence-to-sequence neural machine translation model with only source attention; "Target Attn" refers to our target attention enhanced NMT model.

Table 1. BLEU scores and perplexities of different Neural Machine Translation Models

	BLEU					Perplexities				
	En→Fr	En→De	MT04	MT05	MT06	En→Fr	En→De	MT04	MT05	MT06
RNNSearch	29.93	16.47	34.96	34.57	32.74	4.71	7.36	13.05	11.85	15.03
Target Attn	31.63	17.67	36.71	35.62	33.78	4.19	6.87	12.34	11.24	14.27

One can see that by introducing target attention into conventional NMT model, we can achieve significant improvements on all the three translation tasks. For En→Fr, the gain of BLEU brought by target attention is 1.7; for En→De, we improve BLEU by 1.2; furthermore, the average improvement for Zh→En is 1.28 BLEU point. By applying the statistical significance test for machine translation [11], we get that the results of our target attention mechanism is significantly better than RNNSearch with p-values $p < 0.01$. These results well demonstrate that our target attention algorithm is an effective approach that consistently and significantly improves the performances of different NMT tasks.

After obtaining the translation results, we further process them by the widely used post-processing technique proposed by [8], that is able to replace the "UNK" token in translation sentences. The steps include:

1. Get a word-level translation table T that maps the source language words to target language words; we use the *fastAlign*[6] [5];

---

[6] The script is from https://github.com/clab/fast_align.

2. Given any translated sentence $\hat{y}$, for each of its word $\hat{y}_j$, if $\hat{y}_j$ is UNK, find the corresponding source side word x_i according to the attention weights learnt in NMT model, i.e., $i = \mathrm{argmax}_k \alpha_{kj}$ (refer to the definition in Eq. (4));
3. Look up the table $\mathcal{T}$ to get the corresponding translation for source word x_i.

By applying this technique to En→Fr translation, we can further improve the BLEU to **34.49**, which is a new best result for En→Fr translation conditioned on (i) the model is a single-layer NMT model; (ii) the model is trained with only bilingual data.

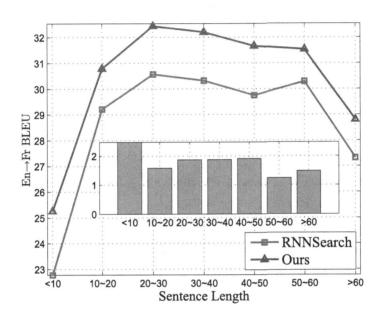

Fig. 2. BLEU w.r.t. input sentence length

We further record the average BLEU scores in En→Fr task for different bilingual sentence pair buckets that are determined by source sentence length and visualize them in Fig. 2. The chart inside Fig. 2 shows the improvements of BLEU score by adding target attention. From this figure we can see that with target attention, the BLEU scores for all buckets clearly increase. Specially,

1. For sentences with $[10, 50)$ words, the longer the sentence is, the more improvement is brought by the target attention. This clearly verifies the effectiveness of target attention in capturing long-range dependency.
2. For sentences with fewer than 10 words, our target attention also brings significant improvements. Note that although the calculation of BLEU is quite sensitive for very short sentences (e.g., p_4 in Eq. (12) is very likely to be zero for sentence-level BLEU), we can still generate better sentences with target attention.

3. Even we remove the sentences with more than 50 words from the training corpus, target attention can achieve improvements on such un-trained subsets containing extremely long sentences (i.e., >50 words), although improvements brought by target attention are slightly less compared to those in the region [10, 50).

In the right part of Table 1 we additionally report the performances of different models using another evaluation measure, i.e., the perplexity, to represent how smooth and natural the translation is[7]. It can be observed that by incorporating target attention, the perplexity scores are decreased by 0.52, 0.49 and 0.69 points for En→Fr, En→De and Zh→En respectively. Such improvements clearly demonstrate the advantages of target attention in generating more smooth translations, mainly out from better dependency modeling within target translation sentence.

3.3 Subjective Evaluation

To better understand and visualize the benefits that target attention brings to NMT, we give three concrete Zh→En examples in Table 2. For each example, we highlight some important parts by bold words to demonstrate either the limitations of baseline model, or the improvements led by incorporating target attention. As discussed in Sect. 1, the three examples respectively show that target attention:

1. Improves long range semantic consistency such as matching the subjects perfectly and avoiding such pattern as *"The founding was the company"*;
2. Eliminates repeated translations such as *"economy could slow down"*[8];
3. Enhances semantic integrity such as successfully translates *"bimen"* and *"xijieweijian gongbu"* in the last example.

Our intuitive explanation for these improvements is that target attention enhances the ability to model a natural sentence, due to which the inconsistent and repeated translations would be assigned with low probabilities and thus not selected. The punishment towards such wrong translation patterns will comparatively improve the possibility of right translation that has not been translated yet, thereby alleviating the semantic inadequacy issue.

3.4 Discussion

In this subsection, we carry out some discussions about our proposed target-target attention framework:

---

[7] For Zh→En, each source sentence x has four references $y(j)$ $j \in \{1, 2, 3, 4\}$. To calculate the perplexities, we simply regard them as four individual sentence pairs $(x, y(j))$.

[8] Although the coverage model in [26] can eliminate repeated translations, it is actually based on source-target attention but not target-target attention. Therefore, [26] can be further combined with our proposed target-target attention. We leave it as a future work.

Table 2. Translation examples. Source, Base, Ours, Ref denote the source sentence, the RNNSearch, target attention and the reference sentence respectively.

Source	*e **tianranqi gongye gufengongsi** chengli yu 1993nian2yue, **shi** shijieshang zuida de tianranqi kaicai gongsi*
Base	***the founding** of the russian gas industry in February 1993 **was** the world's largest natural gas mining company*
Ours	*founded in February 1993, **the russian gas industrial corporation is** the world's largest producer of natural gas mining*
Ref	*Established in February 1993, Gazprom is the largest natural gas exploitation company in the world*
Source	*youyu riyuan shengzhi he pinfu chaju rijian kuoda keneng pohuai jinnian shangbannian xiangyou de nazhong hexie qifen, riben jingji keneng fanghuan sudu, mairu 2005 nian*
Base	*Japan's **economy could slow down** as the japanese **economy could slow down** as the yen appreciated and the disparity between the rich and the poor as a result of a growing gap between the rich and the poor*
Ours	*Japan's economy may slow down in 2005 as the yen's appreciation and the growing gap between the rich and the poor may damage the harmonious atmosphere in the first half of the year*
Ref	*Japan's economy may slow down towards 2005 as yen appreciation and a widening gap between the rich and poor could break the harmonious atmosphere it enjoyed in the first half of this year*
Source	*Xizang liuwang jingshen lingxiu dalai lama de teshi zhengzai beijing he zhongguo guanyuan jinxing **bimen** huiyi, **xijie weijian gongbu***
Base	*the special envoy of the dalai lama, tibet's exiled spiritual leader, was scheduled to meet with chinese officials and officials from the tibetan spiritual leader in beijing and chinese officials*
Ours	*the special envoy of tibet's exiled spiritual leader, the dalai lama, is holding a **closed-door** meeting with chinese officials, and **the details were not disclosed***
Ref	*The special envoy of the Dalai Lama, exiled Tibetan spiritual leader, is currently in Beijing carrying out closed meetings with Chinese officials, but the details have not been released*

1. For decoding speed, our approach indeed takes 17% more time for decoding than previous approach without target attention, considering we have an additional target-target attention model. Such a cost is acceptable considering the BLEU score improvements (i.e., 1.7pts for En→Fr, 1.2pts on En→De and 1.28pts for Zh→En.)

2. A degenerated version of our target-target attention is to use delayed memory. Mathematically, when predicting the t'th word, not only the hidden state s_{t-1} at step $t - 1$, but also the one $s_{t-\tau}$ at step $t - \tau$ are used where τ is a fixed number. Note such the degenerated model might not work well. Take $\tau = 2$ as an example. For the third example in Table 2, "Details were not disclosed"

should be attended to "closed-door", which is far from $t-2$. Delayed memory cannot handle this case since it does not know how many steps to delay. Our model works well due to its adaptive and dynamic attention mechanism.

3. The improvements of target-target attention are not caused by better optimization properties of the new RNN, since its architecture is more complex and more difficult to optimize.

4 Application to Image Captioning

To further verify the effectiveness of incorporating target attention into sequence generation models, we then apply the proposed model to image caption, which targets at describing the content of an image with a natural sentence.

4.1 Settings

We choose a public benchmark dataset, Flickr30k [32], for image caption. In this dataset, every image is provided with 5 reference sentences. We follow the data splitting rule as that used in [9,30] such that the dataset is separated into 28000 training samples, 1000 validation samples and 1000 test samples.

We follow the same model structure as that proposed in [30]. The Oxford VGGnet [24] pretrained on ImageNet is used as the image encoder, which could eventually output 196 features per image, with each feature as a 512 dimensional vector. The decoder is a 512×512 LSTM RNN. Dropout with drop ratio 0.5 is applied to the layer before softmax. The vocabulary is set as the most 10k frequent words in the dataset. Soft source attention is chosen due to better performance in our implementation. In implementation, we base our codes on the open-source project provided by [30][9].

The captions for all images are generated using beam search with beam size 10. To comprehensively evaluate their quality, we adopt several different measures. Following [30], we report the BLEU-3, BLEU-4 without a brevity penalty, which respectively indicates tri-gram and four-gram matching degree with groundtruth caption. Besides, we report the CIDEr, another widely used metric for image caption [27,28]. CIDEr also measures n-gram matching degree, in which each n-gram will be assigned a TF-IDF weighting [18]. For all these metrics, the larger, the better. In addition, we report test perplexities (the smaller, the better) to evaluate whether target attention improve sentence smoothness in image caption task. (Refer to Eq. (13) for the definition of perplexity.)

At the beginning of training process, we warm start our model with a pretrained model with only source attention, which is previously optimized with Adadelta for about one day. Then in the training process, we incorporate target attention mechanism into the initialized captioning model and continue to train it using plain stochastic gradient descent for another one day, with a fixed learning rate as 0.05 selected by validation performance.

---

[9] https://github.com/kelvinxu/arctic-captions.

4.2   Results

We present our results in Table 3. The second row, labeled with "source-attn" represents the performance of baseline captioning model, which approximately matches the numbers reported in [30] (e.g., BLEU-3 as 28.8 and BLEU-4 as 19.1), while the third row labeled with "Ours" records the results after adding target attention into caption generation.

Table 3. Results of image caption with/without target attention. For perplexity, the lower, the better, while for other measures, the higher, the better.

	BLEU-3	BLEU-4	CIDEr	perplexity
source-attn	28.3	19.0	37.6	28.56
Ours	29.6	20.5	40.4	23.06

From Table 3, we can clearly see that target attention achieves significant improvements over baseline in terms of all evaluation measures. In particular, the decrease of perplexity further demonstrates the better probabilistic modelling for captions by adding target attention.

Table 4. Two examples showing different captioning result

source-attn	*A group of people are playing instruments on stage*
Ours	*A band is playing <u>on stage</u> <u>in front of a crowd</u>*
Ref	*(1) Two men , one sitting , one standing , are playing their guitars on stage while the audience is looking on .*
	(2) band doing a concert for people
source-attn	*A black dog is running through the grass*
Ours	*A black dog is running through the grass <u>with a toy in its mouth</u>*
Ref	*(1) A curly brown dog runs across the lawn carrying a toy in its mouth*
	(2) The black dog is running on the grass with a toy in its mouth .

We also list two examples of image caption in Table 4, including the image, its two referenced captions (marked by "Ref") and the captioning results with/without target attention (respectively marked by "source-attn" and "Ours"). It is clearly shown that target attention mechanism generates better captions for both images (see the highlighted underlined words), mainly owing to the enhanced stimulation from already generated words in target side, right brought by target attention.

To better demonstrate how target attention helps to improve the image caption quality, we take the right figure in Table 4 as an example and analyze the target attention weights in the decoder. To be concrete, in the internal green bar chart of Fig. 3, we show the weights of the previously generated words when generating the last word *mouth* in our caption (i.e., *A black dog is running through the grass with a toy in its mouth*). Here we remove the most common words like *a, is* to make the illustration more compact and clearer. Simultaneously, the outer bar chart in Fig. 3 shows the words co-occurrence statistics for the word *mouth*, which is calculated by

$$\text{co-occurrence}(\text{word}_i) = \frac{\text{The number of sentences containing word}_i \text{ and } mouth}{\text{The number of sentences containing } mouth}.$$

From the outer chart, it is clearly observed that *mouth* is highly correlated to relevant words such as *dog* and *black*. The target attention thereby stimulates the generation of word *mouth*, by learning from the significant weights assigned to these two previously decoded words. This clearly shows that target attention mechanism accurately characterizes the semantic relation among the decoded words, thus improves both the coherence and completeness of the caption by attending to the past.

5   Related Work

The attempt of applying attention mechanism with deep learning dates back to several works in computer vision, represented by [4,15] and [30]. Particularly [15] and [30] leverage a "hard" attention mechanism, which in every step attend to only one part of the image. For NMT, [1] first incorporates the "soft" attention mechanism that automatically assign attentive weights to every source side words in decoding target sentence word. Since then many varieties have been designed to improve such a source-target attention mechanism for neural machine translation. For example, [12] proposes a local attention model, which is effective and computationally efficient. In this model, for each decoding step t, the attention weights are only assigned over a subset of source hidden states within a window $[p_t - D, p_t + D]$, where D is the window size, and p_t is the window center determined by a selective mechanism.

There are some other works that target improving the network structure for attention model in neural machine translation. For example, [14] propose an interactive attention model for NMT named as NMT_{IA}. NMT_{IA} can keep track of the interaction between the decoder and the representation of source sentence

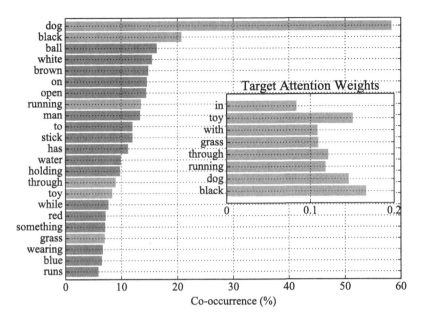

Fig. 3. Analysis of co-occurrence and target attention weights (Color figure online)

during translation by both reading and writing operations, which are helpful to improve translation quality. Similarly, [31] design a recurrent attention model, in which the attention weights are tracked by an LSTM.

As to decoder side, [22] proposes a self-attention model to maintain coherence in longer responses for neural conversation model: the decoded words would be concatenated with words in the source side encoder by their embeddings, and the self-attention model will generate contexts by these "faked" words. Such a proposal does not fit to other sequence-to-sequence tasks when the words in the encoder and decoder are not in the same language, like NMT and image caption. Therefore, such a model based on source-target words concatenation is limited and cannot be generalized to more general scenarios. [2] proposed a similar model like ours but [2] did not focus on sequence generation tasks.

6   Conclusion

In this work, motivated from tailored observations and analysis, we design a target attention model to enhance the dependency within decoder side components and thereby improve performances of sequence generation tasks. We conduct extensive evaluations and analysis on neural machine translation and image caption. Significant better results with the proposed model are observed on these two tasks, which illustrate the effectiveness of target attention.

There are many interesting directions left as future works. First, we aim to adapt target attention into different model structures [29] and different training techniques [6,23]. Second, we plan to study how to make the target attention model more effective by combining it with the advanced source attention mechanisms discussed in related work. Last, target attention will be tested on more tasks such as document summarization, neural dialogue generation and question-answering.

References

1. Bahdanau, D., Cho, K., Bengio, Y.: Neural machine translation by jointly learning to align and translate. In: International Conference on Learning Representations (2015)
2. Cheng, J., Dong, L., Lapata, M.: Long short-term memory-networks for machine reading. In: EMNLP (2016)
3. Cho, K., van Merrienboer, B., Gulcehre, C., Bahdanau, D., Bougares, F., Schwenk, H., Bengio, Y.: Learning phrase representations using RNN encoder-decoder for statistical machine translation. In: EMNLP, pp. 1724–1734 (2014)
4. Denil, M., Bazzani, L., Larochelle, H., de Freitas, N.: Learning where to attend with deep architectures for image tracking. Neural Comput. **24**(8), 2151–2184 (2012)
5. Dyer, C., Chahuneau, V., Smith, N.A.: A simple, fast, and effective reparameterization of IBM model 2. In: Proceedings of the Conference of the North American Chapter of the Association for Computational Linguistics: Human Language Technologies, pp. 644–649 (2013)
6. He, D., Xia, Y., Qin, T., Wang, L., Yu, N., Liu, T., Ma, W.Y.: Dual learning for machine translation. In: Advances In Neural Information Processing Systems, pp. 820–828 (2016)
7. Hochreiter, S., Schmidhuber, J.: Long short-term memory. Neural Comput. **9**(8), 1735–1780 (1997). https://doi.org/10.1162/neco.1997.9.8.1735
8. Jean, S., Cho, K., Memisevic, R., Bengio, Y.: On using very large target vocabulary for neural machine translation. In: The Annual Meeting of the Association for Computational Linguistics (2015)
9. Karpathy, A., Fei-Fei, L.: Deep visual-semantic alignments for generating image descriptions. In: IEEE Conference on Computer Vision and Pattern Recognition, pp. 3128–3137 (2015)
10. Koehn, P.: Pharaoh: a beam search decoder for phrase-based statistical machine translation models. In: Frederking, R.E., Taylor, K.B. (eds.) AMTA 2004. LNCS (LNAI), vol. 3265, pp. 115–124. Springer, Heidelberg (2004). https://doi.org/10.1007/978-3-540-30194-3_13
11. Koehn, P.: Statistical significance tests for machine translation evaluation. In: EMNLP, pp. 388–395. Citeseer (2004)
12. Luong, M.T., Pham, H., Manning, C.D.: Effective approaches to attention-based neural machine translation. arXiv preprint arXiv:1508.04025 (2015)
13. Luong, T., Cho, K., Manning, C.: Tutorial: Neural Machine Translation. ACL 2016 tutorial (2016)
14. Meng, F., Lu, Z., Li, H., Liu, Q.: Interactive attention for neural machine translation. In: COLING (2016)
15. Mnih, V., Heess, N., Graves, A., et al.: Recurrent models of visual attention. In: Advances in Neural Information Processing Systems, pp. 2204–2212 (2014)
16. Papineni, K., Roukos, S., Ward, T., Zhu, W.J.: Bleu: a method for automatic evaluation of machine translation. In: The Annual Meeting of the Association for Computational Linguistics, pp. 311–318 (2002)

17. Pascanu, R., Mikolov, T., Bengio, Y.: On the difficulty of training recurrent neural networks. In: International Conference on Machine Learning, vol. 28, pp. 1310–1318 (2013)

18. Robertson, S.: Understanding inverse document frequency: on theoretical arguments for IDF. J. Documentation **60**(5), 503–520 (2004)

19. Rush, A.M., Chopra, S., Weston, J.: A neural attention model for abstractive sentence summarization. In: EMNLP, pp. 379–389 (2015)

20. Sennrich, R., Haddow, B., Birch, A.: Neural Machine Translation of Rare Words with Subword Units. In: The annual meeting of the Association for Computational Linguistics (2016)

21. Shang, L., Lu, Z., Li, H.: Neural responding machine for short-text conversation. arXiv preprint arXiv:1503.02364 (2015)

22. Shao, L., Gouws, S., Britz, D., Goldie, A., Strope, B., Kurzweil, R.: Generating long and diverse responses with neural conversation models. arXiv preprint arXiv:1701.03185 (2017)

23. Shen, S., Cheng, Y., He, Z., He, W., Wu, H., Sun, M., Liu, Y.: Minimum Risk Training for Neural Machine Translation. In: The Annual Meeting of the Association for Computational Linguistics (2016)

24. Simonyan, K., Zisserman, A.: Very deep convolutional networks for large-scale image recognition. In: International Conference on Learning Representations (2015)

25. Tu, Z., Liu, Y., Shang, L., Liu, X., Li, H.: Neural machine translation with reconstruction. In: AAAI (2017)

26. Tu, Z., Lu, Z., Liu, Y., Liu, X., Li, H.: Modeling coverage for neural machine translation. In: The Annual Meeting of the Association for Computational Linguistics, pp. 76–85 (2016)

27. Vedantam, R., Lawrence Zitnick, C., Parikh, D.: Cider: consensus-based image description evaluation. In: IEEE Conference on Computer Vision and Pattern Recognition, pp. 4566–4575 (2015)

28. Vinyals, O., Toshev, A., Bengio, S., Erhan, D.: Show and tell: a neural image caption generator. In: IEEE Conference on Computer Vision and Pattern Recognition, pp. 3156–3164 (2015)

29. Wu, Y., Schuster, M., Chen, Z., Le, Q.V., Norouzi, M., Macherey, W., Krikun, M., Cao, Y., Gao, Q., Macherey, K., et al.: Google's neural machine translation system: Bridging the gap between human and machine translation. arXiv preprint arXiv:1609.08144 (2016)

30. Xu, K., Ba, J., Kiros, R., Cho, K., Courville, A.C., Salakhutdinov, R., Zemel, R.S., Bengio, Y.: Show, attend and tell: Neural image caption generation with visual attention. In: International Conference on Machine Learning vol. 14, pp. 77–81 (2015)

31. Yang, Z., Hu, Z., Deng, Y., Dyer, C., Smola, A.: Neural machine translation with recurrent attention modeling. arXiv preprint arXiv:1607.05108 (2016)

32. Young, P., Lai, A., Hodosh, M., Hockenmaier, J.: From image descriptions to visual denotations: new similarity metrics for semantic inference over event descriptions. Trans. Assoc. Comput. Linguist. **2**, 67–78 (2014)

33. Zeiler, M.D.: Adadelta: an adaptive learning rate method. arXiv preprint arXiv:1212.5701 (2012)

Wikipedia Vandal Early Detection: From User Behavior to User Embedding

Shuhan Yuan[1], Panpan Zheng[2], Xintao Wu[2(✉)], and Yang Xiang[1]

[1] Tongji University, Shanghai 201804, China
{shxiangyang,4e66}@tongji.edu.cn
[2] University of Arkansas, Fayetteville, AR 72701, USA
{pzheng,xintaowu}@uark.edu

Abstract. Wikipedia is the largest online encyclopedia that allows anyone to edit articles. In this paper, we propose the use of deep learning to detect vandals based on their edit history. In particular, we develop a multi-source long-short term memory network (M-LSTM) to model user behaviors by using a variety of user edit aspects as inputs, including the history of edit reversion information, edit page titles and categories. With M-LSTM, we can encode each user into a low dimensional real vector, called user embedding. Meanwhile, as a sequential model, M-LSTM updates the user embedding each time after the user commits a new edit. Thus, we can predict whether a user is benign or vandal dynamically based on the up-to-date user embedding. Furthermore, those user embeddings are crucial to discover collaborative vandals. Code and data related to this chapter are available at: https://bitbucket.org/bookcold/vandal_detection.

1 Introduction

Wikipedia, as one of the world's largest knowledge bases on the web, heavily relies on thousands of volunteers to make contributions. This crowdsourcing mechanism based on the freedom-to-edit model (i.e., any user can edit any article) leads to a rapid growth of Wikipedia. However, Wikipedia is plagued by vandlism, namely "deliberate attempts to damage or compromise integrity"[1]. Those vandals who commit acts of vandalism damage the quality of articles and spread false information, misleading information, or nonsense to Wikipedia users as well as information systems such as search engines and question-answering systems.

Reviewing millions of edits every month incurs an extremely high workload. Wikipedia has deployed a number of tools for automatic vandalism detection, like ClueBot NG[2] and STiki[3]. These tools use heuristic rules to detect and revert apparently bad edits, thus helping administrators to identify and block vandals. However, those bots are mainly designed to score edits and revert the worst-scoring edits.

---

[1] https://en.wikipedia.org/wiki/Wikidata:Vandalism.
[2] https://en.wikipedia.org/wiki/User:ClueBot_NG.
[3] https://en.wikipedia.org/wiki/Wikipedia:STiki.

© Springer International Publishing AG 2017
M. Ceci et al. (Eds.): ECML PKDD 2017, Part I, LNAI 10534, pp. 832–846, 2017.
https://doi.org/10.1007/978-3-319-71249-9_50

Detecting vandals and vandalized pages from crowdsourcing knowledge bases has attracted increasing attention in the research community [12,15,21]. For example, [12] focused on predicting whether an edit is vandalism. They developed a set of 47 features that exploit both content and context information of users and edits. The content features of an edit are defined at levels of character, word, sentence, and statement whereas the context features are used to quantify users, edited items, and their respective revision histories. [15] focused on predicting whether a user is a vandal based on user edits. The developed VEWS system adopted a set of behavior features based on edit-pairs and edit-patterns, such as vandals make faster edits than benign user, benign users spend more time editing a new page than vandals, or benign users more likely edit a meta-page than vandals. All the above features empirically capture the differences between good edits and vandalism to some extent and there is no doubt that classifiers (e.g., randomforest or SVM) with these features as inputs can achieve good accuracy of detecting vandalism.

Different from the existing approaches that heavily rely on hand-designed features, we tackle the problem of vandal detection by automatically learning user behavior representations from their edit sequences. Each edit in a user's edit sequence contains many attributes such as PageID, Title, Time, Categories, PageType, Revert Status, and Content. We transform each edit sequence into multiple aspect sequences and develop a multi-source long-short term memory network (M-LSTM) to detect vandals. Each LSTM processes one aspect sequence and learns the hidden representation of the corresponding aspect of user edits. The LSTM as a sequence model can represent the user edit sequence with variable-length as fixed-length real vectors, i.e., aspect representations. We then apply the attention model [4,33] to derive the user representation, called user embedding, by combining all aspect representations. The user embedding accurately captures all aspects of a user's edit information. Thus we can use user embeddings as classifier inputs to separate vandals from benign users. To the best of our knowledge, this is the first work to use the deep neural network to represent users as user embeddings which capture the information of user behavior for vandal detection.

Our approach has several advantages over past efforts. First, neither heuristic rules nor hand-designed features are needed. Second, while each user may have a different number of edits and each user may have different edit behavior, we map each user into the same low-dimensional embedding space. Thus user embeddings can be easily used for a variety of data mining tasks such as classification, clustering, outlier analysis, and visualization. Third, by using various aspect information (e.g., article title and categories), our M-LSTM is able to effectively capture hidden relationships between different users. The derived user embeddings can be used to analyze collaborative vandals who commit acts of vandalism together to impose big damages and/or evade detection. Fourth, our M-LSTM can naturally achieve early vandal detection and has great potential to be deployed for dynamically monitoring user edits and conducting real-time vandal detection.

The rest of the paper is organized as follows. Section 2 summarizes the related work about deep neural networks and the vandal detection. Section 3 introduces our M-LSTM model for vandal (early) detection. The experimental results and analysis are discussed in Sect. 4. Section 5 concludes the paper.

2   Related Work

Deep neural networks have achieved promising results in image [11], text [20], and speech recognition [9]. The key ingredient for the successful of deep neural network is because it learns meaningful representations of inputs [5]. For example, in text area, all the words are trained to represent as real-valued vectors called word embeddings which capture the semantic relations among words [20]. Then, a neural network model can further combine the word embeddings to represent the sentences or documents [33]. For image recognition, a deep neural network can learn different levels of image representations on different levels of the neural network [38]. In this work, we propose a M-LSTM model to train the representations of users and further use them to predict vandals.

Most work for vandalism detection extracts features, e.g., content-based features [1,14,21], context features to measure user reputation [2], spatial-temporal user information [31], and then uses those features as classifier inputs to predict vandalism. Moreover, [19] utilizes search engine to check the correctness of user edits. However, it is difficult to apply these approaches based on hand-design features to detect subtle and collaborative vandalism.

Wikipedia vandal detection is related to fake review detection. In [22], different types of behavior features were extracted and used to detect the fake reviews in Yelp. [17] have identified several representative behaviors of review spammers. [32] studied the co-anomaly patterns in multiple review-based time series. [8] proposed approaches to detect fake reviews by characterizing burstiness of review. There has been extensive research on detecting anomaly from graph data [3,28,34] and detecting Web ranking spams [27]. In [35], the authors developed a fraud detection framework that combines deep neural networks and spectral graph analysis. They developed two neural networks, deep autoencoder and convolutional neural network, and evaluated them in a signed graph extracted from Wikipedia data. [23] studied various aspects of content-based spam on the Web and presented several heuristic methods for detecting content based spam. Finding time points at which graph changes significantly given a sequence of graphs has also been studied [24]. Although some of above approaches can be used for vandal detection, they are not able to automatically extract and fuse multiple aspects of user edit behaviors.

Several network embedding methods including DeepWalk [26], LINE [30], Node2vec [10], SNE [36], and DDRW [16] have been proposed. These models are based on the neural language model. Some works learn the network embedding by considering the node attribute or other information from heterogeneous networks [6,29]. Unlike all the embedding works described above, in this paper, we explore to build user embedding from user edit sequence data.

3   Multi-source LSTM for Vandal Early Detection

Our key idea is to adopt multiple LSTMs to transform a variable-length edit sequence into multiple fixed-length aspect representations and further use the attention model to combine all aspect representations into the user embedding. As user embeddings capture all aspects of user edits as well as relationships among users, they can be used for detecting vandals and examining behaviors of vandalism.

3.1   LSTM Revisited

Long short-term memory network [13], as one class of recurrent neural networks (RNNs), was proposed to model the long-range sequences and has achieved great success in natural language processing and speech recognition recently. Figure 1 shows the structure of the standard LSTM for classification. Given a sequence $\mathbf{x} = (\mathbf{x}_1, \ldots, \mathbf{x}_t, \ldots, \mathbf{x}_T)$ where $\mathbf{x}_t \in \mathbb{R}^d$ denotes the input at the t-th step, LSTM maintains a hidden state vector $\mathbf{h}_t \in \mathbb{R}^h$ to keep track the sequence information with the input from the current step $\mathbf{x}_t$ and the previous hidden state $\mathbf{h}_{t-1}$. LSTM is composed by a special unit called memory block in the recurrent hidden layer to compute the hidden state vector $\mathbf{h}_t$. Each memory block contains self-connected internal memory cells and special multiplicative units called gates to control what kinds of information need to be encoded to the internal memory or discarded. Each memory block has an input gate to control the input information into the memory cell, a forget gate to forget or reset the current memory, and an output gate to control the output of cell into the hidden state. Formally, the hidden state $\mathbf{h}_t$ is computed by

$$
\begin{aligned}
\tilde{\mathbf{c}}_t &= \tanh(\mathbf{W}_c \mathbf{x}_t + \mathbf{U}_c \mathbf{h}_{t-1} + \mathbf{b}_c) \\
\mathbf{i}_t &= \sigma(\mathbf{W}_i \mathbf{x}_t + \mathbf{U}_i \mathbf{h}_{t-1} + \mathbf{b}_i) \\
\mathbf{f}_t &= \sigma(\mathbf{W}_f \mathbf{x}_t + \mathbf{U}_f \mathbf{h}_{t-1} + \mathbf{b}_f) \\
\mathbf{o}_t &= \sigma(\mathbf{W}_o \mathbf{x}_t + \mathbf{U}_o \mathbf{h}_{t-1} + \mathbf{b}_o) \\
\mathbf{c}_t &= \mathbf{i}_t \odot \tilde{\mathbf{c}}_t + \mathbf{f}_t \odot \mathbf{c}_{t-1} \\
\mathbf{h}_t &= \mathbf{o}_t \odot \tanh(\mathbf{c}_t)
\end{aligned}
\tag{1}
$$

where σ is the sigmoid activation function; $\odot$ represents element-wise product; $\mathbf{i}_t, \mathbf{f}_t, \mathbf{o}_t, \mathbf{c}_t$ indicate the input gate, forget gate, output gate, and cell activation vectors and $\tilde{\mathbf{c}}_t$ denotes the intermediate vector of cell state; $\mathbf{W}$ and $\mathbf{U}$ are the weight parameters; $\mathbf{b}$ is the bias term. We denote all LSTM parameters ($\mathbf{W}$, $\mathbf{U}$ and $\mathbf{b}$) as $\mathbf{\Theta}_1$.

After the LSTM reaches the last step T, $\mathbf{h}_T$ encodes the information of the whole sequence and is considered as the representation of the sequence. It can then be used as an input of the softmax classifier,

$$
P(\hat{y} = k | \mathbf{h}_T) = \frac{\exp\left(\mathbf{w}_k^T \mathbf{h}_T + b_k\right)}{\sum_{k'=1}^{K} \exp(\mathbf{w}_{k'}^T \mathbf{h}_T + b_{k'})},
\tag{2}
$$

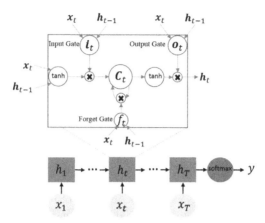

Fig. 1. Standard LSTM for classification

where K is number of classes, $\hat{y}$ is the predicted class of the sequence, $\mathbf{w}_k$ and b_k are the parameters of softmax function for the k-th class, and $\mathbf{w}_k^T$ indicates the transpose of $\mathbf{w}_k$. All softmax parameters $\mathbf{W}_k$ and $\mathbf{b}_k$ over K classes are denoted as $\boldsymbol{\Theta}_2$. The LSTM model aims to optimize $\boldsymbol{\Theta}_1$ and $\boldsymbol{\Theta}_2$ by minimizing the cross entropy loss function,

$$L = -\frac{1}{N} \sum_{i=1}^{N} y_i * log(P(\hat{y}_i)), \tag{3}$$

where y_i is the true class of the i-th sequence, and N is the number of training sequences.

3.2 Multi-source LSTM

We develop a multi-source LSTM model to capture all useful aspects of edits. As different aspects carry different weights for vandal detection, we adopt the attention model [4,33] to dynamically learn the importance of each aspect. The user embeddings are then used as inputs of softmax classifier to separate vandals from benign users.

Formally, for a user u with T edits, his edits can be modeled as a sequence $\mathbf{e}_u = (\mathbf{e}_{u_1}, \ldots, \mathbf{e}_{u_t}, \ldots, \mathbf{e}_{u_T})$ where $\mathbf{e}_{u_t}$ includes all related information about the t-th edit. Please note different users may have different numbers of edits. For each edit sequence $\mathbf{e}_u$, we transform it into M aspect sequences. Its m-th aspect sequence, denoted as $\mathbf{x}^{(m)} = (\mathbf{x}_1^{(m)}, \mathbf{x}_2^{(m)}, \ldots, \mathbf{x}_T^{(m)})$, captures the m-th aspect of edit information and is used as the input of the m-th LSTM in our multi-source LSTM. Figure 2 illustrates our M-LSTM model with M aspect sequences as inputs. The last hidden states $\mathbf{h}_T^{(m)}$ $(m = 1, \cdots, M)$ encode all the aspect

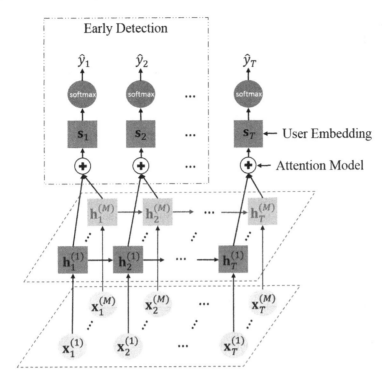

Fig. 2. Multi-source LSTM

information of the user's edit sequence. We apply the attention model as shown in Eqs. 4–6 to combine all aspect information into the user embedding.

$$\mathbf{z}_T^{(m)} = \tanh(\mathbf{W}_a \mathbf{h}_T^{(m)}), \tag{4}$$

$$\alpha_T^{(m)} = \frac{\exp(\mathbf{u}_a^T \mathbf{z}_T^{(m)})}{\sum_{m'=1}^{M} \exp(\mathbf{u}_a^T \mathbf{z}_T^{(m')})}, \tag{5}$$

$$\mathbf{s}_T = \sum_{m=1}^{M} \alpha_T^{(m)} \cdot \mathbf{h}_T^{(m)}, \tag{6}$$

where $\mathbf{W}_a \in \mathbb{R}^{h*h}$ is a trained projection matrix; $\mathbf{u}_a \in \mathbb{R}^h$ is a trained parameter vector. All the parameters, $\mathbf{W}_a$ and $\mathbf{u}_a$, in the attention model are denoted as Θ_3.

In the attention model, we first compute a hidden representation $\mathbf{z}_T^{(m)}$ of the last hidden state $\mathbf{h}_T^{(m)}$ based on a one-layer neural network by Eq. 4. After obtaining all the M hidden representations, $\mathbf{z}_T^{(1)}, \ldots, \mathbf{z}_T^{(M)}$, we apply the softmax function to calculate the weight of each hidden state $\alpha_T^{(m)}$ by Eq. 5. Finally, we compute the user embedding $\mathbf{s}_T$ as the weighted sum of the M hidden states by Eq. 6. The advantage of the attention model is that it can dynamically learn a weight of each aspect according to its relatedness with the user class (e.g., vandal or benign).

We use the user embedding $\mathbf{s}_T$ to predict $P(\hat{y}|\mathbf{s}_T)$, i.e., the probability of user u belonging to each class k based on the softmax function shown in Eq. 2. We adopt the standard cross-entropy loss function (Eq. 3) to train our M-LSTM model.

Algorithm 1 shows the pseudo-code of M-LSTM training. Given a training dataset D that contains edit sequences and class labels of N users, we first construct the M aspects of edit sequences for each user. After initializing the parameters, Θ_1, Θ_2, and Θ_3, in M-LSTM, in each running, we compute the M last hidden states by LSTM networks (Line 8). Then, we adopt the attention model to combine the M hidden states to the user embedding (Line 9). Finally, we update the parameters of M-LSTM by using the user embedding to predict the user label (Line 10). The parameters of M-LSTM are optimized by Adadelta [37] with the back-propagation.

Algorithm 1. Multi-source LSTM Training

    Inputs : $D = \{(\mathbf{e}_u, y_u); u = 1, \cdots, N\}$
                 Maximum training epoch $Epoch$
    Outputs: Well-trained parameters Θ_1, Θ_2, Θ_3

1 **foreach** $user\ u\ in\ D$ **do**
2 | construct M aspect sequences $\mathbf{x}^{(m)}$ $(m = 1, \ldots, M)$ from the edit sequence
      | $\mathbf{e}_u$;
3 **end**
4 initialize parameters Θ_1, Θ_2, Θ_3 in M-LSTM;
5 $j \leftarrow 0$;
6 **while** $j < Epoch$ **do**
7 | **foreach** $user\ u\ in\ D$ **do**
8 | | compute $\mathbf{h}_T^{(m)}$ $(m = 1, \ldots, M)$ on M sequences of aspect vectors;
9 | | compute the user embedding $\mathbf{s}_T$ by the attention model (Eqs. 4, 5, 6) on
    | | M last hidden states;
10 | | optimize the parameter Θ_1, Θ_2, Θ_3 in M-LSTM based on the loss
    | | function shown in Eq. 3 with Adadelta.
11 | **end**
12 | $j \leftarrow j + 1$;
13 **end**

M-LSTM for Vandal Early Detection. Our trained M-LSTM model can then be used to predict whether a new user v is vandal or benign given his edit sequence $\mathbf{e}_v = (\mathbf{e}_{v_1}, \cdots, \mathbf{e}_{v_t}, \cdots)$. The upper-left region of Fig. 2 shows our M-LSTM based vandal early detection. At each step t, we first derive its M aspect sequences from the user's edit sequence till the step t. The hidden states are updated with the new input $\mathbf{e}_{v_t}$. Thus, they are able to capture all user's edit aspects until the t-th step. We then adopt the attention model shown in

Eqs. 4, 5, and 6 (replacing all subscript T with t) to calculate the user embedding $\mathbf{s}_t$. The user embedding $\mathbf{s}_t$ captures all the user's edit information till the step t. Then, we can use the classifier to predict the probability $P(\hat{y}|\mathbf{s}_t)$ of the user to be a vandal based on $\mathbf{s}_t$. We set a threshold τ to evaluate whether the user is vandal. When $P(\hat{y}|\mathbf{s}_t) > \tau$, the user is labeled as vandal.

4   Experiments

We conduct our evaluation on UMDWikipedia dataset [15]. This dataset contains information of around 770K edits from Jan 2013 to July 2014 (19 months) of 17105 vandals and 17105 benign users. We focus on identifying the user behaviors on the Wikipedia articles. We remove those edits on meta pages (i.e., with titles containing "User:", "Talk:", "User talk:", "Wikipedia:") because they do not cause damages.

For each edit, we extract three aspects, article title, article categories, and revert status. We choose these three aspects because both the title and categories capture the topic information of the edited article and revert status (reported by bots) indicates whether the edit is good or bad. It is imperative to use them to derive user embeddings and then predict whether users are vandal or benign.

We represent article titles and categories to their title embeddings and category embeddings based on word embeddings. Specifically, we first map each word in the titles and categories to its word embedding and then adopt average operation over the word embeddings to get the title embeddings and category embeddings, respectively. The title embeddings and category embeddings reflect the hidden features about the pages. We use the off-the-shelf pre-trained word embeddings[4] provided by [25]. These word embeddings are widely used and have been shown to achieve good performances on many NLP tasks. We randomly initialize the words which don't have pre-trained word embeddings. The dimension of the word embeddings is 50. The dimension of the hidden layer of the M-LSTM network is 32. The training epoch is 25.

4.1   Vandal Detection

To evaluate the performance of vandal detection, we split the training and testing dataset chronologically. We use the first 9 months of users as the training dataset and the last 10 months of users as the testing dataset. The training dataset has 8620 users and the testing dataset has 10418 users. We report the mean values of 10 different runs.

Table 1 shows the precision, recall, F1 and accuracy for vandal detection with different thresholds. Precision indicates the ratio of vandals who are correctly detected. Recall indicates the ratios of vandals who are correctly detected from the test dataset. The default threshold for binary classification used in vandal detection is 0.5, where our model achieves the best performance. We can also

---

[4] http://nlp.stanford.edu/projects/glove/.

Table 1. Experimental results on precision, recall, F1, and accuracy of vandal detection with different thresholds

τ	Precision	Recall	F1	Accuracy
0.5	88.35%	96.67%	92.32%	91.33%
0.6	88.69%	96.01%	92.20%	91.24%
0.7	89.31%	94.85%	92.00%	91.10%
0.8	90.36%	92.27%	91.31%	90.52%
0.9	93.13%	74.10%	82.53%	83.09%

observe that the model achieves good performances of vandal detection with different thresholds τ from 0.5 to 0.8. Meanwhile, with increasing τ, the precision increases accordingly while the recall decreases, which indicates with a higher threshold, the model can detect vandal more accurately but can mis-classify the vandals as benign users. Overall, the F1 and accuracy decease significantly at the 10^{-4} level with t-tests while the threshold $\tau > 0.6$.

We further compare our results with the VEWS approach [15]. The VEWS approach uses a set of hand-crafted features to detect vandals. When incorporating the revision status information, the VEWS can achieve around 90% classification accuracy. Our M-LSTM achieves better accuracy on vandal detection. More importantly, our M-LSTM does not need to design dozens of features to predict vandals. Hence our model can be easily extended to identify vandals from other crowdsourcing knowledge bases like Wikidata.

4.2   User Embeddings

As each user has different edits and each edit has many different aspects, it is challenging to derive users' edit patterns. Our M-LSTM derives user embeddings based on user edits. As user embeddings capture user edit behaviors, they can then be used to differentiate between benign users and vandals and detect potential collaborative vandals.

Visualization. We randomly select user embeddings of 3000 users and map them to a two-dimensional space based on t-SNE approach [18]. Figure 3 shows the visualization plot of user embeddings from M-LSTM. We observe that the benign users and vandals are clearly separated in the projected space. This indicates the user embeddings successfully capture the information whether a user is benign or vandal. Hence, they can be used for vandal detection.

Clustering of User Embedding. In this experiment, we adopt the classic DBSCAN algorithm [7] to cluster user embeddings. We set the maximum radius of the neighborhood $\epsilon = 0.05$ and the minimum number of points $minPts = 3$. DBSCAN produces 211 clusters. Among them, 139 clusters contain only vandals and the total number of vandals is 502 whereas 46 clusters contain only benign users and the total number of benign users is 495. It indicates the benign users

Fig. 3. Visualization of 3000 users in the dataset. Color of a node indicates the type of the user. Red: "Vandals", blue: "Benign Users". (Color figure online)

often form large-size clusters. On the contrary, the vandals usually cluster to small groups to damage articles. For the rest 26 clusters that contain mixed vandals and benign users, there are 17 clusters in which the vandals constitute the majority and 9 clusters in which the benign users constitute the majority. Similarly, the 17 (vandal-majority) clusters are small with only 52 vandals whereas there are 3663 benign users in the 9 (benign-majority) clusters. Embeddings of benign users are closer to each other than that of vandals, which can also be observed in Fig. 3. We conclude that "Benign users are much more alike; every vandal vandalizes in its own way."

When setting the maximum radius of the neighborhood $\epsilon = 0$ and the minimum number of points $minPts = 2$, DBSCAN produces 701 groups containing 1687 user embeddings. Note that under this setting all embeddings within the same group are exactly the same. Among them, 575 groups only contain vandals and the total number of vandals is 1396 whereas 68 groups only contain benign users and the total number of benign users is 144. The largest vandal group contains 13 vandals and the largest benign group contains 17 benign users.

Table 2 shows three examples of potential collaborative vandal groups. In Row 1, the group has 8 vandals who attacked the same page consecutively within a short time window. In Row 2, the group has three vandals who attacked one same page on different days. Because all these vandals were blocked after revising the page, these vandals have high chance to be controlled by a malicious user or group and aim to vandalize the specific page. In Row 3, we show a vandal group containing five vandals. All the five vandals edited the same two pages, "Shakugan no Shana" and "The Familiar of Zero", which are both Japanese light novels, consecutively within a short time window. These three examples demonstrate one advantage of our M-LSTM, i.e., detecting potential collaborative vandals with different behavior patterns.

In Table 3, we further show article titles edited by three pairs of vandals. Each pair of vandals are closed to each other in the embedding space. The first row shows that two vandals damage almost the same pages, which indicates vandals who edit the same pages are close to each other. The second row shows pages edited by two vandals have common words in titles although the title

Table 2. Three example of potential collaborative vandal groups. The vandals of each group damage the same page(s). Group 1 damages the page "List of the X factor finalists (U.S. season 2)" in 2013-01-05 within a short time window. Group 2 damages the page "Niels Bohr" on different days. Group 3 damages the two pages, "Shakugan no Shana" and "The Familiar of Zero", consecutively in 2014-04-18 within a short time window.

Group ID	User ID	Page ID	Revision time
Group 1 2013-01-05	4203021	37310371	02:36:32
	4203016		02:42:02
	4203009		02:42:55
	4203006		02:44:58
	4202998		02:45:32
	4203002		02:47:12
	4202988		02:52:12
	4202986		02:56:21
Group 2	4584127	21210	2013-10-04
	4597541		2013-10-23
	4939865		2014-01-08
Group 3 2014-04-18	5063994	2548832	21:33:51
		5982921	21:34:07
	5063996	2548832	21:35:53
		5982921	21:35:53
	5063998	2548832	21:45:06
		5982921	21:45:28
	5064002	2548832	21:47:21
		5982921	21:47:29
	5064006	2548832	21:48:56
		5982921	21:49:01

names are different. This indicates our M-LSTM can discover the semantic collaborative behaviors on pages based on user embeddings. The last row shows that our M-LSTM can further identify vandals who damage the pages with similar subject areas although there are no any common words in the titles. This example shows the usefulness of incorporating page category information in our M-LSTM. All above examples demonstrate that users who are closed in the low-dimensional user embedding space have similar edit patterns. Therefore, analyzing user embeddings can help capture and understand collaborative vandal behaviors.

Table 3. Three pairs of vandals and their edited page titles. Each pair has similar embeddings based on the cosine similarity.

Vandal IDs	Page title	Page title
4266603 and 4498466	Live While We're Young, What Makes You Beautiful, Up All Night (One Direction album), Take Me Home (One Direction album)	Live While We're Young, Best Song Ever (song), What Makes You Beautiful, Up All Night (One Direction album), Take Me Home (One Direction album)
4422121 and 4345947	Super Mario 3D World, Super Mario Galaxy, Sonic Lost World, Pringles, Action Girlz Racing, Data Design Interactive	Super Mario World, Super Mario World 2: Yoshi's Island, Super Mario Bros. 3, Virtual Boy, Nintendo DS, Kirby Super Star, Yogurt
5032888 and 4592537	Matthew McConaughey, Maggie Q, Theo James, Theo James, Dexter (TV series), Laker Girls, Bayi Rockets, Arctic Monkeys, Dulwich College Beijing	Nicolas Cage, Alan Carr, Liam Neeson, Dale Winton, Craig Price (murderer), Manuel Neuer

4.3   Vandal Early Detection

Our vandal early detection is achieved after each edit is submitted. Although our M-LSTM exploits revert status of the edit, we emphasize that the revert status is inspected by the ClueBot NG in a real time manner. Hence, our M-LSTM can be deployed for real time vandal detection. We evaluate the vandal early detection on the 6427 users who have at least two edits in the testing dataset. Table 4 shows the precision, recall and F1 of our M-LSTM on vandal early detection. We vary the threshold τ from 0.5 to 0.9. Similar to the results of vandal detection shown in Table 1, with increasing τ, a classifier with a higher threshold has more confidence about the prediction, resulting in a higher precision. On the contrary, the recall decreases with the increasing of τ because fewer users will be marked as vandals. The F1 score increases significantly at the 0.005 level with $\tau > 0.6$ and reaches the maximum with $\tau = 0.9$. However, comparing with the results of vandal detection, the vandal early detection has a lower precision but much higher recall. This indicates that we lose some precision but achieve big recall when using partial edit information to do early vandal detection.

Table 4 further shows the average number of edits before the vandals were blocked by the administrators and the ratio of vandals who can be early detected over the whole testing dataset. We can observe that the average number of edits and the ratio of early detected vandals both have a significant decreasing while the threshold $\tau = 0.9$. Note that the ratios of early detected vandals with thresholds from 0.5 to 0.8 are only a little lower than the recall values, which indicates that most of the vandals who are correctly detected are early detected. Overall, setting threshold $\tau = 0.8$ will achieve a balance performance between vandal early detection and accurate prediction.

Table 4. Experimental results on precision, recall and F1 of vandal early detection, the average number of edits before the vandals are blocked, and the ratio of vandals who are early detected.

τ	Precision	Recall	F1	# of edits	% of early detected
0.5	84.10%	99.07%	90.97%	3.50	97.35%
0.6	84.96%	98.99%	91.44%	3.48	96.87%
0.7	85.81%	98.82%	91.86%	3.41	95.94%
0.8	86.88%	98.76%	92.44%	3.33	93.34%
0.9	89.89%	98.34%	93.93%	2.48	72.32%

5   Conclusion and Future Work

In this paper, we have developed a multi-source LSTM model to encode the user behaviors to user embeddings for Wikipedia vandal detection. The M-LSTM is able to simultaneously learn different aspects of user edit information, thus user embeddings accurately capture the different aspects of user behaviors. Our M-LSTM achieves the state-of-the-art results on vandal detection. Furthermore, we show that user embeddings are able to identify collaborative vandal groups with various patterns. Different from existing works which require a list of hand-designed features, our M-LSTM can automatically learn user embeddings from user edits. The user embeddings can be used for a variety of data mining tasks such as classification, clustering, outlier analysis, and visualization. Our empirical evaluation has demonstrated its potential for analyzing collaborative vandals and early vandal detection.

In the future, we plan to incorporate into our M-LSTM more information about user edits, e.g., user-user relations and hyperlink relations among articles. These relations are modeled as graphs and can be naturally incorporated into M-LSTM by using network embedding approaches [26,30]. We also plan to conduct comprehensive evaluations on collaborative vandal detection.

Repeatability. Our software together with the datasets used in this paper are available at https://bitbucket.org/bookcold/vandal_detection.

Acknowledgments. The authors would like to thank anonymous reviewers for their valuable comments and suggestions. The authors acknowledge the support from the 973 Program of China (2014CB340404), the National Natural Science Foundation of China (71571136), and the Research Projects of Science and Technology Commission of Shanghai Municipality (16JC1403000, 14511108002) to Shuhan Yuan and Yang Xiang, and from National Science Foundation (1564250) to Panpan Zheng and Xintao Wu. This research was conducted while Shuhan Yuan visited University of Arkansas.

References

1. Adler, B.T., de Alfaro, L.: A content-driven reputation system for the wikipedia. In: WWW, pp. 261–270 (2007)
2. Adler, B.T., de Alfaro, L., Mola-Velasco, S.M., Rosso, P., West, A.G.: Wikipedia vandalism detection: combining natural language, metadata, and reputation features. In: CICLing, pp. 277–288 (2011)
3. Akoglu, L., McGlohon, M., Faloutsos, C.: oddball: Spotting anomalies in weighted graphs. In: Zaki, M.J., Yu, J.X., Ravindran, B., Pudi, V. (eds.) PAKDD 2010. LNCS (LNAI), vol. 6119, pp. 410–421. Springer, Heidelberg (2010). https://doi.org/10.1007/978-3-642-13672-6_40
4. Bahdanau, D., Cho, K., Bengio, Y.: Neural machine translation by jointly learning to align and translate. In: ICLR (2015)
5. Bengio, Y., Courville, A., Vincent, P.: Representation learning: a review and new perspectives. TPAMI **35**(8), 1798–1828 (2013)
6. Chang, S., Han, W., Tang, J., Qi, G.J., Aggarwal, C.C., Huang, T.S.: Heterogeneous network embedding via deep architectures. In: KDD (2015)
7. Ester, M., Kriegel, H.P., Sander, J., Xu, X., et al.: A density-based algorithm for discovering clusters in large spatial databases with noise. In: KDD, pp. 226–231 (1996)
8. Fei, G., Mukherjee, A., Liu, B., Hsu, M., Castellanos, M., Ghosh, R.: Exploiting burstiness in reviews for review spammer detection. In: ICWSM, pp. 175–184 (2013)
9. Graves, A., Mohamed, A.R., Hinton, G.: Speech recognition with deep recurrent neural networks. In: ICASSP, pp. 6645–6649 (2013)
10. Grover, A., Leskovec, J.: node2vec: scalable feature learning for networks. In: KDD (2016)
11. He, K., Zhang, X., Ren, S., Sun, J.: Deep residual learning for image recognition. In: CVPR, pp. 770–778 (2016)
12. Heindorf, S., Potthast, M., Stein, B., Engels, G.: Vandalism detection in wikidata. In: CIKM, pp. 327–336 (2016)
13. Hochreiter, S., Schmidhuber, J.: Long short-term memory. Neural Comput. **9**(8), 1735–1780 (1997)
14. Javanmardi, S., McDonald, D.W., Lopes, C.V.: Vandalism detection in wikipedia: a high-performing, feature-rich model and its reduction through lasso. In: WikiSym, pp. 82–90 (2011)
15. Kumar, S., Spezzano, F., Subrahmanian, V.: Vews: a wikipedia vandal early warning system. In: KDD, pp. 607–616 (2015)
16. Li, J., Zhu, J., Zhang, B.: Discriminative deep random walk for network classification. In: ACL (2016)
17. Lim, E.P., Nguyen, V.A., Jindal, N., Liu, B., Lauw, H.W.: Detecting product review spammers using rating behaviors. In: CIKM, pp. 939–948 (2010)
18. Maaten, L.V.D., Hinton, G.: Visualizing data using t-SNE. JMLR **9**, 2579–2605 (2008)
19. McKeown, K., Wang, W.: Got you!: automatic vandalism detection in wikipedia with web-based shallow syntactic-semantic modeling. In: COLING, pp. 1146–1154 (2010)
20. Mikolov, T., Corrado, G., Chen, K., Dean, J.: Efficient estimation of word representations in vector space. In: ICLR (2013)

21. Mola-Velasco, S.M.: Wikipedia vandalism detection through machine learning: feature review and new proposals. arXiv:1210.5560 [cs] (2012)
22. Mukherjee, A., Venkataraman, V., Liu, B., Glance, N.S.: What yelp fake review filter might be doing? In: ICWSM, pp. 409–418 (2013)
23. Ntoulas, A., Najork, M., Manasse, M., Fetterly, D.: Detecting spam web pages through content analysis. In: WWW, pp. 83–92 (2006)
24. Papadimitriou, P., Dasdan, A., Garcia-Molina, H.: Web graph similarity for anomaly detection. J. Internet Serv. Appl. **1**(1), 19–30 (2010)
25. Pennington, J., Socher, R., Manning, C.D.: Glove: global vectors for word representation. In: EMNLP, pp. 1532–1543 (2014)
26. Perozzi, B., Al-Rfou, R., Skiena, S.: Deepwalk: online learning of social representations. In: KDD, pp. 701–710 (2014)
27. Spirin, N., Han, J.: Survey on web spam detection: principles and algorithms. SIGKDD Explor. Newsl. **13**(2), 50–64 (2011)
28. Sun, J., Qu, H., Chakrabarti, D., Faloutsos, C.: Neighborhood formation and anomaly detection in bipartite graphs. In: ICDM, pp. 1–8 (2005)
29. Tang, J., Qu, M., Mei, Q.: PTE: Predictive text embedding through large-scale heterogeneous text networks. In: KDD (2015)
30. Tang, J., Qu, M., Wang, M., Zhang, M., Yan, J., Mei, Q.: Line: large-scale information network embedding. In: WWW, pp. 1067–1077 (2015)
31. West, A.G., Kannan, S., Lee, I.: Detecting wikipedia vandalism via spatio-temporal analysis of revision metadata? In: EUROSEC, pp. 22–28 (2010)
32. Xie, S., Wang, G., Lin, S., Yu, P.S.: Review spam detection via temporal pattern discovery. In: KDD, pp. 823–831 (2012)
33. Yang, Z., Yang, D., Dyer, C., He, X., Smola, A., Hovy, E.: Hierarchical attention networks for document classification. In: NAACL, pp. 1480–1489 (2016)
34. Ying, X., Wu, X., Barbará, D.: Spectrum based fraud detection in social networks. In: Proceedings of the 27th International Conference on Data Engineering, Hannover, Germany, pp. 912–923. ICDE 2011, 11–16 April 2011 (2011)
35. Yuan, S., Wu, X., Li, J., Lu, A.: Spectrum-based deep neural networks for fraud detection. CoRR abs/1706.00891 (2017)
36. Yuan, S., Wu, X., Xiang, Y.: SNE: signed network embedding. In: Kim, J., Shim, K., Cao, L., Lee, J.-G., Lin, X., Moon, Y.-S. (eds.) PAKDD 2017. LNCS (LNAI), vol. 10235, pp. 183–195. Springer, Cham (2017). https://doi.org/10.1007/978-3-319-57529-2_15
37. Zeiler, M.D.: Adadelta: An adaptive learning rate method. arXiv:1212.5701 [cs] (2012)
38. Zeiler, M.D., Fergus, R.: Visualizing and understanding convolutional networks. In: Fleet, D., Pajdla, T., Schiele, B., Tuytelaars, T. (eds.) ECCV 2014. LNCS, vol. 8689, pp. 818–833. Springer, Cham (2014). https://doi.org/10.1007/978-3-319-10590-1_53

Author Index

Printed in the United States
By Bookmasters